T0291066

Brown Seaweeds (Phaeophyceae) of Britain and Ireland

Brown Seaweeds (Phaeophyceae) of Britain and Ireland

Robert L. Fletcher
University of Portsmouth

Edited by Juliet A. Brodie
Natural History Museum, London

PELAGIC PUBLISHING

Published by Pelagic Publishing
20–22 Wenlock Road
London N1 7GU, UK

www.pelagicpublishing.com

Natural
History
Museum

https://doi.org/10.53061/XDPG8462

British Library Cataloguing in Publication Data

Brown Seaweeds (Phaeophyceae) of Britain and Ireland

Incorporating *Seaweeds of the British Isles,* **a collaborative project
of the British Phycological Society and the Natural History Museum**
Volume 3: Fucophyceae (Phaeophyceae) Part 1

  I. Marine algae – Great Britain
  I. Fletcher, Robert II. British Phycological Society III. Natural History Museum
     589.3941 QK573
     [ISBN 0-565-00992-3]

ISBN 978-1-78427-247-0 Hbk
ISBN 978-1-78427-248-7 ePub
ISBN 978-1-78427-249-4 PDF

Front cover images: *Laminaria digitata* (top), *Ascophyllum nodosum* (bottom)
Back cover images: *Pelvetia canaliculata, Fucus vesiculosus, Fucus serratus* (top row, left to right);
Himanthalia elongata, Saccorhiza polyschides, Ericaria selaginoides (bottom row, left to right).
All photographs © Robert L. Fletcher

Figure I   Diagram of a hypothetical brown algal cell. Reproduced from Bouck,
G.B. 1965. Fine structure and organelle associations in brown algae. *Journal of Cell
Biology* 26: 523–37. By copyright permission of the Rockefeller University Press.

Figure XIII   Bayesian phylogeny of the Phaeophyceae. Reproduced from Silberfeld, T.,
Rousseau, F. & de Reviers, B. 2014. An updated Classification of brown algae (Ochrophyta,
Phaeophyceae). *Cryptogamie, Algologie* 35: 117–56, fig. 1. By copyright permission of the
Division of Scientific Publications du Museum national d'Histoire naturelle, Paris.

The publisher gratefully acknowledges the Trustees of the Natural History Museum,
London for permission to reproduce images from the NHM herbarium.

Typeset by BBR Design, Sheffield

Printed in the UK by Short Run Press Ltd

MIX
Paper | Supporting
responsible forestry
FSC
www.fsc.org   FSC® C014540

Contents

Preface

It was at the first Annual General Meeting of the British Phycological Society, held in London in January 1953, that the initial decision to begin a new British marine algal flora was taken. Work began on a series of volumes entitled *Seaweeds of the British Isles*, as a collaborative project between the British Phycological Society and the Natural History Museum, London. On the newly formed 'Flora Committee', Dr Helen Blackler and Dr Mary Parke were given responsibility for the brown algae. Following the resignation of Dr Blackler from the Committee in 1967, Dr George Russell was asked to prepare the brown algal section of the flora and in 1972 invited Dr Robert Fletcher to assist him in this task.

The present book represents a contributory volume (Vol. 3), indeed the final volume, to the series *Seaweeds of the British Isles*. It is concerned with the brown algae or Phaeophyceae recorded around the coasts of Britain and Ireland and was, originally, to be published in two parts. Part one, published in 1987 with financial support from the Natural Environment Research Council, covered a selected number of families of brown algae, whilst it was proposed that part two would treat the remaining families. However, in view of the extended time lapse between the two parts and recent reclassifications of the Phaeophyceae, based on molecular studies, it was decided to combine both parts into this single volume entitled *Brown Seaweeds (Phaeophyceae) of Britain and Ireland*.

Production of this volume was made possible by support from the British Phycological Society.

The British Phycological Society (BPS) is a learned society and a charity, devoted to the study of algae. Founded in 1952, the BPS was one of the first phycological societies to be established in the world. The Society's aims are to advance education by the encouragement and pursuit of all aspects of the study of algae and to publish the results of this research in journals, and also in other publications.

Membership of the British Phycological Society is open to everybody. https://brphycsoc.org

Counties of Britain and Ireland as covered in this book.

Acknowledgements

I would like to express my sincere thanks to the following: Professor Juliet Brodie for her unfailing and continued support and encouragement, over many years, and for her critical examination, appraisal and editing of all the final drafts; Dr Akira Peters for useful discussions and for his critical comments and appraisal of parts of the manuscript, especially relating to the Ectocarpales; the late Dr Prud'homme van Reine for his appraisal and very helpful comments on the final text of the Sphacelariales; Professor Sung Min Boo for his useful comments on the text of the Ectocarpaceae; Dr W.F. Farnham for his critical appraisal of drafts of all the text; Dr Marie Racault for helpful discussions and for her important research contribution on the molecular phylogeny of several members of the Ectocarpales; Dr Nela Parente for her significant work towards a better understanding of the biology and molecular phylogeny of members of the Lithodermataceae, Nemodermataceae, Chordariaceae and Scytosiphonaceae; Dr Bruno de Reviers for useful discussions.

I would also like to acknowledge the following in relation to part of this volume previously published (Fletcher 1987): Professor E.B. Gareth Jones, for his encouragement, advice, financial support and assistance during my early untenured years at the University of Portsmouth; the Natural Environment Research Council for both full-time and part-time financial support (NERC grants GR3/3500 and GR3/4690 to Professor E.B. Gareth Jones); the late Dr G. Russell for supplying specimens and records, offering much useful advice on the brown algae and contributing to the Introduction; the late Dr D.E.G. Irvine for his detailed critical and valuable examination of early drafts of the manuscript in his capacity as scientific editor of the present volume; the late Mrs L. Irvine for considerable help and advice with nomenclatural problems; members of the British Phycological Society's Flora Committee and in particular the late Mr P. James for all his support, patience and editorial assistance during the final stages of manuscript submission; the late Dr Paul Pedersen for reading the manuscript and making many useful comments: Dr W.F. Farnham for providing specimens, particularly from the sublittoral, and for the generous use of his library; the late Dr Y.M. Chamberlain for her advice and many helpful discussions; the late Mr J. Price and Mr I. Tittley for advice and useful discussions on the brown algae; the staff of the Publications Section at the Natural History Museum (then British Museum (Natural History)) for seeing Part 1 through the press; the technical staff at the Marine Laboratory, University of Portsmouth, especially the late Mr J. Hepburn, Mr N. Thomas and Mrs A. Davis; the late Mr A.E.W. Hawton, Mr C. Derrick and the late Mr K. Purdy for printing some of the photographs; the late Mrs P. Davies for her valuable and helpful assistance in typing out the numerous drafts of the text.

I am also greatly indebted to the following for the loan or gifts of specimens, records, etc.: Dr J. Berryman, the late Dr M.C.H. Blackler, Professor G. Blunden, Dr C.-F. Boudouresque, the late Dr E.M. Burrows, the late Dr Y.M. Chamberlain, the late Dr J.J.P. Clokie, Professor A. Critchley, the late Mrs T. Edelstein, Dr W.F. Farnham, Dr. Ian Evans, Dr P.W.G. Gray, Professor M.D. Guiry, Dr K. Hiscock, Mrs S. Scott, Mr S.I. Honey, Dr R.G. Hooper, the late Miss C. Howson, the late Dr D.E.G. Irvine, the late Mrs L. Irvine, Mr N.A. Jephson, Professor D.M. John, the late Dr W.E. Jones, the late Dr J.M. Kain, the late Dr J. McLachlan, Professor C.A. Maggs, the late Mr O. Morton, the late Dr I. Munda, the late Professor T.A. Norton, the late Dr P. Pedersen, Dr Akira Peters, Mr M.J. Picken, the late Mr H.T. Powell, the late Mr J. Price, the late Dr W. Prud'homme van Reine, Dr O. Ravanko, the late Dr G. Russell, Professor S.C. Seagrief, Professor G.R. South, Dr L. Terry, Mr I. Tittley and the late Dr R.T. Wilce.

I am indebted to the Directors and Curators of the following institutions for working facilities, for permission to examine specimens or borrow material:

Botanical Museum and Herbarium, Copenhagen (Denmark) **C**
Department of Botany, University of St Andrews
Department of Botany, University of Glasgow **GL**
Department of Botany, National University of Ireland, Galway **GALW**
Department of Biology, Memorial University of Newfoundland, St Johns
Dove Marine Laboratory, University of Newcastle upon Tyne, Cullercoats
Ilfracombe Museum, Ilfracombe **ILF**
Laboratoire de Cryptogamie, Museum National d'Histoire Naturelle, Paris (France) **PC**
Marine Biological Association, Plymouth **MBA**
Marine Science Laboratories, University College of North Wales, Menai Bridge **MB**
Marine Biological Station, Port Erin
Millport Laboratory, Cumbrae
Natural History Museum, London **BM**
National Museums Northern Ireland (Ulster Museum), Belfast **BEL**
National Museum Wales **NMW**
Rijksherbarium, Leiden (Netherlands) **L**
Rochester Museum, Rochester
Royal Botanic Gardens, Kew. The algae previously at Kew are now on permanent loan to the Natural History Museum, London **BM-K**
School of Plant Biology, University College of North Wales, Bangor **UCNW**
Scottish Marine Biological Association, Dunstaffnage, Oban
Smithsonian Institution, National Museum of Natural History, Washington (USA) **NMNH**
Wellcome Marine Laboratory, University of Leeds, Robin Hood's Bay
World Museum Liverpool **LIV**

N.B. The Standard Herbarium Abbreviations are given where appropriate.

All the drawings of microscopic details, as well as all the photographs and some of the habit illustrations are by the author (unless otherwise noted). Some of the habit illustrations have been prepared by Mrs S.M. Fletcher and Mr N. Thomas, to whom grateful acknowledgement is given.

I would especially like to acknowledge significant taxonomic advice from Professor Michael Guiry and the website **www.algaebase.org** (referenced as Guiry & Guiry 2023) which provides an invaluable resource of detailed, up-to-date information on the Phaeophyceae, particularly relating to aspects of the synonymy, geographical distribution and literature on species. The website has added substantially to this book.

I would also like to acknowledge the British Phycological Society for providing significant financial support for publication of this volume.

Finally, but most importantly, I would like to thank my wife, Sylvia, for her unfailing patience, encouragement, interest and support during the production and completion of this book.

A note on the figures

Scale bars are given in mm or μm; TS = transverse section; LS = longitudinal section; SP = squash preparation.

Introduction

The brown algae form their own taxonomic class, with four subclasses and twelve orders (see Silberfeld *et al.* 2014) within the Kingdom Chromista (Cavalier-Smith 1981, 1986, 1998), Phylum Ochrophyta, also termed the Heterokontophyta (Cavalier-Smith & Chao 2006 – see also Cavalier-Smith 1995a,b; Guiry & Guiry 2023; de Reviers 2002; Ruggiero *et al.* 2015; Silar 2016). The class has been variously termed the Fucoideae (C. Agardh 1817), Fucophyceae (Christensen 1978), Melanophyceae (Stizenberger 1860), Melanospermeae (Harvey in Mackay 1836) and Phaeophyceae (Kjellman 1891c), the latter being the most widely accepted. The present work accepts the circumscription of the class Phaeophyceae as defined by Bold & Wynne (1985), Christensen (1980), Clayton & King (1990), van den Hoek & Jahns (1978), van den Hoek *et al.* (1995), Kim (2010), de Reviers *et al.* (2015), Round (1973) and Womersley (1987). This introduction provides a summary of the Phaeophyceae and covers aspects such as structure, morphology, anatomy, cytology, reproduction and ecology. It is intended to contribute towards a general understanding of the biology of these species and assist in the use of this brown algal flora of Britain and Ireland. A number of the more critical references are cited; for more general information, the reader is also referred to the treatments of Bold & Wynne (1985), Chapman & Chapman (1973), Christensen (1980), Clayton & King (1981), Fritsch (1945), Graham & Wilcox (2000), van den Hoek *et al.* (1995), Lobban & Wynne (1981), Lüning (1990), Papenfuss (1951), Prescott (1969), de Reviers & Rousseau (1999), de Reviers *et al.* (2015), Russell (1973a), Scagel (1966), Smith (1955) and South & Whittick (1987).

Cell structure

The structure of the principal organelles of the brown algal cell had been elucidated by transmission electron microscopy (TEM) by the early 1970s (see, for example, Bisalputra & Burton 1969; Bouck 1965, 1969; Evans 1966; Evans & Holligan 1972a,b; La Claire & West 1979; Manton 1957, 1959, 1964a,b; Manton & Clarke 1951a,b, 1956; Manton *et al.* 1953; Müller & Falk 1973) and since that time relatively little new information has been added, except for a number of papers on spore/gamete production and structure (Clayton 1984; Katsaros & Salla 1997; Katsaros *et al.* 1991, 1993; Maier 1997a,b; Maier & Müller 1982; Markey & Wilce 1975, 1976a,b; Müller & Falk 1973; O'Kelly & Floyd 1984) and vegetative cell structure, growth, mitosis and cytokinesis (Katsaros & Galatis 1985; Katsaros *et al.* 1983, 1991; La Claire 1982) – see review in van den Hoek *et al.* (1995).

The addition of scanning electron microscopy (SEM) to phycology has given less information on cell structure than for other algal groups, but has provided observations on the spore/zygote germination process (Fletcher 1976, 1977, 1981b; Hardy & Moss 1978, 1979a,b,c; Ramon 1973) and surface structure (Fletcher 1978). These, and other investigations of the brown algal cell, are valuable in pointing to possible evolutionary relationships between taxa but have not brought about any major changes to the overall taxonomy of the group. For a generalized account of the structure of a brown algal cell, reference can be made to Bouck (1965, see Fig. I), Brawley & Wetherbee (1981), Graham & Wilcox (2000), Russell (1973a) and van den Hoek *et al.* (1995).

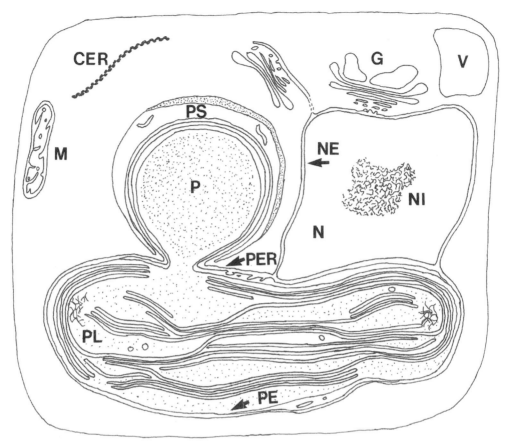

Fig. I. Diagram of a hypothetical brown algal cell (reproduced from Bouck (1965) by copyright permission of the Rockefeller University Press); see text for details of organelle associations. Organelles illustrated include cytoplasmic endoplasmic reticulum (CER), mitochondrion (M), plastid envelope (PE), plastid endoplasmic reticulum (PER), nucleus (N), nuclear envelope (NE), nucleolus (N1), plastid (PL), pyrenoid (P), pyrenoid sac (PS), Golgi bodies (G) and vacuole (V).

Regarding the brown algal cell itself, within the multilayered enclosing wall, the protoplast contains a single, usually central, granular nucleus, variable in shape, on average 5–6 μm in diameter (Russell 1973a) and is usually reported with one, sometimes two, nucleoli. The nucleus is surrounded by a double-membraned, often pored, nuclear envelope in which continuity has been reported with both cytoplasmic endoplasmic reticulum and the plastid endoplasmic reticulum. There are one to several plastids containing the photosynthetic pigments which give the characteristic colour to the brown algal cell. Pyrenoids, which are small hyaline, often globose, electron-dense bodies, are sometimes associated with the plastids. Their composition and biological role is not fully understood, but their high levels of RUBISCO suggest they might play a role in both concentrating CO_2 and in photosynthetic carbon reduction (Jenks & Gibbs 2000; Silberfeld *et al.* 2011). The main food storage material is a soluble polysaccharide, chrysolaminarin, which can constitute up to 34 per cent of the algal dry weight (Craigie 1974; de Reviers *et al.* 2015); also present is mannitol and lipid, with mannitol playing an additional role in osmoregulation (Reed *et al.* 1985). These are distributed in special vacuoles or droplets within the cytoplasm, unlike in green algae where they are stored in grains, or in the red algae in sheaths. One to several variously shaped Golgi bodies are often present, either perinuclear in position with their forming faces adpressed to the nuclear envelope or scattered throughout the cytoplasm. These are the packaging units within the cell, producing vesicles which are widely variable in shape, chemical composition and function. They are particularly active in spore differentiation and settlement (Baker & Evans 1973a,b), sulphated polysaccharide production and secretion (Evans 1974; Evans *et al.* 1973) and cell wall formation (Brawley & Wetherbee 1981; Callow *et al.* 1978; Fletcher 1981b). Other cytoplasmic inclusions include varying numbers of mitochondria, endoplasmic reticula (ER) and a diverse range of vesicles, vacuoles and bodies.

Several of the above-mentioned organelles, including cell walls, plastids, pyrenoids, physodes and iridescent bodies, have proved to be valuable features in the assessment of taxonomic relationships as well as in both generic and specific identification, and are considered in more detail as follows.

The cell wall

With the exception of reproductive spore bodies, all brown algal cells are surrounded by a cell wall. Histochemical and autoradiographic studies have revealed two principal components of the cell walls: a fibrillar, felt-like network which consists of cellulose, arranged in different structural patterns, which is stiffened and strengthened by calcium alginate, and an amorphous component comprising alginates, sulphated polysaccharides, including fucoidan, carboxylated polysaccharides and laminaran. The alginates make up the greater part of the cell wall and, with their colloidal properties, are widely utilized in the food, pharmaceutical and textile industries, with raw material obtained primarily from *Ascophyllum*, *Laminaria*, *Macrocystis* and *Nereocystis* species. In species of the genus *Padina* only, the cell walls are also impregnated with calcium carbonate, giving the thallus a more rigid texture (see van den Hoek *et al.* 1995 for a more detailed summary of the cell wall composition).

The surface cells of all brown algae are covered by a 'cuticle'. The thickness of this cuticle varies for different algae and is also probably determined by inherent characteristics as well as by the surrounding environmental conditions. It is often clearly layered and the outermost layer in some genera, such as *Ascophyllum*, *Fucus* and *Himanthalia*, has been observed to slough off continually – a process considered to reduce the detrimental build up of surface epiphytes (Filion-Myklebust & Norton 1981a,b; Moss 1982; Russell & Veltkamp 1984).

An additional surface and intercellular component of cell walls in many brown algae is mucilage, the principal constituent of which is sulphated polysaccharide. In some genera, such as *Laminaria*, mucilage production is so copious as to render the thallus surface very slimy. In this case, specialized secretory cells are involved in the production of mucilage which is then transported via canals to the thallus surface (Evans *et al.* 1973). In members of the Fucales, mucilage is also synthesized and secreted by active epidermal and outer cortical cells (Evans & Callow 1976). Evans *et al.* (1973) hypothesized that the amount of sulphated polysaccharide retained by the cell of an alga is related to its degree of exposure to the air. For example, *Pelvetia canaliculata* and *Fucus spiralis*, which are found high on the shore, contain large amounts of fucoidan (18–24 per cent on a dry-weight basis) compared to the Laminariales and *Fucus serratus*, members of the lower shore community, which only contain 5 per cent fucoidan on a dry-weight basis. It was suggested that the hygroscopic nature of sulphated polysaccharide served to reduce desiccation.

The intercellular mucilage may play an important role in determining the cohesion of cells and filaments of thalli. The cohesive characteristics of cells and filaments are of considerable importance in specific and generic determination. For instance, the ease of separation of filaments during squash preparations is a very useful taxonomic character for identification of the crustose members of the Ralfsiales and Scytosiphonaceae as well as members of the Chordariaceae. Extracellular and intercellular mucilage might also be responsible for the adhesive nature of the attaching rhizoids (see p. 9).

An additional feature of cell walls observed throughout the brown algae is the presence of pores which permit some degree of cytoplasmic continuity between cells. In adjacent cells the pores are either variously distributed or grouped in distinct pit-fields. Very often cytoplasmic threads (plasmodesmata) pass through the pores, connecting the cytoplasm of adjacent cells. These pores are considered to be secretory in function (van den Hoek *et al.* 1995). They are particularly numerous in cells of members of the Laminariales.

Plastids

All cell types possess one to several prominent plastids, usually located in the peripheral cytoplasm – although in spermatozoa and in hair cells plastids may be very reduced in both number and size. The number and shape of the plastids within a cell are important taxonomic characters and are commonly used to assist in the identification of genera and species (Fig. II, 1–6). Four morphological forms are usually recognized: stellate, discoid, plate-like and ribbon-like. Examples include: one or two axile, stellate plastids in *Bachelotia*, a single, parietal, plate-like plastid in cells of all members of the Scytosiphonaceae, one to several lobed plate-like plastids in various genera such as *Compsonema* and *Myrionema*, and several parietal, ribbon-shaped plastids in *Ectocarpus*. In the great majority of brown algae, however, all cells contain several regularly, or sometimes irregularly shaped, discoid plastids.

The photosynthetic pigments present are chlorophylls a and c, both of which are associated with xanthophylls and carotenoid accessory pigments, the most important being fucoxanthin, violoxanthin, antheraxanthin, neoxanthin, diatoxanthin, diadinoxanthin and β-carotene, which occur in different concentrations in different species (Meeks 1974; Goodwin 1974; van den Hoek *et al.* 1995). It is the abundance of fucoxanthin which gives the Phaeophyceae their characteristic brown colouration, although the depth of pigmentation will be variously influenced by the abundance of the other coloured pigments. For example, brown algae range in colour from olive-brown (*Colpomenia peregrina*, *Punctaria latifolia*), to yellow-brown (*Petroderma maculiforme*, *Corynophlaea crispa*, *Leathesia marina*, *Bifurcaria bifurcata*) to dark brown-black (*Pseudoralfsia verrucosa*, *Battersia mirabilis*, *Halidrys siliquosa*). The colour of an alga will also be determined by its physiological condition; algae in various stages of senescence will often appear green through loss of carotenoids. This effect is particularly well demonstrated in species of *Laminaria* and *Desmarestia*. Moreover, very often fertile regions of some members of the Fucales are distinctly yellow in colour – such as in the male receptacles of *Ascophyllum nodosum* and *Fucus serratus* – possibly reflecting a lower chlorophyll content of these tissues.

Pyrenoids

The pyrenoid is an organelle commonly associated with the plastid in a number of brown algae and is usually clearly discernible under the light microscope as a small hyaline disc (Fig. II, 1, 2 & 6). Under the electron microscope it comprises a granular, dense body, either embedded in the plastid, slightly exserted or projecting from its side on a stalk. It lacks penetrating thylakoids and is bounded by the plastid endoplasmic reticulum, the plastid envelope and an enclosing sheath or cap. In some species with embedded pyrenoids, invaginations of the plastid membrane are present in the pyrenoid. The pyrenoid appears to act as a centre for carbon dioxide (CO_2) fixation and polysaccharide synthesis (Kerby & Evans 1978). The distribution of pyrenoids within the Phaeophyceae was thought to be of phylogenetic importance (Evans 1966; Hori 1971, 1972) and was used as a diagnostic character in ordinal classification (Simon 1954). They were shown to be present in the morphologically simple members of the Ectocarpales but generally absent in the more complex members of orders such as the Dictyotales, Laminariales and Fucales. However, their reported presence in these latter orders, usually as non-pedunculate and rudimentary, either rarely, or confined to certain reproductive

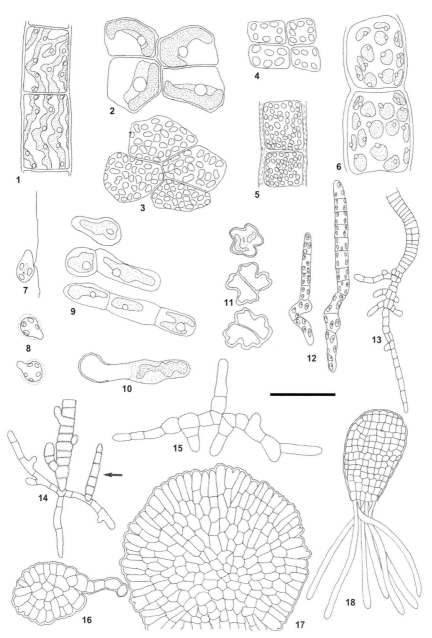

Fig. II. 1–5. Examples of different shaped plastids: ribbon-shaped plastids in *Ectocarpus fasciculatus* (1), plate-like plastids in *Colpomenia peregrina* (2) and discoid plastids in *Desmarestia ligulata* (3), the 'Aglaozonia' phase of *Cutleria multifida* (4) and *Elachista fucicola* (5). Note hyaline, disc-like pyrenoids drawn in 1 and 2. 6. Portion of paraphysis in *Myriactula chordae* showing cells with discoid plastids and small associated pyrenoids. 7. Typical motile brown algal spore, with discoid plastids and laterally inserted flagella of unequal length. 8. Germinating spores of *Hincksia granulosa*. 9. Unipolar germination sequence in *Planosiphon zosterifolius*. 10. Unipolar germination in *Pseudoralfsia verrucosa*, but showing evacuation of contents of original spore cell and part of germ tube. 11. Stellate germination pattern in *Myrionema magnusii*. 12. Young germlings of *Hincksia granulosa* showing growth of erect filaments from original spore cells. 13. Young germling of *Planosiphon filiformis* showing growth of erect blade from the proximal region of the branched filamentous base. 14. Young germling of *Hincksia granulosa* showing secondary shoot production (arrow) from the basal filaments. 15. Young filamentous germling of *Planosiphon zosterifolius*. 16. Young germling of *Petalonia fascia* with terminal disc. 17. Discoid germling produced in cultures of *Stragularia clavata*. 18. Young germling of *Sargassum muticum* with small, multicellular, erect shoot giving rise at the base to 7 rhizoidal filaments (one hidden). Bar = 20 µm (1–4, 6–11), = 50 µm (5, 12, 15–18), = 100 µm (13, 14).

cell types, cast serious doubt on the phylogenetic relevance of this character (Bouck 1965; Bold & Wynne 1985; Chi 1971; Evans 1968; Hori 1972; Prud'homme van Reine & Star 1981). More recent studies have indicated that pyrenoids are poor taxonomic markers and have evolved several times independently in the Phaeophyceae (Silberfeld *et al.* 2010, 2011). Stalked and protruding pyrenoids were, however, used by Rousseau & de Reviers (1999a) as a diagnostic character in their broader concept of the Ectocarpales.

Eyespots

Another notable organelle associated with the plastid in many brown algae, but notably absent in the Laminariales (Henry & Cole 1982a,b), is the eyespot or stigma. Eyespots appear to be confined to the motile reproductive spore and have been described in zoospores of, for instance, *Ectocarpus*, *Pylaiella* and *Scytosiphon* (Lofthouse & Capon 1975; Manton 1957; Markey & Wilce 1975), in sperm of *Cutleria*, *Fucus* and *Ascophyllum* (Bouck 1970; La Claire & West 1979; Manton & Clarke 1951a) and in the female gametangium of *Cutleria* (La Claire & West 1978). They comprise groups of spherical, osmiophilic, carotenoid granules usually projecting slightly from the surface of the plastid and reported to be coloured red or orange under the light microscope (Bouck 1970; Dodge 1973; Manton 1957). These red spots can frequently be seen associated with spores contained within sporangia on the parental thallus. From electron microscope studies there have been reports of an association between the eyespot and a swelling on the smooth posterior flagellum, and together they constitute the photoreceptor apparatus (Berkaloff & Rousseau 1979; Bouck 1970; Dodge 1969; La Claire & West 1978; Manton & Clarke 1951b). They are considered to function as primitive photoreceptors (Bisalputra 1974) and are, therefore, probably involved in phototactic responses.

Flagella

The great majority of brown algae produce motile reproductive spore bodies. Examples of non-motile spores are the tetraspores produced by members of the Dictyotales and the female gametes produced by the oogamously reproducing members of the Laminariales, Desmarestiales and Fucales. Most motile cells are pyriform and possess two laterally inserted flagella with the anterior directed flagellum longer than the posterior (Fig. II, 7). However, variations have been described; for example, a single flagellum is reported for the sperm cells of *Dictyota* (Manton 1959; Manton *et al.* 1953), whilst in the sperm of *Fucus* and *Ascophyllum* and *Laminaria* it is the posterior flagellum which is the longer of the two (Maier & Muller 1982; Manton *et al.* 1953). Both flagella generally end in thin hair-points. A characteristic feature of the anterior flagellum is the presence of variously arranged and orientated hair-like appendages or mastigonemes. These are also present on the single, anterior flagellum of *Dictyota* (Manton *et al.* 1953). The posterior flagellum is devoid of these hairs although they were reported by Loiseaux & West (1970) on *Hincksia* zoospores. Additional ornamentations of the 'hairy' flagella include terminal filaments and spines. In *Fucus*, the sperm have a proboscis-like appendage, extending from the anterior region of the cell, which is considered to play a role in the fertilization process (Manton & Clarke 1956). All flagella have the typical 9 + 2 arrangement of microtubules with a basal body in the cytoplasm (Evans 1974). For a more detailed account of the flagella structure see van den Hoek *et al.* (1995).

Vacuoles and vesicles

A wide variety of vacuoles and vesicles have been described from brown algal cells. Vacuoles are light-coloured, membrane-bound regions, usually centrally positioned within a cell and causing displacement of other cytoplasmic organelles, such as the nucleus and plastids, to the periphery. They are particularly common in the central, more inactive cells of an algal thallus, such as medullary, corticating and hyphal filament cells. Golgi bodies are usually associated with their formation and they are attributed with various functions, particularly storage of water, inorganic ions and organic metabolites. Vacuoles are likely to be important in the responses of the cell to changes in external water potential.

Of the many vesicles described, particularly by electron microscope studies, the physodes or fucosan vesicles are the most prominent. Originally distinguished by a red coloration with vanillin HC1 stains (Kylin 1938) and formed in the perinuclear region of the cell by the Golgi apparatus, these membrane-bound vesicles can be colourless and highly refractile or darkly staining. They contain phlorotannins which comprise a wide variety of different phenolic compounds. These phenolics have multiple roles in the brown algae. They are considered to have antibiotic properties, probably discouraging herbivores and epiphytes, provide protection against excess ultraviolet radiation, are involved in heavy metal resistance, contribute to cell wall formation and early development, and act as adhesives (Clayton 1984b; Ragan 1976; see Schoenwaelder 2002 for review). They are widely distributed in most families of the brown algae but reports of their presence are especially numerous for members of the Dictyotales and Fucales. Further noteworthy cellular inclusions are iridescent bodies which give the thalli of certain brown algae a blue or green iridescence when viewed under water. They have been reported in *Dictyota*, *Ericaria* and *Gongolaria* spp. Electron microscope studies revealed them to be physode-like, membrane-bound vesicles containing electron dense globules (Feldmann & Guglielmi 1972; Pellegrini 1974; Lopez-Garcia et al. 2018). They are proteinaceous with some polysaccharide and probably originate from the Golgi bodies.

Specialized cells

Additional noteworthy morphological/anatomical features of brown algae include hairs, ascocysts and paraphyses (Fig. III).

Hairs

Hairs are specially adapted filaments comprising hyaline cells. Two main types are generally recognized in the brown algae: i) false hairs or pseudohairs, which are usually slightly pigmented, terminate tips of filaments (in the Ectocarpaceae) and lack a distinct basal meristem and sheath; and ii) true hairs (often referred to as 'phaeophycean' and/or 'endogenous' hairs) which are colourless, although sometimes containing vestigial plastids, and have a distinct basal growth zone with or without an enclosing sheath. True hairs, which can be single or grouped, are widely distributed in the Phaeophyceae, arising endogenously or exogenously. Due to their large surface-to-volume ratio, hairs are considered to have a role in the uptake of nutrients and this function has been described for many algal groups including the Cyanobacteria (Sinclair & Whitton 1977), the Chlorophyta (Benson et al. 1983), the Rhodophyta (Rueness et al. 1987) as well as the Phaeophyceae (Hurd et al. 1993; Lichtenberg et al. 2017; Schonbeck & Norton 1979a; Steen 2003a). Hair production appears to be influenced by environmental conditions, including light intensity (Berthold 1882; Oltmanns 1892), light quality, particularly blue light (Dring & Lüning 1975b; Lockhart 1982; Müller & Clauss 1976) and nutrient levels (Lockhart 1979, 1982; Steen 2003a). Additional 'true hair' types include the coloured assimilatory filaments clothing the erect thallus in *Halosiphon* and the coloured apical filaments in *Cutleria*.

Ascocysts

Ascocysts are abnormal, usually enlarged, thick-walled cells, sometimes intercalary, but more often terminal in position, which are initially darkly pigmented and packed with fucosan vesicles containing polymers of phloroglucinol, later often empty and hyaline. They are widely distributed in the Phaeophyceae, including genera such as *Chilionema*, *Pseudolithoderma*,

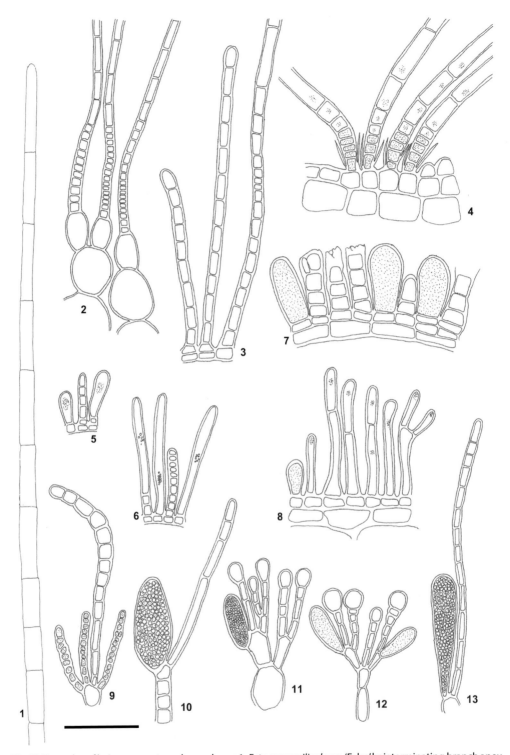

Fig. III. Examples of hairs, ascocysts and paraphyses 1. *Ectocarpus siliculosus*. 'False' hair terminating branch apex. 2–4. Examples of true hairs: 2, *Leathesia marina*; 3, *Chilionema ocellatum*; 4, *Punctaria plantaginea*. 5–8. Ascocysts in *Myrionema magnusii* (5, 6), *Symphyocarpus strangulans* (7) and *Colpomenia peregrina* (8). 9–13. Paraphyses variously associated with both plurilocular (9) and unilocular sporangia (others) in *Microcoryne ocellata* (9), *Stragularia clavata* (10), *Leathesia marina* (11), *Carpomitra costata* (12) and *Myriactula areschougii* (13). Bar = 50 μm (1, 2, 5, 6, 7, 8, 9, 10, 11, 13), = 63 μm (3, 4, 12).

Petroderma, Symphyocarpus, Laminariocolax, Scytosiphon and *Colpomenia*. Discoid brown algae with ascocysts were previously grouped in the genus *Ascocyclus* Magnus, until this was rejected by Loiseaux 1967a (see also Russell 1964b) on the grounds that they represent pathological responses.

Paraphyses

Paraphyses is the name given to a variety of sterile structures that are commonly associated with the reproductive organs, usually in sori. They may be unicellular (e.g. in the Laminariales), multicellular and simple (e.g. in *Pseudoralfsia, Stragularia, Leathesia*) or multicellular and branched (e.g. in *Fucus, Petrospongium* and *Corynophlaea*). In a number of families, such as the Lithodermataceae, Chordariaceae and Scytosiphonaceae, their presence, shape, dimensions, branching pattern, number and shape of constituent cells can be very useful diagnostic characters. They are commonly mucilaginous and perhaps function as protective agents, both from physical damage and desiccation.

The brown algal thallus

Settlement and early development

Although the entrapment of vegetative thalli portions may play an important role in the dissemination and colonization processes of marine algae, including the Phaeophyceae (Clokie & Boney 1980) and may be the only reproductive method of algae at the geographical limits of their distribution, the majority of algae reach new substrata by the production, release and settlement of specialized reproductive spore bodies.

In the great majority of brown algae, motile reproductive spores are produced and settlement commonly involves site selection. The non-motile cells formed in oogamously reproducing algae and members of the Dictyotales (tetraspores) are more at the mercy of the water currents and rely on the more passive process of sinking under gravity. The motile spores appear to be well adapted to their role as substratum colonizers. For instance, they have been reported to remain motile for periods of up to 24 hours in *Petalonia fascia* (Wynne 1969) and *Ectocarpus siliculosus* (Baker & Evans 1973a), to stay viable for up to 5(–14) days in *Undaria pinnatifida* (Forrest *et al.* 2001), to have dispersal ranges of up to 100 m in *Undaria pinnatifida* (Forrest *et al.* 2001) and 200 m in *Laminaria hyperborea* (Fredriksen *et al.* 1995) and to detect and respond to external stimuli such as light (Baker & Evans 1973a; Fletcher 1981b; Flores-Moya *et al.* 2002; Fu *et al.* 2014; Kawai *et al.* 1991; Kinoshita *et al.* 2017a,b; Toth 1976), surface chemistry (Amsler & Neushul 1989, 1990; Kinoshita *et al.* 2017b) and surface topography (Müller 1964a; Russell & Morris 1971).

Following settlement, most brown algal spores initially become attached to a substratum by the release of an adhesive material (Baker & Evans 1973a; Fletcher 1981b; Toth 1976). The spore adhesive has been shown by electron microscope studies to be derived from small vesicles produced by the Golgi bodies. Upon release through the plasmalemma, the glue material, which is probably composed of a polysaccharide-protein complex, spreads out and forms a tenacious attachment to the substratum. Alternative attachment mechanisms of the settling stages recorded in brown algae include direct adhesion of the outer walls, as in *Pelvetia canaliculata* (Hardy & Moss 1979c) and *Halidrys siliquosa* (Hardy & Moss 1978) and the immediate production of adhesive rhizoids, as in *Sargassum muticum* (Fletcher 1980).

Following attachment, the spores undergo a process of germination to form the adult thallus (Fig. II, 8). Usually this occurs immediately, although there can be long periods of delay under adverse environmental conditions (Carney 2011; Kain 1979). Essentially, germination is a dual process: after cell wall formation, the spore initiates the development of two main morphogenetic pathways. These are the formation of (1) a prostrate anchorage system and

(2) the characteristic, erect thallus. During the early stages of this germination process two main patterns of development, unipolar and stellate, have been described (Pedersen 1981a) (Fig. II, 9–11). In unipolar germination a germ tube extends out from the spore and a cross wall is laid down; both cells contain plastids and a nucleus. The younger distal cell then continues growth and division to produce a filamentous germling. Occasionally, two or three germ tubes can be produced from the spore cell, all developing in a similar manner. This germination pattern is very common and distributed throughout most families of brown algae. A slight modification of unipolar germination involves the cell contents of the original spore cell migrating into the germ tube; a cross wall is then laid down leaving the original spore cell empty in the proximal area. This germination pattern has been described in various algae, including *Laminaria* spp. (Kain 1979), *Halosiphon tomentosus* (Toth 1976; as *Chorda tomentosa*) and *Pseudoralfsia verrucosa* (Nakamura 1972; as *Ralfsia verrucosa*).

In the stellate germination pattern (a modification of unipolar germination), the spore forms 3–5 lobes which are later closed by oblique walls. Radial growth of these cells then produces a disc-shaped system. This is particularly characteristic of some epiphytic species and has been reported in various myrionematoid algae (Loiseaux 1967b; Sauvageau 1897) and the chordariacean algae *Cladosiphon zosterae* (Parke 1933; Sauvageau 1924a) and *C. okamuranus* (Shinmura 1974, 1977). These two germination patterns, unipolar and stellate, are not, however, mutually exclusive. Both types have been reported for some species, such as *Pseudoralfsia verrucosa* in which a terminally enlarged unipolar germ tube, rather than the original spore, cleaves into a rosette of four cells (Fletcher 1978) and even for spores derived from a single sporangium – a phenomenon referred to as heteroblasty (see p. 30).

Coupled with this basal developmental process is the formation of the characteristic, erect shoots. Usually the latter are produced directly from the upper, polarised region of the original spore cell, although in many algae the expanding basal system can also produce erect shoots (Fig. II, 12–14). A fundamental aspect of these early stages of development is the extent of growth of the horizontally spreading, primary attachment system. In the least specialized members of the brown algae, development of this basal system is pronounced, either as branched, free filaments or branched, compacted filaments (pseudodisc/disc-like), from various cells of which the erect thallus body is initiated (Fig. II, 15–17). This development of the thallus body in two perpendicular planes is termed heterotrichy (Fritsch 1945) and has been widely reported in such families as the Acinetosporaceae, Chordariaceae and Ectocarpaceae. An extreme example of the heterotrichous condition can be found in some of the very reduced members of the Chordariaceae where, in algae such as *Mikrosyphar*, *Phaeostroma* and *Streblonema*, erect development is absent or reduced to a few short branches or hairs.

Non-heterotrichous thalli are representative of members of the more complex brown algal groups. In the Dictyotales, Fucales and the sporophytes of the Desmarestiales and Laminariales, there is no development of a prostrate vegetative system. The erect thallus develops immediately and attaches only by the basal production of secondary rhizoids (Fig. II, 18). However, traces of the heterotrichous habit persist in gametophytes of the Desmarestiales and Laminariales.

Mode of construction

All brown algae are branched-filamentous, pseudoparenchymatous or parenchymatous in their mode of construction. There are no unicellular, colonial, unbranched-filamentous or siphonaceous forms, such as are variously distributed among the green and red algae. The filamentous algae are common in the families of the Ectocarpales, especially in the Acinetosporaceae and Ectocarpaceae (Fig. IV). They may be little branched, as in *Compsonema*, *Acinetospora*, *Protectocarpus* and *Feldmannia*, or copiously branched, as in various species of *Hincksia*, *Ectocarpus*, *Pylaiella*, *Streblonema* and *Kuckuckia*. Branching may be opposite, as in *Hincksia*

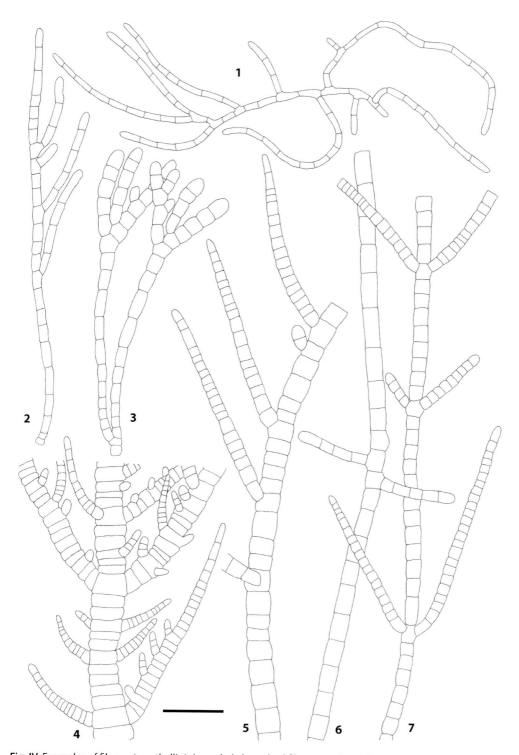

Fig. IV. Examples of filamentous thalli. 1. Irregularly branched filaments of *Streblonema* sp. 2. Single, branched, erect filament of *Protectocarpus speciosus*. 3. Erect, terminally branched filaments of *Compsonema micro-spongium*. 4. Portion of erect axis of *Hincksia granulosa* with opposite laterals. 5. Portion of erect axis of *Hincksia sandriana* with secundly arranged laterals. 6. Portion of main axis of *Acinetospora crinita* with infrequently occurring short, opposite laterals. 7. Portion of main axis of *Pylaiella littoralis* with opposite laterals. Bar = 50 µm (1, 5), = 63 µm (3), = 77 µm (2), = 100 µm (4, 6, 7).

granulosa, *Pylaiella littoralis* and *Feldmannia paradoxa*, irregular and/or alternate, as in *Ectocarpus siliculosus*, *Feldmannia mitchelliae* and *Hincksia sandriana*, or secund, as in *Protectocarpus speciosus* and *Hincksia hincksiae*. Branches can also be closely set, as in *Compsonema*, or widely divergent, as in *Feldmannia paradoxa* and *Acinetospora crinita*. The basal systems of these erect filamentous groups are usually also branched and filamentous, as in *Ectocarpus*; less commonly they have a pseudoparenchymatous base (see below).

In many algae, the mode of construction of the erect thalli is pseudoparenchymatous (sometimes termed haplostichous) in which there is an aggregation of filaments to give rise to a more complex tissue (Fig. V). The derivation of these filaments can be from either a single, initial, uniseriate, erect filament (uniaxial) or from an aggregate of several erect filaments (multiaxial). Examples of pseudoparenchymatous multiaxial thalli include various members of the families Pseudoralfsiaceae, Lithodermataceae, Petrodermataceae, Chordariaceae and Scytosiphonaceae (crustose members only). In members of the Pseudoralfsiaceae, Lithodermataceae and Scytosiphonaceae, the erect filaments are short, quite well compacted and little branched, producing the characteristic crustose type of thalli. These erect filaments can be either firmly united and difficult to separate, as in *Pseudoralfsia*, *Sorapion* and most species of *Pseudolithoderma*, or loosely united and easily separable, as in *Petroderma* and *Microspongium*. In *Myrionema* the erect filaments are very loosely united in a gelatinous matrix and easily separable. In the Chordariaceae and Petrospongiaceae, the erect filaments are much more developed and branched, resulting in cushion forms, as in *Leathesia* and *Petrospongium* and in erect, more macroscopic, cylindrical thalli, as in *Cladosiphon*, *Chordaria* and *Eudesme*.

In many other algae, however, the pseudoparenchymatous form is based on a uniaxial mode of construction. In the chordariacean genera *Mesogloia* and *Spermatochnus*, for instance, there is a single erect axis, and it is the branched derivatives of this which are compacted together to form the solid, gelatinous thallus. The degree of compactness of these chordariacean genera, including both multiaxial and uniaxial types, varies – for example, *Eudesme* is weakly compacted and gelatinous, whilst *Chordaria* is firmly compacted and solid.

A further elaboration of the uniaxial, pseudoparenchymatous mode of construction can be found in the Desmarestiales (Fig. VI, 1–2). In this group both the single erect axis and its lateral branches give rise to rhizoid-like outgrowths which divide and extend outwards and downwards to envelop the axes completely in a solid compacted tissue. The further production of hyphal filaments from the innermost cell layers contributes to the bulk of the thallus.

The third mode of construction of the brown algae is referred to as parenchymatous (sometimes termed polystichous) and involves the cells of a filament undergoing, to a limited or greater extent, longitudinal/periclinal divisions (Fig. VI, 3–10). Unlike pseudoparenchyma, which is the result of filament aggregation, parenchyma is the result of cell division in several planes within a filament. The simplest parenchymatous thalli are represented by species such as *Hecatonema terminale* and *Pylaiella littoralis* in which occasional cells of the erect filaments have divided longitudinally. In algae such as *Compsonema saxicola* and some *Chilionema* spp., the occasional longitudinal divisions are confined to cells of the horizontally spreading filaments. Much more elaborate and complex longitudinal divisions occur in the cells of some members of the Sphacelariales. Indeed, the pattern of internal division is an important and useful taxonomic character. As in the simple parenchymatous types, no significant increase in girth of the thallus is associated with this mode of division.

In many brown algae, however, including representatives of the Scytosiphonaceae, Cutleriaceae, Chordaceae, Laminariaceae, Alariaceae, Dictyotaceae and Fucaceae, parenchymatous divisions in the early formed erect filament/shoot are much more extensive, producing larger and more elaborate thalli with a wide range of forms (Figs VII, VIII). In some algal groups, such as Ectocarpales, Cutleriales, Dictyotales and Laminariales, the thalli are initially polystichous, before cell divisions occur in the three planes to form the parenchyma. This differs from the Fucales and Ascoseirales, in which the apical cells divide in three directions,

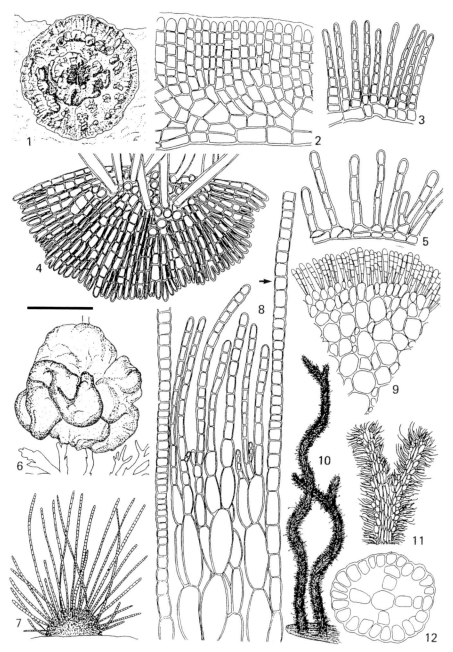

Fig. V. Examples of pseudoparenchymatous thalli. 1. Crustose thallus of *Pseudoralfsia verrucosa*. 2. Vertical section of *Stragularia clavata* crust showing slightly arched, tightly packed, erect filaments. 3. Squash preparation of the *Microspongium gelatinosum* crustose phase of *Scytosiphon lomentaria* showing easily separated erect filaments. 4. Surface view of discoid thallus of young *Myrionema magnusii*. Note erect ascocysts. 5. Squash preparation of *Myrionema strangulans* showing short, loosely associated, erect filaments arising from a monostromatic base. 6. Cushion-like thallus of *Leathesia marina*. 7. Basal hemispherical cushion and exserted filaments of *Elachista fucicola*. 8. Vertical section of *Elachista scutulata* showing closely packed filaments of large cells (which constitute the basal cushion) giving rise terminally to paraphyses and exserted filaments (arrow). 9. Portion of transverse section of *Chordaria flagelliformis* showing compact, multiaxial medulla giving rise peripherally to paraphyses. 10. Branched, cylindrical habit of *Cladosiphon zosterae*. 11. Squash preparation of *Microcoryne ocellata* showing multiaxial medulla of large, elongated cells, giving rise peripherally to paraphyses. 12. Transverse section of *Spermatochnus paradoxus* showing a single, central, axial cell. Bar = 50 μm (2–5, 12), = 10 μm (8, 9), = 0.5 mm (11), = 1.5 mm (7), = 7.5 mm (10), = 15 mm (1), = 30 mm (6).

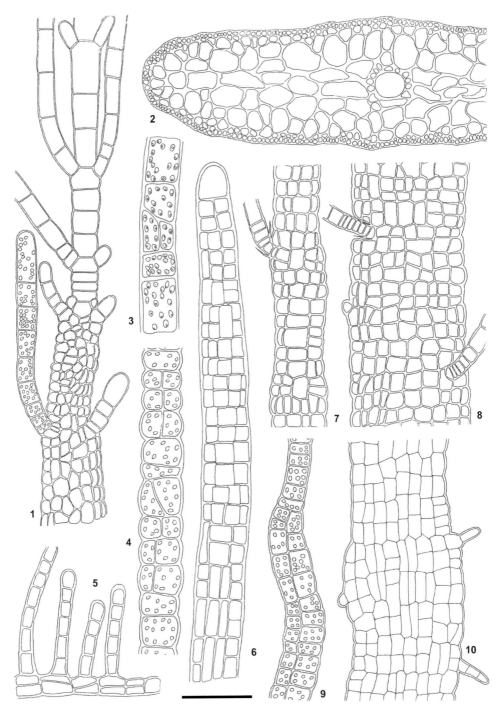

Fig. VI. Examples of both pseudoparenchymatous (1, 2) and parenchymatous thalli (3–10). 1. Apex of *Desmarestia viridis* showing intercalary meristem with upper, branched hair and lower cortical envelope. 2. Transverse section of *Desmarestia ligulata* showing a large, central, axial cell, a broad cortical zone of large colourless cells and a peripheral layer of smaller, pigmented cells. 3. Portion of erect axis of *Pylaiella littoralis* showing a single, longitudinally divided cell. 4. Portion of partly biseriate thallus of *Leblondiella densa*. 5. Portion of *Chilionema ocellatum* showing biseriate basal layer. 6. Apex of *Battersia plumigera* showing segmentation of cells. Note that only a small increase in growth occurs. 7–8. Portions of terete, parenchymatous thalli of *Litosiphon laminariae*. 9. Apical region of *Punctaria tenuissima* showing transformation from uniseriate to biseriate thallus. 10. Portion of young parenchymatous blade of *Punctaria tenuissima*. Bar = 50 μm (3–8, 10), = 63 μm (1, 2, 5, 9).

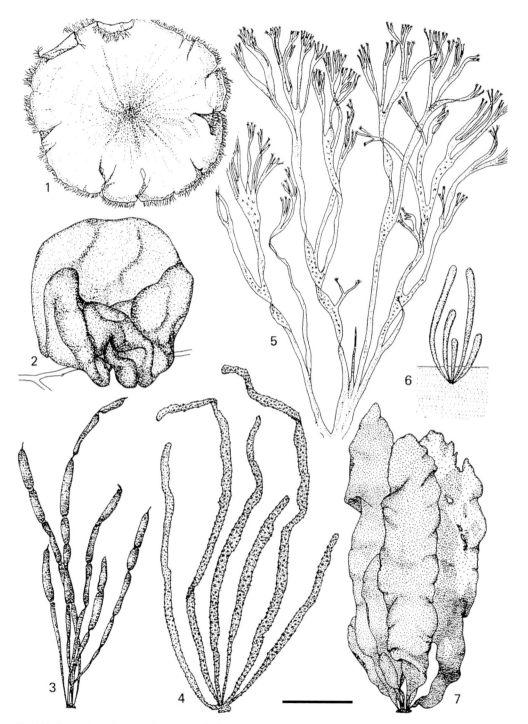

Fig. VII. Examples of parenchymatous thalli 1. *Zanardinia typus*; 2. *Colpomenia peregrina*; 3. *Scytosiphon lomentaria*; 4. *Asperococcus fistulosus*; 5. *Cutleria multifida*; 6. *Leblondiella densa*; 7. *Petalonia fascia.* Bar = 4 mm (6), = 13 mm (1), = 20 mm (5), = 30 mm (2, 4, 7), = 33 mm (3).

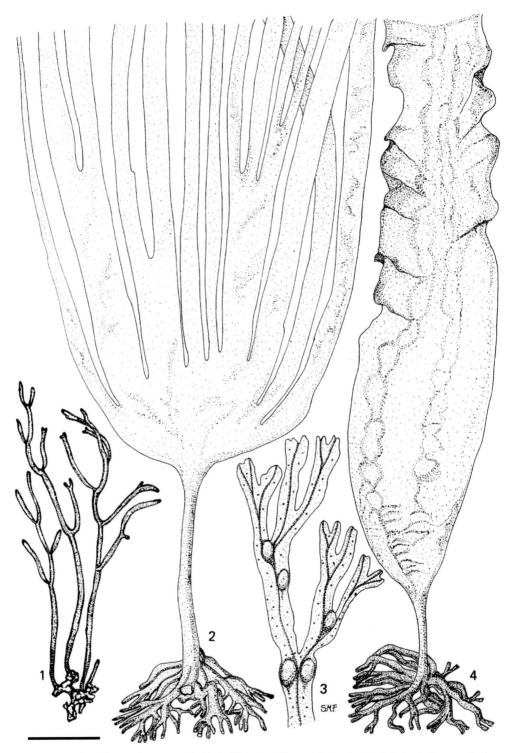

Fig. VIII. Examples of parenchymatous thalli. 1. *Bifurcaria bifurcata*; 2. *Laminaria digitata*; 3. *Fucus vesiculosus* (portion of thallus only); 4. *Saccharina latissima*. Bar = 30 mm.

the segments then undergoing periclinal divisions, a process that is considered to be truly parenchymatous.

Examples of parenchymatous thalli include those that are encrusting/procumbent (*Battersia*, *Zanardinia*, *Aglaozonia* phase of *Cutleria*), saccate (*Colpomenia*), unbranched, solid or hollow, and terete (*Asperococcus*, *Chorda*, *Litosiphon*, *Scytosiphon*), branched and terete (*Bifurcaria*, *Cystoseira*), strap-shaped (*Himanthalia*), unbranched and foliose (*Petalonia*, *Laminaria*, *Punctaria*) or branched and foliose (*Dictyota*, *Cutleria*). Some of these parenchymatous forms, such as the kelps, represent the largest of marine algae, with *Laminaria* spp. growing up to 3(–4) metres in length and the Pacific giant kelp, *Macrocystis pyrifera*, reaching over 50 m.

For both the parenchymatous and pseudoparenchymatous forms, increase in size has usually resulted in considerable structural and anatomical differentiation. For example, in many algae the thallus can be clearly divided into an erect frond/blade, a stipe region and a basal attachment system. The frond consists of the visibly prominent portion of the thallus and contains most of the characteristics/features by which the alga is identified. Apart from the gross morphological components described above, a number of supplementary features are present which characterize the algae. These can include lateral proliferations (*Desmarestia dudresnayi*, *Punctaria latifolia*), surface configurations and bullations (*Saccharina latissima*), the presence of vesicles/bladders used for buoyancy and respiratory functions (*Ascophyllum nodosum*, *Fucus vesiculosus*, *Sargassum muticum*, *Gongolaria baccata*), inflation of thalli (*Scytosiphon lomentaria*, *Colpomenia peregrina*, *Asperococcus bullosus*), the presence of midribs (*Fucus* spp., *Carpomitra*, *Desmarestia ligulata*, *Undaria pinnatifida*) and the presence of leaf-like structures (*Sargassum muticum*). The frond is the main photosynthetic portion of the algal body, with the highest concentration of cells with plastids. It is also responsible for the development of the reproductive organs and, therefore, plays a vital role in the production and dissemination of reproductive bodies. Lastly, the frond is usually the most active portion of the thallus in nutrient uptake and gaseous exchange.

In a number of brown algae, the frond arises from a distinct stalk-like structure, usually referred to as the stipe, which itself arises from a basal attachment system. The extent of development of the stipe varies considerably between algae – it is hardly discernible in *Punctaria* and *Petalonia*, quite well developed in *Fucus*, and especially long and prominent in *Laminaria* spp. It is usually cylindrical in shape, and both strong and flexible, providing a supporting as well as a photosynthetic and translocatory role.

A wide range of attachment systems have been described in the brown algae, the great majority of which are based on the production of secondary rhizoids from vegetative cells on the erect thallus (Fletcher 1977). These rhizoids grow down, usually in large numbers, to the basal substratum. In algae such as *Hincksia* and *Sphacelaria* they also form an enclosing sheath around more distal parts of the thallus and are then termed corticating filaments. In many thalloid algae, such as *Scytosiphon*, *Petalonia* and *Fucus*, they are produced both externally, growing down the outside of the thallus, and internally. At the thallus base the emergent rhizoidal filaments either remain discrete to produce a fibrous mass (*Planosiphon filiformis*, *Halothrix*) or are compacted together to form the pseudoparenchymatous basal holdfast systems which are particularly common in the larger brown algae (e.g. *Fucus*, *Sargassum*, *Punctaria*). The wide range of holdfast morphologies can be of taxonomic importance, as in *Petalonia* spp. (Fletcher 1981a) and *Desmarestia* spp. (Chapman 1972). Two other noteworthy basal systems are the highly specialized finger-like, parenchymatous haptera produced at the base of kelp algae, such as *Laminaria*, and the horizontally spreading, stoloniferous attachment organs produced in the Dictyotales (Fritsch 1945; Richardson 1979). Despite this wide range of attachment systems, all fundamentally depend on individual, microscopic, rhizoidal filaments which have adhesive properties (Fletcher *et al.* 1984).

Concomitant with thallus development is the internal differentiation of the cells into distinct tissue types (Fig. IX). For the majority of algae with thalli in excess of three cells thick, this

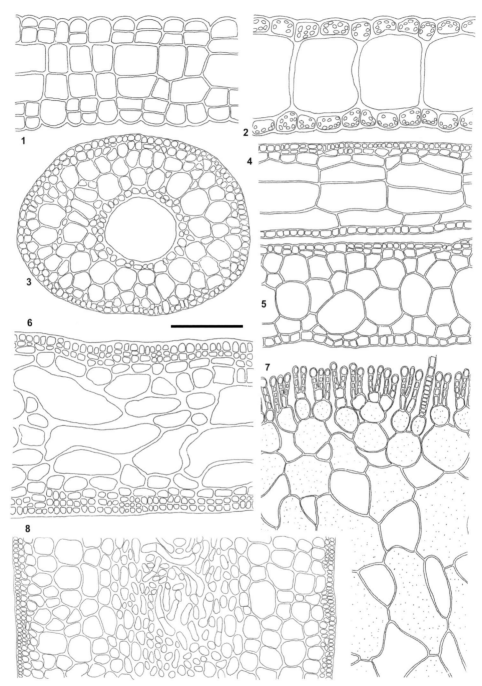

Fig. IX. Examples of thallus structure 1. *Punctaria latifolia*. TS of blade showing no obvious internal differentiation into medulla and cortex. 2. *Dictyota dichotoma*. TS of blade showing a single-celled central medulla enclosed by a single, smaller layer of pigmented cortical cells. 3. *Desmarestia viridis*. TS of thallus showing central axial cell and surrounding broad zone of cortical cells. 4–5. *Zanardinia typus*. VS of thalli showing central region of large, colourless, longitudinally elongate medullary cells enclosed by 1–2 layers of smaller cortical cells. 6. *Petalonia fascia*. TS of thallus showing a central medulla of large, colourless cells enclosed by 1–3 layers of smaller, cortical cells. 7. *Leathesia marina*. VS of thallus showing large, irregularly contorted medullary cells which become smaller peripherally and terminate in short, filamentous paraphyses. 8. *Laminaria digitata*. TS of blade showing central medulla with longitudinal and transverse filaments, surrounded by a cortex of large, more rounded cells and enclosed by 1–2 layers of epidermal cells. Bar = 50 μm (2, 6), = 63 μm (1), = 100 μm (4, 5, 7, 8), = 125 μm (3).

generally comprises an inner 'medulla' and an outer 'cortex'. Medullary cells are usually large, longitudinally elongate, variable in thickness, colourless or with few plastids, and have storage, translocatory and/or mechanical functions. Cortical cells are normally grouped to form a peripheral sub-epidermal layer; they are smaller, fairly thick walled, pigmented and are responsible for much of the photosynthetic activity of the thallus. The relative proportions of medulla and cortex varies for different species. For example, *Dictyota* has a single layer of medullary cells enclosed by a single layer of cortical cells. In *Desmarestia* the medulla is a single, central, axial cell which is surrounded by a broad zone of cortical cells. In *Petalonia*, *Punctaria*, *Leathesia*, *Asperococcus* and many Chordariacean algae, the medulla is a broad zone several cells in thickness, whilst the cortex is only 1–3 cells thick. Very often, further differentiation of cells occurs in these zones; for instance, the broad cortical zone in *Desmarestia* has inner, large, colourless cells and outer smaller, pigmented cells, whilst in *Fucus* and *Laminaria* an inner and outer cortex is generally recognized. In many algae, especially members of the Chordariaceae, some internal differentiation of the medullary tissue is described, with large inner cells and small peripheral cells. However, as for the red algae (Dixon & Irvine 1977), there is fundamentally very little difference between 'medulla' and 'cortex'; the assignation of tissues to one or other of these is frequently arbitrary.

In the structurally more elaborate thalli of the fucoids and the kelps, the internal organization of the tissues is more complex and can, in the case of *Laminaria*, be highly specialized, involving a number of cell types and structural modifications. The epidermal and sub-epidermal regions may be meristematic in function, the products of cell division either contributing to the bulk of the thallus or to its surface area, depending upon the plane of division. The superficial meristem is usually termed a meristoderm. In many of the kelps, distinction is made between primary meristoderm, which is located at the junction of frond and stipe and which is responsible for frond growth and stipe elongation, and secondary meristoderm on the stipe, which is responsible for increase in stipe growth and haptera formation. Cells arising from tangential cell divisions give rise to the cortex from which, in turn, hyphal cells may be borne. These may fuse and interconnect adjacent cells, or spread radially and longitudinally, probably enhancing the conducting and mechanical properties of the thallus, as well as contributing, at the base, to the attachment rhizoidal filaments. Other noteworthy features in *Fucus* and/or *Laminaria* include gelatinization of the medullary cell walls, the formation of trumpet-hyphae and sieve tubes in the medullary cells, the occurrence of mucilage canals, and the formation of growth rings by seasonal secondary meristoderm activity. More details of these tissue types can be obtained from Fritsch (1945).

Despite the increased anatomical specialization of the pseudoparenchymatous and parenchymatous thalli, remnants of a more simple, basic construction can still often be associated with both. For example, many erect, parenchymatous thalli (e.g. *Petalonia*, *Scytosiphon*, *Punctaria*) and pseudoparenchymatous thalli (e.g. *Desmarestia*) clearly originate from the development of a single, erect filament. Moreover, these different modes of construction are not mutually exclusive, and this is well exemplified in those members with a heteromorphic, biphasic life history or with pronounced basal development. The large, structurally elaborate parenchymatous and pseudoparenchymatous sporophytes of the Laminariales and Desmarestiales respectively alternate with simple, branched, filamentous, microscopic gametophytes. The life histories of *Petalonia* and *Scytosiphon* include both erect parenchymatous thalli and pseudoparenchymatous *Stragularia*-like and/or *Microspongium*-like crustose phases. The basal microthalli associated with the parenchymatous thalli of *Punctaria* and *Asperococcus* are discoid and pseudoparenchymatous like those of *Chilionema* and *Hecatonema* spp., whilst branched, fertile *Streblonema*-like, filamentous thalli have also been associated with the less-developed, parenchymatous genera *Myriotrichia* and *Pogotrichum*.

Nevertheless, up until recently, the mode of construction of the brown algae was a fundamentally important taxonomic characteristic. It was the main criterion used in the delimitation of families and, together with aspects of the reproduction and life history, formed the basis of classi-fication at the ordinal level. Members of the Ectocarpaceae were characterized by filamentous thalli, the Lithodermataceae, Chordariaceae, Desmarestiaceae and Arthrocladiaceae were characterized by pseudoparenchymatous thalli, whilst the Scytosiphonaceae, Laminariaceae, Sphacelariaceae, Dictyotaceae and Fucaceae were characterized by parenchymatous thalli.

Thallus growth and longevity

Growth is the result of cell expansion which normally follows cell division. In the brown algae, cell division may be either diffuse or localized (meristematic), with the meristems apical or intercalary in position (Fig. X). Examples of algae with diffuse or random growth include filamentous genera in the Ectocarpaceae, such as *Ectocarpus*, and in the Acinetosporaceae, such as *Hincksia*, as well as pseudoparenchymatous genera, such as *Microcoryne* and *Chordaria*, and erect, thalloid genera, such as *Petalonia*, *Scytosiphon* (Scytosiphonaceae) and *Punctaria* and *Asperococcus* (Chordariaceae). Intercalary growth may be trichothallic, i.e. resulting from cell division at the base of a long, terminal, hair-like filament which contributes cells both above and below; sometimes a single filament is involved, as in *Arthrocladia* and *Desmarestia*; or sometimes a tuft of filaments is involved, as in *Sporochnus*, *Carpomitra*, *Cutleria* and *Zanardinia*. Algae with a tufted, trichothallic growth mechanism are usually pseudoparenchymatous in construction (e.g. *Desmarestia*, *Elachista*); however, in *Cutleria* and *Zanardinia*, fusion of the submeristematic cells, followed by longitudinal divisions, results in a parenchymatous type of construction. The intercalary meristems in other species occupy diverse positions in the thallus: in kelps, such as *Laminaria*, the primary meristem (often referred to as the transition zone) is situated between (and contributes cells to) the blade and the stipe. In *Chorda filum*, also positioned in the Laminariales, a detailed study by Kogame & Kawai (1996) revealed several growth stages, beginning with diffuse growth and accumulating in the development of a well-defined intercalary meristem. In both the kelp and fucoid algae, an additional peripheral meristem (meristoderm) has been demonstrated in the outer cortical tissues which contributes to both blade and stipe growth (usually referred to as primary and secondary meristoderms). In the great majority of brown algae, including filamentous, pseudoparenchymatous and parenchymatous forms, growth of the thallus body is at the apex. Either a single apical cell is involved, as in *Sphacelaria*, *Dictyota* and *Fucus*, or several apical cells are involved. These latter cells are laterally conjoined, as in pseudoparenchymatous genera such as *Leathesia* and *Pseudoralfsia* and parenchymatous genera such as the *Aglaozonia* phase of *Cutleria*, or they are aggregated into a meristem, as in the Australasian genera *Hormosira* and *Notheia*.

In general, diffuse growth is more characteristic of the structurally simple and repro-ductively less elaborate algae and possibly representative of some early divergent brown algal lineages (Graham & Wilcox 2000), although in the case of the Ectocarpales it has been suggested that the simple construction and intercalary growth could be secondarily derived (de Reviers & Rousseau 1999). In contrast, apical growth, particularly involving a single apical cell, followed by cell divisions in various planes is characteristic of the more structurally complex, parenchymatous algae. Algae with trichothallic growth are generally intermediate in structural and reproductive complexity. These growth mechanisms are not mutually exclusive. For instance, diffuse growth is characteristic of the erect thallus portions in heterotrichous ectocarpoid algae, such as *Ectocarpus*, *Hincksia* and *Acinetospora*, but the basal system expands by apical growth; whilst in algae such as *Desmarestia* and *Arthrocladia*, the erect macroscopic sporophyte exhibits trichothallic growth, but the microscopic gametophyte has apical growth. Furthermore, Scytosiphonaceae members (e.g. *Petalonia*, *Scytosiphon*) and Chordariaceae members (e.g. *Punctaria*, *Asperococcus*) have erect blades with diffuse growth but basal systems and/or microthalli with apical growth (*Pseudoralfsia/Microspongium* and

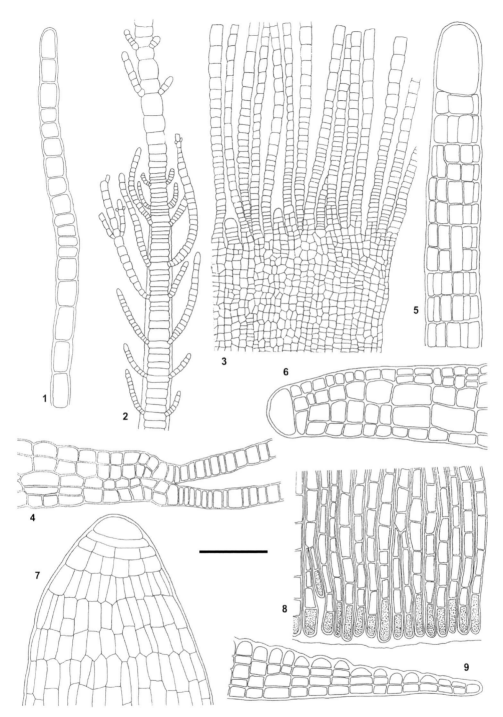

Fig. X. Examples of growth mechanisms 1. *Herponema velutinum*. Terminal portion of erect filament showing an intercalary meristem. 2. *Arthrocladia villosa*. Thallus apex showing trichothallic growth (intercalary meristem with terminal hair). 3. *Zanardinia typus*. Surface view of thallus edge showing trichothallic growth with terminal tuft of filaments. 4. *Zanardinia typus*. VS through thallus edge showing trichothallic growth (note two superimposed extending hair-like filaments). 5. *Battersia plumigera*. Branch apex showing large apical cell. 6. *Cutleria multifida* ('*Aglaozonia*' phase). VS through thallus edge showing large apical cell. 7. *Dictyota dichotoma*. Branch apex showing dome-shaped apical cell. 8. *Chilionema ocellatum*. Surface view of thallus edge showing marginal row of apical cells. 9. *Stragularia clavata*. VS through crust edge showing terminal apical cell. Bar = 50 μm (1, 4–8, 9), = 100 μm (3), = 200 μm (2).

Hecatonema-like respectively). A further example of divergent growth mechanisms is provided by *Cutleria* which has erect thalli with trichothallic growth and prostrate *Aglaozonia*-like thalli with apical growth.

Considerable differences in rates of growth have been reported in brown algae. Growth is particularly slow in crustose algae and very rapid in some of the larger algae, especially members of the Laminariales and Fucales. For example, Dethier (1981) recorded a slow growth rate for *Petalonia fascia* with the crusts only enlarging to 1 cm during early autumn and winter on the Washington coast of north-west America, whilst the present author (unpublished) observed the diameter of *Pseudoralfsia verrucosa* crusts in Kent, south coast of England, to increase by 4–6 mm over a five-month growth period between March and August. These results are in contrast with the maximum growth rates recorded for Atlantic species of *Laminaria* (1.5 cm day^{-1} – Kain 1979), *Saccharina latissima* (2.3 cm day^{-1} – Parke 1948), *Sargassum muticum* (4 cm day^{-1} – Jephson & Gray 1977), *Chorda filum* (4.9 cm day^{-1} – South & Burrows 1967) and *Saccorhiza polyschides* (4.7 cm week^{-1} – Norton 1969).

Great variation is also shown in the longevity of brown algae, ranging from short-lived ephemerals of only a few weeks' lifespan to long-lived perennials, many of which survive for several years and, in some cases, decades. For example, *Laminariocolax tomentosoides* may become fertile at 7 days (Russell 1964b), *Streblonema oligosporum* was reported to reach reproductive maturity in culture after only 9 days at 20°C (Fletcher 1983), whilst reported lifespans include 7–12 months for *Petalonia fascia* (Dethier 1981), up to 12/13 weeks for *Scytosiphon lomentaria* (Clayton 1981), a maximum of one year for *Pylaiella littoralis* (Knight 1923), approximately one year for *Saccorhiza polyschides* (Norton & Burrows 1969a,b), not exceeding 3 years for *Saccharina latissima*, 5–7 years, possibly up to 15 years for *Laminaria hyperborea* (Kain 1963, 1979), whilst *Ascophyllum nodosum* thalli can possibly survive up to 30–35 years (Lüning 1990). However, for many brown algae, determination of longevity of the thalli is complicated by the modifying influence of environmental conditions. These can include, for example, substratum availability; the lifespan of many epiphytic/endophytic algae will be very largely determined by that of the host thallus. This particularly applies to species with only a limited range of possible host organisms; examples include *Ulonema rhizophorum* epiphytic on the erect thalli of *Dumontia contorta*, and *Elachista scutulata*, epiphytic on the thongs of *Himanthalia elongata*. For many algae, and in particular short-lived (ephemeral) species and those with distinct seasonal distribution patterns, the longevity of the thallus is also very largely determined by environmental conditions, such as temperature, light intensity, photoperiod, competition and grazer activity. An increasing number of experimental physiological studies have shown algae able to detect and respond to very subtle changes in a wide range of environmental parameters which very largely determine their lifespan. Further, many of those algae that apparently disappear during unfavourable periods are now known to die back to perennating basal regions/fragments. These are generally hardier than the erect portions and retain the ability to initiate the latter with the approach of more favourable conditions. Examples of these life history stages include perennating *Stragularia*-like thalli in *Petalonia fascia* (Roeleveld *et al.* 1974), *Compsonema*-like thalli in *Planosiphon filiformis* (Fletcher 1981a), *Hecatonema*-like thalli in *Punctaria* and *Asperococcus* spp. (Fletcher 1984; Pedersen 1984), stoloniferous thalli in members of the Dictyotales (Richardson 1979) and basal shoot portions in *Sargassum muticum* (Jephson & Gray 1977).

Environmental influences on thallus establishment and development

Marine macroalgae are subject to the influence of a wide range of environmental conditions which play a major role in aspects of their recruitment, growth, seasonality, distribution, physiology and morphogenesis and, therefore, in their final community structure. Most widely investigated are the effects of light, temperature and salinity on the algae, but attention has also been paid to other aspects such as nutrients, wave action and water movement, sedimentation, grazer activity, the surface properties of substrata and sand/ice scouring. The following sections provide a summary of the literature relating to the effect of these environmental conditions on aspects pertaining to the biology of the Phaeophyceae and act as a supplement to the earlier reviews by Biebl (1937, 1938, 1939, 1958), Den Hartog (1967), Gessner (1970), Gessner & Schramm (1971), Hellebust (1970), Hutchins (1947), Lüning (1981a), and later reviews by Breeman (1988), van den Hoek *et al.* (1990), Lüning (1990), Russell (1987a, 1988) and Wilkinson (1980).

Light and temperature

Table 1 summarizes research papers into the effects of irradiance levels, light quality, daylength/ photoperiod and temperature on brown algal genera.

   Most of these studies involved ecologically dominant brown algae, either wracks from the eulittoral or kelps from the sublittoral, species of economic importance, either as food or as a source of hydrocolloids, or introduced species which may have impacts on the native flora and fauna. With the kelps they included studies both on the sporophyte and gametophyte generations, whilst for the wracks and fucoids they mainly involved embryos and young germlings, sometimes discs or segments of young thalli. Principally, they involved studies on the effect of different temperatures and irradiance levels on germination, growth rates, photosynthesis and photoinhibition which contributed to a better understanding of the ecology of the species, their geographical distribution, depth penetration and distribution in the sublittoral, and their stress/tolerance levels. A number of studies also considered UV light effects on algal pigments, as well as the effect of daylength/photoperiod and light quality on photosynthesis and on growth rates.

   A considerable number of photoperiodic or daylength responses have been described for the Phaeophyceae (see earlier reviews of experimental studies in Breeman 1988; Dring 1984, 1988; Dring & Lüning 1983; Lüning 1980a, 1981a,b, 1990; Lüning & Tom Dieck 1989). In addition to influencing photosynthesis and growth rates, photoperiod/daylength, often in conjunction with temperature and in the presence of adequate nutrient resources, has also been shown to exert a marked influence on erect thallus initiation and development in brown algae with heteromorphic life histories. For instance, they have been reported to determine new frond formation and growth in the kelp species *Laminaria hyperborea* (Lüning 1971, 1986), *L. setchellii* (Tom Dieck 1991) and *Pterygophora californica* (Lüning & Kadel 1993), as well as erect thallus formation in *Petalonia fascia*, *Planosiphon zosterifolius* and *Scytosiphon lomentaria* (Dring & Lüning 1975b; Lüning 1980a), *Punctaria tenuissima* (as *Desmotrichum undulatum*) (Lockhart 1982), *Analipus japonicus* (Nakahara 1984) and *Sphaerotrichia divaricata* (Novaczek & McLachlan 1987). However, the effect of photoperiod on sporophyte production from gametophytes in *Laminaria digitata* was more ambiguous (Ratcliff *et al.* 2017). Photoperiod, combined with temperature, was also reported to influence stipe differentiation and/or erect thallus development/branch elongation in a number of fucoid species including *Sargassum muticum* (Hwang & Dring 2002; Uchida *et al.* 1991), *Sargassum horneri* (Uchida 1993) and *Hizikia fusiformis* (Park *et al.* 1995). In general, the results of these studies were in agreement with the reported seasonal phenology of the species in the field (see Arenas & Fernández 1998; Bartsch *et al.* 2008; Chapman & Craigie 1978; Clayton 1976; DeWreede 1978; Hwang & Dring 2002; Kain 1979; Lüning 1971, 1991; Lüning & Kadel 1993; Novaczek & McLachlan 1987; Parke 1948; Pérez 1969; Schaffelke & Lüning 1994; Tom Dieck 1991, 1994; Toth & Pavia 2002; Uchida 1993; Uchida *et al.* 1991), with short days/low temperatures and long days/high temperatures associated with autumn/winter and spring/summer appearances respectively. Some studies on *Scytosiphon lomentaria* (Lüning 1980a, 1981b) and *Punctaria tenuissima* (as *Desmotrichum undulatum*) (Lockhart 1982) also revealed that genetically different (photoperiodic) ecotypes existed with different developmental responses to temperature and photoperiod which appear to be adapted to their geographical positions. Having different development responses to

Table 1. Summary of studies on the effects of light and temperature on genera of the Phaeophyceae.

Genus	References
Acinetospora	Amsler 1984
Alaria	Fredersdorf *et al.* 2009; Han 1994; Han & Kain 1993, 1996; Lüning & Neushul 1978; Nigan *et al.* 2014; Roleda 2009; Roleda *et al.* 2006
Ascophyllum	Bacon & Vadas 1991; Sheader & Moss 1975; Strömfelt 1978
Chorda	Novaczek *et al.* 1986b
Chordaria	Probyn 1981
Cladosiphon	Fukumoto *et al.* 2018
Costaria	Na *et al.* 2016
Desmarestia	Chapman & Burrows 1970; Fortes & Lüning 1980; Nakahara 1984
Desmotrichum	Lockhart 1982; Rhodes 1970; Rietema & van den Hoek 1981
Eckloniopsis	Komazawa *et al.* 2015
Ectocarpus	Boalch 1961; Bolton 1983; Edwards 1969; Edwards & van Baalen 1970; Schmid & Dring 1992; Schmid & Hillrichs 2001
Egregia	Lüning & Neushul 1978
Eisenia	Maegawa *et al.* 1987; Matson & Edwards 2007
Ecklonia	Altamirano *et al.* 2004; Bolton & Anderson 1987; Bolton & Levitt 1985; Maegawa *et al.* 1987; Mohring *et al.* 2013; Serisawa *et al.* 2001; Staehr & Wernberg 2009; Wood 1987
Fucus	Altamirano *et al.* 2003, 2004; Bird *et al.* 1979; Chapman & Fletcher 2002; Coleman & Brawley 2005; Fortes & Lüning 1980; Hellebust 1970; Huppertz *et al.* 1990; Major & Davison 1998; McLachlan 1974; McLachlan & Bidwell 1983; Munda 1977; Norton *et al.* 1981; Oltmanns 1922; Rohde *et al.* 2008; Steen 2003a; Steen & Scrosati 2004; Strömgren 1978
Glossophora	Hoffmann & Malbrán 1989
Halidrys	Moss & Sheader 1973
Halosiphon	Novaczek *et al.* 1986b
Hedophyllum	Lüning & Neushul 1978
Hincksia	Edwards 1969; Edwards & van Baalen 1970; Fletcher 1981b
Hizikia	Liu *et al.* 2010; Park *et al.* 1995; Wang *et al.* 2017; Zhao *et al.* 2015; Zou *et al.* 2012
Laminaria	Bartsch *et al.* 2008, 2013; Biskupa *et al.* 2014; Boden 1979; Bolton & Lüning 1992; Chapman & Craigie 1978; Cosson 1973a,b, 1975, 1976, 1977; Cosson *et al.* 1976; Delebecq *et al.* 2011, 2016; Fortes & Lüning 1980; Gómez Garreta & Lüning 2001; Han 1993, 1994; Han & Kain 1991; 1992a,b, 1993, 1996; Harries 1932; Kain 1964b, 1965, 1969, 1979; Kain & Jones 1964; Lizumi & Sakanishi 1994; Lüning 1971, 1980b, 1986; Lüning & Neushul 1978; Machalek *et al.* 1996; Martins *et al.* 2017; Mizuta *et al.* 2007; Müller *et al.* 2008; Pereira *et al.* 2011, 2015b; Pérez 1971; Roleda 2009; Roleda *et al.* 2006; Sjøtun & Schoschina 2002; Tom Dieck 1991, 1992, 1993; Toth & Pavia 2002; Ueda 1929; Yhang *et al.* 2013
Lessonia	Hoffmann & Santelices 1982; Martínez 1999; Nelson 2005; Oppliger *et al.* 2011, 2012
Macrocystis	Cie & Edwards 2008; Deysher & Dean 1986; Fain & Murray 1982; Gerard 1986; Graham 1996; Ladah & Zertruche-González 2007; Lüning & Neushul 1978; Reed & Foster 1984; Tala *et al.* 2007
Nereocystis	Vadas 1972
Pelagophycus	Fejtek *et al.* 2011
Pelvetia	Pfetzing *et al.* 2000; Strömfelt 1978
Petalonia	Edwards & Baalen 1970; Hsiao 1970
Pterygophora	Cie & Edwards 2008; Lüning 1991; Lüning & Kadel 1993; Lüning & Neushul 1978; Matson & Edwards 2007

Genus	References
Saccharina	Andersen *et al.* 2013; Bartsch *et al.* 2008; Davison 1987, 1991; Davison *et al.* 1991; Gevaert *et al.* 2003; Han & Kain 1993, 1996; Hanelt *et al.* 1997; Hwang *et al.* 2017; Gerard 1988, 1997; Gerard & Du Bois 1988; Hsiao & Druehl 1971; Lee & Brinkhuis 1988; Liu & Pang 2009; Lüning 1980b; Lüning & Dring 1975; Machalek *et al.* 1996; Norton 1977a; Pereira *et al.* 2011; Roleda 2009; Roleda *et al.* 2006; Sjøtun & Schoschina 2002; Wang *et al.* 2010; Zhang *et al.* 2013
Saccorhiza	Biskupa *et al.* 2014; Norton 1977a; Norton & Burrows 1969a,b; Pereira *et al.* 2011, 2015b
Sargassum	Arai & Miura 1991; Engelen *et al.* 2008; Hales & Fletcher 1989, 1990; Lee & Brinkhuis 1988; Murase *et al.* 2002: Norton 1977b; Pereira *et al.* 2011, 2015; Steen 2003b, 2004; Uchida 1993; Uchida *et al.* 1991; Zhao *et al.* 2013; Zhu & Chen 1997
Scytosiphon	Clayton 1976; Dring 1987; Dring & Lüning 1975a,b; Tom Dieck 1987
Sphaerotrichia	Novaczek & McLachlan 1987
Undaria	Akiyama 1965; Choi *et al.* 2005; Baba 2008; Floc'h *et al.* 1991; Henkel & Hofmann 2008; Kinoshita & Shibuya 1944; Morita *et al.* 2003a,b; Na *et al.* 2016; Pang & Wu 1996; Pang *et al.* 2008; Saito 1956a,b; Segi & Kita 1957

temperature and photoperiod coupled with appearing and competing for resources at different times of the year, as shown by Arenas *et al.* (1995) for the two ecologically dominant, competitive brown algae *Sargassum muticum* and *Gongolaria nodicaulis*, suggests a mutually beneficial strategy.

There are also a number of reports revealing the effect of the above environmental conditions, notably temperature, light intensity, photoperiod/daylength and light quality, acting either single or jointly on the morphogenesis of some brown algae. Environmental influences include:

Light intensity
Light intensity was shown to exert an effect on size (Hellebust 1970; Norton *et al.* 1981) and morphological expression, e.g. blade width, in *Fucus vesiculosus* (Oltmanns 1922) and *Saccharina latissima* (Burrows 1964).

Photoperiod/daylength
A number of studies (see particularly Clayton 1976; Edwards 1969; Fletcher 1978; Nakamura & Tatewaki 1975; Roeleveld *et al.* 1974; Wynne 1969) have also demonstrated that the combination of photoperiod/daylength and temperature not only influences thallus development in members of the Scytosiphonaceae but also determines the morphogenetic response with respect to the production of either the crustose or erect bladed phases. A similar temperature/daylength control of morphological expression (microscopic and prostrate or macroscopic and erect) has also been shown in members of the Chordariaceae (Lockhart 1982; Loiseaux 1969; Rhodes 1970; Rietema & van den Hoek 1981). However, Pedersen (1984) has shown that macrothalli of *Punctaria tenuissima* (as *Desmotrichum undulatum*) may develop from basal systems, irrespective of temperature or photoperiod, implying that genetically different strains varying in their morphogenetic response are present in the North Atlantic.

Temperature
A few investigations have revealed temperature to act independently of photoperiod in brown algal development and morphogenesis. For example, Pedersen (1984) demonstrated that macrothallus production in *Asperococcus fistulosus* was apparently more frequent at low temperatures, irrespective of the photoperiod, whilst different forms of *Streblonema immersum* were produced in low and high temperatures. Rietema & van den Hoek (1981) also reported temperature to be the main modifier of morphology in *Punctaria tenuissima* (as *Desmotrichum undulatum*).

Light quality

A number of reports have revealed light quality, particularly the red and blue wavebands, to exert a control on the morphogenesis of some brown algal species. For example, Dring & Lüning (1975b) and Lüning & Dring (1973) observed that the development of *Scytosiphon lomentaria*, *Petalonia fascia* and *Planosiphon zosterifolius* (as *Petalonia zosterifolia*) in laboratory culture differed in red and blue light. Sporelings developed into *Ralfsia*-like, strongly attached, crustose thalli (profusely branched filamentous thalli in *P. zosterifolius*) in blue light but poorly attached, sparsely branched, filamentous thalli in red light. In addition, the production of erect thalli from the basal systems was restricted to or stimulated by red light only. However, no requirement for either red or blue light for erect blade formation in *Punctaria tenuissima* (as *Desmotrichum undulatum*) was noted by Lockhart (1982). Blue and red light was also observed to induce differential phototrophic effects on the polarity and germination process in *Tilopteris* eggs (Kuhlenkamp 1989) and on haptera growth in *Alaria esculenta* (Nigan *et al.* 2014), to affect growth and photosynthesis and morphogenesis (stipe length and blade dimensions) in *Saccharina latissima* by Lüning & Dring (1975) and Lüning & Markham (1979) respectively, and to influence sporophyte morphology in *Laminaria japonica* (Mizuta *et al.* 2007). Blue light was also shown to be required for optimal growth and differentiation in *Dictyota dichotoma* (Müller & Clauss 1976), whilst Dring (1987) reported that in *Alaria esculenta*, *Fucus* spp. and *Laminaria* spp. the light saturated rate of photosynthesis in blue light was much higher than in red light. A similar requirement for blue light was also shown for photosynthesis and growth in *Fucus* embryos and apices by McLachlan & Bidwell (1983). Finally, blue light was shown to be required for hair formation in *Scytosiphon lomentaria* (Dring & Lüning 1983), branch formation in *Bachelotia antillarum* (Shanab *et al.* 1988), and to enhance recovery/survival of sporophytes of *Laminaria hyperborea* after exposure to damaging UV irradiation (Han & Kain 1991).

Salinity

Investigations of the effects of salinity on survival and growth have included the following brown algal genera: *Asperococcus* (Pedersen 1984); *Chorda* (Russell 1985b, 1988); *Ectocarpus* (Nygren 1975b; Russell & Bolton 1975), *Fucus* (Bird *et al.* 1979; Khfaji & Norton 1979; Munda 1964, 1967a,b, 1977, 1978; Rothäusler *et al.* 2016; Russell 1987a); *Litosiphon* (Nygren 1975b); *Pylaiella* (Bolton 1979; Nygren 1975b; Reed & Barron 1983); *Saccharina* (Peteiro & Sánchez 2012); *Sargassum* (Arai & Miura 1991; Hales & Fletcher 1989; Steen 2004); *Scytosiphon* (Ohno 1969); *Striaria* (Nygren 1975b); *Undaria* (Baba 2008; Peteiro & Sánchez 2012). Most experimental studies involved investigating the effect of different salinities on the more vulnerable germling stages of the algae, and in the case of kelps this included the gametophyte generation. For most of these genera, growth rates and net photosynthesis occurred optimally at salinities around 30 ppt. However, some species and brackish-water ecotypes have been shown to grow better at low salinities, for example *Fucus ceranoides* (Khfaji & Norton 1979), *Fucus serratus* (Bird *et al.* 1979) and *Ectocarpus siliculosus* (Russell & Bolton 1975).

In certain brown algal species, reduced salinity may also be associated with a reduction in thallus size and with other morphological and anatomical changes (Gessner & Schramm 1971; Munda 1978; Norton *et al.* 1981; Pedersen 1984; Wilkinson 1980). This has been particularly well documented in fucoids, where reduced salinity has been linked with modified branching patterns (Jordan & Vadas 1972) and vesiculation (Alexander *et al.* 1935; Den Hartog 1967; Jordan & Vadas 1972). However, it is evident that the morphological impact of reduced salinity is not a simple one as differences in response have been noted between species and between cell types of the same species (Russell 1985a). The influence of reduced salinity on Baltic algal morphogenesis has been discussed by Waern (1952) and Russell (1985a). The ways in which algae respond to changes in external salinity have been reviewed by Russell (1985b, 1986).

Nutrients

The availability of nutrients, particularly of nitrogen, is a key factor influencing growth and productivity of macroalgae (Fong 2008; Hanisak 1983; Lobban & Harrison 1997). Table 2 summarizes research papers into the effects of nutrients on brown algal genera. Most of these studies, using both field and laboratory techniques, were concerned with the effects of nutrients, particularly nitrate and phosphate and their concentrations, on the growth rates of the macroalgae, as well as on photosynthetic rates and physiological/biochemical processes, contributing to a better knowledge and understanding of the biology and seasonal occurrence of the algae, as well as assisting in the mariculture of some species. Nutrient levels have also been linked with the detrimental effects on macroalgae in eutrophicated waters (Bergström *et al.* 2003; Conolly & Drew 1985; Korpinen & Jormalainen 2008a), on the morphogenesis of *Desmotrichum undulatum* and *Cladosiphon zosterae* by Lockhart (1982) and Lockhart (1979) respectively, on hair formation in *Fucus* (Steen 2003a) and blade/crust development in *Petalonia* (Hsiao 1969).

Table 2. Summary of studies on the effects of nutrients on genera of the Phaeophyceae.

Genus	References
Alaria	Lewis *et al.* 2012
Ascophyllum	Asare & Harlin 1983
Chordaria	Probyn 1981, 1984; Probyn & Chapman 1982, 1983
Desmotrichum	Lockhart 1982
Dictyoneurum	Lewis *et al.* 2012
Egregia	Lewis *et al.* 2012
Fucus	Asare & Harlin 1983; Bergström *et al.* 2003; Korpinen & Jormalainen 2008a; Korpinen *et al.* 2007; Steen 2003a; Steen & Scrosati 2004
Laminaria	Biskupa *et al.* 2014; Chapman & Craigie 1977, 1978; Chapman *et al.* 1978; Conolly & Drew 1985; Harries 1932; Kregting *et al.* 2015; Lewis *et al.* 2012; Martins *et al.* 2017; Tom Dieck 1991; Yarish *et al.* 1990; Yoth & Pavia 2002
Lessonia	Hoffmann & Santelices 1982; Hoffmann *et al.* 1984
Macrocystis	Carney 2011; Carney & Edwards 2010; Gerard 1982; Hepburn *et al.* 2007; Lewis *et al.* 2012
Petalonia	Hsiao 1969
Saccharina	Biskupa *et al.* 2014; Chapman *et al.* 1978; Conolly & Drew 1985; Gao 2017; Gerard 1997; Hsiao & Druehl 1973a,b; Nielson *et al.* 2016; Wheeler & Weidner 1983
Sargassum	Gao & Nakahara 1990; Steen 2003b

Degree of exposure to wave action and water movement

In addition to the well-documented effect of wave action and water movement on the distribution and composition of benthic algal communities (Harley & Helmuth 2003; Jones & Demetropoulos 1968; Lewis 1964; Lüning 1990; Scrosati & Heaven 2008; Vadas *et al.* 1990), with wave action strongly influencing the community structure of the eulittoral and currents more influential in the sublittoral, these forces have also been shown to influence both the growth rate and biomass as well as the morphology of algae, including species of Phaeophyceae. Both Andrew & Viego (1998) and Engelen *et al.* (2005) recorded *Sargassum polyceratium* thalli to be much bigger and to achieve higher biomass in more exposed bays compared to sheltered bays. Similar higher values with respect to growth rate and biomass achievements were also recorded in more exposed sites for the kelps *Undaria pinnatifida* and *Laminaria hyperborea* by Peteiro & Freire (2011a) and Sjøtun *et al.* (1998) respectively. Some algae such as *Sargassum muticum*, however, have been shown to be more abundant in sheltered regions (Viejo *et al.* 1995) whilst mean sizes of wave-exposed *Fucus gardneri* increased significantly when transported to protected sites (Blanchette 1997).

Morphological changes in macroalgae as a result of wave exposure include the occurrence of the rather distinctive exposed form of North Atlantic *Fucus vesiculosus* usually identified as *F. vesiculosus* f. *linearis*. Other morphological changes in brown algae as a result of wave action have been reported in *Ascophyllum nodosum* (Cousens 1982), *Ecklonia maxima* (Jarman & Carter 1981), *E. radiata* (Wernberg & Thomsen 2005), several *Laminaria* species, including the *L. cucullata* growth form of *L. digitata* (Svendsen & Kain 1971), *L. longicruris* (Chapman 1973; Egan & Yarish 1990; Gerard & Mann 1979), *L. groenlandica* (Druehl 1967), *L. hyperborea* (Svendsen & Kain 1971), *L. digitata* (Sharp & Carter 1986; Sundene 1962b), *L. setchellii* (Klinger & DeWreede 1988), *L. schinzii* (Jarman & Carter 1981; Sundene 1962b; Molloy & Bolton 1996), *Hedophyllum* (Armstrong 1989), *Chordaria flagelliformis* (Munda 1979), *Nereocystis luetkeana* (Koehl 1986; Koehl & Alberte 1988), *Saccharina latissima* (Gerard 1987; Kain 1976c), *Saccorhiza polyschides* (Norton 1969), *Sargassum cymosum* (Paula & Oliveira 1982) and *Undaria pinnatifida* (Peteiro & Freire 2011b). The effects of wave exposure have been particularly well documented in kelps, with reports of longer, hollower stipes and an increased blade thickness of the thalli in exposed habitats. Relatively higher wave action also often results in narrower and dissected blades, as shown for example in *Laminaria hyperborea*, and for narrower blades in *Saccharina latissima* (Gerard 1987). For many of the above algae, the differences obtained are probably attributable to phenotypic plasticity and not genetically fixed, as proposed by Svendsen & Kain (1971) for 'Laminaria cucullata', by Sundene (1964) for *Laminaria digitata* and by Norton (1969) for *Saccorhiza polyschides*; however, in *Sargassum* and *Chordaria*, genotypic differentiation was suggested to occur with the evolution of distinct ecotypes. In practice, both sources of variation are likely, and in any given thallus, some morphological attributes will prove to be more plastic than others.

In addition to reports of water movement affecting colonization of substrata, either directly by preventing spore attachment (Fletcher & Callow 1992; Norton & Fetter 1981), or indirectly, by influencing silt deposition (see p. 32), it can also influence the morphological (and sometimes anatomical) development of algae. For instance, algal growth can be more luxuriant and enhanced in conditions of strong water movement (Conover 1968), as shown for *L. digitata* (Kregting *et al.* 2015) and *Macrocystis pyrifera* (Hepburn *et al.* 2007; Stephens & Hepburn 2014), probably as a result of the increased uptake of nutrients and/or dissolved gases at the boundary layer of the surface of the macroalgae (Hiscock 1983; Hurd *et al.* 1996; Kregting *et al.* 2015; Lüning 1990).

Areas that have reduced water movement often support free-living populations of algae which are sometimes morphologically different from the attached forms. For instance, in the sublittoral, loose-lying populations of brown algae such as *Halopteris scoparia* (Waern 1952), *Saccorhiza polyschides* (Norton & Burrows 1969a,b), *Pylaiella littoralis* (Russell 1967b; Wilce *et al.* 1982), *Saccharina latissima*, *Asperococcus* spp. and *Desmarestia* spp. (Burrows 1958; Irvine 1974; Tittley *et al.* 1977) have been described. Elsewhere, a number of very distinct, dwarf forms of members of the Fucales have also been observed. For example, saltmarsh areas in the North Atlantic may contain an ecad (i.e. modified by the environment) of *Ascophyllum nodosum*, formerly given taxonomic status as *A. mackayi*, which is free-living, profusely dichotomously branched and either sterile or with reduced fertility (Fritsch 1945; Gibb 1957; Newton 1931; South & Hill 1970). Another small ecad, previously referred to as var. *scorpioides*, contributes to loose-lying communities in sheltered sublittoral regions (Chuck & Mathieson 1976). Widely variant saltmarsh ecads of *Fucus* spp. have also been described (e.g. *F. vesiculosus* ecad *volubilis*, ecad *caespitosus* and ecad *muscoides* [Fritsch 1945]), as well as of *Pelvetia canaliculata* (Chapman 1939; Oliveira & Fletcher 1980). All the above saltmarsh forms of *Ascophyllum*, *Fucus* and *Pelvetia* would appear to be plastic responses to the combined influences of extreme shelter and low salinity, but inherent differences between them and their marine rocky-shore counterparts may also exist (Moss 1971). Interestingly, in the case of the saltmarsh dwarf species *Fucus cottonii*, molecular evidence by Neiva *et al.* (2012a) has revealed that this species does not represent a single genetic entity but has multiple independent origins.

Grazer activity/herbivory

A number of studies, using various field experiments/observations involving clearance techniques, manipulated grazer entry or cage exclusion methods, have highlighted the important role played by grazing activity and herbivory in controlling and regulating macroalgal establishment, biomass and structure of a number of marine benthic communities (Burrows & Lodge 1950; Chapman 1989; Chapman & Johnson 1990; Coleman *et al.* 2007; Creese 1988; Hawkins 1981; Hawkins & Hartnoll 1983, 1985; Hawkins *et al.* 1992; Jenkins *et al.* 2001; Korpinen & Jormalainen 2008a,b; Korpinen *et al.* 2007, 2008; Lubchenco 1978, 1982, 1986; Lubchenco & Gaines 1981; Lubchenco & Menge 1978; Lüning 1990; Menge 1976; Molis *et al.* 2010; Poore *et al.* 2012, 2014; Underwood & Denley 1984; Worm & Chapman 1998).

Soft filamentous and foliose algae are particularly susceptible to grazer activity because they are easy to digest, have rapid growth rates and energy intake, and do not offer strong resistance to herbivory (Littler & Arnold 1982; Littler & Littler 1980; Lotze *et al.* 1999; Steneck & Dethier 1994). However, a wide range of other, more coriaceous algae, such as the wracks and kelps, can also be affected either by direct grazing (Engkvist *et al.* 2000; Lubchenco 1983; Peteiro & Freire 2012a; Schaffelke *et al.* 1995) or indirectly by controlling the abundance of ephemeral algal competitors and removing competitively superior species (see Chapman 1989; Chapman & Johnson 1990; Dethier 1981; Hawkins & Hartnoll 1983; Valentine 2003).

With respect to the Phaeophyceae, especially in temperate regions, sea urchins and, to a lesser extent, some limpets, amphipods, isopods, prosobranchs, crabs and fish, are important grazers of the ecologically important sublittoral kelp forests (Choat & Black 1979; Dean *et al.* 1988; Gagnon *et al.* 2003; Lawrence 1975; Minchin 1992; Molis *et al.* 2010; Oróstica *et al.* 2014; Peteiro & Freire 2012a; Poore *et al.* 2014; Thornber *et al.* 2004; Vadas 1977; Wheeler 1980). Limpets, snails, isopods and amphipods are also considered to play a major role in regulating the abundance of the perennial, eulittoral wrack species (Andrew 1989; Chapman 1989; Engkvist *et al.* 2000; Goecker & Kåll 2003; Jernakoff 1985; Lubchenco 1982, 1983; Lubchenco & Menge 1978; Rothäusler *et al.* 2017; Schaffelke *et al.* 1995; Underwood 1979), although, as proposed by Chapman & Johnston (1990) for the north-west Atlantic, these consumers play two completely different interactive roles. Field observations and experimental studies have revealed some herbivores to be selective grazers and to exhibit strong brown algal preferences (Barker & Chapman 1990; Britton-Simmons *et al.* 2011; Goecker & Kåll 2003; Jones & Long 2017; Jormalainen *et al.* 2001; Malm *et al.* 1999; Prince & LeBlanc 1992; Schaffelke *et al.* 1995; Scheibling & Anthony 2001; Stimson *et al.* 2007; Steinberg 1985; Vadas 1977), this sometimes being associated with the palatability of the algae (Anderson & Velimirov 1982; Moore 1983; Norton 1978; Vadas 1977) and different algal tissues (Loffler *et al.* 2018).

Some studies reveal herbivore resistance in many brown algae. In addition to physical means of resistance such as that demonstrated by the thick, leathery species (Littler *et al.* 1983a,b) and the crustose phases in the life histories of genera such as *Petalonia* and *Scytosiphon* (Dethier 1981; Littler & Kauker 1984; Littler & Littler 1983; Lubchenco & Cubit 1980; Slocum 1980), various Phaeophyceae have been shown to contain and/or release chemical deterrents. These include high concentrations of sulphuric acid in *Desmarestia* species (Anderson & Velimirov 1982), as well as galactolipids and high levels of a range of polyphenolic compounds such as phlorotannins (Deal *et al.* 2003; Denton & Chapman 1991; Denton *et al.* 1990; Flothe *et al.* 2014; Geiselman & Connell 1981; Haavisto *et al.* 2010, 2017; Kurata *et al.* 1990; Littler *et al.*

1983b; Luder & Clayton 2004; Martínez 1996; Nylund *et al.* 2012; Pavia & Toth 2000; Pavia *et al.* 2003, 2012; Ragan 1976; Sawai *et al.* 1994; Schoenwaelder 2002 and references therein; Steinberg 1985; Targett & Arnold 1998; Tugwell & Branch 1989; van Alstyne 1988; Wikström *et al.* 2006). Phlorotannins are usually concentrated in the outer tissues, for example in the outer meristoderm in kelps (Tugwell & Branch 1989). Some studies have shown that the herbivore grazing induces the chemical defence, probably as a response to chemical signals from the grazers (Coleman *et al.* 2007; Pavia & Toth 2000; Rohde *et al.* 2004; van Alstyne 1988), with this involving gene induction and transcriptional reprogramming (Ritter *et al.* 2017) and a subsequent concentration of the polyphenolic compounds around the wound area (Luder & Clayton 2004; Targett & Arnold 1998; van Alstyne 1988).

Particular attention has also been paid to examining the role played by grazing on introduced brown algae, notably on the large and successfully spreading fucoid *Sargassum muticum* and the kelp *Undaria pinnatifida* (Britton-Simmons *et al.* 2011; Cacabelos *et al.* 2010; Enge *et al.* 2017; Engelen *et al.* 2011; Forslund *et al.* 2010; Monteiro *et al.* 2009b; Parker & Hay 2005; Schwartz *et al.* 2017; Strong *et al.* 2009; Thornber *et al.* 2004; Valentine 2003; Valentine & Johnson 2005a,b). The aim of many of these studies, both field based and in the laboratory, was to determine whether grazers showed any preferences towards the native or introduced species and whether the results would provide strategies for controlling the aliens. In general, most experiments revealed local grazers to show a preference for the native species and that the introduced species showed higher resistance, probably due to their possession of more novel chemical weapons (Cacabelos *et al.* 2010; Engelen *et al.* 2011; Monteiro *et al.* 2009a; Schwartz *et al.* 2017; Wikström *et al.* 2006).

Surface properties of substrata

A wide range of surface properties of substrata have been shown to exert an influence on brown algal development and community structure. These include:

Surface free energy
Studies have demonstrated the importance of the surface free energy of a substratum in the attachment process of macroalgae, including members of the Phaeophyceae. The surface free energy can both modify the morphogenesis of the attachment systems and determine their adhesive properties. For example, Hardy & Moss (1979b) revealed that the primary rhizoid of *Fucus* germlings remained long and thin on low energy Teflon surfaces and were unable to form a secure attachment. Fletcher (1976), Fletcher & Baier (1984) and Fletcher *et al.* (1984, 1985) demonstrated that surfaces of different surface energy clearly produced marked differences in the extent of outward growth, morphogenetic appearance and adhesive strength of the rhizoids for a range of algae, including the brown alga *Hincksia granulosa*. The surface energy of a substratum was also shown by Fletcher *et al.* (1985) to influence morphogenesis of the crustose phases in the life histories of *Petalonia* and *Scytosiphon* spp.: high-energy surfaces produced crustose thalli, low-energy surfaces produced knot filamentous thalli. In view of the wide range of substrata available to macroalgae for colonization, and their probable range of surface energy properties, it seems likely that the latter probably play an important role in the colonization process. This might particularly apply to host–epiphyte relationships (see Linskens 1966, and Pedersen's 1984 report of variable morphology in *Mikrosyphar polysiphoniae*). It is also probable that heteroblasty is caused by surface energy effects (see Christensen 1980, p. 135, for reference to heteroblasty).

Surface texture and topography
Experimental field studies and observations have shown that surface texture and topography can have a marked influence on the establishment, survival and density of macroalgal thalli. For example, roughened surfaces generally appear to support denser growths of macroalgae than

smooth surfaces, probably by providing an increased number of surface planes for attachment and reducing the chances of removal by wave and current action (DeNicola & McIntire 1990; Foster 1975; Gao *et al.* 2017; Harlin 1974; Harlin & Lindberg 1977; Malm *et al.* 2003; Neushul *et al.* 1976). Changes in surface topography can also determine local water circulation and the renewal rates of nutrients with implications for photosynthesis and growth (Foster 1975; Neushul 1972), whilst the water-retaining properties of roughened surfaces are also likely to protect the algae from desiccation during unfavourable periods in the eulittoral zone (Brawley & Johnson 1991; Den Hartog 1959; Nienhuis 1969).

There is also evidence that surface topography can influence the *composition* of the attached macroalgae, with some species exhibiting a preference for attachment and growth on roughened surfaces, whilst others preferred smooth surfaces. For instance, experimental field studies by Harlin & Lindberg (1977), using discs on which were cemented different sized quartz particles, revealed different algal communities on the discs, whilst different communities were also observed on different types of natural stone substrata by Luther (1976) and Stephenson (1961). These studies support numerous field reports of different rock types supporting different algal communities, even when in close proximity to each other, for a number of brown algal-dominated shores (de Nicola & McIntire 1990; Den Hartog 1959, 1972; Fletcher 1977b, 1978; Gao *et al.* 2017; Kornmann & Sahling 1977; Lewis 1964; Malm *et al.* 2003; Nienhuis 1969; Rees 1935; Stephenson 1961; Stephenson & Searles 1960; Tittley 1985a,b, 1986; Tittley & Price 1977a,b, 1978; Wynne 1969), this being more evident in the upper eulittoral regions and on sheltered coasts (Nienhuis 1969) and less evident on exposed coasts and in the lower eulittoral regions (Boalch 1957; Foster 1975).

Whilst these floristic distinctions are usually attributed to differences in settlement patterns and survival rates on the different substrata, there is also evidence that they could be related to the attachment processes of the macroalgae. It is well known that friable or fragile rocks offer poor attachment surfaces to larger algae such as laminarians and some fucoids (McLachlan *et al.* 1987; Moss *et al.* 1973). This probably explains the absence of larger kelps on chalk and the restriction of *Pelvetia canaliculata* to lower greensand and not on chalk around the Kent coast of England, and why *Himanthalia elongata* is restricted in distribution on the magnesium limestone on the north-east coasts of England. Some macroalgae appear restricted to certain rock types – for example, Stephenson (1961) reported that the crustose brown alga *Ralfsia* showed a preference for hard rock. A similar preference for hard substrata such as flint has been observed for the crustose brown alga *Petroderma maculiforme* by the present author.

Although there is some evidence that the reproductive stages of algae can select surfaces on which to settle, the subsequent success of the attaching thallus on surfaces is more likely to be dependent on the interaction between the rhizoidal filaments and cells, especially the released adhesive cement, and the surface properties of the substratum. Rhizoids of different morphology have been described for the germlings of various algae, including *Fucus serratus*, on surfaces of different texture (Hardy & Moss 1979b; Moss 1975; Moss *et al.* 1973; Tovey & Moss 1978). Similar reports of differential behaviour of rhizoids according to substratum type, with respect to morphological development and cement release, have also been reported for red macroalgae including *Gelidium* (Seoane-Camba 1989), *Corallina* (Harlin & Lindbergh 1977) and *Polysiphonia* germlings (reported in Fletcher & Callow 1992).

Surface toxicity

Investigations of toxic surfaces, usually panels and ships coated with copper and organo-tin antifouling paints, generally reveal a reduced marine algal flora which includes a small number of brown algae (Christie 1973; Fletcher 1983; Harris 1943, 1946; Igic 1968; Rautenberg 1960). The most frequently reported brown alga is *Ectocarpus*, which is often present in a reduced 'brown mat' form; its presence is in accord with reports of its high copper resistance (Hall *et al.* 1979; Russell & Morris 1970). One curious aspect of algal colonization of these toxic surfaces is the reporting of several brown algae which have become known almost exclusively as epiphytes or epi-endophytes. These include species of *Microspongium*, *Myrionema*, *Porterinema*, *Streblonema* and *Hecatonema* (Fletcher 1983; Harris 1946; Rautenberg 1960).

Sediment loading
Sediment loading in coastal waters, either through natural storm or flood conditions or as a result of human activities, can have mainly detrimental effects on the recruitment, physiology and morphology of macroalgal species and can remove seaweed forests in which brown algae are usually the dominant contributor. For example, Roleda & Dethleff (2011) demonstrated that although short-term burial of sporophyte discs under different sediment types had no negative effects on the physiology and morphology of *Saccharina latissima*, and indeed had a beneficial effect in protecting the algal samples from being bleached and photoinhibited by high PAR and UVR exposure, long-term burial resulted in an adverse smothering effect leading to bleaching, loss of PSII function and tissue decay. A similar adverse effect, with reduced survival, diminished germination and slower growth rates, could probably also apply to both the gametophyte and sporophyte phases of kelps, as demonstrated for *Eisenia bicyclis* by Arakawa (2005), Arakawa & Matuike (1990, 1992), Arakawa & Morinaga (1994), Arakawa *et al.* (2014) and Watanabe *et al.* (2016), for *Macrocystis pyrifera* by Devinny & Volse (1978), for *Saccharina latissima* by Lyngby & Mortensen (1996), for *Saccorhiza polyschides* by Norton (1978) and for *Undaria pinnatifida* by Suzuki *et al.* (1998). Chapman & Fletcher (2002) also demonstrated that sedimentation affected the survival and growth of germlings in *Fucus serratus*, with a similar effect on recruitment being reported for *Fucus vesiculosus* by Eriksson & Johansson (2003), whilst Gao *et al.* (2019) and Bi *et al.* (2016) revealed sedimentation to adversely affect attachment, survival, growth and distribution of the species *Sargassum thunbergii* and *Sargassum horneri*. With all the above examples representing canopy species, it is likely that sedimentation impacts considerably on coastal communities, a situation that will become more problematic with increased human activity.

Sand burial/scouring
Sand burial and scouring will not only select and favour both stress-tolerant strategists and opportunistic strategists and those algae with adaptable/tolerant basal systems (Daly & Mathieson 1977; Littler *et al.* 1983c; Taylor & Littler 1982), but will also favour survival of the prostrate rather than the erect forms of those algae with heteromorphic life histories, such as *Petalonia* and *Scytosiphon* (Fletcher 1974; Wynne 1969).

Ice scouring
Ice scouring in polar regions can severely impact on the biomass, richness and diversity of eulittoral, mainly fucalean-dominated, marine benthic communities as well as the morphology of individual species (Barnes 1999; Gutt 2001; Lüning 1990; Minchinton *et al.* 1997; Scrosati & Heaven 2006, 2007, 2008; Wilce 1959).

Reproduction

Brown algae reproduce vegetatively, asexually and sexually.

Vegetative reproduction
Vegetative reproduction relies upon the release, distribution and re-establishment of vegetative portions of the thallus body. In the majority of brown algae, this represents a means of dispersal, with the dissemination of thallus portions 'released' by environmental agents, such as grazer activity, wave action, water currents, sand scouring and so forth. Using bottle-brush collectors placed in the sea, Clokie & Boney (1980) managed to entrap a large percentage of the local

flora, suggesting that fragmentation may represent an important means of dissemination. This would particularly apply to the filamentous genera of brown algae such as *Acinetospora* (Fritsch 1945, p. 153) and *Feldmannia* (Etherington 1964), which could easily become entrapped and establish firm attachment by secondary rhizoid production. Fragmentation may be of more importance than sexual reproduction in estuarine conditions where spore/gamete production may be limited (e.g. in *Pylaiella* – Russell 1967a).

There are also a few examples of more specialized vegetative reproductive bodies. These include akinetes, described by Sauvageau (1928) on the thalli of *Tilopteris*, and the variously shaped propagules described for species of the genus *Sphacelaria* (Prud'homme van Reine 1982a) (Fig. XI, 1). The latter, which represent modified branches, are very effective dispersal agents and, for many species, make a major contribution to reproduction (Fritsch 1945).

An additional method of vegetative propagation relies upon the production and growth of new erect shoots from horizontally spreading branches or runners which later often become detached from the parental thallus. For example, some epiphytic/endophytic algae, such as *Myriactula clandestina*, *M. stellulata* and *Cylindrocarpus microscopicus*, often proliferate on their respective host thalli by the production of creeping, endophytic threads, whilst stolon/adventitious branch formation has been widely recorded (see particularly Fritsch 1945) in such genera as *Dictyota*, *Padina*, *Cladostephus*, *Bifurcaria* and even *Laminaria*.

Asexual reproduction

In this method of reproduction, spores are produced which are non-sexual in their activity; they settle without sexual fusion and germinate directly to form a thallus. For the great majority of algae, the spores, often referred to as zoospores or swarmers, are motile. Exceptions include the non-motile monospores (produced in monosporangia) of *Acinetospora* and *Tilopteris* (Fig. XI, 2), and the tetraspores (produced in modified unilocular sporangia called tetrasporangia) in members of the Dictyotales (e.g. *Dictyota*, *Padina*) (Fig. XI, 3).

Spores are produced in two principal types of sporangium: unilocular and plurilocular (Fig. XI, 4–16). In unilocular sporangia, the spores (variously termed unispores, unizoids or meiospores) are produced within the sporangial mother cell by nuclear divisions which are not accompanied by the formation of partitioning walls. In the sporangial mother cell of plurilocular sporangia, on the other hand, the spores (often referred to as zoospores, plurispores or plurizoids) are formed individually in small, walled cells (loculi). The nuclear divisions that occur in the formation of plurilocular sporangia are mitotic, and the plurispores are consequently of the same ploidy level as the parental body. Unilocular sporangia differ in that the initial nuclear division is usually meiotic and results in the formation of haploid spores. Although unilocular sporangia can function as the site of meiosis, this is not always the case in many brown algae and reports of apomeiotic unilocular sporangia have been published (for example in *Isthmoplea sphaerophora* – Rueness 1974, and *Ectocarpus* – Bothwell *et al.* 2010). Unilocular sporangia have also been reported on haploid thalli.

Plurilocular sporangia can function either as asexual organs, releasing spores that germinate directly into a thallus with the parental genotype, i.e. an accessory reproductive system, or behave as gametangia releasing gametes (see below). With this dual function they are sometimes referred to as plurilocular zoidangia producing plurizoids. Unilocular sporangia are, therefore, usually associated with the 'sporophyte' thallus (but see Nakamura & Tatewaki 1975), whilst plurilocular sporangia can be associated with either sporophyte or gametophyte.

Sexual reproduction

This mode of reproduction involves the production and the release of reproductive bodies (gametes) which behave sexually. These may be motile and identical in appearance (isogamous

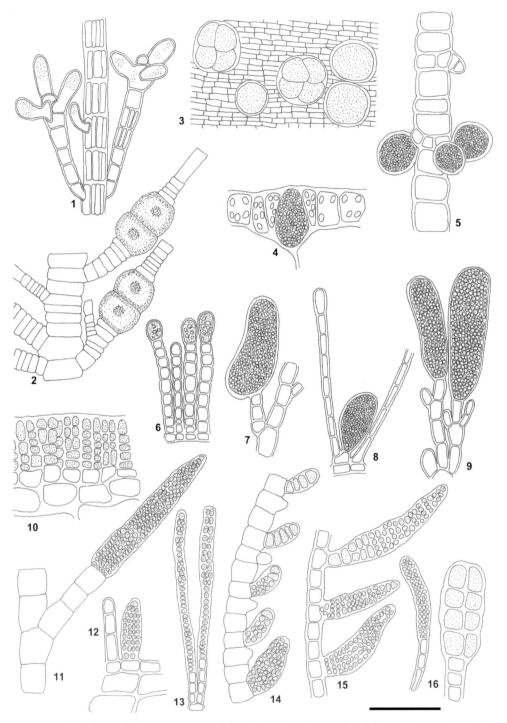

Fig. XI. Examples of reproductive structures. 1. *Sphacelaria cirrosa*. Propagules. 2. *Tilopteris mertensii*. Branches with two-celled chains of monosporangia. 3. *Dictyota dichotoma*. Surface view of tetrasporangia. 4–9. Examples of unilocular sporangia in *Desmarestia ligulata* (4), *Myriotrichia clavaeformis* (5), *Petroderma maculiforme* (6), *Petrospongium berkeleyi* (7), *Myrionema strangulans* (8) and *Elachista scutulata* (9). 10–16. Examples of plurilocular sporangia in *Petalonia fascia* (10), *Ectocarpus fasciculatus* (11), *Asperococcus bullosus* (12), *Petroderma maculiforme* (13), *Hincksia sandriana* (14) and *Hecatonema terminale* (15), and antheridia and oogonia in *Cutleria multifida* (sexual phase) (16). Bar = 20 µm (10), = 25 µm (4), = 50 µm (5–9, 11, 13, 14, 16), = 63 µm (12, 15), = 100 µm (1, 2), = 200 µm (3).

reproduction), as in *Ectocarpus*, motile and differing in size but not in morphology (anisogamous reproduction), as in *Hincksia* and *Cutleria*, or with the female large and immobile and the male small and motile (oogamous reproduction), as in the Laminariales, Desmarestiales and Fucales. In addition, physiological anisogamy can occur in which the gametes look identical but behave differently, immobilized female gametes being fertilized by motile male gametes of similar size for instance (see van den Hoek *et al.* 1995). For the sexual process the isogametes and anisogametes are usually released into the water and fusion occurs after the female gamete has settled (e.g. in *Ectocarpus* – Müller 1967). In an example of isogamete behaviour in *Ascoseira*, Clayton (1987) observed that one gamete typically settled soon after release, whilst the other gamete actively swam for a longer period of time, demonstrating that although morphologically indistinguishable, they were functionally differentiated. Reports of copulating motile gametes have been dismissed as unlikely by Müller (1975). In oogamous reproduction the female gametes are usually fertilized in suspension or when partially emerged from the oogonium (Lüning 1981a; Müller & Luthe 1981; Nakahara 1984). Fertilization of eggs retained on the parental thallus has also been reported in *Sargassum muticum* (Fletcher 1980). There are also some rare reports of some unfused gametes developing into multicellular thalli, for example in *Ascoseira mirabilis* (Clayton 1987) and *Colpomenia peregrina* (Yamagishi & Kogame 1998); in the latter case it was the larger, more pigmented, female gametes which behaved parthenogenetically.

In isogamous and anisogamous reproduction, the gametes are produced in plurilocular gametangia. In oogamous reproduction, the female gametes (termed oospores, ova or eggs) are usually produced singly (except in fucoids) in specialized oogonia, whilst the male gametes (termed antherozoids or sperm) are produced singly or in large numbers in antheridia. Male and female gametangia can either be produced on the same thallus (monoecious), as in *Pelvetia* and *Halidrys* or on different thalli (dioecious), as in many species of *Fucus*, *Ascophyllum* and *Himanthalia*. In some monoecious species, for example *Fucus spiralis*, both sexes are found within the same conceptacle, whilst in others, for example *Sargassum muticum*, they occur in different conceptacles although within the same receptacle. In *Desmarestia* there appears to be a link between gametophyte sexuality and sporophyte morphology, with alternately branched sporophytes having dioecious gametophytes and oppositely branched sporophytes having monoecious gametophytes (Chapman & Burrows 1971, but see Anderson 1982). There have also been a number of reports of unispores behaving as gametes (Loiseaux 1967b; Knight *et al.* 1935; Caram 1972). However, the evidence for this is not altogether satisfactory and was considered to be extremely dubious by Müller (1975).

An increasing number of reports have revealed the release and sexual union of the gametes to be under pheromone control (Fu *et al.* 2014; Kinoshita *et al.* 2016, 2017a; Müller 1981b). This process involves the production of a male-attracting substance by the female gametes and has been described in isogamously reproducing genera, such as *Ectocarpus* (Müller 1978; Müller & Gassmann 1980; Müller & Schmid 1988; Müller *et al.* 1971; Schmid 1993), anisogamously reproducing genera, such as *Cutleria* (Boland *et al.* 1983; Jaenicke *et al.* 1974; Müller 1974), as well as oogamously reproducing genera, such as *Dictyota* (Müller *et al.* 1981a; Phillips *et al.* 1990b), *Ascophyllum* (Müller *et al.* 1982), *Fucus* (Kinoshita *et al.* 2017a; Müller 1972b; Müller & Gassmann 1978; Müller & Jaenicke 1973; Müller & Seferiadis 1977), *Desmarestia* (Müller 1988; Müller & Lüthe 1981), *Hormosira* (Maier *et al.* 1992); *Spermatochnus* (Müller *et al.* 1981b), *Laminaria* (Lüning & Müller 1978; Maier & Müller 1982, 1990; Maier *et al.* 1987; Müller 1988; Müller & Gassmann 1980; Müller *et al.* 1979, 1985), *Saccharina* (Kinoshita *et al.* 2017a) and *Zonaria* (Phillips *et al.* 1990a). Initial contact to the settled gamete is made with the tip of the anterior, mastigoneme-covered flagellum of the motile gamete. Specific glycoproteins on the settled gamete are then detected by receptors on the motile gamete, initially binding the gametes, which is later followed by plasmogamy and karyogamy (Brawley 1992; Maier & Schmid 1995; Schmid *et al.* 1994).

Environmental influences on reproduction

Several environmental parameters have been shown to influence various aspects of the reproductive process in brown algae. These include:

Light quality

Light quality, particularly the red and blue wavebands, exerts an influence on various aspects of reproduction in some brown algae. For example, the blue part of the spectrum plays a critical role in the survival, development and gametogenesis of the gametophyte phases in a number of laminarians, including *Alaria* (Lüning & Neushul 1978), *Egregia* (Lüning & Neushul 1978), *Hedophyllum* (Lüning & Neushul 1978), *Laminaria* (Lüning 1980b; Lüning & Neushul 1978), *Macrocystis* (Lüning & Neushul 1978), *Pterygophora* (Lüning & Neushul 1978) and *Saccharina* (Lüning & Dring 1972, 1975), with the degree of fertility of the gametophytes depending on receiving a critical quantum dose of blue light (Lüning & Dring 1972, 1975). Drew (1983) and Lüning & Neushul (1978) suggested that this might well be limiting at times at certain depths and/or in silty coastal waters, which would seriously impair sporophyte production. Red light has been shown to be inhibitory and blue light promoting of plurilocular sporangia formation in *Scytosiphon* and *Petalonia* in the Scytosiphonaceae (Lüning & Dring 1973) and of gametogenesis in *Dictyota dichotoma* in the Dictyotales (Müller & Clauss 1976). There have also been reports of a blue light requirement for spore/gamete release in the brown algae, e.g. plurispore release in *Punctaria tenuissima* (as *Desmotrichum undulatum*) (Lockhart 1982) and egg release in *Dictyota dichotoma* (Kumke 1973), whilst blue light was reported to inhibit egg release in *Saccharina latissima* (Lüning 1981a).

Light intensity

A few reports have also shown that the level of quantum irradiance can influence aspects of reproduction in some brown algae. Examples include the ratio of unilocular to plurilocular organs in *Ectocarpus siliculosus* (see studies by Boalch 1961; Edwards 1969; Müller 1962), and the critical quantum irradiance levels shown to favour or inhibit fertility in gametophytes of several laminarian and other genera, including *Alaria* (Lüning & Neushul 1978), *Egregia* (Lüning & Neushul 1978), *Hedophyllum* (Lüning & Neushul 1978), *Laminaria* (Lüning 1980b; Lüning & Neushul 1978), *Macrocystis* (Deysher & Dean 1984; Lüning & Neushul 1978), *Nereocystis* (Vadas 1972), *Pterygophora* (Lüning & Neushul 1978), *Saccharina* (Lüning 1980b; Lüning & Dring 1975; Lüning & Neushul 1978), *Sargassum* (Zhao *et al.* 2013) and *Undaria* (Pang *et al.* 2008).

However, reports of gametogenesis inhibition in low white illuminance levels might simply be due to insufficient blue light being present in these conditions (Lüning & Dring 1975).

Photoperiod/daylength

A wide range of photoperiodic responses, related either to short days or long days, are involved in the reproductive processes of many brown algae, usually linked to temperature (see Table 3 for some examples).

The combined action of photoperiod and temperature, probably in association with nutrient levels, also determines the seasonal periodicity of reproductive development in brown algae, which is particularly well documented in many of the perennial wrack species found in Britain and Ireland and throughout the world. Examples include *Ascophyllum nodosum* (Baardseth 1970), *Fucus ceranoides* (Brawley 1992), *Fucus distichus* (Bird & Mclachlan 1975; Pearson & Brawley 1996), *Fucus spiralis* (Niemeck & Mathieson 1976; Subrahmanyan 1961); *Fucus vesiculosus* (Andersson *et al.* 1994; Berndt *et al.* 2002; Pearson & Serrão 2006; Pearson *et al.* 1998), *Halidrys siliquosa* (Moss & Lacy 1963; Moss & Sheader 1973), *Pelvetia canaliculata* (Subrahmanyan 1960) and *Sargassum muticum* (Arenas & Fernández 1998; DeWreede 1978;

Table 3. Photoperiodic responses in the Phaeophyceae.

Regime	Response	References
Long days	Gametangia formation in *Sphacelaria rigidula*	Ten Hoopen 1983; Ten Hoopen *et al.* 1983
	Gametangia formation in *Sphaerotrichia divaricata*	Novaczek & McLachlan 1987
	Gametangia formation in *Myriotrichia clavaeformis*	Peters 1988
Short days	Sporangia formation in *Phyllariopsis brevipes*	Henry 1987a
	Sporangia formation in *Saccharina latissima*	Lüning 1988
	Gametangia formation in *Scytothamnus* spp.	Clayton 1986
	Gametangia formation in *Desmarestia* spp.	Nakamura 1984; Nakahara & Nakamura 1971
	Gametangia formation in *Saccorhiza dermatodea*	Henry 1987b
	Sorus formation in *Laminaria setchelli*	Tom Dieck 1991
	Receptacle initiation in *Ascophyllum nodosum*	Terry & Moss 1980
	Receptacle initiation in *Fucus distichus*	Bird & McLachlan 1976
	Receptacle initiation in *Fucus vesiculosus*	Bäck *et al.* 1991
	Unilocular sporangia formation in *Colpomenia peregrina*	Kogame & Yamagishi 1997
	Unilocular sporangia formation in *Petalonia binghamiae, Petalonia fascia, Planosiphon zosterifolius, Scytosiphon lomentaria*	Nakamura & Tatewaki 1975
	Propagule production in *Sphacelaria furcigera*	Colijn & van den Hoek 1971
Variable daylength/ light:dark ratio	Ratio of unilocular to plurilocular sporangia in *Ectocarpus siliculosus*	Müller 1962
	Gamete release in *Dictyota dichotoma*	Bunning & Müller 1961
	Gamete release in *Pelvetia canaliculata*	Jaffe 1954

Fletcher 1980; Fletcher & Fletcher 1975; Jephson & Gray 1977; Monteiro *et al.* 2009b; Norton 1981b; Okuda 1981) as well as many kelp species, including *Laminaria digitata* (Gayral & Cosson 1973), *Laminaria hyperborea* (Kain 1971), *Saccharina latissima* (Andersen 2013; Lee & Brinkhuis 1986), *Saccorhiza polyschides* (Norton 1970a; Pereira *et al.* 2015a) and *Undaria pinnatifida* (Schaffelke *et al.* 2005).

Lunar/tidal periods

The lunar period is an additional environmental influence on reproduction, probably related to the physiological response to light intensity. Gamete/spore release in various brown algae is periodic and related to the tidal/lunar periods. For instance, lunar rhythms have been observed in *Cystoseira* (Engelen *et al.* 2008), *Dictyota* (Williams 1905; Hoyt 1927; Phillips *et al.* 1990b; Willer 1962), *Pelvetia* (Subrahmanyan 1957), *Sargassum* (Engelen *et al.* 2008; Fletcher 1980; Monteiro *et al.* 2009b; Tahara 1909), *Nemoderma* (Kuckuck 1912) and *Himanthalia* (Gibb 1937). Such a simultaneous release of gametes would certainly increase the probability of fertilization and be clearly advantageous in the completion of the life history.

Salinity

Field reports and experimental studies have indicated that reduced salinity can modify or suppress reproduction in brown algae. For example, Munda (1964, 1967) suggested that reduced salinities influenced the reproductive phenologies of fucoid algae and noted that the rate of receptacle formation in *Ascophyllum nodosum* and *Fucus vesiculosus* was accelerated in diluted seawater. Also Burrows (1964) observed that reproduction is initiated over a wide range of salinities in *Fucus serratus* and *F. ceranoides*, whilst Niemeck & Mathison (1976) reported maturation of receptacles in *Fucus spiralis* during the spring run-off. Other effects reported include sexual reproduction in *Pylaiella* being absent in brackish-water populations (Russell 1971), reduced viability of plurilocular sporangia in *Asperococcus fistulosus* and shorter plurilocular sporangia in *Punctaria tenuissima* (as *Desmotrichum undulatum*) (both reported by Pedersen 1984).

Nutrients

A small number of reports have demonstrated that nutrient availability is an important factor in controlling the reproductive processes of algae, including brown algae (see DeBoer 1981 for a summary).

Life histories

The term 'life history' can be defined as the sum of the reproductive and morphological phases possessed by a species. In the brown algae, the most common life history comprises a cyclic alternation of two chromosomal phases (diploid and haploid) which are linked by meiosis in the diploid phase and fusion of sexual gametes produced by the haploid phase. With the exception of the Fucales, these two phases represent different, free-living generations (diphasic). Superimposed, however, on this cycle are a wide variety of different pathways which have greatly complicated the interpretations of life histories in this group. In general, the simple and cyclical kind of life history is associated with species that possess morphologically complex thalli such as *Fucus* and *Laminaria*. Flexible, more complicated life histories, on the other hand, are usually associated with the simpler thallus forms. However, there are exceptions to both generalizations – the likes of apospory, apogamy and polyploidy have been demonstrated in various Laminariales growing on the Japanese coast (Nakahara 1984), whilst *Laminaricolax*, which has a simple thallus form, possesses a very simple, reduced life history (Russell 1964b).

Numerous attempts have been made to bring order to chaos and classify and code the wide variety of brown algal life histories that have been recorded. One of the earliest terminologies, originally proposed for the Rhodophyta, was that of Svedelius (1915) who recognized two basic life history types: haplobiontic, in which one thallus form occurs, and diplobiontic, in which two thallus forms occur. This basic subdivision is still widely used today, although additional consideration is given to the somatic state of the thallus. Three basic types of life history are usually given as follows:

- *Haplontic* Thalli are haploid, producing gametes which fuse to form a diploid zygote. During zygotic germination, meiosis occurs and resulting spores reform haploid thalli.
- *Diplontic* Thalli are diploid, producing haploid gametes by meiosis. Fusion of gametes re-establishes diploid thalli.
- *Diplohaplontic* Thalli comprise two chromosomal phases – a haploid or gametophytic phase, producing gametes which fuse to form a zygote which germinates into the diploid sporophytic phase. The latter produces haploid spores by meiosis which germinate to reform the haploid phase.

With the exception of the Fucales, to which the second life history applies, all the brown algae have a life history that is basically diplohaplontic. A further subdivision of the latter, proposed by Kylin (1933), recognizes that the morphologies of the two phases can be either similar (isomorphic life history, as in *Padina*, *Zanardinia* and *Dictyota*) or dissimilar (heteromorphic life history, as in *Laminaria* and *Desmarestia*). Indeed, Kylin classified the brown algae according to their life history, recognizing three classes: the Isogeneratae, which possess an alternation of isomorphic generations; the Heterogeneratae, which possess an alternation of heteromorphic generations; and the Cyclosporeae, which lack an alternation of generations (applicable to the Fucales only). Although widely adopted at first, this scheme is no longer supported because of the extreme artificiality of the resulting classification.

Utilizing a combination of Svedelius's (1915) 'morphological' classification and his later (Svedelius 1931) 'somatic state' classification (see Dixon 1973 for details), an attempt was made by Drew (1955) to define more precisely the life history sequences of algae. For example, mono- and dimorphic life histories were terms applied to algae with a single or two distinct thallus forms, respectively. Monomorphic life histories could be haplontic, diplontic or diplo-haplontic, whilst in dimorphic life histories the two thallus forms could either be at the same ploidy level (i.e. dimorphic haplontic or dimorphic diplontic), or at different ploidy levels (i.e. dimorphic, diplohaplontic). Chapman & Chapman's (1961) proposal to extend this termi-nology was generally regarded as too cumbersome and unworkable (Dixon 1973; Russell 1973a). Dixon's (1963, 1970) proposed 'type' method of discrimination between different life histories (e.g. *Ectocarpus*-type life history, *Fucus*-type life history, etc.) was simpler and more flexible but is open to misinterpretation in the brown algae because of the widespread occurrence of different types of life history within the same species.

The present volume uses Drew's (1955) terminology when feasible but it is not imposed on taxa for which it is evidently ill-suited. The chromosomal terminology – e.g. monophasic (either diplontic or haplontic) and biphasic (diplohaplontic) – is applied in conjunction with that of the number of morphological forms (e.g. monomorphic, dimorphic) but only on the basis of cytological data.

In this discussion on life histories of brown algae, problems have arisen even in the inter-pretation of hitherto well-established facts. For instance, members of the Fucales have been regarded as having a monomorphic, diplontic life history (Bold & Wynne 1985) (see below). The free-living thallus body is a diploid gametophyte which undergoes meiosis at the onset of gametogenesis. In this process the first divisions of the oogonium are meiotic and four haploid nuclei are produced which subsequently divide mitotically to produce an eight-nucleate oogonium. Various numbers of these nuclei, according to the different genera, then contribute to egg formation. However, an alternative life history interpretation for the Fucales, revived by Jensen (1974), would be 'dimorphic and diplohaplontic'. In this case the post-meiotic oogonial precursor represents a unilocular sporangium (or megasporangium), and the four nuclei produced by meiosis represent potential reduced gametophytic microthalli (or megagameto-phytes); subsequent mitotic divisions of these nuclei then produce the haploid eggs. The life history could, therefore, be described as dimorphic (more accurately heteromorphic) with both an enclosed, haploid, gametophytic phase and a free-living, diploid, sporophytic phase (i.e. diplohaplontic).

The following examples illustrate the range of life histories described in the brown algae:

Diplontic

This type of life history is usually assigned only to members of the Fucales. Taking *Fucus* as an example (Fig. XII, a), the large diploid thallus produces terminal receptacles on which flask-like conceptacles develop containing male (antheridia) and female (oogonia) gametangia. Production of the male antherozoids and female eggs is preceded by meiosis and fusion of these haploid gametes followed by germination of the resultant zygote re-establishes the diploid macroscopic thallus.

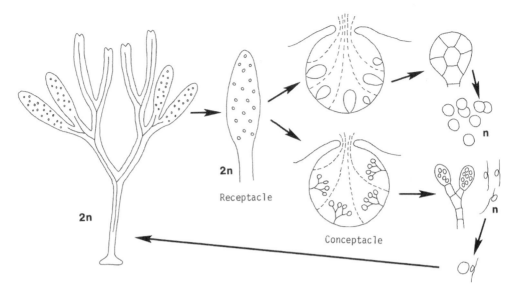

Fig. XII(a). The life history of *Fucus* (monomorphic, diplontic).

Diplohaplontic

Monomorphic

In this life history there is an alternation between morphologically identical haploid and diploid thalli (often termed isomorphic life history) (Fig. XII, b). It has been reported, for example, in many of the simple, erect members of the Ectocarpales (see particularly the now classical studies on *Ectocarpus* by Knight (1929), Papenfuss (1935a) and Müller (1972a) and on *Pylaiella* by Knight (1923)), the filiform parenchymatous Sphacelariales (see particularly van den Hoek & Flinterman's 1968 study of *Sphacelaria furcigera*), and the foliaceous, parenchymatous Dictyotales. The haploid thalli produce gametes by mitotic cell division (from plurilocular sporangia) which fuse. The zygotes germinate into morphologically identical thalli; asexual spore production by meiosis (in the unilocular sporangia, or tetrasporangia in the Dictyotales) then re-establishes the haploid generation. A problem in classifying many of these life histories arises from the fact that they may be slightly dimorphic (often termed heteromorphic), such as in *Ectocarpus* (Müller 1972a), *Hincksia* (Kornmann 1954), *Feldmannia* (Kornmann 1953), *Pylaiella* (Russell 1961a) and *Sphacelaria* (van den Hoek & Flinterman 1968).

Dimorphic (trimorphic, polymorphic, etc.)

In this life history there is an alternation between at least two morphologically dissimilar haploid and diploid thalli (Fig. XII, c–d). In the great majority of algae with this type of life history, there is a regular alternation between a large, diploid thallus (macrothallus) and a microscopic, haploid thallus (microthallus) (Fig. XII, c). Unispores, produced by meiosis in the unilocular sporangia on the diploid thallus, germinate into haploid microthalli; the latter then reproduce sexually and the resultant zygote reforms the macrothallus. This life history is particularly characteristic of members of the Laminariales and Desmarestiales, and has also been reported in various members of the Chordariaceae, for example *Sphaerotrichia* (Ajisaka & Umezaki 1978), *Cladosiphon* (Shinmura 1974), *Striaria* (Caram 1965) and *Litosiphon* (Dangeard 1969).

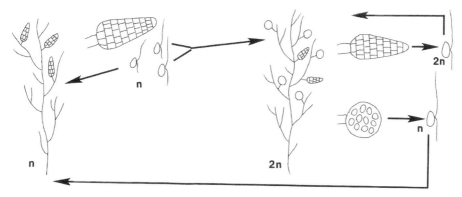

Fig. XII(b). The life history of *Ectocarpus* (monomorphic, diplohaplontic).

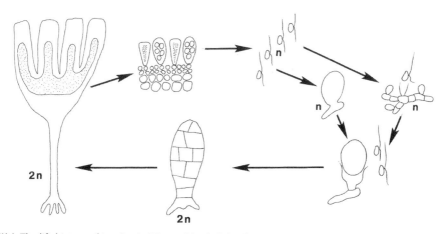

Fig. XII(c). The life history of *Laminaria* (dimorphic, diplohaplontic).

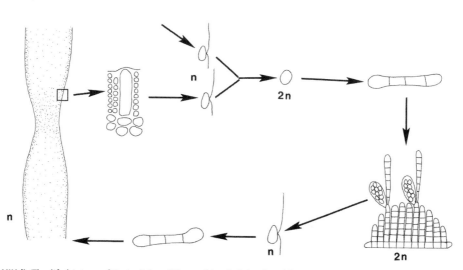

Fig. XII(d). The life history of *Scytosiphon* (dimorphic, diplohaplontic).

In some algae, however, such as *Cutleria* (Cutleriaceae) and members of the Scytosiphonaceae, the macrothallus is haploid and sexual, while the smaller, prostrate thalli are diploid and asexual (Fig. XII, d). In the life history of *Cutleria*, released gametes fuse and germinate into parenchymatous, procumbent, unilocular sporangia-bearing thalli (the '*Aglaozonia*' phase) which is the diploid, asexual generation; unispores formed by meiosis germinate to re-establish the *Cutleria* phase. In certain members of the Scytosiphonaceae, the macrothallus, character-istic of the genera *Colpomenia*, *Petalonia* and *Scytosiphon*, may be haploid, whilst the encrusting *Stragularia*/*Microspongium*-like microthalli are diploid. Often associated with these diplohap-lontic life histories are a number of variations and abbreviations. These include:

Dimorphic, monophasic life histories
An increasing number of species, particularly in the Chordariaceae, are being revealed to have a heteromorphic but asexual life history. There is little evidence that the two distinct, independently existing, morphological growth forms or expressions represent distinct haploid and diploid phases. The life history can, therefore, be described as heteromorphic but 'direct' and monophasic. Such a life history appears to apply to *Asperococcus* and *Punctaria* in which the characteristic, macroscopic, erect, parenchymatous thalli and microscopic, branched, *Hecatonema*-like thalli have been described. Without cytological evidence, it is better to adopt the simple terminology of macrothalli and microthalli for these two growth forms and so avoid commitment to such terms as plethysmothallus (diploid, sporophytic, usually plurilocular sporangia-bearing, self-perpetuating microthallus) and protonema (plethysmothallus which gives rise as a lateral outgrowth to an erect macroscopic sporophyte) as defined by Sauvageau (1932).

Dimorphic, monophasic, asexual life histories have also commonly been described for genera in the Scytosiphonaceae (Fletcher 1974; Roeleveld *et al.* 1974; Wynne 1969); the biphasic life histories appear to be largely confined to Pacific strains of these algae (Clayton 1980; Kogame & Yamagishi 1997; Nakamura & Tatewaki 1975). It is considered that such a life history represents parthenogenetic development of the gametes. A significant aspect of many dimorphic, monophasic life histories is the important role played by environmental conditions in the relationship between the two morphological expressions. Parameters such as temper-ature, photoperiod, light intensity, light quality, salinity, etc. can control the final morpho-logical expression produced (e.g. crustose or erect blade development in the Scytosiphonaceae) and very often the physical connection between the expressions too (e.g. environmental control of erect thallus formation from microthalli in *Punctaria*).

'Vegetative diploidization'
Such a non-sexual, biphasic, life history has been reported in two algae: *Ectocarpus* (Müller 1967) and *Elachista stellata* (Wanders *et al.* 1972), by the direct sprouting of diploid macrothalli from haploid microthalli.

Monomorphic, monophasic life histories
Some brown algae produce spores that germinate directly, without sexual fusion, to repeat the parental thallus form (i.e. a direct, monophasic, monomorphic life history) (Fig. XII, e–f). Such a simple, direct developmental pattern has been observed for plurispores in *Streblonema* (Fletcher 1983), *Hincksia granulosa* (Fletcher 1981b), *Laminariocolax tomentosoides* (Russell 1964b) and *Pseudolithoderma roscoffense* (Loiseaux 1968), as well as for unispores in *Pseudoralfsia verrucosa* (Fletcher 1978; Loiseaux 1968) and *Stragularia spongiocarpa* (as *Ralfsia spongiocarpa* – Fletcher 1981c), suggesting that either the parental thallus is haploid, or if diploid, that apomeiosis has occurred in the unilocular sporangia.

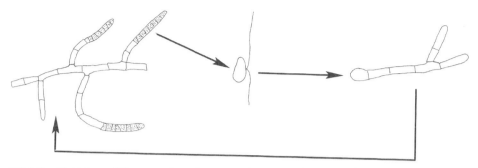

Fig. XII(e). The life history of *Microspongium oligosporum* (monomorphic, monophasic). See Fletcher (1983).

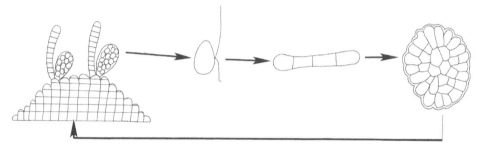

Fig. XII(f). The life history of *Stragularia spongiocarpa* (monomorphic, monophasic).

Community ecology

Since the middle of the twentieth century, intertidal community ecology has developed from a mainly descriptive science to one that is experimental with a variety of field and laboratory techniques. The descriptive phase was marked by much preoccupation with zonation and its causes. Numerous methods for the description of intertidal zonation patterns have been proposed (see Lewis 1964; Russell 1972, 1973a,b,c, 1977, 1980; Stephenson & Stephenson 1949, 1972; Sundene 1953). In the first volume of the series *Seaweeds of the British Isles*, Dixon & Irvine (1977) introduced a zonal system and terminology as shown below. This scheme is directly comparable with that of Lewis (1964) in the way the zones are defined and, by dividing Lewis's 'eulittoral' into upper and lower zones, it is arguably more precise. However, the Dixon and Irvine terminology has not gained widespread acceptance in the same way as Lewis's (1964) system. It also incorporates the term 'midlittoral', which already existed in the literature as a synonym for the eulittoral zone (Stephenson & Stephenson 1949, 1972). The possibility of confusion over what is meant by midlittoral should be avoided. For these reasons, the system of Lewis (1964) is adopted here.

Zonal classification

Lewis 1964		Dixon & Irvine 1977
Littoral	Littoral fringe	Upper littoral
	Eulittoral	Midlittoral + Lower littoral
Sublittoral		Sublittoral

In general, Lewis's three-zone classification has been validated by more objective analytical methods (see Jones *et al.* 1980; Russell 1972, 1973a,b, 1977, 1980) but other patterns are possible.

For instance, estuarine localities frequently lack a sublittoral algal zone (Russell 1973b; Tittley & Price 1977a; Wilkinson 1980).

Zonation has been attributed mainly to tidal (emersion-immersion) factors and differences in adaptation to the associated stresses. Colman (1933) and Doty (1946) took this interpretation to the point of identifying critical tidal levels where zonal boundaries occur. Other workers, such as Lewis (1964), were unable to locate such levels and pointed out that on exposed rocky shores, tidal amplitude was of secondary importance to wave action in determining emersion-immersion times. Nevertheless, the algae of higher shore levels have been shown by Dring (1982) to have superior powers of recovery from desiccation than lower shore species.

The problem of identifying causes of zonation has arisen chiefly from the fact that the physico-chemical factors operating on a rocky shore tend to do so as gradients, whereas zonal boundaries are sharp discontinuities. Therefore, a number of workers have preferred to invoke biological interactive effects as the major factor determining zonal frontiers (Chapman 1974). Others, such as Schonbeck & Norton (1978), have preferred to combine physico-chemical effects with biological interactions in identifying the causes of zonation.

The biological interactive effect of competition on the Phaeophyceae has been investigated and measured in field experiments by a number of workers, for example Dayton (1975), Hawkins & Hartnoll (1980); Lubchenco (1978, 1986); Raffo *et al.* (2009); Wangkulangkul *et al.* (2016); Worm & Chapman (1998) – in many cases by the removal of hypothetical dominants. Particular attention has also more recently been given to the competitive effect of alien species on the biodiversity of native communities (Buschbaum *et al.* 2006; Casas *et al.* 2004; Farrell & Fletcher 2006; Lilley & Schiel 2006; South *et al.* 2016; Valentine & Johnson 2004). Usually the large, brown canopy-forming algae have proved to be dominant in the sense of determining overall algal species composition. Occasionally, however, in stressful situations, ephemeral algae have been found to be capable of competitively excluding perennial forms, with opportunistic filamentous species reducing recruitment success in *Fucus vesiculosus* (Korpinen & Jormalainen 2008a), for example. Annual algae, such as *Ectocarpus*, have also been found to compete successfully, under certain conditions, against other annual algae in laboratory culture (Russell & Fielding 1974), whilst Fletcher (1975) has observed heteroantagonism between crustose species in culture. In general, competition is considered to play a major role in determining the lower growth limits of species, whilst the upper growth limits are determined mainly by desiccation tolerance. Note, however, the important influential role also played by grazers, particularly on exposed shores (see p. 29), extreme environmental events such as storms (Denny *et al.* 2009) and anthropogenic influences such as trampling (Beauchamp & Gowing 1982; Boalch *et al.* 1974; Brosnan & Crumrine 1994; Fletcher & Frid 1996; Keough & Quinn 1998; Micheli *et al.* 2016; Povey & Keough 1991; Schiel & Lilley 2011; Schiel & Taylor 1999).

Field experimentation, particularly involving observations of the recolonization of cleared areas and sometimes the use of test panels, has also played a major role in our understanding of successional changes in macroalgal community structure (see particularly Burrows & Lodge 1950; Dayton 1975; Hawkins 1981; Lodge 1948b; Southward 1956). With respect to elucidating the mechanisms of succession, particular interest has been given to the three 'models' proposed by Connell & Slatyer (1977). These are (1) the 'facilitation' model, which states that only certain pioneering species can colonize and then modify the environment, rendering it more suitable for later successional species; (2) the 'tolerance' model, which suggests that any species capable of living in the habitat as an adult can colonize but that this will not exclude later successional species and so the more environmentally stress-tolerant species will eventually become estab-lished; and (3) the 'inhibition' model, which proposes that any species capable of living in the habitat as an adult can colonize and these will exclude all other species until their death or disturbance creates space. Some workers have since used this system in an attempt to classify and unravel the complexities of succession in various marine ecosystems, particularly with respect to rocky shores (Benedetti-Cecchi 2000; Bertness *et al.* 1999; Farrell 1991; Hawkins

1981; Maggi *et al.* 2011; Sousa 1979; Underwood *et al.* 1983; Van Tamelen 1987). These studies have indicated that no single model will be completely true for all successional stages. For instance, Hawkins's (1981) observation that the early diatom and green algal colonizers are not an essential prerequisite for recruitment and growth of *Fucus* is counter to the facilitation model, whilst Knight & Parke's (1950) observation of *Fucus* germlings growing up through *Enteromorpha* is counter to the inhibition model. In most of the studies undertaken, inhibition appears to be the prevalent mode of succession, with some cases of facilitation being reported – although, as revealed by Maggi *et al.* (2011), the resultant mode can also be density-dependent with respect to the abundance of early colonists and an increase in species richness.

Floristics

Table 4 shows the comparative numbers of brown algal species recorded for North Atlantic regions. It can be seen that, floristically, Britain and Ireland and France hold the largest numbers of brown algae; also relatively rich are Norway (175 species) and Denmark (128), despite the latter having a limited range of shore types. The similarity in the brown algal floras of these regions is not surprising; as Guiry (1978) pointed out, these regions, which would all be classified by van den Hoek (1975) as 'cold-temperate Atlantic-Boreal', are likely to be similar phytogeographically and comprise a fairly rich transition flora. Certainly, it can be seen that passage northwards and southwards to southernmost Greenland and Portugal respectively brings about a corresponding marked diminution in diversity of the brown algal flora.

Table 4. Distribution of brown algae in the North Atlantic.

Country/region	Number of Phaeophyceae	References
Greenland (South)	49	Pedersen 1976
Greenland	67	Pedersen 2011
Faroe Islands	74	Irvine 1982
	83	Nielsen & Gunnarsson 2001
Iceland	72	Caram & Jónsson 1972
	84	Gunnarsson & Jónsson 2002 (checklist revised)
Norway	153	Rueness 2001
Denmark	128	Christensen & Thomsen 1974
Helgoland	100	Bartsch & Kuhlenkamp 2000
Britain	197	Parke & Dixon 1976
	183	Brodie *et al.* 2016
Ireland	161	Guiry 2012
Britain and Ireland	185	Hardy & Guiry 2006
	188	Present study
Roscoff, France Atlantic	160	Feldmann & Magne 1964
France Atlantic + Channel Isles	225	Dizerbo & Herpe 2007
Spain, Atlantic	163	Gallardo García *et al.* 2016
Portugal	98	Ardré 1970
Portugal (North)	67	Araujo *et al.* 2009

Systematics

Table 5 lists the British and Irish brown algae in Harvey (1849), Parke & Dixon (1976), Brodie *et al.* (2016) and in the present study. Except for Harvey, the sequence of the orders in each classification system follows that given by the authors, whilst the families are listed alphabetically within each order. The classification scheme adopted in the present study follows that of the detailed phylogenetic study of Silberfeld *et al.* (2014). This is highlighted in Fig. XIII (pp. 48–49), which shows their ordinal classification and their proposed four new subclasses (all in capitals) listed by order of divergence.

The main groupings of Phaeophyceae listed in Harvey (1849), based almost entirely on morphological details, bear some similarity to the later ordinal classification systems. Differences lie largely in the detail and content of the various classification systems. Increased interest and research on the Phaeophyceae post-Harvey, particularly relating to aspects of their cytology, fine structure, reproduction and life history, resulted in several diverse classification systems and speculation on their phylogeny. These classification systems were very largely based on criteria such as mode of construction (haplostichous, polystichous), mode of growth, reproduction and type of life history (isomorphic, heteromorphic or absence of an alternation of generations), cell structure, number of chloroplasts and whether or not a pyrenoid is present (see de Reviers & Rousseau 1999 for an excellent historical account of the ordinal delineation of the Phaeophyceae). Using combinations of these characteristics, the most notable of these systems were those of Oltmanns (1922), Kylin (1933), Fritsch (1945), Papenfuss (1951), Scagel (1966), Russell (1973a), Parke & Dixon (1976), Wynne (1982), Bold & Wynne (1985), Christensen (1980), Womersley (1987), Clayton (1990) and van den Hoek *et al.* (1995). With respect to the British and Irish flora, the checklist of Parke & Dixon (1976) followed quite closely that of Fritsch (1945), who in turn followed Oltmanns (1922) who had originally conceived the orders. Based on the above-described criteria, the characteristics of and phylogenetic relationships between the orders was expressed diagrammatically by various authors, notably by van den Hoek & Jahns (1978), and this received widespread interest and acceptance, later being modified by Fletcher (1987) and Womersley (1987).

Particularly notable in Table 5 is the broad Fritschian concept of the Ectocarpales, which recognized 16 families and was adopted by Parke & Dixon – it is a concept that was later taken up by Fletcher (1987). This was in contrast to many earlier authors such as Abbott & Hollenberg (1976), Bold & Wynne (1985), Christensen (1980) and Lobban & Wynne (1981), who variously removed and raised a number of the 16 families to ordinal status, resulting in, for example, the Scytosiphonales, Chordariales, Dictyosiphonales, Ralfsiales and Tilopteridales. In the later checklist of British algae by Brodie *et al.* (2016) and in the present study, based on molecular studies that redefined the order (see Rousseau & de Reviers 1999a), only the five families Acinetosporaceae (see Peters & Ramírez 2001), Chordariaceae, Ectocarpaceae, Petrospongiaceae (see Racault *et al.* 2009) and Scytosiphonaceae are now included in the Ectocarpales. The Myrionemataceae, Elachistaceae, Corynophlaceae, Acrotrichaceae, Spermatochnaceae, Striariaceae, Myriotrichaceae, Giraudiaceae, Punctariaceae, Buffhamiaceae and Dictyosiphonaceae have been absorbed into a much broader concept of the Chordariaceae (see Silberfeld *et al.* 2014).

In addition to providing a better understanding of the Ectocarpales, molecular techniques, developed from the 1980s and 1990s onwards, have also given us a much clearer understanding of the ordinal systematics and phylogeny of all the brown algae (see reviews by Draisma 2002; Draisma *et al.* 2003; Druehl *et al.* 1987; de Reviers & Rousseau 1999; Silberfeld *et al.* 2014). These techniques, combined with the morphological and life history data, have afforded a better grasp of the phylogeny of the Phaeophyceae and have provided new phylogeny-based classification schemes that have turned the classical Phaeophyceae phylogenetic tree upside down. For instance, with the exception of the Fucales, algae with a simple filamentous construction

Table 5. The brown algal tribes of Britain and Ireland listed in Harvey (1849), Parke & Dixon (1976), Brodie *et al.* (2016) and in the present study. Orders arranged as in each publication; families arranged alphabetically within each order.

Harvey (1849)	Parke & Dixon (1976)	Brodie *et al.* (2016)	Present study
No subclass	**No subclass**	**No subclass**	**DISCOSPORANGIOPHYCIDAE**
No orders	**Ectocarpales**	**Sphacelariales**	**Discosporangiales**
Chordarieae	Acrotrichaceae	Cladostephaceae	Choristocarpaceae
Dictyoteae	Ectocarpaceae	Sphacelariaceae	**ISHIGEOPHYCIDAE**
Ectocarpeae	Buffhamiaceae	Sphacelodermaceae	**Ishigeales**
Fuceae	Chordariaceae	Stypocaulonaceae	Petrodermataceae
Laminarieae	Corynophlaceae	**Dictyotales**	**DICTYOTOPHYCIDAE**
Sphacelarieae	Dictyosiphonaceae	Dictyotaceae	**Dictyotales**
Sporochnoideae	Elachistaceae	**Ectocarpales**	Dictyotaceae
	Giraudiaceae	Acinetosporaceae	**Sphacelariales**
	Myrionemataceae	Petrospongiaceae	Cladostephaceae
	Myriotrichaceae	Chordariaceae	Lithodermataceae
	Punctariaceae	Ectocarpaceae	Sphacelariaceae
	Ralfsiaceae	Scytosiphonaceae	Sphacelodermataceae
	Scytosiphonaceae	**Discosporangiales**	Stypocaulaceae
	Spermatochnaceae	Choristocarpaceae	**FUCOPHYCIDAE**
	Striariaceae	**Ralfsiales**	**Desmarestiales**
	Tilopteridaceae	Ralfsiaceae	Arthrocladiaceae
	Cutleriales	**Cutleriales**	Desmarestiaceae
	Cutleriaceae	Cutleriaceae	**Ectocarpales**
	Desmarestiales	**Sporochnales**	Acinetosporaceae
	Arthrocladiaceae	Sporochnaceae	Chordariaceae
	Desparestiaceae	**Tilopteridales**	Ectocarpaceae
	Sporochnaceae	Halosiphonaceae	Petrospongiaceae
	Laminariales	Phyllariaceae	Scytosiphonaceae
	Alariaceae	Tilopteridaceae	**Fucales**
	Chordaceae	**Desmarestiales**	Fucaceae
	Laminariaceae	Arthrocladiaceae	Himanthaliaceae
	Sphacelariales	Desmarestiaceae	Sargassaceae
	Choristocarpaceae	**Laminariales**	**Laminariales**
	Cladostephaceae	Alariaceae	Alariaceae
	Sphacelariaceae	Chordaceae	Chordaceae
	Stypocaulaceae	Laminariaceae	Laminariaceae
	Dictyotales	**Fucales**	**Ralfsiales**
	Dictyotaceae	Fucaceae	Pseudoralfsiaceae
	Fucales	Himanthaliaceae	**Sporochnales**
	Cystoseiraceae	Sargassaceae	Sporochnaceae
	Fucaceae		**Tilopteridales**
	Himanthaliaceae		Cutleriaceae
	Sargassaceae		Halosiphonaceae
			Phyllariaceae
			Tilopteridaceae
			Incertae sedis

and showing isogamy, such as members of the Ectocarpales, were previously considered to represent the primitive/ancestral condition, whilst more morphologically complex algae with oogamy were hypothesized to be derived/advanced (Kylin 1933; Papenfuss 1951, 1955; van den Hoek *et al.* 1995). Molecular analyses have now demonstrated this not to be the case and the Dictyotales and Sphacelariales (previously considered advanced) are revealed to be the basal orders (Draisma & Prud'homme van Reine 2001; Rousseau *et al.* 2001) and the Laminariales *s.s.* and Ectocarpales *s.l.* the most derived (Draisma *et al.* 2003).

Molecular studies have also been responsible for the four additional orders, Discosporangiales, Ishigeales, Ralfsiales and Tilopteridales shown in Table 5 and accepted by Brodie *et al.* (2016)

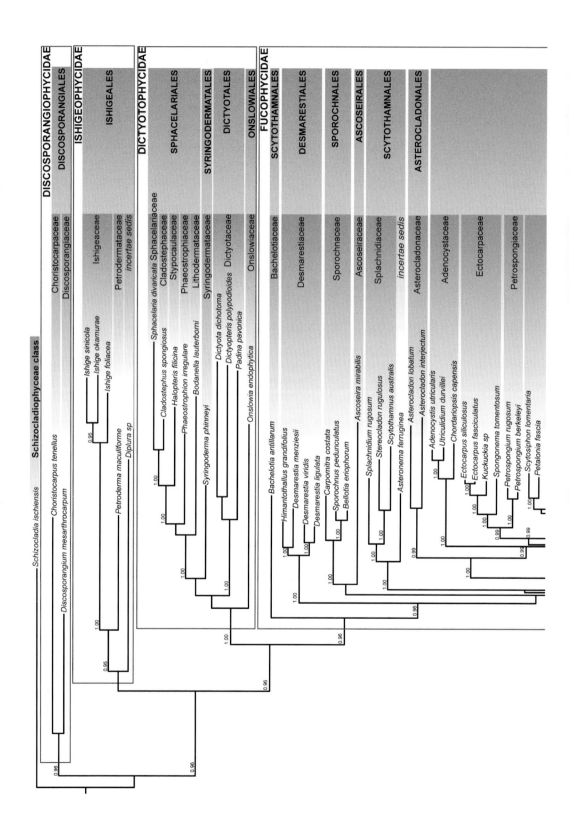

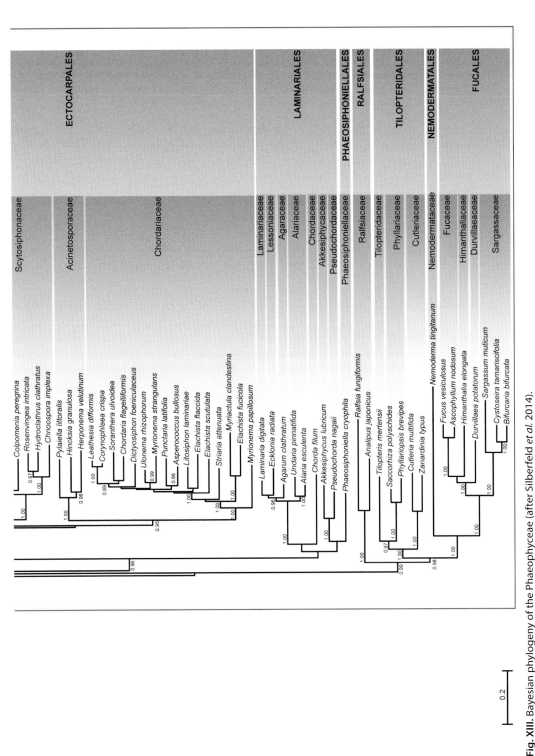

Fig. XIII. Bayesian phylogeny of the Phaeophyceae (after Silberfeld *et al.* 2014).

and in the present treatise. The Discosporangiales (and Discosporangiaceae) originally proposed by Schmidt (1937) and emended and reinstated by Kawai *et al.* (2007), was based on the species *Discosporangium mesarthrocarpum* (Meneghini) Hauck. Molecular techniques used by Kawai *et al.* (2007) revealed the latter species to form a monophyletic clade with *Choristocarpus tenellus*, family Choristocarpaceae and they suggested the latter family should be removed from its usual placement (see Fritsch 1945; Parke & Dixon 1976; Ribera *et al.* 1992) in the Sphacelariales and placed into the Discosporangiales. This found support in a molecular study by Phillips *et al.* (2008a) and has since received widespread acceptance (Brodie *et al.* 2016; Cormaci *et al.* 2012; Gallardo García *et al.* 2016; de Reviers *et al.* 2015; Silberfeld *et al.* 2014).

The Ishigeales was proposed by Cho & Boo in Cho *et al.* (2004) to receive the Ishigeaceae which had been previously placed in the Ectocarpales by Rousseau & de Reviers (1999a). This later found support by the molecular study of Kawai *et al.* (2005). Within the Ishigeales, only the brown crustose family Petrodermataceae, containing the species *Petroderma maculiforme*, is represented in the British and Irish flora. Previously placed in the Lithodermataceae or Ralfsiaceae (see Brodie *et al.* 2016; Fletcher 1978; Parke & Dixon 1976) this species was moved into its own family based on molecular studies and morphological features by Silberfeld *et al.* (2014) which revealed it to have a sister relationship with the Ishigeales.

The Ralfsiales, originally proposed by Nakamura (1972) and later validated by Lim & Kawai in Lim *et al.* (2007), was accepted by a number of authors including Silva & Reviers (2000) and Tanaka & Chihara (1980a). Some other authors, however, such as Nelson (1982), Parke & Dixon (1976), Ribeira *et al.* (1992) and Schneider & Searles (1991), placed all the crustose brown algae within the Ectocarpales, while some, such as Womersley (1987), placed them in the Chordariales. Particularly significant was Tan & Druehl's (1994) molecular study of the type species *Ralfsia fungiformis* revealing it to be distant from the Ectocarpales and, therefore, lending support to the order Ralfsiales. This has received widespread acceptance (Boo *et al.* 2012; Brodie *et al.* 2016; Gallardo *et al.* 2016; Mathieson & Dawes 2017; Silberfeld *et al.* 2014). Following McCauley & Weir's (2007) transfer of the Lithodermataceae – which encompassed *Heribaudiella*, *Lithoderma*, *Pseudolithoderma* and *Bodanella* – from the Ralfsiales to the Sphacelariales, the order then comprised three families: Mesosporaceae, Neoralfsiaceae and Ralfsiaceae (Silberfeld *et al.* 2014), these later being supplemented with the Hapalospongidiaceae León-Alvarez (2017), Pseudoralfsiaceae Parente, Fletcher & G.W. Saunders (2020) and Sungminiaceae Oteng'o *et al.* (2022). Up to very recently only the Ralfsiaceae was represented in the flora of Britain and Ireland, with the species *Ralfsia verrucosa*. Then, on the basis of Parente *et al.*'s (2020) molecular analyses and morpho-anatomical study, *R. verrucosa* was renamed *Pseudoralfsia verrucosa* and transferred from the Ralfsiaceae to the new family Pseudoralfsiaceae, rendering the former family no longer represented in the British and Irish seaweed flora.

The Tilopteridales, originally proposed by Bessey (1907) and emended by Phillips *et al.* (2008a), contains the Cutleriaceae, Halosiphonaceae, Phyllariaceae and Tilopteridaceae in the present study, based on Silberfeld *et al.* (2104), but only the latter three families are included by Brodie *et al.* (2016), with the Cutleriaceae retained in its own order, Cutleriales, as positioned by Parke & Dixon (1976).

Within the Sphacelariales, following the publication of Silberfeld *et al.* (2014), five families are recognized for Britain and Ireland by Brodie *et al.* (2016) and by the present author, viz. Cladostephaceae, Lithodermataceae, Sphacelariaceae, Sphacelodermataceae and Stypocaulaceae, compared to only three families – Cladostephaceae, Sphacelariaceae and Stypocaulaceae – listed in Parke & Dixon (1976) and other authors (e.g. Cormaci *et al.* 2012). Usually positioned within the Sphacelariales, the Cladostephaceae has received quite widespread acceptance (see Coppejans & Kling 1995; Fritsch 1945; Kornmann & Sahling 1977; Newton 1931; Parke & Dixon 1976; Prud'homme van Reine 1982a; Stegenga & Mol 1983; Womersley 1987) although not by some authors, e.g. by Hamel (1931–39) and Rueness (1977) who placed *Cladostephus* within the Sphacelariaceae. However, final recognition of the family

Cladostephaceae was provided in a molecular study by Draisma *et al.* (2010b), overturning their earlier placement of the family within the Sphacelariaceae (Draisma *et al.* 2002). The Sphacelodermataceae is based on *Sphaceloderma caespitulum*, formerly known as *Sphacelaria caespitulum*, and was proposed on the results of a molecular study by Draisma *et al.* (2010b). This has received quite widespread acceptance (see de Reviers *et al.* 2015 and Silberfeld *et al.* 2014).

Apropos the Lithodermataceae, a family usually placed within the Ralfsiales/Ralfsiaceae (Parke & Dixon 1976; Ribera *et al.* 1992), molecular studies by Bittner *et al.* (2008), Lim *et al.* (2007) and McCauley & Wehr (2007) have revealed its taxa to be distant from the Ralfsiales and to form a close relationship with the Sphacelariales, a position that has received more recent support (Silberfeld *et al.* 2014; de Reviers *et al.* 2015).

The only other notable difference between Parke & Dixon (1976) and this and Brodie *et al.*'s (2016) study is the former's recognition of the family Cystoseiraceae within the Fucales, whereas herein, and in Brodie *et al.* (2016), it is absorbed into the Sargassaceae. The inclusion of the Cystoseiraceae within the Sargassaceae was based on molecular studies (see de Reviers & Rousseau 1999) and this later found support in studies by Cho *et al.* (2006b), Harvey & Goff (2006) and Phillips *et al.* (2008a).

Table 6 provides a summary of the main molecular studies carried out on the Phaeophyceae.

Arrangement of the work

The format adopted for the presentation of this volume follows quite closely that of Fletcher (1987). Taxa are arranged according to Silberfeld *et al.* (2014). Prior to the Taxonomic Treatment, two keys – a quick key and a more detailed dichotomous key – are given to the brown algal genera of Britain and Ireland. In the Taxonomic Treatment, full formalized diagnoses are given for each order, family, genus and species, usually followed by additional notes/discussions. The latter can be of an explanatory nature, a summary of the salient features of the taxon, a comparison of closely related taxa, or details of life history features and so forth. Particular emphasis is placed on the latter information, in view of the important role played (or potentially played) by many of the presently included microscopic taxa in the life histories of various macroalgae. The references and proposals for conservation of the family names follow Silva (1980). For each genus, the type species is stated along with the synonymy and, where appropriate, keys to the species are included. For each species, a full synonymy is listed, followed by a detailed description and notes on habitat characteristics, distributional patterns around Britain and Ireland and seasonal aspects of growth and reproduction. Also included is a comprehensive bibliography for each species, relating to aspects of its biology, ecology, distribution, seasonal occurrence, life history and molecular biology; less emphasis is given to aspects of its biochemistry, economic value and mariculture. Where possible, the species description follows the common sequence of morphology, anatomy, cytology and reproduction. Distributional data, which are given on the basis of the pre-1974 county system (see map on page viii), include only verified records and must be considered as conservative. Particular emphasis has been given to the provision of illustrations and at least one compound plate of line drawings and/ or black and white photographs is provided for each species.

Note on county boundaries

The geographical distribution records given for Britain and Ireland are based on the Watson-Praeger vice-counties system used in biological recording.

Table 6. Molecular studies on the Phaeophyceae.

Taxonomic group	References
Class Phaeophyceae (whole class)	Draisma 2002; Draisma & Prud'homme van Reine 2001; Draisma *et al.* 2001, 2003; Kawai *et al.* 1997; Lim *et al.* 1986; McDevit & Saunders 2009; Motonura *et al.* 2010; Phillips *et al.* 2001, 2008a,b; de Reviers & Rousseau 1999; Reviers *et al.* 2007, 2015; Rousseau & de Reviers 1999a; Rousseau *et al.* 2001, Siemer *et al.* 1998, 2001; Silberfeld 2010; Silberfeld *et al.* 2010, 2011, 2014; Snirc *et al.* 2010; Tan 1995; Tan & Druehl 1993; Tan *et al.* 1993.
Orders	
Discosporangiales	Kawai *et al.* 2007.
Ishigeales	Cho *et al.* 2004; Kawai *et al.* 2005; Lee *et al.* 2003, 2009; Peters & Moe 2001.
Dictyotales	Bittner *et al.* 2008; De Clerck *et al.* 2001, 2006; Hoshina *et al.* 2004; Hwang *et al.* 2004a,b; Lee & Bae 2002; Lee & King 1996; Oudot-Le Secq *et al.* 2006; Schultz *et al.* 2015; Silberfeld *et al.* 2013.
Onslowiales	Bittner *et al.* 2008; Draisma & Prud'homme van Reine 2001; Phillips *et al.* 2008a.
Sphacelariales	Bittner *et al.* 2008; Draisma 2002; Draisma *et al.* 2002, 2010b; Heesch *et al.* 2008, 2020; Kawai *et al.* 2005; Keum *et al.* 2005; McCauley & Wehr 2007; Phillips *et al.* 2008a; Reviers *et al.* 2007.
Syringodermatales	Burrowes *et al.* 2003.
Asterocladales	Müller *et al.* 1998; Silberfeld *et al.* 2011; Uwai *et al.* 2005.
Desmarestiales	Kawai & Sasaki 2000; Oudot-Le Secq 2006; Peters *et al.* 1997, 2000; Sasaki *et al.* 2001; Tan & Druehl 1996; Yang *et al.* 2014.
Ectocarpales	Assali & Loiseaux-de Goër 1992; Assali *et al.* 1990, 1991; Boo *et al.* 2011a,b; Burkhardt & Peters 1998; Camus *et al.* 2005; Cho & Boo 2006; Cho *et al.* 2001, 2002, 2003, 2005a,b, 2006b, 2007; Cock *et al.* 2010; Geoffroy *et al.* 2015; Kawai *et al.* 1995; Kim 2010; Kim & Kawai 2002; Kimura *et al.* 2010a; Klochkova *et al.* 2012; Kogame & Masuda 2001; Kogame *et al.* 1999, 2005, 2006, 2015a,b; Lee *et al.* 2002; McCauley & Wehr 2007; McDevit & Saunders 2017; Murúa *et al.* 2018; Oudot-Le Secq *et al.* 2001, 2002; Parente *et al.* 2011b; Peters 2003; Peters & Burkhardt 1998; Peters & Ramírez 2001; Peters *et al.* 2011, 2015; Racault *et al.* 2009; Rousseau & de Reviers 1999a; Rousseau *et al.* 2000; Santiañez & Kogame 2017; Santiañez *et al.* 2018a,b; Siemer & Pedersen 1995; Siemer *et al.* 1998; Silberfeld *et al.* 2011; Stache-Crain *et al.* 1997; Uwai *et al.* 2001, 2002; West *et al.* 2010.
Fucales	Camacho *et al.* 2014; Canovas *et al.* 2011; Cho *et al.* 2006c; Coyer *et al.* 2002a,b,c, 2004, 2006a,b,c; Dixon *et al.* 2012; Draisma *et al.* 2010a; Engelen *et al.* 2001; Hoarau *et al.* 2009; Horiguchi & Yoshida 1998; Kucera & Saunders 2008; LeClerc *et al.* 1998; Lee *et al.* 1999; Lim *et al.* 1986; Lu & Williams 1994; Mattio & Payri 2009; Mattio *et al.* 2008, 2009, 2010; Oak *et al.* 2002; Olsen *et al.* 2002; Oudot-Le Secq *et al.* 2006; Phillips & Fredericq 2000; Phillips *et al.* 2005; Rousseau & de Reviers 1997, 1999b; Rousseau *et al.* 1997, 2001; Saunders & Kraft 1995; Serrão *et al.* 1999; Stiger *et al.* 2000, 2003; Yoshida *et al.* 2000, 2002.
Ralfsiales	Lim *et al.* 2007; Parente *et al.* 2005, 2010, 2011a, 2020; Tan & Druehl 1994.
Scytothamnales	Peters & Clayton 1998; Silberfeld *et al.* 2011.
Nemodermatales	Kawai *et al.* 2016b; Phillips *et al.* 2008a.
Laminariales	Bhattacharya & Druehl 1989; Bhattacharya *et al.* 1991a,b; Billot *et al.* 1998, 1999; Boo & Yoon 2000; Boo *et al.* 1999, 2011a,b; Cho *et al.* 2006a; Coyer *et al.* 1997, 2001; Druehl 1992; Druehl & Saunders 1992; Druehl *et al.* 1997; Kawai & Sasaki 2000; Kawai *et al.* 2001, 2008; Klochkova *et al.* 2017; Kraan & Guiry 2000a,b, 2001; Kraan *et al.* 2000, 2001; Lane & Saunders 2005; Lane *et al.* 2006, 2007; Lim *et al.* 1986; Mayes *et al.* 1992; McDevit & Saunders 2010; Oudot-Le Secq *et al.* 2002; Peters 1998; Peters *et al.* 1997; Robuchon *et al.* 2014; Rousseau & de Reviers 1999a; Sasaki & Kawai 2007; Sasaki *et al.* 2001; Saunders & Druehl 1992, 1993; Saunders & McDevit 2014; Selivanova *et al.* 2007; Stam *et al.* 1988; Tan & Druehl 1993, 1996; Uwai *et al.* 2006a,b, 2007; Yoon & Boo 1999; Yoon *et al.* 2001; Yotsukura *et al.* 1999.
Sporochnales	Kawai & Sasaki 2000; Tan & Druehl 1996; Yee 2007.
Stschapoviales	Kawai *et al.* 2015a,b.
Tilopteridales	Kawai & Sasaki 2000, 2004; Kuhlenkamp *et al.* 1993; Peters 1998; Phillips *et al.* 2008a; Sasaki *et al.* 2001; Siemer *et al.* 1998.

Checklist of Phaeophyceae for Britain and Ireland

On the following pages are listed all the brown algal taxa (Class Phaeophyceae) recorded for Britain and Ireland and covered in the present treatise. The list comprises 4 subclasses, 12 orders, 27 families, 105 genera, 188 species, 1 subspecies, 1 variety and 13 fucoid ecads. It also includes two 'phases' (*Compsonema saxicola* 'phase' of *Petalonia/Scytosiphon* and the *Microspongium gelatinosum* 'phase' of *Scytosiphon lomentaria*). The systematic arrangement followed derives from the molecular study by Silberfeld *et al.* (2014). Subclassal, ordinal and familial names are given with the names of the authorities that described them, along with the dates of publication. The authorities for each order follow that of Silva & de Reviers (2000). All genera are provided with the names of the authors who described them, along with the date of publication and page numbers. All species are provided with the names of the authors who described them.

Largely based on Silberfeld *et al.* (2014), there are a number of differences in the composition of the families when comparing the present treatise and the checklist of British brown algae presented in Brodie *et al.* (2016). These include:

1. The addition of four subclasses in the present treatise.

2. The positioning of *Kuetzingiella* within the Chordariaceae rather than in the Acinetosporaceae.

3. The positioning of *Buffhamia* and *Strepsithalia* in the Chordariaceae rather than in the Chordaceae.

4. The positioning of *Compsonema* (and the species *C. microspongium* and *C. minutum*) in the Chordariaceae rather than in the Scytosiphonaceae. Only *Compsonema saxicola* should be placed in the Scytosiphonaceae in view of its involvement, in some studies, as a phase in the life histories of *Planosiphon* and *Scytosiphon* species.

5. The inclusion of *Microspongium gelatinosum* (with plurilocular sporangia) within the synonymy of *Microspongium globosum*. *Microspongium gelatinosum* with unilocular sporangia is transferred to the Scytosiphonaceae as a 'phase' in the life history of *Scytosiphon lomentaria*.

6. The inclusion of five additional species of *Microspongium*, viz. *M. alariae*, *M. oligosporum*, *M. parasiticum*, *M. radians* and *M. zanardinii*. *Microspongium alariae* and *M. radians* represent new records for Britain and Ireland. *Microspongium oligosporum* (as *Streblonema oligosporum*) was not listed in Brodie *et al.* (2016) and was considered by them (see also Guiry 2012 and Cormaci *et al.* 2012) as a synonymy of *Litosiphon laminariae* (see p. 326 in the present treatise). *Microspongium parasiticum* and *M. zanardinii* are both listed as species of *Streblonema* in Brodie *et al.* (2016).

7. Four species of *Chilionema* – *C. foecundum*, *C. hispanicum*, *C. ocellatum* and *C. reptans* are recognized in the present treatise – compared to only two species (*C. hispanicum* and *C. ocellatum*) in Brodie *et al.* (2016), the latter authors placing *C. foecundum* into the genus *Myrionema* and (probably) placing *Chilionema reptans* within the synonymy of *Asperococcus fistulosus* (see Fletcher 1987 for discussion).

8. *Myrionema orbiculare* is not included in the present treatise, in agreement with Cormaci *et al.* (2012) that it is a Mediterranean species and should be kept distinct from *Myrionema magnusii* which is distributed only in the Atlantic ocean.

9. *Phaeostroma pustulosum* and *Pilinia rimosa* are placed in the Chordariaceae in the present treatise, but placed in the Ectocarpaceae by Brodie *et al.* (2016).

10. Within the Ectocarpaceae, *Pleurocladia lucifuga* and two additional species of *Ectocarpus* (*Ectocarpus crouaniorum* and *E. penicillatus*) are included in the present treatise.

11. The positioning of *Sorapion simulans* as *incertae sedis* in the present treatise (see p. 685) rather than in the Scytosiphonaceae, following Silberfeld *et al.* (2014).

12. *Petroderma* and *Pseudolithoderma* are removed from the Ralfsiaceae and placed elsewhere: *Petroderma* in the Petrodermataceae and *Pseudolithoderma* in the Lithodermataceae, following Silberfeld *et al.* (2014). The Ralfsiales includes the Pseudoralfsiaceae, *Pseudoralfsia* and *Pseudoralfsia verrucosa*, replacing the Ralfsiaceae, *Ralfsia* and *Ralfsia verrucosa*.

13. The family Cutleriaceae is placed in the order Tilopteridales in the present treatise, after Silberfeld *et al.* (2014), but positioned in its own order, Cutleriales, by Brodie *et al.* (2016).

14. Two additional ecads are recognized for *Ascophyllum nodosum* (*scorpioides, minor*).

15. A number of ecads are recognized for some *Fucus* spp. (*F. ceranoides, F. serratus, F. spiralis* and *F. vesiculosus*) and *Pelvetia canaliculata*.

16. The inclusion of two additional genera, *Ericaria* and *Gongolaria*, within the Sargassaceae.

Checklist of the class Phaeophyceae recorded for Britain and Ireland (taxonomic arrangement according to Silberfeld *et al.* 2014)

CLASS PHAEOPHYCEAE Kjellman 1891c

SUBCLASS DISCOSPORANGIOPHYCIDAE Silberfeld, F. Rousseau & de Reviers 2014
ORDER Discosporangiales O.C. Schmidt 1937 *emend.* H. Kawai, Hanyuda, Draisma & Müller 2007
  Choristocarpaceae Kjellman 1891b
    Choristocarpus Zanardini 1860, p. 45
      tenellus Zanardini

SUBCLASS ISHIGEOPHYCIDAE Silberfeld, F. Rousseau & de Reviers 2014
ORDER Ishigeales G.Y. Cho & S.M. Boo in Cho *et al.* 2004
  Petrodermataceae Silberfeld, F. Rousseau & de Reviers 2014
    Petroderma Kuckuck 1897, p. 382
      maculiforme (Wollny) Kuckuck

SUBCLASS DICTYOTOPHYCIDAE Silberfeld, F. Rousseau & de Reviers 2014
ORDER Dictyotales Bory de Saint-Vincent 1828
  Dictyotaceae J.V. Lamouroux ex Dumortier 1822
    Dictyopteris J.V. Lamouroux 1809, p. 332, nom. cons.
      polypodioides (A.P. de Candolle) J.V. Lamouroux
    Dictyota J.V. Lamouroux 1809, p. 38, nom. cons.
      dichotoma (Hudson) J.V. Lamouroux
      spiralis Montagne
    Padina Adanson 1763, p. 13, p. 586, nom. cons.
      pavonica (Linnaeus) Thivy
    Taonia J. Agardh 1848, p. 101
      atomaria (Woodward) J. Agardh

ORDER Sphacelariales Migula 1908
  Cladostephaceae Oltmanns 1922
    Cladostephus C. Agardh 1817, p. xxv
      hirsutus Boudouresque & M. Perret
      spongiosus (Hudson) C. Agardh
  Lithodermataceae Hauck 1883
    Pseudolithoderma Svedelius in Kjellman & Svedelius 1910, p. 175
      extensum (P. Crouan & H. Crouan) S. Lund
      roscoffensis Loiseaux
  Sphacelariaceae Decaisne 1842
    Battersia Reinke ex Batters 1890, p. 59 *emend.* Draisma, Prud'homme & H. Kawai 2010
      mirabilis Reinke ex Batters
      plumigera (Holmes ex Hauck) Draisma, Prud'homme & H. Kawai
      racemosa (Greville) Draisma, Prud'homme & H. Kawai
    Chaetopteris Kützing 1843, p. 293
      plumosa (Lyngbye) Kützing
    Sphacelaria Lyngbye in Hornemann 1818, p. 8
      cirrosa (Roth) C. Agardh
      fusca (Hudson) S.F. Gray
      plumula Zanardini
      rigidula Kützing
      tribuloides Meneghini
    Sphacelorbus Draisma, Prud'homme & H. Kawai 2010b, p. 322
      nanus (Nägeli ex Kützing) Draisma, Prud'homme & H. Kawai
  Sphacelodermaceae Draisma, Prud'homme & H. Kawai 2010
    Sphaceloderma Kuckuck 1894, p. 232
      caespitulum (Lyngbye) Draisma, Prud'homme & H. Kawai
  Stypocaulaceae Oltmanns 1922
    Halopteris Kützing 1843, p. 292
      filicina (Grateloup) Kützing
      scoparia (Linnaeus) Sauvageau
    Protohalopteris Draisma, Prud'homme & H. Kawai 2010b, p. 321
      radicans (Dillwyn) Draisma, Prud'homme & H. Kawai

SUBCLASS FUCOPHYCIDAE Cavalier-Smith 1986
ORDER Desmarestiales Setchell & N.L. Gardner 1925
  Arthrocladiaceae Chauvin 1842
   Arthrocladia Duby 1830, p. 971
     villosa (Hudson) Duby
  Desmarestiaceae Kjellman
   Desmarestia J.V. Lamouroux 1813, p. 24, nom. cons.
     aculeata (Linnaeus) J.V. Lamouroux
     dudresnayi J.V. Lamouroux ex Leman ['dresnayi']
     ligulata (Stackhouse) J.V. Lamouroux
     viridis (O.F. Müller) J.V. Lamouroux

ORDER Ectocarpales Bessey 1907 *emend.* F. Rousseau & de Reviers 1999a
  Acinetosporaceae G. Hamel ex Feldmann 1937
   Acinetospora Bornet 1891, p. 370
     crinita (Carmichael) Sauvageau
   Feldmannia G. Hamel 1931–39, p. 67
     irregularis (Kützing) G. Hamel
     lebelii (J.E. Areschoug ex P. Crouan & H. Crouan) G. Hamel
     mitchelliae (Harvey) H.-S. Kim
     padinae (Buffham) G. Hamel
     paradoxa (Montagne) Hamel
     simplex (P. Crouan & H. Crouan) G. Hamel
   Herponema J. Agardh 1882, p. 55
     solitarium (Sauvageau) G. Hamel
     valiantei (Bornet ex Sauvageau) G. Hamel
     velutinum (Greville) J. Agardh
   Hincksia J.E. Gray 1864, p. 12
     fenestrata (Berkeley ex Harvey) P.C. Silva
     granulosa (Smith) P.C. Silva
     hincksiae (Harvey) P.C. Silva
     ovata (Kjellman) P.C. Silva
     sandriana (Zanardini) P.C. Silva
     secunda (Kützing) P.C. Silva
   Pogotrichum Reinke 1892, p. 61
     filiforme Reinke
   Pylaiella Bory de Saint-Vincent 1823, p. 393, nom. cons.
     littoralis (Linnaeus) Kjellman
  Chordariaceae Greville 1830 *emend.* A.F. Peters & Ramírez 2001
   Acrothrix Kylin 1907, p. 93
     gracilis Kylin
   Asperococcus J.V. Lamouroux 1813, p. 277
     bullosus J.V. Lamouroux
     ensiformis (Delle Chiaje) M.J. Wynne
     fistulosus (Hudson) W.J. Hooker
     scaber Kuckuck
   Botrytella Bory de Saint-Vincent 1822, p. 425
     micromora Bory de Saint-Vincent
   Buffhamia Batters 1895b, pp. 168, 169
     speciosa Batters
   Chilionema Sauvageau 1898, p. 263
     foecundum (Strömfelt) R.L. Fletcher
     hispanicum (Sauvageau) R.L. Fletcher
     ocellatum (Kützing) Kornemann
     reptans (P. Crouan & H. Crouan) Sauvageau
   Chordaria C. Agardh 1817, p. xii, nom. cons.
     flagelliformis (O.F. Müller) C. Agardh
   Cladosiphon Kützing 1843, p. 329
     contortus (Thuret) Kylin
     zosterae (Agardh) Kylin
   Compsonema Kuckuck 1899, p. 58
     microspongium (Batters) Kuckuck

minutum (C. Agardh) Kuckuck
Corynophlaea Kützing 1843, p. 331
  crispa (Harvey) Kuckuck
Cylindrocarpus P. Crouan & H. Crouan 1851, p. 359
  microscopicus P. Crouan & H. Crouan
Dictyosiphon Greville 1830, pp. xliii, 55, nom. cons.
  chordaria J.E. Areschoug
  foeniculaceus (Hudson) Greville
Elachista Duby 1830, p. 972, nom. cons.
  flaccida (Dillwyn) Fries
  fucicola (Velley) J.E. Areschoug
  scutulata (Smith) J.E. Areschoug
  stellaris J.E. Areschoug
Endodictyon Gran 1897, p. 47
  infestans Gran
Eudesme J. Agardh 1882, p. 29
  virescens (Carmichael ex Berkeley) J. Agardh
Fosliea Reinke 1891, p. 45
  griffithsiana (Le Jolis) Hagen ex Athanasiadis
Giraudia Derbès & Solier in Castagne 1851, p. 100
  sphacelarioides Derbès & Solier
Halothrix Reinke 1888, p. 19
  lumbricalis (Kützing) Reinke
Hecatonema Sauvageau 1898, p. 248
  terminale (Kützing) Kylin
Isthmoplea Kjellman 1877, p. 31
  sphaerophora (Carmichael) Goby
Kuetzingiella Kornmann in Kuckuck 1956, pp. 293, 314
  battersii (Bornet ex Sauvageau) Kornmann
  holmesii (Batters) Kornmann
Laminariocolax Kylin 1947, p. 6
  aecidioides (Rosenvinge) A.F. Peters
  tomentosoides (Farlow) Kylin
Leathesia S.F. Gray 1821, pp. 279, 301
  marina (Lyngbye) Decaisne
Leblondiella G. Hamel 1931–39, p. xl
  densa (Batters) G. Hamel
Leptonematella P.C. Silva 1959, p. 63, nom. cons.
  fasciculata (Reinke) P.C. Silva
Litosiphon Harvey 1849, p. 43
  laminariae (Lyngbye) Harvey
Mesogloia C. Agardh 1817, p. xxxvii
  lanosa P. Crouan & H. Crouan
  vermiculata (Smith) S.F. Gray
Microcoryne Strömfelt 1888, p. 382
  ocellata Strömfelt
Microspongium Reinke 1888, p. 20
  alariae (Pedersen) Peters
  globosum Reinke
  immersum (Levring) P.M. Pedersen
  oligosporum (Strömfelt) Fletcher comb. nov.
  parasiticum (Sauvageau) Fletcher comb. nov.
  radians (Howe) Peters
  stilophorae (P. Crouan & H. Crouan) Cormaci & G. Furnari
  zanardinii (P. Crouan & H. Crouan) Fletcher
Mikrosyphar Kuckuck 1895, p. 177
  polysiphoniae Kuckuck
  porphyrae Kuckuck
Myriactula Kuntze 1898, p. 415
  areschougii (P. Crouan & H. Crouan) G. Hamel
  chordae (J.E. Areschoug) Levring

  clandestina (P. Crouan & H. Crouan) Feldmann
  haydenii (Gatty) Levring
  rivulariae (Suhr) Feldmann
  stellulata (Harvey) Levring
 Myriocladia J. Agardh 1841, p. 48
  lovenii J. Agardh
  tomentosa P. Crouan & H. Crouan
 Myrionema Greville 1827, pl. 300
  corunnae Sauvageau
  liechtensternii Hauck
  magnusii (Sauvageau) Loiseaux
  papillosum Sauvageau
  strangulans Greville
 Myriotrichia Harvey 1834, p. 299
  clavaeformis Harvey
 Phaeostroma Kuckuck in Reinbold 1893, p. 43
  pustulosum Kuckuck
 Pilinia Kützing 1843, p. 273
  rimosa Kützing
 Protectocarpus Kornmann 1955, p. 119
  speciosus (Børgesen) Kornmann
 Punctaria Greville 1830, p. 52
  crispata (Kützing) Trevisan
  latifolia Greville
  plantaginea (Roth) Greville
  tenuissima (C. Agardh) Greville
 Sauvageaugloia G. Hamel ex Kylin 1940, p. 32
  divaricata (Clemente) Cremades
 Spermatochnus Kützing 1843, p. 334, nom. cons.
  paradoxus (Roth) Kützing
 Sphaerotrichia Kylin 1940, p. 38
  divaricata (C. Agardh) Kylin
 Stictyosiphon Kützing 1843, p. 301
  soriferus (Reinke) Rosenvinge
  tortilis (Gobi) Reinke
 Stilophora J. Agardh 1841, p. 6, nom. cons.
  nodulosa (C. Agardh) P.C. Silva
  tenella (Esper) P.C. Silva
 Stilopsis Kuckuck 1929, pp. 11, 70
  lejolisii (Thuret) Kuckuck & Nienburg ex G. Hamel
 Streblonema Derbès & Solier in Castagnea 1851, p. 100
  breve (Sauvageau) De Toni
  fasciculatum Thuret in Le Jolis
  helophorum (Rosenvinge) Batters
  intestinum (Reinsch) Batters
 Strepsithalia Bornet ex Sauvageau 1896, p. 64
  buffhamiana (Batters) Batters
 Striaria Greville 1828, p. 44
  attenuata (Greville) Greville
 Ulonema Foslie 1894, p. 131
  rhizophorum Foslie
Ectocarpaceae C. Agardh 1828 *emend.* T. Silberfeld, M. Racault, R.L. Fletcher, A.F. Peters, F. Rousseau & B. de Reviers 2011
 Ectocarpus Lyngbye 1819, p. 130, nom. cons.
  crouaniorum Thuret
  fasciculatus Harvey
  penicillatus (C. Agardh) Kjellman
  siliculosus (Dillwyn) Lyngbye
 Pleurocladia A. Braun 1855, p. 80
  lucifuga (Kuckuck) Wilce

Spongonema Kützing 1849, p. 461
   tomentosum (Hudson) Kützing
Petrospongiaceae M. Racault, R.L. Fletcher, B. de Reviers, G.Y. Cho, S.M. Boo, M. Parente & F. Rousseau 2009
  Petrospongium Nägeli ex Kützing 1858, p. 2
   berkeleyi (Greville) Nägeli ex Kützing
Scytosiphonaceae Ardissone & Straforello 1877
  Colpomenia (Endlicher) Derbès & Solier 1851, p. 95
   peregrina Sauvageau
  Compsonema saxicola 'phase' of *Planosiphon* and *Scytosiphon* spp.
  Microspongium gelatinosum 'phase' of *Scytosiphon lomentaria*
  Petalonia Derbès & Solier in Castagne 1850, p. 265, nom. cons.
   fascia (O.F. Müller) Kuntze
  Planosiphon McDevit & G.W. Saunders 2017, p. 660
   filiformis (Batters) Santiañez & Kogame
   zosterifolius (Reinke) McDevit & Saunders
  Scytosiphon C. Agardh 1820, p. 160, nom. cons.
   dotyi M.J. Wynne
   lomentaria (Lyngbye) Link
  Stragularia Strömfelt 1886, p. 173
   clavata (Harvey) G. Hamel
   spongiocarpa (Batters) G. Hamel
  Symphyocarpus Rosenvinge 1893, p. 896
   strangulans Rosenvinge

ORDER Fucales Bory 1827
 Fucaceae Adanson 1763
  Ascophyllum Stackhouse 1809, pp. 54, 66, nom. cons.
   nodosum (Linnaeus) Le Jolis
   nodosum megecad *limicola* ecad *mackayi*
   nodosum megecad *limicola* ecad *minor*
   nodosum megecad *limicola* ecad *scorpioides*
  Fucus Linnaeus 1753, p. 1158
   ceranoides *Linnaeus*
   ceranoides megecad *limicola* ecad *glomerata*
   ceranoides megecad *limicola* ecad *proliferatus*
   ceranoides megecad *limicola* ecad *ramosissima*
   cottonii M.J. Wynne & Magne
   distichus Linnaeus
   distichus **subsp.** *evanescens* (C. Agardh) H.T. Powell
   macroguiryi Almeida, E.A. Serrão & G.A. Pearson
   serratus Linnaeus
   serratus megecad *limicola* ecad *limicola*
   spiralis Linnaeus
   spiralis megecad *limicola* ecad *nanus*
   vesiculosus Linnaeus
   vesiculosus megecad *limicola* ecad *caespitosus*
   vesiculosus megecad *limicola* ecad *volubilis*
  Pelvetia Decaisne & Thuret 1845, p. 12
   canaliculata (Linnaeus) Decaisne & Thuret
   canaliculata megecad *limicola* ecad *coralloides*
   canaliculata megecad *limicola* ecad *libera*
   canaliculata megecad *limicola* ecad *muscoides*
 Himanthaliaceae (Kjellman) De Toni 1891
  Himanthalia Lyngbye 1819, p. 36
   elongata (Linnaeus) S.F. Gray
 Sargassaceae Kützing 1843
  Bifurcaria Stackhouse 1809, p. 59
   bifurcata R. Ross
  Cystoseira C. Agardh 1820, p. 50, nom. cons.
   foeniculacea (Linnaeus) Greville
   humilis var. *myriophylloides* (Sauvageau) J.H. Price & D.M. John

Ericaria Stackhouse 1809, p. 56
  selaginoides (Linnaeus) Molinari & Guiry
Gongolaria Boehmer in Ludwig 1760, p. 454
  baccata (S.G. Gmelin) Molinari & Guiry
  nodicaulis (Withering) Molinari & Guiry
Halidrys Lyngbye 1819, p. 37, nom. cons.
  siliquosa (Linnaeus) Lyngbye
Sargassum C. Agardh 1820, p. 1, nom. cons.
  muticum (Yendo) Fensholt

ORDER Laminariales Migula 1909
  Alariaceae Setchell & N.L. Gardner 1925
    Alaria Greville 1830, pp. xxxix, 25, nom. cons.
      esculenta (Linnaeus) Greville
    Undaria Suringar 1873, p. 77
      pinnatifida (Harvey) Suringar
  Chordaceae Dumortier 1822
    Chorda Stackhouse 1797, p. xxiv
      filum (Linnaeus) Stackhouse
  Laminariaceae Bory de Saint-Vincent 1827
    Laminaria J.V. Lamouroux 1813, p. 40, nom. cons.
      digitata (Hudson) J.V. Lamouroux
      hyperborea (Gunnerus) Foslie
      ochroleuca Bachelot de la Pylaie
    Saccharina Stackhouse 1809, pp. 53, 65
      latissima (Linnaeus) C.E. Lane, C. Mayes, Druehl & G.W. Saunders

ORDER Ralfsiales Nakamura ex P.-E. Lim & H. Kawai 2007
  Pseudoralfsiaceae Parente, R.L. Fletcher & G.W. Saunders 2020
    Pseudoralfsia Parente, R.L. Fletcher & G.W. Saunders 2020, pp. 15, 16
      verrucosa Parente, R.L. Fletcher & G.W. Saunders

ORDER Sporochnales Sauvageau 1926
  Sporochnaceae Greville 1830
    Carpomitra Kützing 1843, p. 343, nom. cons.
      costata (Stackhouse) Batters
    Sporochnus C. Agardh 1817, p. xii
      pedunculatus (Hudson) C. Agardh

ORDER Stschapoviales H. Kawai 2017
  Halosiphonaceae H. Kawai & H. Sasaki 2000
    Halosiphon Jaasund 1957, p. 211
      tomentosus (Lyngbye) Jaasund

ORDER Tilopteridales Bessey 1907 *emend.* Phillips *et al.* 2008a
  Cutleriaceae J.W. Griffith & Henfrey 1856
    Cutleria Greville 1830, p. 59
      multifida (Turner) Greville
    Zanardinia Nardo ex Zanardini 1841, p. 236
      prototypus (Nardo) P.C. Silva
  Phyllariaceae Tilden 1935
    Saccorhiza Bachelot de la Pylaie 1829, p. 23, nom. cons.
      polyschides (Lightfoot) Batters
  Tilopteridaceae Kjellman 1890
    Haplospora Kjellman 1872, p. 3
      globosa Kjellman
    Tilopteris Kützing 1849, p. 462
      mertensii (Turner) Kützing

  INCERTAE SEDIS
    Sorapion Kuckuck 1894, p. 236
      simulans Kuckuck

Keys to genera

The following keys to the genera include a quick key based on morphological growth forms and a more detailed dichotomous key. For genera such as *Asperococcus* and *Desmarestia*, which can include more than one growth form, identification is sometimes given to the species level.

Quick generic key

Group A

Thallus macroscopic, solitary, tufted, uniseriate branched filaments, usually >1 cm, < 10 cm in length.

Ectocarpus: much branched throughout, ribbon-shaped plastids, sporangia lateral, solitary, stalked, distributed throughout thallus.

Choristocarpus: sparsely branched, mainly pseudodichotomous, cells very light coloured, prominent apical cell, propagules common.

Hincksia: much branched throughout, discoid plastids, sporangia lateral, solitary, sessile, distributed throughout thallus.

Feldmannia (*mitchelliae*): much branched throughout, discoid plastids, plurilocular sporangia distributed throughout thallus, characteristically cigar-like in shape (this is the only species of *Feldmannia* to greatly exceed 1 cm in length, formerly a *Hincksia* but removed from the latter as a result of a molecular study).

Acinetospora: infrequently branched throughout, branches divaricate, discoid to rod-shaped plastids, monosporangia present.

Botrytella: much branched throughout, discoid to rod-shaped plastids, sporangia in dense clusters.

Kuetzingiella: sparsely branched throughout, lobed/plate-like plastids.

Spongonema: much branched throughout; branching densely interwoven; plastids plate-like/ribbon-shaped.

Pylaiella: much branched throughout, discoid plastids, sporangia intercalary in branches.

Group B

Thallus microscopic, usually <1 cm in length, solitary tufts or dense and turf-like, uniseriate branched filaments.

Herponema (*velutinum*): dense turf-like patches on *Himanthalia elongata*.

Pilinia: dense turf-like patches on cave walls.

Feldmannia: branching and sporangia confined to basal region, discoid plastids.

Cylindrocarpus: epiphytic/endophytic on *Gracilaria* spp.

Streblonema (*breve*): epiphytic/endophytic on *Ascophyllum nodosum*.

Herponema: fine turf-like patches on *Ericaria* and *Gongolaria* spp., various Dictyotales and *Himanthalia elongata*.

Hecatonema: mainly epiphytic, forming small tufts of little-branched filaments arising from a monostromatic, in parts distromatic, base with multiseriate plurilocular sporangia borne laterally on the erect filaments.

Group C

Thallus microscopic, mainly epiphytic, forming discs, patches, hemispherical globules, colourations on various hosts

Microspongium: epilithic, epiphytic and endophytic, with a filamentous basal system giving rise to short erect filaments, plurilocular sporangia and/or hairs which, if endophytic, break through the surface of the host.

Mikrosyphar: endophytic, forming discrete patches on *Polysiphonia* spp. and *Porphyra* spp.

Streblonema: endophytic with thallus immersed in host, with only hairs and sporangia sometimes emerging.

Myrionema: mainly epiphytic with a discoid/pseudodiscoid filamentous, monostromatic, uniseriate base giving rise to short, erect, uniseriate filaments with terminal uniseriate, plurilocular sporangia.

Chilionema: epiphytic with a discoid, pseudodiscoid, partly distromatic, base giving rise to short, erect filaments and pluriseriate, plurilocular sporangia.

Ulonema: epiphytic on *Dumontia contorta*, with downward-penetrating rhizoidal base.

Laminariocolax: endophytic, predominantly on kelps, breaking out through the surface as pustules or felt-like patches.

Endodictyon: endozoic in various hydroids and bryozoans.

Phaeostroma: endophytic and epiphytic on various hosts, notably on *Chorda*, forming endophytic threads which give rise to hairs and sporangia, plurilocular sporangia multiseriate and heart-shaped.

Group D

Thallus epiphytic, forming small, brush-like tufts with free, erect, uniseriate, unbranched filaments.

Elachista: tufts with hemispherical, wart-like, pseudoparenchymatous, somewhat gelatinous base, filaments > 1 mm in length and free.

Myriactula: tufts with hemispherical, pseudoparenchymatous wart-like somewhat gelatinous base, with erect short filaments < 1 mm in length and often associated in a gelatinous matrix.

Halothrix: tufts arising from a discoid/pseudodiscoid base or fibrous rhizoidal filaments, erect filaments 25–65 μm in width.

Leptonematella: tufts arising from a discoid/pseudodiscoid base, erect filaments 7–17 μm in width.

Group E

Thallus mainly epilithic, either encrusting or procumbent, pseudoparenchymatous or parenchymatous.

Pseudoralfsia: epilithic or commonly epizoic on limpets and littorinids, dark brown, coriaceous, firmly attached crusts, filaments firmly united, cells with a single plate-like plastid, sporangia in raised gelatinous sori; unilocular sporangia common, at the base of multicellular paraphyses; plurilocular sporangia infrequent, terminal on erect filaments, not accompanied by paraphyses; abundant in eulittoral.

Stragularia: epilithic, epizoic, rarely epiphytic, dark brown, moderately adjoined filaments, cells with a single plate-like plastid and pyrenoid, unilocular sporangia in raised gelatinous discrete sori, borne at base of paraphyses, plurilocular sporangia rare, terminal on erect filaments, without paraphyses, eulittoral.

Petroderma: epilithic, light brown, usually thin crusts, erect filaments gelatinous and easily separated by light pressure, cells with a single plate-like plastid, sporangia terminal, paraphyses absent, eulittoral.

Pseudolithoderma: epilithic, dark brown, firmly attached crusts, filaments strongly adjoined, cells with several discoid plastids, sporangia in slightly raised, gelatinous sori without paraphyses, lower eulittoral and sublittoral.

Sorapion: epilithic on stones and shells, thin crusts, cells with a single, plate-like plastid, terminal unilocular sporangia, paraphyses absent; very rare, only known from Devon, sublittoral.

Symphyocarpus: epilithic, epizoic, soft crust, easily squashed, cells with a single plate-like plastid, terminal, multiseriate plurilocular sporangia; very rare, lower eulittoral, sublittoral.

Battersia mirabilis: epilithic, thin, strongly adherent crust, cells with several discoid plastids, sporangia stalked, in small clumped yellowish sori, projecting from crust surface.

Zanardinia/Aglaozonia: Procumbent, membranous thallus, loosely attached underneath, either with an outer fringe of filaments (*Zanardinia*) or a large apical cell (*Aglaozonia*-asexual phase of *Cutleria*), not pseudoparenchymatous with erect filaments but parenchymatous with internal differentiation of cell types, cells with numerous discoid plastids, sporangia borne in surface sori.

Group F

Thallus pseudoparenchymatous, hemispherical, cushion-like, spongy, easily squashed.

    Leathesia: Thallus epiphytic, globose to irregularly spreading to 5(–8) cm.

    Corynophlaea: Thallus forming hemispherical spots, to 3 mm across on *Chondrus crispus*.

    Petrospongium: Thallus epilithic on bedrock, sometimes epiphytic on *Pseudoralfsia verrucosa*, forming hemispherical, gelatinous cushions to 2 cm in diameter.

Group G

Thallus erect, macroscopic, filiform, pseudoparenchymatous with peripheral, outward-projecting, assimilatory filaments, usually mucilaginous with all constituent filaments separable under pressure.

    Thallus < 5 mm in length.

        = *Microcoryne*.

    Thallus > 5 mm in length.

        Assimilatory filaments only at thallus apices.

            = *Spermatochnus*.

        Assimilatory filaments covering entire thallus, usually greater than 5 cells

            Thallus tough and cartilaginous, with constituent filaments not easily separable under pressure.

            = *Chordaria*.

            Thallus usually soft and mucilaginous, with constituent filaments usually separable under pressure.

                Thallus irregular in diameter.

                    Mesogloia: Thallus irregular in diameter, markedly mucilaginous and easily squashed.

                Thallus regular in diameter, not markedly mucilaginous and easily squashed.

Acrothrix/Myriocladia/Sauvageaugloia: Assimilatory filaments < 8 cells, rarely more than 50 μm in length (= *Acrothrix*), assimilatory filaments < 12 cells, usually 60–150 μm in length (= *Sauvageaugloia*), assimilatory filaments to 11(–14) cells (= *Myriocladia*).

Thallus regular in diameter, markedly mucilaginous and easily squashed.

Eudesme: Terminal regions of thallus giving rise to short, lateral, globular side branches, assimilatory filaments in tufts, plurilocular sporangia rare, epilithic in mid-tide level pools.

Cladosiphon: terminal regions of thallus not giving rise to short, lateral, globular side branches, assimilatory filaments not in tufts, plurilocular sporangia common, epiphytic on *Zostera*.

Assimilatory filaments covering entire thallus, less than 5 cells in length.

Stilopsis: Assimilators papillose, to 3 cells in length.

Sphaerotrichia: Assimilators to 5 cells in length, with inflated apical cell.

Group H

Thallus epiphytic, forming small, solitary, erect shoots lacking primary branches, to 40(–70) mm in height, usually uniseriate below and becoming multiseriate and parenchymatous above, filiform to clavate.

Pogotrichum: erect shoots linear, terete, sporangia formed in surface cells.

Leblondiella: erect shoots terete to slightly clavate, surface cells producing short radial branches which clothe the axis, these branches also branched and usually bearing the sporangia.

Litosiphon: erect shoots filiform, terete, to slightly elongate-clavate, hairs abundant, sporangia formed in surface cells.

Myriotrichia: erect shoots filiform becoming elongate-clavate, producing short, lateral radial branches usually sheathing the main axis, sporangia lateral on main axis or offshoots.

Group I

Thallus forming tufts or turfs, a few mm to 10 cm in height, often epiphytic, the erect shoots/axes/branches often remaining equal in width throughout or are somewhat acuminate, segments clearly parenchymatous and segmented, each segment showing longitudinal divisions (i.e. not uniseriate).

Thallus unbranched, hairs in groups of up to 6 in terminal regions.

= *Giraudia*

Thallus sparsely to frequently branched, lateral branches straddling a cross wall between segments, hairs lacking, solitary or in small groups, propagules absent.

Halopteris: thallus repeatedly branched, often corticated by downward-growing rhizoids, ultimate branches alternate; axes mostly greatly exceeding 50 μm in width, hairs lacking or in groups in axils of branches.

Protohalopteris: branching scattered and irregular, without corticating sheath, axes not exceeding 50 μm in width.

Cladostephus: main axes sparsely to several times branched, with thick corticating sheath, axes mostly greatly exceeding 50 μm in width, ultimate thin branchlets in whorls.

Thallus usually branched, lateral branches arising between the cross walls of a segment, hairs solitary, propagules present or absent.

Sphacelaria: thallus with propagules and without corticating sheath.

Sphacelorbus: thallus without propagules, and without corticating sheath, axes mostly less than 25 µm in width, forming dense carpets in upper eulittoral.

Sphaceloderma: thallus without propagules, and usually without corticating sheath, axes mostly more than 25 µm in width, forming solitary tufts in the sublittoral, usually on *Laminaria* stipes.

Chaetopteris: thallus without propagules, with a corticating sheath, lateral branches feather-like, 75–100 µm in width, to 700 µm with cortex, sporangia usually borne on specialized laterals arising from cortication sheath.

Battersia: thallus without propagules, crust-like, with or without erect filaments, main axes often with a corticating sheath, lateral branches feather-like or scarcely to irregularly branched, sporangia borne on fertile laterals formed on the ultimate feather-like laterals.

Group J

Thallus macroscopic, solitary, tufted, branched filamentous, uniseriate at first becoming pluriseriate and parenchymatous in parts, especially in lower regions.

Isthmoplea: thallus usually > 6 cm in length, unilocular sporangia exserted, often in opposite pairs, monosporangia absent.

Fosliea: Thallus to 14 cm in length, unilocular sporangia formed by the transformation of surface cortical cells; epiphytic on *Palmaria palmata*, rarely on other algae.

Tilopteris: thallus to 21 cm in length, branchlets pinnate, remaining short and often unequal, reproductive organs (monosporangia, oogonia, antheridia) intercalary in monosiphonous parts of thallus.

Haplospora: thallus to 32 cm in length, indeterminate ultimate branches often widely divergent, meiosporangia usually terminal.

Group K

Thallus macroscopic, solitary, tufted, branched or unbranched, pluriseriate throughout, all branches attenuate above to an apical cell or uniseriate filament/hair.

Dictyosiphon: thallus to 60 cm in length, branched, branch apices with an apical cell, lacking a hair.

Striaria: thallus to 44 cm in length, branched, branch apices terminating in a hair, thallus hollow, branching usually opposite, sporangia exserted.

Stictyosiphon: thallus to 1 m in length, branched, branch apices terminating in a hair, thallus hollow only at base, branching not usually opposite, sporangia sunk in thallus.

Buffhamia: thallus to 3 cm in length, simple, linear, fringing *Sauvageaugloia griffithsiana*.

Group L

Thallus parenchymatous, spherical or saccate, membranous, hollow and firm.

= *Colpomenia*.

Group M

Thallus parenchymatous, a flat blade, usually < 30 cm long.

Petalonia: blades unbranched, narrow to fairly broad, flattened, solid, in surface view cells < 18 µm in diameter, cells with a single plate-like plastid with pyrenoid.

Planosiphon: blades unbranched, linear, flattened, hollow or partially hollow, in surface view cells < 18 μm in diameter, cells with a single plate-like plastid with pyrenoid.

Punctaria: blades unbranched, narrow to broad, cells in surface view usually > 18 μm in diameter, cells with several discoid plastids.

Dictyopteris: blades branched, with a central thickened midrib, apical growth by groups of apical cells.

Dictyota: blades branched, no central midrib, growth apical by a single, large cell.

Padina: blades fan-shaped when mature, hairs present in concentric bands.

Taonia: blades irregularly branched/cleft, apices truncate, growth from a marginal row of apical cells; hairs and sporangia in concentric zones/lines across blade.

Cutleria (sexual phase): blades branched, linear, terminal tuft of hair-like filaments.

Group N

Thallus very large, parenchymatous, usually > 1 m in length, coriaceous and clearly divided into a terminal flattened blade, terete or flattened stipe and a root/finger-like attachment base, mainly sublittoral.

Alaria: blades with midrib, stipe cylindrical, frond margin entire or split, lower eulittoral and sublittoral fringe of exposed shores.

Undaria: blades with midrib, stipe flattened, frond margin entire at first, becoming lobed, lower eulittoral pools, sublittoral and sides of floating structures, e.g. marinas.

Saccorhiza: blades without midrib, stipe flattened, holdfast tuberous, inflated and warty.

Laminaria: blade without midrib, stipe cylindrical, blade digitate, smooth edged.

Saccharina: blade without midrib, usually dimpled when older, stipe cylindrical, blade entire, crinkly edged.

Group O

Thallus macroscopic, parenchymatous or pseudoparenchymatous, firm, no obvious differentiation into blade, stipe and holdfast, axes terete.

Thallus hollow/inflated, unbranched.

Chorda: thallus to 6 m in length, fringed by white hairs, sublittoral.

Halosiphon: thallus to 1 m in length, fringed by dark brown, pigmented assimilatory filaments, sublittoral.

Scytosiphon: thallus linear, tubular, constricted, epilithic, to 40 cm in length, intertidal pools

Planosiphon: thallus linear, oval, unconstricted, hollow or partially hollow, epilithic, to 25 cm in length, upper eulittoral.

Asperococcus (*fistulosus*): thallus to 40 cm long, epiphytic, especially on *Fucus* in tidal pools.

Asperococcus (*turneri*): thallus to 30 cm long, epilithic, lower eulittoral, sublittoral.

Thallus solid, branched.

Bifurcaria: thallus cylindrical, dichotomously branched, rhizome-like base, mid-eulittoral pools.

Sargassum: thallus bushy, irregularly branched, air-bladders mainly pedicellate, in leaf axils, holdfast conical, lower eulittoral pools and shallow sublittoral.

Cystoseira: thallus caespitose, arising from a discoid holdfast, primary axis terete or flattened, radially or bilaterally branched, spine-like appendages absent, sometimes slightly iridescent, with intercalary bladders, lower eulittoral pools and shallow sublittoral.

Ericaria: thallus bushy, arising from a conical holdfast, primary axis terete, radially branched, all branches covered with spine-like appendages; strongly iridescent; tophules absent, lower eulittoral pools and sublittoral.

Gongolaria: Thallus non-caespitose, bushy, arising from a discoid/conical holdfast; primary axis flattened and zigzag or terete, branching radial or distichous, occasionally with spine-like appendage, lightly iridescent and with or without scattered aerocysts; tophules present or absent, lower eulittoral pools and sublittoral.

Arthrocladia: thallus axes and branches with regular whorls of hair-like tufts projecting laterally.

Sporochnus: thallus axes and branches giving rise to short, divaricate branches of limited growth, terminated by a tuft of hair-like pigmented filaments.

Desmarestia (*viridis*): thallus without hair-like tufts of filaments, TS shows central large cell.

Carpomitra: thallus with midrib, terminated by a tuft of pigmented hair-like filaments.

Group P

Thallus macroscopic, parenchymatous or pseudoparenchymatous, usually firm, no obvious differentiation into blade, stipe and holdfast, axes dorsiventrally flattened to oval.

Himanthalia: thallus button-shaped, giving rise to long, to 2 m, strap-like, dichotomously branched receptacles, lower eulittoral pools and exposed.

Fucus: thallus dichotomously branched, with midrib and terminally situated receptacles, upper to lower eulittoral.

Pelvetia: thallus dichotomously branched, channelled, without midrib, upper eulittoral.

Ascophyllum: thallus irregularly dichotomously branched, without midrib; large, ovate air-bladders borne at intervals along thallus, receptacles borne on small lateral branches; mid-eulittoral sheltered shores, emersed.

Halidrys: thallus without midrib, air-bladders siliquose and articulate, usually terminal on branches, lower eulittoral pools and shallow sublittoral.

Desmarestia (*dudresnayi, ligulata, aculeata*): thallus firm or papery with or without midrib and veins, opposite or alternate branching, all branches sometimes ending in fascicles of hair-like filaments, central axial cell seen in TS.

Full generic key

1. Whole thallus of uniseriate, heterotrichous filaments; erect filaments if
    present, branched or unbranched, and usually free, not united to form a
    differentiated/undifferentiated pseudoparenchymatous tissue 2
    Thallus of another form 32

2. Filaments of erect systems absent. 3
    Filaments of erect systems (or emergent filaments if thallus endobiotic)
    present, either as assimilatory cells, colourless hairs or both. 4

3. Plurilocular sporangia formed from scarcely modified vegetative cells and
    consisting of 2–4 loculi; thalli endophytic, appearing as brown stains in
    various algae, notably *Porphyra* and *Polysiphonia* spp. . . . . *MIKROSYPHAR*
    Plurilocular sporangia irregular in shape, terminal or intercalary but usually
    < 4 loculi; endozoic in the bryozoans *Alcyonidium, Flustra, Membranopora*
    . . . . *ENDODICTYON*

4. Filaments of erect systems usually profusely branched; thalli usually
    macroscopic (> 5 mm); epilithic or epibiotic but not markedly endobiotic 5
    Filaments of erect system unbranched or sparsely branched; thalli usually
    microscopic (< 5 mm); epilithic, epibiotic or endobiotic 12

5. Plastids ribbon-shaped, branched or unbranched, usually few per cell. 6
    Plastids +/– discoid, usually many per cell 7

6. Plastids usually unbranched; filaments densely interwoven forming a
    branched, string-like frond; crampons present; often very mucilaginous
    . . . . *SPONGONEMA*
    Plastids usually branched; filaments free; crampons absent; thalli not
    particularly mucilaginous . . . . *ECTOCARPUS*

7. Thalli small, rarely > 1.0 cm. 8
    Thalli > 1.0 cm 9

8. Filaments of erect system with diffuse or intercalary growth; usually
    branched only at base; plastids with pyrenoids; reproduction by plurilocular
    sporangia, unilocular sporangia rare; propagules absent. . . . . *FELDMANNIA*
    Filaments of erect system with apical growth from prominent apical
    cell; filaments sparsely branched but throughout length; plastids lacking
    pyrenoids; reproduction by unilocular and plurilocular sporangia;
    propagules of 2 (1–3) distended cells present. . . . . *CHORISTOCARPUS*

9. Erect filaments sparsely branched; branches arising in no obvious pattern;
    angle of branch emergence ±90°; crampons sometimes present; plastids
    irregularly discoid; reproduction by unilocular and plurilocular sporangia,
    and monosporangia . . . . *ACINETOSPORA*
    Erect filaments profusely branched, frequently in +/– orderly succession;
    angle of branch emergence usually < 90°; crampons absent 10

10. All erect filaments terminating in colourless hair with basal growth. . . . . *BOTRYTELLA*
    True hairs absent or occasional. 11

11. Insertion of ultimate branches usually opposite; all filaments ecorticate;
    longitudinal divisions in older vegetative cells quite common; thalli usually
    eulittoral; plurilocular and unilocular sporangia intercalary in branches, latter
    in chains . . . . *PYLAIELLA*

Insertion of ultimate branches rarely opposite; lower parts of main erect filaments commonly corticate; longitudinal cell divisions rare; thalli usually sublittoral; plurilocular and unilocular sporangia sessile or pedicellate but never intercalary in position . . . . *HINCKSIA*

12. Erect filaments present only as colourless hairs; epiphytic on *Zostera*, *Scytosiphon*, *Laminaria* and particularly *Chorda*; procumbent filaments often coalescent; plurilocular sporangia formed from repeated divisions of vegetative cell(s), very irregular in shape, often rounded/heart-shaped with many loculi; unilocular sporangia borne from upper surfaces of procumbent filaments . . . . *PHAEOSTROMA*
Erect assimilatory filaments present, at least in parts 13

13. Thallus epiphytic, forming small discoid or irregularly shaped stains or patches on hosts, with closely packed, short, free, erect filaments often slightly mucilaginous in content 14
Thallus epiphytic or epilithic, forming small tufts, either discrete or forming turf-like patches on substrata, not obviously mucilaginous 18

14. Cells of erect filaments with a single, large, plate-like plastid . . . . *CHILIONEMA* (*C. hispanicum*)
Cells of erect filaments with 1–3 plate-like, lobed plastids. 15

15. Basal cells of erect filaments frequently longitudinally divided (biseriate); erect filaments simple; plurilocular sporangia multiseriate . . . . *CHILIONEMA* (*C. ocellatum, C. foecundum*)
Basal cells of erect filaments uniseriate and not longitudinally divided; erect filaments rarely simple, usually branched; plurilocular sporangia uniseriate 16

16. Erect filaments pseudodichotomously branched, bearing terminal and/or lateral plurilocular sporangia . . . . *MICROSPONGIUM*
Erect filaments unbranched, secundly branched or with short protuberances; plurilocular sporangia usually sessile or shortly stalked on basal layer, more rarely terminal or erect filaments 17

17. Procumbent system a compact disc with no or few downward-growing rhizoids; thalli reported on various hosts, particularly common on *Ulva* spp., rarely epilithic (*Myrionema liechtensternii*) . . . . *MYRIONEMA*
Procumbent system a loose, irregular disc with many downward-extending branched rhizoidal filaments; thalli epiphytic on the red alga *Dumontia contorta* only . . . . *ULONEMA*

18. Erect system greater than prostrate system; thalli epilithic or epibiotic, never endobiotic. 19
Erect systems equal or less than prostrate systems; thalli usually endobiotic, sometimes epiphytic. 27

19. Erect filaments unbranched. 20
Erect filaments frequently branched 23

20. Erect filaments arising from a wart-like, usually hemispherical cushion of compacted, erect filaments of large, colourless cells 21
Erect filaments arising from a discoid/pseudodiscoid base of outward-radiating filaments or from a fibrous network of rhizoidal filaments 22

21. Erect filaments up to 2 mm in length. . . . . *MYRIACTULA*
Erect filaments 3–40 mm in length . . . . *ELACHISTA*

22. Cells of erect filaments 25–65 μm in diameter; plurilocular sporangia clustered, forming densely packed, vertical tiers enclosing vegetative cells; unilocular sporangia unknown . . . . *HALOTHRIX*
    Cells of erect filaments 7–17 μm in diameter; plurilocular sporangia formed by subdivision of vegetative cells; unilocular sporangia rare, borne laterally on the base of erect filaments . . . . *LEPTONEMATELLA*

23. Each vegetative cell contains several discoid plastids; thalli epilithic or epiphytic on *Taonia* and *Padina*; erect filaments sparsely branched; plurilocular sporangia sessile or pedicellate and borne either on procumbent filaments or laterally on erect filaments; unilocular sporangia sessile or pedicellate and borne laterally on erect filaments . . . . *KUETZINGIELLA*
    Each vegetative cell contains 1(–3) plate-like plastids. 24

24. Thallus epilithic forming extensive, confluent tufts of yellow golden-brown compacted filaments in caves; true hairs absent; unilocular sporangia terminal on erect filaments in chains, uniseriate or biseriate; plurilocular sporangia unknown . . . . *PILINIA*
    Thallus epilithic, epiphytic or epizoic, usually forming small discrete tufts of filaments, rarely confluent, upper to lower eulittoral, rarely in caves; true hairs present; unilocular sporangia solitary, terminal or lateral; plurilocular sporangia, if present, multiseriate, lateral. 25

25. Cells of basal layer with longitudinal divisions (biseriate) in part; branching of erect filaments irregularly lateral, with both branches and lateral plurilocular sporangia widely divergent at first, later recurving towards parental filament. . . . . *HECATONEMA*
    Cells of basal layer uniseriate; branching of erect filaments pseudodichotomous or secund, with branches and lateral plurilocular sporangia erect and adpressed closely towards parental filament 26

26. Plurilocular sporangia commonly terminal on erect filaments and frequently branched, usually in a characteristic cock's-comb (secund) fashion . . *PROTECTOCARPUS*
    Plurilocular sporangia rarely terminal on erect filaments and never branched in a cock's-comb fashion . . . . *COMPSONEMA*

27. Growth of emergent filaments less than that internally and frequently reduced to sporangia, hairs and tips of erect filaments . . . . *STREBLONEMA*
    Growth of filaments emergent from host as great as that internally 28

28. Emergent (erect) filaments unbranched or sparsely branched; branching never fastigiate; thalli not mucilaginous. 29
    Emergent (erect) filaments repeatedly branched; branching usually fastigiate; thalli mucilaginous. 31

29. Emergent filaments usually branched; plastids band-shaped, 2 (1–3) per cell; plurilocular sporangia uniseriate, borne terminally or laterally on emergent filaments; unilocular sporangia unknown; thalli usually endophytic in *Laminaria*. . . . . *LAMINARIOCOLAX*
    Emergent filaments rarely branched; plastids discoid, several per cell 30

30. Hairs present, borne singly or in small clusters at apices of erect filaments; plurilocular sporangia uniseriate and borne in small clusters terminally on erect filaments; unilocular sporangia also in terminal clusters or occasionally on prostrate filaments; sporangia sessile, sometimes shortly pedicellate . . . . *MYRIOTRICHIA*

Hairs absent; plurilocular sporangia multiseriate; plurilocular and unilocular sporangia borne singly, never in clusters; sporangia terminal or lateral on erect filaments, sessile or pedicellate . . . . *HERPONEMA*

31. Emergent filaments attenuate; cells cylindrical, length 2 × diam. or more; unilocular sporangia cylindrical; plurilocular sporangia biseriate; thalli endophytic in *Gracilaria* . . . . *CYLINDROCARPUS*
Emergent filaments clavate; cells isodiametric; unilocular sporangia ovoid or spherical; thalli endophytic in *Mesogloia* and *Sauvageaugloia* . . . .*STREPSITHALIA*

32. Thallus with erect filaments united to form a crust. 33
Thallus of another form 39

33. Erect filaments of crust moderately or weakly adjoined and easily separated in squash preparations. 34
Erect filaments of crust tightly adjoined and not easily separated in squash preparations. 37

34. Crusts smooth, firm and slightly subcoriaceous; erect filaments not obviously mucilaginous, moderately adjoined and not readily separated by light pressure under a coverslip; unilocular sporangia in prominent, raised, gelatinous sori which are obvious in surface view . . . .*STRAGULARIA*
Crusts soft, sponge-like; erect filaments markedly mucilaginous, very loosely adjoined and easily separated by light pressure under a cover-slip; unilocular sporangia (if present) in sori not obvious externally. 35

35. Crusts slightly pulvinate; unilocular sporangia up to 100 × 27 μm, arising laterally at the base of multicellular paraphyses; plurilocular sporangia unknown for Britain and Ireland; ascocyst-like cells not present
. . . . *MICROSPONGIUM GELATINOSUM*
PHASE OF *SCYTOSIPHON LOMENTARIA*
Crusts comparatively thin; unilocular sporangia, if present, up to 23 × 14 μm, arising terminally on erect filament, unaccompanied by paraphyses; plurilocular sporangia and ascocyst-like cells common 36

36. Thallus epilithic, common, reported all around Britain and Ireland, littoral fringe to lower eulittoral especially on embedded flint stones; erect filaments up to 25(–37) cells; unilocular sporangia common; plurilocular sporangia uniseriate, rarely biseriate or triseriate, borne terminally on erect filaments, up to 32 loculi; ascocysts long, intercalary or terminal. . . . . *PETRODERMA*
Thallus epiphytic, rare, reported for only a few scattered localities around Britain and Ireland, lower eulittoral and sublittoral; erect filaments up to 3(–11) cells; unilocular sporangia unknown; plurilocular sporangia multiseriate, arranged in 2 or 4 vertical columns, rarely irregularly clustered, to 4–7 loculi, borne terminally on erect filaments; ascocysts elongate, cylindrical or slightly clavate, borne terminally on filaments . . . . . . *SYMPHYOCARPUS*

37. Several plastids per cell, thallus a thin to moderately thick, dark brown to black, coriaceous crust; in vertical section or squash preparation, thallus clearly pseudoparenchymatous in structure, and comprising closely adherent, erect filaments of cells; plurilocular sporangia and unilocular sporangia borne in continuous, mucilaginous sori on thallus surface; locally common in lower eulittoral pools and sublittoral . . . . *PSEUDOLITHODERMA*
One plastid per cell. 38

38. Erect filaments often > 10 cells; thallus a dark brown-black, thick, coriaceous, loose to moderately well attached crust, sometimes confluent and showing concentric growth rings; margin prominent, usually quite easily lifted by a scalpel, central regions verrucose, bullate and often brittle; vertical sections through margin reveal arched erect filaments with no obviously terminal apical cell; unilocular sporangia borne laterally at the base of multicellular paraphyses, obovate to elongate-pyriform; plurilocular sporangia in closely packed vertical, usually uniseriate columns; common all around Britain and Ireland, usually epilithic/epizoic, throughout eulittoral zone, emersed or immersed in pools, frequently epizoic on limpets and barnacles. . . . . *PSEUDORALFSIA*
Erect filaments < 6 cells; thallus a comparatively thin and strongly adherent, smooth crust with a discrete margin not lifted by a scalpel; vertical sections through margin reveal a single, large, apical cell; erect filaments vertical, not arched, near margin; unilocular sporangia pyriform and widest distally, borne apically on erect filaments; secondary assimilatory filaments absent; plurilocular sporangia unknown; thalli epizooic on mollusc shells and only known from the sublittoral zone near Plymouth, Devon . . . . *SORAPION*

39. Thallus pseudoparenchymatous and formed by lateral fusion of axes, branches or both; essentially, the filamentous nature of the thallus is evident and often only a squash preparation under a microscope slide is required to separate constituent filaments; thallus rarely massive and in the form of globose or irregular-shaped cushions, or a branched cylindrical frond, often filiform, +/− hollow, occasionally flattened, +/− mucilaginous, soft or cartilaginous; axial filaments single (uniaxial) or multiple (multiaxial) 40
Thallus truly parenchymatous and formed by repeated transverse and longitudinal cell divisions within axes, branches or both; separate filaments lacking and thallus not easily squashed to separate cells and tissues; thallus variable in form, often massive but simplest species may be subfilamentous 59

40. Thallus globose, hemispherical, cushion or finger-shaped to clavate-like, simple or more rarely branched, mucilaginous, pseudoparenchymatous, easily squashed under a microscope slide into constituent filaments 41
Thallus erect, filiform, much branched, either soft and mucilaginous or tough and cartilaginous, pseudoparenchymatous in structure. 44

41. Thallus epiphytic, forming small (to 5 mm in height) discrete, erect, cylindrical, clavate, more rarely forked, finger-like extensions on hosts . . . . *MICROCORYNE*
Thallus epiphytic or epilithic, forming discrete, less commonly confluent, hemispherical to irregularly globose cushions on substrata 42

42. Thallus epilithic or epiphytic on crusts of the brown alga *Pseudoralfsia verrucosa* . . . . *PETROSPONGIUM*
Thallus epiphytic, reported on various hosts, in particular *Chondrus crispus* and *Corallina officinalis* 43

43. Thallus hemispherical and smooth, light/dark brown in colour, solid throughout, not exceeding 3 mm in diameter, internal tissue consisting of large cylindrical, not stellate colourless cells; terminal paraphyses 5–17 cells in length; epiphytic on *Chondrus crispus* . . . . *CORYNOPHLAEA*

Thallus hemispherical becoming globose, irregular and convoluted later, yellow-brown in colour, solid becoming hollow, to 8 cm in diameter, internal tissue consisting of large stellate, colourless cells; terminal paraphyses 2–5 cells in length; epiphytic on various hosts, especially *Corallina officinalis* ... *LEATHESIA*

44. Thallus uniaxial 45
    Thallus multiaxial 52

45. Thallus constructed from a dense cortication of a single axial filament and its branches; cortication forms a small-celled tissue, axial cell remains distinct in TS; thallus bears prominent assimilatory hairs, especially in juvenile parts (may be absent in winter); often tough and cartilaginous, not mucilaginous; filiform, terete or ligulate; branching opposite or alternate 46
    Thallus constructed by a loose cortication of a single axial filament and its branches; cells of cortication similar in size to those of axis, axial cell becomes indistinct; cortication arises from short lateral assimilatory filaments and may, in turn, bear secondary assimilatory filaments; hairs colourless; thallus soft and often mucilaginous, never cartilaginous; branching pseudodichotomous or irregular 47

46. Thallus filiform, up to 15 cm in length; pigmented hairs borne in branched tufts and arranged in whorls around axes over the entire length; unilocular sporangia in chains on whorled tufts . . . . *ARTHROCLADIA*
    Thallus filiform, terete or ligulate, usually greater than 15 cm in length; no whorled arrangement of pigmented hairs; unilocular sporangia sunk in peripheral layer of thallus . . . . *DESMARESTIA*

47. Assimilatory filaments present only at frond apices; frond surface consisting of a small-celled layer bearing sporangia in sori; sporangia almost always unilocular; sori contain paraphyses and often show a spiral or whorled arrangement; frond pseudodichotomously branched; branches with a prominent, exposed, dome-shaped apical cell . . . . *SPERMATOCHNUS*
    Assimilatory filaments clothing entire frond 48

48. Assimilatory filaments usually > 8 cells in length 49
    Assimilatory filaments usually < 8 cells in length 51

49. Frond filiform, irregular in diameter, becoming distended, markedly mucilaginous . . . . *MESOGLOIA*
    Frond filiform, +/– constant in diameter, not markedly mucilaginous 50

50. Assimilatory filaments 6–8 cells in length; axes profusely branched . . . . . . *ACROTHRIX*
    Assimilatory filaments > 12 cells in length; axes sparsely branched . . . . . *MYRIOCLADIA*

51. Assimilatory filaments 2–3 cells in length, lacking prominent apical cell; frond always becoming hollow but with narrow bore and not distended; unilocular sporangia in sori irregularly distributed over thallus surface; frond texture soft; branches terminate in apical cell . . . . *STILOPSIS*
    Assimilatory filaments approx. 5 cells in length and consisting of narrow cells capped by large +/– spherical cell; thallus solid but may become hollow; unilocular sporangia present but not in very distinct sori; frond texture firm; branches terminate in short filaments . . . . *SPHAEROTRICHIA*

52. Assimilatory filaments clothing axes from apex to base 53
    Assimilatory filaments absent or present only at frond apices 57

53. Morphological boundary between assimilatory filaments and axes very sharp 54
    Morphological boundary between assimilatory filaments and axes +/– indistinct 56

54. Thallus tough, cartilaginous, dark brown, solid, bearing many and long primary branches but relatively few secondary and tertiary branches; unilocular sporangia borne at bases of assimilatory filaments and apparently sunk in peripheral layers of frond; plurilocular sporangia unknown on macrothallus . . . .*CHORDARIA*
Thallus soft and mucilaginous 55

55. Thallus +/– richly branched, branches long; frond 10–30 cm; cells of axes 20 × diam. in length; unilocular sporangia borne at bases of assimilatory filaments; plurilocular sporangia unknown on macrothallus; epilithic, sometimes epiphytic. . . . . *SAUVAGEAUGLOIA*
Thallus +/– sparsely branched, branches short; frond 2–20 cm; cells of axes 4–8 × diam. in length; unilocular sporangia borne at bases of assimilatory filaments; plurilocular sporangia uniseriate or multiseriate, borne laterally on or at the apices of assimilatory filaments; thalli epiphytic on *Zostera* leaves . . . . *CLADOSIPHON*

56. Thallus to 50 cm in length, +/– abundantly branched, dark olive-brown; assimilatory filaments in tufts which are pedicellate on axes; thalli epilithic on stones, often in sandy or pebbly areas; sporangia as in *Cladosiphon* . . . .*EUDESME*
Thallus to 40 cm in length, +/– sparsely branched, yellow-brown; assimilatory filaments single or in small tufts borne directly on axes, not pedicellate; often epiphytic on *Zostera* . . . . *CLADOSIPHON*

57. Thallus apices lacking tuft of hairs; apices attenuate, never swollen; assimilatory filaments present but only on distal parts of branches, peripheral tissue layer of most of frond composed of small epidermal cells; thallus filiform with only 3–5 axial filaments; unilocular and plurilocular sporangia present, latter uniseriate and branched or unbranched, approx. 6 loculi; sporangia in prominent sori scattered throughout frond; sori contain colourless hairs and paraphyses. . . . .*STILOPHORA*
Thallus apices each crowned with a prominent tuft of hairs; apices sometimes swollen; assimilatory filaments absent; thallus filiform or flattened, with many axial filaments. 58

58. Thallus filiform, terete; assimilatory filaments absent but paraphyses present – crowded together with, and just proximal to, tufts of hairs at frond apices; paraphyses branched and with large apical cells; frond alternately or irregularly branched, 10–45 cm, soft but not mucilaginous; unilocular sporangia borne on paraphyses; plurilocular sporangia unknown on macrothallus . . . . *SPOROCHNUS*
Thallus filiform, ligulate; axis forms midrib on which two lateral wings of tissue are produced; assimilatory filaments absent but paraphyses similar to those of *Sporochnus* present at frond apices; frond 10–30 cm, dichotomously branched; unilocular sporangia borne on paraphyses; plurilocular sporangia unknown on macrothallus. . . . . *CARPOMITRA*

59. Thallus a thin, black, closely adherent, coriaceous crust; in VS or squash preparation thallus parenchymatous in structure and not separable into erect filaments of cells; plurilocular and unilocular sporangia borne on short, polysiphonous filaments grouped together in discrete, yellowish brown sori; in shallow, sandy, lower eulittoral pools, and only known from Berwick and Fife in Britain. . . . .*Battersia mirabilis*
Thallus of another form, procumbent or erect 60

60. Thallus procumbent, membranous, loosely attached underneath and easily
    removed intact. 61
    Thallus essentially erect, of another form 62

61. Thallus roundish or reniform, up to 20 cm in diameter; growing edge of
    thallus bears fringe of hairs; older thalli becoming dark and coriaceous with
    ragged edges; upper surfaces of thalli bear colourless hairs and reproductive
    sori of either plurilocular gametangia (oogonia, antheridia) or unilocular
    sporangia . . . . *ZANARDINIA*
    Thallus orbicular at first, becoming irregularly spreading later, 1–2 cm in
    diameter, frequently with overlapping lobes, with entire and rounded margin
    without an outer fringe of hair-like filaments; upper surface of thallus bearing
    colourless hairs and reproductive sori comprising unilocular sporangia only,
    life history phase of *Cutleria multifida* . . . . *AGLAOZONIA*

62. Thallus spherical, hollow and epibiotic. . . . . *COLPOMENIA*
    Thallus of another form 63

63. Thallus filiform, usually < 2(–3) mm diameter, not massive; soft or
    cartilaginous; some species only slightly parenchymatous and clearly
    filamentous in parts, thallus solid, sometimes hollow in older tissues 64
    Thallus greatly variable in form, if filiform usually > 2(–3) mm in diameter,
    can be massive, can be solid or tubular, usually cartilaginous or leathery 84

64. Axes branched throughout length 65
    Axes unbranched or branched only at base 79

65. Thallus with a prominent cell at the apex of every growing branch 66
    Thallus with diffuse growth; no prominent apical cells 73

66. Lateral branches borne between the cross walls of one articulation; thalli
    small, seldom exceeding 5 cm; commonly epibiotic; often showing vegetative
    reproduction by means of propagules. 67
    Lateral branches borne straddling a cross wall between two adjacent
    articulations, thalli usually > 5 cm; usually epilithic; never forming propagules. 71

67. Propagules present . . . . *SPHACELARIA*
    Propagules absent. 68

68. Thallus with a corticating sheath; sporangia borne on specialized laterals/
    branches arising from the terminal laterals or from the outer cells of a rhizoidal cortex. . . 69
    Thallus +/– a corticating sheath; sporangia sessile or shortly stalked on the
    terminal laterals. 70

69. Sporangia borne on densely packed, specialized laterals arising from the
    surface cells of the corticating sheath at base of thallus, more rarely borne on
    branched laterals arising from the ultimate pinnate laterals . . . . *CHAETOPTERIS*
    Sporangia borne on fertile laterals arising from the ultimate laterals. . . . . . . *BATTERSIA*

70. Main axes mainly 15–23 µm wide, rarely swollen in parts to 35 µm; only some
    segments with secondary transverse walls; basal crust only a few cells high;
    forming dense carpets in the upper eulittoral, caves etc. . . . . *SPHACELORBUS*
    Main axes mainly 20–34 µm, swollen in parts to 48 µm; nearly all segments
    with secondary transverse walls; crust base quite thick, with superimposed
    crusts; forming usually solitary tufts in the sublittoral, on rocks and stones
    and especially epiphytic on *Laminaria* stipes . . . . *SPHACELODERMA*

71. Axes not exceeding 50 µm in width, and not corticated. . . . . *PROTOHALOPTERIS*
    Axes exceeding 50 µm in width and corticated 72

72. Thallus axes repeatedly branched; ultimate branches alternate, borne directly
on axes . . . . *HALOPTERIS*
Thallus axes branched once or twice; ultimate branches in whorls around
axes, borne on secondary thickening which encases axes . . . . *CLADOSTEPHUS*

73. Thallus monosiphonous in parts 74
Thallus parenchymatous from apex to base. 78

74. Reproductive structures partly or fully immersed in parenchyma. . . . *STICTYOSIPHON*
Reproductive structures never immersed in parenchyma 75

75. Reproductive structures formed by the transformation of surface cortical cells 76
Reproductive structures either intercalary or terminal on branchlets,
sometimes in short chains 77

76. Thallus up to 6 cm in length, commonly epiphytic on *Plumaria elegans*. . . *ISTHMOPLEA*
Thallus up to 14 cm in length, epiphytic on *Palmaria palmata*, rarely on other
algae . . . . *FOSLIEA*

77. Reproductive structures (monosporangia) pedicellate and laterally borne
on monosiphonous parts of frond, rarely intercalary and then only singly;
branching typically irregular; plurilocular sporangia intercalary. . . . . *HAPLOSPORA*
Reproductive structures (monosporangia) always intercalary in
monosiphonous parts of frond and usually in short chains of 2 or 3; branching
mainly opposite and unequal; plurilocular sporangia intercalary . . . . *TILOPTERIS*

78. Thallus hollow, tubular; branching usually opposite, branches attenuate
at apices and bases; sporangia and hairs in sori which encircle thallus
in equidistant rings; unilocular sporangia exserted on thallus surface;
plurilocular sporangia unknown on macrothallus . . . .*STRIARIA*
Thallus solid though may become tubular in old thalli; branching not usually
opposite; branches seldom attenuate at bases; sporangia and hairs randomly
scattered; unilocular and plurilocular sporangia sunk in thallus; anatomically,
mature thallus in TS resembles young *Dictyosiphon*; epiphytic on various
algae or epilithic, estuarine tolerant. . . . . *STICTYOSIPHON*

79. Axes bearing short uniseriate lateral branches of +/– limited growth 80
Axes lacking laterals of limited growth. 82

80. Lateral branches of constant length (approx. 8 cells); laterals evenly
distributed over axes; thallus up to 3 cm, solitary or gregarious; thallus
around 1 mm diameter, solid; hairs present; plurilocular sporangia cylindrical
or fusiform, uniseriate or biseriate, mixed with, and in length equal to,
laterals of limited growth; unilocular sporangia unknown; epiphytic on
Sauvageaugloia griffithsiana . . . . *BUFFHAMIA*
Lateral branches of variable length 81

81. Lateral branches usually more numerous and longer distally; thallus up to
1 cm, usually gregarious; hairs present; plurilocular sporangia, cylindrical
or conical and usually multiseriate but very variable; unilocular sporangia
present; sporangia usually sessile and borne on axes but occasionally on laterals
and prostrate tissue; epiphytic on *Scytosiphon* and other algae. . . . .*MYRIOTRICHIA*
Lateral branches bearing short branches and evenly and densely distributed
over axes; thallus up to 4 cm, solitary or gregarious; hairs present;
plurilocular sporangia cylindrical uniseriate or biseriate and borne terminally
on laterals; unilocular sporangia borne on laterals; sporangia pedicellate;
epiphytic on *Zostera* . . . .*LEBLONDIELLA*

82. Thallus articulate, attenuate proximally and distally, not exceeding 1.5 cm; each frond terminates in a hair and laterally borne hairs tend to be crowded distally; fronds gregarious; plurilocular sporangia formed by transformation of young (monosiphonous) tissue or borne in sori (sessile) on sides of frond or in clusters (pedicellate) at bases of fronds; plurilocular sporangia uniseriate to multiseriate; unilocular sporangia unconfirmed; epiphytic on *Gongolaria*, other algae, *Zostera* . . . .*GIRAUDIA*
Thallus not articulate, cylindrical, often exceeding 1.5 cm; hairs randomly distributed; fronds solitary or gregarious; plurilocular sporangia sunk in epidermal layer of frond, multiseriate, subspherical; unilocular sporangia also sunk in frond; sporangia not in clusters or sori; epiphytic on *Alaria, Chorda*, other algae 83

83. Erect thallus < 111 μm in width and without laterally produced hairs *POGOTRICHUM*
Erect thallus > 111 μm in width, with hairs commonly arising from surface cells . . . .*LITOSIPHON*

84. Thallus cylindrical, fully or partially hollow 85
Thallus another form 88

85. Frond branched throughout; branching usually opposite, branches attenuate at apices and bases; sporangia and hairs in sori, sori encircling thallus in equidistant rings; unilocular sporangia exserted on thallus surface; plurilocular sporangia unknown on macrothallus . . . .*STRIARIA*
Frond unbranched 86

86. One plastid per vegetative cell; thallus sometimes constricted; plurilocular sporangia uniseriate and forming dense irregular patches on frond surface; sporangia sometimes mixed with cylindrical or pyriform paraphyses consisting of a single cell; unilocular sporangia unknown on macrothallus; epilithic or epizooic, sometimes estuarine 87
Many plastids per vegetative cell; frond unconstricted; sporangia in small sori mixed with multicellular hairs; unilocular and plurilocular sporangia present, sessile or pedicellate, latter sporangia multiseriate; epilithic, epiphytic or epizooic, never estuarine . . . .*ASPEROCOCCUS*

87. Thallus cylindrical, hollow, frequently constricted at intervals; sporangia mixed with cylindrical or pyriform paraphyses consisting of a single cell . . . .*SCYTOSIPHON*
Thallus slightly dorsiventrally flattened to oval, unconstricted, hollow or partially hollow; sporangia without paraphyses . . . . *PLANOSIPHON*

88. Thallus comparatively small, to 20(–40) cm in length, < 10 cells in TS, flattened, delicate and membranous, little differentiated morphologically or internally. 89
Thallus moderate to large, 5 cm to 5 m in length, > 10 cells thick in TS usually tough and coriaceous, rarely membranous, usually markedly differentiated morphologically and internally 95

89. Thallus regularly branched or longitudinally divided 90
Thallus unbranched and not longitudinally divided 94

90. Thallus with a central midrib; growth from a marginal row of meristematic cells; branching dichotomous; asexual thalli bear tetrasporangia in sori on each side of midrib and on both sides of frond; sexual thalli dioecious and bearing oogonia or plurilocular male gametangia in sori; epilithic and sublittoral . . . .*DICTYOPTERIS*
Thallus without a central midrib 91

91. Thallus fan-shaped, primary branches only, deeply divided; < 15 cm in
length; apical edge of thallus incurved . . . .*PADINA*
Thallus ligulate or ribbon-shaped; much dichotomously or irregularly
branched, thallus to 40 cm of more in length; apical edge of frond not incurved 92

92. Apex of thallus fringed by a tuft of hair-like filaments. . . . . *CUTLERIA*
Apex of thallus without a tuft of hair-like filaments 93

93. Thallus with growth from a single apical cell; hairs lacking; branching
dichotomous; thallus sometimes faintly iridescent; tetrasporangia usually
randomly distributed on both sides of the thallus. . . . . *DICTYOTA*
Thallus with growth from a marginal row of meristematic cells; hairs present
in distinct horizontal or concentric bands; branching irregular; thallus not
iridescent; tetrasporangia usually distributed close to the bands of hairs on
both sides of the thallus . . . . *TAONIA*

94. A single, plate-like plastid in each vegetative cell; thallus medium to dark
brown; hairs rare; plurilocular sporangia uniseriate and forming dense
irregular patches on frond surface; unilocular sporangia unknown on
macrothallus; epilithic or epizooic on barnacles; found on open coastal sites
and in estuaries . . . .*PETALONIA*
Many discoid plastids in each vegetative cell; thallus light brown to olive;
hairs common; plurilocular sporangia variable in shape but usually conical
or subspherical, isolated or in sori and partly sunk in thallus; unilocular
sporangia scattered over and partly immersed in frond; epiphytic or epilithic;
found on open coastal sites, absent from estuaries . . . . *PUNCTARIA*

95. Thallus simple or split at distal end but not truly branched; growth
intercalary; usually in sublittoral zone 96
Thallus branched, usually dichotomously; growth apical; usually in eulittoral zone . . 102

96. Thallus cylindrical, chord-like, simple, sometimes very long (up to 8 m),
often gregarious; frond clothed with hyaline hairs or assimilatory filaments,
mucilaginous; unilocular sporangia and clavate unicellular paraphyses
produced in dense layer on surface of thallus; gametangia unknown on
macrothallus; epilithic, usually on small stones. 97
Thallus flattened 98

97. Thallus bearing colourless hairs; frond usually > 30(–100) cm in length;
paraphyses clavate with blunt apices, longer than sporangia . . . . *CHORDA*
Thallus bearing brown assimilatory filaments; frond usually < 30(–100) cm in
length; paraphyses ellipsoid with rounded apices, same length or shorter than
sporangia . . . .*HALOSIPHON*

98. Thallus with cryptostomata and hairs; frond papery, light brown. 99
Thallus lacking cryptostomata and hairs; frond leathery, dark brown,
mucilaginous; stipe tough and holdfast root-like; unilocular sporangia and
clavate unicellular paraphyses produced in irregular patches on frond;
gametangia unknown on macrothallus. 101

99. Thallus without a conspicuous midrib; stipe flattened; holdfast tuberous,
inflated and warty; sporangia borne on blade surface; on sheltered or
moderately exposed shores . . . . *SACCORHIZA*
Thallus with a conspicuous midrib; stipe terete or flattened; holdfast root-like;
unilocular sporangia borne on sporophylls produced as lateral outgrowths of
stipe; on sheltered, moderately exposed or very exposed shores 100

100. Stipe terete; on very exposed shores . . . .*ALARIA*
     Stipe flattened; on sheltered or moderately exposed shores . . . .*UNDARIA*

101. Thallus digitate, smooth . . . . *LAMINARIA*
     Thallus entire, crinkly . . . . *SACCHARINA*

102. Thallus flattened 103
     Thallus cylindrical 107

103. Thallus with midrib; cryptostomata and hairs present; holdfast discoid;
     air-bladders sometimes present; fertile regions (receptacles) are
     metamorphosed branch apices and consist of many flask-shaped cavities
     (conceptacles) containing oogonia and/or antheridia; branching dichotomous;
     8 eggs in each oogonium; monoecious and dioecious. . . . .*FUCUS*
     Thallus lacking midrib. 104

104. Thallus with air-bladders. 105
     Thallus without air-bladders. 106

105. Thallus with large air-bladders borne singly and at intervals along frond;
     branching dichotomous; receptacles on small lateral branches; dioecious;
     four eggs per oogonium. . . . . *ASCOPHYLLUM*
     Thallus with air-bladder compound and articulate in construction, siliquose
     and terminal or subterminal; branching alternate, pinnate . . . . *HALIDRYS*

106. Thallus > 15 cm, comprising long, dichotomously branched receptacles arising
     from the centre of small, button-shaped vegetative tissue; dioecious; one egg
     per oogonium; lower eulittoral and shallow sublittoral . . . .*HIMANTHALIA*
     Thallus < 15 cm; comprising a dichotomously branched and channelled
     frond with terminal receptacles; monoecious; two eggs per oogonium; upper
     eulittoral zone and littoral fringe . . . .*PELVETIA*

107. Thallus distinctly dichotomously branched; holdfast creeping, rhizomatous;
     frond, bearing terminal, unbranched receptacles; monoecious but distal
     conceptacles entirely male and proximal ones all female; one egg per
     oogonium. . . . .*BIFURCARIA*
     Thallus with spiral, alternate, irregular or pinnate branching; holdfast discoid 108

108. Thallus to 3(–5) m in length, never spiny; never iridescent, branching in
     a spiral, alternate fashion, at regular intervals; air-bladders ovoid, single
     or in clusters, borne in leaf axils, usually pedicellate; receptacles simple,
     cylindrical, never branched . . . . *SARGASSUM*
     Thallus rarely exceeding 1(–1.5) m in length, frequently spiny, sometimes
     iridescent; branching radial; air-bladders subspherical and intercalary;
     receptacles simple, branched or with appendages 109

109. Thallus bushy, arising from a conical holdfast, primary axis terete, radially
     branched, all branches covered with spine-like appendages, strongly
     iridescent, tophules absent . . . .*ERICARIA*
     Thallus caespitose or non-caespitose, primary axis terete or flattened,
     occasionally with spine-like appendages, sometimes slightly iridescent 110

110. Thallus non-caespitose, bushy, arising from a discoid/conical holdfast;
     primary axis flattened and zigzag or terete, branching radial or distichous,
     occasionally with spine-like appendage, lightly iridescent and with or
     without scattered aerocysts; tophules present or absent . . . . *GONGOLARIA*
     Thallus caespitose, arising from a discoid holdfast, primary axis terete or
     flattened, radially or bilaterally branched, spine-like appendages absent,
     sometimes slightly iridescent, with intercalary bladders, tophules absent . . .*CYSTOSEIRA*

Other useful keys to the Phaeophyceae are listed below.

Boo, S.-M., Lee, W.J., Hwang, I.K., Keum, Y-S., Oak, J.H. & Cho, G.Y. 2010. *Algal flora of Korea. Volume 2, Number 2. Heterokontophyta: Phaeophyceae: Ishigeales, Dictyotales, Desmarestiales, Sphacelariales, Cutleriales, Ralfsiales, Fucales, Laminariales. Marine brown algae II*, pp. 1–203, figs 1–48, pls 1–2. Incheon: National Institute of Biological Resources.

Coppejans, E. & Kling, R. 1995. *Flora algologique des côtes du Nord de la France et de la Belgique*. pp. [1]–454. Meise: Jardin Botanique National de la Belgique.

Cormaci, M., Furnari, G., Catra, M., Alongi, G. & Giaccone, G. 2012. Flora marina bentonica del Mediterraneo: Phaeophyceae. *Bollettino dell'Accademia Gioenia* 45: 1–508.

Fletcher, R.L. 1980. Catalogue of main marine fouling organisms. Vol. 6 Algae. Office d'Etudes Marines et Atmospheriques, for Comite International Permanent pour la Recherche sur la Preservation des Materiaux en Milieu Marin, Bruxelles (key to Phaeophyta. pp. 19–21).

Gayral, P. 1966. *Les algues des côtes françaises (Manche et Atlantique)*. Paris (see pp. 66–83 for characteristics of orders, families and genera).

Gómez Garreta, A., Barceló Martí, M.C., Gallardo García, T., Pérez-Ruzafa, I.M., Ribera Siguán, M.A. & Rull Lluch, J. (2001c). *Flora Phycologica Iberica. Vol. 1. Fucales*. pp. 192. Universidad de Murcia.

Hiscock, S. 1979. A field key to the British brown seaweeds (Phaeophyta). *Fld. Stud.* 5: 1–44.

Jones, W.E. 1962. A key to the genera of the British seaweeds. *Fld. Stud.* 1: 1–32 (N.B. Revised reprint published in 1964).

Kim, H.-S. & Boo, S.-M. (eds) 2010. *Algal flora of Korea. Volume 2, Number 1. Heterokontophyta: Phaeophyceae: Ectocarpales. Marine brown algae I*, pp. [1–6], 1–195, figs 1–19. Incheon: National Institute of Biological Resources.

Kornmann, P. & Sahling, P.H. 1977. Meeresalgen von Helgoland. Benthische Grün-, Braun- und Rotalgen. *Helgoländer wiss. Meeresunters*. 29: 1–289 (see pp. 9–11 for key to the genera of the brown algae).

Mathieson, A.C. & Dawes, C.J. 2017. *Seaweeds of the Northwest Atlantic*. pp. [i]–x, 1–798, CIX pls. Amherst & Boston: University of Massachusetts Press.

Newton, L. 1931. *A handbook of the British seaweeds*. British Museum (Natural History), London (Phaeophyceae – Key to Genera, pp. 106–12).

Prud'homme van Reine, W.F. 1982. A taxonomic revision of the European Sphacelariaceae (Sphacelariales, Phaeophyceae). *Leiden Botanical Series* 6: x + 1–293 (see pp. 57–8 for key to Sphacelariales and some ecads).

Roberts, M. 1967. Studies on marine algae of the British Isles. 3. The genus *Cystoseira*. *Br. phycol. Bull.* 3: 345–66 (see p. 363 for key).

Schneider, C.W. & Searles, R.B. 1991. *Seaweeds of the Southeastern United States. Cape Hatteras to Cape Canaveral*. pp. xiv + 553, 563 figs, 2 tables. Durham & London: Duke University Press.

South, G.R. & Hooper, R.G. 1980. A catalogue and atlas of the benthic marine algae of the Island of Newfoundland. *Mem. Univ. Nfld. Occas. Pap. Biol.* 3: 1–136 (see pp. 8–13 for key to the genera of brown algae).

Taxonomic treatment

Phaeophyceae Kjellman

PHAEOPHYCEAE Kjellman (1891c), p. 176.

Discosporangiophycidae Silberfeld, F. Rousseau & de Reviers

DISCOSPORANGIOPHYCIDAE Silberfeld, F. Rousseau & de Reviers (2014), p. 124.

Thalli erect, to 4 cm high, tufted, sparsely irregular or subdichotomously branched, comprising colourless, uniseriate filaments with discoid plastids without pyrenoids; growth apical, later probably diffuse; hairs unknown; reproduction by plurilocular sporangia, unilocular sporangia and propagules borne on the erect filaments; plurilocular sporangia sessile, either oval and reported to be of two types, microsporangia and megasporangia reflecting differences in the size of the compartments (*Choristocarpus*), or flattened and plate-like, one-layered and wrapped around the middle of elongated vegetative cells (*Discosporangium*); unilocular sporangia and propagules only known for *Choristocarpus*, both sessile and stalked on the erect filaments.

A subclass of one Order: Discosporangiales.

Discosporangiales O.C. Schmidt

DISCOSPORANGIALES O.C. Schmidt (1937), p. 3.

Thalli epiphytic, epilithic or epizoic, comprising small, erect tufts of sparsely branched, almost colourless, uniseriate, filaments, up to 1(–4) cm high; branching irregular or subdichotomous, with branches arising either at the end (*Choristocarpus*) or middle (*Discosporangium*) of the vegetative cells; growth apical, probably later diffuse; cells with numerous discoid plastids without pyrenoids; hairs unknown; reproduction by plurilocular sporangia, unilocular sporangia, and/or characteristic propagules, all produced laterally on the erect filaments; plurilocular sporangia sessile, either oval and reported to be of two types, microsporangia and megasporangia reflecting differences in the size of the compartments (*Choristocarpus*), or flattened and plate-like, one-layered and wrapped around the middle of elongated vegetative cells (*Discosporangium*); unilocular sporangia with a few large zoospores, c. 16; propagules pedicellate, comprising 1–3 cells.

An order of two families, Discosporangiaceae and Choristocarpaceae, represented by two genera, *Discosporangium* and *Choristocarpus* and two species *D. mesarthrocarpum* and *C. tenellus*. Both genera share similar morphological features and many authors have considered them to be close relatives (Kjellman 1891b; Schmidt 1937; Fritsch 1945). Kjellman (1891b) included both species in the family Choristocarpaceae, although later Schmidt (1937) established a new family Discosporangiaceae and the order Discosporangiales for *D. mesarthrocarpum*. Up to fairly recently, the Choristocarpaceae comprised four genera, *Choristocarpus*, *Discosporangium*, *Onslowia* and *Verosphacela*, and was placed in the Sphacelariales (Fritsch 1945; Henry 1987c; Womersley 1987; Prud'homme van Reine 1993). This was perhaps not surprising in view of their apical mode of growth without transverse division of subapical cells, irregular branching, cell structure and the occurrence of propagules in three of the genera (*Choristocarpus*, *Onslowia* and *Verosphacela*). However, DNA sequence data for *Onslowia* and *Verosphacela* indicated that they were not closely related to *Choristocarpus*, nor indeed to members of the Sphacelariales *sensu stricto* and were consequently transferred to a new family, Onslowiaceae (Draisma & Prud'homme van Reine 2001; Draisma *et al.* 2001, 2003). Clearly, in the absence of longitudinal divisions and a negative test result with Eau de Javelle, it was thought unlikely that *Choristocarpus* belonged to the Sphacelariales (Prud'homme van Reine 1982a, p. 28), and Draisma *et al.* (2003) suggested that *Choristocarpus* may require accommodation in a separate order. In this respect, more recent molecular evidence has grouped *D. mesarthrocarpum* and *C. tenellus* together in a monophyletic clade (Kawai *et al.* 2007) although, based on *rbc*L sequence data, Kawai *et al.* (2007) still considered it reasonable to distinguish between both species at the genus and family levels and proposed the reinstatement of the Discosporangiaceae and Discosporangiales, and the inclusion of Choristocarpaceae in the latter. Molecular phylogenetic analyses of both *C. tenellus* and *D. mesarthrocarpum* (Draisma *et al.* 2001; Burrowes *et al.* 2003; Cho *et al.* 2004; Kawai *et al.* 2007) reveal these species to be the most basal taxa.

The order is represented in Britain by the family Choristocarpaceae.

Choristocarpaceae Kjellman

CHORISTOCARPACEAE Kjellman (1891b), pp. 180, 190.

Thalli erect, tufted, delicate, filamentous, branched, uniseriate, with a large apical cell, without transverse division of subapical cells; growth apical; hairs unknown; cells with numerous discoid plastids without pyrenoids.

Reproduction by plurilocular sporangia, unilocular sporangia, and characteristic 1–3-celled pedicellate propagules, arising laterally on erect filaments; plurilocular sporangia ovoid or flattened and plate-like, one-layered and wrapped around the middle of the elongate vegetative cells.

The family is represented in Britain by the monotypic genus *Choristocarpus*, species *C. tenellus*.

Choristocarpus Zanardini

CHORISTOCARPUS Zanardini (1860), p. 45.

Type species: *C. tenellus* Zanardini (1860), p. 1.

Thalli epiphytic, epilithic and epizoic, comprising small, erect tufts of sparsely branched, almost colourless filaments up to 1 (–2) cm high; branching irregular or subdichotomous; growth apical; cells with numerous discoid plastids without pyrenoids.

Reproduction by plurilocular sporangia, unilocular sporangia, and characteristic propagules, all produced laterally on the erect filament; plurilocular sporangia sessile, and reported to be of two types, microsporangia and megasporangia, reflecting differences in the size of the compartments; unilocular sporangia with about 16 large zoospores; propagules pedicellate, comprising 1–3 cells.

One species in Britain.

Choristocarpus tenellus Zanardini (1860), p. 1.   Fig. 1

Ectocarpus tenellus Kützing (1849), p. 457, nom. illeg.

Thalli epiphytic or epilithic, forming small, flaccid tufts, 1–2 cm high, sparsely branched, comprising erect, monosiphonous, lightly pigmented, almost transparent filaments of cells, attached at the base by downward-growing, secondary rhizoids which clothe the basal cells; branching subdichotomous or irregular; filament cells light-coloured to almost transparent, thin-walled, long, linear, cylindrical (50–)71–130(–142) × 16–20 µm in the terminal regions, (112–)130–172(–190) × (16–)22–28(–31) µm in the mid and basal regions, each with numerous, small, discoid to oblong plastids without obvious pyrenoids; filaments terminated by a prominent, long, usually more darkly pigmented, apical cell responsible for growth, sometimes subtended by a long, cylindrical, subapical cell; all cells without any longitudinal segmentation.

Plurilocular sporangia ovoid, sessile, 30–45 × 20–25 µm, borne laterally on the erect filaments; unilocular sporangia sessile, elongate-cylindrical to elongate-pyriform, 15–35 × 40–60 µm, containing a small number of large zoospores; propagules common, comprising 1–3 cells, pedicellate (one cell), oval, becoming elongate-oval to elongate-pyriform with 1–2 transverse

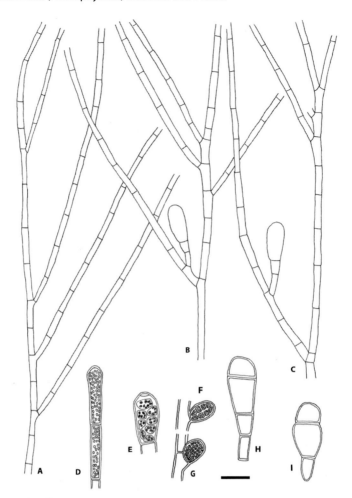

Fig. 1. *Choristocarpus tenellus*
A. Habit of thallus. B, C. Portions of thalli with propagules. D. Apical region of filament showing enclosed discoid plastids. E. Unilocular sporangium. F, G. Portions of erect filaments with plurilocular sporangia. H, I. Propagules. Bar = 80 μm (A–C), = 50 μm (D–I).

walls, 135–146 × 50–54 μm, often associated with the unilocular sporangial thalli. Presumed to have an isomorphic life history.

Reported to be both epilithic (on gravel) and epizoic (on a hydroid), can be locally abundant, sublittoral, to 20 m.

Very rare, and only reported from two localities around Britain: North Cornwall, Isle of Man. Insufficient data to comment on seasonal distribution; records were for July and September.

Boudouresque & Verlaque (1978), p. 265, figs 1–2; Brodie *et al.* (2016), p. 1021; Burrowes *et al.* (2003), pp. 63–73; Cho *et al.* (2004), pp. 921–36; Coppejans & Boudouresque (1976), pp. 195–7; Coppejans (1983), pl. 70; Cormaci *et al.* (2012), p. 82, pl. 17, figs 1–6; Draisma & Prud'homme van Reine (2001), pp. 647–9; Draisma *et al.* (2001), pp. 586–603, (2003), p. 91; Ercegovic (1955), pp. 7–12, fig. 3(a–g); Falkenberg (1878), p. 65; Fletcher in Hardy & Guiry (2003) p. 232; Fritsch (1943), p. 76, (1945), pp. 297–9, fig. 107(A–F); Funk (1955), pp. 1–78; Furnari *et al.* (1999), p. 75; Gallardo García *et al.* (2016), p. 33; Guiry (2012), p. 182; Hamel (1931–39), p. 84, fig. 23(1–4); Hauck (1887), pp. 122–4, pl. 4; Henry (1987c), pp. 182–91; Hiscock & Maggs (1984), p. 85; Kawai *et al.* (2007), pp. 186–94; Kjellman (1891b), pp. 190, 191; Kuckuck (1895a), pp. 290–322, pl. IV, figs 21–35, (1922), p. 316, (1929), p. 13, figs 1–3; Kylin (1917), p. 306; Lodge (1948a), p. 60; Prud'homme van Reine (1993), p. 155; Ribera *et al.* (1992), p. 118; Rousseau *et al.* (2001), pp. 305–19; Sauvageau (1896), p. 247, (1928a), pp. 51–94, (1931), p. 457; Schmidt (1937), pp. 1–4; Varvarigos *et al.* (2007), pp. 253–62; Womersley (1987), pp. 145, 146; Zanardini (1860), p. 45, pl. 1, figs 1–6.

Ishigeophycidae Silberfeld, F. Rousseau & de Reviers

ISHIGEOPHYCIDAE Silberfeld, F. Rousseau & de Reviers (2014), p. 141.

Thalli epilithic or epiphytic, pseudoparenchymatous, either crustose with erect, loosely bound filaments arising from a monostromatic base, or erect, branched, foliose or terete with a central medulla and outer cortex; growth in both forms by apical cells; life history probably isomorphic.

Crustose thalli with a single, lobed plastid with immersed, little protruding pyrenoid in each cell; hairs single, unilocular and plurilocular sporangia terminating the erect filaments; plurilocular sporangia uniseriate to biseriate.

Erect thalli with several discoid plastids without pyrenoids; hairs in clusters growing from cryptostomata; unilocular sporangia transformed from terminal cortical cells; plurilocular sporangia transformed from assimilatory filaments, uniseriate, lacking a terminal cell.

A subclass of one order: Ishigeales.

Ishigeales G.Y. Cho & S.M. Boo

ISHIGEALES G.Y. Cho & S.M. Boo in Cho, Lee & Boo (2004), p. 934 *emend* Silberfeld, F. Rousseau & de Reviers (2014), p. 141.

Thalli epilithic or epiphytic, encrusting or erect and thalloid; growth apical; life history probably isomorphic; encrusting thalli with a monostromatic base giving rise to erect, loosely bound, gelatinous filaments, comprising cells with a single, lobed, plastid with immersed, little protruding pyrenoid; hairs, unilocular and plurilocular sporangia terminal on filaments; plurilocular sporangia mainly uniseriate; hairs single with basal meristem and sheath.

Erect thalli simple or branched, terete or foliose, pseudoparenchymatous with medulla and cortex and outer assimilatory filaments; cortical cells with several discoid plastids without pyrenoids; hairs clustered, growing from cryptostomata; unilocular sporangia terminal, transformed from cortical cells, plurilocular sporangia uniseriate, transformed from assimilatory cells.

Silberfeld *et al.* (2014) recognized two families in this order: Ishigeaceae Okamura in Segawa (1935) and the new family Petrodermataceae Silberfeld, F. Rousseau & de Reviers (2014), p. 141 – only the latter family being represented in Britain and Ireland by the monotypic genus *Petroderma* and the species *Petroderma maculiforme*.

Petrodermataceae Silberfeld, F. Rousseau & de Reviers

PETRODERMATACEAE Silberfeld, F. Rousseau & de Reviers (2014), p. 141.

Thalli epilithic, encrusting, usually thin, rarely slightly pulvinate, sponge-like and gelatinous, firmly adherent to substratum; comprising a monostromatic, discoid base, giving rise to erect, weakly bound, haplostichous filaments; cells with a single, parietal, lobed plastid, each with a single, somewhat enclosed, invaginated pyrenoid; hairs single, terminal on erect filaments, with basal meristem and sheath.

Plurilocular and unilocular sporangia in extensive sori, terminal on erect filaments, without paraphyses; plurilocular sporangia sometimes markedly elongate, uniseriate to biseriate, simple, rarely branched.

Petroderma Kuckuck

PETRODERMA Kuckuck (1897), p. 382.

Type species: *P. maculiforme* (Wollny) Kuckuck (1897), p. 382.

Thalli encrusting, thin or comparatively thick, discrete or extensive, sponge-like and slightly gelatinous, firmly adherent by undersurface to substratum, usually without rhizoids; consisting of a monostromatic discoid base, giving rise to vertical, little branched, mucilaginous filaments loosely bound together except at base and easily separable under pressure; cells with single, lobed, plate-like plastid with a small exserted pyrenoid, not obvious under a light microscope; hairs single, terminal on erect filaments, with basal meristem and sheath.

Plurilocular and unilocular sporangia terminal on erect filaments without accompanying paraphyses, in extensive sori, not obvious externally; unilocular sporangia club-shaped or cylindrical, often present within old sporangial husks; plurilocular sporangia uniseriate to multiseriate, occasionally branched.

One species in Britain and Ireland.

Petroderma maculiforme (Wollny) Kuckuck (1897), p. 382, fig. 1.   Figs 2(A), 3

Lithoderma maculiforme Wollny (1881), p. 31; *Lithoderma lignicola* Kjellman (1883b), p. 256.

Thalli encrusting, ranging from small, thin specks to comparatively thick, confluent extensions up to several centimetres across, light to dark brown, quite firmly attached by undersurface to substratum, rhizoids usually absent; surface smooth, spongy but firm, gelatinous, surface cells rounded, quite closely packed, 7–10 µm in diameter, with prominent single, parietal, ring-shaped plastid without obvious pyrenoid; thallus structure pseudoparenchymatous, consisting of a monostromatic base of laterally united, coherently spreading filaments giving rise to simple or little branched, erect, gelatinous filaments, weakly joined and easily separable under pressure; filaments to 25(–37) cells, 180(–400) µm long, comprising cells 1–3 diameters long, 5–18 × 6–10 µm, quadrate to rectangular, with single, parietal, lobed, plate-like plastid in upper cell region, each with a small exserted pyrenoid not obvious under a light microscope; long intercalary and terminal ascocyst-like cells frequent; hairs single, with basal meristem and sheath, arising from surface cells.

Plurilocular and unilocular sporangia terminal on erect filaments, without paraphyses, crowded in extensive sori on crust surface, not obvious externally; unilocular sporangia club-shaped or cylindrical, 13–23 × 12–14 µm, often formed within empty sporangial husks; plurilocular sporangia uniseriate to multiseriate, occasionally branched to 135 µm (–32 cells) long and 10 µm in diameter.

Epilithic, from littoral fringe to lower eulittoral, usually in shallow pools and channels, although emergent and extensive in littoral fringe under exposed conditions; appears to show a preference for hard substrata, such as flint stones.

Recorded for scattered localities all around Britain (Shetland, Berwick, Yorkshire, Kent, Hampshire, Dorset, Devon, Cornwall, Pembroke) and Ireland (Mayo, Galway); probably under-recorded.

Likely to be perennial, although thalli tend to be most conspicuous during the winter months. Plurilocular and unilocular sporangia recorded October to May.

This is a common and widespread epilithic alga. It was first recorded for Ireland by Cotton (1912) on Clare Island, Mayo, Ireland, but surprisingly was not included in Newton (1931). It is a very closely encrusting alga, occurring as either solitary, fine specks or more usually as expansive confluent crusts several centimetres in diameter. It is encountered on a wide variety of substrata, extending from the littoral fringe to the lower eulittoral region, both exposed and

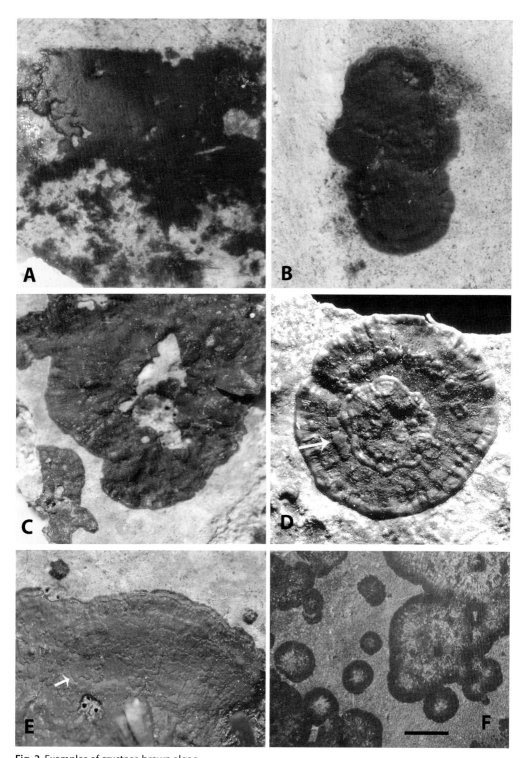

Fig. 2. Examples of crustose brown algae.
A. *Petroderma maculiforme*. B. *Stragularia clavata*. C–E. *Pseudoralfsia verrucosa*. Note the prominent margin and discrete wart-like unilocular sporangial sori (arrow) in D and the more extensive, slightly raised plurilocular sporangial sori in E (arrow). F. *Stragularia spongiocarpa*. Note extensive lighter coloured central area representing the yellow coloured unilocular sporangial sori. Bar in F = 5 mm (B, D, F), = 10 mm (A, C, E).

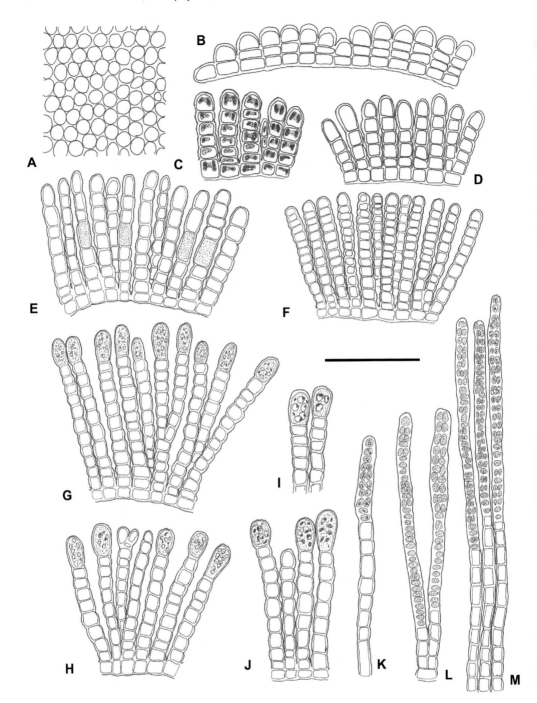

Fig. 3. *Petroderma maculiforme*
A. Surface view of vegetative crust. B. VS of crust margin. C–F. VS of vegetative crusts, showing loosely adjoined erect filaments and intercalary ascocyst-like cells. Note lobed, plate-like plastid drawn in cells in C. G–J. VS of crusts with terminal unilocular sporangia. K–M. VS of crusts with terminal plurilocular sporangia. Bar = 50 μm.

in pools. It is particularly found on hard siliceous flint nodules, often in company with crusts of the red alga *Hildenbrandia*. On the soft chalk substrata of the Kent and Sussex coasts it is very abundant and expansive on the vertical cliff faces in the littoral fringe, often in association with barnacles and *Ulva* spp. Culture studies on Pacific and North Atlantic populations of *P. maculiforme* by Wynne (1969) and Fletcher (1978) respectively, found no evidence of a sexual cycle in the life history, which appeared to be of the 'direct' type. However, in agreement with Wilce *et al.* (1970) more information based on culture studies is required before a further understanding of the life history of this species is possible.

Abbott & Hollenberg (1976), p. 174, fig. 140; Brodie *et al.* (2016), p. 1022; Cotton (1912), pp. 96, 123; Edelstein & McLachlan (1969b), pp. 561–3, figs 1–19; Fletcher (1978), pp. 371–98, figs 8, 9, 20, 21, (1987), pp. 80–2, fig. 1; Fletcher in Hardy & Guiry (2003), p. 284; Gallardo García *et al.* (2016), p. 33; Guiry (2012), p. 157; Kjellman (1883b), p. 256, pl. 26, figs 8–11; Kuckuck (1897), p. 382, figs 9, 10; Mathieson & Dawes (2017), p. 225, pl. LXVIII, figs 16, 17; Pedersen (2011), pp. 99, 100, fig. 104; Peña & Bárbara (2010), pp. 54, 55, fig. 10(A–F); Waern (1949), pp. 663–6, pl. 1, (1952), pp. 141–3, fig. 75; Wilce *et al.* (1970), pp. 119–35, figs 1–5; Wollny (1881), p. 31, pl. II, figs 1–4; Wynne (1969), pp. 9, 10, fig. 3, pl. 3.

Dictyotophycidae Silberfeld, F. Rousseau & de Reviers

DICTYOTOPHYCIDAE Silberfeld, F. Rousseau & de Reviers (2014), p. 140.

Thalli filamentous, pseudoparenchymatous or polystichous at first, later becoming parenchymatous; growth terminal by one or more apical cells; cells with several discoid plastids without pyrenoids; life history isomorphic (Dictyotales, Sphacelariales, Onslowiales) or heteromorphic with reduced gametophytes (Syringodermatales).

A subclass of four Orders (Dictyotales, Onslowiales, Sphacelariales, Syringodermatales), two of which (Dictyotales, Sphacelariales) are represented on the coasts of Britain and Ireland.

Dictyotales Bory

DICTYOTALES Bory (1828), p. 142 ('Dictyoti').

Thalli erect or prostrate, dorsiventrally flattened and membranous, simple, undivided, or cleaved, to regularly dichotomously branched, arising from a fibrous holdfast of secondary rhizoidal filaments, with or without stoloniferous branches; structure parenchymatous, comprising an inner medulla of one to several large, regularly arranged, colourless cells, enclosed by an outer cortex of usually one, sometimes several layers of smaller, photosynthetic, cortical cells; growth apical, from either a single, large, apical cell, or from a marginal row of apical cells; plastids discoid, several per cell, without pyrenoids; hairs scattered in tufts, or in concentric zones on thallus surface; mainly dioecious, life history isomorphic and diplohaplontic; reproductive organs borne singly or in sori arising from surface cells and protruding above thallus surface, with or without an underlying stalk cell. Sporophytes bearing scattered or clustered, sometimes concentrically zoned, unilocular sporangia (usually termed tetrasporangia) bearing 4–8, large, non-motile spores; female gametophytes bearing tightly packed clusters of single celled oogonia; male gametophytes bearing tightly packed clusters of plurilocular sporangia, each compartmentalized.

The Dictyotales represents a distinctive, widely recognized, monophyletic order which finds support in several molecular studies (Rousseau *et al.* 2001; Draisma *et al.* 2001, 2003; Burrowes *et al.* 2003; Cho *et al.* 2004; Kawai *et al.* 2007; Lim *et al.* 2007; Phillips *et al.* 2008a; Silberfeld *et al.* 2014). In some classifications of the brown algae (Womersley 1987; de Reviers & Rousseau 1999; Lee & Hwang 2010; Cormaci *et al.* 2012), the following families have been recognized within the order: Dictyotaceae (Lamouroux ex Dumortier 1922), Dictyotopsidaceae (Allender 1980) and Scoresbyellaceae (Womersley 1987). However, with the transfer of the two families Scoresbyellaceae and Dictyotopsidaceae into the Dictyotaceae, based on molecular studies by Bittner *et al.* (2008) and unpublished data by De Clerck (cited in Bittner *et al.*), the Dictyotaceae now represents the only family within the order (Silberfeld *et al.* 2014). In addition, the subdivision of the Dictyotaceae into the two tribes, Dictyoteae (Greville 1833) and Zonarieae (O.C. Schmidt 1938), based on the number of terminal apical cells, by workers such as Womersley (1987) and Stegenga *et al.* (1997), has also lacked support from molecular studies (Bittner *et al.* 2008; Silberfeld *et al.* 2014) and is not adopted here.

Dictyotaceae J.V. Lamouroux ex Dumortier

DICTYOTACEAE J.V. Lamouroux ex Dumortier (1822), pp. 72, 101.

Thalli epilithic, epizoic or epiphytic, erect or horizontally expanding, dorsiventrally flattened, membranous, blade-like and entire, simple and flabellate, cuneate or reniform, remaining undivided, or becoming cleaved and segmented, or much branched with branching sub-dichotomous or dichotomous and sometimes spirally twisted, arising from a fibrous mass of attaching secondary rhizoidal filaments, with or without a basal horizontal system of rhizome-like adventitious/stoloniferous branches; in structure, parenchymatous, several cells thick in transverse section, with an inner medulla of one to several, large transversely and longitudinally arranged, mainly regularly cuboidal or rectangular, colourless cells, and an outer cortex of usually one, sometimes several, small rectangular to cuboidal pigmented cells; cells with several discoid plastids without pyrenoids; hairs in tufts arising from surface cells, either scattered on blade surface, or in concentric zones, with basal meristem; growth from either a single, large, prominent apical cell, or from a marginal row of apical cells.

Life history isomorphic, diplohaplontic and comprising distinct male, female and sporophytic thalli; reproductive organs borne on both sides of the thalli, arising as transformations of the surface cells; sporophytes bearing scattered or loosely grouped, darker-coloured unilocular sporangia (tetrasporangia) with or without a stalk cell, each tetrasporangium bearing 4 or 8 non-motile spores (tetraspores); gametophytes dioecious, reproduction oogamous with female thalli bearing tightly clustered groups of single celled oogonia and male thalli bearing tightly clustered groups of plurilocular antheridia, sometimes surrounded by single celled paraphyses, sporangial tufts projecting from surface cells and sometimes rupturing the cuticle, with or without a surrounding sterile margin.

This family is represented in Britain and Ireland by four genera: *Dictyopteris*, *Dictyota*, *Padina* and *Taonia*. They are all erect, flat-bladed and parenchymatous, and growth is either by a single, large, apical, meristematic cell which divides equally to form the regularly dichotomously branched *Dictyota*, or by a terminal row of synchronously dividing meristematic cells which can separate fairly regularly and uniformly to form the subdichotomously branched *Dictyopteris*, or irregularly and less frequently to form the flabellate to variously perpendicularly split/dissected *Padina* and *Taonia*. Additional features to distinguish the genera include: the iridescent, blue-green tinge frequently seen on thalli of *Dictyota* and, to a lesser extent, *Dictyopteris* in tidal pools; the white colouring of the thallus surface seen in *Padina*, due to calcification of the surface cells; the occurrence of reproductive organs and/or hairs in concentric lines seen on the blade surface of *Padina* and *Taonia*; the presence of a distinct midrib on the thallus of *Dictyopteris*, as well as its quite strong and pungent smell when freshly collected.

In the present treatment, the genus *Dilophus* is withdrawn and the single species previously recognized for Britain and Ireland, *Dilophus spiralis*, is removed to the older genus *Dictyota*, in agreement with the proposal of Hörnig *et al.* (1992a). J. Agardh (1848) characterized the genus *Dictyota* Lamouroux (1809a) for species with a single apical cell and a single-layer medulla, transferring species with a multilayered medulla into his new genus *Dilophus*. However, the distinction between *Dictyota* and *Dilophus* is not always clear (Setchell & N.L. Gardner 1925; Fritsch 1945): most species of *Dictyota* and *Dilophus* can form a multilayered medulla and this character does not differentiate between the two genera (Womersley 1987; Hörnig *et al.* 1992a).

Dictyopteris J.V. Lamouroux

DICTYOPTERIS J.V. Lamouroux (1809a), p. 332, nom. cons. (Silva 1952, p. 257).

Type species: *D. polypodioides* (A.P. de Candolle) J.V. Lamouroux (1809), p. 332.

Neurocarpus Weber & Mohr (1805), p. 300, nom. rejic.; *Haliseris* Agardh, C. (1820), p. 141.

Thalli consisting of an erect, dorsiventrally flattened, branched blade, arising from a fibrous holdfast of compacted, secondary rhizoidal filaments; erect blades subdichotomously branched, membranous, entire with central thickened midrib and lateral thinner 'wings', sometimes with a traversing network of fine veins; growth apical from groups of meristematic cells at branch tips; in cross section, blades with several tiers of internal, large, colourless, medullary cells enclosed above and below by a single layer of small, cortical, photosynthetic cells, each containing several discoid plastids without pyrenoids; hairs common, solitary or grouped, arising from blade surface.

Dioecious, with tetrasporangia and male and female gametangia all borne on different thalli, on both sides of blade surface, on each side of the midrib; oogonia solitary or in groups; antheridia in distinct sori surrounded by sterile cells; tetrasporangia solitary or grouped in sori, sessile, oblong, protruding and attached by a basal stalk cell.

One species in Britain and Ireland.

Dictyopteris polypodioides (de Candolle) J.V. Lamouroux (1809a), p. 332.   Figs 4, 5

Fucus membranaceus Stackhouse (1795), p. 13; *Fucus polypodioides* Desfontaines (1799), p. 421; *Ulva polypodioides* de Candolle in Lamarck & de Candolle (1805), p. 15; *Fucus ambiguus* Clemente (1807), p. 310; *Polypodoidea membranacea* Stackhouse (1809), p. 97; *Haliseris polypodioides* (de Candolle) C. Agardh (1820), pp. 142, 143; *Ulva membranacea* (Stackh.) J.P. Jones & J.F. Kingston (1829), p. 59; *Dictyopteris membranacea* Batters (1902), p. 54; *Dictyopteris punctata* Noda (1973), p. 3; *Dictyopteris tripolitana* Nizamuddin (1981), p. 18; *Dictyopteris ambigua* (Clemente) Cremades (1990), p. 489.

Thalli epilithic, forming erect, olive-green to dark-brown, dorsiventrally flattened, linear blades, to 30 cm high, 15 mm wide, and attached at the base by an expanded holdfast comprising a fibrous network of secondary, branched rhizoidal filaments; erect blades dichotomously/ subdichotomously branched, membranous, translucent, thin, entire, although frequently torn at the edge, with prominent central midrib, and occasional vein-like striations projecting out across the lamina; midrib to 17 or more cells thick, often persisting at the base of older thalli following erosion of the lamina 'wings' and frequently giving rise to proliferous branches; wing areas of blades gradually narrowing down from the midrib to two cells in thickness, and frequently with microscopic peripheral serrations; in surface view, blade cells initially arranged in both longitudinal and transverse rows and projecting forward from the midrib region, later curving out towards blade margin and becoming more irregularly arranged; blade cells variously shaped, although mainly rectangular and longitudinally elongate to cuboidal at first, more irregularly shaped towards the margin, 4–7 sided, 1–3 diameters long, 22–60 × 18–39 µm; in cross section, midrib region comprises an extensive inner medullary tissue of rounded to angular, longitudinally elongated, colourless cells, usually irregularly arranged or somewhat in radial rows, enclosed within a monostromatic, cortical layer of radially elongated, rectangular to cuboidal, photosynthetic cells containing large numbers of plastids without pyrenoids; in cross section, outer two-layered wing region not differentiated and comprising equal-sized, rectangular to cuboidal, photosynthetic cells containing large numbers of plastids; growth apical, from groups of meristematic cells at branch tips; hairs grouped, and scattered over blade surface, with meristematic basal region.

Dioecious; tetrasporangia and gametangia (antheridia and oogonia) on separate thalli, formed on both sides of the blade surface and either single or grouped in oblong or irregularly shaped sori, either arranged each side of the midrib or extending out onto the laminar blade; gametangia rare; antheridia multiseriate, densely packed in large, irregularly shaped, whitish sori scattered over lamina surface, surrounded by sterile cells; oogonia scattered widely across wing lamina, solitary, sometimes in small groups, rounded, 43–70(–85) µm in diameter, or quite commonly oblong in the direction of the longitudinal rows, 43–70 × 24–50 µm, partially embedded or projecting above surface in cross section; tetrasporangia sometimes solitary, more frequently grouped, in oblong sori usually clustered together in a narrow band each side of the midrib, sessile, rounded in surface view, somewhat misshapen when crowded, 100–160 µm in diameter, rounded to oblong, partially embedded or projecting in vertical section, developing from a single cortical cell, with or without an obvious basal stalk cell.

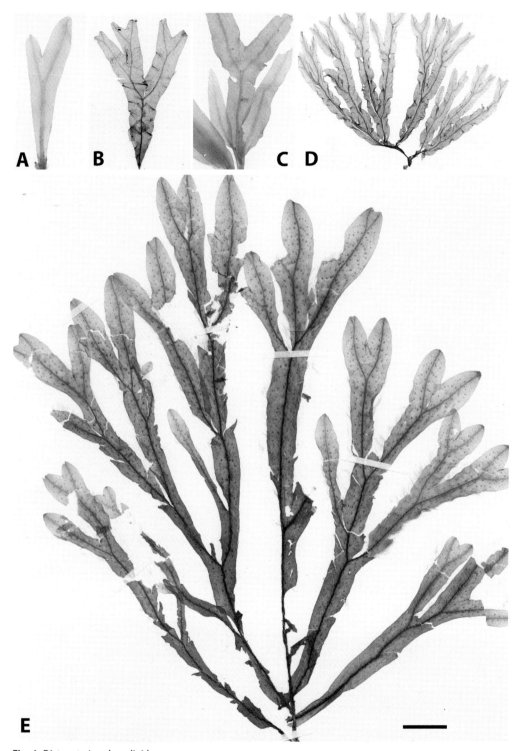

Fig. 4. *Dictyopteris polypodioides*

A–E. Habit of different sized thalli showing the characteristic central midrib and outer translucent wing regions. Note erosion of wing regions at base of the more mature thallus in E. Bar in E = 13 mm (A), = 18 mm (B), = 15 mm (C), = 30 mm (D), = 35 mm (E). (B, D, E, herbarium specimens, NHM.)

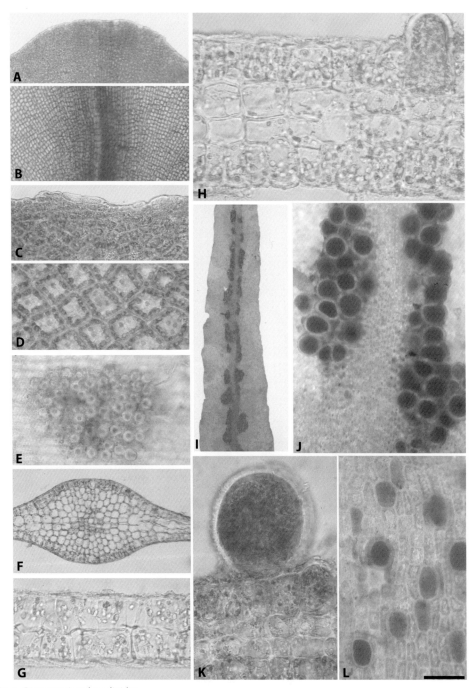

Fig. 5. *Dictyopteris polypodioides*

A. Surface view of thallus apex showing groups of meristematic apical cells. B. Surface view of subapical region of thallus showing thickened central midrib and emerging rows of forward-projecting/curving cells. C. Edge of thallus showing peripheral serrations. D. Surface view of thallus showing cells with several discoid plastids. E. Surface view of thallus showing a group of hair bases. F. TS of thallus showing thickened midrib region with extensive inner medulla of large, colourless cells enclosed by a single layer of smaller, photosynthetic, cortical cells. G. TS of two-cell thick wing area of blade. H. TS of four-cell thick area of blade, close to the midrib, showing a single, young, developing sporangium (top right). I. Surface view of terminal blade region showing tetrasporangial sori arranged each side of the midrib. J. Close-up view of tetrasporangial sori. K. TS of thallus showing tetrasporangium. L. Surface view of developing oogonia. Bar in L = 20 mm (I), = 300 μm (B, J), = 180 μm (A, F), = 90 μm (C, E, L), = 45 μm (D, G, H, K).

Epilithic, lower eulittoral pools, usually in shaded regions, and associated with sand, more common in the sublittoral, down to 25 m.

Common on southern and western shores of Britain and Ireland.

Annual, developing in late winter/spring, reaching maturity in summer, and dying back in late autumn.

Ardré (1970), pp. 267–8; Brodie *et al.* (2016), p. 15; Bunker *et al.* (2017), p. 287; Cabioc'h & Boudouresque (1992), p. 61, fig. 30; Cormaci *et al.* (2012), p. 45, pl. 5, figs 4–6; De Mesquita Rodrigues (1963), p. 44, pl. 4, fig. D, pl. 5, fig. A; Fletcher in Hardy & Guiry (2003), p. 242; Gallardo García *et al.* (2016), p. 33; Gayral (1958), pp. 228, 229, pl. 33, (1966), p. 265, fig. 33A, pl. 42; Georgevitch (1918), pp. 536, 537; Gómez Garreta *et al.* (2002), pp. 153–60; Guiry (2012), p. 132; Hamel (1931–39), pp. 341–3, fig. 56 (i, j); Johnson (1891), pp. 463–70; Misra (1966), pp. 152, 153, fig. 80; Morton (1994), p. 38; Newton (1931), p. 216, fig. 137(A–E); Nizamuddin (1981), pp. 1–112, pl. XXIV; Reinke (1878a), pl. VI, figs 12–20, pl. VII; Ribera *et al.* (1992), p. 119; Schneider & Searles (1991), p. 156, fig. 184; Seoane-Camba (1965), p. 84; Silva *et al.* (1996), pp. 924–7; Stegenga & Mol (1983), p. 117; Stegenga *et al.* (1997a), p. 22; Taylor (1960), p. 227.

Dictyota J.V. Lamouroux

DICTYOTA J.V. Lamouroux (1809a), p. 331.

Type species: *D. dichotoma* (Hudson) J.V. Lamouroux (1809a), p. 42.

Dilophus Agardh, J. (1882), p. 106.

Thalli consisting of erect, dorsiventrally flattened blades, to 20(–40) cm high, 1–20 mm wide, arising from a fibrous base of secondary rhizoidal filaments, with or without stoloniferous outgrowths; blades dichotomously divided, entire, linear, sometimes spirally twisted; growth apical by a single, conspicuous apical cell; in structure, comprising a central medullary of 1(–4) large, colourless cells, enclosed by a single layer of smaller photosynthetic cells; in surface view, cells in distinct longitudinal rows, each with numerous discoid plastids without pyrenoids; hairs in fascicles, scattered on blade surface, with basal meristem.

Dioecious; tetrasporangia and gametangia (antheridia and oogonia) on separate thalli, arising from surface cells over extensive areas on both sides of blade; tetrasporangia single or grouped, arising from a single, basal, stalk cell; antheridia grouped in distinct sori, with or without an involucrum of surrounding elongated sterile cells, arising from a single, basal, stalk cell.

Two species in Britain and Ireland.

KEY TO SPECIES

Terminal branches lanceolate, rounded or acute, thallus base without stoloniferous outgrowths; medulla one cell layered throughout, epilithic and epiphytic . . . .*Dictyota dichotoma*
Terminal branches spatulate, thallus base with stoloniferous outgrowths, medulla sometimes 2–4-celled in parts, epilithic . . . . *Dictyota spiralis*

Dictyota dichotoma (Hudson) J.V. Lamouroux (1809a), p. 42.   Figs 6, 7

Ulva dichotoma Hudson (1762), p. 476; *Fucus zosteroides* J.V. Lamouroux (1805), p. 25, nom. illeg.; *Dictyota implexa* (Desfontaines) J.V. Lamouroux (1809), p. 43; *Dictyota pusilla* J.V. Lamouroux (1809a), p. 43; *Dictyota rotundata* J.V. Lamouroux (1809a), p. 43; *Dictyota implexa* J.V. Lamouroux (1809a), p. 43; *Zonaria dichotoma* (Hudson) C. Agardh (1817), p. xx; *Zonaria rotundata* (Lamouroux) C. Agardh (1817), p. xx; *Fucus dichotomus* (Hudson) Bertolini (1819), p. 314, nom. illeg.; *Zonaria dichotoma* var. *intricata* C. Agardh (1820), p. 134; *Dictyota dichotoma* var. *implexa* (Desfontaines) S.F. Gray (1821), p. 341; *Haliseris dichotoma* (Hudson) Sprengel (1827), p. 328; *Dictyota dichotoma* var. *acuta* Chauvin ex Duby (1830), p. 954; *Dictyota linearis* (C. Agardh) Greville (1830), p. XLIII; *Dictyota dichotoma* (Hudson) Lamouroux var. *intricata* (C. Agardh) Greville (1830), p. 58; *Dictyota setosa* Duby (1830), p. 955; *Dichophyllium dichotomum* (Hudson) Kützing (1843), p. 338; *Dictyota dichotoma* var. *volubilis* Lenormand (1843) (see De Clerck 2003); *Dictyota acuta* Kützing (1845), p. 271; *Dictyota volubilis* Kützing (1849), p. 554; *Dictyota acuta* var. *patens* Kützing (1849), p. 555; *Dictyota dichotoma* var. *rigida* P. Crouan & H. Crouan (1852), p. 69; *Dictyota aequalis* var. *minor* Kützing (1859), p. 9; *Dictyota attenuata* Kützing (1859), p. 6; *Dictyota elongata* Kützing (1859), p. 6; *Dictyota latifolia* nom. illeg. Kützing (1859), p. 6; *Dictyota dichotoma* var. *stenoloba* Hohenacker, nom. inval. (1883), p. 479; *Dictyota dichotoma* var. *elongata* (Kützing) Grunow (1874), p. 24; *Dictyota dichotoma* var. *minor* Kützing (1881), p. 43, nom. inval.; *Dictyota dichotoma* f. *latifrons* Holmes & Batters (1890), p. 86; *Dictyota areolata* Schousboe (1892), p. 227, nom. inval. (see De Clerck 2003); *Dictyota complanata* Schousboe ex Bornet (1892), p. 227, nom. inval. (see De Clerck 2003); *Dictyota dichotoma* f. *attenuata* (Kützing) Vinassa (1892), p. 10; *Dictyota dichotoma* f. *latifolia* (Kützing) Vinassa (1892), p. 110; *Neurocarpus annularis* Schousboe (1892), p. 227; *Neurocarpus areolatus* Schousboe (1892), p. 227; *Dictyota apiculata* J. Agardh (1894), p. 67; *Dictyota vivesii* M.A. Howe (1911), pp. 497–9; *Dictyota dichotoma* f. *elongata* (Kützing) Schiffner (1933) p. 294, nom. inval.; *Dictyota dichotoma* f. *spiralis* Nizamuddin (1981), p. 43.

Thalli forming erect, light golden brown to dark brown, dorsiventrally flattened, ribbon-shaped blades of approximately equal width from base to apex, arising from a fibrous holdfast of secondary rhizoids; erect blades 8–20 cm in length, 1–10 mm broad, regularly dichotomously branched with branches at various angles of divergence up to 90 degrees, linear, sometimes spirally twisted, with slightly narrowing rounded, less commonly pointed, apices terminating in a single, large apical cell; surface cells in distinct longitudinal, less obviously transverse rows, cuboidal to more commonly rectangular in shape, 1–4 diameters long, increasing in size from the apex to the base, 10–70 × 5–23 µm, each containing numerous discoid plastids without pyrenoids; in transverse section, blade thickness 84–164 µm, three celled throughout, with a single, large, central, colourless medullary cell, enclosed above and below by a single layer of small, photosynthetic, cortical cells; hairs in tufts, scattered over surface, each hair with a distinct basal meristematic zone; in pools, some thalli display a blue-green iridescent colouring.

Dioecious; tetrasporangia and gametangia (antheridia and oogonia) on separate thalli, arising from surface cells over extensive areas on both sides of blade; gametangia (antheridia and oogonia) developing from the surface cortical cells and 'packaged' in ovate to longitudinally arranged, raised, oblong sori; in surface view, antheridial sori 180–900 × 140–400 µm, surrounded by a marginal involucrum of elongated sterile cells, with antheridia arranged in longitudinal, less obvious transverse rows, closely packed, each antheridium comprising a cluster of loculi; in cross section, antheridia rectangular to club-shaped, 50–70 × 18–30 µm, comprising vertical columns of tightly packed, cuboidal loculi, usually four columns wide, and up to 18 loculi high, covered by a surface cuticle and arising from a cortical cell; in surface view, oogonial sori 210–600 × 110–280 µm, and comprising closely packed, usually rounded or adpressed oogonia, 40–62 µm in diameter; in cross section, oogonia pear-shaped, 50–90 × 35–58 µm, arising from a cortical cell; in surface view, tetrasporangia either single or irregularly grouped and quite densely distributed over entire blade surface, spherical with a distinct thick wall, 82–130 µm in diameter, developing directly from the cortical cells; in cross section, tetrasporangia rounded to slightly pear-shaped.

Epilithic, and quite commonly epiphytic, in eulittoral pools and emersed, particularly in damp situations on the sides of rocks and gullies, sublittoral to 12 m.

Common and widely distributed all around Britain and Ireland.

Summer annual, becoming conspicuous in February/March, with maximum development and fertility in July–September, although small fragments of thalli can be found throughout the year.

A morphologically highly variable species, with a wide range of growth forms often growing side by side. These include a quite common, densely branched, terminally narrow bladed (to 1–2 mm in width), intricately matted and tangled ecological growth form, variously described in the literature as var. *implexa* or var. *intricata*. Although segregated from *Dictyota dichotoma* by Hörnig & Schnetter (1988), this was not supported by molecular data (Tronholm *et al.* 2010), thus these varieties are herein included within the synonymy of *Dictyota dichotoma*.

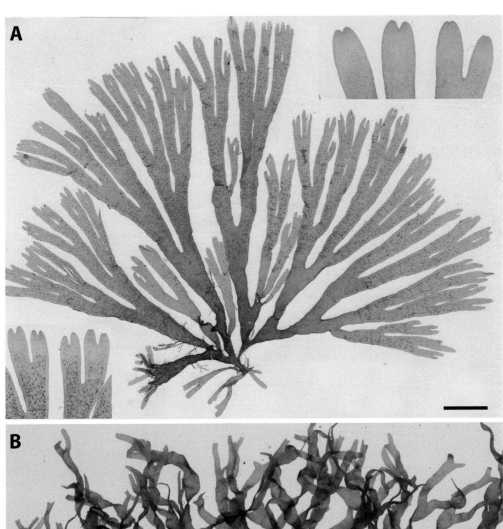

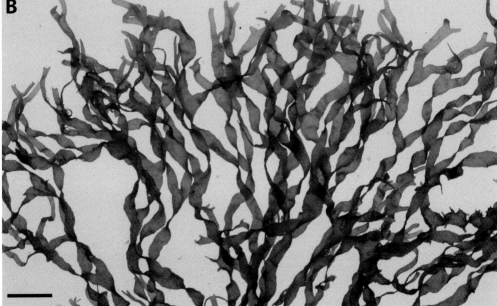

Fig. 6. *Dictyota dichotoma*
A. Habit of relatively broad-bladed form, showing close-ups of vegetative (upper inset) and fertile (lower inset) apical regions. B. Habit of relatively narrow-bladed form (often referred to in the literature as var. *intricata*). Bar in A = 20 mm (= 9 mm top inset, = 12 mm bottom inset), bar in B = 6 mm. (Λ, herbarium specimen, NHM.)

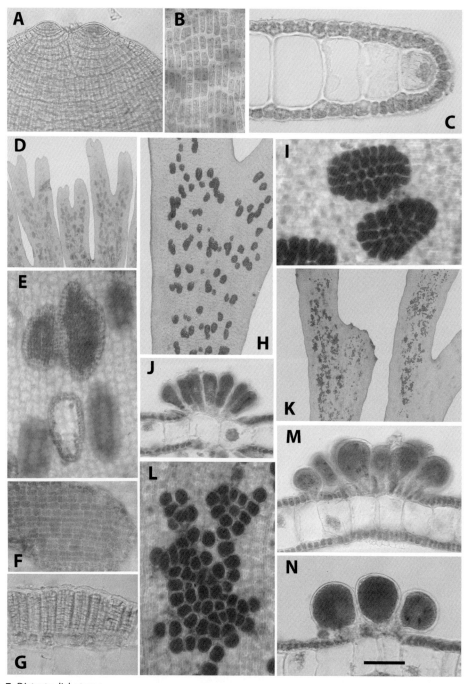

Fig. 7. *Dictyota dichotoma*

A. Surface view of thallus apex showing a recently divided apical cell in the early stages of a dichotomy. B. Surface view of thallus showing the cells in longitudinal (less obviously) transverse rows. C. TS of thallus edge showing the inner, large, colourless, medullary cell and outer, single, row of smaller, photosynthetic, cortical cells. D. Terminal regions of branches showing antheridial sori. E. Close-up view of antheridial sori, with one lighter coloured empty one. F. Close-up of antheridial sorus showing rows of antheridia. G. TS of antheridial sorus showing pluriseriate antheridia. H. Surface view of thallus showing oogonial sori, I. Close-up view of three oogonial sori with closely packed oogonia. J. TS of thallus showing cluster of oogonia. K. Surface view of tetrasporangial thallus with scattered tetrasporangia. L. Close-up view of clustered tetrasporangia. M–N. TS of thallus showing tetrasporangia. Bar in N = 2 mm (H), = 4.5 mm (K), = 6 mm (D), = 300 μm (E, K), = 180 μm (F, I), = 90 μm (A, C, F, J, L, M, N), = 45 μm (B, G).

Ahmad *et al.* (1990), pp. 428–30, fig. 1; Brodie *et al.* (2016), p. 1019; Bunker *et al.* (2017), pp. 284, 285; Børgesen (1926), pp. 84, 85; Cabioc'h & Boudouresque (1992), p. 61, fig. 29; Coppejans & Kling (1995), pp. 188–90, pl. 59(A–H); Cormaci *et al.* (2012), pp. 51, 52, pl. 6, figs 5–7, pl. 7, figs 1–6; De Clerck (2003), pp. 1–205; De Clerck & Coppejans (2003), pp. 275–84; Dizerbo & Herpe (2007), pp. 100–2, pl. 33(2); Dring (1987), p. 302; Feldmann & Gugliemi (1972), pp. 751–4; Fletcher (1980), p. 43, pl. 16, figs 1–3; Fletcher in Hardy & Guiry (2003), p. 244; Flores-Moya (1999) pp. 129–35; Fritsch (1945), pp. 302–4, fig. 108(A–G); Funk (1955), pp. 48, 49; Gaillard & L'Hardy-Halos (1976), pp. 2167–70, (1977), pp. 101–10, (1979), pp. 149–62, (1980), pp. 159–67, (1990), pp. 39–53; Gaillard (1968), pp. 109–15, (1972a), pp. 71–9, (1972b), pp. 145–50; Gaillard *et al.* (1986), pp. 340–57; Gallardo García *et al.* (2016), p. 33; Gayral (1958), pp. 218, 219, pl. 29, fig. 32(D, E), (1966), pp. 254, 255, fig. 32(D–E), pl. XXXVIII; Guiry (2012), pp. 131, 132; Hamel (1931–39), pp. 347–9, fig. 57(IV–VI); Harvey (1846–51), pl. 103; Heil (1924), pp. 119–25; Holden (1913), pp. 1–6; Hörnig & Schnetter (1988), pp. 277–91, fig. 3; Hörnig *et al.* (1992a), pp. 45–62, (1992b), pp. 397–402, (1993), pp. 169–71; Hoyt (1907), pp. 383–92, (1910), pp. 55–7, (1927), pp. 592–619; Hwang *et al.* (2005), pp. 999–1015; Katsaros & Galatis (1985), pp. 263–76; Kornmann & Sahling (1977), p. 161, pl. 88; Kumagae *et al.* (1960), pp. 91–102; Kylin (1947), pp. 35, 36, pl. 3, fig. 9; Lawson & John (1982), p. 141; Lee & Hwang (2010), pp. 34–6, figs 11–12; Lee & Kim (1997), pp. 83–91, figs 1–5; Lewis (1910), pp. 59–64; Lund (1940), pp. 180–94, (1950), pp. 74–7; Mimuro *et al.* (1990), pp. 450–6; Misra (1966), pp. 132, 133, fig. 66(A–D); Morton (1994), p. 38; Müller & Clauss (1976), pp. 461–5; Müller *et al.* (1981a), pp. 1040, 1041; Newton (1931), p. 212, fig. 134(A–D); Nizamuddin (1981), pp. 45, 73–5, pls XV(B), XV(1), XXX(1, 5); Nultsch *et al.* (1984), pp. 217–22, (1987), pp. 93–7; Pybus (1974), pp. 25–7, fig. 2(a); Reinke (1878c), pp. 1–56, pls 1–2; Ribera *et al.* (1992), p. 119; Richardson (1979), pp. 22–6; Rueness (1977), p. 189, pl. 25(2), fig. 12(A, B); Rull Lluch *et al.* (2005), pp. 63–70; Russell (1970), pp. 243–5; Schnetter *et al.* (1987), pp. 193–7; Schreiber (1935), pp. 266–75; Stegenga & Mol (1983), p. 117, pl. 45(1–5); Stegenga *et al.* (1997a), p. 22, (1997b), p. 167, pl. 46(1–3); Taylor (1960), pp. 218, 219, pl. 31(5); Teixeira *et al.* (1990), pp. 87–92; Thuret & Bornet (1878), p. 54, pl. XXVII–XXX; Tronholm *et al.* (2008), pp. 132–44, (2010), pp. 1301–5, fig. 6(A–F); Wenderoth (1933), pp. 185–9; Williams (1897a), pp. 361, 362, (1897b), pp. 545–53, (1903), pp. 184–6, (1904a), pp. 141–60, (1904b), pp. 183–204, (1905), pp. 531–60; Womersley (1987), pp. 194–6, figs 64(H–M), 65(A, B).

Dictyota spiralis Montagne (1846), pp. 29–30.   Figs 8, 9

Dictyota ligulata Kützing (1847), p. 53, nom. illeg.; *Dictyota dichotoma* var. *elongata* P. Crouan & H. Crouan (1867), p. 168; *Dictyota dichotoma* var. *spiralis* (Montagne) P. Crouan & H. Crouan (1867), p. 168; *Dictyota linearis* var. *spiralis* (Montagne) Ardissone (1883), p. 481; *Dictyota fasciola* var. *ligulata* (Kützing) Vinassa (1892), p. 111; *Neurocarpus appendiculatus* Schousboe (1892), p. 227, nom. inval. (see De Clerck 2003); *Neurocarpus complanatus* Schousboe (1892), p. 227, nom. inval. (see De Clerck 2003); *Neurocarpus crispatus* Schousboe (1892), p. 227, nom. inval. (see De Clerck 2003); *Dilophus spiralis* (Montagne) Hamel (1931–39), p. 352; *Dilophus ligulatus* (Kützing) Feldmann (1937), p. 313.

Thalli epilithic, forming erect, olive-green, slightly iridescent, parenchymatous, dorsiventrally flattened, ribbon-shaped blades, arising from a basal system of fibrous, secondary rhizoidal filaments often accompanied by stoloniferous outgrowths; erect blades 5–15(–21) cm in length, 1–5(–8) mm wide, linear, narrowing, thinning and more lightly coloured towards the apex, somewhat sparsely and regularly dichotomously branched, with up to 1–6 dichotomies, with long inter-dichotomies, and often terminating in even longer, spatulate, sometimes somewhat incurved, terminal branches with a rounded apex and a single, large, apical cell; in surface view, cells in distinct longitudinal, less obviously transverse rows, cuboidal to mainly rectangular in shape, 1–5 diameters long, increasing in size down from the apex, 16–70 × 12–27 μm, each containing large numbers of discoid plastids without pyrenoids; in section, blades 95–210(–300) μm in thickness, and predominately with a single, large, central, colourless, cuboidal to rectangular, medullary cell, enclosed above and below by a single layer of much smaller, photosynthetic cortical cells, occasionally, particularly at the margins and in the basal regions, with a localized, multilayered medulla of 2–4 cells; in cross section, inner walls of medullary cells often seen with lomentaceous thickenings; hairs in tufts, scattered over blade surface with distinct, basal, meristematic zone.

Dioecious; tetrasporangia and gametangia (antheridia and oogonia) on separate thalli, borne directly in the central regions of both sides of the blade surface, leaving a prominent, clear

Fig. 8. *Dictyota spiralis*
A, B. Habit of thalli. C, D. Surface view of upper, dichotomously branched regions with scattered tetrasporangia.
Bar in A = 20 mm, bar in B = 12 mm, bar in C = 7.5 mm, bar in D = 15 mm.

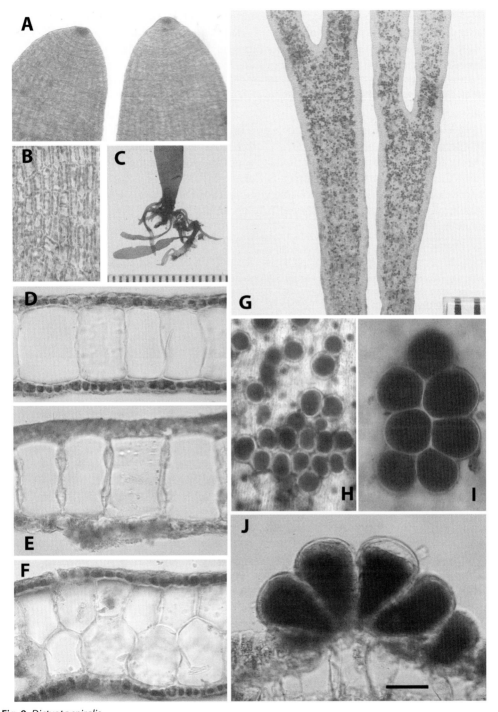

Fig. 9. *Dictyota spiralis*
A. Surface view of two branch apices showing large apical cell. B. Surface view of thallus showing the cells in longitiudinal (less obviously) transverse rows, each with large numbers of discoid plastids. C. Basal region of thallus showing stoloniferous outgrowths and young emergent thalli. D. TS of thallus showing the inner, large, colourless, medullary cell and outer, single row of smaller, photosynthetic, cortical cells. E. TS of thallus showing the inner walls of the medullary cells with lomentaceous thickenings. F. TS of thallus showing a localized multi-layered medulla. G. Terminal branch regions with abundant tetrasporangia. Note the prominent clear margins. H, I. Close-ups of tetrasporangia. J. TS of thallus showing tetrasporangia. Bar in J = 300 µm (H), = 180 µm (A), = 90 µm (D–F, I, J), = 45 µm (B).

margin; gametangia rare and reported to be borne in discrete, raised, ovate to longitudinally arranged, oblong sori very similar to those of *Dictyota dichotoma*; tetrasporangia common, quite densely distributed on thallus surface, sometimes solitary but more usually clustered together in irregularly shaped and variously sized groups, mainly spherical in surface view, 85–120 µm in diameter, with a pronounced thick wall; in transverse section, tetrasporangia rounded to pear-shaped, 132–140 × 78–93 µm, developing directly from the cortical cells and with a small, basal, supporting, stalk cell.

Epilithic, mid to lower eulittoral pools and lagoons and often characteristically in places with intermittent sand scour.

A southern species, apparently confined to south and south-west shores of Britain (Channel Islands, Cornwall, South Devon, Dorset, Isle of Wight) and Ireland (Clare and Galway) with rare historical records elsewhere; probably underrecorded and mistaken for *Dictyota dichotoma*.

A summer annual, mainly observed July to September, more rarely found October to December.

Ardré (1970), p. 269; Børgesen (1926), p. 86; Brodie *et al.* (2016), p. 1019; Bunker *et al.* (2017), p. 286; Cormaci *et al.* (2012), p. 58, pl. 10(1–3); De Clerck (2003), pp. 1–205, (2006), pp. 1271–88; De Mesquita Rodrigues (1963), p. 52, pl. 6(A, D); Dizerbo & Herpe (2007), p. 102, pl. 33(3); Feldmann (1937), pp. 313–17, figs 64(A)–67; Fletcher in Hardy & Guiry (2003), p. 244; Funk (1927), p. 363, fig. 20(C), (1955), p. 50; Gaillard (1968), pp. 109–15; Gallardo García *et al.* (2016), p. 33; Gayral (1958), pp. 220, 221, fig. 32(A–C), pl. 30, (1966), pp. 257–8, fig. 32(A–C), pl. 39; Guiry (2012), p. 132; Hamel (1931–39), pp. 352, 353, fig. 57(VIII, IX); Hörnig & Schnetter (1988), pp. 277–91; Hörnig *et al.* (1992a), pp. 45–62, (1992b), pp. 397–402; Montagne (1846) pp. 29, 30, fig. 10(A–J); Newton (1931), p. 213; Nizamuddin (1981), p. 60, pls XXXVI(42–50), XXXVII; Pybus (1974), pp. 25–7, figs 1, 2(b); Ribera *et al.* (1992), p. 120; Ribera Siguán *et al.* (2011), pp. 205–19; Schnetter *et al.* (1987), pp. 193–7; Tronholm *et al.* (2010), pp. 1301–21, fig. 10.

Padina Adanson

PADINA Adanson (1763), pp. 13, 586.

Type species: *P. pavonica* (Linnaeus) Thivy in Taylor (1960), pp. 234–5.

Thalli consisting of an erect flattened blade, arising from a compacted rhizoidal holdfast and perennating stoloniferous portions; blades coriaceous below, thin and paper-like above, flabellate, entire, or irregularly split, terminally convex, and often with frequently inrolled margins, wedge-shaped below or narrowing to a stipe-like region, both sides lightly calcified and whitish in appearance and with distinct, continuous, concentric zones comprising hairs and reproductive sori; in section, blades several cells thick, with several layers of inner, large, longitudinally elongated, colourless, medullary cells and an enclosing upper and lower row of smaller, cortical, photosynthetic cells containing several discoid plastids, without pyrenoids; growth by a terminal row of meristematic cells.

Monoecious; gametangia (antheridia and oogonia) and tetrasporangia grouped in sori, in concentric rows; gametangia rare, antheridia plurilocular and tiered with basal pedicel, oogonia single-celled and pyriform; tetrasporangia common, pyriform, protruding and surrounded by an indusium.

One species in Britain and Ireland.

Padina pavonica (Linnaeus) Thivy in Taylor (1960), pp. 234–5.   Figs 10, 11

Fucus pavonicus Linnaeus (1753), p. 1162; *Fucus pavonius* Linnaeus (1759), p. 1345, nom. illeg.; *Ulva pavonia* (Linnaeus) Linnaeus (1767), p. 719, nom. illeg.; *Dictyota pavonia* (Linnaeus) Lamouroux (1809a), pp. 39, 40; *Dictyota pavonia* var. *elongata* J.V. Lamouroux (1809a), p. 40, nom. inval.; *Dictyota pavonia* var. *maxima* J.V. Lamouroux (1809a), p. 40, nom. inval.; *Padina pavonia* (Linnaeus) Lamouroux (1816a), p. 304; *Zonaria pavonia* C. Agardh (1820), pp. 125–7; *Padina mediterranea* Bory de Saint-Vincent (1822–31), pp. 589–91.

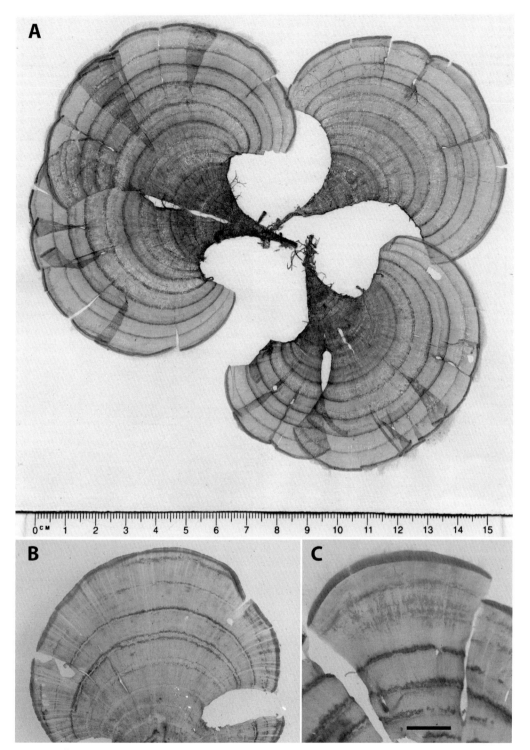

Fig. 10. *Padina pavonica*
A–C. Habit of thalli. Bar in C = 12 mm (B), = 8 mm (C). (Herbarium specimen, NHM.)

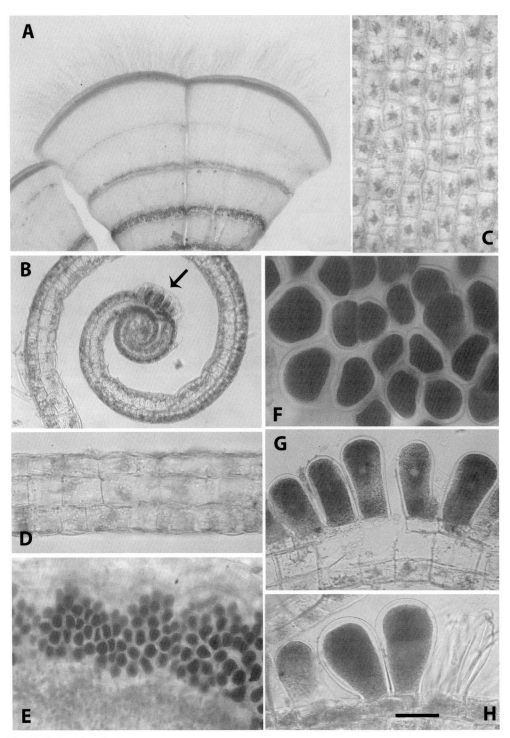

Fig. 11. *Padina pavonica*

A. Terminal portion of thallus, showing outer fringe of hairs, darker inrolled apical region and concentric bands of hairs/sporangia. B. TS of terminal inrolled region. N.B. The young developing fascicle of hairs (arrow). C. Surface view of thallus showing the cells in longitudinal (less obviously) transverse rows. D. TS of thallus showing the three cell layers. E. Surface view of thallus showing part of the concentric band of tetrasporangia. F. Close-up of tetrasporangia. G–H. TS of thallus showing tetrasporangia. Bar in H = 7 mm (A), = 180 μm (B, E), = 90 μm (F, G), = 45 μm (C, D).

Thalli epilithic, solitary or more commonly in groups, forming erect, light brown to white, dorsiventrally flattened blades, up to 11 cm high and 13 cm wide, attached at the base by a fibrous rhizoidal holdfast and often with perennating stoloniferous outgrowths present; erect blades slightly coriaceous below, thin and paper-like above, flabellate and simple when young, becoming deeply and irregularly cleaved later to form wedge-shaped cuneate thalli, with wedges frequently overlapping, and narrowing below to a stipe-like region; blades with margins terminally curved and frequently inrolled down onto the lower face, giving a prominent darker, terminal region, frequently with a subterminal concentric zone of hairs which project out as a fringe from the terminal region; both sides, and particularly the lower side, lightly calcified, and with characteristic concentric lines of hairs/sporangia/gametangia parallel to the margins; growth apical, involving a marginal row of meristematic cells; in surface view, cells arranged in longitudinal but not obviously transverse rows, rectangular to cuboidal in shape, 1–4 diameters long, 30–130 × 26–52 µm, each with large numbers of small discoid plastids, without pyrenoids; in transverse section, thalli usually with three cell layers, 120–130 µm in thickness, with cells of all three layers cuboidal to rectangular in shape and approximately equal in size; hairs in dense concentric zones on the upper abaxial side of blades, multicellular, approximately 18–23 µm in diameter, with basal meristem.

Reproductive sori in continuous concentric zones on one side of the blade surface, usually associated with the bands of hairs; monoecious, gametangia rare, grouped together in small sori; antheridia plurilocular and tiered with basal pedicel; oogonia single celled, 40–50 µm in diameter, and slightly pear-shaped in vertical section; tetrasporangia common, solitary or grouped in linear sori, with conspicuous surrounding indusium, usually rounded in surface view, 80–122 µm in diameter, sometimes mis-shaped when adpressed together, 70–145 × 50–83 µm, pedicellate and pyriform in section, 70–145 × 50–83 µm.

Epilithic, mid to lower eulittoral shallow pools and damp areas, extending into the shallow sublittoral, typically on eroding soft clay and sandstone substrata.

Although more widespread in the past, recent records suggest this species is restricted to the south-west shores of Britain (Channel Islands, Isles of Scilly, Cornwall, South Devon, Dorset, Isle of Wight) with rare, sporadic historical records for the southern North Sea coasts, along the west coast of Britain and in Ireland. Populations on the Isle of Wight probably represent the eastern distributional limit of the species on the south coast of Britain.

Summer annual, April/May to October, although rarely some overwintering fragments have been found. It is doubtful that a full 'alternation of generations' type life history is operating, in view of the rare reports of gametophytes, with most individuals being sporophytic. It is likely that regeneration from basal, perennating, rhizomatous portions plays an important role in the life history and appearance of the thalli.

Alsha & Shameel (2010), pp. 319–40; Ardré (1970), p. 267; Barceló *et al.* (1998), pp. 179–86; Bitter (1899), pp. 255–74; Børgesen (1926), pp. 84, 85; Brodie *et al.* (2016), p. 1019; Bunker *et al.* (2017), p. 288; Carter (1927), pp. 139–59; Coppejans (1983b), pl. 55; Cormaci *et al.* (2012), pp. 64, 66, pl. 11(4–6); De Mesquita Rodrigues (1963), p. 54, pl. 5(F); Dizerbo & Herpe (2007), pp. 102, 103, pl. 33(4); Fagerberg & Dawes (1973), pp. 199–204; Feldmann (1937), pp. 317, 318; Fletcher in Hardy & Guiry (2003), p. 281; Funk (1955), pp. 51, 52; Gallardo García *et al.* (2016), p. 33; Gayral (1958), pp. 230, 231, pl. 34, (1966), pp. 260–2, pl. 40; Gómez Garreta *et al.* (1998), pp. 179–86, (2007), pp. 27–33; Guiry (2012), pp. 132, 133; Hamel (1931–39), pp. 343–5, fig. 57(I–III); Harvey (1846–51), pl. 91; Herbert *et al.* (2016), pp. 1–15, (2018), pp. 112–19; Misra (1966), pp. 154, 155, fig. 81; Newton (1931), pp. 214–16, fig. 136(A–C); Ni-Ni-Win *et al.* (2011), pp. 1193–209; Nizamuddin (1981), pp. 1–122; Oltmanns (1922), p. 179, fig. 433; Price *et al.* (1979), pp. 1–67, 2 figs, 3 plates; Ramon & Friedmann (1966), pp. 183–96; Ribera *et al.* (1992), p. 120; Ribera Siguán *et al.* (2011), pp. 205–19; Seoane-Camba (1965), p. 82, fig. 23(3); Silva *et al.* (1996), p. 606; Taylor (1960), pp. 234, 235; Wenderoth (1933), pp. 185–9; Wolfe (1918), pp. 78–109.

Taonia J. Agardh

TAONIA J. Agardh (1848), p. 101.

Type species: *T. atomaria* (Woodward) J. Agardh (1848), pp. 101–2.

Thalli consisting of an erect flattened blade, arising from a fibrous holdfast of branched rhizoidal filaments; erect blades cuneiform and entire to irregularly divided/vertically split into several wedge-like portions with truncate apices, and narrowing markedly towards the base into a stipe-like region; growth from a marginal row of apical cells; in cross section, thalli up to several cells thick, all of equal size with the inner, medullary cells colourless and the outer cortical cells with several discoid plastids without pyrenoids; hairs and reproductive cells (tetrasporangia and gametangia) either scattered over blade surface or more characteristically in alternately concentric, narrow bands across blade, giving it a zonate appearance.

Dioecious; tetrasporangia, antheridia and oogonia on separate individuals; antheridia in irregularly shaped sori, surrounded by sterile cells, multiseriate and with a basal stalk cell; oogonia solitary or grouped, spherical, without basal stalk cell; tetrasporangia solitary or grouped, spherical with basal stalk cell.

One species in Britain and Ireland.

Taonia atomaria (Woodward) J. Agardh (1848), pp. 101–2.   Figs 12, 13

Ulva atomaria Woodward (1797), p. 53; *Dictyota ciliata* Lamouroux (1809), p. 41; *Dictyota atomaria* (Woodward) Greville (1830), p. xliii; *Padina atomaria* (Woodward) Montagne (1840), p. 146; *Dictyota atomaria* var. *bertolonii* (Woodward) Meneghini (1841), p. 426; *Taonia atomaria* var. *divaricata* Holmes & Batters (1890), p. 86.

Thalli epilithic, forming erect, olive-green to dark brown, dorsiventrally flattened blades to 42 cm in length, with one to several blades arising from a fibrous holdfast of secondary rhizoidal filaments; erect blades cuneiform and flabellate, entire when young, becoming irregularly lobed later, and narrowing markedly at the base, with cuneate, segmented apical regions; blades membranous, translucent and glossy, margins entire or dentate with proliferous truncated branch tips; growth apical, by a marginal row of apical cells; in surface view, cells arranged in longitudinal, less obviously transverse rows, rectangular to cuboidal, 1–3 diameters long, 18–38(–60) × 14–22(–32) μm, each with large numbers of small discoid plastids without pyrenoids; in transverse section, thalli 4–6(–7) cells thick, 165–320 μm, with 2–4 large, quadrate, colourless, inner cells enclosed within an outer single-celled layer of smaller, pigmented, photosynthetic cells; in longitudinal section, all cells longitudinally elongate, with the inner, colourless cells markedly so, 1–4(–6) diameters long; hairs in tufts arising from the surface cortical cells, usually arranged in concentric, wavy bands at intervals across the width of the blade, giving it a characteristic zonate appearance; hairs multicellular, 14–17 μm in diameter, with basal meristem.

Dioecious; tetrasporangia and gametangia (antheridia and oogonia) on separate thalli, arising from surface cells on both sides of blade; gametophytes very rare; antheridia reported to be borne on both sides of the thallus, grouped in scattered, irregularly shaped, light-coloured, slightly protruding sori, surrounded by a narrow band of sterile cells, each gametangium divided into vertical columns of small, cuboidal loculi, approx. 3 μm in diameter; oogonia reported to be solitary or grouped in sori, slightly smaller than the tetrasporangia, to 50 μm in diameter; tetrasporangia common, solitary, irregularly grouped, or arranged in dark concentric zones associated with the bands of hairs, darkly pigmented, oval to elongate spherical/wedge-shaped, 78–100(–130) × 48–62(–79) μm in surface view, club-shaped, 68–95(–105) × 55–68(–78) μm in section, protruding partially or well above blade surface when fully developed, with or without an obvious basal, stalk cell.

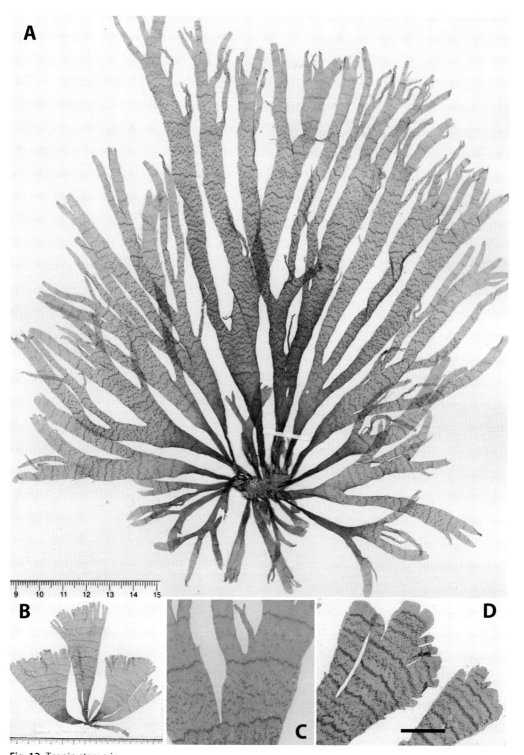

Fig. 12. *Taonia atomaria*

A. Habit of much-branched thallus. B. Habit of cuneate thallus. C–D. Upper regions of thalli showing the dark, concentric, wavy bands of tetrasporangia. Bar in D = 7.5 mm (C) = 9 mm (D). (A, B, D, herbarium specimens, NHM.)

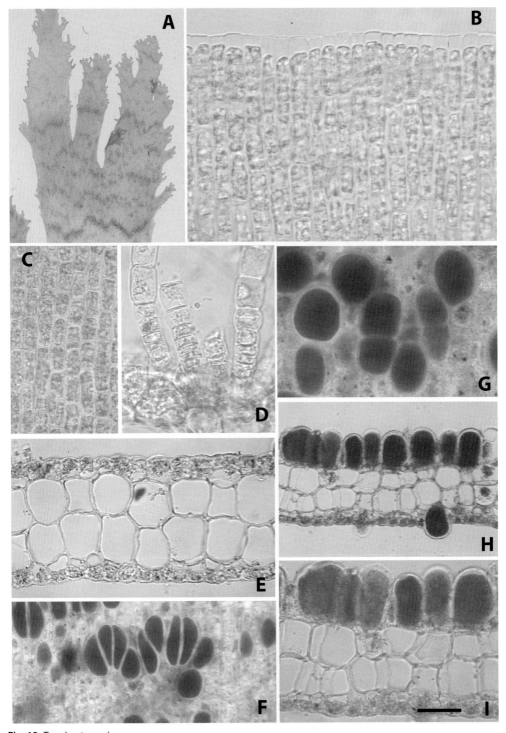

Fig. 13. *Taonia atomaria*

A. Upper regions of thallus showing the dark, concentric, wavy bands of tetrasporangia. B. Apex of thallus showing the terminal cells and thick, hyaline cuticle. C. Surface view of thallus showing the cells in longitudinal (less obviously) transverse rows. D. TS of thallus showing meristematic basal region of a group of hairs. E. TS of thallus showing a two-celled medulla enclosed above and below by a single row of smaller cortical cells. F, G. Surface view of tetrasporangia. H, I. TS of thallus showing tetrasporangia borne on both sides. Bar in I = 10 mm (A), = 180 µm (H), = 90 µm (B, C, E, F, G, I), = 45 µm (D).

Epilithic, eulittoral and sublittoral to 25 m, often characteristically in places with intermittent sand scour.

Locally common and largely restricted to the south and south-west coasts of Britain and Ireland.

Summer annual, from June to October, rarely in November/December.

Ardré (1970), pp. 266, 267; Børgesen (1926), p. 89, figs 34, 35; Brodie *et al.* (2016), p. 1019; Bunker *et al.* (2017), p. 290; Cabioc'h & Boudouresque (1992), p. 60, fig. 28; Coppejans & Kling (1995), p. 190, pl. 60(A–H); Cormaci *et al.* (2012), pp. 74, 75, pl. 14(4–6); De Mesquita Rodrigues (1963), p. 57, pl. 7(A); Dizerbo & Herpe (2007), p. 103, pl. 34(1); Feldmann (1937), pp. 318, 319; Fletcher in Hardy & Guiry (2003), p. 312; Funk (1955), pp. 50, 51; Gaillard (1963), pp. 725, 726; Gallardo García *et al.* (2016), p. 33; Gayral (1958), pp. 232, 233, fig. 34(A), pl. 35, (1966), pp. 262, 263, fig. 33(D), pl. 41; Guiry (2012), p. 133; Hamel (1931–39), pp. 336–8, fig. 56(c–f); Harvey (1846–51), pl. 1; Morton (1994), p. 38; Newton (1931), pp. 213, 214, fig. 135(A–D); Nizamuddin (1981), pls X(A), XIII, XXX, XXXI, fig. 2; Reinke (1878c), pp. 1–56; Ribera *et al.* (1992), p. 120; Ribera Siguán *et al.* (2011), pp. 205–19; Robinson (1932), pp. 113–20; Sauvageau (1897b), pp. 51–5, (1897d), pp. 86–90; Seoane-Camba, J. (1965), p. 82, fig. 23(1); Stegenga & Mol (1983), p. 119, pl. 45(6–8); Stegenga *et al.* (1997a), p. 23; Ubisch (1931), pp. 361–6; Williams (1897a), pp. 361–2, (1897b), pp. 545–53.

Sphacelariales Migula

SPHACELARIALES Migula (1908), p. 237.

Thalli epilithic or epiphytic, erect or crustose; if erect, heterotrichous, tufted, 1 mm to 25 cm in height, attached by monostromatic discs, polystromatic crusts, stolons or endophytic filaments, usually supplemented by secondary rhizoids; if crustose, thin, closely adherent, with or without rhizoids, composed of radiating, laterally spreading, branched filaments, laterally adjoined and disc-like with a marginal row of apical cells, giving rise to vertical files of tightly adjoined filaments, often with superimposed thalli; erect shoots rigid to fairly flaccid, parenchymatous, either foliose or more commonly forming terete, multiseriate segmented filaments, sometimes uniseriate near apices, sparsely to densely branched to several orders, with branching irregular, subdichotomous, whorled or pinnate and opposite, with main axes indeterminate, branches and branchlets determinate or indeterminate, linear or with slight increase in diameter below, ecorticated or corticated owing to limited or extensive meristematic activity of the outer peripheral cells to form downward-growing secondary rhizoidal filaments which can clothe the proximal regions; growth of erect shoots apical, each shoot ending in a prominent apical cell, larger on main axes than on branches; apical cells divide transversely, the daughter cells then dividing transversely and longitudinally, sometimes with additional secondary growth of the secondary segments; in transverse section, erect thalli clearly and variously subdivided to form a parenchymatous tissue, with or without a clear differentiation of cells into inner medullary and outer cortical tissues; branches produced either hypoblastically or acroblastically; all cells of both erect shoots and crusts with several discoid plastids without pyrenoids; hairs with basal meristem and sheath.

Reproduction on erect thalli by plurilocular sporangia of similar or dissimilar (micro and mega) types, unilocular sporangia and/or specialized unicellular or multicellular, deciduous propagules, borne mainly on the terminal laterals, or on special short shoots, sessile or stalked; reproduction on crustose thalli by plurilocular and unilocular sporangia in slightly raised sori on thallus surface, terminating the erect filaments, with or without ascocyst-like cells; sexual reproduction isogamous, anisogamous or oogamous; life history diplohaplontic, with an isomorphic or slightly heteromorphic alternation of generations.

This order was defined by marked heterotrichy, terete, segmented, parenchymatous erect thalli, a prominent apical cell, a mainly isomorphic life history, and a blackening of the cell walls by the bleaching agent Eau de Javelle. Traditionally (Sauvageau 1900–14; Oltmanns 1922;

Fritsch 1945; Newton 1931), and frequently in later texts (Lindauer *et al.* 1961; Stegenga & Mol 1983; Coppejans 1995), three families were included within the Sphacelariales: Sphacelariaceae, Stypocaulaceae and the Cladostephaceae, distinguished mainly by their mode of branching.

1. In members of the Sphacelariaceae, the branches are formed as outgrowths of the secondary segments below the apical cell (hypoblastic branching).

2. In members of the Stypocaulaceae, branching is mainly initiated in the apical cell (acroblastic branching).

3. In members of the Cladostephaceae both patterns of formation of laterals can be observed.

Some authors, however (Hamel 1931–39; Kylin 1947; Taylor 1937a, 1957), only recognized a single family, the Sphacelariaceae. The Choristocarpaceae was included by Fritsch (1945), an addition initially unacceptable to Prud'homme van Reine (1982a), but recognized by Henry (1987c), Womersley (1987), Schneider & Searles (1991) and later by Prud'homme van Reine (1993), although he only provisionally placed the family in the Sphacelariales *sensu lato* under *incertae sedis* pending further studies on this order.

Based on molecular evidence, Draisma *et al.* (2002) merged the monotypic Cladostephaceae with the Sphacelariaceae and cast doubt on the recognition of the family Stypocaulaceae, suggesting that the order might be reduced to a single family. However, a later molecular study by Draisma *et al.* (2010b) resulted in a complete reclassification of the order, with four families recognized for Britain and Ireland:

1. Cladostephaceae. This contains the genus *Cladostephus* with two species, *C. spongiosus* and *C. hirsutus*.

2. Sphacelariaceae. This contains the genus *Sphacelaria* with five species (*S. cirrosa, S. fusca, S. plumula, S. rigidula* and *S. tribuloides*), the reinstated (and emended) genus *Battersia* for the former *Sphacelaria plumigera, S. racemosa* and *S. mirabilis*, the reinstated monotypic genus *Chaetopteris* for the former *Sphacelaria plumosa*, and a new genus *Sphacelorbus* for the former *Sphacelaria nana*.

3. Sphacelodermaceae (Draisma *et al.* 2010b), a new family to incorporate the new genus *Sphaceloderma* for the former *Sphacelaria caespitula*.

4. Stypocaulaceae. This contains the genus *Halopteris* for the two species *H. filicina* and *H. scoparia*, and a new genus *Protohalopteris* for the former *Sphacelaria radicans*.

In agreement with Silberfeld *et al.* (2014), the family Lithodermataceae, previously placed in Ralfsiales, and currently encompassing four crustose genera – *Heribaudiella, Lithoderma, Pseudolithoderma* and *Bodanella* – is also placed within the Sphacelariales based on its close relationship to this order shown by McCauley & Wehr (2007). A similar close relationship between *Pseudolithoderma* and *Heribaudiella* and the Sphacelariales was also revealed in a molecular study by Bittner *et al.* (2008).

European taxa of the Sphacelariaceae received comprehensive treatment by Prud'homme van Reine (1978, 1982a,b) and due acknowledgement to that work is given in the present account of the species in Britain and Ireland.

Cladostephaceae Oltmanns

CLADOSTEPHACEAE Oltmanns (1922), p. 102.

Thalli usually epilithic, solitary or gregarious, with one to several erect, branched axes arising from a thin, crustose base; erect axes firm to spongy in texture, subdichotomously, irregularly branched in parts, to several orders, and clothed in whorls of short, determinate, secondary laterals arising from the superior secondary axial segments, with or without terminal spike-like, determinate laterals; branching of indeterminate main axes and determinate secondary laterals hypacroblastic, branching of determinate laterals acroblastic; secondary axial segments undergoing abundant divisions and enlarging in both length and width (auxocaulic); meristematic activity of the peripheral cells giving rise to a radially arranged secondary rhizoidal cortex, the surface cells of which later undergo abundant anticlinal divisions to form tiers of smaller, isodiametric cells; life history diplohaplontic and isomorphic, with unilocular and plurilocular sporangia borne on specialized, determinate, fertile laterals arising from the surface cortical cells.

Cladostephus C. Agardh

CLADOSTEPHUS C. Agardh (1817), p. xxv.

Type species: *C. spongiosum* (Hudson) C. Agardh (1817), p. xxv.

Thalli epilithic, rarely epiphytic, comprising one to several long main axes, arising from a polystromatic base overlaid by a mass of secondary rhizoidal filaments; main axes subdichotomously or irregularly branched, in parts, to several orders, firm to spongy in texture, indeterminate in length with branching hyparoblastic; indeterminate branches giving rise, from the superior secondary segments, to whorls of short, usually upward-curved, determinate laterals (either densely and overlapping = *Cladostephus spongiosus*, or separately = *C. hirsutus*) which narrow to the apex and base; determinate laterals sometimes terminally branched, giving rise acroblastically to short, determinate, spine-like laterals (branchlets), with or without hairs in their axils; growth apical; central axial cells/segments dividing and enlarging (auxocaulic) to produce a large central medulla of longitudinally elongated, transversely rounded to cuboidal cells, the peripheral cells of which become meristematic to produce an extensive secondary rhizoidal cortex of outward-radiating filaments of cells, further anticlinal divisions of the outermost cells producing compact tiers of isodiametric cells, this secondary cortical growth clothing the base of the determinate laterals.

Plurilocular and unilocular sporangia sessile or stalked, borne on specialized fertile laterals arising from the outermost cortical cells of the axis, between the whorls of determinate laterals.

Prior to the publication of Prud'homme van Reine (1972), most European floristic accounts recognised two species of *Cladostephus*, viz. *C. spongiosus*, in which the whorls of ramuli are very closely packed together and barely distinguishable, and *C. verticillata* in which the whorls of ramuli are more spaced apart and distinct. Prud'homme van Reine (1972), however, recognised these as f. *spongiosus* and f. *verticillatus* respectively of *C. spongiosus*, and this was then adopted in most subsequent taxonomic treatments (see Brodie *et al.* 2016; Cabioc'h *et al.* 1992; Guiry 2012; Mathieson & Dawes 2017; Parke & Dixon 1976), a notable exception being Womersley (1987) who considered that these formae are doubtful taxonomic entities, and the differences between them are more likely to be ecologically based (f. *spongiosus* usually found emersed in the eulittoral and f. *verticillata* usually found in deeper water). Yet a more recent molecular phylogeny and taxonomic reassessment of the genus *Cladostephus* by Heesch *et al.* (2020) indicated that these two previously recognised morphological entities represent two different genetic entities. Differences reported by Heesch *et al.* between the two species include, for example, their geographical distribution (*C. spongiosus* on European Atlantic coasts; *C. hirsutus* in temperate seas of both hemispheres), ecological niche (*C. spongiosus* mid-eulittoral; *C. hirsutus* mid-eulittoral to upper sublittoral), morphology (*C. spongiosus* thallus height 4–10 cm, bushy with numerous overlapping branches, tips rounded, axes width 400–660 µm, not much denuded in winter, whorled branches almost straight to strongly curved; *C. hirsutus* height > 25 cm, thallus with distinct whorls of branches, tips slender to sharply pointed, axes 350–860 µm, denuded in winter, whorled branches forked, strongly curved, sickle-shaped) and reproductive features (*C. spongiosus* unilocular sporangia 45–55 × 25–40 µm, plurilocular gametangia single, opposite, or in pairs, 38–47 × 22–27 µm; *C. hirsutus* unilocular sporangia 55–80 × 35–55 µm, plurilocular gametangia single, opposite, in pairs or unilateral series, 50–100 × 20–35 µm). These differences are broadly supported by the present study, and two species, *Cladostephus spongiosus* (Hudson) C. Agardh and *C. hirsutus* (L.) Boudouresque & M. Perret, should now be recognised for Britain and Ireland. *Cladostephus spongiosus* can be distinguished by its dense clothing of laterals from the verticillate *C. hirsutus* (Heesch *et al.* 2020).

Two species in Britain and Ireland.

KEY TO SPECIES

Thalli rarely exceeding 9 cm in length, the whorled secondary laterals closely packed and overlapping, thalli rarely denuded of secondary laterals in winter, usually mid to low eulittoral, on rocks, frequently associated with mud and sand . . . . *C. spongiosus*
Thalli up to 25 cm in length, the whorled secondary laterals separated into distinct bands at the nodal area of segments, thalli commonly denuded of secondary laterals in winter, lower eulittoral pools and shallow sublittoral, frequently associated with sand . . . . *C. hirsutus*

Cladostephus spongiosus (Hudson) C. Agardh (1817), p. xxvi.   Fig. 14C

Conferva spongiosa Hudson (1762), p. 480; *Fucus bryum* Strøm (1788), p. 348, pl. 1, fig. 2; *Ceramium spongiosum* (Hudson) Wahlenberg (1826), p. 905; *Dasytrichia spongiosa* (Hudson) Bonnemaison (1828), p. 96; *Cladostephus densus* Kützing (1856), pl. 7.

Thalli erect, epilithic, rarely epiphytic, solitary to gregarious, to 5(–9) cm long, dark brown to black, with 1 to several (–22) axes arising from a small, 1–2 cm, crustose base; axes polystichous, growing from conspicuous apical cells, dichotomously, sometimes irregularly branched to several orders, tomentose, terete, fairly rigid and coarse, somewhat spongy, not obviously denuded in winter, except at the extreme base, 185–540 µm wide, elsewhere much wider, 1–2.2 (–3) mm, resulting from the growth of whorls of densely crowded and closely set, overlapping, determinate, secondary laterals; secondary laterals recurved adaxially towards parental axis, 0.8–1.5(–1.9) mm long, 48–66(–80) µm wide, broadening above and either rounded and blunt or spine-like, usually simple, sometimes terminally branched, with 1–3 forward-projecting, spine-like determinate laterals, often with a hair in their axils; secondary laterals segmented, with segments longer than broad below, shorter than broad above, with secondary segments longitudinally divided by 1–6 walls, with up to 7 segments/cells seen is surface view, each segment further divided transversely into up to 4 cuboidal to rectangular, photosynthetic cells; in transverse section, secondary laterals reveal a large single-celled medulla enclosed within a single layer of small cortical cells; in transverse section, axes reveal an inner, central medulla of closely packed, rounded to polygonal cells enclosed by a cortex of radially elongated, colourless cells, enclosed within an epidermal-like layer of smaller, pigmented photosynthetic cells containing densely packed, discoid plastids without pyrenoids; older, basal, denuded axes showing further meristematic activity, with abundant anticlinal divisions producing radial rows of small, quadrate to rectangular, darkly pigmented cells.

Plurilocular and unilocular sporangia occurring on separate thalli, borne laterally on erect, densely packed, brush-like tufts of specialised, fertile laterals arising from the axes between the whorls of secondary laterals over much of the thallus length, sometimes unaccompanied by secondary laterals; fertile laterals shorter than the secondary laterals, stiff, erect, simple, 450–910 µm in length, 31 45 µm in width, widening terminally with an acute, slightly curved apex, segmented, thick-walled, with basal segments longer than wide, upper segments

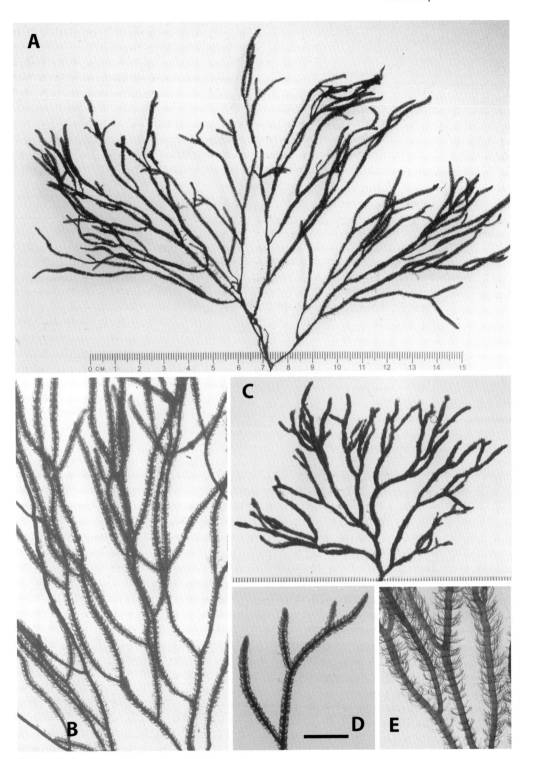

Fig. 14. *Cladostephus* spp.
Cladostephus hirsutus: A, B. Habit of thalli. D, E. Close-up of axes showing distinct whorled arrangements of laterals. *Cladostephus spongiosus*: C. Habit of thallus with indistinct whorls of laterals which are more closely packed and overlapping. Bar in D = 7 mm (B), = 8 mm (D), = 4 mm (E).

usually shorter than wide, with each primary segment divided longitudinally by 1–3 walls and commonly further divided by transverse walls; plurilocular sporangia solitary or grouped, on 1–4 celled, adaxially recurved pedicels, cylindrical, 33–48×22–26 µm (after Kormann & Sahling, 1977); unilocular sporangia sessile or grouped, on 1–4 celled, adaxially recurved pedicels, oval to ovoid, 45–56×31–34 µm.

Mid to lower littoral pools, rocks in damp places, frequently associated with sand and mud. Common and widely distributed all around Britain and Ireland.

Perennial with maximum growth occurring in the summer; unilocular and plurilocular sporangia recorded in winter, particularly January and February although the specialist fertile laterals have been observed at other times (September).

Ardré (1970), pp. 262; Batters (1890b), p. 285; Brodie *et al.* (2016), p. 1018; Bunker *et al.* (2017), p. 292; Cabioc'h & Boudouresque (1992), p. 58, fig. 25; Coppejans & Kling (1995), p. 232, pl. 85(A–F); De Mesquita Rodrigues (1963), pp. 38–40, pl. IVB; Dizerbo & Herpe (2007), p. 99; Draisma *et al.* (2001), pp. 586–603, (2010b), pp. 308–26; Feldmann (1937), pp. 301, 302; Fletcher in Hardy & Guiry (2003), p. 233; Gallardo *et al.* (2016), p. 34; Geyler (1866), p. 523, pl. 36, figs 30–33; Guiry (2012), p. 128; Hamel (1931–39), pp. 268, 269; Harvey (1846–51), pl. 138; Heesch *et al.* 2020, pp. 426–43; Knight & Parke (1931), p. 67; Kornmann & Sahling (1977), pp. 156, 159, pl. 86A–D; Kylin (1947), pp. 31,32, pl. 1(5); Lund (1950), pp. 62–3; Magnus (1873), p. 135, pl. 2(2–32); Mathieson & Dawes (2017), p. 232, pl. XXXV(1–3); Morton (1994), pp. 37, 38; Nelson (2013), pp. 56, 57; Newton (1931), p. 195; Prud'homme van Reine (1972), pp. 139–44; Rueness (1977), p. 188; Sauvageau (1906), pp. 69–94, (1907), pp. 921, 922, (1908), pp. 695–7, (1971), pp. 581–91, 602, 603, figs 126–8; Schreiber (1931), p. 235; Seoane-Camba (1965), pp. 77,78, fig. 21(1); Stegenga & Mol (1983), p. 116, pl. 44(5–7); Stegenga *et al.* (1997a), p. 22.

Cladostephus hirsutus (Linnaeus) Boudouresque & M. Perret-Boudouresque ex Heesch *et al.* (2020), p. 438.   Figs 14(A,B,D,E), 15.

Fucus hirsutus Linnaeus (1767), p. 134; *Conferva verticillata* Lightfoot (1777), p. 984; *Conferva myriophyllum* Roth, nom. illeg. (1801), p. 335 (as '*Conferua Myriophyllum*'); *Cladostephus verticillatus* (Lightfoot) Lyngbye (1819), p. 102; *Cladostephus myriophyllum* Bory, nom. illeg. (1823), p. 182 (as '*Myriophyllum*'); *Cladostephus spongiosus* f. *verticillatus* (Lightfoot) Prud'homme (1972), p. 142.

Thalli erect, epilithic, rarely epiphytic, solitary to gregarious, to 25 cm long, light brown when young, almost black when mature, with one to several (20+) axes arising from a large, several cm, crustose base; axes polystichous, growing from conspicuous apical cells, dichotomously, sometimes irregularly branched to several orders, smooth, frequently denuded in winter, to 0.4(–0.8) mm wide, terete, fairly rigid, elsewhere tomentose, somewhat coarse and spongy, much wider, to 1.1(–1.9) mm, resulting from the growth of distinct, separate whorls of determinate, secondary laterals, 0.4–0.8 mm apart; secondary laterals recurved adaxially towards parental axis, to 1.5 mm long, 58–67 µm wide, broadening above and either rounded and blunt or spine-like, usually simple, sometimes terminally branched, with 1–3 forward-projecting, spine-like determinate laterals, often with a hair in their axils; secondary laterals parenchymatous, segmented, with segments longer than broad below, shorter than broad above, with secondary segments longitudinally divided by 1–6 walls, with up to 7 segments/cells seen is surface view, each segment further divided transversely into up to 4 cuboidal to rectangular, photosynthetic cells; in transverse section, secondary laterals reveal a large single-celled medulla enclosed within a single layer of small cortical cells; in transverse section, axes reveal an inner, central medulla of closely packed, rounded cells enclosed by an extensive outer secondary "rhizoidal" cortex of radially elongated, colourless cells, enclosed within an epidermal-like layer of smaller, pigmented photosynthetic cells containing densely packed, discoid plastids without pyrenoids; older, basal, denuded axes showing further meristematic activity, with abundant anticlinal divisions producing radial rows of small, quadrate to rectangular, darkly pigmented cells.

Plurilocular and unilocular sporangia occurring on separate thalli, more rarely observed on the same thalli, borne laterally on erect, densely packed, brush-like tufts of specialised,

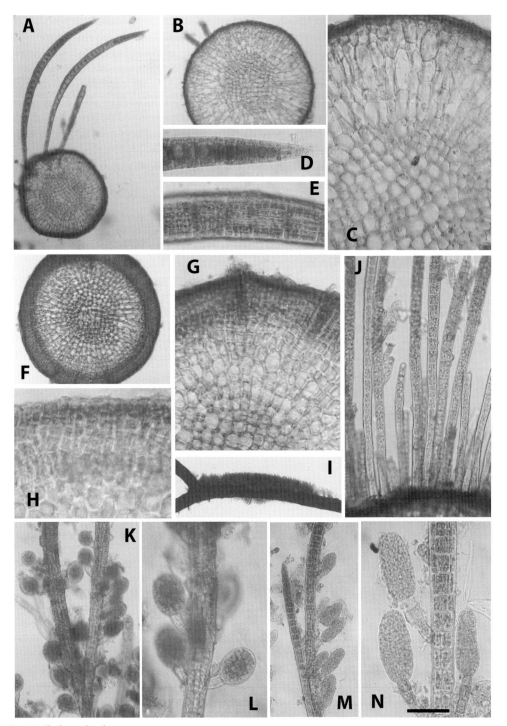

Fig. 15. *Cladostephus hirsutus*
A. TS of axis showing three emergent determinate laterals. B, C. TS of axes showing central medulla and outer secondary cortex of radially elongated cells. D, E. Terminal and mid-region portions of determinate laterals revealing a large, single-celled medulla and an outer single layer of smaller cortical cells. F–H. TS of older axes showing radial rows of small cuboidal cells extending out from the secondary cortex. I. Portion of axis lacking determinate laterals, showing dense tufts of fertile laterals. J. Close-up of fertile laterals. K, L. Portions of fertile laterals with unilocular sporangia. M, N. Portions of fertile laterals with plurilocular sporangia. Bar in N = 4 mm (I), = 300 μm (A), = 180 μm (B, F), = 90 μm (C, G, J, K, M), = 45 μm (D, E, H, L, N).

fertile laterals arising from the axes between the whorls of secondary laterals over much of the thallus length, sometimes unaccompanied by secondary laterals; fertile laterals shorter than the secondary laterals, stiff, erect, simple, to 580 μm in length, 11–30 μm in width, widening terminally with an acute, slightly curved apex, segmented, thick-walled, with basal segments longer than wide, upper segments usually shorter than wide, with each primary segment divided longitudinally by 1–3 walls and commonly further divided by transverse walls; plurilocular sporangia solitary or grouped, on 1–4 celled, adaxially recurved pedicels, cylindrical, 35–66 × 16–28 μm; unilocular sporangia sessile or grouped, on 2–4 celled, adaxially recurved pedicels, oval to ovoid, 38–58 × 28–42 μm.

Epilithic, sometimes epiphytic, lower eulittoral pools, frequently associated with sand, sublittoral to 10 m.

Common and widely distributed all around Britain and Ireland.

Thalli similar in anatomy to those of *C. spongiosus*, but generally larger, up to 25 cm high vs. 4–10 cm in *C. spongiosus*. The most obvious morphological difference is that in *C. hirsutus* the laterals are well-spaced and verticillate (see Fig. 14A, B, D, E) whereas in *C. spongiosus* they are very close together resulting in a spongy texture (see Fig. 14C). Laterals are slightly thicker than in *C. spongiosus* (50–70 μm vs. 45–65 μm). It has a much wider geographical distribution than *C. spongiosus*, being found in temperate seas of both hemispheres, possibly due to introductions, whereas *C. spongiosus* is currently known only from the European Atlantic. A detailed comparison of the two species is provided by Heesch *et al.* (2020).

Ardré (1970), pp. 262–3; Batters (1890b), pp. 285, 286; De Mesquita Rodrigues (1963), pp. 38–40, fig. IVc; Børgesen (1926), pp. 75, 76; Boudouresque *et al.* (1984); Brodie *et al.* (2016), p. 1018; Bunker *et al.* (2017), p. 292; Cabioc'h & Boudouresque (1992), p. 58; Cormaci *et al.* (2012), pp. 419, 420, pl. 144(1–4); De Mesquita Rodrigues (1963), pp. 40, 41, pl. 4C; Dizerbo & Herpe (2007), p. 99; Draisma *et al.* (2001), pp. 586–603, (2010b), pp. 308–26; Feldmann (1937), pp. 301, 302; Fletcher in Hardy & Guiry (2003), p. 233; Funk (1955), p. 45; Gallardo *et al.* (2016), p. 34; Gayral (1958), pp. 208–10, fig. 28A–C, pl. XXV, (1966), pp. 241, 242, fig. 28(A–C), pl. 32; Geyler (1866), p. 523, pl. 36, figs 30–33; Guiry (2012), p. 128; Hamel (1931–39), pp. 266–8, fig. 48(IX–XII); Harvey (1846–51), pl. 33; Heesch *et al.* 2020, pp. 426–43; Knight & Parke (1931), p. 67; Kornmann & Sahling (1977), p. 160, pl. 87A, B; Kylin (1947), p. 31, pl. 1(4); Lindauer *et al.* (1961), pp. 175–7, fig. 24; Lund (1950), pp. 63–4; Magnus (1873), p. 135, pl. 2(2–32); Mathieson & Dawes (2017), p. 232; Mazariegos-Villareal *et al.* (2010), pp. 153–7; Morton (1994), p. 38; Newton (1931), p. 195, fig. 122A–D; Prud'homme van Reine (1972), pp. 139–44; Reinke, J. (1891b), p. 18, pl. 6(1–3); Rueness (1977), p. 188, pl. XXV(1); Sauvageau (1906), pp. 69–94, (1907), pp. 921, 922, (1908), pp. 695–7, (1971), pp. 488–581, 601, 602, figs 93–125; Schreiber (1931), p. 235; Seoane-Camba (1965), pp. 77, 78, fig. 21(1); Stegenga & Mol (1983), p. 116, pl. 44(5–7); Stegenga *et al.* (1997a), p. 22; Taylor (1957), p. 124, pl. 17(9–11); Womersley (1987), pp. 185, 186, figs 60(D), 62(E–G).

Lithodermataceae Hauck

LITHODERMATACEAE Hauck (1883), pp. 318, 402 [as Lithodermaceae].

Thallus epilithic, streblonematoid or heterotrichous and ralfsioid, discrete or expansive, closely adherent to substratum, usually without rhizoids; thallus structure filamentous and/or pseudoparenchymatous, composed of prostrate, radiating, branched filaments, these either remaining free or becoming laterally adjoined and disc-like with a marginal row of apical cells; prostrate filaments either without erect filaments (streblonematoid) or giving rise behind to erect, straight or slightly curved, strongly united, firm, little branched or unbranched, compacted filaments to form a firm, subcoriaceous to coriaceous, haplostichous crust (ralfsioid); vegetative cells with several discoid plastids, without obvious pyrenoids; hairs present or absent; ascocyst-like cells present or absent, intercalary or terminal, on erect filaments.

Plurilocular and unilocular sporangia only known for the crustose genera and occurring in discrete or expansive sori, slightly raised on thallus surface, terminating erect filaments, formed by either direct transformation or extension of surface cells, with or without ascocyst-like cells.

In many earlier treatments of the crustose brown algae (e.g. Feldmann 1937; Taylor 1937a, 1957; Parke 1953; Pankow 1971) up to three families were generally recognized; these were the Lithodermataceae Hauck, the Ralfsiaceae Hauck and the Nemodermataceae Feldmann. The Lithodermataceae were usually characterized by terminally borne unilocular and pluri-locular sporangia and several plastids in each cell; the Ralfsiaceae had lateral unilocular sporangia, intercalary plurilocular sporangia and a single, plate-like plastid in each cell; the monotypic Nemodermataceae had intercalary unilocular sporangia, lateral plurilocular sporangia and several plastids in each cell. However, these differences were not always considered clear-cut and, for example, in many treatments, the Lithodermataceae were included in the circumscription of the Ralfsiaceae (Hollenberg 1969; Abbott & Hollenberg 1976; Parke & Dixon 1976; Rueness 1977). Newton (1931), Fritsch (1945) and Loiseaux (1967a) went further and included all the crusts together with the discoid myrionematoid and hecatonematoid algae within the single, large family Myrionemataceae Foslie. Within this family Loiseaux grouped all the crustose genera in a tribe 'Ralfsiées'.

Recognition of the three families, Lithodermataceae, Ralfsiaceae and Nemodermataceae, distinguished on the basis of the position of the reproductive organs, was resurrected by Nakamura (1972). He placed these families in a new order, Ralfsiales (failing to give a Latin description), members of which were characterized by an *Ectocarpus*-type (isomorphic) life history, a discal type of spore germination pattern and a single, parietal, plate-like plastid which lacks a pyrenoid. This new order was not recognized by Russell & Fletcher (1975), Parke & Dixon (1976), South (1976) or Nelson (1982) but was accepted by Bold & Wynne (1978) and Tanaka & Chihara (1980a, 1982) who erected and included in the order a new family, Mesosporaceae.

Whilst molecular work, notably studies by Tan & Druehl (1994) and Lim *et al.* (2007), lent support for the Ralfsiales, particularly significant here was the molecular study by McCauley & Weir (2007) who removed the Lithodermataceae, encompassing the genera *Heribaudiella*, *Lithoderma*, *Pseudolithoderma* and *Bodanella*, from the Ralfsiales, as the family was found to have a close relationship with the Sphacelariales. A close relationship between *Pseudolithoderma* and *Heribaudiella* and the Sphacelariales was also revealed in a molecular study by Bittner *et al.* (2008). Fletcher (1987) included four genera in the Lithodermataceae (*Petroderma*, *Pseudolithoderma*, *Sorapion* and *Symphyocarpus*). Since then, based on molecular evidence and 'several structural features', Silberfeld *et al.* (2014) removed *Petroderma* from the Ralfsiaceae/ Lithodermataceae and placed it within a new family, Petrodermataceae, positioned as a sister to an emended Ishigeales. Silberfeld *et al.* (2014) also repositioned *Symphyocarpus* within the Scytosiphonaceae based on the genus possessing a single plate-like plastid with a pyrenoid in each cell, and positioned *Sorapion* as *incertae sedis* based on the lack of clarity regarding the presence and type of pyrenoid associated with the plate-like plastid (Bruno de Reviers, pers. comm.). These changes are accepted in the present treatise pending further studies. Of the four genera, viz. *Bodanella*, *Heribaudiella*, *Lithoderma* and *Pseudolithoderma*, included within the Lithodermataceae by Silberfeld *et al.* (2014), only the freshwater genus *Heribaudiella* and the marine genus *Pseudolithoderma* are present in Britain and Ireland.

Pseudolithoderma Svedelius

PSEUDOLITHODERMA Svedelius in Kjellman & Svedelius (1911), p. 175.

Type species: *P. fatiscens* Svedelius in Kjellman & Svedelius (1911), p. 176.

Thallus encrusting, thin to moderately thick, discrete or extensive, light brown to black, firm, smooth, subcoriaceous to coriaceous, firmly attached to the substratum by the whole undersurface, normally without rhizoids; thallus structure pseudoparenchymatous, base monostromatic and discoid, giving rise to erect, little-branched, very tightly bound filaments, covered by a surface cuticle, with or without terminal ascocyst-like cells; cells with several small discoid plastids without obvious pyrenoids; hairs unknown in Britain and Ireland.

Plurilocular sporangia terminal on erect filaments, with or without unicellular paraphysis-like cells, in slightly raised mucilaginous sori, extensive on crust surface, uniseriate to multiseriate, with straight or oblique cross walls; unilocular sporangia present or absent, terminal on erect filaments.

Two species of the genus *Pseudolithoderma* are presently recognized for Britain and Ireland: *P. extensum* and *P. roscoffense*. It is possible that other species of this genus might also be present but await identification. To date three other species have been described in the North Atlantic: *Pseudolithoderma rosenvingei* (Waern) Lund, known from the Swedish and Finnish Baltic coasts, East and West Greenland, the White Sea and north-east Canada (Waern 1949, 1952; Lund 1959; Sears & Wilce 1973); *P. subextensum* (Waern) Lund, known only from the Swedish Baltic coast (Waern 1949, 1952) and *P. paradoxum* Sears & Wilce, known from north-east America (Sears & Wilce 1973). For further information on these species and comparison with crusts of *P. extensum* reference should be made to the literature cited.

Two species in Britain and Ireland.

KEY TO SPECIES
Plurilocular sporangia uniseriate, rarely biseriate; loculi with oblique cross walls . . .*P. extensum*
Plurilocular sporangia biseriate to multiseriate; loculi with straight cross walls . . . .*P. roscoffense*

Pseudolithoderma extensum (P. Crouan & H. Crouan) S. Lund (1959), p. 84, fig. 2.   Fig. 16

Ralfsia extensa P. Crouan & H. Crouan (1867), p. 166. *Lithoderma fatiscens* sensu Kuckuck (1894), p. 238, non Areschoug (1875), p. 22; *Pseudolithoderma fatiscens* (Kuckuck) Svedelius in Kjellman & Svedelius (1910), p. 176; *Lithoderma extensum* (P. Crouan & H. Crouan) Hamel (1931–39), p. 110.

Thallus encrusting, light brown to black, thin to thick, usually confluent, to 10 cm or more in diameter, firm, coriaceous, very firmly attached by whole undersurface to substratum, rhizoids usually absent; surface cells polygonal, closely packed together, 6–10 × 4–6 μm, containing up to six discoid, peripherally placed plastids; thallus structure pseudoparenchymatous, consisting of a monostromatic discoid base giving rise to erect rows of little branched, very firmly adjoined filaments, to 22 cells, 165 μm long, covered by a thick surface cuticle; cells usually markedly wider than high below, quadrate or subquadrate above, 4–11 × 6–20 μm, with discoid, terminally placed plastids without obvious pyrenoids; hairs not observed.

Plurilocular and unilocular sporangia terminal on erect filaments, in very slightly raised, mucilaginous sori extensive on crust surface; plurilocular sporangia common, uniseriate, rarely biseriate, to eight loculi, 40 μm long × approx. 9 μm in diameter, attenuate at apex with characteristic oblique cross walls, frequently with accompanying club-shaped or cylindrical ascocyst-like cells, to 33 μm long × approx. 7 μm in diameter; unilocular sporangia rare, slightly globose, to 33 long × approx. 20 μm in diameter.

Epilithic on stones and bedrock, in lower eulittoral pools, more common and extensive in the sublittoral, to a depth of at least 20 m.

Abundant and widely distributed around Britain and Ireland.

Perennial, sporangia recorded during winter (November to March). This is the only crustose brown alga that is commonly and perennially found in the sublittoral around Britain and Ireland; other crustose species are only rarely reported and are usually confined to the littoral and shallow sublittoral. It occurs on bedrock, boulders, shells, various artificial substrata, mainly contributing to the underflora of *Laminaria* spp. and often in association with crustose coralline and non-coralline red algae.

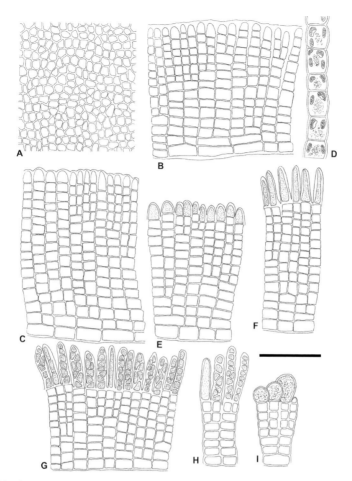

Fig. 16. *Pseudolithoderma extensum*
A. Surface view of vegetative crust. B, C. VS of vegetative crusts showing closely adjoined erect filaments. D. A single erect filament showing cells with 1–4 discoid plastids. E, F. VS of crusts with terminal immature plurilocular sporangia. G, H. VS of crusts with terminal plurilocular sporangia. I. VS of crust with terminal unilocular sporangia. Bar = 20 μm (D), = 50 μm (A–C, E–I).

Culture studies by Kuckuck (1912) and Peters (1989) indicate that an isomorphic alternation of generations type of life history occurs, with both the gametophytes and sporophytes being crustose. Peters observed that released motile isogametes from plurilocular gametangia fused, and the zygotes developed into crustose thalli. Zoospores released from unilocular sporangia on field-collected crusts also developed into crusts.

Batters (1895a), p. 275; Diverbo & Herpes (2007), p. 75, pl. 23(4); Fletcher (1978), pp. 371–98, (1987), pp. 83, 84, fig. 2; Fletcher in Hardy & Guiry (2003), p. 288; Gallardo García *et al.* (2016), p. 33; Hamel (1931–39), pp. 110, 111, fig. 26(D–E); Kuckuck (1894), pp. 238–40, fig. 11(A–B), (1912), pp. 167–76, figs 2–4, pl. 7(1–18); Lund (1938), pp. 1–18, figs 4–6, (1959), p. 84; Mathieson & Dawes (2017), p. 226, pl. LVIX(4–6); Morton (1994), p. 33; Newton (1931), p. 127, fig. 75(A–C); Pankow (1971), pp. 172, 173, fig. 206; Peters (1989), pp. 195–205; Ribera Siguán *et al.* (1992), p. 113; Rueness (1977), p. 126, fig. 51(A–B); Stegenga & Mol (1983), pp. 81, 82; Stegenga *et al.* (1997), p. 18; Taylor (1957), p. 137; Waern (1949), pp. 658, 659, fig. 2 (i–j).

Pseudolithoderma roscoffense Loiseaux (1968), p. 308 (as *P. roscoffensis*).   Fig. 17

Thallus encrusting, thin and light brown to moderately thick and black, firm, smooth, subcoriaceous, circular and discrete or irregular and confluent to 3 cm or more in extent, very firmly attached to the substratum by the undersurface, usually without rhizoids, surface cells polygonal, closely packed and irregularly arranged, 6–13 × 5–9 µm, containing up to ten discoid, peripherally placed plastids; thallus structure pseudoparenchymatous, consisting of a monostromatic discoid base giving rise to erect rows of little-branched, firmly adjoined filaments, to 30 cells, 190 µm long, enclosed by a thick surface cuticle, rarely with terminal, long (to 200 µm) unicellular, colourless, cylindrical ascocysts present; erect filament cells usually subquadrate, rarely rectangular, 0.25–2.5 diameters long, 4–12 × 4–6 µm with 3–6(–10) small discoid plastids, without obvious pyrenoids; hairs not observed.

Plurilocular sporangia terminal, without paraphyses, in slightly raised, mucilaginous sori, extensive on crust surface; in surface view, rounded or irregular in shape, with four, rarely two, loculi; in vertical section, clavate to slightly pyriform in shape, biseriate to multiseriate, 28–45 × 11–18 µm, to 8 loculi long, with straight dividing walls and covered by a thick surface cuticle; plurilocular sporangial initials look like paraphyses or unilocular sporangia, dark-coloured, elongate to slightly pyriform, 15–24 × 6–10 µm; unilocular sporangia unknown.

Epilithic in the lower eulittoral, covering upper (and frequently lower) surface of small flint stones in pools and exposed, in open sheltered places.

Largely restricted to the south coast of England (Sussex, Hampshire, Dorset, Devon and Cornwall), with a more recent find by the present author in Norfolk; a single report for the east coast of Ireland (Dublin).

Plurilocular sporangia have been recorded throughout the year on the south coast of England.

This alga was first described by Loiseaux (1968) based on material collected from Roscoff in Brittany. It was discovered in Britain at Bembridge, Isle of Wight in 1973 and subsequently reported in Parke & Dixon (1976) and Fletcher (1987). It possibly represented a new addition to the south coast, although because of its restricted distribution, it may have been overlooked. It is a eulittoral species and can be distinguished from the other *Pseudolithoderma* species (*P. extensum*) in having the plurilocular sporangia grouped in two or four vertical columns rather than one, and in having straight rather than oblique cross walls. Since its initial discovery it has been found to be very common in the Solent region of the south coast of England, where it dominates the exposed flint stones in the lower eulittoral in the sheltered muddy harbours of Langstone and Portsmouth. It has subsequently been found by the author at other localities in Sussex, Hampshire, Dorset, Devon, Cornwall and more recently in Norfolk – while, to date, there has been a single report for Co. Dublin (Fletcher & Maggs 1985). Despite regular field searches it has not been found on the Kent coast. It appears, therefore, to be largely restricted to southern shores, although it is likely to occur more extensively on the west coasts of both Britain and Ireland.

Culture studies by Loiseaux (1968) and Fletcher (1978) revealed that the life history is of the 'direct' type. Both authors reported the plurispores to germinate directly, without sexual fusion, to repeat the parental crustose thallus; Loiseaux also reported the development of plurilocular sporangia on the crusts. The above description of *P. roscoffense* appears to show a marked similarity to that of *Pseudolithoderma* (*Lithoderma*) *adriaticum* Hauck given by Hamel (1931–39, p. xxxi, fig. 62-19). This species, originally described for the Adriatic Sea and widely distributed in the Mediterranean, was also reported by Hamel (1931–39, p. xxxi & 111) for the North Atlantic (Tatihou and Iles Chausey). It seems possible, therefore, that *P. adriaticum* Hauck *sensu* Hamel is conspecific with *P. roscoffense* Loiseaux, although such a connection was not discussed by Loiseaux (1968). However, the synonymy between these two taxa was challenged by Verlaque (1988), based on differences in thallus thickness and vegetative cell dimensions. More detailed studies, particularly molecular phylogenetic analyses, are clearly required.

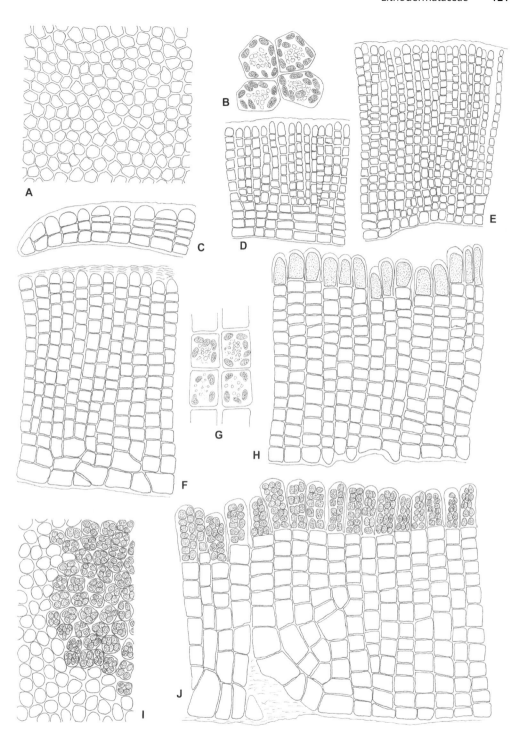

Fig. 17. *Pseudolithoderma roscoffense*
A. Surface view of vegetative crust. B. Surface view of crust cells with several discoid plastids. C. VS of crust margin. D–F. VS of vegetative crusts. G. Portion of two erect filaments showing enclosed discoid plastids. H. VS of crust showing terminal immature plurilocular sporangia. I. Surface view of fertile crust showing plurilocular sporangia. J. VS of fertile crust showing terminal plurilocular sporangia. Bar = 20 μm (B, G), = 50 μm (A, C–F, H–J).

Bartsch & Kuhlenkamp (2002), p. 168; Brodie *et al.* (2016), p. 1022; Cormaci *et al.* (2012), pp. 415, 416, pl. 143(5, 6); Dizerbo & Herpe (2007) p. 76; Fletcher (1978), pp. 371–98, figs 18, 19, 28, (1987), pp. 84, 86, 87, fig. 3; Fletcher & Maggs (1985), pp. 523–6, fig. 2(A–D); Fletcher in Hardy & Guiry (2003), p. 288; Gallardo García *et al.* (2016), p. 33; Guiry (2012), p. 157; Hamel (1931–39), pp. xxxi, 111, fig. 62(19); Loiseaux (1968), pp. 308–11, fig. 6; Peña & Bárbara (2010), p. 56, fig. 12(A–C); Verlaque (1988), pp. 190, 191, figs 17–22.

Sphacelariaceae Decaisne

SPHACELARIACEAE Decaisne (1842), pp. 329, 341.

Thalli epilithic or epiphytic, crustose or erect; if crustose, thin, laterally spreading, comprising vertical files of tightly adjoined filaments, often with superimposed crusts; if erect, heterotrichous, tufted, either solitary or in confluent mats/turfs, 1–130 mm long with one to several main axes, attached by monostromatic discs, polystromatic crusts or stolons, sometimes with endophytic penetration, with or without an overlying mass of secondary rhizoidal filaments; erect shoots rigid or flaccid, terete, sparsely to densely branched, to several orders, with branching irregular or pinnate and opposite, with axes and branches/branchlets determinate or indeterminate, linear or with slight increase in diameter below, narrowing to apex, multiseriate, sometimes with terminal uniseriate branchlets, parenchymatous, and distinctly and usually regularly segmented, with cells longitudinally elongate or cuboidal in surface view, largely isodiametric in transverse section, ecorticate or more or less extensively corticated due to meristematic activity of outer cortex, usually then additionally clothed with downward-growing, rhizoidal filaments; growth apical, all axes terminated by a large, prominent apical cell which divides transversely and longitudinally to form the segmented thallus, secondary segments remaining the same width as the apical cell (leptocaulous); some additional secondary growth of the secondary segments, in length and width, may also occur; branches produced hemiblastically; all cells with several discoid plastids without pyrenoids; hairs with basal meristem and sheath.

Reproduction by specialized propagules, plurilocular sporangia or unilocular sporangia (the plurilocular sporangia similar or dimorphic (micro and mega)), borne mainly on the terminal laterals, or on special short shoots, sessile or stalked; sexual reproduction isogamous or anisogamous.

On the basis of several collaborative molecular studies (see Draisma & Prud'homme van Reine 2010, Draisma *et al.* 2002, 2003, 2010b, 2014), the genus *Sphacelaria* was shown to be polyphyletic and comprised a number of phylogenetic lineages which were then distributed (see Draisma *et al.* 2010b) into the following seven new, revised or reinstated genera: *Battersia, Chaetopteris, Herpodiscus, Protohalopteris, Sphaceloderma, Sphacelorbis* and *Sphacelaria*. Except for *Herpodiscus*, all these genera are present in Britain and/or Ireland.

Battersia Reinke ex Batters

BATTERSIA Reinke ex Batters (1890a), p. 279.

Type species: *B. mirabilis* Reinke ex Batters (1889a), p. 279.

Thalli epilithic, either as extensive, thin, coriaceous, often superimposed, polystromatic crusts lacking erect thalli, or as erect solitary or gregarious tufts of one to several thalli, to 9 cm in height, arising from a discoid monostromatic, later polystromatic base; erect thalli sparsely to much branched to 3(–4) orders, irregularly radial or distichous and opposite, indeterminate, or with terminal, pinnate, opposite laterals of determinate growth; thalli ecorticate, but with proximal regions of thalli usually covered with loosely bound, downward-growing, secondary, rhizoidal filaments; all erect axes and laterals multiseriate, polysiphonous and segmented; growth apical, with longitudinal division of the secondary segments primarily being periclinal, and usually with secondary transverse cell divisions; hairs present or absent; cells with several discoid plastids without pyrenoids.

Unilocular and plurilocular sporangia formed on short unicellular or multicellular, simple or branched, specialized filaments borne along the length of the main axes and laterals of the erect thalli, or in discrete sori on the surface of the crustose thalli.

Three species found in Britain, one of them also occurring in Ireland.

KEY TO SPECIES

1. Thallus a thin, strongly adherent, polystromatic crust only . . . . *B. mirabilis*
    Thallus erect, branched, to several cm in height 2
2. Thallus sparingly and irregularly branched throughout; sporangia borne in
    raceme-like clusters . . . . *B. racemosa*
    Thallus much branched, with branches irregularly or regularly pinnate,
    opposite and feather-like in appearance; in transverse section, clearly
    periclinal; sporangia borne on densely packed, fertile branches distributed
    along the length of the determinate laterals . . . . *B. plumigera*

Battersia mirabilis Reinke ex Batters (1890), p. 279.   Fig. 18

Sphacelaria mirabilis (Reinke ex Batters) Prud'homme van Reine (1982a), p. 168.

Thalli epilithic, forming thin, smooth, coriaceous, hard, very adherent, polystromatic, black crusts which are frequently confluent and reach several centimetres in diameter, usually without rhizoids; in central thicker regions, surface cells polygonal in outline, somewhat irregularly arranged, 10–20 × 8–14 µm, in peripheral regions, surface cells subquadrate to rectangular, arranged in outwardly radiating rows, 10–13 × 5–12 µm, all cells with several discoid, peripherally placed plastids, without pyrenoids; crusts at first monostromatic, with a basal layer of tightly adjoined filaments with an outer, terminal, large apical cell; upward growth of the basal cells gives rise to the polystromatic crust, comprising vertical files of very tightly coalesced cells which become progressively shorter in diameter towards the apex; individual crusts 6–9 cells high, 100–120(–150) µm thick, with crusts often superimposed and up to 300–700 µm or more in height; basal cells rectangular in vertical section, usually much wider than high, 8–20 × 19–46 µm, mid and upper cells usually subquadrate, quadrate or rectangular, 11–18 × 6–28 µm, each with several discoid plastids without pyrenoids, hairs absent.

Plurilocular and unilocular sporangia formed in small, clumped, yellowish sori scattered over the crust surface; both sporangial types borne on short, erect, monosiphonous, or in part polysiphonous, simple or branched, stalks up to 140 µm (8 cells) or more in length; stalk cells usually quadrate to rectangular, (9–)13–26(–33) × (8–)12–18(–22) µm; unilocular sporangia spherical to obovate, (25–)30–45(–63) × (18–)22–35(–43) µm; plurilocular sporangia reported by Prud'homme van Reine (1982a), based on material in culture, to be elongate-cylindrical to somewhat conical, 40–65 × 20–30 µm; propagules unknown.

Epilithic, lower eulittoral, shallow pools and shallow sublittoral; often associated with sand and areas of shade.

Only known from Northumberland (Berwick-upon-Tweed), Fife (St Andrews) and Shetland (Dales Voe). At St Andrews, it occupied shallow pools at the base of the north-facing cliffs, associated with *Sphacelaria radicans*, *S. plumigera* and *Pseudoralfsia verrucosa*. In Shetland, it was collected at 2–3 m depth.

Thalli perennial, with unilocular sporangial sori observed in January and February.

N.B. *Sphacelaria mirabilis* differs from the other species of *Sphacelaria* in being entirely crustose, with no erect axes.

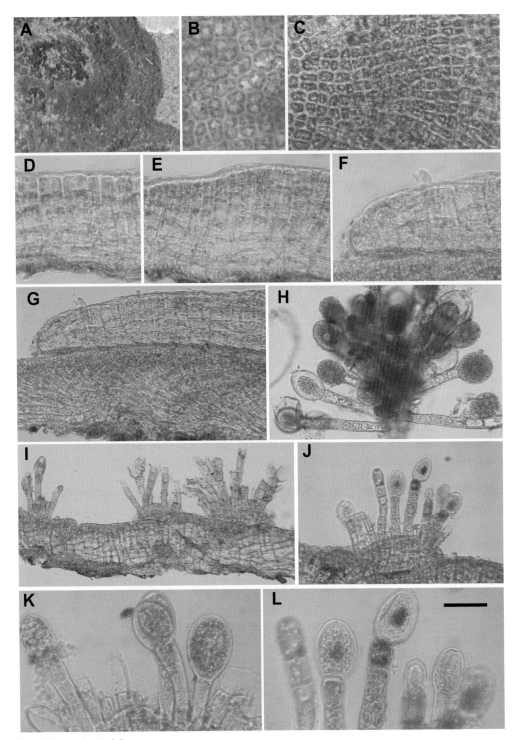

Fig. 18. *Battersia mirabilis*
A. Habit of thallus. B. Surface view of mid region of vegetative crust showing isodiametric cells with peripherally distributed discoid plastids. C. Surface view of edge region of vegetative crust showing more rectangular cells. D, E. VS of vegetative crusts showing closely adjoined erect filaments. F. VS of crust margin showing large apical cell. G. VS of superimposed crusts. H. Squash preparation of reproductive sorus showing terminal unilocular sporangia. I–L. VS of crust showing raised reproductive sori with unilocular sporangia. Bar in L = 3 cm (A), = 180 µm (I), = 90 µm (G, H, J), = 45 µm (B, C–F, K, L).

Batters (1890a), p. 59, pl. 9(1–4), (1890b), p. 279, pl. 9(1–4); (1902), p. 38; Brodie *et al.* (2016), p. 1019; Draisma *et al.* (2010b), pp. 308–26; Fletcher in Hardy & Guiry (2003), p. 297; Holmes & Batters (1890), p. 81; Mathieson & Dawes (2017), p. 234, pl. XXXV(8–11); Newton (1931), pp. 191, 192, fig. 119(A, B); Oltmanns (1922), fig. 373(4); Prud'homme van Reine (1974), p. 170, (1981), pp. 135–9, (1982a), pp. 168–78, figs 380–421, pl. 2(A–C); Reinke (1890), p. 205, fig. 2, (1891a), p. 4, pl. 1(1–6); Sauvageau (1900), p. 224, (1903), p. 80, (1971), pp. 12–17, 249–52, figs 1–2.

Battersia plumigera (Holmes ex Hauck) Draisma, Prud'homme van Reine & H. Kawai (2010), p. 322. Figs 19, 20

Sphacelaria plumigera Holmes ex Hauck (1884), p. 348; *Sphacelaria racemosa* f. *pinnata* Reinke (1892), p. 66; *Sphacelaria plumigera* var. *patentissima* Sauvageau (1903), p. 50.

Thalli epilithic, erect, tufted, up to 9 cm tall, light olive-brown, either solitary or gregarious; main axes arising initially from a monostromatic disc, which later becomes polystromatic, crustose-like and overgrown with attaching, secondary, rhizoidal filaments derived from downward growths of corticating filaments which all arise in the plane of branching and cover the proximal regions of the main axes; branching to 3(–4) orders, irregularly pinnate on both the main axes and branches which have indeterminate growth, giving rise over much of their length to determinate, linear, regularly pinnate, complanate, feather-like, unbranched, opposite laterals, sometimes terminally arranged in a corymbose manner; main axes sometimes denuded at base, bearing short stubs of dehisced laterals; all axes and laterals polysiphonous and segmented except at the extreme apices which are monosiphonous just below the darker-coloured, apical cell; main axes and branches up to 90 μm in width, much more with corticating sheath, each secondary segment usually comprising 1–2 tiers of up to 8–9 mainly rectangular cells seen in surface view and of the periclinal type; determinate laterals to 3 mm in length, 50–60 μm in width, with 4–6 cuboidal to rectangular cells seen in surface view; all cells with several discoid plastids without pyrenoids; hairs common, produced on ultimate laterals, with basal meristem and sheath.

Plurilocular and unilocular sporangia usually occurring on separate individuals, occasionally on the same individual, borne terminally or laterally on ephemeral, forward-projecting, short, narrow, quite densely packed specialized, fertile laterals arising pinnately from the ultimate laterals of determinate growth; fertile laterals to 300(–450) μm in length, 15–22 μm in width, monosiphonous or partly biseriate, mainly simple, occasionally with 1–4-celled, usually monosiphonous stalks; unilocular sporangia common, spherical (40–62 μm in diameter) or ovoid, 38–62(–70) × 26–43(–52) μm; plurilocular sporangia rare in European waters, reported by Prud'homme van Reine (1982a) to be obovate to cylindrical and measuring 45–80 × 30–40 μm; propagules unknown.

Epilithic in lower eulittoral, shady pools and shallow sublittoral, often associated with sand. Rare, but can be locally abundant and widely distributed around Britain and Ireland. Perennial.

Bartsch & Kuhlenkamp (2000), p. 168; Batters (1890a), p. 63, pl. 10(1–3), (1890b), p. 283, pl. X(1–3), (1902), p. 39; Brodie *et al.* (2016), p. 1019; De Haas-Niekerk (1965), pp. 159, 160, figs 64, 66, 67–72; Dizerbo & Herpe (2007), p. 96; Draisma *et al.* (2001), pp. 586–603, (2010), pp. 308–26; Fletcher in Hardy & Guiry (2003), p. 298; Guiry (2012), p. 129; Hamel (1931–39), p. 250; Hauck (1885), p. 348; Holmes (1883), p. 14, (1885), p. 61; Holmes & Batters (1890), p. 8; Irvine (1956), p. 40; Kitayama (1994), pp. 47–51, figs 3–6; Kylin (1947), p. 30, pl. 2(6); Lewis & Taylor (1933), p. 151, pl. 274(1, 4); Lund (1950), pp. 50–5, fig. 11(A–C); Mathieson & Dawes (2017), p. 234, pl. XXXVI(1); Newton (1931), p. 191; Oltmanns (1922), fig. 375(1–3); Prud'homme van Reine (1968), pp. 114–17, fig. 1(a–d), (1974), p. 174, (1978), p. 303, (1982a), pp. 131–46, figs 264–317; Reinke (1890), p. 209, (1891a), p. 12, (1892a), p. 66, pl. 47(1–5); Rueness (1977), pp. 184, 185, fig. 108; Sauvageau (1901), pp. 115, 116, (1903), p. 50, (1971), pp. 94–9, 233, fig. 22; Stegenga & Mol (1983), p. 113, pl. 43(1–5); Stegenga *et al.* (1997a), p. 22; Taylor (1937), p. 132, (1957), p. 122; Tokida (1931), pp. 215–20, figs 1–4, (1954), p. 75; Waern (1945), p. 405, fig. 6, (1952), pp. 99, 100, figs 40, 47(a).

Fig. 19. *Battersia plumigera*
A–C. Habit of thalli. D, E. Close-up of terminal portions of a main axis showing the secondary segments and opposite pinnate branching. Scale in A represents mm. Bar in B = 5 mm (B), = 3 mm (C), = 180 μm (D), = 470 μm (E).

Fig. 20. *Battersia plumigera*

A. Terminal region of laterals showing dark-coloured apical cells. B. Portion of mid-region of main axis showing the regularly pinnate, opposite branching pattern. C. Close-up of main axis showing cells of secondary segments and peripherally distributed discoid plastids. D–I. Portions of upward-pointing fertile laterals borne on the determinate laterals and bearing unilocular sporangia. Bar in I = 300 µm (B), = 180 µm (D, E), = 90 µm (A, F–H), = 45 µm (C, I).

N.B. Sterile specimens of *Battersia plumigera* differ from sterile specimens of *Chaetopteris plumosa* in having corticating filaments only arising in the plane of branching, while in *C. plumosa* all peripheral cells of the axes and branches can grow to form a thick cortex.

Battersia racemosa (Greville) Draisma, Prud'homme van Reine & H. Kawai (2010), p. 322.

Fig. 21

Sphacelaria racemosa Greville (1824a), pl. 96.

Thalli epilithic, erect, forming small scattered dense tufts of one to several shoots, 1–2 cm in height, sometimes solitary, more commonly gregarious, and spreading to several centimetres, light to dark brown in colour, arising from a monostromatic and discoid, later polystromatic and crustose base; main axes sometimes enclosed within downward-growing, closely or loosely attached, mainly uniseriate to biseriate, secondary rhizoidal filaments, which can sometimes develop into stolons upon substratum contact; cortication of axes usually not present; main axes sparingly and irregularly branched to 2(–3) orders, with branches slightly narrowing proximally and closely appressed to parental axes and of indeterminate growth; all axes and laterals multiseriate, polysiphonous and clearly segmented, to 70 μm in diameter, with each secondary segment divided into 1–5(–6) narrow segments seen in surface view, which can also divide by 1–2 secondary transverse walls; in transverse section, older secondary segments variously divided to form an inner medulla of 1–2 large, quadrate cells enclosed by a layer of smaller peripheral cells; all cells with numerous discoid plastids without pyrenoids; hairs rare, with basal meristem and sheath.

Unilocular and plurilocular sporangia borne on separate thalli, produced as clusters at the end of specialized, short, multicellular, branched, uniseriate to biseriate laterals borne on the main axes and branches; plurilocular sporangia rare, with a single, uncertain record (originally reported by Prud'homme van Reine 1982a, p. 167) based on a slide of material collected at Berwick-upon-Tweed by Batters and deposited in BM; the plurilocular sporangia are elongate-cylindrical, sometimes with an irregular outline, and measured 53–108 × 28–50 μm (70–86 × 34–39 μm given by Prud'homme van Reine); unilocular sporangia common, borne in racemose, globular clusters, oval to subspherical, with dimensions of 33–62 × 21–46 μm (48–80 × 40–75 μm given by Prud'homme van Reine 1982a, p. 162); propagules unknown.

Epilithic in lower eulittoral sandy pools; appears to be sporadic in its occurrence.

Rare and apparently confined to northern shores of Britain with records for Buteshire, Fife, Midlothian and Northumberland; one reported record for Pembrokeshire is doubtful according to Prud'homme van Reine (1982a).

Probably perennial; unilocular and plurilocular sporangia occurring in winter and spring.

Abbott & Hollenberg (1976), pp. 218, 219, fig. 180; Bartsch & Kuhlenkamp (2000), p. 168; Batters (1890a), p. 61, (1890b), pp. 281, 282, (1902), p. 38; Brodie *et al.* (2016), p. 1019; Draisma *et al.* (2010b), pp. 308–26; Fletcher in Hardy & Guiry (2003), p. 300; Greville (1824a), pl. 96, (1824b), p. 314; Harvey (1846–51), pl. 349; Holmes (1887), p. 80; Irvine (1956), pp. 39, 40; Lund (1950), pp. 46–8; Newton (1931), p. 189; Oltmanns (1922), figs 370, 375(4); Prud'homme van Reine (1974), p. 174, (1978), p. 303, (1982a), pp. 160–8, figs 353–79; Reinke (1989b), p. 40; Rueness (1977), p. 186; Sauvageau (1901), pp. 137–43, (1903), p. 75, (1971), pp. 99–105, fig. 23; Setchell & N.L. Gardner (1925), pp. 393, 394; Waern (1945), pp. 409–11, fig. 1, (1952), p. 101, fig. 42.

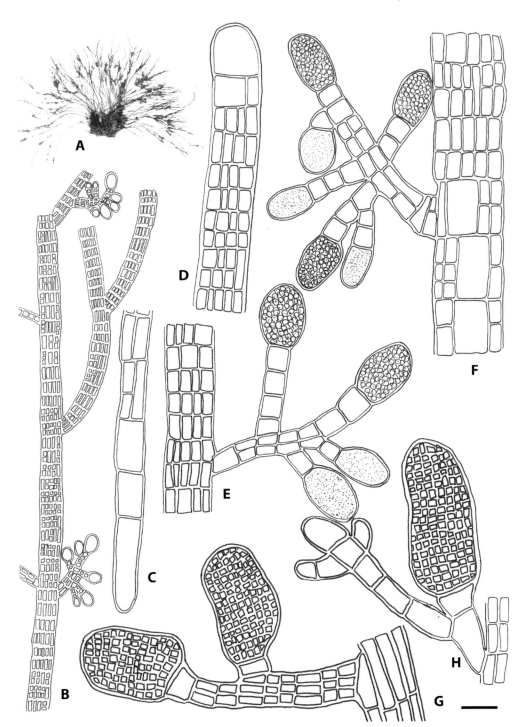

Fig. 21. *Battersia racemosa*
A. Habit of thalli. B. Portion of thallus showing branching and raceme-like clusters of unilocular sporangia.
C. Terminal region of rhizoidal filament. D. Apex of thallus showing large apical cell. E, F. Parts of erect thalli
showing unilocular sporangia. G, H. Parts of erect thalli showing plurilocular sporangia. Bar = 7.5 mm (A),
= 100 μm (B), = 18 μm (C–H). (A, herbarium specimen, NHM.)

Chaetopteris Kützing

CHAETOPTERIS Kützing (1843), p. 293.

Type species: *C. plumosa* (Lyngbye) Kützing (1843), p. 93.

Thalli mainly epilithic, erect, tufted, much branched with one to several axes arising from a perennial crustose base; branching of main axes hemiblastic, irregular, to four orders; axes parenchymatous, and distinctly multiseriate/segmented, largely isodiametric (leptocaulous), increase in girth provided in the proximal regions by a cortication of downward-growing, compact, secondary rhizoidal filaments; axes of indeterminate growth giving rise to opposite, pinnate, determinate, unbranched, ephemeral laterals.

Unilocular and plurilocular sporangia borne on specialized fertile laterals arising from the pinnate laterals or the surface corticating cells.

One species in Britain and Ireland.

Chaetopteris plumosa (Lyngbye) Kützing (1843), p. 93.   Figs 22, 23, 24

Ceramium pennatum Hornemann, nom. illeg. (1813), pl. 1481; *Sphacelaria plumosa* Lyngbye (1819), p. 103; *Sphacelaria plumosa* var. *divaricata* Lyngbye (1819), p. 103; *Cladostephus plumosus* (Lyngbye) Fries (1835), p. 313; *Sphacelaria heteronema* Postels & Ruprecht, nom. nud. (1840), p. II.

Thalli epilithic, less often epiphytic, tufted, usually solitary, sometimes gregarious, forming one to several erect, dark brown main axes up to 6(–10) cm high, arising from a black, spreading, perennial, discoid, later crustose, base overlaid by a thick mat of downward-growing and spreading rhizoidal filaments derived from the erect axes; main axes irregularly branched to 3(–4) orders, giving rise to indeterminate laterals; both main axes and indeterminate laterals sheathed over the greater part of their length by a thick, dark cortication of downward-growing, compact, secondary rhizoids which, proximally, considerably increase the girth of the axes; in spring/summer, terminal region of main axes and indeterminate laterals regularly and oppositely, pinnately branched to give rise to forward-projecting, relatively narrow, linear, determinate, simple or similarly branched, light brown to olive laterals giving the thallus a feather-like appearance and an ovate to lanceolate outline; these are later shed in winter, giving a denuded appearance to the main axes and branches; all axes and laterals originate from a small, relatively inconspicuous, light-coloured apical cell, and are clearly polysiphonous and segmented, with each secondary segment dividing longitudinally, increasing from a uniseriate/partly biseriate apex, 13–22 μm in diameter, to a multiseriate axis, to 80(–115) μm, and 13–14 segments in diameter, before cortication; corticated axes up to 500(–800) μm in diameter; secondary segment cells usually longer than wide, sometimes in part equal in length and width, or shorter than wide, depending on the distance from the apex, 10–20(–32) × 5–15(–22) μm, often with oblique or straight transverse walls; in transverse section, corticated regions reveal an inner, colourless core/axis of secondary segments, variously divided/differentiated to form a larger-celled central medulla enclosed within many small peripheral cells, and an outer, extensive, darkly pigmented, rhizoidal cortex of secondary rhizoids derived as outgrowths of the outer peripheral cells; secondary segments with several discoid plastids without pyrenoids; hairs present or absent, solitary or grouped, arising from the terminal laterals, with basal meristem and sheath.

Plurilocular and unilocular sporangia usually borne on separate thalli, rarely recorded on the same thallus, and formed terminally and laterally on 1–5(–7) celled, divaricate stalks borne on extensive, densely packed, specialized, divaricate, somewhat recurved, unbranched, uniseriate, or in part longitudinally segmented, laterals up to 500(–800) μm in length, 16–25(–30) μm in diameter, arising as a dense felt from the surface cells of the rhizoidal cortex on the bare,

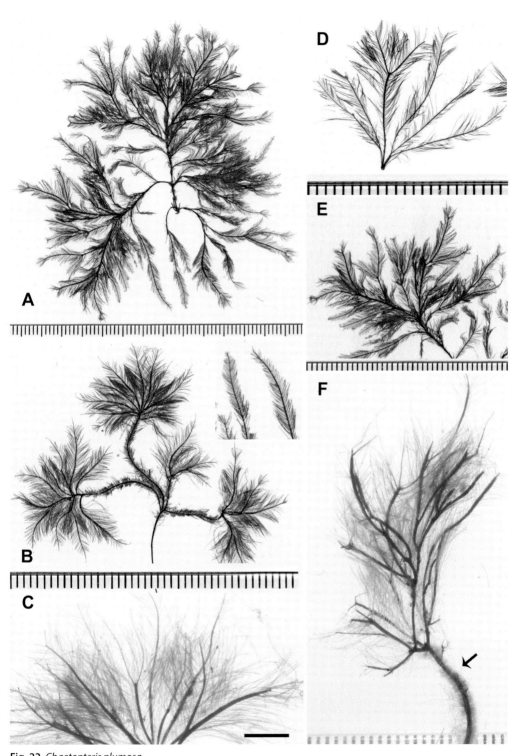

Fig. 22. *Chaetopteris plumosa*
A–F. Habit of thalli. Note specialized fertile laterals (arrow). Bar in C = 2.5 mm.

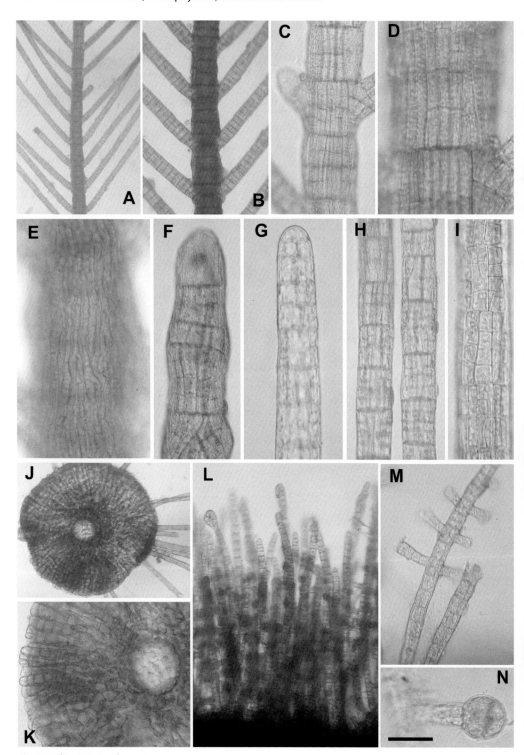

Fig. 23. *Chaetopteris plumosa*

A–I. Portions of erect axes showing opposite branching (A, B), variously sized secondary segments (C, D, H, I), terminal regions (F, G) and cortication of axes (E). J, K. TS of axes showing a small central medulla of colourless cells enclosed by a more extensive, dark-coloured rhizoidal cortex. L. TS of axis showing the specialized fertile laterals. M, N. Portions of fertile laterals with oppositely produced stalks and possible young undeveloped plurilocular sporangium (N). Bar in N = 470 µm (A), = 180 µm (J, L), = 90 µm (B, C, F, H, K, M), = 45 µm (D, E, G, I, N).

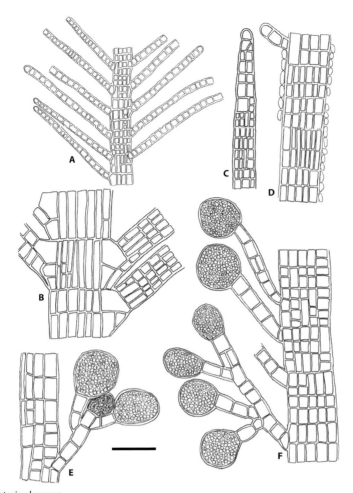

Fig. 24. *Chaetopteris plumosa*
A. Portion of erect thallus showing pinnate branching. B. Portion of the segmented erect thallus. C. Apex of young lateral. D. Portion of erect thallus showing outer corticating filaments. E–F. Unilocular sporangia on fertile laterals arising from the pinnate laterals. Bar = 200 µm, = 50 µm (B–F).

basal regions of the main filaments, more rarely borne terminally on specialized fertile, short, branched, upward-curved, multicellular, uniseriate, or in part, longitudinally segmented laterals arising from the ultimate pinnate laterals; unilocular sporangia ovate to subspherical, 30–60 µm in diameter; plurilocular sporangia reported by Prud'homme van Reine (1982a) to be cylindrical to ovate, 35–67 × 25–42 µm; propagules unknown.

Epilithic on sand-covered bedrock, stones, shells, occasionally epiphytic on larger algae, occurring in lower eulittoral pools and sublittoral to 11 m.

Thalli common and widely distributed around Britain and Ireland.

Perennial, with determinate pinnate laterals formed in spring/summer, then shed in winter. Reproductive sporangia observed in spring and summer.

Bartsch & Kuhlenkamp (2000), p. 168; Batters (1890a), p. 64, pl. 10(4–6), (1890b), pp. 284, 285, pl. 10(4–6), (1902), p. 39; Brodie *et al.* (2016), p. 1019; Dizerbo & Herpe (2007), p. 96; Draisma *et al.* (2010b), pp. 308–26; Fletcher in Hardy & Guiry (2003), p. 299; Guiry (2012), p. 129; Holmes (1883), p. 142; Irvine (1956), pp. 40, 41; Jaasund (1965), p. 72; Kornmann & Sahling (1977), p. 156, pl. 85(A–H); Kylin (1947), p. 31, pl. 2(7); Lund (1950), pp. 55–8; Magnus (1873), p. 133, pl. 1(15–20); Mathieson & Dawes (2017), p. 235, pl. XXXVI(2–5); Morton (1994), pp. 36, 37; Newton (1931), p. 193, fig. 121(A–H); Oltmanns (1922), fig. 378; Pankow (1971), p. 165; Pedersen (2011), p. 87, fig. 93; Prud'homme van Reine (1974), p. 174, (1982a), pp. 116–31, figs 212–63; Reinke (1889b), p. 41, (1892), p. 69, pls 49(1–8), 50(1–4); Rueness (1977), p. 185, pl. XXIV(5), fig. 109; Sauvageau (1901), p. 144, (1971), pp. 106–12, fig. 24(A–H); Schreiber (1931), p. 236; Setchell & N.L. Gardner (1925), p. 298; Taylor (1957), pp. 122, 123; Waern (1945), p. 404.

N.B. For differences between sterile material of *Battersia plumigera* and *Chaetopteris plumosa* see text on the former species.

Sphacelaria Lyngbye in Hornemann

SPHACELARIA Lyngbye in Hornemann (1818), pl. 1600.

Type species: *S. cirrosa* (Roth) C. Agardh (1824), pp. 164–5. See Draisma & Prud'homme van Reine (2010) pp. 1891–2.

Thalli epilithic or epiphytic, tufted, either solitary or in confluent mats/turfs, 1 mm to 15 cm in length, attached by monostromatic discs, polystromatic crusts or stolons, sometimes with endophytic penetration; erect shoots terete, rigid or flaccid, sparsely to much branched to several orders, irregular or pinnate and opposite, determinate or indeterminate, parenchymatous and distinctly multiseriate/segmented, largely isodiametric, except when covered with corticating filaments, with a large prominent apical cell; cells with several discoid plastids without pyrenoids; growth apical; hairs with basal meristem and sheath.

Reproduction by specialized propagules, plurilocular sporangia (micro and mega) and unilocular sporangia, borne mainly on the terminal laterals, and stalked; life history isomorphic.

Five species in Britain and Ireland.

KEY TO SPECIES

1. Propagules with radiating arms... 2
    Propagules wedge-shaped 4

2. Branching pinnate, opposite, alternate, secund or whorled with many laterals
    of determinate growth; propagule arms obviously constricted at base;
    terminal hair present on propagule.................................*S. cirrosa*
    Branching irregular; propagule arms not constricted at base; terminal hair
    absent on propagule.. 3

3. Main filaments usually >50 μm wide....................................*S. fusca*
    Main filaments usually <50 μm wide....................................*S. rigidula*

4. Branching infrequent, irregular ....*S. tribuloides*
    Branching dense, pinnate ....*S. plumula*

Sphacelaria cirrosa (Roth) C. Agardh (1824), pp. 164–5.   Figs 25, 26

Conferva pennata Hudson (1762), p. 486; *Conferva cirrosa* Roth (1800), pp. 214–16; *Ceramium cirrosum* (Roth) C. Agardh (1811), p. 21; *Sphacelaria pennata* Lyngbye (1819), p. 105; *Sphacelaria pennata* var. *gracilis* Lyngbye (1819), p. 105; *Sphacelaria cirrosa* [*cirrhosa*] var. *aegagropila* C. Agardh (1824), p. 165; *Sphacelaria cirrosa* [*cirrhosa*] var. *patentissima* Greville (1827b), pl. 317; *Sphacelaria plumosa* var. *gracilis* (Lyngbye) C. Agardh (1828), p. 25; *Sphacelaria cirrosa* [*cirrhosa*] var. *gracilis* (Lyngbye) Hornemann (1837), p. 694; *Sphacelaria irregularis* Kützing (1845), p. 239; *Sphacelaria rhizophora* Kützing (1849), p. 463; *Stypocaulon bipinnatum* Kützing (1855), p. 28; *Sphacelaria cirrosa* var. *pennata* (Lyngbye) Hauck (1884), p. 345; *Sphacelaria bipinnata* (Kützing) Piccone (1884), p. 54; *Sphacelaria cirrosa* var. *subsecunda* Grunow in Piccone (1884), p. 53; *Sphacelaria cirrosa* [*cirrhosa*] var. *irregularis* (Kützing) Hauck (1885), p. 345; *Sphacelaria hystrix* Suhr ex Reinke (1890), p. 208; *Sphacelaria cirrosa* [*cirrhosa*] f. *mediterranea* Sauvageau (1902), pp. 410, 415; *Sphacelaria cirrosa* [*cirrhosa*] f. *septentrionalis* Sauvageau (1902), pp. 409, 415; *Sphacelaria cirrosa* [*cirrhosa*] f. *meridionalis* Sauvageau (1902), pp. 409, 415; *Sphacelaria bipinnata* (Kützing) Sauvageau (1902), p. 393 (*cirrosa*); *Sphacelaria cirrosa* [*cirrhosa*] f. *patentissima* (Greville) Reinke (1889) p. 40; *Sphacelaria pennata*

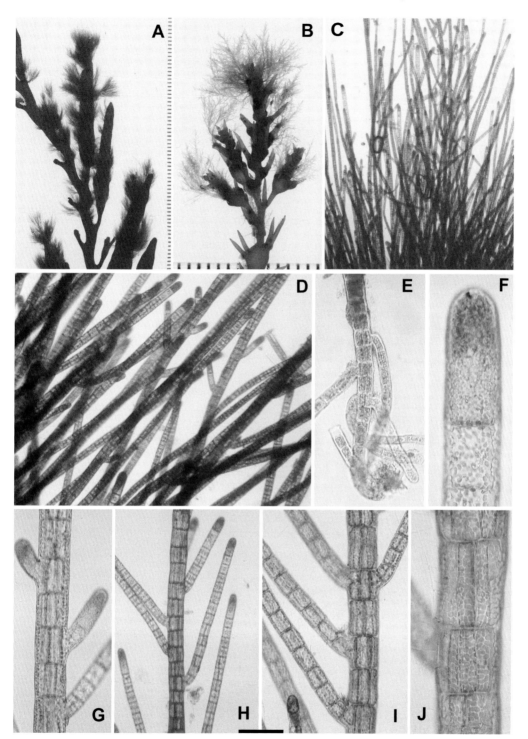

Fig. 25. *Sphacelaria cirrosa*

A. Habit of thalli on *Halidrys*. B. Habit of thalli on *Ericaria*. C, D. Portions of erect axes. E. Base of erect axis showing production of secondary rhizoidal filaments. F. Apex of erect axis. G–I. Portions of erect branched axes showing secondary segments. J. Close-up of secondary segments showing longitudinal septation and discoid to irregularly shaped plastids. Bar in H/I = 600 μm (C), = 300 μm (D), = 180 μm (H), = 90 μm (E, G, I), – 45 μm (F, J).

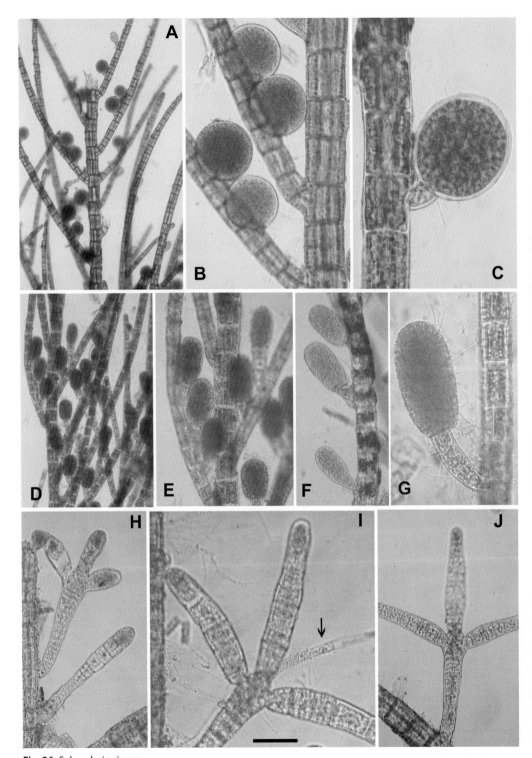

Fig. 26. *Sphacelaria cirrosa*

A–C. Portions of erect axes with unilocular sporangia. D–G. Portions of erect axes with plurilocular sporangia. H–J. Portions of erect axes with propagules. Note terminal cell of stalk between arms giving rise to a hair (arrow). Bar in I = 600 μm (A), = 180 μm (D), = 90 μm (B, E, F, H, J), = 45 μm (C, G, I).

f. *septentrionalis* (Sauvageau) D.E.G. Irvine (1956), p. 31; *Sphacelaria pennata* f. *meridionalis* (Sauvageau) D.E.G. Irvine (1956), p. 32; *Sphacelaria pennata* f. *patentissima* (Greville) D.E.G. Irvine (1956), p. 32; *Sphacelaria pennata* var. *fusca* (Hudson) D.E.G. Irvine (1956), p. 32. See also *Sphacelaria fusca* (Hudson) S.F. Gray.

Thalli epiphytic, less commonly epilithic, forming erect, fairly rigid, solitary to gregarious, olive to dark brown, penicillate tufts or hemispherical cushions on hosts, to 5(–10) mm high, arising from a small, monostromatic disc, later polystromatic crust, sometimes covered by a fibrous mat of branched, secondary rhizoidal filaments, which can be partly endophytic; main erect axes numerous, densely packed, to 70 μm in diameter at the base, branched to 1–3 orders, branching opposite, alternate, secund or whorled, produced either irregularly distichous or from all around the axis, with branches of mainly determinate, sometimes indeterminate growth and borne quite closely adpressed to parental axis, all axes and branches progressively decreasing in diameter above and terminating in a prominent, dark-coloured apical cell; all axes and branches polysiphonous and clearly segmented with each secondary segment divided longitudinally into up to eight narrow cells seen in surface view, very rarely further subdivided by transverse secondary walls; secondary segment cells longer than wide, more rarely quadrate or subquadrate, 15–48(–70) × 5–12(–20) μm; in transverse section, older secondary segments divided by radial walls without any obvious differentiation into a central medulla and outer peripheral cell layers; cortication cover absent on main axes, basal regions giving rise to irregularly produced, downward-growing, monosiphonous, in part parenchymatous, secondary rhizoids; cells with several discoid plastids without pyrenoids; hairs common, with basal meristem, with or without sheath, 12–15 μm in diameter.

Propagules, plurilocular and unilocular sporangia usually borne on separate thalli, rarely on the same thallus; propagules common, borne on upper regions of thalli, comprising a straight, largely parenchymatous, segmented stalk, to 110–240 μm in length, 15–50 μm in width, and distinctly broadening towards the tip, each segment divided longitudinally into 1–4 secondary segments; terminal regions of stalk giving rise to 3(–6) arms; arms uniseriate at first, becoming parenchymatous with 1–4 secondary segments, straight or more commonly outward-curved, of approximately equal length, sausage-shaped, to 100–380 × 25–48 μm, and constricted towards base; terminal cell of stalk often giving rise to a hair between the basal regions of the arms, less frequently it gives rise to a new propagule; plurilocular sporangia irregularly borne on the terminal branches, cylindrical to ellipsoid, solitary or in small groups, on 1–2(–4) celled stalks, 60–85(–103) × 38–48(–58) μm and dimorphic with shorter and wider megazoidangia (= megagametangia) with large loculi and slightly longer and narrower microzoidangia (= microgametangia) with small loculi; unilocular sporangia borne on the adaxial sides of terminal branchlets, solitary or in short series, globular, to 65(–80) μm in diameter and with one stalk cell.

Epiphytic on various algae, in particular *Halidrys siliquosa* and *Gongolaria baccata*, in lower eulittoral pools and shallow sublittoral.

Common and widely distributed around Britain and Ireland.

Fertile thalli usually found throughout the year.

Prud'homme van Reine (1982a) recognized a number of ecads of *S. cirrosa* which he distinguished on the basis of such characters as attached/unattached, the degree of endophytism, the host species, the presence/absence of a corticating layer of rhizoids around the base of the erect shoots, and the occurrence/abundance of the reproductive propagules and sporangia.

Ardré (1970), pp. 258, 259; Bartsch & Kuhlenkamp (2000), p. 168; Batters (1890b), pp. 282, 283, (1902), pp. 38, 39; Børgesen (1926), pp. 72–5; Brodie *et al.* (2016), p. 1019; Bunker *et al.* (2017), p. 279; Chemin (1922), p. 244; Clint (1927), pp. 5–23; Coppejans & Kling (1995), pp. 234, 235, pl. 86(A–D); Cormaci *et al.* (2012), pp. 425, 426, pl. 145(3–8); De Haas-Niekerk (1965), pp. 156, 158; Dizerbo & Herpe (2007), p. 95, pl. 31(2); Draisma *et al.* (2010b), pp. 308–26; Ducreux (1977), pp. 163–84, (1983), pp. 415–29, figs 1–37, (1984), pp. 447–54, figs 1–29; Feldmann (1937), pp. 300, 301; Fletcher in Hardy & Guiry (2003), p. 296; Funk (1955), p. 43, pl. 4(8); Gallardo García *et al.* (2016), p. 34; Goodband (1971), pp. 957–80, (1973), pp. 175–9; Guiry (2012), p. 129; Hamel (1931–39), pp. 257–60, fig. 48(I, III, IV); Harvey (1846–51), pl. 178; Irvine (1956), pp. 29–34; Jaasund (1965), pp. 70–2, fig. 20; Kylin (1947), pp. 29, 30, fig. 24(E); Lindauer *et al.* (1961), pp. 159, 160, fig. 16; Lund (1950), pp. 32–42, figs 6, 7; Mathieson & Dawes (2017), p. 236, pl. XXXVI(6–10); Morton (1994), p. 36; Newton (1931), pp. 189, 190, fig. 118(A–H); Oltmanns (1922), fig. 371; Papenfuss (1934), pp. 437–44; Prud'homme van Reine (1974), pp. 173, 174, (1982a), pp. 225–58, figs 565–646, pl. 6; Reinke (1889b), pp. 39, 40, (1892a), pp. 65, 66, pls 42, 43; Ribera *et al.* (1992), p. 119; Rueness (1977), pp. 182–4, fig. 107(B); Sauvageau (1898b), p. 1672, (1902), pp. 349, 393, 415, (1903), pp. 45–53, (1971), pp. 191, 193–205, 211–37, figs 39–41, 42(A–D), 44–6; Stegenga & Mol (1983), p. 113, pl. 42(1–2); Stegenga *et al.* (1997b), p. 22; Taylor (1957), p. 121, pl. 17(106); Wittrock (1884), pp. 93–5; Womersley (1987), pp. 164–6, figs 51(c), 53(H–K).

Sphacelaria fusca (Hudson) S.F. Gray (1821), p. 333.   Fig. 27

Conferva fusca Hudson (1762), p. 486; *Ceramium fuscum* (Hudson) Roth (1800), p. 487; *Delisella fusca* (Hudson) Bory (1823), p. 540; *Sphacelaria cirrosa* [*cirrhosa*] var. *fusca* (Hudson) P. Crouan & H. Crouan (1852), exsicc. No. 35; *Sphacelaria cirrosa* [*cirrhosa*] f. *fusca* (Hudson) Holmes & Batters (1890), p. 81; *Sphacelaria subfusca* Setchell & N.L. Gardner (1924), p. 1; *Sphacelaria pennata* var. *fusca* D.E.G. Irvine (1956), p. 32.

Thalli epilithic or epiphytic forming small, dark brown, fairly rigid, somewhat matted, brush-like tufts on rocks and larger algae, solitary or gregarious, to 2(–3) cm in height, arising from a tightly adherent, irregularly shaped, monostromatic, discoid base; main erect axes numerous, densely packed, 23–58 µm in diameter, linear, usually uniform in diameter throughout their length and terminating in a prominent, darkly pigmented, apical cell, sparingly to numerously irregularly branched, to 2(–3) orders, with branches fairly closely adpressed to widely divaricate, greatly variable in length, some indeterminate and indistinguishable from primary axes; primary axes and laterals polysiphonous and segmented with each secondary segment divided longitudinally into up to 5(–6) cells seen in surface view; secondary segment cells longer than wide, more rarely quadrate or subquadrate, 22–63 × 8–25 µm, occasionally divided by transverse walls; in transverse section, older secondary segments divided by radial walls without any obvious differentiation into a central medulla and outer peripheral cell layers (radial type); cortication sheath absent on main axes, and basal regions without downward-growing secondary rhizoids; cells with several discoid plastids lacking pyrenoids; hairs common, formed on the terminal branchlets, with basal meristem and sheath.

Propagules common, borne in upper regions of thalli, comprising a forward-projecting, sometimes widely divaricate stalk giving rise terminally to 2–3 widely divaricate arms; stalk to 350(–570) µm long, 20–31 µm wide, straight, linear or slightly broadening terminally, largely parenchymatous, segmented, each segment divided longitudinally into 1–3 secondary segments; terminal region of stalk giving rise to a small lenticular cell between the arms; arms to 200(–280) in length × 18–25 µm in diameter, usually of approximately equal length, linear, and cylindrical, largely parenchymatous, segmented, each primary segment divided longitudinally into 1–3 secondary segments; plurilocular and unilocular sporangia unknown.

Thalli epilithic or epiphytic on various hosts, lower eulittoral pools and sublittoral.

Rare and only recorded from a few scattered localities around Britain and Ireland, yet likely to be more common and widespread than records suggest.

Appears to be perennial; insufficient records to comment on seasonality of propagule occurrence.

Ardré (1970), pp. 259, 260; Batters (1902), p. 39; Børgesen (1926), p. 74; Brodie *et al.* (2016), p. 1019; Coppejans & Kling (1995), p. 236, pl. 87(A–C); Cormaci *et al.* (2012), pp. 426–8, pl. 146(1–2); Dawes & Mathieson (2008), p. 132, pl. 15(16–17); De Haas-Niekerk (1965), pp. 148–50, figs 1–33; De Mesquita Rodrigues (1963), p. 31, pl. 1(A–C); Dizerbo & Herpe (2007), p. 95; Draisma *et al.* (2010b), pp. 308–26; Fletcher in Hardy & Guiry (2003), p. 297; Gallardo García *et al.* (2007), p. 95, (2016), p. 34; Gayral (1958), p. 201, fig. 26(A–B), pl. 22, (1966), p. 235, fig. 26, pl. 28; Goodband (1971), pp. 957–80; Guiry (2012), p. 130; Hamel (1931–39), pp. 260, 261; Harvey (1846–51), pl. 149; Holmes & Batters (1890), p. 81; Irvine (1956), pp. 32, 33; Keum (2010), pp. 81–3, fig. 3; Keum *et al.* (2005), pp. 6–13, figs 13–22; Knight & Parke (1931), pp. 66, 111, 112; Morton (1994), p. 36; Newton (1931), p. 190; Prud'homme

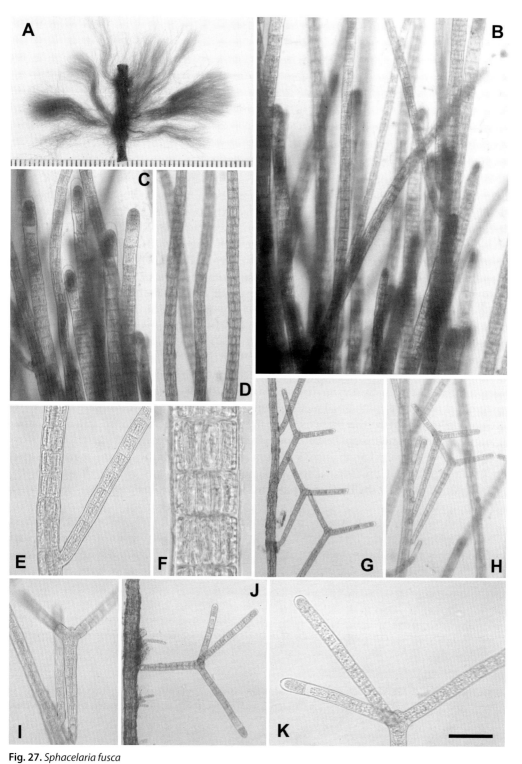

Fig. 27. *Sphacelaria fusca*

A. Habit of thallus. B–E. Portions of erect axes. F. Close-up of main axis showing secondary segments. G–K. Portions of erect axes showing propagules. Bar in K = 300 μm (B), = 180 μm (C, D, G, H, J), = 90 μm (E, I, K), = 45 μm (F).

van Reine (1974), p. 174, (1978), p. 303, (1982a), pp. 220–4, figs 555–64; Ribera *et al.* (1992), p. 119; Sauvageau (1902), p. 399, fig. 43(A–D), (1971), pp. 206–11, fig. 43(A–D); Taylor (1957), p. 120, (1960), p. 210; Womersley (1987), pp. 168, 169, figs 51(E), 54(H–J).

Sphacelaria plumula Zanardini (1864), p. 9.   Figs 28, 29

Sphacelaria pseudoplumosa P. Crouan & H. Crouan (1867), p. 164; *Sphacelaria pectinata* Reinsch (1875), p. 102; *Sphacelaria plumula* var. *cervicornis* Sauvageau (1901), p. 109; *Sphacelaria plumula* var. *patentissima* Sauvageau (1903), p. 49.

Thalli epilithic or epiphytic, forming small, light to dark brown, usually solitary tufts up to 2(–4) cm high, comprising one to several, erect main axes arising from a small, discoid holdfast of compacted, secondary, rhizoidal filaments; a few descending rhizoidal filaments originate as outgrowths from the proximal regions of the axes; erect axes irregularly branched, to 2(–3) orders, with the main, indeterminate branches giving rise distally to regularly pinnate and opposite, linear, unbranched laterals of determinate length which are often shed below and shorter above giving the thalli a feather-like appearance and a frequently lanceolate/triangular shape; main axes to 125 μm in diameter at the base, terminal laterals 32–63 μm in diameter and narrowing towards the apex, 15–20 μm in diameter; all axes are multiseriate, polysiphonous and clearly segmented, except for the apices of some branches which are monosiphonous, with all branches and axes terminated by a large apical cell; secondary segments divided longitudinally into 3–7(–10) cells seen in surface view; secondary segment cells mostly longer than wide, 28–72 × 7–20 μm, rarely divided by transverse walls; in transverse section, secondary segments are subdivided by radial and semi-radial walls to form larger central cells and many smaller peripheral cells; cells with several discoid plastids without pyrenoids; hairs rare, produced on terminal branchlets, with basal meristem and sheath.

Propagules lateral on the branchlets, at first globular or club-shaped, later becoming triangular in shape, multiseriate except for terminal cells of the triangular corners, to approximately 160(–190) μm in length and 75–130 μm in width; unilocular sporangia, not reported for Britain and Ireland; plurilocular sporangia unknown.

Epilithic on bedrock, shells etc., or epiphytic on various algae, deep lower eulittoral pools and sublittoral.

Thalli rare, with records mainly for the south, south-west and west coasts of Britain (Channel Islands, Cornwall, Devon, Dorset, Sussex, Shetland, Argyll and Isle of Man) and Ireland (Donegal and Galway) with an absence of records for the east and south-east coasts.

Thalli probably perennial; propagules observed on thalli collected in June, August and September.

Sphacelaria plumula is very similar to *Battersia plumigera*. According to Prud'homme van Reine (1982a), the former differs from the latter in having almost no transverse walls in the secondary segments. The angle of axils of laterals arising from the main axis is also larger in *S. plumula* than in *Battersia plumigera*, while the laterals themselves are usually much smaller in *S. plumula* than in *S. plumigera*.

Ardré (1970), pp. 257, 258; Bartsch & Kuhlenkamp (2000), p. 168; Batters (1902), p. 39; Brodie *et al.* (2016), p. 1019; Cormaci *et al.* (2012), p. 428, pl. 146(3–4); P. Crouan & H. Crouan (1867), p. 164, pl. 25(161); Dizerbo & Herpe (2007), p. 95, pl. 31(4); Feldmann (1937), p. 299; Fletcher in Hardy & Guiry (2003), p. 299; Funk (1955), p. 44, pl. 4(4); Gallardo García *et al.* (2016), p. 34; Gayral (1966), p. 235, pl. 29; Guiry (2012), p. 130; Hamel (1931–39), pp. 251–3, fig. 47(6–8); Irvine (1956), p. 37; Keum *et al.* (2003), pp. 114–23; Lund (1950), pp. 44–6, fig. 9; Mathieson & Dawes (2017), p. 237, pl. XXXVI(13, 14); Morton (1994), pp. 36, 37; Newton (1931), p. 191; Oltmanns (1922), fig. 379(6); Prud'homme van Reine (1974), p. 173, (1978), p. 303, (1982a), pp. 192–202, figs 473–507, pl. 5; Reinke (1891b), p. 10, (1892a), p. 67, pl. 48(1–7); Ribera *et al.* (2012), p. 119; Rueness (1977), pp. 186, 187, fig. 110(A, B, C); Sauvageau (1901), pp. 94–111, figs 18–20, (1971), pp. 90–4, figs 18–21; Stegenga & Mol (1983), pp. 113–16, pl. 42(5, 6); Stegenga *et al.* (1997b), p. 22; Zanardini (1860b), p. 139, pl. 33, (1864), pp. 9–12, pl. 33(1–4).

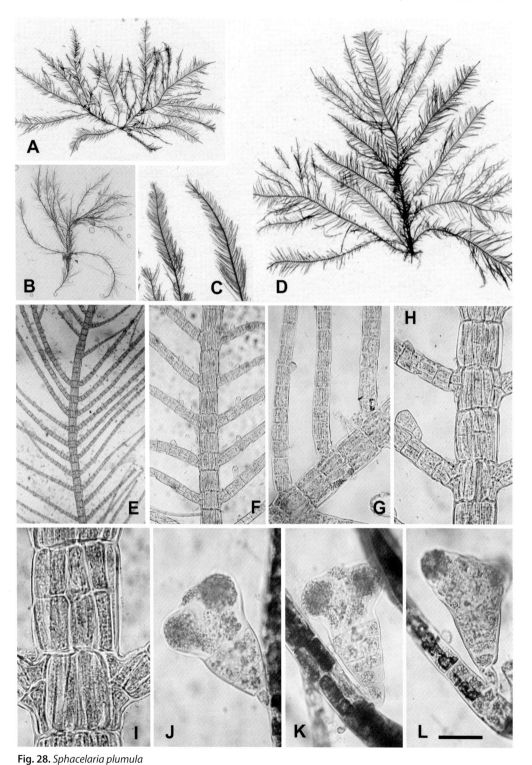

Fig. 28. *Sphacelaria plumula*
A–D. Habit of thalli. E–I. Portions of erect axes showing branching pattern. J–L. Portions of erect axes showing propagules (NHM herbarium and microslides). Bar in L = 11 mm (A), = 12 mm (B), = 3 mm (C), = 2.5 mm (D), = 300 μm (E), = 180 μm (F), = 90 μm (G, H), = 45 μm (I–L).

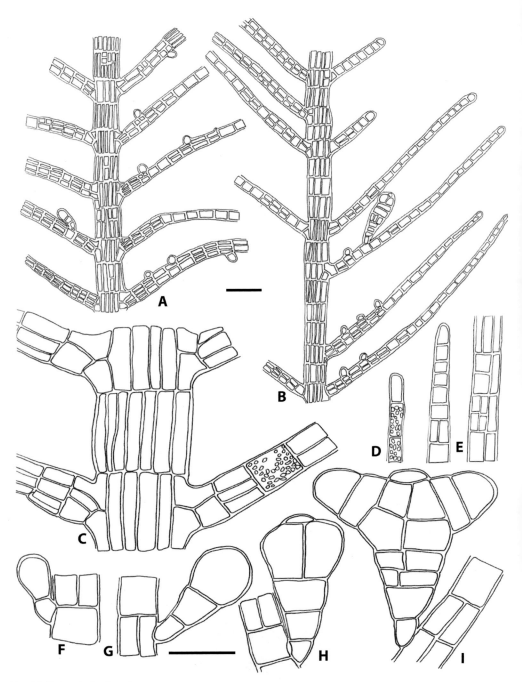

Fig. 29. *Sphacelaria plumula*

A, B. Portions of erect thalli showing young developing propagules. C. Portion of erect thallus showing opposite branching. D, E. Apical and subapical portions of young laterals. F–I Portions of erect thalli showing propagules at various stages of development. Small bar = 20 μm (A, B), large bar = 50 μm (C–I).

Sphacelaria rigidula Kützing (1843), p. 292.   Fig. 30

Sphacelaria divaricata Montagne (1849), p. 62; *Sphacelaria furcigera* Kützing (1855), p. 27; *Sphacelaria variabilis* Sauvageau (1901), pp. 412, 414; *Sphacelaria iridaeophytica* Nagai (1932), p. 145; *Sphacelaria caespitosa* Takamatsu (1943), p. 167; *Sphacelaria linearis* Takamatsu (1943), p. 181; *Sphacelaria apicalis* Takamatsu (1943), p. 179; *Sphacelaria expansa* Noda (1969), p. 31; *Sphacelaria iwagasakensis* Noda (1970), p. 31.

Thalli epilithic, sometimes epiphytic, forming small, reddish to dark brown, fairly stiff, dense, brush-like tufts, to approximately 1(–1.5) cm in height, solitary, sometimes confluent to form mats several cm in extent, arising from a thin monostromatic disc, polystromatic crust or creeping stolons and rhizoids; epiphytic thalli sometimes with endophytic penetration into host; erect axes numerous, 20(–28)-40(–50) μm in diameter, uniform in diameter throughout their length, or only slightly wider towards the base, sparsely branched to 2(–3) orders, with branches irregularly to helically produced, sometimes widely divergent, more usually closely adpressed to parental axis, and indeterminate in growth, not uncommonly exceeding the height of the main axes; all axes terminated by an elongated, more darkly pigmented, apical cell; proximal region of primary axes with downward-growing laterals which function as rhizoidal filaments and spread out to assist in attachment; corticating secondary rhizoids not observed; all primary axes and laterals polysiphonous and segmented, with each secondary segment divided longitudinally into 1–4(–6) segments/cells seen in surface view; secondary segment cells usually longer than wide, sometimes in part equal in length and width, or shorter than wide, 18–50 μm in height, sometimes with straight or oblique transverse divisions; in transverse section, older secondary segments clearly radially/sub-radially divided to form a parenchymatous tissue without obvious differentiation into medullary or cortical/peripheral cell types (radial type); secondary segment cells with several discoid plastids without pyrenoids; hairs present or absent, arising from the terminal laterals, with or without obvious basal meristem and sheath.

Propagules usually abundant, occurring on terminal laterals and comprising a narrow, linear, cylindrical, or slightly widening, forward-projecting stalk, 230–340(–440) μm in length, 15–20(–25) μm in width, giving rise terminally to 1–3(–4) (usually 2) divaricate, narrow, straight, simple or bifurcate, cylindrical arms of equal, less frequently unequal, length, 120–260(–310) μm in length, 14–18(–20) μm in width; both stalk and arms segmented with primary segments variously subdivided into secondary segments; stalk apex often forming a small, lenticular apical cell; terminal region of arms sometimes giving rise to new arms.

Unilocular sporangia rare, borne on erect axes and laterals, on unicellular stalks, quite strongly recurved towards parental axis, either single or grouped, at first clavate, later becoming globular, ovoid or obovoid, 38–50(–70) μm in diameter; plurilocular sporangia have not been observed on material collected around Britain and Ireland but have been reported from Hoek van Holland (Prud'homme van Reine 1982a, pp. 206–7) to be stout-cylindrical or spherical and borne on 1–3-celled stalks, and of two types viz. megazoidangia (= megagametangia) (34–)57–126(–200) μm in length × (25–)37–58 μm in width, with large loculi, and microzoidangia (= microgametangia), (43–)50–72(–91) μm in length × 33–47(–68) μm in width, with small loculi.

Epilithic, or epiphytic on various algae, lower eulittoral, usually in pools, and shallow sublittoral, sometimes emersed in damp situations and contributing to sand-binding communities on rocks and wooden breakwaters.

Common and widely distributed around Britain and Ireland.

Appears to be perennial; propagules have been found on material collected from spring to autumn, while unilocular sporangia have only been found on collections made in winter, viz. 'winter' collections made on the Isle of Man by Knight and Parke (1931) (as *S. cirrosa* var. *fusca*), collections from October to December made at St Andrews by Blackler (1966) (as *S. pennata* var. *fusca*) and January collections made at West Runton, Norfolk by the present author.

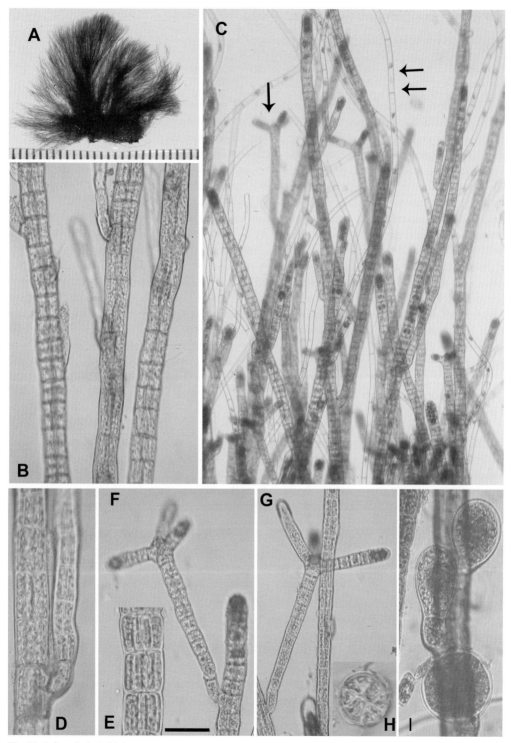

Fig. 30. *Sphacelaria rigidula*

A. Habit of thallus. B. Portions of branched erect axes. C. Portion of erect axes showing propagules (single arrow) with terminal arms and hairs (double arrow). D. Portion of axis showing closely adpressed branch. E. Portion of erect axis showing secondary segments. F, G. Portions of erect axes showing propagules with terminal arms. H. TS of axis (radial type). I. Portion or erect axis showing unilocular sporangia. Bar in F = 180 μm (C), = 90 μm (B, F, G), = 45 μm (D, E, H, I).

Aisha & Shameel (2012), pp. 247–64; Bartsch & Kuhlenkamp (2000), p. 168; Børgesen (1926), p. 72; Brodie *et al.* (2016), p. 1019; Coppejans (1983), pls 80, 81; Cormaci *et al.* (2012), pp. 428, 429, pl. 146(5–10); Dawes & Mathieson (2008), p. 133, pl. 16(1); De Haas-Niekerk (1965), p. 150, fig. 33; Dimitriadis *et al.* (2001), pp. 23–34; Dizerbo & Herpe (2007), p. 96; Fletcher in Hardy & Guiry (2003), p. 301; Gallardo García *et al.* (2016), p. 34; Gayral (1958), pp. 201–3, pl. XXII(26A, B); Goodband (1971), pp. 957–80; Guiry (2012), p. 130; Hamel (1931–39), p. 255, fig. 47(16); Irvine (1956), pp. 28, 29; Karyophyllis *et al.* (2000a), pp. 25–33, figs 1–12, (2000b), pp. 195–203, figs 1–23; Keum (2010), p. 90, fig. 11; Keum *et al.* (2005), pp. 1–13; Kitayama (1994), pp. 72–85, figs 22–3; Kornmann & Sahling (1977), p. 153, pl. 82(A–C); Lawson & John (1982), p. 134, pl. 13(1); Lund (1950), pp. 29–32, fig. 5; Mathieson & Dawes (2017), p. 237, pl. XXXVII(1–4); Newton (1931), p. 189; Norris (2010), pp. 112–14, fig. 51; Oltmanns (1922), figs 377(6–7), 379(3–5); Prud'homme van Reine (1974), p. 174, (1978), p. 303, (1982a), pp. 203–20, figs 508–54; Reinke (1890), p. 208, (1891a), p. 14, pl. 5(5–13); Ribera (1992), p. 119; Rueness (1977), p. 184; Saunders & McDevit (2013), pp. 1–23; Sauvageau (1900), p. 221; (1901), p. 379, (1902), p. 394, (1971), pp. 145–56, 207, fig. 35(A–K); Setchell & N.L. Gardner (1925), pp. 396–7, pl. 37(29); Stegenga & Mol (1983), p. 116, pl. 44(3, 4); Stegenga *et al.* (1997a), pp. 159, 162, pl. 44(1–2), (1997b), p. 22; Takamatsu (1943), p. 167, fig. 7; Tamura *et al.* (1996), pp. 63–8; Taylor (1957), p. 119, (1960), pp. 210, 211, pl. 29(5); Ten Hoopen (1983), pp. 1–87; Ten Hoopen *et al.* (1983), pp. 285–9; Van den Hoek & Flinterman (1968), pp. 193–242; Womersley (1987), pp. 166–8, figs 51(D), 54(A–G).

Sphacelaria tribuloides Meneghini (1840), p. 2.   Figs 31, 32

Sphacelaria rigida Hering in Krauss (1846), p. 213; *Sphacelaria tribuloides* var. *radicata* De Notaris (1846), p. 69; *Sphacelaria mexicana* W.R. Taylor (1945), p. 86.

Thalli epilithic or epiphytic, forming erect, dense, usually solitary tufts or hemispherical cushions, 1–2 cm high, olivaceous to dark brown in colour, arising from horizontally spreading stolons which probably overlay a small basal disc; erect primary axes numerous, fairly stiff, unbranched or sparsely irregularly, secundly or oppositely branched, to 3(–4) orders, with branches either adpressed to parental axes or divaricate, and of indeterminate but approximately equal growth; all primary axes and branches linear, uniform in diameter throughout their length, polysiphonous and segmented, 18–50(–70) μm in diameter, varying from 0.5 to 2 diameters long, with each secondary segment divided longitudinally into 1–3(–4) cells seen in surface view, these rarely with transverse walls; secondary segment cells usually longer than wide, sometimes shorter than wide, 28–70 × 7–20 μm, each containing several discoid plastids without pyrenoids; hairs common arising from terminal laterals, with basal meristem, with or without sheath.

Propagules common, borne on the terminal laterals, characteristically inverted triangular in shape when mature, 50–90 × 30–120 μm including 'horns', with one arm attached basally by a cylindrical, monosiphonous, 1–4-celled stalk, up to 125 μm in length, 14–36 μm in diameter, broadening above, with the two remaining upper arms usually short and horn-like, with a small lenticular cell situated between; arms of propagule sometimes giving rise to a new propagule.

Plurilocular and unilocular sporangia unknown from material in Britain and Ireland, but reported for various European countries by Prud'homme van Reine (1982a); unilocular sporangia rare, spherical, on a unicellular stalk, solitary or in a unilateral series, 65–80 μm in diameter (dimensions from Sauvageau (1971), p. 237, fig. 47); plurilocular sporangia rare, ellipsoid to cylindrical, shortly stalked and often in unilateral series, dimorphic, sporangia with large loculi measuring 55–100 × 29–55 μm, sporangia with small loculi measuring 34–59 × 24–32 μm (dimensions from Prud'homme van Reine (1982a), p. 182; see also Sauvageau (1971), p. 131, fig. 29(L)).

Epilithic, in lower eulittoral rock pools.

Rare, only recorded from a few localities around Britain (East Lothian, Northumberland, South Devon and Dorset), but likely to also occur in Ireland (see Guiry 2012).

Appears to be a summer annual (June to October).

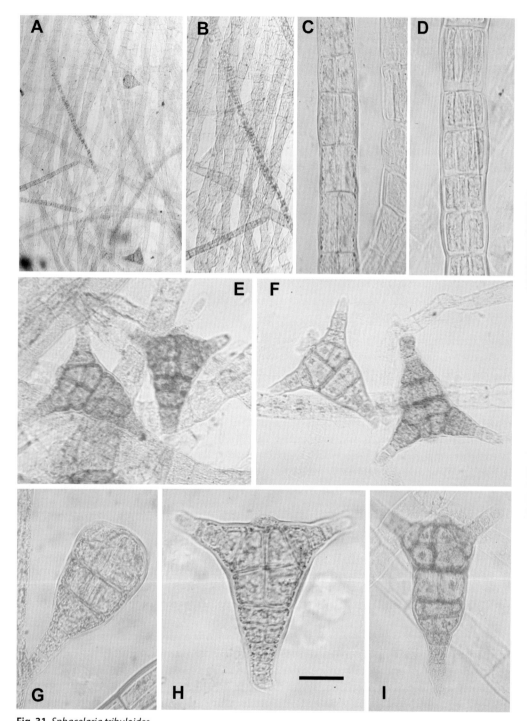

Fig. 31. *Sphacelaria tribuloides*

A, B. Portions of erect axes. C, D. Portions of erect axes showing secondary segments. E–I. Propagules at various stages of development. Bar in H = 180 μm (A, B), = 90 μm (E, F), = 45 μm (C, D, G, H, I).

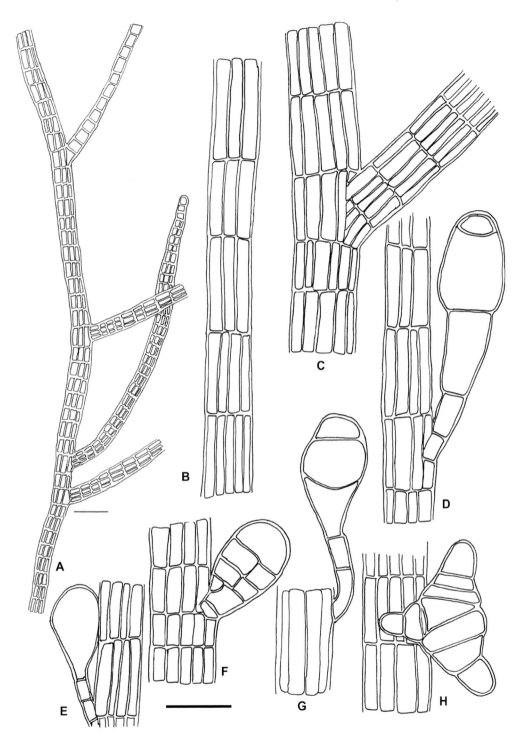

Fig. 32. *Sphacelaria tribuloides*
A–C. Portions of erect thalli. D–H. Portions of erect thalli showing propagules at various stages of development. Small bar = 100 μm (A), large bar = 50 μm (B–H).

Aisha & Shameel (2012), pp. 247–64; Ardré (1970), p. 260; Batters (1890b), p. 282, (1902), p. 38; Børgesen (1926), p. 72, (1941), pp. 41, 42, fig. 18(a–c); Brodie *et al.* (2016), p. 1019; Coppejans (1983), pls 85, 86; Cormaci *et al.* (2012), pp. 429, 430, pl. 147(1–5); Dawes & Mathieson (2008), p. 133, pl. 16(2); Dizerbo & Herpe (2007), p. 96; Draisma *et al.* (2010b), pp. 308–26; Feldmann (1937), pp. 299, 300; Fletcher in Hardy & Guiry (2003), p. 301; Funk (1955), p. 43; Galatis *et al.* (1977), pp. 139–51, figs 1–24; Gallardo García *et al.* (2016), p. 34; Hamel (1931–39), pp. 253, 254, fig. 47(9–12); Holmes & Batters (1890), p. 81; Irvine (1956), p. 36; Katsaros *et al.* (1983), pp. 16–30; Keum *et al.* (2003), pp. 113–24; Kitayama (1994), pp. 59–63, figs 12–15; Lawson & John (1982), pp. 134, 135, pl. 13(4); Lindauer *et al.* (1961), pp. 158, 159, fig. 16; Lund (1950), pp. 42–4, fig. 8; Misra (1966), pp. 127–9, fig. 64(A–F); Newton (1931), p. 189; Norris (2010), pp. 116, 117, fig. 53; Oltmanns (1922), figs 377(2–5), 379(1–2); Pringsheim (1874), p. 166, pl. 8(7–23); Prud'homme van Reine (1982a), pp. 179–88, figs 422–54; Ribera *et al.* (1992), p. 119; Rodrigues (1963), p. 34, pl. 1(J); Sauvageau (1901), pp. 233–42, (1903), pp. 53–5, (1971), pp. 123–32, 237–9, figs 28, 29, 47; Taylor (1945), p. 86, pl. 3(1–8), (1960), p. 211, pl. 29(6); Womersley (1987), pp. 160, 162, figs 45(G), 52(A–C).

Sphacelorbus Draisma, Prud'homme van Reine & H. Kawai

SPHACELORBUS Draisma, Prud'homme van Reine & H. Kawai (2010), p. 322.

Type species: *S. nanus* (Nägeli ex Kützing) Draisma, Prud'homme van Reine & H. Kawai (2010), p. 322.

Thalli gregarious, forming dense, turf-like patches of erect, sparsely branched, somewhat flaccid thalli arising from either a monostromatic disc or polystromatic crust and giving rise at the base to an extensive, spreading system of simple or branched stoloniferous laterals capable of disc/crust formation upon substratum contact, and the initiation of new erect shoots; erect thalli leptocaulous and fairly uniform in diameter, parenchymatous and segmented throughout; secondary laterals indeterminate, irregularly formed, solitary, unbranched, closely adpressed to parental axis, and hypacroblastic; cells with several discoid plastids without pyrenoids; growth apical; hairs with basal meristem and sheath.

   Unilocular and plurilocular sporangia born on basal regions of primary and secondary laterals, also on stoloniferous laterals, solitary or in small groups, and shortly stalked.

One species in Britain and Ireland.

Sphacelorbus nanus (Nägeli ex Kützing) Draisma, Prud'homme van Reine & H. Kawai
(2010), p. 322.   Fig. 33

Sphacelaria nana Nägeli ex Kützing (1855), p. 26; *Sphacelaria olivacea* Pringsheim (1874), nom. illeg. (see Sauvageau 1901, p. 54 and note below); *Sphacelaria saxatilis* (Kuckuck) Kuckuck ex Sauvageau (1900), p. 217; *Sphacelaria britannica* Sauvageau (1901), p. 53.

Thalli epilithic, sometimes epiphytic, olive to dark-brown, soft and flaccid, forming turf-like patches of erect, gregarious and quite dense thalli, 2–7(–10) mm in height and spreading over substratum to several centimetres in extent; thalli little branched, with branches closely adpressed to parental axis and indeterminate, often exceeding the height of the main axes, all axes terminating in a small apical cell; thalli at first attached by a monostromatic and discoid base, later becoming polystromatic and crustose, usually irregular in outline and less than 500 μm in diameter with peripheral growth by terminal apical cells, later by an extensive intertwining system of laterally spreading, sometimes coalesced, creeping stolons which commonly re-attach at intervals by disc formation and regularly initiate new erect shoots; erect thalli linear, fairly uniform in diameter throughout their length although sometimes with irregular swellings, polysiphonous and segmented, 15–23(–35) μm in diameter, with each secondary segment divided longitudinally into 2–4(–5) narrow segments seen in surface view, these sometimes with transverse secondary walls; secondary cells usually longer than wide, becoming quadrate or subquadrate by 1–3 transverse divisions, 12–25(–30) × 6–9(–12) μm; in

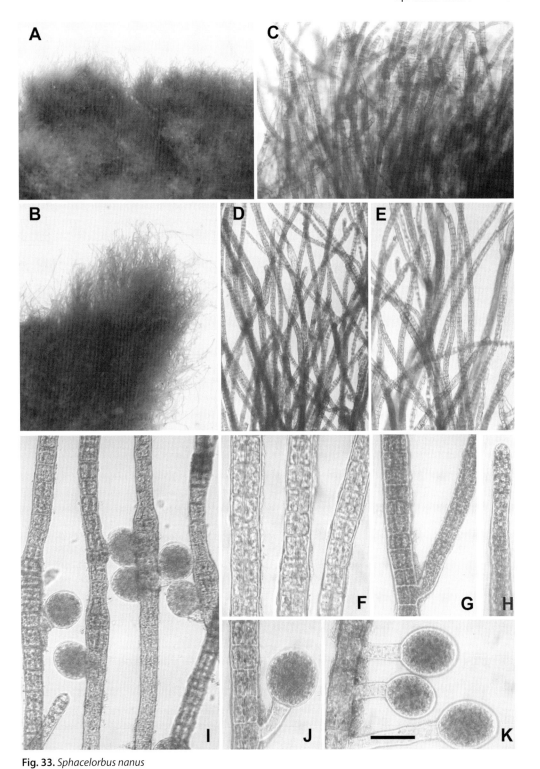

Fig. 33. *Sphacelorbus nanus*

A, B. Habit of thalli. C–H. Portions of erect axes. I–K. Portions of erect axes with unilocular sporangia. Bar in K = 6 µm (A), = 3.5 µm (B), = 470 µm (C), = 180 µm (C, D), = 90 µm (I), = 45 µm (F–H, J, K).

transverse section, older secondary segments clearly divided by radial and transverse walls, without any obvious differentiation into a central medulla and outer peripheral cell layers; base of thalli giving rise to lateral branches which often become stolons and also function as attaching rhizoids; corticating sheaths not observed; cells with several discoid plastids without pyrenoids; hairs rarely seen, solitary with basal meristem and usually with sheath.

Plurilocular and unilocular sporangia borne on separate individuals; unilocular sporangia common, solitary, or in small groups, produced at the base of erect shoots and branches, also reported on the stolons and discs/crusts, ovoid to ellipsoid, 34–50(–65) × 32–44(–58) µm, and borne on monosiphonous, later in part parenchymatous, segmented stalks, usually 1–6 segments in length, occasionally to 17 segments and usually recurved towards parental axis; plurilocular sporangia rare, only reported for Britain (St Andrews, Fifeshire) by Blackler & Jackson (1966), solitary or in small groups, borne in basal regions of erect shoots/branches, also on the discs/crusts and stolons, multiseriate, elongate-cylindrical, linear or irregular in outline, 88–124 × 33–41 µm, and borne on monosiphonous, in parts parenchymatous, 2–4-celled segmented stalks; propagules unknown.

Commonly found on vertical cliff faces and walls, and boulders, in the upper eulittoral, particularly in north facing/shaded habitats; a common inhabitant of the walls and ceilings of marine caves, sometimes associated with *Pilinia rimosa* and *Rhodochorton purpureum*, forming dense and extensive carpets; also observed to form extensive mats among halophytes on the soil of salt marshes.

Thalli common and widely distributed around Britain and Ireland; can be locally abundant in some habitats.

Perennial, and reported throughout the year; appears to be winter fertile with unilocular sporangia observed from November to February (personal observation), and plurilocular sporangia reported for January by Blackler & Jackson (1966).

N.B. The *Sphacelaria olivacea* as described by Pringsheim (1874) is a mixture of *Sphacelorbus nanus* and *Sphacelaria solitaria* (Pringsheim) Kylin. The latter species has not been found in Britain and Ireland.

Bartsch & Kuhlenkamp (2000), p. 168; Batters (1902), p. 38; Blackler & Jackson (1966), p. 85; Brodie *et al.* (2016), p. 1019; De Haas-Niekerk (1965), p. 152, figs 41–3; Dizerbo & Herpe (2007), pp. 94, 95, pl. 31(3); Draisma *et al.* (2010b), pp. 308–26; Fletcher in Hardy & Guiry (2003), p. 298; Gallardo García *et al.* (2016), p. 34; Greville (1824b), p. 314; Guiry (2012), p. 130; Hamel (1931–39), p. 251; Irvine (1956), pp. 26–8; Jaasund (1965), pp. 69, 70; Kuckuck (1894), p. 232, fig. 5(A–B); Kylin (1947), pp. 27, 28, fig. 24(B), pl. 1(3); Lund (1950), pp. 22–9, figs 3–4, (1959), p. 95, pp. 22–9, fig. 54; Mathieson & Dawes (2017), p. 238, pl. XXXVII(5–7); Newton (1931), p. 189; Oltmanns (1922), figs 373(3), 379(8); Pedersen (2011), p. 87; Prud'homme van Reine (1974), p. 174, (1978), p. 303, (1982a), pp. 93–115, figs 152–211, pl. 3(b), 4; Ribera *et al.* (1992), p. 119; Sauvageau (1900), p. 217, (1901), p. 61, (1971), pp. 5, 53, 66–76, figs 16(A–L), 17(A–K); Stegenga & Mol (1983), p. 113, pl. 42(3–4); Stegenga *et al.* (1997a), p. 22; Taylor (1957), p. 119; van den Hoek (1958a), pp. 188–95; Waern (1945), pp. 402–4, pl. 2, (1952), pp. 95, 96.

Sphacelodermaceae Draisma, Prud'homme van Reine & H. Kawai

SPHACELODERMACEAE Draisma, Prud'homme van Reine & H. Kawai (2010), p. 321.

Thalli tufted, with one to several erect shoots arising from a discoid or crustose base; erect shoots simple or sparsely branched, terete, parenchymatous, segmented and leptocaulous; branches indeterminate, produced hypacroblastically; growth apical; corticating sheath and basally produced rhizoidal filaments absent; hairs absent; cells with several discoid plastids without pyrenoids.

Reproduction by plurilocular sporangia and unilocular sporangia borne on the lower regions of mature shoots, rarely on the basal crust (unilocular sporangia), usually solitary on 1–4-celled stalks; propagules unknown.

Sphaceloderma Kuckuck

SPHACELODERMA Kuckuck (1894), p. 232.

Type species: *S. helgolandicum* Kuckuck (1894), p. 323.

Thalli tufted, either solitary, or confluent, with one to several erect main shoots arising from monostromatic discs, later polystromatic crusts which are often superimposed and tiered; erect shoots fairly rigid, terete, uniform in diameter throughout (leptocaulous), simple or sparsely branched, to 1(–4) orders, with branching irregular, sometimes secund with branches indeterminate; branches produced hypacroblastically; thalli with prominent apical cell, parenchymatous, segmented, secondary segments divided by radial and transverse walls without obvious division into a medulla and peripheral cells; corticating sheath and basally produced rhizoidal filaments absent; hairs absent; cells with several discoid plastids without pyrenoids.

Reproduction by plurilocular sporangia and unilocular sporangia borne on the lower regions of mature thalli, rarely on the basal crust (unilocular sporangia), usually solitary on 1–4-celled stalks; propagules unknown.

One species in Britain and Ireland.

Sphaceloderma caespitulum (Lyngbye) Draisma, Prud'homme van Reine & Kawai (2010), p. 321.   Fig. 34

Sphacelaria caespitula Lyngbye (1819), pp. 105–6; *Sphaceloderma helgolandica* Kuckuck (1894), p. 232; *Sphacelaria helgolandica* (Kuckuck) Waern (1945), p. 399.

Thalli epiphytic or epilithic forming small, erect, fairly rigid, light brown tufts, to 2.5(–3) mm in height, usually solitary, sometimes confluent and spreading up to 0.5–1 cm in extent, arising initially from an outward-growing, perennial, monostromatic disc, later from a polystromatic crust up to 5(–8) cells, 100 µm high, often several of which overgrow each other to form tiers of crusts up to 450 µm in thickness; erect axes numerous, 20–34(–48) µm in diameter, fairly uniform in diameter throughout their length, simple or sparsely, and irregularly, sometimes secundly, branched, to 1–4 orders, with branches usually closely adpressed to parental axis and indeterminate in growth, often exceeding the height of the main axis; all primary axes and laterals originating from a colourless, elongated apical cell, polysiphonous and clearly segmented with each secondary segment divided longitudinally into 1–4(–6) narrow cells seen in surface view, with nearly all segments further dividing by 1–4 transverse secondary walls to form smaller cells; secondary segment cells longer than wide, quadrate or subquadrate, 8–20(–26) × 5–8(–11) µm; in transverse section, older secondary segments clearly divided by radial and transverse walls without any obvious differentiation into a central medulla and outer peripheral cell layers; secondary cells with several discoid plastids without pyrenoids; axes without obvious production of downward-growing secondary rhizoids and corticating sheath; hairs not observed.

Unilocular and plurilocular sporangia occurring on separate thalli, borne proximally on the main axes and lower laterals, usually solitary, on 1–3(–4) celled stalks, and quite strongly recurved towards parental axis, rarely on basal crust (unilocular sporangia); plurilocular sporangia subspherical to ellipsoid, 85–113(–130) × 48–85(–100) µm; unilocular sporangia not observed by the present author, reported by Prud'homme van Reine (1982a) for material collected at St Andrews, Fifeshire, to be spherical to subspherical, 80–110 × 60–100 µm; propagules unknown.

Commonly epiphytic on the stipes and holdfasts of *Laminaria hyperborea*, also reported on the stipes of *Saccorhiza polyschides*, as well as epilithic on rocks and stones, sublittoral.

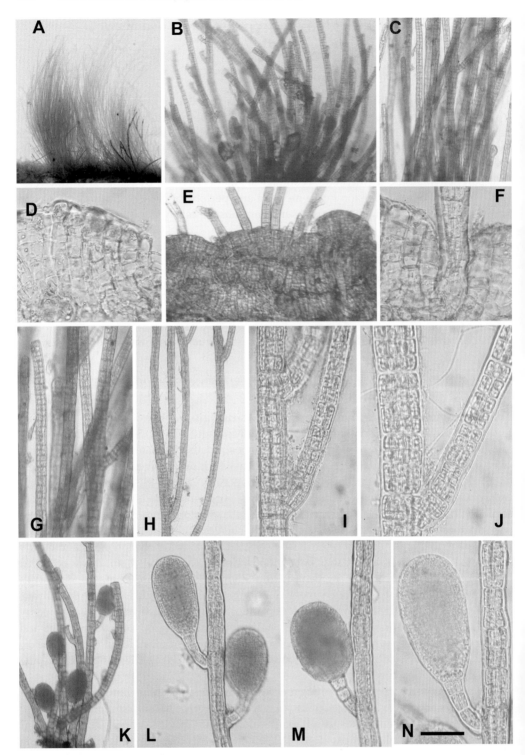

Fig. 34. *Sphaceloderma caespitula*
A. Habit of thalli. B, C. Portions of erect axes. D. VS of crustose base showing vertical tiers of cells. E. VS of super-imposed crusts showing emergent erect axes. F. VS of basal crust showing emergent erect axis. G–J. Portions of erect axes showing secondary segments. K–N. Portions of erect axes with plurilocular sporangia. Bar in N = 2 mm (A), = 300 μm (B), = 180 μm (C, E, G, H, K), = 90 μm (D, L, M), = 45 μm (F, I, J, N).

A rare species, reported to be widely distributed around Britain (North Devon, Sussex, Anglesey, Isle of Man, Ross and Cromarty, Northumberland) and Ireland (Cork, Clare, Galway, Waterford).

Appears to be perennial.

Bartsch & Kuhlenkamp (2000), p. 168; Batters (1884), p. 353, (1890a), p. 59, pl. 9(5–8), (1890b), p. 279, (1902), p. 39; Brodie *et al.* (2016), p. 1019; Dizerbo & Herpe (2007), p. 95; Draisma *et al.* (2010b), pp. 308–26; Fletcher in Hardy & Guiry (2003), p. 296; Guiry (2012), p. 130; Irvine (1956), pp. 37, 38; Keum *et al.* (1994), pp. 137–44; Kornmann & Sahling (1977), p. 156, pl. 84(A–H); Kuckuck (1912), p. 178, pl. 19(4); Kylin (1947), pp. 26, 27, fig. 24(A); Lund (1950), pp. 12–17, fig. 1; Mathieson & Dawes (2017), p. 238, pl. XXXVII(8–10); Newton (1931), pp. 190, 191; Prud'homme van Reine (1974), p. 174, (1982a), pp. 80–93, figs 118–51; Reinke (1891b), p. 13, pl. 4(1–4); Rueness (1977), p. 183, fig. 107(A).

Stypocaulaceae Oltmanns

STYPOCAULACEAE Oltmanns (1922), p. 95.

Thalli erect, tufted, much branched, arising from a small basal attachment disc overlaid by rhizoids growing down from basal cells of the erect filaments; erect axes terete, fairly rigid, parenchymatous and segmented throughout, except in terminal branchlets, produced by prominent apical cells with no obvious increase in width (leptocaulous); branching radial or distichous and alternate, to several orders, with branches determinate and arising by acrohet-eroblasty; in transverse section, comprising large, inner medullary cells surrounded by layers of smaller corticating cells, with or without additional layers of corticating rhizoidal filaments towards the base; growth apical, with or without secondary growth of segments; hairs single or clustered, arising from axils of branches; cells with numerous discoid plastids without pyrenoids.

Life history isomorphic and diplohaplontic, with asexual unilocular sporangia and sexual gametangia arising from the axils of terminal branchlets; sexual reproduction anisogamous or oogamous, with gametangia borne on the same thalli and of two types, plurilocular micro-sporangia (antheridia) and plurilocular megasporangia (oogonia); propagules unknown.

Some authors, including Newton (1931), Knight & Parke (1931), Guiry (1978), South & Tittley (1986) recognized two genera for this family in Britain and Ireland, viz. *Halopteris* Kützing and *Stypocaulon* Kützing. Differences between the genera included *Halopteris* having leptocaulous growth, four large central cells in transverse section, acroheteroblastic, regular, alternate, pinnate (distichous) branching, an absence of pericysts, and sporangia borne singly in the axils of secondary branchlets, while *Stypocaulon* has leptocaulous growth, large numbers of cells in transverse section, acroheteroblastic, often rather irregular, radial branching, pericysts in the main filaments, oogamous gametangia and asexual sporangia borne in groups in the axils of laterals. Other authors, including Sauvageau (1903), Hamel (1931–39), Fritsch (1945), Tokida (1954), Taylor (1957), Gayral (1966), Ardré (1970), Parke & Dixon (1976), Prud'homme van Reine (1978) and Womersley (1987) combined the genera under *Halopteris*. Support for maintaining two separate genera was later given by Prud'homme van Reine (1991, 1993) and Kawai & Prud'homme van Reine (1998), based on the differences in reproduction and branching pattern. However, support for the single genus *Halopteris* has since been provided by the molecular study of Draisma *et al.* (2010b) and this is accepted in the present treatise.

Halopteris Kützing

HALOPTERIS Kützing (1843), p. 292.

Type species: *H. filicina* (Grateloup) Kützing (1843), p. 293.

Thalli erect, tufted, much branched, arising from a crustose base, secondarily attached by rhizoids originating from the basal cells of the erect axes; erect axes terete, fairly rigid, parenchymatous and clearly segmented in the main axes, originating from a large prominent apical cell with no obvious increase in width over the original apical cell (leptocaulous); branching distichous, alternate or radial, and determinate, to several orders with branches arising by acroheteroblasty; terminal branchlets monosiphonous; in transverse section, comprising an inner medulla of large cells surrounded by a layer of smaller corticating cells, with or without additional layers of corticating rhizoidal cells towards the base; cells with several discoid plastids, without pyrenoids: hairs present or absent, clustered, arising from axils of branches.

Unilocular sporangia and gametangia borne in clusters, in the axils of the terminal branchlets; gametangia are plurilocular microsporangia (antheridia) and unilocular or plurilocular megasporangia (oogonia).

Two species in Britain and Ireland.

KEY TO SPECIES

Tufts olive-brown, fairly dense and feather-like; branching regularly alternate, pinnate (distichous) throughout; pericysts absent; sporangia borne singly or in pairs in the axils of secondary branchlets. . . . . *H. filicina*
Tufts dark brown to reddish brown, very dense and fasciculate, frequently in inverse cone-like clusters; branching regularly alternate above, becoming irregularly radial below; pericysts present; sporangia borne in groups in axils of laterals . . . . *H. scoparia*

Halopteris filicina (Bonnemaison) Geyler 1866, p. 507.   Figs 35, 36

Ceramium filicinum Grateloup (1806), p. 33; *Sphacelaria disticha* Vahl ex Lyngbye (1819), p. 104; *Sphacelaria filicina* (Grateloup) Agardh (1824), p. 22; *Halopteris filicina* var. *sertularia* (Bonnemaison) Geyler (1866), pp. 507, 508.

Thalli erect, usually gregarious, 2–10 cm long, light or yellowish brown in colour, much branched and feather-like in appearance and rigid, arising from a basal fibrous holdfast of compacted rhizoidal filaments overlying an attachment disc; axes terete, fairly rigid, regularly (usually every second segment) and alternately branched (except basal region of terminal laterals), in the lower parts clothed over much of the length by downward-growing, tomentose, corticating, branched filaments, which spread out basally as attachment rhizoids; upper laterals densely packed, distichous and multipinnate, with terminal laterals erect and closely adpressed to parental branch, lower branches often dehisced leaving branch scars; except for greater part of terminal branchlets which are uniseriate, axes clearly polysiphonous and segmented, each secondary segment 1–1.5 diameters long, 25–170 × 25–140 μm, and divided longitudinally into up to 7(–13) narrow cells seen in surface view, these usually then further divided by 1–3 secondary transverse walls; cells of secondary segments quadrate to rectangular, 1.5–5 diameters long, 28–100 × 12–32 μm, each with densely packed discoid plastids without pyrenoids; all axes derived from the activity of a large, pronounced, terminal apical cell; corticating rhizoidal filaments largely uniseriate, comprising cells 2–4 diameters long, 60–150 × 25–55 μm each with discoid plastids; in transverse section, axes and laterals reveal a central medulla of several, largely quadrate to rectangular, colourless cells (periclinal type) surrounded by an outer peripheral layer of 1–2 smaller, photosynthetic cells, the outer

Fig. 35. *Halopteris filicina*
A. Habit of thallus. B–E. Portions or erect axes showing branching pattern and secondary segments. Bar in E = 1.9 mm (B), = 0.7 mm (C), = 470 µm (E), = 180 µm (D).

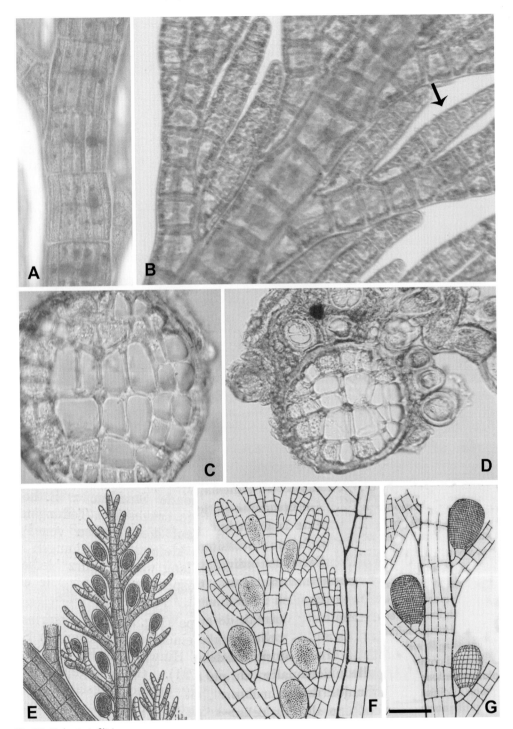

Fig. 36. *Halopteris filicina*

A. Portion of erect axis showing longitudinally and transversely divided secondary segments. B. Portion of terminal thallus region showing regular alternate branching pattern but characteristically with an additional branch formed in each axil (arrow). C. TS of axis showing central medulla of large colourless cells enclosed by a peripheral layer of 1–2 smaller cells. D. TS of axis showing outer attached corticating filaments. E–F. Portions of terminal regions of erect axes showing unilocular sporangia. G. Portion of terminal region of erect axis showing plurilocular sporangia (E from Reinke (1891), F & G from Sauvageau (1900–14)). Bar in G = 180 μm (B), = 90 μm (A, D), = 55 μm (F), = 85 μm (G), = 105 μm (E), = 45 μm (C).

thickened wall usually showing large numbers of attached, corticating filaments; hairs rare, arising from the axils of the terminal laterals, with basal meristem and sheath.

Unilocular sporangia and plurilocular gametangia occurring on different thalli, borne in the axils of the terminal branchlets; gametangia ovoid, 80–100 × 45–60 µm, sessile or pedicellate, of two types, orange-red microsporangia (antheridia) and colourless megasporangia (oogonia), with different-sized locules; unilocular sporangia ovoid, elongate ovoid, 60 × 40–50 µm, pedicellate (dimensions according to Hamel 1931–39).

Epilithic, in pools in the lower littoral, and sublittoral to 25 m.

Mainly reported for southern and western shores of Britain, extending eastwards to Hampshire and northwards to the Isle of Man, with a single record for Durham; more widely distributed in Ireland.

Perennial; unilocular sporangia recorded in spring and July–September.

Ardré (1970), pp. 260, 261; Batters (1902), p. 40; Brodie *et al.* (2016), p. 1019; Bunker *et al.* (2017), p. 278; Cabioc'h & Boudouresque (1992), p. 58, fig. 24; Cormaci *et al.* (2012), p. 434, pl. 148(1–4); Dawes & Mathieson (2008), p. 134, pl. 16(3); De Mesquita Rodrigues (1963), p. 35, pl. 16(B); Dizerbo & Herpe (2007), p. 98, pl. 32(3); Draisma *et al.* (2010b), pp. 308–26; Feldmann (1937), p. 301; Fletcher in Hardy & Guiry (2003), p. 255; Funk (1955), p. 44, pl. 4(3–6); Gallardo García *et al.* (2016), p. 34; Gayral (1958), pp. 204, 205, fig. 27(B), pl. 24, (1966), p. 239, fig. 27(B), pl. 31; Guiry (2012), p. 131; Hamel (1931–39), pp. 261–3, fig. 48(5–6); Harvey (1846–51), pl. 142; Katsaros & Galatis (1986), pp. 358–70, (1990), pp. 63–74, figs 1–21; Keum (2010), pp. 93, 94, fig. 12(A–F); Keum *et al.* (1995) pp. 137–44; Mathias (1935), pp. 25–8, figs 1–10; Moore (1951), pp. 265–78; Morton (1994), p. 37; Newton (1931), p. 196, fig. 123; Oltmanns (1922), fig. 380(1–5); Prud'homme van Reine (1993), pp. 153–5; Ribera *et al.* (1992), p. 119; Sauvageau (1903), pp. 378–423, figs 55–63, (1931a), pp. 33–50, (1971), pp. 294–332, figs 55–63; Seoane-Camba (1965), p. 76, fig. 21(3–4).

Halopteris scoparia (Linnaeus) Sauvageau (1904), pp. 349, 377.   Figs 37, 38

Conferva scoparia Linnaeus (1753), p. 1165; *Ceramium scoparium* (Linnaeus) de Candolle (1805), p. 263; *Sphacelaria scoparia* (Linnaeus) Lyngbye (1819), p. 104; *Sphacelaria scoparioides* Lyngbye (1819), p. 107; *Sphacelaria scoparia* var. *aestivalis* J. Agardh (1842), p. 29; *Sphacelaria scoparia* var. *hiemalis* J. Agardh (1842) p. 29; *Stypocaulon scoparium* (Linnaeus) Kützing (1843), p. 293; *Sphacelaria scoparia* f. *aestivalis* (J. Agardh) Meneghini (1850), p. 164; *Sphacelaria scoparia* f. *hiemalis* (J. Agardh) Areschoug (1850), p. 164; *Stypocaulon scoparium* f. *spinulosum* (Lyngbye) Kjellman (1890), p. 66; *Stypocaulon scoparium* var. *scoparioides* (Lyngbye) Holmes & Batters (1890), p. 82; *Stypocaulon scoparium* f. *patentissimum* (Sauvageau) Lund (1950), p. 58; *Halopteris scoparia* var. *scoparioides* (Lyngbye) Prud'homme van Reine (1978), p. 303, nom. inval.

Thalli erect, solitary or gregarious, to 16 cm high, dark to reddish-brown in colour, much branched to several orders, becoming densely tufted and fasciculate with age, frequently in obconic clusters, arising from a large, fibrous holdfast of densely compacted, rhizoidal filaments; main axes terete, fairly rigid and clearly regularly and alternately branched above, more irregularly branched and often naked below, all axes and laterals arising from a single, large, dark-coloured, apical cell; all main axes and laterals parenchymatous, and clearly segmented throughout, covered over the greater part by tomentose, downward-growing, corticating filaments which form the fibrous holdfast; terminal branchlets distichous, pinnate, spine-like and parenchymatous; secondary segments 0.5 to 2(–3) diameters long, 46–120 × 90–170 µm and divided longitudinally into up to 10(–16) narrow segments seen in surface view, these then further divided by 1–3 secondary transverse walls; cells of secondary segments quadrate to rectangular, 1–3 diameters long, 10–35(–50) × 6–15(–20) µm, each with densely packed, peripherally placed discoid plastids without pyrenoids; corticating rhizoidal filaments largely multiseriate and segmented, segments 3–6 cells wide, 40–95 × 55–105 µm; in transverse section, axes and laterals reveal a central medulla of numerous, regularly arranged, often in rows, large quadrate to rectangular, colourless cells surrounded by an outer peripheral layer of 1(–2) smaller photosynthetic cells (periclinal type), the outer thickened wall usually with large numbers of attached, corticating filaments; hairs solitary or grouped, with basal meristem

Fig. 37. *Halopteris scoparia*
A. Habit of thalli in the field. B–E. Habit of thalli. F. Terminal region of erect axes showing dark-coloured apical cells. Bar in E = 7 cm (A), = 12 mm (C), = 5 mm (D), = 26 mm (E), = 470 μm (F).

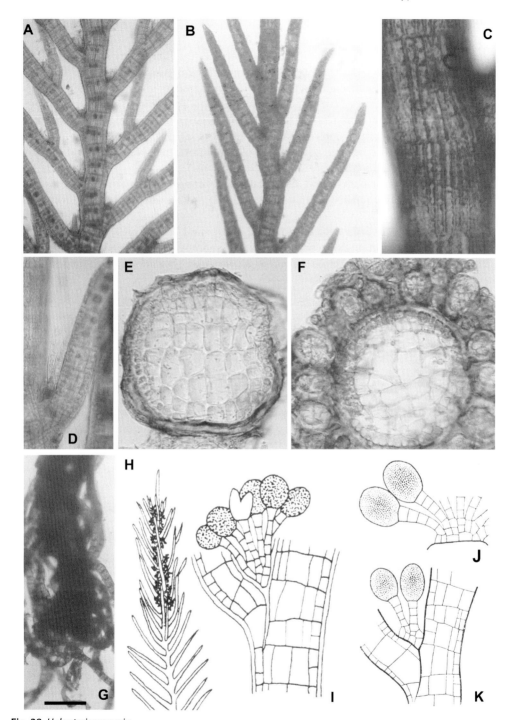

Fig. 38. *Halopteris scoparia*

A, B. Portions of terminal regions of erect axes showing alternate branching pattern and acuminate determinate laterals. C. Close-up of longitudinally and transversely divided secondary segments showing peripheral distribution of plastids in the cells. D. Axis of terminal branch showing axially positioned hair. E. TS of axis showing inner medulla of large colourless cells enclosed within a peripheral layer of 1–2 smaller cells. F. TS of axis towards the base showing outer attached corticating filaments. G. Basal region of axis showing secondary corticating attachment rhizoids. H–K. Terminal regions of thallus showing position of unilocular sporangia in the branch axils. (H from Reinke 1891, I–K from Sauvageau, 1971.) Bar in G = 470 μm (A, G), = 300 μm (B), = 90 μm (C, D, E, F), = 8 cm (H), = 75 μm (I), = 65 μm (J), = 85 μm (K).

and sheath, arising from axils of terminal laterals; each secondary segment has four pericysts, which may grow into new branches.

Unilocular sporangia and gametangia borne on separate thalli, arising in small clusters from a placenta-like mound of cells in the axils of the terminal laterals; unilocular sporangia terminal on simple, multicellular stalks, globular, 60–80 μm in diameter; gametangia rare, borne on simple, multicellular stalks, of two types, microsporangia (antheridia) and megasporangia (oogonia); microsporangia oval, 100–110 × 90–100 μm; megasporangia oval to pyriform, 90–105 × 75–80 μm (sporangial dimensions according to Hamel 1931–39).

Epilithic, occasionally epiphytic, usually in pools, midlittoral to lower littoral, and sublittoral to 9 m, often in sandy situations.

Recorded throughout Britain and Ireland; quite common on the south-west coasts of England becoming rarer further north and on the east coast; not recorded for Orkney and Shetland.

Perennial, sporangia very rarely reported for Britain and Ireland and insufficient data to comment on phenology; a single record of unilocular sporangia in autumn for the Isle of Man (Knight & Parke 1931) and a single record of plurilocular sporangia in September for the Channel Islands (Dixon 1961).

Ardré (1970), p. 261; Børgesen (1926), p. 75; Brodie *et al.* (2016), p. 1019; Bunker *et al.* (2017), p. 293; Cabioc'h & Boudouresque (1992), p. 56, fig. 23; Cormaci *et al.* (2012), p. 436, pl. 149(1–3); De Mesquita Rodrigues (1963), p. 36, pls 4(A), 14(A), 15(9); Dizerbo & Herpe (2007), pp. 98, 99, pl. 32(4); Draisma *et al.* (2010b), pp. 308–26; Feldmann (1937), p. 301; Fletcher in Hardy & Guiry (2003), p. 311; Funk (1955), pp. 44, 45, pl. 4(1–2); Gallardo García *et al.* (2016), p. 34; Gayral (1958), p. 204, fig. 27A, pl. XXIII, (1966), pp. 237–9, fig. 27(A, B), pl. 30; Geyler (1866), p. 481, pl. 34(1–13); Guiry (2012), p. 131; Hamel (1931–39), pp. 263–6, fig. 48(VII, VIII); Harvey (1846–51), pl. 37; Higgins (1931), p. 345; Kawai & Prud'homme van Reine (1998), pp. 265–9; Kützing (1843), p. 293, pl. 18(II); Kylin (1947), pp. 32, 33, figs 25, 26; Lawson & John (1982), p. 135; Lund (1950), pp. 58–61, fig. 12; Lyngbye (1819), p. 104, pl. 31(4); Magnus (1873), p. 139, pl. 2(38–41); Mathieson & Dawes (2017), p. 240, pl. XXXVII(11–14); Morton (1994), p. 37; Newton (1931), p. 197, fig. 124(A–G); Novaczek *et al.* (1989), pp. 183–93; Oltmanns (1922), figs 372, 381–3; Reinke (1891b), p. 24, pl. 7(1–5); Ribera *et al.* (1992), p. 119; Rueness (1977), pp. 187, 188, fig. III; Sauvageau (1904), p. 349, (1907), p. 506, (1908), p. 62, (1909), p. 44, figs 1–10, (1971), pp. 349–85, figs 69–75; Seoane-Camba (1965), p. 77, fig. 20(3); Taylor (1937), p. 134, (1957), p. 134; Tokida (1954), p. 76; Waern (1952), pp. 108–10, figs 45, 46, 47c.

Protohalopteris Draisma, Prud'homme & H. Kawai

PROTOHALOPTERIS Draisma, Prud'homme & H. Kawai (2010), p. 321.

Type species: *P. radicans* (Dillwyn) Draisma, Prud'homme & H. Kawai (2010), p. 321.

Thalli tufted, comprising densely packed erect axes arising from monostromatic discs, later polystromatic crusts; erect axes sparsely branched, with branching mainly hypacroblastic, sometimes acroblastic; erect axes leptocaulous, parenchymatous and segmented, with hairs (grouped) and pericysts; secondary segments with transverse walls and pericysts; growth apical; cells with many discoid plastids without pyrenoids.

Plurilocular and unilocular sporangia formed on separate thalli, borne on the erect axes, solitary or grouped, unilocular sporangia sessile or shortly (1–2 cells) stalked, plurilocular sporangia stalked (1–4 cells); propagules unknown.

One species in Britain and Ireland.

Protohalopteris radicans (Dillwyn) Draisma, Prud'homme & H. Kawai (2010), p. 321.

Fig. 39

Conferva radicans Dillwyn (1809), p. 57, pl. c; *Sphacelaria olivacea* (Dillwyn) Greville (1824), p. 96; *Sphacelaria radicans* (Dillwyn) C. Agardh (1824), p. 165; *Sphacelaria cirrosa* [*cirrhosa*] var. *simplex* C. Agardh (1828), p. 29; *Sphacelaria olivacea* var. *radicans* (Dillwyn) J. Agardh (1848), p. 31; *Sphacelaria radicans* f. *aegagropila* Hylmo (1916), exs. No. 42, nom. illeg.

Thalli epilithic, forming solitary, more commonly confluent, irregularly spread, densely crowded, light to dark brown, fairly stiff, carpet-like tufts, to several cm in diameter, 3–10(–20) mm in height, arising from a monostromatic, later polystromatic, discoid base; monostromatic discs spreading marginally by apical growth, becoming polystromatic by upward growth of the basal cells, 3–5 cells, 20–40 μm thick, little branched, giving rise terminally to numerous erect thalli; erect thalli linear, fairly uniform in diameter throughout their length, parenchymatous and clearly segmented, 25–38(–49) μm in diameter, with each secondary segment divided longitudinally into 1–4(–8) narrow cells seen in surface view, these sometimes seen with transverse walls; secondary segment cells usually longer than wide, becoming quadrate or subquadrate by 1–3 transverse divisions, 8–22(–34) × 6–10(–13) μm; thalli little branched with branching mainly hypacroblastic, sometimes acroblastic, irregular, sometimes bifurcate, to three orders, with branches usually closely adpressed to parental axis at first, later becoming more divergent, and indeterminate, often exceeding the height of the main axis, all axes and branches terminating in a large, prominent, often dark-pigmented, apical cell; in transverse section, older secondary segments clearly divided by radial and transverse walls, without any obvious differentiation into a central medulla and outer peripheral cell layers (radial type); in the superior secondary segments usually one cell extends from the centre to the periphery of the transverse section; the larger cells function as pericysts to develop new branches; in living specimens the pericysts have more transparent contents than other cells but in dried algal specimens most pericysts become very dark in colour; base of shoot sometimes giving rise to divaricate, downward-growing, parenchymatous, rhizoidal filaments, which form discs upon surface contact; corticating sheaths not observed; hairs present or absent, single or grouped, with basal meristem and sheath.

Plurilocular and unilocular sporangia formed on separate thalli; plurilocular sporangia rare, shortly stalked (1–4 segments), solitary or grouped, cylindrical to obovate, 60–83 × 36–52 μm; unilocular sporangia common, sessile or shortly (1–2 segments) stalked, solitary or grouped in twos or fours, spherical or club-shaped, 42–60(–78) × 38–46(–53) μm, formed on erect shoots, sometimes reported on rhizoids; propagules unknown.

Epilithic, more rarely epiphytic, lower eulittoral pools, extending to upper eulittoral in shaded situations, sublittoral to 5 m, often associated with sand.

Common and widely distributed around Britain and Ireland.

Perennial; reproductive sporangia found during the winter (January, February).

Ardré (1970), p. 257; Bartsch & Kuhlenkamp (2000), p. 168; Batters (1890b), p. 280, (1902), p. 38; Brodie *et al.* (2016), p. 1019; Coppejans & Kling (1995), p. 236, pl. 87(E, F); De Haas-Nieker (1956), pp. 158, 159, figs 51–63; Dizerbo & Herpe (2007), p. 96, pl. 32(1); Draisma *et al.* (2010b), pp. 308–26; Fletcher in Hardy & Guiry (2003), p. 300; Gallardo García *et al.* (2016), p. 34; Guiry (2012), p. 131; Hamel (1931–39), p. 249, fig. 4(13); Harvey (1846–51), pl. 189; Holmes (1887a), pp. 79–82; Irvine (1956), pp. 38–40; Jaasund (1965), pp. 68, 69, (1977), p. 187; Kornmann & Sahling (1977), p. 154, pl. 83(A–D); Kuckuck (1894), pp. 229–32, fig. 4(A–L); Kylin (1947), p. 27, fig. 24(C); Lund (1950), pp. 17–22, fig 2(A–F); Mathieson & Dawes (2017), p. 239, pls XXXVII(15), XXXVIII(1, 2); Morton (1994), p. 37; Newton (1931), pp. 188, 189; Oltmanns (1922), figs 374(1), 377(8–9); Pankow (1971), p. 162, figs 187, 188; Pedersen (2011), p. 87; Prud'homme van Reine (1974), p. 174, (1978), p. 303, (1982a), pp. 62–80, figs 60–117, pl. 3(a); Reinke (1890), p. 207, (1891a), p. 8, pl. 3(1); Rodrigues (1963), p. 33; Rueness (1977), p. 187; Saunders & McDevit (2013), pp. 1–23; Sauvageau (1901), p. 34, (1909a), p. 64, figs 10, 11, (1971), pp. 63–5, figs 14–15; Sundene (1953), p. 157; Stegenga & Mol (1983), p. 116, pl. 44, figs 1–2; Stegenga *et al.* (1997b), p. 22; Taylor (1957), pp. 119, 120; Waern (1945), p. 400, (1952), pp. 96–9, fig. 38.

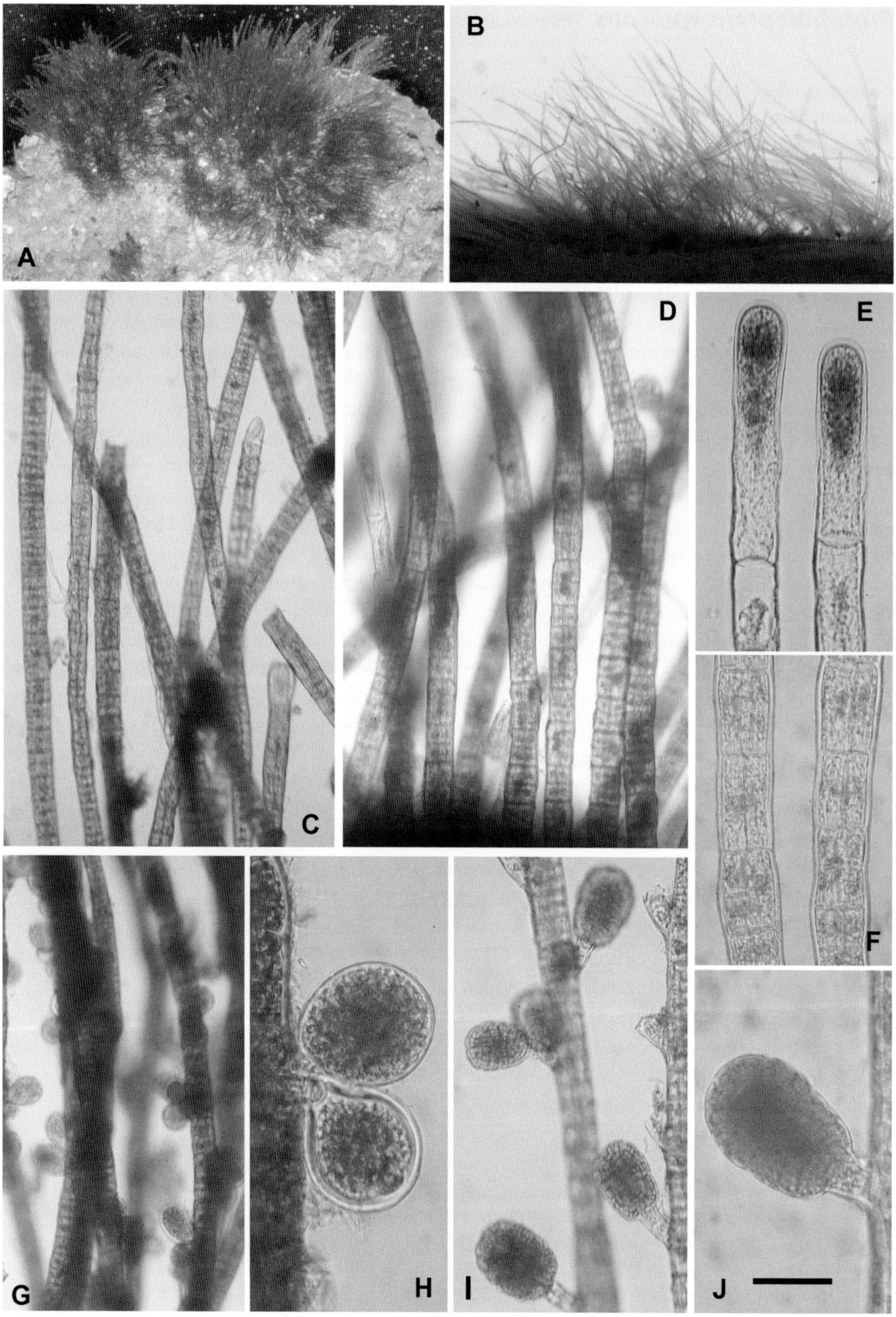

Fig. 39. *Protohalopteris radicans*

A, B. Habit of thalli. C, D. Portions of erect axes. E. Two apical cells. F. Portions of erect axes showing longitudinally divided secondary segments. G, H. Portions of erect axes showing unilocular sporangia. I, J. Portions of erect axes showing plurilocular sporangia. Bar in J = 5 mm (A), = 3.5 mm (B), = 180 μm (C, D, G), = 90 μm (E, I), = 45 μm (F, H, J).

Fucophycidae Cavalier-Smith

FUCOPHYCIDAE Cavalier-Smith (1986), p. 341.

A large subclass of a dozen orders, of which eight are present in Britain and Ireland (see Silberfeld *et al.* 2014).

Desmarestiales Setchell & N.L. Gardner

DESMARESTIALES Setchell & N.L. Gardner (1925), p. 554.

Thalli erect, terete, slightly compressed to blade-like, branched or unbranched, branching distichous, alternate or opposite, to several orders, with or without terminal fascicles of hair-like filaments and whorls of 3–4 short, laterally projecting fascicles of hair-like filaments; growth trichothallic, structure pseudoparenchymatous and solid with central axial filament surrounded by a cortex of large colourless cells, enclosed by 1–3 layers of small, pigmented cells.

   Unilocular sporangia in chains, fasciculate, borne laterally at the base of the whorled, hair-like filaments or scattered over blade surface in little modified surface cells; unispores germinating into filamentous, monoecious, microscopic gametophytes bearing antheridia and oogonia.

Two families are represented in Britain and Ireland: Arthrocladiaceae Chauvin and Desmarestiaceae Kjellman.

Arthrocladiaceae Chauvin

ARTHROCLADIACEAE Chauvin (1842) p. 66 [as Arthrocladieae].

Thalli erect, terete, much branched, branching opposite, sometimes irregular or alternate, all axes with whorls of 3–4 short, laterally projecting fascicles of hair-like, branched, rarely simple filaments; growth trichothallic, structure pseudoparenchymatous, uniaxial with broad cortex of compacted, colourless cells.

   Unilocular sporangia in chains, fasciculate, borne laterally at the base of the whorled, hair-like filaments; unispores germinating into filamentous, monoecious, microscopic gameto-phytes bearing antheridia and oogonia.

Only one genus of the Arthrocladiaceae, *Arthrocladia*, occurs in Britain and Ireland.

Arthrocladia Duby

ARTHROCLADIA Duby (1830), p. 971.

Type species: *A. villosa* (Hudson) Duby (1830), p. 971.

Thalli consisting of erect, much-branched, narrow, terete fronds arising from a small, discoid holdfast; thalli with opposite, less frequently irregular or alternate, branches; all axes bearing regular whorls of 3–4 fascicles of branched, rarely simple, laterally projecting, hair-like filaments; growth trichothallic, structure pseudoparenchymatous, comprising a large, central axial cell, surrounded by a broad cortex of longitudinally elongate, transversely rounded or irregular, colourless cells, becoming smaller outwards and enclosed by 1–2 layers of thick-walled, pigmented cells; surface cells in longitudinal rows, much longer than wide, usually with pointed ends; plastids discoid, several per cell, without pyrenoids.

Unilocular sporangia in moniliform, fasciculate chains, lateral at the base of the whorled, hair-like filaments. Germination of unispores gives rise to filamentous, monoecious, microscopic gametophytes bearing oogonia and antheridia.

One species in Britain and Ireland.

Arthrocladia villosa (Hudson) Duby (1830), p. 971.   Figs 40, 41

Conferva villosa Hudson (1778), p. 603; *Chordaria villosa* (Hudson) C. Agardh (1817), p. 14; *Sporochnus villosus* (Hudson) C. Agardh (1824), p. 266; *Arthrocladia australis* Kützing (1845), p. 275; *Arthrocladia villosa* f. *australis* (Kützing) Hauck (1884), p. 381.

Thalli forming erect, much-branched, terete fronds, solitary or more commonly gregarious, light yellow to mid-brown, arising from a small discoid holdfast; erect thalli with a distinct main axis, firm and cartilaginous, becoming soft and flaccid terminally; solid, to 40(–100) cm long, 0.5(–0.7) mm wide, branching opposite, more rarely irregular, to 2(–3) orders with branches well spaced, divaricate and terminating in a fine, hair-like filament; all axes, except in some older basal regions, with regular whorls of 3–4 fascicles of laterally extending, hair-like filaments up to 2(–4) mm long, uniseriate, simple or branched, with branching whorled, opposite, irregular or unilateral, comprising cells 13–80 × 8–26 µm, fairly short and broad below, becoming long and narrow above; surface cells of thalli in longitudinal rows, much longer than wide, 1–5 diameters long, rectangular or more usually with pointed ends, 16–42 × 7–14 µm; in section, thalli with a large, central axial cell, surrounded by a broad cortex of thick-walled, longitudinally elongate, transversely rounded or irregular, colourless cells becoming smaller outwards and enclosed by 1–2 layers of thick-walled, pigmented cells; all cells with numerous discoid plastids without pyrenoids.

Unilocular sporangia borne in place of branches in the lower regions of the whorled, hair-like filaments, usually unilateral, more rarely opposite or irregular, sessile or on 1–3-celled stalks, in moniliform/compressed moniliform, simple, straight or slightly curved, fasciculate chains, to 36 sporangia long (–250 µm); individual sporangia usually much shorter than wide, 4–9 × 11–15 µm, releasing spores via a lateral pore.

Epilithic, particularly on small stones, gravel, shells etc., more rarely epiphytic, lower eulittoral and sublittoral to at least 8 m.

Generally distributed around Britain and Ireland, although there are more records from south-western and western coasts.

Thalli annual, June to September, although most records for August and September.

The only reported culture studies of *A. villosa* have been on Mediterranean isolates (Sauvageau 1931; Müller & Meel 1982 – as *A. villosa* f. *australis*). Müller & Meel (1982) revealed the occurrence of a pronounced heteromorphic life history with macroscopic sporophytes alternating with monoecious, microscopic gametophytes bearing antheridia and oogonia. Fertilization, however, was not observed and the sporophyte originated apomictically. Culture studies are required on North Atlantic isolates to determine if a similar apomeiotic and parthenogenetic life history occurs. The culture results gave support to the placement of *Arthrocladia* in the Desmarestiales, but not to Sauvageau's (1931) proposal to establish a separate order, Arthrocladiales, for this genus.

Bartsch & Kuhlenkamp (2000), p. 165; Brodie *et al.* (2016), p. 1022; Bunker *et al.* (2017), p. 299; Cormaci *et al.* (2012), pp. 34–6, pl. 1(1–4); Dizerbo & Herpe (2007), p. 108, pl. 35(4); Fletcher (1987), pp. 272–4, figs 78, 79, pl. 12; Fletcher in Hardy & Guiry (2003), p. 224; Fritsch (1945), p. 180, figs 60(D), 62(B, D, F, I, M); Gallardo García *et al.* (2016), p. 34; Guiry (2012), p. 161; Hamel (1931–39), pp. 286, 287, fig. 49(q); Harvey (1884), pl. 64; Hauck (1884), p. 381; Mathieson & Dawes (2017), pp. 132, 133, pl. LVIII(2–5); Morton (1994), p. 39; Müller & Meel (1982), pp. 419–25, figs 1–15; Newton (1931), p. 167, fig. 104; Ribera *et al.* (1992), p. 117; Rosenvinge & Lund (1947), pp. 55–7; Rueness (1977), p. 171, (2001), p. 14; Sauvageau (1931), pp. 95–121, figs 17–22; Taylor (1957), pp. 152, 153, pls 13(2), 17(7–8); Womersley (1987), pp. 266, 267, figs 94(C), 96(F–H), (2003), p. 499.

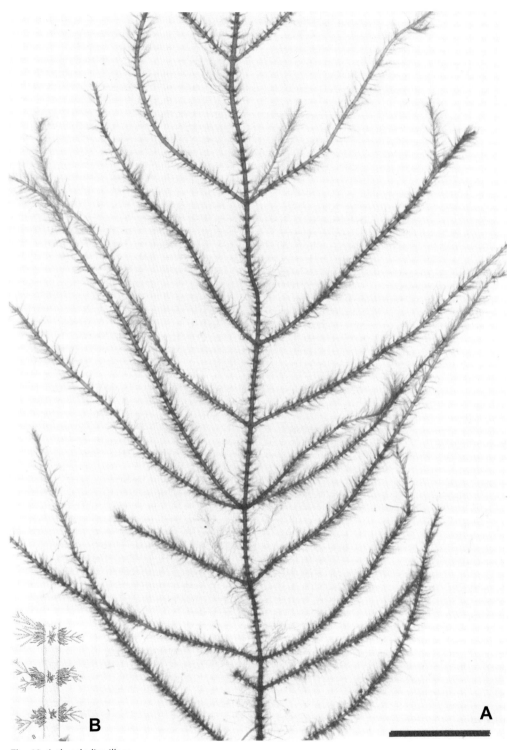

Fig. 40. *Arthrocladia villosa*
A. Habit of thallus. B. Portion of thallus showing whorls of branched filaments. Bar = 10 mm (A), = 0.65 mm (B).

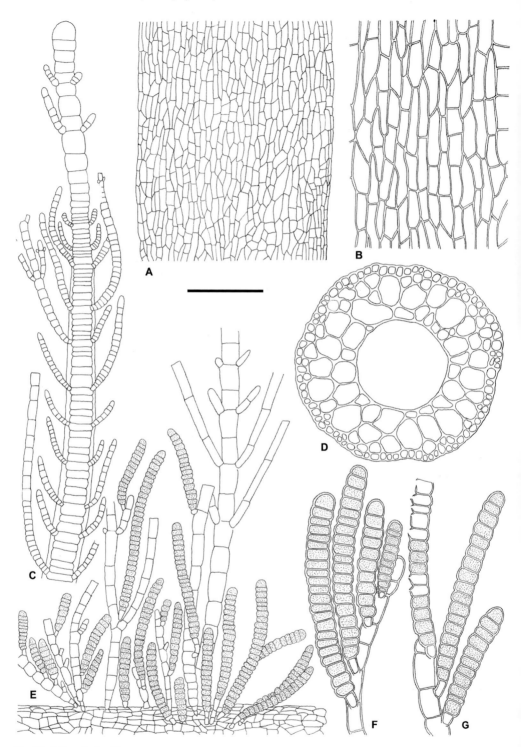

Fig. 41. *Arthrocladia villosa*

A. Portion of young thallus. B. Surface view of cells. C. Terminal region of branch. D. TS of thallus. E. Unilocular sporangia arising from whorls of branched hair-like filaments. F. Unilocular sporangia. Note some sporangia have released contents. Bar = 100 μm (A, D, E), = 200 μm (C), = 50 μm (B, F).

Desmarestiaceae Kjellman

DESMARESTIACEAE Kjellman (1880), p. 10.

Thalli erect, unbranched or branched, terete, slightly compressed to blade-like, branching distichous, alternate or opposite, to several orders, with terminal fascicles of hair-like filaments; growth trichothallic, structure pseudoparenchymatous and solid with central axial filament surrounded by a cortex of large, colourless cells, enclosed by 1–3 layers of small pigmented cells; cells with numerous discoid plastids without pyrenoids.

Unilocular sporangia scattered over blade surface in little modified surface cells; unispores germinating into microscopic, monoecious or dioecious, branched, filamentous gametophytes bearing oogonia and antheridia.

Only one genus of the Desmarestiaceae, *Desmarestia*, occurs in Britain and Ireland.

Desmarestia J.V. Lamouroux

DESMARESTIA J.V. Lamouroux (1813), p. 43, nom. cons.

Type species: *D. aculeata* (Linnaeus) J.V. Lamouroux (1813), p. 25.

Thalli erect, much branched, terete, slightly compressed or distinctly flattened and blade-like; branching distichous, alternate or opposite, branches of limited growth, terminal branches sometimes bearing fascicles of hair-like, pigmented filaments; growth trichothallic, structure pseudoparenchymatous with a central, axial filament surrounded by a solid cortex of large, thick-walled, colourless cells enclosed by 1–2 layers of small, pigmented cells; all cells with numerous discoid plastids without pyrenoids.

Unilocular sporangia discrete, scattered, formed in little modified surface cells, slightly immersed; released unispores germinating into separate, microscopic, filamentous, monoecious or dioecious, gametophytes bearing oogonia and antheridia.

Desmarestia is a notoriously polymorphic genus and within the two major filiform and ligulate groups, species delimitation is often considered to be markedly difficult. For example, in major revisions of members of both these groups recorded for the Northern Hemisphere, using numerical techniques, Chapman (1972a,b) considerably reduced the number of recognized taxa. In the present work, four species of *Desmarestia* are recognized for Britain and Ireland which are easily distinguished on morphological features. These include *D. dudresnayi*, retained here as a separate species and not included within the synonymy of *D. ligulata* var. *firma* (C. Agardh) J. Agardh as proposed by Chapman (1972b).

Culture studies on several species of *Desmarestia* around the world (see Schreiber 1932; Abe 1938; Kornmann 1962a; Chapman 1969; Chapman & Burrows 1970; Nakahara & Nakamura 1971; Müller & Lathe 1981; Anderson 1982) have all revealed a life history comprising a heteromorphic alternation between a macroscopic sporophyte and a microscopic gametophyte. As a general rule, oppositely branched species were found to be monoecious while alternately branched species were dioecious; the only contradiction to this was Anderson's report of the oppositely branched *D. firma* being dioecious. Notable aspects of these culture studies on *Desmarestia* species include reports of the influence of environmental conditions such as temperature, light and nutrients on gametophyte maturation (Kornmann 1962a; Nakahara & Nakamura 1971; Anderson 1982; Nakahara 1984), the involvement of hormonal interactions in gametic discharge and attraction (Müller & Lathe 1981) and the occurrence of parthenogenesis in the life history of *D. viridis* (Nakahara 1984).

Desmarestia thalli rapidly deteriorate following collection due to the presence of high levels of free sulphuric acid within the cells. The thalli quickly turn green, emit a characteristic pungent smell and if not removed, acid released from the thalli will quickly spoil accompanying algae.

Of the four species recorded for Britain and Ireland, *D. dudresnayi*, *D. ligulata* and *D. viridis* are summer annuals which probably overwinter mainly as microscopic gametophytes; *D. aculeata* is perennial.

Four species in Britain and Ireland.

KEY TO SPECIES

1. Thalli filiform and terete 2
   Thalli flattened and membranous 3
2. Thalli oppositely branched, main axis cartilaginous. . . . .*D. viridis*
   Thalli alternately or irregularly branched, main axis coarse and stiff . . . . *D. aculeata*
3. Thalli much branched; axes not more than 7 mm wide . . . .*D. ligulata*
   Thalli unbranched, or rarely with marginal proliferations; axes up to 6 cm
   wide. . . . . *D. dudresnayi*

Desmarestia aculeata (Linnaeus) J.V. Lamouroux (1813), p. 25.   Figs 42, 43

Fucus muscoides Hudson (1762), p. 470, nom. illeg.; *Fucus aculeatus* Linnaeus (1763), p. 1632; *Fucus virgatus* Gunnerus (1766), p. 45; *Sphaerococcus aculeatus* (Linnaeus) Stackhouse (1797), p. xxiv; *Fucus aculeatus* var. *muscoides* Turner (1802), p. 122; *Hippurina aculeata* (Linnaeus) Stackhouse (1809), p. 89; *Sporochnus aculeatus* (Linnaeus) C. Agardh (1817), p. 10; *Desmia aculeata* (Linnaeus) Lyngbye (1819), p. 34, pl. 44 B(1); *Ectocarpus densus* Lyngbye (1819), p. 133, pl. 44B; *Sporochnus aculeatus* var. *complanatus* C. Agardh (1820), p. 152; *Sporochnus aculeatus* var. *plumosus* Suhr (1834), pl. 2132; *Desmarestia aculeata* var. *plumosa* (Suhr) Oersted (1844), p. 48; *Spinularius aculeata* (Linnaeus) Ruprecht (1850), p. 183; *Desmarestia aculeata* f. *subnervis* E.S. Sinova (1921), p. 40; *Desmarestia aculeata* f. *typica* E.S. Sinova (1921), p. 140, nom. inval.; *Desmarestia aculeata* f. *viridescens* A.D. Zinova (1951), p. 105.

Thalli forming erect, solitary, sometimes gregarious fronds, 0.5–1(–2) m in length, light brown, cartilaginous when young becoming dark, coarse and stiff later, much branched, main axis fairly distinct, oval, to 3 mm wide at base, slightly more compressed, 1 mm or less above, arising from a small, 2–3 cm long, terete stipe and attached by a bulbous holdfast; branching regularly or 2–3 times alternate, more rarely opposite towards base, ultimate branches short and spine-like; in spring/early summer branches bearing opposite fascicles of 1–3 light brown, oppositely branched, uniseriate filaments 2–4 mm long, comprising cells 14–120 × 13–28 µm, generally 1(–2) diameters long below, 3–8 diameters long above, bearing large numbers of discoid plastids without pyrenoids; surface cells of branches irregularly shaped and placed, 8–18 × 5–12 µm, each with numerous, small, discoid plastids; in section, structure pseudoparenchymatous with a large central axial cell, surrounded by a broad cortex of variously sized and irregularly distributed, thick-walled, colourless cells enclosed by an outer layer of 1–2 small, usually taller than wide, pigmented cells.

Unilocular sporangia scattered over both sides of thallus surface, oval in shape and slightly darker than surrounding vegetative cells; in section, partially immersed, oval, taller than wide, 15–23 × 9–13 µm.

Epilithic in shaded lower eulittoral pools, or more commonly in the sublittoral to at least 15 m.

Generally distributed around Britain and Ireland.

Perennial; spring and early summer thalli clothed in opposite fascicles of short, branched, hair-like filaments; these also terminate all branches and are shed by late summer, giving thalli a serrated appearance.

Culture studies by Schreiber (1932), Chapman (1969), Chapman & Burrows (1970) and Mailer & Lathe (1981) revealed that North Atlantic isolates of *D. aculeata* have a heteromorphic life history with the macroscopic sporophyte thalli alternating with a microscopic, dioecious gametophyte. Gamete formation in the gametophytic thalli showed a requirement for blue

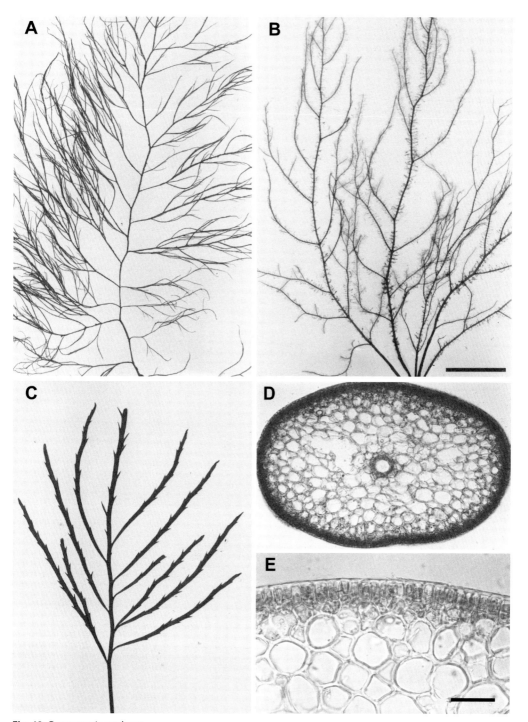

Fig. 42. *Desmarestia aculeata*
A. Portion of thallus (winter form). B. Portion of thallus (summer form). C. Portion of thallus (winter form) with prominent serrations. D. TS of thallus. E. TS of thallus margin. Bar in B = 24 mm (A), = 10 mm (B), = 14 mm (C). Bar in E = 25 μm (D), = 8 μm (E).

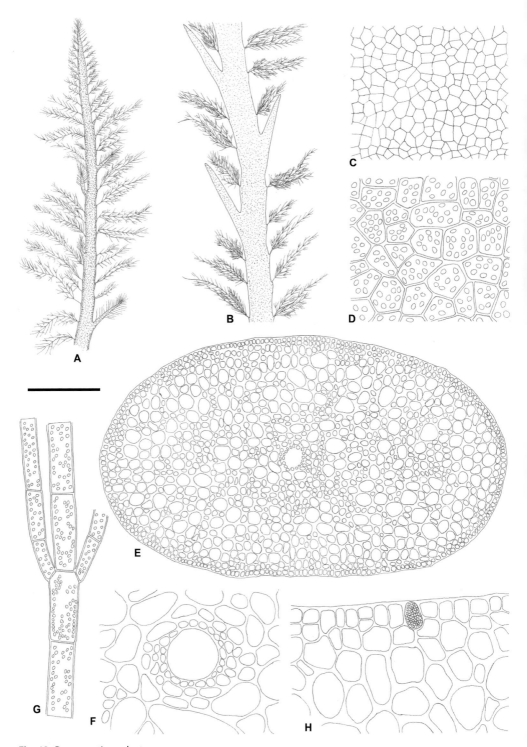

Fig. 43. *Desmarestia aculeata*
A. Terminal branch region showing opposite fascicles of filaments. B. Portion of thallus showing fascicles of filaments and short spine-like protuberances. C, D. Surface view of thallus showing cells with numerous discoid plastids. E. TS of thallus. F. Central axial filament in TS. G. Portion of branched hair-like filament. H. Section of thallus margin showing unilocular sporangium. Bar = 2.5 mm (A), = 2 mm (B), = 200 μm (E), = 50 μm (C, F–H), = 20 μm (D).

light, while discharge and subsequent attraction of the spermatozoids to the egg was controlled by a hormonal interaction (Müller & Lüthe 1981).

Bartsch & Kuhlenkamp (2000), p. 165; Brodie *et al.* (2016), p. 1022; Bunker *et al.* (2017), p. 301; Chapman & Burrows (1970), pp. 103–8, figs 1–2, (1971), pp. 63–76, figs 1–15; Cormaci *et al.* (2012), pp. 38, 40, pl. 2(1–2); Dizerbo & Herpe (2007), p. 105, pl. 34(4); Fletcher (1980), p. 44, pl. 16(6), (1987), pp. 276–9, figs 80, 81, pl. 13; Fletcher in Hardy & Guiry (2003), p. 240; Gallardo García *et al.* (2016), p. 34; Gayral (1966), p. 288, pl. LII; Guiry (2012), p. 161; Hamel (1931–39), p. 283, fig. 49(m–p); Hauck (1884), p. 378, fig. 163; Kjellman in Englert & Prantl (1897), p. 210, fig. 147(A–B); Kornmann & Sahling (1977), p. 140, fig. 75(A–D); Mathieson & Dawes (2017), pp. 133, 134, pls LVIII(6, 7), LXII(4–6); Morton (1994), p. 39; Müller & Lüthe (1981), pp. 351–6, figs 1–14; Newton (1931), p. 164, fig. 103; Pedersen (2011), p. 92, fig. 98(A, B); Perestenko & Zakhodnova (2008), pp. 1112–17; Ribera *et al.* (1992), p. 116; Rosenvinge & Lund (1943), pp. 51–4, fig. 19; Rueness (1977), p. 172, pl. 23(3); Saunders & McDevit (2013), pp. 1–23; Schreiber (1932), pp. 561–82, figs 1–12; Stegenga *et al.* (1997a), p. 22; Taylor (1957), p. 101, pl. 13(4, 5), pl. 14(7).

Desmarestia dudresnayi J.V. Lamouroux ex Léman (1819), p. 105.   Figs 44, 45

Thalli forming erect, solitary, light to yellow-brown, becoming olive-brown, dorsiventrally flattened blades arising from a small, 2–4 cm long, distinct, terete stipe and attached by a conical holdfast; erect blades simple, rarely with marginal proliferations, thin, solid, somewhat translucent, membranous, papery, flaccid, delicate, frequently torn, slightly lubricous, linear-lanceolate to strap-shaped, with smooth or scalloped margins, to 35(–55) cm long, 6 cm wide, narrowing below; blades with distinct midrib extending from stipe, bearing opposite primary veins which branch into secondary veins; primary veins of young thalli only extending to blade margin and terminating in projecting trichothallic, branched, hair-like filaments; in surface view, cells irregularly arranged, variable in shape, although approximately rectangular, 14–32 × 9–17 µm, each with large numbers of discoid plastids without pyrenoids; in section, blades to 130(–180) µm thick and pseudoparenchymatous in structure, comprising a central axial filament, usually intact although not uncommonly disrupted and indiscernible, surrounded by a cortex of 1(–2) layers of large, thick-walled, colourless cells, enclosed by an outer layer of small, pigmented cells.

Unilocular sporangia scattered over both sides of blade surface, slightly darker and larger than accompanying vegetative cells, rounded or oval in surface view, 13–21 × 12–16 µm, partly immersed, usually oval, 16–21 × 10–16 µm in vertical section.

Epilithic on small stones and shells embedded in gravel in the sublittoral, in areas of moderate to strong water current, to 30 m.

Only recorded in Britain for Devon, Cornwall and Argyll, and in Ireland for Antrim, Donegal, Galway and Kerry.

Annual, May to September.

Thalli are frequently colonized by the endophytic red alga *Colacodictyon reticulatum* (Batters) J. Feldmann.

Altamirano *et al.* (2014), pp. 61–3; Blackler (1961), p. 87; Brodie *et al.* (2016), p. 1022; Cormaci *et al.* (2012), p. 40, pl. 2(3–4); Dizerbo (1965), p. 504; Dizerbo & Herpe (2007), p. 106, pl. 35(1); Drew & Robertson (1974), pp. 195–200, figs 1–6; Fletcher (1987), pp. 279–82, figs 82, 83; Fletcher in Hardy & Guiry (2003), p. 240; Gallardo García *et al.* (2016), p. 34; Guiry (2012), p. 161; Hamel (1931–39), pp. 284, 285; Morton (1994), p. 39; Newton (1931), p. 166; Ribera *et al.* (1992), p. 117; Sauvageau (1925), pp. 1–13; Yang *et al.* (2014a), pp. 149–66, (2014b), p. 771.

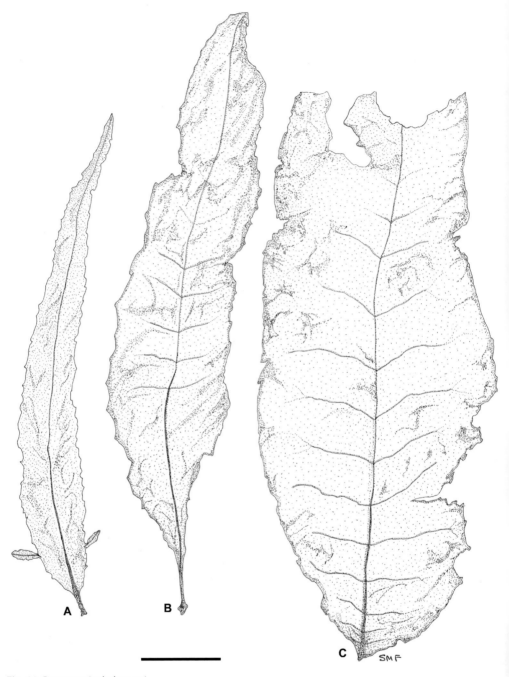

Fig. 44. *Desmarestia dudresnayi*
A–C. Habit of thalli. A–B. Young thalli. C. Portion of older thallus. Bar = 30 mm.

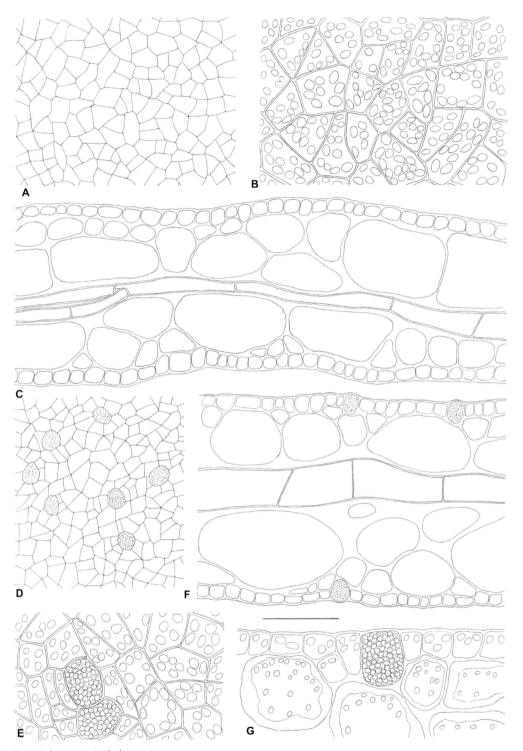

Fig. 45. *Desmarestia dudresnayi*
A, B. Surface view of thallus showing cells with several discoid plastids. C. TS of thallus. Note central axial filament. D, E. Surface view of fertile thallus showing unilocular sporangia. F, G. TS of fertile thalli showing unilocular sporangia associated with surface cells. Bar = 50 μm (A, C, D, F), = 20 μm (B, E, G).

Desmarestia ligulata (Stackhouse) J.V. Lamouroux (1813), p. 25.   Figs 46, 47

Fucus ligulatus Lightfoot (1777), p. 946, nom. illeg.; *Fucus herbaceus* Hudson (1778), p. 252, nom. inval.; *Herbacea ligulata* Stackhouse (1809), p. 89; *Fucus ligulatus* var. *angustior* Turner (1809), p. 74; *Fucus ligulatus* var. *dilitatus* Turner (1809), p. 74; *Laminaria ligulata* (Lightfoot) C. Agardh (1817), p. xiii; *Desmia ligulata* (Lightfoot) Lyngbye (1819), p. 33, pl. 7: fig. B; *Sporochnus ligulatus* (Lightfoot) C. Agardh (1820), p. 158; *Desmarestia ligulata* var. *angustior* (Turner) Batters (1902), p. 23; *Desmarestia ligulata* var. *dilatata* (Turner) Batters (1902), p. 23; *Desmarestia jordanii* N.L. Gardner (1940), p. 269, pl. 32; *Desmarestia linearis* N.L. Gardner in G.M. Smith (1944), p. 120, pl. 18, fig. 2; *Desmarestia adriatica* Ercegovic (1948), p. 113, figs 15–19.

Thalli forming erect, solitary, cartilaginous, much-branched, dorsiventrally flattened fronds, to 0.5–1(–2) m long, 2–3(–7) mm wide, light olive-brown in colour, arising from a small, terete stipe and lobed holdfast; main axis usually prominent, giving rise to opposite, distichous branches of limited growth similarly branched to several orders; all branches attenuate at base and apex, main branches with distinct midrib; ultimate branches short and spine-like, with or without terminal tufts of hair-like filaments; filaments to 1–2 mm long, uniseriate, oppositely branched, comprising cells 28–80 × 20–28 μm, approximately quadrate below, rectangular to 3(–4) diameters long above, containing numerous discoid plastids without pyrenoids; in surface view, thallus cells irregularly shaped and placed, 8–18 × 5–15 μm, each with several

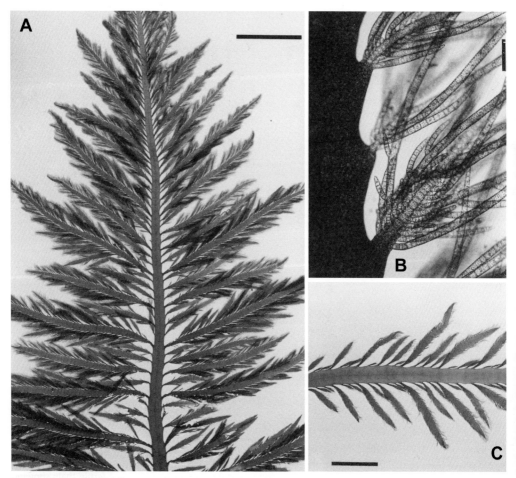

Fig. 46. *Desmarestia ligulata*
A. Portion of thallus. Bar = 20 mm. B. Edge of branch showing tufts of filaments. Bar = 50 μm. C. Portion of branch. Bar = 10 mm.

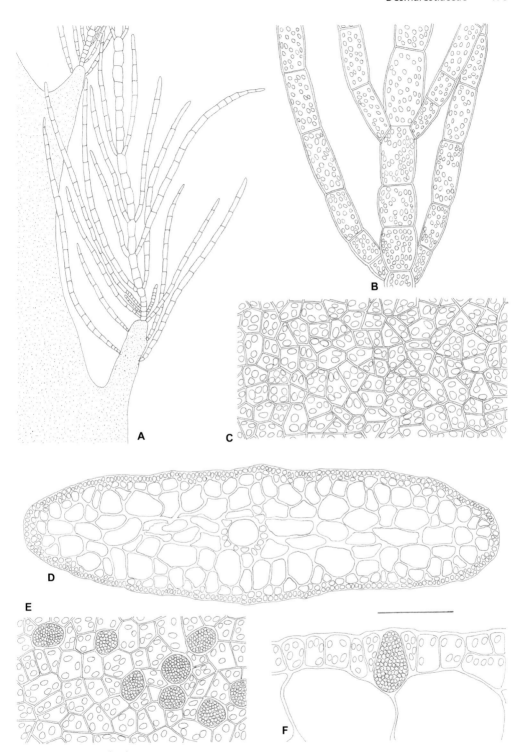

Fig. 47. *Desmarestia ligulata*

A. Thallus margin showing tufts of branched filaments. B. Portion of branched filament. C. Surface view of vegetative thallus showing cells with numerous discoid plastids. D. TS of vegetative thallus. Note central axial cell. E. Surface view of fertile thallus showing unilocular sporangia. F. Unilocular sporangia associated with surface cells. Bar = 200 μm (A), = 50 μm (B), = 20 μm (C, E–F), = 100 μm (D).

discoid plastids; in section, thallus pseudoparenchymatous with a large central axial cell surrounded by a broad cortex of large, thin-walled, colourless cells enclosed by 1–2 layers of small, pigmented cells.

Unilocular sporangia scattered over both sides of blade surface, rounded or oval in shape, slightly darker than surrounding vegetative cells; in vertical section, oval in shape, taller than wide, partly immersed, 15–21 × 9–12 μm.

Epilithic, lower eulittoral pools, more commonly sublittoral to at least 12 m.

Generally distributed around Britain and Ireland, although appears to be much more common on southern and western shores.

Annual, spring and summer; in spring and early summer ultimate branches bear tufts of hair-like branched filaments; these are shed during the summer. Drift thalli commonly washed ashore during the late summer/autumn period.

Culture studies by Nakahara (1984) revealed Japanese isolates of *D. ligulata* to have a heteromorphic life history with the macroscopic sporophyte alternating with a microscopic, monoecious gametophyte.

Abbott & Hollenberg (1976), pp. 222, 224, 225, fig. 185; Adams (1994), p. 92, pl. 28; Anderson (1985), pp. 437–47; Brodie *et al.* (2016), p. 1022; Bunker *et al.* (2017), p. 302; Baardseth (1941), pp. 121, 122; Chapman (1972b), pp. 1–20; Coppejans & Kling (1995) pp. 176, 177, pl. 5(A, B); Cormaci *et al.* (2012), pp. 40, 42, pl. 3(1–3); Dizerbo & Herpe (2007), p. 106, pl. 35(2); Fletcher (1987), pp. 282–5, figs 84, 85, pl. 14; Fletcher in Hardy & Guiry (2003), p. 241; Gallardo García *et al.* (2016), p. 34; Gayral (1958), p. 234, pl. 37, (1966), p. 287, pl. 51; Guiry (2012), p. 162; Hamel (1931–39), pp. 283, 284; Harvey (1846–51), pl. 115; Lee & Hwang (2010), pp. 60, 61, fig. 33; Moe & Silva (1977), pp. 159–67; Morton (1994), p. 39; Nakahara (1984), pp. 102–8, figs 14–21, pls IV(H–L), V(A–H); Newton (1931), pp. 164, 165; Oates & Cole (1990), pp. 529–32; Ribera *et al.* (1992), p. 117; Rueness (1977), p. 172, pl. 23(2); Sato *et al.* (1976), pp. 51–6; Scrosati (1989), pp. 89–98; Womersley (1987), pp. 264–6, fig. 96(A–E); Yang *et al.* (2014a), p. 771, (2014b), pp. 149–66.

Desmarestia viridis (O.F. Müller) J.V. Lamouroux (1813), p. 45.   Figs 48, 49

Fucus viridis O.F. Müller (1782), p. 5, pl. DCCCXXXVI; *Iridea fluitans* Stackhouse (1816), p. xii; *Chordaria viridis* (O.F. Müller) C. Agardh (1817), p. 14; *Gigartina viridis* (O.F. Müller) Lyngbye (1819), p. 44; *Sporochnus viridis* (O.F. Müller) Greville (1830), p. 39; *Dichloria viridis* (O.F. Müller) Greville (1830), pp. xl, 39; *Spinularius viridis* (O.F. Müller) Ruprecht (1850), p. 183; *Desmarestia pacifica* Setchell & N.L. Gardner (1924), p. 6; *Desmarestia media* var. *tenuis* Setchell & N.L. Gardner (1924), p. 7; *Krobylopteris oltmannsii* Schmidt (1942), fig. 7A, pl. II, fig. 1.

Thalli forming erect, solitary, light golden brown, much-branched, terete thalli, to 0.5 m long, arising from a small, 2–3 cm long, terete stipe and attached by a bulbous or flattened holdfast; thalli cartilaginous below, becoming soft and flaccid above, slightly lubricous, main axis distinct to 1(–2) mm wide, slightly compressed, giving rise to regularly opposite, distichous, divaricate, narrowing branches of limited growth, similarly branched to several orders, with or without terminal tufts of branched, hair-like filaments; terminal filaments to 2(–3) mm long, uniseriate, oppositely branched, comprising cells 18–75 × 15–30 μm, 1(–1.5) diameters long below, 2–4 diameters long above, each enclosing numerous discoid plastids without pyrenoids; surface cells of axes irregular in shape, irregularly arranged, although in longitudinal rows in terminal branches, 12–20 × 7–15 μm, each enclosing numerous discoid plastids without pyrenoids; in section, thallus structure pseudoparenchymatous, with a large, central, axial cell surrounded by a broad cortex of large, colourless cells, enclosed by 1–2 layers of small, pigmented cells.

Unilocular sporangia scattered over thallus surface, slightly larger and darker than accompanying vegetative cells, oval, partly immersed in vertical section, approximately 10–15 μm in diameter.

Epilithic, occasionally epiphytic on larger algae, in shaded lower eulittoral pools and in the sublittoral to at least 10 m.

Fig. 48. *Desmarestia viridis*
Habit of thallus. Bar = 20 mm.

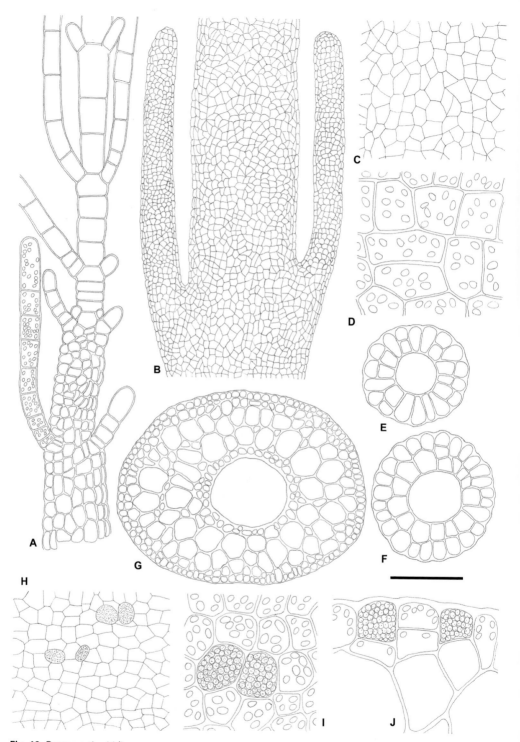

Fig. 49. *Desmarestia viridis*
A. Terminal branch region. B. Portion of oppositely branched thallus. C, D. Surface view of thallus showing cells with numerous discoid plastids. E–G. TS of different sized axes. H, I. Surface view of thallus with unilocular sporangia. J. TS of fertile thallus showing unilocular sporangia. Bar = 50 µm (A, C, E, F, H), = 100 µm (B, G), = 20 µm (D, I, J).

Generally distributed around Britain and Ireland.

Annual, spring and summer; in young, developing thalli terminal fascicles of filaments are present, rendering the thalli softer and lubricous; these fascicles are usually shed by late summer.

Culture studies by Abe (1938), Kornmann (1962a) and Nakahara (1984) on Japanese isolates of *D. viridis* revealed a heteromorphic life history with the macroscopic sporophyte alternating with a microscopic, monoecious, gametophyte. Nakahara also revealed parthenogenetic development of unfertilized eggs to produce haploid sporophytic thalli. Zoospores were produced from the latter without meiosis in the unilocular sporangia.

Abbott & Hollenberg (1976), p. 225, fig. 187; Abe (1938), pp. 475–82; Bartsch & Kuhlenkamp (2000), p. 165; Brodie *et al.* (2016), p. 1022; Bunker *et al.* (2017), p. 303; Chapman (1972a), pp. 225–31, figs 1–4, 10; Cormaci *et al.* (2012), p. 42, pl. 4(1–3); Dizerbo & Herpe (2007), pp. 106–8, pl. 35(3); Fletcher (1980), p. 44, pl. 16(4–5), (1987), pp. 285, 286, 288, figs 86, 87, pl. 15; Fletcher in Hardy & Guiry (2003), p. 241; Guiry (2012), p. 162; Hamel (1931–39), p. 282; Kornmann (1962a), pp. 287–92; Kornmann & Sahling (1977), p. 142, pl. 76(A–D); Kremer (1973), pp. 609, 610; Lamouroux (1813), p. 45; Lee & Hwang (2010), pp. 61–4, fig. 35(A–H); Mathieson & Dawes (2017), pp. 134, 135, pl. LVIII(10–13); Morton (1994), p. 39; Nakahara (1984), pp. 92–102, figs 4–13, pl. IV(A–G); Newton (1931), p. 164; Oates & Cole (1990), pp. 529–32; Pedersen (2011), p. 91, fig. 97; Ribera *et al.* (1992), p. 117; Rosenvinge & Lund (1947), pp. 54, 55; Rueness (1977), p. 172, pl. 23(3); Stegenga & Mol (1983), p. 105, pl. 116(4); Stegenga *et al.* (1997a), p. 22; Taylor (1957), pp. 160, 161, pl. 13(3).

Ectocarpales Bessey

ECTOCARPALES Bessey (1907), p. 288.

Chordariales Setchell & N.L. Gardner (1925), p. 570; *Dictyosiphonales* Setchell & N.L. Gardner (1925), p. 586; *Punctariales* Kylin (1933), p. 93; *Scytosiphonales* J. Feldmann (1949), p. 112.

Thallus an unmodified or modified heterotrichous filament.

Unmodified thalli erect, free, solitary or gregarious and tufted/turf-like, from less than a few mm to several cm in length, to over 1 m in sheltered conditions, usually attached at the base, although sometimes loose-lying and/or free-floating; thalli branched, filamentous, uniseriate, rarely biseriate/multiseriate as a result of some longitudinal divisions, sparingly or copiously branched, with or without corticating filaments; basal system of creeping, branched, filaments derived from both primary and secondary rhizoids, either spreading on host or substratum surface or partly or entirely endophytic in host thalli; growth apical, diffuse or intercalary, with or without obvious meristematic zones, often occurring at the base of branches; hairs usually present, either false or true, with basal meristem; plastids discoid or ribbon-shaped, simple or branched, with pedunculate pyrenoids; reproduction by unilocular sporangia, uniseriate to pluriseriate plurilocular sporangia and gametangia, or monosporangia, borne laterally, terminally or, less frequently, intercalary; gametangia isomorphic, producing isogametes or heteromorphic producing anisogametes; life history diplohaplontic, isomorphic or slightly heteromorphic, involving the production of haploid zoospores by meiosis in the unilocular sporangia, and haploid gametes in the plurilocular gametangia, plus direct, asexual life histories involving germination of zoospores released from the plurilocular sporangia.

Modified thalli include:

1. *Discoid/pulvinate/globose/irregularly lobed forms* resulting from either a regular or irregular lateral fusion of the basal filaments, the constituent cells remaining uniseriate, more rarely becoming biseriate, typically giving rise – by vertical growth – to erect, generally short, branched or unbranched, uniseriate filaments, approximately equal in height and often loosely associated in a mucilaginous matrix; filaments sometimes differentiated into an inner medulla of large, long, usually cylindrical, branched, colourless cells which give rise distally to an outer, more compact cortex of usually branched

filaments of small, more pigmented cells; true hairs common, borne on the terminal cortical filaments, also from the outer medullary cells; growth apical, subapical, intercalary, diffuse or trichothallic; plastids discoid, with pedunculate pyrenoids; reproduction by unilocular sporangia and uniseriate plurilocular sporangia borne laterally or terminally on the ultimate filaments, with or without paraphyses; gametangia borne on a filamentous, microscopic gametophyte generation, uniseriate, producing isogametes; life history diplohaplontic, heteromorphic.

2. *Upright, pseudoparenchymatous forms* comprising an inner core of one or more erect, central, axial filaments of unlimited growth giving rise to lateral branches of limited growth; all filaments monosiphonous and loosely or firmly bound in a gelatinous or cartilaginous matrix; thalli usually macroscopic, cylindrical, simple or branched, solid or hollow, with the central, axial filament(s) forming a medulla of large, usually elongated, colourless cells, giving rise peripherally to a simple or branched, filamentous, outward-radiating cortex of few to many shorter, smaller, in some genera well compacted, pigmented cells, sometimes referred to as assimilatory threads; attachment by a holdfast of compacted rhizoidal filaments; true hairs present or absent; growth of axial filaments monopodial or sympodial, sometimes trichothallic, with an intercalary, subapical growth zone; cells with discoid plastids and pedunculate pyrenoids; reproduction by unilocular sporangia, usually borne at the base of the terminal branches, and by uniseriate plurilocular sporangia borne laterally or terminally in the terminal branches; life history diplohaplontic and heteromorphic, the macroscopic sporophyte generation alternating with a microscopic, gametophyte generation bearing gametangia; gametes isogamous, rarely anisogamous; direct development from plurilocular sporangial spores and parthenogenetic development of gametes also reported.

3. *Upright, parenchymatous forms* resulting from more or less abundant periclinal and anticlinal cell divisions within an erect, heterotrichous filament producing macrothalli of diverse form and structure, which can range from simple or branched, terete to flattened, clavate or globular, solid or hollow, with or without a terminal uniseriate filament giving rise to a hair, and with, or without, internal differentiation into a medulla and cortex; growth apical, subapical or by intercalary cell divisions, with or without, solitary or clustered, true hairs; cells with discoid plastids and pedunculate pyrenoids; unilocular and plurilocular sporangia solitary or in sori, embedded or superficial, with or without accompanying paraphyses and hairs; reproduction by unilocular and plurilocular sporangia, usually borne on the thallus surface.

In general, thallus growth in the discoid, pulvinate and pseudoparenchymatous forms is localized at or near the apex, and growth in the parenchymatous forms more diffuse. Hairs can be present or absent, either false or true with basal meristem, while plastids are mainly discoid, more rarely ribbon-shaped, with pedunculate pyrenoids. Reproductive organs are terminal, lateral or intercalary. Asexual reproduction is by zoospores from unilocular and plurilocular sporangia or, rarely, by monospores. Sexual reproduction is usually isogamous, occasionally anisogamous by gametes from plurilocular gametangia. Observations on life histories in culture have shown that certain parenchymatous forms may possess alternative crustose or filamentous microthalli which may be sexual or asexual in function; the microthallus may be persistent, given appropriate environmental conditions, or induced to metamorphose into the macrothallus. The pseudoparenchymatous and the largest of the heterotrichous forms may also possess filamentous microthalli that may be sexual (prothalli) or asexual (plethysmothalli) in function.

The circumscription of the Ectocarpales adopted in the present treatise is similar to that proposed by Fritsch (1945). Many workers preferred to distinguish at ordinal rank the crustose (Ralfsiales: Nakamura 1972), pseudoparenchymatous (Chordariales: Setchell &

N.L. Gardner 1925) and parenchymatous forms (Dictyosiphonales: Setchell & N.L. Gardner 1925; Punctariales: Kylin 1933; Scytosiphonales: Feldmann 1949). However, the relationships between the main thallus forms, as revealed by life-history studies, and the ease with which many forms may be interchanged, suggested that these classifications were not natural. Consequently, a broader circumscription of the Ectocarpales was required, incorporating all the above forms, as originally favoured by Fritsch.

This broader concept of the Ectocarpales and the artificial nature of previous classifications based on thallus organization and life histories, has since found support in a number of molecular studies (Tan & Druehl 1993; Siemer *et al.* 1998; Rousseau & de Reviers 1999a; de Reviers & Rousseau 1999; Peters & Ramírez 2001; Draisma *et al.* 2003; Cho *et al.* 2004; Phillips *et al.* 2008a; Silberfeld *et al.* 2014). Notable was the pioneering work of Tan & Druehl (1993), who separated the brown algae with stalked pyrenoids from all the other brown algae. In a later reinvestigation into the concept of the Ectocarpales, using molecular analyses and a combination of diagnostic features (one to several plastids with one to several stalked and exserted pyrenoids, isogamy and anisogamy, never oogamy, and a haplodiplontic life history), Rousseau & de Reviers (1999a) merged all the families of the orders Chordariales, Dictyosiphonales (including the Punctariales), Ectocarpales and Scytosiphonales, with the exception of the family Ralfsiaceae, into an emended Ectocarpales *sensu lato*. A previous study by Tan & Druehl (1994), of the type species *Ralfsia fungiformis* of the Ralfsiales (proposed by Nakamura 1972 but validated by Lim *et al.* 2007) revealed it should be placed outside the Ectocarpales. Molecular support for this concept of the Ectocarpales and its monophyletic lineage, using more extensive DNA sequence data sets was further provided by Siemer *et al.* (1998), Peters & Ramírez (2001), Draisma *et al.* (2003), Cho *et al.* (2004), Phillips *et al.* (2008a) and Silberfeld *et al.* (2014). Siemer *et al.* (1998) additionally confirmed the exclusion of the Tilopteridales, lacking a pyrenoid, from this broadened Ectocarpales, while Peters & Ramírez (2001), using a combination of new DNA sequence data and non-molecular characters such as the life history and plastid structure, proposed that all the genera, previously distributed in the 23 formerly accepted families within the Ectocarpales be placed in the following five families: Acinetosporaceae, Adenocystaceae (erected by Rousseau *et al.* 2000), Chordariaceae, Ectocarpaceae and Scytosiphonaceae. This found support in a study by Phillips *et al.* (2008b). A further family, Petrospongiaceae, was later added by Racault *et al.* (2009). This reduction in the number of Ectocarpales families to six was later supported in a comprehensive review article by Silberfeld *et al.* (2014).

It is also noteworthy that while Fritsch, and later authors such as van den Hoek (1995), considered the Ectocarpales *sensu stricto* to be the most primitive/ancestral, molecular studies (Tan & Druehl 1993, 1994) have revealed this not to be the case. The Ectocarpales that possess pyrenoids are the sister taxon of those without pyrenoids or with a reduced pyrenoid, which form the 'crown' radiation of brown algae (see Cho *et al.* 2004).

With the removal of the crustose forms into other orders/families and the addition of the new family Petrospongiaceae, the order Ectocarpales comprises five families in Britain and Ireland:

1. Acinetosporaceae
2. Chordariaceae
3. Ectocarpaceae
4. Petrospongiaceae
5. Scytosiphonaceae

The richest in genera is the Chordariaceae (47) followed by the Acinetosporaceae (6), Scytosiphonaceae (6), Ectocarpaceae (3) and the Petrospongiaceae (1).

Acinetosporaceae G. Hamel ex J. Feldmann

ACINETOSPORACEAE G. Hamel ex J. Feldmann (1937), p. 250.

Thalli erect, free, solitary or gregarious, densely tufted, a few mm to 1 m in length, simple or branched, arising from either a branched filamentous or discoid base; erect thalli either filamentous and uniseriate throughout, or becoming pluriseriate and parenchymatous above, either in localized areas or more extensively to form erect shoots of diverse form, with or without internal differentiation. Growth of prostrate filaments apical and/or diffuse, that of the upright filaments intercalary, diffuse or in short meristematic regions. True hairs with basal meristem absent, with filaments sometimes terminating in colourless pseudohairs with reduced numbers of plastids. Plastids disc-shaped, ovoid, or slightly elongated, each with conspicuous pyrenoids.

Reproduction is mainly by means of swarmers from unilocular sporangia and plurilocular sporangia and gametes from gametangia, the latter sometimes with different-sized loculi. Asexual reproduction is by means of zoospores, or (rarely) non-motile monospores. Sexual reproduction is by isogametes or anisogametes. Reproductive structures may be terminal, lateral or intercalary in position, the latter by transformation of vegetative cells.

Life histories are diplohaplontic, isomorphic or slightly heteromorphic. Meiosis occurs at initiation of unilocular sporangia which are, therefore, found only on diploid thalli, usually accompanied by plurilocular sporangia. Haploid thalli bear plurilocular gametangia only. Sometimes direct life histories are present. Asexual reproduction is widespread and common in both ploidies. Sexual reproduction is affected by production of sexual attractants by female gametes.

Based on molecular studies (Siemer *et al.* 1998; Peters & Ramírez 2001), six genera, *Acinetospora, Feldmannia, Geminocarpus, Hincksia, Pogotrichum* and *Pylaiella*, were included within this family, while later, an additional two genera, *Internoretia* and *Herponema*, were included by Silberfeld *et al.* (2014), the latter genus as a result of molecular studies by Silberfeld *et al.* (2011). Six of these genera (*Acinetospora, Feldmannia, Herponema, Hincksia, Pogotrichum* and *Pylaiella*) are present on the coasts of Britain and Ireland.

Acinetospora Bornet

ACINETOSPORA Bornet (1891), p. 370.

Type species: *A. pusilla* (A.W. Griffiths ex Harvey) De Toni (De Toni 1895), p. 566 (= *A. crinita*), (Carmichael) Sauvageau (1899a), p. 118.

Thalli epilithic, more usually epiphytic, erect, filamentous, branched, monosiphonous, usually in the form of dense interwoven tufts; branching irregular, more frequent below, much more sparse and widely spaced above, usually giving rise to short, few-celled branchlets produced at right angles to the parental filament; some branchlets slightly curved ('crampons') which probably assist in attachment to other algae; growth intercalary with frequent, reasonably distinctive meristematic zones of short, tightly packed cells; plastids discoid to oblong, with pyrenoids; false hairs sometimes terminating branches, especially towards the base of thalli.

Plurilocular sporangia, unilocular sporangia and monosporangia irregularly scattered on erect filaments, usually isolated, sometimes in small groups, sessile or pedicellate on 1–5+ celled stalks, given off at right angles to parental filament; plurilocular sporangia often curved, obovate to conical, with large loculi; unilocular sporangia spherical to subspherical; monosporangia ovoid to spherical.

One species in Britain and Ireland.

Acinetospora crinita (Carmichael) Sauvageau (1899a), p. 118.   Fig. 50

Ectocarpus crinitus Carmichael (1833), p. 326; *Ectocarpus pusillus* Harvey (1841), p. 41; *Ectocarpus vidovichii* Meneghini ex Zanardini (1843), p. 41; *Acinetospora pusilla* (A.W. Griffiths ex Harvey) Bornet (1891), p. 18; *Haplospora vidovichii* (Meneghini) Bornet (1891), p. 363; *Acinetospora pusilla* (A.W. Griffiths ex Harvey) Buffham (1893), pp. 88, 89; *Acinetospora pusilla* (A.W. Griffiths ex Harvey) De Toni (1895), p. 566; *Ectocarpus pusillus* var. *riparia* Sauvageau (1895b), pp. 274–87; *Acinetospora vidovichii* (Meneghini) Sauvageau (1898a), p. 1581; *Acinetospora pusilla* var. *crinita* (Carmichael) Batters (1902), p. 53.

Thalli epiphytic, occasionally epilithic, forming flaccid, branched filaments, sometimes single and arising from a basally attached rhizoid cell, or much more commonly forming dense, interwoven, fleece-like, filmy tufts arising from a basal system of outward-spreading, branched, filamentous rhizoids, some of the latter being secondarily produced as downward-growing extensions of basal cells of the erect filaments; tufts usually up to 5–10 cm in height, reaching up to 1 m or more in very sheltered areas or as free-floating populations; filaments distinctly light brown in colour, entirely uniseriate throughout, more densely branched below, much more sparingly above with extensive unbranched regions; branching irregular, with branches sometimes clustered, more usually scattered, variable in length, ranging from short, bud-like extensions of the parental axial cell to longer multicellular filaments, and frequently in the form of straight or curved, short, spine-like branchlets ('crampons'), 2–3 cells long with obtuse end cells and characteristically given off at right angles to the parental filaments; basal branches usually longer, and also curved, with the curvature probably assisting in binding the filaments both together and to other algae; vegetative cells 15–160 × 18–40 μm, quadrate to mainly rectangular in shape, 0.5–5 diameters long; intercalary growth occurring frequently along filaments, consisting of fairly densely packed, short, quadrate to sub-quadrate, photosynthetic cells; plastids sparsely distributed in long cells, more noticeably packed in short cells, usually discoid or oblong, more rarely elongate-oblong/ribbon-shaped, with or without branches, and with associated pyrenoids; cells often showing centrally situated clusters of physodes; basally positioned branches sometimes tapering into pseudohairs.

Thalli bearing plurilocular sporangia, unilocular sporangia and/or monosporangia; plurilocular sporangia common, irregularly positioned and scattered, although more usually distributed towards basal regions of filaments, more sparse above, sessile or pedicellate on 1–5-celled stalks, arising perpendicular to the parental filament, straight or often curved, ovate-oblong to conical, 75–128 × 24–42 μm with large loculi; unilocular sporangia infrequent, sessile or pedicellate, spherical to subspherical, arising perpendicular to parental filament, 25–50 μm in diameter; monosporangia common, solitary or in groups, borne at right angles to parental filament, sessile or pedicellate, ovoid to spherical, 30–48 × 25–35 μm, containing a single large, non-motile spore.

Lithophytic, or more commonly epiphytic, on various hosts, upper to lower eulittoral, exposed or in pools, sublittoral, particularly under sheltered conditions; extensive growths have been observed growing out from the sides of floating pontoons in some very sheltered yachting marinas.

Common and widely distributed around Britain and Ireland.

Annual; mainly spring and early summer but also found in autumn.

The main distinguishing characteristics of this species include the fleece-like, intricately matted habit of mature thalli with no obvious polarity of development, the light brown colour of the filaments, the considerable length of the constituent cells (> 160 μm), the scarcity of branch formation over much of the length of the filaments, with the branches usually limited in growth to a few cells only and emerging at right angles to the parental filament, the large loculi of the plurilocular sporangia and the occurrence of monosporangia.

The taxonomic affinity of this species is uncertain. Based on the presence of monosporangia, Sauvageau (1928a), included it within the family Tilopteridaceae, order Tilopteridales, along

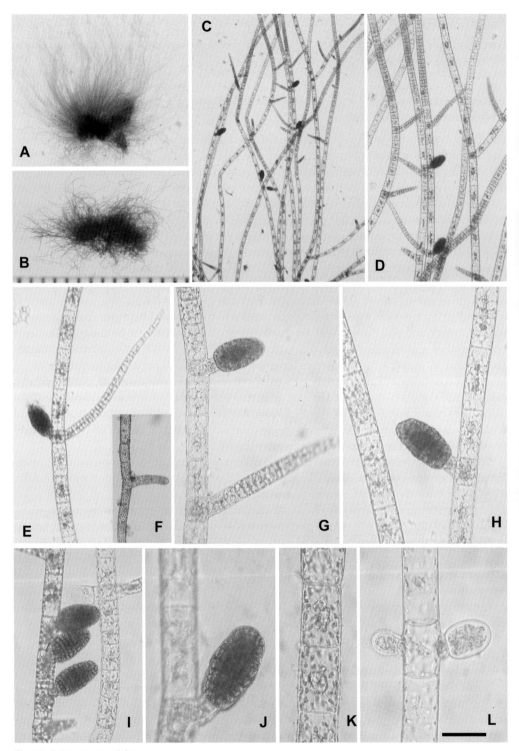

Fig. 50. *Acinetospora crinita*
A, B. Loose tangled habit of thalli. C–E. Filaments with plurilocular sporangia and small lateral branchlets. F. Part of filament with slightly curved lateral branchlet. G–J. Part of filaments with plurilocular sporangia. K. Portion of erect filament showing enclosed discoid plastids. L. Portion of erect filament with developing monosporangia.
Bar in L = 9 mm (A), = 0.6 mm (C), = 300 μm (D), = 180 μm (E, F), = 90 μm (G–I), = 45 μm (J–L).

with species of the genera *Haplospora* and *Tilopteris*. This was accepted by some early workers (e.g. Newton 1931; Knight & Parke 1931). However, subsequently most authors placed the species withing the family Ectocarpaceae following Kornmann's (1953) suggestion that it might be the sporophytic phase in the life history of a *Feldmannia* species (see also Knoepffler-Péguy 1974, 1977). Based on molecular work (Siemer *et al.* 1998; Peters & Ramírez 2001), this species was subsequently placed into a separate family, Acinetosporaceae, within the Ectocarpales.

Amsler (1984), pp. 377–82, figs 4–5; Ardré (1970), pp. 242, 243, pl. 35(8); Bartsch & Kuhlenkamp (2000), p. 164; Batters (1885), p. 537; Blomquist (1955), pp. 46–9, figs 1–10; Børgesen (1926), pp. 30–5, figs 15–17; Bornet (1892a), pp. 356–8, pls 7(1–5), 8(1–3); Brodie *et al.* (2016), p. 1019; Buffham (1893), pp. 88, 89; Cardinal (1964), pp. 69–72, fig. 37(A–F); Carmichael ex Harvey in Hooker (1833), p. 326, pl. 330(112–13); Clayton (1974), pp. 749–51, figs 1(a–f); Cormaci *et al.* (2012), pp. 87, 88, pl. 18(1–5); P. Crouan & H. Crouan (1852), pl. 15; Dizerbo & Herpe (2007), p. 60, pl. 17(1); Fletcher (1980), p. 37, pl. 10(1–3); Fletcher in Hardy & Guiry (2003), p. 223; Gallardo García *et al.* (2016), p. 34; Guiry (2012), p. 134; Hamel (1931–39), pp. 75–80, fig. 22(1–5); Harvey (1846–51), pl. 153; Kim (2010), pp. 38–41, fig. 14(A–D); Knoepffler-Péguy (1972), pp. 101–4, (1973), pp. 171–89, (1974), pp. 43–72, figs 1–7, pl. 1, (1977), pp. 111–28; Kornmann (1953), pp. 205–24, figs 1–14; Kornmann & Sahling (1977), p. 115, fig. 60(A–H); Morton (1994), p. 31; Müller (1986), pp. 219–24; Newton (1931), pp. 210, 211, fig. 133(A–C); Pedersen & Kristiansen (2001), pp. 209–18; Peters & Ramírez (2001), p. 196; Ribera *et al.* (1992), p. 111; Rosenvinge & Lund (1941), pp. 65, 66, fig. 35; Rueness *et al.* (2001), p. 14; Sauvageau (1895b), pp. 274–87, figs 1–6, (1899), pp. 107–26, figs 1–4; Schmidt (1940), pp. 23–8; Schneider & Searles (1991), pp. 111, 112, figs 112, 113; Siemer *et al.* (1998), pp. 1038–48; Stegenga & Mol (1983), p. 72, pl. 18(1–2); Stegenga *et al.* (1997a), p. 15, (1997b), pp. 135–7, pl. 32(1–4); Taylor (1960), p. 214; Van den Hoek (1958b), p. 193, fig. 2; Womersley (1987), pp. 46, 47, figs 10(A), 11(A–E).

Feldmannia Hamel

FELDMANNIA Hamel (1931–39), p. xi.

Type species: *F. lebelii* (P. Crouan & H. Crouan) Hamel (1939), p. xvii.

Thalli epiphytic, with basal regions endophytic, less commonly epilithic and epizoic, small, less than 5 cm in height, densely tufted, filamentous, monosiphonous, arising from a much-branched, basal system of rhizoidal filaments, penetrating quite deeply and spreading in host species if epiphytic; erect filaments long, infrequently branched, with basal intercalary meristem, uniform in diameter throughout, and often narrowing terminally into long, false hairs; terminal laterals sometimes with additional meristematic zone; branches usually confined to lower region of main filaments, with branching irregular, alternate, unilateral, spiral or opposite; plastids discoid or slightly elongate, with associated pyrenoids.

Plurilocular sporangia common, multiseriate, pedicellate, more rarely sessile, largely basally positioned occurring below meristematic growth zones on erect filaments; unilocular sporangia uncommon, discoid to spherical, usually associated with the plurilocular sporangia.

Feldmannia is quite similar in appearance to *Hincksia*, but differs in most species (an exception being *F. mitchelliae*) by the tufts usually being much smaller, much less obviously branched, with branching confined to the base of the tufts only. The resultant erect filaments are long, linear, simple with a distinct, basal, intercalary, meristematic region and a terminal, well-developed, colourless pseudohair. The sporangia are solitary, rarely in series, usually pedicellate and borne at the base of the erect filaments below the meristematic regions.

As described for *Hincksia*, some species of *Feldmannia* (*F. lebelii*, *F. mitchelliae*, *F. padinae*) have been reported with plurilocular sporangia comprising different-sized loculi. These have been variously described as (neutral) plurilocular sporangia (sometimes wrongly termed meiosporangia), megasporangia and microsporangia, the latter two representing female and male gametangia respectively and indicating that the basic life history is isomorphic, involving anisogamy.

Five species in Britain and Ireland, one species in Britain only. One species (*F. mitchelliae*) represents a transfer from the genus *Hincksia*, based on molecular studies by Kim (2014).

KEY TO SPECIES

1. Erect filaments little branched, to 1(–2) orders; sporangia and branching of
    erect filaments mainly confined to the basal/lower regions of thalli 2
    Erect filaments much branched, to several orders; sporangia and branching of
    erect filaments distributed throughout the thalli........................*F. mitchelliae*

2. Branching irregular, alternate, rarely opposite; plurilocular sporangia
    irregularly formed, elongate cylindrical/conical, usually recurved upwards
    towards parental filament 3
    Branching mainly opposite; plurilocular sporangia frequently opposite, or
    opposite a branch, globose, usually divaricate and at right angles to parental
    filament... *F. paradoxa*

3. Thalli forming fairly discrete, erect tufts of filaments, on various hosts, to
    1–2(–5) cm high... 4
    Thalli forming continuous, turf-like growths on hosts, 2–3(–4) mm high............ 5

4. Thalli mainly epiphytic on *Ericaria selaginoides* and *Sargassum muticum*,
    emerging from old conceptacle ostioles; branching alternate, sometimes
    opposite; upper filament cells up to 30 μm in diameter......................*F. lebelii*
    Thalli epiphytic on various hosts; branching irregular or pseudodichotomous;
    upper filament cells up to 20 μm in diameter.........................*F. irregularis*

5. Thalli epiphytic on *Codium* spp., with extensive rhizoidal system penetrating
    down between the host's utricles; plurilocular and unilocular sporangia often
    occurring on same thalli; plurilocular sporangia on 1–3-celled stalks, conical
    to elongate-cylindrical/lanceolate....................................... *F. simplex*
    Thalli epiphytic on *Padina pavonica*, with extensive system of rhizoidal
    filaments spreading horizontally between the layers of 3 cells; plurilocular
    sporangia with a single-celled stalk, ovate to elongate-ovate; unilocular
    sporangia unknown .... *F. padinae*

Feldmannia irregularis (Kützing) Hamel (1939), p. xvii.   Fig. 51

Ectocarpus irregularis Kützing (1845), p. 234; *Ectocarpus arabicus* Figari & De Notaris (1853), p. 169; *Ectocarpus arabicus* Kützing (1855), p. 21, nom. illeg.; *Ectocarpus guadelupensis* P. Crouan & H. Crouan in Schramm & Mazé (1865); *Ectocarpus simpliciusculus* var. *vitiensis* Askenasy (1888), p. 20; *Ectocarpus mucronatus* De A. Saunders (1898), p. 152; *Ectocarpus coniger* Børgesen (1935), p. 31; *Ectocarpus nanus* Levring (1938), p. 18; *Ectocarpus coniger* var. *arabicus* Nasr (1941), pp. 61, 62; *Ectocarpus izuensis* Segawa (1941), p. 254; *Ectocarpus lebelii* var. *agigensis* Celan (1964), p. 33; *Giffordia irregularis* (Kützing) Joly (1965), pp. 72, 73; *Hincksia irregularis* (Kützing) Amsler in Schneider & Searles (1991), p. 120.

Thalli epiphytic, forming small, up to 5 mm high, fairly dense, light to yellow-brown, solitary tufts of several, long, erect, uniseriate, linear filaments arising from a basal system of outward-spreading, or deeply host-penetrating, intertwined, branched, attaching, rhizoidal filaments derived as basal extensions of the erect filaments; rhizoid cells colourless, irregularly shaped, somewhat contorted, (25–)35–86(–110) × (10–)14–18(–23) μm; erect filaments rarely and irregularly or pseudodichotomously branched, with branching largely confined to the basal regions, below a meristematic zone, with all filaments terminating in colourless, false hairs; growth by a single, intercalary meristematic zone of shorter, darker coloured, compacted cells, usually positioned singly at the base of the principal filaments and laterals; mid to upper filament cells cylindrical, 1–3 diameters long, (23–)30–45(–54) × (15–)17–19(–20) μm, lower filament cells cylindrical to barrel-shaped, 1–3 diameters long, (18–)36–72(–92) × (13–)18–25(–32) μm; hair

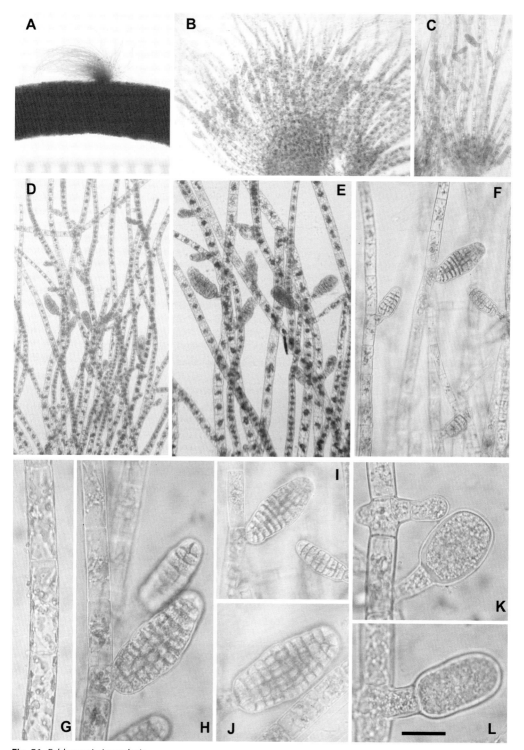

Fig. 51. *Feldmannia irregularis*
A. Habit of thallus on host. B. SP of thallus showing densely packed, erect filaments terminating in colourless hairs. C–F. Portions of erect thalli showing erect filaments with basally positioned lateral plurilocular sporangia. G. Portion of erect filament showing cells with discoid plastids. H–J. Plurilocular sporangia. K, L. Unilocular sporangia. Bar in L = 300 μm (B, D), = 180 μm (C, E), = 90 μm (F), = 45 μm (G, H, I, J, K, L).

cells cylindrical, 1–5(–9) diameters long, (22–)40–110(–153) × 13–23 µm; all cells with several discoid plastids with pyrenoids.

Plurilocular and unilocular sporangia occurring on the same individuals, borne at the base to mid region of the erect filaments below the meristematic zones; plurilocular sporangia common, sessile or more usually with a single-celled pedicel, sometimes terminal on branches, isolated, occasionally in small groups, ovate to elongate-lanceolate, (47–)63–88(–113) × (18–)25–38(–53) µm; unilocular sporangia rare, only observed on one occasion, usually with a single-celled pedicel, isolated, occasionally in groups or opposite, ovoid, (38–)45–55(–60) × (23–)25–30(–34) µm.

Epiphytic on various hosts, especially *Codium* spp., lower eulittoral pools and shallow sublittoral.

Widely distributed but rarely recorded around Britain (Channel Islands, Sussex, Ayr) and Ireland (Cork).

Summer annual.

Abbott & Hollenberg (1976), p. 136, fig. 99; Aisha & Shameel (2011), pp. 134–6, fig. 3(a–d); Ardré (1970), p. 241; Batters (1902), p. 31; Børgesen (1926), p. 25, figs 12–14, (1941), p. 23, figs 9, 10; Bornet (1892b), p. 245; Brodie *et al.* (2016), p. 15; Cardinal (1964), pp. 54, 55, fig. 29(A–K); Clayton (1974), pp. 777–9, figs 19(A–G), 20(A–H); Coppejans (1983), pl. 60; Cormaci *et al.* (2012), pp. 94, 95, pl. 19(3–7); Dawes & Mathieson (2008), p. 137, pl. 16(13); Dizerbo & Herpe (2007), p. 65, pl. 19(1); Ercegovic (1955), pp. 51–7, figs 24, 25; Feldmann (1937), p. 248; Fletcher in Hardy & Guiry (2003), p. 248; Gallardo García *et al.* (2016), p. 34; Guiry (2012), p. 134; Hamel (1931–39), pp. xvii, 45–7, figs 13, 61(F), (1939), p. 67; Joly (1965), pp. 72, 73, pl. VIII(111–19); Kim (2010), pp. 46–8, fig. 18(A–E); Kim & Lee (1994), pp. 154–9, figs 1(A–B), 2(A–C), 3(A–K), 4(A–G); Knoepffler-Péguy (1970), pp. 173, 174; Kuckuck (1963), pp. 371–3, fig. 6(A–F); Kützing (1845), p. 234, (1855), pl. 62(1); Lindauer *et al.* (1961), pp. 148, 149, fig. 8; Mathieson & Dawes (2017), p. 136, pl. XLII(5, 6); Miranda (1936), p. 369, fig 4E; Misra (1966), pp. 78, 79, fig. 33; Müller & Frenzer (1993), pp. 37–44; Nasr (1941), pp. 4–6, figs 3, 4; Newton (1931), p. 119; Norris (2010), pp. 167, 168, fig. 80; Oltmanns (1922), p. 9, fig. 294; Ribera *et al.* (1992), p. 111; Robledo *et al.* (1994), pp. 247–51, figs 1–3; Rosenvinge & Lund (1941), pp. 50–3, figs 23(A–D), 24(A–D); Rueness *et al.* (2001), p. 14; Sauvageau (1933a), pp. 101–11, figs 24–7; Schneider & Searles (1991), pp. 120–2, figs 130, 131; Stegenga & Mol (1996), p. 107, pl. II(2–3); Stegenga *et al.* (1997a), p. 15, (1997b), pp. 141, 142, pl. 34(2); Womersley (1987), p. 42, figs 6(D), 8(A–C).

Feldmannia lebelii (Areschoug ex P. Crouan & H. Crouan) Hamel (1939), p. 67.   Fig. 52

Ectocarpus lebelii J.E. Areschoug ex P. Crouan & H. Crouan (1867), p. 163; *Giffordia lebelii* (Areschoug ex P. Crouan & H. Crouan) Batters (1893a), p. 86; *Feldmannia caespitula* var. *lebelii* (J.E. Areschoug ex P. Crouan & H. Crouan) Knoepffler-Péguy (1970), p. 160.

Thalli epiphytic, forming small, erect tufts of uniseriate filaments up to approximately 1 cm in height, arising from a well-developed base of branched, contorted, rhizoidal filaments penetrating host tissue; rhizoidal cells 21–50 × 10–18 µm; erect filaments sparingly branched, with branching confined to the lower regions, alternate or opposite, giving rise to long, simple, terminal regions which usually end in a false hair; growth intercalary, occasionally by diffuse divisions, more usually by a single meristematic growth zone positioned towards the base of the long filaments, sometimes subterminal in position, comprising a series of shorter, darker cells; basal cells of erect filaments 1–3 diameters long, (28–)40–70(–112) × (21–)26–33(–38) µm; terminal cells of erect filaments 1–3 diameters long, (22–)35–48(–68) × (13–)19–24(–30) µm, each with several discoid plastids with pyrenoids; hair cells 4–8 diameters long, (85–)120–145(–163) × (18–)21–24(–26) µm.

Plurilocular sporangia common, solitary, confined to the basal region, always below the meristematic growth zones, sessile or with a single-celled, more rarely multi-celled, pedicel, recurved towards the parental filament, globular, ovoid to cylindrical, (45–)56–73(–118) × (24–)30–38(–45) µm and reported by various authors (Sauvageau 1897c; Hamel 1931–39), although not observed by the present author, to be of two types: neutral plurilocular sporangia and microsporangia (termed 'antheridies' by Hamel), which can be found on the same or

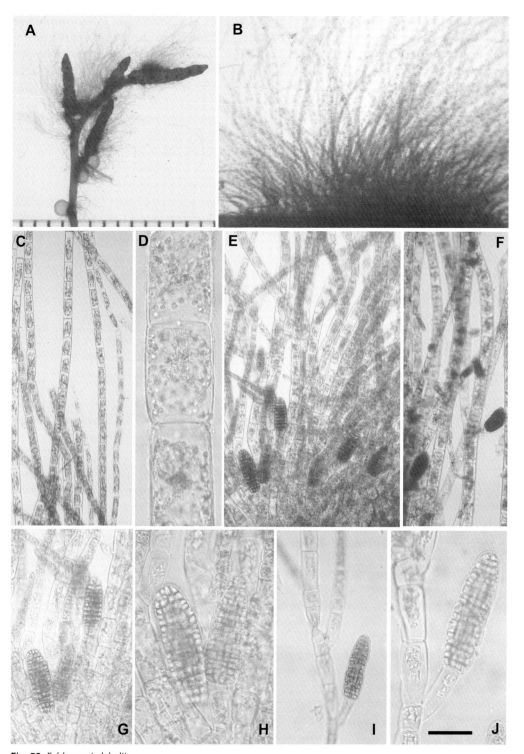

Fig. 52. *Feldmannia lebelii*

A, B. Habit of thalli epiphytic on receptacles of *Sargassum muticum*. C. Portions of erect filaments. D. Portion of erect filament showing cells with enclosed discoid plastids. E–J. Portions of erect filaments with basally positioned, lateral plurilocular sporangia. Bar in J = 400 μm (B), = 180 μm (C, E, F), = 90 μm (G, I), = 45 μm (D, H, J).

separate individuals, and distinguished by the number, size and colour of the component locules, which are more numerous, much smaller in dimensions and greyish in the microsporangia compared to the larger and darker brown locules of the plurilocular sporangia; unilocular sporangia unknown, although Hamel (1931–39) described globular unilocular sporangia-like cells, with fine granular contents in what he considered to be parasitized plurilocular sporangia.

Epiphytic on various hosts but notably on *Ericaria selaginoides*, emerging from the cryptostomata.

Rarely recorded and apparently restricted to the south coasts of England (Devon and Dorset) and Ireland (Wexford).

Probably a summer annual, with records for August and September.

Although unilocular sporangia are unknown in this species, on the basis of cultures studies, both Kornmann (1953) and Knoepffler-Péguy (1974, 1977) suggested that *F. lebelii* is the gametophyte phase of an *Acinetospora* sp.

Batters (1893a), p. 86, (1902), p. 34; Brodie *et al.* (2016), p. 1019; Cormaci *et al.* (2012), p. 96, pl. 20(1–4); P. Crouan & H. Crouan (1867), p. 163; Feldmann (1937), p. 247, fig. 35(D, E); Fletcher in Hardy & Guiry (2003), p. 249; Gallardo García *et al.* (2016), p. 34; Guiry (2012), p. 134; Hamel (1931–39), pp. 41–3, fig. 11(A–C); Holmes & Batters (1890), p. 79; Kim (2010), pp. 48–50, fig. 20; Knoepffler-Péguy (1970), pp. 137–88, (1974), pp. 43–72, (1977), pp. 111–18; Kornmann (1953), pp. 205–24; Mathieson & Dawes (2017), p. 136, pl. XLII(4); Miranda (1931), p. 24; Newton (1931), p. 123; Norton (1970b), p. 261; Ribera *et al.* (1992), p. 111; Sauvageau (1897c), pp. 5–14, figs 2–6; Womersley (1987), p. 44, fig. 9(E–J).

Feldmannia mitchelliae (Harvey) H.-S. Kim (2010), p. 51, figs 21–2.   Fig. 53

Ectocarpus mitchelliae Harvey (1852), pp. 142–3, pl. XII.G, figs 1–3; *Ectocarpus virescens* Thuret in Sauvageau (1896c), p. 124; *Ectocarpus mitchelliae* f. *brevicarpum* Børgesen (1939), pp. 76–8; *Giffordia mitchelliae* (Harvey) Hamel (1931–39), p. xiv; *Hincksia mitchelliae* (Harvey) P.C. Silva in Silva *et al.* (1987), p. 73.

Thalli epiphytic, more rarely epilithic and epizoic, forming dense, very flaccid, feathery, solitary, more rarely confluent tufts of uniseriate filaments up to 8(–10) cm long, light brown in colour becoming paler, olive-green when dried and arising from a spreading basal system of irregularly branched, rhizoidal filaments derived from the basal regions of the main axes; tufts much branched and sometimes terminally clustered, with branches irregular, spiral, alternate, sometimes secund, and gradually attenuating to either an acute tip, or commonly to a colourless, false hair; main filaments often covered at the base in a sheath of downward-growing, rhizoidal filaments; meristems either indistinct or appearing intercalary, scattered, or more usually positioned at the base of branches, and distinguished as a series of darker cells; cells usually longer than wide throughout, except in the growth zones, cylindrical to barrel-shaped, thick walled, 1–3 diameters long, 43–140 × 38–50 μm in the basal regions, 8–58 × 10–28 μm in the middle to upper regions, terminal cells acute, 4–9 diameters long, 20–65 × 5–12 μm; cells with several discoid, sometimes slightly irregular/elongate plastids and associated pyrenoids, densely packed in lower cells, more thinly distributed in upper cells rendering the cells almost colourless.

Plurilocular sporangia borne on the terminal branchlets, common, sessile, or sometimes one-celled pedicellate, either solitary or more usually borne in series, on the upper adaxial sides of branches, characteristically cigar-like in shape, ellipsoidal or linear cylindrical, usually with broadly rounded/obtuse, sometimes tapering ends, measuring 68–120 × 21–30 μm; unilocular sporangia rare, occurring with plurilocular sporangia, scattered, sessile, rarely pedicellate, ovoid, 35–50 × 15–25 μm.

Usually found in lower eulittoral pools and shallow eulittoral; also found on the sides of floating pontoons.

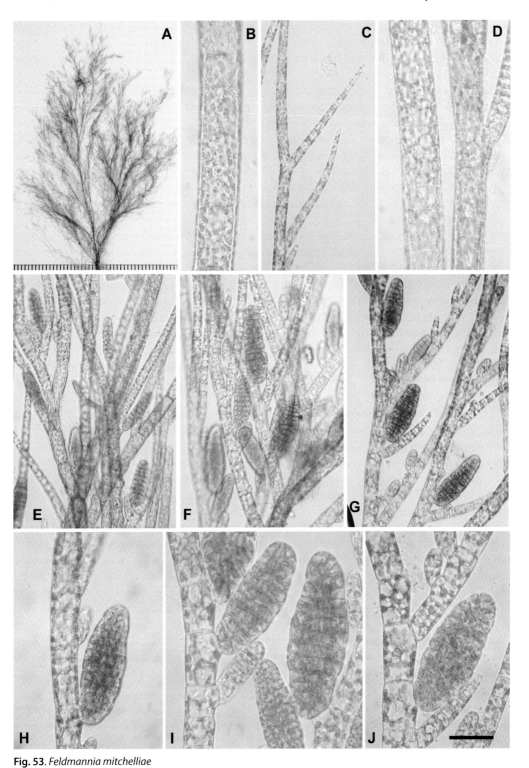

Fig. 53. *Feldmannia mitchelliae*
A. Habit of thallus. B. Portion of erect filament showing cells with enclosed discoid plastids. C, D. Portions of erect filaments showing lateral branching. E–J. Portions of erect filaments showing branching pattern and lateral plurilocular sporangia. Bar in J = 180 μm (E), = 90 μm (C, F, G), = 45 μm (B, D, H, I, J).

Rarely recorded in Britain and Ireland with a widespread distribution but probably more common than records suggest.

Appears to be a spring/summer annual, mainly May to October, rarely recorded at other times.

Several authorities have described different types of plurilocular sporangia in *F. mitchelliae*, usually referred to as microsporangia and megasporangia, comprising small and large locules respectively. These were reported by Sauvageau (1896), Hamel (1931–39) and Cardinal (1964) for the Atlantic French coast, Setchell & N.L. Gardner (1925) and Abbott & Hollenberg (1976) for the Californian coast, Børgesen (1926) for the Canary Islands, Ercegovic (1955) for the Adriatic, Clayton (1974) for southern Australia, and Müller (1969) for North Carolina. In the present study, only one type of plurilocular sporangium has been observed in collections from Britain and Ireland (cf. Schneider & Searles 1991; Edwards 1969; Womersley 1987) which is assumed to be sporangial.

Culture studies on *F. mitchelliae* have been carried out by Sauvageau (1933a), Edwards (1969) and Clayton (1974), with all three reporting a similar mode of development. Particularly noteworthy was the study by Sauvageau (1933a) who provided a much better insight into the life history of this alga, which is complicated by the different types of plurilocular 'sporangia' that have been described. Three types of plurilocular sporangia have been described for *H. mitchelliae*, based on the different sizes of the compartments; these are microsporangia and megasporangia, bearing small and large loculi respectively, which probably represent gametangia, and meiosporangia, made up of locules intermediate in size between those of the two gametangia, which probably represent an accessory means of asexual reproduction, and would be better termed plurilocular sporangia (= neutral plurangia according to Abbott & Hollenberg 1976, p. 143).

Unilocular sporangia have been rarely reported (Børgesen 1939; Sauvageau 1933a) and these probably play a minor role in the life history. Sauvageau (1933a) found thalli with meiosporangia and unilocular sporangia. He reported that in culture, swarmers from the unilocular sporangia germinated into thalli bearing megasporangia and microsporangia. Swarmers from the megasporangia were pigmented and sluggish in movement, while swarmers from the microsporangia were colourless and more rapid in movement. No sexual union was observed to form unilocular sporangia-bearing thalli and completing a likely isomorphic alternation of generations-type life history, involving anisogamy. As Hamel (1939, p. xv) pointed out, the microspores have the same form and dimensions as spores released from meiosporangia, but they are very pale, yellowish and with very small locules. The meiosporangia probably represent an accessory means of asexual reproduction occurring on both sporophyte and gametophyte generations; this is supported by reports of unilocular and meiosporangia occurring on the same thalli (Børgesen 1939) and of thalli bearing megasporangia or meiosporangia both possessing the same number of chromosomes (Svedelius 1928). Additional accessory reproduction is likely to be provided by the direct germination of both the megaspores and microspores (Sauvageau 1933). Vegetative propagation by fragmentation is also likely to play a role in the life history of the alga (Clayton 1974).

Abbott & Hollenberg (1976), p. 143, fig. 105; Aisha & Shameel (2011), pp. 140–2, fig. 6(a–c); André (1970), pp. 239, 240; Børgesen (1914), p. 162, figs 129, 130, (1926), pp. 18–23, figs 9, 10, (1930), p. 105, fig. 8; Brodie *et al.* (2016), p. 1019; Cardinal (1964), pp. 45–7, fig. 23(A–H); Clayton (1974), pp. 779–82, figs 21(A–H), 22(A–E); Coppejans (1983), pl. 61; Cormaci *et al.* (2012), p. 109, pl. 24(8–11); Dizerbo & Herpe (2007), p. 67, pl. 20(2); Edwards (1969), pp. 59–114, fig. 13; Ercegovic (1955), pp. 22–6, figs 10(a–e), 11(a–d); Feldmann (1937), p. 246, fig. 35(C); Flahault (1988), p. 383; Fletcher (1980), p. 40, pl. 13(5, 6); Fletcher in Hardy & Guiry (2003), p. 261; Gallardo García *et al.* (2016), p. 34; Guiry (2012), p. 134; Hamel (1931–39), pp. xiv–xv, fig. 61(C, D), pp. 29–32, figs 7, 8(I, II); Harvey (1852), pp. 142, 143, pl. 12(G), figs 1–3; Kim (2010), pp. 51–4, figs 2, 22; Kim & Lee (1992b), pp. 246–8, fig. 3(A–D); Kuckuck (1963), pp. 374–7, figs 7, 8; Mathieson & Dawes (2017), p. 138, pl. LXII(10, 11); Misra (1966), pp. 91–3, fig. 43(A–G); Müller (1969), p. 220, fig. 1; Newton (1931), p. 119; Norris (2010), pp. 170, 171, fig. 82; Ribera *et al.*

(1992), p. 112; Saunders (1898), p. 153, pl. 21; Sauvageau (1896c), pp. 98–107, figs 1–6, (1933a), pp. 67–79, figs 15, 16; Schneider & Searles (1991), p. 122, figs 132–4; Setchell & N.L. Gardner (1925), p. 428; Silva *et al.* (1987), p. 73; Stegenga & Mol (1983), p. 73, pls 20(4), 115(4); Stegenga *et al.* (1997a), p. 16; Svedelius (1928), p. 289, figs 1–4; Taylor (1957), p. 111, (1960), p. 206, pl. 29(1, 2); Womersley (1987), p. 52, figs 10(D), 12(E–G).

Feldmannia padinae (Buffham) Hamel (1939), p. 67.   Fig. 54

Giffordia padinae Buffham (1893), pl. 185, p. 89; *Ectocarpus padinae* (Buffham) Sauvageau (1896a), p. 268.

Thalli epiphytic on both sides of mature blades of host, forming small, isolated tufts or turf-like growths of long, erect, branched, linear, uniseriate filaments up to approximately 3–4 mm in height, attached at the base by downward-growing, secondary, branched, rhizoidal filaments which narrow and penetrate through the host surface, and then spread out extensively between the host cells; erect filaments irregularly and sparsely branched, with branching confined to the basal regions; growth of principal filaments and laterals intercalary from a single, basally positioned, short, meristematic growth zone of shorter, darker, compacted cells; erect filaments giving rise above to false hairs; upper cells of erect filaments 2–5 diameters long, (40–)60–80(–90) × (18–)22–32(–38) µm, basal cells of erect filaments 1–2 diameters long, (22–)35–66(–77) × (24–)30–38(–42) µm, each with numerous discoid or slightly irregularly elongate plastids with pyrenoids; hair cells 4–9 diameters long, 73–92 × 9–16 µm, and colourless with few plastids.

Plurilocular sporangia borne laterally at the base of the erect filaments, below the meristematic regions, occasionally terminally on short, emergent, erect filaments, pluriseriate, usually with a single-celled pedicel, ovate to elongate-ovate, (48–)56–83(–105) × (28–)38–56(–78) µm; unilocular sporangia unknown.

In agreement with Sauvageau (1897), Hamel (1931–39) and Cardinal (1964) (but not observed by Buffham (1893), Cotton (1907) or Newton (1931)), plurilocular sporangia on one thallus appeared to be of different types, varying in the size of the constituent loculi, and can perhaps be provisionally identified as microgametangia and megagametangia pending further studies (see figs 54(H) and 54(I) respectively). Ovoid, pedicellate monosporangia reported on thalli by Cardinal (1974) have not been observed on material collected in Britain and Ireland.

Epiphytic on the blades of *Padina pavonica*, in mid to lower eulittoral pools.

Rare and only recorded in Britain for the Channel Islands, Dorset, south Devon and Cornwall.

A summer annual, only recorded for August, September and October, in keeping with the main seasonal distribution of the host species.

Culture studies by Sauvageau (1897b, 1920a) using megaspores produced thalli in culture with similar features to those of *Acinetospora crinita*, with numerous meristematic zones, crampon-like short lateral branches and similar plurilocular sporangia. Cardinal's (1974) later reports of the presence of monosporangia on specimens of *F. padinae* add further support to a life history connection between the two species, with *Acinetospora* possibly representing the sporophyte phase and *F. padinae* representing the gametophyte phase. Pending further studies, the two species are kept separate in the present treatise.

Batters (1902), p. 34; Brodie *et al.* (2016), p. 1019; Buffham (1893), pp. 88, 89, pl. 185(5–7); Cardinal (1964), pp. 60, 61, fig. 33(A–F); Cormaci *et al.* (2012), pp. 96, 98, pl. 21(1–3); Cotton (1907), pp. 371, 372; Dizerbo & Herpe (2007), p. 65, pl. 19(2); Fletcher in Hardy & Guiry (2003), p. 249; Gallardo García *et al.* (2016), p. 34; Guiry (2012), p. 180; Hamel (1931–39), pp. xvii, 43–5, fig. 12(A–D); Newton (1931), p. 123; Ribera *et al.* (1992), p. 111; Sauvageau (1896a), p. 268, (1897b), pp. 24–34, figs 7–10, (1920a), pp. 1041–4, (1931b), p. 972.

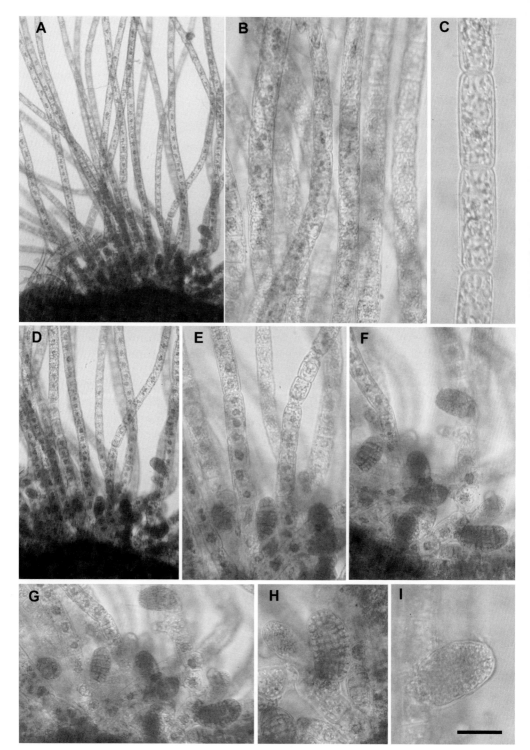

Fig. 54. *Feldmannia padinae*
A. Habit of thallus showing long, unbranched erect filaments and basally positioned plurilocular sporangia.
B. Portions of erect filaments. C. Portion of erect filament showing cells with enclosed discoid plastids.
D–I. Portions of erect filaments showing basally positioned plurilocular sporangia. Note different sized loculi
in the plurilocular sporangia figured in H and I, possibly representing megagametangia and microgametangia
respectively. Bar in I = 300 μm (A), = 180 μm (D), = 90 μm (B, E, F, G), = 45 μm (C, H, I).

Feldmannia paradoxa (Montagne) Hamel (1939), p. xvii.   Fig. 55

Ectocarpus paradoxus Montagne (1839), p. 175; *Ectocarpus caespitulus* J. Agardh (1842), p. 26; *Ectocarpus globifer* Kützing (1843), pp. 289, 290; *Ectocarpus pusillus* Kützing (1849), p. 450; *Ectocarpus insignis* P. Crouan & H. Crouan (1852), p. 450; *Ectocarpus globifer* var. *rupestris* Batters (1902), p. 31; *Feldmannia globifera* (Kützing) Hamel (1931–39), p. xvii; *Feldmannia caespitula* (J. Agardh) Knoepffler-Péguy (1970), p. 146; *Feldmannia paradoxa* var. *donatiae* (Ercegovic) Antolic & Span (2010), p. 24; *Feldmannia paradoxa* var. *caespitula* (J. Agardh) Cormaci & G. Furnari (2012), p. 98.

Thalli epiphytic, more rarely epilithic or epizoic, forming small, erect, filamentous, uniseriate, medium to dark brown, fairly dense, solitary tufts, up to 1 cm high, arising from a basal system of outward-spreading, sometimes host-penetrating, branched, filamentous, multicellular, attaching rhizoids arising from the basal cells of the main axes; rhizoid cells thick walled, often irregular and contorted, 2–8 diameters long, 36–115 × 9–24 µm; erect tufts branched predominantly towards the base, giving rise to long, linear, unbranched filaments with a single, basally positioned, meristematic region of compacted, markedly shorter than wide, almost lenticular-like cells; branching irregular, opposite or unilateral, with principal filaments and laterals often terminating in long, colourless false hairs; basal cells of erect filaments quadrate to rectangular, 1–3 diameters long, (50–)57–63(–100) × (28–)33–45(–55) µm with thickened walls, mid and upper region cells of erect filaments quadrate to rectangular, 0.5–3 diameters long, (27–)34–72(–100) × (29–)38–47(–52) µm, each with numerous, small, discoid or slightly rod-shaped plastids and associated pyrenoids; hair cells 3–8 diameters long, (80–)98–140(–173) × (20–)23–27(–31) µm.

Plurilocular and unilocular sporangia largely confined to the basal regions of erect filaments arising below the meristematic zones; plurilocular sporangia common, usually on one-celled pedicels, solitary, sometimes in opposite pairs, or opposite a branch, somewhat perpendicular to parental filament, pluriseriate, globose to ovate, (48–)60–90(–128) × (32–)46–60(–78) µm; unilocular sporangia rare, not observed on material from Britain and Ireland, reported by Kuckuck (1958) for individuals collected at Rovigno, Northern Adriatic to be sessile or on 1–4-celled pedicels, ovoid, 70–90 × 50–70 µm arising in a manner similar to that of the plurilocular sporangia.

Lower eulittoral pools and shallow sublittoral, sometimes occurring on the sides of floating structures.

Rarely recorded but widely distributed around Britain (Cornwall, south Devon, Hampshire, Kent, Northumberland) and Ireland (Antrim, Down, Mayo).

Appears to be a spring/summer annual, April to September.

Abbott & Hollenberg (1976), pp. 134, 135, fig. 97; Ardré (1970), pp. 241, 242; Batters (1902), p. 31; Børgesen (1926), pp. 43–52, figs 22–6; Bornet (1892a), pp. 358–61, pl. VII(6–7), (1892b), p. 245; Brodie *et al.* (2016), p. 1019; Cardinal (1964), pp. 57–60, figs 31(A–G), 32(A–H); Clayton (1974), pp. 754, 763–4, figs 3, 11(A–D); Coppejans (1983), pls 58, 59; Cormaci *et al.* (2012), pp. 98–102, pls 21(4–6), 22(1–3); Cotton (1912), p. 94; P. Crouan & H. Crouan (1867), p. 168, (1852), pl. 14; De Mesquita Rodrigues (1963), p. 18, pl. 1(F); De Toni (1895), pp. 539, 540; Derbès & Solier (1856), p. 49, pl. 14(9–11); Dizerbo & Herpe (2007), pp. 63–5, pl. 18(3, 4); Ercegovic (1955a), pp. 30–2, 40–8, figs 13, 18, 19; Feldmann (1937), p. 247, (1954), p. 31; Fletcher in Hardy & Guiry (2003), p. 250; Funk (1955), p. 32; Gallardo García *et al.* (2016), p. 34; Guiry (1978), pp. 190, 194, (2012), pp. 134, 135; Hamel (1931–39), pp. xvii–xviii, fig. 61(G), pp. 47–50, figs 14, 15(E); Harvey (1851), p. 32; Holmes (1887), pp. 161, 162, pl. 274(1a–c); Kim (2010), pp. 42–4, fig. 16(A–D); Kim & Lee (1994), pp. 159–64, figs 1(C–D), 2(D–E), 5(A–F); Knoepffler-Péguy (1970), pp. 139–60, 174–6, figs 1–8, pls I–III; Kuckuck (1928), p. 125, figs 5–10, (1958), pp. 179–92, figs 5–10; Kützing (1843), pp. 289, 290, (1845), p. 232, (1849), p. 450, (1855), V, 49; Mathieson & Dawes (2017), p. 137, pl. XLII(7); Morton (1994), pp. 8, 31; Newton (1931), p. 119; Oltmanns (1922), p. 10, fig. 295; Ribera *et al.* (1992), p. 111; Rosenvinge & Lund (1941), pp. 53–6, figs 25(A–D), 26(A–D), 27(A–C); Rueness *et al.* (2001), p. 14; Sauvageau (1933), pp. 93–101, figs 22–3; Setchell & N.L. Gardner (1925), pp. 438, 439; Stegenga & Mol (1996), pp. 105–7, fig. 1; Stegenga *et al.* (1997a), p. 15; Taylor (1957), p. 109; Womersley (1987), pp. 42–4, figs 6(E, F), 9(A–D).

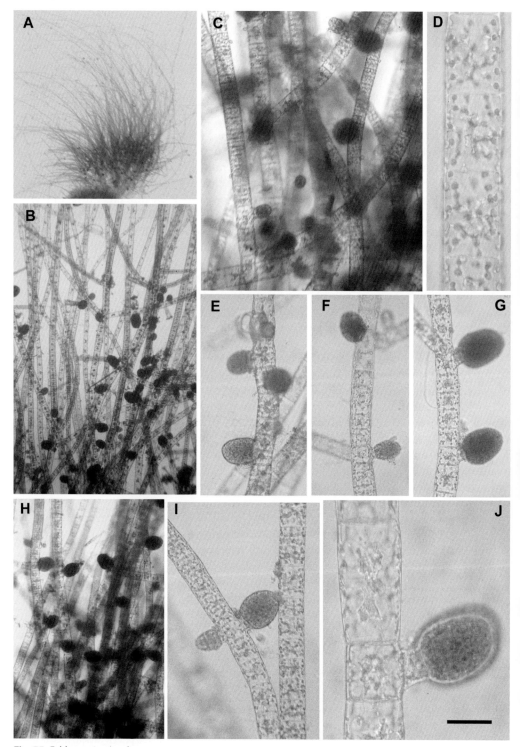

Fig. 55. *Feldmannia paradoxa*

A. Habit of thallus. B, C. Portions of erect filaments with plurilocular sporangia. D. Portion of erect filament showing cells with enclosed discoid plastids. E–J. Portions of erect filaments showing plurilocular sporangia with rounded ends, some oppositely positioned. Bar in J = 2.5 mm (A), = 470 μm (B), = 300 μm (H), = 180 μm (C), = 90 μm (E, F, G, I), = 45 μm (D, J).

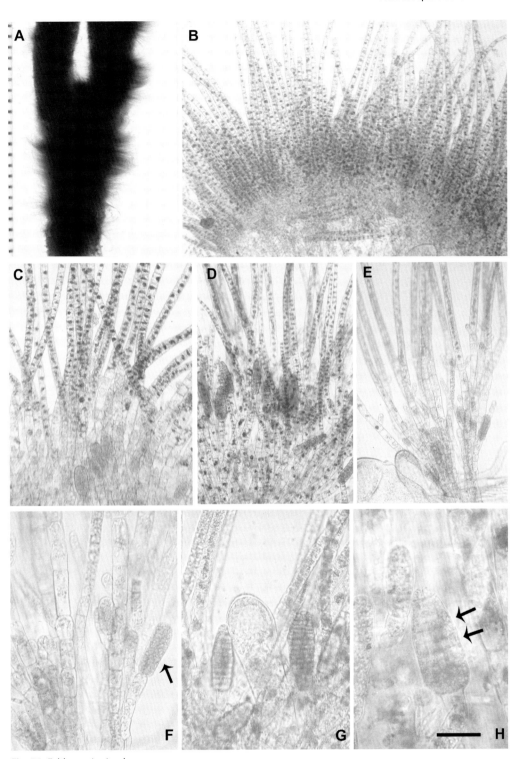

Fig. 56. *Feldmannia simplex*
A. Habit of thallus on *Codium*. B. VS of thallus showing closely packed, erect filaments arising through the *Codium* host. C–H. SP of thalli showing much-branched basal region, upper, long, largely unbranched filaments, and basally positioned unilocular (F, single arrow) and plurilocular (H, double arrows) sporangia. Note large, central utricle of *Codium* in G. Bar in H – 300 μm (B), = 180 μm (C, E), = 90 μm (D, F), = 45 μm (G, H).

Feldmannia simplex (P. Crouan & H. Crouan) Hamel (1939), p. xviii.   Fig. 56

Ectocarpus simplex P. Crouan & H. Crouan (1867), p. 163; *Ectocarpus cylindricus* De A. Saunders (1898), p. 150; *Ectocarpus flocculiformis* Setchell & N.L. Gardner (1922), p. 409; *Ectocarpus cylindricus* var. *codiophilus* Setchell & N.L. Gardner (1922), p. 415; *Ectocarpus socialis* Setchell & N.L. Gardner (1922), p. 412; *Feldmannia cylindrica* (De A. Saunders) Hollenberg & I.A. Abbott (1966), p. 19 [29]; *Feldmannia paradoxa* var. *cylindrica* (De A. Saunders) Kim & Lee (1994), pp. 163, 164, nom. inval.

Thalli epiphytic, partly endophytic, on the host alga, forming small, 2–3 mm in height, confluent and extensive, densely packed, dark brown, turf-like growths of uniseriate filaments arising from colourless, branched, twisted and irregularly contorted, rhizoidal filaments which penetrate down between the utricles of the host; cells of rhizoidal filaments 38–115 × 10–20 µm; emergent erect filaments quite densely packed, and pseudodichotomously to irregularly branched below, with principal filaments and laterals giving rise to long, unbranched, linear filaments with tapering apices when young, with or without terminal hairs; hair cells up to 5(–7) diameters long, 46–122 × 12–19 µm; growth by distinct meristematic growth zones at the base of both the principal filaments and laterals, supplemented by occasional areas of diffuse divisions; upper cells of erect filaments 1–5 diameters long, (18–)30–65(–80) × (13–)16–19(–23) µm, basal cells of filaments 1–3.5 diameters long, (13–)28–63(–70) × (14–)16–20(–23) µm, each with densely packed large discoid, to irregularly elongate plastids, becoming lighter in colour with fewer plastids at the filament apices, each plastid with one or more pyrenoids.

Plurilocular and unilocular sporangia occurring on the same individuals, basally positioned on the erect filaments below the meristematic zones, solitary, sometimes opposite to another sporangium or branch; plurilocular sporangia common, pluriseriate, pedicellate on 1–3-celled stalks, conical to elongate-cylindrical, (63–)85–115(–130) × (18–)30–38(–42) µm; unilocular sporangia rare, pedicellate, ovoid to cylindrical, (35–)65–80(–90) × (17–)21–32(–43) µm.

A common epiphyte on *Codium* spp., completely discolouring the host thallus when abundant, upper to lower eulittoral pools and shallow sublittoral.

Common and widely distributed around Britain and Ireland.

A summer annual.

Abbott & Hollenberg (1976), pp. 132, 134, fig. 96; André (1970), pp. 240, 241; Batters (1902), p. 30; Brodie *et al.* (2016), p. 1019; Cardinal (1964), pp. 55–7, fig. 30(A–K); Clayton (1974), pp. 754–61, figs 3(A–D), 4(A–C), 5(A–E); P. Crouan & H. Crouan (1852), pl. 13; Dizerbo & Herpe (2007), p. 65, pl. 19(3); Fletcher in Hardy & Guiry (2003), p. 250; Gallardo García *et al.* (2016), p. 34; Guiry (2012), p. 135; Hamel (1931–39), pp. xviii, 50–1, figs 61(H), 15(A–D); Holmes (1887), p. 161, pl. 274, fig. 2(A–C); Knight & Parke (1931), p. 61, pl. IX; Knoepffler-Péguy (1970), pp. 176–8; Kützing (1843), pp. 289, 290, (1845), p. 232, (1849), p. 450; Morton (1994), p. 8; Müller *et al.* (1996), pp. 61–3, figs 1–6; Müller & Frenzer (1993), pp. 37–44; Newton (1931), p. 118; Ribera *et al.* (1992), p. 112; Robledo *et al.* (1994), pp. 247–51, figs 1–3; Rueness *et al.* (2001), p. 14; Saunders (1898), p. 150, pl. 16; Sauvageau (1933), pp. 80–92, figs 17–21; Setchell & N.L. Gardner (1925), pp. 438, 439; Stegenga *et al.* (1997a), p. 15.

Herponema J. Agardh

HERPONEMA J. Agardh (1882), p. 55.

Type species: *H. velutinum* (Greville) J. Agardh (1890), p. 56.

Thalli epiphytic, forming small (< 4–5 mm), tufts or low-lying, turf-like growths of filaments on host surface, arising from a limited or extensive, deeply penetrating endophytic basal system of branched, tortuous filaments spreading through host tissue; erect filaments uniseriate, simple or rarely branched, with branching irregular; growth intercalary, with distinctive meristematic zones of short cells; hairs either false or true with basal meristem; cells containing several discoid plastids with associated pyrenoids.

Unilocular sporangia and plurilocular sporangia either terminal on short, emergent, erect filaments, or lateral, sessile or pedicellate at the base of longer filaments; plurilocular sporangia multiseriate, conical – one species (*H. solitarium*) reported to be heteromorphic, with locules of different dimensions, suggesting they represent gametangia; unilocular sporangia ovoid or spherical; species generally host specific.

Two species in Britain and Ireland, one species in Britain only.

KEY TO SPECIES

1. Thalli arising from galls on *Ericaria selaginoides* and *Gongolaria baccata* . . . . *H. valiantei*
    Thalli not associated with galls on the above algae, and reported on other host species . . . 2
2. Thalli forming sparse, microscopic tufts of erect filaments epiphytic
    predominantly on members of the Dictyotales . . . . *H. solitarium*
    Thalli forming prominent velvety, felt-like patches of erect filaments on
    Himanthalia elongata . . . .*H. velutinum*

Herponema solitarium (Sauvageau) Hamel (1939), p. xix.   Fig. 57

Ectocarpus solitarius Sauvageau (1892), pp. 97, 126; *Streblonema solitarium* (Sauvageau) De Toni (1895), p. 576.

Thalli epiphytic, forming solitary to confluent, turf-like patches on host surface, particularly prominent along the margins but not visible to the unaided eye; comprising a basal system of endophytic, branched filaments spreading just below the surface of the host, giving rise at intervals to emergent erect filaments; erect filaments uniseriate, simple, to 1–(1.5) mm (20–30 cells) in length, comprising lightly pigmented cells, cuboidal to rectangular, sometimes slightly barrel-shaped, (5)14–33(50) × 9–18 µm, each with a small number of discoid to small, plate-like plastids with associated pyrenoids; terminal region of filaments usually seen extending into a colourless hair, comprising cells 2–12 diameters long, 30–118 × 9–17 µm; meristematic-like regions of small numbers of subquadrate to quadrate cells observed subtending the terminal hair (possibly trichothallic).

Plurilocular sporangia terminal on short, few-celled emergent erect filaments, or more commonly lateral on the erect filaments; lateral sporangia sessile or more commonly one-celled pedicellate, ovate to globular, multiseriate, (50)59–78(93) × 22–33(–38) µm and usually strongly recurved towards the parental filament.

Cardinal (1964) reported the plurilocular sporangia to be heteromorphic, one type being colourless and made up of a large number of small locules, the other type being more pigmented and made up of larger locules, suggesting they represent microgametangia and megagametangia respectively; only one type, with relatively large locules, has been observed by the present author; unilocular sporangia unknown.

Epiphytic on older thalli of various Dictyotales, notably *Dictyota dichotoma*, *D. spiralis*, *Dictyopteris polypodioides* and *Taonia atomaria*, more rarely reported from other host algae, e.g. *Desmarestia ligulata*; lower eulittoral pools.

Rare, and only recorded for a few localities, mainly on the south and west coasts of Britain (south Devon, Dorset, Anglesey) and south and west coasts of Ireland (Waterford and Mayo).

A spring/summer annual, in keeping with the seasonal distribution of the main host species.

Anon (1952), p. 23; Aziz & Humm (1962), p. 61, figs 2–5; Batters (1902), p. 30; Brodie *et al.* (2016), p. 1019; Cardinal (1964), p. 64, fig. 34(K–N); Cormaci *et al.* (2012), p. 258, pl. 73 (figs 4–5); Dizerbo & Herpe (2007), p. 70, pl. 21(1); Feldmann (1954), p. 32; Fletcher in Hardy & Guiry (2003), p. 258; Gallardo García *et al.* (2016), p. 34; Guiry (1978), p. 202, (2012), p. 135; Hamel (1931–39), pp. xii, xix, 58, fig. 18(C); Hardy & Guiry (2003), p. 258; Kuckuck (1956), pp. 300, 301; Morton (1994), p. 8; Newton (1931), p. 118; Ribera *et al.* (1992), p. 112; Sauvageau (1892), pp. 97–100, 126, pl. 3(24–7); Schneider & Searles (1991), pp. 117, 118, figs 123, 124; Stegenga (1996), pp. 201, 203, pl. III(1–2); Stegenga *et al.* (1997a), p. 15, (2007), pp. 125–43; Van Heurck (1908), p. 24.

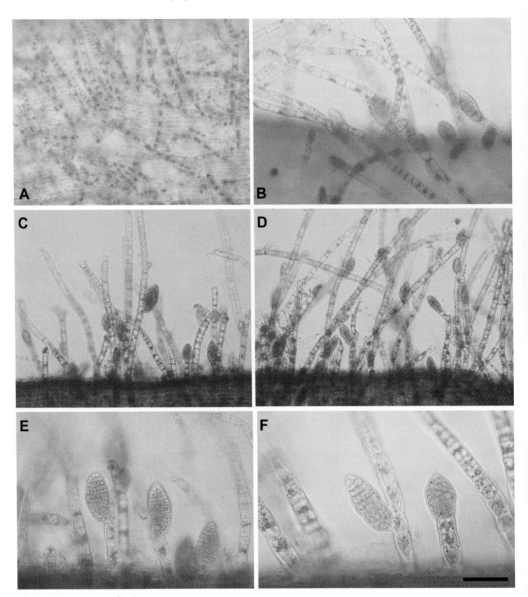

Fig. 57. *Herponema solitarium*

A. Habit of thalli showing the erect filaments of *Herponema* arising from between the cells of the host *Dictyota dichotoma*. B–F. Habit of thalli showing the erect filaments and both terminal and lateral plurilocular sporangia. Bar in F = 200 μm (A), = 110 μm (B–D), = 70 μm (E, F).

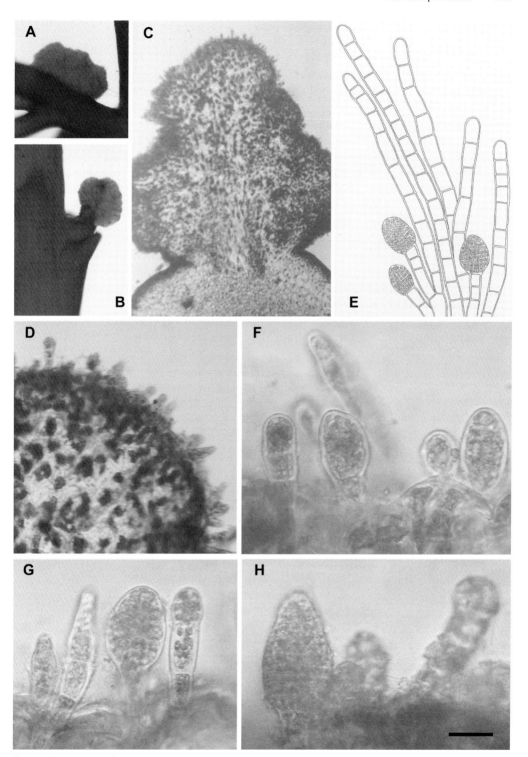

Fig. 58. *Herponema valiantei*
A, B. Gall-like growths on the host *Ericaria selaginoides* formed as a result of the *Herponema*. C. VS of a single gall-like growth. D. Close-up of the edge of the gall shown in C. E–H. VS through edges of gall-like growths showing emergent filaments and plurilocular sporangia. Bar in H = 1 mm (A), = 2 mm (B), = 470 μm (C), = 120 μm (D), = 55 μm (E), = 45 μm (F–H).

Herponema valiantei (Bornet) Hamel (1939), p. xix.   Fig. 58

Ectocarpus valiantei Bornet ex Sauvageau (1892), pp. 57, 124; *Streblonema valiantei* (Bornet) De Toni (1895), p. 576.

Thalli epi-endophytic, forming small, erect tufts of filaments, arising from a hemispherical, knobbly, gall-like growth on the host alga formed as a result of the presence of the epi-endophyte; comprising an extensive, endophytic basal system of irregularly to dichoto-mously branched, downward-penetrating and spreading filaments of cylindrical or irregularly contorted, pigmented cells, 1–6 diameters long, 8–40 × 8–20 μm, giving rise at the surface of the gall-like structure to a fine turf of erect filaments; erect filaments to 0.4(–1) mm in height, fairly rigid, uniseriate, simple or pseudodichotomously branched at the base, and sometimes terminated by a false hair, comprising quadrate to rectangular cells, 1–3 diameters long, (12–)18–46(–60) × (8–)11–19(–21) μm, each containing several discoid plastids with associated pyrenoids; hairs colourless, comprising cells 3–5 diameters long, 32–60 × 10–12 μm.

   Plurilocular sporangia terminal or lateral on the erect filaments, usually basally situated and associated with short, newly emergent filaments, sessile or pedicellate, multiseriate, oval to globular, (33–)43–60(–68) × (19–)28–35(–40) μm; unilocular sporangia unknown.

   Epi-endophytic, forming hemispherical gall-like structures on *Ericaria selaginoides* and *Gongolaria baccata*, lower eulittoral pools and shallow sublittoral.

   Very rare with records only for the Channel Islands and the Isles of Scilly.

   A spring/summer annual, recorded from March to September.

Batters (1902), p. 29; Brodie *et al.* (2016), p. 1019; Cormaci *et al.* (2012), pp. 258, 259, pl. 74(1–3); Dixon (1961), p. 73; Dizerbo & Herpe (2007), p. 70, pl. 21(2); Feldmann (1937), p. 249, (1954), p. 32; Fletcher in Hardy & Guiry (2003), p. 258; Gallardo García *et al.* (2016), p. 34; Guiry (2012), p. 181; Hamel (1931–39), pp. xii, xix, 60, fig. 18(F); Hardy & Guiry (2003), p. 258; Kuckuck (1956), pp. 301, 302, fig. 3(A–C); Lyle (1920), p. 8; Miranda (1931), p. 25; Newton (1931), p. 116; Ribera *et al.* (1992), p. 112; Sauvageau (1892), pp. 57–9, 76, 124, pl. II(8–10); Van Heurck (1908), p. 24.

Herponema velutinum (Greville) J. Agardh (1882), p. 56.   Fig. 59

Sphacelaria velutina Greville (1828), p. 42, pl. 350; *Linkia velutina* Carmichael in Harvey in Hooker (1833), p. 325; *Ectocarpus velutinus* (Greville) Kützing (1845), p. 236; *Streblonema velutinum* (Greville) Thuret in Le Jolis (1863), p. 73; *Ectocarpus velutinus* var. *laterifructus* Batters (1902), p. 30.

Thalli epiphytic, forming turf-like, oval to confluent, velvety patches on host, covering the surface to several cm in length, drying to a yellow-gold colour; comprising a basal attachment system of irregularly branched filaments of contorted, usually elongated, slightly pigmented cells, deeply penetrating and ramifying throughout the host tissue and emerging, usually much branched, through the surface to give rise to short, closely packed, erect, uniseriate filaments; erect filaments unbranched, or very rarely pseudodichotomously branched, variable in length, to 1.5 mm high and, in the summer months, terminated by long, colourless false hairs; erect filament cells, quadrate to rectangular cells, 1–2(–3) diameters long, (16–)22–42(–50) × (13–)15–17(–18) μm, thick walled, each containing several discoid plastids with associated pyrenoids; growth intercalary with distinctive meristematic zones of short cells usually positioned terminally just below the false hairs, comprising a closely packed series of short, subquadrate to quadrate cells, 0.5–1 diameters long, 5–13 × 13 μm; hair cells rectangular, 3–9 diameters long, 51–125 × 14–16 μm.

   Unilocular sporangia common, either terminal on short, erect filaments, or lateral, sessile or pedicellate towards the basal region of longer filaments, ovoid to elliptical, (47–)57–82(–96) × (31–)36–48(–52) μm; plurilocular sporangia not observed by the present author, but reported for St Andrews, Scotland by Laverack & Blackler (1974) and for Helgoland material by Kuckuck (1956), with Kuckuck describing them as occupying the same position as the unilocular sporangia, and as multiseriate, short-lanceolate with a slightly curved apex.

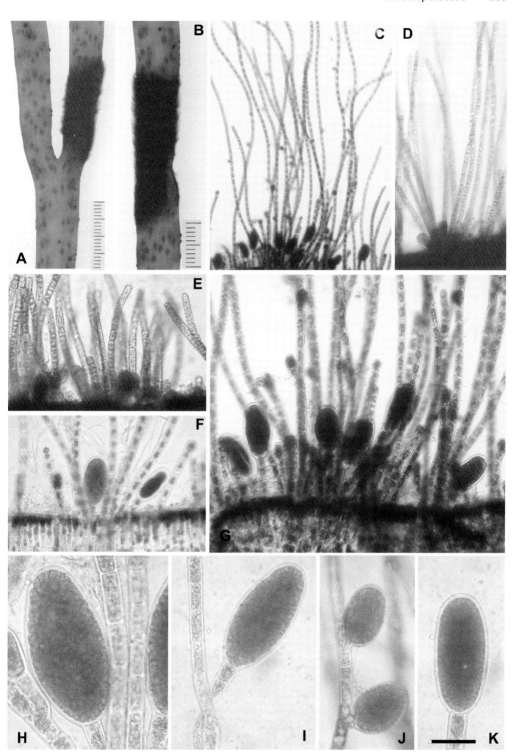

Fig. 59. *Herponema velutinum*

A, B. Turf-like habit of thalli on receptacles of *Himanthalia elongata*. C–G. VS of thalli showing erect, largely unbranched, erect filaments and largely basally positioned unilocular sporangia. H–K. Close-ups of terminal and lateral unilocular sporangia. Bar in K = 300 μm (C), = 180 μm (D), = 90 μm (E, F, G), = 45 μm (H–K).

Commonly epiphytic on the receptacles of *Himanthalia elongata*, lower eulittoral pools and exposed.

Fairly widespread and common around Britain and Ireland, in accordance with the distribution pattern of the host species.

Appears on young *H. elongata* in the spring/early summer, with maximum development during September and October; can be found throughout the year and on overwintering host thalli; unilocular sporangia observed throughout the year; plurilocular sporangia reported by Blackler (1974) for St Andrews material from January to March.

Anon (1952), p. 23; André (1970), p. 245; Bartsch & Kuhlenkamp (2000), p. 259; Batters (1890), p. 274, (1902), p. 30; Blackler (1974), p. 242; Brodie *et al.* (2016), p. 15; Cardinal (1964), pp. 64, 65, fig. 34(A–E); Cormaci *et al.* (2012), p. 259, pl. 74(4–6); De Mesquita Rodrigues (1963), p. 25; Dixon (1961), p. 73; Dizerbo & Herpe (2007), p. 70, pl. 21(3); Ercegovic (1955), p. 63, fig. 28; Feldmann (1954), p. 32; Fletcher in Hardy & Guiry (2003), p. 259; Funk (1955), p. 33; Gallardo García *et al.* (2016), p. 34; Gayral (1966), p. 227, fig. 23; Guiry (1978), p. 202, (2012), p. 135; Hamel (1931–39), pp. XII, XIX, 60–1, fig. 18(A); Harvey (1846–51), pl. 28(B); Knight & Parke (1931), p. 61, pl. X(15); Kuckuck (1956), pp. 294–300, figs 1, 2; Le Jolis (1863), p. 73; Morton (1994), p. 32; Newton (1931), p. 117; Printz (1926), pp. 150–3, pls IV–V; Ribera *et al.* (1992), p. 112; Rueness (1977), p. 116, (2001), p. 15; Russell & Wareing (1979), p. 128; Sauvageau (1892), pp. 55–7, pl. 1(7); Stegenga & Mol (1983), p. 76, pl. 22(1); Stegenga *et al.* (1997a), p. 16; Van Heurck (1908), p. 24.

Hincksia J.E. Gray

HINCKSIA J.E. Gray (1864), p. 12.

Type species: *H. hincksiae* (Harvey) P.C. Silva (1987), p. 130.

Thalli epiphytic, epilithic or epizoic, heterotrichous, forming erect, dense tufts of filaments up to 10(–25) cm long, arising from a basal system of branched, outward-spreading, secondary, rhizoidal filaments formed at the base of the erect shoots; erect filaments uniseriate, much and variously branched, to several orders, the ultimate branches usually attenuated to thin, light-coloured cells and/or false hairs; growth intercalary, either diffuse or with ill-defined scattered meristems; cells with numerous discoid or short, band-shaped plastids with one or more pyrenoids.

Plurilocular sporangia common, multiseriate, usually ovoid to short-conical, rarely cylindrical, usually broadly based, asymmetrical, recurved upwards towards parental filament, usually sessile, rarely stalked, borne singly or grouped, sometimes in unilateral, seriate, adaxially arranged rows on terminal, erect filaments, sometimes with up to three different sized loculi, variously described as meiosporangia/neutral sporangia (probably producing accessory, asexual plurispores), microsporangia and megasporangia, the latter representing male and female gametangia producing anisogametes from small and large loculi respectively; unilocular sporangia uncommon, usually sessile, borne singly on the erect filaments.

The genus *Hincksia* is principally characterized by the following combination of characters: erect, filamentous tufts, variously branched to several orders, the ultimate branchlets attenuating or giving rise to false hairs; growth intercalary and either diffuse or by scattered meristems; cells with mainly several discoid plastids, with pyrenoids; reproduction by usually sessile unilocular sporangia and plurilocular sporangia, usually positioned on terminal branchlets, rarely reported on prostrate rhizoidal filaments (*H. ovata*; Cardinal 1964 (as *H. intermedia*); *H. hincksiae* Sauvageau 1897b, 1933); plurilocular sporangia solitary, opposite or in series, usually adaxially positioned, broad-based, asymmetrical and upward-curved and, in some species, comprising different-sized locules, and variously identified as asexual plurilocular sporangia, megasporangia (female gametangia) and microsporangia (male gametangia). Most culture studies have revealed a direct life history, with apogamous development of the megagametes, along with swarmers from the neutral sporangia. However, an isomorphic alternation involving sexual fusion of the gametes might be normal in the field. Unilocular sporangial thalli are rare and they might have different seasonal or local habitat preferences.

Hincksia bears a close resemblance to *Feldmannia* and generally differs in the following respects; *Hincksia* is much branched to several orders above, growth is intercalary and either diffuse or by scattered meristems, the ultimate branches either attenuating or giving rise to false hairs, whereas in *Feldmannia*, branching is largely basally positioned, giving rise to long, ultimate, unbranched filaments with a distinct, basally positioned meristem and terminal sterile hair; in *Hincksia*, the plurilocular sporangia are usually sessile, ovoid to short-conical, broadly based, asymmetrical, upward-curved, adaxially positioned and distributed largely on the ultimate branchlets, whereas in *Feldmannia*, they are usually pedicellate, cylindrical, symmetrical and usually occur in the basal regions of the tufts below the meristematic growth zones.

It was the marked anisogamy shown by the then recognised type species, *H. secunda* (as *Ectocarpus secunda*), which prompted Batters (1893c) to remove it from *Ectocarpus* to a new genus *Giffordia*. However, the validity of this was questioned (see Sauvageau 1896b; Fritsch 1945, p. 121) and subsequently, the majority of authors such as Setchell & N.L. Gardner (1925), Newton (1931), Hamel (1931–39), Knight & Parke (1931) continued to use the genus *Ectocarpus*. Hamel (1931–39) later redefined and enlarged the genus to include those species of *Ectocarpus* with 'growth zones commonly more distally placed, giving rise to short filaments or false hairs, discoid plastids, sessile plurilocular sporangia, which are often arranged in series, four different types of sporangia ("unilocular sporangia", "antheridia", "megasporangia", and "meiosporangia") and the occurrence of a sexual life history involving anisogametes in some species (*G. secunda*, *G.* (= *Feldmannia*) *mitchelliae*)'. This rendered the genus *Giffordia* much more acceptable and was adopted extensively by such workers as Parke & Dixon (1964, 1968, 1976), Ercegovic (1955), Taylor (1957, 1960) Kuckuck (1961, 1963), Cardinal (1964), Clayton (1964), Abbott & Hollenberg (1976), Rueness (1977), Kornmann & Sahling (1977), Stegenga & Mol (1983), Womersley (1987) and Coppejans (1995). Notable exceptions, however, were Rosenvinge & Lund (1941), Smith (1944) and Lindauer *et al.* (1961) who continued to use the genus *Ectocarpus*. The species are now placed in the genus *Hincksia* which has priority over *Giffordia* (Silva *et al.* 1987).

Five species in Britain and Ireland, one species in Britain only.

KEY TO SPECIES

1. Branching opposite, at least in part 2
   Branching alternate, pseudodichotomous or secund 3

2. Branching opposite in main filaments, sub-secund or secund in ultimate branches; plurilocular sporangia borne on adaxial sides of ultimate branches, sessile, ovoid, not in series; thalli up to 25 cm in length . . . .*H. granulosa*
   Branching partly opposite but mainly alternate; plurilocular sporangia borne on main filaments and often in opposite pairs or clusters, sessile, narrowly ovoid; thalli <3 cm in length . . . .*H. ovata*

3. Branching alternate, irregular or pseudodichotomous . . . . *H. fenestrata*
   Branching secund, at least in distal portions of thallus. 4

4. Plurilocular sporangia conical, sessile and borne in long series on adaxial sides of ultimate branches, thus giving the filaments a serrated appearance; usually epiphytic on *Saccorhiza polyschides* . . . .*H. hincksiae*
   Plurilocular sporangia not markedly conical or arranged in a long, serrated series; epiphytic on a range of species. 5

5. Plurilocular sporangia ovoid, sometimes curved, length <2 × breadth and up to 60 μm; branching consistently secund; thalli up to 2(–5) cm in length . . . . . *H. secunda*
   Plurilocular sporangia spindle-shaped, length >2 × breadth and up to 90 μm; some alternate branching in main filaments; thalli up to 20 cm in length . . . . *H. sandriana*

N.B. Forms of *H. granulosa* exist with alternate rather than opposite branching in main filaments. These can be distinguished from *H. sandriana* by the greater diameter (>50 μm) of the main filaments, but can be easily confused with *H. secunda*.

Hincksia fenestrata (Harvey ex Berkeley) P.C. Silva (1987), p. 130.   Fig. 60

Ectocarpus fenestratus Harvey ex Berkeley (1846–51), pl. 257; *Giffordia fenestrata* (Harvey ex Berkeley) Batters (1893c), p. 86.

Thalli epilithic or epiphytic, olive-brown, heterotrichous, forming fine, filamentous tufts up to 5(–7) cm in length arising from a branched, filamentous, basal system; main axes little branched with branches well-spaced and pseudodichotomous, alternate or irregular, with ultimate branches quite flaccid, delicate and simple; cells cuboidal to rectangular, 1–4 diameters long, (18–)30–75(–90) × (11–)16–30(–38) μm, each containing numerous discoid plastids with associated pyrenoids.

Plurilocular sporangia solitary, pedicellate (single-celled, rarely two-celled), at first clavate, later becoming elliptic-oblong to fusiform, recurved towards parental filament, (46–)58–85(–142) × (19–)26–43(–58) μm; unilocular sporangia unknown.

Epilithic or epiphytic, lower eulittoral pools.

Very rare with only historical records for Britain (Cornwall, South Devon, Sussex and Northumberland).

Summer annual, with records from May and June.

Anon (1952), p. 21; Batters (1893a), p. 86, (1902), p. 34, (1994b), p. 116; Bornet (1892c), pp. 148–50, fig. 1(A); Brodie *et al.* (2016), p. 1019; Dizerbo & Herpe (2007), p. 65; Fletcher in Hardy and Guiry (2003), p. 260; Guiry (2012), p. 180; Harvey (1846–51), pl. 257; Holmes & Batters (1890), p. 79; Merrifield (1863) p. 519; Newton (1931), p. 123; Silva *et al.* (1987), p. 130; Tregelles (1933), p. 301.

Hincksia granulosa (Smith) P.C. Silva (1987), p. 130.   Fig. 61

Conferva granulosa Smith (1790–1814), pl. 2351; *Ectocarpus granulosus* (Smith) C. Agardh (1828a), p. 45; *Ectocarpus oviger* Harvey (1862), p. 167; *Ectocarpus granulosus* var. *tenuis* Farlow (1881), p. 70; *Ectocarpus granulosus* var. *refractus* Batters (1902), p. 34, nom. inval. (as 'refracta'); *Giffordia granulosa* (Smith) Hamel (1931–39), p. xv; *Giffordia granulosa* (Smith) Hamel var. *laeta* (C. Agardh) Kuckuck (1961), p. 121; *Giffordia granulosa* (Smith) Hamel var. *seriata* Kuckuck (1961), p. 124; *Giffordia granulosa* (Smith) Hamel var. *eugranulosa* Kuckuck (1961), p. 126; *Ectocarpus recurvatus* Kuckuck (1961), p. 132; *Giffordia recurvata* Kuckuck ex Cardinal (1964), p. 51; *Giffordia oviger* (Harvey) Hollenberg & I.A. Abbott (1966), p. 18 [628]; *Hincksia recurvata* (Kuckuck) P.C. Silva in Silva *et al.* (1987), p. 130; *Hincksia recurvata* (Cardinal) Athanasiadis (1996), p. 142.

Thalli epiphytic, epizoic or epilithic, forming dense, slightly rigid/coarse, medium to dark brown, much-branched, filamentous tufts up to 5–10(–25) cm in height, solitary or confluent with several main axes arising from a branched, outward-spreading, filamentous rhizoidal base; principal filaments sometimes intertwined into rope-like mats, and usually clothed below by a sheath of downward-growing, corticating, secondary, rhizoidal filaments which spread out and form the attachment base; main axial filaments usually oppositely branched, sometimes whorled, terminal branchlets either opposite, alternate or unilateral and secund/ comb-like, all filaments usually tapering from base to apex; basal cells quadrate to cylindrical, usually longer than broad, 1–2 diameters long, 65–113 × 53–70 μm, middle region cells subquadrate to cylindrical, 0.5–2.5 diameters long, 28–80 × 33–62 μm, terminal branchlet cells subquadrate to quadrate, less commonly cylindrical, 0.3–1(–1.5) diameters long, 9–34 × 10–45 μm; growth intercalary, either diffuse or by scattered meristems, these more notably positioned at the base of terminal branches; false hairs common, terminal on the erect filaments of ultimate branches; cells with numerous densely packed, discoid plastids, each with an associated pyrenoid, and centrally positioned cluster of physodes.

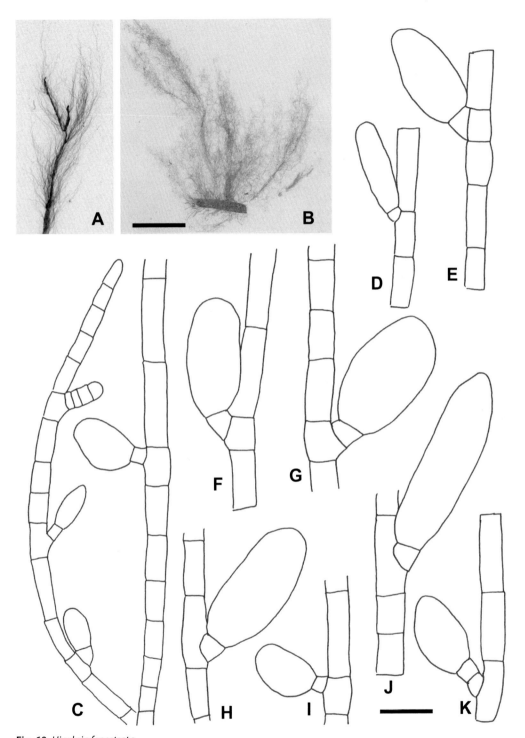

Fig. 60. *Hincksia fenestrata*

A, B. Habit of thalli. C–K. Portions of erect filaments with plurilocular sporangia. Photo bar = 1.7 mm (A, B), line drawings bar = 40 µm (C–K).

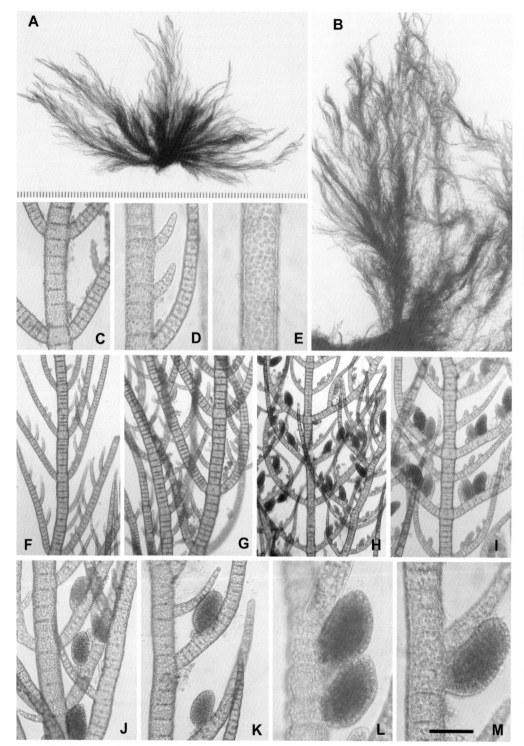

Fig. 61. *Hincksia granulosa*

A, B. Habit of thalli. C, D. Filaments with opposite and lateral branching, respectively. E. Portion of filament showing enclosed discoid plastids. F, G. Filaments showing opposite branching. H–K. Filaments showing opposite branching and sessile plurilocular sporangia. L, M. Close-ups of plurilocular sporangia. Bar in M = 0.8 mm (B), = 300 μm (H), = 180 μm (F, G, I), = 90 μm (C, J, K), = 45 μm (E, L, M).

Plurilocular sporangia borne adaxially on the terminal branchlets, very common and numerous, sessile, conical or more usually ovoid in shape, broad-based, 48–85(–110) × 28–60(–80) μm, solitary or borne unilaterally in a short, secund series, asymmetrically and slightly recurved towards parental filament, suspected by Knight & Parke (1931) to be of two types, differing in the size of the compartments; unilocular sporangia infrequent, on thalli bearing plurilocular sporangia, sessile, ovoid, measuring 50–70 × 40–55 μm.

Commonly epiphytic on various algae, in particular *Laminaria* spp., also epilithic and epizoic particularly on limpets, in eulittoral pools, and shallow sublittoral. Frequently reported as a common fouling species at, or just below, the water line on floating structures, such as buoys, pontoons etc., particularly under wave-washed conditions.

Abundant and widely distributed around Britain and Ireland.

Recorded throughout the year, but particularly luxuriant during spring and summer.

In some populations, the thalli are predominantly unilaterally branched and resemble *Hincksia secunda*. These thalli were described as separate forms by Kuckuck (1961). However, their varietal status was not accepted by Cardinal (1964) on the basis of intergrades being observed between the different varieties, and this is accepted in the present work. As proposed by Clayton (1974), probably the only way of separating *H. granulosa* and *H. secunda* is by observing the reproductive structures which comprise plurilocular sporangia and hetero-morphic gametangia in the latter species. In the present study, only one type of plurilocular sporangium has been observed in *Hincksia granulosa*, contrary to the two types suspected by Knight & Parke (1931).

Culture studies by Fletcher (1981c) revealed an asexual 'direct' type of life history for populations on the south coast of England, with spores from the plurilocular sporangia germi-nating to repeat the parental phase. A similar life history was described by Clayton (1974) for Australian material.

Abbott & Hollenberg (1976), pp. 140–1, fig. 103; Ardré (1970), pp. 237, 238; Bartsch & Kuhlenkamp (2000), p. 166; Brodie *et al.* (2016), p. 1019; Cardinal (1964), pp. 39–41, 51–3, figs 19(A–J), 20(A–J), 27(A–G), 28(A–J); Clayton (1974), pp. 770–3, figs 15(A–F), 16(A–G); Coppejans (1980), figs 120–4, (1982), p. 234, (1995), p. 202, pl. 67(A–E); Cormaci *et al.* (2012), pp. 106, 107, pl. 23(7–12); De Mesquita Rodrigues (1963), p. 20, pl. 2(C); Dizerbo & Herpe (2007), pp. 65, 67, pl. 19(4); Ercegovic (1955), pp. 27–30, fig. 12(d, e); Feldmann (1937), p. 246; Fletcher (1980), p. 39, pl. 12(1–4), (1981), pp. 211–21; Fletcher in Hardy & Guiry (2003), p. 260; Gallardo García *et al.* (2016), p. 34; Guiry (2012), p. 135; Hamel (1931–39), pp. xv, 37, 38, figs 8(III), 61(E); Harvey (1846–51), pl. 200; Kim (2010), pp. 57–9, fig. 25(A–D); Kim & Lee (1992), pp. 252–4, fig. 7(A–D); Knight & Parke (1931), p. 63; Kornmann & Sahling (1977), p. 109, pl. 55(A–E); Kuckuck (1961), pp. 119–37, figs 1–9; Kylin (1947), p. 11, fig. 4; Lindauer *et al.* (1961), pp. 145, 146, fig. 4; Mathieson & Dawes (2017), p. 138, pl. XLII(8–9); Morton (1994), pp. 31, 32; Newton (1931), p. 123; Norris (2010), p. 170, fig. 81(A–B); Ribera *et al.* (1992), p. 112; Rosenvinge & Lund (1941), p. 45, fig. 19; Rueness (1977), p. 114, fig. 35, (2001), p. 15; Sauvageau (1933), pp. 59–66, figs 13, 14; Schneider & Searles (1991), pp. 119, 120, figs 125–8; Setchell & N.L. Gardner (1925), pp. 426, 427; Smith, G.M. (1944), pp. 81, 82, pl. 11(1–2); Smith, J.E. (1811), pl. 2351; Stegenga & Mol (1983), p. 73, pl. 20(1–3); Stegenga *et al.* (1997a), p. 16, (1997b), p. 142, pl. 34(3–5); Taylor (1957), pp. 112, 113, pl. 7(7–9); Womersley (1987), p. 54, figs 10(E), 13(A–D).

Hincksia hincksiae (Harvey) P.C. Silva (1987), p. 130.   Fig. 62

Ectocarpus hincksiae Harvey (1841) p. 40; *Hincksia ramulosa* J.E. Gray (1864), p. 12, nom. illeg.; *Ectocarpus hincksiae* var. *irregulans* Børgesen (1902), p. 412; *Giffordia hincksiae* (Harvey) Hamel (1931–39), p. xv; *Giffordia hincksiae* var. *californica* Hollenberg & I.A. Abbott (1968), p. 1238.

Thalli epiphytic, forming very fine and delicate, light brown tufts of uniseriate filaments up to 30(–60) mm in length; filaments gregarious and often intertwined into rope-like mats and corticated in the basal regions of larger thalli, with downward-growing, secondary, rhizoidal filaments which spread out over the host, either irregularly or in a pseudodiscoid manner, sometimes penetrating the host tissue; erect filaments branched, with branching infrequent and irregular or spirally arranged near the base, markedly unilateral and comb-like above with

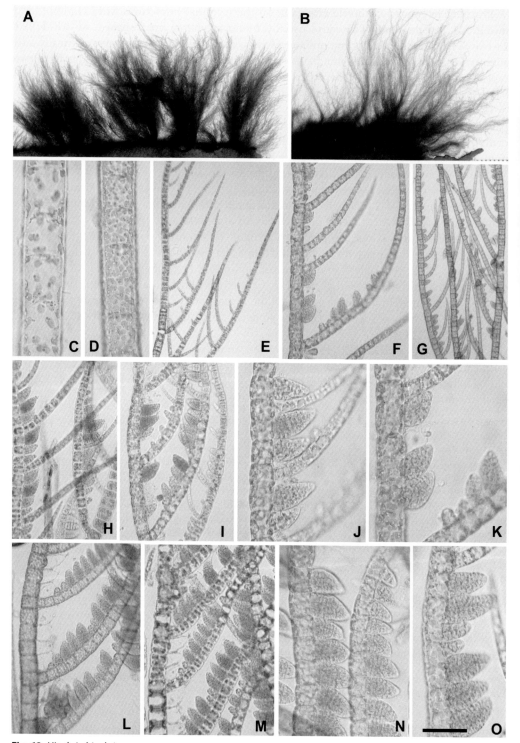

Fig. 62. *Hincksia hincksiae*

A, B. Habit of thalli on host. C, D. Portions of filaments showing cells with sparsely and densely packed discoid plastids, respectively. E. Principal filament showing unilateral branching. F–M. Erect filaments showing unilateral branching and adaxially arranged plurilocular sporangia. N, O. Close-ups of plurilocular sporangia. Bar in O = 10.5 mm (A, B), = 180 μm (E, G), = 90 μm (F, H, I, L, M), = 45 μm (D, J, K, N, O).

branches tapering from base to apex and often markedly recurved towards parental filament; ultimate branchlets narrowing with cells distinctly longer than wide, lighter in colour, or more commonly giving rise to false hairs, the cells of which are markedly longer than wide and lightly coloured due to the small number of plastids present; intercalary meristems scattered but more usually observed at the base of distal branches; cells usually longer than broad towards the base, 0.3–1(–2) diameters long, cylindrical, sometimes pronouncedly thick walled, 15–80 × 35–58 µm, usually becoming more shorter than broad above, 0.3–1 diameters long, 12–45 × 28–58 µm, with terminal branchlet cells 0.6–2 diameters long and much narrower, 8–18 × 8–18 µm; all cells with a prominent central nucleus surrounded by numerous densely packed discoid or irregularly elongated plastids with associated pyrenoids.

Plurilocular sporangia common, sessile, rarely stalked, conical to ovoid in shape, broadly based, 30–50 × 24–30 µm, slightly pointed distally and recurved towards the parental filament and borne in a closely packed, secund series (serrated appearance) along the inner, adaxial faces of the ultimate branches and branchlets and short recurved lower branches; unilocular sporangia rare, scattered, sessile, spherical to elongate-spherical, 30–55 µm in diameter.

Commonly found epiphytic on *Saccorhiza polyschides*, usually in great local abundance, less commonly recorded on *Laminaria* spp. and other algae, lower littoral pools and shallow sublittoral, 1–5 m.

Common and widely distributed around Britain and Ireland, except for south-eastern and eastern England where it has probably just been overlooked.

Appears to be mainly a spring/summer annual, March to September, but commonly reported throughout the year.

Ardré (1970), pp. 238, 239, pl. 34(10); Bartsch & Kuhlenkamp (2000), p. 166; Børgesen (1902), pp. 412, 413, fig. 72; Brodie *et al.* (2016), p. 1019; Cardinal (1964), pp. 44, 45, fig. 22(A–O); Cormaci *et al.* (2012), pp. 108, 109, pl. 24(3–7); Dizerbo & Herpe (2007), p. 67, pl. 20(1); Etherington (1965), pp. 472–7, figs 1(a–d), 2(a–g); Fletcher (1980), p. 39, pl. 12(4–6); Fletcher in Hardy & Guiry (2003), p. 261; Gallardo García *et al.* (2016), p. 34; Guiry (2012), pp. 135, 136; Hamel (1931–39), pp. xv–xvii, 38–41, fig. 10(A–C); Harvey (1846–51), p. 85, pl. xxii; Knight & Parke (1931), p. 63; Kornmann & Sahling (1977), p. 112, pl. 58(A–E); Morton (1994), p. 32; Newton (1931), p. 122; Parodi & Müller (1994), pp. 213–17, figs 1–17; Ribera *et al.* (1992), p. 112; Rueness (1977), p. 114, fig. 36(A, B), (2001), p. 15; Sauvageau (1897b), pp. 66–71, fig. 11, (1933), pp. 23–49, figs 5–12; Stegenga (1996), p. 199, pl. I(1–3); Stegenga *et al.* (1997a), p. 16.

Hincksia ovata (Kjellman) P.C. Silva (1987), p. 130.   Fig. 63

Ectocarpus ovatus Kjellman (1877b), p. 35; *Ectocarpus ovatus* var. *arachnoideus* Reinke (1889a), pl. 20; *Ectocarpus ovatus* var. *elongatus* Rosenvinge (1893), p. 888; *Ectocarpus holmii* Rosenvinge (1893), p. 889; *Ectocarpus ovatus* var. *holmii* (Rosenvinge) Rosenvinge (1898), p. 78; *Ectocarpus ovatus* var. *tenuis* Rosenvinge (1898), p. 80; *Ectocarpus ovatus* var. *intermedius* Rosenvinge (1941), p. 49; *Giffordia ovata* (Kjellman) Kylin (1947), p. 9; *Giffordia ovata* var. *arachnoidea* (Reinke) Kylin (1947), p. 10; *Giffordia fuscata* (Zanardini) Kuckuck in Kornmann (1954), p. 41; *Giffordia intermedia* (Rosenvinge) S. Lund (1959), p. 48; *Hincksia intermedia* (Rosenvinge) P.C. Silva in P.C. Silva, E.G. Meñez, & Moe (1987), p. 130.

Thalli forming small, erect, epiphytic or epizoic, light brown, solitary tufts, usually less than 1–3 cm in height, attached at the base by branched, filamentous rhizoids originating as corticating extensions of the basal cells; erect filaments uniseriate, usually with a distinct, somewhat sparsely branched, main axis, with branching either irregular and alternate, or more commonly opposite with widely divergent branches which often remain short, unequal in length, and markedly attenuate; growth intercalary, diffuse or by ill-defined meristems on the main axes, more commonly seen on terminal laterals/branchlets and usually subtending acute, hair-like extensions comprising cells up to 11 diameters long, 49–66 × 6–10 µm; basal cells subquadrate to rectangular, 0.5–1(–2) diameters long, 18–85 × 29–47 µm, middle-region cells usually wider than long, 0.5–1(–2) diameters long, 10–38 × 18–30 µm, terminal branchlet cells subquadrate

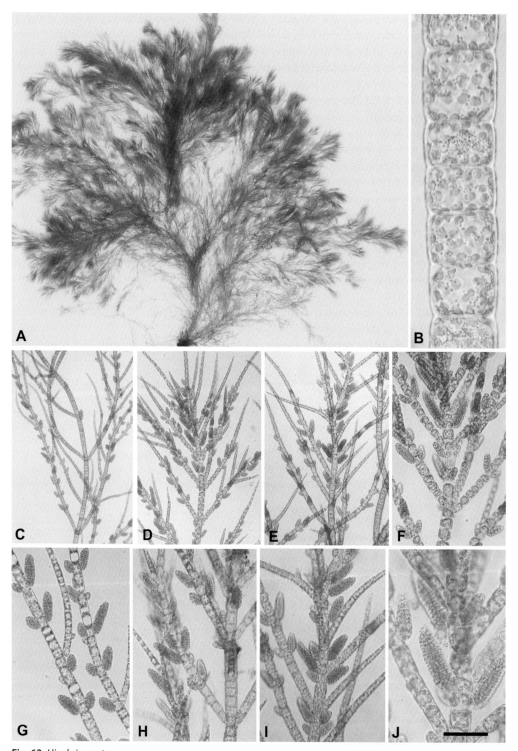

Fig. 63. *Hincksia ovata*
A. Habit of thallus. B. Portion of filament showing cells with enclosed discoid plastids. C–J. Portions of filaments showing branching patterns and frequently oppositely arranged, plurilocular sporangia. Bar in J = 350 μm (A), = 300 μm (C), = 180 μm (D, E), = 90 μm (F–H), = 66 μm (J), = 27 μm (B).

to rectangular, 0.5–1(–3) diameters long, 8–38 × 8–18 µm, each with several discoid plastids and associated pyrenoids.

Plurilocular sporangia abundant, sometimes scattered and solitary, or in short series, more usually in opposite pairs, opposite a branch, in whorls, or clustered on small branchlets, sessile, rarely pedicellate, ovoid to elongate-ovoid/conical, more rarely cylindrical in shape, strongly recurved towards parental filament, measuring 30–85(–100) × 16–23(–30) µm; unilocular sporangia rare, scattered or more often in opposite pairs, sessile, ovoid, measuring 20–40 × 20–35 µm.

Rare, epiphytic and epizoic on various hosts in the lower eulittoral and shallow sublittoral.

Only recorded from a small number of widely distributed localities around Britain and Ireland.

Appears to be a winter/spring annual, mainly November to May, more rarely reported in the summer months.

This is a small species, characterized by the abundant ovoid to conical plurilocular sporangia, which are usually sessile and arranged in opposite pairs or opposite a branch.

Brodie *et al.* (2016), p. 1019; Cardinal (1964), pp. 49–51, fig. 26(A–L), (1987), p. 47, figs 24, 25; Clayton (1974), p. 767, figs 13, 14; Coppejans & Kling (1995), p. 202, pl 68(A–G); Cormaci *et al.* (2012), pp. 109–10, pl. 25(1–3); Dizerbo & Herpe (2007), p. 67; Fletcher in Hardy & Guiry (2003), p. 262; Guiry (2012), p. 136; Kim (2010), pp. 61–3, fig. 28(A–D); Kim & Lee (1992b), p. 252, figs 6, 8(A–D); Knight & Parke (1931), p. 63, pl. xii(30); Kornmann (1953b), p. 17, (1954), pp. 41–52, figs 1–5; Kornmann & Sahling (1977), p. 112, fig. 57(A–D); Kuckuck (1961), pp. 140–52, pls 12(A–C), 16(A–E); Kylin (1947), pp. 9, 10, fig. 3(A, B); Lund (1959), pp. 43–8, fig. 4(A–E); Mathieson & Dawes (2017), p. 139, pl. XLII(12, 13); Misra (1966), pp. 95, 96, fig. 46(A, B); Morton (1994), p. 32; Newton (1931) pp. 121, 122; Pedersen (1979a), pp. 57–65, figs 1–5, (2011), pp. 67, 68, fig. 69(A, B); Ribera *et al.* (1992), p. 112; Rosenvinge (1898), pp. 77–80, fig. 16(A–C); Rosenvinge & Lund (1941), pp. 46–9, figs 20–2; Rueness *et al.* (2001), p. 15; Schneider & Searles (1991), p. 125, fig. 140; Taylor (1957), pp. 110, 111; Womersley (1987), pp. 54, 56, fig. 14(A–C).

Hincksia sandriana (Zanardini) P.C. Silva (1987), p. 130.   Fig. 64

Ectocarpus sandrianus Zanardini (1843), p. 41; *Ectocarpus sandrianus* Zanardini in Kützing (1849), p. 451; *Ectocarpus elegans* Thuret (1863), p. 77; *Ectocarpus sandrianus* var. *balticus* Reinke (1889b), p. 43; *Ectocarpus sandrianus* f. *implexus* Gran (1893), p. 31, fig. 8; *Ectocarpus granulosoides* Setchell & N.L. Gardner (1922), p. 410; *Ectocarpus parksii* Setchell & N.L. Gardner (1924), p. 1; *Giffordia sandriana* (Zanardini) Hamel (1931–39), p. xiv; *Giffordia granulosoides* (Setchell & N.L. Gardner) Hollenberg & I.A. Abbott (1966), p. 18 [628]; *Giffordia elegans* (Thuret) Knoepffler-Péguy (1974), p. 64; *Hincksia sandriana* var. *baltica* (Reinke) Athanasiadis (1996), p. 142.

Thalli epiphytic, sometimes epilithic, forming dense, usually flaccid and delicate, light brown, much-branched filamentous tufts up to 3–8 cm in length, solitary or confluent, sometimes covered at the base in downward-growing, corticating secondary rhizoids which spread out over the host thallus/substratum; main axes irregularly, subdichotomously or alternately branched below, unilateral and comb-like (secund) above, with branches somewhat divergent and tapering from base to apex, terminal cell acute without obvious false hairs; cells quadrate to rectangular, cylindrical to barrel-shaped, 1–3 diameters long, quite thick walled, 40–128 × 33–52 µm in the basal regions, 1–2 diameters long, 20–47 × 25–32 µm in the upper branch regions, gradually becoming more elongate and narrower terminally, 1–3 diameters long, 9–25 × 8–16 µm, each containing several discoid plastids with pyrenoids; growth intercalary, diffuse.

Plurilocular sporangia very common and abundant, sessile, rarely on one-celled stalks, elongate-ovoid to slightly conical in shape, narrowing to a point, distinctly recurved towards parental filament and adaxially arranged, often in a long secund series, on the upperside of the ultimate branchlets, 25–78(–90) × 13–20(–30) µm; unilocular sporangia rare, 15–30(–35) × 50–60 µm, oval in shape, sessile and positioned adaxially on the parental filament.

Usually found epiphytic on various algae and *Zostera* spp., sometimes epilithic, in eulittoral pools and shallow sublittoral, particularly in sheltered places; not uncommonly found on floating structures and associated with marinas and docks.

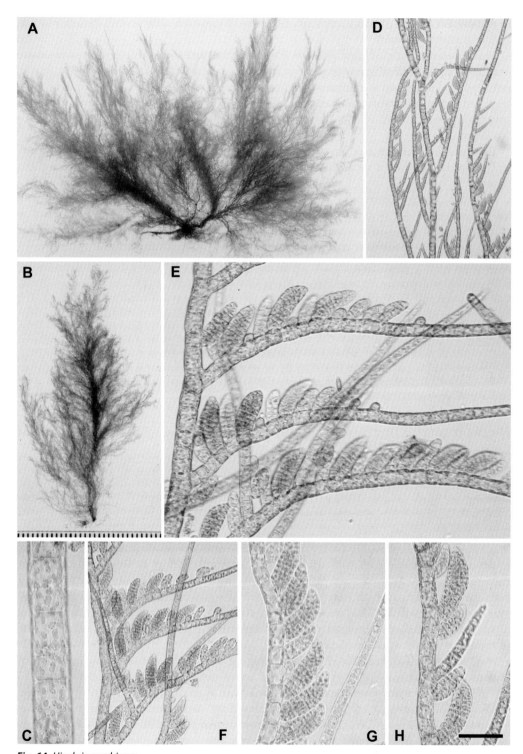

Fig. 64. *Hincksia sandriana*

A, B. Habit of thalli. C. Portion or filament showing cells with discoid plastids. D. Portions of erect filaments showing unilateral branching pattern above. E, F. Portions of erect filaments showing secundly arranged plurilocular sporangia on the upper sides of branches. G, H. portions of filaments showing unilaterally arranged plurilocular sporangia. Scale bar (mm) in B also applies to A. Bar in H = 180 μm (D), = 90 μm (E, F), = 45 μm (C, G, H).

Common, and widely distributed around Britain and Ireland.

Appears to be a spring/summer annual, mainly March to June, more rarely reported at other times.

This species is characterized by its very delicate, flaccid habit and the abundant plurilocular sporangia arranged adaxially on the branches, often in a secund series.

Abbott & Hollenberg (1976), pp. 145, 146, fig. 106; André (1970), pp. 236, 237; Bartsch & Kuhlenkamp (2000), p. 166; Brodie *et al.* (2016), p. 1019; Cardinal (1964), pp. 37, 38, fig. 18(A–J); Clayton (1974), p. 782, figs 23–4; Cormaci *et al.* (2012), pp. 110–12, pl. 25(4–8); De Mesquita Rodrigues (1963), p. 23, pl. 2(D); Dizerbo & Herpe (2007), pp. 67, 70, pl. 20(3); Ercegovic (1955), p. 26, fig. 12; Fletcher (1980), p. 40, pl. 13(1–4); Fletcher in Hardy & Guiry (2003), p. 262; Gallardo García *et al.* (2016), p. 34; Guiry (2012), p. 136; Hamel (1931–39), pp. 28, 29, fig. 6(I–II); Hollenberg & I.A. Abbott (1966), p. 18[628], as *Giffordia granulosoides* (Setchell & N.L. Gardner) Hollenberg & I.A. Abbott; Kim (2010), p. 65, fig. 30(A–D); Kim & Lee (1992), pp. 248–50, fig. 4(A–D); Kornmann & Sahling (1977), p. 109, pl. 56(A–E); Kützing (1849), p. 451; Kylin (1947), p. 10, fig. 3(C–D); Le Jolis (1863), p. 77, pl. II; Mathieson & Dawes (2017), p. 139, pl. XLII, figs 12, 13; Misra (1966), pp. 94, 95, fig. 45(A–C); Morton (1994), p. 32; Newton (1931), p. 119; Ribera *et al.* (1992), p. 112; Rosenvinge & Lund (1941), p. 44, fig. 18; Setchell & N.L. Gardner (1922), p. 410, pl. 45, figs 7, 8, (1925), p. 431; Smith (1944), p. 82, pl. 11, figs 3, 4; Stegenga & Mol (1983), pp. 73, 75, 76, pl. 21, figs 1–3; Stegenga *et al.* (1997a), p. 16; Taylor (1957), pp. 111, 112, (1960), p. 207; Womersley (1987), pp. 50–2, fig. 12(A–D); Zanardini (1865), p. 143, pl. 74b.

Hincksia secunda (Kützing) P.C. Silva (1987), p. 130.   Fig. 65

Ectocarpus secundus Kützing (1847), p. 54; *Giffordia secunda* (Kützing) Batters (1893), p. 85.

Thalli epiphytic, forming dense, light brown, uniseriate, filamentous tufts up to 2(–5) cm in length; principal filaments usually clothed at the base by a corticating sheath of downward-growing filaments which spread out across the host surface as branched, attaching, rhizoidal filaments; main axes sparsely branched below, more copiously branched above; branching alternate or unilaterally and secund, rarely opposite, sometimes recurved, with branches tapering from base to apex, sometimes into a recognizable, light-coloured false hair; ultimate branches more distinctly unilateral and adaxially branched; cells usually wider than long throughout, occasionally quadrate, becoming more pronouncedly longer than wide in the ultimate branchlet tips; basal cells 0.5–1(–1.5) diameters long, 28–88 × 40–72 μm, middle cells 0.4–1 diameters long, 20–55 × 48–59 μm, ultimate branchlets cells 0.5–1(–3) diameters long, 10–50 × 10–22 μm, each with numerous discoid plastids with pyrenoids; growth intercalary, diffuse.

Plurilocular sporangia common, sessile, sometimes solitary, but more often in rows of 2–3, secund and adaxial in arrangement on filaments, 50–92 × 30–60 μm, ovate to obtuse, broad-based, incurved and reported to be of three types: megasporangia comprising large locules with dark-pigmented contents probably functioning as oogonia, microsporangia comprising much smaller locules with yellow- to orange-coloured contents, probably functioning as antheridia, and neutral plurilocular sporangia with compartments intermediate in size between those of the micro- and megasporangia, producing zoospores asexual in behaviour; unilocular sporangia unknown.

Epiphytic on various algae, particularly on *Saccorhiza polyschides*, in lower eulittoral pools and shallow sublittoral.

Frequent and widely distributed around Britain and Ireland.

Found throughout most of the year, spring to autumn, more rarely found at other times.

In habit and microscopic characteristics, especially with respect to the plurilocular sporangia, this species bears a close resemblance to *Hincksia granulosa*. However, it is smaller, and unlike the latter species, it has unilateral branching and never displays opposite branching. Unlike *H. granulosa*, *H. secunda* has also been reported with dimorphic plurilocular sporangia, although these have not been observed by the present author.

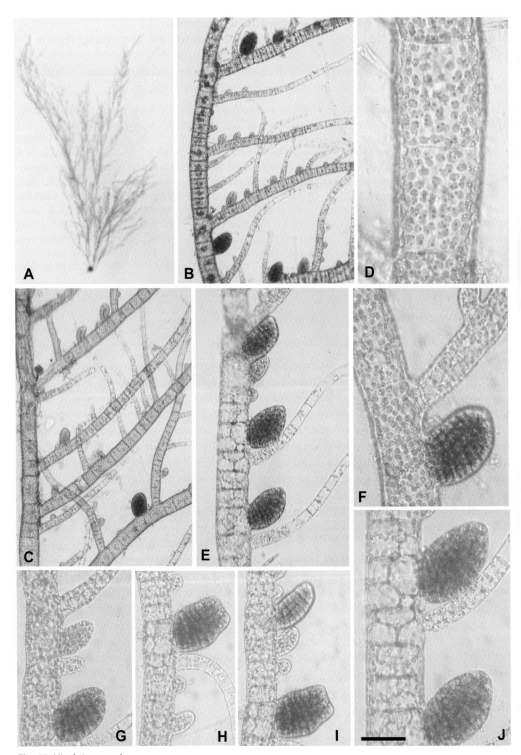

Fig. 65. *Hincksia secunda*
A. Habit of thallus. B, C. Portions of erect filaments showing immature and mature plurilocular sporangia and unilaterally arranged branches terminating in hairs. D. Portion of filament showing cells with enclosed discoid plastids. E–J. Portions of filaments showing lateral branches with terminal hairs and sessile, unilaterally arranged plurilocular sporangia. Bar in J = 800 µm (A), = 180 µm (B, C), = 90 µm (E), = 45 µm (D, F, G, H, I, J).

It was the occurrence of male and female gametangia, reported by Bornet (1892a) that prompted Batters (1893) to transfer the species from *Ectocarpus* to a new genus, *Giffordia*. The occurrence of a sexual life history, with observed fusion of the male and female anisogametes, was later shown by Sauvageau (1896). Supplementary reproduction was provided by parthenogenetic development of the megagametes and production of asexual zoospores from the neutral plurilocular sporangia.

Ardré (1970), p. 238; Batters (1893), pp. 85, 86; Bornet (1892a), pp. 353–6, pl. 6(1–8); Brodie *et al.* (2016), p. 1019; Cardinal (1964), pp. 42–4, fig. 21(A–H); Coppejans & Kling (1995), p. 204, pl. 69(A–C); Cormaci *et al.* (2012), p. 112, pl. 25(9–12); Dizerbo & Herpe (2007), p. 70, pl. 20(4); Fletcher in Hardy & Guiry (2003), p. 263; Gallardo García *et al.* (2016), p. 34; Guiry (2012), p. 136; Hamel (1931–39), pp. xv, 35–7, fig. 9; Kim & Lee (1992), p. 244, fig. 1; Kuckuck (1961), pp. 136–40, figs 10, 11; Mathieson & Dawes (2017), p. 140, pl. XLII(15); Morton (1994), p. 32; Newton (1931), p. 123; Ribera *et al.* (1992), p. 112; Sauvageau (1896b), p. 260, fig. 1, (1896d), pp. 388–98, fig. 1, (1931), pp. 971, 972, (1933), pp. 50–8; Stegenga & Mol (1983), p. 76, pls 21(3), 116(1); Stegenga *et al.* (1997a), p. 16; Taylor (1957), p. 112.

Pogotrichum Reinke

POGOTRICHUM Reinke (1892), p. 61.

Type species: *P. filiforme* Reinke (1892), p. 62.

Thalli epiphytic or epilithic, forming erect tufts attached by basally produced rhizoidal filaments, sometimes arising from a small basal disc; erect thalli simple, filiform, cylindrical or slightly elongate-clavate, uniseriate becoming pluriseriate and parenchymatous above; in section, outer cells usually small and pigmented, inner cells usually large and colourless, sometimes without size differentiation; cells with several discoid plastids and associated pyrenoids; hairs unknown.

Unilocular and plurilocular sporangia borne on the same or different thalli, developing from surface vegetative cells, single or more commonly clumped; unilocular sporangia ovate, or elliptical, formed directly in vegetative cells; plurilocular sporangia formed directly in vegetative cells by subdivision.

The genus *Pogotrichum* was originally described by Reinke (1892) based on *Pogotrichum filiforme*. The species was later transferred to the genus *Litosiphon* by Batters (1902) and renamed *Litosiphon filiformis*. The latter was widely accepted in the literature until Pedersen (1978a) revived *Pogotrichum*, recognising two species, *P. filiforme* and *P. setiforme* (Rosenvinge) Pedersen, and placed it together with the monotypic genus *Omphalophyllum* and species *O. ulvaceum*, into a new family Pogotrichaceae.

In Britain and Ireland only one species within this family, *P. filiforme*, is represented; *O. ulvaceum* is a deep water, Arctic species originally described from Greenland while *P. setiforme* has been described only from Greenland and Denmark. In addition, *P. setiforme* is doubtfully distinct as it appears to be included within the circumscription of *P. filiforme* although maintained as a separate species by Guiry & Guiry (2023).

According to Pedersen the characteristic features of the genus *Pogotrichum* which distinguish it from the genus *Litosiphon*, as typified with *L. pusillus* (Carm. ex Hook.) Harvey, include (a) the absence of hairs, (b) the extensive confluent development of sporangia which can leave greater parts of the thallus empty following spore release and (c) plurilocular sporangia are formed directly in surface vegetative cells, following simple subdivision, without preceding vegetative divisions.

Examination of British material by Fletcher (1987) confirmed these differences and support was given to Pedersen's (1978a) resurrection of Reinke's (1892) old name *Pogotrichum* to include the entity usually referred to as *Litosiphon filiforme* (Reinke) Batters. Fletcher also recognized the family Pogotrichaceae.

However, later molecular studies by authors such as Siemer (1989), Siemer *et al.* (1998), Peters & Ramírez (2001) and Rousseau & de Reviers (1999a) revealed a much broader concept of the Ectocarpales, to include the families Acinetosporaceae, Chordariaceae, Ectocarpaceae and Scytosiphonaceae, and that *Pogotrichum* lies positioned within the Acinetosporaceae, together with the genera *Acinetospora*, *Feldmannia*, *Hincksia*, *Pylaiella* and *Geminocarpus*. This placement is accepted in the present treatise, with the Acinetosporaceae taking priority over Pogotrichaceae Pedersen (1978a) and Pilayellaceae Pedersen (1984).

One species in Britain and Ireland.

Pogotrichum filiforme Reinke (1892), p. 62.   Fig. 66

Litosiphon filiformis (Reinke) Batters (1902), p. 25; *Pogotrichum filiformis* f. *gracilis* Batters (1892b), p. 18.

Thalli epiphytic, forming erect, light to dark brown tufts on hosts; tufts discrete, not confluent, commonly gregarious and densely fringing host surface, each comprising large numbers of erect, densely packed shoots arising from the centre of a discoid base; basal disc one cell layer, comprising outward-radiating, branched, closely packed filaments of cells, not firmly attached to host surface and often transversely divided with straight and oblique cross walls to form plurilocular sporangia; erect shoots 10–20(–50) mm long, 16–60(–115) μm wide, filiform, terete, solid, cylindrical, uniseriate, often becoming pluriseriate and parenchymatous above by frequent longitudinal and transverse divisions; cells subquadrate, quadrate or rectangular in uniseriate regions, 0.5–1(–2) diameters long, 14–55 × 18–35 μm, more irregularly shaped in polysiphonous regions, 13–34 × 12–27 μm, in transverse but not longitudinal rows below, irregularly placed above, each with several discoid plastids with pyrenoids; in section, inner cells colourless, outer cells pigmented, approximately equal in size or slightly differentiated into an inner medulla and outer cortex; hairs not present.

Unilocular and plurilocular sporangia borne on different, less commonly similar shoots, single, or more commonly grouped, formed from surface cells; unilocular sporangia rare, ovate, elliptical, bulging slightly, 19–29 × 16–21 μm; plurilocular sporangia common, formed directly from single cortical cells following subdivisions, often in continuous sori, 14–26 × 12–16 μm.

Epiphytic on various hosts, in particular *Saccharina latissima* and *Laminaria digitata*, lower eulittoral pools and shallow sublittoral.

Common and widely distributed around Britain and Ireland.

Annual, spring and summer (April to September).

Culture studies on Greenland material by Pedersen (1978a), using a wide variety of culture conditions, indicate that a 'direct' life history is in operation, without a sexual process; swarmers from the plurilocular sporangia on both the prostrate and erect parts of the thalli germinated without copulation to repeat the fertile parental phase. A similar 'direct' life history was reported by Kuckuck (1917) and Kylin (1937).

Bartsch & Kuhlenkamp (2000), p. 167; Batters (1892b), pp. 18, 19, (1902), p. 25; Brodie *et al.* (2016), p. 1019; Coppejans & Kling (1995), p. 177, pl. 52(A–F); Dizerbo & Herpe (2007), p. 115; Gallardo García *et al.* (2016), p. 34; Fletcher (1987), pp. 163–5, fig. 34(A–E); Guiry (2012), p. 137; Hamel (1931–39), pp. 219, 220, fig. 43(8–10); Jaasund (1965), pp. 85–7, figs 24, 25; Kornmann & Sahling (1977), p. 128, fig. 67; Kuckuck (1917), pp. 557–66, figs 1–3; Kylin (1947), p. 75, fig. 60(B–D); Lund (1959), pp. 150–2, fig. 33; Mathieson & Dawes (2017), p. 140, pl. XLII(16–18); Newton (1931), p. 181; Pedersen (1978a), pp. 61–8, figs 1–11, (2011), pp. 71, 72, fig. 74; Reinke (1892a), p. 62, figs A, B, pl. 41(12–25); Rosenvinge & Lund (1947), pp. 20, 21, fig. 4; Rueness (1977), p. 162, fig. 94, (2001), p. 16; Stegenga & Mol (1983), p. 97, pl. 23(2–6); Stegenga *et al.* (1997a), p. 20; Taylor (1957), pp. 162, 163.

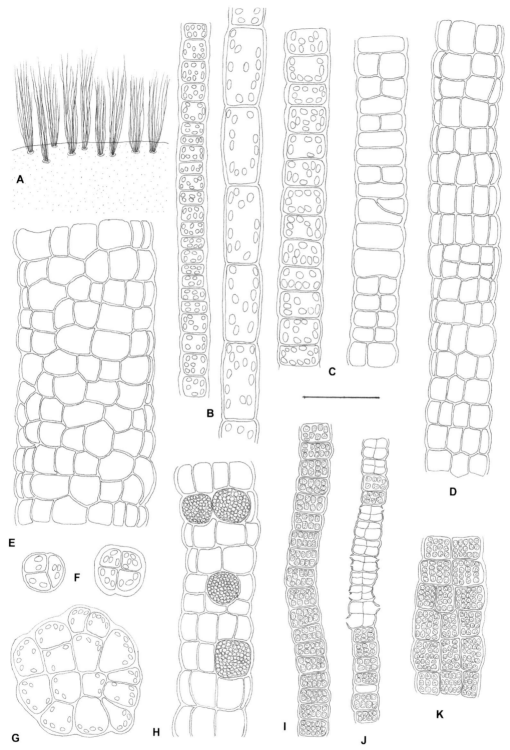

Fig. 66. *Pogotrichum filiforme*

A. Habit of thalli epiphytic on *Laminaria*. B–E. Portions of erect thalli showing different stages of development. F, G. TS of three thalli at different stages of development. H. Portion of erect thallus with unilocular sporangia. I–K. Portions of erect thalli at different stages of development, with plurilocular sporangia. Note dehisced sporangia with released contents. Bar = 12 mm (A), 50 μm (others).

Pylaiella Bory Saint-Vincent

PYLAIELLA Bory de Saint-Vincent (1823), p. 393.

Type species: *P. littoralis* (Linnaeus) Kjellman (1872), p. 99, nom. cons.

Thalli epilithic, epiphytic and epizoic, sometimes loose-lying/floating in sheltered conditions, erect, tufted, to 10–20 cm, exceptionally to 1 m in length, filamentous, monosiphonous, with occasional longitudinal divisions, much branched; branching opposite, usually unequal, or irregular, with branches terminating in false hairs; true hairs with basal meristem absent; growth diffuse; plastids discoid with pyrenoids.

Plurilocular and unilocular sporangia intercalary, more rarely terminal, on ultimate branchlets, derived by transformation of vegetative cells.

One species in Britain and Ireland.

Pylaiella littoralis (Linnaeus) Kjellman (1872), p. 99, nom. cons.   Figs 67, 68

Pylaiella littoralis var. *brachiata* Batters (1902), p. 34; *Pylaiella littoralis* var. *longifructus* (Harvey) Batters (1902), p. 34; *Pylaiella littoralis* var. *ramellosa* Kuckuck; *Conferva compacta* (Roth) Weber & Mohr; *Ceramium tomentosum* var. *ferrugineum* Roth (1800b), p. 470; *Conferva littoralis* Linnaeus (1753), p. 1165; *Ceramium compactum* Roth (1806), p. 148; *Ceramium littorale* (Linnaeus) Dillwyn (1809), pl. 31; *Ceramium ferrugineum* (Roth) C. Agardh (1811), p. 18; *Ceramium brachiatum* C. Agardh (1817), p. 67; *Ceramium ferrugineum* var. *compactum* C. Agardh (1817), p. 65; *Ectocarpus littoralis* (Linnaeus) Lyngbye (1819), p. 130; *Conferva ferruginea* Lyngbye (1819), p. 159; *Ectocarpus littoralis* var. *protensus* Lyngbye (1819), pl. 42 C; *Ectocarpus littoralis* var. *ruber* Lyngbye (1819), p. 131; *Ectocarpus ferrugineus* C. Agardh (1824), p. 163; *Ectocarpus brachiatus* (C. Agardh) C. Agardh (1824), p. 163; *Ectocarpus ferrugineus* var. *compactus* (C. Agardh) C. Agardh (1824), p. 163; *Ectocarpus siliculosus* var. *protensus* (Lyngbye) C. Agardh (1824), p. 162; *Ectocarpus compactus* (Roth) C. Agardh (1828), p. 41; *Ectocarpus siliculosus* var. *firmus* C. Agardh (1828), p. 38; *Ectocarpus littoralis* var. *brachycarpus* C. Agardh (1828), p. 40; *Ectocarpus compactus* var. *densissimus* C. Agardh (1828), p. 42; *Lyngbya littoralis* (Linnaeus) Dillwyn ex Gaillon (1828), p. 394, nom. inval.; *Ulothrix compacta* (Roth) Kützing (1833), p. 520; *Ectocarpus subverticillatus* Kützing (1845), p. 255; *Ectocarpus ramellosus* Kützing (1845), p. 236; *Ectocarpus firmus* (C. Agardh) J. Agardh (1848), p. 23; *Ectocarpus longifructus* Harvey (1849), pl. CCLVIII; *Ectocarpus littoralis* var. *brachiatus* (C. Agardh) Areschoug (1850), p. 402; *Ectocarpus littoralis* var. *compactus* (C. Agardh) Areschoug (1850), p. 402; *Ectocarpus firmus* f. *vernalis* Areschoug (1861–72), p. 105; *Ectocarpus firmus* var. *rupincola* Areschoug (1861–72), nr. 113; *Ectocarpus littoralis* f. *pumilus* Areschoug (1861–72); *Pylaiella littoralis* f. *compacta* Kjellman (1872), p. 105; *Pylaiella littoralis* f. *vernalis* Kjellman (1872), p. 100; *Pylaiella littoralis* f. *typica* Gobi (1878), p. 59, nom. inval.; *Pylaiella littoralis* f. *tilopterioides* Gobi (1878), p. 60; *Pylaiella litoralis* f. *rupincola* (Areschoug) Kjellman (1890), p. 84; *Pylaiella littoralis* var. *firma* (C. Agardh) Kjellman (1890), p. 84; *Pylaiella littoralis* var. *opposita* Kjellman (1890), p. 84; *Pylaiella littoralis* var. *divaricata* Kjellman (1890), p. 85; *Pylaiella littoralis* f. *aegagropila* Kjellman (1890), p. 85; *Pylaiella littoralis* f. *praetorta* Kjellman (1890), p. 85; *Pylaiella littoralis* f. *subsalsa* Kjellman (1890), p. 85; *Pylaiella littoralis* f. *olivacea* Kjellman (1890), p. 84; *Pylaiella littoralis* f. *parvula* Kjellman (1890), p. 84; *Pylaiella littoralis* f. *crassiuscula* Kjellman (1890), p. 84; *Pylaiella littoralis* f. *elongata* Kjellman (1890), p. 84; *Pylaiella littoralis* f. *nebulosa* Kjellman (1890), p. 84; *Ectocarpus littoralis* subsp. *divaricatus* (Kjellman) Kuckuck (1891), p. 12; *Ectocarpus littoralis* subsp. *firmus* (C. Agardh) Kuckuck (1891), p. 9; *Ectocarpus littoralis* subsp. *oppositus* (Kjellman) Kuckuck (1891), p. 8; *Ectocarpus littoralis* subsp. [*divaricatus*] f. *ramellosus* Kuckuck (1891), p. 12; *Ectocarpus littoralis* subsp. [*firmus*] f. *lividus* Kuckuck (1891), p. 10; *Ectocarpus littoralis* subsp. [*firmus*] f. *subglomeratus* Kuckuck (1891), p. 10; *Ectocarpus littoralis* subsp. [*oppositus*] f. *rectangulans* Kuckuck (1891), p. 9; *Ectocarpus littoralis* subsp. [*firmus*] f. *pachycarpus* Kuckuck (1891), p. 11; *Ectocarpus littoralis* subsp. [*oppositus*] f. *subverticillatus* (Kützing) Kuckuck (1891), p. 8; *Ectocarpus littoralis* var. *divaricatus* (Kjellman) Rosenvinge (1893), p. 881; *Ectocarpus littoralis* var. *firmus* (C. Agardh) Rosenvinge (1893), p. 881; *Ectocarpus littoralis* var. *oppositus* (Kjellman) Rosenvinge (1893), p. 881; *Pylaiella littoralis* f. *rectangularis* De Toni (1895), p. 532, nom. illeg.; *Pylaiella ramellosa* (Kützing) Laing (1927), p. 136; *Pylaiella rupincola* (Areschoug) Kylin (1937), p. 5; *Pylaiella kylinii* Du Rietz (1941), p. 6.

Thalli epilithic, epiphytic or epizoic, sometimes loose-lying or floating in sheltered conditions, pale yellowish-brown to mid-brown or red-brown in colour, erect, tufted, much branched throughout, filamentous, uniseriate, with occasional longitudinal divisions in older somatic

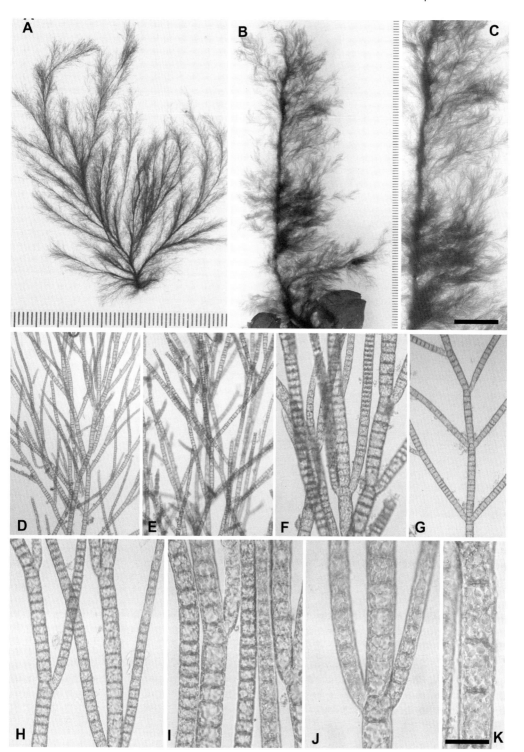

Fig. 67. *Pylaiella littoralis*
A–C. Habit of thalli. D–J. Portions of erect filaments showing both unilateral and opposite branching. K. Portion of filament showing cells with enclosed discoid plastids. Bar in C = 12 mm. Bar in K = 300 μm (D), = 180 μm (G), = 90 μm (E, F, H, I), = 45 μm (J, K).

Fig. 68. *Pylaiella littoralis*

A–E. Portions of filaments with rows of intercalary unilocular sporangia. F–H. Portions of filaments with plurilocular sporangia. Bar in D = 180 μm (A, B, C), = 90 μm (D, E, G), = 45 μm (F, H).

cells, to 10(–20), exceptionally to over 90 cm in length, the principal filaments often covered in the lower regions by a sheath of corticating filaments which spread out at the base as branched, filamentous, secondary, attaching rhizoids; principal filaments usually plaited to form a cable; branching opposite, usually unequal, sometimes irregular, with branches terminating in long, colourless, false hairs; true hairs with basal meristem absent; growth diffuse, with no obvious intercalary, meristematic regions; basal cells subquadrate to rectangular, 0.5–1(–3) diameters long, (28–)40–80(–108) × (31–)36–40(–43) μm, upper cells subquadrate to rectangular 0.3–1(–2) diameters long, (13–)20–38(–48) × (20–)25–43(–48) μm; all cells with large numbers of discoid plastids and associated pyrenoids; hairs colourless, with few plastids, terminal on filaments and sporangia, comprising cells 1–4 diameters long, 20–80 × 11–25 μm.

Plurilocular and unilocular sporangia common, intercalary in position, more rarely terminal, formed by direct transformation of one or more vegetative cells; plurilocular sporangia irregularly cylindrical, to 540 μm long, 35–56 μm in diameter, releasing zoospores via one or more lateral pores; unilocular sporangia sometimes solitary, but usually in series of up to 33, barrel-shaped, uniseriate, occasionally biseriate, 30–35 µm in diameter, releasing zoospores through a single lateral pore.

Common and widespread, over the whole eulittoral zone, sometimes extending into the littoral fringe and sublittoral, both emersed and in tidal pools, occurring lithophytically, epiphytically and epizoically, on a wide range of hosts and substrata, and on a wide variety of shores. Particularly common as an epiphyte of *Fucus* species. It is one of the few branched, filamentous, brown algae occurring exposed on rocks. It is very common in estuaries, salt marshes, ditches and other habitats of unstable salinity. It is also a common contributor to loose-lying communities developing in sheltered eutrophic environments, adopting a different morphology with widely divergent branches in very irregular patterns. The existence of euryhaline genotypes is known.

Abundant and widely distributed around Britain and Ireland. *P. littoralis* is probably the most commonly occurring and successful, small, brown 'ectocarpoid' species on British and Irish rocky shores.

A wide seasonal distribution pattern, usually being found abundant at all times of the year.

A highly morphologically variable species, with wide variation in branching positions linked with thallus age and environments.

Abbott & Hollenberg (1976), p. 148, fig. 109; Asensi (1966), pp. 1–14; Assali & Loiseaux-de Goër (1992), pp. 209–13; Assali *et al.* (1990), pp. 307–15; Bartsch & Kuhlenkamp (2000), p. 167; Brodie *et al.* (2016), p. 1019; Cabioc'h & Boudouresque (1992), p. 55, fig. 18; Cardinal (1964), pp. 11–13, fig. 1(A–L); Coppejans & Kling (1995), p. 208, pl. 71(A–F); Cormaci *et al.* (2012), p. 116, pl. 26(5–9); Dalmon & Loiseaux (1981), pp. 241–51; Dizerbo & Herpe (2007), pp. 71, 73, pl. 22(2); Fletcher (1980), pp. 41, 42, pl. 14(1–3); Fletcher in Hardy & Guiry (2003), p. 291; Gachon *et al.* (2007), pp. 395–403; Gallardo García *et al.* (2016), p. 35; Gayral (1966), p. 223, pl. XXVII; Geoffroy *et al.* (2015), pp. 480–9; Gross & Cheney (1993), p. 8, (1994), p. 13; Guiry (2012), p. 137; Hamel (1931–39), pp. 11–14, fig. 2(A–C); Hauck (1885), p. 339, fig. 42; Knight (1923), pp. 343–60, pls 1–6; Kornmann & Sahling (1977), p. 105, pl. 53(A–E); Kuckuck (1891), pp. 3–6, 33–9; Kylin (1947), p. 5, fig. 1(A–C); Littlauer (2010), pp. 1–2; Loiseaux & Mache (1977), p. 110; Loiseaux & Rozier (1978), pp. 333–40; Loiseaux *et al.* (1979), pp. 619–29, (1980), pp. 381–8; Longtin & Scrosati (2009), pp. 535–8; Maier *et al.* (1998), pp. 213–20, figs 1–23; Markey & Wilce (1975), pp. 219–41, (1976a), pp. 147–73; Martin *et al.* (1993), pp. 111–13; Mathieson & Dawes (2017), pp. 141–2, pl. XLIII(1–3); Misra (1966), pp. 70, 71, fig. 27(A–C); Morton (1994), p. 32; Müller & Stache (1989), pp. 71–8; Newton (1931), p. 124, fig. 71; Pankow (1971), pp. 145, 146, figs 159–61; Pedersen (1984), pp. 13–18, figs 1–7; Ribera *et al.* (1992), p. 113; Rosenvinge & Lund (1941), pp. 5–11; Rueness (1977), p. 120, fig. 53; Russell (1958), pp. 34, 35, (1961a), pp. 30, 31, (1961b), p. 101, (1963), pp. 469–83, (1964a), pp. 322–6; Saunders & McDevit (2013), pp. 1–23; Setchell & N.L. Gardner (1925), pp. 402–4, pl. 37(32); Smith (1944), p. 93, pl. 12, fig. 3; Stegenga & Mol (1983), p. 78, pl. 23(1–4); Stegenga *et al.* (1997a), p. 17; Taylor (1957), pp. 102, 103, pl. 9(1–3); Tokida (1954), pp. 77, 78; Waern (1952), pp. 110–13, fig. 49; Wilce *et al.* (1982), pp. 336–54; Womersley (1987), pp. 38–40, fig. 7(G–J).

Chordariaceae Greville

CHORDARIACEAE Greville (1830), p. 44.

Thalli epilithic, epiphytic, epizoic or largely endophytic, with or without obvious heterotrichy; prostrate/attaching filaments uniseriate, branched, endophytic and spreading in host tissue, or on a substratum surface, free or uniting to form a pseudoparenchyma, or larger holdfast-like base; erect filaments remaining monosiphonous throughout or becoming parenchymatous by both transverse and longitudinal divisions; monosiphonous filaments simple or branched, free and tufted, microscopic or macroscopic, or weakly/strongly coalesced to form pulvinate, globose, cylindrical or flattened, cartilaginous or gelatinous, pseudoparenchymatous thalli up to 50 cm in height and originating from either a single axial filament (uniaxial) or by several axial filaments (multiaxial) forming a medullary core; erect parenchymatous thalli limited in development with few longitudinal divisions or extensive in development to form variously sized and shaped, solid or hollow, globose, terete or flattened macroscopic thalli; both pseudo-parenchymatous and parenchymatous thalli often internally differentiated into a central medullary region of large colourless cells, with or without accessory secondary assimilatory/ hyphal filaments, which give rise to an outer cortex-like region of smaller, pigmented cells; true hairs with basal meristem; cells with discoid or lobed, plate-like plastids with or without (pedunculate) pyrenoids; growth intercalary, diffuse, apical, subapical or trichothallic.

Reproduction by plurilocular (neutral) sporangia, plurilocular gametangia and unilocular sporangia; life history haplodiplontic and heteromorphic with a microscopic, plurilocular gametangia-bearing gametophyte generation developing from spores released from unilocular sporangia on the more macroscopic sporophyte phase.

The Chordariaceae is now recognized as a large, morphologically diverse and rich family with 95 genera listed by Silberfeld *et al.* (2014). On the basis of several molecular studies (see, in particular, Siemer *et al.* 1998; Rousseau & de Reviers 1999a; Peters & Ramírez 2001; Draisma *et al.* 2003; Cho *et al.* 2004; de Reviers *et al.* 2007), it now incorporates all genera traditionally placed within the Chordariales, Dictyosiphonales and Punctariales and, correspondingly, the great majority of families and genera which had been previously assigned within the broader, monophyletic concept of the Ectocarpales originally proposed by Fritsch (1945) and accepted by various authors such as Parke & Dixon (1976). The families Myrionemataceae, Elachistaceae, Corynophlaeaceae, Chordariaceae, Acrotrichaceae, Spermatochnaceae, Striariaceae, Myriotrichiaceae, Giraudiaceae, Punctariaceae, Buffhamicaeae and Dictyosiphonaceae, and over half the genera placed within the Ectocarpaceae, listed in Parke & Dixon (1976), are now placed within the Chordariaceae, giving a total of 36 genera and approximately 89 species present in Britain and Ireland.

Acrothrix Kylin

ACROTHRIX Kylin (1907), p. 93.

Type species: *A. gracilis* Kylin (1907), p. 93.

Thalli macroscopic, erect, to 40 cm high and 1 mm wide, arising from a small discoid holdfast; thalli cylindrical, at first solid, quickly becoming tubular, much branched, branching regularly alternate, occasionally irregular in parts, all branches tapering gradually above and terminating in a hair; thalli haplostichous with a central, uniseriate, axillary filament of markedly elongated cells from which whorls of primary laterals are produced which branch to form a surrounding medulla of 1(–3), large, thin-walled, longitudinally elongate, transversely rounded or irregularly shaped, colourless, perimedullary cells which become progressively smaller, more pigmented and cortex-like towards the periphery; primary laterals (and later the cortical cells) also giving rise to free, multicellular, primary assimilators and true hairs, with basal meristem, which cover the thallus surface; primary assimilators thinning out below apex and replaced by the later-formed secondary assimilators; growth apical and trichothallic, followed by subterminal, pseudoparenchymatous growth of the primary laterals; cells with numerous discoid plastids, each with a pyrenoid.

Unilocular sporangia produced at the base of the secondary assimilators, sessile, ovoid to slightly pyriform; plurilocular sporangia unknown on macrophyte.

One species in Britain and Ireland.

Acrothrix gracilis Kylin (1907), p. 93.   Figs 69, 70

Acrothrix novae-angliae Taylor (1928b), p. 577; *Acrothrix norvegica* Levring (1937), p. 62.

Thalli macroscopic, erect, solitary, to 40 cm high, to 0.6(–1) mm wide on main axes, light/ medium brown to pale brown in colour, firm, somewhat straw-like and not gelatinous, arising from a small discoid holdfast; thallus cylindrical, at first solid, quickly becoming tubular, much

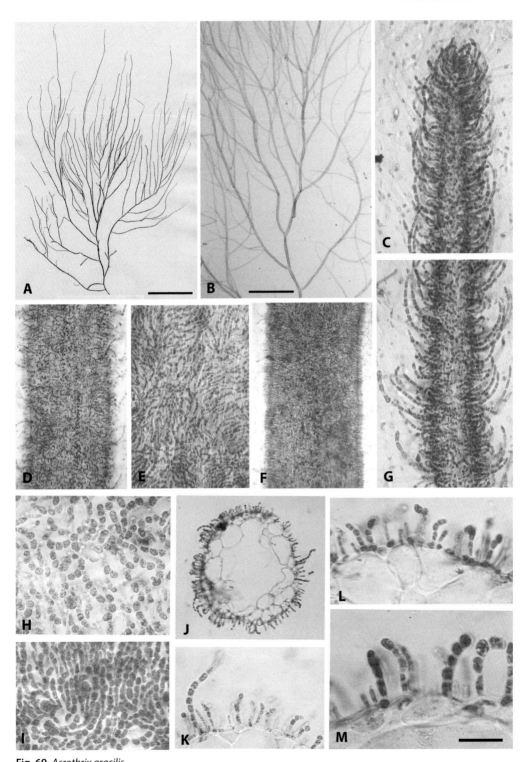

Fig. 69. *Acrothrix gracilis*

A, B. Habit of thalli. C. Apex of thallus showing recurved assimilatory filaments. D. Portion of thallus, subapical region, showing free primary assimilators. E–I. Surface views of portions of thalli showing thinly packed assimilators. Note unilocular sporangia in H and I. J–M. TS of thalli showing secondary assimilators. Bar in A = 20 mm, bar in B = 12 mm, bar in M = 235 µm (D, F), = 90 µm (C, E, G), = 45 µm (H–M).

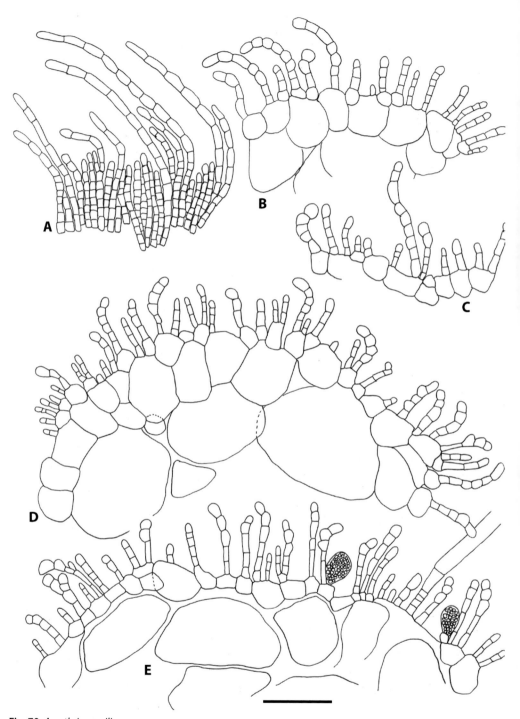

Fig. 70. *Acrothrix gracilis*

A. SP of thallus near apex showing recurved primary assimilators. B, C. TS of thallus edge showing a mixture of primary and secondary assimilators. D, E. TS of thallus edge showing central lacuna and an outer subcortex region of large, colourless cells, the outer cells giving rise to a fringe of secondary assimilators, unilocular sporangia and hairs. Bar = 50 µm.

branched, to 3–4 orders, especially towards the base, branching regularly alternate, occasionally irregular in parts, with distinct primary axis and long, widely divergent, secondary branches, all branches tapering gradually above and terminating in a hair, more clearly seen on young branches; growth apical and trichothallic, by means of a short, transversely dividing, uniseriate meristem with terminal true hair, followed by subterminal, pseudoparenchymatous growth of the primary laterals; in structure, thalli haplostichous, with terminal apical meristem and hair (usually shed in older specimens) and a central, sometimes laterally displaced, uniseriate, axillary filament of markedly elongated cells (not always discernible in section) lying within a large central cavity, from which whorls of primary laterals are produced which branch to form a surrounding medulla of 1(–3), large, thin-walled, longitudinally elongate, transversely rounded or irregularly shaped, colourless, perimedullary cells which become progressively smaller and cortex-like towards the periphery, constituting a subcortex region; in the subterminal region, primary laterals also giving rise to free, long, upward-curved, unbranched, multicellular, slightly mucilaginous, primary assimilators and true hairs with basal meristem, both assimilators and hairs forming an overarching involucre around the apical meristem and subsequently covering the thallus surface, thinning out below and being replaced by the later-formed secondary assimilators; primary assimilators to 195 µm (13 cells) in length, with cells quadrate below to cylindrical and bead-like in appearance above, (8–)11–18(–21) × 5–8 µm, each with numerous discoid plastids with pyrenoids; hair cells colourless, to 15 diameters in length, 40–145 µm; secondary assimilators also produced directly from the primary laterals, are usually distributed over the entire body of the thallus and are shorter than the primary assimilators, (4–)5–7(–8) cells long, usually simple, rarely branched at the base, somewhat club-shaped, curved, comprising rectangular cells below, (6–)10–14(–18) × 3–6 µm, the terminal cells more isodiametric, curved and swollen, usually on one side, (5–)8–11(–13) × 6–8(–9) µm.

Unilocular sporangia produced at the base of the secondary assimilators, sessile, ovoid to slightly pyriform, 26–35(–50) × (13–)16–24(–30) µm; plurilocular sporangia unknown on macrophyte.

Epilithic on stones/pebbles, also reported to be epiphytic on various algae, shallow sublittoral to 10 m, occasionally found in lower eulittoral pools.

Rare, reported for northern shores of Britain (Inverness, Isle of Man, Argyll, Orkney and Shetland) and scattered localities in Ireland (Clare, Cork, Galway and Wexford).

A summer annual, June to September.

Culture studies on Japanese material by Ajisaka & Kawai (1986) revealed a 'direct' type of life history, with released zoospores germinating directly into a prostrate system of microscopic branched filaments which either formed plurilocular sporangia or issued the characteristic erect thalli with unilocular sporangia to complete the life history. Spores from the plurilocular sporangia also behaved asexually and developed in the same manner as the zoospores. Environmental conditions, notably water temperature and photoperiod, rather than any possible differences in ploidy level, influenced the life history process.

Ajisaka & Kawai (1986), pp. 129–36, figs 1–4; Brodie *et al.* (2016), p. 1019; Cormaci *et al.* (2012), pp. 120–2, pl. 27(1–2); Cotton (1912), p. 124; Fletcher in Hardy & Guiry (2003), p. 223; Forward & South (1985), pp. 347–59; Fritsch (1945), pp. 87–9, fig. 24(A–E); Guiry (2012), p. 137; Jaasund (1965), p. 67; Kawai (1983), pp. 167–72, figs 1–4; Kuckuck (1929), pp. 65–7, figs 87–91; Kylin (1907), p. 93, figs 22, 23, pl. 2, (1933), fig. 25, (1940), p. 45, figs 22, 23, (1947), pp. 60, 61, figs 52, 53; Levring (1937), pp. 62–4, fig. 9(A–B), pl. 1(1–2); Mathieson & Dawes (2017), pp. 143, 144, pl. LXIII(7–10); Morton (1994), p. 35; Parke (1933), pp. 13, 33–5, text fig. 16, pls II(7), IX(57); Rosenvinge & Lund (1943), pp. 36, 37, fig. 13(A–E); Rueness (1977), pp. 149, 150, pl. 20(1); Taylor (1928b), pp. 577–83, (1957), p. 150, pl. 13(1); Verlaque (2001), p. 38, fig. 21.

Asperococcus J.V. Lamouroux

ASPEROCOCCUS J.V. Lamouroux (1813), p. 277.

Type species: *non designatus*.

Thalli consisting of erect tubular fronds arising from a small stipe and basal disc or holdfast; fronds simple, solid or hollow, cylindrical, filiform, or elongate-clavate, inflated or dorsi-ventrally compressed, with or without dark brown punctations; comprising an inner medulla of large, colourless medullary cells enclosed by an outer cortex of smaller pigmented cells; cells with several discoid plastids and pyrenoids; hairs solitary or grouped arising superficially, with basal meristem and sheath.

Plurilocular and/or unilocular sporangia borne on fronds in discrete or continuous sori, arising from surface cells, and associated with multicellular, paraphysis-like filaments; unilocular sporangia spherical or pyriform; plurilocular sporangia cylindrical or conical, protruding, uniseriate or multiseriate.

The genus *Asperococcus* is characterized by an erect, simple, usually hollow thallus internally differentiated into an inner medulla of large, colourless cells and an outer cortex of small, pigmented cells; additional characters include cells with several discoid plastids with pyrenoids, hairs with a basal meristem and sheath and sporangia arising from surface cells usually associated with short, multicellular, paraphysis-like filaments. The four species present in Britain and Ireland are quite easily distinguished on morphological characters.

Although European isolates of three of these species (*A. bullosus*, *A. ensiformis* and *A. fistulosus*) have been investigated in culture, their life histories are still not fully understood and often conflicting interpretations have been made by various research workers. In general it seems likely that a 'direct' type of life history is operating, which is heteromorphic (comprising both macro- and microthalli) and monophasic; reported observations of copulation between plurispores derived from plurilocular sporangia on the microthalli (Knight *et al.* 1935 for *A. fistulosus* and *A. bullosus* (as *A. turneri*)) and between unispores derived from unilocular sporangia on the macrothalli (Knight *et al.* 1935 and Dangeard 1968c for *A. fistulosus*) must be treated with caution (see Pedersen 1984, pp. 19–20).

A notable aspect of the life history studies was the resemblance between the microthalli produced in the cultures of *Asperococcus* and morphology of species in the genus *Hecatonema*. Similar characteristics very often included a discoid/pseudodiscoid thallus, short upright, simple or little branched filaments, with siliquose, multiseriate plurilocular sporangia borne on the basal filaments or lateral/terminal on the erect filaments, cells with several discoid plastids and pyrenoids and hairs with basal meristem and enclosing sheath (see especially Kylin (1933), fig. 14; Dangeard (1968c) pl. 5(A); Pedersen (1984) figs 8–9). Pedersen identified the microthalli produced in his cultures of *A. fistulosus* with one or more of four species of *Hecatonema* depending on the light level under which they were grown (see p. 239 for discussion). As an adjunct to these studies, Loiseaux (1969) and Fletcher (1984) reported *Asperococcus*-like macrothalli developing in cultures of field-collected thalli of *Hecatonema terminale* (see p. 304).

Three species in Britain and Ireland, one species in Britain only.

KEY TO SPECIES
1. Thallus solid and less than 3 mm long and 0.5 mm wide. *A. scaber*
   Thallus hollow and more than 5 cm long and 2 mm wide. 2
2. Thallus distinctly dorsiventrally flattened . . . .*A. ensiformis*
   Thallus distinctly inflated 3
3. Thallus cylindrical and linear, gradually attenuate towards base, flaccid but
   fairly strong, up to 20 mm wide; paraphyses to 190 μm and 7 cells long . . . . *A. fistulosus*

Thallus inflated and bulbous, sharply attenuate towards base, soft and
delicate, easily torn, up to 40 mm wide; paraphyses to 120 μm and 4 cells long
.... *A. bullosus*

Asperococcus bullosus J.V. Lamouroux (1813), p. 277.   Figs 71(D), 72

Ulva turneri Dillwyn ex J.E. Smith (1814), pl. 2570; *Gastridium opuntium* Lyngbye (1819), p. 71; *Encoelium bullosum*
(J.V. Lamouroux) C. Agardh (1820), p. 146; *Physidrum bullosum* (J.V. Lamouroux) Delle Chiaje (1829), pl. 97;
Asperococcus turneri (Dillwyn ex Smith) W.J. Hooker (1833), p. 277; *Encoelium macgregori* Suhr ex Kützing (1859),
p. 4; *Asperococcus turneri* var. *profundus* (Feldmann) Ballesteros (1983), nom. inval.

Thalli forming erect, solitary, or more commonly gregarious, olive to light brown fronds arising
from a discoid holdfast; thalli simple, membranous, very soft and flaccid, easily torn, hollow,
inflated and bulbous, cylindrical, 10–30 cm long × 5–40 mm wide, attenuate sharply below to
a narrow stipe-like region, obtuse above, surface with or without small, dark-coloured punctations; surface cells quadrate or rectangular, in rows, 23–62 × 14–50 μm, each with numerous
discoid plastids and pyrenoids; in section, thalli distinctly hollow with an outer cortex of one,
rarely two, small, plastid-bearing cells enclosing an inner medulla of 1–2(–3) large, longitudinally elongate, colourless, often ruptured cells; hairs abundant, arising from surface cells,
solitary, later often grouped and associated with the reproductive sori, with a dark brown
basal meristem and enclosing sheath, 19–23(–28) μm in diameter.

Plurilocular and unilocular sporangia borne on the same or different thalli, crowded in
discrete, irregularly shaped, punctiferous sori, widely scattered over blade surface, produced
from the surface cells, with associated hairs and paraphysis-like filaments; unilocular sporangia
abundant, rounded in surface view, 28–55 μm in diameter, pyriform and sessile in transverse
section, 60–82 × 52–72 μm; plurilocular sporangia rare, ellipsoidal in section, sessile or on
1–3-celled stalks, multiseriate, 34–56 × 13–25 μm; paraphyses usually uncommon, erect, multicellular, simple, rigid, thick walled, dark pigmented (especially the terminal cell), linear or
slightly elongate-clavate, to 1–3 cells (120 μm) long, comprising quadrate to rectangular cells
1–2 diameters long, 32–55 × 26–32 μm.

Epilithic, less commonly epiphytic on various algae, upper to lower eulittoral pools and
sublittoral to 12 m.

Common and widely distributed around Britain and Ireland.

Annual, summer (April to October).

Culture studies have been carried out on this species by Sauvageau (1929), Knight *et al.*
(1935), Kylin (1918, 1933) and Pedersen (1984). Knight *et al.* (1935) reported a heteromorphic,
haplodiplontic life history, with zoospores from the unilocular sporangia on the macroscopic
phase germinating directly to form microscopic, creeping, filamentous gametophytes bearing
plurilocular sporangia, the gametes of which copulated to reform the *Asperococcus* thalli. A
'direct' diplophasic life history, however, was reported by Sauvageau (1929) and Kylin (1933) in
which zoospores from either unilocular or plurilocular sporangia formed creeping microscopic
filaments, with or without plurilocular sporangia, which could directly form new *Asperococcus*
thalli. Successive generations of this microscopic phase could also be produced by the direct
development, without copulation, of the zoospores from the plurilocular sporangia. The
culture results of Pedersen (1984) are in approximate agreement with those of Sauvageau and
Kylin. The life history appeared to be of the 'direct' type, with macrothalli developing directly
from the microthalli. Copulation between spores was not observed and the life history was
interpreted as apparently monophasic and heteromorphic. The present author agrees with this
proposed life history for *A. bullosus* and express doubt on the haplodiplontic, heteromorphic
cycle reported by Knight *et al.* (1935).

As shown for *Asperococcus fistulosus*, noteworthy here is the similarity between the plurilocular sporangia-bearing microthalli produced in the cultures of *A. bullosus* by Kylin and

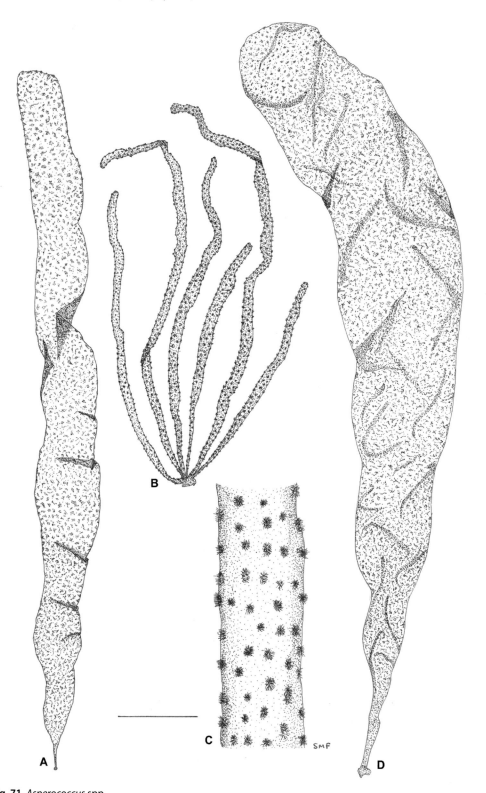

Fig. 71. *Asperococcus* spp.
A. *Asperococcus ensiformis*. Habit of thallus. B, C. *Asperococcus fistulosus*. B. Habit of thallus. C. Portion of erect thallus. D. *Asperococcus bullosus*. Habit of thallus. Bar = 3 cm (A, B), = 1.5 cm (D), = 1.2 mm (C).

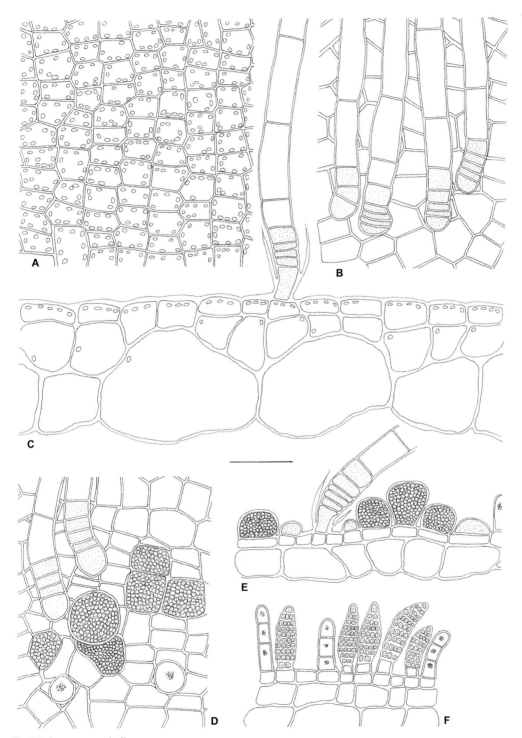

Fig. 72. *Asperococcus bullosus*

A. Surface view of thallus showing cells with numerous plastids. B. Surface view of thallus with hairs. C. Section of vegetative thallus showing outer 1–2 layers of small cortical cells (and emergent hair with basal sheath) and inner 1–2 layers of larger medullary cells enclosing central cavity. D. Surface view of fertile thallus showing unilocular sporangia. E. Section of fertile thallus with unilocular sporangia and hair arising from surface cells. F. Section of fertile thallus with plurilocular sporangia and associated paraphyses arising from surface cells. Bar = 50 μm.

Pedersen and the genus *Hecatonema* (see especially figure 14 in Kylin and figure 36E in Pedersen).

Ardré (1970), p. 277; Brodie *et al.* (2016), p. 1020; Bullhorn (1891), pp. 321–3, pl. 314(1–3); Bunker *et al.* (2017), p. 64; Cormaci *et al.* (2012), pp. 123, 124, pl. 27(3–5); Dizerbo & Herpe (2007), pp. 111, 114, pl. 37(3); Feldmann (1937), p. 154, fig. 55; Fletcher (1987), pp. 187–9, figs 40D, 44; Fletcher in Hardy & Guiry (2003), p. 226; Fritsch (1945), p. 107, fig. 37; Funk (1955), pp. 41, 42; Gallardo García *et al.* (2016), p. 35; Guiry (2012), p. 138; Hamel (1931–39), pp. 223–5; Harvey (1846–51), pl. 11; Knight *et al.* (1935), pp. 91–7, figs 1–3; Kylin (1933), pp. 38–44, figs 12–16, (1947), pp. 75, 76, fig. 60(e–f), pl. 11(38); Lamouroux (1813), p. 62, pl. 6(5); Lindauer *et al.* (1961), pp. 260, 261, pl. IV; Morton (1994), p. 40; Newton (1931), p. 172, fig. 107; Ohmori (1973), pp. 87–95; Oltmanns (1922), p. 63, fig. 355(5); Pedersen (1984), pp. 56, 57, figs 35, 36E; Ribera *et al.* (1992), p. 117; Rosenvinge & Lund (1947), pp. 41–3, fig. 14; Rueness (1977), pp. 159, 160, pl. 21(3), (2001), p. 14; Sauvageau (1929), pp. 385–96, figs 19–20; Womersley (1987), pp. 320, 322, figs 114(C), 116(A, B), 117(A); Yamada (1936), pp. 135–40.

Asperococcus ensiformis (Delle Chiaje) M.J. Wynne (2003), p. 474.   Figs 71(A), 73

Laminaria ensiformis Delle Chiaje (1829), p. 82; *Asperococcus compressus* Griffiths ex W.J. Hooker (1833), p. 278; *Haloglossum griffithsianum* Kützing (1843), pl. 52 I; *Haloglossum compressum* (Griffiths ex Hooker) Hamel (1931–39), p. 225.

Thalli forming erect, gregarious, olive-green, becoming light brown, thalli arising from a discoid holdfast; thalli simple, flaccid, hollow, but dorsiventrally compressed, elongate-clavate, to 40(–80) cm long, 8–20 mm wide, covered throughout with dark brown punctations; in section, thalli distinctly hollow with an inner medulla of 1–3 large, colourless, sometimes ruptured cells, enclosed by an outer cortex of 1–2 smaller pigmented cells; surface cells quadrate or rectangular, usually in rows, 20–30(–40) × 13–20(–30) μm, each with several discoid plastids and pyrenoids; hairs common, arising from surface cells and grouped in discrete punctate sori, with basal meristem and enclosing sheath, 11–14 μm in diameter.

Unilocular sporangia common, crowded in discrete, irregularly shaped, punctiferous sori, produced from surface cells, with associated hairs and paraphysis-like filaments; unilocular sporangia rounded in surface view, 27–46 μm in diameter, globose or pyriform in section, usually sessile, occasionally stalked, 40–60 × 26–43 μm; paraphyses erect, simple, colourless, linear or clavate, 1–3(–4) celled, 58–87 × 13–23 μm, arising from surface cells; plurilocular sporangia rare, not observed on British or Irish material, reputed to be on separate thalli, in small sori, cylindrical, conical, sessile or stalked, to 35 μm long × 9–15 μm wide.

Epiphytic, epilithic in eulittoral pools and sublittoral to 10 m.

Rare but widely distributed around Britain and Ireland.

Annual, spring and summer, April to September.

Reinke (1878a) and Sauvageau (1929) reported a 'direct' type of life history in this species with unispores from the unilocular sporangia germinating to produce creeping, filamentous, sometimes plurilocular sporangia-bearing microthalli from which new, erect macrothalli developed. Fertile microthalli, described as hecatonemoid were also reported in cultures of this species by Pedersen (1984).

Bartsch & Kuhlenkamp (2000), p. 168; Brodie *et al.* (2016), p. 1020; Cabioc'h *et al.* (1992), p. 62; Cormaci *et al.* (2012), p. 124, pl. 28(1–3); Dizerbo & Herpe (2007), p. 111, pl. 37(1); Fletcher (1987), pp. 179–81, figs 40A, 41; Fletcher in Hardy & Guiry (2003), p. 226; Gallardo García *et al.* (2016), p. 35; Guiry (2012), p. 138; Hamel (1931–39), pp. 225, 226; Knight *et al.* (1935), pp. 86, 87; Kuckuck (1912), p. 178, pl. 8(11); Newton (1931), p. 172; Pedersen (1984), p. 57, fig. 3(6F); Reinke (1878a), pp. 268, 269; Rosenvinge & Lund (1947), pp. 47, 48; Rueness (1977), p. 158, (2001), p. 14; Sauvageau (1895), pp. 336–8, fig. 1, (1929), pp. 366–81, figs 15–17; Stegenga *et al.* (1997b), p. 186; Womersley (1987), pp. 318, 320, figs 114(B), 115(E–F).

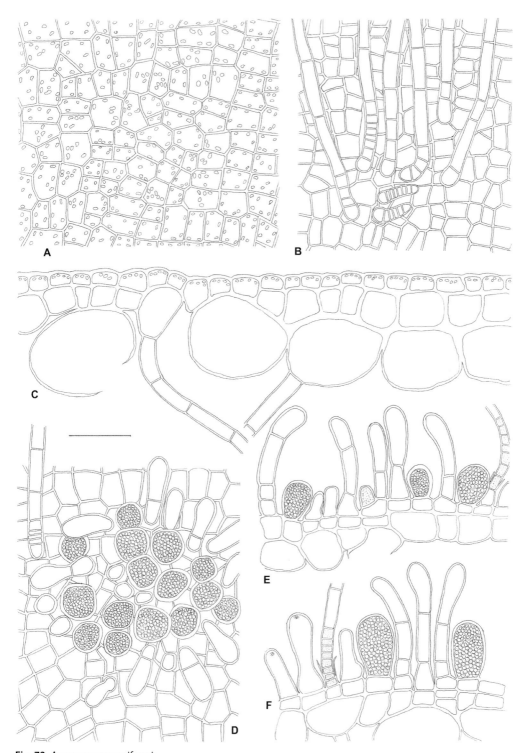

Fig. 73. *Asperococcus ensiformis*

A. Surface view of thallus showing cells with numerous discoid plastids. B. Surface view of thallus with hairs. C. Section of vegetative thallus showing outer 1–2 layers of small cortical cells and inner layers of large, medullary cells enclosing central cavity. Note rhizoidal extensions of medullary cells. D. Surface view of fertile thallus showing unilocular sporangia and associated 1–2-celled paraphyses. E, F. Sections of fertile thalli showing unilocular sporangia and paraphyses arising from surface cells. Bar = 50 μm.

Asperococcus fistulosus (Hudson) W.J. Hooker (1833), p. 277.

Figs 71(B,C), 74, 75, 76(E,F), 77

Ulva fistulosa Hudson (1778), p. 569; *Conferva echinata* Mertens ex Roth (1806), p. 1701; *Scytosiphon fistulosus* (Hudson) C. Agardh (1811), p. 24; *Asperococcus rugosus* Lamouroux (1813), p. 62; *Chordaria filum* var. *fistulosa* (Hudson) C. Agardh (1817), p. 14; *Asperococcus echinatus* (Mertens ex Roth) C. Agardh (1817), p. xxi; *Encoelium echinatum* (Mertens ex Roth) C. Agardh (1820), p. 145; *Scytosiphon filum* var. *fistulosus* (Hudson) C. Agardh (1820), p. 163; *Ectocarpus reptans* P. Crouan & H. Crouan (1867), p. 151, pl. 24; *Ectocarpus reptans* P. Crouan & H. Crouan (1867), p. 151, pl. 24; *Ascocyclus reptans* (P. Crouan & H. Crouan) Reinke (1889), p. 44; *Ectocarpus repens* Reinke (1889), p. 21, pl. 19, figs 5, 6; *Scytosiphon lomentaria* f. *fistulosus* (Hudson) Foslie (1890), p. 98; *Phycocelis reptans* (P. Crouan & H. Crouan) Kjellman (1890), p. 81; *Myrionema reptans* (P. Crouan & H. Crouan) Foslie (1894), p. 130; *Chilionema reptans* (P. Crouan & H. Crouan) Sauvageau (1897), p. 268, fig. 25; *Hecatonema reptans* Sauvageau (1897), p. 273 but see Pedersen (1984), p. 60; *Hecatonema maculans* Sauvageau (1897), p. 248, première forme; *Hecatonema reptans* Kylin (1907), p. 41, nom. illeg.; *Asperococcus echinatus* f. *villosa* Kylin (1907), p. 77; *Asperococcus fistulosus* f. *villosus* Kylin (1907), p. 77; *Hecatonema kjellmani* Nordstedt (1912), p. 237; *Phycocelis crouaniorum* Athanasiadis (1996), p. 178, nom. illeg.

Asperococcus phase

Thalli forming erect, gregarious, light olive to dark brown thalli arising from a discoid holdfast; thalli simple, flaccid, hollow, cylindrical, linear or slightly elongate-clavate, 5–40 cm long × 2–5(–20) mm wide, gradually attenuate below, more sharply attenuate and obtuse above, surface unevenly inflated and contoured, slightly roughened in texture, covered throughout with dark brown punctations; surface cells quadrate or rectangular, usually in rows, 10–21(–44) × 8–19(–26) µm, each with large numbers of discoid plastids with associated pyrenoids; in section, thalli distinctly hollow with an outer cortex of 1(–2) small, plastid-bearing cells enclosing an inner medulla of 1–3 large, longitudinally elongate, colourless, sometimes ruptured cells; hairs abundant, grouped in discrete, punctate sori, with dark brown basal meristem and enclosing sheath, 13–18 µm in diameter.

Plurilocular and unilocular sporangia borne on the same or different thalli, crowded in discrete, irregularly shaped, abundant and extensive, punctiferous sori, produced from cortical cells, with associated hairs and paraphysis-like filaments; unilocular sporangia abundant, rounded in surface view, 40–70 µm diameter, globose or pyriform in section, 30–85 × 20–70 µm, sessile; plurilocular sporangia rare, only reported for the Isle of Man, dimensions uncertain; paraphyses erect, simple, or rarely branched, rigid, thick walled, dark pigmented, especially the terminal cell, linear or slightly elongate-clavate, multicellular 2–5(–7) cells, 80–120(–190) µm long, arising directly from surface cells, cells subquadrate, quadrate or rectangular, 18–47 × 18–36 µm.

Epiphytic on various algae, especially *Fucus* spp. and *Gongolaria* spp., more rarely epizoic, in upper to lower eulittoral pools and sublittoral to 18 m.

Common and widely distributed around Britain and Ireland.

Annual, summer (May to September) less commonly reported throughout the year.

Hecatonemoid phase

Thalli epiphytic, forming thin, closely adherent spots on hosts, usually solitary and circular, 1–4 mm in diameter, occasionally confluent and irregular; in surface view, comprising a monostromatic peripheral region of outward-spreading, laterally adjoined (discoid), branched filaments of quadrate to rectangular cells, 1–2 diameters long, 6–21 × 8–16 µm, each with 1–4 plate-like plastids with pyrenoids, an outer central region of irregularly spaced, thick-walled, rounded cells, 5–8 µm in diameter and an inner central region which is raised, turf-like, similar to a thin carpet pile; in section, or squash preparation, consisting of a monostromatic, later distromatic, basal layer of cells giving rise, in the central region, to erect filaments, hairs and/or plurilocular sporangia; erect filaments loosely associated, simple, linear, slightly mucilaginous, 10–15(–28) cells, 220(–390) µm long, occasionally arising within sporangial husks, comprising cells quadrate or rectangular, 1–2(–3) diameters long, 7–31 × 8–12 µm, each with 1–2 lobed,

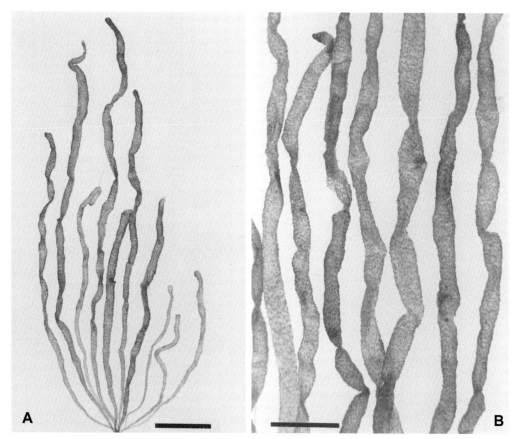

Fig. 74. *Asperococcus fistulosus* (*Asperococcus* phase)
A. Habit of thalli. Bar = 20 mm. B. Portions of erect thalli. Bar = 10 mm.

plate-like plastids; hairs common, terminal on erect filaments with basal meristem and sheath, approx. 19 μm in diameter; rhizoids not observed.

Plurilocular sporangia common, usually sessile on the basal layer or on 1–3-celled stalks, more rarely terminal or lateral on the erect filaments, simple, cylindrical, biseriate, later often triseriate, 58–91 × 14–22 μm, to 15 loculi, with straight, occasionally oblique cross walls; unilocular sporangia unknown.

Epiphytic on various hosts, in particular *Fucus serratus*, lower eulittoral, in pools and shallow sublittoral.

Rare, but widely distributed around Britain (Shetland, Essex, Kent, Dorset and Argyll) and probably Ireland as well.

Recorded June to November.

European isolates of the macroscopic thalli have been the subject of laboratory culture studies by Sauvageau (1929), Knight & Parke (1931), Kylin (1934), Knight *et al.* (1935), Rosenvinge & Lund (1947), Dangeard (1968c, 1969) and Pedersen (1984). In the majority of these studies the life history has generally been interpreted to conform to the haplodiplophasic, heteromorphic type, with unispores derived from the unilocular sporangia on the diploid macroscopic phase producing creeping, microscopic, filaments bearing plurilocular sporangia which are considered to represent the gametophyte phase. However, copulation of the plurispores from these plurilocular sporangia was only reported by Knight & Parke (1931) and Knight *et al.* (1935); usually the zoospores behaved asexually to produce several generations of fertile

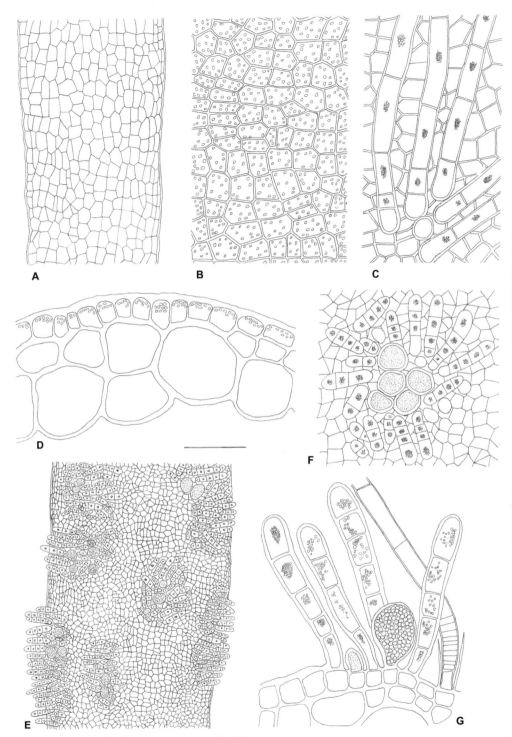

Fig. 75 *Asperococcus fistulosus* (*Asperococcus* phase)

A. Portion of erect vegetative thallus. B. Surface view of vegetative thallus showing cells with several discoid plastids. C. Surface view of vegetative thallus with hairs. D. Section of vegetative thallus showing outer small cortical cells and inner large medullary cells enclosing central cavity. E. Portion of fertile thallus showing unilocular sporangial sori. F. Surface view of fertile thallus showing unilocular sporangia and associated paraphyses. G. Section of fertile thallus showing paraphyses, a single hair with basal sheath and unilocular sporangia. Bar = 50 μm (A–D, G), = 100 μm (F), = 650 (E).

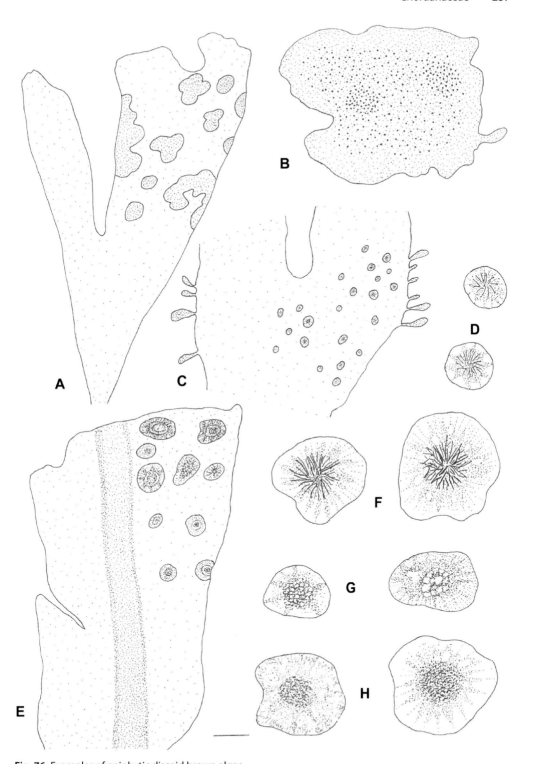

Fig. 76. Examples of epiphytic discoid brown algae

A, B. *Chilionema foecundum*. Habit of thalli on *Palmaria palmata*. C, D, F. *Chilionema ocellatum*. Habit of thalli on *Palmaria palmata*. E, G, H. *Asperococcus fistulosus* (*Hecatonema reptans* phase). Habit of thalli on *Fucus serratus*.
Bar = 10 mm (A), = 0.4 mm (B, D, F–H), = 30 mm (C), = 5 mm (E).

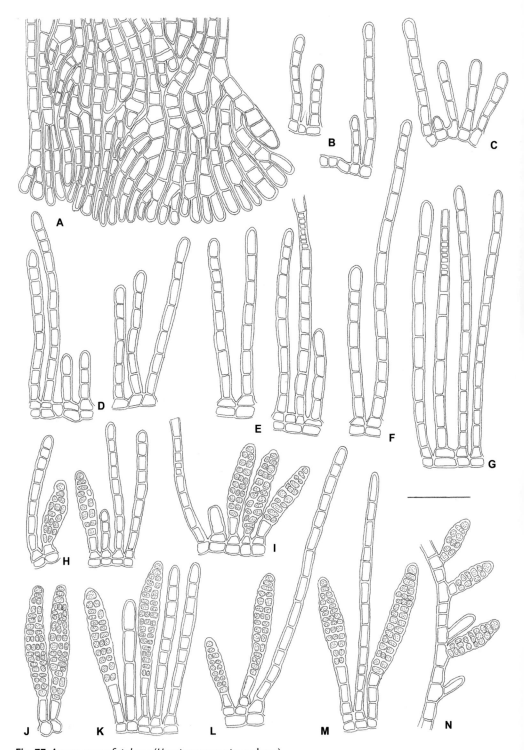

Fig. 77. *Asperococcus fistulosus* (*Hecatonema reptans* phase)
A. Peripheral region of thallus showing outward-spreading filaments. B–G. SP of vegetative thallus showing erect filaments and hairs arising from a distromatic base. H–N. SP of fertile thallus showing erect filaments and plurilocular sporangia. Bar = 50 μm.

microscopic filaments. An additional feature of the life history reported by Knight *et al.* (1935) and Dangeard (1968c) was the copulation of zoospores from the unilocular sporangia, thus apparently directly repeating the diploid macroscopic phase (diplophasic life history).

From the results of a later laboratory culture study, Pedersen (1984) proposed that a 'direct' type of life history was probably operating in this species. He questioned the validity of reports of spore fusions and thought it unlikely that the fertile microthalli represented gametophytes. He therefore interpreted the life history as monophasic and heteromorphic, with the microthalli serving a useful function as a resting phase during unfavourable environmental conditions. Agreement is expressed with Pedersen by the present author in this interpretation of the life history.

A noteworthy feature of the life history of the macrothalli in these studies reported by Kylin (1934), Dangeard (1969) and Pedersen (1984) was the hecatonemoid growth form of the fertile microthalli. Kylin noted the resemblance of the microthalli to *Hecatonema maculans* (Collins) Sauvageau and *Ectocarpus terminalis* Kützing, while Pedersen identified the microthallus as either *Hecatonema reptans* Sauvageau and *Ectocarpus repens* Reinke or *Hecatonema maculans* (Collins) Sauvageau and *Hecatonema terminale* (Kützing) Kylin, depending on the culture conditions of illumination at 4°C. *Hecatonema terminale* was also linked in life history with an *Asperococcus* species (probably *A. fistulosus*) by Loiseaux (1979) (probably misidentified as *Myriotrichia*) and Fletcher (1984). Both authors reported a monophasic, heteromorphic life history which lacked sexuality and that there was a facultative relationship between the *Hecatonema*-like microthalli and the thalloid macrothalli which was governed by environmental conditions, particularly temperature and photoperiod. *Hecatonema*-like thalli were produced under conditions of 18/20°C (Loiseaux 1979) and 15/20°C, 16–8 h light/dark photoregime (Fletcher 1984), while *Asperococcus*-like thalli were formed at 12°C (Loiseaux) and 10°C, 8–16 h light/dark photoregime (Fletcher 1984). All released spores behaved asexually and no evidence was obtained of microthalli functioning as gametophytes or of an obligate alternation between the two expressions due to a difference in ploidy level. These results add support to Pedersen's (1984) interpretation of the life history of European isolates of *A. fistulosus*.

In view of many of the above-mentioned cultures studies, and in agreement with Guiry & Guiry (2023), Cormaci *et al.* (2012), Gallardo García *et al.* (2016) and Mathieson & Dawes (2017), it seems probable that *Hecatonema reptans* represents the gametophyte/microthallus in the life history of *Asperococcus fistulosus* and should be included in the latter's synonyms. It is possible that references to the life history connection between *Hecatonema terminale* (usually as *Hecatonema maculans*) and *Asperococcus fistulosus* relate to the former's quite close morpho-logical similarity to the 'Première forme' of *Hecatonema terminale* described by Sauvageau (1897) and that the 'Deuxième forme' and 'Troisième forme' of *Hecatonema terminale* described by Sauvageau might represent the gametophyte/microthallus of *Punctaria latifolia* (including *P. tenuissima*). More work is obviously required on the *Hecatonema terminale*/*Hecatonema reptans* complex of possible genotypes to determine their life history connections with *Asperococcus* and *Punctaria* and the possible occurrences of independent, isomorphic life histories as shown by some authors (e.g. Clayton 1974; Pedersen 1984). It would also be interesting to undertake culture studies and molecular studies on the morphologically similar species of *Chilionema*.

Bartsch & Kuhlenkamp (2000), p. 165; Brodie *et al.* (2016), p. 1020; Cabioc'h *et al.* (1992), p. 64, fig. 34; Coppejans & Kling (1995), p. 180, pl. 54(A–E); Cormaci *et al.* (2012), pp. 126, 128, pl. 29(1–6); Cotton (1912), p. 21; Crouan frat. (1867), p. 161, pl. 24, fig. 158(3–4); Dangeard (1969), pp. 75–6, fig. 3(A), pls XIV–XV; Dizerbo & Herpe (2007), p. 111, pl. 37(2); Fletcher (1984), p. 193, (1987), pp. 181, 183–5, figs 40(B, C), 42, pl. 5; Gallardo García *et al.* (2016), p. 35; Guiry (2012), pp. 138, 147; Hamel (1931–39), p. 97, fig. 25(VI), pp. 222, 223, fig. 43(VII); Harvey (1848), pl. 194; Knight & Parke (1931), pp. 59, 105–10; Knight *et al.* (1935), pp. 87–90; Kylin (1934), pp. 13–15, figs 7, 8, (1947), pp. 15, 16, fig. 12; Mathieson & Dawes (2017), p. 144, pl. XLIII(11, 12); Morton (1994), p. 40; Newton (1931), pp. 156, 157, 172, fig. 98; Pedersen (1984), pp. 18–20, figs 8, 9, 36A, 37; Reinke (1889a), pp. 7, 19, pls 14, 15(3–6) (as *Ascocyclus reptans*), (1889b), pp. 44, 53; Ribera *et al.* (1992), p. 117; Rosenvinge & Lund (1947), p. 43, fig. 15;

Rueness (1977), pp. 132, 159, figs 57, 92, (2001), p. 14; Sauvageau (1897a), pp. 268–73, fig. 25, (1929), pp. 381–5, fig. 18; Stegenga & Mol (1983), p. 99, pl. 34; Taylor (1957), pp. 130, 169–70, pls 15(7), 16(5–6); Womersley (1987), pp. 32, 34, figs 114(D), 116(D).

Asperococcus scaber Kuckuck (1899a), p. 18.   Fig. 78

Thalli erect, solitary or gregarious, up to 13 mm long and 0.25 mm wide, arising from a small pseudo-discoid base of outward-spreading, compacted rhizoids; thalli simple, filiform, cylindrical, linear or elongate-clavate, gradually attenuating above and below, rarely obtuse above, solid, parenchymatous; surface cells quadrate or rectangular, in longitudinal less obviously transverse rows, 10–27 × 10–21 μm, each with numerous discoid plastids with pyrenoids; in section, thalli solid with 4–6 large, colourless, central cells, with or without a small central cavity, enclosed by 1–2 smaller, pigmented cortical cells; hairs common, single, arising from surface cells, with basal meristem and sheath.

Plurilocular sporangia common, borne in crowded extensive sori often occupying greater length of erect shoot, arising and protruding laterally from surface cells, cylindrical or more commonly conical, sessile or stalked, uniseriate to multiseriate, 22–55 × 10–18 μm, rarely with associated short, 2–3-celled, paraphysis-like filaments; unilocular sporangia not observed on British material, reputed to be on the same or separate thalli, uncommon, spherical or pyriform, to 45 μm long × 30–35 μm wide.

Epilithic on stones, in pools, in the lower eulittoral.

Rare, recorded only for Britain in Dorset and Argyll.

Collections made in April; information inadequate for comment on seasonal distribution.

The records for Britain are based on collections made by E.A.L. Batters in April 1892 and 1897. Slides of this material are deposited in the Natural History, London. No collections have been made by the present author.

Batters (1902), p. 28; Brodie *et al.* (2016), p. 1020; Cormaci *et al.* (2012), pp. 128, 129, pl. 30(1–3); Fletcher (1987), pp. 185–7, fig. 43; Fritsch (1945), fig. 37(B–F); Gallardo García *et al.* (2016), p. 35; Hamel (1931–39), pp. 224, 225; Knight *et al.* (1935), pp. 90, 91; Kuckuck (1899a), pp. 14–20, figs 1–4, pl. II; Newton (1931), pp. 171, 172; Ribera *et al.* (1992), p. 117.

Botrytella Bory de Saint-Vincent

BOTRYTELLA Bory de Saint-Vincent (1822), p. 425.

Type species: *B. micromora* Bory (1822), p. 425.

Sorocarpus Preingsheim (1862), p. 9; *Polytretus* Sauvageau (1900), p. 218.

Thalli epiphytic, forming dense, bushy, tufts of erect, monosiphonous filaments, arising from a basal system of outward-growing, radially arranged, branched filaments, which are laterally adjoined and discoid in appearance; main filaments much branched, with branching irregular, alternate, spiral or secund; basal regions of erect filaments giving rise to secondary rhizoids, which form a loose cortication; true, colourless hairs with basal meristem common, terminal or lateral on branches; growth diffuse; cells containing several, discoid to rod-shaped plastids without obvious pyrenoids.

Plurilocular sporangia conical, or ovoid, multiseriate, occurring singly, or in small to large, irregularly shaped to globular, densely crowded clusters (sori), resembling a bunch of grapes, and arising on the adaxial side of the ultimate branchlets, sessile or on short branches, often terminating the branchlets at the base of a hair; unilocular sporangia unknown for Atlantic material.

One species in Britain and Ireland.

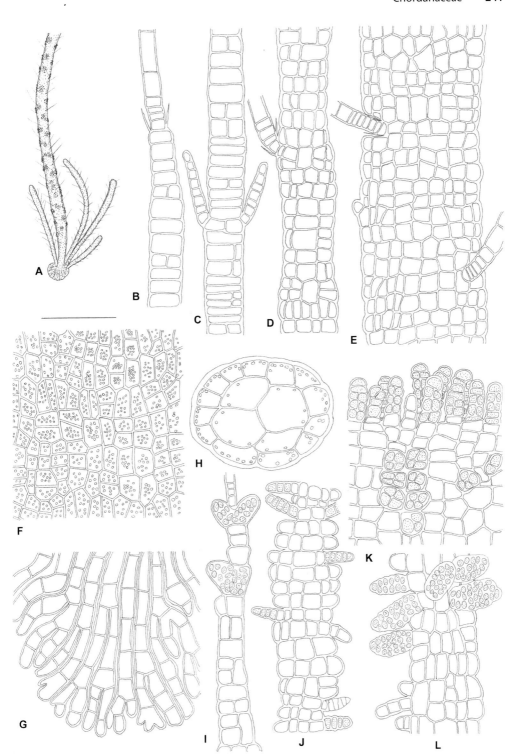

Fig. 78. *Asperococcus scaber*
A. Habit of thallus. B. Portions of erect thalli at various stages of development. Note lateral hairs. C. Surface view of thallus showing cells with numerous discoid plastids. D. Peripheral region of basal attachment disc. E. TS of vegetative thallus. F. Portions of fertile thalli showing plurilocular sporangia. Bar = 2 mm (A), = 50 μm (others).

Botrytella micromora Bory (1822–31), p. 425.   Fig. 79

Ectocarpus siliculosus var. *uvaeformis* Lyngbye (1819), p. 132; *Sorocarpus uvaeformis* (Lyngbye) Pringsheim (1862), p. 9; *Sorocarpus micromorus* (Bory de Saint-Vincent) Silva (1952), p. 256; *Botrytella uvaeformis* (Lyngbye) Kornmann & Sahling (1988), pp. 1–12.

Thalli epiphytic, forming dense, bushy, tufts of erect, monosiphonous filaments, to 5 cm in length, arising from a basal system of outward-growing, radially arranged, branched filaments, which are laterally adjoined and discoid/pseudodiscoid in appearance; basal regions of erect filaments, giving rise to secondary rhizoids, which form a loose cortication; main filaments much branched, with branching irregular, alternate or spiral, ultimate branches irregular or secund; branching sympodial; basal cells cylindrical, quite thick walled, 1–2.5 diameters long, (20–)50–76(–106) × (32–)44–65(–76) μm, becoming shorter, more barrel-shaped, of equal length and width or distinctly shorter than broad, (11–)15–26(–30) × (12–)14–26(–33) μm in the upper regions; cells with several, discoid to rod-shaped plastids without obvious pyrenoids; growth intercalary and diffuse; true, colourless hairs with basal meristem common, terminal or lateral on branches.

   Plurilocular sporangia conical, or ovoid in shape, multiseriate, (15–)20–30(–34) × (10–)13–22(–25) μm, occurring singly, or in small to large, to 83 × 55 μm, irregularly shaped to globular, densely crowded clusters (sori), resembling a bunch of grapes, and arising on the adaxial side of the ultimate branchlets, sessile or on short branches, often occurring terminally on the branchlets, at the base of a hair; unilocular sporangia unknown for Atlantic material.

   Epiphytic on various algae, lower eulittoral pools and shallow sublittoral.

   Rare but widely distributed around Britain (Dorset, Renfrew and Argyll) and Ireland (Galway, Mayo and Waterford).

   Thalli reported for spring and summer but insufficient data to comment on seasonality with any certainty.

   Culture studies on *Botrytella micromora* by Clayton (1974), Kornmann & Sahling (1984) and Pedersen (1974a) revealed an asexual, 'direct' type life history with plurispores from the plurilocular sporangia germinating directly without copulation to produce a branched, filamentous basal system which gave rise to upright filaments resembling those of the parent thallus. In contrast, Abe (1935) working on Pacific material reported an isomorphic alternation of generations in this species with assumed copulation between anisogametes and meiosis occurring in some of the plurilocular sporangia. Unilocular sporangia (and pyrenoids) have also been described for Japanese material (Tatamatsu 1936). It is possible, therefore, that either Atlantic populations and Japanese populations are not conspecific or, as Kylin (1937), Lund (1959) and Pedersen (1974a) propose, Atlantic material also has an isomorphic alternation of generations but with one of the generations suppressed.

Abe (1935), pp. 329–37; Bartsch & Kuhlenkamp (2000), p. 165; Batters (1902), p. 34; Brodie *et al.* (2016), p. 20; Clayton (1974), p. 799, fig. 32; Cormaci *et al.* (2012), pp. 129, 130, pl. 30(4–6); Cotton (1912), p. 95; Guiry (2012), p. 138; Hamel (1931–39), p. 64; Holmes & Batters (1890), p. 80; Kim (2010), p. 72, fig. 32(A–D); Kjellman (1891a), pp. 177, 178; Kornmann & Sahling (1977), p. 115, pl. 59(A–D), (1984), pp. 87–101, pls 5–7, pp. 1–12, pls 1, 3, 4; Kylin (1947), p. 15, fig. 9(A–C); Lund (1959), pp. 55–9, fig. 8(A–M); Mathieson & Dawes (2017), p. 145, pl. XLIII(13, 14); Munda (1964), pp. 535, 536; Newton (1931), pp. 130, 131, fig. 77(A–C); Oliveira & Bisalputra (1978), pp. 439–45, pls 1–4; Pankow (1971), p. 153, fig. 171; Pedersen (1974), pp. 57–61, figs 1–3, (2011), p. 64, fig. 65; Pringsheim (1862), p. 9, pl. 3A(1–8); Rosenvinge & Lund (1941), pp. 58, 59, fig. 30(A–H); Rueness (1977), p. 120, (2001), p. 16; Schneider & Searles (1991), pp. 113, 114, fig. 116; Silva (1950b), p. 256; Stegenga & Mol (1983), pp. 78–81, pl. 24(1–2); Stegenga & Mol (1996), p. 105, pl. 1(3–4); Taskin (2008), p. 175, fig. 2; Tatamatsu (1936), p. 90, pls 4–5; Taylor (1957), p. 117, pl. 9(6), (1960), pp. 208, 209; Womersley (1987), pp. 36–8, figs 6(B), 7(E–F).

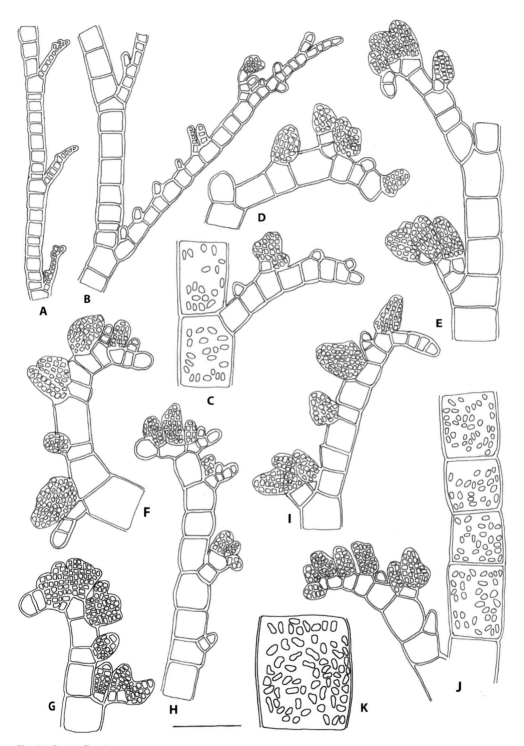

Fig. 79. *Botrytella micromora*

A, B. Portions of erect filaments showing branching patterns. C–J. Portions of erect filaments showing single and clusters of plurilocular sporangia. K. Cell with discoid to rod-shaped plastids. Bar = 50 μm.

Buffhamia Batters

BUFFHAMIA Batters (1895b), pp. 168, 169.

Type species: *B. speciosa* Batters (1895b), p. 311.

Thalli erect, macroscopic, to 3 cm in length, 0.5(–1) mm wide, simple, cylindrical, linear, attenuate above and below, arising from a holdfast of secondary rhizoidal filaments; in structure, thallus parenchymatous, solid, with a central medulla of large, thick-walled cells, enclosed by a single peripheral cortical layer of smaller, pigmented cells containing numerous discoid plastids with pyrenoids; hairs common, arising from surface cells, with basal meristem, but without obvious sheath.

   Plurilocular sporangia occurring in extensive sori on surface, formed as outgrowths from surface cells, on 2–3-celled stalks, uniseriate to biseriate, linear-oblong to spindle shaped, accompanied by multicellular, cylindrical to clavate, paraphyses; unilocular sporangia unknown.

One species in Britain.

Buffhamia speciosa Batters (1895b), pp. 169, 311.   Fig. 80

Thalli erect, macroscopic, solitary or gregarious, to 1(–3) cm in length, to 0.5(–1) mm wide, olive-brown, arising from a holdfast base of outward-spreading and host-penetrating, compacted secondary rhizoids derived as downward extensions of the basal cells; thalli simple, filiform, cylindrical, linear, gradually attenuating above and below to a few cells, solid, soft, fairly flaccid, parenchymatous; surface cells quadrate, in longitudinal and transverse rows each with numerous discoid plastids and associated pyrenoids; in transverse section, thalli solid, with a central medulla of large, thick-walled, angular to rounded, colourless, central cells, enclosed by a single, peripheral, cortical layer of smaller, pigmented cells; hairs common, arising from surface cells and from terminal cells on young thalli, with basal meristem but without obvious sheath.

   Plurilocular sporangia common, borne in extensive sori covering the greater length of the mature thallus surface, arising and protruding laterally from surface cells, accompanied by paraphyses which are cylindrical to clavate, multicellular, simple or dichotomously branched, 3–7 cells long with cells approximately isodiametric, 10–18 × 9 μm; plurilocular sporangia arising from the cortical cells, on 2–3-celled stalks which can be branched, linear-oblong to spindle shaped, obtuse or pointed, uniseriate to biseriate, to 50(–75) μm in length, approximately 15 μm in diameter; unilocular sporangia unknown.

   Epiphytic, fringing erect thalli of *Sauvageaugloia divaricata*, in lower eulittoral pools.

   Very rare, only recorded for Britain in Dorset and Hampshire on the south coast of England.

   Collections made in September; information inadequate for comment on seasonal distribution. Not found by the present author; the only material available for examination was that deposited by Batters as dried specimens in the algal herbarium at the Natural History Museum, London.

Batters (1895b), pp. 307–11, pl. XI(1–10), (1895c), pp. 168, 169, (1902), p. 45; Brodie *et al.* (2016), p. 1022; Fletcher in Hardy & Guiry (2003), p. 229; Guiry (2012), p. 181; Hamel (1931–39), p. 203; Newton (1931), p. 144, fig. 89.

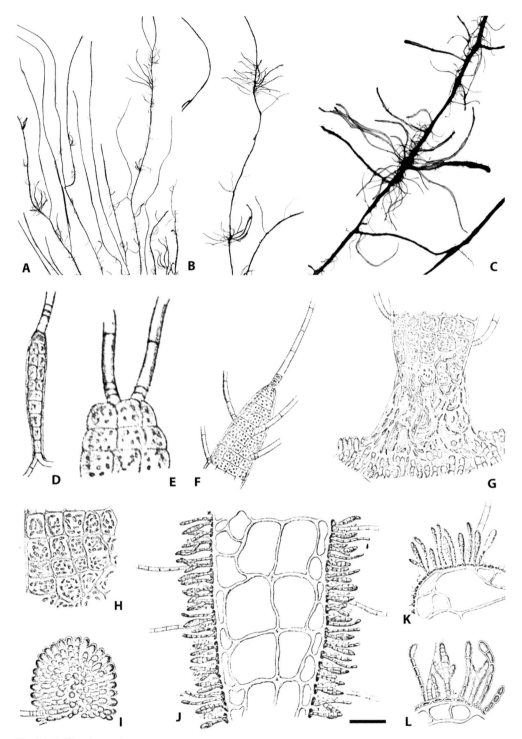

Fig. 80. *Buffhamia speciosa*

A–C. Habit of thalli epiphytic on *Sauvageaugloia divaricata*. D–F. Apex of thalli showing terminal hairs with basal meristem. G. Base of thallus showing downward-growing and spreading secondary rhizoids. H. Surface view of portion of thallus showing cells with discoid plastids. I. SP of portion of thallus. J–L. LS of thallus showing central, large, cortical cells and a peripheral layer of small cortical cells giving rise to paraphyses and stalked plurilocular sporangia. Bar in J = 26 mm (A, B), = 12 mm (C), = 45 μm (D), = 11 μm (E), = 24 μm (F, H), = 23 μm (G), = 110 μm (I), = 63 μm (J), = 51 μm (K), = 38 μm (L). (A–C, herbarium specimens, NHM; D–L, from Batters 1895b.)

Chilionema Sauvageau

CHILIONEMA Sauvageau (1897a), p. 263.

Lectotype species: *Chilionema nataliae* Sauvageau (1898), p. 263 [= *Chilionema ocellatum* (Kützing) Kornmann 1953, p. 325]

Note: Two species were included in the original description of *Chilionema* by Sauvageau. A type does not appear to have been designated (Guiry & Guiry 2023) and so is designated here.

Thallus epiphytic, forming small spots on hosts, consisting of a layer of prostrate filaments, laterally adjoined and disc-like, with or without prominent, peripheral apical cells, monostromatic or distromatic in central parts, giving rise, densely or irregularly, to erect filaments, hairs, ascocysts and/or sporangia; erect filaments usually of equal height, remaining short, simple, rarely branched, comprising cells with 1–3 plate-like plastids with pyrenoids; hairs common, usually terminal or erect filaments, rarely arising from basal layer, with basal meristem and sheath; ascocysts in some species usually borne on basal layer, more rarely terminal on erect filaments.

   Plurilocular sporangia borne on basal layer, sessile or on many-celled stalks, more rarely terminal or lateral on erect filaments; unilocular sporangia unknown.

   Chilionema is an epiphytic genus characterized by a mainly distromatic basal disc which gives rise, somewhat irregularly, to short, erect filaments of approximately equal size which are simple or rarely branched near the base. Hairs (with basal meristem and enclosing sheath) and multiseriate, plurilocular sporangia are also borne on the disc and are sessile or shortly stalked. Unilocular sporangia are unknown. It is closely related to (and possibly synonymous with) *Hecatonema*, type species *H. terminale* (Kützing) Kylin, which is based on *Phycocelis maculans* described by Collins (1896) for the north-east American coast. In his original paper, Sauvageau described and figured three 'formes' of *H. terminale* which represented progressive stages of increasing development of the erect filaments. The 'Première forme' (form one) was described with short, simple, erect filaments and plurilocular sporangia borne on the basal disc, while the 'Deuxième' and 'Troisième' forms had well-developed, erect, branched filaments on which the plurilocular sporangia were mainly borne, either terminally or laterally. Fletcher (1987) recognized *H. terminale* (and hence *Hecatonema*) in terms of forms 2 and 3 described by Sauvageau. He therefore excluded form 1 which he transferred to the genus *Chilionema*, in particular to the species *C. ocellatum* (Kützing) Kuckuck. He also proposed that two species of *Hecatonema* which have been reported for Britain and Ireland, viz. *H. foecundum* (Strömfelt) Loiseaux and *H. hispanicum* (Sauvageau) Loiseaux, would also appear to be more suitably placed in *Chilionema*. They form small, epiphytic, discoid basal thalli bearing both short, erect, predominantly simple, filaments and sessile or stalked, multiseriate, plurilocular sporangia. Fletcher thus proposed the following two new combinations: *Chilionema foecundum* (Strömfelt) Fletcher and *Chilionema hispanicum* (Sauvageau) Fletcher.

   Four species of *Chilionema* were recognized by Fletcher (1987) for Britain and Ireland: *C. foecundum*, *C. hispanicum*, *C. ocellatum* and *C. reptans*. Differences between the latter two species are very small and they may well be conspecific. *C. foecundum* and *C. hispanicum* are characterized by the frequent presence of ascocyst-like cells arising from the basal layer. *C. hispanicum* differs from the other species of this genus in having a monostromatic basal layer and cells with a single, plate-like plastid. In some respects, it shows resemblance to *Symphyocarpus strangulans* Rosenvinge, family Scytosiphonaceae (see p. 521). However, the species is retained in *Chilionema* here pending further investigation.

   Any proposals that *Chilionema* comprises a disparate group of taxa would find support in the life history variations which have been reported (usually under the genus *Hecatonema*). For example, *C. foecundum* and *C. hispanicum* (as *Hecatonema foecundum* and *H. hispanicum*, respectively) were both reported by Loiseaux (1967b and 1966/1967b respectively) to have a haplodiplontic, heteromorphic lifecycle with a haploid gametophytic microthallus and a diploid sporophytic macrothallus. An additional 'short-circuit' life history with a fusion of unispores was reported in *C. hispanicum*. *Chilionema*-like thalli also appear to have been reported as microthalli in the life histories of species of the macroalgal genus *Asperococcus*. For instance, material referable to the 'Première forme' of *Hecatonema maculans* described by Sauvageau (1897a) and included within the synonymy of *Chilionema ocellatum* by Fletcher (1987), was reported to be a phase in the life history of an *Asperococcus* sp. by Loiseaux (1969) and Fletcher (1984). Furthermore, Pedersen (1984) considered *Hecatonema reptans* Sauvageau and *Ectocarpus repens* Reinke, both of which were included within the synonymy of *Chilionema reptans* by Fletcher (1987), as phases in the life history of *Asperococcus fistulosus*. Clearly the situation is unsatisfactory and more extensive studies, particularly involving culture work, are necessary to ascertain fully the taxonomic relationship between *Chilionema* and *Hecatonema* and their respective involvement as prostrate stages in the life histories of *Asperococcus* and *Punctaria* species. However, in view of the above culture studies by Fletcher (1987), Loiseaux (1969) and Pedersen (1984), there appears to be good support for concluding that *Chilionema reptans* (= *Hecatonema reptans*) probably represents the gametophyte in the life history of *Asperococcus fistulosus*, as proposed by Mathieson & Dawes (2017) and the former is included in the synonymy of the latter in the present treatise.

In view of the above discussion, *Chilionema* is represented in Britain and Ireland by three species, *C. foecundum*, *C. hispanicum* and *C. ocellatum*.

Two species in Britain and Ireland, one species in Britain only.

KEY TO SPECIES

1. Thalli without ascocyst-like cells . . . . *C. ocellatum*
    Thalli with ascocyst-like cells 2
2. Basal cells clearly distromatic in parts; plastids multilobed . . . . *C. foecundum*
    Basal cells monostromatic throughout; plastids plate-like. . . . .*C. hispanicum*

Chilionema foecundum (Strömfelt) R.L. Fletcher (1987), p. 191.   Figs 76(A,B), 81

Phycocelis foecunda Strömfelt (1888), p. 383; *Ascocyclus foecundus* (Strömfelt) Reinke (1889), p. 46; *Myrionema foecundum* (Strömfelt) Sauvageau (1897a), p. 170; *Ascocyclus sphaerophorus* Sauvageau (1897a), p. 280; *Ascocyclus islandicus* Jónsson (1903), p. 149; *Ascocyclus foecundus* (Strömfelt) Cotton (1912), p. 122; *Ascocyclus saccharinae* Cotton (1912), p. 122; *Chilionema borgensenii* Printz (1926), p. 137; *Ascocyclus distromaticus* Taylor (1937b), p. 228; *Hecatonema foecundum* (Strömfelt) Loiseaux (1967a), p. 338.

Thallus epiphytic, forming loosely adherent light to dark brown spots on hosts, usually solitary and circular 1–5 mm in diameter, occasionally irregular and confluent; central region dark brown, turf-like, similar to a thick carpet pile, composed of densely packed vertical filaments; peripheral region of light-coloured, branched, radiating filaments, composed of rectangular cells, 1–3 diameters long, 9–18 × 5–8 µm, with a single multilobed plastid; terminal apical cells not noticeably coalesced; in section, or squash preparation, consisting of loosely united, radiating, monostromatic, frequently distromatic filaments, easily separable under pressure, giving rise, in central regions to erect, slightly gelatinous, loosely associated, occasionally branched, erect filaments of 6–9(–13) cells, 80–134 µm high; cells cylindrical, 1–2 diameters long, 8–21 × 8–13 µm, with a single large multilobed plastid with 1–3 pyrenoids; ascocyst-like cells common, either produced from basal cells or terminal on erect filaments; hairs frequent, arising from basal layer, with basal meristem and sheath.

Plurilocular sporangia usually abundant, either on basal cells or terminal on erect filaments, biseriate or (more rarely) triseriate, subcylindrical, attenuate towards apex, 26–59 × 9–18 µm, to 12 loculi long, frequently formed within empty sporangial husks; unilocular sporangia unknown.

Epiphytic, on *Palmaria palmata* and *Saccharina latissima* blades, lower eulittoral and shallow sublittoral.

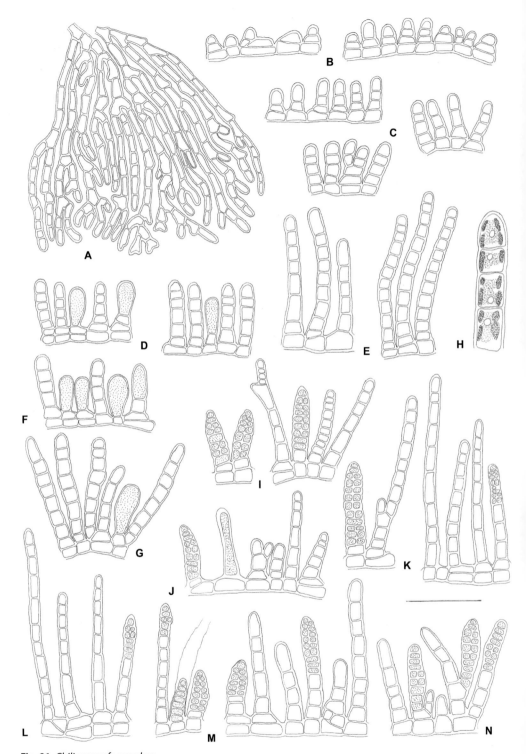

Fig. 81. *Chilionema foecundum*

A. Peripheral region of thallus showing outward-spreading branched vegetative filaments. B–G. SP of vegetative thallus showing erect filaments arising from a monostromatic/distromatic base. Note enlarged ascocysts. H. Portion of erect filament showing lobed plate-like plastid. I–N. SP of fertile thallus showing erect filaments, plurilocular sporangia and ascocysts. Bar = 50 µm (A–G, I–N), = 20 µm (H).

Recorded for scattered localities around Britain (Shetland, Kent, Hampshire, Dorset, Devon, Anglesey and Argyll) and Ireland (Wexford and Mayo).

Summer annual, commonly recorded in August and September.

Chilionema foecundum has been investigated in laboratory culture by Loiseaux (1964, 1967b) (as *Ascocyclus sphaerophorus* Sauvageau). The life history was reported to be heteromorphic and haplodiplophasic. Plurispores released from plurilocular sporangia on the diploid, discoid thallus germinated directly and, by the process of heteroblasty, formed two morphologically different thalli: ectocarpoid, filamentous growths and stellate, later discoid growths. Both growth forms produced plurilocular sporangia, the plurispores from which directly repeated the diploid ectocarpoid and/or discoid growths. In addition, a sexual cycle was observed with unilocular sporangia developing on the diploid ectocarpoid growths, the unispores from which germinated directly to form filamentous ectocarpoid, gametophytes. Released plurispores from 'gametangia' on the thalli were observed to fuse and the zygote reformed the ectocarpoid, diploid thalli.

Bartsch & Kuhlenkamp (2000), p. 165; Batters (1892a), p. 174, (1892b), p. 20; Børgesen (1902), p. 427, fig. 82; Brodie *et al.* (2016), p. 1020; Cotton (1912), pp. 122, 123, pl. X(4–9); Dizerbo & Herpe (2007), p. 76, pl. 24(3); Fletcher (1987), pp. 191–3, figs 45(A, B), 46; Fletcher in Hardy & Guiry (2003), p. 286; Gallardo García *et al.* (2016), p. 36; Guiry (2012), p. 148; Hamel (1931–39), p. 100, fig. 25(VII); Jaasund (1957), pp. 228–30, figs 9, 10, (1965), p. 60; Levring (1937), pp. 51, 52, fig. 7(C–E); Loiseaux (1967a), pp. 329–47, figs 3, 5(D), 6, pl. 1; Lund (1959), pp. 109–11, fig. 22; Morton (1994), p. 8; Newton (1931), p. 159; Peña & Bárbara (2010), pp. 55, 56, fig. 11; Printz (1926), pp. 137–9, pl. 2(12–17); Rueness (1977), p. 133, (2001), p. 14; Sauvageau (1897a), pp. 280–5, figs 28, 29; Taylor (1957), pp. 156, 157, pl. 11(7–12).

Chilionema hispanicum (Sauvageau) R.L. Fletcher (1987), p. 195.   Fig. 82

Ascocyclus hispanicus Sauvageau (1897a), p. 274; *Hecatonema hispanicum* (Sauvageau) Loiseaux (1967a), p. 338; *Phycocelis hispanicus* (Sauvageau) Athanasiadis (1996), p. 182.

Thallus epiphytic, forming dark brown-black spots on hosts, solitary and circular, to 5 mm or more in diameter, becoming confluent and irregular, several centimetres across; comprising a monostromatic base of laterally adjoined, firmly attached, outward-spreading filaments of cells, giving rise to erect filaments, plurilocular sporangia, ascocysts and/or hairs; basal cells rounded, quadrate or rectangular in section, usually longer than wide, $8–21 \times 10–16$ µm; erect filaments loosely associated, rarely branched, to 10 cells (–91 µm), cells subquadrate, quadrate or rectangular, 1–2 diameters long, $5–14 \times 5–11$ µm, each with a single terminal, plate-like plastid; ascocysts common, usually sessile on basal layer or on 1–2-celled stalks, conical in shape, $27–50 \times 13–20$ µm; hairs infrequent, arising from the basal layer, with basal meristem and sheath, approx. 8–9 µm in diameter.

Plurilocular sporangia common, terminal on erect filaments, or on 1–3-celled stalks, biseriate, frequently triseriate, subcylindrical, $40–50 \times 11–17$ µm; unilocular sporangia unknown.

Epiphytic on the brown algae *Saccorhiza polyschides* and *Saccharina latissima*, probably present on other hosts, lower eulittoral and shallow sublittoral.

Only recorded for south Devon and Dorset by E.A.L. Batters but probably more widely distributed.

Recorded in September; insufficient data to comment on seasonal distribution.

Chilionema hispanicum has been investigated in culture by Loiseaux (1966; 1967b) (as *Ascocyclus hispanicus* Sauvageau). The life history was reported as heteromorphic and haplodiplophasic. A 'short-circuit' lifecycle was also reported to occur by the fusion of unispores from the unilocular sporangia of the sporophyte.

Batters (1902), p. 42; Brodie *et al.* (2016), p. 1020; Dizerbo & Herpe (2007), p. 76; Fletcher (1987), pp. 195, 196, fig. 47; Fletcher in Hardy & Guiry (2003), p. 230; Gallardo García *et al.* (2016), p. 35; Guiry (2012), pp. 138, 139; Hamel (1931–39), p. 99, fig. 25(VIII); Loiseaux (1966), pp. 68–71, figs 1, 2, (1967a), p. 338; Newton (1931), p. 159; Ribera *et al.* (1992), p. 115; Sauvageau (1897a), pp. 274–80, figs 26, 27.

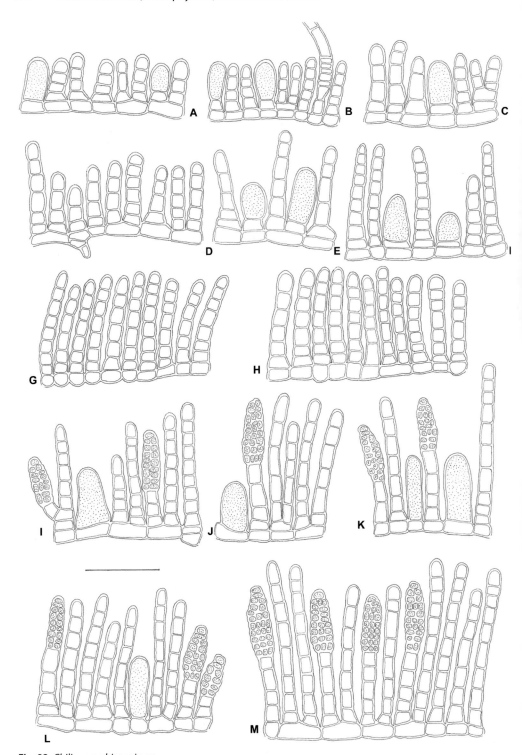

Fig. 82. *Chilionema hispanicum*

A–H. VS of vegetative thallus showing erect filaments, hairs and ascocysts arising from a monostromatic base. I–M. VS of fertile thallus showing erect filaments, ascocysts and plurilocular sporangia. Bar = 50 μm. (Microscope slide preparation: Batters, NHM.)

Chilionema ocellatum (Kützing) Kornmann (1953), p. 325.

Figs 76(C,D), 83

Phyllactidium ocellatum Kützing (1843), p. 295; *Myrionema ocellatum* (Kützing) Kützing (1849), p. 540; *Ascocyclus ocellatus* (Kützing) Reinke (1889b), p. 44; *Chilionema nathaliae* Sauvageau (1897a), p. 263; *Phycocelis ocellatus* (Kützing) Athanasiadis (1996), p. 181.

Thallus epiphytic, forming light brown spots on hosts, usually solitary and circular, 1–3 mm in diameter, occasionally confluent and irregular; in surface view, comprising a light brown, peripheral region of branched, outward-radiating, monostromatic filaments, quite firmly united and discoid in structure, with peripheral row of dark-coloured apical cells enclosed within outer cuticle, cells quadrate or rectangular in shape, 1–2 diameters long, 12–30 × 9–20 µm with 3–5 plate-like plastids with pyrenoids; outer central region of scattered, rounded, thick-walled cells, 13–21 µm in diameter, inner central region turf-like, similar to a sparse carpet pile; in squash preparation, consisting of an outer monostromatic, inner distromatic base, from most cells in the central region of which arise erect filaments, hairs and/or plurilocular sporangia; basal cells thick walled, 1–3 diameters long, 12–26 × 7–13 µm; erect filaments, short, linear, simple, slightly mucilaginous, to 11(–17) cells, 180(–280) µm high, comprising thick-walled, quadrate or more usually rectangular cells, 1–2(–3.5) diameters long, 9–26(–42) × 9–13 µm with 2–3 plate-like plastids with pyrenoids; hairs common, stalked and arising from the basal cells, or more usually terminal on the erect filaments, 13–16 µm in diameter with basal meristem and sheath; rhizoids not observed.

Plurilocular sporangia common, usually on 1–6-celled stalks arising from the basal layer, biseriate or more commonly multiseriate, simple, ovate-lanceolate or subconical, 40–65 × 14–23 µm, to 10 loculi; unilocular sporangia unknown.

Epiphytic on the red alga *Palmaria palmata*, in the lower eulittoral and shallow sublittoral.

Recorded for scattered localities in Britain (Channel Islands, Dorset, Devon, Pembroke, Anglesey and Isle of Man) and Ireland (Wexford and Mayo). Probably more common and widely distributed.

Summer annual; June to September.

Fletcher (1987), placed *Hecatonema terminale* (Collins) Sauvageau 'Première forme' within the synonymy of *Chilionema ocellatum* in view of their close morphological and anatomical similarity. There seems little distinction between the former and *Hecatonema reptans* (including *Chilionema reptans*) and both are herein placed within the synonymy of *Asperococcus fistulosus* on the basis of culture work revealing hecatonematoid microthalli in the life history of the *Asperococcus*. It may well be that *Chilionema ocellatum* should also be included within the synonymy of *Asperococcus fistulosus*. However, pending further culture studies it is retained herein as a separate taxon.

Batters (1893a), p. 23, (1900), pp. 371, 372; Brodie *et al.* (2016) p. 1020; Cotton (1912), p. 95; Dizerbo & Herpe (2007), p. 78, pl. 24(3); Fletcher (1987), pp. 195, 197–9, figs 45(C, D), 48; Fletcher in Hardy & Guiry (2003), p. 230; Guiry (2012), p. 139; Hamel (1931–39), pp. 96–8, fig. 25(IV–V) (as *Chilionema nathaliae* and *C. ocellatum*); Kuckuck (1953), pp. 325–34, figs 4–8; Newton (1931), pp. 157, 158, fig. 98 (as *Chilionema nathaliae* and *C. ocellatum*); Reinke (1889a), pp. 19, 22, pls 1(1–2), 19(5–6), (1889b), p. 44; Rueness (1977), p. 131, fig. 56, (2001), p. 14; Sauvageau (1897), pp. 263–8, figs 23–4.

Chordaria C. Agardh

CHORDARIA C. Agardh (1817), p. xii.

Type species: *C. flagelliformis* (O.F. Müller) C. Agardh (1817), pp. xii, 12.

Thalli macroscopic, erect, much branched, cylindrical, filiform, solid, cartilaginous, arising from a small discoid holdfast; branching to 1, rarely to 2(–3) orders, with subsimple primary axis and numerous long secondary branches; in structure, thallus haplostichous, multiaxial, with a solid

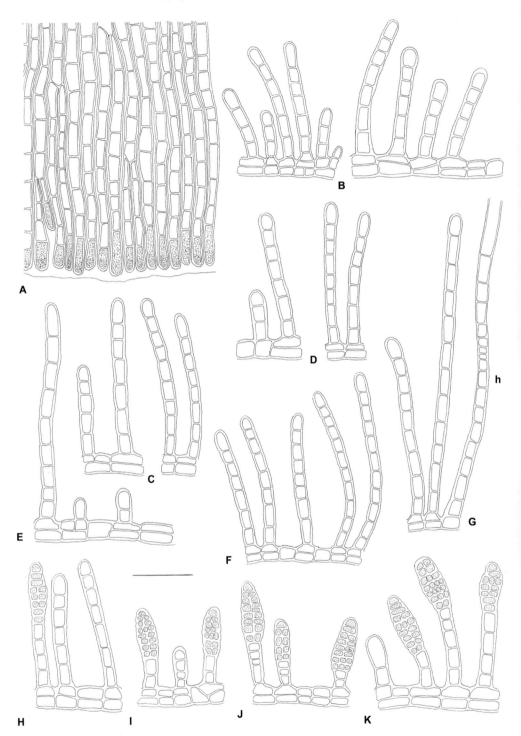

Fig. 83. *Chilionema ocellatum*

A. Peripheral region of thallus showing outward-spreading, laterally united, branched filaments. B–G. SP of vegetative thallus showing erect filaments and hair (h) arising from a distromatic base. H–K. SP of fertile thallus showing plurilocular sporangia. Bar = 50 μm.

medulla of colourless, longitudinally elongated, thick-walled cells, comprising both primary filament cells and narrow hyphae-like cells; outer medullary cells giving rise to a cortex of densely packed, short, primary assimilatory filaments comprising cells with numerous discoid plastids, each with an associated pyrenoid; hairs abundant, clothing the thallus surface, with basal meristem, arising from the outer medullary cells amid the assimilators.

Unilocular sporangia borne at the base of the assimilators, ovate to pyriform.

One species in Britain and Ireland.

Chordaria flagelliformis (O.F. Müller) C. Agardh (1817), pp. xii, 12.   Figs 84, 85

Conferva elongata Gunnerus (1772), p. 116, nom. illeg.; *Conferva flagelliformis* Gunnerus (1772), p. 105; *Fucus corneus* Zoega (1772), nom. illeg.; *Fucus flagelliformis* O.F. Müller (1775), pl. 650; *Chordaria flagelliformis* var. *minor* C. Agardh (1817), p. 13; *Fucus flagelliformis* var. *minor* Wahlenberg (1826), p. 891; *Chordaria flagelliformis* f. *densa* Farlow (1881), p. 84; *Chordaria flagelliformis* f. *typica* Kjellman (1883b), p. 249, nom. inval.; *Dichosporangium chordariae* Wollny (1886), p. 127; *Chordaria flagelliformis* var. *firma* Kjellman (1890), p. 37; *Streblonema chordariae* (Wollny) Cotton ex Newton (1931), p. 128, nom. illeg.

Thalli macroscopic, erect, to 80 cm in length, 0.2–0.5(–0.8) mm in width, linear throughout with uniform thickness, solitary or gregarious, dark brown to black, arising from a discoid holdfast; thallus terete, cylindrical, filiform, solid, firm and compact, cartilaginous, slippery, both young and older thalli much branched throughout, except at the base, to 1, rarely 2(–3) orders; branching alternate to irregular and widely divergent, with the long, secondary, whiplash-like branches, especially those produced in the upper regions, usually exceeding the length of the relatively short, subsimple, primary axis in older thalli, remaining very short and almost bottlebrush-like in young thalli; thallus haplostichous in structure, multiaxial with a broad, solid filamentous medulla of colourless, longitudinally elongated, transversely rounded or irregular, thick-walled cells comprising both primary filament cells intermingled with their outgrowths of narrower hyphae-like cells; outer medullary cells smaller and giving rise directly to a cortex of densely packed, short, primary, assimilatory filaments bound in a gelatinous matrix; assimilatory filaments to 5–8(–10) cells, 65–115 μm in length, frequently abaxially branched at the base, sometimes simple, club-shaped in younger parts, elongate-clavate in older parts, with usually colourless, cylindrical lower cells 3–5 diameters long, 10–28 × 4–10 μm, giving rise to generally shorter, quadrate to rectangular, somewhat barrel-shaped, more pigmented upper cells, 6–27 × 5–10 μm, sometimes with an enlarged, ovoid terminal cell; cells with numerous discoid plastids and associated pyrenoids; hairs abundant, clothing the thallus surface, with basal meristem, arising from the outer medullary cells amid the assimilators, sometimes terminal on the assimilators.

Unilocular sporangia borne at the base of the assimilators, sessile, rarely pedicellate, ovoid to pyriform, 38–60 × 13–23 μm; plurilocular sporangia not known on the macrophyte but reported on the microscopic *Streblonema*-like gametophyte phase.

Epilithic on stones and rock, usually in shallow pools, upper to lower eulittoral, shallow sublittoral, frequently associated with sand.

Common and widely distributed around Britain and Ireland.

Annual, May to September, small thalli occasionally found throughout the year.

Bartsch & Kuhlenkamp (2000), p. 165; Batters (1902), pp. 43, 44; Brodie *et al.* (2016), p. 1020; Bunker *et al.* (2017), p. 300; Caram (1955), pp. 18–36, figs 1–9; Cotton (1912), p. 120; Dizerbo & Herpe (2007), p. 88, pl. 28(3); Fletcher in Hardy and Guiry (2003), p. 231; Fritsch (1945), pp. 82–5, fig. 22(A, C–F); Gayral (1966), pp. 272, 273, pl. XLV; Guiry (2012), p. 139; Hamel (1931–39), pp. 172, 173; Jaasund (1963), pp. 1–3, fig. 1; Kim & Kawai (2002), pp. 328–39; Kornmann (1962c), pp. 276–9; Kornmann & Sahling (1977), p. 124, fig. 65(A–C); Kylin (1940), p. 40, fig. 21(A, B), (1947), pp. 59, 60, fig. 51(A, D); Lindauer *et al.* (1961), pp. 232, 233, fig. 54(1–2); Mathieson & Dawes (2017), p. 146, pls XLIV(3–5), LXII(8); Morton (1994), p. 35; Munda (1979), pp. 567–91; Newton (1931), pp. 128, 146; Pankow (1971), pp. 184–6, figs 232, 233; Pedersen (2011), p. 94, fig. 100; Pedersen *et al.* (1958), pp. 57–60;

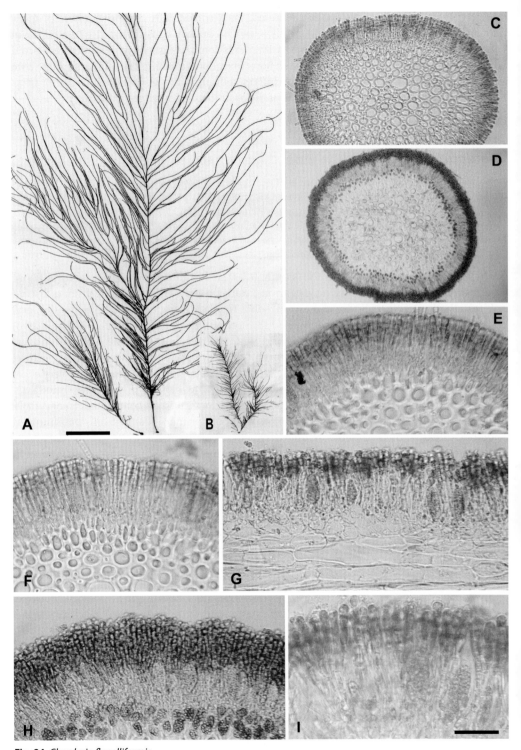

Fig. 84. *Chordaria flagelliformis*
A. Habit of mature thallus, showing long, whiplash-like branches. B. Bottlebrush-like habit of young thallus.
C–F. TS of thalli showing a broad, solid, central medulla giving rise to a smaller, peripheral cortex of closely packed, erect, assimilatory filaments. G–I. TS of fertile thalli showing unilocular sporangia borne at the base of the assimilators. Bar in A = 4 cm (A), = 6 cm (B). Bar in I = 300 µm (C, D), = 90 µm (E, F, G, H), = 45 µm (I). (A, B, herbarium specimens, NHM.)

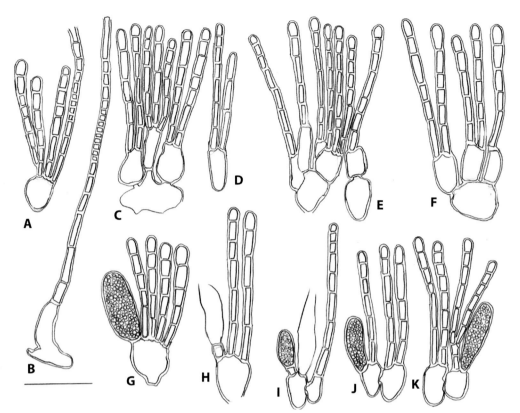

Fig. 85. *Chordaria flagelliformis*
A–K. SP of outer cortex region showing multicellular assimilators, unilocular sporangia and hairs. Bar = 50 μm.

Probyn (1981), pp. 339–44, (1984), pp. 271–5; Probyn & Chapman (1982), pp. 129–33, (1983), pp. 243–71; Reinke (1889b), pp. 74, 75, fig. 7, (1892a), p. 58, pl. 39(1–7); Rice & Chapman (1982), pp. 107–11; Rosenvinge & Lund (1943), pp. 34–6; Rueness (1977), p. 146, fig. 77, (2001), p. 14; Saunders & McDevit (2013), pp. 1–23; Sauvageau (1929), p. 266, fig. 2; Setchell & N.L. Gardner (1925), pp. 572, 573, pl. 39(44); Stegenga & Mol (1983), p. 84, pl. 32(1–2); Stegenga *et al.* (1997a), p. 17; Taylor (1957), p. 148, pls 12(6), 14(4); Wollny (1886), pp. 127–30, pls i(1–5), ii(3).

Cladosiphon Kützing

CLADOSIPHON Kützing (1843), p. 329, pl. 251.

Type species: *C. mediterraneus* Kützing (1842), p. 329.

Thalli erect, macroscopic, cylindrical, filiform, at first solid, later becoming tubular, gelatinous, mucilaginous, sparsely to much branched, to 3(–4) orders, attenuate above and below, arising from a discoid holdfast; thalli multiaxial in structure, pseudoparenchymatous, haplostichous with a central core of closely bound, medullary filaments of large, colourless, elongated cells surrounded by a primary subcortex of branched filaments which give rise peripherally to a primary cortex of assimilatory filaments; compactness of subcortical and cortical regions increased by downward-growing filamentous extensions of the basal cells (to form the secondary subcortex and cortex respectively); medullary filaments and subcortical filaments terminated above by sheathed hairs; assimilatory filaments derived from both primary and secondary cortex, usually simple, multicellular, curved, elongate-clavate, with cylindrical basal cells and moniliform terminal cells; assimilatory cells with several discoid plastids each with a pyrenoid; hairs common, with basal meristem and sheath; growth trichothallic with a short intercalary meristem terminating all filaments, with subtending colourless hair.

Plurilocular sporangia common, occurring as uniseriate, filiform, outgrowths of the distal cells of both the primary and secondary assimilators; unilocular sporangia rare, formed at the base of assimilatory filaments, sessile, ovoid.

Two species in Britain and Ireland.

KEY TO SPECIES

Frond usually > 10 cm length, branched; assimilatory filaments branched, fasciculate; plurilocular sporangia in clusters . . . . *C. contortus*
Frond usually < 10 cm length, simple or with a few short branches; assimilatory filaments simple or branched once; plurilocular sporangia single. . . . .*C. zosterae*

Cladosiphon contortus (Thuret) Kylin (1940), p. 27.   Fig. 86

Castagnea contorta Thuret in Le Jolis (1863), p. 86.

Thalli erect, macroscopic, to 25–40 cm long, 1–2(–3) mm wide, yellowish brown, cylindrical, filiform, at first solid, later tubular, gelatinous, mucilaginous, sparsely to much branched, to 3(–4) orders, attenuate above and below, arising from a small discoid holdfast; main axes sometimes indistinct and exceeded in length by long irregularly, pseudoparenchymatously or trichotomously produced, sparsely branched, widely divergent, secondary branches; terminal branch regions often with short, very thin, branchlets; in structure, thalli multiaxial, pseudoparenchymatous, haplostichous with a central core of quite closely bound medullary filaments of large, colourless, elongated cells which branch out laterally to form an enclosing, short, compact primary subcortex of branched filaments, from the terminal regions of which arise clusters of outward-radiating, peripheral assimilatory filaments (primary cortex) with associated unilocular sporangia, plurilocular sporangia and hairs; compactness of subcortical and cortical regions increased by downward-growing filamentous extensions of the basal cells, those of the subcortical regions comprising large, distended cells which compact around the outer medullary tissue (secondary subcortex), while those of the cortical regions being much narrower and rhizoid-like (secondary cortex); medullary filaments and subcortical filaments terminated above by sheathed hairs; assimilatory filaments derived from both primary and secondary cortex, simple, or sometimes branched above, 150–200 μm, 15–20 cells long, curved, elongate-clavate, multicellular, basal cells cylindrical, 2–5 diameters long, 8–10 μm wide, terminal cells moniliform, length equal to diameter, 12–16 μm wide with outer walls distended; assimilatory cells with several discoid plastids each with a pyrenoid; hairs common, with basal meristem and sheath; growth trichothallic with short intercalary meristem terminating all filaments, with a subtending colourless hair.

Plurilocular sporangia and unilocular sporangia occurring on the same or separate thalli; plurilocular sporangia common, occurring as uniseriate, filiform, filamentous outgrowths of the distal cells of both primary and secondary assimilators, to 3–6 loculi, 40 μm long, 12 μm in diameter; unilocular sporangia rare, formed at the base of assimilatory filaments, sessile, ovoid, 90–100 × 50–65 μm.

Epiphytic on both sides of the leaves and on rhizomes of *Zostera* spp., shallow sublittoral.

Rare and reported for a few scattered localities around Britain (Channel Islands, Dorset, and Ayrshire) and one locality in Ireland (Donegal).

A summer annual, June to September.

Batters (1902), p. 45; Brodie *et al.* (2016), p. 1020; Cormaci *et al.* (2012), pp. 133, 134, pl. 31(1); Dizerbo & Herpe (2007), p. 88, pl. 28(4); Fletcher in Hardy & Guiry (2003), p. 232; Gallardo García *et al.* (2016), p. 35; Guiry (2012),

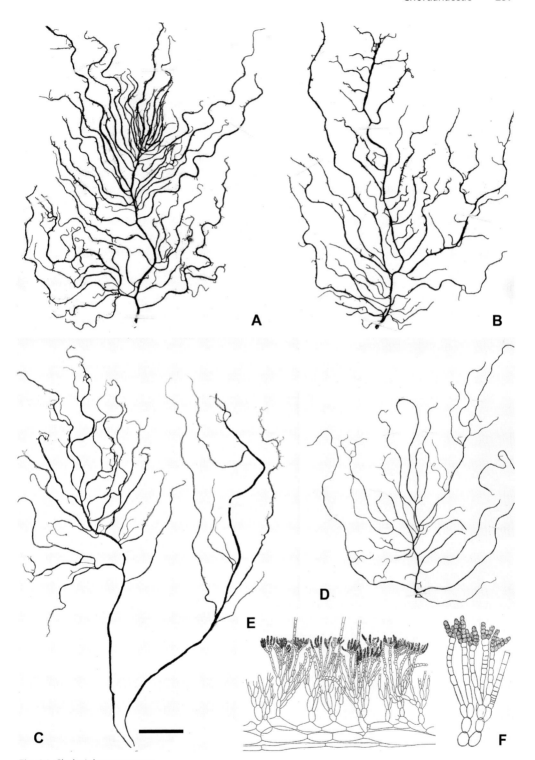

Fig. 86. *Cladosiphon contortus*
A–D. Habit of thalli. E, F. SP of portions of thalli showing the central, medullary filaments of large, elongated cells, branching out to form the subcortex, which gives rise to the peripheral assimilatory filaments and plurilocular sporangia. (E from Kuckuck 1929; F from Hamel 1931–39.) Bar in C = 28 mm (A, B), = 27 mm (C), = 70 mm (D), = 130 μm (E), = 159 μm (F). (A–D, herbarium specimens, NHM.)

p. 139; Hamel (1931–39), pp. 163, 164, fig. 36(d); Kuckuck (1929), p. 50, fig. 60; Newton (1931), p. 147; Parke (1933), pp. 18–20, pl. I, text fig. 6, pls I(2), VII(42–4); Ribera *et al.* (1992), p. 114; Sansón *et al.* (2006), pp. 529–45, figs 1–14.

Cladosiphon zosterae (J. Agardh) Kylin (1940), p. 28.   Figs 87, 88

Myriocladia zosterae J. Agardh (1841), p. 49; *Mesogloia virescens* var. *zostericola* Harvey (1846–51), pp. 1, 82; *Castagnea zosterae* (J. Agardh) Thuret in Le Jolis (1863), p. 95; *Eudesme zosterae* (J. Agardh) Kylin (1907), p. 85.

Thalli erect, macroscopic, 8–15(–20) cm long, 2–3 mm wide in the main axes, 1–2 mm in the branches, olive-green to light brown, cylindrical, filiform, solid, gelatinous, very mucilaginous, somewhat contorted, simple or sparsely and shortly branched above, with branches widely divergent, arising from a small discoid holdfast; in structure, thalli multiaxial, pseudoparenchymatous, haplotrichous with a central core of closely bound medullary filaments of colourless, elongated cells, 20–70 µm in diameter, 1–5(–9) diameters long, which branch out laterally to form a short subcortex of branched filaments of more globose cells, from the terminal regions of which arise clusters of outward-radiating, peripheral, assimilatory filaments with associated unilocular sporangia, plurilocular sporangia and hairs; compactness of medulla and subcortex increased by downward-growing filamentous extensions of the basal cells; medullary filaments and subcortical filaments terminated above by sheathed hairs; assimilatory filaments densely crowded, simple, or sometimes branched above, to 300(–450) µm, 25 cells long, fairly rigid, curved, elongate-clavate, multicellular; basal to mid-region cells of the assimilatory filaments cylindrical, 1–3(–3.5) diameters long, 8–32 × 6–9 µm, terminal cells typically moniliform, 0.5–2 diameters long, 8–19 × 8–13 µm; assimilatory cells with several discoid plastids each with a pyrenoid; hairs common, with basal meristem without obvious sheath, hair cells 5–11 diameters long, 60–165 × 13–16 µm; growth trichothallic with short intercalary meristem terminating all filaments, with subtending colourless hair.

Plurilocular and unilocular sporangia occurring on the same or separate thalli; plurilocular sporangia common, occurring as uniseriate, rarely biseriate, siliquose, quite densely tufted, transformations/outgrowths of cells in the terminal branched regions of the assimilators, to 3(–6) loculi long, to 40 µm in length, 12 µm in diameter; unilocular sporangia rare, formed at the base of assimilators, sessile, globular to ovoid, 38–30 × 18–40 µm.

Epiphytic on both sides of the leaves of *Zostera* spp., usually around the apex; lower eulittoral and shallow sublittoral.

Reported for a few scattered localities on southern and western coasts of Britain (Isles of Scilly, Cornwall, North Devon, Dorset, Isle of Man, Ayrshire and Argyll) and Ireland (Cork, Galway and Mayo).

A summer annual June to September.

Life history unknown, but based on culture studies of *Cladosiphon okamuranus* (Shinmura 1977) probably heteromorphic, with the macroscopic, diploid sporophyte alternating with a microscopic, discoid gametophyte bearing uniseriate gametangia and releasing isogametes. It is also likely that the sporophyte phase reproduces directly via spores from the plurilocular sporangia. Spore germination is stellate, which is considered to be an adaptation to an epiphytic habit, and similarities have been drawn between this mode of germination and that of the myrionemoid alga *Myrionema magnusii* (Loiseaux 1964), which possibly implicates this species in the life history of *Cladosiphon zosterae*.

Batters (1902), p. 45; Bornet (1892b), p. 236; Brodie *et al.* (2016), p. 1020; Cormaci *et al.* (2012), p. 140, pl. 33(4–5); Cotton (1912), p. 96; Dizerbo & Herpe (2007), p. 88, pl. 29(1); Fletcher in Hardy & Guiry (2003), p. 233; Gallardo García *et al.* (2016), p. 35; Guiry (2012), p. 139; Hamel (1931–39), pp. 160–2, figs 36 (b, c), 37(b); Kuckuck (1929), p. 47, pl. 4(15); Kylin (1907), p. 85, (1933), pp. 56–64, figs 26–9, (1940), p. 28, pl. 4(9), (1947), pp. 57, 58, figs 49, 50; Mathieson & Dawes (2017), p. 148, pl. XLIV(6–9); Newton (1931), p. 147; Oltmanns (1922), fig. 310(2); Parke (1933), p. 20, pls II(6), VII(38–41); Ribera *et al.* (1992), p. 114; Rueness (1977), p. 147, pl. XVIII(5), fig. 78, (2001), p. 14; Sauvageau (1924), p. 1381, (1927), pp. 369–431, figs 1–12; Taylor (1957), p. 146, pls 10(10, 11), 12(2).

Fig. 87. *Cladosiphon zosterae*

A–C. Habit of thalli on *Zostera*. D, E. SP of portions of thalli. F, G. SP of portions of vegetative thalli showing the assimilatory filaments. H–J. SP of portions of fertile thalli showing plurilocular sporangia transformed directly from the upper portion of the assimilatory filaments. Bar in A = 4 cm, bar in B = 12 mm, bar in I = 470 μm (D), = 300 μm (E), = 90 μm (F, G, H, I), = 45 μm (J).

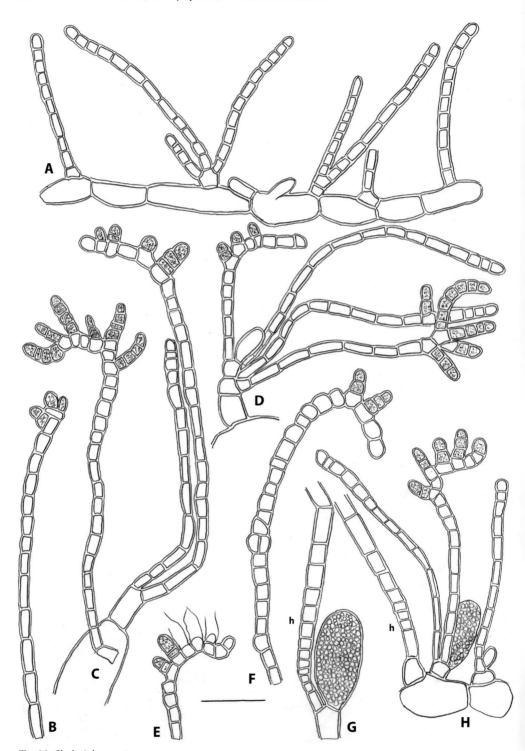

Fig. 88. *Cladosiphon zosterae*
A. SP of vegetative thallus showing the peripheral assimilatory filaments. B–F, H. SP. of fertile thallus showing the peripheral, assimilatory filaments transformed terminally into plurilocular sporangia. G, H. SP of fertile thallus showing unilocular sporangia at the base of the assimilatory filaments and hairs (h). Bar = 50 μm.

Composonema Kuckuck

COMPSONEMA Kuckuck (1899a), p. 58.

Type species: *C. gracile* Kuckuck (1899a), p. 58 (= *C. minutum* (Agardh) Kuckuck).

Thalli epiphytic or epilithic, forming either small hemispherical cushions or dense carpet, pile-like tufts; comprising a prostrate system of branched, outward-radiating, monostromatic filaments, usually remaining free although sometimes adjoined laterally into discs, giving rise to short, erect, simple or little-branched, gelatinous, later fibrous filaments usually of approximately equal length; plastids single, plate-like and lobed with pyrenoids; hairs with basal meristem and sheath.

Plurilocular and unilocular sporangia lateral or terminal on erect filaments, more rarely borne directly on basal filaments; plurilocular sporangia multiseriate.

The genus *Composonema*, established by Kuckuck in 1899a, is cosmopolitan and characterized by the following combination of features: a monostromatic basal layer of branched, outward-spreading filaments, which can either be free or laterally united into a discoid structure; fairly short, simple or little-branched, free, erect filaments; cells with one or two plate-like plastids; laterally situated sporangia, which in the case of the plurilocular sporangia are multiseriate. With the exception of Kuckuck (1953), Parke & Dixon (1964), Jaasund (1965) and Dangeard (1970) who placed the genus in the Ectocarpaceae, it is usually placed in the Myrionemataceae. It resembles the other myrionemoid genus *Myrionema* and *Microspongium* in having a monostromatic base and the genera *Chilionema* and *Hecatonema* in having multiseriate plurilocular sporangia.

However, in agreement with Abbott and Hollenberg (1976) it is a 'poorly defined genus' and has, over the years, included a disparate group of taxa, such as species for which only unilocular sporangia are known, species with sporangia borne on the basal layer rather than positioned on the erect filaments, species with variable numbers and shapes of plastids and, more significantly, species that have been reported with a distromatic rather than a monostromatic base.

Particularly pertinent to any discussion about the status of the genus *Composonema* have been the results of some life history studies. In three investigations, a direct, monophasic life history has been shown. The present author (unpublished) and Dangeard (1970) observed a repetition of the parental phase via plurispores in both *C. microspongium* (collected from the south coast of England) and *C. minutum* (collected from the Bay of Biscay), while a similar life history via unispores was reported by Pedersen (1981c) for *C. saxicola* (collected from Denmark). In two other investigations, however, *Composonema* species have been implicated as phases, probably diploid, in the life histories of members of the erect bladed family Scytosiphonaceae. Loiseaux (1970b) cultured the Californian species *C. sporangiiferum* Setchell & N.L. Gardner and produced a small, erect, *Scytosiphon*-like blade, which she tentatively identified as *S. pygmaeus* Reinke, while later there were reports of *C. saxicola* collected from the south coast of England occurring as a phase in the life histories of *Petalonia* and/or *Scytosiphon* species (Fletcher 1981a, and unpublished).

Such differences in life histories raise questions about the status of the genus *Composonema* and its relationship with Scytosiphonaceae members, but the indications are that the revealed heteromorphic life histories may not be typical for all members of this genus. Both *C. saxicola* and the closely related *C. sporangiiferum* differ from other *Composonema* spp. in possessing a unique combination of characters – i.e. a single plate-like plastid with large pyrenoid, partly distromatic base and basally positioned unilocular sporangia. Indeed, out of 18 species of *Composonema* described from California by Setchell & N.L. Gardner (1925), only *C. sporangiiferum* possessed a distromatic base. On the basis of these results, *C. saxicola* was transferred to the Scytosiphonaceae by Fletcher (1987), following the proposal of Pedersen (1981c) (see p. 486), while *Compsonema* was retained as an autonomous genus within the Chordariaceae pending further studies.

One species in Britain and Ireland, one species in Britain only.

KEY TO SPECIES

Thalli epiphytic, forming hemispherical cushions up to 1 mm high on
Pseudoralfsia verrucosa, in late summer, autumn . . . .*C. microspongium*
Thalli epilithic, forming confluent, pile-like tufts on bedrock, up to 0.5 mm high,
in winter . . . . *C. minutum*

Compsonema microspongium (Batters) Kuckuck (1953), p. 347.   Figs 89, 90

Ectocarpus microspongius Batters (1897), p. 436.

Thalli epiphytic, forming small, yellow-brown, hemispherical cushions on host, usually solitary and rounded, 1–4 mm wide to 1 mm high, occasionally oblong or confluent and netlike; consisting of an endophytic base of branched, outward-spreading, monostromatic filaments ramifying through host tissue, giving rise to numerous erect filaments; emergent filaments free, loosely aggregated, slightly gelatinous at first, later becoming more fibrous in texture, frequently pseudodichotomously branched; lower branches of older thalli spreading, matted and rhizoid-like, comprising cells irregularly nodose, 2–4 diameters long, 12–32 × 5–11 µm, upper branches erect, often secund, comprising quadrate to rectangular cells, 1(–2) diameters long, 8–18 × 9–16 µm; plastids single, plate-like and lobed, with pyrenoids, usually occupying upper cell region; hairs common, terminal on erect filaments with basal meristem and sheath.

Plurilocular sporangia common on upper branches, biseriate to multiseriate, solitary, or more often in short secund series, sessile or on 1–3-celled stalks, usually erect and closely adpressed to parental filament, oblong-lanceolate in shape, 65–90 × 10–20 µm; unilocular sporangia unknown.

Epiphytic, on crusts of *Pseudoralfsia verrucosa*, in the upper eulittoral.

Only recorded for the south-west coasts of Britain (south Cornwall and north and south Devon) and Mayo in Ireland.

Probably more widely distributed than records suggest. Appears to be a summer annual, June to September, more rarely found in October, November.

Laboratory culture studies by Fletcher (unpublished) revealed the plurispores to behave asexually and germinate directly to repeat the fertile, parental thallus. The life history appears, therefore, to be of the 'direct' type.

Batters (1897), pp. 436, 437; Brodie *et al.* (2016), p. 35; Cotton (1912), p. 94; Fletcher (1987), pp. 95–7, fig. 6; Fletcher in Hardy & Guiry (2003), p. 234; Gallardo García *et al.* (2016), p. 35; Guiry (2012), p. 155; Kuckuck (1953), pp. 347–50, fig. 15(A–B); Newton (1931), p. 117.

Compsonema minutum (C. Agardh) Kuckuck (1953), p. 341.   Fig. 91

Ectocarpus minutus C. Agardh (1827), p. 639; *Ectocarpus monocarpus* C. Agardh (1820–28), p. 48; *Compsonema gracile* Kuckuck (1899a), p. 56.

Thalli epilithic, forming small, light brown tufts to 0.5 mm high, densely crowded and carpet pile-like in appearance, spreading to 5 cm or more in extent; consisting of a basal layer of outward-spreading, irregularly branched, monosiphonous, free filaments, not obviously laterally adjoined and disc-like, giving rise to erect filaments, hairs and/or sporangia; in squash preparation, basal layer cells rectangular or more commonly irregularly nodose, usually 1–2 diameters long, 9–13 × 5–9 µm, with erect growths arising from mid-cell wall region; erect filaments short, to 400 µm, 35 cells long, simple or with 1(–3) branches, linear or slightly

Fig. 89. *Compsonema microspongium*
A, B. Erect vegetative filaments emerging from the host *Pseudoralfsia verrucosa*. C, D. Erect filaments with laterally produced plurilocular sporangia. Bar in B = 17 μm (A–D).

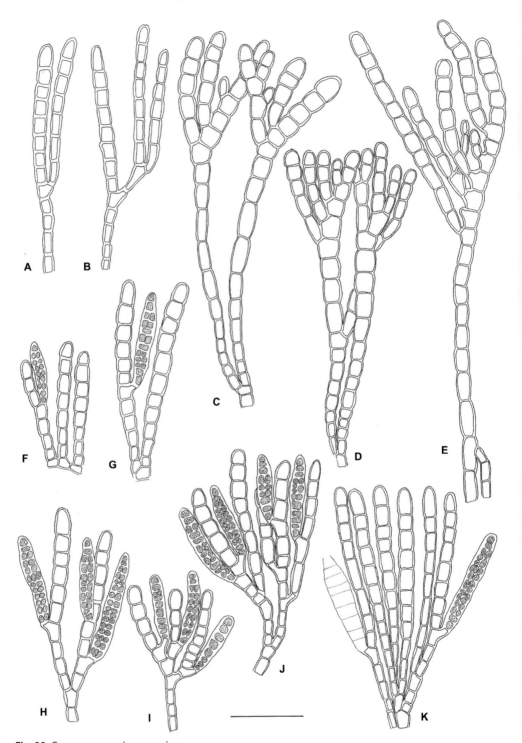

Fig. 90. *Compsonema microspongium*

A–E. Terminal portions of erect, vegetative filaments. F, G. Short, erect filaments arising from a monostromatic base with lateral plurilocular sporangia. H–K. Terminal portions of erect, fertile filaments with lateral plurilocular sporangia. Bar = 50 µm.

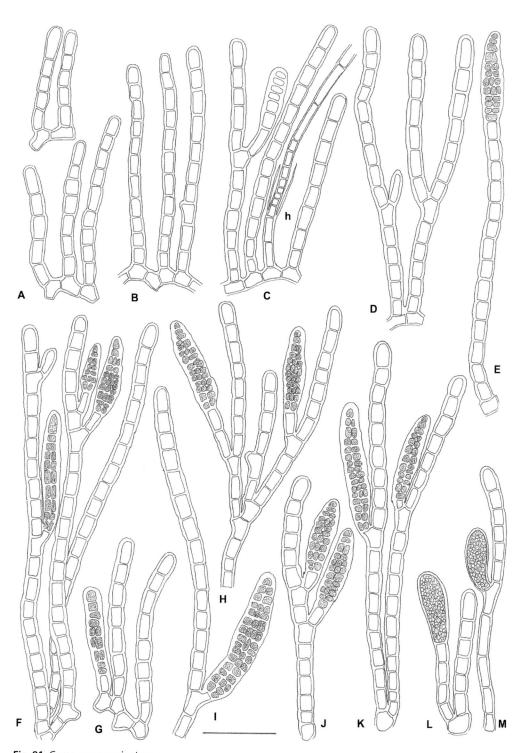

Fig. 91. *Compsonema minutum*

A–D. Erect, vegetative filaments arising from a monostromatic base. Note terminal hair with sheath (h). E–K. Erect, fertile filaments with terminal and lateral plurilocular sporangia. L, M. Erect, fertile filaments with terminal and lateral unilocular sporangia. Bar = 50 μm.

attenuate towards base, free or loosely aggregated in a mucilaginous matrix; branching pseudodichotomous, more rarely unilateral, to 1(–2) orders, usually restricted to the upper region, with branches strongly recurved towards main axis; erect filament cells rectangular, or slightly barrel-shaped, 1(–2) diameters long, 9–25 × 9–16 μm, with thick hyaline walls and enclosing a single, irregularly lobed, plate-like plastid with pyrenoids; outer cell wall regions often with short, closely set, upward-pointing, spine-like protuberances (possibly remains of outer cell wall); hairs common, usually terminal on erect filaments and branches, approx. 6–7 μm in diameter, with basal meristem and sheath.

Sporangia borne on erect filaments, more rarely arising from basal cell layer; plurilocular sporangia common, usually lateral and midway on erect filaments, sessile or on 1–2-celled stalks and strongly recurved towards parental filament, more rarely terminal on erect filaments or arising from basal cell layers, lanceolate in shape, margin irregular, biseriate to multiseriate, (50–)65–100 × 15–21 μm; unilocular sporangia infrequent, usually associated with the plurilocular sporangia, lateral or terminal on erect filaments, elongate-pyriform in shape, 40–45 × 15–20 μm.

Epilithic, on bedrock in the upper eulittoral.

Only recorded for Hampshire (Isle of Wight) in Britain; probably more widely distributed on the south coast of Britain.

Collections made in December and January; insufficient data to comment on seasonal distribution.

Laboratory culture studies by Dangeard (1970) on *C. minutum* collected at Mimizan, in the Bay of Biscay, revealed the occurrence of a 'direct' type life history with the plurispores behaving asexually and germinating directly to repeat the fertile parental thallus.

Børgesen (1926), pp. 59–62, fig. 31; Brodie *et al.* (2016), p. 1021; Cormaci *et al.* (2012), pp. 277, 278, pl. 80(1–4); Dangeard (1970), pp. 63–5, figs 1–13, pls 1–2; Fletcher (1987), pp. 97–9, fig. 7; Fletcher in Hardy & Guiry (2003), p. 235; Gallardo García *et al.* (2016), p. 35; Guiry (2012), p. 182; Kuckuck (1899a), pp. 56–60, figs 6–9, (1953), pp. 341–3, fig. 13; Ribera *et al.* (1992), p. 115; Schiffner (1916), p. 154, figs 70–4.

Corynophlaea Kützing

CORYNOPHLAEA Kützing (1843), p. 331.

Type species: *C. umbellata* (Agardh) Kützing (1843), p. 331.

Thallus epiphytic, forming small, gelatinous hemispherical cushions on hosts, less than 3 mm in diameter; comprising a prostrate base of outward-spreading, pseudodichotomously branched, monostromatic filaments, free or irregularly associated and pseudodiscoid, each cell giving rise to erect filaments; erect filaments short, dichotomously, sometimes trichotomously branched, loosely compacted in a gelatinous matrix, comprising large, thin-walled, colourless cells, elongated, cylindrical and narrow below becoming broader and more oval above, terminating in narrow thread-like photosynthetic filaments (paraphyses), hairs and/ or sporangia; paraphyses multicellular, gelatinous, simple or branched, slightly recurved with terminal moniliform cells; plastids single, lobed, plate-like or several and discoid, with pyrenoids; hairs with basal meristem but lacking obvious sheath.

Unilocular and plurilocular sporangia borne at the base of the terminal filaments; unilocular sporangia elongate-pyriform; plurilocular sporangia unknown for Britain and Ireland, reported elsewhere to be cylindrical, uniseriate, rarely biseriate.

One species in Britain and Ireland.

Corynophlaea crispa (Harvey) Kuckuck (1929), p. 42.   Figs 92, 93

Leathesia crispa Harvey (1857), p. 201; *Leathesia concinna* Kuckuck (1897), p. 387.

Thalli forming small light brown, hemispherical spots on host, usually circular and solitary, occasionally irregular and confluent, to 1–3 mm across; consisting of a prostrate system of branched, pseudodiscoid filaments giving rise to erect, dichotomously branched filaments loosely compacted together in a gelatinous matrix and easily separable under pressure; erect filament cells, thin walled, colourless, except for 1–4 small plate-like/discoid plastids, elongate below, 75–150 × 10–35 µm, 3–6(–10) diameters long, becoming more swollen above, 40–120 × 25–50 µm, 1–2 diameters long, finally becoming globular terminally, 18–34 × 15–30 µm, 1–2 diameters long, and giving rise to numerous paraphyses-like assimilatory filaments, hairs and/or unilocular sporangia; assimilatory filaments simple or branched, multicellular, 9–17 cells, 65–140 µm long, curled, comprising cells rectangular or barrel-shaped near base, 1–3 diameters long, 6–18 × 5–9 µm, approximately quadrate and crenate in appearance near apex, to 14 µm long × 8–12 µm in diameter; cells usually with one, more rarely two, parietal, lobed, plate-like plastids, with pyrenoids; hairs common, arising from cells at base of the assimilators, 8–9 µm in diameter, with basal meristem, but without obvious sheath.

Unilocular sporangia common, pyriform, 40–70 × 16–25 µm, solitary or in small groups at base of assimilators; plurilocular sporangia unknown in Britain and Ireland, reported elsewhere to be rare, usually simple and uniseriate, more rarely biseriate, occasionally branched, terminal or lateral on the assimilatory filaments.

Epiphytic on blades, especially ones yellow and senescent, of the red alga *Chondrus crispus*, upper eulittoral to shallow sublittoral.

South and west coasts of Britain and Ireland, as far north as Argyll and as far east as Dorset; in Ireland, Mayo, Clare, Wicklow.

Annual, summer and autumn, April to September.

Without microscopic examination this species can easily be confused with young thalli of *Leathesia difformis*. However, the latter are usually lighter in colour, firmer and gelatinous, with surface cells much more closely packed. Squash preparations of *C. crispus* often reveal endophytic *Streblonema*-like thalli.

Bartsch & Kuhlenkamp (2000), p. 165; Batters (1906), p. 2; Brodie *et al.* (2016), p. 1020; Cormaci *et al.* (2012), pp. 143, 144, pl. 35(1–3); Cotton (1908a), p. 329; Dizerbo & Herpe (2007), p. 83; Fletcher (1987), p. 141, fig. 24, pl. 2(e); Fletcher in Hardy & Guiry (2003), p. 236; Gallardo García *et al.* (2016), p. 35; Guiry (2012), p. 140; Hamel (1931–39), pp. 142, 143, fig. 32(F, G); Harvey (1857), pp. 201, 202, pl. XII(1A–3A); Kuckuck (1897), p. 387, fig. 12, (1929), p. 42, figs 46–8; Newton (1931), p. 141; P.M. Pedersen (1983), pp. 3, 4, fig. 6(a–d); Pybus (1975), pp. 153–5, fig. 1; Racault *et al.* (2009), pp. 111–23; Taskin (2006), pp. 217–25, figs 1–4.

Cylindrocarpus P. Crouan & H. Crouan

CYLINDROCARPUS P. Crouan & H. Crouan (1851), p. 359.

Type species: *C. microscopicus* P. Crouan & H. Crouan (1851), p. 359.

Thalli epiphytic, forming small (up to approximately 0.5 mm in diameter), felt-like tufts on host; comprising a basal system of branched, endophytic filaments which give rise to an emergent, medulla-like region of erect, dichotomously, more rarely trichotomously branched, uniseriate filaments of large, elongate-cylindrical, colourless cells, which become progressively more globose above and a terminal, cortex-like region of filamentous assimilators comprising narrower, rectangular, pigmented cells; medulla-like cells sometimes giving rise to downward-growing rhizoidal filaments; true hairs present, with basal meristem; cells with a single, lobed, plate-like plastid with pyrenoids.

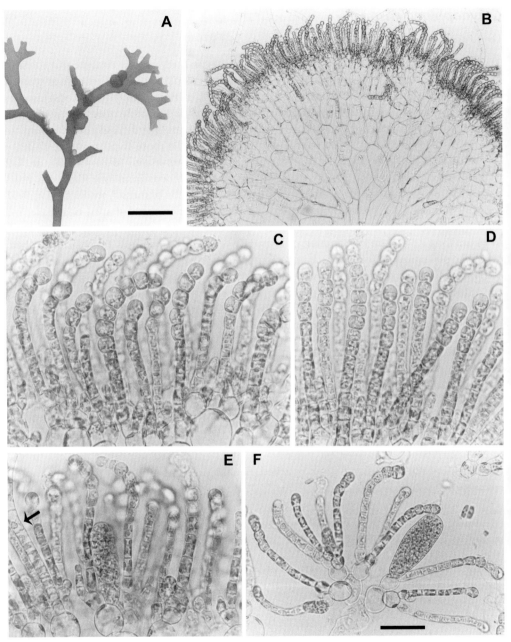

Fig. 92. *Corynophlaea crispa*

A. Habit of thalli epiphytic on host *Chondrus crispus*. B. VS of thallus showing the extensive medulla-like region of dichotomously branched filaments of large colourless cells giving rise terminally to an outer fringe of paraphysis-like, assimilatory filaments. C–F. SP of outer fringe-like region showing the paraphysis-like filaments, basally produced unilocular sporangia and a single hair (arrow). Bar in A = 11 mm, bar in F = 180 μm (B), = 45 μm (C–F).

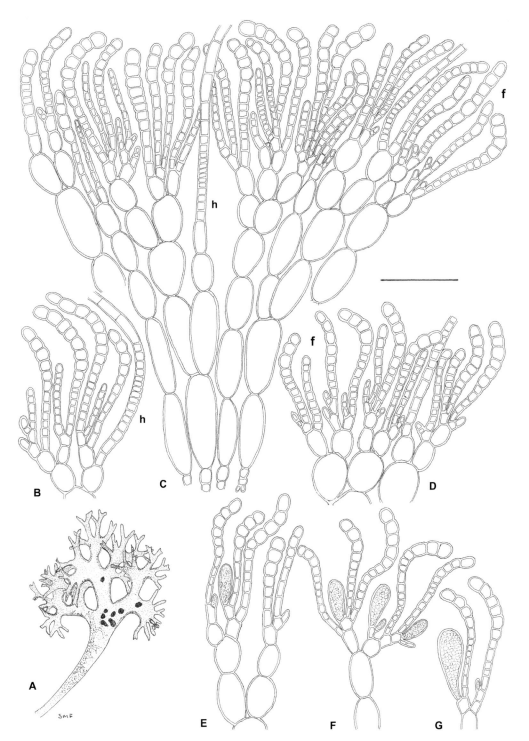

Fig. 93. *Corynophlaea crispa*

A. Habit of thalli epiphytic on *Chondrus crispus*. B–D. SP of vegetative thalli showing terminal assimilatory filaments (f) and occasional hairs (h). E–G. Portions of fertile thalli showing assimilatory filaments and unilocular sporangia. Bar = 30 mm (A), = 50 μm others.

Plurilocular sporangia rare, cylindrical, simple, uniseriate to biseriate, terminal on short, emergent, erect filaments, or borne laterally on the terminal assimilators; unilocular sporangia common, ovate to elongate-cylindrical, terminal on short, emergent, erect filaments or more commonly borne laterally on the upper medulla-like cells.

One species in Britain.

Cylindrocarpus microscopicus P. Crouan & H. Crouan (1851), p. 359.   Figs 94, 95

Ectocarpus investiens Hauck (1885), p. 325; *Streblonema investiens* (Hauck) Ardissone (1886), p. 79; *Ectocarpus microscopicus* (P. Crouan & H. Crouan) Batters (1902), p. 80.

Thalli epiphytic, forming small, solitary, sometimes confluent, light brown, felt-like tufts on host, 1–2 mm in extent and just visible to the unaided eye, arising from an endophytic base of narrow, branched, contorted, colourless, rhizoidal filaments spreading between the outer cortical tissue of the host; erect filaments constricted at point of entry into host surface, uniseriate, up to 500–600 µm in height, loosely aggregated in a slightly gelatinous matrix which is easily spread on a microscope slide under the light pressure of a coverslip, and distinctly differentiated into a sparsely branched, medulla-like, basal region of large, colourless, elongate-cylindrical cells which become progressively shorter and more globose above, and a more profusely branched, terminal, cortex-like region comprising narrower, rectangular, pigmented cells; medulla-like region dichotomously, more rarely trichotomously branched, terminal region dichotomously, more rarely irregularly or secundly branched; endophytic cells to 50 µm in length, 3–5 µm in diameter, medullary cells 1–5(–8) diameters long, 16–123 × 15–34 µm, cortical cells 2–4 diameters long, 12–27 × 6–8 µm; true hairs common, delicate, borne on both medullary and cortical cells, comprising cells to 15 diameters (–135 µm) long, 9–11 µm in diameter, with basal meristem; medulla-like cells often giving rise to downward-extending, narrow, branched, multicellular filaments; cells with a single, branched, plate-like plastid with pyrenoids.

Plurilocular sporangia rare, terminal on short emergent filaments or, more rarely, borne on the cortical filaments, cylindrical, simple, uniseriate sometimes biseriate, 6 to 9 loculi, 31–40 × 9–12 µm, sometimes associated with unilocular sporangia on the same thallus; unilocular sporangia common, either terminal on short emergent filaments, or more commonly borne laterally in the upper medullary region, if lateral then also usually sessile, ovoid, 65–105 × 26–55 µm.

Commonly occurring as an epi-endophyte on *Gracilaria bursa-pastoris*, but also reported on other *Gracilaria* spp. (e.g. *G. multipartita*); if present in the eulittoral zone, then only in rock pools, otherwise in the upper sublittoral and in conditions of shelter.

Restricted to the south coast of Britain (Isle of Wight, Dorset) and thus has a narrower range than its host species.

Summer annual, May to September.

Batters (1902), p. 30; Brodie *et al.* (2016), p. 1020; Cormaci *et al.* (2012), pp. 148, 149, pl. 37(1); P. Crouan & H. Crouan (1851), pp. 359–61, pls 16(1–11), (1867), p. 160; Dizerbo & Herpe (2007), p. 83, pl. 26(4); Feldmann (1937), pp. 285, 286; Fletcher in Hardy & Guiry (2003), p. 237; Gallardo García *et al.* (2016), p. 35; Guiry *et al.* (2012), p. 181; Hamel (1931–39), pp. 145–7, fig. 33(C); Hauck (1884), p. 325, fig. 135; Holmes (1883), p. 141; Holmes & Batters (1890), p. 78; Kuckuck (1899), pp. 49–55, figs 1–5, pl. VI; Newton (1931), pp. 116, 117; Oltmanns (1922), p. 25, fig. 315; Ribera *et al.* (1992), p. 114; Sauvageau (1892), pp. 41–3, pl. 1(6).

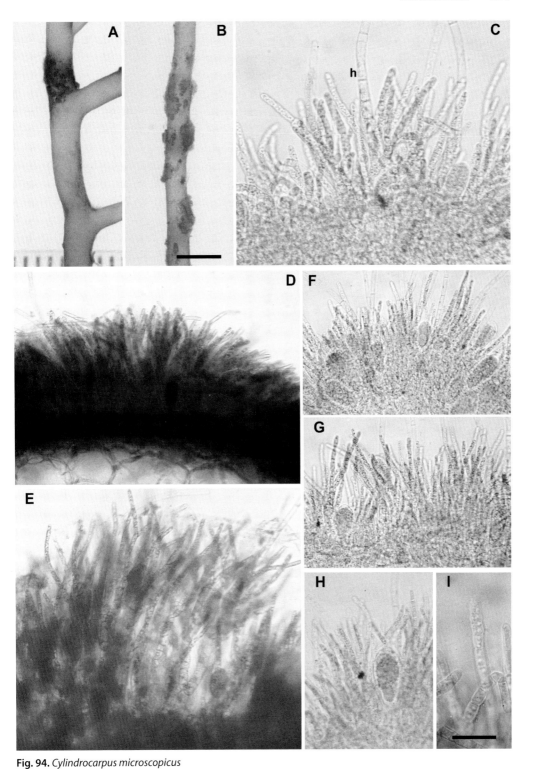

Fig. 94. *Cylindrocarpus microscopicus*
A, B. Habit of thalli epiphytic on *Gracilaria*. C–H. TS of thallus showing loosely aggregated erect filaments, basally situated unilocular sporangia and terminal hairs. I. Terminal plurilocular sporangium. Bar in B = 4 mm, bar in I = 180 µm (D, E), = 90 µm (C, F, G, H), = 45 µm (I).

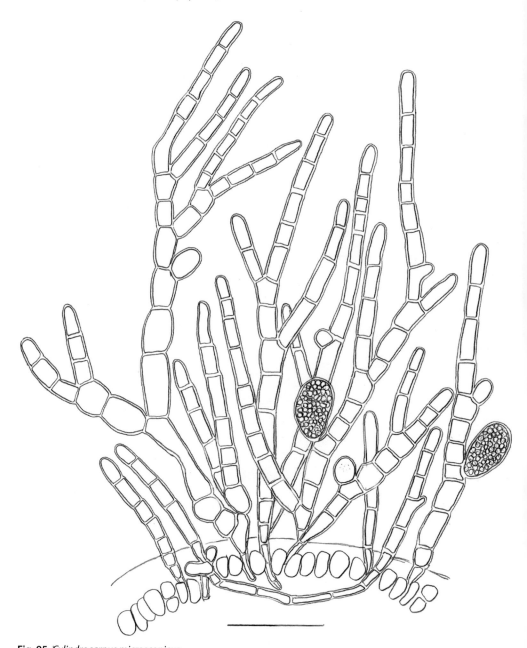

Fig. 95. *Cylindrocarpus microscopicus*
TS of thallus epiphytic on host showing the loosely associated erect filaments and basally positioned unilocular sporangia. Bar = 50 μm.

Dictyosiphon Greville

DICTYOSIPHON Greville (1830), p. XLIII, 55.

Type species: *D. foeniculaceus* (Hudson) Greville (1830), p. 56.

Thalli macroscopic, erect, sparingly to much branched, cylindrical, filiform, parenchymatous, solid at first in the younger parts, sometimes remaining so in the basal regions, becoming hollow in older parts, arising from a holdfast of compacted, secondary rhizoids; main axis prominent, branching opposite, irregular or alternate, with branches sometimes attenuate, with terminal apical cell, capped by a hair in young specimens; in structure, thallus clearly pluriseriate throughout, comprising an inner medulla of four large, longitudinally elongated, colourless cells in the younger parts which later collapse to form the hollow cavity, surrounded by a cortex of several layers of progressively smaller, colourless cells with a peripheral, epidermis-like layer of smaller, pigmented cells, each with numerous discoid plastids with associated pyrenoids; hairs common, arising directly from surface cells, usually solitary, with basal meristem but without obvious sheath; initial growth apical, then diffuse.

   Unilocular sporangia scattered over thallus surface, formed in enlarged outer cortical cells, oval to ellipsoidal and partly immersed; plurilocular sporangia unknown on macrophyte.

Two species in Britain and Ireland.

KEY TO SPECIES

Branching mainly primary, few secondary branches produced; branches attenuate
at base . . . . *D. chordaria*
Secondary, tertiary and other branches produced in profusion; branches not
attenuate at base. . . . . *D. foeniculaceus*

Dictyosiphon chordaria Areschoug (1847), p. 372.   Fig. 96

Scytosiphon chordarius (Areschoug) Fries (1845), p. 124; *Dictyosiphon mesogloia* Areschoug (1861–79), Exs. no. 106, (1875), p. 33; *Coilonema chordarium* (Areschoug) Areschoug (1872), no. 323; *Dictyosiphon chordarius* var. *bahusiensis* Areschoug (1873), p. 170; *Dictyosiphon chordarius* var. *simpliciusculus* Areschoug (1873), p. 170; *Cladosiphon balticus* Gobi (1874), p. 12; *Dictyosiphon chordarius* f. *simpliciusculus* (Areschoug) Kjellman (1877b), p. 40; *Chordaria baltica* (Gobi) Gobi (1879), p. 89; *Coilonema chordarium* f. *bahusiense* (Areschoug) Kjellman (1880), p. 10; *Coilonema mesogloia* (Areschoug) Kjellman (1880), p. 10; *Dictyosiphon finmarkicus* Foslie (1881), p. 6; *Coilonema chordarium* f. *simpliciusculum* (Areschoug) Strömfelt (1884), p. 12; *Gobia baltica* (Gobi) Reinke (1889b), p. 65; *Dictyosiphon balticus* (Gobi) Du Rietz (1930), p. 427; *Dictyosiphon chordarius* f. *balticus* (Gobi) Häyren (1940), p. 58.

Thalli macroscopic, erect, to 36 cm long, 0.5–1.5 mm wide, solitary, dark olive to light brown, soft, sparse to much branched, cylindrical, parenchymatous, at first solid in the younger parts, becoming hollow in older parts, arising from a holdfast of compacted, secondary rhizoids derived from downward-growing hyphal extensions of the inner cortical cells; main axis usually prominent and much branched to 2(–3) orders, primary laterals usually sparsely branched or simple, branching irregular, branches fairly uniform in diameter, although sometimes tapering to a fine point; in surface view, thallus clearly pluriseriate throughout, with cells in longitudinal, less obvious transverse rows, especially in younger regions, less obvious in older regions, approximately cuboidal to polygonal, isodiametric, each with numerous discoid plastids with associated pyrenoids; in both transverse and longitudinal sections, young parts clearly differentiated into an inner medulla of four large, longitudinally elongated, colourless cells, which later collapse to form the hollow cavity, surrounded by a cortex of several layers of progressively smaller, colourless cells, the peripheral ones dividing tangentially to form an epidermis-like layer of smaller, pigmented cells; hairs common, arising directly from surface cells, usually solitary, with basal meristem but without obvious sheath; growth apical, additionally followed by diffuse.

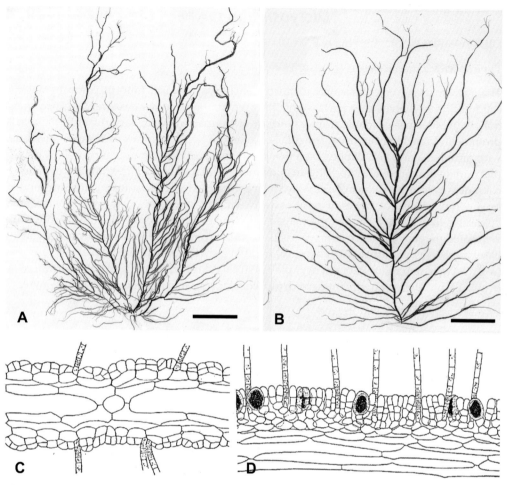

Fig. 96. *Dictyosiphon chordaria*

A, B. Habit of thalli. C. LS of young vegetative thallus showing the central elongated medullary cells giving rise to a cortex of smaller cells enclosed by an epidermis-like layer. D. LS of fertile thallus showing partly immersed unilocular sporangia. Bar in A = 24 mm (A), bar in B = 24 mm (B), = 38 μm (C), = 48 μm (D). (C, D after Kuckuck 1929; A, B, herbarium specimens, NHM.)

Unilocular sporangia scattered over thallus surface, formed in slightly enlarged outer cortical cells, larger and darker than accompanying vegetative cells, oval to ellipsoidal, with long axis parallel to the surface, partly immersed in vertical section, 20–40 μm in diameter; plurilocular sporangia unknown on macrophyte.

Epilithic, on rocks and stones, mid to lower eulittoral pools and shallow sublittoral.

Rare, with a northerly distribution in Britain (Yorkshire, Northumberland, Fifeshire, Orkney, Shetland, Isle of Cumbrae, Bute, Ayrshire) and reported for various localities in Ireland (Clare, Down, Antrim).

Summer annual, mainly collected June to August, rarely in October.

Areschoug (1875), p. 33; Batters (1885), p. 537, (1890), p. 271, (1895a), p. 275, (1902), p. 24; Brodie *et al.* (2016), p. 1020; Du Rietz (1940), pp. 35–46, figs 10, 11; Fletcher in Hardy & Guiry (2003), p. 243; Guiry (2012), p. 140; Holmes & Batters (1890), p. 77; Kylin (1947), p. 79; Levring (1940), pp. 56–8; Mathieson & Dawes (2017), p. 151, pl. XLV(7A, 7B, 8); Morton (1984), p. 42; Newton (1931), p. 169; Pankow (1971), pp. 202, 203, figs 268, 269; Pedersen (2011), p. 83, fig. 88; Reinke (1889b), p. 65, fig. 3; Ribera *et al.* (1992), p. 117; Rueness (1977), pp. 165, 166, fig. 98, (2001), p. 14; Taylor (1957), p. 174.

Dictyosiphon foeniculaceus (Hudson) Greville (1830), p. 56. Figs 97, 98

Conferva foeniculacea Hudson (1762), p. 479; *Scytosiphon foeniculaceus* (Hudson) C. Agardh (1811), p. 25; *Ulva foenic-ulacea* (Hudson) C. Agardh in Liljeblad (1816), p. 604; *Halymenia foeniculacea* (Hudson) C. Agardh (1817), p. 38; *Chordaria flagelliformis* var. *intricata* C. Agardh (1817), p. 13; *Scytosiphon tomentosus* Hornemann (1818), pl. 1594; *Scytosiphon hippuroides* Lyngbye (1819), p. 63; *Scytosiphon foeniculaceus* var. *intricatus* (C. Agardh) Lyngbye (1819), p. 63; *Scytosiphon foeniculaceus* var. *membranaceus* Lyngbye (1819), p. 63; *Sphaerococcus plicatus* var. *hippuroides* (Lyngbye) C. Agardh (1822), p. 314; *Sphaerococcus hippuroides* (Lyngbye) C. Agardh (1824), p. 235; *Ilea foeniculacea* (Hudson) Fries (1835), p. 321; *Scytosiphon ramellosus* J. Agardh (1836), p. 16; *Gigartina plicata* var. *hippuroides* (Lyngbye) Örsted (1844), p. 52; *Chordaria flagelliformis* var. *hippuroides* (Lyngbye) J. Agardh (1848), p. 66; *Dictyosiphon fragilis* Harvey in Kützing (1849), p. 485; *Dictyosiphon hippuroides* (Lyngbye) Kützing (1856), pl. 52 II; *Dictyosiphon foeniculaceus* subsp. *flaccidus* Areschoug (1873), p. 169; *Dictyosiphon foeniculaceus* var. *flaccidus* (Areschoug) Areschoug (1875), p. 31; *Dictyosiphon hippuroides* var. *fragilis* Kjellman (*c.* 1877); *Dictyosiphon hippuroides* f. *typicus* Kjellman (*c.* 1877), nom. illeg.; *Dictyosiphon foeniculaceus* subsp. *hispidus* Kjellman (1877b), p. 39; *Dictyosiphon hispidus* (Kjellman) Kjellman (1877b), p. 47; *Dictyosiphon foeniculaceus* f. *flaccidus* (Areschoug) Kjellman (1880), p. 10; *Dictyosiphon corymbosus* Kjellman (1883b), p. 330; *Dictyosiphon corymbosus* f. *abbreviatus* Kjellman (1883b), p. 330; *Dictyosiphon foeniculaceus* f. *typicus* Kjellman (1883b), p. 333; *Dictyosiphon foeniculaceus* var. *hispidus* (Kjellman) Collins (1937b), p. 226; *Dictyosiphon foeniculaceus* f. *hippuroides* (Lyngbye) Levring (1940), p. 55.

Thalli macroscopic, erect, to 30(–60) cm long, to 0.5 mm in width, solitary or gregarious, pale olive-brown to medium brown, firm and membranous, usually much branched and bushy, cylindrical, filiform, very fine, parenchymatous, at first solid in the younger parts and remaining so in the basal regions, becoming hollow, sometimes partly inflated in older parts, arising from a holdfast of compacted, secondary rhizoids; main axis prominent and branched to 2(–3) orders, branching opposite, alternate or irregular with short to long primary branches, and ultimate branches variable in length, somewhat curved and bushy, tapering to a fine point with a prominent, dome-shaped apical cell, capped by a hair in young specimens; in surface view, thalli pluriseriate throughout, with cells in longitudinal and, less obviously, transverse rows, especially in younger regions, more irregularly placed below, approximately cuboidal to polygonal, isodiametric, measuring approximately 5–14 × 5–12 μm, each containing numerous discoid plastids with associated pyrenoids; in both transverse and longitudinal sections, young parts of the thalli are clearly differentiated into an inner medulla of four longitudinally elongated, colourless cells, to 30(–45) μm in diameter which later collapse to form the hollow cavity, surrounded by a cortex of several layers of progressively smaller, colourless cells, the peripheral cells dividing tangentially to form an epidermis-like layer of smaller, pigmented cells measuring approximately 5–10 μm in depth; hairs common, arising directly from surface cells, usually solitary, with basal meristem but without obvious sheath, the hair cells to 20 diameters long, averaging approximately 8 μm in diameter; growth apical, followed by diffuse.

Unilocular sporangia scattered over thallus surface, formed in slightly enlarged outer cortical cells, larger and darker than accompanying vegetative cells, oval to ellipsoidal, with long axis parallel to the surface, partly immersed in vertical section, 18–60 μm in diameter in surface view; plurilocular sporangia unknown on macrophyte.

Epiphytic on various algae, particularly on *Chordaria flagelliformis*; other host species include *Scytosiphon lomentaria*, *Chorda filum* and *Gongolaria* spp. Occasionally epilithic, mid to lower eulittoral pools, sometimes emersed, and shallow sublittoral, to 12 m, in conditions of sheltered or moderate exposure; locally common; some tolerance of reduced salinity.

Common and widely distributed around Britain and Ireland.

Summer annual, mainly collected June to September, rarely found at other times.

Culture studies by Sauvageau (1917) and Peters & Müller (1985) revealed a diplohaplontic, heteromorphic life history, with the macroscopic, diploid generation which bore unilocular sporangia alternating with a microscopic, filamentous, dioecious, gametangia-bearing, haploid generation. Zygotes, and unfused gametes from the haploid generation then developed into the macroscopic diploid thalli.

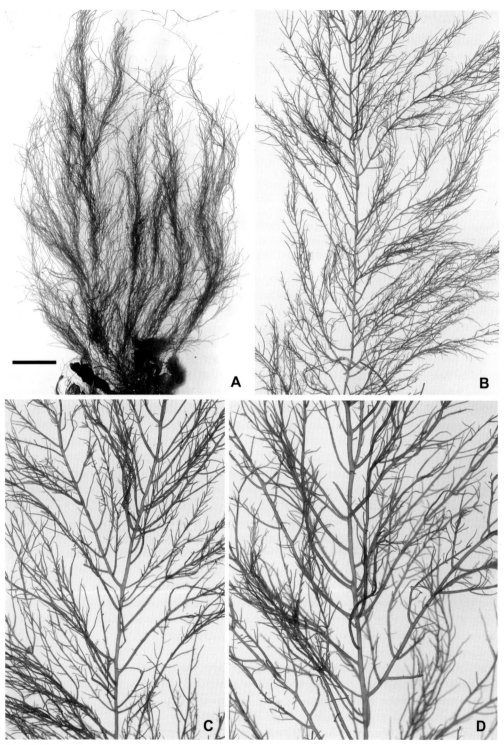

Fig. 97. *Dictyosiphon foeniculaceus*
A–D. Habit of thalli. Bar in A = 2.5 cm (A), = 12 mm (B), = 6 mm (C, D).

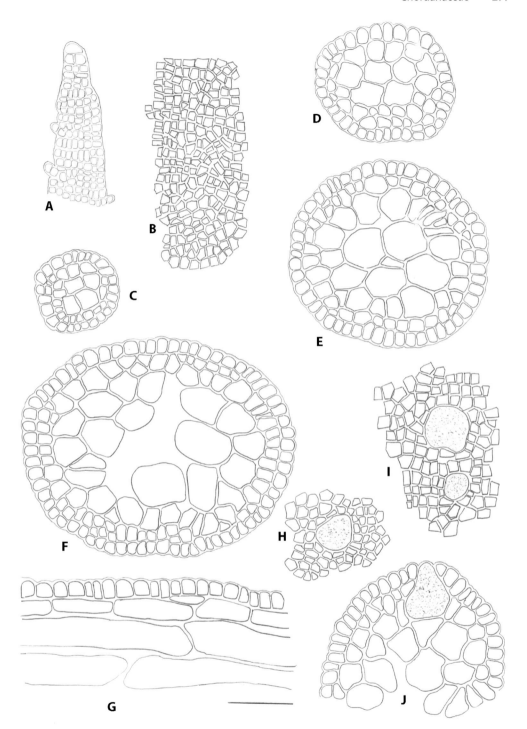

Fig. 98. *Dictyosiphon foeniculaceus*

A. Apex of branch showing dome-shaped apical cell. B. Surface view of thallus showing irregular arrangement of cells. C–F. TS of progressively thicker thalli. Note central cavity in F. G. LS. of thallus edge showing elongated, cortical cells enclosed by an epidermis-like layer of cells. H, I. Surface view of fertile thalli showing scattered unilocular sporangia. J. TS of fertile thallus showing partly immersed unilocular sporangium. Bar = 50 µm.

Areschoug (1873), pp. 161–71; Bartsch & Kuhlenkamp (2000), p. 165; Batters (1890b), p. 270, (1902), p. 23; Brodie *et al.* (2016), p. 1020; Bunker *et al.* (2010), p. 304; Cotton (1912), p. 94; Dizerbo & Herpe (2007), p. 115, pl. 38(3); Fletcher in Hardy & Guiry (2003), p. 243; Fritsch (1945), p. 114, fig. 41(A–J); Guiry (2012), pp. 140, 141; Hamel (1931–39), pp. 270–2, fig. 40(a–i); Holmes (1883), p. 140; Jaasund (1965), pp. 94, 95, fig. 29; Kornmann & Sahling (1977), p. 278; Kylin (1947), pp. 78, 79, pl. 13(42, 43); Levring (1940), pp. 55, 56; Mathieson & Dawes (2017), p. 152, pl. XLVI(1); Morton (1994), p. 42; Newton (1931), p. 168, fig. 105(A–C); Pankow (1971), p. 202, figs 265–7; Pedersen (2011), p. 82, fig. 87; Peters & Müller (1985), pp. 441–7; Reinke (1889b), p. 63; Rueness (1977), p. 167, pl. XVIII(3), fig. 99(A, B), (2001), p. 14; Saunders & McDevit (2013), pp. 1–23; Sauvageau (1917), pp. 829–31; Taylor (1957), pp. 172, 173, pls 12(4), 14(2).

Elachista Duby

ELACHISTA Duby (1830), p. 972 (nom. cons.).

Type species: *E. scutulata* (Smith) Areschoug (1843), p. 262.

Thalli epiphytic, forming small, brush-like tufts on hosts, with or without basal hemispherical cushions, arising from a prostrate, monostromatic layer of outward-radiating, branched filaments, free or laterally united and discoid/pseudodiscoid; filaments of tufts erect, free, linear, flaccid, multicellular, monosiphonous, simple or branched at base, acuminate sharply below, more gradually above, with basal meristem, comprising thick-walled cells with numerous discoid plastids without obvious pyrenoids; hemispherical cushions, if present, solid, firm and cartilaginous or slightly soft and gelatinous, pseudoparenchymatous, comprising vertical, compact, dichotomously branched filaments of large, colourless, thin-walled cells, giving rise terminally to the free erect filaments, paraphyses and/or sporangia, basally to downward-growing rhizoidal filaments which often penetrate host tissue; paraphyses multicellular, simple, linear or curved, photosynthetic; hairs unknown.

Unilocular and plurilocular sporangia usually formed at the base of the free erect filaments, usually lateral and/or terminal on the paraphyses, more rarely borne on the upper regions of the erect filaments; unilocular sporangia pyriform, plurilocular sporangia filiform, uniseriate, simple.

Species of the genus *Elachista* can generally characterized by an epiphytic habit and small brush-like tufts of free, erect, simple filaments which arise initially from a discoid/pseudodiscoid basal layer, later from a solid, cushion-like, pseudoparenchymatous thallus comprising vertical rows of compacted, branched, filaments of large colourless cells. The reproductive sporangia are usually borne at the base of the free, erect filaments, associated with paraphysis-like filaments.

In Britain and Ireland this genus is represented by four fairly distinct species: *Elachista flaccida, E. fucicola, E. scutulata* and *E. stellaris. Elachista stellaris* has previously been placed into two separate genera *Symphoricoccus* and *Areschougia* by Kuckuck (1929) and Meneghini (1844) respectively, on the bases of sporangial development on the erect filaments and the ability of the erect filaments to act as stolons (see Wanders *et al.* 1972 and Dangeard 1968a for discussion on this subject). However, in agreement with Wanders *et al.* (1972), these are not considered sufficient criteria for placing *E. stellaris* in a separate genus.

Three species in Britain and Ireland, one species in Britain only.

KEY TO SPECIES

1. Thallus epiphytic on the thongs of the brown alga *Himanthalia*; basally produced unilocular sporangia elongate-cylindrical, on 1–4-celled pedicels . . *E. scutulata*
    Thallus not epiphytic on *Himanthalia*; basally produced unilocular sporangia pyriform, not pedicellate. 2

2. Erect filaments commonly 50–135 µm in diameter . . . . *E. flaccida*
    Erect filaments usually less than 50 µm in diameter. 3

3. Epiphytic on *Fucus* spp., rarely other algae; paraphyses obvious and
    abundant; unilocular sporangia borne at base of erect filaments only;
    plurilocular sporangia not known on macrothallus . . . . *E. fucicola*
    Not recorded on *Fucus* spp.; paraphyses not obvious; unilocular sporangia
    borne laterally on the basal and upper regions of the erect filaments;
    plurilocular sporangia recorded on macrothallus . . . . *E. stellaris*

Elachista flaccida (Dillwyn) Fries (1843), p. 262.   Figs 99(A,B), 100

Conferva flaccida Dillwyn (1802–09), p. 52; *Conferva breviarticulata* Suhr (1836), p. 384; *Elachista breviarticulata* (Suhr) Areschoug (1842), p. 234.

Thalli epiphytic, forming light to yellow-brown, brush-like tufts on hosts, to 1(–2) cm in height, single or gregarious, arising at the base from a hemispherical, solid, wart-like cushion, 2–4 mm in diameter; cushions firm and cartilaginous or slightly gelatinous, comprising vertical rows of dichotomously branched, closely packed filaments of large, thin-walled, colourless cells, 6–30 μm in diameter, elongate-cylindrical below becoming more globose above, penetrating below into host tissue, giving rise terminally to the erect brush-like tufts of filaments, paraphyses and/or sporangia; erect filaments 7–10(–16) mm long, free, linear, attenuate at base and apex, monosiphonous, comprising thick-walled cells, 0.5–1(–2) diameters long, 70–125 × 55–160 μm, each with large numbers of small discoid plastids without obvious pyrenoids; paraphyses abundant, closely packed, multicellular, simple, to 160(–280) μm long, 10(–15) cells, slightly curved, markedly clavate, with upper cells becoming moniliform, 13–47 × 15–60 μm; hairs unknown.

Unilocular sporangia lateral at the base of the paraphyses, pyriform, 70–90 × 23–33 μm; plurilocular sporangia unknown.

Epiphytic on *Gongolaria* spp., more rarely on *Halidrys siliquosa*, in lower eulittoral pools and shallow sublittoral.

Generally distributed around Britain and Ireland, but apparently absent from south-east England (Sussex and Kent) and much of eastern England and eastern Ireland. Recorded on *Gongolaria* spp. for the Channel Islands, Isles of Scilly, eastwards to the Isle of Wight, northwards to Mull, in Ireland northwards to Donegal, eastwards to Wexford. Recorded for scattered localities on *Halidrys* (Isle of Man; Outer Hebrides, Orkney, Shetland, Berwick).

Annual; spring and summer, April to September, rarely recorded at other times.

Ardré (1970), p. 250; Areschoug (1842), p. 234, figs 4, 5; Brodie *et al.* (2016), p. 1020; Cormaci *et al.* (2012), pp. 150, 152, pl. 37(2–3); Cotton (1912), p. 95; De Mesquita Rodrigues (1963), pp. 60, 61, pl. VI(b); Dizerbo & Herpe (2007), pp. 80, 82, pl. 25(4); Fletcher (1987), p. 125, fig. 18, pl. 2(c); Fletcher in Hardy & Guiry (2003), p. 246; Gallardo García *et al.* (2016), p. 35; Gayral (1958), p. 237, fig. 35, (1966), p. 267, fig. 34; Guiry (2012), p. 141; Hamel (1931–39), pp. 121, 122, fig. 27(c); Harvey (1846–51), pl. 260; Kuckuck (1929), p. 23; Morton (1994), p. 34; Newton (1931), p. 134; Ribera *et al.* (1992), p. 115; Stegenga & Mol (1983), p. 92; Stegenga *et al.* (1997a), p. 18.

Elachista fucicola (Velley) Areschoug (1842), p. 235.   Figs 99(C,D), 101

Conferva fucicola Velley (1795), pl. 4; *Conferva fucorum* Roth (1797), p. 190; *Conferva flaccida* Hornemann (1827), pl. 1906, nom. illeg.; *Elachista globosa* Örsted (1844), p. 50; *Elachista lubrica* Ruprecht (1850), p. 196; *Elachista grevillei* Arnott in Harvey (1857), p. 202; *Elachista fucicola* var. *lubrica* (Ruprecht) Rosenvinge (1893), p. 878; *Elachista fucicola* f. *lubrica* (Ruprecht) Printz (1926), p. 161; *Elachista fucicola* f. *grevillei* Hamel (1931–39), p. 119; *Myriactula lubrica* (Ruprecht) Jaasund (1960), p. 101.

Thalli epiphytic, forming light to dark brown, brush-like tufts on hosts, to 2(–4) cm long, single or gregarious, arising either directly from a monostromatic basal layer of outward-radiating filaments, laterally adjoined and discoid/pseudodiscoid, or from a basal hemispherical, solid, wart-like cushion; basal cushions, if present, cartilaginous or slightly mucilaginous, comprising

Fig. 99. *Elachista* spp.
A, B. *Elachista flaccida*. Habit of thalli epiphytic on *Gongolaria nodicaulis*. C, D. *Elachista fucicola*. Habit of thalli epiphytic on *Fucus serratus* and *F. vesiculosus*. E. *Elachista stellaris*. Habit of thalli on *Gongolaria nodicaulis*. Bar in E = 40 mm (A), = 15 mm (B), = 4 cm (C), = 3 cm (D), = 1 cm (E).

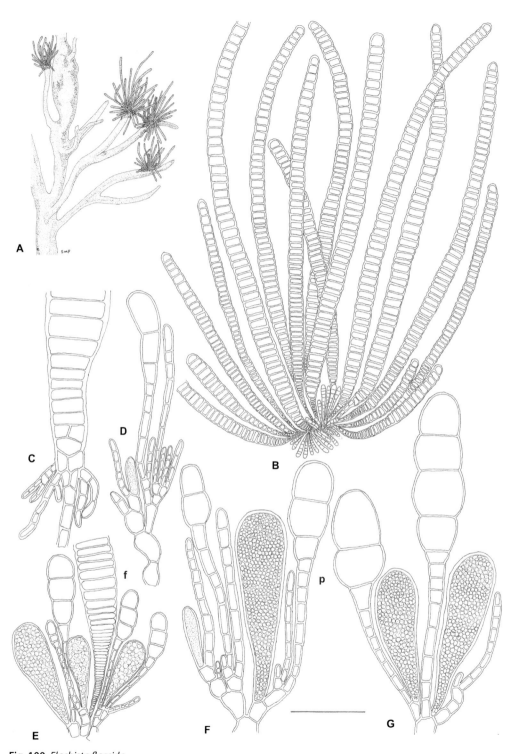

Fig. 100. *Elachista flaccida*

A. Habit of thalli epiphytic on *Gongolaria*. B. SP of thallus showing erect filaments. C. Basal region of erect filament. D–G. SP of fertile thallus showing erect filaments (f), paraphyses (p) and large, pyriform unilocular sporangia. Bar = 18 mm (A), = 0.5 mm (B), = 50 μm (C–G).

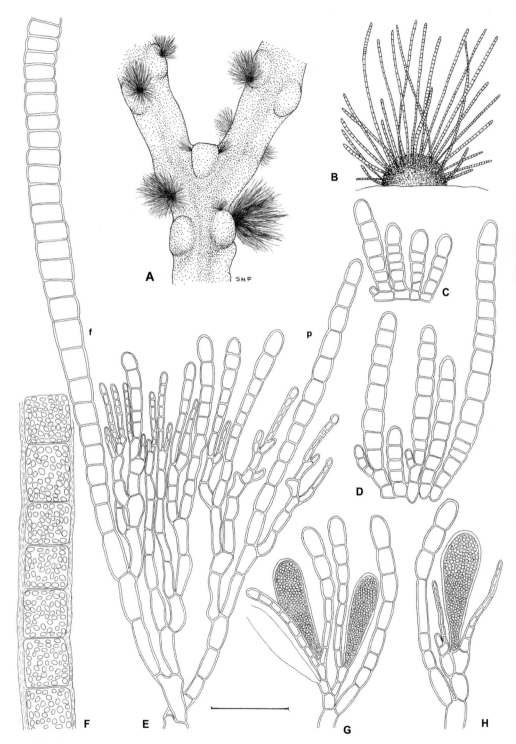

Fig. 101. *Elachista fucicola*

A, B. Habit of thalli, epiphytic on *Fucus vesiculosus*. C, D. SP of young thalli. E. SP of thallus showing erect filaments (f). F. Portion of erect filament showing thick-walled cells with large numbers of discoid plastids. G, H. SP of thallus showing unilocular sporangia arising at the base of paraphyses (p). Bar = 36 mm (A), = 5 mm (B), = 50 μm (C, D, F), = 100 μm (E, G, H).

vertical rows of closely packed, dichotomously branched filaments of large, thin-walled, colourless cells, elongate-cylindrical below and penetrating host tissue, gradually becoming globose above and terminating in the brush-like tufts of free, erect filaments, paraphyses and/or unilocular sporangia; young thalli without obvious basal cushion and erect filaments, paraphyses and/or unilocular sporangia arising directly from monostromatic basal layer; erect filaments to 4 cm long, simple, linear, with basal meristem, acute below, acuminate above, uniseriate, comprising thick-walled cells, quadrate to rectangular, 10–43 × 23–40 µm, each with numerous discoid plastids without obvious pyrenoids; paraphyses multicellular, simple, clavate, slightly curved, to 120 µm long, 7–11 cells, basal cells rectangular, terminal cells moniliform to 18 µm in diameter; hairs unknown.

Unilocular sporangia common, lateral at the base of the paraphyses, pyriform, 52–90 × 24–37 µm; plurilocular sporangia not known on macroscopic thalli (see below).

Epiphytic on *Fucus* spp., especially *F. vesiculosus*, more rarely on *F. spiralis*, *F. serratus* and various other algae, eulittoral.

Common and widely distributed all around Britain and Ireland.

Recorded throughout the year, with unilocular sporangia, although usually more common midsummer to early winter. Young thalli develop in spring and early summer, producing unilocular sporangia during autumn and winter as the brush-like tufts of erect filaments are gradually shed from the basal cushion. It is possible that the basal cushions can survive the winter period and produce new erect filaments in the spring.

Culture studies have been carried out on European isolates of *E. fucicola* by Kylin (1937), Kornmann (1962b), Blackler & Katpitia (1963) and Koeman & Cortel-Breeman (1976). In the culture studies of Kylin, Kornmann and Blackler & Katpitia, the unispores were observed to grow directly into microscopic branched filaments from which new macroscopic thalli developed. Kornmann and Blackler & Katpitia additionally reported the presence of plurilocular sporangia in the cultures, which Kornmann specifically described on the microscopic filaments; plurispores from these sporangia also germinated directly, without sexual fusion and repeated the parental phase. The species was therefore attributed with a 'direct' type of life history with apparently apomeiotic unilocular sporangia. Essentially similar results were obtained by Koeman & Cortel-Breeman, although they did provide additional information about the influence of environmental factors on the life history and about the ploidy level of the two heteromorphic phases (microthallus and macrothallus). For example, higher temperatures (16 and 20°C) promoted the direct budding of the typical erect, macroscopic tufts from the microthallus, while under low temperatures (4°C) the microthalli developed plurilocular sporangia. They also observed a haploid number of chromosomes in a few young unispore-derived germlings only, which suggested that meiosis may take place in the unilocular sporangia and that vegetative diploidization could occur during the development of macrothalli from the microthalli as described in *Elachista stellaris* (Wanders *et al.* 1972). However, they concluded that the relationship between the two phases is not an obligate alternation of ploidy levels but a facultative one, and more likely to be controlled by environmental conditions (e.g. temperature and daylength).

Abbott & Hollenberg (1976), pp. 178, 179, fig. 145; Areschoug (1842), p. 235, pl. 8(6–7); Bartsch & Kuhlenkamp (2000), p. 166; Batters (1883), p. 110; Blackler & Katpitia (1963), pp. 392–5; Brodie *et al.* (2016), p. 1020; Coppejans & Kling (1995), p. 168, pl. 46; Cormaci *et al.* (2012), pp. 152, 154, pl. 37(4–6); Cotton (1912), p. 95; De Mesquita Rodrigues (1963), pp. 61, 62, pl. VI(c); Deckert & Garbary (2005), pp. 363–8; Dizerbo & Herpe (2007), p. 82, pl. 26(1); Fletcher (1987), pp. 127–9, fig. 19; Fletcher in Hardy & Guiry (2003), p. 246; Gallardo García *et al.* (2016), p. 35; Guiry (2012), p. 141; Hamel (1931–39), pp. 117, 118, fig. 27(b); Harvey (1849), pl. 240, (1857), p. 202, pl. 12B(1–3); Jaasund (1960), p. 105, (1965), pp. 64, 65; Koeman & Cortel-Breeman (1973), p. 213, (1976), pp. 107–17, figs 1–22; Kornmann (1962b), pp. 293–7, figs 1–4; Kornmann & Sahling (1977), p. 121, fig. 64; Kuckuck (1929), pp. 22, 23; Kylin (1947), p. 51, fig. 44(A); Mathieson & Dawes (2017), pp. 153, 154, pl. XLVI(5, 6); Morton (1994), p. 34; Newton (1931), p. 133, fig. 80; Pedersen (1979), pp. 151–9, figs 3–7, (2011), p. 90, fig. 96; Printz (1926), pp. 159–61; Ribero *et al.* (1992)), p. 115; Rosenvinge (1935), pp. 19–24; Rueness (1977), pp. 139, 140, fig. 67, (2001), p. 14; Saunders & McDevit (2013); Setchell & N.L. Gardner (1925), pp. 503, 504, pl. 38(33–5); Stegenga & Mol (1983), p. 92, pl. 29(1); Stegenga *et al.* (1997a), p. 18; Sundene (1953), pp. 63–164; Taylor (1957), p. 140, pl. 10(1–3); Uwai *et al.* (2000), pp. 267–80; Waern (1952), pp. 153–5.

Elachista scutulata (Smith) Areschoug (1843), p. 262.   Figs 102, 103

Conferva scutulata Smith (1790–1814), pl. 2311; *Elachista scutulata* (Smith) Duby (1830), p. 972.

Thalli epiphytic, forming dark brown-black, low-lying, closely packed tufts on host, to 5(–10) mm in height, usually discrete and circular, commonly completely enveloping host tissue, sometimes gregarious and irregularly spreading to 2(–3) cm, arising from a well-developed, hemispherical, solid cushion; cushions fairly soft and gelatinous, quite easily squashed, composed of horizontally and vertically spreading, dichotomously to trichotomously branched, closely packed filaments of large, thin-walled colourless cells, usually elongate-cylindrical, occasionally moniliform and barrel shaped, 30–70 × 15–45 µm, giving rise below to downward- or backward-extending rhizoidal filaments penetrating conceptacle host tissue, giving rise above to the erect sparsely packed, brush-like tufts of filaments, paraphyses and/or sporangia; erect filaments to 5(–10) mm long, free, linear, uniseriate, comprising thick-walled cells, 1–2(–3) diameters long, 16–48 × 16–26 µm, each with numerous discoid plastids without obvious pyrenoids; paraphyses abundant, simple, linear to elongate-clavate, multicellular, to 195(–325) µm long, comprising 10(–14) cells, 20–47 × 8–21 µm, basal cells elongate-rectangular, 4–6 diameters long, terminal cells rectangular, barrel-shaped or slightly moniliform, 1(–1.5) diameters long; hairs unknown.

Unilocular sporangia common, borne at the base of the paraphyses, on 1–4-celled pedicels, elongate-cylindrical, 80–120 × 30–55 µm; plurilocular sporangia rare, terminal on modified paraphyses, filiform, uniseriate, simple, to 120 long × approx. 7–10 µm wide, 5–30 loculi.

Epiphytic on the thongs (receptacles) of *Himanthalia elongata*, emerging from conceptacles, upper to lower eulittoral, exposed or in pools, and shallow sublittoral.

Common and widely distributed all around Britain and Ireland.

Annual, June to October, although infrequently observed throughout the year. Sporangia present in late summer/autumn with the unilocular sporangia occurring first, followed by the plurilocular sporangia. The plurilocular sporangia were first reported by Katpitia & Blackler (1962) and are only known for Fife and Devon.

André (1970), p. 251; Brodie *et al.* (2016), p. 1020; Cotton (1912), p. 95; Dizerbo & Herpe (2007), p. 82, pl. 26(2); Fletcher (1987), pp. 129–31, fig. 20, pl. 2(d); Fletcher in Hardy & Guiry (2003), p. 247; Gallardo García *et al.* (2016), p. 35; Guiry (2012), p. 141; Hamel (1931–39), pp. 119, 120, fig. 28(A–B); Harvey (1846–51), pl. 323; Katpitia & Blackler (1962), pp. 173, 174; Lee (2000), pp. 9–21; Morton (1994), p. 34; Newton (1931), pp. 134, 135; Ribera *et al.* (1992), p. 115; Rueness (1977), p. 140, (2001), p. 14; Stegenga & Mol (1983), p. 92, pl. 29(2–4); Stegenga *et al.* (1997a), p. 18; Thuret (1850), pp. 236, 237, pl. 25(1–3).

Elachista stellaris J.E. Areschoug (1842), p. 233.   Figs 99(E), 104, 105

Areschougia stellaris (Areschoug) Meneghini (1844), p. 293; *Phycophila stellaris* (Areschoug) Kützing (1845–71), pl. 97; *Symphoricoccus radians* Reinke (1888), p. 17; *Leptonema fasciculatum* var. *flagellare* Reinke (1889b), p. 51; *Elachista fracta* H. Gran (1893), p. 28; *Symphoricoccus stellaris* (Areschoug) Kuckuck (1929), p. 34.

Thalli epiphytic, forming light brown, usually discrete, brush-like tufts on hosts, 8–12 mm across, comprising fairly densely packed, free, erect filaments arising from a small, solid, hemispherical to elongated dome-shaped, wart-like, basal cushion; basal cushions slightly soft and gelatinous, comprising an extensive, medullary zone of vertical to curved rows of closely packed, dichotomously or trichotomously branched filaments of large, elongate-cylindrical, thin-walled, colourless cells, 85–325 × 22–110 µm, giving rise terminally to a narrow, more

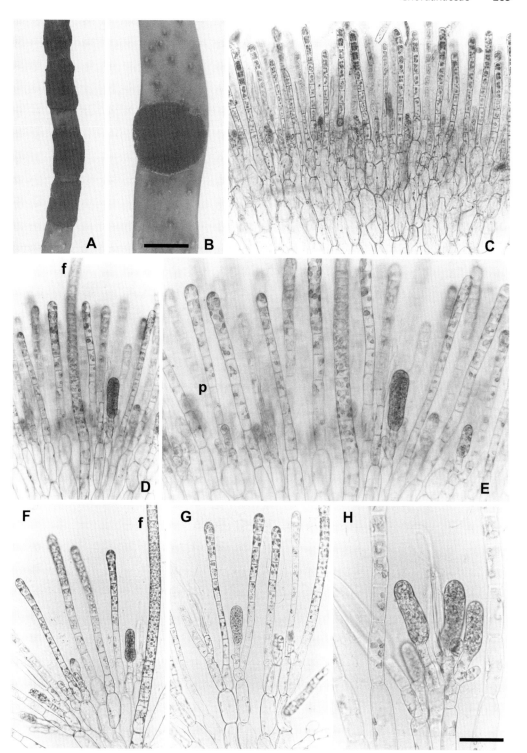

Fig. 102. *Elachista scutulata*

A, B. Habit of thalli on *Himanthalia elongata*. C–G. VS of thalli showing large, colourless cells of basal cushion giving rise terminally to erect, exserted filaments (f), fairly densely packed paraphyses (p) and basally situated unilocular sporangia. H. Unilocular sporangia at various stages of development. Bar in B = 20 mm (A), = 12 mm (B), bar in H = 100 μm (C), = 90 μm (D, F, G), = 45 μm (E, H).

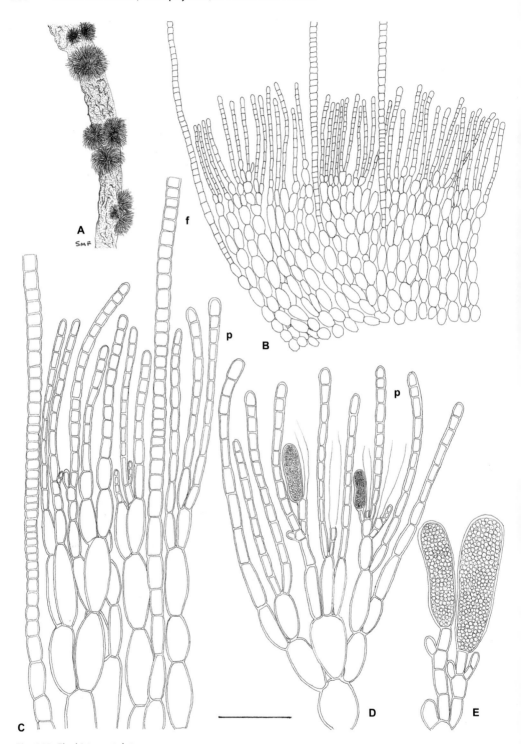

Fig. 103. *Elachista scutulata*

A. Habit of thalli epiphytic on *Himanthalia elongata*. B, C. VS of vegetative thallus showing basal cushion of large, thin-walled cells giving rise terminally to exserted filaments (f) and paraphyses (p). D, E. VS of fertile thallus showing paraphyses (p) and unilocular sporangia. Bar = 15 mm (A), = 200 μm (B), = 100 μm (C–D), = 50 μm (E).

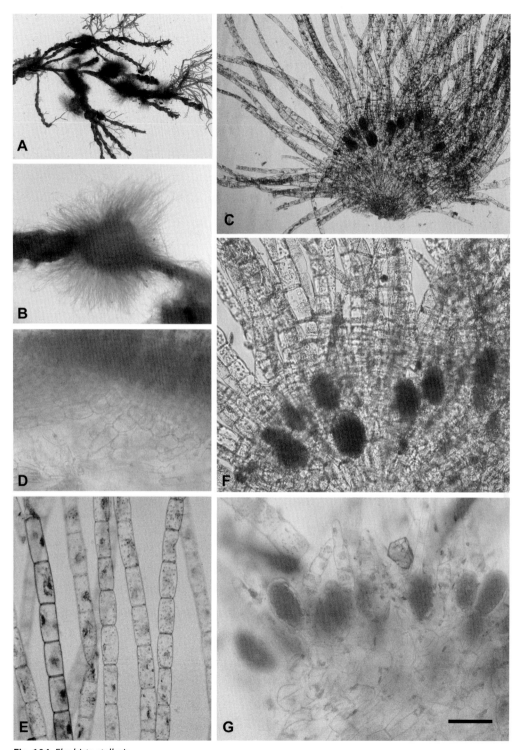

Fig. 104. *Elachista stellaris*

A, B. Habit of thalli epiphytic on *Gongolaria nodicaulis*. C. Habit of thallus showing the basal cushion giving rise to the erect filaments and dark-coloured unilocular sporangia. D. SP of basal cushion showing the constituent large, colourless medullary cells. E. Portions of erect filaments. F, G. SP of portions of thalli showing unilocular sporangia and erect filaments arising from the peripheral cortex of the basal cushion. Bar in G = 24 mm (A), = 6 mm (B), − 300 μm (C), = 150 μm (D), = 125 μm (E, G), = 120 μm (F).

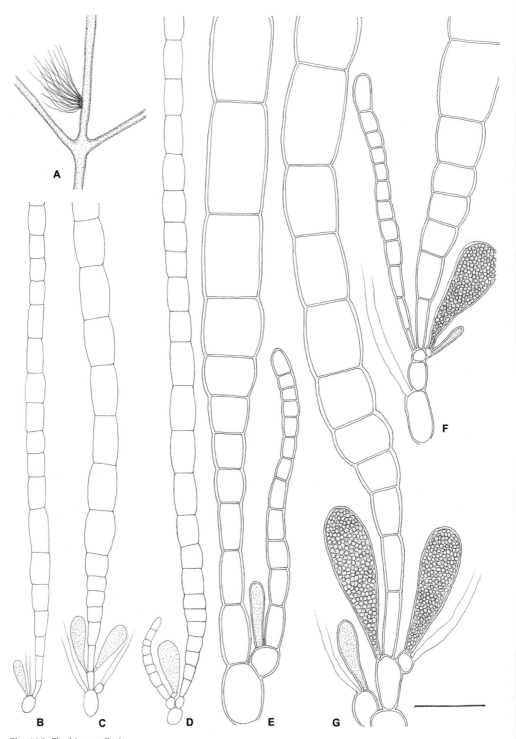

Fig. 105. *Elachista stellaris*
A. Habit of thallus epiphytic on *Arthrocladia villosa*. B–G. SP of fertile thalli showing a single, erect filament with basal unilocular sporangia. Bar = 8 mm (A), = 100 μm (B–D), = 50 μm (E–G).

loosely packed photosynthetic cortex of oval to barrel-shaped cells, 35–80 × 25–77µm; lower medullary cells producing downward-growing rhizoidal filaments which penetrate host tissue, terminal cortical cells producing free, erect filaments and sporangia; erect filaments to 8 mm long, linear, simple, uniseriate, usually tapering towards the base and apex, with basal meristematic zone, comprising fairly thick-walled, rectangular to barrel-shaped cells, 1–3(–4) diameters long, 45–122(–145) × 20–50(–75)µm, each with numerous plastids without pyrenoids; hairs unknown.

Unilocular and plurilocular sporangia lateral at the base of the erect free filaments, unaccompanied by obvious paraphyses, more rarely reported in mixed sori on the upper parts of the erect filaments; basally positioned unilocular sporangia pyriform, 78–140 × 28–90 µm, plurilocular sporangia lateral or terminal on small branches, densely crowded, filiform, uniseriate, to 50 µm long (6 loculi) × approx. 5–8 µm wide; unilocular sporangia on upper parts of erect filaments, globular, lateral, sessile or on one-celled stalks, plurilocular sporangia filiform or conical, lateral.

Epiphytic, on various algae in the lower eulittoral.

Only recorded for Dorset, South Devon and Anglesey.

Spring and summer annual.

Life history investigations of *Elachista stellaris* have been conducted by Kylin (1934, 1937 – European isolates) and Wanders *et al.* (1972 – Mediterranean isolates). Kylin concluded that this species has an apomeiotic, 'direct' type of life history; zoospores from both unilocular and plurilocular sporangia produced microscopic, fertile (with plurilocular sporangia) filaments from which new macroscopic thalli sprouted directly. The culture study by Wanders *et al.* (1972), however, was more significant. They demonstrated a heteromorphic life history (with both a microthallus and a macrothallus) which was determined by environmental conditions of photoperiod, temperature and nutrients, rather than by a sexual process. The diploid macroscopic thallus (showing the morphology of *E. stellaris*) could reproduce directly by large zoospores from plurilocular sporangia or produce haploid microscopic thalli from smaller zoospores formed (probably by meiosis) in the unilocular sporangia. The microscopic phase could also directly repeat itself via zoospores from plurilocular sporangia, the latter smaller than those developed on the macroscopic thallus. New macroscopic thalli could then develop directly from the microscopic phase, under certain culture conditions, a phenomenon attributed to the process of 'spontaneous diploidization'. Their culture results suggest that *E. stellaris* thalli would produce plurilocular sporangia throughout the vegetative growth period in the spring and summer, while unilocular sporangia would be produced in the autumn. The microthalli produced from the released unispores would then serve as overwintering phases from which would grow new macrothalli in the following spring.

Brodie (2016), p. 1020; Cormaci *et al.* (2012), p. 156, pls 40(7), 41(1–3); Dangeard (1968a), pp. 87–94, pl. 1(1–9); Dizerbo & Herpe (2007), p. 82; Fletcher (1987), pp. 131–3, fig. 21; Fletcher in Hardy & Guiry (2003), p. 247; Gallardo García *et al.* (2016), p. 35; Guiry (2012), p. 141; Hamel (1931–39), pp. 124, 125, fig. 29(a–b); Kuckuck (1929), figs 26–31; Kylin (1934), pp. 9–13, figs 4–6, (1937), pp. 14, 15, fig. 5, (1947), pp. 49, 50, fig. 43; Mathieson & Dawes (2017), p. 154, pls XLVI(7, 8), CVII(4); Müller & Schmidt (1988), pp. 153–8; Newton (1931), p. 133; Pankow (1971), p. 180, figs 219–21; Reinke (1889a), p. 3, pl. 2, p. 13 (s), pl. 10(10–11); Ribera *et al.* (1992), p. 115; Rosenvinge (1935) pp. 24–6; Rueness (1977), pp. 141, 142, fig. 69, (2001), p. 14; Sauvageau (1933b), pp. 179–88; Stegenga & Mol (1983), p. 92, pl. 30(1); Stegenga *et al.* (1997), p. 18; van den Hoek *et al.* (1972), pp. 57–63, figs 3–5; Wanders *et al.* (1972), pp. 458–91, pls I–VI.

Endodictyon Gran

ENDODICTYON Gran (1897), p. 47.

Type species: *E. infestans* Gran (1897), p. 47.

Thalli endozoic, forming brown spots on hosts; comprising a network of irregularly to oppositely branched, sometimes anatomising, often recurved and concentric filaments ramifying through host tissues; filaments usually free, sometimes adjoined and pseudoparenchymatous; plastids plate-like with pyrenoids.

Plurilocular sporangia terminal or intercalary on the filaments, solitary or in series, pluriseriate, irregularly globular in shape; unilocular sporangia unknown.

One species in Britain.

Endodictyon infestans Gran (1897), p. 47.   Fig. 106

Streblonema infestans (H. Gran) Batters (1902), p. 29.

Thalli entirely endozoic, forming a profusely branched network of sometimes anastomosing filaments ramifying through the tissues of the host species visible to the unaided eye as small brown patches; filaments profusely and irregularly branched, sometimes opposite, often recurved to form concentric whorls, comprising rectangular or irregularly shaped cells, 1–4 diameters long, 13–44 × 5–16 µm; filaments free, or not uncommonly laterally united in parts to form a pseudoparenchyma; cells with 1–2 plate-like, lobed plastids with pyrenoids; hairs not observed.

Plurilocular sporangia usually borne near the host surface, terminal or intercalary on the filaments, solitary or in series, pluriseriate, irregularly globular in shape, to 20 µm in length, 25 µm in diameter; unilocular sporangia unknown.

Endozoic in various invertebrate hosts, including the hydroids *Dynamena pumila* and *Obelia geniculata*, and the bryozoans *Alcyonidium hirsutum*, *Flustrella hispida* and *Membranipora membranacea*. Specimens more frequently reported in cast-up hosts.

Rarely recorded in Britain (Hampshire, Isle of Man, Shetland), but probably more widely distributed around Britain and Ireland.

Insufficient data to comment on seasonality. Plurilocular sporangia reported for April on the Isle of Man (Knight & Parke 1931).

Russell in Irvine *et al.* (1975) tentatively proposed that *Streblonema infestans* may simply be a plethysmothallus of an *Ectocarpus* sp., possibly *E. fasciculatus*.

Batters (1900), p. 372, (1902), p. 29; Brodie *et al.* (2016), p. 1020; Dizerbo & Herpe (2007), p. 63, pl. 18(2); Fletcher in Hardy & Guiry (2003), p. 308; Gran (1897), p. 47, pl. I(12–17); Guiry (2012), p. 182; Hamel (1931–39), pp. 72, 73, fig. 21c; Irvine *et al.* (1975), pp. 68–71, fig. 8(A–E); Jaasund (1965), p. 51; Knight & Parke (1931), p. 60, pl. xix(78); Kylin (1947), p. 24; Levring (1935b), p. 456, (1937), p. 37, (1945), p. 189; Mathieson & Dawes (2017), p. 187, pl. LV(4); Newton (1931), p. 128; Printz (1926), p. 147; Rosenvinge & Lund (1941), p. 64; Rueness (2001), p. 14; Russell (1975), p. 79.

Eudesme J. Agardh

EUDESME J. Agardh (1882), p. 29.

Type species: *E. virescens* (Carmichael ex Berkeley) J. Agardh (1882), p. 31.

Thalli erect, macroscopic, cylindrical, solid, later in old parts hollow, gelatinous, sparingly branched, arising from a small, thin, basal disc; in structure, thalli clearly haplostichous, multiaxial with a central medulla of erect rows of numerous, easily separable filaments of elongated, colourless cells, which branch out laterally to form a subcortex of branched

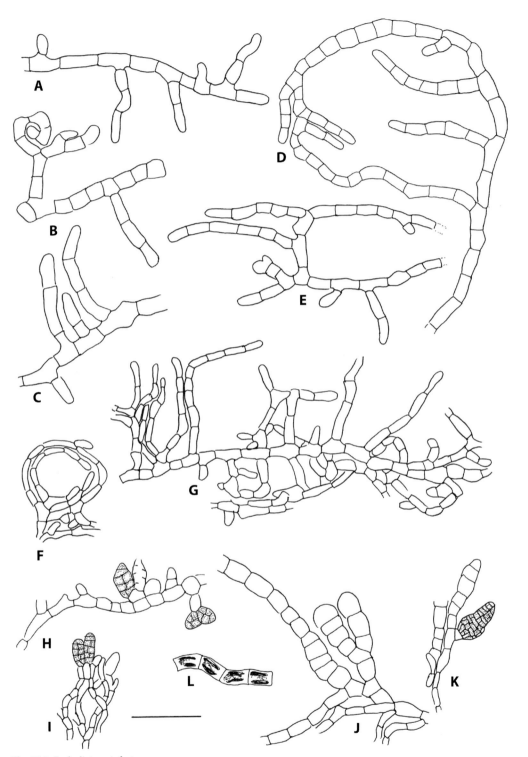

Fig. 106. *Endodictyon infestans*
A–G. Surface view of vegetative thalli showing the irregularly branched and spreading filaments. H, I. Fertile filaments with plurilocular sporangia. J, K. Young stages of *Ectocarpus* sp. on *Flustrella hispida*, showing marked similarity to *Endodictyon infestans*. L. Portion of vegetative filament showing plate-like plastids. Bar = 50 μm.

filaments which turn out to a radiating, peripheral, cortical region comprising assimilatory filaments, unilocular sporangia, plurilocular sporangia and/or hairs; both medullary and subcortex filaments giving rise to secondary tissues, the former as downward-growing, rhizoid-like extensions, the latter as upward growing secondary filaments which give to rise to the secondary assimilators; assimilators clustered, simple, rarely branched, clavate, multicellular, each cell containing several discoid plastids with associated pyrenoids; growth trichothallic with short meristematic zone of compacted cells at the apex of the medullary filaments, subtended by a colourless hair; hairs common, with basal meristem, lacking a sheath.

Plurilocular sporangia rare, formed by transformation of the terminal cells of the paraphyses, unilocular sporangia common, borne at the base of the assimilators, sessile, globular to ovoid.

One species in Britain and Ireland.

Eudesme virescens (Carmichael ex Berkeley) J. Agardh (1882), p. 31.   Figs 107, 108

Rivularia zosterae Weber & Mohr (1810), p. 367; *Linckia zosterae* Lyngbye (1819), p. 194; *Mesogloia virescens* Carmichael ex Berkeley (1833), p. 44; *Aegira zosterae* (Mohr) Fries (1835), p. 325; *Helminthocladia virescens* (Carmichael) Harvey (1841), p. 46; *Mesogloia zosterae* (Lyngbye) Areschoug (1842), p. 228; *Castagnea virescens* (Carmichael ex Harvey) Thuret in Le Jolis (1863), p. 85; *Castagnea zosterae* (Mohr) Thuret in Le Jolis (1863), p. 85, nom. illeg.; *Castagnea virescens* f. *vernales* Areschoug (1875), p. 19; *Castagnea virescens* f. *aestivales-autumnales* Areschoug (1875), p. 19; *Aegira virescens* (Carmichael ex Harvey in Hooker) Setchell & N.L. Gardner (1925), p. 547.

Thalli epilithic, erect, either with a prominent main axis or tufted, to 50 cm long, 0.5–2(–4) mm wide, usually solitary, sometimes gregarious, yellowish brown to deep olive, cylindrical, linear, filiform, at first solid, later with a central cavity, soft, flaccid, extremely gelatinous, usually sparingly branched, arising from a small, thin, basal disc; branching irregular, to 1(–2) orders, with branches long, flexuose and widely divergent, not uncommonly giving rise in the terminal regions to short, lateral, globular, side branchlets; thallus clearly haplostichous in structure, multiaxial with a central medulla of erect rows of numerous, large, loosely bound and easily separable filaments of mainly elongated colourless cells, 40–200(–250) × 28–65 μm, which branch out laterally to form a subcortex of narrower, vertically and laterally orientated and associated, branched filaments and hyphal filaments, which turn outwards to a radiating, peripheral, cortical region comprising assimilatory filaments, unilocular sporangia, plurilocular sporangia and/or hairs; central medullary cells terminating at the thallus apices in colourless hairs; central medullary cells also giving rise to narrow, branched, parallel running, hyphae-like filaments comprising cells up to 70 μm or more in length and 8–12 μm in diameter, the basal ones rhizoidal and contributing to the basal attachment disc; assimilators clustered, straight to slightly curved, often almost enclosing the unilocular sporangia, falcate, simple, rarely branched, clavate, multicellular, to 8–15(–18) cells, 265 μm in length, comprising mainly quadrate to rectangular cells below, 1–2.5 diameters long, 5–23 × 6–9 μm, mainly more bead-like to slightly globose cells above, 1–2 diameters long, 12–16 × 8–12 μm, each containing several discoid plastids, with associated pyrenoid; downward-growing branched, rhizoidal filaments commonly produced from the medullary filaments, comprising cells 86–180 × 10–12 μm and contributing below to the discoid attachment base; growth trichothallic with meristematic zones of short compacted cells at the apex of the medullary filaments subtended by a hair; hairs common, arising at the base of the assimilators, with a basal meristem of up to 13 subquadrate cells and lacking a sheath.

Unilocular sporangia common, and widely distributed over the thallus body, borne at the base of the assimilators, sessile, rarely on a 1–3-celled pedicel, globular to ovoid, (42–)50–68(–72) × (30–)34–52(–60) μm; plurilocular sporangia rare, formed by enlargement and transformation of the terminal cells of the assimilators, in a series of 3–6 sporangia, appearing first as small bulges on the upper sides of the cells, later dividing transversely and/or obliquely

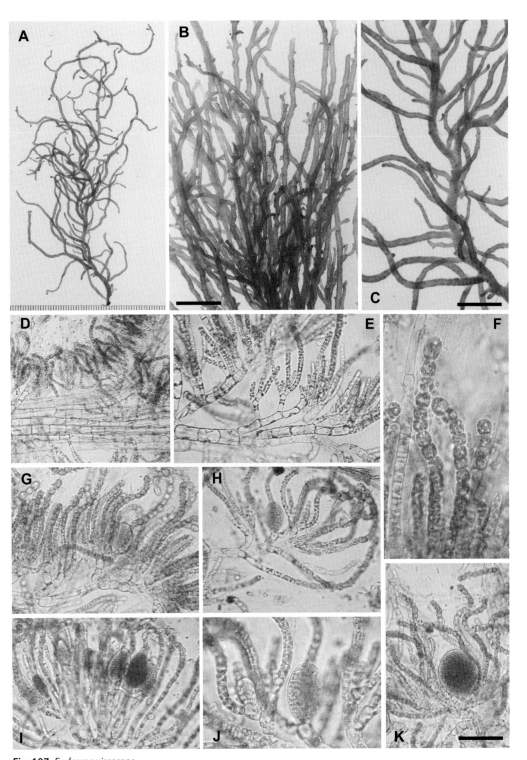

Fig. 107. *Eudesme virescens*

A–C. Habit of thalli. D, E. SP of vegetative thalli showing central, elongated, medullary cells branching out to form a subcortex which turns outwards to a radiating cortex of photosynthetic assimilatory filaments. F. Assimilatory filaments. G–K. SP of portions of fertile thalli showing assimilatory filaments with basally positioned unilocular sporangia. Bar in B = 12 mm, bar in C = 12 mm, bar in K = 180 μm (D), = 90 μm (E, G, H, I), = 45 μm (F, J, K).

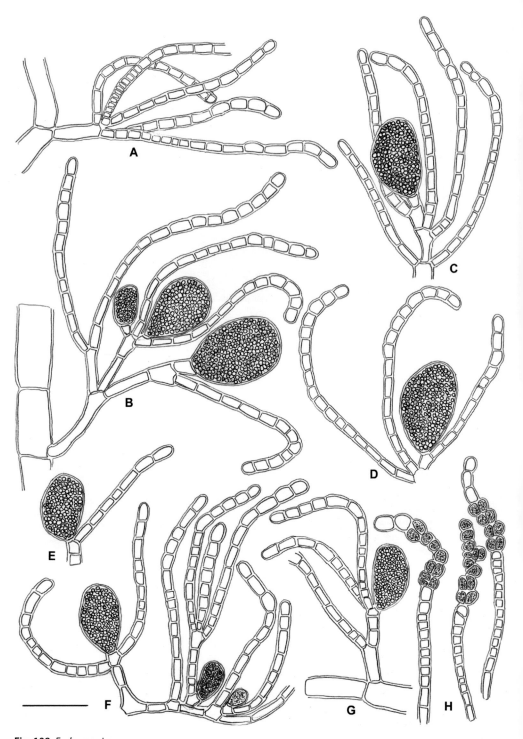

Fig. 108. *Eudesme virescens*

A. Portion of thallus showing the outer, assimilatory filaments and accompanying hair. B–G. SP of fertile thalli showing the assimilatory filaments and unilocular sporangia. H. Plurilocular sporangia developing terminally on the assimilatory filaments. Bar = 50 μm.

mainly into two, at first hemispherical, later globular, loculi slightly protruding and darker than the vegetative cells.

Epilithic, on rock in upper to lower eulittoral, usually in very shallow, often sandy pools and runnels; sublittoral to 3 m.

Can be locally abundant, reported for scattered localities around Britain and Ireland.

A summer annual, May to September, with the populations being quite short-lived.

Culture studies by Kawai (1986) revealed a 'direct', asexual life history with unispores from the unilocular sporangia germinating into microscopic, branched, filamentous microthalli from which new erect thalli developed bearing unilocular sporangia. However, a sexual life history was reported by Cole (1967) with the microthallus producing gametes which fused, although the zygotes merely repeated the microthallus phase which was not observed to form macrothalli.

Bartsch & Kuhlenkamp (2000), p. 166; Batters (1902), pp. 44, 45; Brodie *et al.* (2016), p. 1020; Cole (1967), pp. 665–73; Cormaci *et al.* (2012), p. 158, pl. 40(4–5); Cotton (1912), p. 36; Dizerbo & Herpe (2007), p. 88, pl. 29(2); Fletcher in Hardy & Guiry (2003), p. 248; Funk (1955), p. 34; Gallardo García *et al.* (2016), p. 35; Gayral (1966), pp. 274, 275, pl. XLVI; Guiry (2012), p. 141; Hamel (1931–39), pp. 158–60, fig. 36(a); Harvey (1846), pl. 82; Hooker (1833), p. 387; Kawai (1986), pp. 203–8; Kuckuck (1929), p. 47, fig. 55; Kylin (1947), pp. 56, 57, fig. 48(B); Levring (1940), pp. 49, 50; Lund (1959), pp. 119–21, fig. 25(A–D); Mathieson & Dawes (2017), p. 156, pl. XLVI(11–13); Morton (1994), p. 35; Newton (1931), pp. 146, 147, fig. 91; Oltmanns (1922), fig. 310(1); Pankow (1971), p. 186, fig. 234; Parke (1933), pp. 6, 15–18, pls 1(3), 5(25, 27), 6(30–7); Pedersen (2011), p. 95, fig. 102; Peters (1987), pp. 223–64; Reinke (1889b), p. 76, fig. 8; Ribera *et al.* (1992), p. 114; Rosenvinge & Lund (1943), pp. 28–30, fig. 10; Rueness (1977), p. 147, fig. 79, (2001), p. 14; Sauvageau (1929), p. 281; Setchell & N.L. Gardner (1925), pp. 547, 548, pl. 42(59, 60); Stegenga & Mol (1983), p. 94, pl. 32(3–4); Stegenga *et al.* (1997), p. 17; Taylor (1957), pp. 145, 146, pl. 12(3); Välikangas (1928), pp. 59, 60.

Fosliea Reinke

FOSLIEA Reinke, 1891, p. 45.

Type species: *Fosliea curta* (Foslie) Reinke (1891), p. 45.

Thalli epiphytic, macroscopic, erect, to 14 cm in length, at first solid, later becoming hollow in older parts, arising from a base of compacted, spreading, secondary rhizoidal filaments, with or without downward penetration into host tissue; thalli uniseriate above, pluriseriate below as the result of regular longitudinal divisions; older thalli often with short opposite branchlets following longitudinal divisions in a cell to produce four cells, the outer two functioning as branch initials; branching mainly opposite, sometimes whorled or quaternate, and usually associated with the pluriseriate regions, with branches quite widely spreading, terminating in a uniseriate, hair-like filament, lacking a basal meristematic zone; in section, pluriseriate thallus regions with one or more central, elongated cells surrounded by a ring of peripheral cells; cells with numerous discoid plastids each with pyrenoids; growth diffuse and intercalary.

Unilocular sporangia formed directly in surface cortical cells, scattered, solitary or confluent in sori, sometimes associated with the nodes of branches, slightly protruding and darker than the vegetative cells; plurilocular sporangia, if present, intercalary.

One species in Britain and Ireland

Fosliea griffithsiana (Le Jolis) Hagen ex Athanasiadis 2021: 7, 756.   Fig. 109

Ectocarpus brachiatus Harvey (1833), p. 327, nom. illeg.; *Ectocarpus griffithsianus* Le Jolis (1863), p. 37; *Phloeospora brachiata* Bornet (1878), p. 16, footnote; *Stictyosiphon griffithsianus* (Le Jolis) Holmes & Batters (1890), p. 78 (as 'Griffithsianus'); *Stictyosiphon brachiatus* (Bornet) Hygen & Jorde (1935), p. 29.

Thalli epiphytic, macroscopic, erect, to 5(–14) cm long, light brown, to greenish brown when dried, soft and delicate in texture, densely branched and tufted, filiform, at first uniseriate, later becoming pluriseriate and parenchymatous in parts; thalli solid at first, becoming

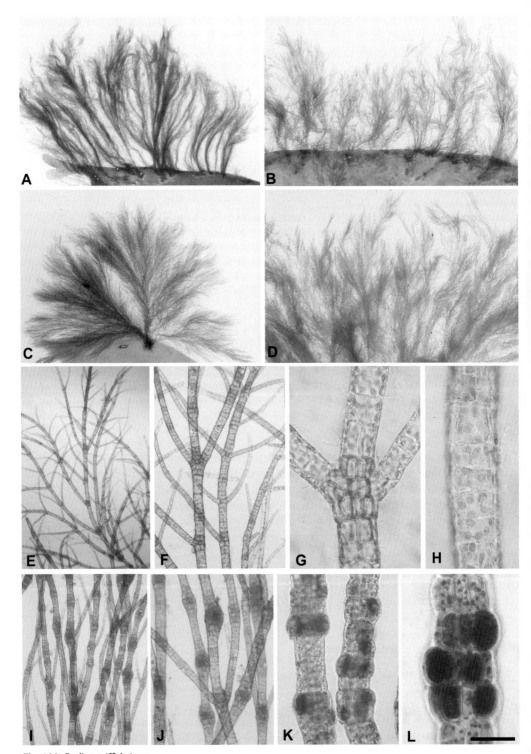

Fig. 109. *Fosliea griffithsiana*

A–D. Habit of thalli on host (*Palmaria*). E–H. Portions of erect thalli showing opposite branching and pluriseriate structure. I–L. Portion of fertile thallus showing the dark protruding unilocular sporangia associated with the pluriseriate region. Bar in L = 15 mm (A, C), = 18 mm (B), 5 mm (D), = 240 μm (E, I), = 180 μm (F, J), = 90 μm (K), = 45 μm (G, H, L).

hollow in older, basal regions, arising from a small, filamentous, rhizoidal holdfast with rhizoids penetrating down into host tissue; branching mainly opposite, sometimes whorled or quaternate, usually associated with the pluriseriate regions, with branches quite widely spreading, terminating in a uniseriate, hair-like colourless filament comprising cells 65–100 × 9–15 μm; thalli at first uniseriate, later becoming irregularly pluriseriate; cells in uniseriate regions mainly subquadrate, 0.5–1 diameters long, 21–38 × 41–53(–74) μm, in pluriseriate regions these cells dividing internally to produce rows of quadrate to rectangular cells, 18–35 × 9–30 μm, 2–6(7) seen in surface view; in transverse section, pluriseriate regions with a central cell surrounded by a number of pericentral cells; cells with numerous discoid plastids each with pyrenoids; growth intercalary and diffuse.

Unilocular sporangia scattered on upper branches, usually associated with the pluriseriate regions, solitary or in small groups, frequently opposite, borne in enlarged cortical cells, darker and more protruding than vegetative cells, resulting in localized swellings along branch length; plurilocular sporangia not known on macrophyte.

Epiphytic on *Palmaria palmata*, more rarely on other algae, mid to lower eulittoral pools and emersed, sublittoral to 27 m.

Rare but widely distributed around Britain and Ireland, although more common on southern and western coasts of England and Ireland. In Britain, reported for the Channel Islands, Isles of Scilly, Cornwall, South Devon, North Devon, Dorset, Hampshire, Kent, Isle of Man, Berwickshire, Wigtownshire, Aberdeenshire, West Lothian, Fife, Inverness, Orkney, Shetland and Renfrewshire; in Ireland reported for Dublin, Antrim, Waterford, Wexford, Dublin, Mayo and Donegal.

Summer annual, May to September.

Adams (1908), p, 48; Athanasiadis (2021), p. 7; Batters (1902), p. 25; Brodie *et al.* (2016), p. 1021; Burel *et al.* (2019), p. 21; Cotton (1912), p. 94; Dizerbo & Herpe (2007), p. 110; Fletcher in Hardy & Guiry (2003), p. 303; Guiry (1977), p. 82, (2012), pp. 150, 151; Hamel (1931–39), p. 209, fig. 43(III); Harvey (1846–51), pl. 4; Holmes & Batters (1890), p. 78, fig. 186; Jaasund (1965), p. 81; Knight & Parke (1931), p. 57; Kuckuck (1929), pp. 80, 82, 83, figs 125–8; Le Jolis (1861), pp. 36, 37, (1863), pp. 78, 79; Lyle (1920) p. 7; Mathias (1935b), pp. 1–25; Mathieson & Dawes (2017), p. 184, pl. LIV(10, 11); Morton (1994), p. 40; Munda (1979a), p. 461; Naylor (1958a), p. 15, (1958b), pp. 20, 21; Nielsen & Gunnarsson (2001), p. 81; Newton (1931), p. 187, fig. 117(A, B); Parke & Dixon, (1976), p. 560; Rueness (1977), p. 154, fig. 85, Rueness *et al.* (2001), p. 17; Taylor (1937a), 178, (1957), p. 157, pl. 4.

Giraudia Derbès & Solier

GIRAUDIA Derbès & Solier in Castagne (1851), p. 100.

Type species: *G. sphacelarioides* Derbès & Solier (1851), p. 101.

Thalli forming minute, brush-like tufts, comprising several, erect, simple, rarely basally branched shoots, at first monosiphonous, later parenchymatous, cylindrical, segmented and tapering above; in transverse section, thalli solid, with large colourless, central, medullary cells enclosed by 1–2 smaller, pigmented, cortical cells; growth both trichothallic with subterminal meristem and terminal, hyaline, multicellular hair, and intercalary with a basal meristem of short, closely packed, lenticular cells; hairs common, gregarious and densely tufted in terminal regions, more solitary elsewhere, arising from surface cells, with basal meristem and sheath; cells with numerous discoid plastids without obvious pyrenoids.

Plurilocular sporangia in extensive, closely packed, embedded or projecting sori, formed from surface cells in the terminal regions of the erect thalli, and/or arising terminally or laterally on short, branched, lateral filaments at base of thalli, oblong-lanceolate, multiseriate; unilocular sporangia unknown.

One species in Britain and Ireland.

Giraudia sphacelarioides Derbès & Solier (1851), p. 101.   Figs 110, 111

Thalli epiphytic, more rarely reported to be epilithic and epizoic, forming minute, erect, densely packed, gregarious, yellow to olive-brown, tufts of shoots, up to 10(–15) mm high, and arising from a basal system of branching, rhizoidal filaments aggregated into monostromatic pseudodiscs, formed as lateral outgrowths of the basal cells of the shoots; basal rhizoids reported to produce new erect shoots; erect shoots simple, rarely branching at the base to produce new erect thalli, terete, filiform, cylindrical and linear throughout, at first monosiphonous, later becoming polysiphonous and distinctly segmented, except at the tapering apex, to 40(–53) μm in diameter; segments to 40 μm in height, comprising tiers of up to 10(–15) rectangular, vertically elongate cells, up to 2–5 seen in surface view; basal regions with short, monosiphonous branches; growth both trichothallic with subterminal meristem and terminal, hyaline, multicellular hair, and intercalary with a basal meristem of short, closely packed, uniseriate, lenticular cells; hairs common, gregarious and densely tufted, in groups of up to six, in terminal regions, more solitary elsewhere, arising from surface cells, with basal meristem and sheath; hair cells 3–6 diameters long, 30–80 × 11–14 μm; cells with numerous discoid plastids without obvious pyrenoids.

Plurilocular sporangia of two types; one type formed by multiple cell divisions and transformation of surface cells in the terminal regions of the erect thalli, forming discrete or extensive, compacted, intercalary loculi in surface view, either embedded or projecting, to 40 μm in length, 10–15 μm in width, uniseriate to biseriate and partly or wholly encircling thalli; the other type borne terminally or laterally on short filaments arising at the base of the thalli, below the meristematic region, oblong-lanceolate, uniseriate to multiseriate, simple or branched, to 120 μm in length, 10–15 μm in diameter; unilocular sporangia unknown.

Epiphytic on various algae, especially *Gongolaria* spp. and *Zostera* spp., lower eulittoral pools and shallow sublittoral.

Thalli rare with only a few records for Britain (Hampshire and Dorset) and Ireland (Clare, Cork and Donegal); probably more widespread than records suggest.

Appears to be a spring and summer annual.

Culture studies by Sauvageau (1927b) and Skinner & Womersley (1984) have revealed an asexual 'direct' type of life history with spores from the plurilocular sporangia germinating directly to form a plurilocular sporangium-bearing microthallus which gave rise to further fertile macrothalli. Spores released from the microthallus developed in a similar manner to those derived from the macrothallus. The microthalli developed in both a filamentous and stellate manner, to become myrionematoid and ascocyclus-like in appearance.

Ardré (1970), p. 274; Batters (1902), p. 37; Brodie *et al.* (2016), p. 1020; Coppejans (1983), pls 50–2; Cormaci *et al.* (2012), pp. 160–2, pl. 41(1–4); Cotton (1912), p. 95; Derbès & Solier (1851), p. 101, (1852–56), p. 49, pl. 14(12–16); Dizerbo & Herpe (2007), pp. 110, 111, pl. 36(4); Feldmann (1937), p. 290; Fletcher in Hardy & Guiry (2003), p. 254; Funk (1955), p. 39; Gallardo García *et al.* (1985), p. 35; Guiry (2012), p. 142; Hamel (1931–39), pp. 189–92, fig. 42(1–8); Hauck (1884), p. 335, fig. 139; Holmes (1883), p. 141; Holmes & Batters (1890), p. 81; Knight & Parke (1931), p. 64, pl. X(18); Kuckuck (1929), pp. 28–32, figs 19–25; Kylin (1947), pp. 66, 67, fig. 56(A–C); Mathieson & Dawes (2017), p. 157, pl. XLVI(14–17); Morton (1994), p. 8; Newton (1931), pp. 135, 136, fig. 81(A–D); Pankow (1971), p. 198, figs 256–8; Ribera *et al.* (1992), p. 117; Rosenvinge & Lund (1947), p. 55, fig. 19; Rueness (1977), pp. 157, 158, fig. 91, (2001), p. 15; Sauvageau (1927b), pp. 1–74, figs 1–18, (1928b), p. 269; Skinner & Womersley (1984), p. 164, figs 1–10; Stegenga & Mol (1983), p. 97; Stegenga *et al.* (1997a), p. 19; Taylor (1957), p. 144; Womersley (1987), pp. 306–8, fig. 110(D–H).

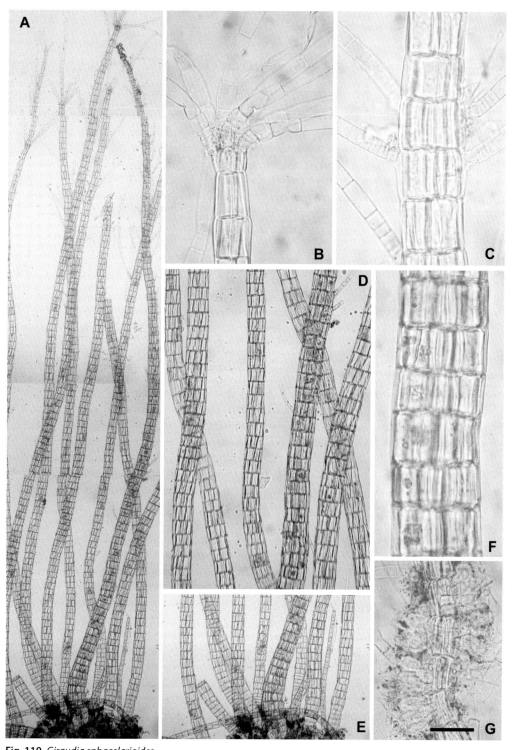

Fig. 110. *Giraudia sphacelarioides*
A. Habit of thalli. B. Apex of thallus showing terminal hairs. C. Subterminal region of thallus showing laterally produced hairs. D. Portions of upper regions of thalli. E. Portions of basal regions of thalli. F. Portion of thallus. G. Portion of terminal region of thallus showing the outward-projecting plurilocular sporangia. Bar = 125 µm (A), = 18 µm (B), = 38 µm (C), = 94 µm (D), − 107 µm (F), = 25 µm (F), = 20 µm (G).

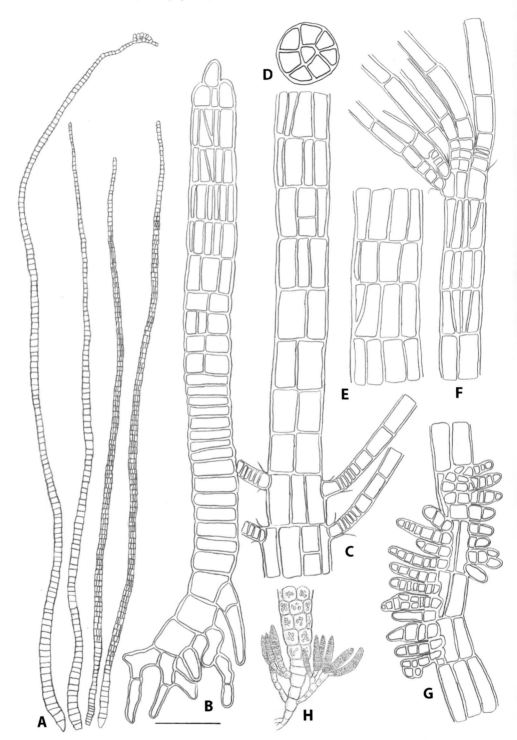

Fig. 111. *Giraudia sphacelarioides*
A. Habit of thalli. B. Basal region of thallus showing growth of secondary rhizoidal filaments. C. Mid region of thallus showing the polysiphonous segment and lateral hairs. D. TS of thallus showing a central, medullary cell enclosed by seven peripheral cells. E. Portion of thallus. F. Apex of thallus showing cluster of terminal hairs. G. Portion of fertile thallus showing projecting tiers of plurilocular sporangia. H. Basal region of fertile thallus showing plurilocular sporangia terminal on short, branched lateral filaments. Bar = 190 μm (A), = 50 μm (B–F), = 35 μm (G), = 90 μm (H). H from Newton (1931).

Halothrix Reinke

HALOTHRIX Reinke (1888), p. 19.

Type species: *H. lumbricalis* (Kützing) Reinke (1888), p. 19.

Thalli epiphytic, forming small, brush-like tufts on host, consisting of erect, simple, or rarely branched, linear, uniseriate filaments arising from a basal system of compacted, branched, rhizoids; growth by an intercalary meristem, usually basally positioned; cells with numerous discoid plastids without obvious pyrenoids.

    Plurilocular sporangia intercalary or terminal on erect filaments, formed by subdivision of vegetative cells, densely packed, multiseriate, clustered; unilocular sporangia unknown.

One species in Britain.

Halothrix lumbricalis (Kützing) Reinke (1888), p. 19.   Figs 112, 113

Ectocarpus lumbricalis Kützing (1845), p. 233; *Elachista lumbricalis* (Kützing) Hauck (1883–85), p. 354; *Halothrix rectiuscula* Y.-P. Lee (2001), p. 40.

Thalli epiphytic, forming small (1–2 mm high), solitary, brush-like tufts on hosts, just visible to the unaided eye; consisting of a basal fibrous network of densely packed, irregularly branched, thin, colourless, rhizoidal filaments, 6–8 µm in diameter, penetrating host tissue below, giving rise above to numerous erect filaments; erect filaments predominantly linear throughout, narrowing

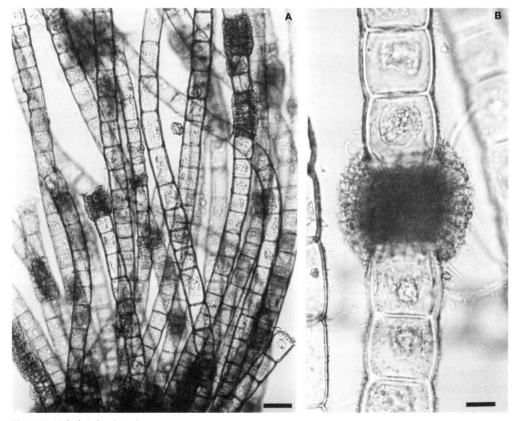

Fig. 112. *Halothrix lumbricalis*
A. Erect filaments of thalli with intercalary plurilocular sporangia. B. Portion of erect filament with intercalary plurilocular sporangium. Bar in A = 55 µm, bar in B = 27 µm.

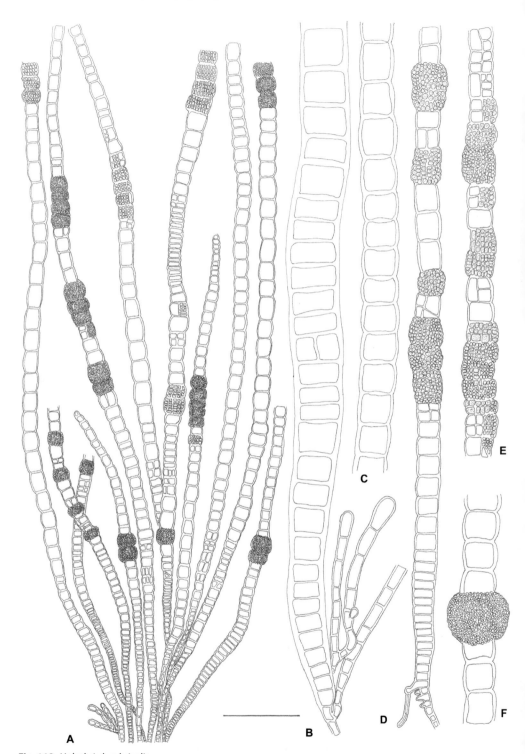

Fig. 113. *Halothrix lumbricalis*

A. Habit of thalli showing erect filaments and intercalary plurilocular sporangia. B. Basal region of erect filament. C. Portion of erect filament. D. Portions of erect filament showing plurilocular sporangia. Bar = 200 μm (A), = 50 μm (B, C), = 100 μm (D–F).

towards base, simple or rarely branched towards base, comprising cells 1–2(–2.5) diameters long, barrel-shaped to rectangular, 20–85 × 25–65 μm, light coloured, thick walled, with numerous, well-scattered, small discoid plastids without obvious pyrenoids; growth zone intercalary, usually near base, recognizable by a series of distinctly subquadrate, almost lenticular cells; hairs unknown.

Plurilocular sporangia common, formed by internal divisions of erect filament cells, solitary or more commonly in 1–6 cell chains, slightly enlarged to bulbous, dark coloured, to 88 μm wide, comprising compacted loculi in surface view, 3–5 × 3–4 μm, becoming multiseriate, clustered and projecting later, rounded in surface view; unilocular sporangia unknown.

Epiphytic on *Zostera* spp. in the lower eulittoral and shallow sublittoral.

Rare but widely distributed around Britain (Argyll, Isle of Man and Dorset).

Summer annual, recorded April to August.

Culture studies by Pedersen (1979) revealed *Halothrix lumbricalis*, collected from Denmark, to have a 'direct' type of life history. Zoospores from the plurilocular sporangia germinated directly, without sexual fusion, to form prostrate systems from which arose new erect filaments bearing plurilocular sporangia. Unilocular sporangia were not observed either on field or cultured material.

Batters (1892a), p. 174; Brodie *et al.* (2016), p. 1020; Cormaci & Furnari (1979), pp. 221–50; Cormaci *et al.* (2004), pp. 153–202, (2012), pp. 162, 163, pl. 41(5–8); Dizerbo & Herpe (2007), p. 83; Fletcher (1987), pp. 134–6, fig. 22, pl. 3; Fletcher in Hardy & Guiry (2003), p. 256; Guiry (2012), p. 181; Hamel (1931–39), p. 126, fig. 29(c–d); Kuckuck (1929), pp. 26–8, figs 15–18; Mathieson & Dawes (2017), p. 158, pl. XLVII(4, 5); Newton (1931), pp. 132, 133, fig. 79; Oltmanns (1922), fig. 325; Pankow (1971), p. 180, figs 223–5; Pedersen (1979b), pp. 151–9, figs 1–2; Reinke (1889a), p. 1, pl. 1; Ribera *et al.* (1992), p. 115; Rosenvinge (1935), pp. 37, 38, figs 36, 37; Rueness (1977), p. 141, fig. 68, (2001), p. 15; Taylor (1957), p. 143; Womersley (1987), p. 82.

Hecatonema Sauvageau

HECATONEMA Sauvageau (1897a), p. 256.

Type species: *H. terminale* (Kützing) Kylin (1937), p. 8.

Thalli epiphytic or epilithic, forming small solitary or confluent tufts up to 1 mm high, consisting of a basal layer of outward-spreading, dichotomously branched filaments, either free or laterally associated and pseudodiscoid, monostromatic, sometimes partly distromatic, giving rise to short, erect, free, unbranched or little branched filaments (usually all of equal height), plurilocular sporangia and/or hairs; cells with several discoid or 1–3 plate-like plastids with pyrenoids; hairs common, lateral or terminal on erect filaments with basal meristem and sheath.

Plurilocular sporangia common, borne on basal layer or terminal, more usually lateral, on erect filaments, multiseriate; unilocular sporangia unknown for British and Irish material.

In the present treatment, *Hecatonema* is characterized by a filamentous or discoid basal layer, which is monostromatic at first, later becoming distromatic and which gives rise to short, erect, branched filaments, hairs with a basal meristem and sheath, and multiseriate, plurilocular sporangia. The sporangia are usually lateral or terminal on the erect filaments, rather than arising directly from the basal layer. It very closely resembles and is not easily distinguished from the genus *Chilionema*, and the two genera might be synonymous. However, here they are maintained separately pending the results of further studies.

Particular interest has been given to reports of the involvement of *Hecatonema* species and *Hecatonema*-like microthalli in the life histories of species of the macroscopic genera *Asperococcus*, *Desmotrichum* and *Punctaria*. These include investigations by Sauvageau (1929), Kylin (1933, 1937), Dangeard (1963a), Loiseaux (1969), Clayton & Ducker (1970), Rhodes (1970), Clayton (1974), Rietema & van den Hoek (1981), Lockhart (1982), Fletcher (1984) and Pedersen (1984). In general, there appears to be a facultative relationship between the *Hecatonema*-like microthalli and the thalloid macrothalli rather than one based on differences in ploidy level. Indeed, the two types of morphological expression were often shown to be produced in response to environmental conditions.

It is partly because of these life history connections that *Hecatonema* was removed from the Myrionemataceae and placed in the Punctariaceae by Fletcher (1987), thus following the procedure of Christensen (1980). However, much more work is required, particularly culture studies, to elucidate fully the extent of involvement and relationship of *Hecatonema*-like thalli and described taxa of *Hecatonema* (and *Chilionema*) with macroscopic members of the previously recognized Punctariaceae, now placed in the Chordariaceae (see Silberfeld 2014). In this respect, it is pertinent to note culture studies that have revealed *Hecatonema* thalli to behave independently and do not develop macrothalli in culture (see Edelstein *et al.* 1971; Clayton 1974; Pedersen 1984).

Four species of the genus *Hecatonema* were listed for Britain and Ireland by Parke & Dixon (1976): *H. foecundum* (Strömfelt) Loiseaux, *H. hispanicum* (Sauvageau) Loiseaux, *H. liechtensternii* (Hauck) Batters and *Hecatonema terminale* (Kützing) Kylin. Following the transfer of the former two species to the genus *Chilionema* and *H. liechtensternii* to the genus *Myrionema* (see Fletcher 1987), only *Hecatonema terminale* is represented in the British and Irish flora.

One species in Britain and Ireland.

Hecatonema terminale (Kützing) Kylin (1937), p. 8.   Fig. 114

Ectocarpus terminalis Kützing (1845–71), p. 236; *Ascocyclus major* Foslie (1891), p. 15; *Myrionema majus* Foslie (1894), p. 15, pl. III, figs 18, 19; *Phycocelis maculans* Collins (1896), p. 459; *Hecatonema maculans* (Collins) Sauvageau (1897a), p. 256, deuxième forme and troisième forme.

Thalli forming small tufts of filaments, up to 1 mm high, solitary or confluent and spreading, similar to a sparse carpet pile, to several centimetres in diameter; comprising a basal network of outward-spreading, dichotomously branched filaments giving rise to erect filaments, pluri-locular sporangia and/or hairs; in squash preparation, basal filaments free or sometimes irregularly and laterally associated and pseudodiscoid in appearance, monostromatic, or in parts distromatic, comprising cells variable in shape, although usually longer than wide 13–25 × 13–17 µm; erect filaments to 30(–60) cells long, 0.5–1 mm, linear, free, unbranched or more commonly sparingly and laterally branched with branches characteristically widely divergent at first, later recurving towards parental filament, comprising cells 1–2 diameters long, 13–34 × 12–17 µm, with thick hyaline walls and occasionally divided longitudinally; cells with several discoid or 1–3 plate-like, lobed plastids with pyrenoids; hairs common, terminal on erect filaments or branches, 9–12 µm diameter, with basal meristem and outer sheath.

Plurilocular sporangia common, sessile or stalked on basal layer, or more commonly terminal or lateral on erect filaments; laterally positioned sporangia widely divergent at first, later recurved towards parental filament, sessile or stalked, solitary or more commonly in unilateral groups; sporangia biseriate, or more commonly multiseriate, elongate, ovate-lanceolate in shape, occasionally bifurcate at tip, 65–90(–135) × 20–28 µm, to 30 loculi; unilocular sporangia unknown.

Epiphytic on various algae, more rarely epilithic, in the eulittoral and shallow sublittoral.

Rare and widely distributed around Britain and Ireland.

Recorded throughout the year.

A small number of culture studies have linked *Hecatonema terminale* (as *H. maculans*) in the life history of macroscopic genera formerly positioned within the Punctariaceae. In three studies, *H. terminale* thalli were connected with *Punctaria tenuissima* (as *Desmotrichum undulatum*) and *Punctaria latifolia* (Clayton & Ducker 1970; Clayton 1974; Fletcher 1984 – Sauvageau's 'Deuxième forme' only), while in one study (Pedersen 1984) it was linked with *Asperococcus fistulosus* (see, however, this species life history connection with *H. reptans* and the 'Première forme' of *H. maculans* described by Sauvageau 1897). There are also reports of isolates of *H. terminale* possessing an independent life history (Edelstein *et al.* 1971; Clayton 1974). Support is thus given to the proposal by Clayton (1974) and Guiry (2102) that *H. terminale* probably represents a mixture of genotypes that are morphologically indistinguishable and, pending further studies, it is maintained as a distinct taxon in the present treatise.

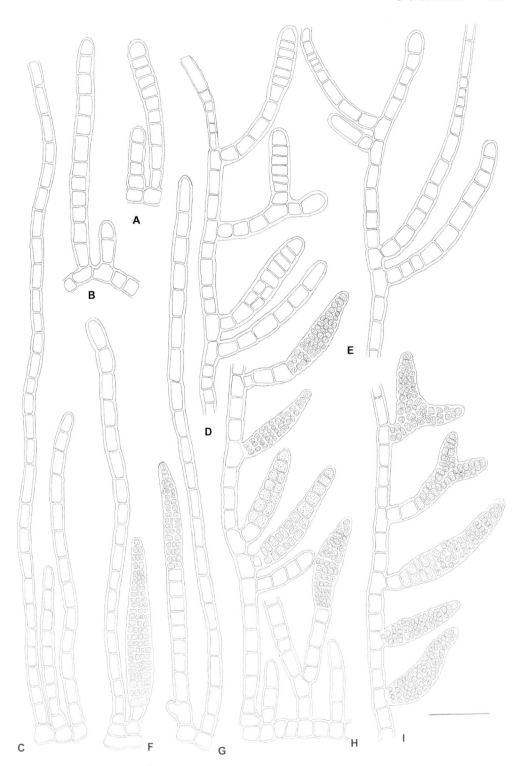

Fig. 114. *Hecatonema terminale*

A–C. Erect filaments arising from basal layer. D, E. Portions of branched, erect filaments. F–I. SP of fertile thallus showing erect filaments arising from a monostromatic/distromatic base with lateral/terminal plurilocular sporangia. Bar = 50 μm.

Bartsch & Kuhlenkamp (2000), p. 166; Batters (1900), p. 371; Børgesen (1926), pp. 52–5, figs 27, 28; Cardinal (1964), pp. 76, 77, fig. 40; Clayton (1974), pp. 790–3, figs 28, 29; Collins (1896), pp. 459, 460, pl. 278(1–5); Cotton (1912), p. 95; Dizerbo & Herpe (2007), p. 78; Edelstein *et al.* (1971), pp. 1248, 1249, figs 4, 11, 12; Fletcher (1987), pp. 202–4, fig. 50; Fletcher in Hardy & Guiry (2003), p. 257; Gallardo García *et al.* (2016), p. 35; Guiry (2012), p. 142; Hamel (1931–39), pp. 51, 52, fig. 17(A); Kuckuck (1953), pp. 319–25, pls 1–3; Kylin (1937), p. 8, (1947), p. 15, figs 10, 11; Mathieson & Dawes (2017), p. 159, pl. XLVII(6, 7); Ribera *et al.* (1992), p. 115; Rueness (1977), pp. 132, 133, fig. 58, (2001), p. 15; Stegenga *et al.* (1997a), p. 20; Taylor (1957), pp. 153, 154; Womersley (1987), pp. 315, 316, fig. 115(A, B); Wynne (1998), p. 101.

Isthmoplea Kjellman

ISTHMOPLEA Kjellman (1877b), p. 31.

Type species: *I. sphaerophora* (Carmichael) Kjellman ex Gobi (1878), pp. 12, 58.

Thalli epiphytic, forming tufts of densely branched filaments arising from a discoid base; filaments oppositely branched, less often alternately branched, comprising pluriseriate areas below and uniseriate areas in the upper branchlets; growth intercalary and diffuse; cells with numerous discoid plastids.

   Plurilocular sporangia formed by transformation of a series of vegetative cells in the secondary, ultimate branches; unilocular sporangia globose, sessile, developing as enlargements of vegetative cells in pluriseriate regions, often in opposite pairs or opposite a branch, more rarely solitary.

One species in Britain and Ireland.

Isthmoplea sphaerophora (Carmichael) Kjellman ex Gobi (1878), pp. 12, 58.   Figs 115, 116

Capsicarpella sphaerospora (Carmichael ex Harvey) Bory (1823), p. 178; *Ectocarpus sphaerophorus* Carmichael (1833), p. 326.

Thalli epiphytic, forming dense, flaccid, filamentous, olive-brown tufts up to 6 cm high, attached below by branched, rhizoidal filaments aggregated into outward-spreading discs, supplemented by corticating filaments arising from, and sheathing, the basal cells; erect

Fig. 115. *Isthmoplea sphaerophora*
A, B. Habit of thalli. Bar in A = 1 cm, bar in B = 0.7 cm. (Herbarium specimens, NHM.)

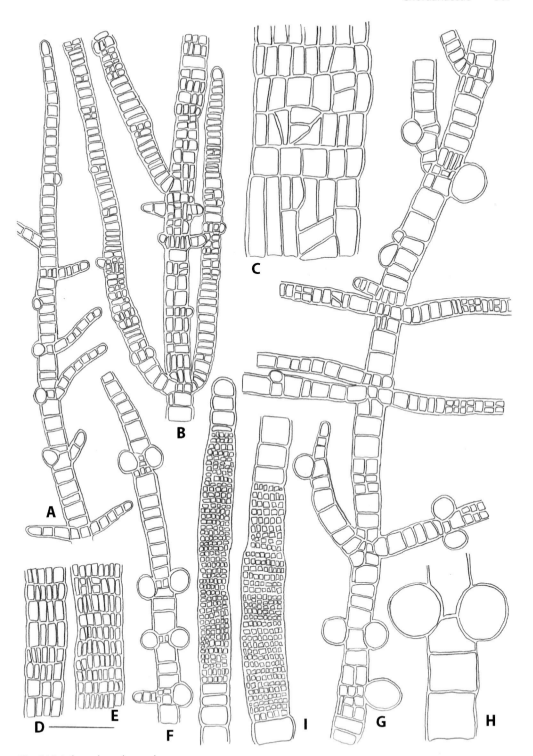

Fig. 116. *Isthmoplea sphaerophora*
A, B. Portions of thalli. C–E. Surface views of thalli. F–H. Portions of fertile thalli showing unilocular sporangia.
I. Portions of fertile thalli showing intercalary plurilocular sporangia. Bar = 50 μm.

filaments branched, with branching usually opposite or whorled, less often alternate, uniseriate above, comprising quadrate to cylindrical cells, 1–2 diameters long, 19–32 × 18–28 μm above, becoming 0.5–1 diameters long, (14–)32–53(–75) × (23–)32–52(–64) μm below and then dividing longitudinally and transversely, occasionally obliquely, to become pluriseriate, to 80(–90) μm and up to eight cells in width towards the base, the cells mainly rectangular in surface view; growth intercalary and diffuse; true hairs absent; cells with several discoid plastids.

Plurilocular sporangia intercalary and immersed, formed by transformation of the outer vegetative cells in the terminal branchlets, usually in a long continuous series, up to 650 μm in length, 32–50 μm (to 9 loculi) in diameter; unilocular sporangia spherical, (25–)40–60(–70) μm in diameter, either sessile and projecting or immersed in branches, solitary or more usually in opposite pairs, occasionally whorled or opposite a branch, formed by the longitudinal subdivision of a cell into three cells, the outer two developing into sporangia.

Epiphytic on various small algae, in particular *Gaillona hookeri*, *Plumaria plumosa* and *Cladophora rupestris*, lower eulittoral pools and shallow sublittoral.

Common and widely distributed around Britain and Ireland.

Appears to be a spring and summer annual, recorded between April and June.

Culture studies by Edelstein *et al.* (1971), Pedersen (1975) and Jónsson (1977) revealed a heteromorphic life history with spores from unilocular sporangia on the macrothallus developing asexually into streblonematoid microthalli which gave rise to new unilocular sporangia-bearing macrothalli and, in some races, also produced plurilocular sporangia.

Bartsch & Kuhlenkamp (2000), p. 166; Batters (1890b), pp. 277, 278, (1902), p. 35; Brodie *et al.* (2016), p. 1020; Coppejans & Kling (1995), p. 186, pl. 57(A–C); Cotton (1912), p. 95; Dizerbo & Herpe (2007), p. 110; Edelstein *et al.* (1971), pp. 1247, 1248, figs 1–3; Fletcher in Hardy & Guiry (2003), p. 263; Gallardo García *et al.* (2016), p. 35; Guiry (2012), p. 142; Hamel (1931–39), pp. 231, 232, fig. 45(3–5); Harvey (1833), p. 326; Jaasund (1960b), pp. 176–81, figs 4(A), 5(A–E), (1961b), pp. 215–22, (1965), p. 77; Jónsson (1903), pp. 162–5, figs 18, 19, (1977), pp. 433–5; Knight & Parke (1931), p. 64, pl. X(18); Kornmann & Sahling (1977), pp. 124, 125, pl. 66(A–H); Kylin (1937), p. 11, (1947), pp. 67, 68, fig. 56(D–E); Lund (1959), pp. 39, 40; Mathieson & Dawes (2017), p. 161, pl. XLVII(13–15); Morton (1994), p. 40; Newton (1931), pp. 182, 183, fig. 114(A–C); Pedersen (1975), pp. 165–8, (2011), p. 69, fig. 71; Rueness (1974), pp. 323–8, (2001), p. 15; Stegenga & Mol (1983), p. 97, pl. 3, fig. 1; Stegenga *et al.* (1997a), p. 20; Taylor (1957), p. 156, pl. 9(4, 5).

Kuetzingiella Kornmann

KUETZINGIELLA Kornmann (1956), p. 293.

Type species: *K. battersii* (Bornet ex Sauvageau) Kornmann (1956), p. 314.

Thalli epiphytic or epilithic, forming short, wool-like, intricately matted tufts of erect filaments arising from a basal system of outward-spreading, branched, monosiphonous filaments, either loosely associated or discoid/pseudodiscoid in appearance, more rarely penetrating host species when epiphytic; erect filaments free, monosiphonous, delicate, simple or sparsely and irregularly branched, sometimes giving rise to short, few-celled branchlets produced at right angles to parental filament; growth diffuse or from distinct meristematic regions subtending hairs (= trichothallic growth); hairs false, terminating filaments and branchlets; cells with several discoid/plate-like plastids, with pyrenoids.

Plurilocular sporangia and unilocular sporangia borne on the basal layer and/or on the erect filaments, terminal or lateral, sessile or shortly stalked; plurilocular sporangia multiseriate, unilocular sporangia ovate.

Two species in Britain and Ireland.

KEY TO SPECIES

Thalli epilithic; sporangia (mainly unilocular) borne only on erect filaments,
never from the prostrate system . . . .*Kuetzingiella holmesii*
Thalli epiphytic; sporangia (mainly plurilocular) borne on prostrate as well as on
erect filaments. . . . . *Kuetzingiella battersii*

Kuetzingiella battersii (Bornet ex Sauvageau) Kornmann (1956), p. 314.   Fig. 117

Ectocarpus battersii Bornet ex Sauvageau (1895a), p. 351; *Feldmannia battersii* (Bornet ex Sauvageau) Hamel (1931–39), p. xvii.

Thalli epiphytic, sometimes in part endophytic, forming small, light-coloured, fleece-like tufts up to 2 mm high on host surface; comprising an initial monostromatic basal system of branched, outward-spreading, monosiphonous filaments, irregularly arranged, or more usually, laterally adjoined and pseudodiscoid/discoid in appearance, especially around the edge, with cells 1–2 diameters long, 6–16 × 6–8 μm in diameter, giving rise below to occasional, downward-penetrating, and spreading endophytic, rhizoidal filaments, and later above to numerous, erect filaments; erect filaments, short, to 2 mm in height, free, quite intricately matted, unbranched or sparingly and irregularly branched, and usually terminating in long, colourless hairs; hair cells 5–13 diameters long, 73–170 × 11–19 μm; erect filament cells cuboidal to rectangular, 1–2(–3) diameters long, early formed filament cells 6–20 × 5–10 μm, later on more developed erect filaments (18–)26–40(–45) × (11–)14–18(–20) μm; growth trichothallic, with erect filaments producing terminal hairs from intercalary, meristematic zones of short, closely packed, cuboidal cells; cells with several discoid to plate-like plastids, each with a pyrenoid.

Plurilocular and unilocular sporangia borne on the same or different thalli, occurring on both the basal layer and erect filaments; plurilocular sporangia common, ovoid to elongate-ovoid, multiseriate, (25–)28–32(–35) × (14–)16–19(–22) μm, commonly borne on the basal layer, prior to their development on the erect filaments, sessile or on 1–3-celled pedicels, recurved, solitary or in secund series on the erect filaments; unilocular sporangia infrequent, subglobose, sessile or shortly stalked, 30–38 × 22–30 μm.

Epiphytic on older thalli of *Taonia atomaria*, although reported elsewhere on other hosts (e.g. *Dictyota*; Cardinal 1964), lower eulittoral pools and shallow sublittoral.

Very rare and only reported for South Devon in Britain and Mayo in Ireland. Probably under-recorded and more widely distributed in keeping with the distribution of the host species.

Appears to be a summer annual, in keeping with the seasonal distribution of the host species, with fertile thalli observed June to August.

Kuetzingiella battersii usually initially appears on *T. atomaria* as slightly darker coloured, irreg-ularly shaped streaks on the surface of the blades, comprising a monostromatic, filamentous to pseudodiscoid layer which can give rise to the sporangia directly. Later, this layer gives rise to the erect tufts of filaments that can be seen more clearly along the edge of the host.

Ardré (1970), p. 242; Batters (1902), p. 30; Børgesen (1926), pp. 36–43, figs 18–21; Brodie *et al.* (2016), p. 1019; Cardinal (1964), pp. 65–7, fig. 35(A–H); Cormaci *et al.* (2012), pp. 113, 114, pl. 26(1–3); Cotton (1912), p. 94 (as *Ectocarpus battersii*); De Toni (1895), pp. 540, 541; Dizerbo & Herpe (2007), p. 71, pl. 21(4); Ercegovic (1955), pp. 32–4, fig. 14; Fletcher in Hardy & Guiry (2003), p. 264; Gallardo García *et al.* (2016), p. 35; Guiry (2012), p. 136; Hamel (1931–39), pp. XVII, 52–4, fig. 16(A, B); Holmes & Batters (1890), p. 79; Kuckuck (1956), pp. 314–19, pls 10(A, B), 11(A–J), 12(A–K); Misra (1966), pp. 79, 80, fig. 34; Morton (1994), p. 8; Newton (1931), pp. 117, 118; Ribera *et al.* (1992), p. 112; Sauvageau (1895a), pp. 351–9, 362, 363, figs 1–5; Stegenga (1996), p. 204, pl. III(3–4); Stegenga *et al.* (1997), p. 16.

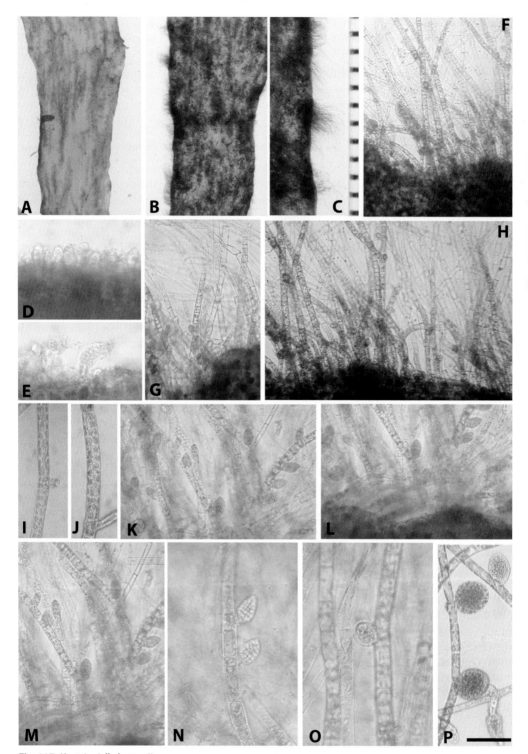

Fig. 117. *Kuetzingiella battersii*

A–C. Habit of thalli on *Taonia*. A. Early formed prostrate, monostromatic habit. B, C. Later formed tufted habit. D, E. Plurilocular sporangia borne on the basal layer of thalli. F–J. Erect vegetative filaments of thalli. K–N. Portions of fertile erect filaments with plurilocular sporangia. O, P. Portions of fertile erect filaments with unilocular sporangia. Bar in P = 4.5 mm (A, B), = 189 μm (H), = 90 μm (F, G, K, L, M), = 45 μm (D, E, I, J, N, O, P).

Kuetzingiella holmesii (Batters) G. Russell in Dixon & Russell (1964), pp. 279–80.   Fig. 118

Ectocarpus holmesii Batters (1888), p. 454.

Thalli epilithic, forming confluent, yellowish-brown (when dried), wool-like, intricately matted tufts of filaments spreading over substrata, commonly to several cm in extent, exceptionally to over 1 m; comprising a basal system of outward-spreading, irregularly branched, loosely compacted filaments of irregularly shaped, sometimes contorted cells, measuring 7–21 × 5–7 μm, which give rise to quite densely packed, fairly delicate and soft, erect filaments of varying length, but not exceeding 2–3(–7) mm; erect filaments monosiphonous, simple over much of their length or occasionally branched with branches often widely divergent and short, few celled; basal cells giving rise to downward-growing, secondary rhizoids which can contribute to attachment system; filament cells commonly rectangular or quadrate, and straight-walled, never barrel-shaped, 0.5–1(–3) diameters long, 6–48 × 10–18 μm, each with several, plate-like plastids; growth probably diffuse, although meristem-like regions of closely packed, short, quadrate to subquadrate cells often seen on branches, usually subtending hairs (possibly trichothallic growth); hairs terminal on branches, false, comprising non-pigmented cells, 4–7 diameters long, 40–90 × 10–15 μm.

Unilocular and plurilocular sporangia borne on the same or on separate thalli; both sorts of sporangia infrequent, usually arising laterally on the parental filament, irregularly positioned and scattered, although more usually distributed towards basal regions of filaments, sometimes terminating some branches; if lateral, sporangia either sessile or pedicellate, and usually arising perpendicular to the parental filament, sometimes quite strongly recurved towards it; plurilocular sporangia multiseriate, conical, 27–60 × 18–29 μm; unilocular sporangia ovoid to cylindrical, 38–50(–80) × 30–52 μm.

Epilithic on rocks, ledges, woodwork, stakes etc.; also reported on the sides of caves, in the upper eulittoral, usually in the shade; locally fairly common.

Rare, but widely distributed around Britain (Anglesey, Dorset, Kent, Norfolk, North and South Devon, Northumberland, Somerset) and Ireland (Cork, Mayo).

Collected between January and August, but data insufficient to comment on seasonal distribution.

Kuetzingiella holmesii was first described (as *Ectocarpus holmesii*) by Batters (1888), based on material collected at Minehead, Somerset by Miss I. Gifford. Batters subsequently reported it from a number of localities around Britain. The material had previously been identified by Dr Walker-Arnott as *Ectocarpus crinitus* (= *Acinetospora crinita*) to which it bears a close similarity with respect to its habitat (exposed, high eulittoral), tufted habit, branching pattern (sparse and irregular, with widely divergent branches and sporangia) and cytology (discoid plastids). However, there are some notable differences which help to separate the two species. These include:

- *Seasonal differences*: *K. holmesii* appears to be a winter to late summer species, whereas *A. crinita* is largely a spring/early summer annual.

- *Habitat differences*: *K. holmesii* is epilithic, whereas *A. crinita* is usually epiphytic, more rarely epilithic.

- *Habit differences*: *K. holmesii* has a very small thallus, barely exceeding 3–6 mm in height, whereas *A. crinita* can reach up to 10 cm or more in length.

- *Thallus differences*: the erect filaments of *K. holmesii* are much smaller in dimensions than those of *A. crinita*, the cells ranging from 6–25 μm in length and not exceeding 10–14 μm in diameter, whereas the cells of *A. crinita* range from 32–135 μm in length and reach 18–29 μm in diameter.

- *Reproductive differences*: monosporangia have only been reported for *A. crinita*.

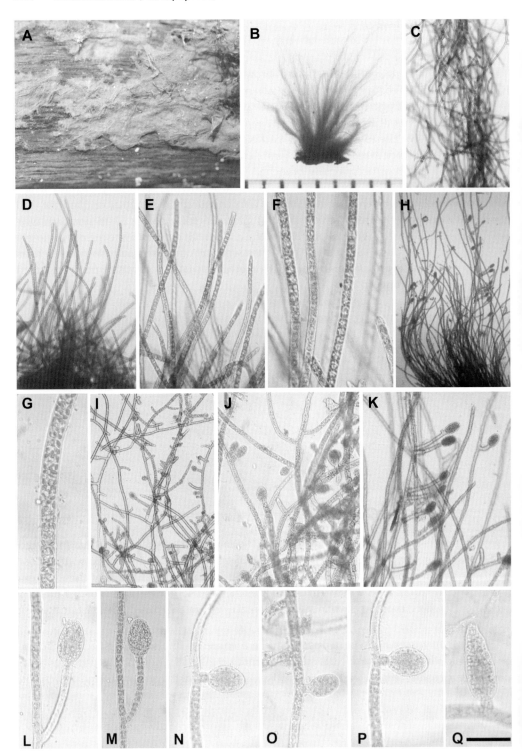

Fig. 118. *Kuetzingiella holmesii*

A. Fleece-like habit of thalli on wooden substrata. B. Habit of thalli. C–F. Portions of erect filaments. G. Portion of erect filament showing cells with discoid plastids. H–P. Portions of fertile filaments with unilocular sporangia. Q. Plurilocular sporangium. Bar in Q = 15 mm (A), = 1 mm (C), = 1.1 mm (H), = 180 µm (D, E, I), = 90 µm (F, J, K), = 45 µm (G, L–Q).

Bartsch & Kuhlenkamp (2000), p. 166; Batters (1888), pp. 450–7, pl. 18(7–16), (1889a), p. 55, pl. 8(7–16), (1890b), p. 275, pl. VIII(7–16), (1902), p. 31; Brodie *et al.* (2016), p. 1019; Cotton (1912), p. 94 (as *Ectocarpus holmesii*); Cotton (192), p. 94; Dixon & Russell (1964), pp. 279, 280; Fletcher in Hardy & Guiry (2003), p. 265; Gallardo García *et al.* (2016), p. 35; Guiry (2012), p. 136; Hamel (1931–39), p. 56; Kornmann & Sahling (1977), p. 118, pl. 61(A–D); Kuckuck (1897a), pp. 378–80, fig. 5(A–L); Morton (1994), p. 8; Newton (1931), pp. 118, 119; Printz (1926), pp. 149, 150, fig. 7; Rees (1935), p. 128; Rueness (1997), p. 117, (2001), p. 15.

Laminariocolax Kylin

LAMINARIOCOLAX Kylin (1947), p. 6.

Type species: *L. tomentosoides* (Farlow) Kylin (1947), p. 6.

Thalli epiphytic/endophytic, forming either spreading, fine, felt-like growths or solitary/ confluent pustules on host surface, arising from an extensive endophytic network of branching filaments; erect filaments usually short, simple or sparsely and irregularly/secundly branched; growth diffuse; plastids plate-like, 1–3 per cell, each with a pyrenoid; true hairs, with basal meristem, absent or present.

Plurilocular sporangia common, terminal or lateral on erect filaments, sessile or shortly pedicellate, cylindrical, uniseriate, simple or rarely branched; unilocular sporangia rare or absent, sessile or shortly pedicellate on erect filaments, or lateral at the base of short filaments; ascocysts sometimes present.

Two species in Britain and Ireland.

KEY TO SPECIES
Thalli solitary to confluent pustules on host surface; erect filaments absent or a few cells, < 40 μm in height, simple; plurilocular and unilocular sporangia on the same or different individuals; ascocysts absent . . . .*L. aecidioides*
Thalli extensive, felt-like growths covering host surface; erect filaments to 500 μm in height, simple or sparsely branched; plurilocular sporangia only; ascocysts common. . . . . *L. tomentosoides*

Laminariocolax aecidioides (Rosenvinge) A.F. Peters (1998), p. 689.   Fig. 119

Ectocarpus aecidioides Rosenvinge (1893), p. 894; *Phycocelis aecidioides* (Rosenvinge) Kuckuck (1894), p. 234; *Myrionema aecidioides* (Rosenvinge) Sauvageau (1897a), p. 177; *Entonema aecidioides* (Rosenvinge) Kylin (1947), p. 21; *Streblonema aecidioides* (Rosenvinge) Foslie ex Jaasund (1963), p. 3; *Gononema aecidioides* (Rosenvinge) Pedersen (1981a), p. 270.

Thalli endophytic, consisting of a network of branched, uniseriate filaments either deeply penetrating and widely spreading in host tissue or aggregated and/or packed together into a localized, monostromatic layer of cells between the outer cortex and epidermis, giving rise to erect filaments, sporangia and hairs which are emergent as small pustules on host surface; in surface view, pustules discrete and circular, occasionally elongate, 90–140(–180) μm in diameter, comprising, in surface view, sparsely packed, thick-walled, rounded cells, 4–8 μm in diameter; in transverse section, deeply penetrating filaments quite frequently branched, tortuous and intricate, spreading between walls of both cortical and medullary tissue, comprising cells of variable shape, some simple and cylindrical, some bulbous and contorted, 7–21 × 2–8 μm; monostromatic, pustule area cells irregularly shaped, 5–13 μm wide × 5–8 μm high, sometimes obviously connecting down with deeply penetrating filaments; erect filaments infrequent on plurilocular sporangia-bearing thalli, more commonly present on unilocular sporangia-bearing thalli, up to 40 μm (5 cells) high, comprising cuboidal, more commonly, cylindrical cells, 1–2 diameters long, 5–11 × 4–5 μm, each with 1(–2) plate-like plastids, without obvious pyrenoids; hairs frequent, with basal meristem (and possibly sheath), 8–10 μm wide.

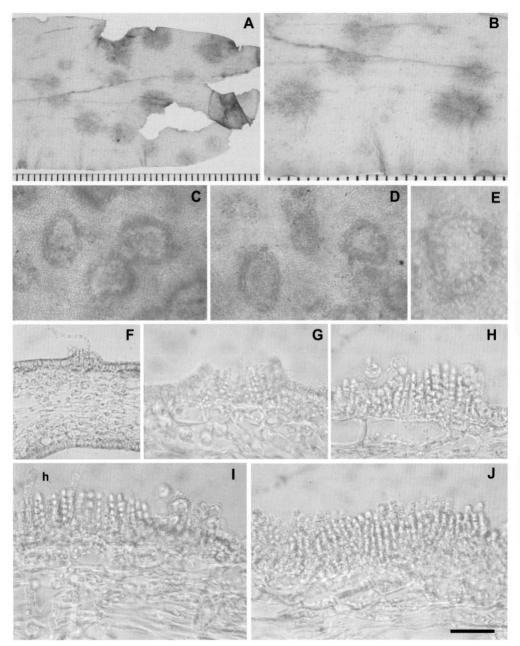

Fig. 119. *Laminariocolax aecidioides*
A, B. Habit of thalli forming spots on host (*Undaria pinnatifida*). C–E. Habit of thalli forming pustules on *U. pinnatifida*. F–H. TS of host (*U. pinnatifida*) showing pustule forming habit of thallus. I, J. TS of host (*U. pinnatifida*) showing uniseriate plurilocular sporangia and hairs (h). Bar in J = 90 μm (C, D, F), = 45 μm (E, G–J).

Plurilocular and unilocular sporangia either in separate or in the same pustules, usually sessile, more rarely stalked on basal layer; plurilocular sporangia abundant, closely packed, uniseriate, simple, cylindrical, 28–47 × 5–8 μm, to 9 loculi; spore loculi subcuboidal, often lenticular, 3–5 × 5–7 μm, with straight, rarely oblique cross walls; unilocular sporangia rare, ovate to pyriform, 18–36 × 15–23 μm, sometimes lateral at the base of erect filaments.

Endophytic in *Laminaria digitata*, less commonly *Saccharina latissima*, *Saccorhiza polyschides* and *Undaria pinnatifida*, in lower eulittoral pools and shallow sublittoral.

Recorded for scattered localities, probably more common and more generally distributed throughout Britain and Ireland.

Recorded between spring and autumn, probably present throughout the year.

Culture studies by Pedersen (1981a), using material collected from Greenland, revealed a 'direct' life history with swarmers from the plurilocular sporangia repeating the parental phase.

Abbott & Hollenberg (1976), p. 150, fig. 112; Bartsch & Kuhlenkamp (2000), p. 167; Batters (1895a), p. 276, (1902), p. 41; Brodie *et al.* (2016), p. 1020; Burkhardt & Peters (1998), p. 689, figs 8, 9; Ellertsdóttir & Peters (1997), pp. 135–43; Fletcher in Hardy & Guiry (2003), p. 267; Foslie (1894), pp. 136–9, pl. 1(7–10); Gallardo García *et al.* (2016), p. 35; Guiry (2012), p. 142; Hamel (1931–39), p. 92, fig. 25(I, II); Jaasund (1963), pp. 3, 4, fig. 3(B), (1965), pp. 47, 48, fig. 14; Kuckuck (1894), pp. 234–5, fig. 8(A–C); Kylin (1947), pp. 21, 22, fig. 18(A, B); Levring (1937), pp. 49, 50, fig. 6(A–C); Mathieson & Dawes (2017), p. 162, pl. XLVIII(1, 2); Newton (1931), p. 151; Parke & Dixon (1976), p. 559; Pedersen (1981a), pp. 263–70, figs 18–22, (2011), p. 103, fig. 109; Peters & Schaffelke (1996), pp. 111–16; Rosenvinge (1893), p. 894, fig. 27, (1894), pp. 118–20, fig. 27(A–D); Rueness (1977), p. 13, fig. 61, (2001), p. 15; Sauvageau (1897a), pp. 175–7; Stegenga & Mol (1983), p. 85, pl. 26(1); Stegenga *et al.* (1997a), p. 15; Taylor (1957), p. 115; Veiga *et al.* (1997), pp. 155, 156; Yoshida & Akiyama (1979), pp. 219–23.

Laminariocolax tomentosoides (Farlow) Kylin (1947), p. 6.　　　　　Fig. 120

Ectocarpus tomentosoides Farlow (1889), p. 11; *Streblonema tomentosoides* (Farlow) De Toni (1895), p. 573.

Thalli epiphytic, forming, yellow to dark brown, solitary or confluent, fine, felt-like, irregularly shaped patches on both sides of host surface, up to several cm in extent; comprising an extensive network of irregularly branched, much contorted, endophytic filaments, ramifying throughout host tissue, composed of long, narrow, sometimes distorted cells, 4–10 diameters long, 3–5 μm wide each containing up to 5–6 plate-like plastids with associated pyrenoids; endophytic filaments more copiously branched just before emerging through host surface to give rise to numerous erect, free, simple or irregularly/secundly branched filaments to 5 mm in height, collectively forming the surface covering growths; branching of erect filaments to one order only, with branches emerging from filaments at wide angles (45–90 degrees); erect filament cells cylindrical, 1–5 diameters long, 3–9 μm in diameter wide; growth diffuse but most active at branch apices, especially in endophytic filaments; all cells with 2 (1–3) parietal, plate-like plastids each with a pyrenoid; true hairs and pseudohairs absent.

Plurilocular sporangia common, borne terminally or laterally on the erect filaments, sessile, sometimes shortly pedicellate, cylindrical, uniseriate, simple, sometimes branched, 24–250 × 6–8 μm, emerging from filaments at wide angles; unilocular sporangia unknown, but frequent occurrence of ascocysts may lead to misidentification of these structures.

Commonly found as an epi-endophyte on the lamina of *Laminaria digitata*, but also found on the kelps *L. hyperborea*, *Saccharina latissima*, *Saccorhiza polyschides* and *Alaria esculenta*, as well as the red alga *Palmaria palmata*, lower eulittoral pools and shallow sublittoral.

Common and widely distributed around Britain and Ireland.

Emergent filaments found September–June with maximum thallus size and greatest abundance in March. Reproductive thalli observed October–June, with maximum development in March. Fertile thalli rare or absent in July and August. Species probably over-summers as an endophyte.

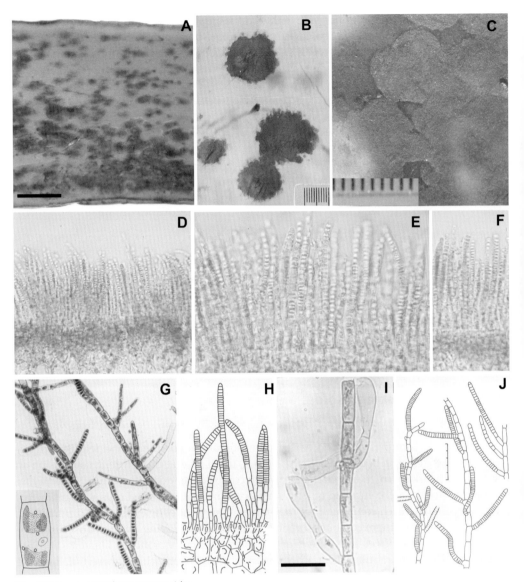

Fig. 120. *Laminariocolax tomentosoides*

A–C. Habit of thalli forming dark, irregularly shaped, felt-like spots on host. D–H, J. Columns of uniseriate pluri-locular sporangia emerging from host. Inset in G showing cell with two plate-like plastids. I. Portion of erect filament showing ascocysts. Bar in A = 40 mm, bar in I = 90 μm (D, F), = 45 μm (E), = 50 μm (G, J), = 60 μm (H), = 20 μm (I), = 13 μm (inset in G).

Bartsch & Kuhlenkamp (2000), p. 167; Batters (1892b), p. 20; Brodie *et al.* (2016), p. 1020; Cardinal (1964), pp. 67–9, fig. 36(A–H); Cotton (1912), p. 94; Dangeard (1934), pp. 98–102; Dizerbo & Herpe (2007), p. 71, pl. 22(1); Farlow (1889), pp. 11, 12, pl. LXXXVII(4); Fletcher in Hardy & Guiry (2003), p. 267; Gallardo García *et al.* (2016), p. 35; Gran (1893), pp. 1–15, figs 1–8; Guiry (2012), pp. 142, 143; Hamel (1931–39), pp. XXIV–XXVI, fig. 62(1); Hygen & Jorde (1935), pp. 1–60; Jónsson (1903), p. 154; Knight & Parke (1931), p. 61, pl. VIII(1–2); Kornmann & Sahling (1977), p. 103, pl. 52(A–D); Kuckuck (1894), p. 234, (1899), pp. 370–2, figs 5–7; Kylin (1937), pp. 7, 8, fig. 1(A–D), (1947), pp. 6, 7, fig. 1(D–E); Lund (1959), pp. 51–3, fig. 7; Mathieson & Dawes (2017), pp. 162, 163, pl. XLVIII(4–6); Morton (1994), p. 32; Newton (1931), p. 117; Pedersen (1984), pp. 42, 43, fig. 29, (2011), p. 68, fig. 70; Printz (1926), p. 149; Rosenvinge (1894), p. 116, fig. 25; Rosenvinge & Lund (1941), pp. 43, 44, fig. 7(A–E); Rueness (1977), p. 118, fig. 40, (2001), p. 15; Russell (1964b), pp. 601–12, figs 1(A–E), 2(A–C), 3, pl. 1(1–9); Stegenga & Mol (1983), p. 76, pl. 22(2–4); Stegenga *et al.* (1997a), p. 17; Taylor (1957), pp. 108, 109; Villalard-Bohnsack & Harlin (2001), p. 376, tab. 2.

Leathesia S.F. Gray

LEATHESIA S.F. Gray (1821), p. 301.

Type species: *L. tuberiformis* S.F. Gray (1821), p. 301 (= *Leathesia difformis* Areschoug (1847), p. 376).

Thalli epiphytic or epilithic, globose, gelatinous and soft, quite easily squashed under pressure, smooth or irregularly convoluted, solid when young, becoming hollow in mature specimens; comprising erect rows of dichotomously–trichotomously branched compacted filaments of large, irregularly shaped and contorted, colourless cells, elongate-cylindrical below, progressively becoming globose towards the periphery and producing short, 2–5-celled, clavate, pigmented, tightly packed, terminal branches (paraphyses); paraphyses' cells with several discoid plastids with pyrenoids; hairs single or in fascicles, arising at base of paraphyses, with basal meristem and sheath.

Plurilocular and unilocular sporangia arising at base of paraphyses; plurilocular sporangia simple, filiform and uniseriate; unilocular sporangia globose or pyriform.

One species in Britain and Ireland.

Leathesia marina (Lyngbye) Decaisne (1842), p. 370.   Figs 121, 122

Tremella difformis Linnaeus (1755), p. 429; *Rivularia tuberiformis* Smith (1809), p. 1956, nom. inval.; *Nostoc marinum* C.A. Agardh (1810–12), p. 45; *Chaetophora marina* Lyngbye (1819), p. 193; *Leathesia tuberiformis* Gray (1821), p. 301; *Clavatella nostoc-marina* Bory (1823), p. 197, nom. illeg.; *Corynephora marina* (C. Agardh) C. Agardh (1824), p. 2, nom. inval.; *Nostoc mesentericum* C.A. Agardh (1824), p. 21; *Clavatella difformis* (Linaeus) Fries (1835), p. 316; *Leathesia difformis* Areschoug (1847), p. 376; *Leathesia difformis* var. *tingitana* Schousboe ex Bornet (1892), p. 237; *Leathesia nana* Setchell & N.L. Gardner (1924), p. 3; *Leathesia amplissima* Setchell & N.L. Gardner (1924), p. 3.

Thalli epiphytic, more rarely epilithic, light to yellow-brown in colour, firm fleshy and mucilaginous in texture, solid, globose and smooth when young, hollow, irregular and convoluted later, either solitary and discrete or confluent and expanding to 5(–8) cm in diameter; in squash preparation, thalli constructed of upward and outward-radiating, dichotomously–trichotomously branched filaments of cells, loosely packed together and fairly easily separable under pressure, solid throughout at first becoming irregularly hollow later; comprising large, elongate-cylindrical or irregularly contorted and almost stellate, colourless cells, becoming smaller and more globose towards the periphery, terminating in short 2–5-celled clavate, simple, densely crowded, pigmented filaments (paraphyses), hairs, unilocular and/or plurilocular sporangia; cells of paraphyses 2–3 diameters long, elongate-cylindrical below, globose terminally, with 1–3 plate-like plastids with pyrenoids; hairs and sporangia arising at the base of the paraphyses; hairs single or in fascicles with basal meristem and sheath.

Unilocular sporangia globose or pyriform, 20–35 × 17–23 μm; plurilocular sporangia filiform, simple, uniseriate, 15–33 × 4–7 μm, to 5–7 loculi.

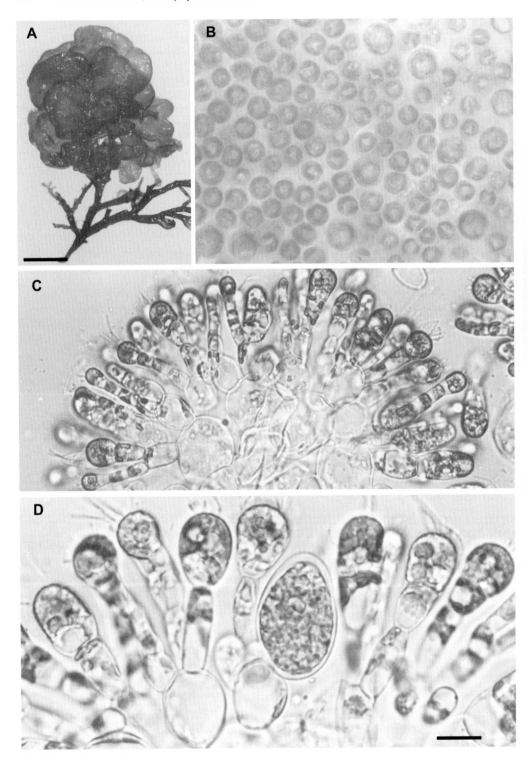

Fig. 121. *Leathesia marina*
A. Habit of thallus. B. Surface view of thallus showing loosely associated cells in gelatinous matrix. C. SP of thallus showing medulla-like region of large, colourless cells giving rise terminally to short filaments. D. SP of peripheral region showing short terminal filaments and a single unilocular sporangium. Bar in A = 1.5 cm, bar in D = 10 μm (B–D).

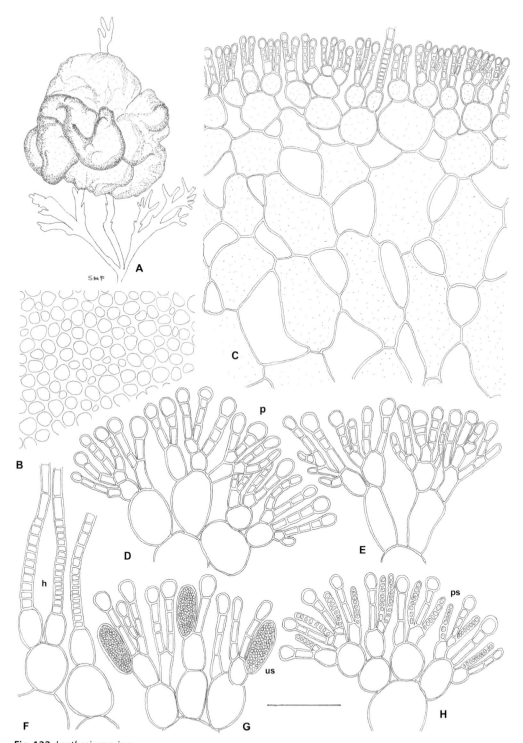

Fig. 122. *Leathesia marina*
A. Habit of thallus. B. Surface view of thallus showing loosely associated cells in gelatinous matrix. C. VS of thallus showing large, irregularly shaped central cells giving rise terminally to short filaments (paraphyses) and a single hair. D, E. Portions of peripheral thallus region. F. Portion of peripheral thallus region showing hairs (h). G. Portion of peripheral thallus showing unilocular sporangia (us). H. Portion of peripheral thallus showing plurilocular sporangia (ps). Bar = 17 mm (A), = 50 µm (B–H), = 100 µm (C).

Epiphytic on diverse algae, especially the red alga *Corallina officinalis*, upper to lower eulittoral, in pools or emergent.

Commonly distributed around Britain and Ireland.

Annual; spring and summer.

European isolates of *Leathesia marina* (as *L. difformis*) have been studied in culture by Sauvageau (1925, 1928, 1929), Dammann (1930), Kylin (1933) and Dangeard (1965c, 1969). It is generally considered that the life history is heteromorphic, comprising a diploid, sporophytic macrothallus which alternates with a haploid, gametophytic microthallus. The life history is further complicated, however, by a direct recycling of the diploid sporophyte via plurispores from plurilocular sporangia on the macrothallus. A similar direct recycling of the haploid gametophyte via plurispores from plurilocular sporangia on the microthallus (by partheno-genesis) also occurs which is in addition to their reported gametic sexual role in returning the sporophyte. Dangeard additionally reported a 'short-circuit' life history with the development of the diploid macrothallus via sexual fusion of unispores from the unilocular sporangia (but note Müller 1975 cast doubt on reports of unispores fusing). A noteworthy feature of the culture studies (commented upon by Sauvageau 1925) was the morphological resemblance between the microthalli of *L. marina* and *Myrionema*. This adds further support to Pedersen's (1984) hypothesis that this genus does not represent a natural phylogenetic unit.

Abbott & Hollenberg (1976), pp. 176, 177, fig. 142; Ardré (1970), pp. 253, 254; Bartsch & Kuhlenkamp (2000), p. 167; Brodie *et al.* (2016), p. 1020; Bunker *et al.* (2017), p. 297; Coppejans & Kling (1975), pp. 168, 170, pl. 47(A–C); Cormaci *et al.* (2004), p. 166, (2012), p. 166, pl. 42(6–8); Cotton (1912), p. 96; Dammann (1930), p. 11, fig. 3, pl. 1(1); Dangeard (1965c), pp. 5–43, pls 1–5, (1969), pp. 79–81, fig. 4; Dizerbo & Herpe (2007), pp. 83, 85, pl. 27(1); Fletcher (1987), pp. 143–5, fig. 25; Fletcher in Hardy & Guiry (2003), p. 268; Gallardo García *et al.* (2016), p. 35; Gayral (1966), p. 269, fig. 35, pl. XLIII; Guiry (2012), p. 143; Hamel (1931–39), pp. 138–40, fig. 32(A–D); Harvey (1846–51), pl. 324; Kim (2010), pp. 135–7, fig. 61; Kuckuck (1929), p. 43, figs 49–52, (1930), p. 44, figs 49–52; Kylin (1933), pp. 64–6, fig. 30, (1947), pp. 53, 54, fig. 46; Lindauer *et al.* (1961), pp. 220, 221, fig. 44; Mathieson & Dawes (2017), p. 163, pl. XLVIII(7, 8); Morton (1994), p. 34; Newton (1931), p. 141, fig. 87; Oltmanns (1922), figs 318, 319; Pedroche *et al.* (2008), p. 60; Poza *et al.* (2017), pp. 579–89; Ribera *et al.* (1992), p. 115; Rosenvinge & Lund (1943), p. 8, fig. 1(A, B); Rueness (1977), pp. 142, 143, fig. 70, (2001), p. 15; Sauvageau (1925), pp. 1632–5, (1928), p. 268, (1929), p. 403; Schneider & Searles (1991), p. 134, fig. 150; Setchell & N.L. Gardner (1925), pp. 511, 512, pl. 40(52), pl. 43(65, 66); Stegenga & Mol (1983), pp. 92, 93, pl. 31; Stegenga *et al.* (1997a), p. 18, (1997b), p. 151, pl. 41(5–6); Taylor (1957), p. 149, pls 12(5), 14(8); Verlaque *et al.* (2015), p. 69; Womersley (1987), p. 102, figs 27(C), 29(D–F).

Leblondiella G. Hamel

LEBLONDIELLA G. Hamel (1931–39), p. xl.

Type species: *L. densa* (Batters) Hamel (1931–39), p. xl.

Thalli epiphytic, forming small, dark brown-black, flaccid tufts on host, solitary or gregarious, arising from a small, discoid, attachment base; comprising up to several erect, filiform shoots, uniseriate, simple and sharply tapering below, multiseriate and radially branched above with obtuse apex; branches either sparse and tufted or confluent and expansive, clothing central parenchymatous axis, short, dichotomously or unilaterally branched, strongly recurved, comprising cylindrical to moniliform cells, each with several discoid plastids; hairs uncommon with basal meristem, lacking sheath.

Unilocular and plurilocular sporangia borne on the same or different shoots; unilocular sporangia spherical or ovoid, sessile or stalked, at base of radial branches; plurilocular sporangia lateral on primary axis below, lateral and terminal on radial branches above, solitary or clustered, sessile or stalked, cylindrical to conical, uniseriate or biseriate.

Leblondiella is a monotypic genus, based on *Myriotrichia densa* Batters (1895b). It differs from *Myriotrichia* in the following features: it has a discoid rather than a filamentous base; the mature, branched, erect shoots are filiform and linear, sharply attenuate at base and dark brown-black in colour rather than elongate-clavate, with a long tapering base and light/mid brown in colour; the short radial branches are dichotomously and/or unilaterally branched with branches strongly recurved, rather than irregularly or radially branched with divergent branches; the constituent cells are more moniliform in *Leblondiella* than in *Myriotrichia* and remain uniseriate and filamentous rather than becoming multiseriate and parenchymatous; hairs are also infrequent in *Leblondiella* rather than abundant as in *Myriotrichia*. It is possible that the above differences will prove insufficient to maintain *Leblondiella* as an autonomous genus. However, it is retained here pending the results of future culture studies.

One species in Britain and Ireland.

Leblondiella densa (Batters) Hamel (1931–39), p. xl.   Fig. 123

Myriotrichia densa Batters (1895b), p. 313.

Thalli forming erect tufts on host, single or gregarious, each comprising 2–5 shoots arising from a disc-like base; erect shoots to 25(–40) mm long, 0.2(–0.3) mm wide, simple, dark brown-black, filiform, more or less linear, soft and flaccid in texture, obtuse above, sharply tapering to base; basal disc to 300(–600) µm in diameter, comprising an outward spreading network of dichotomously/irregularly branched filaments of thin-walled, colourless, often irregularly shaped cells, weakly joined together in a mucilaginous matrix; erect shoots uniseriate and simple below comprising cells 0.5–1 diameters long, 16–31 × 27–36 µm becoming multiseriate above by longitudinal and transverse divisions to form a narrow terete thallus, with quadrate or irregularly shaped surface cells which produce short, radial branches, either sparsely to form irregular tufts or more often abundantly to clothe the primary axis; lateral filaments dichotomously or unilaterally branched, with branches strongly recurved and slightly attenuate, to 130 µm long (8 cells) with cells moniliform, subcylindrical to cylindrical, 0.5–1(–2) diameters long, 8–34 × 8–34 µm, each with several discoid plastids without obvious pyrenoids; hairs uncommon, with basal meristem and lacking sheath, 10–16 µm in diameter.

Unilocular and plurilocular sporangia more commonly reported on different shoots, sometimes mixed; unilocular sporangia less common, sessile or shortly stalked, at base of branched filaments, spherical or ovoid, 34–60 µm in diameter; plurilocular sporangia common arising directly from primary axis, more commonly on lateral branches, single or clustered, sessile or stalked, cylindrical, subcylindrical, lanceolate or conical, to 10 loculi long, 25–48(–60) × 8–13 µm, uniseriate, more commonly biseriate with straight or oblique cross walls.

Epiphytic on decaying leaves of *Zostera* spp., lower eulittoral pools and shallow sublittoral.

Only recorded for a few scattered localities around Britain (Hampshire, Dorset, south Devon, Cornwall, Isle of Man, Argyll) and Mayo in Ireland.

Annual; summer (June to August).

Batters (1895b), pp. 311–13, pl. XI(11–13), (1895c), p. 169; Brodie *et al.* (2016), p. 1020; Buffham (1891), pp. 322, 323, figs 4–12; Dizerbo & Herpe (2007), p. 108; Fletcher (1987), pp. 167–9, fig. 35; Fletcher in Hardy & Guiry (2003), p. 268; Gallardo García *et al.* (2016), p. 35; Guiry (2012), p. 143; Hamel (1931–39), pp. 239, xl–xli, fig. 46(V); Kuckuck (1899a), p. 75; Mathieson & Dawes (2017), p. 164, pl. XLVIII(9–11); Morton (1994), p. 8; Newton (1931), p. 174; Rueness (1977), p. 157, (2001), p. 15; Sauvageau (1931), pp. 76, 77.

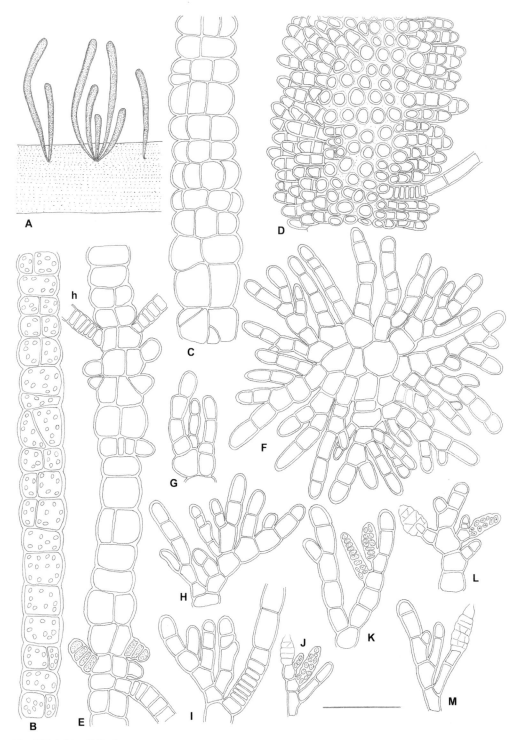

Fig. 123. *Leblondiella densa*
A. Habit of thalli epiphytic on *Zostera*. B–D. Portions of erect thalli showing increased anatomical development. E. Portion of thallus showing hairs (h) and short, lateral plurilocular sporangia (shaded). F. TS of well-developed thallus showing branched lateral filaments. G–I. SP of three portions of vegetative thallus showing branched lateral filaments. J–M. SP of four portions of fertile thallus showing plurilocular sporangia. Bar = 4 mm (A), = 50 μm (others).

Leptonematella Silva

LEPTONEMATELLA Silva (1959), p. 63.

Type species: *L. fasciculata* (Reinke) P.C. Silva (1959) p. 63.

Leptonema Reinke (1888), p. 19, non *Leptonema* A. de Jussieu.

Thalli epiphytic, epilithic or epizoic, forming small dense tufts of erect filaments arising from a pseudodiscoid base of outward-radiating branched filaments; erect filaments simple, or little branched at the base, uniseriate, acuminate, with basal intercalary meristem; cells with several plate-like plastids, without obvious pyrenoids in the North Atlantic; hairs unknown.

Unilocular sporangia ovate or elongate-pyriform, sessile or stalked, lateral at the base of the erect filaments; plurilocular sporangia intercalary or terminal on erect filaments, in series, formed by simple subdivision of vegetative cells.

One species in Britain and Ireland.

Leptonematella fasciculata (Reinke) P.C. Silva (1959), p. 63.   Fig. 124

Leptonema fasciculatum Reinke (1888), p. 19; *Leptonema fasciculatum* var. *majus* Reinke (1889a), p. 13, pl. 10; *Leptonema fasciculatum* var. *flagellare* Reinke (1889a), p. 13, pl. 10, figs 10, 11; *Leptonema fasciculatum* var. *uncinatum* Reinke (1889a), p. 13, pl. 9; *Elachista fasciculata* (Reinke) H. Gran (1893), p. 29; *Leptonema fasciculatum* var. *subcylindricum* Rosenvinge (1893), p. 879; *Leptonema neapolitanum* Schussnig (1930), p. 174; *Leptonematella neapolitana* (Schussnig) Cormaci & G. Furnari in Gallardo García (1992), p. 324.

Thalli usually epiphytic, less commonly epizoic and epilithic, forming erect, dense, light brown, flaccid tufts, solitary, not confluent, arising from the centre of a small basal disc; disc comprising outward-radiating, branched filaments, one-layered or with superimposed layers in parts, either free or closely compacted and pseudodiscoid in structure, cells variable in shape, frequently irregularly distended, 1–4 diameters long, 6–18 × 4–6 µm, with branches often arising from mid cell wall region only; erect tufts to 12 mm long, uniseriate throughout, linear, acuminate, simple or branched towards base, with basal, intercalary meristem, comprising cells largely quadrate or barrel shaped below, rectangular above, 1–3(–4) diameters long, 11–53 × 7–17 µm, each with several, plate-like plastids without obvious pyrenoids in the North Atlantic; hairs unknown.

Unilocular and plurilocular sporangia borne on the same thalli; unilocular sporangia rare, borne laterally at the base of erect filaments, sessile or more commonly on one to several celled stalks, ovate or elongate-pyriform, 80–120 × 26–40 µm; plurilocular sporangia common, intercalary or terminal on erect filaments, formed by subdivisions of vegetative cells, usually four sporangia per cell produced, 7–9 × 8–18 µm, sometimes laterally protruding to 50 µm in diameter, 4–6(–8) loculi per sporangium seen in surface view, loculi 3–5 × 3–4 µm.

Epiphytic, epizoic on shells, lower eulittoral pools and sublittoral to 10 m.

Recorded for scattered localities around Britain (Shetland, Argyll, Isle of Man, Devon) and Ireland (Cork, Down); probably more widely distributed.

Annual, spring and summer.

Leptonematella fasciculata has been investigated in culture by Dangeard (1966b, 1968d, France isolate), Wynne (1969, Washington State, USA isolate) and Pedersen (1978a, West Greenland isolate). All three authors reported the zoospores from the plurilocular sporangia to germinate directly, without sexual fusion, into more or less disc-shaped prostrate systems of uniseriate branched filaments, which developed lateral plurilocular sporangia and sprouted new erect filaments. Plurilocular sporangia, similar to those reported on field collected material, later developed on the erect filaments. Spores from the plurilocular sporangia, on both the erect and creeping filaments, repeated the above sequence of development. Only Pedersen observed

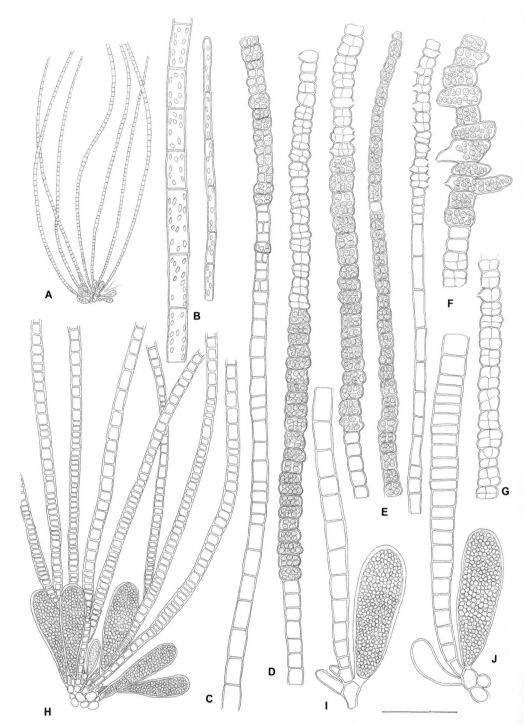

Fig. 124. *Leptonematella fasciculata*

A. Habit of thallus. Note basal unilocular sporangia. B. Portions of erect filament showing cells with discoid plastids. C–G. Portions of fertile erect filaments with mature and dehisced plurilocular sporangia. H, I, J. Basal regions of erect filaments showing unilocular sporangia. Bar = 0.5 mm (A), = 50 µm (B–G, I, J), = 100 µm (H).

unilocular sporangia in culture which developed only under short day conditions. Released unispores behaved in an identical manner to the plurispores. The life history appears, therefore, to be asexual and of the 'direct' type. However, detailed chromosomal studies on material in culture is obviously essential before the exact relationship between the microthalli and macrothalli can be determined.

Bartsch & Kuhlenkamp (2000), p. 167; Batters (1892a), p. 174, (1892b), p. 20, (1894), p. 91; Brodie *et al.* (2016), p. 1020; Cormaci *et al.* (2012), p. 169, pl. 44(1–4); Cotton (1912), p. 95; Dangeard (1966b), pp. 1692–4, pls I–II, (1968d), pp. 117–30, pls I–II, (1969), pp. 86–7; Dizerbo & Herpe (2007), p. 83; Fletcher (1987), pp. 136–8, fig. 23; Fletcher in Hardy & Guiry (2003), p. 269; Gallardo García *et al.* (2016), p. 35; Guiry (2012), p. 143; Hamel (1931–39), pp. 127, 128; Jaasund (1965), p. 64; Kim (2010), pp. 100–2, fig. 47; Knight & Parke (1931), pp. 66, 110–11, pl. 13(35, 36, 39); Kornmann & Sahling (1977), p. 212, fig. 63; Lund (1959), pp. 113–16, fig. 24; Mathieson & Dawes (2017), pp. 164, 165, pl. XLVIII(12–14); Morton (1994), p. 34; Newton (1931), p. 131, fig. 78(A, D, not 78B, C); Oltmanns (1922), p. 32, fig. 323; Pankow (1971), pp. 181, 182, figs 226, 227; Pedersen (1976), pp. 40, 41, (1978a), pp. 61–8, figs 12–16, (1984), pp. 43, 44, (2011), p. 89, fig. 95; Printz (1926), p. 159, pl. 3(28); Reinke (1889a), pp. 13, 14, pls 9(1–15), 10(1–11); Ribera *et al.* (1992), p. 115; Rueness (1977), pp. 133, 134, fig. 59, (2001), p. 15; Saunders & McDevit (2013), pp. 1–23; Silva (1959), p. 63; Stegenga & Mol (1983), p. 83, pl. 25(3–6); Stegenga *et al.* (1997a), p. 18; Taylor (1957), p. 143; Wynne (1969), pp. 11–13, fig. 4.

Litosiphon Harvey

LITOSIPHON Harvey (1846–51), p. 43.

Type species: *L. pusillus* (Carmichael) Harvey (1849), p. 43 (= *L. laminariae* (Lyngbye) Harvey (1849), p. 43).

Thalli epiphytic, gregarious, forming tufts of erect shoots on hosts, attached by downward-penetrating rhizoidal filaments; shoots simple, solid, parenchymatous, filiform/cylindrical or elongate-clavate; in section, comprising an outer cortex of 1–2 small, pigmented cells enclosing a central medulla of large, elongate, colourless cells; cells with several discoid plastids with pyrenoids; hairs common with basal meristem, lacking sheath.

Unilocular and plurilocular sporangia borne on the same or different shoots, developing from surface cells, single or grouped; unilocular sporangia ovate or elliptical, formed directly from a single vegetative cell; plurilocular sporangia formed indirectly from vegetative cells following vegetative division.

Previously placed in the Punctariaceae by various authors such as Parke & Dixon (1976), Rueness (1977) and Kornmann & Sahling (1977), the genus *Litosiphon* was later removed by Pedersen (1984) to the Myriotrichiaceae. By comparing *Litosiphon pusillus* and *Punctaria plantaginea* (both the type species of their respective genera), Pedersen concluded that they differed in a number of important characteristics. These include: the occurrence of sympodial branching in *L. pusillus* but not *P. plantaginea*; sheathed hairs are absent in *L. pusillus* but present in *P. plantaginea*; the embryospore occasionally shows immediate differentiation in *L. pusillus* but not in *P. plantaginea*; the microthalli are *Streblonema*-like in *L. pusillus* but *Hecatonema*-like in *P. plantaginea*. In particular, the occurrence of the two characteristics, sympodial branching and immediate differentiation of the embryospore (which are shared with *Myriotrichia*, type genus of the *Myriotrichiaceae*) persuaded Pedersen to transfer *Litosiphon* from the Punctariaceae to the Myriotrichiaceae. However, molecular studies (see Rousseau & de Reviers 1999a; de Reviers & Rousseau 1999; Silberfeld *et al.* 2014) do not support the recognition of either the Myriotrichiaceae or the Punctariaceae as distinct families and place *Litosiphon* within the family Chordariaceae within the order Ectocarpales.

Of the three species of *Litosiphon* recognized for Britain and Ireland by Parke & Dixon (1976) viz. *L. filiformis* (Reinke) Batters, *L. pusillus* (Carmichael ex Hooker) Harvey and *L. laminariae* (Lyngbye) Harvey, only *L. laminariae* is accepted in the present text. *L. filiformis* is removed to the genus *Pogotrichum* Reinke, following the proposal of Pedersen (1978a), while *L. pusillus* is included within the synonymy of *L. laminariae*. Certainly there appears to be little justification for maintaining a distinction between *L. laminariae* and *L. pusillus* based on previously accepted criteria such as host specificity (*L. laminariae* on *Alaria esculenta*, *L. pusillus* on *Chorda filum*), thallus colour (dark brown thalli in *L. laminariae*, light brown thalli in *L. pusillus*), thallus size (longer, broader thalli in *L. pusillus* compared to *L. laminariae*) and basal attachment system (pustule formation on host surface by *L. laminariae* only). The discovery of a wider range of hosts as well as intermediate forms of these two species indicate their probable conspecificity.

One species in Britain and Ireland.

Litosiphon laminariae (Lyngbye) Harvey (1849), p. 43.   Figs 125, 126

Bangia laminariae Lyngbye (1819), p. 84; *Bangiella laminariae* (Lyngbye) Gaillon (1833), p. Appendix [5], nom. illeg.; *Asperococcus pusillus* Carmichael in W.J. Hooker (1833), p. 277; *Encoelium pusillum* (Carmichael ex W.J. Hooker) J. Agardh (1836), p. 15; *Streblonema thuretii* Sauvageau (1936), p. 199 (as 'thureti'); *Litosiphon tenuis* Levring (1937), p. 66; *Scytosiphon pusillus* (Carmichael ex W.J. Hooker) Fries (1845), p. 124; *Dictyosiphon pusillus* (Carmichael ex W.J. Hooker) Areschoug (1847), p. 149; *Streblonema danicum* Kylin (1947), p. 46; *Asperococcus laminariae* (Lyngbye) J. Agardh (1848), p. 79; *Litosiphon pusillus* (Carmichael ex W.J. Hooker) Harvey (1849), p. 43; *Punctaria laminariae* (Lyngbye) P. Crouan & H. Crouan (1867), p. 167; *Pogotrichum hibernicum* Johnson (1892), p. 6; *Litosiphon hibernica* (Johnson) Batters (1902), p. 25; *Pilocladus danicus* (Kylin) Kornmann (1954), p. 117.

Thalli epiphytic, forming erect, light to dark brown, commonly gregarious, dense tufts on hosts, each tuft solitary, not confluent, comprising up to 40(–70) erect shoots, closely associated and attached at the base by an intricate mat of downward-growing secondary rhizoids; rhizoids sometimes penetrating and producing small (usually less than 1 mm) swellings on host surface; erect shoots 10–35(–70) mm long, 50–180(–280) µm wide (–22 cells), simple, filiform, terete, firm, solid, linear or more commonly, slightly elongate-clavate, fairly smooth, occasionally tortuous in parts, parenchymatous throughout except at extreme base of young shoots, with obtuse apex; comprising surface cells in straight, more commonly spirally twisted, longitudinal, less obviously, transverse rows, cells 1–2(–4) diameters long, 8–32 × 8–21 µm, quadrate or rectangular in shape, each with several discoid plastids with pyrenoids; in transverse section, shoots comprising an outer cortex of 1–2 small, pigmented cells enclosing a central medulla of large, slightly elongated, colourless cells; hairs abundant, arising singly from surface cells, 13–16 µm in diameter, multicellular with short basal meristem and lacking sheath.

Unilocular and plurilocular sporangia usually borne on different, occasionally the same shoots, discrete or confluent, occurring over greater length of shoot, formed in surface cells; unilocular sporangia variable in shape, usually oval or elliptical, 16–35 × 15–27 µm; plurilocular sporangia not formed in subdivided surface cells but in daughter cells following transverse and longitudinal vegetative divisions, variable in shape, 11–14 × 6–11 µm in surface view, slightly protruding, each comprising 2–4(–6) loculi.

Epiphytic on various hosts, especially *Chorda filum* and *Alaria esculenta* but also commonly occurring on *Scytosiphon lomentaria*, *Saccorhiza polyschides* and *Laminaria* spp., lower eulittoral pools and shallow sublittoral.

Common and widely distributed around Britain and Ireland.

Annual; summer and autumn.

Culture studies by Sauvageau (1929, 1933a) and Dangeard (1965a) (Atlantic coast of France), Kylin (1934) and Nygren (1975a,b, 1979) (Sweden) and Pedersen (1978a, 1981c, 1984) (Denmark) indicate that a 'direct' life history is operating. Released swarmers directly, without copulation, develop into thalli of similar morphology. Fertile prostrate systems (microthalli) are also commonly reported (Nygren 1975a; Sauvageau 1933a; Pedersen 1978a, 1981c, 1984) which may exist alone under various suboptimal conditions of lower salinity and reduced light intensity

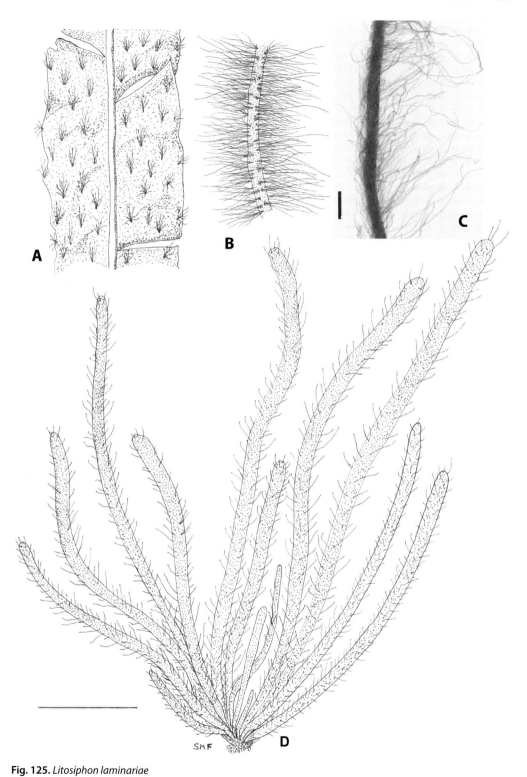

Fig. 125. *Litosiphon laminariae*

A. Habit of thalli on *Alaria esculenta*. B, C. Habit of thalli epiphytic on *Chorda filum*. D. Habit of young thallus. Bar in D = 34 mm (A), = 16 mm (B), = 1.2 mm (D), bar in C = 5 mm.

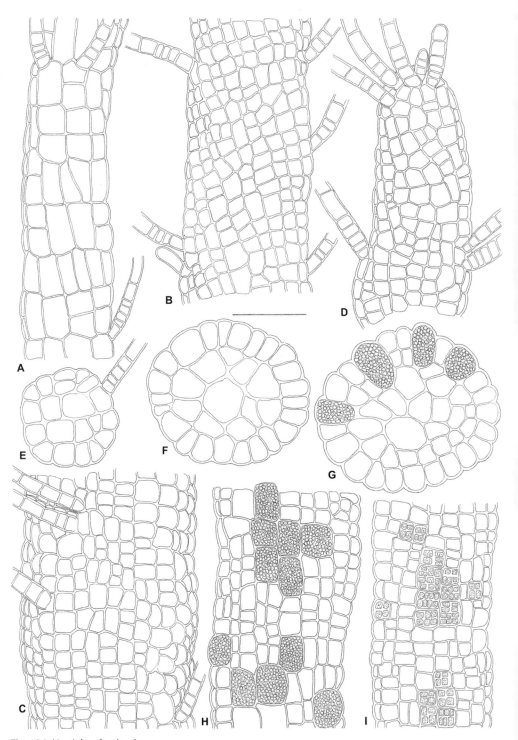

Fig. 126. *Litosiphon laminariae*
A–C. Portions of erect thalli with lateral hairs. D. Terminal portion of thallus. E, F. TS of vegetative thalli. G. TS of fertile thallus with unilocular sporangia. H. Portion of fertile thallus with unilocular sporangia. I. Portion of fertile thallus with plurilocular sporangia. Bar = 50 µm.

(Pedersen 1981c, 1984). Their morphology has been described as streblonematoid and similarities have been drawn between the *Litosiphon* microthalli and species of *Streblonema*, such as *S. thuretii* Sauvageau, *S. volubile* (P. Crouan & H. Crouan) Thuret in Le Jolis, *S. danicum* Kylin and *Microspongium oligosporum* (Strömfelt).

Ardré (1970), p. 276; Bartsch & Kuhlenkamp (2000), p. 167; Batters (1892b), p. 19; Brodie *et al.* (2016), p. 1020; Cormaci *et al.* (2012), pp. 170, 172, pl. 44(5–6); Cotton (1912), p. 94; Dangeard (1965), p. 47, pls 11–13, (1969), pp. 72–4, fig. 2(B); Fletcher (1987), pp. 170–3, figs 36, 37, pl. 4(b); Fletcher in Hardy & Guiry (2003), p. 270; Gallardo García *et al.* (2016), p. 35; Guiry (2012), p. 144; Hamel (1931–39), pp. 217–19, fig. 43(4–6); Harvey (1846–51), pl. 270, (1849), p. 43, pl. 8(D); Jaasund (1957), pp. 218–22, fig. 7(c–g), (1965), pp. 83–5; Johnson (1892), pp. 5, 6; Kylin (1933), pp. 25–33, figs 5–7, (1947), p. 74, fig. 60(A); Levring (1937), pp. 66–8; Mathieson & Dawes (2017), p. 165, pl. XLIX(1–5); Morton (1994), p. 41; Newton (1931), p. 180, fig. 113; Nygren (1975a), pp. 131–41, figs 16–21, (1975b), pp. 143–7, fig. 4; Pedersen (1978a), pp. 61–8, figs 17–19, (1981c), pp. 194–217, (1984), pp. 26–32, figs 14–19; Rosenvinge & Lund (1947), pp. 15–19, fig. 3; Rueness (1977), p. 162, fig. 95, (2001), p. 15; Sauvageau (1929), pp. 350–62, (1933a), pp. 5–22, figs 1–4; Stegenga & Mol (1983), p. 99; Stegenga *et al.* (1997a), p. 19.

Mesogloia C. Agardh

MESOGLOIA C. Agardh (1817), p. 126.

Type species: *M. vermiculata* (Smith) S.F. Gray (1821), p. 320.

Thalli macroscopic, erect, pale olive to yellowish-brown, cylindrical, worm-like, solid, slightly cartilaginous, strongly mucilaginous and easily squashed under light pressure, much branched throughout, polymorphic, irregular in diameter, arising from a small, discoid holdfast; in structure, thallus multiaxial, haplostichous with a central medulla of erect, loosely bound, parallel, longitudinal rows of filaments of large, markedly elongated, colourless cells which branch out laterally and radially to form a subcortex of dichotomously to trichotomously branched, loosely compacted filaments of irregularly shaped and swollen cells and which give rise peripherally to an outer cortex comprising assimilatory filaments, unilocular sporangia, plurilocular sporangia and hairs; growth subapical, by an intercalary meristematic zone on the central, primary, axial filament giving rise terminally to a long, free, multicellular filament; assimilatory filaments curved, multicellular, simple or branched, straight or elongate-clavate, largely of moniliform cells above; hairs common, with basal meristem, but without basal sheath; plastids sparsely distributed in medullary and subcortical cells, numerous in the assimilators, discoid or elongate-discoid, each with a pyrenoid.

Unilocular and plurilocular sporangia borne at the base of the assimilators; unilocular sporangia sessile, occasionally on 1–3-celled stalks, ovoid to spheroid; plurilocular sporangia borne terminally on short, 1–5-celled pedicels, pluriseriate.

A number of authors have speculated about the relationship between *Mesogloia* and *Liebmannia*, as anatomically the genera are very similar, the only distinction being the presence of plurilocular sporangia in the latter genus. As early as 1842, Meneghini transferred the type species *Liebmannia leveillei* to *Mesogloia*, which found later support by Brodie *et al.* (2016), Cormaci *et al.* (2012), Kuckuck (1929), and Parke (1933). This was not universally accepted, with authors such as Dizerbo & Herpe (2007), Gallardo *et al.* (2016), Guiry (2012), Hamel (1931–39), Kylin (1940), Parke & Dixon (1976), and Siberfeld *et al.* (2014) keeping the two genera separate. However, of particular significance, is the recent molecular phylogenetic study by Kawai *et al.* (2019) using mitochondrial *cox*1 and *cox*3 and chloroplast *atp*B, *psa*A, *psb*A and *rbc*L gene sequences which found *Liebmannia leveillei* to be genetically indistinguishable from *Mesogloia vermiculata*. On the basis of this study, *Liebmannia leveillei* is, herein, placed within the synonymy of *Mesogloia vermiculata*.

Two species in Britain and Ireland.

KEY TO SPECIES

Thallus smooth in appearance; assimilators clavate, comprising cells increasing in diameter towards the apex, often with terminal, swollen, globose cells . . . . *M. vermiculata*
Thallus woolly and velutinous in appearance; assimilators cylindrical, comprising cells of equal diameter throughout and lacking terminal, swollen, globose cells . . . *M. lanosa*

Mesogloia lanosa P. Crouan & H. Crouan (1867), p. 166.   Figs 127(A), 128

Thalli erect, to 25 cm long, to 3–8 mm wide, pale yellow-brown, cylindrical, worm-like, woolly and velutinous in appearance, solid, more or less linear, strongly mucilaginous and slimy and easily separable on a slide under the weight of a coverslip, arising from a small discoid holdfast; main axis sometimes hardly distinguishable and much branched throughout with branching irregular, opposite or bifurcate, to 2(–4) orders, incurved or widely divergent, with main axes and branches often irregular in diameter, cylindrical or slightly flattened in parts, constricted at base, and bearing short, simple, tertiary branchlets; thallus multiaxial, haplostichous, with a central medulla, derived from radial branching, followed by upward elongation, of the initial central medullary filament, of erect, parallel rows of large, dichotomously branched, loosely bound and easily separable filaments of elongate-cylindrical to elongate barrel-shaped, colourless cells, which branch out laterally and radially to form a subcortex of dichotomously branched, loosely compacted filaments of irregularly shaped, colourless cells which gives rise in the outer peripheral regions to a cortex comprising assimilatory filaments, unilocular sporangia, and hairs; apex with distinct, primary, single, axial filament with intercalary, meristematic zone of short, dividing cells giving rise terminally to a long, multicellular, elongate-clavate filament; medulla and subcortex with additional secondary, branched, hyphae-like filaments running horizontally and vertically throughout thallus; assimilators straight to curved, cylindrical, multicellular, simple, to 250(–310) μm, 15(–18) cells in length, comprising subquadrate/quadrate

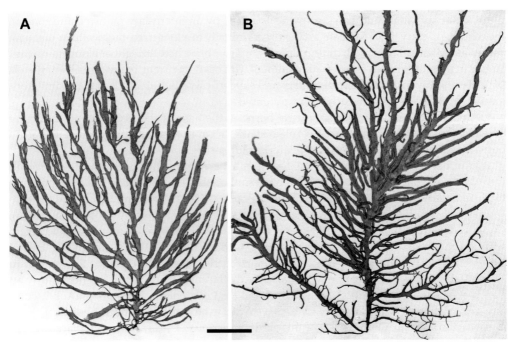

Fig. 127. *Mesogloia* spp.
Habit of thalli. A. *Mesogloia lanosa*. B. *Mesogloia vermiculata*. Bar = 35 mm. (Herbarium specimens, NHM.)

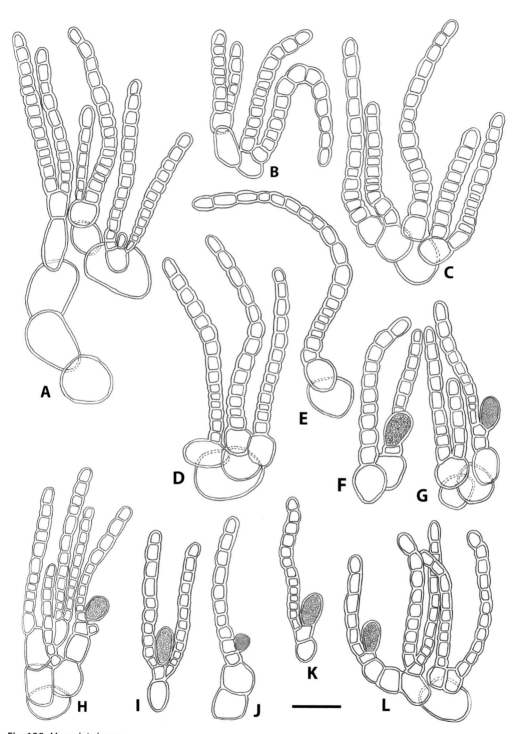

Fig. 128. *Mesogloia lanosa*

A–L. SP of thalli showing the outer, peripheral, assimilatory filaments arising from the large, subcortex cells, some with basally produced unilocular sporangia. Bar = 50 μm.

cells below, more cylindrical to moniliform cells above, approximately equal in length and width, (9–)14–26(–30) × (10–)13–18(–20) µm; plastids few in number in medullary and subcortical cells, numerous in the assimilators, discoid or elongate-discoid, each with a pyrenoid; growth apical with intercalary meristematic zone; hairs common with basal meristem but without basal sheath, comprising cells 1–4 diameters long, 6–70 × approx. 20 µm.

Unilocular sporangia common, distributed all over the thallus, borne at the base of the assimilators, sessile, ovoid to spherical, (33–)40–58(–63) × (19–)21–35(–38) µm; plurilocular sporangia unknown.

Epilithic on rock, also reported epiphytic on *Zostera* spp., lower eulittoral pools or sublittoral to 10 m.

Rare and only recorded for a few localities around Britain (Channel Islands, Dorset, Isle of Man, Argyllshire, Shetland) and for Mayo in Ireland.

A summer annual, May to August.

Brodie *et al.* (2016), p. 1020; Cormaci & Furnari (1988), p. 220, figs 15, 16; Cormaci *et al.* (2012), pp. 173, 174, pl. 45(1–3); Cotton (1912), pp. 96, 123; P. Crouan & H. Crouan (1867), pl. 26(166); Dizerbo & Herpe (2007), p. 90; Fletcher in Hardy & Guiry (2003), p. 270; Guiry (2012), p. 144; Hamel (1931–39), p. 170; Morton (1994), p. 8; Newton (1931), p. 149; Ribera *et al.* (1992), p. 114; Rosenvinge & Lund (1943), pp. 19, 20, fig. 6; Taskin & Öztürk (2007), p. 184, fig. 15.

Mesogloia vermiculata (Smith) S.F. Gray (1821), p. 320.　　Figs 127(B), 129, 130, 131, 132

Rivularia vermiculata J.E. Smith in Smith & Sowerby (1808), pl. 1818; *Mesogloia vermicularis* C. Agardh (1817), p. 126; *Trichocladia vermicularis* (C. Agardh) Harvey in Mackay (1836), p. 184; *Helminthocladia vermicularis* Harvey (1838), pl. 397; *Liebmannia leveillei* J. Agardh (1842), p. 35; *Mesogloia leveillei* (J. Agardh) Meneghini (1842–43), p. 283; *Liebmannia major* P.L. Crouan & H.M. Crouan (1867), p. 166.

Thalli erect, to 10–40(–76) cm long, to 4(–10 in squashed specimens) mm wide, olive to yellow brown, cylindrical above, becoming somewhat dorsiventrally compressed below, worm-like, solid, smooth in appearance, slimy, slightly to strongly mucilaginous, soft, flaccid, easily squashed under light pressure, much branched throughout, highly polymorphic, non-linear and irregular in diameter and unequally distended and constricted, arising from a small discoid holdfast; main axis usually prominent and distinguishable, uniform to irregular in diameter or tapering below, much branched, branching irregular, occasionally subdichotomous, to 2(–4) orders, primary branches long, particularly from the middle regions of the main axis, often irregular in diameter, cylindrical or slightly flattened in parts, somewhat tapering above and below, and bearing short, simple, somewhat tooth-like, branchlets; in structure, thallus multiaxial, haplostichous. pseudoparenchymatous with a central medulla, derived from radial branching, followed by upward elongation, of the initial central medullary filament, of erect, parallel rows of dichotomously branched, large, loosely bound and easily separable filaments of elongate-cylindrical to elongate barrel-shaped, colourless cells up to 300(–600) µm or more in length, 30–55(–150) µm in width, which branch out laterally and radially to form a subcortex of dichotomously to trichotomously branched, loosely compacted filaments of rounded to irregularly shaped, colourless cells which gives rise in the outer peripheral regions to a cortex comprising assimilatory filaments, unilocular sporangia, plurilocular sporangia and hairs; apex with distinct, primary, single, axial filament with an intercalary, meristematic zone of short, dividing cells giving rise terminally to a long, multicellular, elongate-clavate filament up to 400 µm (30+ cells) in length (only seen in young apices); medulla and subcortex with additional secondary, branched, hyphae-like filaments running horizontally and vertically throughout thallus, strengthening the thallus and contributing below as rhizoidal components of the basal disc, comprising cylindrical cells 40–80 × 8–11 µm; assimilatory filaments densely packed, determinate, clavate, curved, multicellular, simple, rarely branched at the base, (110–)140–275(–305) µm, (9–)13–18(–21) cells long, comprising quadrate to rectangular cells below, 8–19 × 7–12 µm, which become more cylindrical, moniliform to globose above, 13–36 × 10–30 µm; plastids few in number in medullary and subcortical

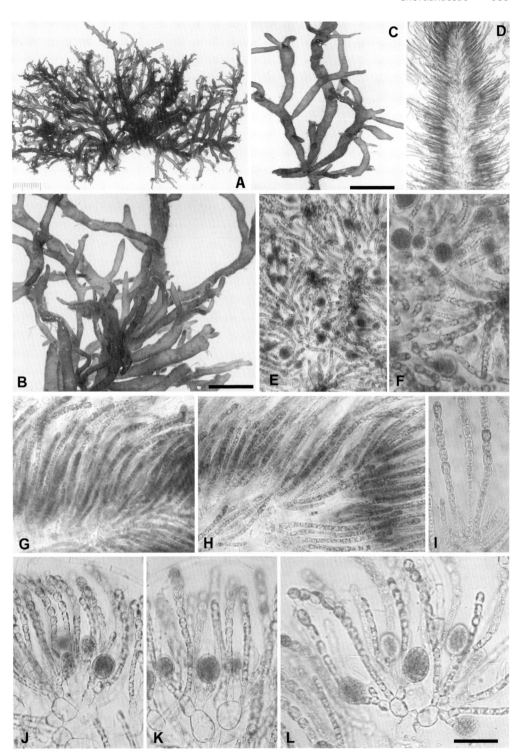

Fig. 129. *Mesogloia vermiculata*

A–C. Habit of thalli. D. SP of thallus apex, showing the outer, assimilatory filaments. E, F. Surface view of thallus showing unilocular sporangia among the loosely packed, assimilatory filaments. G–I. SP of thalli showing the outer assimilatory filaments. J–L. SP of thalli showing unilocular sporangia at the base of the assimilatory filaments. Bar in B = 12 mm, bar in C – 20 mm, bar in L = 320 µm (D), = 180 µm (F), = 90 µm (F), = 85 µm (G, H), = 72 µm (I, J, K, L).

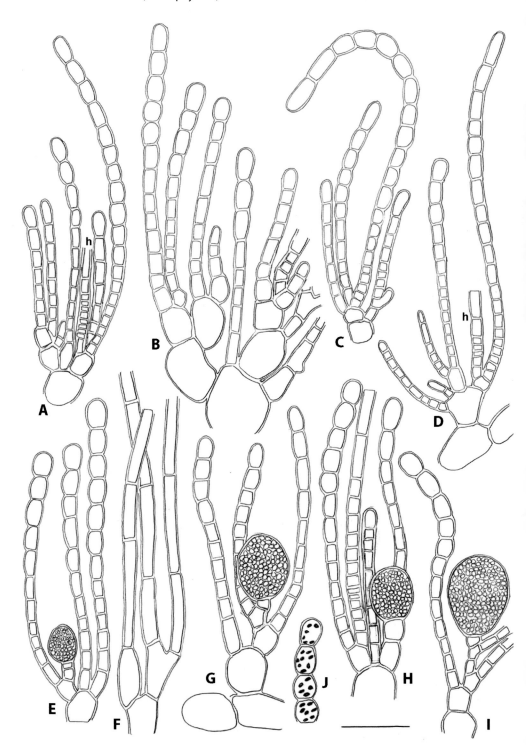

Fig. 130. *Mesogloia vermiculata*
A–I. SP of thalli showing the outer, peripheral, assimilatory filaments, basal regions of hairs (h) and basally positioned unilocular sporangia. J. Portion of an assimilatory filament showing cells with discoid plastids. Bar = 50 μm.

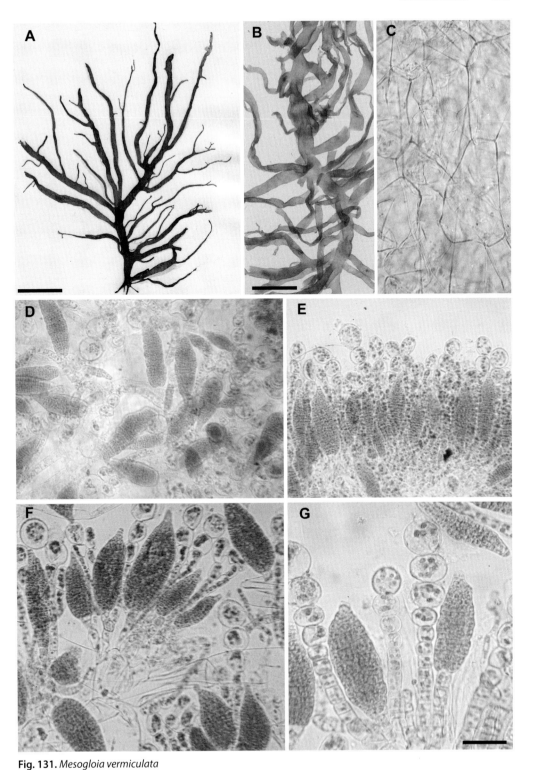

Fig. 131. *Mesogloia vermiculata*

A, B. Habit of thalli. C. SP of thallus showing central, branched medullary filaments. D–G. SP of thalli showing pedicellate, pluriseriate, plurilocular sporangia associated with the assimilators. Bar in A = 36 mm, bar in B = 20 mm, bar in G = 90 μm (C–F), = 45 μm (G). (A, herbarium specimen, NHM.)

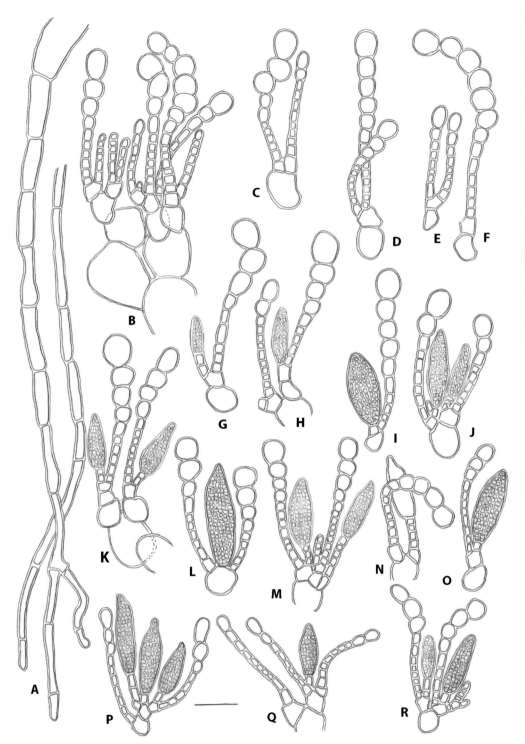

Fig. 132. *Mesogloia vermiculata*
A. Downward-growing hyphae-like filaments. B–R. SP of thalli showing the assimilatory filaments and pluri-seriate, plurilocular sporangia. Bar = 50 μm.

cells, numerous in the assimilators, discoid or elongate-discoid, each with a pyrenoid; primary growth subapical by the intercalary meristematic zone at the base of the main axial filament; hairs common in younger regions, more rarely seen in mature regions, with basal meristem but without basal sheath, comprising cells 4–9 diameters long, 38–90 × 8–11 μm.

Unilocular and plurilocular sporangia borne on the same or separate thalli: unilocular sporangia abundant on older thalli, regularly distributed throughout the whole thallus, usually borne singly as an outgrowth of the basal cell of the assimilators, sessile, more usually on a 1–2(–3) celled pedicel, ovoid to spheroid, 50–75 × 40–50 μm; plurilocular sporangia more rarely observed, developed in place of an assimilator, on pedicels of 1–5 cells, pluriseriate, siliquose, lanceolate and ectocarpoid, simple, rarely bifid, (50–)68–110(–133) × (14–)19–42(–47) μm.

Epilithic on rock and stones, sometimes epiphytic on various algae, mid to lower eulittoral pools or exposed, shallow sublittoral, often associated with sand, in fairly sheltered regions.

Common, sometimes locally abundant, reported from all around Britain and Ireland.

A summer annual, first appearing at the end of May, declining early September.

Mesogloia vermiculata has been investigated in culture by a number of authors, including Knight & Parke (1931), Kylin (1933) and Parke (1933). The evidence points towards the occurrence of a heteromorphic life history, with the unilocular sporangia-bearing macrothalli alternating with microscopic, filamentous, plurilocular sporangia-bearing microthalli.

Ardré (1970), p. 256; Bartsch & Kuhlenkamp (2000), pp. 167, 179; Batters (1895a), p. 275; Brodie *et al.* (2016), p. 1020; Bunker *et al.* (2017), p. 277; Coppejans (1983b), pl. 32; Cormaci *et al.* (2012), pp. 174, 175, pls 45(4–6), 46(1–2); Cotton (1912), p. 96; Dizerbo & Herpe (2007), p. 90, pl. 29(3,4); Feldmann (1937), p. 287; Fletcher in Hardy & Guiry (2003), pp. 269, 271; Gallardo *et al.* (2016), p. 35; Gayral (1966), pp. 278, 279, pl. XLVIII; Guiry (2012), pp. 143, 144; Hamel (1931–39), pp. 166–169, fig. 38(B, C); Harvey (1846–51), pl. XXXI; Hauck (1884), pp. 363, 365, figs 154, 155; Knight & Parke (1931), pp. 70, 113; Kuckuck (1929), pp. 54–56, figs 67–71; Kylin (1940), pp. 5, 7, figs 1–2(A, B), (1947), p. 54, fig. 47(A); Maggs *et al.* (1983), p. 262; Martín *et al.* (2002), pp. 87–98; Morton (1994), pp. 35, 36; Oltmanns (1922), figs 307, 311; Newton (1939), pp. 148, 149, fig. 92(A, B); Parke (1933), pp. 7, 22–29, text figure 13, pls I(1,5), III(10–16), IV(17–23), V(24, 26, 28, 29); Ribera *et al.* (1992), p. 114; Rosenvinge & Lund (1943), pp. 16–19, fig. 5; Rueness (1977), pp. 147, 148, fig. 80, (2001), p. 15; Rueness *et al.* (1977), p. 15; Sauvageau (1929), pp. 272–81, figs 3–4; Stegenga & Mol (1983), pp. 94, 95; Stegenga *et al.* (1997), p. 18; Taskin & Öztürk (2007), p. 184, fig. 16.

Microcoryne Strömfelt

MICROCORYNE Strömfelt (1888), p. 382.

Type species: *M. ocellata* Strömfelt (1888), p. 382.

Thalli epiphytic, erect, minute, globose, cylindrical or clavate, simple or shortly forked/branched, solid, soft and gelatinous, protruding from hosts; constructed of vertical and outward-radiating rows of closely packed gelatinous, dichotomously to trichotomously branched filaments, easily separable under pressure, comprising colourless cells, predominantly elongate-cylindrical, becoming more globose towards the periphery, terminating in paraphyses, plurilocular sporangia and/or hairs; paraphyses multicellular, simple, curved comprising cells with several discoid plastids and pyrenoids; hairs with basal meristem without obvious sheath.

Plurilocular sporangia in dense fascicles at base of paraphyses, cylindrical, simple, uniseriate or partly biseriate; unilocular sporangia unknown.

One species in Britain and Ireland.

Microcoryne ocellata Strömfelt (1888), p. 382.   Figs 133, 134

Thalli epiphytic, forming small, erect, solitary, sometimes clustered, extensions from host surface, globose, cylindrical or more commonly clavate, simple or once forked, more rarely with short, radiating branches, solid, soft and gelatinous, easily squashed, to 4(–5) mm long × 1.5 mm wide;

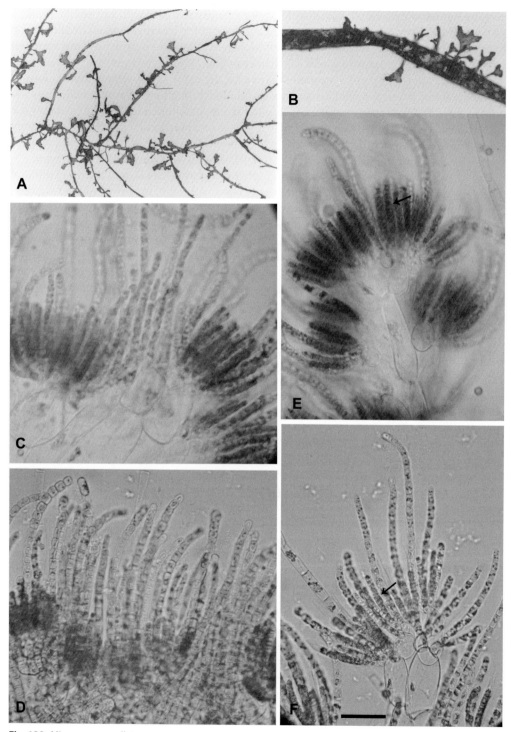

Fig. 133. *Microcoryne ocellata*
A, B. Habit of thalli on *Sauvageaugloia griffithsiana* and *Chorda filum*. C, E, F. SP of thallus showing the large, elongated cells of the central filaments and the outer rows of paraphyses-like assimilatory filaments and dense, dark clusters of plurilocular sporangia (arrow in E). D. SP of thallus showing outer projecting assimilatory filaments. Bar = 26 mm (A), = 5 mm (B), = 45 μm (C, E), = 50 μm (D), = 43 μm (F). (A, B, herbarium specimens, NHM.)

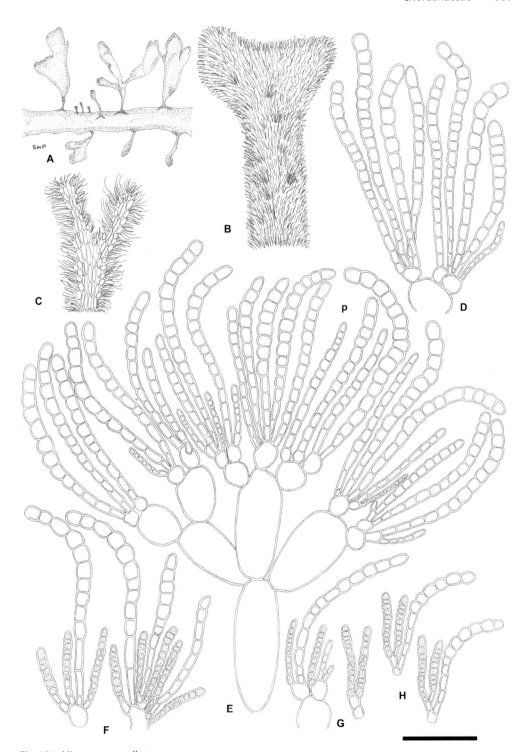

Fig. 134. *Microcoryne ocellata*

A. Habit of thalli, epiphytic on *Sauvageaugloia griffithsiana*. B. Habit of young thallus. C. SP of portion of young thallus showing elongated cells of central filaments. D, E. Terminal thallus portions showing paraphyses-like assimilatory filaments (p). F, G. Portions of fertile thalli showing plurilocular sporangia arising at base of the assimilatory filaments. Bar = 4 mm (A), = 0.5 mm (B, C), = 50 μm (D–G).

in squash preparation, consisting of upward and outward-radiating, dichotomously to trichotomously branched filaments, loosely packed together and easily separable under pressure, comprising elongate-cylindrical, colourless cells up to 350+ µm in length × approx. 18–25 µm in diameter, becoming shorter, more globose towards the periphery, 32–112 × 17–35 µm, terminating in paraphyses-like assimilatory filaments, plurilocular sporangia and/or hairs; paraphyses quite densely crowded, multicellular, simple, slightly curved, to 20(–26) cells, 130–(180–210)–250 µm long, pigmented, comprising cells rectangular below, becoming barrel-shaped and/or moniliform above, 7–21 × 6–13 µm, each with several discoid plastids and pyrenoids; hairs colourless, 30–180 × 10–14 µm with distinct basal meristem, enclosing sheath not observed.

Plurilocular sporangia common, in dense fascicles at the base of the paraphyses, arising directly from large colourless cells, either sessile or on branched 1–2-celled stalks, to 16 loculi long, 32–75 × ~6–11 µm, simple, cylindrical, uniseriate, sometimes partly biseriate, with straight, sometimes oblique cross walls, comprising quadrate, more commonly subquadrate or lenticular loculi, 3–5 × 6–8 µm; unilocular sporangia unknown.

Epiphytic on various algae including *Chorda filum*, *Dasya hutchinsiae*, *Palmaria palmata*, *Sauvageaugloia griffithsiana* and *Gongolaria nodicaulis* in lower eulittoral pools and shallow sublittoral.

Recorded only for Dorset and Anglesey in Britain and Mayo in Ireland; probably more widely distributed.

Summer annual, May to September.

Microcoryne ocellata was originally described as an epiphyte on *Chorda filum* in Norway by Strömfelt (1888). It was subsequently reported for Britain, at Weymouth, Dorset, by E.A.L. Batters (Batters 1892c) and later at Portland by E.M. Holmes (reported in Batters 1893b), epiphytic on various host algae. It was also reported for Clare Island, County Mayo, Ireland by Cotton (1912). Despite regular and widespread searches throughout Britain by the present author, it was only recently rediscovered, epiphytic on *Gongolaria nodicaulis* in mid-tidal pools, on Anglesey, North Wales. The present description and ecological notes are based on the examination of material deposited in the NHM, London, by E.A.L. Batters, E.M. Holmes and A.D. Cotton (Fig. 133A,B and Fig. 134) and on the material collected from Anglesey (Fig. 133C–F).

Batters (1892c), pp. 51, 52, (1893b), p. 51; Brodie *et al.* (2016), p. 1020; Cormaci *et al.* (2012), pp. 178, 179, pls 46(3–4), 47(1–4); Cotton (1912), p. 96; Fletcher (1987), p. 146, fig. 26, pl. 4(a); Fletcher in Hardy & Guiry (2003), p. 271; Gallardo García *et al.* (2016), p. 35; Guiry (2012), p. 144; Hamel (1931–39), pp. 143, 144; Kuckuck (1929), pp. 45, 46, fig. 53; Kylin (1907), pp. 81–3, figs 18, 19, (1947), pp. 52, 53, fig. 45; Morton (1994), p. 8; Newton (1931), pp. 139, 140, fig. 85; Rosenvinge & Lund (1943), pp. 11–16, figs 2–4; Rueness (1977), p. 143, fig. 71, (2001), p. 15; Stegenga & Mol (1983), pp. 92, 93; Stegenga *et al.* (1997a), p. 18; Strömfelt (1888), p. 382, pl. 3(2–3); Taskin *et al.* (2010), pp. 892–6.

Microspongium Reinke

MICROSPONGIUM Reinke (1888a), p. 20.

Type species: *M. gelatinosum* Reinke (1888a), p. 20 (pro. part. with plurilocular sporangia).

Thalli epiphytic, endophytic, more rarely epilithic; if epiphytic, thallus crustose, pulvinate or globular, sponge-like and gelatinous in texture, consisting of a monostromatic base of outward-spreading filaments, laterally united and discoid, giving rise to short, erect, densely crowded, branched filaments, easily separable under pressure; if epilithic, thallus forming a monostromatic base of irregularly spreading, fairly compacted filaments giving rise to short, erect filaments; if endophytic, thallus forming an irregularly spreading network of filaments within host tissue, giving rise to erect filaments which may just emerge from host surface; hairs frequent, with basal meristem, with or without, basal sheath; cells with 1–3 plate-like plastids.

Plurilocular sporangia terminal or lateral, with or without stalk, on erect filaments, just projecting through host surface if thallus endophytic, cylindrical to subcylindrical, simple, uniseriate, sometimes partly biseriate; unilocular sporangia unknown.

Eight species of *Microspongium* are currently recognized for Britain and Ireland, viz. *M. alariae*, *M. globosum*, *M. immersum*, *M. oligosporum*, *M. parasiticum*, *M. radians*, *M. stilophorae* and *M. zanardinii*. *Microspongium alariae* and *M. radians* represent new records for Britain and Ireland. In agreement with Pedersen (1984), the heterotrichous nature of the thallus shown in culture by *M. immersum* supports its transfer from the genus *Streblonema*. Similarly, and with the occurrence of uniseriate, plurilocular sporangia, it also seems appropriate to transfer the species *Streblonema oligosporum*, *S. parasiticum* and *S. zanardinii* into the genus *Microspongium* (see Pedersen 1984; Fletcher 1973; Peters 2003). *Microspongium radians* and *M. stilophorae* (including *M. tenuissimum*) have also been transferred from the genus *Streblonema*, and *M. alariae* from the genus *Gononema*, based on molecular studies by Peters (2003).

However, the considerable overlap in the descriptions of some of these species, and the over-reliance on the host species for separating them, casts some doubt on their taxonomic status and further culture and molecular studies are required to determine the extent of the species diversity of this genus.

Four species in Britain and Ireland, four species in Britain only.

KEY TO SPECIES

1. Thalli endophytic or lithophytic 2
    Thalli entirely epiphytic, forming globular or hemispherical cushions on hosts
    . . . .*M. globosum*

2. Thalli lithophytic, forming prostrate, closely adherent, branched, filamentous
    growths . . . .*M. oligosporum*
    Thalli endophytic 3

3. Thalli forming large, circular to confluent brown spots on *Grateloupia turuturu* . . .*M. radians*
    Thalli forming small brown flecks just visible to the unaided eye on other hosts 4

4. Thalli forming brown flecks on the sporophylls and senescent blades of *Alaria*
    esculenta . . . . *M. alariae*
    Thalli forming brown flecks on other hosts 5

5. Hairs with basal meristem, without obvious basal sheath 6
    Hairs with basal meristem, with basal sheath 7

6. Erect filament cells mainly 4–5 μm in diameter with a plate-like plastid,
    plurilocular sporangia to 62 μm in length . . . .*M. parasiticum*
    Erect filament cells mainly 7–10 μm in diameter with discoid plastids,
    plurilocular sporangia to 40 μm in length . . . .*M. stilophorae*

7. Thalli endophytic in *Chylocladia* . . . . *M. zanardinii*
    Thalli endophytic in other hosts 8

8. Erect filament cells 3–5 μm in diameter . . . .*M. immersum*
    Erect filament cells 3–9 μm (–13 μm in other descriptions) in diameter . . .*M. oligosporum*

Microspongium alariae (P.M. Pedersen) A.F. Peters (2003), p. 301.   Fig. 135

Phycocelis alariae Norum (1913), p. 150; *Myrionema alariae* (Norum) Printz (1926), p. 136; *Streblonema aecidioides* f. *alariae* Jaasund (1963), p. 4; *Entonema alariae* Jaasund (1965), p. 49; *Gononema alariae* Pedersen (1981a), p. 269.

Thalli epi-endophytic, forming dense, small circular to oblong, light to dark brown spots on host, just visible to the unaided eye; comprising either a network of endophytic filaments spreading throughout host tissue between host cells, re-emerging at the surface as small pustules, or forming spreading, epiphytic colonies on host surface; thallus forming a monostromatic surface layer either between the outer cortex and epidermis within a pustule,

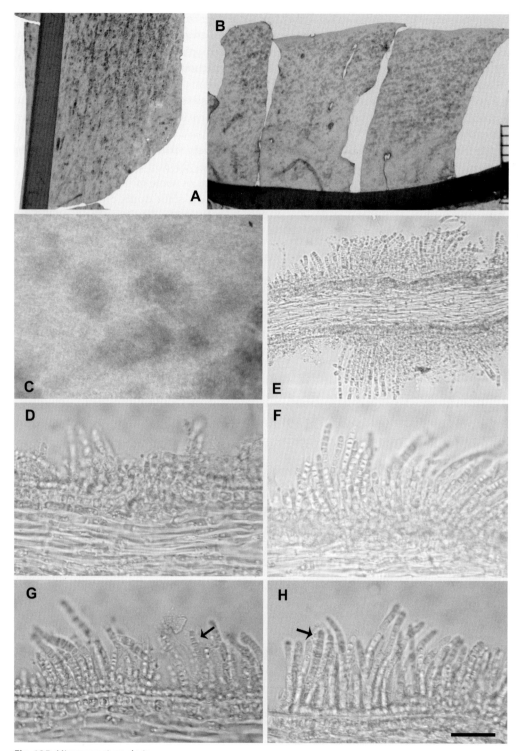

Fig. 135. *Microspongium alariae*

A–C. Habit of thalli on *Alaria* blades. D. VS of *Alaria* blade showing young emerging pustule. E–H. TS of *Alaria* blades showing closely packed erect filaments of thalli, many with terminal uniseriate plurilocular sporangia (arrows). Note epiphytic habit of thalli in G, H with erect filaments arising from a monostromatic basal layer. Scale bar in B applies to A and B. Bar in H = 300 μm (C), = 90 μm (E), = 45 μm (D–H).

or spreading on top of the host epidermis; surface layer cells rounded or slightly taller than wide in vertical sections, 3–5 × 5–8 µm, and giving rise to hairs, erect filaments and/or pluri-locular sporangia; erect filaments to 115 µm, closely packed, simple, very rarely branched, comprising quadrate to rectangular cells, 1–2(–2.5) diameters long, 8–17 × 4–6 µm, each with 1–2 lobed, plate-like plastids; hairs with basal meristem rarely formed.

Plurilocular sporangia terminal on the erect filaments, or shortly stalked arising from the basal layer, to 50 µm in length, uniseriate, cylindrical, simple, comprising up to 16 rows of quadrate, subquadrate to lenticular loculi, 2–4 × 3–5 µm, with straight to oblique cross walls; unilocular sporangia unknown.

Epi-endophytic on both sides of aged sporophylls and senescent blades of *Alaria esculenta*.

Originally recorded by the present author for Berwick-upon-Tweed, Northumberland (18 February 2009) and Anglesey (3 May 1985), and later for Aberdeenshire and Argyll in October 2014 (Murúa *et al.* 2018), this species represents a new record for Britain.

Records suggest no obvious seasonality.

In the material examined, the thalli rarely formed pustules and were mainly epiphytic on the epidermis of the host. It is possible the pustule forming and epiphytic thalli represent two different entities; however, they are grouped together in this treatise pending further studies.

Culture studies by Pedersen (1981a), using material collected from Greenland, revealed a 'direct' life history with swarmers from the plurilocular sporangia repeating the parental phase. Peters (2003), working on material collected on the coast of Maine, reported that the ITS1 sequence of this species was similar to that of *Microspongium* spp. and not to *Laminariocolax* spp. or to that of *Gononema pectinatum* (Skottsberg) Kuckuck & Skottsberg reported by Burkhardt & Peters (1998), which is the type species of *Gononema*. As proposed by Peters (2003), *Microspongium alariae* is therefore included in *Microspongium*.

Jaasund (1963), pp. 4, 5, fig. 2(C), (1965), p. 49; Mathieson & Dawes (2017), p. 155, pl. XLVI(9); Murúa *et al.* (2018), pp. 343–54; Norum (1913), p. 150, pl. II; Pedersen (1981a), pp. 263–70, figs 1–18, (2011), pp. 64, 65, fig. 66; Peters (2003), pp. 293–302; Printz (1926), p. 136, pl. II(1–10); Sears (1998), p. 163.

Microspongium globosum Reinke (1888), p. 20. Figs 136, 137, 138

Microspongium gelatinosum Reinke (1888a), p. 20 (pro. part. with plurilocular sporangia); *Ascocyclus globosus* Reinke (1889a), p. 20; *Myrionema globosum* (Reinke) Foslie (1894), p. 130; *Phycocelis globosa* (Reinke) De Toni (1895), p. 582; *Myrionema polycladum* Sauvageau (1897a), p. 233; *Ectocarpus pulvinatus* H. Gran (1897), p. 45; *Hecatonema globosum* (Reinke) Batters (1902), p. 41; *Hecatonema globosum* var. *nanum* Cotton (1907), p. 370; *Myrionema subglobosum* Kylin (1907), p. 37.

Thalli epiphytic, forming light to dark brown spots on hosts, usually solitary, globular or circular, 2–3(–4) mm in diameter, occasionally confluent, slightly pulvinate, smooth or radially ridged; mid region obviously thickened, surface cells rounded, densely packed, 6–11 µm in diameter, marginal region thin, of loosely adjoined, frequently branched, outward-radiating filaments, comprising rectangular cells 1–2 diameters long, 10–16 × 8–10 µm, each with 1–3 plate-like multilobed plastids and associated pyrenoids; in squash preparation, consisting of a loosely attached, monostromatic base giving rise to erect filaments, hairs and/or plurilocular sporangia; erect filaments to 75(–150) µm, 4–9(–18) cells long, branched, linear or sometimes distinctly clavate with enlarged, terminal, more darkly staining cell, mucilaginous, comprising mainly rectangular cells, 1–2(–5) diameters long, 5–15(–20) × 5–7 µm (terminal cell to 22 × 12 µm) with 1–3 plate-like, lobed plastids and pyrenoids; hairs frequent, arising from basal cells or terminal on erect filaments, with basal meristem and sheath.

Plurilocular sporangia common, terminal or lateral on erect filaments, frequently formed within empty sporangial husks, uniseriate, subcylindrical, to 35(–50) µm, 10(–15) loculi long × 5–7 µm in diameter, with occasional oblique cross wall; unilocular sporangia unknown.

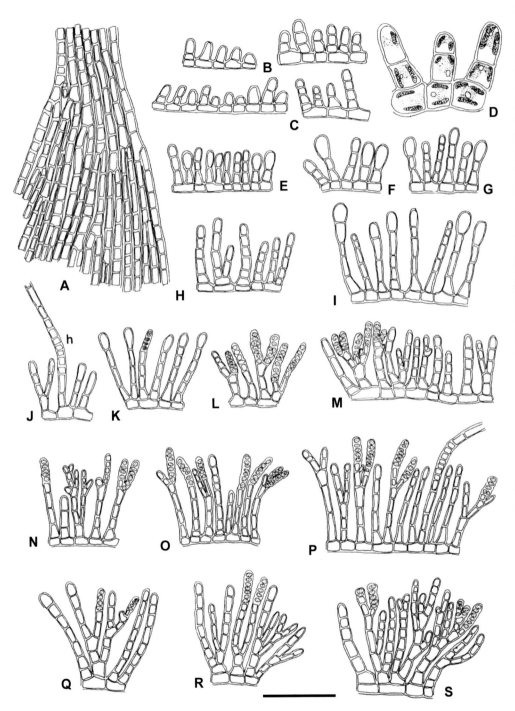

Fig. 136. *Microspongium globosum*

A. Surface view of peripheral thallus region showing branched outward-spreading, closely associated vegetative filaments. B–J. SP of vegetative thallus showing erect filaments arising from a monostromatic base. Note occasional hair (h) and enlarged terminal cells. K–S. SP of fertile thallus showing more extensively branched, erect filaments and both terminal and lateral plurilocular sporangia. Bar = 20 μm (D), = 50 μm (others).

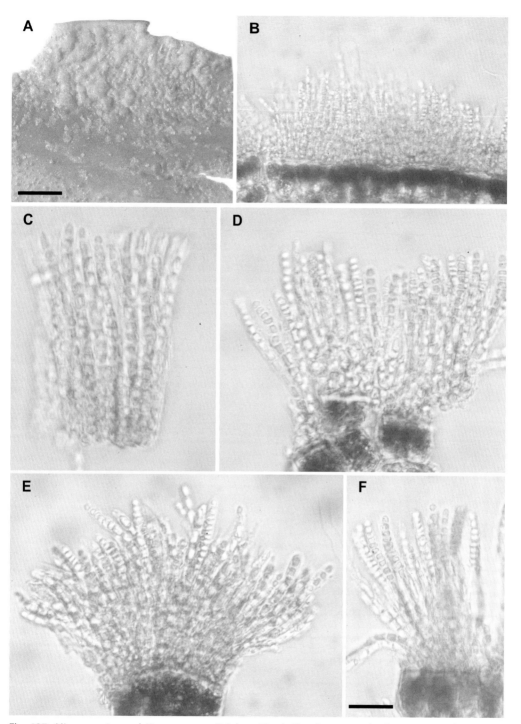

Fig. 137. *Microspongium gelatinosum sensu* Reinke with plurilocular sporangia. Dorset material, on *Fucus serratus*.

A. Pulvinate habit of thallus on *F. serratus*. B. TS of thallus on *F. serratus*. C. SP of vegetative thallus. D–F. SP of fertile thalli showing terminal plurilocular sporangia. Bar in A = 6 mm, bar in F = 50 µm (B), = 35 µm (C–F).

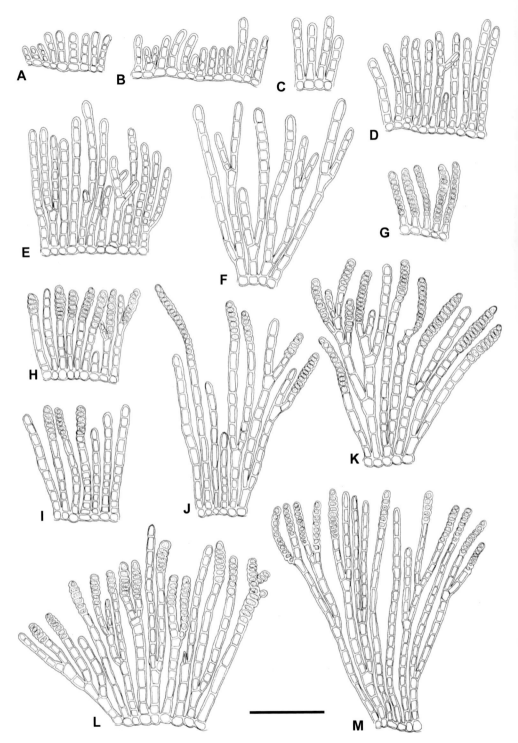

Fig. 138. *Microspongium gelatinosum sensu* Reinke with plurilocular sporangia. Dorset material, on *Fucus serratus.*

A–F. SP of vegetative thallus showing erect, branched filaments. G–M. SP of fertile thallus showing terminal and lateral plurilocular sporangia. (Form typical of thalli epiphytic on *Fucus* spp.) Bar = 50 μm.

Epiphytic on various hosts including *Fucus* spp., *Palmaria palmata*, *Chaetomorpha* spp., *Cladophora* spp. and *Polysiphonia* spp., eulittoral in pools and sublittoral to 5 m.

Recorded for scattered localities around Britain (East Lothian, Northumberland, Hampshire, Dorset, Devon, Cheshire, Argyll) and for Mayo in Ireland.

Summer annual, April to September.

Noteworthy here is the material observed on *Fucus* spp., which has been collected from two localities:

1. County Mayo, Ireland by A.D. Cotton in 1909; Natural History Museum slide nos 2642–4; on *F. vesiculosus*.

2. Peveril Point, Swanage, Dorset, by the present author, 21 April 1985; on *F. serratus*.

The host algae, and the squash preparations of the Dorset material (see Fig. 137) showing the length of the erect filaments and plurilocular sporangia, reveal close similarities to Reinke's *M. gelatinosum* described with plurilocular sporangia; however, the latter was described by Reinke with a distromatic base. This is not in agreement with the present study and with the authors Kylin (1947), Rautenberg (1960) and Pankow (1971), who described *M. gelatinosum* with a monostromatic base and it is possible that Reinke was mistaken on this point. There would, therefore, be no real criteria left to distinguish the plurilocular sporangia-bearing material of *M. globosum* and *M. gelatinosum* and, pending further studies, they are combined in the present treatise.

Bartsch & Kuhlenkamp (2000), p. 167; Batters (1892b), p. 21, (1900), p. 371, (1902), p. 41; Børgesen (1902), pp. 419–21, figs 76, 77; Brodie *et al.* (2016), p. 1020; Cormaci *et al.* (2012), p. 180, pl. 47(5–6); Cotton (1907), pp. 369, 370, (1912), p. 95; Dizerbo & Herpe (2007), p. 78; Fletcher (1987), pp. 103, 104, figs 8, 9; Fletcher in Hardy & Guiry (2003), p. 272; Gallardo García *et al.* (2016), p. 35; Guiry (2012), p. 145; Hamel (1931–39), pp. 92, 93, 96; Jaasund (1951), pp. 134, 135, fig. 4(a–e), (1965), pp. 60, 61; Jónsson (1903), pp. 146–8, figs 5–7; Kornmann & Sahling (1983), p. 50, figs 30, 31; Kristiansen (1960), pp. 251–4; Kristiansen & Pedersen (1979), p. 55; Kylin (1907), pp. 37, 38, fig. 8, (1947), pp. 39, 41–2, figs 31, 35(A–B), 36(A–C); Levring (1937), pp. 50, 51, (1940), pp. 39, 40, fig. 7(A–C); Lund (1959), pp. 111–13, fig. 23; Newton (1931), pp. 151, 156; Pankow (1971), pp. 170–2, figs 204, 205; Parke & Dixon (1976), p. 559; Printz (1926), p. 135; Rautenberg (1960), p. 135, figs 7, 8; Reinke (1888), pp. 16, 20, (1889a), pp. 11, 12, 20, pls 7(1–10, 20), 17(1–7), (1889b), p. 46; Rosenvinge (1898), pp. 86–9, figs 19, 20; Rueness (1977), pp. 134, 135, fig. 60, (2001), p. 15; Sauvageau (1897a), pp. 233–7, fig. 13A–K; South & Hooper (1980), p. 134; Stegenga *et al.* (1997a), p. 19; Sundene (1953), p. 161; Taskin *et al.* (2006), pp. 135–42, figs 1–4, (2008), p. 60; Taylor (1937), pp. 152, 53, 154–5, (1957), pp. 128, 131.

Microspongium immersum (Levring) P.M. Pedersen (1984), p. 36.   Fig. 139

Streblonema immersum Levring (1937), p. 39; *Entonema immersum* (Levring) Waern (1950), p. 33.

Thalli endophytic, forming small localized patches of branched, more or less horizontally spreading filaments just within the host surface with downward-penetrating rhizoidal filaments and emerging filaments, hairs and/or plurilocular sporangia; endophytic filaments irregularly branched, comprising quadrate to rectangular, sometimes globose cells 5–16 × 4–8 µm; erect filaments free, sometimes clustered together, simple, occasionally dichotomously branched, to 95 µm in height, comprising quadrate to rectangular cells 1–3 diameters long, 4–10(–15) × 4–5 µm, each with a single plastid; hairs approximately 6 µm in diameter, with basal meristem and (according to Levring 1937 and Pedersen 1980) with sheath.

Plurilocular sporangia formed by transformation of the erect filament cells, sessile or shortly stalked, single or clustered, to 36 µm long, 5–8 µm wide, simple, sometimes branched, linear, cylindrical, uniseriate, to partly biseriate, with straight or oblique cross walls, and comprising rows of up to 13 quadrate to lenticular loculi.

Endophytic and first reported for Britain by Pedersen (1980) in cultures obtained from the red algal host *Harveyella mirabilis* (Reinsch) Reinke; since then it has more commonly been recorded from eulittoral pools in the red alga *Dumontia contorta*, a common host reported for this endophyte in the literature (Levring 1937; Jaasund 1965).

Only recorded for Britain on the Isle of Man, Fifeshire and the Outer Hebrides; probably more widespread and common than records suggest.

Probably annual, based on the seasonal distribution of the host species.

Described by Waern (1952, p. 129) as 'one of the smallest brown algae heard of', he referred to it as a small biotype of *Streblonema oligosporum*. Similarly, Fletcher (1983) suggested that *Streblonema immersum* may be a reduced form of *Streblonema oligosporum*.

Culture studies on the isolate by Pedersen (1984) revealed a 'direct' life history, with spores released from the plurilocular sporangia germinating directly into branched, filamentous microthalli which again bore plurilocular sporangia. The heterotrichous organization of the culture material with erect, subdichotomously branched filaments morphologically different from the prostrate filaments, suggested to Pedersen (1984, p. 36) that the species should be removed from the genus *Streblonema* and transferred to *Microspongium*; this has been accepted in the present treatise.

Brodie *et al.* (2016), p. 1020; Fletcher in Hardy & Guiry (2003), p. 273; Guiry (2012), p. 18; Jaasund (1965), p. 44; Levring (1937), p. 39, fig. 3(H–I); Mathieson & Dawes (2017), p. 168, pl. XLIX(13, 14); Pedersen (1980b), pp. 247, 248, figs 4, 5, (1984), pp. 35, 36, fig. 23; Rueness (1977), p. 122; Waern (1950), pp. 128, 129, fig. 57, (1952), pp. 128, 129.

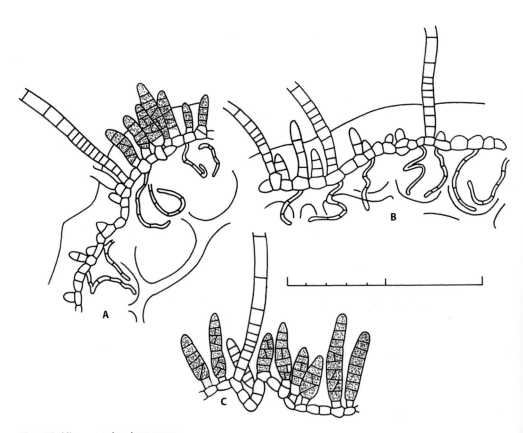

Fig. 139. *Microspongium immersum*

A–C. Portions of thalli spreading just below host surface showing protruding plurilocular sporangia and hairs. Bar = 100 μm.

Microspongium oligosporum (Strömfelt) R.L. Fletcher comb. nov. Fig. 140

Streblonema oligosporum Strömfelt (1884), p. 15; *Entonema oligosporum* (Strömfelt) Kylin (1947), p. 20; *Clathrodiscus oligosporus* (Strömfelt) Waern (1950), p. 33.

Thalli lithophytic or endophytic, forming small, initially circular spots, 0.5–1 mm in diameter, later confluent and spreading to several mm on substratum/host; comprising a basal system of outward-spreading filaments, giving rise to erect filaments, hairs and/or plurilocular sporangia; basal filaments monosiphonous, frequently and irregularly branched with branches often arising from the mid-cell wall region only, closely compacted although not regularly and laterally adjoined into a discoid structure, comprising cells 1–2(–2.5) diameters long, 4.2–13 × 3–7 μm, narrow and cylindrical distally, wider, more rounded and distended proximally, each with one, more rarely two, plate-like parietal plastids each with a pyrenoid; erect filaments common, usually short, to 9 cells (38 μm) in length, simple, linear, slightly attenuate, comprising cells 1–2 diameters long, 3.1–9.5 × 3.1–6.2 μm, each containing one plate-like plastid with pyrenoid occupying upper cell region; hairs infrequent, produced from basal cells, approximately 3 μm in diameter, with basal meristem and sheath.

Plurilocular sporangia usually abundant, arising from the basal cells, commonly on 1–2-celled stalks, occasionally sessile or terminal on erect filaments, to 9(–12) loculi (50 μm) long, slightly swollen, 3–6.2 μm wide, simple, uniseriate with dividing straight, occasionally oblique, cross walls; loculi usually shorter than wide, 0.5–1.0 diameters long, 2.0–3.5 × 3.5–6.2 μm; unilocular sporangia not observed.

Only recorded for Britain by Fletcher (1983a) on a copper-based, antifouled panel immersed from a test raft in Langstone Harbour, Hampshire.

Only known in Britain for Hampshire but likely to be much more common and widespread, particularly as an endophyte.

Appears to be a summer/autumn annual, recorded commonly between July and October, becoming increasingly rare in November/December.

The present record of this species refers to material collected from a toxic antifouling coating. Similarly, this species was recorded on toxic surfaces by Rautenberg (1960). Usually the species is recorded as an endophyte, although Taylor (1957) recorded it spreading over, as well as within, its host species. Culture studies, by Fletcher (1983b) revealed a 'direct' life history with spores from the plurilocular sporangia germinating directly, without sexual fusion, to resemble the parental thallus. However, Pedersen (1981c, 1984) considered *S. oligosporum* to be a microthallus phase in the life history of *Litosiphon laminariae* (Lyngbye) Harvey (as *L. pusillus* (Carmichael ex Hooker) Harvey) and included the former within the synonymy of the latter, a position accepted by Cormaci *et al.* (2012, p. 266) and more recently by Brodie *et al.* (2016) and Guiry in Guiry & Guiry (2023). However, in view of the 'direct' life history shown for material of this species by the present author, and pending further life history and molecular studies, it is retained here as an independent species. The transfer from the genus *Streblonema* to *Microspongium* was made on the basis of a heterotrichous thallus in culture (Fletcher 1983b) and the presence of uniseriate plurilocular sporangia. Pedersen (1984) had used the same argument to transfer *Streblonema immersum* to *Microspongium immersum*.

Fletcher (1983a), pp. 415–23, figs 1, 2; Jaasund (1951), p. 136, fig. 5(a), (1965), p. 45; Kristiansen (1978), p. 210; Kylin (1947), pp. 20, 21, fig. 16; Levring (1935b), pp. 19, 20, fig. 3(A–F), (1937), p. 38, (1940), pp. 27, 28, fig. 3(A–E); Pankow (1971), pp. 156, 157, fig. 177; Pedersen (1984), pp. 26–32, 66, figs 14–19; Rautenberg (1960), p. 134, figs 1–3; Ribera *et al.* (1992), p. 113; Rueness (1977), p. 122, fig. 47; Schneider & Searles (1991), pp. 127, 128, fig. 145; Strömfelt (1884), pp. 133–5, pl. 1(4, 5); Sundene (1953), p. 156; Svedelius (1901), p. 104; Taskin *et al.* (2008), p. 58; Taylor (1957), p. 114, pl. 11(6); Waern (1952), pp. 126–9, figs 56–7.

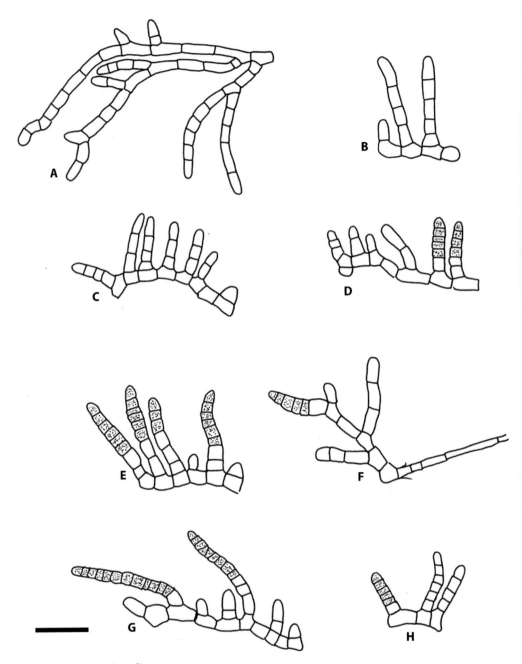

Fig. 140. *Microspongium oligosporum*

A–H. Portions of thalli showing monostromatic base and erect filaments, hairs and plurilocular sporangia (shaded). Bar = 25 µm.

Microspongium parasiticum (Sauvageau) Fletcher comb. nov.   Fig. 141

Ectocarpus parasiticus Sauvageau (1892), pp. 92, 125, 126; *Streblonema parasiticum* (Sauvageau) De Toni (1895), p. 575; *Entonema parasiticum* (Sauvageau) Hamel (1931–39), p. xxvi.

Thalli endophytic and discernible as rounded to irregular brown spots on host, forming a branched, filamentous system ramifying through the outer, cortical tissues of host and emergent as discrete pustules on the surface; endophytic filaments procumbent in parts, irregularly branched, sometimes widely spreading, enveloping and sometimes penetrating host cells, giving rise above to erect filaments which break through the host surface and emerge as short filaments, hairs and/or sporangia; endophytic cells variously sized and irregularly shaped, largely colourless or with a small plate-like plastid, 1–7 diameters long, (8–)13–28(–36) × (2–)4–7(–10) μm; emergent filaments short, usually simple, sometimes dichotomously branched, quite crowded, obtuse to acuminate, to 6(–9) cells, to 90 μm in length, comprising quadrate to rectangular cells, 1–5 diameters long, (5–)8–13(–17) × 3–4.5(–8) μm, each with a single, plate-like plastid; terminal cell of filament rounded, or giving rise to a hair; hairs infrequent, with basal meristem without obvious sheath.

Plurilocular sporangia borne terminally or laterally at the base of the emergent filaments, often densely packed, sessile or on 1–2-celled stalks, uniseriate or sometimes biseriate, often with oblique cross walls, subcylindrical, 40–62 × 3–6(–10) μm; unilocular sporangia unknown.

Endophytic in various algae, including *Palmaria palmata*, *Ceramium virgatum*, *Cystoclonium purpureum*, *Gracilaria gracilis* and *Dictyota dichotoma*.

Rarely recorded for Britain (Channel Islands, Dorset, Sussex, Northumberland, Bute, Isle of Man, Anglesey) and Ireland (Cork, Waterford) but probably more widespread and common.

Summer annual.

Batters (1892c), p. 51, (1902), p. 29; Brodie *et al.* (2016), p. 1021; Cormaci *et al.* (2012), p. 267, pl. 76(5); Dizerbo & Herpe (2007), p. 75; Fletcher in Hardy & Guiry (2003), p. 308; Guiry (2012), p. 152; Hamel (1931–39), pp. 56–8, fig. 18(B); Knight & Parke (1931), p. 60, pl. XIII(33, 37); Levring (1937), pp. 38, 39; Mathieson & Dawes (2017), p. 187, pl. LXII(1); Newton (1931), p. 115; Ribera *et al.* (1992), p. 113; Rueness (1977), p. 122; Sauvageau (1892), pp. 92–6, 125, 126, pl. III(20–3); Taskin & Öztürk (2007), p. 180, fig. 10; Taskin *et al.* (2008), p. 58; Taylor (1957), p. 116.

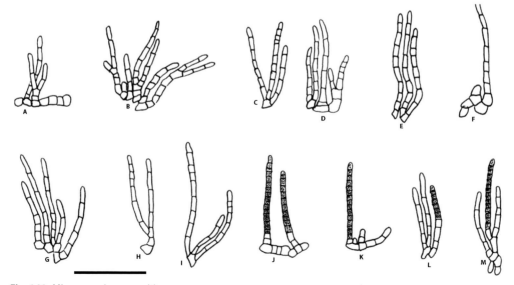

Fig. 141. *Microspongium parasiticum*
A–M. Portions of thalli showing monostromatic base giving rise to erect filaments and plurilocular sporangia (shaded). Bar = 50 μm.

Microspongium radians (Howe) A.F. Peters (2003), p. 301   Fig. 142

Streblonema radians M. Howe (1914), p. 47.

Thalli endo-epiphytic, forming brown patches on host surface, particularly in the basal regions of more mature blades, initially solitary and circular to 1(–2) cm in diameter, later often confluent to several cm in extent and causing localized green discolouration and loss of host tissue to form holes; comprising an extensive, endophytic network of widely spreading, irregularly and infrequently branched, pigmented filaments, which later break through the host surface, mainly in the central regions of the infection, as slightly protruding, short, erect filaments and/or plurilocular sporangia; endophytic cells rectangular to slightly contorted, 1–4(–8) diameters long, 11–49 × 4–12 µm, each with 1–2, centrally situated, plate-like plastids, with associated pyrenoid; emergent filaments commonly pseudodichotomously, dichoto-mously or trichotomously branched below, with branches closely packed and fastigiated, simple, rarely branched above, to 1–2 orders, to 50(–75) µm (–20 cells) in length, comprising cells 0.5–1(–2.5) diameters long, 3–9 × 3–5 µm, each with a single, plate-like plastid with pyrenoid; hairs not observed.

Plurilocular sporangia usually terminal on erect filaments, simple, cylindrical, uniseriate, slightly tortuose, comprising up to 10 quadrate, subquadrate to lenticular loculi, 13–25(–40) × 5–6 µm in length; unilocular sporangia unknown.

A common endo-epiphyte on the blades of the introduced red alga *Grateloupia turuturu* (see Farnham 1980 for a review of the introduction of this alga).

Only recorded for Britain (Kent and Hampshire) but probably more widely distributed in accordance with the geographical distribution of the host species. *Grateloupia turuturu* is a major contributor to the waterline fouling communities on the sides of floating structures, particularly marina pontoons, on the Hampshire coast and is frequently colonized by the epiphyte/endophyte. Indeed, most records of the species relate to host material collected from marinas; intertidal populations of the host are rarely epiphytized.

Appears to be mainly a winter species and has been recorded between October and May. However, more information is required on its seasonality.

This is the first report of *Microspongium radians* for Britain. It was first described and illustrated by Howe (1914) (as *Streblonema radians*) for Peru, endo-epiphytic in 'Grateloupia', most likely *G. doryphora*, which was one of two species recorded for this country. A further description was given by Dawson *et al.* (1964).

Peters (2003), working on material collected in South Africa, reported that the ITS1 sequence of this species was similar to that of *Microspongium* spp. The endo-epiphyte is, therefore, included in this genus in the present treatise. The host alga, *Grateloupia turuturu*, was first recorded as an alien species for Britain by Farnham in 1969, on Southsea beach, Hampshire (Farnham & Irvine 1973) and it seems likely that *Microspongium radians* was introduced with it.

Howe (1914), pp. 47, 48, pls 6(B), 11(8–13); Dawson *et al.* (1964), pp. 17, 18, pl. 13(E–F); Burkhardt & Peters (1998), pp. 682–91; Villalard-Bohnsack & Harlin (2001), pp. 372–80; Peters (2003), pp. 293–301.

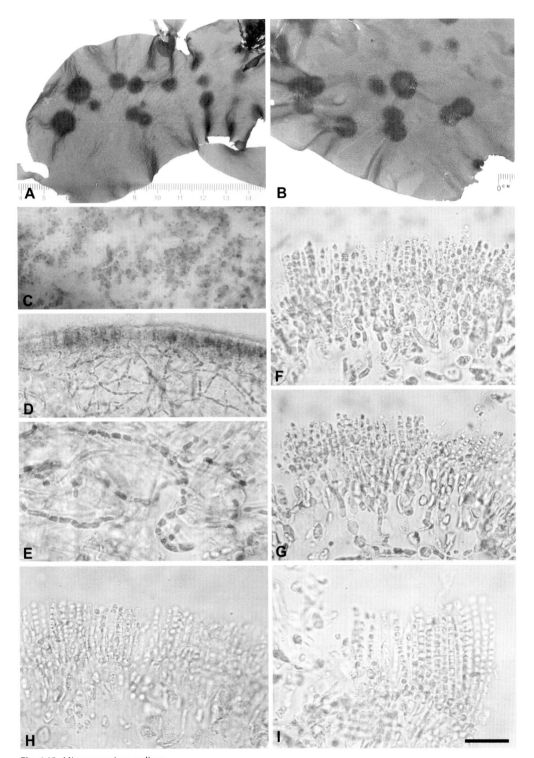

Fig. 142. *Microspongium radians*

A, B. Habit of thalli on host *Grateloupia turuturu*. C–E. Surface view of host showing endophytic filaments of the *Microspongium* F–I. TS of host showing emergent filaments and uniseriate plurilocular sporangia of the *Microspongium*. Bar in I = 90 μm (D), = 45 μm (C, E–I).

Microspongium stilophorae (P. Crouan & H. Crouan) Cormaci & G. Furnari (2012), p. 180.

Figs 143, 144

Ectocarpus stilophorae P. Crouan & H. Crouan (1867), p. 161; *Streblonema tenuissimum* Hauck (1884), p. 323; *Ectocarpus stilophorae* f. *caespitosus* Rosenvinge (1893), p. 892; *Streblonema stilophorae* var. *caespitosum* (Rosenvinge) De Toni (1895), p. 574; *Streblonema stilophorae* (P. Crouan & H. Crouan) De Toni (1897), p. 574; *Streblonema stilophorae* (P. Crouan & H. Crouan) Kylin (1908), p. 5; *Streblonema stilophorae* f. *caespitosum* (Rosenvinge) Kylin (1908), p. 5; *Ectocarpus stilophorae* var. *caespitosus* (Rosenvinge) Lily Newton (1931), p. 115; *Microspongium tenuissimum* (Hauck) Peters (2003), p. 301.

Thalli endophytic, not obviously visible to the unaided eye, forming a discrete and localized, basal network of irregularly branched filaments ramifying through the surface layers of the host tissue, mainly towards the base, but sometimes more vigorous and extending to the growing apices of the host thallus, giving rise terminally and laterally to short, emergent, erect branchlets bearing hairs and/or plurilocular sporangia; procumbent filament cells irregularly shaped, sometimes slightly bulbous or barrel-shaped, 1–4 diameters long, (5–)13–20(–28) × (3.5–)5–8(–11) μm; erect filaments to 60 μm in length, to 5(–8) cells, simple or dichotomously branched, uniseriate, comprising subquadrate to rectangular cells, 1–3 diameters long, (6–)8–15(–17) × (5–)7–10(–12) μm, each with discoid plastids; hairs common, usually terminal on the erect filaments, with basal meristem, without obvious basal sheath, hair cells 4–8 diameters long, 20–50 × 5–6 μm.

Plurilocular sporangia terminal or lateral on erect filaments, sessile or shortly stalked, to 40 μm, 8 loculi in length, single or terminally clustered and fasciculate, usually simple, sometimes branched, usually linear, cylindrical, uniseriate, sometimes partly biseriate, with straight or obliques cross walls, (14–)21–40(–70) × (5–)6–8(–11) μm, comprising rows of up to 22 quadrate to lenticular compartments; unilocular sporangia unknown.

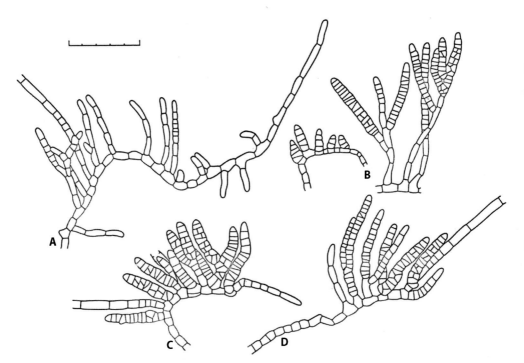

Fig. 143. *Microspongium stilophorae*

A–D. Portions of thalli showing branched endophytic filaments (A) giving rise to erect filaments, hairs and uniseriate to biseriate plurilocular sporangia. Bar = 50 μm.

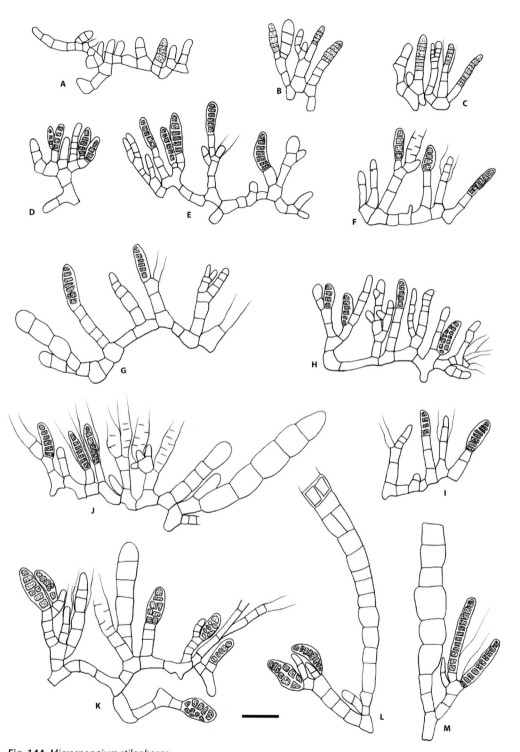

Fig. 144. *Microspongium stilophorae*
A–M. Portions of thalli giving rise to erect filaments with terminal plurilocular sporangia (shaded). Note that some thalli appear to be giving rise to young erect thalli of *Myriotrichia* (J–M). Bar = 25 μm.

Endophytic in various hosts in the eulittoral including *Nemalion elminthoides*, *Sauvageaugloia griffithsiana*, *Stilophora tenella* and *Stilopsis lejolisii*.

Rare, only reported for a few scattered localities in Britain (Cornwall, Devon, Dorset, Anglesey, the Isles of Scilly) and for Cork in Ireland; likely to be more widely distributed.

Summer/Autumn annual with fertile thalli observed in September.

Molecular studies by Siemer *et al.* (1998) and Peters (2003) on various isolates of this species (identified as *S. tenuissima*) revealed it to be closely related to *Microspongium globosum* Reinke and *M. gelatinosum* Reinke, and based on this evidence and the occurrence of uniseriate, plurilocular sporangia, Peters (2003) transferred it to the genus *Microspongium*; this has been accepted in the present treatise.

Some squash preparations of *Microspongium stilophorae* revealed erect thalli resembling young *Myriotrichia clavaeformis*, also often present as an epiphyte on the *Stilophora*. The *Myriotrichia*-producing microthalli differed only from *Microspongium stilophorae* in that the plurilocular sporangia occurred more commonly in dense fascicles and were often biseriate. Of note *Streblonema*-like thalli have also been described in the life history of *Myriotrichia clavaeformis* by Pedersen (1978b) and it is possible, therefore, that *Microspongium stilophorae* represents a microthallus phase in the life history of the *Myriotrichia*.

Batters (1894a), p. 91, (1902), p. 29; Brodie *et al.* (2016), p. 1020; Burkhardt & Peters (1998); Cormaci *et al.* (2012), p. 180, pl. 47(7, 8); P. Crouan & H. Crouan (1867), p. 161; Dizerbo & Herpe (2007), p. 75; Fletcher in Hardy & Guiry (2003), p. 309; Gallardo García *et al.* (2016), p. 35; Guiry (2012), p. 145; Hamel (1931–39), pp. XXII, 70, fig. 20(e); Hauck (1885), p. 323; Levring (1935), p. 19, fig. 3(L, M), (1937), p. 39; Mathieson & Dawes (2017), p. 168, pl. LXII(9, 10); Newton (1931), pp. 115, 130; Pankow (1971), pp. 177, 178, fig. 214; Peters (2003), pp. 293–301; Printz (1926), pp. 148, 149; Reinke (1889a), p. 21, pl. 19(1–4), (1889b), p. 42, fig. 3(I–M); Ribera *et al.* (1992), p. 113; Rosenvinge (1893), p. 892, fig. 26, (1894), fig. 26(A–C); Rosenvinge & Lund (1941), p. 62, fig. 33; Rueness (1977), p. 124; Siemer *et al.* (1998), pp. 1038–48.

Microspongium zanardinii (P. Crouan & H. Crouan) Fletcher comb. nov.   Fig. 145

Ectocarpus zanardinii P. Crouan & H. Crouan (1867), p. 161; *Streblonema zanardinii* (P. Crouan & H. Crouan) De Toni (1885), p. 572.

Thalli endophytic, forming small, usually discrete, circular spots on host, mainly towards the base, but sometimes extending to the growing apices; comprising a horizontally spreading, procumbent network of pseudodichotomously to irregularly branched filaments of cells growing over the surface of the outer layer of cortical cells and within the gelatinous sheath of the host, giving rise below to downward-penetrating, branched filaments spreading between the cortical cells, and above to emergent erect filaments, plurilocular sporangia and/or true hairs; procumbent filaments comprising rectangular to subglobose/globose/irregularly distorted cells, 1–3 diameters long, $4–12 \times 3–8$ μm; in surface view, erect filaments cells rounded, 5–9 μm in diameter; in section, filaments quite closely adpressed, simple, sometimes branched at the base, up to 12 cells, 48 μm in height, comprising rectangular to cuboidal cells, $5–9 \times 5–7$ μm, the upper cells extending beyond the gelatinous sheath of the host; procumbent and erect cells with 1–3 small plate-like/discoid plastids; hairs common, with basal meristem of 4–7 cells, with basal sheath, comprising colourless cells, 5–10 diameters long, $30–70 \times 5–19$ μm.

Plurilocular sporangia terminal on erect filaments, simple, sometimes bifurcate, to 22(–40) μm in height, uniseriate, cylindrical, comprising up to 9 cuboid/subcuboid loculi, $3–5 \times 3–5$ μm, with straight sometimes oblique cross walls; unilocular sporangia unknown.

Endophytic in *Chylocladia verticillata*, forming brown spots on the host.

Rare, reported for widely scattered localities around Britain (Cornwall, South Devon, Dorset, Hampshire, Channel Islands, Bute, Isle of Man) and Ireland (Cork, Dublin, Waterford).

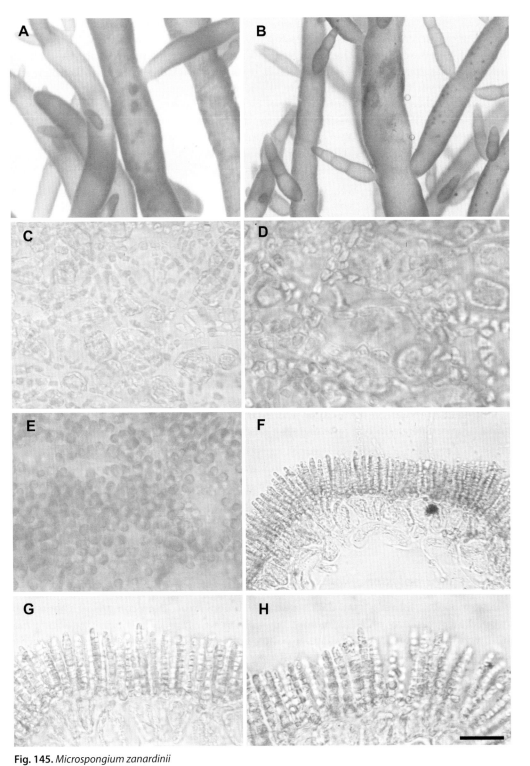

Fig. 145. *Microspongium zanardinii*

A, B. Habit of thalli on host *Chylocladia*. C, D Endophytic filaments. E. Surface view of thalli on host F–H. TS of host showing closely packed erect filaments and plurilocular sporangia of the *Microspongium*. Bar = 7 mm (A, B), = 90 μm (F), = 45 μm (C–E, G, H).

Spring and summer annual, with plurilocular sporangia reported in April and May.

The above description is based on endophytic material from *C. verticillata*, the only host observed with the endophyte by the present author, although Cotton (1906) also described it from *Champia parvula*. Only cylindrical, uniseriate sporangia have been observed by the present author. However, biseriate to multiseriate plurilocular sporangia, up to 20 μm and 5 loculi in diameter have been described for this species by some authors (Cotton 1906; Hamel 1931–39; Knight & Parke 1931). It is possible, therefore, that the two sporangial types represent two different endophytic taxa within the host *C. verticillata*. However, pending further studies, the material with uniseriate, plurilocular sporangia described above is, herein, attributed to *Streblonema zanardinii*.

Batters (1902), p. 29; Brodie *et al.* (2016), p. 1020; Cotton (1906), pp. 296, 297, pl. 12(6); P. Crouan & H. Crouan (1867), p. 161; Dizerbo & Herpe (2007), p. 75; Fletcher in Hardy & Guiry (2003), p. 310; Guiry (2012), p. 144; Hamel (1931–39), p. 54, fig. 17(D, E); Knight & Parke (1931), pl. 12, figs 28, 32; Levring (1937), p. 37, fig. 3(A–G); Newton (1931), p. 129.

Mikrosyphar Kuckuck

MIKROSYPHAR Kuckuck (1895b), p. 177.

Type species: *M. zosterae* Kuckuck (1895b), p. 177.

Thalli endophytic, forming small, brown spots on host algae, comprising a creeping network of branched filaments within hosts, either free or pseudodiscoid; cells with 1–2, plate-like plastids with pyrenoids; true hairs, if present, with basal meristem and emerging through host surface.

Plurilocular sporangia terminal or intercalary on filaments, positioned just below host surface, uniseriate; unilocular sporangia not observed on material from Britain and Ireland.

Two species in Britain and Ireland.

KEY TO SPECIES
Thalli irregularly shaped, without central radiating points; endophytic in
Polysiphonia spp. and other algae....................................*M. polysiphoniae*
Thalli more or less circular in shape with central radiating points; endophytic in
Porphyra spp. .... *M. porphyrae*

Mikrosyphar polysiphoniae Kuckuck (1897b), p. 355.   Fig. 146

Thalli endophytic, forming small, dark brown colourations on host alga; comprising a horizontal network of irregularly branched filaments, creeping haphazardly just below cuticle of the host, without obvious central radiating points; filaments either free, or laterally associated and pseudodiscoid in appearance, monosiphonous throughout, although occasionally with periclinal cell divisions present; cells 1–3(–3.5) diameters long, 5–18 × 5–8 μm and usually containing one parietal plastid with pyrenoid; growth both apical and diffuse.

Plurilocular sporangia intercalary on filaments, either solitary or confluent into sori, irregular in shape, comprising up to four or more loculi and 10 μm or more in diameter; unilocular sporangia not observed, reported by Taylor (1957, p. 116) to be formed on the ends of very short, 1–2-celled branchlets.

Endophytic in the cell walls of host species, usually *Polysiphonia stricta*, but also observed in *Vertebrata fucoides*, *Gaillona seposita*, *Ceramium* spp., *Halurus flosculosus*, *Hildenbrandia rubra*, *Cladophora rupestris* and *Chaetomorpha melagonium*; eulittoral in pools and exposed, shallow sublittoral.

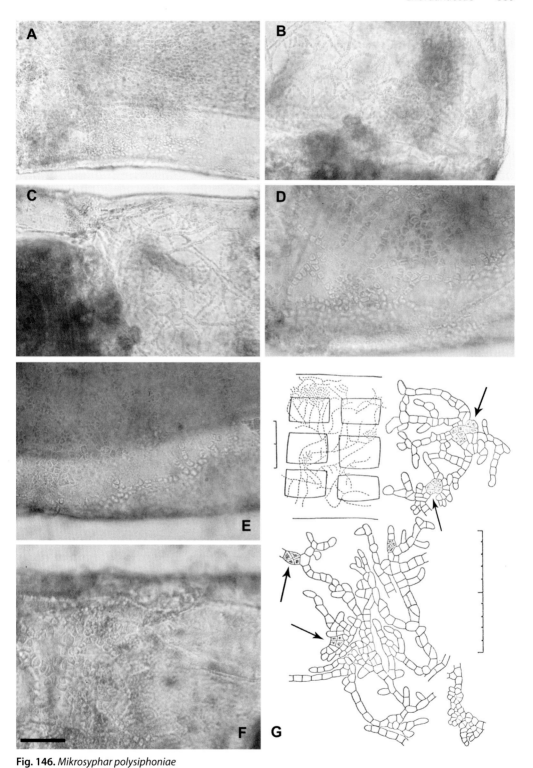

Fig. 146. *Mikrosyphar polysiphoniae*

A–F. Surface view of infected host showing the radiating filaments/pseudodiscs of the *Mikrosyphar*. G. Filaments/
pseudodiscs of the *Mikrosyphar* with intercalary plurilocular sporangia (shaded). Bar in F = 45 µm (A–F), the two
bars in G = 100 µm. Arrows in G indicate plurilocular sporangia.

Rare, only recorded for a few scattered localities around Britain (Isle of Man, Northumberland, Kent) and for Cork in Ireland; probably more common than records suggest.

Probably a spring/summer annual, but insufficient data to comment on its seasonal distribution.

Ardré (1970), p. 246; Bartsch & Kuhlenkamp (2000), pp. 167, 179; Batters (1897), p. 436; Brodie *et al.* (2016), p. 1020; Coppejans & Kling (1995), p. 204, pl. 70(A–C); Cormaci *et al.* (2012), p. 182, pl. 47(9); Dizerbo & Herpe (2007), p. 71; Fletcher in Hardy & Guiry (2003), p. 273; Gallardo García *et al.* (2016), p. 35; Guiry (2012), p. 145; Jaasund (1965), p. 50; Knight & Parke (1931), pp. 56, 105; Kornmann & Sahling (1977), p. 274, pl. 157(A, B); Kuckuck (1897a), p. 381, fig. 7(A, B), (1897b), pp. 349–58, pl. X(1–9); Kylin (1908), p. 4, (1947), p. 24, fig. 20(A); Levring (1935), p. 18; Lund (1959), p. 146; Mathieson & Dawes (2017), p. 109, pl. L(2, 3); Morton (1994), p. 8; Newton (1931), p. 125; Pankow (1971), p. 158, fig. 182; Pedersen (1976), p. 35, fig. 15, (1984), pp. 36–40, figs 24(A–F), (2011), p. 115; Rueness (1977), p. 118, fig. 41, (2001), p. 15; Stegenga & Mol (1983), p. 78; Stegenga *et al.* (1997a), p. 17; Sundene (1953), p. 156; Taskin & Öztürk (2007), p. 169, fig. 4; Waern (1952), pp. 121, 122, fig. 53.

Mikrosyphar porphyrae Kuckuck (1897a), p. 381.   Fig. 147

Thalli endophytic, forming small, brown spots on host, sometimes locally common and confluent; comprising a horizontal network of outward-radiating, free or sometimes laterally adjoined, irregularly branched, monosiphonous filaments, which spread between and around host cells, and extend to both sides of the host; growth apical or diffuse; cells 1–3(–4) diameters long, 6–16 × 4–9 μm, containing 1–2, parietal, plate-like plastids, with pyrenoids; hairs sometimes present, emergent from host and with a basal meristem.

Plurilocular sporangia formed at the branch tips, just below the cuticular host surface, single or grouped, uniseriate, cylindrical, consisting of 2–4 loculi, 10–16 × 5–6 μm; unilocular sporangia not reported for Britain and Ireland, elsewhere reported as terminal on short branchlets, and are spherical, approximately 6–7 μm in diameter.

Endophytic in old blades of *Porphyra* spp.; eulittoral, emersed, more conspicuous when *Porphyra* fronds are sun-bleached.

Rare, only recorded for a few scattered localities around Britain (Isles of Scilly, South Devon, Kent, Fife, East Lothian, Isle of Man) and for Wexford in Ireland; probably more common than records suggest.

Appears to be a summer annual, but insufficient data to comment on seasonal distribution.

This species may be a plethysmothallus of another alga. For example, it has been implicated in the life history of a myrionematoid species by Ravanko (1970), and its radial symmetry supports that view. It is possible that more than one genotype may be involved in the *M. porphyrae* phenotype.

Bartsch & Kuhlenkamp (2000), p. 167; Batters (1897), p. 436; Brodie *et al.* (2016), p. 1020; Coppejans & Kling (1995), p. 204; Cormaci *et al.* (2012), p. 182; Fletcher in Hardy & Guiry (2003), p. 274; Gallardo García *et al.* (2016), p. 36; Guiry (2012), p. 145; Hamel (1931–39), p. 73, fig. 21(d, e); Jaasund (1965), p. 49; Kuckuck (1897a), p. 381, fig. 6(A, B), (1897b), pp. 351–3, pl. IX(4–12); Kylin (1907), p. 47, (1947), p. 23, fig. 20(B–E); Levring (1935), p. 18, (1937), p. 36; Mathieson & Dawes (2017), p. 169, pl. L(4–5); Morton (1994), p. 8; Newton (1931), p. 124, fig. 72; Pedersen (1984), pp. 36–40, fig. 25(A–I); Rosenvinge & Lund (1941), p. 65; Rueness (1977), p. 118, (2001), p. 15; Stegenga & Mol (1983), p. 78; Stegenga *et al.* (1997a), p. 17; Sundene (1953), p. 146; Taylor (1957), p. 116, pl. 11(3–5).

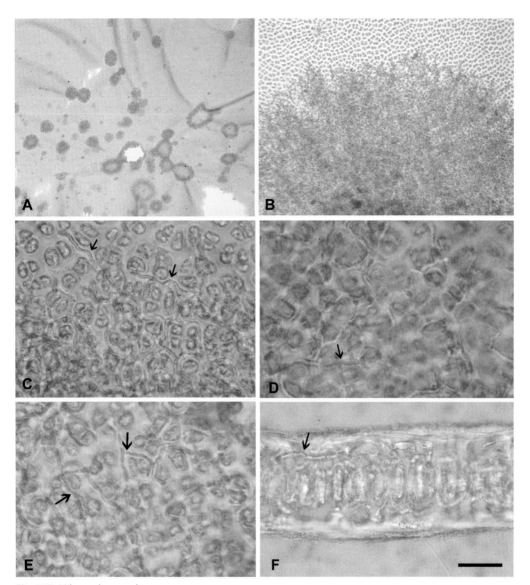

Fig. 147. *Mikrosyphar porphyrae*
A, B. Habit of thalli, forming dark spots of outward-radiating filaments on host. C–E. Surface view of host, showing radiating filaments (arrows) of the *Mikrosyphar* ramifying between the host cells. F. TS of host, showing part of a filament (arrow) of the endophytic *Mikrosyphar*. Bar in F = 6 mm (A), = 3 mm (B), = 23 μm (C–F).

Myriactula Kuntze

MYRIACTULA Kuntze (1891–98), p. 415.

Type species: *M. pulvinata* (Kützing) Kuntze (1891–98), p. 415 (= *M. rivulariae* (Suhr in Areschoug) J. Feldmann).

Myriactis Kützing (1843), p. 330; *Gonodia* Nieuwland (1917), p. 30.

Thalli epiphytic, sometimes partly endophytic, forming minute pustules, cushions or tufts on hosts, solitary or gregarious, discrete or sometimes confluent and spreading; comprising a basal system of branched filaments, with or without downward-extending endophytic/

rhizoidal filaments, giving rise to erect, dichotomously to trichotomously branched filaments terminating in paraphyses, hairs and/or sporangia; erect filaments remaining short and little differentiated or vertically extensive and becoming closely packed to form a soft, gelatinous, somewhat cartilaginous pseudoparenchyma of large, thin-walled, colourless cells, elongate-cylindrical below giving rise to downward-growing rhizoidal filaments, becoming more globose above; paraphyses loosely associated, gelatinous, multicellular, simple, linear, clavate or fusiform, comprising cells with several discoid plastids and pyrenoids; hairs with basal meristem without obvious sheath.

Unilocular and plurilocular sporangia borne at the base of the paraphyses, sessile or shortly stalked; unilocular sporangia pyriform; plurilocular sporangia clustered, cylindrical, simple, uniseriate.

Five species in Britain and Ireland, one species in Britain only.

KEY TO SPECIES

1. Thalli epiphytic on the thongs of the brown alga *Himanthalia elongata*; paraphyses commonly less than 14 cells in length . . . . *M. areschougii*
   Thalli not epiphytic on *Himanthalia elongata*; paraphyses commonly greater than 14 cells in length. 2

2. Thalli epiphytic on *Fucus* spp.; paraphyses up to 250 μm in length; cells of paraphyses less than 10 μm in diameter . . . . *M. clandestina*
   Thalli not epiphytic on *Fucus* spp.; paraphyses commonly greater than 250 μm in length; cells of paraphyses greater than 10 μm in diameter 3

3. Paraphyses distinctly fusiform; cells of paraphyses larger than 25 μm in diameter; thalli epiphytic on *Gongolaria* spp. and *Sargassum muticum*, more rarely on *Halidrys siliquosa* and other algae. . . . . *M. rivulariae*
   Paraphyses clavate, linear or slightly fusiform; cells of paraphyses less than 25 μm in diameter; thalli not reported on *Gongolaria* spp., *Sargassum muticum* and *Halidrys siliquosa* 4

4. Cells of paraphyses less than 16 μm in diameter; thalli epiphytic on *Scytosiphon lomentaria* only. . . . . *M. haydenii*
   Cells of paraphyses commonly greater than 16 μm in diameter; thalli recorded on several hosts, only occasionally on *Scytosiphon lomentaria* 5

5. Thalli epiphytic on *Dictyota dichotoma* only; paraphyses up to 285 μm in length . . . . *M. stellulata*
   Thalli not reported on *Dictyota dichotoma*; paraphyses up to 470 μm in length *M. chordae*

Myriactula areschougii (P. Crouan & H. Crouan) Hamel (1935), p. xxxii.   Figs 148, 149

Elachista areschougii P. Crouan & H. Crouan (1867), p. 160; *Myriactis areschougii* (P. Crouan & H. Crouan) Batters (1892a), p. 173; *Gonodia areschougii* (P. Crouan & H. Crouan) Hamel (1931–39), p. 132.

Thalli epiphytic, forming small, slightly raised pustules on host, circular or occasionally oblong, solitary or rarely grouped, to 2 mm in diameter and just visible to the unaided eye; in surface view, thalli turf-like, similar to a carpet pile, slightly projecting above, and surrounded by raised rim of host material; comprising a soft, gelatinous, somewhat cartilaginous cushion occupying host conceptacle, consisting of upward-growing, closely packed, dichotomously branched filaments of large, thin-walled, colourless cells, elongate-cylindrical below and occasionally giving rise to branched, multicellular rhizoidal filaments ramifying down into host tissue, gradually becoming smaller, more globose above and terminating in paraphyses, unilocular sporangia and/or hairs; paraphyses erect, simple, multicellular, slightly elongate-clavate, curved, usually

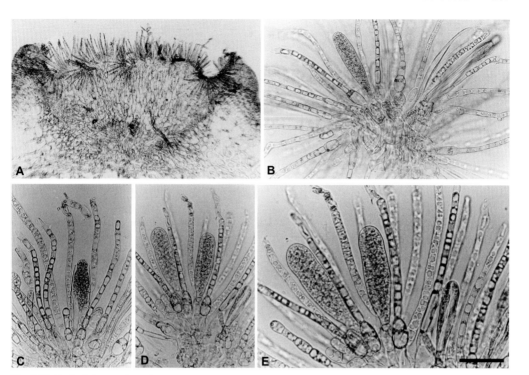

Fig. 148. *Myriactula areschougii*
A. VS of thallus embedded within host. B–E. SP of thalli showing terminal paraphyses, hairs and unilocular sporangia. Bar in E = 470 μm (A), = 90 μm (B–E).

8–14 cells, 180–250 μm long, comprising predominantly elongate-rectangular, thin-walled cells throughout, 2–5 diameters long, 13–40 × 5–10 μm becoming less elongate towards the apex, thick walled, rectangular, barrel-shaped or somewhat irregularly swollen, 1–2 diameters long, 10–21 × 7–13 μm often with terminal cells ruptured; all cells with numerous discoid plastids, with pyrenoids; hairs infrequent, with basal meristem and sheath, 8–10 μm in diameter.

Unilocular sporangia borne at base of paraphyses, elongate-pyriform, 65–105 × 18–28 μm; plurilocular sporangia unknown.

Epiphytic, occupying conceptacles of *Himanthalia elongata*, in lower eulittoral pools.

Only recorded for a few widely scattered localities around Britain (Devon, Bute, East Lothian, Northumberland, Shetland) and Ireland (Antrim, Mayo, Galway).

Thalli annual, June to October.

Batters (1885), p. 537, (1892a), p. 173; Brodie *et al.* (2016), p. 1020; Cotton (1912), p. 95; P. Crouan & H. Crouan (1867), p. 160, pl. 24(157-3, 4); Dizerbo & Herpe (2007), p. 85; Fletcher (1987), p. 150, fig. 27; Fletcher in Hardy & Guiry (2003), p. 274; Gallardo García *et al.* (2016), p. 36; Guiry (2012), p. 145; Hamel (1931–39), pp. 132–4, fig. 30(B, C); Morton (1994), p. 34; Newton (1931), p. 142; Sauvageau (1892), pp. 36–41: Kuckuck (1929), p. 37.

Myriactula chordae (Areschoug) Levring (1937), p. 57.   Fig. 150

Elachista stellaris var. *chordae* Areschoug (1875), p. 18; *Elachista stellaris* f. *chordae* (Areschoug) Kjellman (1880), p. 10; *Elachista chordae* (Areschoug) Kylin (1907), p. 61; *Gonodia chordae* (Areschoug) Hamel (1931–39), p. 137; *Gonodia pulvinata* f. *chordae* (Areschoug) Rosenvinge (1935), p. 31.

Thalli epiphytic, forming small, hemispherical cushions on hosts, to 0.5(–1) mm in diameter, soft and lubricous, easily squashed under pressure; consisting of a small, gelatinous, pseudopa-renchymatous base of short, erect, closely packed, dichotomously to trichotomously branched

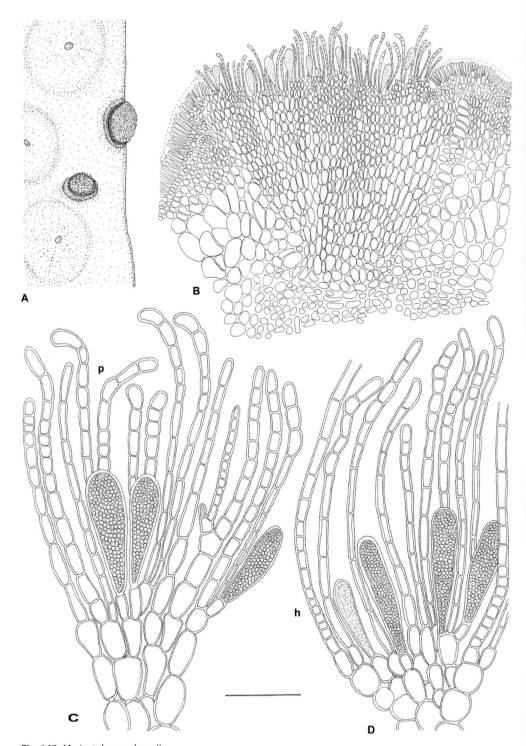

Fig. 149. *Myriactula areschougii*
A. Habit of thallus on *Himanthalia elongata*. B. VS of thallus embedded within host tissue. C, D. Portions of thalli showing terminal paraphyses (p), hairs (h) and unilocular sporangia. Bar = 2 mm (A), = 200 μm (B), = 50 μm (C, D).

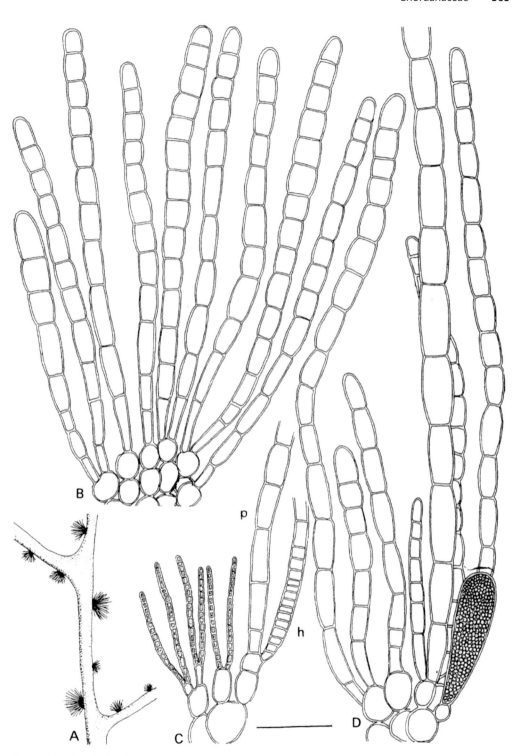

Fig. 150. *Myriactula chordae*
A. Habit of thalli on *Sauvageaugloia griffithsiana*. B. SP of thallus showing erect paraphyses arising from basal cushion of cells. C. Portion of thallus showing paraphyses (p), a hair (h) and plurilocular sporangia. D. Portion of thallus showing paraphyses and a unilocular sporangium. Bar = 2 mm (A), = 50 μm (B–D).

filaments of large, thin-walled, colourless cells, becoming smaller above and giving rise terminally to paraphyses, sporangia and/or hairs; paraphyses erect, multicellular, elongate-clavate, becoming linear, simple, to 800 μm long (–22 cells), comprising elongate-cylindrical or rectangular cells 28–52 × 14–29 μm, each with numerous discoid plastids with pyrenoids; hairs common, borne at base of paraphyses, with basal meristem, without obvious sheath, 10–16 μm in diameter.

Unilocular and plurilocular sporangia recorded on the same or different thalli, borne at the base of the paraphyses; unilocular sporangia pyriform, 70–94 × 23–33 μm; plurilocular sporangia in dense clusters, sessile or on 1–3-celled stalks, simple, cylindrical, uniseriate, to 91 μm long (–19 loculi) × approx. 6–8 μm wide.

Epiphytic on various algae, in lower eulittoral pools, and shallow sublittoral.

Recorded for a few scattered localities around Britain (Orkney, Hampshire, Dorset).

Summer annual; May to September.

Brodie *et al.* (2016), p. 1020; Dizerbo & Herpe (2007), p. 85; Feldmann (1945), pp. 222–9; Fletcher (1987), p. 152, fig. 28; Fletcher in Hardy & Guiry (2003), p. 275; Gallardo García *et al.* (2016), p. 36; Hamel (1931–39), pp. 137, 138, fig. 30(F); Kylin (1947), pp. 47, 48, fig. 41; Mathieson & Dawes (2017), p. 170, pl. L(7, 8); Newton (1931), p. 133; Rosenvinge (1935), pp. 31–7, figs 29–35, (1943), p. 7; Rueness (1977), p. 145, fig. 72, (2001), p. 16; Taylor (1957), p. 142.

Myriactula clandestina (P. Crouan & H. Crouan) Feldmann (1945), p. 223.   Fig. 151

Elachista clandestina P. Crouan & H. Crouan (1867), p. 160; *Ectocarpus clandestinus* (P. Crouan & H. Crouan) Sauvageau (1892), p. 13; *Gonodia clandestina* (P. Crouan & H. Crouan) Hamel (1931–39), p. 134; *Entonema clandestinum* (P. Crouan & H. Crouan) Hamel (1931–39), p. xxvi.

Thalli primarily endophytic, emerging from host tissue as small rounded pustules, 0.5–1 mm in diameter, solitary or more often gregarious; comprising a network of branched, multicellular, filaments, 3–6 μm in diameter ramifying through host tissue, particularly just below surface cells, emerging to the surface, at first forming small swellings or raised bumps, later rupturing through the surface as small pustules; central region turf-like, similar to a carpet pile, spongy in texture, consisting of a basal layer of emergent, branched, endophytic filaments, the cells of which give rise to dense, erect tufts of paraphyses, unilocular sporangia and/or hairs hardly extending above raised rim of host material; paraphyses gelatinous, easily separable under pressure, simple, multicellular, linear or slightly elongate-clavate, straight or slightly curved, to 180(–250) μm long, 13–22(–30) cells, comprising colourless or only faintly pigmented, thick-walled cells throughout, 0.5–1(–2) diameters long, rectangular, rarely moniliform, 5–9(–18) × 6–10 μm in upper and mid regions, 3–5 diameters long, elongate-rectangular, 12–20 × 3–5 μm towards the base; cells with 2–3 small discoid plastids, without obvious pyrenoids; hairs common, produced from basal filaments, sometimes terminal on paraphyses, with basal meristem, without obvious sheath.

Unilocular sporangia formed directly from basal cells, emerging at base of paraphyses, pear-shaped to elongate-cylindrical, 55–90 × 14–32 μm, often on a single-celled pedicel; plurilocular sporangia unknown.

Epiphytic on *Fucus* spp., particularly *F. serratus*, over whole length of thallus, although more commonly near base, in the upper to lower eulittoral, in pools or emergent.

Recorded for widely scattered localities all around Britain (Shetland, Fife, Berwick, Yorkshire, Kent, Sussex, Hampshire, Dorset, Devon, Cornwall, Pembroke, Isle of Man) and for Mayo in Ireland; probably more common and widespread than records suggest.

No obvious seasonal distribution pattern; recorded throughout the year with unilocular sporangia.

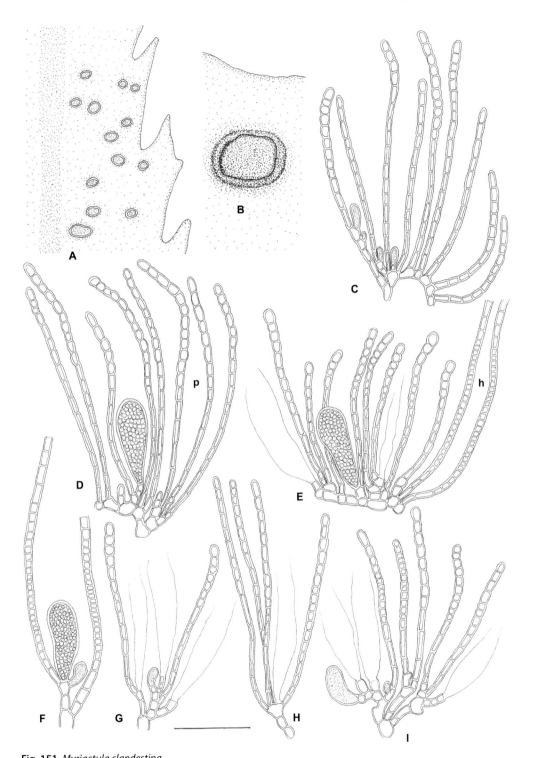

Fig. 151. *Myriactula clandestina*

A, B. Habit of thalli, forming pustules on *Fucus serratus*. C–I. SP of thalli showing paraphyses (p), hairs (h) and unilocular sporangia (some dehisced) Bar = 4 mm (A), = 1 mm (B), = 50 μm (C–I).

Batters (1895a), p. 276; Brodie *et al.* (2016), p. 1020; Cotton (1912), p. 94 (as *Ectocarpus clandestinus*); P. Crouan & H. Crouan (1867), p. 160, pl. 24(157-1, 2); Dangeard (1965b), pp. 16–18, (1968b), pp. 81–6, pl. 1; Dizerbo & Herpe (2007), p. 85; Feldmann (1945), pp. 222–9; Fletcher (1987), pp. 152–4, fig. 29; Fletcher in Hardy & Guiry (2003), p. 275; Guiry (2012), p. 146; Hamel (1931–39), p. 134, fig. 30(D–E); Mathieson & Dawes (2017), p. 171, pl. L(9); Morton (1994), p. 8; Newton (1931), p. 115; Pedersen (2011), p. 88; Sauvageau (1892), pp. 38–41.

Myriactula haydenii (Gatty) Levring (1937), p. 57.   Figs 152, 153

Elachista haydenii Gatty (1863), p. 162; *Elachista moniliformis* Foslie (1894), p. 120; *Myriactis haydenii* (Gatty) Batters (1902), p. 36; *Myriactis moniliformis* (Foslie) Kylin (1910), p. 13; *Gonodia moniliformis* (Foslie) Setchell & N.L. Gardner (1924), p. 722; *Myriactula moniliformis* (Foslie) Feldmann (1945), p. 223.

Thalli epiphytic, forming minute tufts on host, to 0.5 mm in height, usually confluent and spreading to 2–3 mm in extent; comprising a basal system of outward-spreading, fairly tightly packed, branched filaments on host surface, comprising cells irregular in shape, 7–13 × 4–7 µm, from which arise short, erect, branched filaments bearing paraphyses, sporangia and/or hairs; erect filaments to 9 cells long, cells rectangular, 6–16 × 4–9 µm with 1–2(–3) plate-like plastids; paraphyses erect, simple, multicellular, to 400 µm (–23 cells) long, linear or slightly clavate, tapering gradually below, comprising quadrate, barrel-shaped or more commonly rectangular cells, 1–2(–2.5) diameters long, 8–30 × 10–16 µm, each with several discoid plastids without obvious pyrenoids and sometimes with central, large, hyaline vacuole/oil vesicle present; hairs abundant, arising directly from erect filaments or terminal on paraphyses, with basal meristem but without obvious sheath, 9–12 µm in diameter.

Unilocular and plurilocular sporangia arising at base of paraphyses, terminal or lateral on erect filaments; plurilocular sporangia abundant, densely crowded, sessile or stalked, simple or rarely branched, cylindrical, uniseriate, to 70 µm long (–11 loculi) × approx. 5–7 µm in diameter, often arising within old sporangial wall husks, comprising subquadrate loculi, 3–5 × 5–6 µm with straight, rarely oblique cross walls; unilocular sporangia pyriform, not observed for Britain and Ireland, rarely reported elsewhere in Europe.

Epiphytic on *Scytosiphon lomentaria*, in eulittoral pools.

Only recorded for a few scattered localities around Britain (Northumberland, Hampshire, Cheshire, Isle of Man) and for Cork and Mayo in Ireland; probably more widespread and common than records suggest.

Summer annual, June to September.

Brodie *et al.* (2016), p. 1020; Cotton (1912), p. 95; Fletcher (1987), pp. 154–6, fig. 30; Fletcher in Hardy & Guiry (2003), p. 276; Foslie (1894), pp. 120–3, pl. 1(3–6); Guiry (2012), p. 146; Jaasund (1965), p. 65; Kuckuck (1929), pp. 37, 38, figs 35, 36; Kylin (1947), p. 48, fig. 42(B); Levring (1937), p. 57; Morton (1994), p. 8; Newton (1931), p. 143; Rueness (1977), p. 145, fig. 74(A), (2001), p. 16; Womersley (1987), p. 92, figs 25(E–G).

Myriactula rivulariae (Suhr ex Areschoug) J. Feldmann (1937), p. 274.   Figs 154, 155

Elachista rivulariae Suhr ex Areschoug (1842), p. 235; *Phycophila rivulariae* (Suhr ex Areschoug) Kützing; *Myriactis pulvinata* Kützing (1843), p. 330; *Elachista attenuata* Harvey (1846–51), pl. 28A; *Elachista pulvinata* (Kützing) Harvey (1846–51), pl. 17; *Myriactula pulvinata* (Kützing) Kuntze (1898), p. 415; *Gonodia pulvinata* (Kützing) Nieuwland (1917), p. 30.

Thalli epiphytic, forming small, hemispherical cushions on hosts, to 2 mm in diameter, 1 mm in height, just visible to the unaided eye, soft and lubricous easily squashed under pressure; consisting of a soft and gelatinous, somewhat cartilaginous, pseudoparenchymatous lower region of erect, closely packed, dichotomously to trichotomously branched filaments, emergent from host tissue, comprising large, thin-walled, colourless cells, containing only a few discoid plastids, which gradually become smaller above and give rise terminally to paraphyses, sporangia and/or hairs; paraphyses erect, closely packed but gelatinous and easily separated

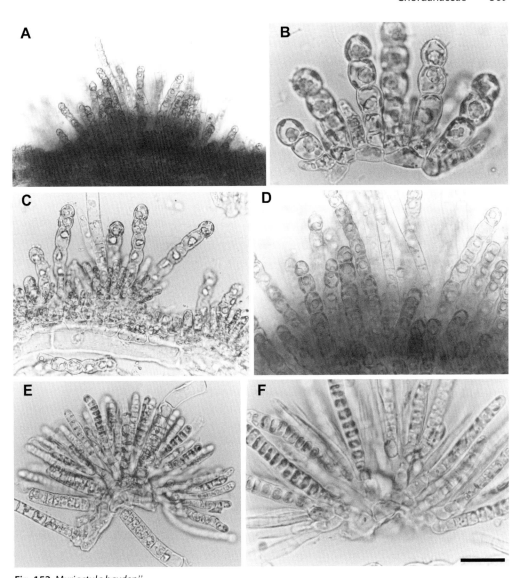

Fig. 152. *Myriactula haydenii*

A. Habit of thalli on *Scytosiphon*. B–D. SP of thalli showing erect paraphyses-like filaments and hairs. E, F. SP of thalli showing closely packed tiers of plurilocular sporangia. Bar in F = 180 μm (A), = 90 μm (C–E), = 45 μm (B, F).

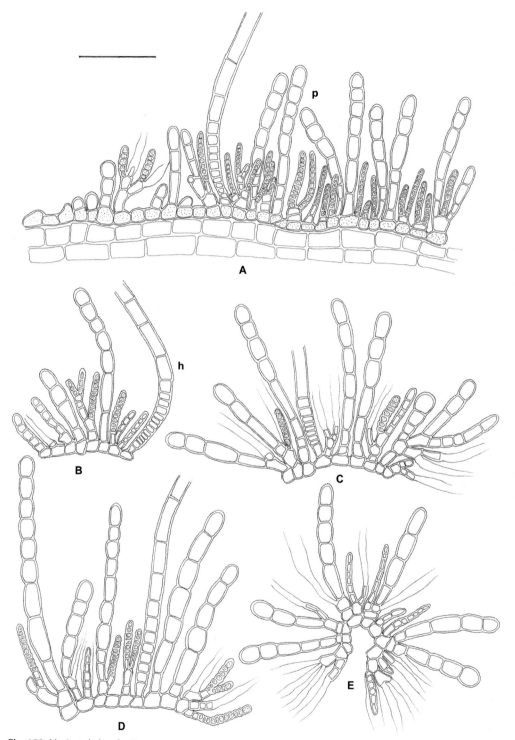

Fig. 153. *Myriactula haydenii*
A–E. SP of thalli showing paraphyses (p), hairs (h) and plurilocular sporangia (some dehisced) arising from the basal layer. Bar = 50 μm.

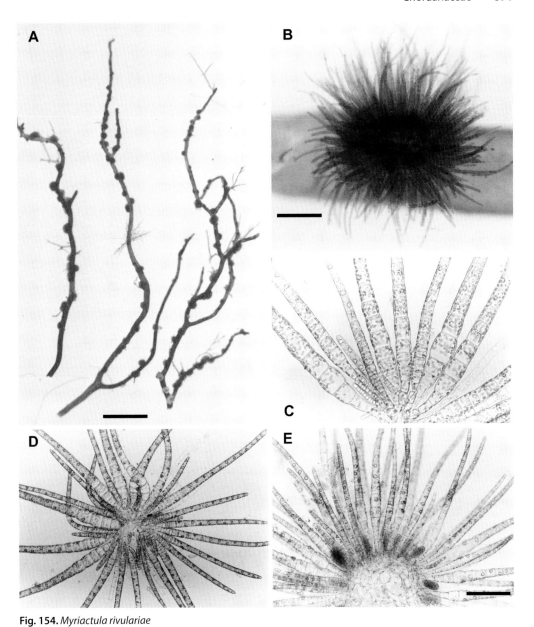

Fig. 154. *Myriactula rivulariae*

A. Globular habit of thalli on host. B. Habit of thallus on host. C, D. SP of vegetative thalli. E. SP of fertile thallus showing unilocular sporangia arising at the base of the paraphyses-like assimilatory filaments. Bar in A = 6 mm, bar in B = 470 μm, bar in E = 90 μm (C), = 180 μm (D, E).

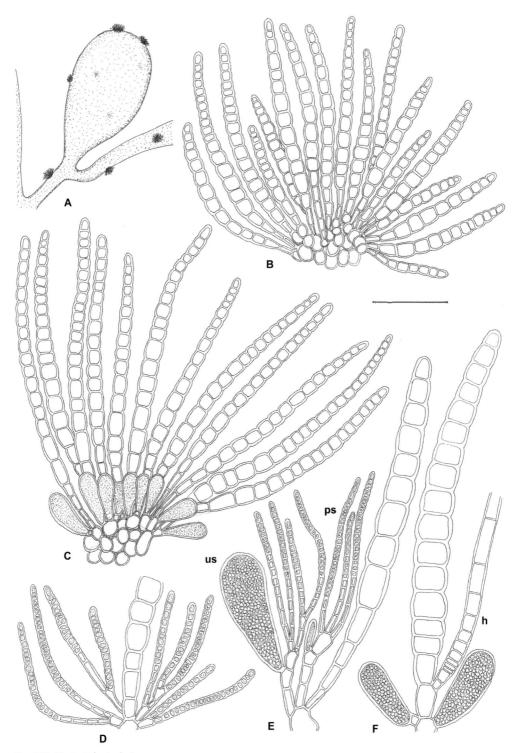

Fig. 155. *Myriactula rivulariae*
A. Habit of thalli epiphytic on an air-bladder of *Sargassum muticum*. B. SP of vegetative thallus showing erect, paraphyses-like assimilatory filaments. C–F. SP of fertile thalli showing unilocular (us) and plurilocular sporangia (ps) arising at base of assimilatory filaments. Note hair (h). Bar = 2.5 mm (A), = 100 μm (B–C), = 50 μm (D–F).

under pressure, fusiform, multicellular, simple, slightly curved, to 620 µm long (–31 cells), tapering gradually towards apex, more sharply towards base, comprising thick-walled cells, mainly barrel-shaped throughout, less commonly rectangular and quadrate, especially at base, later becoming enlarged and swollen in mid-lower region, 4–14 diameters long, 16–50 × 25–43 µm remaining much smaller and less obviously swollen in mid-terminal region, 1–14 diameters long, 14–32 × 16–26 µm, all cells with numerous discoid plastids with pyrenoids; hairs common, borne at base of paraphyses, with basal meristem, without obvious sheath, approx. 9 µm in diameter.

Unilocular and plurilocular sporangia recorded on the same or different thalli, borne at the base of the paraphyses; unilocular sporangia sessile or on 1-celled stalks, pyriform, 54–90 × 22–40 µm; plurilocular sporangia on 1–2-celled stalks, simple, cylindrical, uniseriate, to 60 µm long (–10 loculi) × approx. 7–9 µm wide.

Epiphytic on *Gongolaria* spp. and *Sargassum muticum*, less commonly on *Halidrys siliquosa*, usually emergent from conceptacles or cryptostomata. More rarely recorded on various other algae including *Scytosiphon lomentaria* and *Laminaria* spp., upper and lower eulittoral pools and shallow sublittoral.

Common on *Gongolaria* spp. on south-west shores of Britain (Hampshire, Channel Islands, Devon, Cornwall, Isles of Scilly) and Ireland (Wexford, Waterford, Cork, Kerry, Clare, Galway, Mayo). Recorded on *Halidrys siliquosa*, from widely scattered localities in Britain (Argyll, Isle of Man, Yorkshire), and on other algae from Dorset; probably more widely distributed on these hosts.

Summer annual, July to September.

Brodie *et al.* (2016), p. 1020; Cormaci *et al.* (2012), p. 192, pl. 50(4–50); Cotton (1912), p. 95 (as *Myriatis pulvinata*); Dizerbo & Herpe (2007), p. 85, pl. 27(2); Feldmann (1937), p. 274, fig. 46(A–C); Fletcher (1987), pp. 156–8, fig. 31, pl. 2(f); Fletcher in Hardy & Guiry (2003), p. 276; Gallardo García *et al.* (2016), p. 36; Guiry (2012), p. 146; Hamel (1931–39), pp. 135, 136, fig. 31(1, 2); Kuckuck (1929), pp. 39, 40, figs 40, 41; Morton (1994), p. 8; Newton (1931), p. 142 (as *Myriatis pulvinata*); Ribera *et al.* (1992), p. 115; Rosenvinge (1935), p. 28, figs 27, 28; Stegenga & Mol (1983), p. 94, pl. 30(2–3); Stegenga *et al.* (1997a), p. 18; Taskin *et al.* (2008), p. 59, pl. 51(1).

Myriactula stellulata (Harvey) Levring (1937), p. 57.   Figs 156, 157

Conferva stellulata Harvey (1841), p. 132; *Elachista stellulata* (Harvey) A.W. Griffiths (1843), p. 261; *Myrionema stellulatum* (Harvey) J. Agardh (1848), p. 49; *Myriactis stellulata* (Harvey) Batters (1892a), p. 174; *Gonodia stellulata* (Harvey) Hamel (1931–39), p. 132.

Thalli epiphytic, forming minute, solitary, brush-like tufts on host, 0.5–1 mm in diameter; comprising endophytic, branched filaments of largely colourless or faintly pigmented cells ramifying mainly between the cortical and medullary cell layers of host, occasionally spreading down between medullary cells, which break through to the surface forming small pustules; emergent filaments remaining fairly short, erect, dichotomously to trichotomously branched, closely packed and cushion forming, although gelatinous and easily separated under pressure, comprising large, somewhat swollen, thin-walled, colourless cells, becoming smaller above and terminating in paraphyses, sporangia and/or hairs; paraphyses erect, gelatinous, multicellular, to 12(–17) cells, 155–285 µm long, simple, rarely branched, at first linear or elongate-clavate later becoming slightly fusiform comprising cells elongate-rectangular below, 1–3 diameters long, 16–25 × 7–12 µm becoming more rectangular, barrel-shaped or slightly moniliform above, 0.5–2.5 diameters long, 7–27 × 10–24 µm each containing several discoid plastids with pyrenoids; hairs common, borne at base of paraphyses, with basal or sub-basal meristem, without obvious sheath, approx. 9–11 µm in diameter.

Unilocular and plurilocular sporangia recorded on the same or different thalli, borne at the base of the paraphyses; unilocular sporangia common, sessile or shortly stalked,

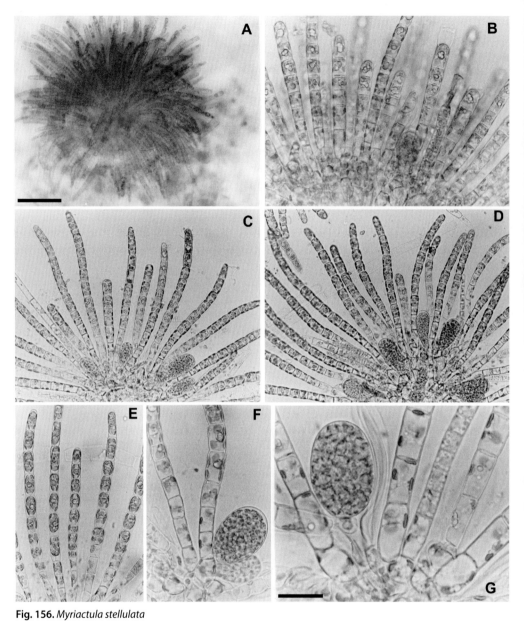

Fig. 156. *Myriactula stellulata*

A. Habit of thalli epiphytic on *Dictyota dichotoma*. B–G. SP of thalli showing the erect paraphyses-like assimilatory filaments and basal unilocular sporangia. Bar in A =180 μm, bar in G = 90 μm (B–E), = 45 μm (F, G).

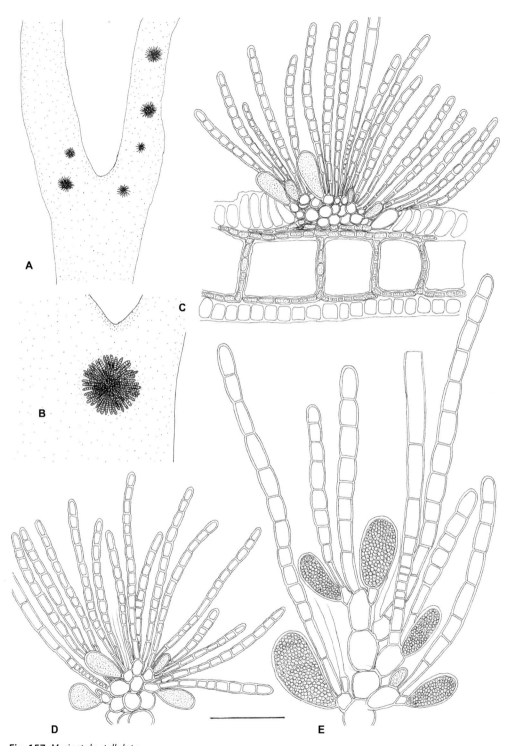

Fig. 157. *Myriactula stellulata*

A, B. Habit of thalli epiphytic on *Dictyota dichotoma*. C. VS of fertile thallus emerging from host tissue. D, E. SP of thalli showing the paraphyses-like assimilatory filaments, hairs and unilocular sporangia. Bar= 2 mm (A), = 0.5 mm (B), = 100 μm (C, D), = 50 μm (E).

elongate-pyriform, 36–72 × 20–33 µm; plurilocular sporangia rare, in dense clusters, cylindrical, simple, uniseriate, to 80 µm long (13 loculi) × approx. 8–10 µm wide.

Epiphytic on *Dictyota dichotoma*, usually near the base, in pools in the eulittoral, and shallow sublittoral.

Southern and western shores of Britain (Channel Islands and extending eastwards to Hampshire and northwards to Ayr) and Ireland (eastwards to Wexford and northwards to Mayo).

Thalli annual, June to September and usually fertile; plurilocular sporangia only recorded for the Isle of Man but probably more widely distributed.

Brodie *et al.* (2016), p. 1020; Coppejans (1983), pls 33, 34; Cormaci *et al.* (2012), pp. 192, 194, pl. 51(2, 3); Cotton (2012), p. 95; Dizerbo & Herpe (2007), p. 86, pl. 27(3); Feldmann (1937), pp. 272–4, fig. 45(A–C); Fletcher (1987), pp. 158–60, fig. 32; Fletcher in Hardy & Guiry (2003), p. 277; Gallardo García *et al.* (2016), p. 36; Guiry (2012), p. 146; Hamel (1931–39), p. 132, fig. 30(A); Kuckuck (1929), p. 34, figs 32–4; Morton (1994), p. 8; Newton (1931), p. 142, fig. 88; Ribera *et al.* (1992), p. 115; Rueness (1977), p. 146, fig. 76, (2001), p. 16; Sauvageau (1892), pp. 6–10, pl. 1(1, 2); Schneider & Searles (1991), pp. 135, 136, fig. 151.

Myriocladia J. Agardh

MYRIOCLADIA J. Agardh (1841), p. 48.

Type species: *M. lovenii* J. Agardh (1841), p. 48.

Thalli erect, macroscopic, cylindrical, filiform, terete, solid, becoming hollow in older parts, flaccid, slightly gelatinous, sparingly and mainly irregularly branched, to 1(–3) orders, arising from a small basal disc; in structure, thallus at first uniaxial, becoming multiaxial, haplostichous, with usually one projecting assimilatory filament at apex; initial axial filament cells giving rise to secondary medullary filaments to form a small, central, multiaxial, region; axial cells and secondary filament cells giving rise directly to an outer cortical region of assimilatory filaments, sporangia and/or hairs; growth probably intercalary with a meristematic zone at base of apical assimilatory filament; assimilatory filaments not obviously packed in a gelatinous substance, except below, simple, rarely branched, linear, comprising cylindrical cells; hairs with basal meristem but no obvious sheath; assimilatory cells with numerous discoid plastids with pyrenoids.

Unilocular sporangia borne at the base of the assimilatory filaments, sessile, sometimes reported to be stalked, oblong to pyriform; plurilocular sporangia reported to arise by transformation of the middle cells of the assimilatory filaments.

The genus *Myriocladia* was originally distinguished and separated from *Mesogloia* by J. Agardh (1841) partly on the basis of an incomplete mucilaginous covering of the thallus. Essentially only the basal region of the assimilators is embedded in mucilage and the middle and upper regions of the filaments are free. This was not, however, accepted as a valid characteristic by Sauvageau (1927). Parke (1933) also discussed the question of whether the thallus is solid or hollow. According to Sauvageau (1927), quoting J. Agardh, the thallus is hollow, and this seems to be supported by the illustration of *Myriocladia tomentosa* given by P. Crouan & H. Crouan (1867). Kuckuck (1929) stated that the thallus was solid.

Many authors have kept *Myriocladia* as a distinct genus and have retained it in the family Chordariaceae (e.g. Hamel 1931–39; Rosenvinge & Lund 1943; Rueness 1977; South & Tittley 1986). Oltmanns (1922, p. 34), however, established a new subfamily (Myriocladieae) for it under the family Spermatochnaceae, although as Setchell & N.L. Gardner (1925, p. 555) pointed out, *Spermatochnus* differs in having a distinct apical cell so its placement in this family, as accepted by Newton (1931), appears to be inappropriate. Setchell & N.L. Gardner (1925) referred it, along with *Myriogloia*, to the family Myriogloiaceae while Kuckuck (1929, p. 63), erected a new family, the Myriocladaceae, to accommodate it which was accepted by such authors as Christensen (1980–94).

In structure, *Myriocladia* is uniaxial with one distinct, central, persistent filament observed at the apex of all axes; however, secondary growth of the axial cell produces columns of enclosing filaments giving the thallus a multiaxial appearance in squash preparations. This central medulla region remains relatively undeveloped and gives rise directly to the outer 'cortex' of assimilatory filaments, with no real intermediate subcortex region. The assimilatory filaments comprise cells which are cylindrical in shape, not becoming moniliform above as found in many related genera.

Two species in Britain.

KEY TO SPECIES
Axis > 1 mm diam.; assimilators cylindrical, branched at base. . . . . *M. lovenii*
Axis < 1 mm diam.; assimilators attenuate at apices, unbranched . . . . *M. tomentosa*

Myriocladia lovenii J. Agardh (1841), p. 48.   Figs 158, 159

Thalli erect, macroscopic, to 9 cm long, to 0.5(–1) mm wide in the main axes, olive-brown, cylindrical, terete, filiform to slightly dorsiventrally flattened, solid, flaccid, slightly gelatinous, sparingly branched, arising from a small basal disc of outward-spreading, rhizoidal filaments; branching bifurcate or irregular, to one, rarely two, orders, with branches scattered and often remaining short; in structure, thallus uniaxial, haplostichous, with a central uniaxial filament of large, mainly rectangular, colourless cells, 2–4 diameters long, 50–110 × 15–50 μm, terminating at the apex in one or two, projecting assimilatory filaments; central axial filament cells in the terminal regions producing filamentous, multicellular offshoots (sometimes termed cortical threads) which cover and grow upwards, parallel with the axial filament, the constituent cells of which enlarge, giving a central medulla-like, seemingly multiaxial region; terminal regions of both the axial cells and the offshoot filament cells turning abruptly outwards, and somewhat distichously, to form a cortex-like region of single or tufts/fascicles of fairly long, multicellular, pigmented, assimilatory filaments of limited growth, hairs and unilocular sporangia; additional, laterally emerging assimilatory filaments also formed adventitiously from the medulla-like region; basal cells of offshoots and assimilatory filaments giving rising to downward-growing, multicellular rhizoidal filaments comprising quite elongated cells; growth probably intercalary with a meristematic zone of dividing cells at the base of the apical assimilatory filament; assimilatory filaments not obviously packed in a gelatinous substance, except below, simple, sometimes dichotomous at the base, linear, uniseriate, to 14(–18) cells, 275(–470) μm and comprising cylindrical cells throughout, markedly longer than wide, 2–5 diameters long, (10–)15–25(–30) × (5–)6–8(–11) μm; hairs with a well-developed basal meristem but no obvious sheath, comprising cells 5–14 diameters long, 60–145 × 7–11 μm; assimilatory cells with numerous discoid plastids with pyrenoids.

Unilocular sporangia occurring over much of the length of the thallus, borne at the base of the assimilatory filaments, sessile, sometimes reported to be stalked, oblong to pyriform, (40–)52–70(–75) × (20–)25–35(–40) μm; plurilocular sporangia not observed on material collected from Britain and Ireland, reported elsewhere (Kjellman 1891b) to arise by transformation of the middle cells of the assimilatory filaments, and to occur in short globular rows distinctly wider than their parental vegetative cells.

Epilithic, on stones and boulders, also elsewhere reported (Kylin 1947) to be epiphytic on the blades of *Laminaria hyperborea*; sublittoral, to 18 m.

Very rare, and widely distributed around Britain (Sussex, Isle of Man, Inner Hebrides).

Appears to be a summer annual, but insufficient information to comment on seasonality.

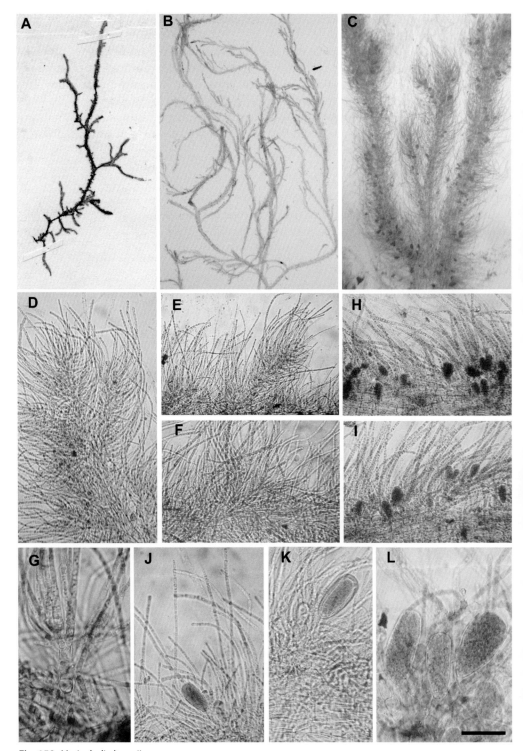

Fig. 158. *Myriocladia lovenii*
A, B. Habit of thalli. C–F. Portions of thalli showing the outer fringe of assimilatory filaments. G. Apex of thallus showing two projecting assimilatory filaments. H–L. Unilocular sporangia borne at the base of the assimilatory filaments. Bar in L = 8 mm (A), = 10 mm (B), = 470 μm (C), = 10 μm (E, F), = 90 μm (D, G–K), = 45 μm (L). (A is herbarium material, NHM.)

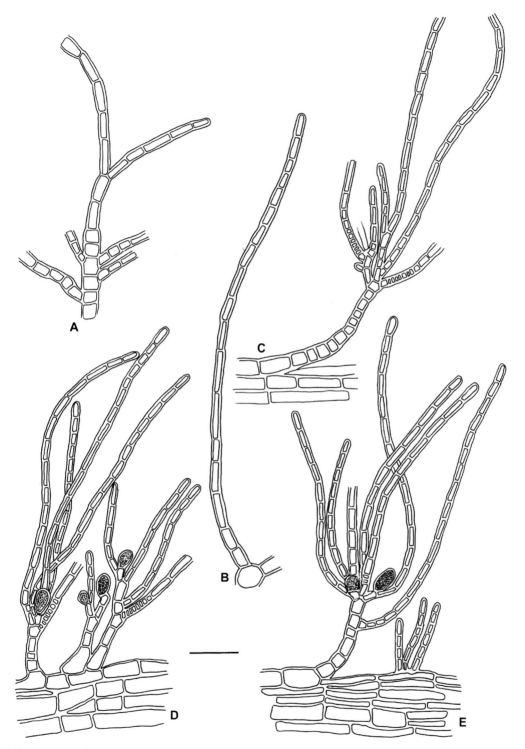

Fig. 159. *Myriocladia lovenii*
A. Apex of branch showing two projecting assimilatory filaments. B–E. SP of thalli showing the origin of the assimilatory filaments, hairs and unilocular sporangia from the medullary filaments. Bar = 50 µm.

According to Parke & Dixon (1964, p. 525; 1976, p. 560), British records of *M. lovenii* were considered to be misidentifications of *M. tomentosa*, However, G. Russell reported no significant differences between these specimens and the type material of *M. lovenii* (see Maggs *et al*. 1983).

Agardh (1841), pp. 48, 49; Brodie *et al*. (2016), p. 1020; Dizerbo & Herpe (2007), p. 90; Fletcher in Hardy & Guiry (2003), p. 277; Fritsch (1945), p. 79, fig. 19(B–D); Guiry (2012), p. 181; Kjellman (1890), p. 39; Kuckuck (1929), p. 65, figs 85, 86; Kylin (1907), pp. 88–90, fig. 21(a–c), (1933), pp. 52–6, figs 21–4, (1947), pp. 55, 56, fig. 47(B); Levring (1937), pp. 60–2, fig. 8(A, B), pl. II(3); Maggs *et al*. (1983), p. 262; Mathieson & Dawes (2017), p. 172, pl. L(11–13); Newton (1931), p. 160, fig 100(A, B); Oltmanns (1922), pp. 37, 38, figs 328, 329; Parke (1933), pp. 31–3, text figs 14–15; Parke & Dixon (1964), p. 525, (1976), pp. 560, 563; Rosenvinge & Lund (1943), pp. 20–2, fig. 7(A, B); Rueness (1977), p. 148, fig. 81, pl. XIX(3), (2001), p. 16.

Myriocladia tomentosa P. Crouan & H. Crouan (1867), p. 165. Fig. 160

Thalli erect, macroscopic, to 30 cm long, to 1–2 mm wide in the main axes, olive-brown, cylindrical, filiform, terete, tomentose, solid, flaccid, slightly gelatinous and sparingly branched along main axes; branching irregular, sometimes bifurcate or opposite, to two, rarely three, orders, with branches scattered, sometimes remaining short, widely divergent and markedly attenuate in tertiary branchlets; in structure, thallus reported to be uniaxial, haplostichous, with a central uniaxial filament of large, rectangular, colourless cells, terminating at the apex in one or two projecting assimilatory filaments; central axial filament cells in the terminal regions producing filamentous, multicellular offshoots (sometimes termed cortical threads) which clothe and grow upwards, parallel with the axial filament, the constituent cells of which enlarge, giving a central medulla-like, seemingly multiaxial region, both the axial cells and the offshoot filament cells also giving rise directly and later outwardly to an outer cortex-like region of single or tufts/fascicles of short, multicellular, pigmented, primary and secondary assimilatory filaments respectively of limited growth, hairs and unilocular sporangia; basal cells of offshoots and assimilatory filaments giving rising to downward-growing, multicellular rhizoidal filaments comprising quite elongated cells; growth probably intercalary with a meristematic zone of dividing cells at the base of the apical assimilatory filament; assimilatory filaments not obviously packed in a gelatinous substance, except below, simple, linear, recurved, uniseriate, attenuate towards the apex, to 10(–12) cells, and comprising cylindrical cells, 2–3 diameters long; hairs common, with basal meristem but no obvious sheath; assimilatory cells with numerous discoid plastids with pyrenoids.

Unilocular sporangia borne at the base of the assimilatory filaments, sessile, spherical to ovoid; plurilocular sporangia unknown.

Epiphytic on the leaves of *Zostera* spp., sublittoral.

Very rare and only known for Dorset in Britain.

Appears to be a summer annual, with collections made in September, but insufficient information to comment on seasonality.

N.B. Levring (1937, p. 61) considered *M. tomentosa* possibly conspecific with *Myriocladia ekmani* (Areschoug.) Kylin.

Brodie *et al*. (2016), p. 1020; P. Crouan & H. Crouan (1867), p. 165, pl. 26, fig. 165; Dizerbo & Herpe (2007), p. 90; Fletcher in Hardy & Guiry (2003), p. 278; Gallardo García *et al*. (2016), p. 36; Guiry (2012), p. 181; Hamel (1931–39), pp. 173, 174; Newton (1931), p. 160.

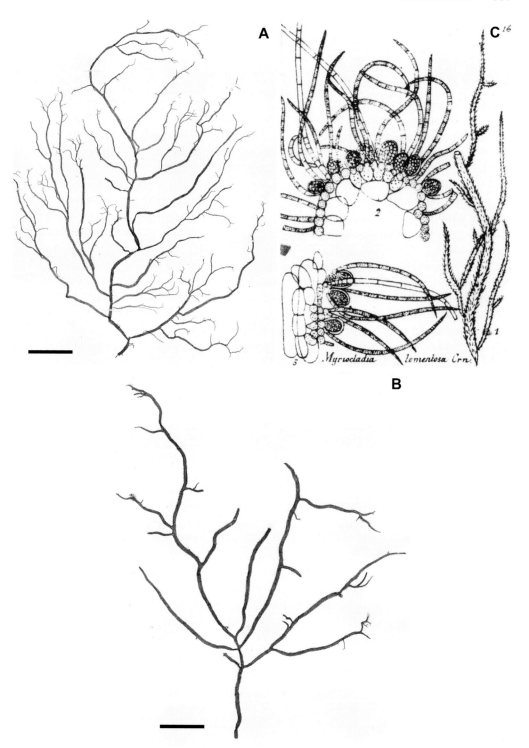

Fig. 160. *Myriocladia tomentosa*

A, B. Habit of thalli (herbarium material, NHM). C. Habit of thalli and SP showing the unilocular sporangia at the base of the assimilatory filaments (C from P. Crouan & H. Crouan 1867). Bar in A = 4 cm, bar in B = 2 cm.

Myrionema Greville

MYRIONEMA Greville (1827a), pl. 300.

Type species: *M. strangulans* Greville (1827a), pl. 300.

Thalli epiphytic or rarely epilithic, forming small circular, less frequently irregular spots on hosts; consisting of a basal prostrate layer of outward-spreading, pseudodichotomously branched, monostromatic, filaments of cells, laterally adjoined and disc-like in appearance, often with a distinct marginal row of synchronously growing apical cells, less frequently irregularly associated and pseudodiscoid; basal cells without rhizoids, each giving rise, except near the periphery to 1(–2) erect filaments, hairs, ascocysts and/or sporangia; erect filaments commonly short, simple, more rarely branched or with short protuberances, loosely compacted, often in a mucilaginous matrix, comprising cells with 1–3 plate-like plastids and pyrenoids; hairs, ascocysts and sporangia usually sessile or stalked on basal layer, less commonly terminal or lateral on erect filaments; hairs with basal meristem and sheath.

Plurilocular and unilocular sporangia arising from basal layer, sessile or shortly stalked, or terminal/lateral on erect filaments; plurilocular sporangia common, simple or rarely branched, uniseriate or rarely biseriate; unilocular sporangia uncommon, ovate or pyriform.

The genus *Myrionema* is characterized by a monostromatic discoid base, with short, unbranched filaments arising from the central cells. Culture studies on a small number of species reveal this discoid morphology to originate in a 'stellate' spore germination pattern. Pedersen (1984) pointed out that such a similar 'stellate' germination pattern followed by disc formation are characteristic of the early developmental processes in some members of the families Chordariaceae and Giraudiaceae (see particularly Sauvageau 1924b; Shinmura 1977; Lockhart 1979). This suggests that *Myrionema* species do not form a natural phylogenetic group and that at least some species may represent, or have represented, microthallial phases of various algae presently included in other diverse families. Further studies relating to life history and developmental patterns are required before the taxonomic identities of these algae can be fully assessed.

Five species of *Myrionema* are recognized for Britain and Ireland: *M. corunnae*, *M. liechtensternii*, *M. magnusii*, *M. papillosum* and *M. strangulans*. Meanwhile, *M. aecidioides* and *M. polycladum*, recognized by Parke & Dixon (1976), are transferred to other genera. *M. aecidioides* was to be transferred to *Gononema* in agreement with Pedersen (1981a) but, herein, is transferred to *Laminariocolax* based on molecular evidence from nrDNA ITS sequences by Burkhardt & Peters (1998). *M. polycladum* is included within the circumscription of *Microspongium globosum*. Out of the five species of *Myrionema* recognized here, *M. corunnae* appears anomalous and is only retained in this genus pending further studies. It differs from the other species of *Myrionema* in having a more irregularly structured, pseudodiscoid rather than discoid basal layer, slenderer basal and erect filaments, hairs with reduced basal meristem lacking sheath, and the occurrence of plurilocular sporangia only. *M. liechtenstermii* is transferred from the genus *Hecatonema* and only retained here pending further studies; British and Irish material of this species does come rather close in description to *Microspongium globosum*.

Three species in Britain and Ireland, two species in Britain only.

KEY TO SPECIES

1. Thallus epilithic . . . . *M. liechtensternii*
   Thallus epiphytic 2
2. Thalli with ascocysts . . . . *M. magnusii*
   Thalli without ascocysts 3
3. Erect filaments with short, lateral protuberances . . . .*M. papillosum*
   Erect filaments without short, lateral protuberances 4

4. Erect filaments and plurilocular sporangia clavate; erect filaments usually
    > 5 μm diameter; plurilocular sporangia uniseriate, frequently biseriate, up
    to 50 μm long; unilocular sporangia common; epiphytic mainly on species of
    Ulva, rarely on *Laminaria* blades . . . .*M. strangulans*
    Erect filaments and plurilocular sporangia linear; erect filaments usually
    < 5 μm diameter; plurilocular sporangia uniseriate rarely biseriate, up to
    125 μm long; unilocular sporangia unknown; epiphytic on *Laminaria* blades
    only . . . .*M. corunnae*

Myrionema corunnae Sauvageau (1897a), p. 237.   Fig. 161

Thalli epiphytic, forming small spots on hosts, usually solitary and circular, 0.5–1.0 mm across, occasionally confluent to 5 mm or more in extent; in surface view, peripheral region composed of a monostromatic, pseudodiscoid layer of branched, outward-radiating, irregularly associated filaments comprising quadrate, rectangular or irregularly shaped cells, 1–3 diameters long, 5–13 × 4–6 μm, each with 1(–3) plate-like plastids, outer central region composed of closely packed, rounded cells, 5–8 μm in diameter, inner central region turf-like, similar to a carpet pile; in squash preparations, central basal cells usually 1(–2) diameters long, 5–12 × 5–8 μm each giving rise to one, rarely two, erect filaments, hairs and/or plurilocular sporangia; erect filaments short, linear, loosely associated and gelatinous, multicellular, to 13 cells (150 μm), simple or very rarely branched, comprising quadrate or more commonly rectangular cells, 1–3(–4) diameters long, 4–21 × 4–5 μm with 1(–2) plate-like plastids and pyrenoids; hairs common, usually arising from basal cells, rarely terminal on erect filaments, without sheath and with reduced basal meristem, approx. 6 μm in diameter; rhizoids not observed.

Plurilocular sporangia abundant, simple or rarely branched, sessile or more usually terminal on erect filaments, uniseriate, rarely biseriate, with straight, sometimes oblique cross walls, slightly tortuous, to 25 loculi (125 μm) long × 5–8 μm in diameter; unilocular sporangia unknown.

Epiphytic on *Laminaria* blades, lower eulittoral, in pools and shallow sublittoral.

Recorded for a few scattered localities around Britain (Berwick, Kent, Hampshire, Dorset, Devon, Cornwall, Channel Islands) and for Wexford in Ireland; probably more common than the records suggest.

No data on seasonal growth. Fertile thalli have been recorded throughout the year.

A number of authors have linked *M. corunnae* as a stage in the life history of *Ectocarpus fasciculatus* Harvey. Intimate association of the two species on the host alga *Laminaria* was noted by both Jaasund (1961, 1965) in Norway and Sauvageau (1897b) in Spain. *M. corunnae* was also proposed as a myrionemoid variant of *E. fasciculatus* (collected from Canada and Britain) based on laboratory culture studies by Baker & Evans (1971). The evidence for such a life history connection is not, however, convincing and more extensive culture studies are required; the myrionemoid variants may represent contaminant organisms as proposed by Clayton (1972). Culture studies of *M. corunnae* from California (Loiseaux 1970a) revealed a 'direct' life history to be operating, with plurispores directly repeating the fertile parental thallus.

Abbott & Hollenberg (1976), p. 158, fig. 123; André (1970), p. 249; Børgesen (1902), p. 426, fig. 80; Brodie *et al.* (2016), p. 1020; Coppejans & Kling (1995), pp. 170–2, pl. 48(A–G); Cotton (1912), p. 95; Dizerbo & Herpe (2007), p. 78; Fletcher (1987), pp. 105–7, fig. 10; Fletcher in Hardy & Guiry (2003), p. 278; Gallardo García *et al.* (2016), p. 36; Guiry (2012), p. 146; Hamel (1931–39) p. 91, fig. 24(16–18); Jaasund (1957), pp. 223–8, fig. 8, (1961), pp. 239–41, fig. 1, (1965), pp. 56, 57; Kylin (1947), pp. 38, 39, fig. 30; Levring (1937), p. 49, fig. 7(A–B); Loiseaux (1970a), p. 251; Mathieson & Dawes (2017), p. 173, pl. L(15); Morton (1994), p. 8; Newton (1931), p. 151; Rueness (1977), p. 137, fig. 62, (2001), p. 16; Sauvageau (1897a), pp. 237–42, fig. 14; Setchell & N.L. Gardner (1925), p. 458; Stegenga & Mol (1983), p. 85, pl. 26(2); Stegenga *et al.* (1997a), p. 19; Taylor (1957), p. 155.

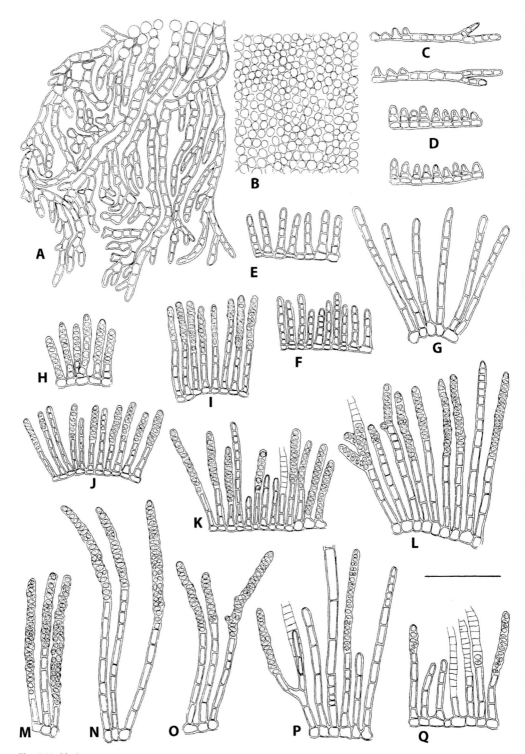

Fig. 161. *Myrionema corunnae*
A. Surface view of peripheral thallus region showing irregularly spreading, vegetative filaments. B. Surface view of central thallus region (vegetative). C–G. SP of vegetative thallus showing erect filaments arising from a monostromatic base. H–Q. SP of fertile thalli showing terminal plurilocular sporangia. Bar = 50 μm.

Myrionema liechtensternii Hauck (1877), p. 185.   Fig. 162

Ascocyclus lichtensteinii (Hauck) Holmes & Batters (1890), p. 82; *Hecatonema liechtensternii* (Hauck) Batters (1902), p. 42; *Phaeosphaerium liechtensternii* (Hauck) De Toni

Thalli epilithic, forming thin, closely adherent, light/olive-brown patches on substrata, either solitary and circular, to 1 mm in diameter, or confluent and irregular to several millimetres in extent; in surface view, peripheral region monostromatic, discoid/pseudodiscoid composed of outward-radiating, closely packed filaments, central region of rounded cells or squashed pile-like in appearance; in squash preparations, consisting of a basal layer of uniseriate cells, 6–14 × 4–8 μm, usually quadrate, rectangular or irregular in shape, sometimes with short, 2–3-celled, rhizoidal filaments, giving rise in the central regions to one, rarely two, erect filaments, hairs and/or plurilocular sporangia; erect filaments short, linear or elongate-clavate, simple or dichotomously branched to 1(–2) orders, loosely associated and slightly mucilaginous to 7(–12) cells (50–106 μm) long, comprising cells often subquadrate at the base, but quadrate to rectangular above, 1–3.5 diameters long, 4–13(–18) × 4–10 μm often with enlarged, sometimes pyriform and ascocyst-like apical cell, 16–26 × 6–16 μm; cells with 1(–2) multi-lobed, plate-like plastids with pyrenoids; hairs infrequent arising from basal layer, approximately 10 μm in diameter, with basal meristem without obvious sheath.

Plurilocular sporangia common, sessile or stalked on basal layer, more commonly terminal or lateral on erect filaments, frequently formed in old sporangial husks, simple, cylindrical, uniseriate to 7(–14) loculi, 24–48 × 4–7 μm; loculi subquadrate, less commonly quadrate, 3–6 × 5–7 μm with straight, occasionally oblique cross walls; unilocular sporangia unknown.

Epilithic on bedrock, in upper eulittoral and littoral fringe, associated with algae such as *Planosiphon filiformis*, *Rivularia* spp. and *Phymatolithon lenormandii*.

Only reported for Berwick-upon-Tweed, Northumberland in Britain.

Material collected in October; information insufficient to comment on seasonal distribution.

The above description and notes are based upon examination of herbarium specimens deposited in the Natural History Museum, London and originally collected by E.A.L. Batters in 1889. The material comes very close in description to the epiphytic alga *Microspongium globosum* (in particular, the material formally described as *Myrionema polycladum* Sauvageau) and is only retained here as an independent taxon pending further studies. It differs from *M. globosum* in being epilithic (British material only), has longer erect filaments and slightly larger vegetative cells.

Athanasiadis (1985), p. 465, fig. 16; Brodie *et al.* (2016), p. 1020; Coppejans and Dhondt (1976), pp. 112–17, figs 1–9; Cormaci *et al.* (2012), pp. 197, 198, pl. 52(6–10); Fletcher (1987), pp. 107–9, fig. 11; Fletcher in Hardy & Guiry (2003), p. 279; Gallardo García *et al.* (2016), p. 36; Guiry (2012), p. 182; Hauck (1877), pp. 185, 186, figs 1–2, (1883–85), pp. 321, 322; Newton (1931), p. 157; Ribera *et al.* (1992), p. 116; Verlaque and Boudouresque (1981), p. 147, figs 12, 13.

Myrionema magnusii (Sauvageau) Loiseaux (1967a), p. 338, nom. inval.   Figs 163, 164

Ascocyclus magnusii Sauvageau (1927b), p. 13.

Thalli epiphytic, forming light to dark brown spots on host, usually solitary and circular, 150–500 μm in diameter, occasionally confluent to 2 mm or more in extent; in surface view, peripheral region light brown, monostromatic, discoid, composed of branched, outward-radiating, quite firmly united filaments enclosed within outer cuticle, comprising rectangular cells 1–3(–4) diameters long, 6–13 × 3–7 μm, each with a single, multilobed plastid and pyrenoid, central region dark brown, of rounded cells, 4–9 μm in diameter, later with colourless ascocysts emerging; structure consisting of a basal layer of uniseriate cells giving rise in the central regions to erect filaments, hairs, ascocysts and/or sporangia; erect filaments

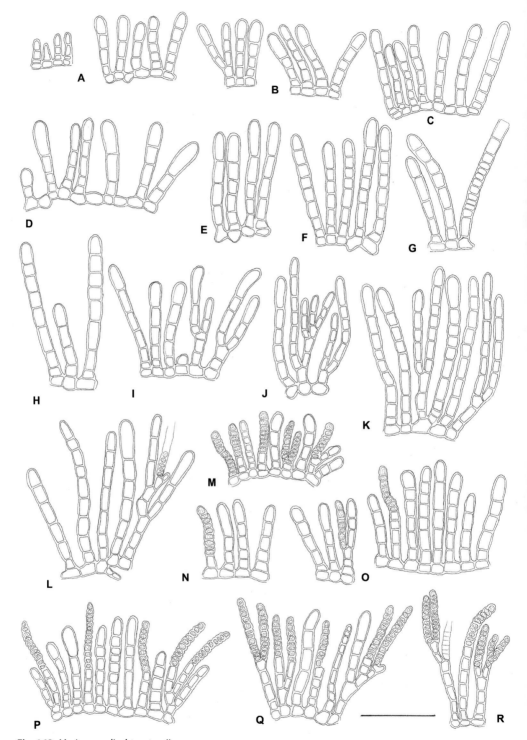

Fig. 162. *Myrionema liechtensternii*
A–K. SP of vegetative thallus showing erect filaments arising from a monostromatic base. L–R. SP of fertile thallus showing terminal plurilocular sporangia. Bar = 50 μm.

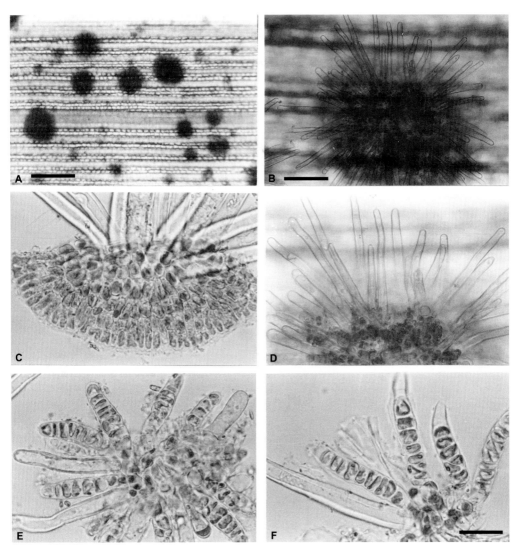

Fig. 163. *Myrionema magnusii*
A, B. Habit of thalli on *Zostera*. C, D. Thalli with abundant ascocysts. E, F. SP of thalli showing plurilocular sporangia and ascocysts. Bar in A = 450 μm, bar in B = 100 μm, bar in F = 20 μm (C), = 40 μm (D), = 35 μm (E), = 35 μm (F).

short, simple, loosely associated and slightly mucilaginous, to 4–12 cells (40–90 μm) long, comprising cells quadrate to rectangular, usually 1–2(–3) diameters long, 5–20 × 7–10 μm each with a single, multilobed plastid with pyrenoid; ascocysts extremely common, either produced from basal cells or on short 1–3-celled stalks, darkly pigmented and slightly globose when young becoming linear, elongated, 45–105(–120) × 8–11 μm, thick walled and colourless later; hairs frequent, arising singly from basal layer with basal meristem and sheath, 8–10 μm in diameter.

Unilocular and plurilocular sporangia borne on the same thalli; plurilocular sporangia abundant, uniseriate, rarely biseriate, simple, cylindrical, 34–58 × 7–9 μm, to 7–13 loculi, on 1–4-celled stalks, loculi subquadrate, often lenticular, 3–5 × 5–9 μm, with straight, occasionally oblique cross walls and quite distinct red eye spot; unilocular sporangia rare, sessile or shortly stalked on basal layer, ovate to pyriform, 28–33 × 14–16 μm.

Epiphytic, on *Zostera* leaves, in lower eulittoral, and shallow sublittoral.

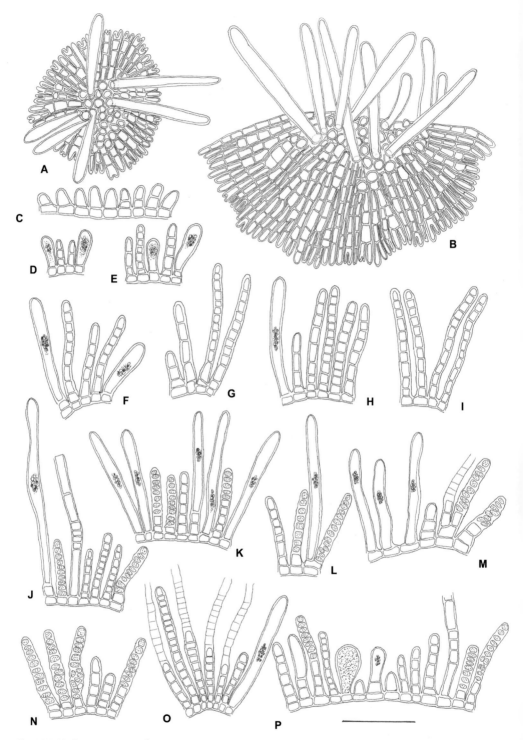

Fig. 164. *Myrionema magnusii*

A, B. Surface view of young thalli. Note central elongated ascocysts. C–I. SP of vegetative thalli showing erect filaments and ascocysts arising from a monostromatic base. J–O. SP of fertile thalli showing ascocysts and plurilocular sporangia. P. SP of fertile thallus showing ascocysts, plurilocular sporangia and a single unilocular sporangium. Bar = 50 μm.

Generally distributed around Britain and Ireland.

Summer annual, May to September, with sporangia.

Culture studies by Loiseaux (1964) revealed *M. magnusii* (recorded as *Ascocyclus magnusii*) to have a 'direct' type of life history. The plurispores behaved asexually and, by the process of heteroblasty, germinated directly to form two morphologically different thalli: ectocarpoid filamentous growths and stellate, later discoid growths. Both growth forms produced plurilocular sporangia, the plurispores from which germinated to repeat the above described developmental processes. A 'direct' life history was also shown for a Swedish isolate by Kylin (1933) (as *Ascocyclus orbicularis*). To date, no evidence supports Parke's idea (1933) that *M. magnusii* (as *A. orbicularis*) is connected in life history with *Cladosiphon contortus* (as *Castagnea contorta*).

Agreement is expressed with Cormaci *et al.* (2012, pp. 198, 199) that *Myronema orbiculare* and *M. magnusii* should be considered as separate species, the former being confined to the Mediterranean and the latter confined to the Atlantic.

Abbott & Hollenberg (1976), p. 158, fig. 124; Brodie *et al.* (2016), p. 1021; Christensen (1958), pp. 129–32, pl. 4; Dixon & Russell (1964), pp. 281, 282; Dizerbo & Herpe (2007), p. 78, pl. 25(1); Fletcher (1987), pp. 109–11, fig. 12, pl. 2(a); Fletcher in Hardy & Guiry (2003), p. 279 (as *Myronema orbiculare*); Gallardo García *et al.* (2016), p. 36; Guiry (2012), p. 147 (as *Myronema orbiculare*); Hamel (1931–39), pp. 100, 101; Kylin (1907), p. 39, fig. 9, (1933), pp. 22–5, fig. 4, (1947), p. 40, figs 32, 33; Levring (1940), pp. 40–3, fig. 8; Loiseaux (1964), pp. 2903–5, figs 1–2, (1967a), pp. 329–47, figs 1, 2a, 4, 5b, pl. 1, (1967b), p. 567; Mathieson & Dawes (2017), pp. 173, 174, pl. LI(3–4); Morton (1994), p. 8; Newton (1931), p. 159, fig. 99; Price *et al.* (1978), pp. 137, 138; Rueness (1977), p. 127, fig. 63, (2001), p. 16; Sauvageau (1927b), p. 13; Schneider & Searles (1991), p. 133; Stegenga & Mol (1983), p. 85, pl. 25, fig. 7; Stegenga *et al.* (1997a), p. 19; Taylor (1957), p. 158; Waern (1952), pp. 149–51, fig. 68(A, B, E).

Myronema papillosum Sauvageau (1897a), p. 242.   Figs 165, 166

Thalli epiphytic, forming light brown spots on hosts, usually solitary and circular, 0.5–1.0 mm across; in surface view, peripheral region consisting of laterally united, outward-radiating, branched, filaments of quadrate, more usually rectangular, cells, 1–3 diameters long, 8–20 × 5–10 μm, each with 1–3 plate-like plastids with pyrenoids, outer central region composed of closely packed, rounded cells, 6–12 μm in diameter, inner central region turf-like, similar to a carpet pile; in squash preparations, comprising a monostromatic basal layer of quadrate to rectangular cells, 1–5 diameters long, 8–22 × 4–7 μm, each giving rise to 1–2 erect filaments, hairs and/or sporangia; erect filaments short, linear or slightly elongate-clavate, loosely associated and gelatinous, multicellular, to 12 cells (170 μm) long, simple, or rarely branched, later commonly giving rise to numerous 1–2(–3)-celled lateral protuberances, comprising cells 1–3(–4) diameters long, 5–21 × 5–8 μm, each with 1–3 plate-like plastids and associated pyrenoids; hairs rare, arising from basal layer, with basal meristem and sheath; rhizoids not observed.

Plurilocular and unilocular sporangia often associated on same thallus; plurilocular sporangia common, terminal or lateral on erect filaments, simple, cylindrical or slightly moniliform, uniseriate, to 5–9 loculi (22–42 μm) long × 7–9 μm wide, with straight, occasionally oblique cross walls; unilocular sporangia infrequent, lateral on erect filaments, sessile, ovate to pyriform, 28–65(–90) × 17–34(–45) μm.

Epiphytic, on *Saccharina latissima* blades, lower eulittoral pools and sublittoral to 5 m.

Only recorded for Dorset in Britain and Cork and Mayo in Ireland but probably of more widespread occurrence.

Summer annual, July to October.

Batters (1900), p. 371; Brodie *et al.* (2016), p. 1021; Cotton (1912), p. 95; Dizerbo & Herpe (2007), p. 80, pl. 25(2); Fletcher (1987), p. 112, fig. 13; Fletcher in Hardy & Guiry (2003), p. 280; Guiry (2012), p. 147; Hamel (1931–39), pp. 91, 92, fig. 24(11–15); Morton (1994), p. 8; Newton (1931), p. 151; Sauvageau (1897a), pp. 242–7, figs 15–17.

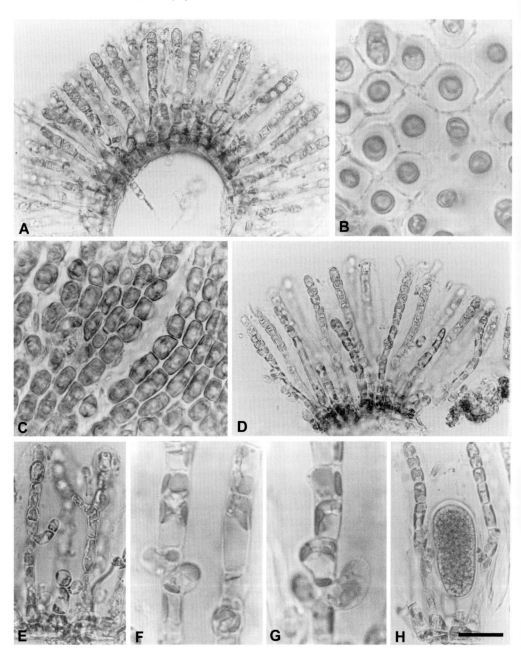

Fig. 165. *Myrionema papillosum*
A. SP of thallus showing the erect filaments. B. Surface view of the erect filaments showing the gelatinous matrix surrounding the cells. C. Surface view of a disc edge, showing the filaments in longitudinal rows. D–G. SP of thalli showing the lateral protuberances on the erect filaments. H. Unilocular sporangium borne at the base of the erect filaments. Bar in H = 40 μm (A, D, E), = 45 μm (B, C, H), = 16 μm (F, G).

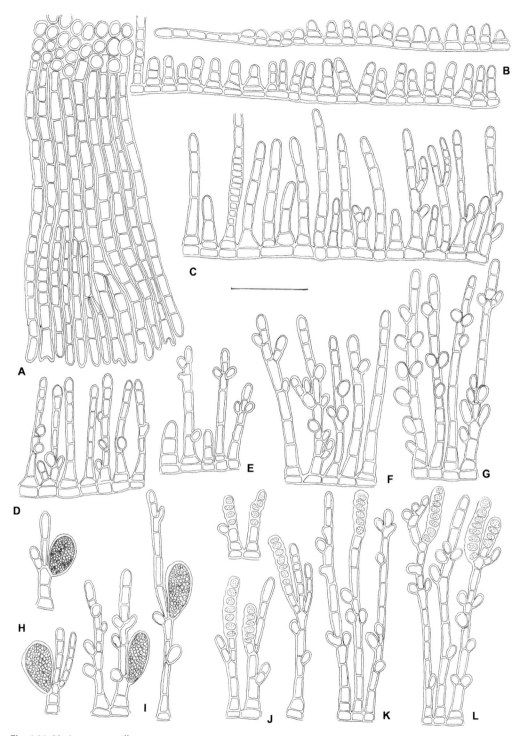

Fig. 166. *Myrionema papillosum*

A. Surface view of peripheral thallus region showing outward-spreading, closely united filaments. B–G. SP of vegetative thallus showing erect filaments arising from a monostromatic base. Note lateral, usually one-celled, protuberances on erect filaments. H, I. SP of fertile thallus showing erect filaments with lateral unilocular sporangia. J–L. SP of fertile thallus showing erect filaments with lateral and terminal plurilocular sporangia. Bar = 50 μm.

Myrionema strangulans Greville (1827a), pl. 300.   Figs 167, 168, 169

Linckia punctiforme Lyngbye (1819), p. 195, nom. inval.; *Myrionema punctiforme* (Lyngbye) Harvey (1833), p. 391; *Myrionema maculiforme* Kützing (1845), p. 264; *Myrionema leclancherii* Harvey (1846), pl. XLI(A); *Myrionema punctiforme* (Lyngbye) Areschoug (1847), p. 380, nom. illeg.; *Myrionema vulgare* Thuret in Le Jolis (1863), p. 82; *Myrionema vulgare* var. *maculaeformis* (Kützing) Piccone (1889), p. 22; *Phaeosphaerium punctiforme* (Lyngbye) Kjellman (1890), p. 41; *Myrionema intermedium* Foslie (1894), p. 13; *Myrionema strangulans* var. *punctiforme* (Harvey) Lyle (1920), p. 10.

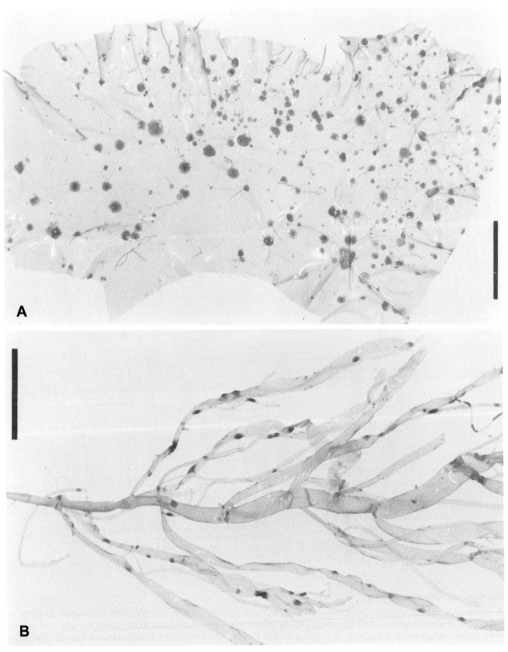

Fig. 167. *Myrionema strangulans*
Habit of thalli on *Ulva* spp. Bars = 10 mm.

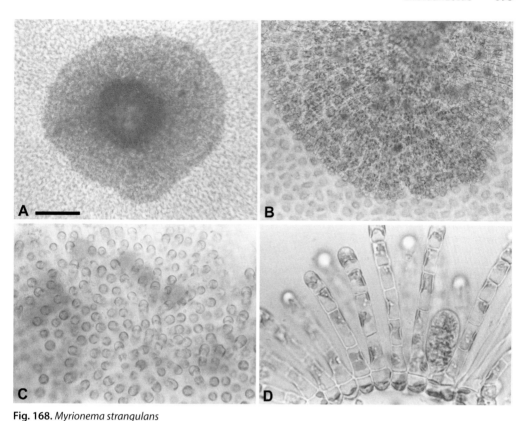

Fig. 168. *Myrionema strangulans*

A, B. Discoid habit of thalli on host. C. Surface view of thalli showing loosely associated erect filaments. D. SP of thallus showing the erect filaments and basal unilocular sporangia. Bar in A = 0.25 mm (A), = 0.15 mm (B), = 50 μm (C), = 20 μm (D).

Thalli epiphytic, forming light brown spots on hosts, usually solitary and circular, 0.5–1 mm in diameter, occasionally confluent to 2–3 mm; in surface view, peripheral region light brown and disc-like in appearance, composed of monostromatic, frequently branched, outward-radiating, loosely united filaments, with terminal apical cells often enclosed within outer cuticle, central region dark brown, pile-like, composed of rounded cells, variously packed, 7–11 μm in diameter, later becoming turf-like, similar to a carpet pile; in structure, consisting of a basal layer of uniseriate cells, 1–4 diameters long, 11–24 × 5–6 μm, each giving rise over the greater part of the thallus to 1–2, erect filaments, hairs and/or sporangia; erect filaments short, simple, loosely associated and gelatinous, elongate-clavate or elongate-cylindrical, 3–7(–9) cells, 55–85 μm long, comprising rectangular cells, 1–5 diameters long, 7–24 × 4–9 μm, each with 1–3 discoid plastids and associated pyrenoids; hairs frequent, arising singly from basal layer, with basal meristem and sheath, approx. 6–8 μm in diameter.

Unilocular sporangia abundant, spherical to pyriform, 39–60 × 22–30 μm, sessile or stalked on basal layer or more commonly lateral at the base of vegetative filaments; plurilocular sporangia less frequent, cylindrical, uniseriate, occasionally biseriate, sessile on basal layer or terminal on erect filaments, 20–45 × 7–11 μm, to 8 loculi long.

Epiphytic on various hosts, in particular *Ulva* spp., littoral fringe to lower eulittoral, in pools or (more rarely) emergent.

Common and widely distributed throughout Britain and Ireland.

Summer annual, found occasionally throughout the year.

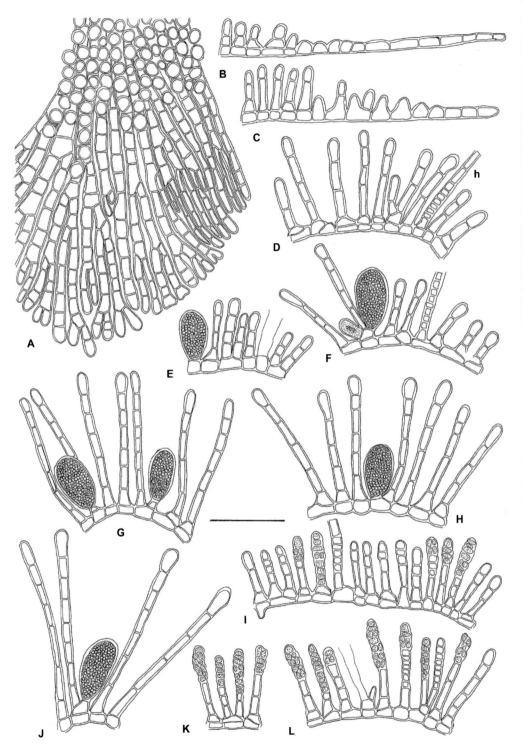

Fig. 169. *Myrionema strangulans*

A. Surface view of peripheral thallus region showing outward-spreading, closely united filaments. B–D. SP of vegetative thallus showing erect filaments and hair (h) arising from a monostromatic base. E–H, J. SP of fertile thallus showing unilocular sporangia sessile or shortly stalked on basal layer. I, K, L. SP of fertile thallus showing plurilocular sporangia. Bar = 50 μm.

M. strangulans has been investigated in culture by Kylin (1934), Loiseaux (1967b) and Pedersen (1981c). A heteromorphic life history appears to be operating with a diploid macrothallus alternating with a haploid, filamentous microthallus. Plurispores released from the microthallus were reported to be sexual and fused to reform the sporophyte (Loiseaux 1967b). The life history was further supplemented by asexual reproduction of unfused gametes and sexual reproduction by the fusion of unispores (Loiseaux 1967b). Heteroblasty was also reported with stellate and filamentous germlings produced from spores originating in the same sporangium; Pedersen (1981c) suggested that this phenomenon might be determined by the initial degree of germling/surface contact.

Abbott & Hollenberg (1976), pp. 158, 159, fig. 125; Ardré (1970), p. 249; Bartsch & Kuhlenkamp (2000), p. 167; Brodie *et al.* (2016), p. 1021; Coppejans & Kling (1995), pp. 172–4, pl. 49(A–N); Cormaci *et al.* (2012), p. 201, pl. 54(1–6); Cotton (1912), p. 95; Dawes & Mathieson (2017), p. 174, pl. LI(7–9); Dizerbo & Herpe (2007), p. 80, pl. 25(3); Fletcher (1987), pp. 112, 114, 115, fig. 14, pls 1, 2(b); Fletcher in Hardy & Guiry (2003), p. 280; Gallardo García *et al.* (2016), p. 36; Guiry (2012), p. 147; Hamel (1931–39), pp. 88–90, fig. 24(1–10); Harvey (1846–51), pl. 280, pl. 41(A–B); Jaasund (1965), p. 57, fig. 17; Kim (2010), pp. 78–80, fig. 37; Kornmann & Sahling (1983), pp. 46–50, figs 27–9; Kylin (1934), pp. 5–9, figs 1–3, (1947), p. 36, fig. 28; Lindauer *et al.* (1961), pp. 205, 206, fig. 35; Loiseaux (1967a), pp. 329–47, figs 2(B), 5(A), (1967b), pp. 529–76; Morton (1994), p. 33; Newton (1931), p. 150, fig. 93; Pedersen (1981c), pp. 194–217, fig. 5.1(D–G); pl. 40, fig. 51; Ribera *et al.* (1992), p. 116; Rueness (1977), pp. 137, 138, fig. 64(A, B), (2001), p. 16; Sauvageau (1897a), pp. 185–229, figs 1–11; Schneider & Searles (1991), pp. 133, 134; Setchell & N.L. Gardner (1925), p. 471, pl. 35(12); Smith (1955), pp. 245, 246, fig. 141(A–C); Stegenga & Mol (1983), p. 85, pl. 26(3–5); Stegenga *et al.* (1997a), p. 19; Taskin *et al.* (2008), p. 60; Taylor (1957), p. 156, pl. 11(13, 14); Womersley (1987), pp. 62, 64, figs 15(A), 16(A–C).

Myriotrichia Harvey

MYRIOTRICHIA Harvey (1834), p. 299.

Type species: *M. clavaeformis* Harvey (1834), p. 300.

Thalli epiphytic forming small, erect tufts on hosts arising from an endophytic, or more commonly, epiphytic base of branched, rhizoidal filaments; tufts comprising several erect, flaccid shoots, uniseriate, becoming multiseriate; cells with several discoid plastids without obvious pyrenoids; hairs abundant, with basal meristem but lacking sheath.

Unilocular and plurilocular sporangia occurring on the same or different shoots, borne on basal filaments, or more commonly on erect shoots; unilocular sporangia sessile or stalked, solitary or grouped, lateral on primary axis or at base of radial branches, spherical; plurilocular sporangia sessile or stalked, solitary or grouped, lateral on primary axes or more commonly lateral and terminal on radial branches, simple or branched, uniseriate to biseriate, cylindrical to conical.

One species in Britain and Ireland.

Myriotrichia clavaeformis Harvey (1834), p. 300.   Figs 170, 171, 172

Myriotrichia filiformis Harvey (1841), p. 44; *Myriotrichia harveyana* Nägeli (1847), p. 149; *Ectocarpus sphaericus* Derbès & Solier in Castagne (1851), p. 100; *Myrionema irregulare* Jaasund (1951), p. 131; *Streblonema sphaericum* (Derbès & Solier in Castagne) Thuret in Le Jolis (1863), p. 73; *Myriotrichia repens* Hauck (1879), p. 242; *Myriotrichia clavaeformis* var. *minima* Holmes & Batters in Holm. Fasc. No. 167; *Dichosporangium repens* (Hauck) Hauck (1883–85), p. 339; *Myriotrichia clavaeformis* f. *filiformis* (Harvey) Kjellman (1890), p. 47; *M. clavaeformis* var. *subcylindrica* Batters (1895b), p. 312.

Thalli forming erect tufts on host, usually visible to the unaided eye, solitary, rarely confluent, comprising several erect shoots attached at the base by spreading, filamentous, branched, rhizoidal filaments; erect shoots filiform, linear, often slightly nodose, becoming

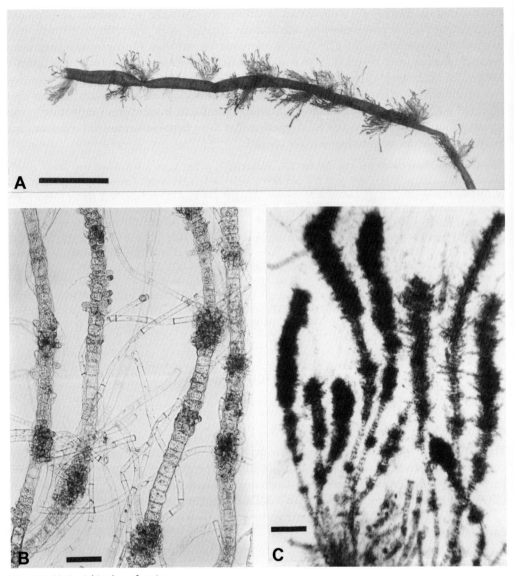

Fig. 170. *Myriotrichia clavaeformis*
A. Habit of thalli on *Scytosiphon lomentaria*. B, C. Portions of erect thalli at different stages of development. Bar in A = 10 mm, bar in B = 80 μm, bar in C = 200 μm.

elongate-clavate, to 6–30(–40) mm long, 0.2–0.5 mm wide, simple or with short radial branches, soft and flaccid in texture; shoots uniseriate and filamentous throughout, or more commonly uniseriate below, multiseriate and parenchymatous above; parenchymatous regions discrete and shoots nodose, sometimes becoming confluent and expansive; surface cells of parenchymatous regions giving rise to lateral branches, hairs and/or sporangia; lateral branches radially produced either irregular and discrete or confluent and closely packed, sheathing the primary axis, short, uniseriate, becoming multiseriate, simple or irregularly and alternatively branched, usually longer in terminal regions contributing to clavate appearance of shoots, later terminated by plurilocular sporangia; cells quadrate, cylindrical or barrel-shaped, 0.5–1(–2) diameters long, 8–36 × 13–20(–40) μm, each with several discoid plastids without obvious

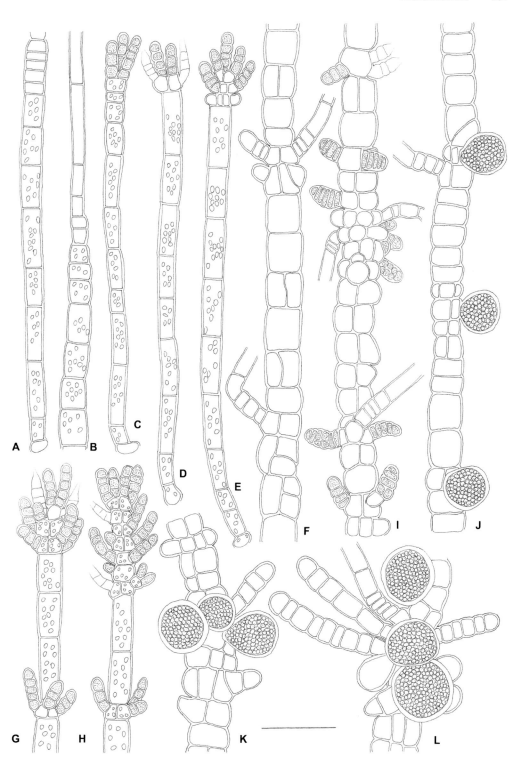

Fig. 171. *Myriotrichia clavaeformis*

A. Short erect filament with terminal hair. B. Terminal portion of erect filament with hair. C–E. Three short filaments with terminal clusters of plurilocular sporangia. F. Portion of erect thallus. G–I. Portions of erect thalli with lateral plurilocular sporangia. J–L. Portions of erect thalli with lateral branches and unilocular sporangia. Bar = 50 μm.

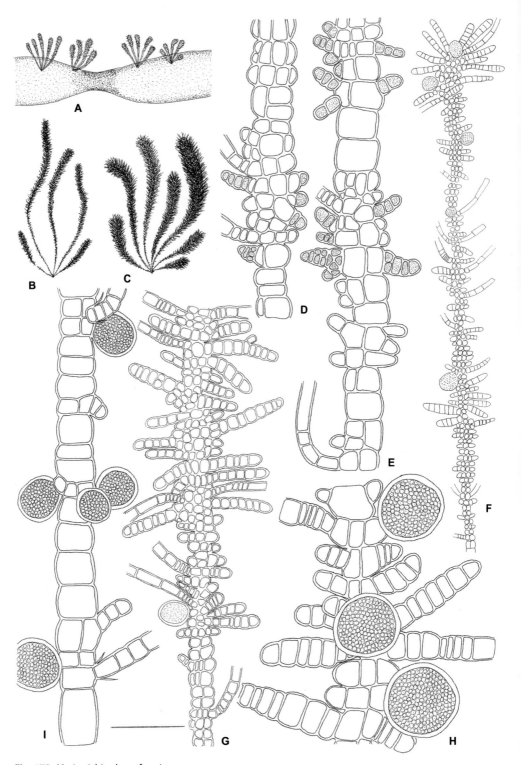

Fig. 172. *Myriotrichia clavaeformis*

A. Habit of thalli on *Scytosiphon lomentaria*. B, C. Habit of thalli. D, E. Portions of erect thalli with plurilocular sporangia. F–I. Four portions of erect thalli with unilocular sporangia. Bar = 6 mm (A), = 2 mm (B, C), = 50 μm (D, E, G, I), = 200 μm (F), = 100 μm (H).

pyrenoids; hairs abundant, 10–13 μm in diameter, colourless except for short, 2–4-celled, pigmented basal meristem, without enclosing sheath.

Unilocular and plurilocular sporangia commonly borne on different thalli, occasionally mixed; unilocular sporangia sessile or stalked on primary axis or lateral branches, single or more commonly clumped in sori, spherical, 20–40(–52) μm diameter; plurilocular sporangia terminal or lateral on lateral branches, single or more commonly clustered in dense sori, simple or branched, uniseriate to biseriate with straight and oblique cross walls, cylindrical, subcylindrical or conical, 4–12 loculi, 20–58 × 7–13 μm.

Endophytic or more commonly epiphytic on various algae and *Zostera* spp., much more commonly recorded on brown algae, in particular *Scytosiphon lomentaria* and species of *Asperococcus*, *Punctaria* and *Stictyosiphon*; in eulittoral pools and shallow sublittoral.

Common and widely distributed around Britain and Ireland.

Annual; summer.

The present treatment agrees with Pedersen (1978b) that the three entities described as *M. repens*, *M. filiformis* and *M. clavaeformis* represent progressive stages in development of the same species, for which *M. clavaeformis* has priority. The form described under *M. repens* is a microscopic epi/endophyte, occurring on various hosts particularly under suboptimal light and salinity conditions. It comprises a basal system of outward-spreading and penetrating rhizoidal filaments giving rise to short simple, erect filaments which are predominantly uniseriate, partly biseriate, bearing hairs and single, or more usually, clustered, unilocular and/or plurilocular sporangia. Occasionally both sporangia can be observed on the basal filaments and in many respects this growth form has been likened to *Streblonema sphaericum* (see Pedersen 1978b for discussion).

With increasing length of the erect shoots and more regular development of parenchymatous knots of tissue along the terminal region of the uniseriate filament, the form described under *M. filiformis* is produced. Very often short branches, unilocular and plurilocular sporangia are associated with these node-like regions on the erect shoot. Finally, with the greater part of the erect shoot parenchymatous and clothed in the lateral branches and sporangia, the form described under *M. clavaeformis* arises. All the above described forms may be observed either separately or in association.

Although *M. clavaeformis* has been investigated in laboratory culture by various authors such as Karsakoff (1892), Sauvageau (1931), Kylin (1933), Dangeard (1965d) and Pedersen (1978b) (as *M. clavaeformis*, *M. filiformis* and/or *M. repens*), most of these studies were often incomplete and the life history is still not fully understood. Evidence points towards the occurrence of a direct, asexual, life history with streblonematoid microthalli. Loiseaux's (1969) report of a *Myriotrichia* sp. (provisionally identified as *M. clavaeformis*) being linked to the life history of the *Hecatonema terminale* microthallus was undoubtedly incorrect (see Pedersen 1978b, p. 290 and Fletcher 1984, p. 193).

Ardré (1970), pp. 276, 277; Batters (1893a), p. 23, (1895b), pp. 312, 313; Brodie *et al.* (2016), p. 1021; Cormaci *et al.* (2012), pp. 204–6, pl. 55(1–7); Cotton (1912), p. 95; Dangeard (1965d), pp. 79–98, pls XVI–XIX, (1969), pp. 78, 79; Dizerbo & Herpe (2007), p. 108, pl. 36(1); Fletcher (1987), pp. 174–7, figs 38, 39, pl. 4(c–e); Fletcher in Hardy & Guiry (2003), p. 281; Funk (1955), p. 41; Gallardo García *et al.* (2016), p. 36; Guiry (2012), p. 147; Hamel (1931–39), pp. 233–9, fig. 46(III, IV, VI); Harvey (1834), p. 300, pl. 138, (1841), p. 44, (1846–51), pls 101, 156; Hauck (1879), p. 242, pl. 4(1–2); Jaasund (1965), pp. 78, 79, fig. 22 (as *Myriotrichia repens*); Karsakoff (1892), pp. 433–44, figs I–II, pl. XIII(1–10); Knight & Parke (1931), p. 64, pl. XI(21–2); Kuckuck (1899a), pp. 21–30, 37–41, pl. 3(1–10); Kylin (1933), pp. 33–6, figs 8–10, (1947), pp. 70–2, fig. 58; Mathieson & Dawes (2017), p. 175, pl. LI(10–12); Morton (1994), p. 39; Newton (1931), pp. 174, 175, fig. 109; Oltmanns (1922), p. 14, fig. 301; Pedersen (1978b), pp. 281–91, figs 1–34; Ribera *et al.* (1992), p. 117; Rosenvinge & Lund (1947), pp. 48–55, figs 16–18; Rueness (1977), pp. 155–7, figs 89, 90, (2001), p. 18; Sauvageau (1897b), p. 36, fig. 1, (1931), pp. 51–77, figs 10–12; Taskin *et al.* (2008), p. 60; Taylor (1957), pp. 161, 162, pl. 10(4–7), Waern (1952), pp. 159–62, figs 71–3; Womersley (1987), p. 304, fig. 110(C).

Phaeostroma Kuckuck in Reinbold

PHAEOSTROMA Kuckuck in Reinbold (1893), p. 43.

Type species: *P. pustulosum* Kuckuck in Reinbold (1893), p. 43.

Thalli epiphytic and endophytic, forming small, brown spots on host less than 1 mm wide and barely visible to the unaided eye; comprising an initial, monostromatic, prostrate basal system of irregularly branched, monosiphonous filaments of rectangular to irregularly shaped cells, either free or laterally associated/united and pseudodiscoid/discoid in appearance, which later can become superimposed and several layers thick; prostrate layer giving rise to hairs and sporangia; true hairs with basal meristem and elongated basal cell, arising from prostrate cells or borne on stalks; cells with several discoid plastids without obvious pyrenoids.

Plurilocular sporangia and unilocular sporangia formed by direct transformation of vegetative cells, lateral or terminal on the filaments; plurilocular sporangia multiseriate.

One species in Britain and Ireland.

Phaeostroma pustulosum Kuckuck in Reinbold (1893), p. 43.   Figs 173, 174

Phaeocladia prostrata Gran (1893a), p. 32; *Streblonema aequale* Oltmanns (1894), p. 214; *Phaeostroma prostratum* (Gran) Kuckuck (1895), p. 185; *Phaeostroma aequale* (Oltmanns) Kuckuck (1897), p. 385; *Entonema aequale* (Oltmanns) Kylin (1947), p. 21.

Thalli epiphytic and endophytic, forming small brown spots on hosts less than 1 mm wide and barely visible to the unaided eye; comprising a prostrate, basal system of irregularly branched, spreading filaments, either free or pseudodiscoid/discoid depending on the host, usually monostromatic sometimes becoming distromatic to tristromatic by longitudinal cell divisions; basal system giving rise to hairs and sporangia; prostrate cells rectangular to irregularly shaped, 1–3.5 diameters long, 9–36 × 5–16 µm; hairs arising from the basal layer, usually comprising 1–4 basal cells, the lowermost cells being noticeably elongate and cylindrical, the uppermost cells being more barrel shaped to bulbous in appearance, the hair cells being 2–9 diameters long, 22–70 × 6–9 µm; cells with several discoid plastids.

Plurilocular sporangia and unilocular sporangia formed by direct transformation of the vegetative cells or hairs; plurilocular sporangia usually on 1–3-celled stalks, multiseriate, irregular in shape but sometimes heart-shaped, 18–40 × 18–38 µm; unilocular sporangia usually laterally placed on the prostrate filaments, sessile, or on 1–2-celled pedicels, oval, reported by Hamel (1931–39, p. 105) to measure 43 × 24 µm.

Endo-epiphytic on various hosts, including *Zostera* spp., *Laminaria* spp., *Scytosiphon lomentaria*, *Porphyra umbilicalis* and particularly on *Chorda filum*, in pools or emersed, lower eulittoral and shallow sublittoral.

Rarely recorded in Britain (South Devon, Northumberland, Fifeshire, Argyllshire, Anglesey) and Ireland (Antrim, Cork, Mayo) but probably more common and more widely distributed than records suggest.

Appears to be summer/early autumn annual.

A morphologically variable species, depending on the host. On firm hosts it is epiphytic and forms pseudoparenchymatous discs, whereas on softer hosts it can be endophytic and forms filamentous thalli. On host species such as *Chorda filum*, the thalli are found both on the surface of the host and ramifying through the outer cortical layers, and give rise to erect, stalked sporangia arising through the reproductive sori of the host and taking on a shape very similar to their surroundings, as noted by Jaasund (1965) for the host *Chordaria*.

Culture studies by Pedersen (1981b) revealed a direct life history with spores released from both types of sporangia germinating directly to repeat the parental phase.

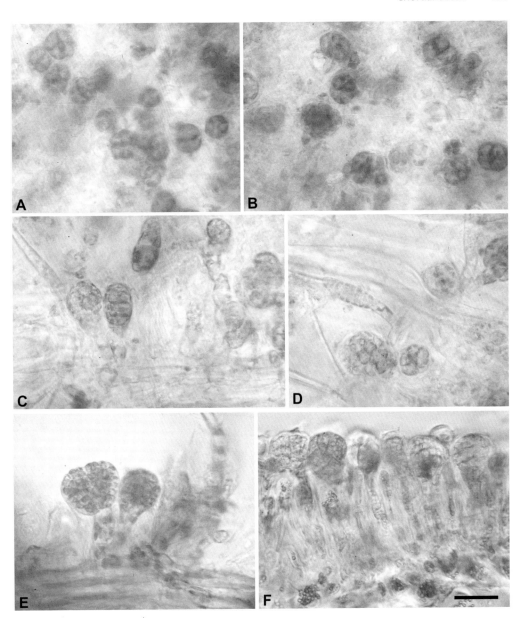

Fig. 173. *Phaeostroma pustulosum*

A–D. Surface view of host (*Chorda*) showing plurilocular sporangia, prostrate filaments and hairs. E, F. TS of host (*Chorda*) showing the plurilocular sporangia of the *Phaeostroma* terminal on filaments. Bar in F = 35 μm (A–F).

Bartsch & Kuhlenkamp (2000), p. 167; Batters (1902), p. 29; Blackler (1974), p. 243; Brodie *et al.* (2016), p. 1021; Dizerbo & Herpe (2007), p. 71; Fletcher in Hardy & Guiry (2003), p. 285; Gallardo García *et al.* (2016), p. 36; Guiry (2012), p. 154; Hamel (1931–39), pp. 69, 70, 103, 104, fig. 20(f); Jaasund (1951), p. 137, (1963), pp. 5, 6, figs 3(A–F), 5(C–E), (1965), pp. 51, 52; Kuckuck (1895b), p. 185, pl. 7(1–12), (1897a), pp. 385, 386, fig. 11(A–D); Kylin (1907), pp. 48, 49, (1947), pp. 21, 25, figs 17(A–D), 22(A–B); Levring (1940), pp. 24–6, 30, fig. 2(A–F); Lund (1959), pp. 63, 64; Mathieson & Dawes (2017), pp. 177, 178, pl. LII(6, 7); Morton (1994), p. 32; Newton (1931), pp. 125, 126, 128, fig. 73; Oltmanns (1894), p. 214, pl. III(14–16); Pankow (1971), pp. 158, 159, figs 178, 179, 185; Pedersen (1976), pp. 36, 37, figs 17, 18, (1981d), pp. 271–6, figs 24, 25, (2011), p. 102; Printz (1926), p. 147, pl, 3, figs 20–4; Rienbold (1893), p. 43; Rosenvinge (1898), pp. 68–70, fig. 15, (1910), pp. 117, 118, Rosenvinge & Lund (1941), pp. 61, 65; Rueness (1977), pp. 119, 122, figs 42, 45, (2001), p. 16; Sauvageau (1897c), pp. 181–3; Sundene (1953), p. 156.

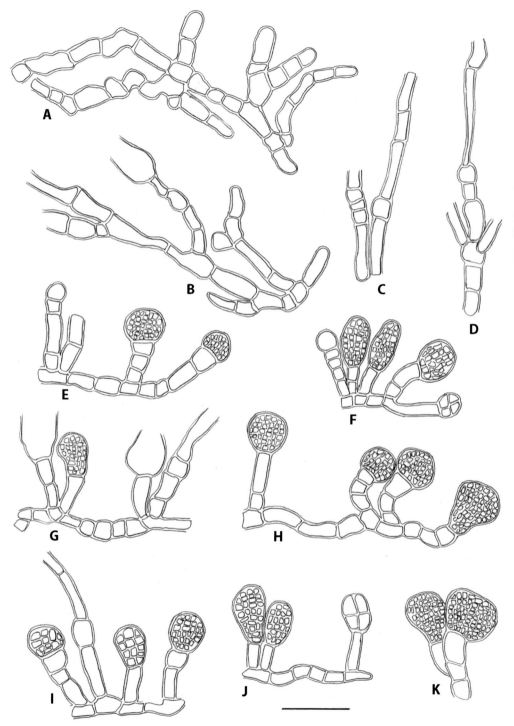

Fig. 174. *Phaeostroma pustulosum*
A–D. Surface view of thallus showing branched, prostrate filaments and hairs. E–K. SP of fertile thalli showing erect filaments, hairs and terminal plurilocular sporangia. Bar = 50 μm.

Pilinia Kützing

PILINIA Kützing (1843), p. 273.

Type species: *P. rimosa* Kützing (1843), p. 273.

Thalli epilithic, forming confluent, turf-like growths, comprising an attachment system of basally spreading, branched filaments, giving rise to short, simple or sparsely branched, erect filaments; branching pseudodichotomous, irregular, sometimes unilateral with branches recurved towards parental filament; cells small, < 10 μm in diameter with 1–2 plate-like plastids with a single pyrenoid.

Plurilocular sporangia unknown; unilocular sporangia terminal on erect filaments, formed by direct transformation of vegetative cells, uniseriate or biseriate, usually arranged in a secund series and laterally protruding in a cock's-comb fashion, each with four zoospores.

One species in Britain and Ireland.

Pilinia rimosa Kützing (1843), p. 273. Fig. 175

Leptonema lucifugum Kuckuck (1897a), p. 364; *Waerniella lucifuga* (Kuckuck) Kylin (1947), p. 26; *Leptonematella lucifuga* (Kuckuck) P.C. Silva (1959), p. 63; *Waerniella lucifuga* var. *australis* A.B. Cribb (1965), p. 271.

Thalli epilithic, forming solitary to confluent and expansive, light yellow-brown, firm, spongy, densely tufted, turf-like patches, to several cm (-m) in extent but less than 1 mm in height; comprising an attaching basal region of outward-spreading, branched, uniseriate filaments giving rise to closely packed, erect filaments; basal filament cells irregularly shaped and contorted to rounded, thick-walled, 11–15(–28) × (6–)8–11(–15) μm; erect filaments to 400 μm high, uniseriate, free, simple or more commonly sparingly branched, especially at the base, with branching pseudodichotomous to irregular, sometimes unilaterally, with branches recurved adaxially towards parental filament and frequently exceeding length of parental axis, comprising cells quadrate, subquadrate to more commonly rectangular to barrel-shaped, 1–2(–3.5) diameters long, 8–13(–18) × 5–10 μm, each containing 1–2 plate-like plastids with a single pyrenoid; basal cells of erect filaments giving rise to secondary rhizoids; growth probably diffuse; hairs not observed.

Unilocular sporangia formed in rows, terminal or intercalary on erect filaments, formed by direct transformation of vegetative cells, uniseriate to biseriate, sometimes arranged in a cock's-comb fashion, with straight or oblique cross walls, to 13 μm in diameter, each sporangium containing only four zoospores.

Epilithic, on the walls of caves, also on pilings, boulders in shaded areas high up in the eulittoral, usually forming a distinct zone above *Rhodochorton purpureum*, *Sphacelorbus nanus* and *Gaillona hookeri*.

Rare, but widely distributed all around Britain (South Devon, Cheshire (Hibre Island), Bute, Hampshire, Kent, Northumberland, Fifeshire) and Ireland (Cork, Galway, Mayo, Wexford); can be locally abundant.

Perennial, and probably winter fertile with unilocular sporangia observed in October, November, January and March.

Pilinia rimosa was originally placed in the green algae (Kützing 1843), and included in floras as such (e.g. in Taylor 1957; Newton 1931; Burrows 1991). Hooper *et al.* (1987), on the basis of light microscopy and later confirmed by electron microscopy by O'Kelly (1989), showed it be a brown alga and identical to the species *Waerniella lucifuga* (Kuckuck) Kylin. The reproductive structures are unique. Long thought to be plurilocular sporangia, they were shown in an electron microscope study by O'Kelly (1989) to be very small unilocular sporangia, containing only four spores.

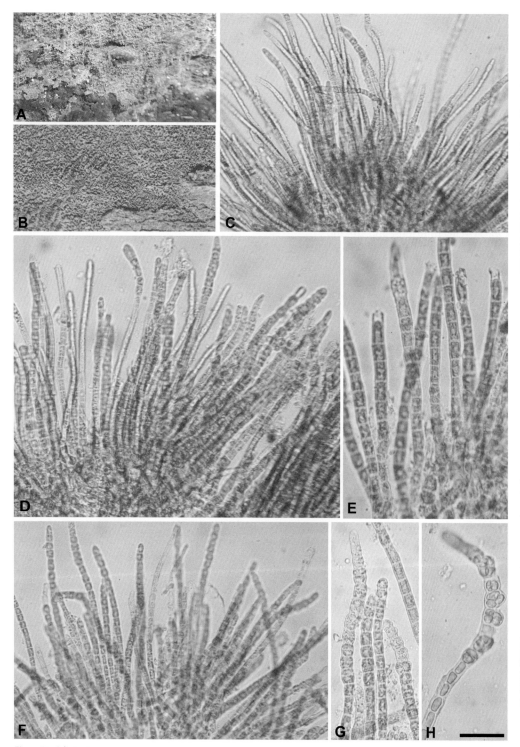

Fig. 175. *Pilinia rimosa*

A, B. Turf-like habit of thalli on the walls of caves. C–F. SP of thalli showing the densely packed, erect filaments. G, H. Early and late stages in the transformation of the terminal cells of the filaments into unilocular sporangia. Bar in H = 6 cm (A), = 2 cm (B), = 55 μm (C, D), = 45 μm (E, G), = 80 μm (F) = 55 μm (H).

Bartsch & Kuhlenkamp (2000), pp. 167, 179; Batters (1906), p. 2; Blackler (1956), p. 52, (1974), p. 245; Børgesen (1902), pp. 415, 416, fig. 74; Brodie *et al.* (2016), p. 1021; Cormaci *et al.* (2012), p. 263, pl. 75(5–8); Dizerbo & Herpe (2007), p. 75, pl. 23(3); Fletcher in Hardy & Guiry (2003), p. 286; Gallardo García *et al.* (2016), p. 36; Guiry (2012), p. 154; Hamel (1931–39), pp. 128, 129, fig. 29(e); Hooper *et al.* (1987), pp. 439, 440; Kuckuck (1897a), pp. 362–4, pl. 7(20–4), (1897b), p. 387; Kylin (1947), p. 26, fig. 23; Levring (1940), pp. 47, 48; Mathieson & Dawes (2017), p. 193, pl. LV(13, 14); Morton (1994), p. 8; O'Kelly (1989), pp. 369–74, figs 1–18; Pankow (1971), pp. 159, 160, figs 183, 184; Rosenvinge & Lund (1935), p. 40, fig. 40; Rueness (1977), p. 124, fig. 49, (2001), p. 16; Silva *et al.* (1996), p. 717; Skuja (1928), pp. 39–44, pl. 1(1–10); Stegenga & Mol (1983), p. 81; Stegenga *et al.* (1997a), p. 17; Waern (1936), pp. 329–42, (1940), pp. 1–6, (1952), pp. 118–20, fig. 52.

Protectocarpus Kornmann

PROTECTOCARPUS Kornmann (1955), p. 119.

Type species: *Non designatus.*

Thalli epilithic or epizoic, forming small, gregarious tufts up to 1(–3) mm high; consisting of a monostromatic basal layer of outward-spreading, pseudodichotomously branched filaments which are laterally adjoined and discoid or irregularly associated and pseudodiscoid in appearance, most cells of which, except at the periphery, give rise to erect filaments, hairs and/or sporangia; erect filaments simple or branched to the first order, with branches often in unilateral rows and recurved towards parental filament; cells with 1–2 plate-like, lobed plastids with pyrenoids; hairs terminal or lateral on erect filaments, with basal meristem and sheath.

Plurilocular sporangia uniseriate to multiseriate, either sessile or stalked on basal layer, more commonly terminal or lateral, often in unilateral rows, on the erect filaments, simple or with unilateral, branch-like extensions; unilocular sporangia sessile or shortly stalked on the basal layer, or lateral on the erect filaments.

One species in Britain and Ireland.

Protectocarpus speciosus (Børgesen) Kornmann (1955), p. 119.   Figs 176, 177

Myrionema speciosum Børgesen (1902), p. 421; *Myrionema faeroense* Børgesen (1902), p. 424; *Hecatonema diffusum* Kylin (1907), p. 39; *Hecatonema speciosum* (Børgesen) Cotton (1912), p. 121; *Ectocarpus speciosus* (Børgesen) Kuckuck in Oltmanns (1922), p. 13; *Compsonema speciosum* (Børgesen) Setchell & N.L. Gardner (1922), p. 356; *Hecatonema faeroense* (Børgesen) Levring (1937), p. 47.

Thalli epiphytic, more rarely epilithic and epizoic, forming small solitary or confluent tufts, just visible to the unaided eye; comprising a monostromatic base of outward-spreading, frequently branched filaments, more or less laterally adjoined and discoid/pseudodiscoid in appearance, the uniseriate cells of which give rise in the central regions to erect filaments, hairs and/or sporangia; in surface view, outer basal cells rectangular 1–2(–3) diameters long, 6–22 × 5–10 µm, in squash preparation, inner basal cells, rectangular, quadrate less frequently ovate, slightly longer than wide, 8–13 × 4–7 µm; erect filaments free, loosely associated, slightly gelatinous, simple or more commonly sparingly branched, to 570 µm high (38 cells), comprising uniseriate cells 1–3(–5) diameters long, 7–39 × 6–10 µm, each with a single, terminal, parietal, plate-like plastid with pyrenoid; branching usually to one order only, with branches short, pinnate and secund, later often being transformed into plurilocular sporangia; hairs common, arising from basal cells or more usually terminal or lateral on erect filaments and branches, with basal meristem and sheath.

Plurilocular sporangia abundant, sessile or stalked on basal cells, more commonly terminal and lateral on erect filaments; lateral sporangia sessile or stalked, usually arranged in terminal regions of filament; sporangia elongate-cylindrical, tapering upwards, simple or branched,

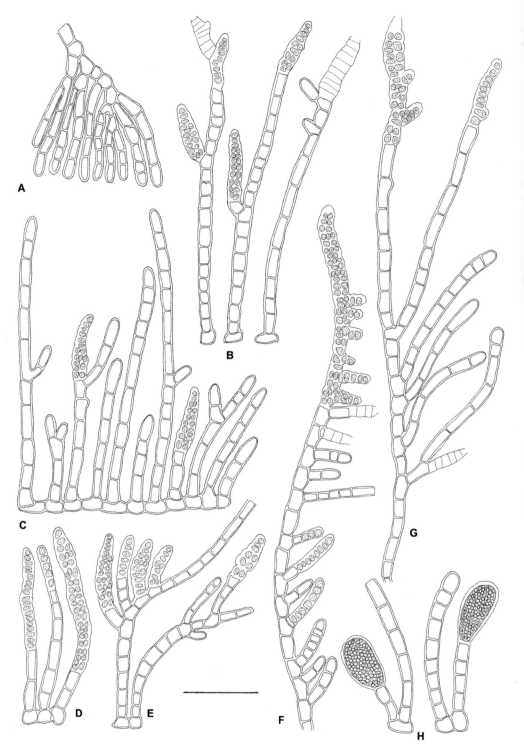

Fig. 176. *Protectocarpus speciosus*
A. Surface view of portion of peripheral thallus region showing outward-spreading basal filaments. B–G. SP of fertile thallus showing erect filaments arising from a monostromatic base, with terminal and lateral plurilocular sporangia. H. SP of fertile thallus showing unilocular sporangia. Bar = 50 μm.

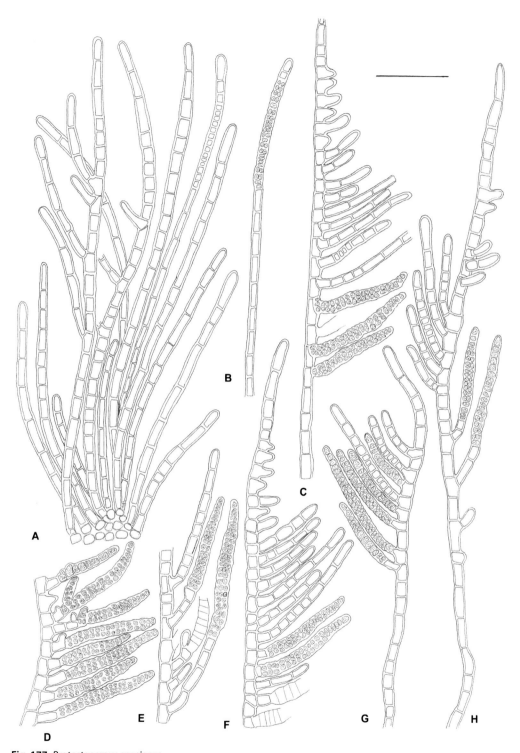

Fig. 177. *Protectocarpus speciosus*

A. SP of vegetative thallus. B–H. SP of fertile thalli showing portions of erect filaments with plurilocular sporangia. Bar = 50 μm.

with branches usually short, secund, widely divergent and arranged in a cock's-comb fashion, uniseriate or more commonly biseriate to multiseriate, with straight or oblique cross walls, 32–100(–130) × 9–16 μm; unilocular sporangia common, on basal layer or more commonly lateral at the base of erect filaments, usually stalked, elongate-ovate, 40–60 × 22–25 μm.

Epiphytic on various algae, less commonly epizoic on shells and epilithic on stones, eulittoral and shallow sublittoral.

Recorded for scattered localities around Britain (Shetland, Orkney, East Lothian, Essex, Hampshire, Channel Islands, Isles of Scilly, Inner Hebrides, Outer Hebrides) and Ireland (Cork, Wexford, Mayo).

Spring and summer annual.

Bartsch & Kuhlenkamp (2000), p. 167; Børgesen (1902), pp. 421–4, fig. 78, (1926), pp. 55–9, figs 29–30; Brodie *et al.* (2016), p. 1021; Cardinal (1964), p. 77; Cormaci *et al.* (2012), pp. 214, 216, pl. 58(2, 3); Cotton (1912), p. 121; Dizerbo & Herpe (2007), p. 80; Fletcher (1987), pp. 116–19, figs 15, 16; Fletcher in Hardy & Guiry (2003), p. 287; Gallardo García *et al.* (2016), p. 36; Guiry (2012), p. 148; Jaasund (1965), pp. 53–5, fig. 16; Kim (2010), pp. 81–3, fig. 38; Kornmann & Sahling (1977), p. 275, fig. 158; Kuckuck (1955), pp. 120–35, figs 1–8; Kylin (1907), pp. 39–41, (1947), pp. 17, 18, fig. 13; Mathieson & Dawes (2017), p. 179, pl. LIII(1–3); Morton (1994), p. 8; Newton (1931), p. 157; Oltmanns (1922), p. 11, fig. 300; Ribera *et al.* (1992), p. 116; Rueness (1977), pp. 138, 139, fig. 65, (2001), p. 16; Stegenga & Mol (1983), p. 89, pl. 27(1–3); Stegenga *et al.* (1997a), p. 19; Tanaka (1986), pp. 287–92; Taskin *et al.* (2008), p. 60.

Punctaria Greville

PUNCTARIA Greville (1830), p. 52.

Type species: *P. plantaginea* (Roth) Greville (1830), p. 53.

Diplostromium Kützing (1843), p. 298; *Phycolapathum* Kützing (1843), p. 299; *Desmotrichum* Kützing (1845), p. 244; *Homoeostroma* J. Agardh (1896), p. 7; *Nematophlaea* J. Agardh (1896), p. 12; *Rhadinocladia* Schuh (1900), p. 3.

Thalli consisting of erect flattened blades arising from a small stipe and discoid holdfast; blade simple, entire or with marginal proliferations, solid, flaccid or subcoriaceous, 2–4(–9) cells thick, central medullary cells if present, large, thick-walled, colourless, longitudinally elongate, transversely rounded, outer cortical cells smaller, pigmented; in surface view, cells large, in rows, each with numerous discoid plastids and associated pyrenoids; hairs single or grouped, with basal meristem and enclosing sheath.

Unilocular and plurilocular sporangia on the same or different blades, extensive on both surfaces, solitary or grouped, formed in surface cells; unilocular sporangia infrequent or rare, ovate; plurilocular sporangia common, quadrate or polygonal in surface view, conical, sometimes bifurcate, protruding above surfaces in transverse section.

The genus *Punctaria* is principally characterized by an erect, simple, flattened blade, variable in dimension and shape, usually becoming several cells thick and with only slight internal differentiation into an inner medulla of large colourless cells and an outer cortex of 1–3 small photosynthetic cells. The cells contain several discoid plastids with pyrenoids, hairs have a basal meristem and sheath and the multiseriate sporangia develop within the cortical cells. These characteristics should enable the genus to be distinguished from other morphologically similar genera such as *Petalonia* and *Laminaria*.

The early stage of development comprises an erect uniseriate filament terminated by one or several hairs. Transverse and longitudinal divisions then produce a laterally expanded blade mainly two cells thick. These cells then divide parallel to the blade surface to produce a thallus four cells thick, the outer cells of which then assume the main photosynthetic role and become heavily pigmented. Further transverse and longitudinal divisions of these cells are also responsible for the increased thickness of the blades seen in some species and the formation of the sporangia that frequently project above the surface of the thallus.

The extent of the development of the erect blades is an important character used in the recognition of the species. In *P. tenuissima* (which includes *Desmotrichum undulatum* in the present treatment), lateral development of the blades appears to be limited (to less than 12 mm) and they remain ribbon-shaped. The thickness of the blades also rarely exceeds 60 μm and usually comprises two, sometimes four cell layers. Some very reduced forms of this species have also been described (formerly under *Desmotrichum*) with very narrow blades, which can be uniseriate in parts. Much wider and thicker blades, on the other hand, are characteristic of the other *Punctaria* species found in Britain and Ireland. However, some doubt must be expressed on the validity of these characters in species identification. Considerable overlap does seem to occur amongst British and Irish material. The usefulness of such characters as blade dimension, shape, colour and thickness, the presence or absence of hairs, whether hairs are solitary or grouped, present on the blade surface or in the margin, the dimensions and extent of protrusion of the sporangia, is undoubtedly questionable and may well have to be abandoned. A particularly noteworthy paper by Rietema & van den Hoek (1981) has revealed many of these characters to be modified by environmental conditions of temperature, daylength and light intensity.

Life history studies of a number of species of *Punctaria* have indicated that a 'direct' monophasic life history is operating with unispores and/or plurispores producing small microthalli, from which budded thalli identical with the parents, i.e. with apparently apomeiotic unilocular sporangia. This has been shown for *P. latifolia* in both the North Atlantic and Australia (Sauvageau 1929, 1933a; Dangeard 1963a, 1966a (as *P. crouanii*); Clayton & Ducker 1970; Pedersen 1984), for *P. plantaginea* and *P. orbiculata* Jao collected in Newfoundland and Washington respectively (South 1980) and for *P. tenuissima* (as *Desmotrichum undulatum*) on the west coast of Sweden (Kylin 1933), the north-east coast of North America (Rhodes 1970; Lockhart 1982), the Netherlands (Rietema & van den Hoek 1981) and Denmark (Pedersen 1984). Fertile microthalli (with plurilocular sporangia) were frequently observed in the cultures that Pedersen (1984) identified with both the genus *Hecatonema* and the species *Streblonema effusum* Kylin. It is also of note that cultures of *Hecatonema maculans* (= *H. terminale*) were reported by Clayton (1974) and Fletcher (1984) to give rise to macrothalli resembling *Punctaria* and *Desmotrichum* (= *Punctaria* herein). The relationship between these two morphological forms (i.e. microthallus and macrothallus) appeared to be facultative and influenced by environmental conditions (particularly temperature, daylength and light intensity) rather than differences in ploidy level. The only report of a sexual life history in *Punctaria* was given by Knight (1929), in which she claimed to have observed copulation of unispores in *P. plantaginea*: in view of the results of later publications this now needs confirmatory studies.

Three species in Britain and Ireland, one species in Britain only.

KEY TO SPECIES

1. Thallus dark brown, 6–8 cells in thickness, firm and subcoriaceous,
   plurilocular sporangia rare or absent. 2
   Thallus light/olive-brown, 2–4 (rarely 6) cells in thickness, soft and flaccid,
   plurilocular sporangia common 3

2. Thallus lanceolate becoming oblong or orbicular, to 70 × 24 cm, with ruffled
   margins; restricted to the Isles of Scilly . . . . *P. crispata*
   Thallus lanceolate becoming obovate, to 30 × 5 cm with smooth margins;
   widely recorded around Britain and Ireland . . . . *P. plantaginea*

3. Thallus to 8 cm in width, ovate, linear-lanceolate or oblong in shape; 2–4
   (rarely 6) cells in thickness. . . . . *P. latifolia*
   Thallus to 2 cm in width, elongate-lanceolate to ribbon shaped; 1–2 (rarely 4)
   cells in thickness . . . . *P. tenuissima*

Punctaria crispata (Kützing) Trevisan (1849), p. 428. Fig. 178

Phycolapathum crispatum Kützing (1843), p. 299; *Punctaria laminarioides* P. Crouan & H. Crouan (1867), p. 167.

Thalli forming erect, dark brown, dorsiventrally flattened blades arising from a short, cylindrical stipe and discoid holdfast; erect blades solitary or gregarious, simple, often irregularly split, solid, firm and subcoriaceous, oblong, orbicular or lanceolate, to 30(–70) cm long, 24 cm broad, with ruffled margins, tapering sharply below; surface cells either in rows or irregularly placed, markedly thick walled, quadrate, rectangular or irregular in shape, 15–39 × 13–26 µm, each with numerous discoid plastids with pyrenoids; in transverse section, blades to 180(–220) µm thick, 6–8 cells, slightly differentiated into inner, large, colourless cells and outer (1–2 cells) smaller, pigmented cells; hairs not observed.

Unilocular sporangia scattered on both sides of blade, slightly immersed, formed from within surface cells, orbicular or ovate in surface view, 26–35 × 22–30 µm, hemispherical or quadrilateral, usually slightly protruding, more rarely sunken in transverse section, 28–40 × 23–30 µm; plurilocular sporangia unknown.

Epiphytic on *Zostera* spp., lower eulittoral pools.

Only known for the Isles of Scilly in Britain and Co. Clare in Ireland.

Annual; summer (May to October).

The distinguishing features of this species include the large, dark brown, subcoriaceous, peripherally ruffled blades, easily mistaken for *Saccharina latissima*, the apparent absence of hairs, the relatively small, inconspicuous unilocular sporangia which are sometimes sunken and the absence of plurilocular sporangia.

Batters (1900), pp. 372, 372; Brodie *et al.* (2016), p. 1021; Dizerbo & Herpe (2007), p. 114; Fletcher (1987), p. 206, fig. 51; Fletcher in Hardy & Guiry (2003), p. 289; Guiry (2012), p. 149; Hamel (1931–39), pp. 216, 217, fig. 44(6); Mathieson & Dawes (2017), p. 180, pl. LIII(4, 5); Newton (1931), p. 186.

Punctaria latifolia Greville (1830), p. 52. Figs 179(A), 180(A), 181

Punctaria plantaginea var. *crouanii* Thuret in Le Jolis (1863), p. 70; *Punctaria crouaniorum* (Thuret) Bornet & Thuret (1876–80), p. 15; *Punctaria hiemalis* Kylin (1907), p. 70; *Homoeostroma latifolium* Okamura (1916), p. 8.

Thalli forming erect, light or greenish brown, membranous, dorsiventrally flattened blades arising from a small discoid holdfast; erect blades simple, sometimes proliferous around margins, solitary or more commonly gregarious, flaccid, slightly lubricous, solid, elongate-ovate, linear-lanceolate or oblong, to 30(–45) × 2–4(–8) cm, with flat or ruffled margins, tapering below fairly sharply to a short, narrow stipe, 2–4(–8) mm in length; surface cells in rows, quadrate to rectangular, 17–30(–50) × 11–26(–35) µm, each with large numbers (up to 30) of discoid plastids and associated pyrenoids; in transverse section, blades 60–120(–160) µm thick, 2–4(–6) cells, either with quadrate or subquadrate cells (in 2–4-celled thick thalli) or with large, colourless, thick-walled, rounded, central cells (usually elongate-rectangular in longitudinal section) and small, quadrate, subquadrate to vertically elongate, pigmented, surface cells (in 4–6-celled thick thalli); hairs abundant, arising from surface or subsurface cells, single or more commonly grouped, 10–16 µm in diameter, fairly flaccid, with short basal meristem and enclosing sheath.

Plurilocular and unilocular sporangia borne on the same or different blades, widely distributed on both surfaces, formed within surface cells; plurilocular sporangia common, solitary and scattered, or grouped in extensive patches, formed by direct subdivision of surface cells, more commonly formed in 2–4 daughter cells following vegetative divisions, quadrate to rectangular, 14–42(–59) × 12–26(–42) µm in surface view, subquadrate or more commonly conical and projecting, sometimes bifurcate, superficial or immersed, 25–46 × 13–26 µm in

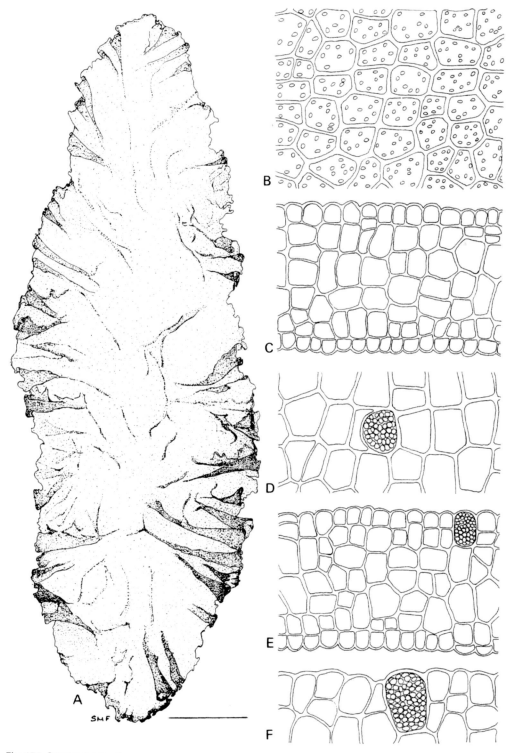

Fig. 178. *Punctaria crispata*
A. Habit of thallus. B. Surface view of vegetative thallus showing cells with several discoid plastids. C. VS of vegetative thallus. D. Surface view of fertile thallus showing unilocular sporangia. E, F. VS of fertile thalli showing unilocular sporangia. Bar = 4 cm (A), = 50 μm (B, D, F), = 100 μm (C, E).

Fig. 179. *Punctaria* spp.
Habit of thalli. A. *Punctaria latifolia*. B, C. *Punctaria plantaginea*. Bar in C = 2 cm (A), = 5 cm (B), = 2 cm (C).

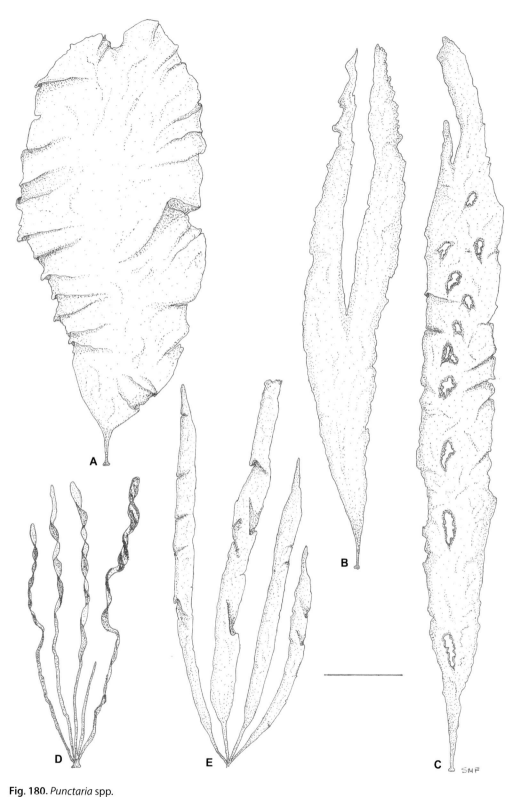

Fig. 180. *Punctaria* spp.

Habit of thalli. A. *Punctaria latifolia*. B, C. *Punctaria plantaginea*. D, E. *Punctaria tenuissima*. Bar = 22 mm (A), = 21 mm (B), = 19 mm (C), = 30 mm (D, E).

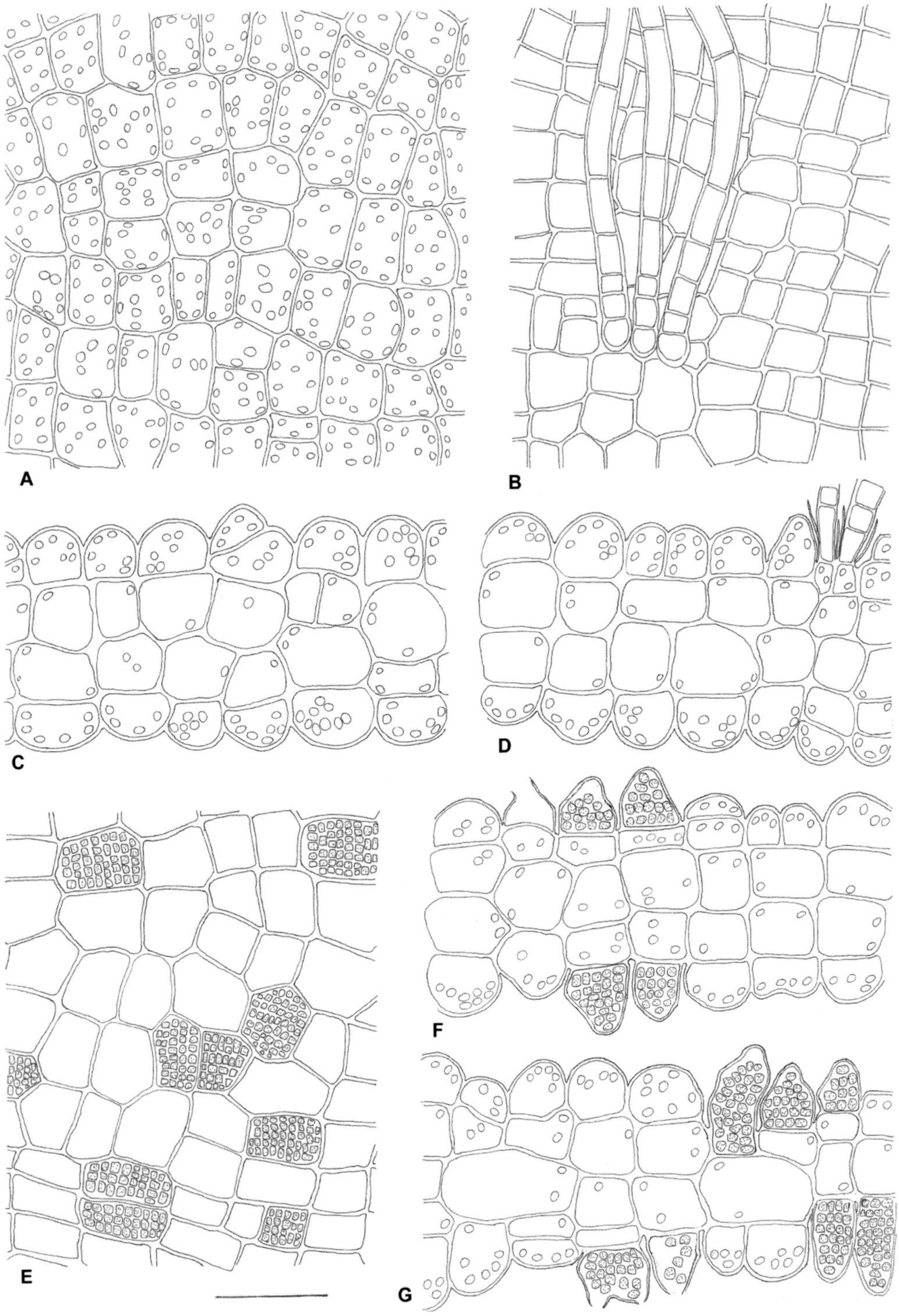

Fig. 181. *Punctaria latifolia*
A. Surface view of thallus showing cells with numerous discoid plastids. B. Surface view of thallus showing hairs. C, D. VS of vegetative thalli. E. Surface view of fertile thallus showing plurilocular sporangia. F, G. VS of fertile thalli showing plurilocular sporangia. Bar = 50 μm.

transverse section; unilocular sporangia rare, orbicular or ovate, 40–50 × 36–43 μm in surface view, ovate or vertically elongate in transverse section.

Epiphytic on a wide variety of algae, less commonly epilithic, upper to lower eulittoral pools and sublittoral to 15 m.

Common and widely distributed around Britain and Ireland.

Annual; summer.

Included here within the synonymy of *P. latifolia* is the morphologically and anatomically similar *P. crouaniorum* which, according to Hamel (1931–39), is primarily distinguished by the occurrence of elongate plurilocular sporangia. However, such erect, protruding, plurilocular sporangia have been observed on three species of British and Irish material and, therefore, the taxonomic importance of this character is questionable. They are very common on *P. latifolia* in agreement with French material of this species described by Dangeard (1963a) and Australian material described by Clayton & Ducker (1970).

Particularly interesting are the laboratory culture studies of Clayton (1974) and Fletcher (1984b) in which they revealed a life history connection between *Punctaria latifolia* and *Hecatonema terminale*, and more specifically the deuxième forme and troisième forme of Sauvageau (1897). However, in view of reports of *H. terminale* also having an independent life history (Edelstein *et al.*, 1971; Clayton, 1974), pending further life history and molecular studies, the latter species is retained as a separate taxon here.

Ardré (1970), p. 275, pl. 37(1); Bartsch & Kuhlenkamp (2000), p. 168; Brodie *et al.* (2016), p. 1021; Cabioc'h (1976), pp. 23–6, fig. 1; Clayton & Ducker (1970), pp. 293–300, figs 1–10; Coppejans & Kling (1995), pp. 182–4, 186, pls 55(A–E), 56(A–D); Cotton (1912), p. 94; Dangeard (1963a), pp. 205–24, pls 1–3, (1966a), pp. 157–67, pls 1–3, (1969), pp. 70–2, fig. 2A; Dizerbo & Herpe (2007), pp. 114, 115, pl. 38(2); Fletcher (1987), pp. 209–11, figs 52(A), 53, pl. 6(a); Fletcher in Hardy & Guiry (2003), p. 289; Fritsch (1945), p. 99, fig. 31(C–H, L); Funk (1955), p. 39; Gallardo García *et al.* (2016), p. 36; Guiry (2012), p. 149; Hamel (1931–39), pp. 211–13, fig. 44(1–2); Harvey (1846–51), pl. 8; Jaasund (1965), p. 91; Kornmann & Sahling (1977), p. 278, fig. 16; Kylin (1907), pp. 70–2, fig. 17; Lindauer *et al.* (1961), pp. 254, 255, fig. 64; Mathieson & Dawes (2017), p. 180, pl. LIII(9–11); Morton (1994), p. 41; Newton (1931), pp. 184, 185, fig. 116; Pedersen (1984), pp. 21, 22, fig. 10; Rhodes (1970), pp. 312–14, figs 1–11; Ribera *et al.* (1992), p. 117; Rueness (1977), p. 164, fig. 96(A, B), (2001), p. 16; Sauvageau (1929), pp. 334–49, figs 11–14, (1933a), pp. 1–4; Schneider & Searles (1991), pp. 144, 145, fig. 165; Stegenga & Mol (1983), pp. 99–103, pls 35, 36; Stegenga *et al.* (1992a) p. 20; Taskin *et al.* (2008), p. 60; Taylor (1957), p. 166, pl. 15(5).

Punctaria plantaginea (Roth) Greville (1830), p. 53. Figs 180(B,C), 182

Ulva plantaginea Roth (1800), p. 243; *Phycolapathum plantagineum* (Roth) Kützing (1849), p. 483; *Laminaria plantaginea* (Roth) C. Agardh (1817), p. 20; *Ulva rubescens* Lyngbye (1819), p. 27; *Zonaria plantaginea* (Roth) C. Agardh (1820) p. 138; *Asperococcus plantagineus* (Roth) Fries (1835), p. 307; *Homoeostroma plantagineum* J. Agardh (1896), p. 11; *Punctaria rubescens* (Lyngbye) J. Agardh (1896), p. 6.

Thalli forming erect, light to dark brown, dorsiventrally flattened blades, arising from a small discoid holdfast; erect blades solitary or more commonly gregarious, simple, solid, subcoriaceous, entire or often irregularly or longitudinally split, lanceolate or obovate, gradually tapering below to a short stipe, to 15(–30) × 1–3(–5) cm; surface cells quadrate to rectangular, in longitudinal less obvious transverse rows, 20–50 × 14–39 μm each with large numbers of discoid plastids and associated pyrenoids; in section, blades 120–180(–210) μm thick, 4–6(–9) cells, inner cells large, colourless, thick walled, quadrate or elongate-rectangular in longitudinal section, rounded in transverse section, outer cells smaller, pigmented, quadrate to vertically elongate; hairs abundant, arising from surface cells, single or more commonly grouped, thick walled, fairly stiff, 15–20(–25) μm in diameter, with extensive basal meristem and enclosing sheath.

Plurilocular and unilocular sporangia borne on the same or different blades, widely distributed on both surfaces, formed within surface cells; plurilocular sporangia solitary or grouped, formed by direct subdivision of surface cells, or following 2–4 vegetative divisions, quadrate to rectangular in surface view, 20–40 × 11–32 μm, conical, often bifurcate, projecting

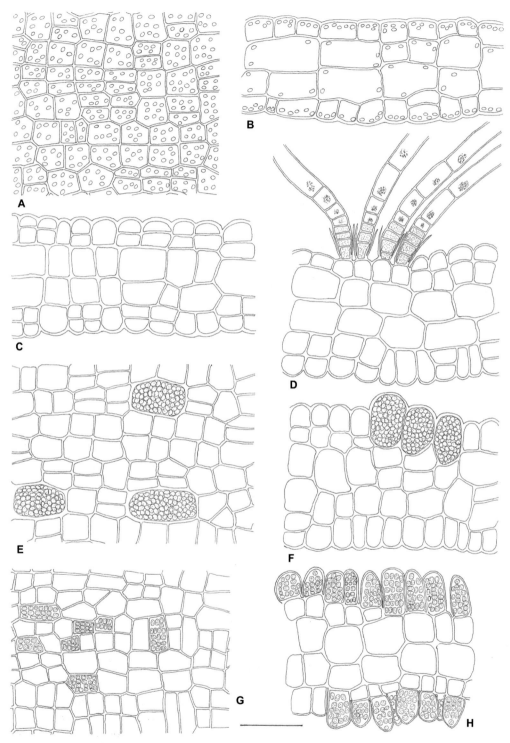

Fig. 182. *Punctaria plantaginea*

A. Surface view of vegetative thallus showing cells with numerous plastids. B–D. VS of vegetative thalli with lateral hairs (D). E. Surface view of fertile thalli with unilocular sporangia. F. VS of fertile thallus with unilocular sporangia. G. Surface view of fertile thallus with plurilocular sporangia. H. VS of fertile thallus with plurilocular sporangia. Bar = 50 μm.

from the surface in section, 29–52 × 14–29 µm; unilocular sporangia orbicular in surface view 45–85 × 31–40 µm, ovate in section, slightly projecting, 32–52 × 29–43 µm.

Epilithic, epizoic, upper to lower eulittoral pools, sublittoral to 18 m.

Only recorded in a few scattered localities around Britain and Ireland.

Annual; spring and summer.

North Atlantic isolates of *P. plantaginea* have been studied in culture by Knight (1929), South (1980) and Pedersen (1984). Both South and Pedersen reported a 'direct' asexual life history with unispores germinating into sterile microthalli from which further generations of unilocular sporangia-bearing erect thalli were produced. In the culture study by Knight a sexual life history was reported, with unispores fusing together. Very few details, however, were given in Knight's paper and her claim must remain suspect. Also requiring further investigation is Pedersen's (1984) suggestion that thalli of *P. plantaginea* reported with plurilocular sporangia may represent a cold-water form of *P. latifolia*.

Bartsch & Kuhlenkamp (2000), p. 168; Brodie *et al.* (2016), p. 1021; Cotton (1912), p. 94; Fletcher (1987), pp. 211–13, figs 52(B, C), 54, pl. 6(b); Fletcher in Hardy & Guiry (2003), p. 290; Gallardo García *et al.* (2016), p. 36; Guiry (2012), p. 149; Hamel (1931–39), pp. 213–15, fig. 44(3); Harvey (1946–51), pl. 128; Jaasund (1965), pp. 89–91, fig. 27; Kornmann & Sahling (1977), p. 128, fig. 68(A–C); Kylin (1947), p. 73; Lund (1959), pp. 133–5; Mathieson & Dawes (2017), p. 181, pl. LIII(12–14); Morton (1994), p. 41; Pankow (1971), p. 194, pl. 21(2); Pedersen (1984), pp. 20–2, (2011), pp. 77, 78, fig. 82; Ribera *et al.* (1992), p. 117; Rosenvinge & Lund (1947), pp. 11–14, fig. 2; Rueness (1977), p. 64, pl. 21(4), fig. 97, (2001), p. 16; South (1980), pp. 266–72, figs 1–5; Stegenga & Mol (1983), p. 103; Stegenga *et al.* (1997a), p. 20; Taskin *et al.* (2008), p. 61; Taylor (1957), pp. 166, 167, pls 15(4), pl. 16(4).

Punctaria tenuissima (C. Agardh) Greville (1830), p. 54.   Figs 180(D,E), 183

Ulva plantaginifolia Wulfen (1803), n. 3; *Zonaria tenuissima* C. Agardh (1824), p. 268; *Punctaria undulata* J. Agardh (1836), p. 15; *Diplostromium tenuissimum* (C. Agardh) Kützing (1843), p. 298; *Diplostromium tenuissimum* Kützing (1849), p. 483; *Desmotrichum balticum* Kützing (1849), p. 470; *Desmotrichum scopulorum* Reinke (1888), p. 18; *Desmotrichum undulatum* (J. Agardh) Reinke (1889b), p. 57; *Diplostromium balticum* (Kützing) J. Agardh (1896), p. 18; *Desmotrichum balticum* f. *paradoxum* Gran (1897), p. 37; *Punctaria baltica* (Kützing) Batters (1902), p. 26; *Streblonema effusum* Kylin (1907), p. 49; *Desmotrichum repens* Kylin (1907), p. 66; *Desmotrichum scopulorum* f. *fennicum* Skottsberg (1911), p. 5; *Entonema effusum* (Kylin) Kylin (1947), p. 20.

Thalli forming erect, olive to light brown, dorsiventrally flattened blades, arising from a small discoid holdfast and short stalk; erect blades solitary or more commonly gregarious, simple, solid, linear, elongate-lanceolate, tapering to base and apex, occasionally obtuse at apex, soft and flaccid, frequently spirally twisted with crisp margins, entire or with small serrations, to 10(–20) cm × 2–6(–12) mm; surface cells quadrate or rectangular, in rows, 14–48 × 13–27 µm each with numerous, small discoid plastids with associated pyrenoids; in transverse section, blades 32–66 µm thick, 1–2(–4) cells, cells rectangular, higher than wide, 30–38 × 13–17 µm, in one cell thick regions, quadrate, subquadrate 15–21 × 15–20 µm in two cells thick regions and with large, colourless, longitudinally elongate, transversely rounded, central cells and smaller, pigmented, quadrate or wider than high surface cells in regions 4 cells thick; hairs abundant, single or grouped, arising from blade surface or margin, 10–12 µm in diameter with small, basal meristem and enclosing sheath.

Plurilocular and unilocular sporangia borne on the same or different blades, single or grouped on both surfaces, formed from surface cells; plurilocular sporangia common, quadrate or rectangular in surface view, 24–36 × 20–27 µm, conical and sometimes bifurcate and projecting above thallus surface in section, 26–42 × 14–23 µm; unilocular sporangia rare, ovoid or quadrate in surface view, 30–40 µm in diameter.

Epiphytic on various algae, upper to lower eulittoral pools and shallow sublittoral.

Widely distributed around Britain and Ireland.

Annual; spring and summer.

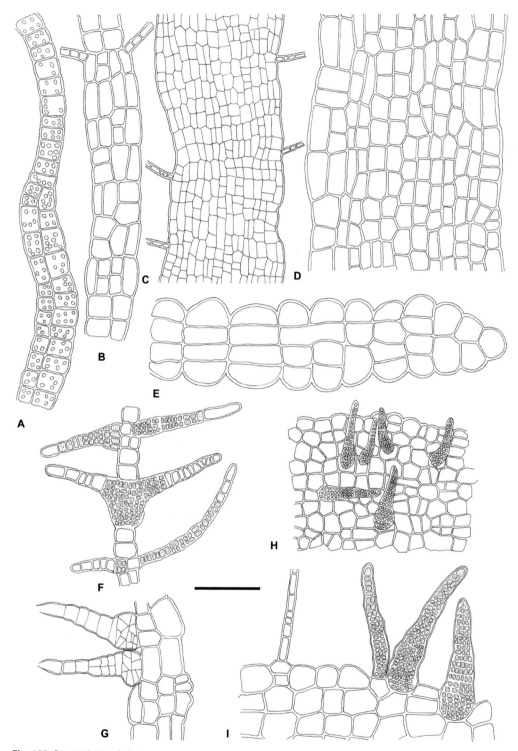

Fig. 183. *Punctaria tenuissima*
A–D. Portions of erect thalli showing enlargement of blade. E. VS of thallus margin. F, G. Portions of narrow thalli with lateral mature and dehisced plurilocular sporangia. H, I. Edge portions of erect thalli showing protruding plurilocular sporangia. Bar = 100 μm (C), = 50 μm (A, B, D, H), = 25 μm (E–G, I).

Culture studies by Kylin (1933), Rhodes (1970), Lockhart (1982), Rietema & van den Hoek (1981), Pedersen (1984) and Parente *et al.* (2010a) on *Punctaria tenuissima* (sometimes as *Desmotrichum undulatum*) reveal isolates of this species to have both direct and heteromorphic, asexual life histories. In all cultures, spores developed directly into microthalli capable of producing new erect blades. For many isolates, the formation of the blades was influenced by environmental factors including temperature, light intensity, light quality, photoperiod and nutrient content of medium. In some isolates, and in some culture conditions, the microthalli became fertile with plurilocular sporangia and these were thought to resemble *Hecatonema reptans*, *Hecatonema terminale* and *H. maculans* (the latter two species joined in synonymy herein) and *Streblonema effusum* Kylin by Pedersen (1984), and to resemble *Hecatonema terminale* (as *H. maculans*) by Parente *et al.* (2010a). Other studies (Clayton & Ducker 1970; Clayton 1974; Fletcher 1984b), have also linked *Punctaria tenuissima* (usually as *Desmotrichum undulatum*) with *Hecatonema terminale* (as *H. maculans*). Clearly the situation is unsatisfactory, as material identified as *Hecatonema reptans* has also been implicated in the life history of *Asperococcus fistulosus* (see pp. 229 and 234–39), and the taxonomic status of *Hecatonema* needs investigation. Some isolates of *Punctaria tenuissima* from different geographical regions varied in their response to various environmental conditions, suggesting that this species is represented by various genotypes. There is also support for Parente *et al.*'s (2010) suggestion, based on culture and molecular studies, that *Punctaria tenuissima* should be a synonym of *Punctaria latifolia*. However, pending further research, the latter two species are maintained here as separate taxa: *Hecatonema reptans* (incorporating the Première forme of *Hecatonema maculans* sensu Sauvageau) is linked to the life history of *Asperococcus fistulosus* and *Hecatonema terminale* is maintained as a separate species.

Batters (1892b), pp. 17, 18; Brodie *et al.* (2016), p. 1021; Cormaci *et al.* (2012), pp. 216–18, pl. 59, figs 1–5; Cotton (1912), p. 94; Dizerbo & Herpe (2007), p. 115; Fletcher (1987), pp. 213–15, figs 52(D, E), 55; Fletcher in Hardy & Guiry (2003), p. 290; Gallardo García *et al.* (2016), p. 36; Guiry (2012), p. 149; Harvey (1846–51), pl. 248; Jaasund (1957), pp. 216–18, fig. 6, (1965), pp. 87–9, fig. 26; Kylin (1907), pp. 66–70, fig. 16, (1933), pp. 36–8, figs 11, 12, (1947), pp. 72, 73, fig. 59(A, B); Mathieson & Dawes (2017), pp. 181, 182, pls LIII(16), LIV(1, 2); Morton (1994), p. 41; Newton (1931), pp. 185, 186; Pankow (1971), pp. 192–4, fig. 249, pl. 22(1); Parente *et al.* (2010a) pp. 223–31; Reinke (1889a), p. 15, pls 11–13; Rhodes (1970), pp. 312–14, figs 1–11; Ribera *et al.* (1992), p. 117; Rietema & van den Hoek (1981), pp. 321–35, figs 1–10; Rosenvinge & Lund (1947), pp. 6–11, fig. 1; Rueness (1977), pp. 160, 164, fig. 93, (2001), p. 16; Schneider & Searles (1991), pp. 145, 146, figs 166–8; Stegenga *et al.* (1997a), p. 20; Taylor (1957), p. 165, pls 15(6), 16(1, 2); Womersley (1987), pp. 316, 318, figs 114(A), 115(D).

Sauvageaugloia Hamel ex Kylin

SAUVAGEAUGLOIA Hamel ex Kylin (1940), p. 32.

Type species: *S. griffithsiana* (Greville) Hamel ex Kylin (1940), p. 33.

Thalli epilithic, erect, cylindrical, filiform, linear, soft and gelatinous, frequently and irregularly branched along the axes to 1(–3) orders, solid becoming tubular in older specimens, arising from a small, flattened, discoid holdfast; thallus haplostichous, comprising a central medulla of several erect rows of sympodially branched filaments of elongate-cylindrical, colourless cells, the outer cells of which give rise directly and laterally to a relatively short, peripheral, pigmented, cortical region comprising assimilators, plurilocular sporangia, unilocular sporangia and/or hairs; assimilators straight to slightly curved, multicellular, simple, rarely branched at the base, each containing several discoid plastids; hairs common, with basal meristem, lacking a sheath.

Plurilocular sporangia formed by transformation of several of the terminal cells of the assimilators, uniseriate to biseriate; unilocular sporangia borne on the same or different thalli, at the base of the paraphyses, sessile, ovoid to pyriform.

The characteristic features of this genus and the criteria by which Hamel (1931–39) separated it from the genus *Mesogloia*, based on *M. griffithsiana* (the generic diagnosis later modified by Kylin 1940), are the hollow, central medulla and the absence of a well-developed subcortex, with the cortex of assimilatory filaments originating directly from the outer, elongated, medullary cells.

One species in Britain and Ireland.

Sauvageaugloia divaricata (Clemente) Cremades in Cremades & Pérez-Cirera (1990), p. 492.   Figs 184, 185

Ulva divaricata Clemente (1807), p. 320; *Mesogloia griffithsiana* Griffiths ex Harvey (in W.J. Hooker) (1833), p. 387; *Helminthocladia griffithsiana* (Griffiths ex W.J. Hooker) Harvey (1841), p. 46; *Myriocladia chordariaeformis* P. Crouan & H. Crouan (1867), p. 165; *Cladosiphon chordariaeformis* (P. Crouan & H. Crouan) J. Agardh (1882), p. 42; *Castagnea griffithsiana* (Griffiths ex Harvey) J. Agardh (1882), p. 38; *Castagnea chordariaeformis* (P. Crouan & H. Crouan) Thuret in Flahault (1888), p. 381; *Sauvageaugloia griffithsiana* (Greville ex Harvey) Hamel ex Kylin (1940), p. 33; *Sauvageaugloia chordariiformis* (P. Crouan & H. Crouan) Kylin (1940), p. 33.

Thalli epilithic, erect, to 20 cm long, linear and uniformly 0.5–1(–1.5) mm in diameter throughout with slightly attenuate apices, yellow-brown to olive-green, cylindrical, filiform, linear, soft to slightly firm, gelatinous, slimy, slippery, frequently and irregularly branched along the axes, solid becoming tubular in older specimens, arising from a small, flattened, discoid holdfast; in transverse section or squash preparation, thallus haplostichous, and comprising moderately compacted filaments in a gelatinous matrix and quite easily separable under pressure, differentiated into a central medulla of several erect rows of sympodially branched filaments of elongate-cylindrical, colourless cells, to ×20 diameters long, less elongated around the edge, the outer cells of which give rise terminally and laterally to a relatively short, peripheral cortical region comprising assimilators, plurilocular sporangia, unilocular sporangia and/ or hairs; assimilators straight to slightly curved, multicellular, simple, rarely branched at the base, to 4–9(–12) cells long, (50–)60–150(–170) µm, comprising mainly cylindrical cells below, 2–4 diameters long, (18–)22–30(–32) × (5–)6–8(–9) µm, which become more ovoid to isodiametric, often slightly enlarged on one side with occasional transverse divisions in the upper regions, 0.5–2 diameters long, (10–)13–20(–23) × (9–)12–22(–26) µm; all cells with several discoid plastids; hairs common, with basal meristem, lacking a sheath, with cells 5–15 diameters long, 55–110 × 5–14 µm.

Plurilocular sporangia not observed by the present author, but described, for example, by Cormaci *et al.* (2012) to be siliquose, uniseriate to biseriate, unilateral outgrowths of several of the terminal cells of the assimilators; unilocular sporangia borne on the same or different thalli, at the base of the assimilators, sessile, ovoid to pyriform, (35–)42–73(–80) × (25–)29–40(–48) µm.

Epilithic on rock and stones, in shallow, mid to lower eulittoral pools, locally fairly abundant.

Occasional, reported for widely scattered localities around Britain (Isles of Scilly, Channel Islands, Cornwall, South Devon, Dorset, Hampshire, Sussex, Argyllshire, Ayrshire, Anglesey) and Ireland (Cork, Galway, Mayo).

A summer annual, June to September.

Culture studies (Caram 1961) revealed that *Sauvageaugloia divaricata* (as *S. griffithsiana*) had a heteromorphic life history, with the macroscopic sporophyte generation alternating with a microscopic, gametophyte generation bearing uniseriate to biseriate, plurilocular sporangia (gametangia).

Ardré (1970), pp. 255, 256; Bartsch & Kuhlenkamp (2000), pp. 168, 179; Bornet (1892b), p. 235; Brodie *et al.* (2016), p. 1021; Caram (1961), pp. 594–6, figs 1–10, (1965), pp. 43–68 (figs 1–13), pp. 69–85, figs 1–8, pls I–II; Cormaci *et al.* (2012), pp. 218, 220, pl. 60(1–5); Cremades & Pérez-Cirera (1990), p. 492, fig. 1(i–j); P. Crouan & H. Crouan (1867), p. 165; Dizerbo & Herpe (2007), p. 91, pl. 30(2); Fletcher in Hardy & Guiry (2003), p. 293; Gallardo García *et al.* (2016), p. 30, Gayral (1966), p. 277, pl. 47(36A, B); Guiry (2012), p. 149; Hamel (1931–39), pp. 162, 163, 171,

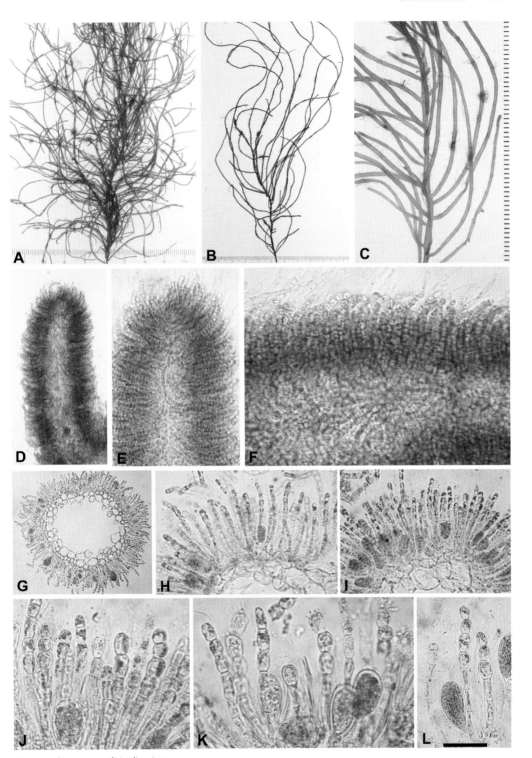

Fig. 184. *Sauvageaugloia divaricata*

A–C. Habit of thalli. D, E. SP of branch apices, showing the outer cortical fringe of assimilatory filaments. F. SP of subapical branch region showing the densely packed peripheral assimilatory filaments. G–I. TS of thalli showing the central cavity and large rounded colourless, outer cells of the medulla giving rise to the assimilatory filaments and unilocular sporangia. J–L. Assimilatory filaments and unilocular sporangia. Bar in L = 300 μm (G), = 180 μm (D), = 90 μm (E, F, H, I), = 45 μm (J–L).

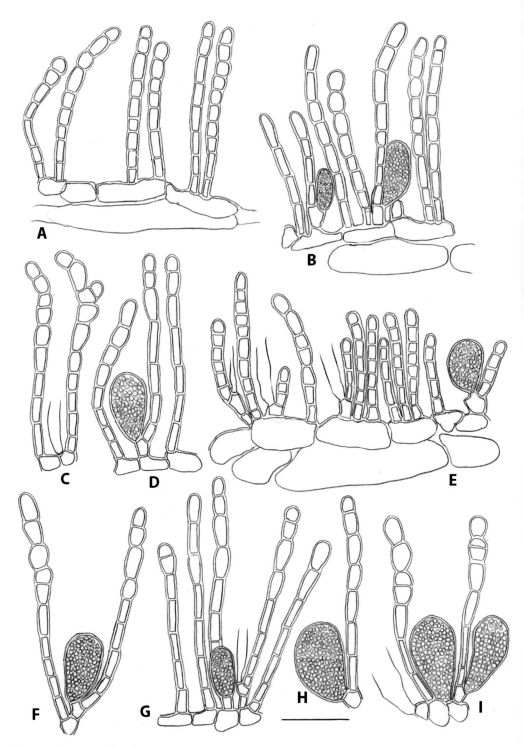

Fig. 185. *Sauvageaugloia divaricata*
A–I. SP of thalli showing the large, elongated, medullary cells giving rise laterally to the assimilatory filaments, some bearing unilocular sporangia at their base. Bar = 50 μm.

pl. XXXVIII, fig. 36(e); Harvey (1846–51), pl. 138; Kim (2010), p. 122; Kuckuck (1912), p. 177, pl. VIII(14), (1929), p. 53; Kylin (1940), p. 33, fig. 16(B–C); Morton (1994), p. 8; Newton (1931), p. 149; Parke (1933), p. 40, pl. I(4); Ribera *et al.* (1992), p. 114; Sauvageau (1897), pp. 13, 45, figs 4–5.

Spermatochnus Kützing

SPERMATOCHNUS Kützing (1843), p. 334, nom. cons.

Type species: *S. paradoxus* (Roth) Kützing (1843), p. 96.

Thalli macroscopic, to 40 cm long, to 1(–2) mm wide, arising from a small, discoid holdfast; thallus cylindrical, solid becoming tubular in older parts, much branched, branching regularly dichotomous, in parts irregular, all branches gradually tapering above; in structure, thallus haplostichous, uniaxial with central axial filament of cells giving rise laterally, from their upper region, to a radiating, petal-like whorl of 4–5 elongated, separated, primary lateral cells, making the medulla partly hollow, which branch radially and laterally to form a peripheral pseudoparenchymatous, 3–4-celled cortex; growth initially by a single apical cell on axial filament, followed by lateral subapical, pseudoparenchymatous growth; apices regularly giving rise to projecting whorls of 4–5, short, pigmented, simple or branched below filaments (= primary assimilators) with hairs; hairs, with basal meristem; cells with several discoid plastids each with a pyrenoid.

    Unilocular and plurilocular sporangia borne on the same or different thalli, occurring on older parts of the thallus, in scattered, or sometimes somewhat striated, punctuate, tufted sori, with secondary assimilators and numerous hairs; secondary assimilators simple or branched at the base, pigmented, usually slightly curved, 4–8 cells in length and clavate; unilocular sporangia common, sessile, ovoid to pyriform; plurilocular sporangia reported to be rare, pedicellate, siliquose, uniseriate, 4–8 locules long.

One species in Britain and Ireland.

Spermatochnus paradoxus (Roth) Kützing (1843), p. 96.   Fig. 186

Conferva paradoxa Roth (1806), p. 33; *Chordaria rhizodes* var. *paradoxa* (Roth) C. Agardh (1817), p. 15; *Scytosiphon paradoxus* (Roth) Hornemann (1818), pl. 1595, fig. 2; *Chordaria paradoxa* (Roth) Lyngbye (1819), p. 53; *Sporochnus rhizodes* var. *paradoxus* (Roth) C. Agardh (1824), p. 226; *Stilophora lyngbyei* J. Agardh (1841), p. 6; *Stilophora paradoxa* (Roth) Areschoug (1845), p. 124.

Thalli macroscopic, to 40 cm long, to 0.5–1(–2) mm wide, erect, olive to light brown, arising from a small, discoid holdfast; thallus cylindrical, filiform, solid becoming tubular in older parts, fairly course, slightly gelatinous, much branched, branching regularly dichotomous, in parts irregular, all branches gradually tapering above; thallus haplostichous in structure, uniaxial with the constituent, central, relatively narrow, markedly elongated, axial filament cells giving rise laterally, from their upper region, to a radiating, petal-like whorl of 4–5 elongated, separated, well-spaced apart, primary, lateral cells, which branch radially and laterally to form cortical threads of up to 3–4 cells, which become progressively smaller and closely bound to form a peripheral, smooth surfaced cortex of angular cells seen in surface view; growth apical, by a prominent, exposed, dome-shaped apical cell terminating the axial filament, followed by lateral subapical, growth; subapical regions of young thalli reveal projecting whorls of 4–5, short, pigmented, filaments (= primary assimilators), usually accompanied by a hair, developing from cortical cells lying immediately beyond each primary lateral, which, as a result of vertical elongation of the inner cortical cells, later appear as evenly spaced areas/patches scattered over thallus surface; primary assimilators curved towards the apex, simple, or branched below, 3–6(–8) cells long, 40–80 µm, comprising cells somewhat moniliform

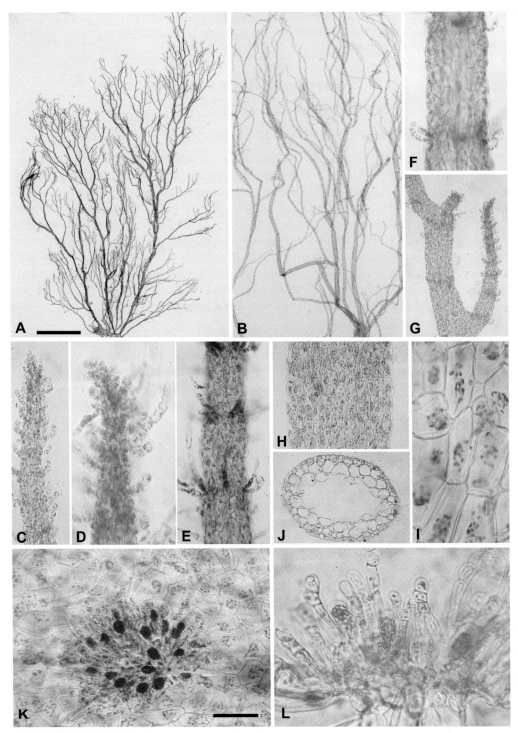

Fig. 186. *Spermatochnus paradoxus*

A, B. Habit of thalli. C–G. Apical regions of young thalli showing the whorled arrangement of the primary assimilators. H, I. Surface views of cortex showing irregularly arranged angular cells and enclosed discoid plastids. J. TS of thallus showing central hollow cavity and outer cortex. K, L. Surface views of mature thalli showing the punctuate reproductive sori comprising secondary assimilators, hairs and unilocular sporangia. Bar in A = 5 cm (A), = 3.5 cm (B), bar in K = 180 μm (G), = 90 μm (C, E, F, H, J, K), = 45 μm (D, I), = 30 μm (L). (A, herbarium specimen, NHM.)

above, more compressed, shorter than wide below, (5–)10–15(–18) × (10–)12–18(–21) μm; hairs common, especially in the branch apices, usually associated with the assimilatory filaments, with basal meristem, lacking sheath, comprising cells 3–9 diameters long, 22–115 × 8–16 μm; cells with several small, discoid plastids each with a pyrenoid.

Unilocular and plurilocular sporangia borne on the same or different thalli, occurring on older parts of the thallus, in scattered, or sometimes somewhat striated, punctuate, tufted sori, which initially develop around, and sometimes spread out from, the clumps of primary assimilators, accompanied by secondary assimilators and numerous hairs; secondary assimilators simple or branched at the base, pigmented, usually slightly curved, 4–8 cells, 40–90 μm in length, clavate, with basal cells cylindrical, 2–3 diameters long (11–)15–25(–28) × (4–)5–10(–12) μm, terminal cells more globular, usually distended on one side, and overlapping; unilocular sporangia common, sessile, ovoid to pyriform, (26–)28–34(–42) × (15–)17–21(–24) μm; plurilocular sporangia not observed by the present author, reported by Sauvageau (1936, p. 123, fig. 1F, G), to be rare, pedicellate, siliquose, uniseriate, 4–8 locules, 35–50 μm long, 7–8 μm in diameter.

Epilithic and epiphytic, lower eulittoral, sublittoral, 5–20 m, usually associated with sand or mud and sheltered habitats such as bays and estuaries; tolerant of reduced salinity.

Rare but widely distributed around Britain (Channel Islands, Cornwall, Dorset, Argyllshire, Ayrshire, Orkney, Shetland) and Ireland (Cork, Down, Mayo, Galway, Kerry, Dublin).

A summer annual; May to September.

Culture studies by Papenfuss (1935b), Caram (1968) and Müller (1981a) revealed *S. paradoxus* to have a heteromorphic, diplohaplontic life history with the macroscopic, diploid sporophyte alternating with a microscopic, filamentous haploid gametophyte bearing uniseriate gametangia. Müller (1981a) also showed that temperature affected life history expression: at 20°C the microthalli behaved asexually and propagated themselves, while at 9°C the microthalli behaved as gametophytes and fusion of the isogametes resulted in the diploid macrothallus.

Brodie *et al.* (2016), p. 1021; Bunker *et al.* (2017), p. 306; Caram (1968), pp. 1828–30; Cormaci *et al.* (2012), p. 222, pl. 61(1–5); Cotton (1912), p. 96; Dizerbo & Herpe (2007), p. 86, pl. 27(4); Fletcher in Hardy & Guiry (2003), p. 295; Fritsch (1945), pp. 90–3, fig. 27(A–H); Gayral (1966), p. 281, fig. 36(D); Guiry (2012), p. 150; Hamel (1931–39), pp. 182–4, figs 40(IX), 41; Harvey (1846–51), pl. 237; Kylin (1940), p. 49, fig. 27, (1947), pp. 61, 62, fig. 54(A–D), pl. 6(20); Morton (1994), p. 35; Müller (1981a), pp. 384–9, figs 1–4; Müller *et al.* (1981b), p. 478; Newton (1939), pp. 161, 162, fig. 101(A–C); Oltmanns (1922), p. 37, fig. 333; Pankow (1971), pp. 186–8, fig. 235; Papenfuss (1935b), pp. 1–4, figs 1–10; Peters *et al.* (1987), pp. 223–64; Reinke (1889b), pp. 66–70, fig. 4(A–C), (1892a), pp. 53, 54, pls 33–5; Ribera *et al.* (1992), p. 116; Rosenvinge & Lund (1943), pp. 38–40, fig. 14; Rueness (1977), pp. 150, 151, fig. 83(A, B), pl. XX(2), (2001), p. 16; Sauvageau (1931), pp. 128–32, figs 24, 25; Womersley (1987), pp. 132–4, figs 39(C), 41(A–D).

Sphaerotrichia Kylin

SPHAEROTRICHIA Kylin (1940), p. 38.

Type species: *S. divaricata* (C. Agardh) Kylin (1940), p. 38.

Thalli macroscopic, erect, solid becoming partly hollow later, cartilaginous, gelatinous and slimy, terete, much branched, covered with long gelatinous hairs, arising from a small, basal disc; branching mainly dichotomous/pseudodichotomous or irregular, to several orders, gradually attenuating above, with branches often long, flexuose and widely divaricate; in structure, thallus multiaxial, haplostichous, pseudoparenchymatous with central filamentous medulla of longitudinally elongate, transversely rounded, colourless cells interspersed with thinner secondary rhizoidal filaments, surrounded by a cortex of smaller, pigmented cells from which radially project assimilators, hairs, and/or unilocular sporangia bound in a gelatinous matrix; assimilators multicellular, short, simple, clavate, with swollen terminal cell; cells with numerous discoid plastids each with associated pyrenoids; hairs with basal meristem but without obvious sheath; growth apical, by a large, subspherical cell capping the primary central axial filament of limited growth.

Unilocular sporangia formed in extensive sori covering the entire surface, borne on the cortical cells at the base of the assimilators, sessile, or stalked, ovate to pyriform; plurilocular sporangia not known on macrophyte.

One species in Britain and Ireland.

Sphaerotrichia divaricata (C. Agardh) Kylin (1940), p. 38.   Fig. 187

Chordaria divaricata C. Agardh (1817), p. 12; *Castagnea divaricata* (C. Agardh) J. Agardh (1882), p. 37; *Nemacystus divaricatus* (C. Agardh) Hygen (1934), p. 185; *Sphaerotrichia japonica* Kylin (1940), p. 38; *Sphaerotrichia chordarioides* Yamada (1944) p. 167; *Sphaerotrichia chordarioides* var. *gracilis* Yamada (1944), p. 167; *Sphaerotrichia divaricata* f. *typica* Inagaki (1954), p. 11, nom. inval.; *Sphaerotrichia divaricata* f. *gracilis* (Yamada) Inagaki (1954), p. 13.

Thalli macroscopic, erect, to 50(–70) cm long, to 0.5(–1) mm wide, pale to olive-brown, usually solitary, sometimes gregarious, solid becoming hollow later, cartilaginous, gelatinous and slimy to the touch, thread-like, terete, much branched, clothed with long gelatinous hairs, arising from a small, basal disc; main axis variable in length and degree of branching, with primary branches often long, flexuose, widely divaricate, and sparsely branched along their length, more copiously and intricately branched towards the apex with short, somewhat forked, branchlets, branching mainly dichotomous/pseudodichotomous or irregular, sometimes in parts secund or alternate, to several orders, gradually attenuating above; in structure, multiaxial, haplostichous, pseudoparenchymatous with a broad, extensive, largely solid, laterally adjoined, filamentous medulla of longitudinally elongate, transversely rounded, colourless cells with interspersed, thinner, secondary rhizoidal filaments, surrounded by a cortex of smaller, pigmented cells from which project short assimilators, hairs, and/or unilocular sporangia bound in a gelatinous matrix; central axis gradually becoming hollow to form a large central cavity; assimilatory filaments 3–5 cells in length, straight, simple, or rarely branched, clavate, comprising mainly cylindrical cells along their length, 1–2 diameters long, 8–25 × 8–16 µm and terminated by a large inflated, ovoid to globose apical cell, 22–45 × 20–40 µm, each with numerous discoid plastids with associated pyrenoids; hairs common, clothing the axes, with basal meristem but without obvious sheath; growth apical, by a large, subspherical cell capping the primary central axial filament of limited growth, which later becomes obscured, and enclosed within, the upward extending and dividing primary laterals constituting the medullary tissue.

Unilocular sporangia formed in extensive sori covering the entire surface, borne on the cortical cells at the base of the assimilators, sessile, or stalked, ovate to pyriform, 40–65 × 20–40 µm; plurilocular sporangia not known on macrophyte.

Epilithic, lower eulittoral pools, on moderately exposed shores.

Rare, recorded for a few widely scattered localities around Britain (Cornwall, Isle of Man, Ayrshire, Bute, Fair Isle, Inner Hebrides) and for Antrim in Ireland, usually as drift.

Appears to be a late summer/early autumn annual with collections made from August to October.

Culture studies by Hygen (1934), Ajisaka & Umezaki (1978) and Peters *et al.* (1987) revealed a heteromorphic life history, with zoospores from the diploid macrophyte germinating directly into streblonematoid, filamentous, haploid, dioecious gametophytes bearing uniseriate plurilocular sporangia, the contents of which behaved sexually with the resultant zygote developing into macrophytes bearing unilocular sporangia; an asexual, direct life history was also reported by Peters *et al.* (1987) for some strains.

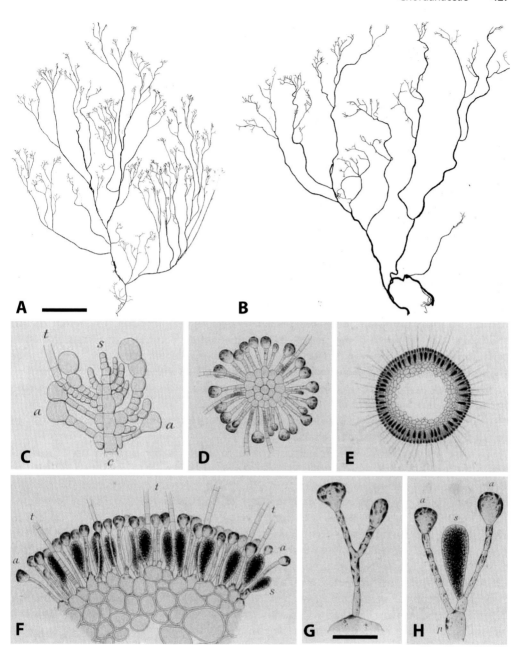

Fig. 187. *Sphaerotrichia divaricata*

A, B. Habit of thalli. C. Apical region of branch. D. TS of young thallus showing the central, medullary cells surrounded by the cortical assimilatory filaments. E, F. TS of mature thalli showing the central cavity, enclosing medullary filaments, the outer cells giving rise to the cortex of short, assimilatory filaments and unilocular sporangia. G. Assimilatory filament. H. Assimilatory filaments and unilocular sporangium. (*t* = hair, *a* = assimilatory filament, *c* = central axial filament, *s* = subspherical cell). Bar in A = 10 cm (A), = 6 cm (B), bar in G = 60 μm (C), = 0.5 mm (E),= 75 μm (F), = 35 μm (G, H). (A, B, herbarium specimens, NHM; C–H, from Reinke 1892a.)

Ajisaki & Umezaki (1978), pp. 53–9, figs 1–5; Brodie *et al.* (2001), p. 1021; Dizerbo & Herpe (2007), p. 91; Fletcher in Hardy & Guiry (2003), p. 302; Guiry (2012), p. 150; Harvey (1846–51), pl. 17; Hygren (1934), pp. 242–68, figs 1–10, pls I–XVI; Kim (2010), p. 119; Kim *et al.* (2003), pp. 183–92; Kuckuck (1929), pp. 70, 72, fig. 100; Kylin (1940), p. 38, fig. 20(C–D), (1947), pp. 58, 59, fig. 52(C); Mathieson & Dawes (2017), pp. 183, 184, pl. CVII(1, 2); Morton (1994), p. 36; Newton (1931), p. 146, fig. 90(A–C); Novaczek & McLachlan (1987); Peters & Müller (1986b), pp. 69–73, figs 11–15; Peters (1987), pp. 223–64; Peters *et al.* (1987), pp. 457–66, figs 1–16, (1993), pp. 31–6; Reinke (1892a), pp. 57, 58, pl. 39(8–16); Ribera *et al.* (1992), p. 114; Riouall (1985), pp. 83–6; Rosenvinge & Lund (1943), pp. 31–4, figs 11(A–B), 12; Rueness (1977), p. 149, fig. 82, (2001), p. 17; Taylor (1957), p. 147, pls 12(1), 14(1); Verlaque (2001), p. 32.

Stictyosiphon Kützing

STICTYOSIPHON Kützing (1843), p. 301.

Type species: *S. adriaticus* Kützing (1843), p. 301.

Kjellmania Reinke (1888b), p. 241.

Thalli macroscopic, erect, 1 cm-1 m in length, fine to coarse in texture, much branched, cylindrical, filiform, very fine, parenchymatous, at first solid, later becoming hollow in older parts, arising from a holdfast of compacted, spreading, secondary rhizoidal filaments; young thalli and branch apices uniseriate, terminating in a hair, older tissues pluriseriate; in section, thallus showing some degree of differentiated, with an inner medulla of usually four large, colourless cells, surrounded by a cortex of 1–2 layers of smaller, pigmented cells; corticating hyphae sometimes present; cells with numerous discoid plastids each with pyrenoids; hairs common, arising directly from surface cells, solitary, sometimes in whorls, and terminating branches, with basal meristem but without obvious sheath; growth probably trichothallic, followed by diffuse.

Unilocular and/or plurilocular sporangia formed directly in surface cells, scattered, solitary or confluent in slightly protruding sori.

Two species of *Stictyosiphon* are currently recognized for Britain and Ireland, *S. soriferus* and *S. tortilis*. Pending further studies, *S. subarticulatus* (Areschoug) Hauck and *S. adriaticus* Kützing are omitted from the present treatise. The former species was first reported for Britain by Holmes (1883, p. 140), listed in Batters (1902, p. 25) and Cotton (1912, p. 94), and described in Newton (1931, pp. 179–80). It is currently listed as a synonym of *S. laxus* (J. Agardh) Athanasiadis and is unlikely to occur around Britain and Ireland (Hardy & Guiry 2003). It seems likely that the material identified and described by Newton (1931) represents either *S. soriferus* or *S. tortilis*. Note that Naylor (1958a) included *S. subarticulatus* (as *Phloeospora subarticulata*) within the synonymy of *S. tortilis*.

Material identified as *S. adriaticus* was newly reported for Ireland by Cotton (1912, pp. 119, 120), based on specimens dredged in Clew Bay, Clare Island, County Mayo. Initially identified as *Kjellmania sorifera* Reinke (= *Stictyosiphon soriferus*), the specimens were later renamed *S. adriaticus* by Kuckuck as he considered Reinke's species to be synonymous with the latter. *Stictyosiphon adriaticus* was not included in Newton (1931), and the only subsequent record of this species appears to be for the Isle of Man (Parke 1933, p. 36). It is likely that these records of *S. adriaticus* represent *S. soriferus* (see discussion under the latter species).

Two species in Britain and Ireland.

KEY TO SPECIES
1. Thalli soft, fragile; yellow-brown in colour; branch apices uniseriate for many
    cells (usually until emergence of lateral branches or beyond); medullary
    cells rounded, of equal length and breadth; plurilocular sporangia forming
    conspicuous mosaic-like sori on frond or as intercalary or terminal structures
    on uniseriate parts of frond . . . . *S. soriferus*

Thalli robust, cartilaginous; dark brown-black in colour; branch apices
uniseriate for only few cells (longitudinal cell divisions before emergence
of lateral branches); medullary cells elongated, length 3–10 × breadth;
plurilocular sporangia forming confluent areas or scattered in subapical parts
of frond, never found in uniseriate parts of frond . . . . *S. tortilis*

Stictyosiphon soriferus (Reinke) Rosenvinge (1935), p. 9.   Fig. 188

Kjellmania sorifera Reinke (1889a), p. 5; *Kjellmania striarioides* Gran (1896), p. 38; *Stictyosiphon corbieri* Sauvageau (1929), p. 298.

Thalli macroscopic, erect, to 30 cm long, to 0.3–0.35 mm wide, light brown, very fine, extremely flaccid in texture, much to sparsely branched, cylindrical, filiform, parenchymatous, solid, possibly becoming hollow in older parts, arising from a well-developed holdfast consisting of a fibrous mat of secondary, downward-extending rhizoidal filaments; branching irregular, alternate or opposite, sometimes widely divaricate, with branches frequently with long, terminal, uniseriate regions and capped by a true hair; in surface view, much of the older thallus regions clearly pluriseriate, to 10–20 cells wide; cells more or less in transverse, less clearly in longitudinal rows especially in younger regions, less obvious in older regions; surface cells angular in shape, approximately isodiametric, measuring 10–50 × 12–30 μm, each with numerous, small, discoid plastids without obvious pyrenoids; cells in terminal, uniseriate regions usually much shorter than wide and lenticular, 0.2–2 diameters long, 5–19 × 7–28 μm; in both transverse and longitudinal section, thallus clearly differentiated into an inner medulla of usually four spherical to subspherical, fairly uniformly sized, colourless cells, to 100 μm in diameter surrounded by a cortex of 1–2 layers of smaller, pigmented cells; corticating hyphae sometimes reported to be present; hairs common, arising directly from surface cells and often seen terminating branches, with basal meristem but without obvious sheath; growth probably trichothallic, followed by diffuse.

Plurilocular and unilocular sporangia formed by direct transformation of surface vegetative cells and slightly protruding; plurilocular sporangia common, solitary, or more commonly grouped in irregularly distributed sori in the parenchymatous regions of the thallus, square, rectangular or irregularly shaped, 28–55 × 20–40 μm in surface view; plurilocular sporangia also formed in extensive rows in the terminal monosiphonous regions of the branches, forming long uniseriate chains, darker in colour and obviously more protruding than the vegetative cells; unilocular sporangia rare, solitary, sometimes in small sori, roundish in surface view, 16–30 × 15–29 μm.

Epilithic, rarely epiphytic on various algae, lower eulittoral pools and shallow sublittoral, 6–10 m, usually in sheltered, calm areas; also observed on the sides of seawater tanks.

Rare, recorded for a few widely scattered localities around Britain (South Devon, Hampshire, Isle of Man, Inner Hebrides, Shetland) and for Mayo in Ireland.

Spring/Summer annual, April to September; plurilocular sporangia in spring and summer; unilocular sporangia (a single, new record for Britain and Ireland) reported in April.

The characteristic features of this species include the branched, somewhat delicate, parenchymatous thallus with extensive uniseriate terminal branch regions, an inner medulla of usually four rounded, colourless cells which are not longitudinally elongated in section, and sporangia which are formed by direct transformation of the surface cells in the parenchymatous regions and which also occur in uniseriate rows in the terminal uniseriate regions. *S. soriferus* is very similar to *S. adriaticus* and further studies are required to determine the limits of the species (see Naylor 1958 and Rosenvinge 1935 for discussions on this topic).

Culture studies on *S. soriferus* by Rosenvinge & Lund (1935), Sauvageau (1929) (as *S. corbierei* Sauvageau), South & Hooper (1976), Pedersen (1984) and Kawai (1991) (as both *S. soriferus* from the Atlantic and *Kjellmania arasakii* Yamada from Japan) have revealed a direct life history with a replication of the parental thallus by asexually behaving zoospores. Differences in

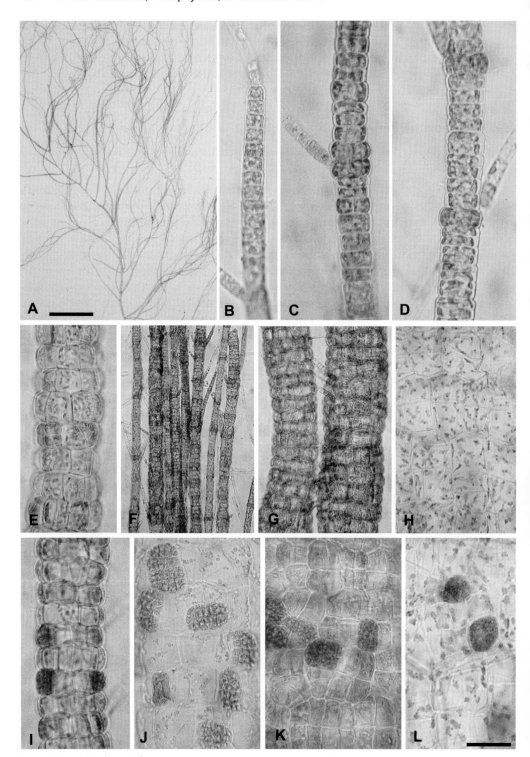

Fig. 188. *Stictyosiphon soriferus*

A. Habit of thalli. B. Terminal region of thalli capped by a hair. C–G. Portion of thalli showing at first uniseriate and then, later, pluriseriate regions. H. Surface view of portion of thallus showing cells with numerous discoid plastids. I–K. Portions of fertile thalli with plurilocular sporangia. L. Portion of fertile thallus with unilocular sporangia. Bar in A = 4 cm, bar in L = 180 μm (F), = 90 μm (G), = 45 μm (B–E, H, I–L).

temperature–reproduction relationships between geographically separated isolates indicated that intraspecific variation may occur (see Pedersen 1984, p. 53).

Bartsch & Kuhlenkamp (2000), pp. 169, 181; Brodie *et al.* (2016), p. 17; Coppejans & Kling (1995), pp. 186–8, pl. 58(A–D); Cormaci *et al.* (2012), pp. 228–30, pl. 63(5–10); Cotton (1912), p. 119; Dizerbo & Herpe (2007), p. 110; Feldmann (1937), pp. 296–9, figs 57–9; Fletcher in Hardy & Guiry (2003), p. 304; Gallardo García *et al.* (2016), p. 36; Gran (1896), pp. 38–40, pl. 1(8–9); Guiry (2012), p. 150; Hamel (1931–39), pp. 205, 206, fig. 42(10–14); Kawai (1991), pp. 319–28, figs 1–34; Kornmann & Sahling (1977), p. 275, pl. 159(A–D); Kuckuck (1929), pp. 81, 82, figs 118–24 (as *S. adriaticus*); Mathieson & Dawes (2017), p. 185, pl. LIV(12–14); Morton (1994), p. 8; Naylor (1958a), pp. 16–20, text fig. 3(C, D); Pankow (1971), pp. 200, 201, fig. 264; Reinke (1889a), p. 5, pl. 3(1–13), (1989b), pp. 59, 60; Ribera *et al.* (1992), p. 117; Rosenvinge & Lund (1935), pp. 9–18, figs 9–19; Rueness (1977), p. 154, fig 86(A–C), (2001), p. 17; Saunders & McDevit (2013), pp. 1–23; Sauvageau (1929), pp. 298–315, figs 5–7 (as *S. corbierei*); South & Hooper (1976), pp. 24–9, figs 3–8; Stegenga & Mol (1996), pp. 109, 110, pl. 4(1–4); Stegenga *et al.* (1997a), p. 20; Taskin *et al.* (2008), p. 61; Womersley (1987), p. 314, figs 109(C), 113(E–J).

Stictyosiphon tortilis (Gobi) Areschoug ex Reinke (1889b), p. 55.

Figs 189, 190

Scytosiphon tortilis Ruprecht (1850), p. 373 nom. illeg.; *Dictyosiphon tortilis* Gobi (1874), p. 15; *Phloeospora tortilis* (Gobi) Areschoug (1876), p. 34; *Phloeospora pumila* Kjellman (1877a), p. 45.

Thalli macroscopic, erect, to 1 m long, 20–280 µm wide, olive, yellowish to dark brown-black, coarse/wiry and cartilaginous in texture, not adhering to paper, copiously branched throughout, and densely intertwined in the mid to lower regions, cylindrical, filiform, very fine, parenchymatous, at first solid, becoming hollow in older parts, arising from a holdfast of compacted, spreading, secondary rhizoids; branching mainly opposite, but also alternate or irregular, with branches terminated by a hair and pluriseriate throughout, rarely with a short, uniseriate region; in surface view, thallus clearly pluriseriate, to 10–15(–25) cells wide, cells in longitudinal, less obvious transverse rows, especially in younger regions, less obvious in older regions, longitudinally elongate to approximately isodiametric, 9–33 × 9–19 µm, and each with numerous discoid plastids with pyrenoids; in both transverse and longitudinal sections, thallus clearly differentiated into an inner medulla of four large, longitudinally elongated, colourless cells, to 75(–500) µm in length, 25–50 µm in width, surrounded by a cortex of 1–2 layers of smaller, pigmented cells; hairs common, arising directly from surface cells, solitary, sometimes in whorls, and terminating branches, with basal meristem but without obvious sheath; growth probably trichothallic, followed by diffuse.

Plurilocular sporangia formed in pluriseriate region only, by a tangential division of a cortical cell to form a sporangial initial cell which enlarges slightly and protrudes above surface; sporangia isolated but more commonly grouped into confluent patches, rounded in surface view, 18–30 µm in diameter; unilocular sporangia unknown.

Epilithic, on rock, stones and boulders, frequently epiphytic, mid to lower eulittoral pools and shallow sublittoral, locally abundant; tolerant of some sand cover and variation in salinity.

Widely distributed around Britain and Ireland, but mainly with a northern distribution and more common on the coasts of Scotland and the north of England.

Summer annual, mainly May to September, but some individuals can be found throughout the year, overwintering as a small macrothallus or perennating from the holdfast system.

Culture studies on *Stictyosiphon tortilis* by Kuckuck (1912), Rosenvinge & Lund (1935), Naylor (1958a), Dangeard (1966) and Caram (1966) were not conclusive, with different interpretations being offered based on whether the sporangia on the macrothalli were considered to represent either plurilocular sporangia (Kuckuck 1912; Rosenvinge & Lund 1935; Pedersen 1984) or unilocular sporangia (Naylor 1958a; Caram 1966), as their correct identification has remained in some doubt (see Rosenvinge & Lund 1935, p. 4; Hamel 1931–39, p. 206). As discussed by Pedersen (1984), the macrothalli could be interpreted as haploid, with the swarmers from the plurilocular sporangia functioning either as gametes or zoospores, or the

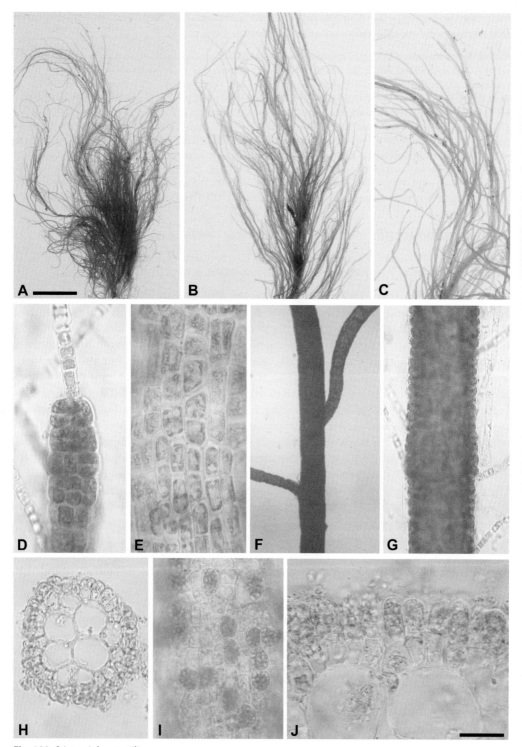

Fig. 189. *Stictyosiphon tortilis*

A–C. Habit of thalli. D. Apex of branch showing terminal hair and immediate pluriseriate development. E. Surface view of thallus showing cells in longitudinal, less obvious, transverse rows. F, G. Portion of erect thalli, showing branch formation and lateral hair formation. H. TS of thalli showing large, inner medullary cells and outer cortex of 1–2 smaller cells. I. Plurilocular sporangia seen in surface view. J. TS of thallus showing plurilocular sporangia. Bar in A = 4 cm (A), = 1.5 cm (B), = 1 cm (C), bar in J = 300 μm (F), = 90 μm (G, H), = 45 μm (D, E, I, J).

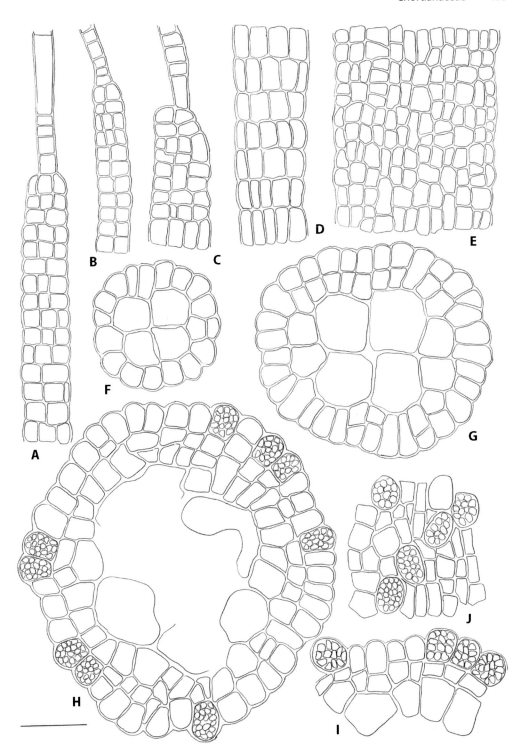

Fig. 190. *Stictyosiphon tortilis*

A–C. Apices of branches showing terminal hair and immediate pluriseriate development. D, E. Portions of thalli showing surface cells. F, G. TS of vegetative thalli showing inner, large, medullary cells and surrounding, smaller, cortical cells. H, I. TS of fertile thalli showing plurilocular sporangia. J. Surface view of thallus showing plurilocular sporangia. Bar = 50 μm.

macrothalli could be interpreted as diploid, and after meiosis the unilocular sporangia produce haploid swarmers which develop into microscopic gametophytes.

Abbott & Hollenberg (1976), pp. 186, 187, fig. 150; Bartsch & Kuhlenkamp (2000), p. 169; Brodie *et al.* (2016), p. 1021; Caram (1966), pp. 2333–5; Cormaci *et al.* (2012), p. 230, pl. 64(1–4); Cotton (1912), p. 94; Dangeard (1965), p. 15, (1966), pp. 65–78, (1968a), pp. 131–9, pl. I(1–14); Dizerbo & Herpe (2007), p. 110, pl. 36(2); Fletcher in Hardy & Guiry (2003), p. 304; Gobi (1874), pp. 15, 16, pl. 2(12–16); Guiry (2012), p. 151; Hamel (1931–39), pp. 206–8, fig. 42(9); Jaasund (1965), pp. 82, 83; Kylin (1947), p. 68; Lund (1959), pp. 130–2; Mathieson & Dawes (2017), p. 185, pl. LIV(15, 16); Morton (1994), p. 8; Naylor (1958a), pp. 3–22, text figs 1–4, pls I–III; Naylor (1958a), pp. 8–14, text figs 1(A, B), 2(B, F), (1958b), pp. 1–22, text figs 1–4, pl. I–III; Newton (1931), p. 180; Oltmanns (1922), fig. 346(2); Pankow (1971), pp. 199, 200, figs 259–61; Pedersen (2011), p. 84, fig. 90; Reinke (1891a), pp. 47–52, pls 31(1–19), 32(1–13); Ribera *et al.* (1992), p. 117; Rosenvinge & Lund (1935), pp. 3–9, figs 1–8; Rueness (1977), pp. 154, 155, fig. 87, pl. XXI(1), (2001), p. 17; Saunders & McDevit (2013), pp. 1–23; Smith (1944), p. 133, pl. 20(6–7); Taylor (1957), p. 158; Waern (1952), pp. 155–9, fig. 70; Wynne (1972b), pp. 134–7, figs 17–22.

Stilophora J. Agardh

STILOPHORA J. Agardh (1841), p. 6 nom. cons.

Type species: *S. tenella* (Esper) P.C. Silva (1996), p. 624.

Thalli macroscopic, erect, cylindrical, filiform, mainly solid but sometimes becoming hollow, firm, non-gelatinous to slightly slimy, much branched, arising from a discoid holdfast; branching dichotomous/subdichotomous to irregular, all axes and branches usually tapering above; thalli haplostichous in structure, multiaxial with a central medullary region of 4–5, closely packed, columns/axial filaments of colourless, vertically elongated cells, surrounded by a small, narrow, cortex of 4–5, smaller, tightly packed, firm cells; growth apical, each medullary filament with a terminal apical cell; apical regions surrounded by an involucrum of enclosing, simple, curved, pigmented filaments of up to 16+ cells long (primary assimilators) which later appear in localized to extensive areas/patches on thallus surface; hairs common, with basal meristem, mainly associated with primary assimilators; cells with several discoid plastids, each with a pyrenoid.

Unilocular and plurilocular sporangia occurring in scattered, raised, punctuate sori, associated with both the primary and newly formed, secondary assimilators (paraphyses); secondary assimilators usually short, 4–13 cells long, simple or branched, linear or more commonly club-shaped, curved, sometimes straight; plurilocular sporangia borne at the base or upper regions of the secondary assimilators, sessile or pedicellate, cylindrical, filiform, uniseriate, 4–6(–10) locules; unilocular sporangia borne at the base of the secondary assimilators, sessile, clavate to ovoid.

Two species in Britain and Ireland.

KEY TO SPECIES
Thallus mainly < 20 cm long, dark brown drying to black; primary and secondary assimilators (paraphyses) and reproductive sori in discrete punctuate patches scattered over surface . . . . *S. nodulosa*
Thallus mainly > 20 cm long, light brown; primary and secondary assimilators (paraphyses) and reproductive sori in extensive vertically elongated, punctuate sori leaving few areas of the smooth thallus surface showing . . . . *S. tenella*

Stilophora nodulosa (C. Agardh) P.C. Silva (1996), p. 929.   Fig. 191

Ceramium tuberculosum Hornemann (1816), pl. 1546, nom. illeg.; *Chaetophora nodulosa* C. Agardh (1817), p. 127; *Chordaria tuberculosa* Lyngbye (1819), p. 65; *Chordaria nodulosa* (C. Agardh) C. Agardh (1820), p. 165; *Castagnea tuberculosa* (Hornemann) J. Agardh (1882), p. 36; *Stilophora tuberculosa* (Hornemann) Reinke (1889b), p. 72; *Stilophora tuberculosa* f. *corniculata* Reinke (1889b), p. 73; *Stilophora tuberculosa* f. *gracilior* Reinke (1889b), p. 73.

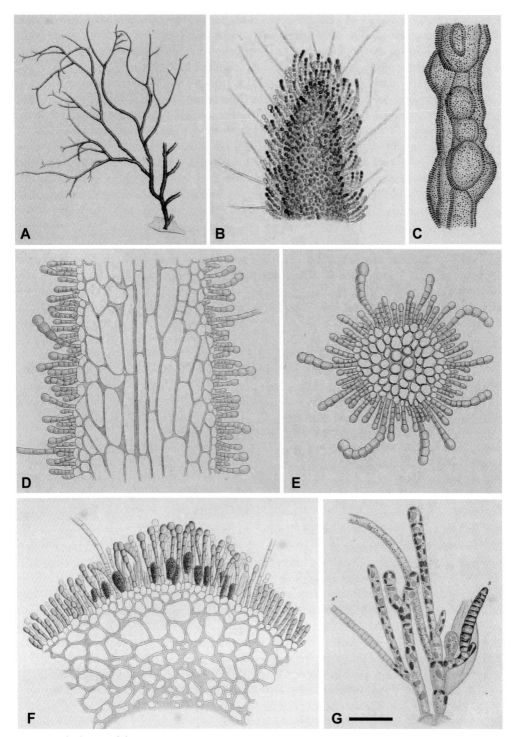

Fig. 191. *Stilophora nodulosa*
A. Habit of thallus. B. Apex of thallus. C. Portion of thallus with sori. D. LS of thallus showing central medulla of longitudinally elongated cells and peripheral assimilators. E. TS of thallus. F. TS of thallus showing sorus with unilocular sporangia at the base of the assimilators and two projecting hairs. G. Portion of sorus showing an immature unilocular sporangium and an immature (s1) and mature (S) uniseriate plurilocular sporangium. Bar in G =15 mm (A), = 150 μm (B), = 300 μm (C), = 70 μm (D), = 72 μm (E), = 75 μm (F), = 25 μm (G). (All figures from Reinke 1892a.)

Thalli macroscopic, 10–20(–30) cm long, to 1.5 mm wide, erect, light to dark brown-black in colour, arising from a discoid holdfast; erect thalli subcylindrical, filiform, mainly solid but sometimes becoming hollow, firm, fairly coarse, somewhat slippery, much branched, branching subdichotomous to dichotomous, all axes and branches usually tapering above, sometimes cylindrical, tufted in appearance and noticeably thicker at the bases; in structure, thalli haplostichous, multiaxial with a central medullary region of 4–5 closely packed, terminally free, columns/axial filaments of colourless, vertically elongated cells, surrounded by a narrow cortex of 4–5, smaller, tightly packed, firm, cortical cells which become smaller and more pigmented towards the periphery; growth apical, each medullary filament with a terminal apical cell; apical regions of young thalli surrounded by an involucrum of enclosing curved, short, pigmented, filaments (= primary assimilators) which persist and are added to, forming extensive, localized patches covering most of the surface, leaving few areas of the smooth thallus surface showing; primary assimilators as described for *S. tenella*; hairs, with basal meristem, frequent, especially in the branch apices, usually associated with the assimilatory filaments; cortical cells with several discoid plastids, each with a pyrenoid.

Unilocular and plurilocular sporangia borne on the same or different thalli, occurring on all but the younger regions of the thalli, associated with the primary assimilators in extensive, closely packed, vertically elongated, raised, punctuate, tufted sori; secondary assimilators (paraphyses) also reported to be associated with the sori, the centrally positioned ones longer than those at the edge of the sori, giving the thallus the characteristic rough, warty appearance; secondary assimilators associated with the plurilocular sporangia reported to be longer and more branched than those of *Stilophora tenella* and the upper cells only slightly swollen and not noticeably different in diameter to those further down; plurilocular sporangia borne laterally on the secondary assimilators, in the mid to upper regions, sessile or shortly stalked, cylindrical, filiform, uniseriate, to 10 locules, 45 μm long × 7 μm wide; unilocular sporangia borne at the base of the secondary assimilators, sessile, obovoid-clavate, 50–70 × 20–30 μm.

Reported to be epiphytic on *Fucus* spp., and locally abundant.

Rare, only reported for the south and west coasts of Britain (Channel Islands, Swanage in Dorset and the Isle of Cumbrae in Argyll) and for Kerry in Ireland.

A summer annual.

This species was first reported for Britain by Batters (1893b), based on material collected by E.M. Holmes from Swanage in Dorest. It was subsequently reported for the Isle of Cumbrae (Batters 1902), and Co. Kerry in Ireland (Cullinane 1974; Guiry 1978). It has been reported to differ from *Stilophora rhizodes* in the following aspects: it is a smaller species and has a thicker, firmer and darker thallus; all cortical cells of the surface of the thallus, either immediately or later develop primary assimilators, so that very few smooth areas remain; the surface is almost completely covered in reproductive sori, except for the younger regions, which are often longitudinally extended in the direction of the thallus; while the secondary filaments (paraphyses) associated with the unilocular sporangia appear similar to those in *S. tenella*, those associated with the plurilocular sporangia are longer and more branched with the upper cells only slightly swollen; the plurilocular sporangia often occur high up on the paraphyses, in addition to their basal position. However, some doubt was expressed by Rosenvinge & Lund (1943) about the validity of some of these distinguishing characteristics and they, along with authors such as Reinke (1889b) and Peters & Müller (1986a), question the validity of maintaining them as separate species. Pending further studies, and examination of further collections, however, they are described herein as distinct species.

Agardh, J. (1882), p. 36, pl. 6(22); Batters (1893), pp. 23, 24, (1902), p. 43; Cullinane (1974), p. 67; Guiry (1978), p. 218, (2012), p. 151; Hamel (1931–39), p. 186, fig. 40(VII); Hornemann (1816), pl. 1546; Kylin (1907), p. 95, (1947), pp. 63, 64; Newton (1931), p. 162; Pankow (1971), pp. 188, 189, figs 240, 241; Peters & Müller (1986), pp. 417–23, figs 1–14; Reinke (1892a), pp. 55, 56, pl. 37(1–7), (1889b), pp. 72, 73; Ribera *et al.* (1992), p. 116; Rosenvinge & Lund (1943), pp. 45–7, fig. 16(A C); Rueness (1977), p. 151; Taskin *et al.* (2008), p. 61.

Stilophora tenella (Esper) P.C. Silva (1996), p. 624.   Figs 192, 193

Fucus tenellus Esper (1800), p. 197; *Ceramium tuberculosum* Roth (1800), pp. 162–4; *Conferva gracilis* Wulfen (1803), n. 23; *Ceramium rhizodes* C. Agardh (1811), p. 18; *Chordaria rhizodes* (C. Agardh) C. Agardh (1817), p. 15; *Chordaria rhizodes* var. *simplex* C. Agardh (1817), p. 16; *Fucus rhizodes* Ehrhart ex Turner (1811–19), pl. 235, nom. illeg.; *Sporochnus rhizodes* (C. Agardh) C. Agardh (1820), p. 156; *Zonaria papillosa* C. Agardh (1820), p. 135; *Dictyota papillosa* J.V. Lamouroux ex C. Agardh (1820), p. 135, nom. inval.; *Dictyota rhizodes* (Turner) J.V. Lamouroux (1824), p. 484; *Fucus rhizodes* var. *tenuior* Wahlenberg (1826), p. 893; *Dictyota papillosa* (C. Agardh) Greville (1830), p. xliii; *Stilophora rhizodes* (C. Agardh) J. Agardh (1841), p. 6, nom. illeg.; *Spermatochnus papillosus* (C. Agardh) Kützing (1843), p. 335; *Stilophora rhizodes* f. *papillosa* (C. Agardh) Hauck (1885), p. 385.

Thalli macroscopic, to 20(–60) cm long, 1(–3) mm wide, erect, yellow to light brown in colour, arising from a discoid holdfast; erect thalli cylindrical, filiform, solid at first, becoming hollow, firm, fairly rigid and coarse, non-gelatinous, much branched, branching dichotomous/subdichotomous to irregular, all axes and branches tapering slightly above; in structure, thalli haplostichous, multiaxial with a central medullary region of 4–5 closely packed columns/axial filaments of colourless, vertically elongated cells, surrounded by a small, narrow cortex of 4–5, smaller, tightly packed, firm, cortical cells which become slightly smaller and more pigmented towards the periphery; growth apical, each medullary filament with a terminal apical cell; apical regions of young thalli surrounded by involucra of enclosing curved, pigmented, filaments (= primary assimilators) which, as a result of vertical elongation of the inner cortical cells, later appear in localized areas/patches scattered over the remaining smooth, thallus surface; in surface view, thallus cells not arranged in obvious transverse or longitudinal rows, angular and generally elongated in the longitudinal axis, 15–80 × 12–21 µm; primary assimilators abundant, closely packed and strongly recurved towards apex at first, more sparsely distributed and less recurved below, simple, to 17 cells (–210 µm) long, slightly elongate-clavate, comprising large, ovoid to barrel-shaped, almost moniliform, upper cells, (10–)14–22(–26) × (12–)14–17(–19) µm, smaller and more quadrate/subquadrate below, (4–)6–9(–11) × (5–)8–12(–14) µm; hairs common, especially in the branch apices and associated with the assimilatory filaments, with basal meristem, lacking a sheaths, hair cells 3–11 diameters long, 35–90 × 8–12 µm; cortical cells and primary assimilator cells with several discoid plastids, each with a pyrenoid.

Unilocular and plurilocular sporangia borne on the same or different thalli, occurring extensively throughout thallus, associated with the primary assimilators in scattered, raised, punctuate, tufted sori, occurring also at the base of closely packed, newly formed, secondary assimilators; secondary assimilators simple or branched at the base, pigmented, curved, 3–9 cells, to 90 µm in length, comprising cells narrow, elongate-rectangular, below, to 2–5 diameters long, 10–18 × 4–6 µm, becoming large, barrel-shaped to moniliform, not infrequently unilaterally swollen above, 6–15 × 9–14 µm; plurilocular sporangia cylindrical, filiform, uniseriate, comprising 4–6(–12) locules, 30–50 µm long, 5–8 µm in diameter; unilocular sporangia sessile, clavate to ovoid, 25–60 × 12–28 µm.

Epilithic and epiphytic on larger algae, lower eulittoral pools and shallow sublittoral, frequently associated with sandy areas; some tolerance of low salinity; occasionally infected by *Streblonema stilophorae*.

Widespread and common around Britain and Ireland, but reported mainly from south-western/western shores.

A summer annual; June to September.

Culture studies on *Stilophora tenella* (as *S. rhizodes*) throughout the North Atlantic (see Thuret 1850; Kylin 1918, 1933; Sauvageau 1931c; Dangeard 1968b; Peters & Müller 1986a; Novaczek *et al.* 1986b) revealed that zoidangia/meiospores and plurispores from the macroscopic thalli gave rise to filamentous, streblonematoid microthalli, the latter, under certain environmental conditions, acting as dioecious gametophytes releasing isogametes. *Stilophora tenella* can, thus,

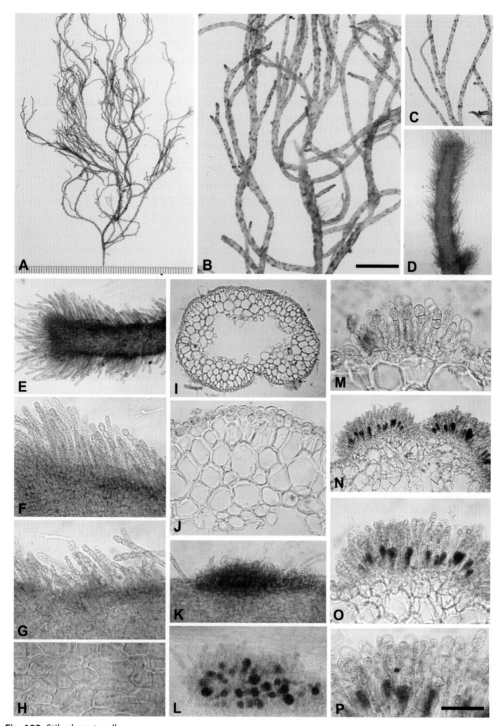

Fig. 192. *Stilophora tenella*

A–C. Habit of thalli. D, E. Branch apices showing involucre of primary assimilatory filaments. F–G. Edges of thalli showing primary, assimilatory filament. H. Surface view of thallus showing irregularly arranged angular cells. I, J. TS of mature thalli showing central hollow cavity surrounded by large medullary cells and outer cortex of 1–2 layers of smaller photosynthetic cells. K, L. Surface view of fertile thalli showing the punctuate sori. M–P. TS of fertile thallus showing the punctuate sori comprising secondary assimilatory filaments and unilocular sporangia. Bar in B = 15 µm (B), = 20 mm (C), bar in P = 300 µm (D, I), = 180 µm (E, N), = 90 µm (F, G, J, K, O), = 45 µm (H, M, P).

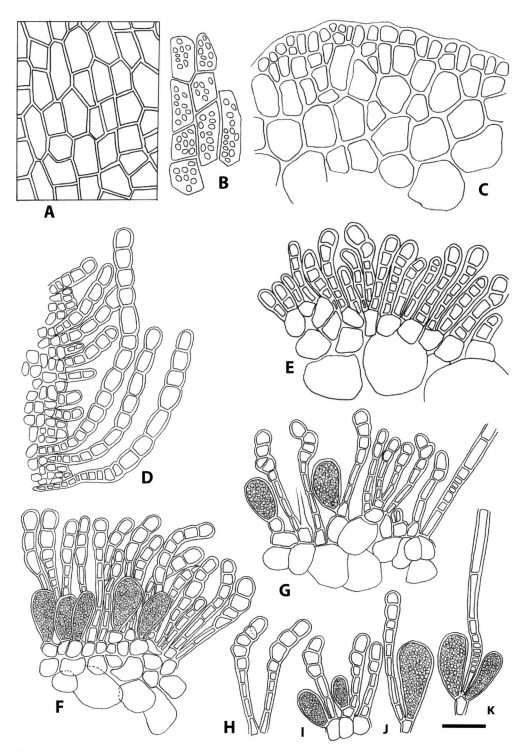

Fig. 193. *Stilophora tenella*

A, B. Surface view of thalli showing the irregularly arranged angular cells with discoid plastids. C. TS of portion of thallus showing the inner, large medullary cell and smaller cortical surface cells. D. Edge of thallus showing recurved primary assimilatory filaments. E–K. SP of fertile thalli showing the multicellular secondary assimilatory filaments and the basally borne unilocular sporangia. Bar = 50 μm.

be considered to have a heteromorphic, diplohaplontic type life history with a macroscopic, diploid sporophyte alternating with a microscopic, haploid gametophyte.

Brodie *et al.* (2016), p. 17; Bunker *et al.* (2017), p. 306; Coppejans (1983b), pls 40, 41; Cormaci *et al.* (2012), pp. 231, 232, pl. 65(1–3); Cotton (1912), p. 96; Dawes & Mathieson (2008), pp. 145, 146, pl. XVII(17–18); Dizerbo & Herpe (2007), p. 86, pl. 28(1); Fletcher in Hardy & Guiry (2003), p. 305; Fritsch (1945), p. 94, fig. 28; Gallardo García *et al.* (2016), p. 36; Gayral (1966), p. 283, pl. XLIX; Guiry (2012), p. 151; Hamel (1931–39), pp. 184, 185, fig. 40(V), VI; Harvey (1846–51), pl. 70; Hauck (1884), p. 385, fig. 166; Kylin (1907), p. 96, (1918), pp. 12, 13, figs 4–5, (1933), pp. 66–9, figs 31, 32, (1940), p. 51, fig. 29; Mathieson & Dawes (2017), p. 186, pl. LV(1–2); Morton (1994), p. 35; Newton (1939), p. 162, fig. 102(A–E); Novaczek *et al.* (1986b), pp. 407–16, figs 1–19; Oltmanns (1922), pp. 39, 40, fig. 334(1, 2); Pankow (1971), p. 188, figs 236–9; Peters (1987), pp. 223–63; Peters & Müller (1986), pp. 417–23, figs 1–14; Reinke (1889b), pp. 70, 71, (1892), p. 55, pl. 36; Ribera *et al.* (1992), p. 116; Rosenvinge & Lund (1943), pp. 42–4, fig. 15(A–E); Rueness (1977), p. 151, fig. 84, pl. XX(3), (2001), p. 17; Schneider & Searles (1991), pp. 139, 140, fig. 159; Silva *et al.* (1996), pp. 624, 928, 929; Taskin *et al.* (2008), p. 61; Taylor (1957), p. 151, pl. 13(6), 14(6); Womersley (1979), pp. 128–30, figs 39(A), 40(A–E).

Stilopsis Kuckuck

STILOPSIS Kuckuck (1929), pp. 11, 70.

Type species: *S. lejolisii* (Thuret) Kuckuck (1931–39), p. 180.

Thalli macroscopic, 20–30(–50) cm long, to 1(–2) mm wide, erect, filiform, solid, much branched, branching dichotomous to subdichotomous, with axes and branches tapering above and below, attached by a small, discoid holdfast; structure haplostichous, uniaxial, with central axial filament branching laterally to form an enclosing envelope of large, elongated, colourless cells, which by further divisions form a cortex up to 4–5 cells thick, with cells gradually decreasing in diameter towards the periphery; outer cortical cells giving rise to short, 1–3-celled, filaments of limited growth, short, 3–6-celled paraphyses, unilocular sporangia, plurilocular sporangia and/or abundant hairs; growth apical; hairs with basal meristem, without obvious sheath; filaments of limited growth forming a continuous, papillose covering over surface.

Unilocular and plurilocular sporangia occurring on the same or different thalli, in scattered sori on thallus surface, at the base of the paraphyses; plurilocular sporangia uniseriate; unilocular sporangia ovoid to pyriform.

One species in Britain and Ireland.

Stilopsis lejolisii (Thuret) Kuckuck (1931–39), p. 180.   Figs 194, 195

Stilophora lejolisii Thuret in Le Jolis (1863), p. 89; *Spermatochnus lejolisii* (Thuret) Reinke (1889b), p. 60.

Thalli macroscopic and erect, 20–30(–50) cm long, to 1(–2) mm wide, yellow to pale brown, and attached by a small, discoid holdfast; thalli filiform, firm, solid, gelatinous and somewhat slippery, and much branched, all axes and branches tapering above, and less obviously, below; branching dichotomous to subdichotomous; in structure, thalli haplostichous, uniaxial with central axial filament, surrounded by a cortex of large, elongated, colourless cells, up to 4 cells thick, with cells gradually decreasing in diameter towards the periphery; growth by an apical cell; outer cortical cells giving rise to short filaments of limited growth, short paraphyses, unilocular sporangia, plurilocular sporangia and/or abundant hairs; short filaments forming a continuous papillose covering over surface, 1–2(–3) cells high, 15–45 × 10–16 µm, appearing rounded in surface view; hairs with basal meristem, without sheath, colourless, 5–12 µm in diameter; cells with several discoid plastids each with a pyrenoid.

Unilocular and plurilocular sporangia occurring on the same or different thalli, produced at the base of the paraphyses in discrete to continuous, irregularly shaped, punctate sori extending to 200–300 µm in diameter on thallus surface; plurilocular sporangia not observed

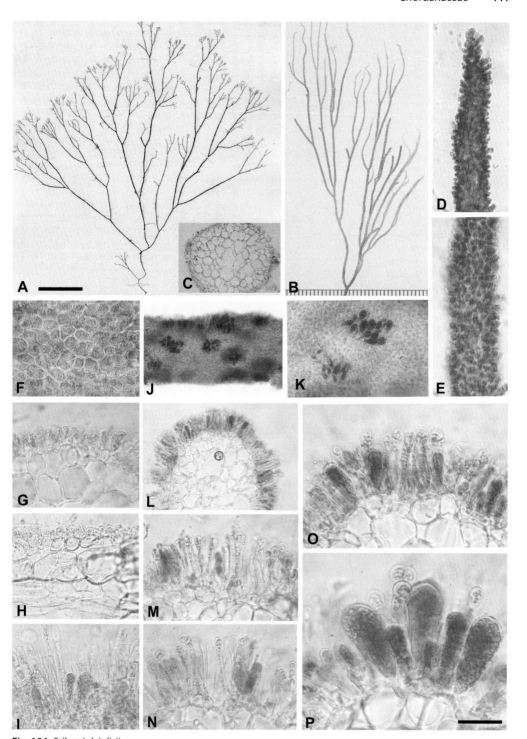

Fig. 194. *Stilopsis lejolisii*

A, B. Habit of thalli. C. TS of thallus. D. Apical portion of branch. E. Subapical portion of branch. F. Surface view of thallus. G, H. TS and LS of thalli respectively, showing short filaments of limited growth. J, K. Surface view of thalli showing the punctuate sori. I, L–P. TS of thalli showing the unilocular sporangia borne at the base of the paraphyses. Bar in A = 12 mm, bar in P = 300 μm (J), = 90 μm (C–E, G, L, I), = 45 μm (F, H, M–P). (A, herbarium specimen, NHM.)

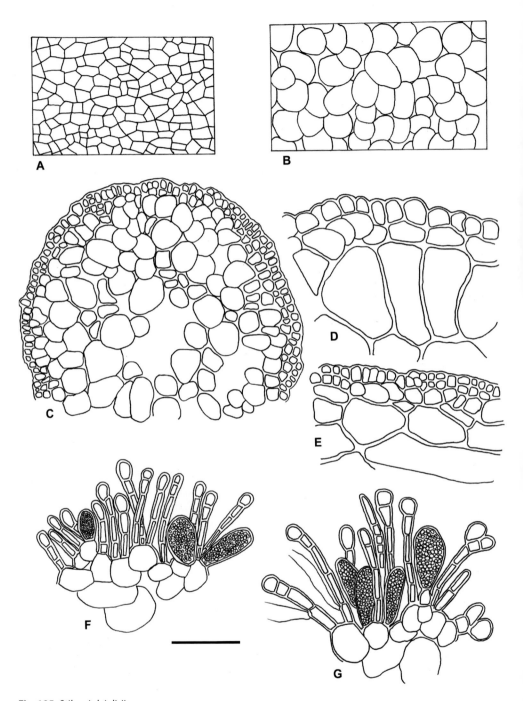

Fig. 195. *Stilopsis lejolisii*

A. Surface view of young branch tip. B. Surface view of more mature thallus. C–E. TS of thalli showing the outer short filaments. F, G. TS of thalli showing the unilocular sporangia at the base of the paraphyses. Bar = 50 μm.

by the present author, reported by Hamel (1931–39) to be uniseriate, 30–4 × 10 µm; unilocular sporangia pyriform, (33–)40–55(–60) × (15–)18–25(–28) µm; paraphyses comprising (2–)3–5(–6) cells, 32–90(–120) µm in length, with cylindrical basal cells, 2–6 diameters long, (12–)15–25(–31) × (4–)5–8(–10) µm, and bulbous terminal cells, 17–22 × 10–16 µm.

Epilithic, on pebbles, or epiphytic on larger algae, usually *Gongolaria* spp., in sheltered areas and usually associated, more or less strongly, with sand; lower eulittoral pools and shallow sublittoral.

Rare, reported for scattered localities around Britain (Cornwall, Sussex, Ayrshire, Bute, Inverness) and for Wexford in Ireland.

Summer annual; June to September.

Brodie *et al.* (2016), p. 17; Dizerbo & Herpe (2007), p. 88, pl. 28(2); Fletcher in Hardy & Guiry (2003), p. 305; Guiry (2012), p. 151; Hamel (1931–39), pp. 180–2, fig. 40(III, IV); Kuckuck (1929), pp. 11, 70, figs 96–8; Kylin (1940), p. 50, fig. 28; Le Jolis (1863), p. 89; Morton (1994), p. 8; Newton (1931), p. 162; Rosenvinge & Lund (1943), pp. 40, 41.

Streblonema Derbès & Solier

STREBLONEMA Derbès & Solier (1851), p. 100

Type species: *S. sphaericum* (Derbès & Solier) Thuret (1863), p. 73.

Thalli primarily endophytic, comprising a limited to extensive network of irregularly branched, colourless to pigmented filaments spreading throughout host tissue, sometimes penetrating deeply into the cortical zone, with or without limited emergence and development onto the host surface; true colourless hairs, with basal meristem, present or absent; growth diffuse; cells usually with either 1–3 plate-like or several discoid plastids, usually without obvious pyrenoids.

Unilocular and plurilocular sporangia terminal or lateral on the endophytic filaments, sometimes on emergent filaments, sessile or shortly pedicellate; plurilocular sporangia common, linear, cylindrical, simple or branched, or ovate to conical and multiseriate when mature, solitary or in fascicles; unilocular sporangia rare, usually globose.

Species of the genus *Streblonema* are characterized by a very simple thallus of free, branched filaments which usually occur as endophytes in a range of red or brown host algae. Their primary characteristic is that they lack heterotrichy i.e. their erect filaments are morphologically the same as those of the prostrate filaments. The greater part of the thallus is usually endophytic, forming a branched network of irregularly spreading filaments, giving rise to hairs, unilocular and/or plurilocular sporangia which can protrude through the host surface. The plurilocular sporangia are multiseriate when mature. However, it is a polyphyletic, i.e. artificial, genus and the species do not form a natural grouping. While previously Parke & Dixon (1976) listed 12 species, culture and molecular studies (see references below) have reduced the number of species within *Streblonema* to five for Ireland (Guiry 2012) and 6 for Britain (Brodie *et al.* 2016). While some species/isolates, for example *S. immersum* (see Pedersen 1984) and *S. oligosporum* (see Fletcher 1983b), have been shown to be independent entities, many others have been revealed to be microscopic phases in the life histories of several macroscopic genera. For example, Pedersen (1984) revealed *S. immersum* to be an independent taxon and Fletcher (1983b) revealed a direct life history for *S. oligosporum*. In contrast to this, Pedersen (1978b) reported that *S. sphaericum* was probably the microthallus of *Myriotrichia clavaeformis*, Ajisaka & Umezaki (1978) and Ajisaka (1979) reported *Streblonema*-like microthalli in the life histories of *Sphaerotrichia divaricata* (C. Agardh) Kylin and *Acrothrix pacifica* Okamura & Yamada respectively, while Pedersen (1984) considered *Streblonema thuretii* Sauvageau, *S. volubile* (P. Crouan & H. Crouan) Thuret, *S. danicum* Kylin and *S. oligosporum* Strömfelt to be similar to the microthalli of *Litosiphon laminariae* (as *L. pusillus*). These *Streblonema*-like microthalli have been variously interpreted as gametophytes, protonemata, prostrate systems and plethysmothalli (Pedersen

1984), with the relationship between the microthalli and macrothalli based on either ploidy differences, giving rise to a biphasic life history, and/or environmental conditions.

As the type species of *Streblonema* is *S. volubile*, which has multiseriate plurilocular sporangia, and in agreement with Peters (2003) that sporangial structure is a conservative trait, those species recorded for Britain and Ireland with uniseriate plurilocular sporangia are, herein, removed from *Streblonema* and transferred to *Microspongium* following the proposal of Peters. The following *Streblonema* species are therefore transferred into *Microspongium*: *S. immersum* (see Pedersen 1984), *S. oligosporum* (see Fletcher 1983b), *S. parasiticum*, *S. stilophorae* (including *S. tenuissima*) and *S. zanardinii*. With the additional inclusion of some *Streblonema* species within the synonymy of others (*S. sphaericum* within *Myriotrichia clavaeformis*, *S. aequale* within *Phaeostroma pustulosum*, *S. effusum* within *Punctaria tenuissima*, *S. chordaria* within *Chordaria flagelliformis* and *S. volubilis* within *Litosiphon laminariae* – see relevant sections in the present treatise), fewer species remain within *Streblonema* as compared to the numbers given by Newton (1931) and Parke & Dixon (1976). In the present treatise, four species of *Streblonema* are recognized for Britain and Ireland but this is likely to change pending further culture and molecular studies on this difficult group.

Two species in Britain and Ireland, two species in Britain only.

KEY TO SPECIES

1.   Thallus endophytic in *Ascophyllum nodosum*. . . . . *S. breve*
    Thallus endophytic in other hosts 2

2.   Thallus endophytic in the crustose 'Petrocelis cruenta' phase of *Mastocarpus*
    stellatus . . . . *S. helophorum*
    Thallus endophytic in other hosts 3

3.   Thallus endophytic in the red alga *Vertebrata byssoides*; plurilocular sporangia
    solitary . . . . *S. intestinum*
    Thallus endophytic mainly in gelatinous hosts (*Eudesme virescens*, *Helminthora divaricata*, *Sauvageaugloia griffithsiana*, *Stilophora tenella*) and *Ulva* spp.; plurilocular sporangia sometimes solitary, more usually in dense clusters. . . . . *S. fasciculatum*

Streblonema breve (Sauvageau) De Toni (1895), p. 576.   Figs 196, 197

Ectocarpus brevis Sauvageau (1892), p. 76; *Entonema breve* (Sauvageau) Hamel (1931–39), p. xxvi.

Thalli endophytic, forming very fine, yellowish, confluent turfs, up to several cm in extent, on *Ascophyllum nodosum*, particularly obvious when the host has dried; comprising an extensive endophytic network of quite deeply penetrating, and horizontally spreading, irregularly branched filaments of often contorted cells, 1–5 diameters long, 8–29 × 5–14 µm, which become perpendicularly oriented and the cells more linear and rectangular as they approach and eventually narrow and break through the host surface as short, closely packed, mainly simple, although occasionally irregularly branched to one order, rigid, erect filaments; erect filaments obtuse, sometimes with characteristic dome-shaped apices, sometimes with downward-growing multicellular, secondary rhizoids arising from basal cells, and up to 20 cells, 350 µm in height, and comprising fairly thick-walled, quadrate to rectangular, sometimes slightly barrel-shaped cells, 0.5–1 diameters long, 6–16 × 7–12 µm, each containing several discoid plastids; growth probably apical; hairs not observed.

Plurilocular sporangia and unilocular sporangia terminal or lateral on erect filaments, quite commonly produced on short, newly emergent filaments; if lateral, then sessile or on 1–2-celled pedicels; plurilocular sporangia common, oval to ovate-lanceolate in shape, multiseriate, 21–60 × 14–27 µm, enclosing a relatively small number of loculi; unilocular sporangia rare, ovate, 20–27 × 17–22 µm.

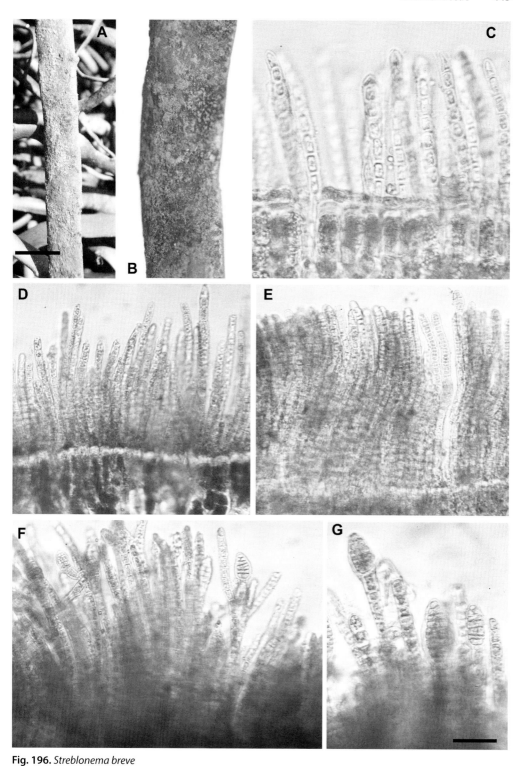

Fig. 196. *Streblonema breve*

A, B. Turf-like habit of thalli on host *Ascophyllum nodosum*. C–E. TS of *A. nodosum* showing emerging, closely packed filaments of the *Streblonema*. F, G. Portions of erect filaments with plurilocular sporangia. Bar in A = 8 mm (A), = 4 mm (B), bar in G = 90 μm (D–F), = 45 μm (C, G).

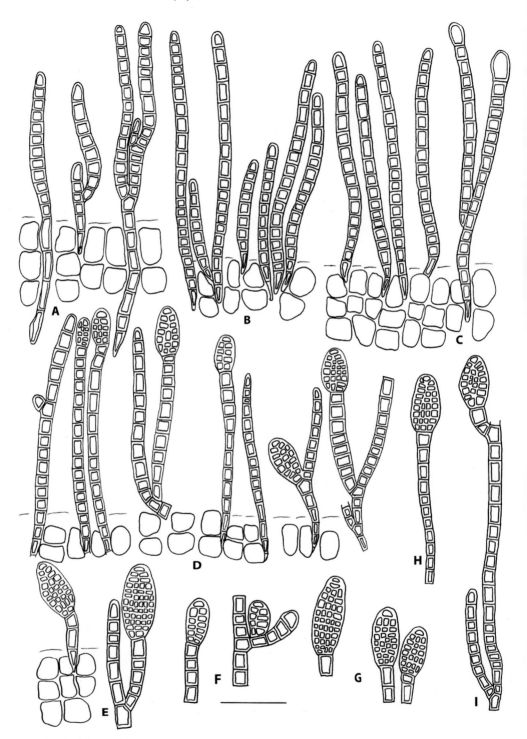

Fig. 197. *Streblonema breve*

A–C. TS of *Ascophyllum nodosum* showing emergent filaments of the *Streblonema*. D–I. Portions of erect filaments with plurilocular sporangia. Bar = 50 μm.

On older thalli of *Ascophyllum nodosum*, particularly at the base of the fronds.

Rare and recorded at only a few localities around Britain (Northumberland, South Devon, Isle of Man) and Ireland (Cork, Dublin); probably more common than records suggest.

Recorded throughout the year (January, March–June, October), always with plurilocular sporangia, with unilocular sporangia in April.

Batters (1892), pp. 50, 51, (1902), p. 29; Brodie *et al.* (2016), p. 1021; Dizerbo & Herpe (2007), p. 73; Fletcher in Hardy & Guiry (2003), p. 307; Guiry (2012), p. 152; Hamel (1931–39), pp. 59, 60, fig. 18(E); Knight & Parke (1931), p. 60, pl. XIX(79); Newton (1931), p. 116; Sauvageau (1892), pp. 76, 77, pl. II(11).

Streblonema fasciculatum Thuret (1863), p. 73. Fig. 198

Ectocarpus pringsheimi Reinke (1889b), p. 42; *Streblonema fasciculatum* var. *simplex* Batters (1902), p. 29; *Streblonema fasciculatum* f. *typica* Kuckuck in Kuckuck & Kornmann (1954) nom. inval.

Thalli endophytic, forming a network of branched filaments spreading among the outer cortical cells of the host, emerging at the surface, and just visible to the unaided eye as small, discrete to confluent, dark-coloured spots with dense, outward-projecting hairs; primary filaments procumbent, irregularly to pseudodichotomously branched, comprising cylindrical to irregularly globose cells, 1–3 diameters long, 14–21(–26) × 8–12(–19) µm; secondary and fertile branches erect, with more regular, rectangular cells; all cells containing 1–2 small, plate-like plastids; terminal regions of branchlets giving rise to hairs and/or plurilocular sporangia; hairs colourless, with short basal meristematic zone, lacking a sheath; hair cells 4–11 diameters long, 48–135 × 10–15 µm.

Plurilocular sporangia common, uniseriate to pluriseriate, terminal or lateral on short branchlets, solitary or more usually in dense clusters and fasciculate, and secundly arranged on parental filaments, cylindrical, simple, less commonly branched, conical to ovoid-lanceolate/ spindle in shape, 40–130 × 10–15(–20) µm.

Endophytic, most commonly in *Eudesme virescens*, but also reported for *Helminthora divaricata*, *Sauvageaugloia griffithsiana*, *Stilophora tenella* and *Ulva* spp. during the summer, particularly in older host tissue, mid to lower eulittoral pools.

Rarely recorded but widely distributed around Britain (Isles of Scilly, South Devon, Dorset, Channel Islands, Anglesey, Isle of Man, Lewis, Aberdeenshire, Northumberland) and Ireland (Wexford, Clare).

Summer annual, with host species.

N.B. Formae *typica* and *simplex* are sometimes used by authors and refer to material with branched and simple plurilocular sporangia respectively. Note that Pedersen (1976) considered material he identified as *Streblonema fasciculatum* var. *simplex* from southernmost Greenland to resemble the microthalli of *Delamarea attenuata* pictured in Pedersen (1974b).

Aisha & Shameel (2011), pp. 145, 146, fig. 9(a, b); Bartsch & Kuhlenkamp (2000), p. 169; Batters (1902), p. 29; Brodie *et al.* (2016), p. 1020; Dizerbo & Herpe (2007), p. 75; Fletcher in Hardy & Guiry (2003), p. 307; Guiry (2012), p. 152; Hamel (1931–39), p. 69, fig. 20(C); Jaasund (1951), pp. 135, 136, fig. 4(f), (1965), p. 41, fig. 12(A); Knight & Parke (1931), p. 60; Kuckuck & Kornmann (1954), pp. 109, 110, fig. 4(A–C); Mathieson & Dawes (2017), p. 187, pl. LV(3); Newton (1931), p. 128; Pankow (1971), p. 177, fig. 215; Pedersen (1974b), pp. 313–18, fig. 6, (1976), pp. 33, 34, figs 12, 14(a–b), (2011), p. 115; Reinke (1889b), p. 42; Rosenvinge (1894), p. 118; Rosenvinge & Lund (1941), pp. 60, 61, figs 31, 32; Rueness (1977), p. 122, fig. 46, (2001), p. 17; Taskin *et al.* (2008), p. 58; Taylor (1957), pp. 114, 115; Womersley (1987), p. 57, fig. 13(D).

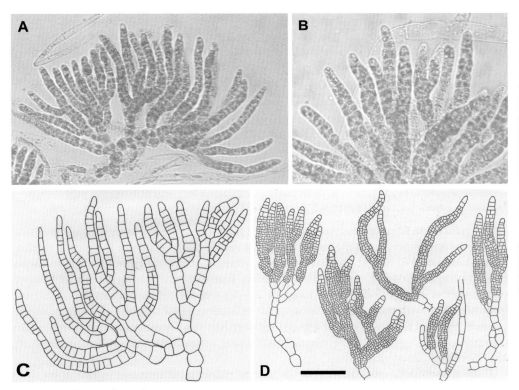

Fig. 198. *Streblonema fasciculatum*
A–D. Habit of thalli showing the terminal fascicles of uniseriate to pluriseriate plurilocular sporangia. Bar in D = 66 μm (A), = 30 μm (B, D), = 54 μm (C).

Streblonema helophorum (Rosenvinge) Batters (1902), p. 29.   Fig. 199

Ectocarpus helophorus Rosenvinge (1898), pp. 82–5.

Thalli entirely endophytic in host, forming a procumbent, horizontal network of branched filaments ramifying through host tissue, giving rise to erect filaments extending up to host surface; procumbent filaments of irregularly flexuose cells, much longer than wide, 2–4(–5) μm in diameter; erect filaments to 200 μm or more in height, elongate-clavate, branched, especially towards the apex, comprising long slender cells below, to 5(–11) diameters long, to 21(–58) × 3–4 μm, and shorter cells above, to 2(–3) diameters long, to 13(–18) × 5(–7) μm, usually with an enlarged clavate to subclavate terminal cell, 4–6 μm in diameter, each containing a single, plate-like plastid; hairs not observed.

Unilocular sporangia terminal on erect filaments, sometimes lateral and sessile, clavate, 9–22 × 9–10 μm; plurilocular sporangia unknown.

Endophytic in the red crustose 'Petrocelis cruenta' phase of *Mastocarpus stellatus*.

Rare, only reported for Northumberland (Berwick-on-Tweed) and Bute (Isle of Cumbrae) in Britain (see Batters 1902), and not collected by the present author.

Collections of this species confined to the winter months, January for Berwick-on-Tweed and November for the Isle of Cumbrae, and insufficient to comment on seasonal distribution.

No detailed description was given by Batters (1900), or subsequently by Newton (1931), and the above description and illustrations have been derived for Rosenvinge (1889). A slide labelled *Ectocarpus helophorus* deposited by Batters in the Natural History Museum, collected on 27 January 1887 from Berwick-on-Tweed, revealed erect branched filaments up to 250 μm

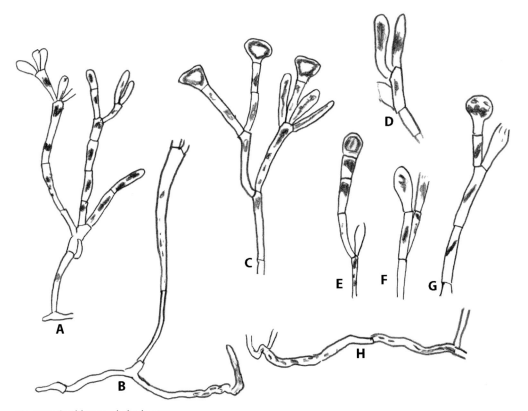

Fig. 199. *Streblonema helophorum*
A–H. Procumbent filaments giving rise to erect, branched filaments and terminal unilocular sporangia (after Rosenvinge 1898).

long arising through the host material, but lacking in sporangia and with cellular dimensions of 39–45 × 18–27 μm, which are much larger than those in the description and illustrations given by Rosenvinge, and with no obvious markedly elongated lower cells on the filaments. Pending further studies, this record for Britain must remain suspect.

Batters (1900), p. 372, (1902), p. 29; Brodie *et al.* (2016), p. 1021; Fletcher in Hardy & Guiry (2003), p. 41; Jónsson (1904), p. 38; Lund (1959), pp. 69, 70; Newton (1931), p. 129; Pedersen (1976), p. 73; Rosenvinge (1898), pp. 82–5, figs 17, 18.

Streblonema intestinum (Reinsch) Batters (1891), p. 525.   Fig. 200

Entonema intestinum Reinsch (1875), p. 5.

Thalli entirely endophytic, except for emergent plurilocular sporangia; forming a network of branched filaments ramifying between the pericentral cells of the host; filaments much branched, branching irregular or alternate, sometimes bifurcating at the tip, comprising cells 1–2 diameters long, (7–)10–20(–23) × (7–)9–11(–13) μm, usually more elongate below, more quadrate above, sometimes with terminal cells of branchlets swollen and hyaline; plastid characteristics not discernible and hairs not observed.

Plurilocular sporangia emergent from the host surface, simple, pluriseriate, acutely ovoid, 54–63 × 15–23 μm

Endophytic in the red alga *Vertebrata byssoides*.

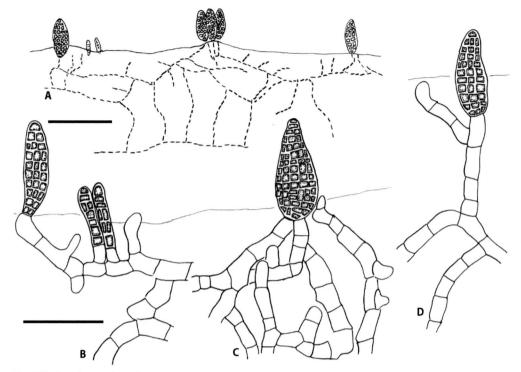

Fig. 200. *Streblonema intestinum*

A–D. Edge view of host showing irregularly spreading endophytic filaments giving rise to emergent, pluriseriate, plurilocular sporangia. Bar in A = 100 μm, bar in B = 50 μm (B–D).

Very rare, only known for Dorset (Weymouth) in Britain but probably more widely distributed.

Collection period given as April.

Batters (1902), p. 29; Brodie *et al.* (2016), p. 1021; Cotton (1906), pp. 295, 296, pl. 12(5); Fletcher in Hardy & Guiry (2003), p. 41; Holmes & Batters (1891), p. 525; Newton (1931), p. 128; Reinsch (1875), p. 5, pl. VIA(2a).

Strepsithalia Bornet ex Sauvageau

STREPSITHALIA Bornet ex Sauvageau (1896d), p. 64.

Type species: *S. curvata* Sauvageau (1896d), p. 53.

Thalli endophytic/epiphytic, microscopic, forming an endophytic network of branched, filaments which give rise to erect, short, large-celled, medullary filaments, in turn giving rise to cortical filaments of determinate growth, protruding through host surface; cortical filaments simple or branched, to 7–10 cells and embedded in a gelatinous matrix; hairs with basal meristem arising from medullary filaments; cortical cells with several discoid plastids, each with a pyrenoid.

Unilocular and plurilocular sporangia borne at the base of the cortical filaments; unilocular sporangia sessile, spherical to oval; plurilocular sporangia uniseriate to biseriate.

One species in Britain.

Strepsithalia buffhamiana (Batters) Batters (1902), p. 28.   Fig. 201

Streblonema buffhamianum Batters (1896), p. 386.

Thalli endophytic/epiphytic, microscopic, forming an endophytic network of branched, laterally spreading filaments giving rise to erect, medullary filaments, which in turn give rise to small, pulvinate tufts of cortical, paraphysis-like filaments protruding through host surface; endophytic filaments spreading, much branched, ramifying throughout cortical tissues of host, comprising variously shaped, colourless cells, some slightly nodose, 60–80 × 9–10 μm; medullary cells few in number, colourless, larger than the parental filaments; cortical filaments clustered, determinant in length, simple, occasionally branched, to 7–10 cells, clavate, with basal cells measuring up to 30 μm in length, 9–15 μm in diameter, upper cells 30–45 μm in length, to 30 μm in diameter.

Unilocular sporangia borne at the base of the cortical filaments, sessile, spherical to oval, 30–45 μm long, to 30 μm in diameter; plurilocular sporangia unknown.

Endophytic, ramifying throughout the cortical tissues of *Sauvageaugloia griffithsiana* and *Mesogloia vermiculata*, lower eulittoral pools.

Very rare, only reported for Britain (Cornwall, Dorset).

A summer annual, as for the hosts.

This species was first described by Batters (1896) based on material collected at Falmouth in Cornwall, sent to him by T.H. Buffham who thought it might represent a new species of *Strepsithalia*. However, as the erect, protruding cortical filaments were not embedded in jelly, Batters (1896) initially assigned the material to *Streblonema*, but indeed later transferred it to *Strepsithalia* (Batters 1902, p. 28). It differs from *Streblonema* spp. in showing some differentiation of the endophytic filaments, notably with respect to the production of erect, short medullary filaments of larger cells, and the pulvinate, gelatinous features of the emergent cortical/paraphyses-like filaments. No material has been found by the present author and the description is based on slides deposited by Batters in the algal herbarium (BM) at the Natural History Museum, London with due acknowledgement to Newton's (1931) description and illustrations. Unfortunately, no illustrations were given in Batters's (1896) original description.

Batters (1896), p. 386, (1902), p. 28; Brodie *et al.* (2016), p. 1022; Fletcher in Hardy & Guiry (2003), p. 42; Newton (1931), p. 152, fig. 94(A–C).

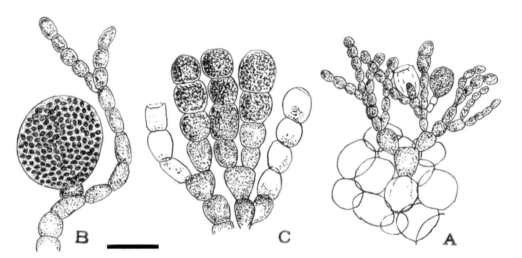

Fig. 201. *Strepsithalia buffhamiana*
A–C. SP of thalli on host showing emergent pulvinate tufts of branched, cortical filaments with basally formed unilocular sporangia. Drawings from Newton (1931).

Striaria Greville

STRIARIA Greville (1828), p. 44.

Type species: *S. attenuata* Greville (1828), p. 44.

Thalli macroscopic, erect, arising from a small discoid holdfast; thalli cylindrical, at first terete and solid, later becoming flaccid, somewhat compressed and tubular, much branched, branching regularly opposite, sometimes alternate or in whorls of up to 4, to 2–3 orders, frequently with occasional adventitious branches, all branches tapering sharply below, more gradually above, and ending in a uniseriate filament with a terminal true hair; in structure, thallus parenchymatous, with a large central cavity surrounded by a medulla of 1(–2), large, thin-walled, longitudinally elongate, transversely rounded or irregularly shaped, colourless cells surrounded by a peripheral, single layered cortex of smaller, pigmented cells; true hairs, with basal meristem and without obvious sheath, single or grouped, sometimes in whorls around thallus, arising from surface cells; growth trichothallic, followed subapically by longitudinal divisions; cells with numerous discoid plastids without obvious pyrenoids.

Unilocular sporangia projecting, produced directly from surface cells, either scattered or more abundantly grouped in irregular-shaped sori, sometimes characteristic whorled giving a striated/banded appearance to the mature, thallus surface, with unicellular, clavate paraphyses and hairs, ovoid to slightly pyriform; plurilocular sporangia unknown on macrophyte.

One species in Britain and Ireland.

Striaria attenuata (Greville) Greville (1828), p. 44.   Figs 202, 203

Carmichaelia attenuata Greville (1827a), pl. 288; *Striaria fragilis* J. Agardh (1841), p. 5; *Solenia crinita* J. Agardh (1842), p. 41; *Striaria crinita* (C. Agardh) J. Agardh (1842), p. 41; *Encoelium ramosissimum* Kützing (1843), p. 336; *Striaria attenuata* var. *crinita* (C. Agardh) Kützing (1845), p. 270; *Asperococcus ramosissimus* (Kützing) Zanardini (1863), p. 275; *Striaria attenuata* f. *ramosissima* (Kützing) Hauck (1884), p. 377; *Striaria attenuata* var. *crassa* Ardissone (1887), p. 137; *Striaria attenuata* f. *typica* Kjellman (1890), p. 54, nom. illeg.; *Striaria attenuata* f. *fragilis* (J. Agardh) Kjellman (1890), p. 54; *Striaria attenuata* f. *crinita* (J. Agardh) Kylin (1907), p. 74; *Striaria attenuata* var. *ramosissima* (Kützing) Schiffner in Schiffner & Vatova (1938), p. 198; *Striaria attenuata* f. *tenuissima* Kylin (1947), p. 69.

Thalli macroscopic, erect, solitary, to 32(–44) cm high, to 3–4 mm wide on main axes, yellow to light/medium brown in colour, arising from a small discoid holdfast; thalli cylindrical, at first terete and solid, later becoming flaccid, slightly compressed and tubular, much branched, branching regularly opposite, sometimes alternate or in whorls of up to 4, to 2–3 orders, not uncommonly with occasional adventitious branches, all branches tapering sharply below, more gradually above, and ending in a uniseriate filament with terminal true hair; in surface view, surface cells quadrate to rectangular, in longitudinal less obvious transverse rows, 10–28 × 6–20 μm, each with large numbers of small discoid plastids; in structure, thallus parenchymatous, with a large central cavity surrounded by a medulla of 1(–2), large, thin-walled, longitudinally elongate, transversely rounded or irregularly shaped, colourless cells surrounded by a peripheral, single layered cortex of smaller, pigmented cells; growth trichothallic, by means of a transversely dividing, hair-capped, uniseriate meristem of subcuboidal cells, which is followed subapically by longitudinal divisions; true hairs solitary or grouped, sometimes in whorls around thallus, opposite in apical regions, arising from surface cells, with basal meristem and no obvious sheath; cells with numerous discoid plastids without obvious pyrenoids.

Unilocular sporangia produced directly from surface cells and exserted, either single or more usually grouped in abundant, irregularly shaped sori, which can be scattered or characteristically whorled, giving a striated/banded appearance to the mature thallus surface;

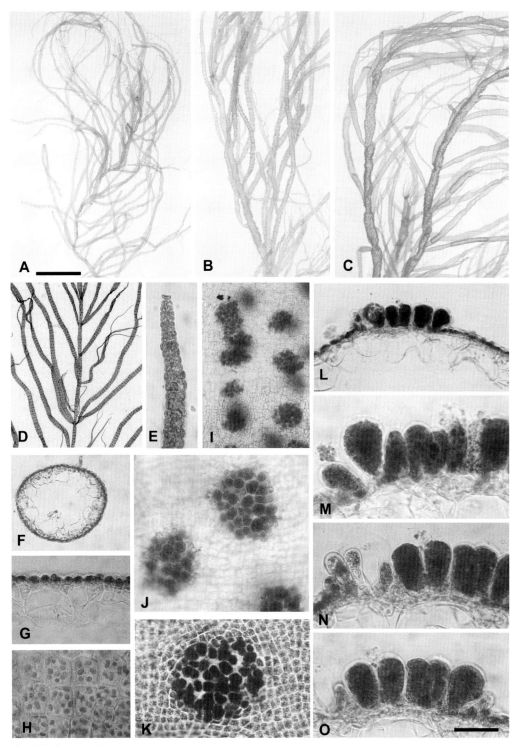

Fig. 202. *Striaria attenuata*

A–D. Habit of thalli. E. Apex of branch. F, G. TS of thalli showing large, central cavity with surrounding large medullary cells enclosed by a single layer of photosynthetic cortical cells. H. Surface view of thallus showing cortical cells in rows and containing discoid plastids. I–K. Surface view of fertile thalli showing reproductive sori. L–O. TS of fertile thalli showing unilocular sporangia. Bar in A = 2 cm (A, B), = 1 cm (C), bar in O = 1 mm (F), = 300 µm (I), = 180 µm (J), = 90 µm (E, K, L), = 45 µm (G, H, M–O). (D, herbarium specimen, NHM.)

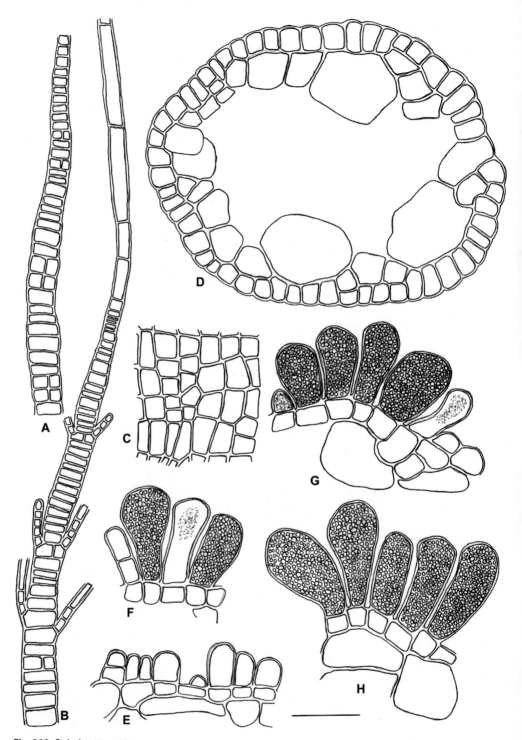

Fig. 203. *Striaria attenuata*

A, B. Apices of branches showing uniseriate filament and terminal hair. C. Surface view of thallus showing cells in rows. D. TS of thallus showing central cavity, large colourless medullary cell and an outer single layer of cortical cells. E–H. TS of sori showing unilocular sporangia being formed from the surface cortical cells. Bar = 50 μm.

sporangia ovoid to slightly pyriform, (5–)60–85(–93) × (28–)32–41(–45) μm, with occasional, unicellular, clavate paraphyses and hairs; plurilocular sporangia unknown on macrophyte.

Epilithic, epiphytic on various algae, shallow sublittoral to 12 m, occasionally reported in lower eulittoral, exposed or in tidal pools/lagoons, usually occasional, sometimes luxuriant; also found as loose-lying populations in extreme sheltered conditions.

Rare but widely distributed around Britain (Cornwall, North Devon, South Devon, Channel Islands, Hampshire, Sussex, Isle of Man, Bute, Ayrshire, Argyllshire, Orkney) and Ireland (Antrim, Cork, Donegal, Dublin, Down, Galway, Mayo).

A summer annual, commonly May/June with collections made from March to September.

Striaria attenuata has been investigated in culture by a number of authors including Kylin (1934), Caram (1965, 1966), Caram & Nygren (1970), Kornmann & Sahling (1973), Nygren (1975a) and Pedersen (1984). Although the results have differed, most evidence points towards *S. attenuata* having a monophasic, heteromorphic life history, with the unilocular sporangia-bearing macrothallus alternating with a microscopic, filamentous, plurilocular sporangia-bearing microthallus. Variations on this basic life history include the occurrence of apomeiotic unilocular sporangia, the production of unilocular sporangia on the microthallus, and zooids developing into both microthalli and macrothalli, as well as behaving as gametes.

Bartsch & Kuhlenkamp (2000), p. 169; Batters (1902), p. 25; Brodie *et al.* (2016), p. 1021; Bunker *et al.* (2017), p. 306; Caram (1964), pp. 2495–7, (1965), pp. 101, figs 1–12; Caram & Nygren (1970), pp. 130–5; Cormaci *et al.* (2012), pp. 234–6, pl. 66(1–3); Cotton (1912), p. 94; Dizerbo & Herpe (2007), p. 110, pl. 36(3); Fletcher in Hardy & Guiry (2003), p. 310; Fritsch (1945), pp. 103, 104, fig. 35(A–D); Gallardo García *et al.* (2016), p. 36; Greville (1827a), p. 44, pl. 288; Guiry (2012), pp. 152, 153; Hamel (1931–39), pp. 228–31, fig. 45(1–2); Harvey (1846–51), pl. 25; Hauck (1883–85), p. 377, fig. 162; Kornmann & Sahling (1973), pp. 14–25, figs 1–9, (1977), p. 280, pl. 161; Kylin (1934), pp. 15–17, fig. 9(A–E), (1947), pp. 69, 70, fig. 57(C–E), pl. 9(31); Lindauer *et al.* (1961), pp. 265, 266, fig. 69; Mathieson & Dawes (2017), p. 188, pl. LV(7–9); Morton (1994), p. 40; Nelson & Maggs (1996), pp. 449–53; Newton (1931), pp. 172, 173, fig. 108(A–C); Nygren (1975a), pp. 135, figs 11–15; Ohmori (1973), pp. 87–95; Pedersen (1984), pp. 54, 55; Peters (1991), pp. 261–9; Reinke (1891a), p. 51, figs A–C; Ribera *et al.* (1992), p. 118; Rosenvinge & Lund (1947), p. 59, fig. 21; Rueness *et al.* (1977), p. 155, pl. XXI(2), (2001), p. 17; Schneider & Searles (1991), p. 142, fig. 183; Stefanov *et al.* (2000), pp. 141–5; Stegenga *et al.* (1997a), pp. 20, 21; Taskin *et al.* (2008), p. 61; Taylor (1957), p. 159; Womersley (1987), p. 312, figs 109(B), 113(A–D).

Ulonema Foslie

ULONEMA Foslie (1894), p. 131.

Type species: *U. rhizophorum* Foslie (1894), p. 131.

Thalli epiphytic, forming small circular spots on host; consisting of a monostromatic basal layer of outward-spreading, pseudodichotomously branching filaments, not obviously laterally adjoined and discoid but irregular and diffuse, giving rise in the central regions to erect filaments, hairs and/or sporangia and downward-penetrating, branched, multicellular, rhizoidal filaments; erect filaments short, linear or clavate, simple or more frequently secundly branched to one order, with branches remaining short, 1–3-celled; cells with 1–3 plate-like plastids with pyrenoids; hairs common, with basal meristem and sheath.

Plurilocular sporangia either sessile or shortly stalked on basal layer, uniseriate, simple; unilocular sporangia commonly lateral at the base of erect filaments, ovate to pyriform.

The genus *Ulonema* is very closely related to *Myrionema* from which it differs in having a diffuse and irregularly spreading, filamentous base with deeply penetrating, branched rhizoidal filaments rather than a discoid base without rhizoids. However, these may be insufficient criteria for separating the genera. Some *Myrionema* species have been reported with rhizoids, while the diffuse, filamentous base of *Ulonema* might merely be an adaptation to the surface type of the host alga *Dumontia contorta*. In his original diagnosis for *Ulonema*, Foslie (1894) stated that plurilocular sporangia were absent. However, plurilocular sporangia have been described in *Ulonema* by Sauvageau (1897a, p. 233, fig. 12D) and Hamel (1931–39, p. 34). These observations lend support to the inclusion of *Ulonema* as a synonym of *Myrionema*. Pending confirmatory experimental studies the autonomy of *Ulonema* is retained here.

One species in Britain and Ireland.

Ulonema rhizophorum Foslie (1894), p. 132.   Fig. 204

Thalli epiphytic, forming light brown spots on hosts, usually solitary and circular, 0.25–0.75 mm in diameter, occasionally confluent and irregular; in surface view, peripheral region thin and prostrate comprising outward-radiating, frequently pseudodichotomously branched, diffused filaments, not laterally cohered into discs, with inner region raised, hemispherical and pile-like comprising cells variously packed in surface view, 7–12 µm in diameter; in squash preparations, consisting of a monostromatic basal layer of uniseriate cells, 1–4 diameters long, 8–21 × 7–9 µm, all cells, except at periphery, giving rise to erect filaments, hairs and/or sporangia, with many cells giving rise below to single or multicellular, occasionally branched, rhizoidal filaments; erect filaments short, loosely associated in a gelatinous matrix, simple or sparingly branched, linear or clavate, to 4–7 cells (60–85 µm) long, comprising cells 2–6 diameters long, 16–23 × 4–7 µm each enclosing 1–3 plate-like plastids with pyrenoids; hairs frequent, arising singly from basal layer, with basal meristem and sheath, approx. 7–8 µm in diameter.

Unilocular sporangia abundant, either sessile on basal layer or on 1-celled stalks and usually lateral at the base of the erect filaments, spherical to pyriform in shape, 26–46 × 17–25 µm; plurilocular sporangia rare, sessile or shortly stalked on basal layer, to 25 µm long, 8 loculi, uniseriate.

Epiphytic on the red alga *Dumontia contorta*, especially on mature thalli, littoral fringe to lower eulittoral pools.

Common and widely distributed around Britain and Ireland.

Annual; mainly spring and early summer.

In many floristic studies, *U. rhizophorum* has either been included within the circumscription of *Myrionema strangulans* (Levring 1937; Jónsson 1903) or authors have commented on their possible conspecificity (Jaasund 1951, p. 130; Parke & Dixon 1976, p. 562; South & Hooper 1980, p. 35). However, agreement is expressed with Jaasund (1951, p. 131) that *Ulonema rhizophorum* does appear to differ from *Myrionema strangulans* and its autonomy is thus retained in the present work pending further experimental studies.

Batters (1895a), p. 275, pl. III; Brodie *et al.* (2016), p. 1021; Cotton (1912), p. 95; Dizerbo & Herpe (2007), p. 80; Edelstein & McLachlan (1969a), pp. 555–7, figs 5–12; Fletcher (1987), pp. 120–2, fig. 17; Fletcher in Hardy & Guiry (2003), p. 313; Foslie (1894), pp. 131–4, pl. III(11–17); Guiry (2012), p. 153; Hamel (1931–39), p. 94, fig. 24(19–21); Jaasund (1951), pp. 130, 131, fig. 1(e–f), (1965), pp. 58, 59, fig. 18(A, B); Knight & Parke (1931), p. 68, pl. IX(7, 10, 13); Kylin (1947), pp. 40, 41, fig. 34; Mathieson & Dawes (2017), pp. 188, 189, pl. LV(10–12); Newton (1931), pp. 154, 155, fig. 96; Rueness *et al.* (2001), p. 17; Sauvageau (1897a), pp. 229–33, fig. 12; Stegenga & Mol (1983), p. 89, pl. 27(4–5); Stegenga *et al.* (1997a), p. 19.

Ectocarpaceae C. Agardh.

ECTOCARPACEAE C. Agardh (1828), p. 9.

Thalli epiphytic, lithophytic or epizoic, and heterotrichous; prostrate filaments branched, free, and never united to form a pseudodisc, disc or crust; erect filaments free, usually tufted, sometimes intertwined and rope-like, usually much branched, especially above, to 10–20 cm in length, exceptionally up to 1 m in sheltered conditions, and monosiphonous throughout; branching mainly alternate to pseudodichotomous, more rarely subsecund, helical, irregular and opposite, to several orders, with laterals indeterminate, the ultimate branchlets often tapering gradually into either colourless, false hairs (e.g. in *Ectocarpus* and *Herponema*), or true

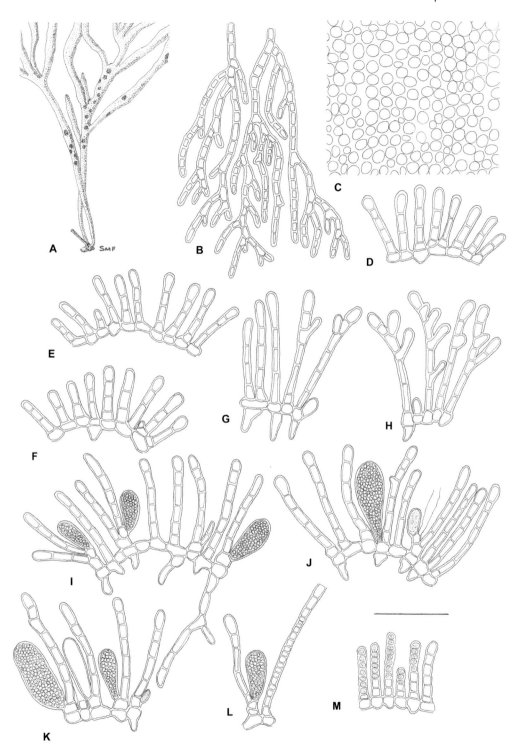

Fig. 204. *Ulonema rhizophorum*

A. Habit of thallus, forming small spots on the red alga *Dumontia contorta*. B. Surface view of peripheral thallus region showing irregularly spreading basal filaments. C. Surface view of central thallus region showing terminal cells of erect filaments. D–H. SP of vegetative thallus showing erect, later branched filaments arising from a monostromatic base. Note rhizoids emerging from basal layer. I–L. SP of fertile thallus showing unilocular sporangia. M. SP of fertile thallus showing plurilocular sporangia. Bar = 15 mm (A), = 50 μm (others).

hairs with basal meristem (*Kuckuckia*); corticating filaments sometimes covering basal regions of main axes, giving rise below to a rhizoidal base; basal attachment filaments totipotent and can periodically give rise to further erect filaments; growth apical or diffuse on basal system, intercalary and diffuse, rarely apical on erect system, without definite meristematic zones; cells with either (sometimes branched) elongate, ribbon-shaped or 1(–2) plate-like plastids, with or without obvious pyrenoids.

Reproduction by plurilocular sporangia, gametangia and unilocular sporangia; plurilocular sporangia/gametangia uniseriate or multiseriate, terminal, lateral or intercalary on the erect filaments, sessile or pedicellate, sometimes borne below a terminal false hair; unilocular sporangia terminal or lateral on the erect filaments, sessile or pedicellate.

Life history probably haplodiplontic and isomorphic to slightly heteromorphic, as shown for *Ectocarpus* in detailed studies by Kornmann (1956), Müller (1964b, 1966, 1967, 1972, 1975, 1976a,b, 1977, 1979) and Kornmann & Sahling (1977). Müller's studies revealed the life history to be complex in *Ectocarpus*: meiosis is always associated with unilocular sporangium initiation and, as a consequence, both diploid and haploid thalli occur, the former with unilocular and plurilocular sporangia and the latter with plurilocular gametangia only; sexual reproduction is by isogametes and assisted by the production of sexual attractants by female gametes; there may be some incompatibility between geographically isolated populations (1976b); asexual reproduction is widespread and common in both phases.

Although the family comprised four genera in the late nineteenth century, it became centred on *Ectocarpus*. Based on life history, morphological and cytological evidence, Sauvageau (1897, 1900), Hamel (1931–39), Kylin (1947) and Kornmann in Kuckuck (1955, 1956, 1958) separated several new genera from *Ectocarpus* (e.g. *Feldmannia*, *Hincksia* (*Giffordia*), *Kuckuckia*) and these have generally been accepted by most authors (e.g. Taylor 1957; Cardinal 1964; Parke & Dixon 1968; Abbott & Hollenberg 1976; Stegenga & Mol 1983; Womersley 1987; Schneider & Searles 1991; Kim & Lee 1992; Coppejans 1995). Notable exceptions include Lindauer *et al.* (1961), Chapman & Aiken (1961) and Chapman (1963). Molecular evidence (Peters & Ramírez 2001) revealed these genera to be polyphyletic and grouped the data into two clades which represented the Acinetosporaceae comprising all genera with discoid plastids, and the Ectocarpaceae, comprising all genera with ribbon-shaped plastids. North Atlantic members of the Ectocarpaceae were represented by *Ectocarpus* and *Kuckuckia*, the former characterized by false hairs and the latter by true phaeophycean hairs. Of these two genera, only *Ectocarpus* has been reported for Britain and Ireland, although it is likely that *Kuckuckia* is also present but has been overlooked due to its close similarity to *Ectocarpus*. However, based on molecular evidence, notably that of Silberfeld *et al.* (2011, 2014) and McCauley & Wehr (2007), the Ectocarpaceae was expanded to include *Spongonema* and *Pleurocladia*, both of which are represented in the marine flora of Britain and Ireland. Three genera viz *Ectocarpus*, *Pleurocladia* and *Spongonema* are therefore included within the Ectocarpaceae for Britain and Ireland.

Ectocarpus Lyngbye

ECTOCARPUS Lyngbye (1819), p. 130.

Type species: *E. siliculosus* (Dillwyn) Lyngbye (1819), p. 131.

Thalli epiphytic, lithophytic or epizoic, heterotrichous, forming erect, branched, filamentous tufts up to 10(–20) cm, exceptionally to 1 m in length in sheltered situations, arising from a branched, filamentous, outward-spreading, rhizoidal basal system; erect filaments monosiphonous throughout, much and variously branched, with lateral branches indeterminate, and produced alternately, pseudodichotomously, irregularly, helically or secundly, rarely oppositely, to several orders, and often tapering into a false hair; true hairs with basal meristem absent;

narrow corticating filaments sometimes covering basal regions of main axes giving rise below to a rhizoidal base; basal system serves mainly for attachment but also gives rise to further erect filaments; growth apical on basal filaments, intercalary and diffuse on erect filaments; cells with (sometimes branched) elongate, ribbon-shaped plastids, each with several pyrenoids.

Reproduction by plurilocular and unilocular sporangia; plurilocular sporangia common, multiseriate, terminal on erect filaments, or sometimes borne below a terminal false hair, sessile or pedicellate and borne laterally and adaxially, sometimes terminally on erect filaments; unilocular sporangia uncommon, ovoid to ellipsoid, sessile, rarely pedicellate, on terminal laterals; life history haplodiplontic, isomorphic to slightly heteromorphic, and complex (Kornmann 1956; Müller 1964b, 1966, 1967, 1972; Kornmann & Sahling 1977). Plurilocular sporangia can occur on both the gametophyte and sporophyte phase, on the former as gametangia, producing isogametes, and on both phases as 'neutral' sporangia which directly, and asexually, reproduce the parental phase.

With the exception of *Kuckuckia*, the conspicuously ribbon-shaped plastids distinguish this genus from all other erect, branched, filamentous, brown algae.

A large genus with nearly 500 species (Guiry & Guiry 2023), showing considerable phenotypic plasticity. Hamel's (1931–39) division of *Ectocarpus* into two species complexes, *Ectocarpi siliculosi* and *Ectocarpi fasciculati*, based on branching pattern and sporangium shape, was later vindicated by Russell (1966) based on branching pattern only. These two species complexes were generally accepted and often referred to as separate species, each with numerous subspecies/varieties (Caram & Jónsson 1972; Hardy & Guiry 2003; Kim & Lee 1992; Parke & Dixon 1974; Ribera 1992; Schneider & Searles 1991; South & Tittley 1986; Stachen Crain *et al.* 1997; Stegenga *et al.* 1997; Womersley 1987), although sometimes additional species were also recognized by some of these authors, for example by Kim & Lee (1992), Ribera *et al.* (1992), Schneider & Searles (1991) and Stegenga *et al.* (1997). Further support for two species complexes was provided by the molecular studies of Stache-Crain *et al.* (1997) and Peters (2004).

Later work by Peters *et al.* (2010), on isolates collected from the coasts of Brittany and Normandy, based on five molecular markers from the nuclear, mitochondrial and plastid genomes, and unpublished studies by Kim & Boo (2014) on isolates of several species collected on the coast of Korea, based on *rbc*L sequence phylogeny, have provided evidence for the resurrection of some species within the *Ectocarpi siliculosi* complex, two of which, *Ectocarpus crouaniorum* and *E. penicillatus*, now require recognition within the flora of Britain and Ireland.

Three species in Britain and Ireland, one species in Britain only.

KEY TO SPECIES

1. Thalli with no discernible main axis, much branched throughout with branches evenly distributed; terminal branches alternate or irregular, never in subsecund series; plurilocular sporangia formed by transformation of laterals, scattered and never in subsecund series, usually with a terminal hair . . . . . . *E. siliculosus*
   Thalli with a fairly distinct main axis, often devoid of branches below; terminal branches alternate or becoming unilateral, subsecund and fasciculate/densely branched; plurilocular sporangia sessile, or with short or long pedicel, usually without terminal hairs 2

2. Thalli not fasciculate above, with branches irregularly produced and well-spaced apart; plurilocular sporangia sessile or shortly pedicellate above, on long pedicels below. . . . .*E. crouaniorum*
   Thalli fasciculate above, with branches and plurilocular sporangia dense, unilateral, subsecund and borne in series on adaxial side of terminal branchlets 3

3. Plurilocular sporangia 44–146 μm in length. . . . .*E. fasciculatus*
   Plurilocular sporangia 50–240 μm in length. . . . .*E. penicillatus*

Ectocarpus crouaniorum Thuret (1863), p. 75.   Fig. 205

Ectocarpus confervoides var. *crouanii* (Thuret) A. Cardinal (1964), p. 21; *Ectocarpus siliculosus* var. *crouaniorum* (Thuret) Gallardo (1992), p. 325.

Thalli epiphytic, less commonly epilithic or epizoic, erect, heterotrichous, to 2(–5) cm in length, light to yellow-brown in colour, filamentous, monosiphonous, much branched, to 1–2 orders, attached by secondary rhizoids produced at the base of the main axes; main axes pseudodichotomously branched with branches indeterminate; lateral branches regularly and alternatively branched, with branches spaced apart above, often terminating in a false hair of markedly elongated, colourless cells, 2–10 diameters long, 33–100 × 9–21 µm; true hairs absent; growth diffuse with no obvious meristematic regions; basal cells of main axes cuboidal to rectangular, barrel-shaped or half barrel-shaped, 0.5–1 diameters long, thick walled and often slightly constricted at septa, comprising cells 29–56(–70) × 50–66(–70) µm; upper cells of main axes/laterals more rectangular, 1–3 diameters long, (10–)13–59(–90) × (14–)18–25(–31) µm; plastids ribbon-shaped, linear or irregularly branched and usually lying parallel with the long axes of cells, each with several pyrenoids.

Plurilocular sporangia spaced apart, sessile or with short pedicels in terminal regions and long pedicels below, rarely terminal on hairs, elongate-conical to narrow-linear in shape, 75–115(–140) × 18–30(–38) µm; unilocular sporangia not observed for material from Britain and Ireland; studies by Peters *et al.* (2010) revealed a slightly heteromorphic life history for this species in Brittany and Normandy, with a larger, epiphytic, gametophyte generation bearing plurilocular sporangia in spring-early summer, and a smaller, epilithic or epizoic, sporophyte generation bearing unilocular and plurilocular all year round.

Gametophyte thalli occurring in shallow, upper, eulittoral pools, epiphytic on various algae, mainly *Scytosiphon lomentaria*, sometimes on *Bifurcaria bifurcata*; according to Peters *et al.* (2010), sporophyte thalli are also found in a similar habitat, but epilithic and epizoic.

Recorded for a small number of scattered localities around Britain (Cornwall, South Devon, Essex, Argyllshire); probably more widely distributed than records suggest.

Annual, spring and early summer.

Peters *et al.* (2010) noted that this species is morphologically very similar to *Ectocarpus siliculosus* and collections of the two species can be hard to distinguish, with evidence of viable hybridization occurring between some populations. These authors also observed that basally formed plurilocular sporangia of both species can have long pedicels, a diagnostic character originally described by Le Jolis (1863) and Hamel (1931–39) for *E. crouaniorum* only.

Differences between *Ectocarpus crouaniorum* and *Ectocarpus siliculosus* include:

1. Mature thalli of *E. crouaniorum* are much smaller than those of *E. siliculosus* with gameto-phytes seldom exceeding 2–3(–5) cm in length, and sporophytes seldom exceeding 3 cm in length, whereas those for *Ectocarpus siliculosus* can reach up to 50+ cm in length.

2. The plurilocular sporangia of mature gametophyte thalli of *E. crouaniorum* do not have terminal hairs whereas they are usually commonly found on the plurilocular sporangia of *E. siliculosus*.

3. Mature gametophyte thalli of *E. crouaniorum* appear to be restricted to shallow upper eulittoral pools, whereas *E. siliculosus* is usually restricted to mid- to lower eulittoral pools and also occurs in the sublittoral.

4. Mature gametophyte thalli of *E. crouaniorum* are epiphytic only and found predomi-nantly on the thalli of *Scytosiphon lomentaria* commonly attached to limpets, more rarely on other macroalgae, whereas *E. siliculosus* can be epiphytic, epilithic and epizoic, occurring on a wide range of hosts and substrata.

5. Mature gametophyte thalli of *E. crouaniorum* are spring, early summer annuals, whereas *E. siliculosus* can be found throughout the year.

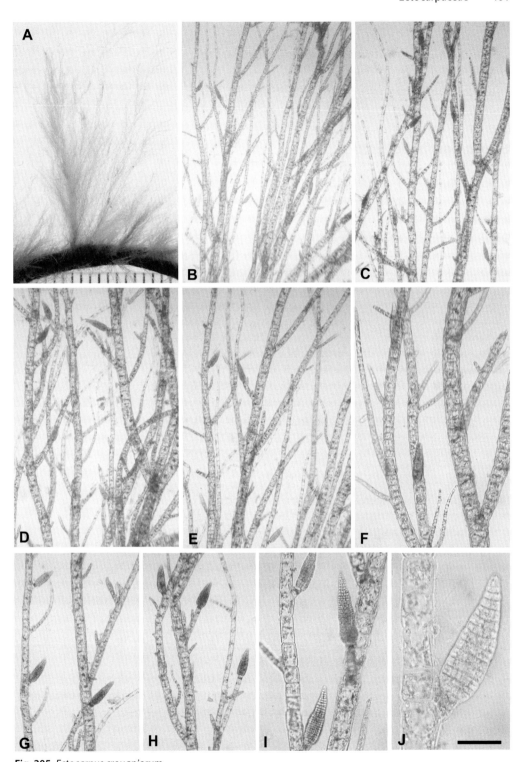

Fig. 205. *Ectocarpus crouaniorum*

A. Habit of thalli on *Scytosiphon lomentaria*. B–F. Portions of filaments showing branching. G–J. Portions of filaments showing laterally and terminally positioned plurilocular sporangia. Scale in A = mm; bar in J = 470 μm (B), = 300 μm (C, E), = 180 μm (D, F, G, H), = 90 μm (I), = 45 μm (J).

Ardré (1970), p. 230; Batters (1902), p. 31; Cardinal (1964), pp. 21, 22, fig. 8(A–K); Chalon (1905), p. 104; Cormaci *et al.* (2012), pp. 241, 242, pl. 67(1–4); P. Crouan & H. Crouan (1852), pl. 28, (1867), p. 161; De Toni (1895), p. 56; Dizerbo & Herpe (2007), p. 60, pl. 17(3); Feldmann (1954), p. 29; Hamel (1931–39), p. 25, fig. 2(F); Kim (2010), pp. 22–4, figs 5–6; Kim & Lee (1992a), pp. 226–30, figs 1–2; Kuckuck (1964), p. 12, 3 figs; Newton (1931), p. 119; Peters *et al.* (2010), pp. 157–70; Ribera *et al.* (1992), p. 111; Rodrigues (1963), p. 14, pl. 1(D, H); Taskin *et al.* (2008), p. 57, (2013), p. 147; Thuret in Le Jolis (1863), p. 75, figs 5, 6; Tsiamis *et al.* (2013), p. 147; Van Heurck (1908), p. 27.

Ectocarpus fasciculatus Harvey (1841), p. 40.   Fig. 206

Ectocarpus landsburgii Harvey (1849), pl. 233, fig. 1; *Ectocarpus fasciculatus* var. *draparnaldioides* P. Crouan & H. Crouan (1867), p. 162; *Ectocarpus draparnaldioides* (P. Crouan & H. Crouan) Kjellman (1872), p. 87; *Ectocarpus pycnocarpus* Rosenvinge (1893), p. 886, fig. 23; *Ectocarpus fasciculatus* var. *pycnocarpus* (Rosenvinge) Cardinal (1964), p. 29, fig. 14.

Thalli epiphytic, more rarely epilithic and epizoic, heterotrichous, erect, filamentous, monosiphonous, 2–5(–15) cm in length, medium to dark brown in colour, much branched and attached at the base by outward-spreading, and partly penetrating, rhizoidal filaments derived as lateral outgrowths of the basal vegetative cells, sometimes in sufficient quantity to produce a cushion of cells; rhizoidal cells cylindrical to irregularly shaped, approximately 6 μm in diameter, sometimes giving rise to sporangia; main axes fairly distinct in mature thalli, and often devoid of branches below; branching in main axes alternate, sometimes helical, occasionally adventitious, with branches sometimes recurved, branching in ultimate branches subsecund, with branches (and sporangia) often borne in a series on the adaxial surface of the short, filamentous branches and crowded in fascicles, sometimes with additional branches oppositely or suboppositely formed, the combination of alternate and subsecund branching giving a characteristic zig-zag appearance of the thallus; all branches terminating in long, colourless, attenuating, false hairs; growth diffuse with no obvious intercalary meristematic regions; cells of principal filaments frequently constricted at septa. i.e. barrel-shaped, usually 1–3 diameters long below, 42–70(–111) × 23–33(–40) μm, 0.3–1(–2) diameters long above, (–15) 38–52(–68) × 18–50(–73) μm; cells of ultimate laterals 1–4 diameters long, 8–25(–32) × 7–11(–13) μm; hairs attenuating, comprising colourless cells, 1–7 diameters long, 28–70 × 5–12 μm; plastids ribbon-shaped and irregularly branched, with numerous pyrenoids.

Plurilocular sporangia common, conical-lanceolate, 44–110(–146) × 14–22(–30) μm, sometimes sessile, more usually with a short pedicel (1–2 cells), or apical on branches of varying length, rarely with apical hairs and often characteristically arranged in subsecund series on adaxial faces of lateral branches; unilocular sporangia ovoid, 30–36(–42) × 20–26(–30) μm, sessile or pedicellate, borne on adaxial faces of lateral branches, sometimes in subsecund series or helically arranged around axial filaments.

Mainly epiphytic on larger algae, particularly common on *Laminaria digitata* and *Himanthalia elongata*; also found epiphytic on the eel grass *Zostera angustifolia* and epizoic on the limpet *Patella vulgata* and the topshell snail *Gibbula cineraria*.

Upper eulittoral, down to 5 m in the sublittoral; restricted to rock pools in upper part of vertical range and absent from estuaries; best developed in areas of moderate wave action.

Abundant and widely distributed around Britain and Ireland.

According to Russell (1966), small thalli appear in December or January; growth is rapid in March, April and May; maximum size and number occurs in midsummer after which the species becomes increasingly scarce; minimum thalli numbers observed in October and November. Unilocular sporangia observed in April–June; plurilocular sporangia all year. In the free-living, drifting state this species forms short, spiky and widely divergent, lateral branches and these morphological forms were previously considered to be distinct species, viz. *Ectocarpus distortus* and *E. landsburgii*.

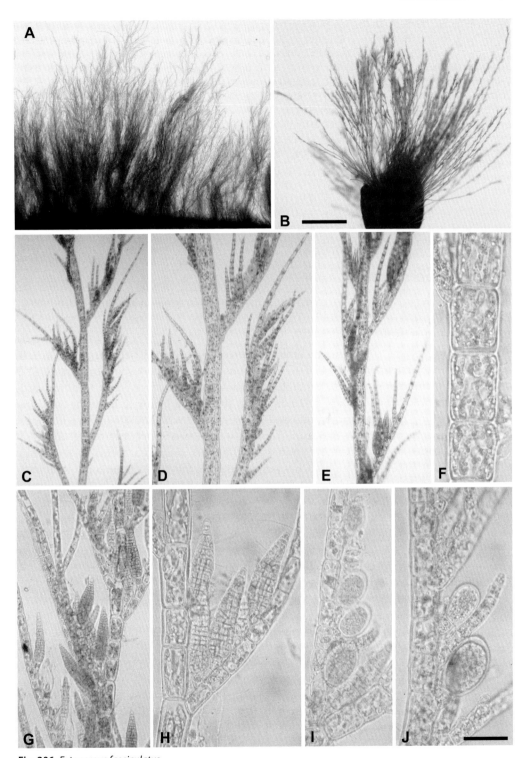

Fig. 206. *Ectocarpus fasciculatus*
A, B. Habit of thalli. C–E. Portions of erect filaments showing fasciculate branching pattern. F. Portion of erect filament showing cells with ribbon-shaped plastids. G, H. Upper regions of filaments showing secund arrangements of plurilocular sporangia on adaxial faces on lateral branches. I, J. Portions of filaments showing subsecund arrangement of unilocular sporangia. Bar in B = 3.5 mm (B), = 1 cm (A), bar in J = 300 μm (C), = 180 μm (D, E), = 90 μm (G), = 45 μm (F, H–J).

André (1970), pp. 231–3, pl. 32, figs 3, 4, pl. 33(3); Baker & Evans (1971), pp. 73–80, (1972), pp. 99, 100, (1973a), pp. 1–13, (1973b), pp. 181–9; Brodie *et al.* (2016), p. 1021; Cardinal (1964), pp. 26–30, figs 12–14; Coppejans & Kling (1995), pp. 196–200, pl. 65(A–C), pl. 66(A–F); Cormaci *et al.* (2012), pp. 242–4, pls 67(5–7), 68(1–2); De Mesquita Rodrigues (1963), pp. 15, 16, pl. 1(L); Dixon *et al.* (2000), pp. 258–63; Dizerbo & Herpe (2007), p. 63, pl. 18(1); Feldmann (1937), p. 244, fig. 35(A–B); Fletcher (1980), p. 38, pl. 11(5, 6); Fletcher in Hardy & Guiry (2003), p. 245; Gallardo García *et al.* (2016), p. 36; Gayral (1958), p. 197, fig. 25(A–C), (1966), p. 229, fig. 24(A–C); Guiry (2012), p. 153; Hamel (1931–39), pp. 25–7, fig. 5(1–5); Harvey (1846–51), pl. 273; Jaasund (1961a), pp. 239–41, (1965), pp. 33–6, figs 7, 8; Knight & Parke (1931), p. 62, pl. XI(23, 25); Kornmann & Sahling (1977), p. 97, pls 48(A–F), 49(A–C); Kylin (1947), p. 9, fig. 2(D); Maier *et al.* (1997), pp. 838–44, (2002), pp. 227–32; Mathieson & Dawes (2017), p. 190, pl. LVI(4, 5); Morton (1994), p. 31; Müller (1972), pp. 11–13; Müller & Eichenberger (1994), pp. 219–25, (1995), pp. 173–6, (1997), pp. 79–81; Newton (1931), p. 121; Parodi & Müller (1994), pp. 113–17; Pedersen (2011), pp. 63, 64, fig. 64; Ribera *et al.* (1992), p. 111; Rosenvinge & Lund (1941), pp. 24, 35–41, figs 6, 11–14; Rueness (1977), p. 110, fig. 31; Russell (1966), pp. 267–94, (1967a), pp. 233–50; Sauvageau (1892), p. 37, pl. IV; Schneider & Searles (1991), p. 115, figs 119, 120; Sengco *et al.* (1996), pp. 73–8; Stache-Crain *et al.* (1997), pp. 152–68; Stegenga *et al.* (1997a), p. 15, (1997b), pp. 138, 139, pl. 34(1); Suneson (1939), pp. 53–6; Taylor (1957), pp. 110, 111; Womersley (1987), p. 34, fig. 5(F–H).

Ectocarpus penicillatus (C. Agardh) Kjellman (1890), p. 76.   Figs 207, 208

Ectocarpus siliculosus var. *penicillatus* C. Agardh (1824), p. 162; *Ectocarpus confervoides* f. *penicillatus* Kjellman (1872), pp. 80–3; *Ectocarpus confervoides* sensu Waern (1952), p. 113; *Ectocarpus siliculosus* (*confervoides* type) sensu Kim & Lee (1992a), p. 230; *Ectocarpus siliculosus* sensu Lee (2008), p. 7.

Thalli mainly epiphytic, sometimes epilithic and epizoic, erect, tufted, heterotrichous, to 5 cm in length, light to yellow-brown in colour, filamentous, monosiphonous and branched to 2–3 orders, with main axes somewhat entangled at times and giving rise below to downward-growing and sheathing secondary rhizoids which spread out below to form the attaching system; branching pseudodichotomous and sparse below, with laterals fairly wide apart and indeterminate, alternately and more copiously branched above, becoming unilateral and secund, fasciculate and adaxially arranged in the short, somewhat recurved, terminally acute branchlets, sometimes with branches terminating in a false hair of markedly elongated, colourless cells, 4–11 diameters long, 40–90 × 8–14 µm; true hairs absent; growth diffuse with no obvious meristematic regions; basal cells of principal filaments subquadrate, quadrate to rectangular, 0.75–1(–2) diameters long, (30–)52–86(–108) × (42–)49–58(–63) µm, upper and terminal region cells subquadrate, quadrate to rectangular. 1–2 diameters long, (12–)22–36(–42) × (8–)12–18(–23) µm; plastids ribbon-shaped, linear or irregularly branched and usually lying parallel with long axes of cell, each with several pyrenoids.

Plurilocular sporangia conical-lanceolate in shape, (50–)90–185(–240) × (16–)22–38(–42) µm, shortly (1–4 cells) pedicellate, rarely sessile, arranged adaxially on the terminal branchlets, or lateral on the main laterals; unilocular sporangia unknown.

Epiphytic on various algae and *Zostera* spp., less commonly epilithic and epizoic, lower eulittoral pools and shallow sublittoral.

Recorded for a small number of widely scattered localities around Britain (South Devon, Dorset, Argyll) and for Wexford in Ireland; probably more widely distributed than records suggest.

Recorded throughout the year with no obvious seasonal periodicity.

Agardh, C. (1824), p. 162; André (1970), pp. 234, 235, pl. 33, figs 1, 2; Cardinal (1964), p. 24, fig. 11; Dizerbo & Herpes (2007), p. 63; Feldmann & Magne (1964), p. 11; Furnari *et al.* (1999), p. 86; Hamel (1931–39), p. 25, fig. 61(A); Jaasund (1965), p. 29; Kim (2010), pp. 24–6, fig. 7(A–F); Kim & Lee (1992a), p. 230, figs 3(A–C), 5(A–D) (*Ectocarpus siliculosus confervoides* type), pp. 234–8, figs 9(A–D), 11(A–D) (*Ectocarpus penicillatus*); Kjellman (1872), pp. 80–3, (1880), p. 76, (1890), p. 76, figs 7, 8; Kuckuck (1891), p. 98, fig. 5(A–B); Kylin (1907), p. 54, (1947), p. 8; Lakowitz (1929), p. 217, fig. 303(A–B); Lee (2008), p. 7, 5 figures; Newton (1931), p. 121; Pankow (1971), p. 147, figs 162, 163: Yoshida *et al.* (1995), p. 123; Ribera *et al.* (1992), p. 111; Rosenvinge &

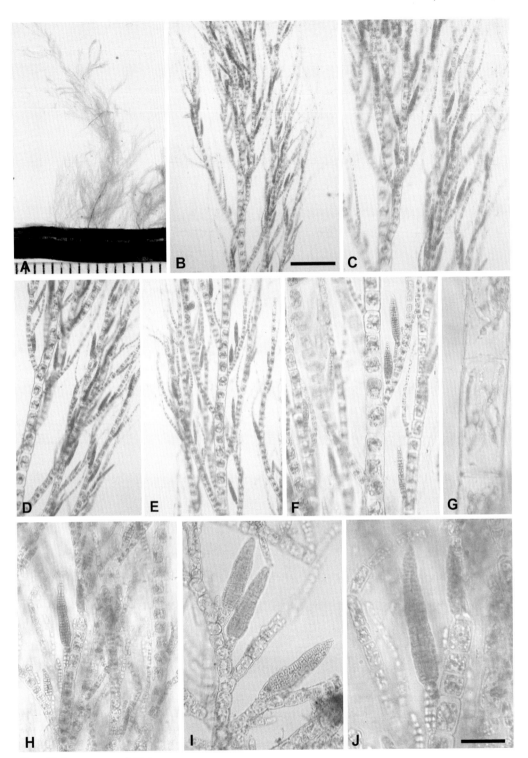

Fig. 207. *Ectocarpus penicillatus*
A. Habit of thallus. B–F. Terminal portions of filaments showing alternate branching pattern. G. Portion of filament showing cells with ribbon-shaped plastids. H–J. Terminal portions of erect filaments showing lateral arrangement of plurilocular sporangia. Bar in B = 66 μm, bar in J = 180 μm (C–E), = 90 μm (F, H, I), = 45 μm (G, J).

Fig. 208. *Ectocarpus penicillatus*
A. Habit of thallus. B–G. Terminal regions of filaments showing laterally produced, stalked plurilocular sporangia.
Bar in G = 180 µm (B, C), = 90 µm (D–F), = 45 µm (G).

Lund (1941), pp. 34, 35, fig. 11; Russell (1966), pp. 267–94; Saunders (1898), p. 155, pl. 21(3–4); Taskin *et al.* (2008), p. 57, (2013), p. 147; Taylor (1957), p. 107; Tsiamis *et al.* (2013), p. 148; Waern (1952), pp. 113–18, fig. 51; Zinova (1953), p. 69, fig. 60(A–B).

Ectocarpus siliculosus (Dillwyn) Lyngbye (1819), p. 131.   Fig. 209

Ceramium confervoides Roth (1797), p. 191, nom. illeg.; *Ceramium densum* Roth (1800), pp. 473, 474; *Ceramium confervoides* var. *ferrugineum* Roth (1806), p. 148; *Conferva siliculosa* Dillwyn (1809), p. 69; *Ceramium siliculosum* (Dillwyn) C. Agardh (1811), p. 18; *Ceramium siliculosum* var. *atrovirens* C. Agardh (1817), p. 65; *Ceramium siliculosum* var. *ferrugineum* (Roth) C. Agardh (1817), p. 66; *Ceramium siliculosum* var. *nebulosum* C. Agardh (1817), p. 66; *Ectocarpus siliculosus* var. *atrovirens* (C. Agardh) C. Agardh (1824), p. 162; *Ectocarpus siliculosus* var. *caespitosus* C. Agardh (1824), p. 162; *Ectocarpus siliculosus* var. *ferrugineus* (Roth) C. Agardh (1824), p. 162; *Ectocarpus siliculosus* var. *nebulosus* (C. Agardh) C. Agardh (1824), p. 162; *Ectocarpus arctus* Kützing (1843), p. 289; *Ectocarpus confervoides* Le Jolis (1863), p. 75; *Ectocarpus hiemalis* (P. Crouan & H. Crouan ex Kjellman) Kjellman (1867), p. 162; *Ectocarpus confervoides* f. *arctus* (Kützing) Kjellman (1872), p. 71; *Ectocarpus confervoides* f. *siliculosus* (Dillwyn) Kjellman (1872), p. 73; *Ectocarpus confervoides* f. *spalatinus* (Kützing) Kjellman (1872), p. 76; *Ectocarpus confervoides* var. *siliculosus* (Dillwyn) Farlow (1881), p. 71; *Ectocarpus siliculosus* f. *nebulosa* (C. Agardh) Kjellman (1890), p. 78; *Ectocarpus hiemalis* f. *spalatinus* (Kützing) Kjellman (1890), p. 78; *Ectocarpus siliculosus* var. *confervoides* (Roth) Kjellman (1890), p. 78; *Ectocarpus siliculosus* f. *arctus* (Kützing) Kuckuck (1891), pp. 40, 67–9; *Ectocarpus siliculosus* var. *hiemalis* (P. Crouan & H. Crouan ex Kjellman) Gallardo (1891), p. 127; *Ectocarpus confervoides* var. *hiemalis* (Foslie) Farlow (1891), p. 67; *Ectocarpus hansteeni* Foslie (1894), p. 22; *Ectocarpus confervoides* var. *brumalis*

Fig. 209. *Ectocarpus siliculosus*
A, B. Habit of thalli on hosts. C. Portion of filament showing enclosed narrow, ribbon-shaped plastids.
D–G. Portions of filaments showing branching pattern and long plurilocular sporangia with terminal hairs.
H. Terminal region of filaments showing unilocular sporangia. I. Terminal region of filaments showing pluri-
locular and unilocular sporangia on the same thallus. J. Close-up of unilocular sporangium. Bar in B = 1 cm (B),
bar in I = 300 μm (D), = 180 μm (E), = 90 μm (F–I), = 45 μm (C, J).

Holden (1899), no. 576; *Ectocarpus confervoides* f. *irregularis* Collins (1906), p. 107; *Ectocarpus confervoides* var. *arctus* (Kützing) Lily Newton (1931), p. 120; *Ectocarpus siliculosus* f. *fluviatilis* (Kützing) Kuylenstierna (1990), p. 45, nom. inval.

Thalli epilithic, epiphytic or epizoic, heterotrichous, sometimes loose-lying or floating in sheltered conditions, erect, filamentous, monosiphonous, to 30(–50+) cm in length, light to yellowish brown in colour, much branched throughout length with no discernible main axis, the basal regions of main axes often clothed in a sheath of corticating filaments derived as outgrowths of the vegetative cells, which spread out below to form the attaching system of secondary, rhizoidal filaments; principal filaments often entangled into rope-like mats; branching subdichotomous below, regularly alternate to irregular above; branches indeterminate and usually evenly distributed over thallus length, tapering gradually at the apex, frequently to a long, false hair characteristically of colourless, elongate-rectangular, narrow cells, 3–8 diameters long, 25–65 × 6–11 µm, containing few plastids; true hairs, with basal meristem, absent; growth diffuse with no obvious meristematic regions; basal cells of principal filaments, thick walled, quadrate to rectangular, sometimes barrel-shaped, 1–1.5 diameters long, 40–60(–70) × 38–42(–46) µm, mid-region cells quadrate to rectangular, 1–3 diameters long, 32–55(–70) × 23–33(–37) µm, terminal branch cells quadrate to rectangular, 1–3 diameters long, 8–18(–23) × 5–14(–17) µm; plastids sparsely distributed, ribbon-shaped, finely linear, irregularly branched, and usually lying parallel with the long axes of the cells, each with several pyrenoids.

Plurilocular and unilocular sporangia borne on the same or different thalli; plurilocular sporangia common, formed directly on terminal laterals of variable length and usually large in size, cylindrical, narrow-linear to elongate-lanceolate, (60–)90–190(–230) × 18–26(–32) µm, usually terminating in an apical hair, sometimes hairless and intercalary; unilocular sporangia rare, ovoid or broadly ellipsoid, sessile or shortly pedicellate, 34–52(–58) × 20–28(–35) µm, sometimes adpressed to axes.

Upper to lower eulittoral and sublittoral to 7 m; restricted to rock pools in upper part of vertical range and sometimes found in estuaries; best developed in sheltered areas, commonly epiphytic on larger algae, chiefly *Chorda filum*, *Asperococcus fistulosus* and *Scytosiphon lomentaria*, various artificial substrata, particularly floating structures such as buoys, pontoons, boats etc., and epizoic on the limpet *Patella vulgata*, European lobster *Homarus gammarus* and the crab *Macropodia longirostris*; also euryhaline.

Common and widely distributed all around Britain and Ireland.

According to Russell (1966), seasonal growth of *E. siliculosus* is accompanied by changes in vertical distribution. From October until February most thalli are to be found in rock pools in the lower littoral fringe or upper sublittoral; from March until September more thalli are to be found in the lower eulittoral, either in pools or on rock and host spp.; from April until September, sublittoral thalli are common. Plurilocular sporangia occur all year round; unilocular sporangia occur in spring and early summer.

Branching pattern varies less than in *E. fasciculatus* but size range of thalli is greater; the largest thalli (over 50 cm) are to be found during summer in sheltered, sublittoral areas.

E. siliculosus is typically a less specialized species than *E. fasciculatus*. It has a wider host and substrata range, a wider vertical range and a wider geographical range.

Abbott & Hollenberg (1976), pp. 128, 129, fig. 90; Amsler *et al.* (1999), pp. 239–44; André (1970), pp. 229, 230, pl. 31(1–4); Bartsch & Kuhlenkamp (2000), p. 166; Bolton (1983), pp. 131–8; Bräutigam *et al.* (1995), pp. 823–7; Brodie *et al.* (2016), p. 1021; Busch & Schmid (2001), pp. 61–70; Cardinal (1964), pp. 14–19, figs 2–5; Charrier *et al.* (2008), pp. 319–32; Coppejans & Kling (1995), p. 196, pl. 64(A–D); Cormaci *et al.* (2012), pp. 254, 256, pl. 72(4–7); Dixon & Russell (1964), pp. 282, 283; Dizerbo & Herpe (2007), pp. 60, 62, pl. 17(4); Ercegovic (1955), p. 3, figs 1–5; Fletcher in Hardy & Guiry (2003), p. 245; Funk (1955), p. 31; Gallardo García *et al.* (2016), p. 36; Gayral (1966), p. 230; Geller & Müller (1981), pp. 53–66; Guiry (2012), p. 153; Hall (1980), pp. 73–8; (1981), pp. 223–8; Hall *et al.* (1979), pp. 195–9; Hamel (1931–39), pp. 21–3, fig. 3(A, B); Harvey (1846–51), pl. 162(1–5); Hillrichs & Schmid

(2001), pp. 71–80; Katsaros *et al.* (1991), pp. 87–92, (1993), pp. 787–97; Kawai *et al.* (1990), pp. 292–7; Kim (2010), pp. 26–8, fig. 9(A–H); Kinoshita *et al.* (2016), pp. 139–48; Knight (1929), pp. 307–22, pls 1–6; Kornmann (1956), pp. 32–43; Kornmann & Sahling (1977), p. 94, pl. 46(A–F); Kreimer *et al.* (1991), pp. 268–76; Kuckuck (1891), pp. 65–71, figs 1, 3, (1892), pp. 256–9; Kuhlenkamp & Müller (1994), pp. 525–30; Kylin (1943), pp. 295–8, (1947), p. 7, fig. 2(A, B); Le Bail *et al.* (2008), pp. 1269–81; Lindauer *et al.* (1961), pp. 143, 144, fig. 2; Lyngbye (1819), p. 131, pl. 43(C); Maier (1997a), pp. 241–53, (1997b), pp. 255–66; Mathieson & Dawes (2017), p. 191, pl. LVI(6–9); Montecinos *et al.* (2017), pp. 17–31; Morton (1994), p. 31; Müller (1964b), p. 1402, (1966), pp. 57–68, (1967a), pp. 39–54, (1967b), pp. 30–64, (1972), pp. 87–98, (1975), pp. 315–21, (1976a), pp. 89–94, (1976b), pp. 252–4, (1977), pp. 131–6, (1978), pp. 371–7, (1979), pp. 312–18, (1980), pp. 3–15, (1988), pp. 469–76, (1991a), pp. 101–2, (1991b), pp. 739–43; Müller & Eichenberger (1994), pp. 219–25, (1995), pp. 173–6; Müller & Falk (1973), pp. 313–22; Müller & Frenzer (1993), pp. 37–44; Müller & Kawai (1991), pp. 151–5; Müller & Schmid (1988), pp. 647–53; Müller *et al.* (1990), pp. 72–82; Norris (2010), p. 185; Pankow (1971), pp. 147–9, fig. 166; Papenfuss (1933), pp. 390, 391, (1935), pp. 421–46; Pedersen (2011), p. 63, fig. 63; Peters *et al.* (2004a), pp. 235–42, (2004b), pp. 1079–88, (2008), pp. 1503–12; Prud'homme van Reine *et al.* (1973), pp. 93–6; Ribera *et al.* (1992), p. 111; Rosenvinge & Lund (1941), pp. 14–23, 27–35, figs 1–5, 10–11; Rueness (1977), p. 110, fig. 32(A, B), (2001), p. 14; Russell (1966), p. 275, fig. 4a, (1967a), pp. 233–50; Russell & Bolton (1975), pp. 91–4; Russell & Morris (1970), pp. 288, 289; Saez *et al.* (2015), pp. 425–9; Sauvageau (1896), pp. 431–3; Schmid (1993), pp. 437–43; Schmid & Dring (1992), p. 100; Schmid & Hillrichs (2001), pp. 257–64; Schneider & Searles (1991), pp. 115–17, figs 121, 122; Sengco *et al.* (1996), pp. 73–8; Setchell & N.L. Gardner (1925), pp. 410, 411; Silva *et al.* (1996), p. 923; Stache (1989), pp. 173–86; Stache-Crain *et al.* (1997), pp. 152–68; Stegenga & Mol (1983), p. 72, pl. 19(2–4); Stegenga *et al.* (1997a), p. 15, (1997b), p. 141; Taskin (2008), p. 57, (2013), p. 147; Taylor (1957), p. 105, pl. 8(4, 5); Thomas & Kirst (1990a), p. 97, (1990b), pp. 26–36, (1991), pp. 1–9; Tsiamis *et al.* (2013), p. 147; Womersley (1987), pp. 33, 34, figs 2(D), 5(A–E); Woolery & Lewin (1973), pp. 131–8; Yoshida *et al.* (1985), p. 64.

Pleurocladia A. Braun in Rabenhorst

PLEUROCLADIA A. Braun in Rabenhorst 1855: no. 441.

Type species: *P. lacustris* A. Braun (1855), p. 80.

Thalli heterotrichous, of densely packed, short, to 1 mm in height, erect filaments arising from a spreading, filamentous basal system; erect filaments somewhat gelatinous, monosiphonous, sparingly branched above, with branching irregular, pseudoparenchymatous, unilateral to opposite, each cell containing a single, lobed, plate-like plastid with no obvious pyrenoid; hairs with basal meristem and sheath.

Unilocular sporangia terminal or lateral on erect filaments, sessile or shortly stalked, one cell sometimes initiating and supporting more than one sporangium; plurilocular sporangia produced terminally and intercalary on the erect filaments, uniseriate to biseriate in chains.

One species in Britain.

Pleurocladia lucifuga (Kuckuck) Wilce (1966), p. 64.   Fig. 210

Ectocarpus lucifugus Kuckuck (1897), 359, pl. XI–XII; *Kolderupia lucifuga* (Kuckuck) Lund (1959), p. 54.

Thalli epilithic, forming small, solitary to confluent, pale brown, gelatinous tufts of erect filaments, to 1 mm in height and several mm in extent; comprising a basal attachment system of uniseriate, outward-spreading, irregularly branched, free, monosiphonous filaments giving rise to fairly closely packed, short, erect filaments loosely associated in a gelatinous matrix; basal filament cells variable in shape, often rounded, somewhat contorted, thick walled, 15–36 × 5–22 µm; erect filaments monosiphonous, uniseriate, free, to 900 µm in height, sparingly branched above, to 1(–2) orders, relatively bare below, with branching irregular, pseudodichotomous, more commonly unilateral; basal to mid region cells of erect filaments rectangular, 2–4 diameters long, (18–)24–35(–42) × (6–)7–9(–11) µm, upper cells of erect filaments quadrate to more commonly rectangular, 1–3 diameters long, (14–)20–31(–35) × (9–)12–15(–17) µm, each

Fig. 210. *Pleurocladia lucifuga*

A. Fine tuft-like habit of thalli on rock (arrow). B, C. SP of thalli showing erect filaments. D–L. SP of thalli showing branched, erect filaments and terminal unilocular sporangia. Bar in A = 2 cm (A), = 300 μm (B), = 90 μm (C, E–H, J), = 45 μm (D, I, K, L).

containing a single large, plate-like plastid without obvious pyrenoids; all erect filaments surrounded by a gelatinous sheath; erect filament growth probably diffuse; hairs not observed.

Unilocular sporangia terminal on erect filaments and laterals, sometimes lateral, sessile or shortly stalked, one cell sometimes initiating and supporting more than one sporangium, cylindrical to elongate pyriform and can be curved apically, (28–)35–56(–66) × (13–)16–20(–23) μm; plurilocular sporangia not observed by the present author, but reported by Kuckuck (1897a) for Helgoland to be associated with unilocular sporangia, and produced terminally and intercalary on the erect filaments by subdivisions of the cells to form fairly short, 2–5 locules, uniseriate to biseriate chains.

Epilithic, high upper eulittoral, exposed on cliff faces, associated with, and sometimes emerging through, gelatinous colonies of Chrysophyceae, and frequently mixed with the fleece-like growths of the green alga *Chaetomorpha tortuosa*; this is the highest shore height recorded for brown algae and is probably only reached by the highest spring tides and/or seawater spray.

Rarely recorded and only collected in Britain by the present author from the chalk cliff faces and cuttings of the Kent (Thanet) coast, and from the sandstone cliff faces of St Andrews, Scotland; likely to be more widespread.

Collections have been made January to May, with fertile material bearing unilocular sporangia observed January to March; possibly perennial.

A molecular study by McCauley & Wehr (2007) of the freshwater *Pleurocladia lacustris* revealed it to be related to the Ectocarpaceae, a position also adopted in the present treatise for the marine *Pleurocladia lucifuga* pending further studies.

Athanasiadis (1996), p. 142; Bartsch & Kuhlenkamp (2000), pp. 166, 178; Hooper *et al.* (1987), pp. 439, 440; Kuckuck (1897), pp. 259–361, 363–4, pls xi(1–13), xii(14–19); Lund (1959), p. 54; Nielsen & Gunnarsson (2001), pp. 45–108; McCauley. & Wehr (2007), pp. 429–39.

Spongonema Kützing

SPONGONEMA Kützing (1849), p. 461.

Type species: *S. tomentosum* (Hudson) Kützing (1849), p. 461.

Thalli of densely branched, intricately matted, monosiphonous, filaments arising from a monostromatic, pseudodiscoid/discoid, in part endophytic, base; thallus initially fleece-like, light to yellow-brown in colour and spreading over host surface, later forming erect, dark brown, branched, rope-like chords of entwined filaments; filaments monosiphonous, irregularly branched, with branches widely divergent, ultimate branchlets characteristically recurved and hook-like; cells with 1–2 plate-like to linear-like plastids with pyrenoids.

Plurilocular sporangia terminal, lateral or intercalary on erect filaments, sessile or shortly stalked, multiseriate, light brown coloured, and strongly recurved towards parental filament; unilocular sporangia globose, terminal or lateral, sessile or shortly stalked on erect filaments.

One species in Britain and Ireland.

Spongonema tomentosum (Hudson) Kützing (1849), p. 461.   Fig. 211

Conferva tomentosa Hudson (1762), p. 480; *Ceramium tomentosum* (Hudson) Roth (1800), pp. 468–70; *Ectocarpus tomentosus* (Hudson) Lyngbye (1819), p. 132; *Ectocarpus tomentosus* var. *clavatus* C. Agardh (1824), p. 163; *Chordaria tomentosa* (Hudson) Fries (1845), p. 124; *Scytosiphon tomentosus* (Hudson) J. Agardh (1848), p. 127, nom. illeg.; *Ectocarpus luteolus* Sauvageau (1892), p. 25; *Ectocarpus minimus* Nägeli in Sauvageau (1892), p. 77; *Streblonema luteolum* (Sauvageau) De Toni (1895), p. 575; *Streblonema minimum* (Nägeli) De Toni (1895); *Ectocarpus ellipticus* De A. Saunders (1898), p. 149; *Herponema luteolum* (Sauvageau) Hamel (1931–39), p. xix; *Feldmannia elliptica* (De A. Saunders) Hollenberg & I.A. Abbott (1966), p. 18.

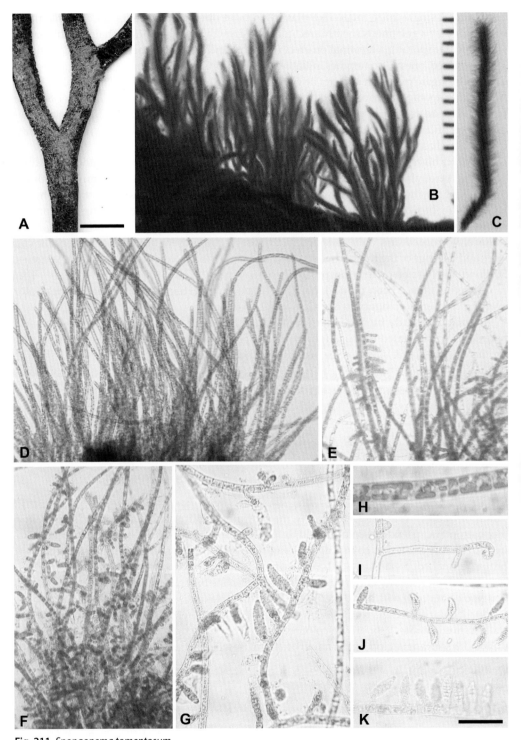

Fig. 211. *Spongonema tomentosum*

A. Early stage fleece-like habit (light coloured areas) of thalli on *Fucus*. B. Later-formed branched, chord-like habit of thalli on *Fucus*. C. Chord-like thallus showing outer fringe of free filaments. D. Outer free filaments of tomentose-like thallus. E–G. Close-up of filaments showing terminal hairs and laterally produced plurilocular sporangia. H. Portion of filament showing enclosed plate-like plastids. I. Portion of filament showing hook-like branchlets. J, K. Portions of filaments showing plurilocular sporangia. Bar in A = 1 cm (A), = 2.4 mm (C), bar in K = 180 μm (D–F), = 90 μm (G, J), = 45 μm (H, I, K). Scale in B = mm.

Thalli epiphytic, forming erect, densely matted, light yellow-brown to dark brown-black, fleece-like tufts or rope-like chords of filaments on host surface; tufts yellow-brown at first, especially when seen dried, relatively low lying and fleece-like, spreading several cm over host surface, later becoming more consolidated, structured and erect, extending up from the host surface as irregularly branched, soft, spongy, rope-like chords up to 5(–10) cm in height; basal system of branched, monosiphonous, outward-spreading filaments, in parts pseudodiscoid/discoid in appearance, often accompanied by some considerable endophytic penetration amongst the cortical cells of the host; erect filaments monosiphonous throughout, much branched and intricately matted together, with branching dichotomous/subdichotomous, alternate or irregular, and often divaricate and widely spreading, sometimes characteristically producing short, hook-like determinate branchlets ('crampons') at right angles to parental filament; filaments usually terminated by a long pseudohair, comprising colourless cells, 3–7 diameters long, $25–60 \times 9–11$ μm; cells of erect filaments quadrate to rectangular, 1–3(–4) diameters long, $(7–)13–30(–42) \times (8–)10–12(–13)$ μm, each containing one, rarely two, plate-like to linear plastids, with 1–2 pyrenoids; growth diffuse, or by a fairly indistinct intercalary meristem of short, compacted cells usually positioned below a terminal hair.

Plurilocular sporangia and unilocular sporangia borne on the erect filaments, usually together; plurilocular sporangia common, pale brown in colour, terminal, intercalary, or more commonly lateral, sessile or shortly pedicellate, often hook-like and recurved towards parental filament, solitary or in a secund series, simple, sometimes branched, multiseriate, ovoid to lanceolate, $(22–)28–58(–78) \times (12–)13–15(–17)$ μm; unilocular sporangia less common, pale brown in colour, borne on the erect filaments, terminal or lateral, sessile or stalked, ovoid to globular, $25–50 \times 20–30$ μm.

Commonly epiphytic, particularly on *Fucus* spp., more rarely epilithic, mid-eulittoral pools and exposed.

Common and widely distributed around Britain and Ireland.

Initially observed as fine, felt-like growths in late winter/early spring, with maximum development and formation of the rope-like strands in late spring/summer, dying down in September/October.

Abbott & Hollenberg (1976), pp. 148, 149, fig. 111; Aisha & Shameel (2011), pp. 143–5, fig. 8(a, b); Ardré (1970), pp. 235, 236, pls 34(1–9), 50(4); Bartsch & Kuhlenkamp (2000), p. 168; Batters (1902), p. 30; Brodie *et al.* (2016), p. 1021; Cardinal (1964), pp. 34–6, fig. 17(A–M); Coppejans (1983), pp. 1–5; Coppejans & Kling (1995), pp. 208, 209, pls 72(A–D), 73(A–Z); Cormaci *et al.* (2012), pp. 263, 264, pl. 76(1–3); Dizerbo & Herpe (2007), p. 73, pl. 22(3); Fletcher in Hardy & Guiry (2003), p. 302; Gallardo García *et al.* (2016), p. 36; Guiry (2012), p. 154; Hamel (1931–39), pp. XI–XII, 32–4, 58–9, figs 8(IV, V), 18(D); Harvey (1846–51), pl. 182; Jaasund (1965), p. 37, fig. 9; Knight and Parke (1931), pl. XI(19, 20, 24); Kornmann & Sahling (1977), p. 101, pl. 50(A–D); Kuckuck (1960), pp. 93–113, figs 1–6; Kylin (1933), pp. 20–2, fig. 3(A–F), (1947), p. 12, pl. 1(2); Levring (1937), pp. 40, 41, fig. 4; Mathieson & Dawes (2017), pp. 194, 195, pl. XLVIII(3); Morton (1994), p. 33; Pankow (1971), pp. 151, 152, fig. 170; Printz (1926), pp. 156, 157, fig. 8; Ribera *et al.* (1992), p. 113; Rosenvinge & Lund (1941), pp. 41–3, fig. 15; Rueness *et al.* (2001), p. 17; Sauvageau (1892), pp. 25–7, pl. II(14–19), (1895c), pp. 1–14, (1928c), pp. 121–35, figs 1–2; Stegenga & Mol (1983), p. 81, pls 24(3–6), 116(3); Stegenga *et al.* (1997a), p. 17; Taskin (2008), p. 58; Taylor (1957), p. 108, pl. 8(6–8).

Petrospongiaceae Racault *et al.*

PETROSPONGIACEAE M. Racault, R.L. Fletcher, B. de Reviers, G.Y. Cho, S.M. Boo, M. Parente & F. Rousseau (2009), p. 121.

Thalli epiphytic or epilithic, hemispherical, spongy, cushion-like with broad radiating lobes; comprising a prostrate system of radially spreading, branched filaments which give rise above to an extensive medullary region of erect, dichotomously to trichotomously branched filaments of large, elongate-cylindrical to globular cells, giving rise below (if epiphytic) to downward-growing rhizoidal filaments, and above to a determinate cortical region of filaments comprising smaller, more barrel-shaped cells, each containing one plate-like or 1–2 discoid plastids, each with an exserted pedunculate pyrenoid. Hairs arising from the cortical cells, with basal meristem and sheath.

Unilocular sporangia common, laterally attached at the base of the terminal cortical filaments.

The Petrospongiaceae is represented in Britain and Ireland by a single genus *Petrospongium* and by a single species, *P. berkeleyi* (Greville) Nägeli ex Kützing (1858), based on *Chaetophora berkeleyi* described by Greville (1833) in Berkeley. It was transferred to *Cylindrocarpus* (type species *C. microscopicus* P. Crouan & H. Crouan 1851) by P. Crouan & H. Crouan (1851), which was accepted by most authors (Kuckuck 1929; Hamel 1931–39; Feldmann 1937; Gayral 1966; Parke & Dixon 1976; Dizerbo & Herpe 2007). Fletcher (1987), however, concluded that they represent unrelated taxa. *C. microscopicus* differs from *C. berkeleyi* in a number of features. It has an extensive network of endophytic filaments in the host red alga *Gracilaria bursa-pastoris*, which emerge as erect microscopic tufts of branched filaments rather than gelatinous cushions; the erect filaments are also not closely packed into a pseudoparenchyma and, although the basal and mid-region cells are slightly swollen, there is no obvious differentiation into a medulla-like region of large, thin-walled colourless cells. Based on these differences, *C. berkeleyi* was transferred to *Petrospongium* while *C. microcopicus* was removed from the Corynophlaeaceae pending further studies revealing its true taxonomic identity. While Batters (1902), Setchell & N.L. Gardner (1925) and Newton (1931) linked *C. microscopicus* with the Ectocarpaceae, it seemed more suitably placed in the Chordariaceae emend A.F. Peters & Ramírez where it is now placed (see Silberfeld *et al.* 2014).

Fletcher (1987) retained *Petrospongium* in the family Corynophlaeaceae, in keeping with many other authors (see Kuckuck 1929; Hamel 1931–39; Abbott & Hollenberg (1976); Russell & Fletcher 1975; Rueness 1977; Ribera *et al.* 1992; Christensen 1994; Stegenga & Mol 1983). Other authors, however, placed the genus in the Leathesiaceae (see Rosenvinge 1935; Taylor 1957; Womersley 1987; Yoshida *et al.* 1990). *Petrospongium* was usually grouped together in these families with genera such as *Corynophlaea*, *Leathesia*, *Microcoryne* and *Myriactula* which all occur in Britain and Ireland. Morphologically these genera comprise small globular, hemispherical or finger-like cushions which are almost exclusively epiphytic on other algae; only *Petrospongium* is sometimes reported as a lithophyte. In Britain and Ireland these genera also tend to be host-specific or have been reported as epiphytes on a limited range of algae. For instance, *Corynophlaea* is apparently confined to the host species *Chondrus crispus*, *Petrospongium* is confined to the brown crustose alga *Pseudoralfsia verrucosa*, while three species of *Myriactula* are each confined to a single host species or genus. In structure, they all comprise a pseudo-parenchymatous thallus of erect, closely packed, branched filaments of large, colourless cells (sometimes referred to as the 'medulla'), giving rise peripherally to short, multicellular, usually simple, pigmented filaments (termed paraphyses in view of their similarity to these structures in *Elachista*), sporangia and/or hairs (sometimes referred to as the 'cortex'). Unlike *Elachista*, however, they do not additionally form the long, free, filaments contributing to the brush-like tufts. All thalli are soft and lubricous and easily squashed under light pressure.

Molecular studies have revealed that orders, such as the Chordariales, Dictyosiphonales and Punctariales as well as some families within the Ectocarpales, including the Corynophlaeaceae and Leathesiaceae, are unjustified, and they were all subsumed, along with their genera, within a broader concept of the Chordariaceae s.l. (Peters & Ramírez 2001) and within an enlarged Ectocarpales. However, phylogenetic relationships between these cushion-forming genera remained uncertain until Cho & Boo (2006), using *rbc*L and *psa*A sequence data, revealed *Petrospongium* to be clearly separate from *Leathesia* and other members of the Chordariaceae. This was supported by Racault *et al.* (2009) who, based on LSU, *rbc*L and *psa*A sequence data, revealed the type species of *Petrospongium*, *P. berkeleyi* and its Pacific vicariant *P. rugosum* formed an independent monophyletic group which was a sister clade to the Ectocarpaceae.

Differences in the morphology and plastid types separated *Petrospongium* from members of the Ectocarpaceae and a new family, Petrospongiaceae, was proposed to accommodate the genus *Petrospongium* (Racault *et al.* 2009). This proposal has been accepted in the present treatise.

The family Petrospongiaceae is represented by a single genus, *Petrospongium*.

Petrospongium Nägeli in Kützing

PETROSPONGIUM Nägeli in Kützing (1845–71), p. 2.

Type species: *P. berkeleyi* (Greville in Berkeley) Nägeli in Kützing (1845–71), p. 2.

Thalli epiphytic or epilithic, forming solid, hemispherical, gelatinous cushions; comprising a prostrate system of outward-spreading, branched filaments giving rise to erect, loosely compacted, dichotomously to trichotomously branched filaments, easily separable under pressure; filaments infrequently branched below, comprising large, elongate-cylindrical to globular colourless cells, often with downward-growing host-penetrating (if epiphytic), multicellular, branched, rhizoidal filaments, becoming more frequently branched and compacted above, comprising quadrate, rectangular to barrel-shaped, smaller, pigmented cells; hairs and sporangia associated with terminal, pigmented filaments; hairs with basal meristem and sheath.

Unilocular sporangia laterally inserted at the base of the terminal, pigmented filaments, sessile or shortly stalked, elongate-cylindrical becoming deformed; plurilocular sporangia reported to be terminal on pigmented filaments, multiseriate.

One species in Britain and Ireland.

Petrospongium berkeleyi (Greville in Berkeley) Nägeli in Kützing (1845–71), p. 2.

Figs 212, 213

Chaetophora berkeleyi Greville in Berkeley (1833), p. 5; *Leathesia berkeleyi* (Greville in Berkeley) Harvey (1846–51), p. 176; *Cylindrocarpus berkeleyi* (Greville in Berkeley) P. Crouan & H. Crouan (1851), p. 363.

Thalli epiphytic or epilithic, forming small, light to dark brown, hemispherical cushions, discrete and circular, or confluent and irregularly spreading to 10(–20) mm in diameter, 2–3(–4) mm high; in squash preparation, consisting of erect, multicellular, branched filaments, loosely compacted in a mucilaginous matrix; lower and mid region filaments sparsely dichotomously branched, comprising large, colourless cells, elongate-cylindrical below, becoming more globose above, 1–4 diameters long, 25–65 × 15–30 µm, frequently producing from the basal region downward-growing, branched, cylindrical or irregularly contorted, colourless, rhizoidal filaments; rhizoidal cells 28–50 × 6–12 µm, reputed to penetrate below into host tissue; upper regions of thallus more frequently dichotomously to trichotomously branched forming closely compacted pigmented, straight, thread-like filaments, comprising cells quadrate, rectangular, or more frequently barrel-shaped in appearance, 1(–11) diameters long, 9–16 × 8–13 µm, each with a single, plate-like plastid with pyrenoid occupying upper cell region; hairs produced at base of terminal thread-like filaments, with basal meristem and sheath, 8–10 µm in diameter.

Unilocular sporangia common, lateral on thread-like filaments, sessile or shortly stalked, elongate-cylindrical or slightly deformed and characteristically wider below, basally, sometimes laterally, attached, 50–81 × 17–32 µm; plurilocular sporangia rare, reported to be terminal on the thread-like filaments, pod-like, multiseriate, 46–113 × approx. 11 µm.

Epilithic on bedrock, less obviously epiphytic on crusts of the brown alga *Pseudoralfsia verrucosa*, in the upper to lower eulittoral, especially in shaded situations.

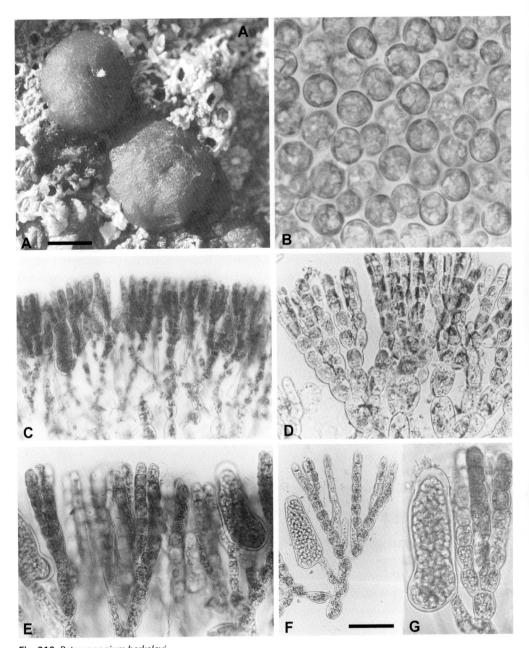

Fig. 212. *Petrospongium berkeleyi*
A. Globose habit of two thalli epiphytic on *Pseudoralfsia verrucosa*. B. Surface view of thallus showing loosely associated cells. C, D. SP of thalli showing terminal filaments. E. SP of terminal filaments with associated unilocular sporangia F, G. Portions of fertile thalli showing laterally produced unilocular sporangia. Bar in A = 1.5 mm, bar in F = 70 μm (C), = 35 μm (E, G), = 25 μm (D), = 10 μm (B, G).

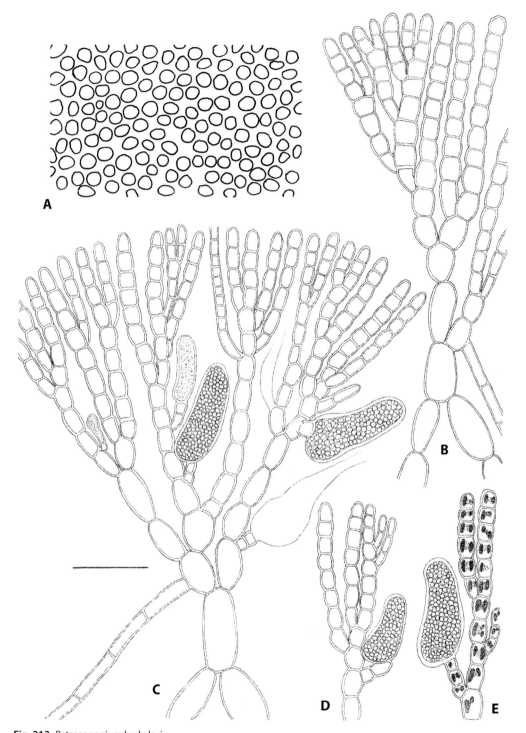

Fig. 213. *Petrospongium berkeleyi*
A. Surface view of thallus showing loosely associated cells. B. Portion of vegetative thallus. C–E. Portions of fertile thalli showing unilocular sporangia. Bar = 50 μm.

Southern and western shores of Britain and Ireland, extending northwards to Orkney and eastwards to Dorset; in Ireland, northwards to Donegal, eastwards to Wexford, Antrim.

Thalli annual, summer and autumn.

Caram (1957) reported *P. berkeleyi* (as *Cylindrocarpus berkeleyi*) to have an isomorphic type of life history. Meiosis was assumed to occur in the unilocular sporangia to form haploid zoospores which developed into the plurilocular sporangia-bearing haploid thalli. She also reported a 'short-circuit' life history involving direct development of the diploid thallus from zygotes produced by sexual fusion of the unispores.

Bartsch & Kuhlenkamp (2000), p. 167; Brodie *et al.* (2016), p. 1019; Caram (1957), pp. 440–3; Cormaci *et al.* (2002), pp. 270, 271, pl. 78(1–3); P. Crouan & H. Crouan (1851), p. 364, pl. 17(12, 13), (1852), pl. 25(159); Dangeard (1969), pp. 81, 82; Dizerbo & Herpe (2007), p. 83; Fletcher (1987), pp. 160–2, fig. 33; Fletcher in Hardy & Guiry (2006), p. 284; Gallardo García *et al.* (2016), p. 37; Gayral (1966), p. 271, pl. XLIV; Guiry (2012) p. 137; Hamel (1931–39), pp. 147, 148, fig. 33(a, b); Hanna (1899), pp. 461–4; Hauck (1884), p. 358, fig. 151; Kuckuck (1929), p. 46, fig. 54; Miranda (1931), p. 26, fig. 1; Morton (1994), p. 34; Newton (1931), p. 140, fig. 86; Racault *et al.* (2009), pp. 111–23; Ribera *et al.* (1992), p. 115.

Scytosiphonaceae Farlow

SCYTOSIPHONACEAE Farlow (1881), pp. 15, 62 [as Scytosiphoneae].

Chnoosporaceae Setchell & N.L. Gardner (1925), p. 552.

Thalli erect or prostrate, epilithic, epizoic or epiphytic, arising from either a fibrous mat of rhizoids, discs or crusts.

Erect thalli macroscopic, thalloid, saccate, cylindrical or dorsiventrally flattened, simple, hollow or entire; parenchymatous in structure, with inner medulla of large, thick-walled, colourless cells and outer cortex of small, pigmented cells; cells with a single, large, plate-like plastid with 1–2 pyrenoids; growth initially apical, later diffuse; hairs with basal meristem with or without obvious sheath; plurilocular sporangia and unilocular sporangia borne in local or extensive sori in terminal thallus region, arising from, or mixed together, with the outer cortical cells; plurilocular sporangia in closely packed, vertical columns, uniseriate to multiseriate, with or without associated ascocysts; unilocular sporangia associated with multicellular paraphyses.

Prostrate thalli microscopic or macroscopic, densely tufted, resembling a carpet pile, slightly pulvinate or crustose; comprising outward-spreading, branched filaments either free, with or without longitudinally divided cells, or compacted and pseudodiscoid/discoid in structure, giving rise to erect, short, simple or little-branched filaments, loosely aggregated and gelatinous or laterally adjoined, weakly or quite firmly, and pseudoparenchymatous in structure; cells with a single, plate-like plastid with pyrenoid; hairs solitary or grouped, arising superficially or immersed, with basal meristem but lacking obvious sheath; thalli with or without ascocysts, occurring on the basal cells or terminal on erect filaments; unilocular sporangia common, irregularly distributed on basal filaments, sessile or stalked, more commonly grouped in immersed or raised mucilaginous sori, terminal on compacted erect filaments, with or without associated, further projecting, multicellular paraphyses arising from the supporting cell; plurilocular sporangia rare, terminal on compacted erect filaments, usually grouped in sori without accompanying paraphyses, uniseriate to multiseriate, without terminal sterile cells.

Life history less commonly isomorphic, mainly heteromorphic with erect, gametophyte thalli alternating with prostrate, sporophyte thalli, or abbreviated and direct, sexual or asexual; gametes isogamous or anisogamous.

Circumscription of the family has been principally based on the parenchymatous erect thalli, in particular a single plate-like plastid with pyrenoid in each cell and the occurrence of plurilocular sporangia only on the erect thalli, which are in closely packed, mainly uniseriate, vertical columns, with or without accompanying paraphysis-like cells. These features were even considered by Feldmann (1949) to be sufficiently important to erect the Scytosiphonales. Reaction to this order was mixed; it was accepted by Jaasund (1965), Wynne (1969), Wynne & Loiseaux (1976), Nakamura (1972), Abbott & Hollenberg (1976), Rueness (1977), Bold & Wynne (1978) and Christensen (1980) but rejected by Russell & Fletcher (1975), Parke & Dixon (1976), Kornmann & Sahling (1977) and South & Hooper (1980). Even the family Scytosiphonaceae did not receive full support, with genera such as *Petalonia* and *Scytosiphon* variously included in the Punctariaceae or Asperococcaceae (Taylor 1957; Lindauer *et al.* 1961; Pankow 1971).

As Bold & Wynne (1978) pointed out, an additional significant criterion for the delimitation of the group, is the involvement of prostrate, principally crustose, thalli bearing unilocular sporangia in the life histories of many of the members. This heteromorphic life history was in contrast to the isomorphic life history originally suggested by Feldmann (1949). The relationship between the two morphologically different thalli is not fully understood, but an extensive study of several genera in Japan by Nakamura & Tatewaki (1975) first indicated that the basic life history probably involves a heteromorphic alternation between an erect bladed gametophytic (i.e. sexual) phase and a prostrate, usually crustose, sporophytic (i.e. asexual) phase. The sexual thalli are isomorphic and heterothallic releasing either isogametes or anisogametes from the plurilocular sporangia. Gamete fusion results in the development of the sporophytic, prostrate phase bearing unilocular sporangia in which meiosis occurs. Release and germination of the unispores completes the lifecycle with the formation of the erect gametophytic thallus. In addition to this sexual, diphasic, lifecycle, an abbreviated asexual life history has often been reported, via parthenogenetic development of the gametes. Depending on the culture conditions of temperature and daylength used, these parthenogametes could either repeat the parental expression or form unilocular sporangia bearing crustose thalli. The unilocular sporangia in these haploid crusts do not undergo meiosis and the released unispores behave asexually and develop similarly to the parthenogametes. Both sexual and abbreviated asexual life histories were reported by Nakamura & Tatewaki (1975) in five species of the Scytosiphonaceae; *Petalonia fascia* differed in possessing only an abbreviated life history.

Nakamura & Tatewaki's (1975) report of the occurrence of a heteromorphic life history in members of the Scytosiphonaceae has received support from a number of culture studies throughout the world, with the erect blade alternating with a prostrate, usually crustose thallus, the latter sometimes bearing unilocular sporangia (Blackler 1981; Clayton 1976a,b, 1979, 1980a,b, 1982; Edwards 1969; Fletcher 1974b, 1987; Kogame 1996, 1997a,b; Kogame & Masuda 2001; Nakamura 1965; Parente *et al.* 2003, 2011; Rhodes & Connell 1973; Roeleveld *et al.* 1974; Tatewaki 1966; Wynne 1969), more rarely reported bearing plurilocular sporangia (Blackler 1981; Clayton 1978, 1979; Kogame 1997a, 2001; Kogame & Yamagishi 1997; Parente *et al.* 2011; Sauvageau 1927; Toste *et al.* 2003; Wynne 1972). In a number of these studies, the life history was abbreviated and asexual with the plurizoids released from the erect blades germinating into basal systems that either directly produced erect blades or developed into discs/crusts which bore unilocular sporangia. Reports of sexual reproduction in the Scytosiphonaceae have been less common and confined to populations in Japan (Kogame 1996, 1997b, 1998; Kogame *et al.* 2005; Kunieda & Arasaki 1947; Nakamura & Tatewaki 1975; Shannon *et al.* 1988; Tatewaki 1966) and Australia (Clayton 1979, 1980, 1982). A number of these studies further revealed that temperature and daylength played a notable role in whether erect blades or discoid/crustose thalli were produced in the cultures (see Dring & Lüning 1975b, 1983; Edwards 1969; Fletcher 1974b, 1978; Nakamura & Tatewaki 1975; Rhodes & Connell 1973; Roeleveld *et al.* 1974; Takewaki 1966; Wynne 1969) and these environmental cues accounted for the seasonality of the gametophyte thalli in the yearly cycle. Temperature and daylength were also reported to influence unilocular and plurilocular sporangia formation on the basal systems, rather than the formation of the erect thalli (Kogame 1996, 1997a; Kogame & Yamagishi 1997), which would also control the seasonal occurrence of the erect blades. It would appear that the erect bladed thalli represent potential gametophytes, in which sexuality can be absent, rare or requires specific environmental conditions, for example daylength and temperature, for this phase to occur. It has also been suggested that age might also be a factor regulating gametogenesis (Clayton 1981).

A notable feature of the heteromorphic life histories revealed in this family is the range of prostrate forms described. These have frequently been linked in identity with thalli of such genera as *Streblonema*, *Chilionema*, *Compsonema*, *Myrionema*, *Microspongium*, *Stragularia* (usually identified as *Ralfsia*) and *Hapterophycus*. In the North Atlantic, connections have been established between the parenchymatous, erect bladed genera and *Stragularia* species as well as connections with material identified as *Compsonema saxicola* and the unilocular sporangia-bearing material of *Microspongium gelatinosum*. For this reason, the genus *Stragularia* is now commonly placed in the Scytosiphonaceae, with the species often placed within the synonymy of the erect bladed species, while *Compsonema saxicola* and the unilocular sporangia bearing material of *Microspongium gelatinosum* are removed from the Chordariaceae (in which they were unsuitably placed) and usually provisionally interpreted as 'phases' in the life histories of *Scytosiphon* and/or *Petalonia* spp. Indeed, the possession of a single, plate-like plastid in each cell would alone justify the inclusion of the above prostrate algae in the family Scytosiphonaceae (Christensen 1980; Pedersen 1981c, 1984).

While a number of phylogenetic studies have supported Rousseau & de Reviers' (1999b) merger of the Chordariales, Dictyosiphonales and Scytosiphonales within a broadened concept of Ectocarpales (Burkhardt & Peters 1998; Peters & Burkhardt 1998; Peters & Clayton 1998; Kogame *et al.* 1999; Siemer *et al.* 1998), and have highlighted a single evolutionary origin for the family Scytosiphonaceae (Cho *et al.* 2001, 2006; Kogame *et al.* 1999; Peters & Ramírez 2001), these studies and others have also highlighted a number of genus level discrepancies and considerable taxonomic uncertainty which question the traditional basis of classification of the group. Species assigned to the Scytosiphonaceae characteristically feature a single, plate-like plastid with a pedunculate pyrenoid and a heteromorphic life history and have traditionally been distributed in the following genera: *Colpomenia*, *Endarachne* (now synonymized with *Petalonia*, see Vinogradova 1973 and Kogame *et al.* 1999, p. 496), *Hydroclathrus*, *Iyengaria*, *Jolyna*, *Petalonia*, *Rosenvingea* and *Scytosiphon*. However, the basic concept of a heteromorphic life history dramatically changed with the later additions to the family of *Myelophycus*, based on a suggestion by Tanaka & Chihara (1984) and a molecular study by Cho *et al.* (2003), and of *Melanosiphon*, based on a molecular study by McDevit (2010b). These two genera have an isomorphic life history with unilocular sporangia on the erect thalli. In addition, *Chnoospora* was moved from the Chnoosporaceae Setchell & N.L. Gardner (1925) into the Scytosiphonaceae, and the family synonymized with the Scytosiphonaceae, based on a molecular study by Kogame *et al.* (1999).

There have also been a number of ambiguities in the demarcation of several genera in the Scytosiphonaceae, with members having intermediary traits in both morphology and structure (Kogame *et al.* 1999; Wynne 1972; Wynne & Norris 1976). This uncertainty has been addressed by a number of molecular studies on several genera within the family (Cho *et al.* 2001, 2006, 2007; Kogame *et al.* 1999; McDevit & Saunders 2017; Parente *et al.* 2011; Santiañez *et al.* 2018a,b), particularly with respect to *Colpomenia*, *Petalonia* and *Scytosiphon*. For instance, using molecular markers Kogame *et al.* (1999) revealed that these genera were not monophyletic and that their species were variously distributed between two large clades, one distributed in warm temperate to tropical regions and the other mostly distributed in cold-temperate regions. *Colpomenia peregrina* occupied a position outside the two clades. The cool-temperate clade also exhibited two subclades, each also comprising a mixture of various species and genera. These results cast considerable doubt on the monophyly of the genera *Colpomenia*, *Petalonia* and *Scytosiphon* and the use of morphological criteria in an understanding of their phylogeny. For example, whether thalli are hollow or solid does not distinguish between *Scytosiphon* and *Petalonia* (Kogame *et al.* 1999). These studies also revealed that, rather than the morphology and structure of the erect gametophytes, it was the morphological character-istics of the prostrate sporophytes which were congruent with the molecular phylogeny and offered more important taxonomic criteria for the classification of both genera and families. Hence, close molecular relationships were found between genera and species with the same type of prostrate thalli (for example *Composonema*-like or *Stragularia*-like) and whether or not they produced plurilocular sporangia on prostrate thalli. These results found support in the more extensive molecular studies by Cho *et al.* (2001, 2003, 2006), who reported two similarly composed clades, each separated according to their distribution patterns and repro-ductive organs: the warm-temperate species produced plurilocular and unilocular sporangia on the sporophytes while the cool-temperate species produced only unilocular sporangia. Moreover, in agreement with Kogame *et al.* (1999), *Scytosiphon* was separated into two clades. Finally, McDevit & Saunders (2017), have also provided a better understanding of the genera *Scytosiphon* and *Petalonia*, which has resulted in the new genus *Planosiphon*.

This family is represented in Britain and Ireland by six genera: the two pseudoparenchy-matous (syntagmatic) crustose genera (*Symphyocarpus, Stragularia*) and four parenchymatous, thalloid genera (*Colpomenia, Petalonia, Planosiphon* and *Scytosiphon*). Also included here is the microscopic tufted species *Composonema saxicola* and the unilocular sporangia-bearing material of the crustose alga *Microspongium gelatinosum*, both referred to as 'phases'. The prostrate thalli can be distinguished using a combination of morphological and anatomical features. *C. saxicola* forms confluent tufts while the *M. gelatinosum* phase, *Stragularia* and *Symphyocarpus* form crustose thalli. Unlike *Stragularia* and *Symphyocarpus, M. gelatinosum* forms slightly pulvinate thalli comprising erect filaments which are only loosely associated and easily glide apart under slight pressure, while *Stragularia* and *Symphyocarpus* have erect or slightly curved erect filaments which are quite firmly adjoined but separable under pressure.

The four thalloid genera included here are more easily distinguished: *Colpomenia* forms inflated saccate thalli, *Petalonia* forms expanded, flattened, solid thalli, *Planosiphon* forms linear, flattened, unconstricted tubular thalli which are hollow or partially hollow, and *Scytosiphon* forms linear, tubular, hollow, constricted thalli.

Colpomenia (Endlicher) Derbès & Solier

COLPOMENIA (Endlicher) Derbès & Solier (1851), p. 95.

Type species: *C. sinuosa* (Mertens ex Roth) Derbès & Solier (1851), p. 95.

Thalli spherical or saccate, membranous, hollow, furrowed, firm and entire or more frequently collapsed and irregularly split, attached at the base by rhizoidal filaments; surface cells irregu-larly placed, each with a single plate-like, parietal plastid and pyrenoid; comprising an inner medulla of large, thick-walled, colourless cells and an outer cortex of smaller, pigmented cells; hairs tufted, grouped, arising from depressions, with basal meristem, with or without obvious sheath.

Plurilocular sporangia crowded, discrete or extensive on thallus surface, in closely packed vertical columns, uniseriate to multiseriate, with ascocyst-like cells; unilocular sporangia unknown on thallus surface, reported to be associated with a ralfsioid phase in the life histories of Pacific isolates only.

Two species of *Colpomenia* have been recorded in the North Atlantic: *C. sinuosa* (Roth) Derbès & Solier, which is restricted to warm-temperate waters and only extends as far north as the Spanish coast in the Bay of Biscay (Ardré 1970; Hamel 1928; Miranda 1931) and *C. peregrina* (Sauvageau) Hamel which is mainly cool-temperate in distribution and extends from Morocco to Norway. The two species are distinguished mainly on features such as surface configuration, hair origin, wall thickness and sorus shape. However, many of these were rejected by Clayton

(1975) in her comparative study of populations of these two species in Australia. Only the shape of the sori (irregular/extensive in *C. peregrina*, punctate in *C. sinuosa*) and the presence of a cuticle on the plurilocular sporangia (in *C. sinuosa* only) were considered by her to be effective.

A number of *Colpomenia* species have been investigated in laboratory culture throughout the world; these include *C. bullosa* Yamada (Blackler 1981; Kogame & Masuda 2001; Nakamura & Tatewaki 1975), *C. peregrina* (Blackler 1981; Clayton 1979; Dangeard 1963b; Kogame & Yamagishi 1997; Sauvageau 1927a) and *C. sinuosa* (Blackler 1981; Kogame 1997; Kunieda & Suto 1938; Parente *et al.* 2011; Toste *et al.* 2003; Wynne 1972a – as f. *deformans* Setchell & N.L. Gardner). In all these investigations the life history was shown to be heteromorphic with an erect, thalloid gametophyte alternating with a prostrate crustose or filamentous (*Ectocarpus* or *Chilionema*-like) sporophyte. In some studies, the prostrate thalli became fertile with plurilocular and/or unilocular sporangia and this appeared to be under the influence of temperature and daylength; Kogame & Yamagishi (1997) proposed that this regulatory control of fertility might well be responsible for the annual cycle of the gametophytes. While in most studies, the life histories shown were asexual, in some, for example in Japan (Kogame & Masuda 2001; Kogame & Yamagishi 1997; Nakamura & Tatewaki 1975), Australia (Clayton 1979) and the Azores (Toste *et al.* 2003) the life histories was shown to be sexual. The culture studies also revealed that geographical isolates of some species have different life histories, while morphological differences were also shown between the sporophytes of *Colpomenia* species, indicating that the genus is polyphyletic, this finding supported by the molecular study of Kogame *et al.* (1999).

One species in Britain and Ireland.

Colpomenia peregrina Sauvageau (1927), p. 321.   Figs 214, 215

Colpomenia sinuosa (Mertens ex Roth) Derbès & Solier var. *peregrina* Sauvageau (1927), p. 321.

Thalli consisting of erect, spherical or saccate, hollow thalli, 3–7(–9) cm across, entire and fleshy when young, frequently furrowed, collapsed and irregularly torn later, broadly attached at the base by localized patches of rhizoidal filaments produced from the outer cortical cells; saccate portion membranous, smooth and slightly lubricous outside, roughened inside, olive-brown, drying to green, surface cells polygonal, irregularly displaced, 8–16 × 5–12 μm, each with a single, parietal, plate-like plastid and pyrenoid; hairs frequent, immersed in pits, with basal meristem, with or without obvious sheath; in section, thalli with an outer cortex of 1–3 small, pigmented cells and an inner medulla of 1–4 layers of large, thick-walled, frequently ruptured and collapsed colourless cells.

Plurilocular sporangia crowded, in extensive or discrete dark patches on thallus surface, in closely packed vertical columns arising from outer cortical cells, to 12 loculi (–55 μm) long, uniseriate to multiseriate, 5–11 μm in diameter, usually with abundant scattered ascocyst-like cells present, 17–35 × 7–13 μm; unilocular sporangia unknown on thallus surface and not reported in the life history of Atlantic isolates.

Epiphytic, rarely epilithic, upper to lower eulittoral pools and damp rock, and sublittoral to 3 m.

Generally distributed around Britain and Ireland, but more common on south-western shores.

Thalli annual, appearing in early summer and dying back during late autumn and early winter, rarely recorded in winter and spring and probably lying dormant as basal filaments. At Bembridge on the Isle of Wight, thalli are abundant in May and June, less frequent during July and August with a noticeable increase in numbers again in September and October.

Colpomenia peregrina represents an introduced species to the North Atlantic, probably originating from the north-west Pacific (see Jones 1974; Farnham 1980; Cho *et al.* 2005). It was first

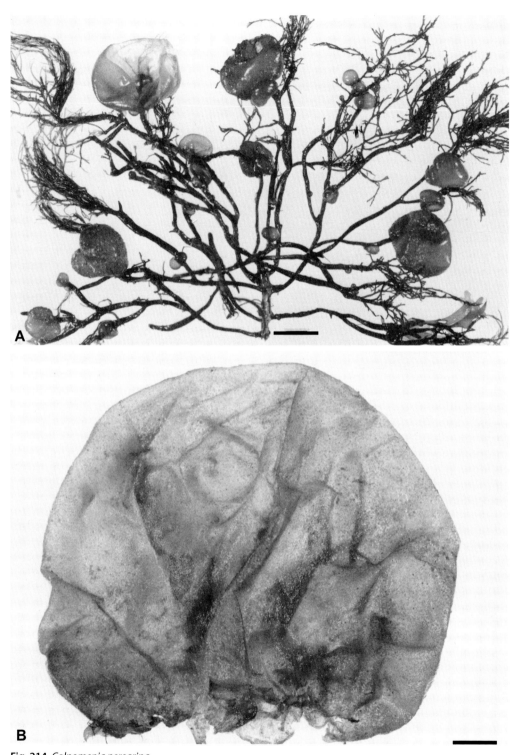

Fig. 214. *Colpomenia peregrina*
A. Habit of thalli on *Gongolaria*. Bar = 20 mm. B. Single large thallus. Bar = 12 mm.

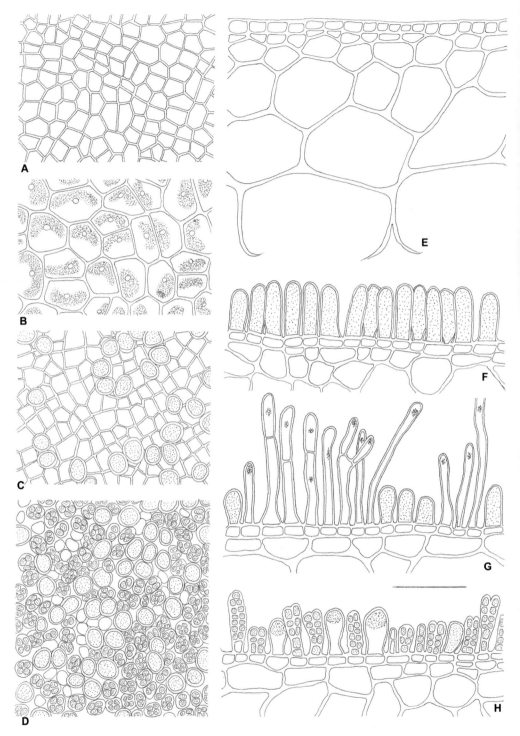

Fig. 215. *Colpomenia peregrina*
A, B. Surface view of thallus. Note single plate-like plastid with pyrenoid. C. Surface view of thallus showing ascocysts. D. Surface view of thallus showing ascocysts and plurilocular sporangia with 1–4 loculi. E. VS of vegetative thallus showing 1–2 layers of small, cortical cells and 3–4 layers of large, medullary cells. Note inner cells, enclosing central cavity, are ruptured. F, G. VS of thalli showing variously shaped ascocysts. H. VS of fertile thallus showing ascocysts and plurilocular sporangia arising from surface cells. Bar = 20 μm (B), = 50 μm (others).

reported at the beginning of the twentieth century (1906) in the Gulf of Morbihan on the west coast of France (Sauvageau 1918) and has since spread as far north as Norway (Rueness *et al.* 2001). It was first reported for Britain by Cotton (1908b,c, 1911) but was omitted from Newton (1931). It was later reported that it had been introduced into the north-west Atlantic (Villalard-Bohnsack 2002).

Culture studies on Atlantic isolates by Sauvageau (1927a), Dangeard (1963b) and Blackler (1981) revealed the occurrence of a direct, abbreviated, asexual life history. Plurispores from plurilocular sporangia on the saccate blades germinated directly to form filamentous microthalli bearing ectocarpoid plurilocular sporangia, from which a further generation of saccate thalli developed. There was no evidence of a sexual cycle and the life history could, therefore, be described as heteromorphic and monophasic. A similar life history was also reported by Blackler (1981) for isolates obtained from the Mediterranean, Australia and California. This life history is in contrast to the heteromorphic, biphasic life history described for this species in Australia by Clayton (1979) and in Japan by Kogame & Yamagishi (1997). Clayton revealed that the saccate macrothalli are potential gametophytes which alternate with filamentous to pseudodiscoid, unilocular sporangia-bearing sporophytes. Sexual reproduction was described as anisogamous, with large female gametes and small male gametes. However, functional gametophytes were described as rare, with a restricted winter occurrence; usually the life history is 'direct', with the plurispores (which could be either zoospores or gametes) behaving asexually and germinating directly into similar thalli, not unlike the sequence described for European isolates by Sauvageau and Dangeard.

In the field and culture study of Japanese isolates of *C. peregrina* (Kogame & Yamagishi 1997), naturally occurring sporophytes were reported for the first time, although these could not be identified with any known genera. Morphologically similar fertile prostrate thalli were also described in culture by these authors and noted to be similar to those described in Clayton's cultures but differed in having ascocysts and ectocarpoid plurilocular sporangia in addition to unilocular sporangia. They were reported to resemble species of *Myrionema*, *Chilionema* and *Symphyocarpus*. They were therefore unlike previously described sporophytes in the Scytosiphonaceae which resembled species of the genera *Streblonema*, *Compsonema*, *Microspongium*, *Stragularia* and *Hapterophycus*. Kogame & Yamagishi (1997) suggested that either the Australian *C. peregrina* has a different life history or that the culture conditions were not inducive to plurilocular sporangia formation on the prostrate thalli. In this connection, they reported that culture conditions, such as photoperiod and temperature, determined whether plurilocular or unilocular sporangia were formed on the prostrate sporophytes and this response to these parameters was considered by the authors to determine the seasonality of the gametophytes.

More culture studies on European isolates of this species throughout its seasonal occurrence, and using a range of temperature and photoperiod regimes, are thus required to determine if a sexual life history, with a unilocular and plurilocular sporangia-bearing prostrate phase is present, similar to that reported for Australian and Japanese thalli.

Abbott & Hollenberg (1976), pp. 1–10, figs 1–10; Adams (1994), p. 76, pl. 18; Bird & Edelstein (1978), pp. 181–7; Blackler (1964), p. 50, (1967), pp. 5–8, (1981), p. 133; Boo (2010), pp. 173, 174, figs 12(A, B), 13; Brodie *et al.* (2016), p. 1021; Bunker *et al.* (2017), p. 296; Cho *et al.* (2005), pp. 103–11; Clayton (1975), pp. 187–95, figs 1–14, (1979), pp. 1–10, figs 1–10, (1980b), pp. 113–18, figs 1–3; Coppejans & Kling (1995), pp. 228, 229, pl. 83(A, B); Cormaci *et al.* (2012), pp. 274–6, pl. 79(1–3); Cotton (1980b), pp. 73–7, (1908c), pp. 82, 83, (1911), pp. 153–7; Dangeard (1963b), pp. 66–73, pls XII–XIII, XX–XXI; Dizerbo & Herpe (2007), pp. 115, 116, pl. 38(4); Domingo (1957), pp. 92–8; Farnham (1980), p. 877; Feldmann (1949), pp. 103–15, fig. 2; Fletcher (1987), pp. 219–21, figs 56, 57, pl. 7(a, b); Fletcher in Hardy & Guiry (2003), p. 234; Gallardo García *et al.* (2016), p. 37; Gayral (1966), p. 254, pl. XXXIII; Green *et al.* (2012), pp. 643–7; Guiry (2012), p. 154; Hamel (1931–39), pp. 201, 202; Jones (1974), p. 107; Kogame & Yamagishi (1997), pp. 337–44; Kogame *et al.* (1999), pp. 496–502; Lund (1945), pp. 1–16; Mathieson & Dawes (2017), pp. 195, 196, pl. LVI(14–17); Mathieson *et al.* (2016), pp. 276–305; Matta & Chapman (1991), pp. 303–13; Minchin (1991), pp. 380, 381; Morton (1994), p. 42; Oates (1985), pp. 109–19, (1988), pp. 57–63, (1989), pp. 475–8;

Ouiz (2012), p. 154; Parsons (1982), pp. 295–7, figs 6–7, 13; Pedroche *et al.* (2008), p. 64; Ribera *et al.* (1992), p. 118; Rosenvinge & Lund (1947), p. 37, figs 12, 13; Rueness (1977), p. 167, pl. XXII(5), (2001), p. 14; Sauvageau (1918), pp. 1–28, (1927a), pp. 308–53, figs 1–8; Stefanov *et al.* (1996), pp. 475–8; Stegenga & Mol (1983), p. 103, pl. 37(1–2); Stegenga *et al.* (1997a), p. 21; Vandermeulen (1986), pp. 138–44; Vandermeulen & De Wreede (1986), pp. 31–47, (1987), pp. 91–8; Vandermeulen *et al.* (1984), pp. 325, 326, figs 5–8; Villalard-Bohnsack (2002), pp. 130–2; Womersley (1987), p. 298, figs 107(B), 108(G, H); Yamagishi & Kogame (1998), p. 217.

Compsonema saxicola 'phase'

Compsonema saxicola 'phase' of *Planosiphon/Scytosiphon*   Figs 216, 217

Myrionema saxicola Kuckuck (1897), p. 381; *Compsonema saxicolum* (Kuckuck) Kuckuck (1953), p. 343.

Thalli epilithic or epizoic, forming small, light brown tufts not visible to the unaided eye, discrete and circular or more commonly confluent (resembling a carpet pile) and spreading to approximately 1 mm in diameter; in squash preparation, consisting of a basal layer of branched, outward-spreading, irregularly contorted, loosely associated filaments (pseudo-discoid) comprising mainly rectangular cells, 7–13 × 5–8 µm, frequently longitudinally divided (biseriate), many of which, especially in the central regions, give rise to erect filaments, hairs and/or unilocular sporangia; erect filaments free and not laterally adjoined, loosely associated in a gelatinous matrix and easily separable under pressure, simple or rarely branched, to 280 µm long, 14–25(–35) cells, slightly tapering above, comprising cells mainly quadrate, sometimes rectangular, 5–11 × 5–7 µm, each with a single, plate-like and parietal plastid occupying the upper cell region, with pyrenoid; hairs fairly frequently arising from basal layer or terminal on erect filaments, with basal meristem and sheath.

Unilocular sporangia common, oval to pyriform, 40–70 × 19–35 µm, usually arising directly from basal cells, sessile or on 1–3-celled stalks, less frequently lateral or terminal on the erect filaments; plurilocular sporangia unknown.

Epilithic on bedrock, or epizoic on limpets, in the upper eulittoral and littoral fringe.

Only recorded for the south coast of England (Kent, Dorset, Devon) and west coast of Ireland (Mayo); probably more widely distributed.

Probably a summer annual; recorded from April to October.

This species is probably common and widely distributed around Britain and Ireland. It appears to favour colonization of limpet shells in the littoral fringe, forming small confluent tufts just visible under a low power binocular microscope. Algae frequently associated with *C. saxicola* include *Blidingia* spp., *Petroderma maculiforme*, *Hecatonema* spp., *Pseudoralfsia verrucosa*, *Ulva* spp., young fucoid germlings and blue-green algae. It is characterized by the unbranched erect filaments, cells with a single, plate-like plastid, basally positioned unilocular sporangia and a parenchymatous prostrate system with frequent biseriate cells.

C. saxicola has been investigated in laboratory culture by Pedersen (1981c) and the present author (unpublished information). Pedersen's (1981c) work revealed the occurrence of a 'direct', monophasic life history for Danish material; the unispores behaved asexually and germinated directly into fertile thalli, similar to the parental field material. Culture studies by the present author, using material collected from Dorset on the south coast of England, revealed a very similar sequence of development; however, an additional erect thallus arose from the basal filaments. These erect thalli were poorly developed in culture but resembled *Planosiphon* (identified as *Petalonia*) or *Scytosiphon*. The involvement of *Compsonema*-like thalli in the life history of a member of the Scytosiphonaceae finds support in the investigations of Loiseaux (1970b), Fletcher (1981a) and Kogame (1998). Loiseaux established a life history connection between the Californian species *Compsonema sporangiiferum* Setchell & N.L. Gardner (which she noted to be very similar to *C. saxicola*) and a minute *Scytosiphon*, close to *S. pygmaeus* Reinke. Fletcher observed fertile microthalli in cultures of *Planosiphon filiformis*

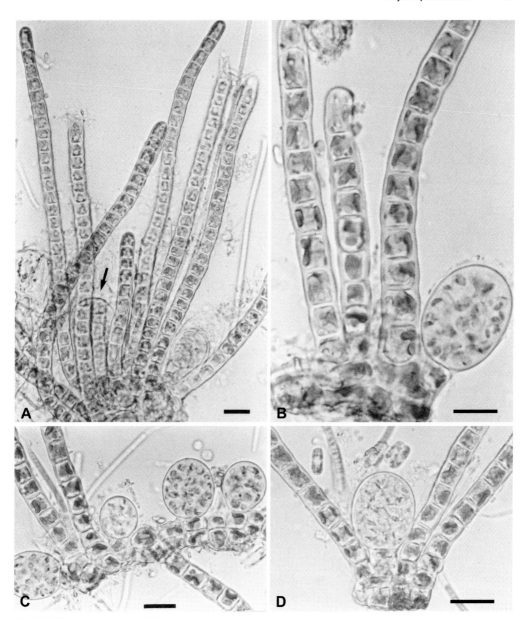

Fig. 216. *Compsonema saxicola* 'phase'.
A. SP of thallus showing erect filaments and basal unilocular sporangia (arrow). B–D. SP of thallus portions showing distromatic base giving rise to erect filaments and unilocular sporangia. Bar in A = 25 μm, bar in B = 12 μm, bar in C = 15 μm, bar in D = 20 μm.

(as *Petalonia filiformis*) collected from the south-east coast of England, which showed a marked similarity to *C. saxicola*, while Kogame revealed *Compsonema*-like sporophytic thalli in the life history of *Scytosiphon gracilis* in Japan.

There is evidence that environmental conditions, particularly temperature and photoregime, play a role in the relationship between the *Compsonema* microthalli and the erect macrothalli. Studies by the present author (unpublished) revealed erect thallus production to be enhanced in low temperature/short day conditions, while fertile *Compsonema*-like thalli showed maximum development in warm temperature/long day conditions. These results agree with

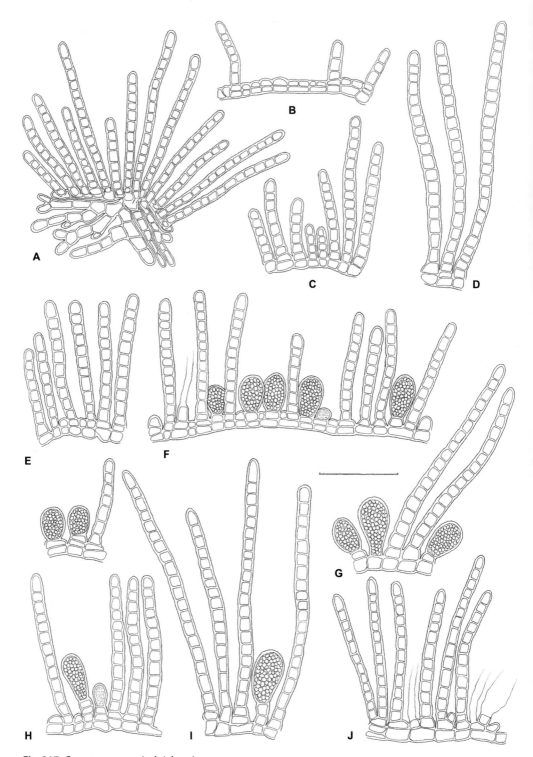

Fig. 217. *Compsonema saxicola* 'phase'.
A–E. SP of vegetative thalli showing erect filaments arising from a distromatic base. F–J. SP of fertile thalli showing unilocular sporangia. Bar = 50 μm.

the mainly winter/spring seasonal occurrence of *Planosiphon* and *Scytosiphon* spp. around Britain and Ireland and the generally summer/autumn field occurrence of the *C. saxicola* thalli. Kogame (1998) also reported that, in the development of plurizoids of *Scytosiphon gracilis*, erect thalli arose more abundantly and were longer in lower temperatures and short day conditions. More culture studies are required to determine the occurrence of *Composonema*-like thalli in the life histories of genera and species of the Scytosiphonaceae and their exact relationship.

Composonema saxicola was originally described from Helgoland by Kuckuck (1897) as *Myrionema saxicola*. It was subsequently reported for the Britain at Swanage, Dorset by Batters (1900, p. 371) and described in Newton (1931, p. 151). Kuckuck (1953) transferred the species to the genus *Composonema* Kuckuck (1899a), removing the latter from its original place in the Myrionemataceae to the Ectocarpaceae. The placement of *Composonema* into the Ectocarpaceae was accepted by Parke (1953) and Parke & Dixon (1964) although later (Parke & Dixon 1968, 1976) the genus was transferred back to the Myrionemataceae, probably as a result of the work by Loiseaux (1967a). On the basis of the culture results it is included here in the Scytosiphonaceae following the proposal of Pedersen (1981c). Other features consistent with its inclusion in this family include the frequent occurrence of longitudinal divisions in the basal cells and the presence of a single, plate-like plastid with pyrenoid in each cell.

Bartsch & Kuhlenkamp (2000), p. 165; Batters (1900), p. 371; Brodie *et al.* (2016), p. 1021; Cotton (1912), p. 121; Fletcher (1981a), pp. 103, 104, (1987), pp. 221–4, fig. 58, pl. 8(a, b); Fletcher *et al.* (1988), pp. 1–8, fig. 2; Fletcher in Hardy & Guiry (2003), p. 235; Guiry (2012), p. 155; Knight & Parke (1931) p. 68, pl. XII(27–9), non *Myrionema saxicola*; Kuckuck (1897), pp. 381, 382, fig. 8, (1953), pp. 343–7, fig. 14; Mathieson & Dawes (2017), p. 196, pl. LVII(1); Parente *et al.* (2011), p. 43; Pedersen (1981c), pp. 213, 214, fig. 55.

Microspongium gelatinosum 'phase'

Microspongium gelatinosum 'phase' of *Scytosiphon lomentaria*   Figs 2(A), 218

Microspongium gelatinosum Reinke (1888), p. 16; *Microspongium gelatinosum* Reinke (1889b), p. 46 (unilocular sporangial material only).

Thalli crustose or slightly pulvinate and spongy, medium to dark brown, to approximately 0.5 mm thick, discrete and circular or more commonly confluent and spreading, to 30 mm or more in extent, firmly attached to substratum by undersurface, usually without rhizoids; in surface view, central vegetative cells rounded, quite closely packed, 7–11 µm in diameter, edge region cells rectangular or quadrate 8–18 × 5–10 µm; consisting of a monostromatic base of coherently spreading, laterally united filaments with a marginal row of apical cells giving rise, over the greater part, to simple, or little branched, erect, slightly gelatinous filaments only loosely adjoined and quite easily separable under pressure, except basal 1–3 cells; filaments to 25 cells (–430 µm) long, comprising mainly rectangular cells, 5–10 µm in diameter each with a single, parietal, plate-like plastid occupying upper cell region, with pyrenoid.

Unilocular sporangial sori extensive over central thallus region, not visible in surface view; unilocular sporangia elongate-pyriform or elongate-cylindrical, 48–100 × 16–27 µm, sessile or more usually on 1–3-celled pedicels, borne at the base of loosely associated, gelatinous, multicellular paraphyses; paraphyses linear or elongate-clavate, 95–170 × 5–7 µm, mostly of 6–8 cells which are 1–2.5 diameters long above, narrower and up to 4(–9) diameters long below; plurilocular sporangia not recorded on crustose thalli in Britain and Ireland, only reported on Danish specimens by Lund (1966, figure 3N) forming terminal, uniseriate clusters on branches of the erect filaments; plurilocular sporangia usually confined to the erect bladed *Scytosiphon lomentaria* phase.

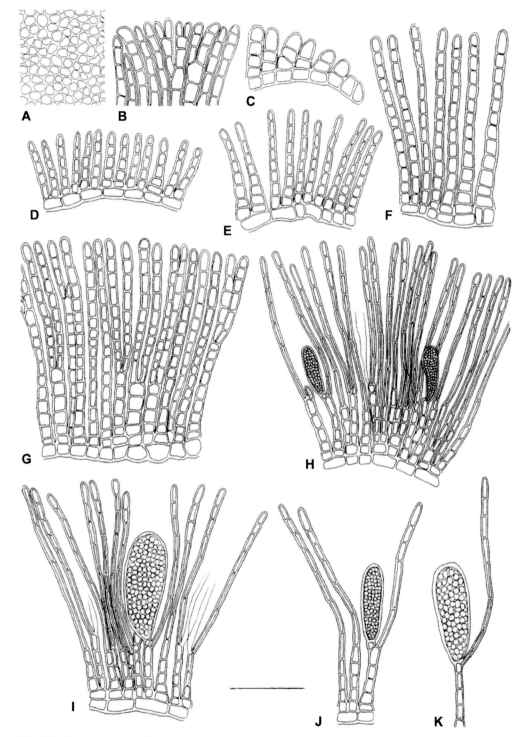

Fig. 218. *Microspongium gelatinosum* 'phase'
A. Surface view of central crust region. B. Surface view of peripheral crust region. C. SP of crust edge showing terminal apical cell. D–G. SP of vegetative thalli showing loosely adjoined erect filaments. H–K. SP of fertile thalli showing unilocular sporangia arising at base of paraphyses. Bar = 50 μm.

Epilithic and epizoic, in pools and emergent, upper to lower eulittoral, usually in association with other crustose brown algae such as *Petroderma maculiforme*, *Stragularia clavata* and *Pseudoralfsia verrucosa*.

Infrequent, recorded for widely scattered localities around Britain (Fife, East Lothian, Berwick, Northumberland, Yorkshire, Kent, Sussex, Hampshire, Dorset, Devon, Pembroke, Argyll) and Ireland (Down).

Vegetative thalli recorded throughout the year and probably perennial; unilocular sporangia recorded September to April.

The above-described crustose phase of *Scytosiphon lomentaria* represents the unilocular sporangia-bearing thalli of *Microspongium gelatinosum* originally discovered by Reinke (1889b) in the Baltic. It is now recognized that these thalli represent a different taxon to the thalli of *M. gelatinosum* he had earlier described with plurilocular sporangia (Reinke 1888a) and they must consequently be separated (see pp. 508–10). The unilocular sporangia-bearing material (herein termed the *M. gelatinosum* 'phase') is very similar in structure to the genus *Stragularia*, from which it differs in having slightly more pulvinate, spongy crusts, unilocular sporangial sori not clearly distinguishable from the vegetative thallus in surface view and erect vegetative filaments which are loosely associated (except near the base) and easily separated, forming almost fan-shaped squash preparations. In addition, the paraphyses in the *M. gelatinosum* phase are less clearly discernible from the erect vegetative filaments. Using the above characteristics, field material of the *M. gelatinosum* 'phase' can easily be distinguished from *Stragularia* spp., even the similar *S. clavata*. This negates the use of the term '*Ralfsia*-like' for the unilocular sporangial thalli of the *M. gelatinosum* phase as proposed by Kristiansen & Pedersen (1979), despite the nomenclatural benefits that would accrue.

Field and laboratory culture studies of the *M. gelatinosum* phase in the North Atlantic have shown it to be a life history phase of the macroscopic alga *Scytosiphon lomentaria* (see Fletcher 1978 – south coast of England; Lund 1966 – Denmark; McLachlan *et al.* 1971 – Nova Scotia; Parente *et al.* 2003a – Azores). Unispores from the unilocular sporangia on the crusts germinated directly, without sexual fusion, to form uniseriate filamentous, knot-filamentous and/or '*Microspongium*'-like prostrate systems. In addition to repeating the parental phase and developing unilocular sporangia, these thalli gave rise to erect, parenchymatous, tubular fronds identified as *S. lomentaria*. The evidence indicates that both the crustose and erect thalli probably have the same ploidy level and the life history can therefore be described as heteromorphic and monophasic. Fletcher revealed a facultative relationship between the erect fronds and the 'basal' growths which was influenced by environmental conditions. For example, frond initiation appeared to be enhanced in low temperature/short day conditions while basal growths, particularly crustose development was enhanced in warm temperature/long day conditions. The involvement of crustose thalli in the life history of *S. lomentaria* is supported by plurispore culture studies of the latter alga throughout the world, although there is some conflicting evidence from widely different geographical isolates as to whether these are '*Microspongium*'-like or '*Ralfsia*'-like (see p. 511).

On the basis of the culture studies of North Atlantic isolates and the occurrence of a single, plate-like plastid in each cell, *M. gelatinosum* Reinke (unilocular sporangia material only) is removed from the Myrionemataceae and herein included in the Scytosiphonaceae.

Fletcher (1974a), pp. 78–91, figs 13–16, pls 23–5, (1978), pp. 371–98, figs 10, 22, 31, 32, 33, 35, (1987), pp. 224–8, figs 59(A), 60, pl. 8(c); Fletcher & Maggs (1985), pp. 523–6, fig 1(A–H); Fletcher *et al.* (1988), pp. 1–8, fig. 1(df); Kristiansen & Pedersen (1979), pp. 31–56; Lund (1966), pp. 70–8, figs 1–4; McLachlan *et al.* (1971), pp. 82–7, figs 1–10; Parente *et al.* (2003), pp. 353–9, (2011), p. 43; Peña & Bárbara (2010), pp. 56, 57, fig. 13(A–E); Reinke (1888a), p. 16, (1889a), pp. 11, 12, pl. 8, (1889b), pp. 46, 47 (in both references, unilocular sporangial thalli only); Tanaka & Chihara (1980c), pp. 339–41, fig. 1(B) (as *Ralfsia bornetii*).

Petalonia Derbès & Solier nom. cons.

PETALONIA Derbès & Solier (1850), p. 265.

Type species: *P. debilis* (C. Agardh) Derbès & Solier (1850), p. 266 (= *P. fascia* (O.F. Müller) Kuntze (1891–98), p. 419).

Ilea Fries (1835), p. 321; *Phyllitis* Kützing (1843), p. 342; *Endarachne* J. Agardh (1896), p. 26.

Thalli consisting of erect, light to dark brown, bilaterally flattened blades arising from an expanded encrusting base or small discoid holdfast; blades single or gregarious, simple, solid or with small central cavities, extremely variable in shape, although usually linear to broadly lanceolate, tapering sharply or slowly to base, less obviously to apex, sometimes spirally twisted in narrow specimens, entire and without proliferations, rarely longitudinally indented/split, with or without ruffled, delicate margins, delicate and flaccid to subcoriaceous in texture, often terminally eroded on mature specimens; in surface view, cells small, rectangular to polygonal, irregularly arranged or in longitudinal, less obviously transverse rows, each containing a single, parietal, plate-like plastid with 1–2 pyrenoids; in section, comprising an inner medulla of large, thick-walled, longitudinally elongate, transversely rounded, colourless cells, solid throughout or with small intercellular cavities, surrounded by an outer cortex of 1–3 smaller, pigmented cells; hairs single or tufted arising from surface or depressions, with or without obvious basal meristem and sheath.

Plurilocular sporangia in crowded, closely packed sori, extensive in terminal blade region, in vertical columns, arising by division of cortical cells, uniseriate or biseriate with subquadrate to rectangular loculi; unilocular sporangia not known on blade surface but in some species reported on crustose *Stragularia*-like thalli.

Until recently, *Petalonia fascia*, *P. filiformis* and *P. zosterifolius* have been reported for Britain and Ireland (Fletcher 1987). They differed with respect to their width, which decreased in the order *P. fascia*, *P. zosterifolius* and *P. filiformis*, their shape, being oval in *P. filiformis*, flattened in *P. fascia* and *P. zosterifolius*, and their basal systems, being a crust or holdfast in *P. fascia*, and filamentous and rhizoidal in *P. filiformis* and *P. zosterifolius*. Fletcher (1987) also reported differences in life history between the species; in their heteromorphic life history, the prostrate phase in *P. filiformis* was *Compsonema*-like, in *P. zosterifolius* it was branched filamentous while in *P. fascia* it was crustose and *Stragularia*-like. There was no evidence of a sexual cycle and the life histories were described as heteromorphic and monophasic. These results were in general agreement with other studies on these species throughout the world. Notable differences, however, included reports of the prostrate thalli in *P. fascia* being *Streblonema*-like, *Myrionema*-like as well as *Stragularia*-like, and the prostrate thalli being *Stragularia*-like in a culture study of *P. zosterifolius* in Japan (Nakamura & Tatewaki 1975). Notable differences also included the reports of a sexual cycle in *P. zosterifolius* (Nakamura & Tatewaki 1975) and *P. fascia* (Kogame 1997). Clearly the situation seemed unsatisfactory and using the above morphological characters combined with a DNA barcode molecular study, McDevit & Saunders (2017) uncovered considerable genetic variation in the genus *Petalonia* which supported the transfer of species with flattened thalli that were hollow to partially hollow, and possessed *Compsonema*-like prostrate sporophytic thalli that bore only unilocular sporangia, to a new genus *Planosiphon*. *Petalonia zosterifolius* was duly transferred to the new genus, followed by the transfer of *P. filiformis* (Santiañez & Kogame 2017). As a result of these studies, both are transferred to the genus *Planosiphon* here, leaving *Petalonia* represented in Britain and Ireland by a single species, *P. fascia*.

Petalonia fascia blades may be confused with those of species of *Punctaria* Greville and young *Laminaria* Lamouroux. However, the latter two genera may be distinguished from *Petaloniu* by the following anatomical and cytological details: in surface view, the cells of

Punctaria and *Laminaria* blades are much larger and contain several discoid plastids; in section *Punctaria* shows no pronounced differentiation into cortical and medullary cells, the cells often being approximately equal in size, while in *Laminaria* the blades are three-layered with an outer, epidermis-like meristoderm of small pigmented cells, enclosing a cortical zone of large, thin-walled, colourless cells and a central medullary region; while *Petalonia* is attenuate towards the base, *Laminaria* and sometimes *Punctaria* have a distinct stipe.

One species in Britain and Ireland.

Petalonia fascia (O.F. Müller) Kuntze (1891–98), p. 419.   Figs 219, 220

Fucus fascia O.F. Müller (1771–82), pl. 768; *Laminaria fascia* (O.F. Müller) C. Agardh (1817), p. 19; *Ulva fascia* (O.F. Müller) Lyngbye (1819), p. 28; *Ulva fascia* var. *tenuior* Lyngbye (1819), p. 28; *Ilea fascia* (O.F. Müller) Fries (1835), p. 321 (pro parte); *Punctaria caespitosa* J. Agardh (1836), p. 14; *Phyllitis fascia* (O.F. Müller) Kützing (1843), p. 342; *Phycolapathum cuneatum* Kützing (1845–71), pl. 4911; *Laminaria caespitosa* (J. Agardh) J. Agardh (1848), p. 73; *Phyllitis fascia* var. *curvata* Kützing (1849), p. 567; *Phyllitis debilis* Kützing (1849), p. 567; *Punctaria debilis* (Kützing) Trevisan (1849), p. 428; *Petalonia debilis* (Kützing) Derbès & Solier (1850), p. 266; *Petalonia debilis* (C. Agardh) Derbès & Solier (1850), p. 266; *Phyllitis caespitosa* Le Jolis (1863), p. 68; *Scytosiphon fascia* (O.F. Müller) P. Crouan & H. Crouan (1867), p. 169; *Phyllitis fascia* var. *caespitosa* (J. Agardh) Farlow (1881), p. 62; *Phyllitis fascia* (O.F. Müller) Kützing var. *debilis* (C. Agardh) Hauck (1883–85), p. 391; *Phyllitis fascia* var. *latior* Hariot (1889), p. 46; *Saccharina caespitosa* (J. Agardh) Kuntze (1891), p. 915; *Saccharina fascia* (O.F. Müller) Kuntze (1891), p. 915; *Phyllitis fascia* var. *tenuissima* Batters (1902), p. 27; *Ilea caespitosa* (J. Agardh) Nordstedt (1912), p. 239; *Ralfsia californica* Setchell & N.L. Gardner (1924), p. 2; *Ilea fascia* f. *debilis* (Kützing) Setchell & N.L. Gardner (1924), p. 12; *Ilea fascia* f. *caespitosa* (J. Agardh) Setchell & N.L. Gardner (1924), p. 12; *Petalonia fascia* var. *debilis* (Kützing) Hamel (1937), p. 198; *Ilea fascia* f. *crispa* (A.D. Zinova) A.D. Zinova (1953), p. 122.

Thalli forming erect, light to dark brown, single or more commonly gregarious, dorsiventrally flattened blades arising from a thin expanding crust or small discoid holdfast; erect blades simple, solid, thin and flaccid, becoming thicker and subcoriaceous, linear-lanceolate to broadly lanceolate to 15(–40) cm × 5–15(–40) mm, tapering slowly or sharply to base, usually rounded, less frequently tapering above, entire and without proliferations, more rarely longitudinally indented above and often terminally eroded, with flat, in parts occasionally ruffled, margins; in surface view, cells irregularly placed or in longitudinally less obviously transverse rows, rectangular to polygonal, small, 6–13 × 4–8 μm, each enclosing a single, plate-like, parietal plastid with conspicuous pyrenoid; in section, blades to 85 μm (vegetative) and 210 μm (reproductive) thick comprising an inner medulla of 3–6 large, thick-walled, longitudinally elongate, transversely rounded, often collapsed and ruptured colourless cells, enclosed by an outer cortex of 1–3 small, pigmented cells; hairs rarely observed on field-collected material, single or grouped arising from surface cells or depressions, with basal meristem, lacking obvious sheath.

Plurilocular sporangia in dark brown sori extensive in terminal blade region, in closely packed vertical columns arising from cortical cells, uniseriate, occasionally biseriate to 60 μm (–14 loculi) long × 3–6 μm in diameter, with subquadrate to rectangular loculi, without paraphyses; unilocular sporangia not known on blade surface but recorded on a *Stragularia*-like crustose thallus.

Epilithic, or more rarely epiphytic, upper to lower eulittoral, in open shallow pools and particularly channels in both sheltered and moderately exposed localities; tolerant of some sand cover and some variation in salinity.

Very common and generally distributed around Britain and Ireland.

Blades annual; found throughout the year but especially common during late winter and spring; thalli probably persist throughout remainder of year as perennial *Stragularia*-like basal crusts.

Isolates of *P. fascia* investigated in laboratory culture in the North Atlantic (Caram 1965; Dangeard 1962b, 1963b; Edwards 1969; Fletcher 1974b; Hsiao 1969; Kylin 1933; Reinke 1878a;

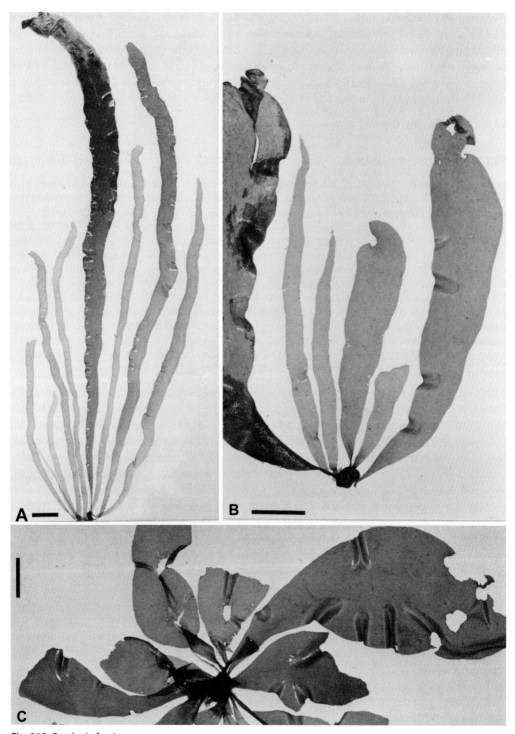

Fig. 219. *Petalonia fascia*
A–C. Habit of thalli showing variations in blade morphology. Bars = 10 mm.

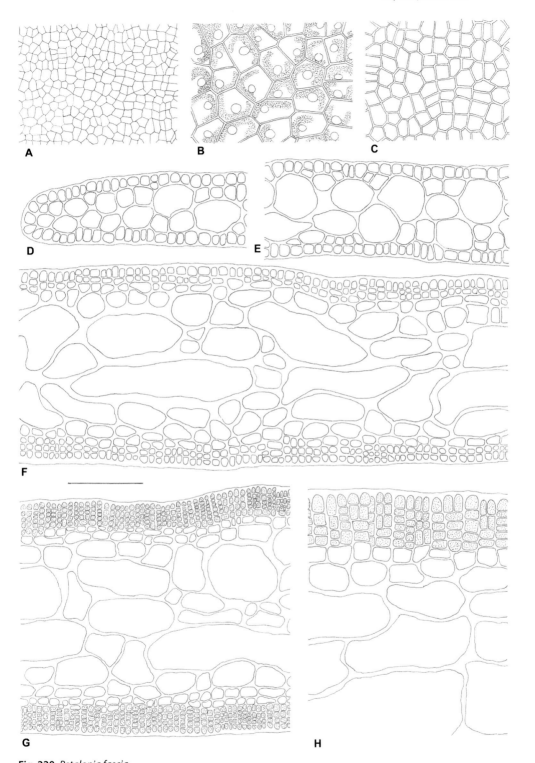

Fig. 220. *Petalonia fascia*

A, B. Surface view of vegetative thallus. Note cells with single plate-like plastid and pyrenoid. C. Surface view of fertile thallus showing plurilocular sporangia. D. TS of thallus margin. E, F. TS of vegetative thalli. Note outer small cortical cells and inner large medullary cells. G, H. TS of fertile thallus showing outer vertical rows of plurilocular sporangia. Bar = 20 µm (B, C, H), = 50 µm (others).

Rhodes & Connell 1973; Roeleveld *et al.* 1974; Sauvageau 1929 – as *Phyllitis debilis* Kützing), the Pacific coast of North America (Wynne 1969, 1972a,b) and Japan (Kogame 1997; Kunieda & Arasaki 1947; Nakamura & Tatewaki 1975; Yendo 1919) have largely shown a 'direct' type of life history. Plurispores released from the erect blades germinated directly without sexual fusion to form prostrate systems, from which arose further generations of erect blades. Only the latter two studies reported the occurrence of a sexual, biphasic cycle with a diploid prostrate sporophyte alternating with a haploid, bladed gametophyte releasing isogametes, this bearing out earlier reports of the copulation between plurispores from erect blades of *P. fascia* in Japan by Kunieda & Arasaki (1947).

Considerable variation was shown in the morphology of these prostrate systems, which were *Streblonema*-like, *Myrionema*-like and/or *Stragularia*-like. There have also been reports of *Stragularia*-like crusts collected in the field which, when investigated in laboratory culture, produced erect blades identifiable as *P. fascia* (Edelstein, *et al.* 1970; Fletcher 1974b, 1978; Kogame 1997; Rhodes & Connell 1973; Roeleveld *et al.* 1974; Wynne 1969). Three species of '*Stragularia*' have to date been implicated in the life history of *P. fascia*, viz. *S. californica* Setchell & N.L. Gardner (Wynne 1969), *S. clavata* (Edelstein *et al.* 1970; Fletcher 1974b, 1978; Roeleveld *et al.* 1974) and *S. bornetii* Kuckuck (Edelstein *et al.* 1970). This suggests that at least two entities are involved in our concept of *P. fascia* each with its own distinct '*Stragularia*' phase. For now, however, the three crustose taxa are considered here as variants of a single species for which the name *S. clavata* would have priority. *P. fascia*, therefore, represents a range of genotypes with slightly differing crustose bases. Support for such a genetic divergence was provided by Parente *et al.*'s (2011) finding that collections of *P. fascia* from the Canary Islands were genetically divergent from all other Scytosiphonaceae. *P. fascia* is polymorphic with several 'formae' and 'varieties' previously attributed to it. Work by Wynne (1969) further indicated that these morphological variants are probably genotypically determined. He found that only the narrow, astipitate blades (forma *fascia*) were capable of producing crusts in culture bearing unilocular sporangia. Blades of forma *caespitosa* produced only sterile crusts whereas in cultures of forma *debilis* no thickened crusts were seen at all. Also, the prostrate systems produced in cultures of *P. fascia* by Edwards (1969) were almost exclusively filamentous.

A significant feature of the 'direct' type life histories was the facultative relationship observed between the prostrate and erect bladed expressions. Plurispores derived from the plurilocular sporangia on the erect bladed expression and unispores derived from unilocular sporangia on the crustose expression could inherently undergo identical processes of germination and development and produce the same range of morphological expressions. Indeed, environmental conditions such as the combination of different temperatures and photoperiods (Edwards 1969; Fletcher 1974b, 1978; Nakamura & Tatewaki 1975; Roeleveld *et al.* 1974; Wynne 1969) and the nutrient level (Hsiao 1969) determined the ratio between the erect bladed and crustose thalli. In general, low temperatures and short-day photoperiods favoured the formation of erect blades while high/warm temperatures and long day photoperiods favoured prostrate thallus development. These studies generally lend support to the predominantly winter/spring seasonal occurrence of the erect blades in the North Atlantic and suggest a summer seasonal growth of the basal systems (i.e. crusts). Studies by Wynne (1969) and Nakamura & Tatewaki (1975) further suggest that a period of low temperatures and short days is then necessary for unilocular sporangia development on the crusts; this would agree with the winter fertility observed on crusts of *S. clavata* around Britain and Ireland. The importance of temperature and photoperiod in controlling unilocular sporangia formation rather than erect thallus formation also found support in the study by Kogame (1997), which suggested that this accounted for the seasonal occurrence of the erect thalli.

It is likely that the 'direct' life history revealed for *P. fascia* in many of the above culture studies is 'abbreviated' and that the basic life history is sexual and involves a heteromorphic alternation between an erect bladed (i.e. *Petalonia*-like) gametophyte and a crustose (i.e. *Stragularia*-like) sporophyte. Such a heteromorphic, haplodiplontic, biphasic life history might remain undetected because of its rarity (perhaps requiring specific environmental conditions) or geographical isolation.

Abbott & Hollenberg (1976), p. 200, fig. 163; Adams (1994), p. 77, pl. 19; Bartsch & Kuhlenkamp (2000), p. 167; Batters (1883), p. 113; Boo (2010), pp. 164, 165, fig. 3; Brodie *et al.* (2016), p. 1021; Bunker (2017), p. 289; Caram (1965), pp. 86–98, figs 1–8; Coppejans & Kling (1995), p. 230, pl. 84(A–C); Cormaci *et al.* (2012), pp. 282–4, pl. 81(3–9); Dangeard (1962b), pp. 3290–2, pl. 1, (1963b), pp. 48–57, 84, pl. VII–VIII, (1969), pp. 60–70, fig. 1(B), pl. XI–XII; Dizerbo & Herpe (2007), p. 116, pl. 39(1); Edelstein *et al.* (1970); Edwards (1969), pp. 85–7, figs 38–42; Feldmann (1949), pp. 103–15, fig. 1(C–G); Fletcher (1974a), pp. 92–108, figs 17–19, pls 26–30, (1978), pp. 371–98, (1980), p. 42, pl. 15(1–4), (1987), pp. 230–4, figs 61(A), 62, pl. 9(a–c); Fletcher in Hardy & Guiry (2006), p. 282; Flores-Moya & Fernandez (1998), pp. 125–30; Funk (1955), p. 40; Gallardo García *et al.* (2016), p. 37; Gayral (1958), p. 212, fig. 29, pl. XXVI, (1966), p. 249, pl. XXXV; Guiry (2012), p. 155; Hamel (1931–39), pp. 197, 198; Hsiao (1969), pp. 1611–16, figs 1–4, (1970), pp. 1359–61, figs 1–2; Katoh & Ehara (1990), pp. 439–47; Kogame (1997), pp. 389–94; Kogame *et al.* (1999), pp. 496–502; Kornmann & Sahling (1977), pp. 134, 137, figs 71(A–C), 72(A–F); Kylin (1933), pp. 44–7, fig. 17; Lee *et al.* (2003), pp. 333–9; Lindauer *et al.* (1961), pp. 255, 256, fig. 64; Mathieson & Dawes (2017), pp. 197, 198, pl. LVII(2–4); Matsumato *et al.* (2014), pp. 127–36; Morton (1994), p. 42; Nakamura (1965), pp. 109, 110, fig. 5; Nakamura & Tatewaki (1975), pp. 72–6, figs 11–14; Newton (1931), p. 176, fig. 110; Parente *et al.* (2011), p. 43; Pedersen (2011), p. 76, fig. 80; Reinke (1878a), pp. 262–7, pl. 11(1–2); Rhodes & Connell (1973), pp. 211–15, figs 1–10; Ribera *et al.* (1992), p. 118; Roeleveld *et al.* (1974), pp. 410–26, figs 1–7; Rosenvinge & Lund (1947), pp. 31–4, fig. 10; Rueness (1977), p. 168, fig. 100, pl. XXII(2); Saunders & McDevit (2013), pp. 1–23; Sauvageau (1929), pp. 316–30, figs 8–10 – as *Phyllitis debilis* Kützing; Stegenga & Mol (1983), p. 105, pl. 38, figs 1–5; Stegenga *et al.* (1997a), p. 21, (1997b), pp. 184, 185, pl. 52(3–4); Taylor (1957), pp. 167, 168, pls 14(5), 15(3); Williams & Herbert (1989), pp. 515–22; Womersley (1987), p. 292, figs 106(A), 108(A, B); Wynne (1969), pp. 17–31, figs 6–8, pls 6–13, (1972a) pp. 133–6, (1972b), pp. 137–41, figs 24–7; Yendo (1919), pp. 171–84, pl. II(1–14); Yoneshigue-Valentin & Pupo (1994), pp. 489–96.

Planosiphon McDevit & G.W. Saunders

PLANOSIPHON McDevit & G.W. Saunders (2017), p. 657.

Type species: *P. complanatus* (Rosenvinge) McDevit & G.W. Saunders (2017), p. 660, fig. 5(a–c).

Thalli forming erect, simple, linear, slightly dorsiventrally flattened to oval blades arising from a fibrous mat of rhizoidal filaments; erect blades unconstricted, hollow or partially hollow; surface cells rectangular to polygonal, irregularly arranged or arranged in longitudinal, less obviously transverse rows, each enclosing a single plate-like plastid with a prominent pyrenoid; in section, blades with an inner medulla of large, longitudinally elongated, transversely rounded, sometimes irregularly collapsed, colourless cells enclosed by an outer cortex of 1–3 layers of small pigmented cells.

Plurilocular sporangia in extensive sori in terminal blade region, in closely packed, uniseriate, vertical columns arising from the cortical cells, without paraphyses; unilocular sporangia not known on blade surface but recorded on prostrate *Compsonema*-like thalli.

Based on a recent taxonomic and molecular phylogenetic study, McDevit & Saunders (2017) proposed a new genus, *Planosiphon* to accommodate species of *Scytosiphon* and *Petalonia* with flattened thalli that are hollow to partially hollow, lack paraphyses and have a life history in culture which includes *Compsonema*-like prostrate sporophytic thalli that bear only unilocular sporangia, rather than *Stragularia*-type sporophytic thalli (Fletcher 1987; Brophy & Murray 1989; Kogame 1998; Kogame *et al.* 1999; Parente *et al.* 2003; Wynne 1969). They noted that such characteristics have been described for *Petalonia zosterifolius* and they duly transferred this species into *Planosiphon*. Shortly afterwards, Santiañez & Kogame (2017) noted that similar morphological and life history characteristics had also been described by Fletcher (1987) for *Petalonia filiformis* and they also proposed transferring this species to the genus *Planosiphon*.

Two species in Britain and Ireland.

KEY TO SPECIES

Blades 100–200(–400) μm broad, not markedly dorsiventrally flattened, often oval
in cross section and frequently spirally twisted . . . .*P. filiformis*
Blades more than 0.5 mm broad, dorsiventrally flattened and not spirally twisted
. . . . *P. zosterifolius*

Planosiphon filiformis (Batters) Santiañez & Kogame (2017), p. 2.   Figs 221(A,B), 222

Phyllitis filiformis Batters (1888), p. 451; *Petalonia filiformis* (Batters) Kuntze (1891–98), p. 419.

Thalli forming erect, light to yellow-brown, gregarious, slightly dorsiventrally flattened to oval blades arising from a fibrous mat of rhizoidal filaments; erect blades simple, solid or with small central cavities, thin and flaccid, narrow, filiform and ribbon-like, entire, without proliferations or obviously ruffled margins, frequently spirally twisted, to 4–6(–8) cm long × 0.1–0.35 (0.4) mm wide, tapering slowly to base, usually terminally eroded; in surface view, cells small, rectangular to polygonal, 4–16 × 4–10 μm arranged in straight or spiral longitudinal, less obviously transverse rows, each enclosing a single, plate-like, parietal plastid with conspicuous pyrenoid; in section, blades to 75 μm (vegetative) and 130 μm (reproductive) thick, comprising an inner medulla of 2–4(–6) large, thick-walled, longitudinally elongated, transversely rounded sometimes irregularly collapsed, colourless cells enclosed by an outer cortex of 1–3 layers of small, pigmented cells; hairs not observed on field material.

Plurilocular sporangia in slightly darker coloured sori extensive in terminal blade region, in closely packed vertical columns arising from cortical cells, uniseriate, to 30 μm (–7 loculi) long × 3–6 μm in diameter, with subquadrate to rectangular loculi, without paraphyses; unilocular sporangia not known on blade surface but recorded on *Composonema*-like thalli.

Epilithic, exposed on the sides of vertical cliff faces, large boulders etc., in the littoral fringe.

Apparently confined to south-eastern and eastern shores of England and Scotland extending west to Sussex and north to Fife (Fife, Northumberland, Yorkshire, Essex, Kent, Sussex); in Ireland a single record for Co. Down.

Blades annual, January to May, with plurilocular sporangia.

P. filiformis is fairly common in the upper eulittoral region, where it forms isolated, sometimes dense stands on the vertical faces of cliffs, harbour walls, breakwaters, large boulders and similar substrates in moderately exposed areas. It occurs more commonly in north-facing positions, and can often be found at the entrance to caves, crevices etc. Collections have been made from a wide variety of substrata including chalk (Kent), sandstone (Northumberland, Fife) and both concrete and brick-built artificial sea walls (Kent). The gregarious, light to yellow-brown thalli contribute to a generally low-lying sward of algae which can include *Ulothrix flacca*, *Urospora penicilliformis*, *Blidingia minima*, *Ulva* spp., *Pylaiella littoralis*, *Aglaothamnion hookeri*, *Gelidium pusillum* and *Polysiphonia stricta*, often with underlying crusts of *Petroderma maculiforme* and *Pseudoralfsia verrucosa* present. The only reports of this species occurring outside Britain and Ireland have been by Munda (1979) for Iceland, Kuckuck (1897) and Bartsch & Kuhlenkamp (2000) for Helgoland and Mathieson & Dawes (2017) for the north-west Atlantic.

Culture studies by Fletcher (1981a) indicated that a 'direct' life history was operating; the plurispores from the erect blades behaved asexually and germinated into branched, filamentous or knot-filamentous bases, from which sprouted a further generation of blades. These prostrate systems further developed unilocular sporangia, which were usually accompanied by erect, multicellular, paraphysis-like filaments; these fertile growths were considered by Fletcher to show a marked similarity to the brown alga *Composonema saxicola* (see p. 486). This facultative relationship between the erect bladed (i.e. *Planosiphon*) form and the microthallial (i.e. *Composonema*) form was also revealed by Fletcher (unpublished)

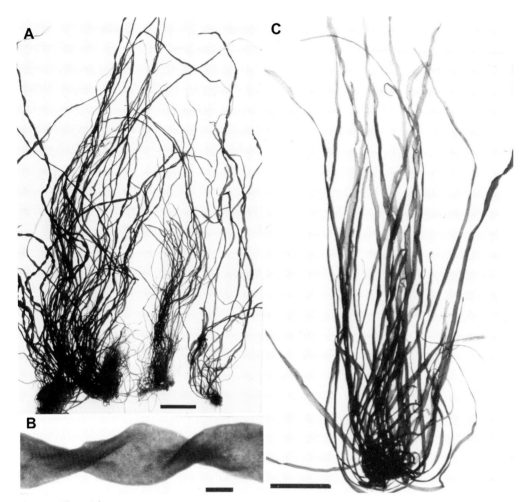

Fig. 221. *Planosiphon* spp.

A, B. *Planosiphon filiformis*. A. Habit of thalli. B. Portion of erect thallus showing spiral twisting. C. *Planosiphon zosterifolius*, habit of thalli. Bar in A = 5 mm, bar in B = 180 μm, bar in C = 10 mm.

to be influenced by environmental conditions. For example, blade production and development was more enhanced in low temperature/short day conditions while microthallus development was enhanced in warm temperature/long day conditions. These results agree with the observed winter/spring seasonal occurrence of the erect blades and the survival of the thalli as *Compsonema*-like microthalli during the more unfavourable summer and autumn periods (Fletcher 1974a). The life history of *P. filiformis* can, therefore, be described as asexual, heteromorphic and monophasic with both morphological forms probably having the same ploidy level. This may be, however, an abbreviated life history. More culture studies of isolates are required to determine if a sexual cycle is also operating, albeit of rare occurrence.

Bartsch & Kuhlenkamp (2000), p. 167; Batters (1888), pp. 451, 452, pl. 18(1–6); Brodie *et al.* (2016), p. 1021; Brophy & Murray (1989), pp. 6–15; Fletcher (1974a), pp. 109–27, figs 20–3, (1981a), pp. 103, 104, (1987), pp. 234–6, figs 61(b), 63, pl. 10(a, b); Fletcher in Hardy & Guiry (2003), p. 283; Guiry (2016), p. 155; Kogame (1998), pp. 39–46; Kogame & Kawai (1993), pp. 29–37; Kogame *et al.* (1999), pp. 496–502; Mathieson & Dawes (2017), p. 198, pl. CIII(6–8); Morton (1994), p. 42; Newton (1931), p. 176; Santiañez & Kogame (2017), pp. 1–3; Saunders & McDevit (2013), pp. 1–23.

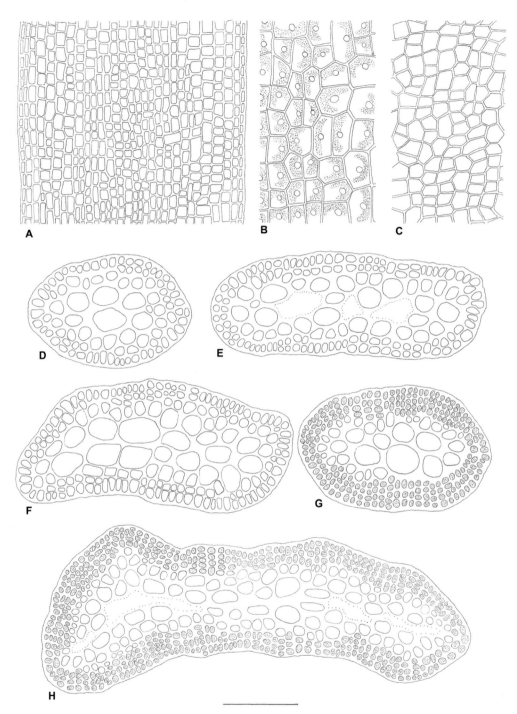

Fig. 222. *Planosiphon filiformis*
A. Portion of narrow blade. B. Surface view of vegetative thallus showing single plate-like plastid with pyrenoid.
C. Surface view of fertile thallus showing plurilocular sporangia. D–F. TS of vegetative thalli showing outer small cortical and inner large medullary cells. Note central cavities. G, H. TS of fertile thalli showing outer vertical rows of plurilocular sporangia. Bar= 20 μm (B, C), = 50 μm (others).

Planosiphon zosterifolius (Reinke) McDevit & G.W. Saunders (2017), p. 659.

Figs 221(C), 223

Phyllitis zosterifolia Reinke (1889b), p. 61; *Petalonia zosterifolia* (Reinke) O. Kuntze (1891–98), p. 419; *Ilea zosterifolia* (Reinke) Norstedt (1912), p. 239; *Petalonia fascia* var. *zosterifolia* (Reinke) W.R. Taylor (1937), p. 230; *Ilea zosterifolia* f. *typica* A.D. Zinova (1951), p. 403, nom. inval.; *Ilea zosterifolia* f. *filiformis* A.D. Zinova (1951), p. 404, figs 3, 4; *Ilea zosterifolia* f. *lata* A.D. Zinova (1951), p. 403, figs 1, 2.

Thalli forming erect, medium to dark brown, single or more commonly gregarious, dorsiventrally flattened blades arising from an intricate mat of basally produced, branched, rhizoidal filaments; erect blades simple, solid or with central cavities, fairly thin and flaccid, more or less linear, entire, without proliferations or ruffled margins, to 15(–25) cm long × 0.5–2(–4) mm wide, tapering slowly to base, usually terminally eroded; in surface view, cells small, rectangular to polygonal, 6–18 × 4–11 µm, arranged in longitudinal, less obviously transverse rows, each enclosing a single, plate-like parietal plastid with conspicuous pyrenoid; in section, blades to 140 µm (vegetative) and 210 µm (reproductive) thick comprising an inner medulla of 4–6(–8) large, thin-walled, longitudinally elongate, transversely rounded, often collapsed, colourless cells enclosed by an outer cortex of 1–3 small, pigmented cells; hairs common, single, arising from surface cells with basal meristem and sheath.

Plurilocular sporangia in dark brown sori extensive in terminal blade region, in closely packed vertical columns arising from cortical cells, uniseriate, to 40 µm (–10 loculi) long × 3–6 µm wide, with subquadrate to rectangular loculi, without paraphyses; unilocular sporangia not recorded for Britain and Ireland, reported by Dangeard (1962a, 1963b) in France on basal microthalli in culture.

Epilithic, upper eulittoral to littoral fringe, on the sides of vertical cliff faces, platforms, large boulders in moderately to very exposed areas. Also frequent on the upper wave-washed regions of exposed floating structures such as buoys.

Recorded for scattered localities all around Britain and Ireland but more common on the south-west and west coasts of Britain (Northumberland, Hampshire, Dorset, Devon, Cornwall, Pembroke, Anglesey, Isle of Man).

Blades annual, December to March with plurilocular sporangia.

Several authors have studied the life history of *Planosiphon zosterifolius* (as *Petalonia zosterifolia*) in culture, including Boone & Kapraun (1989), Dangeard (1962a, 1963a), Fletcher (1974b), Kawai (1993), Kogame & Kawai (1993), Nakamura & Tatewaki (1975) and Wynne (1969). Fletcher revealed that a 'direct' type life history was operating in material from Britain. The plurispores released from the blades behaved asexually and germinated directly into branched, filamentous microthalli from which arose a further generation of erect macrothalli. *Stragularia*-like crustose thalli similar to those present in the life history of *P. fascia* were not observed. A similar life history was reported for the French North Atlantic coast by Dangeard (1962a, 1963b) although in his cultures the microthalli became fertile with both plurilocular and unilocular sporangia. Wynne (1969) questioned the identification of Dangeard's cultures with plurilocular sporangia as the accompanying filaments revealed cells with several plastids inside, while Kogame & Kawai doubted Dangeard's report of ralfsioid prostrate thalli in the life history, leading all authors to suspect the presence of contaminants. Dangeard's illustrations of branched, filamentous, prostrate thalli bearing unilocular sporangia (Dangeard 1963, Plates XIV, XV) do, however, seem representative of this phase in the species life history. A similar life history with unilocular sporangia bearing filamentous prostrate thalli was also described by Boone & Kapraun (1989) for North Carolina material. The life history of these three sets of isolates can, therefore, be described as heteromorphic and monophasic. In contrast to these studies, Nakamura & Tatewaki's (1975) study of Japanese isolates reported a sexual heteromorphic and biphasic life history with an erect bladed (i.e. *Planosiphon*) gametophyte bearing plurilocular sporangia, alternating with a *Stragularia* (*Ralfsia*)-like sporophyte bearing

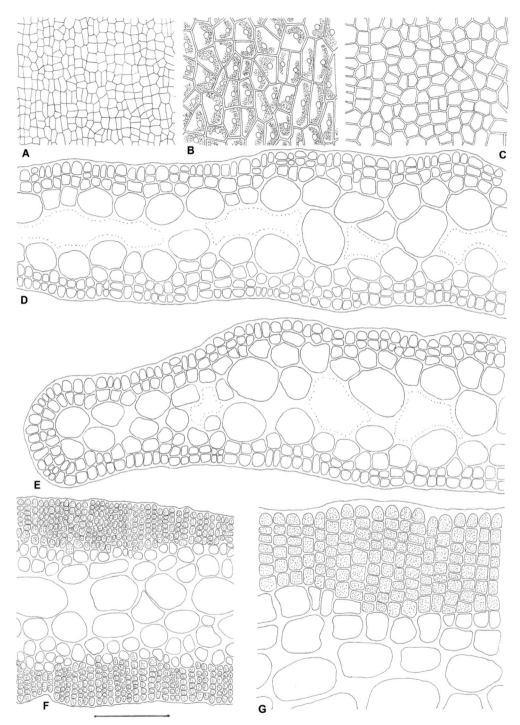

Fig. 223. *Planosiphon zosterifolius*

A, B. Surface view of vegetative thallus. C. Surface view of fertile thallus showing plurilocular sporangia. D, E. TS of vegetative thallus showing outer small cortical cells and inner large medullary cells. Note central cavities. F, G. TS of fertile thallus showing outer vertical rows of plurilocular sporangia. Bar = 20 μm (B–C, G), = 50 μm (others).

unilocular sporangia. But Kogame & Kawai (1993) cast doubt on the identification of Nakamura & Tatewaki's material, which they determined as a *Scytosiphon* sp. later identified as *S. tenellus* (Kogame 1998). They then reported on the findings of their life history study of *Planosiphon zosterifolius* from Japan, revealing it to be similar to that reported by Fletcher (1974) and Boone & Kapraun (1989) i.e. direct, asexual, monophasic and heteromorphic, with the filamentous prostrate thalli identified as *Compsonema saxicola*.

Bárbara & Cremades (1990), pp. 492–4; Bartsch & Kuhlenkamp (2000), p. 167; Boo (2010a), pp. 165–7, figs 4, 5; Brodie *et al.* (2016), p. 1021; Cho *et al.* (2002), pp. 135–44; Dangeard (1962a), pp. 1895, 1896, pl. 1, (1963b), pp. 12–47, pls 14–16, (1969), pp. 60–70, fig. 1(A); Dizerbo & Herpe (2007), p. 117; Fletcher (1980), pp. 42, 43, pl. 15(5), (1987), pp. 236–9, figs 64, 65, pl. 10(c); Fletcher in Hardy & Guiry (2003), p. 283; Gallardo García *et al.* (2016), p. 37; Guiry (2012), p. 155; Hamel (1931–39), pp. 198, 199; Kogame & Kawai (1993), pp. 29–37; Kornmann & Sahling (1977), p. 137, fig. 73; Mathieson & Dawes (2017), pp. 198, 199, pl. LVII, figs 5–7; McDevit & Saunders (2017), pp. 653–71; Morton (1994), pp. 42, 43; Nakamura & Tatewaki (1975), pp. 65–71, figs 6–10, pl. 2; Newton (1931), p. 176; Nordstedt (1912), pp. 237–9; Pedersen (2011), p. 77, fig. 81; Ribera *et al.* (1992), p. 118; Rosenvinge & Lund (1947), pp. 34–7, fig. 11; Rueness (1977), pp. 168, 169; Stegenga & Mol (1983), p. 105, pl. 38(6–7).

Scytosiphon C. Agardh

SCYTOSIPHON C. Agardh (1820), p. 160, nom. cons.

Type species: *S. filum* var. *lomentaria* (Lyngbye) C.A. Agardh (= *Chorda lomentaria* Lyngbye; *S. lomentaria* (Lyngbye) Link (1833), p. 233 typ. cons.); see ICBN (1961), p. 213.

Thalli consisting of erect, yellow to dark brown, tubular thalli arising from a slightly expanded encrusting base, small discoid holdfast or fibrous mat of rhizoidal filaments; thalli single or gregarious, simple, hollow, tubular, cylindrical, with or without regular constrictions on mature thalli, tapering slowly to base, more sharply to apex; in surface view, cells small, rectangular to polygonal, irregularly arranged, each with a single, plate-like parietal plastid with 1–2 pyrenoids; in section, thalli parenchymatous, hollow, comprising an inner medulla of 2–5 layers of large, thick-walled, longitudinally elongate, transversely rounded or irregularly shaped colourless cells, enclosed by an outer cortex of 1–3 layers of small, pigmented cells; hairs single or grouped arising from surface cells or depressions.

Plurilocular sporangia in continuous sori, extensive in terminal thallus region, in closely packed vertical columns, arising by division of cortical cells, uniseriate to biseriate, rarely multiseriate, with subquadrate to rectangular loculi, with or without unicellular ascocyst-like cells; unilocular sporangia not known on thallus surface but reported on crustose *Microspongium*, *Stragularia* or *Compsonema*-like thalli.

The genus *Scytosiphon* is characterized by an erect, simple, hollow, sometimes constricted thallus with small surface cells, each with a single plate-like plastid with pyrenoids, a parenchymatous mode of construction with an inner medulla of large colourless cells and an outer cortex of small, pigmented cells, and plurilocular sporangia only which occur in densely packed, vertical columns extensive in the terminal regions. In Britain and Ireland this genus is similar in habit to *Asperococcus* (in particular *A. fistulosus*), and small thalli of *Chorda filum*. However, it can easily be distinguished from these genera using a combination of the above described characteristics.

Scytosiphon has been the subject of life history studies throughout the world (Camus *et al.* 2005; Clayton 1976a,b, 1980a,b; Dangeard 1963b; Fletcher 1974b; Frye & Phifer 1930; Kogame 1996; Kogame *et al.* 2005; Kristiansen & Pedersen 1979; Kristiansen *et al.* 1991; Nakamura & Tatewaki 1975; Parente *et al.* 2003a; Pedersen 1980a; Tatewaki 1966; Wynne 1969). Species investigated include *S. canaliculatus*, *S. dotyi*, *S. gracilis*, *S. lomentaria* and *S. tenellus* and a wide variety of both asexual and sexual life history patterns have been described. These include asexual, 'direct' monophasic, monomorphic type life histories, in which the plurispores from the erect thalli germinate into a small prostrate thallus from which new erect thalli developed similar to the parental thallus (for example, see Camus *et al.* 2005; Dangeard 1963b; Fletcher 1974b; Kristiansen *et al.* 1991; Pedersen 1980; Wynne 1969), as well as asexual, monophasic, heteromorphic type life histories in which there is much greater development of the prostrate system which could be considered a separate phase in the life history, not only capable of giving rise to erect '*Scytosiphon*' thalli, but can also develop unilocular sporangia (for example, see Camus *et al.* 2005; Dangeard 1963b; Dring & Lüning 1975a,b; Fletcher 1974b; Kogame 1996; Kristiansen & Pedersen 1979; Kristiansen *et al.* 1991, Parente *et al.* 2003a, 2011; Pedersen 1980; Rhodes & Connell 1973; Wynne 1969). In this situation, both the prostrate and the erect systems possess the same ploidy level.

Considerable variation has been shown in the morphology of these prostrate phases and they have been described as *Streblonema*-like, *Stragularia*-like, *Compsonema*-like, *Microspongium*-like, *Ralfsia*-like or, in the case of *S. canaliculata*, identified as the crustose genus *Hapterophycus*. For many *Scytosiphon* isolates investigated, culture conditions, in particular temperature and photoperiod, appeared to determine which morphological expression was produced. More often, low temperatures and short days favoured erect thallus development while warm temperatures and long days favoured prostrate development and this was likely to determine the seasonality of the two phases (Correa *et al.* 1986; Dring & Lüning 1975b, 1983; Rhodes & Connell 1973; Tatewaki 1966; Tom Dieck 1987; Wynne 1969). In some reports, however, these culture conditions also played a role in unilocular sporangia development on the prostrate phase and this was considered to determine the seasonality of the gametophyte phase.

In apparent contrast to this monophasic type life history, there have been reports of a sexual life history in some populations of *S. lomentaria*, in *S. canaliculatus*, *S. gracilis* and *S. tenellus* from Japan (Kogame 1996, 1998; Nakamura 1965; Nakamura & Nakahara 1977; Nakamura & Tatewaki 1975; Tatewaki 1966) and in *S. tenellus* from Northern Chile (Camus *et al.* 2005). The erect thalli represent the gametophyte phase, producing isogametes or anisogametes, while the prostrate thalli represent the sporophyte phase bearing unilocular sporangia in which meiosis occurs. An asexual life history also occurred in some of these studies by the development of parthenogametes liberated from plurilocular sporangia on the erect thalloid phase. This parthenogenetic life history is in agreement with the monophasic life history reported for isolates in other geographical regions. The basic life history of *Scytosiphon* species is, therefore, considered to be heteromorphic and biphasic but in some populations, or under certain environmental conditions or geographical locations, this is not fully expressed but abbreviated.

Scytosiphon is a widely distributed genus in temperate and cold seas represented by eight species (Guiry & Guiry 2023). However, the delimitation of taxa within the genus has long been problematical, not only due to the range of infraspecific taxa described, which probably represent environmental induced variants, but also to the considerable morphological variation within the genus and the relatively few useful taxonomic characteristics available. While some studies, notable those of Clayton (1976, 1978) and Kogame (1998), have provided more taxonomic characters such as thallus size, thallus form, structural differences in the plurilocular zoidangial sori, the presence or absence of ascocysts and the morphology of the sporophytic thalli, much more recent progress in species identification has been obtained by the increasing number of molecular analyses carried out (see Camus *et al.* 2005; Cho *et al.* 2001, 2002, 2006, 2007; Kogame *et al.* 1999, 2005, 2015a,b; McDevit & Saunders 2017; Santiañez & Kogame 2017). These have revealed considerable genetic variation in *Scytosiphon* and the occurrence of many cryptic species, particularly in studies of material identified as *S. lomentaria* (Cho *et al.* 2007; Kogame *et al.* 2005, 2015b; McDevit & Saunders 2017; Parente *et al.* 2011) and clearly more taxonomic work is required on this genus.

Two species are recorded for Britain and Ireland: *S. dotyi* Wynne and *S. lomentaria* (Lyngbye) Link. *S. dotyi* Wynne, a Pacific species, represents a relatively new addition to the marine flora of Britain. *S. dotyi* is distinguished from *S. lomentaria* in the following aspects (see Wynne 1969, p. 36): the thalli are relatively short and narrow, lacking constrictions; hairs are tufted, emerging from depressions; paraphyses are absent; it occurs in the littoral fringe, exposed on vertical rock faces; the life history is 'direct' without a crustose expression. However, the reliability of these characters is questionable and the two species may well be conspecific. Interestingly, McDevit & Saunders (2017) pointed out that in Lyngbye's (1819) description of *Scytosiphon lomentaria*, no mention was made of the presence of paraphyses and they suggest this attribute may have been later wrongly assigned to this species. In view of this, they suggest that as paraphyses are also absent in *S. dotyi*, then the latter might be a juvenile form or smaller individual of *S. lomentaria*. Clearly more studies are required, particularly using molecular tools.

One species in Britain and Ireland, one species in Britain only.

KEY TO SPECIES

Thalli not exceeding 14 cm in length, fairly rigid in texture; mature thalli unconstricted at intervals, with grouped hairs and without ascocysts present; recorded in littoral fringe only, exposed on rock; restricted to the east coast of England (Yorkshire and Kent) . . . . *S. dotyi*
Thalli to 40 cm long, flaccid in texture; mature thalli constricted at intervals, with single hairs and ascocysts present; widely distributed in the littoral, usually in pools; recorded throughout Britain and Ireland . . . . *S. lomentaria*

Scytosiphon dotyi Wynne (1969), p. 34.   Figs 224, 225

Scytosiphon lomentaria f. *complanatus* minor Setchell & N.L. Gardner (1925), p. 534; *Scytosiphon lomentaria* f. *tortilis* Yamada (1935), p. 12; *Scytosiphon lomentaria* f. *cylindricus nanus* Tokida (1954), p. 105.

Thalli forming erect, gregarious, mid to dark brown, tubular thalli arising from a small discoid holdfast; erect thalli simple, hollow, cylindrical or more usually slightly dorsiventrally flattened, flaccid to slightly rigid in texture, without constrictions although sometimes twisted, to 14 cm long × 0.25–1 mm wide, attenuate above and below, sometimes with curled terminal region; in surface view, cells small, irregularly arranged, 4–13 × 4–8 μm each with a single plate-like plastid with pyrenoid; in section, thalli parenchymatous, with a large central cavity, enclosed by an inner medulla of 1–5 large, thick-walled, longitudinally elongate, transversely rounded, colourless cells and an outer cortex of 1–3 small, pigmented cells; hairs mainly grouped, arising from cortical cells sunken in fertile regions.

   Plurilocular sporangia in dark-coloured continuous sori, extensive in terminal thallus region, in closely packed vertical columns, arising by division of cortical cells, uniseriate, sometimes terminally biseriate, to 35 μm (–7 loculi) long, with subquadrate to rectangular loculi, without accompanying ascocyst-like cells; unilocular sporangia unknown.

   Epilithic, littoral fringe, exposed on the sides of vertical cliff faces, large boulders etc. particularly with a north-facing aspect, in moderately exposed areas.

   Only reported for Britain on the east coast of England (Yorkshire, Kent) and in Ireland for Co. Down; probably more widely distributed.

   Thalli annual; February to May.

   Culture studies by Wynne (1969) revealed *S. dotyi* to have a 'direct', asexual life history. Under a range of culture conditions, plurispores from the erect thalli germinated directly to produce small polystromatic crusts of only limited extent, from the centre of which developed a further generation of erect thalli. A distinct heteromorphic development of the spores into either unilocular sporangia bearing crusts and plurilocular sporangia bearing blades depending on the culture conditions, as revealed for *S. lomentaria*, was not observed.

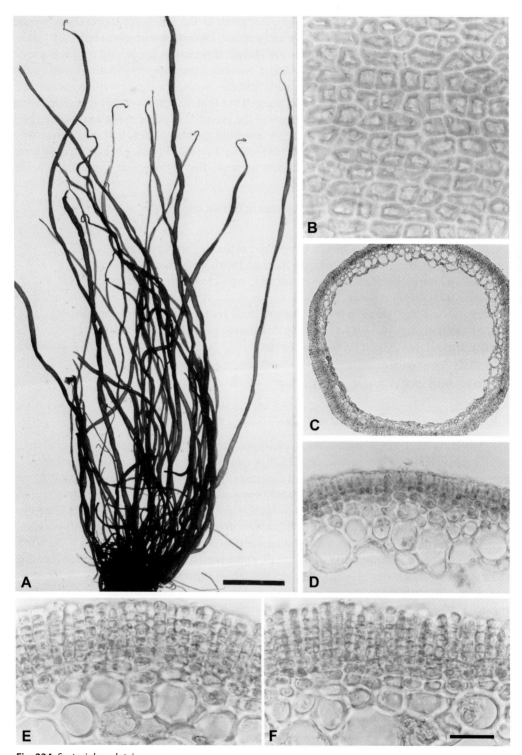

Fig. 224. *Scytosiphon dotyi*
A. Habit of thalli. B. Surface view of vegetative thallus showing cells with a single plate-like plastid. C. TS of a mature thallus showing large, central cavity. D. TS through edge of young thallus showing outer small cortical cells. E, F. TS through edge of mature thallus showing terminal rows of plurilocular sporangia. Bar in A = 10 mm, bar in F = 200 μm (C), = 90 μm (D), = 45 μm (B, E, F).

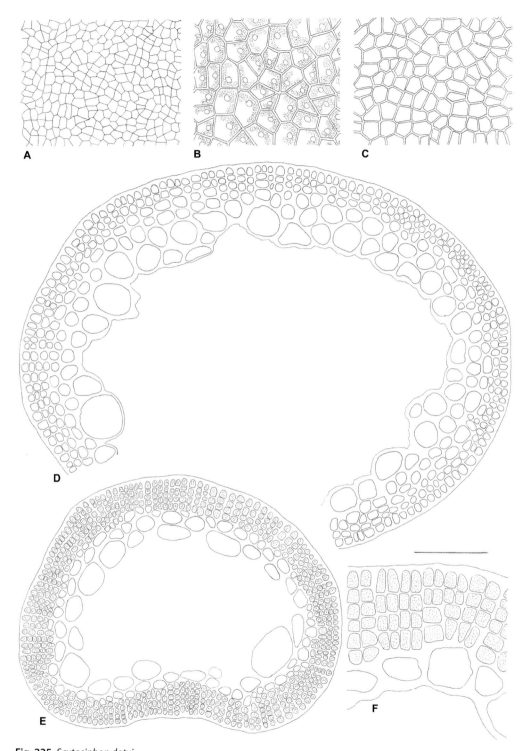

Fig. 225. *Scytosiphon dotyi*

A, B. Surface view of vegetative thallus. Note single, plate-like plastid with pyrenoid. C. Surface view of fertile thallus showing plurilocular sporangia. D. TS of young thallus showing outer small cortical cells, inner large medullary cells and large, central cavity. E. TS of small fertile thallus showing terminal rows of plurilocular sporangia. F. Plurilocular sporangia. Bar = 20 μm (B–C, F), = 50 μm (others).

Abbott & Hollenberg (1976), p. 198, fig. 161; Bárbara *et al.* (2012), p. 25, fig. 5(F); Brodie *et al.* (2016), p. 1021; Cormaci *et al.* (2012), p. 287, pl. 82(4–6); Fletcher (1987), pp. 245–8, figs 67(A), 68, pl. 11(c); Fletcher in Hardy & Guiry (2003), p. 293; Gallardo García *et al.* (2016), p. 37; Guiry (2012), p. 156; Mathieson & Dawes (2017), p. 201, pl. LVII(10, 11); McDevit & Saunders (2017), pp. 653–71; Perez-Cirera *et al.* (1991), pp. 135–8, 2 figs; Ribera *et al.* (1992), p. 118; Setchell & N.L. Gardner (1925), pl. 44, fig. 74; Taskin (2012), pp. 33–5; Verlaque *et al.* (2015), p. 99; Wynne (1969), pp. 34–9, fig. 9, pls 18–19.

Scytosiphon lomentaria (Lyngbye) Link (1833), p. 233.   Figs 226, 227

Ulva simplicissima Clemente (1807), p. 320, nom. rejic.; *Chorda lomentaria* Lyngbye (1819), p. 74; *Scytosiphon filum* var. *lomentarius* (Lyngbye) C. Agardh (1820), p. 162; *Fucus lomentarius* (Lyngbye) Sommerfelt (1826), p. 184; *Asperococcus castaneus* W.J. Hooker (1833), p. 277; *Chlorosiphon shuttleworthianus* Kützing (1843), p. 301; *Chorda fistulosa* Zanardini (1843), p. 37; *Chorda autumnalis* Areschoug (1862), no. 95; *Chorda lomentaria* var. *autumnalis* (Areschoug) Areschoug (1862); *Scytosiphon lomentaria* var. *zostericola* Thuret (1863), p. 67, nom. illeg.; *Scytosiphon lomentaria* f. *castaneus* (Carmichael) Kleen (1874), p. 39; *Chorda lomentaria* f. *autumnalis* (Areschoug) Areschoug (1875), p. 16; *Chordaria attenuata* Foslie (1887), p. 176; *Scytosiphon pygmaeus* Reinke (1888), p. 18; *Scytosiphon lomentaria* f. *fistulosus* Reinke (1907), p. 32, nom. illeg.; *Scytosiphon lomentaria* f. *typicus* Setchell & N.L. Gardner (1925), p. 533, nom. inval.; *Scytosiphon simplicissimus* (Clemente) Cremades (1990), p. 492.

N.B. Although *Ulva simplicissima* predates the basionym *Chorda lomentaria* and Cremades & Perez-Cirera (1990) renamed *S. lomentaria* as *S. simplicissimus*, this was rejected in view of the widespread use of the former (see Mathieson & Dawes 2017 for discussion).

Thalli forming erect, single or more commonly gregarious, yellow to dark brown, tubular thalli arising from a small encrusting base and attached by a slightly swollen discoid holdfast; erect thalli simple, fairly flaccid, hollow, cylindrical, either collapsed or inflated, usually characteristically constricted at intervals when mature, to 300(–400) mm long × 4–6(–8) mm wide, tapering slowly to base usually more sharply delimited at apex; in surface view, cells small, usually irregularly arranged, more rarely in longitudinal rows, 5–13 × 3–9 µm, each with a single plate-like parietal plastid with 1–2 pyrenoids; in section, thalli parenchymatous, with a large central cavity enclosed by an inner medulla of 2–5 layers of large, thick-walled, longitudinally elongate, transversely rounded, colourless cells and an outer cortex of 1–3 layers of small, pigmented cells; hairs single, more rarely in tufts, arising from outer cortical cells, sunken in fertile thalli, with basal meristem, lacking obvious sheath, approx. 6–8 µm in diameter.

Plurilocular sporangia in dark-coloured continuous sori, extensive in terminal thallus region, in closely packed vertical columns arising by divisions of cortical cells, uniseriate to multiseriate, to 90 µm (–20 loculi) long, 4–8 µm wide with subquadrate to rectangular loculi, often with accompanying scattered, pyriform, ascocyst-like cells, 18–33 × 11–16 µm; unilocular sporangia unknown on erect thallus but reported on crustose *Microspongium gelatinosum* phase.

Epilithic, in pools/channels and emergent, littoral fringe to lower eulittoral and shallow sublittoral; tolerant of some sand cover.

Common and generally distributed around Britain and Ireland.

Annual; reported throughout the year but most common in spring/early summer; probably persists throughout the remainder of year as perennial basal crust.

S. lomentaria has been investigated in laboratory culture throughout the world including the North Atlantic (Dammann 1930; Dangeard 1962b, 1963b; Dring & Lüning 1975a,b; Fletcher 1974b; Frye & Phifer 1930; Kristiansen & Pedersen 1979; Kuckuck 1898; Kylin 1933; Lüning 1980a; Lüning & Dring 1973; Reinke 1878a; Rhodes & Connell 1973; Roberts & Ring 1972; Sauvageau 1929), Mediterranean (Berthold 1881; Reinke 1878a), California (Wynne 1969) and Japan (Abe 1935; Nakamura 1965; Nakamura & Tatewaki 1975; Tatewaki 1966). In the great majority of these studies, a 'direct' asexual life history was revealed with the liberated plurispores from the erect thalli germinating into filamentous, discoid or crustose prostrate systems from which were produced further generations of thalli. For many isolates, environmental

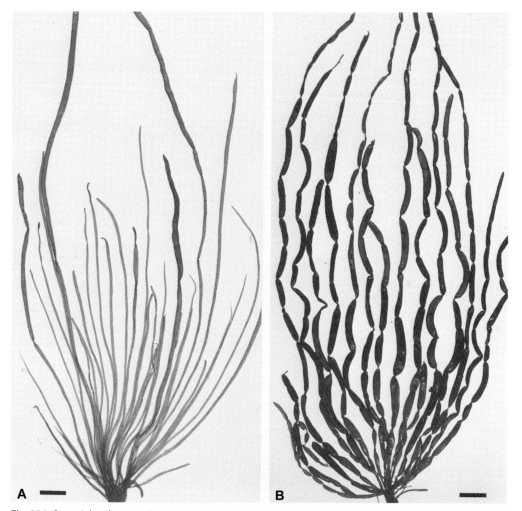

Fig. 226. *Scytosiphon lomentaria*
Habit of thalli. A. Narrow form with few constrictions. B. broad form with prominent constrictions. Bar = 20 mm (A), = 12 mm (B).

conditions appeared to play an important role in determining the relative occurrence of the prostrate and erect systems. For instance, low temperatures and short days generally favoured erect thallus development while warm temperatures and long days favoured prostrate (usually crust) development (Correa *et al.* 1986; Dring & Lüning 1975b, 1983; Rhodes & Connell 1973; Tatewaki 1966; Tom Dieck 1987; Wynne 1969). Moreover, red light was shown to enhance erect thallus formation while white and blue light favoured crust formation (Lüning & Dring 1973). Some isolates, however, developed independently of environmental conditions of temperature and daylength (Kristiansen & Pedersen 1979; Pedersen 1980). The life history can, therefore, be interpreted as typically heteromorphic but monophasic with both the prostrate and erect systems possessing the same ploidy level.

Culture studies in Japan by Nakamura (1964), Tatewaki (1966) and Nakamura & Tatewaki (1975) revealed the occurrence of an additional sexual cycle, with an alternation of heteromorphic, biphasic generations. The erect thalli represented the sexual haploid generation releasing plurispores which behaved as gametes, while the crustose thalli represented the asexual diploid generation bearing unilocular sporangia. Chromosomal studies by Nakamura

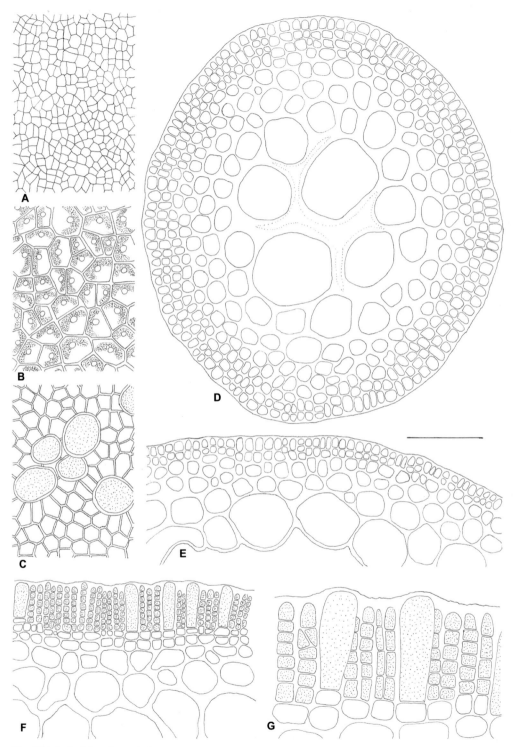

Fig. 227. *Scytosiphon lomentaria*
A, B. Surface view of vegetative thallus. C. Surface view of fertile thallus showing plurilocular sporangia and large ascocysts. D. TS of young thallus showing outer small cortical cells and inner large medullary cells. E. TS of portion of thallus showing outer cortical cells, inner medullary cells and central cavity. F, G. TS of fertile thalli showing terminal rows of plurilocular sporangia and large ascocysts. Bar = 20 μm (B–C, G), = 50 μm (others).

& Tatewaki (1975) further confirmed the occurrence of meiosis in the formation of the unispores and the alternation of nuclear phases in the lifecycle. Nakamura & Tatewaki (1975) also reported asexual development of plurispores released from the erect thalli, the final morphological expression obtained (i.e. crusts or erect thalli) being dependent on environmental conditions; as Nakamura & Tatewaki pointed out, this parthenogenetic development of the 'gametes' is in accordance with the 'direct' life history widely reported for this species.

The above reports of sexual reproduction in *S. lomentaria* give some credence to earlier worldwide observations of plurispores behaving as gametes (Abe 1935; Berthold 1881; Kuckuck 1912). The possibility of a sexual life history occurring, albeit of rare occurrence in North Atlantic waters, needs reinvestigation.

A noteworthy feature of the life history of *S. lomentaria* is the wide range of reported prostrate systems and their resemblance to previously described taxa. This has been augmented by field and laboratory studies of the prostrate systems which have shown them to be connected to the life history of *S. lomentaria*. Such 'prostrate' taxa include *Streblonema* (Dangeard 1963b), *Myrionema* (Dangeard 1963b), *Compsonema* (Loiseaux 1970b, as *Scytosiphon pygmaeus*; Fletcher 1987, p. 222), *Microspongium* (Fletcher 1978; Lund 1966; Kristiansen & Pedersen 1979; McLachlan *et al.* 1971) and *Ralfsia* (Nakamura 1965; Nakamura & Tatewaki 1975; Tatewaki 1966; Wynne 1969). It is possible that *S. lomentaria* represents various genotypes with morphologically different basal expressions. Alternatively, the basal expressions could be due to different degrees of surface contact achieved by the germlings (see Lüning 1980a) or to the use of different culture conditions. For example, the crustose expressions implicated in the life history of *S. lomentaria* have been reported to be *Microspongium gelatinosum*-like in the North Atlantic and *Ralfsia*-like in the Pacific, supporting the argument that different entities are involved (Lund 1966). This is further complicated by reports of *Ralfsia*-like phases also occurring in the life history of Atlantic material of *S. lomentaria* (Rhodes & Connell 1973), although the evidence is not altogether convincing, as the illustrations provided by these authors (figs 11, 12) appear to resemble *Microspongium* more than '*Ralfsia*'. It is further possible that the *Ralfsia*-like crusts reported in the life history of Pacific *S. lomentaria* are also based on misidentifications. Illustrations provided by many of the authors (Nakamura 1965, figs 2, 3; Nakamura & Tatewaki 1975, figs 3B, E, pl. I(D, E); Tatewaki 1966, figs 6, 7) strongly resemble *Microspongium* thalli. Only the illustrations provided by Wynne (1969, pl. 15) of cultured unilocular sporangial crusts of *S. lomentaria* are convincingly *Ralfsia*-like. It is the author's view here that the crustose thallus involved in the life histories of Atlantic and Pacific populations of *S. lomentaria* is *Microspongium*-like rather than *Ralfsia*-like and can be more specifically identified as Reinke's *M. gelatinosum* bearing unilocular sporangia. Further studies on a worldwide basis are required to determine if the distinction between the *Microspongium*-like and *Ralfsia*-like phases, based primarily on differences in the cohesion of the erect filaments, can be justifiably maintained.

Several molecular studies (Camus *et al.* 2005; Cho *et al.* 2007; Kogame *et al.* 2005, 2015a; McDevit & Saunders 2017) report different prostrate expressions in culture studies (see Fletcher 1987), and reports of varied morphogenetic responses (erect thallus or prostrate thallus formation) to temperature and photoperiod shown in some culture studies (see Kristiansen & Pedersen 1979; Kristiansen *et al.* 1991; Lüning & Dring 1975; Pedersen 1980; Tom Dieck 1987) all indicate that, throughout the world, *Scytosiphon lomentaria* represents a complex of genetically isolated groups or species. Kogame's (2015a) finding of at least four species passing under the name *S. lomentaria* in the north-east Atlantic and Mediterranean is particularly pertinent. It is likely, then, that more than one species is passing under the name of *Scytosiphon lomentaria* in the marine flora of Britain and Ireland and further studies, particularly using molecular tools similar to those used by McDevit & Saunders (2017) in their study of Canadian material, are required. McDevit & Saunders (2017) amended the diagnosis for *S. lomentaria* referring to a lack of unicellular paraphyses in this species, but this is in conflict with the description given in the present treatise.

Abbott & Hollenberg (1976), p. 198, fig. 162; Adams (1994), p. 75, pl. 18; Bartsch & Kuhlenkamp (2000), p. 168; Boo (2010a), pp. 168–70, fig. 8; Brodie *et al.* (2016), p. 1021; Bunker *et al.* (2017), p. 298; Camus *et al.* (2005), pp. 931–41; Cho *et al.* (2007), pp. 657–65; Clayton (1976a), pp. 187–98, figs 1–9, (1976b), pp. 201–8, figs 1–7, (1980a), pp. 105–18, figs 1–11, (1980b), pp. 113–18, figs 1–3; Clayton & Beakes (1983), pp. 4–16; Contreras *et al.* (2007), pp. 1320–8; Coppejans & Kling (1995), p. 230, pl. 84(D–E); Cormaci *et al.* (2012), pp. 287–9, pl. 83(1–4); Correa *et al.* (1986), pp. 469–75; Dangeard (1963b), pp. 57–64, pls IX–XI, (1969), pp. 60–70, fig. 1(B); Dizerbo & Herpe (2007), p. 118, pl 39(2); Dring (1984), pp. 159–92, figs 1, 4; Dring & Lüning (1975a), pp. 107–17, (1975b), pp. 25–32; (1983), pp. 545–68; Du Rietz (1941), pp. 1–11; Feldmann (1949), pp. 103–15, fig. 1(A–B); Fletcher (1978), pp. 371–98, fig. 35, (1980), pp. 41, 42, pl. 14(4–6), (1987), pp. 248–52, figs 67(B), 69, pl. 11(a, b); Fletcher in Hardy & Guiry (2003), p. 294; Flores-Moya *et al.* (2002), pp. 134–40; Frye & Phifer (1930), pp. 234–45, figs 1–39; Fujita *et al.* (2005), pp. 159–67; Funk (1955), p. 40; Gallardo García *et al.* (2016), p. 37; Gayral (1966), p. 247, pl. XXXIV; Guiry (2012), p. 156; Hamel (1931–39), pp. 194–6, fig. 431; Kato *et al.* (2006), pp. 65–71; Kawai *et al.* (1995), pp. 306–11; Kimura *et al.* (2010b), pp. 143–52; Kogame *et al.* (2005), pp. 313–22, (2015a), pp. 367–74; Kornmann & Sahling (1977), pp. 137–9, fig. 74; Kristiansen (1981), pp. 321–6, (1984), pp. 719–24, fig. 1; Kristiansen & Pedersen (1979), pp. 31–56; Kristiansen *et al.* (1991), pp. 375–83, (1994) pp. 444–54; Kylin (1933), pp. 47–9, fig. 18; Lindauer *et al.* (1961), pp. 256, 257, fig. 65; Littler & Littler (1983), pp. 425–31; Lubchenco & Cubit (1980), pp. 676–87; Lund (1966a), pp. 67–78, (1966b), p. 155; Lüning (1980a), pp. 920–30, figs 2, 3; Lüning & Dring (1973), pp. 333–8, figs 13–18; McDevit & Saunders (2017), pp. 653–71; Miller & Connell (2012), pp. 105–13; Morton (1994), p. 43; Munda (1984), pp. 371–6; Nagasato *et al.* (2000), pp. 339–48, (2003), pp. 1172–80, (2004), pp. 266–72; Nakamura (1965), pp. 109, 110, fig. 5, (1972), pp. 148, 149, fig. 1; Nakamura & Tatewaki (1975), pp. 59–65, figs 1–5, pl. 1; Newton (1931), pp. 178, 179, fig. III; Oltmanns (1922), p. 57, fig. 348(b); Parente *et al.* (2003), pp. 353–9, figs 2–4; Pedersen (1980), pp. 391–8, (2011), p. 73, fig. 75; Pedersen & Kristiansen (1994), p. 645; Reinke (1878a), pp. 267, 268, pl. 11, figs 13, 14; Rhodes & Connell (1973), pp. 211–15, figs 1, 11–14; Ribera *et al.* (1992), p. 118; Rosenvinge & Lund (1947), pp. 27–31, fig. 9; Rueness (1977), p. 169, fig. 101, pl. XXII(4); Sauvageau (1929), pp. 331–3; Setchell & N.L. Gardner (1925), p. 531, pl. 44, figs 72, 74; Stefanov *et al.* (1996), pp. 475–8; Stegenga & Mol (1983), p. 105, pl. 37, figs 3–5; Stegenga *et al.* (1997a), p. 21, (1997b), p. 185, pl. 53(1–2); Tatewaki (1966), pp. 62–6, figs 1–10; Taylor (1957), pp. 168, 169, pls 15(2), 16(3); Tom Dieck (1987), pp. 307–21; Wallentinus (1974), pp. 81–8; Womersley (1987), pp. 294, 295, pl. 3(3), figs 106(B, C), 108(C, D); Wynne (1969), pp. 32–4, pls 14–17; Yamano *et al.* (1996), pp. 155–9.

Stragularia Strömfelt

STRAGULARIA Strömfelt (1886a), p. 173.

Type species: *S. adhaerens* Strömfelt (1886a), p. 173 (= *S. clavata* (Harvey in Hooker) Hamel (1931–39), p. xxxi).

Thalli encrusting, discrete and orbicular becoming confluent and irregular, epilithic, epizoic, rarely epiphytic, light yellow-brown to dark brown, fairly smooth and firm, sometimes sponge-like, comparatively thin, firmly adherent by undersurface to substratum, usually without rhizoids; structure pseudoparenchymatous, consisting of a monostromatic discoid base of outward-spreading, frequently branched, quite firmly united filaments, with terminal row of synchronously growing apical cells, giving rise behind, to vertical or slightly curved, simple or little branched, fairly well-united filaments of cells containing a single, plate-like, parietal plastid with pyrenoid; hairs infrequent, terminal on erect filaments, arising from surface or depressions, with basal meristem, lacking obvious sheath.

Unilocular and plurilocular sporangia borne on the same or different thalli, terminal on erect filaments, grouped in sori; unilocular sporangial sori common, indefinite and extensive over central thallus region, obvious in surface view, slightly raised, sponge-like, light to yellow-brown, gelatinous, comprising ovate, obovate-pyriform or elongate-cylindrical sporangia arising laterally at the base of paraphyses, more rarely terminal on erect filaments with or without paraphyses; paraphyses multicellular, simple, linear or clavate, loosely associated and gelatinous; plurilocular sporangial sori present or absent, if present rare, not obvious externally, arising as terminal extensions of the erect filaments, cylindrical, uniseriate, rarely biseriate in parts.

The genus *Stragularia* is revived here to include two species previously assigned to the closely related genus *Ralfsia*. *Stragularia* differs from *Ralfsia* in having vertically directed filaments arising from the basal layer which are not initially almost prostrate, and having expansive, rather than discrete unilocular sporangial sori. Other features that distinguish *Stragularia* from *Ralfsia* include: crusts are comparatively thin, more closely encrusting with the margin gradually, rather than sharply, delimited and almost impossible to raise intact; young thalli are light yellow-brown rather than olive-brown, matt rather than glossy; older thalli are fairly soft, smooth and firmly adherent throughout rather than firm, coriaceous, verrucose, bullate with rust-red underside; erect filaments are less firmly coherent and can be squashed apart with relative ease; there is single plate-like plastid with pyrenoids; the plurilocular sporangia are grouped in less conspicuous sori and are not constructed of closely packed vertical columns each with a terminal sterile cell; thalli have a more restricted local (and in most species geographical) distribution, more or less restricted to tidal pools and runnels.

Crustose algae with *Stragularia* features have commanded particular attention in recent years in view of their life history connections with macroscopic, erect bladed members of the Scytosiphonaceae. In most investigations of these heteromorphic life histories the relationship between the crustose and erect fronded expression was shown to be asexual, facultative and determined by environmental conditions. In a few cases, a sexual relationship was observed with the erect blades representing the sexual, gametophytic phase and the crustose thalli representing the asexual, sporophytic phase. The so-called *Stragularia* stages have been implicated in the life histories of such species as *Petalonia fascia* (Edelstein *et al.* 1970; Fletcher 1974b, 1978; Nakamura 1965; Rhodes & Connell 1973; Roeleveld *et al.* 1974; Wynne 1969), *Colpomenia bullosa* (Nakamura & Tatewaki 1975), *C. peregrina* (Clayton 1979), *Petalonia binghamiae* (Nakamura & Tatewaki 1975), *Scytosiphon tenellus* (Kogame 1998) and *S. lomentaria* (Clayton 1980a; Fletcher 1974b, 1978; Nakamura 1965; Nakamura & Tatewaki 1975; Tatewaki 1966; Wynne 1969). To date, only a small number of these '*Stragularia*' taxa have been identified, including *S. clavata* (as *Ralfsia clavata*), *R. bornetii*, *R. californica* and *Microspongium gelatinosum*. This is further complicated by species such as *R. californica* reported to be linked in life history with both *S. lomentaria* and *P. fascia* (Wynne 1969). Also, species such as *R. clavata* include geographical isolates which appear to have a direct, monomorphic life history (Kylin 1934; Loiseaux 1968). More work, particularly culture studies, is therefore required to identify and define all the '*Stragularia*'-type crustose algae and the extent of their involvement with the erect, thalloid members of the Scytosiphonaceae.

One species in Britain and Ireland, one species in Britain only.

KEY TO SPECIES
Erect filament cells of crust not exceeding 13 μm in diameter; unilocular sporangial sori indefinite and irregularly spreading on crust surface, light brown; paraphyses distinct from crust filaments, not exceeding 7 cells, with cells, particularly towards base, markedly longer than wide, 1–7 diameters long . . . . *S. clavata*
Erect filament cells of crust up to 19 μm in diameter; unilocular sporangial sori indefinite and regularly spreading over whole central crust region, yellow-brown; paraphyses not markedly different from crust filaments, usually of 8–12 cells, with cells not exceeding 1–2 diameters long . . . . *S. spongiocarpa*

Stragularia clavata (Harvey in Hooker) Hamel (1931–39), p. xxxi.   Figs 2(B), 228

Myrionema clavatum Harvey in Hooker (1833), p. 391; *Ralfsia clavata* (Carmichael ex Harvey) P. Crouan & H. Crouan (1867), p. 166; *Ralfsia disciformis* P. Crouan & H. Crouan (1867), p. 166; *Myrionema henschei* Caspary (1871), p. 142; *Stragularia adhaerens* Strömfelt (1886), p. 49; *Ralfsia bornetii* Kuckuck (1894), p. 245; *Ralfsia clavata* f. *laminariae* Collins in Collins, Holden & Setchell (1907), no. 1309, nom. inval.; *Ralfsia tenuis* Kylin (1947), p. 45.

Thalli mainly epilithic, encrusting, comparatively thin, discrete and orbicular becoming confluent and irregular to 5 cm or more in extent, frequently with central regions eroded away, light to dark brown-black, firmly adherent to substratum by underside, usually without rhizoids; peripheral region thin, sometimes monostromatic and discoid, light brown, firm, subcoriaceous, closely encrusting and impossible to raise intact; central region dark brown-black, smooth, soft, slightly spongy, comprising rounded or polygonal cells quite closely packed and irregularly arranged in surface view, 5–13 μm in diameter, each with a single, parietal, plate-like plastid with pyrenoid; thallus pseudoparenchymatous in structure, comprising a monostromatic basal layer of outward-spreading, frequently branched, quite firmly united filaments of mainly rectangular cells, 1–2 diameters long, to 20 × c. 8 μm in diameter; basal layer regularly discoid, with marginal row of distinct, synchronously dividing apical cells, giving rise behind, immediately or belatedly, to vertical or slightly curved, unbranched or little branched compacted filaments; erect filaments quite firmly adjoined, although separable under pressure, to 180 μm (–20 cells) long, comprising quadrate to rectangular cells, 8–12 × 5–13 μm each containing a single, plate-like, parietal, yellow-brown plastid with pyrenoid, occupying upper cell region; hairs infrequent, terminal on erect filaments arising superficially or from depressions, with basal meristem but lacking obvious sheath.

Unilocular and plurilocular sporangia borne on the same or different thalli, terminal on erect filaments, grouped in sori; unilocular sporangial sori common, indefinite and irregularly spreading over crust surface, slightly raised and obvious in surface view, light brown, soft, sponge-like, very gelatinous comprising obovate, pyriform, elongate-pyriform or elongate-cylindrical unilocular sporangia, 45–80(–120) × 16–28 μm, sessile and lateral at the base of paraphyses; paraphyses clearly distinguishable from the erect filament, simple, elongate-clavate to elongate-cylindrical, multicellular, 2–5(–7) cells, 60–165 μm long, comprising quadrate, more commonly rectangular cells, 1–2 diameters long above, 7–16 × 7–9 μm, markedly elongate below, cylindrical or elongate-clavate, 1–7 diameters long, 24–47 × 4–7 μm, frequently with ruptured outer cell walls; plurilocular sporangia rare, terminal on erect filaments, without associated paraphyses, uniseriate or partly biseriate, to 20 μm (–7 loculi) long.

Epilithic especially on soft substrata, epizoic, more rarely epiphytic, upper to lower eulittoral pools/channels, sometimes emergent in damp situations especially under cover of fucoids, in sheltered to moderately exposed areas; tolerant of some sand cover.

Common and widely distributed around Britain and Ireland.

Thalli probably perennial, unilocular sporangia recorded November to May; plurilocular sporangia recorded November (single record). Crusts becoming recognizable on the shore from October onwards, reaching maximum development and fertility in winter/spring and then dying back during the summer or remaining as small pieces of crust edge.

This is the most common and widely distributed species of *Stragularia* present in Britain and Ireland. Although often recorded on the sloping sides of standing shallow pools (rarely in deep pools) it shows a particular preference for runnels and channels associated with running seawater. Associated algae usually include the *Microspongium gelatinosum* phase of *Scytosiphon lomentaria*, *Pseudoralfsia verrucosa*, *Petroderma maculiforme* and *Hildenbrandia rubra* and, in particular, the erect bladed *Petalonia fascia*. The crusts are frequently subjected to a periodic cover of sand. The species is seldom emergent on rock except under cover of fucoids, in particular *F. serratus*. It has been recorded on a wide variety of natural and artificial substrata although more commonly on softer materials such as chalk, limestone, shale etc.

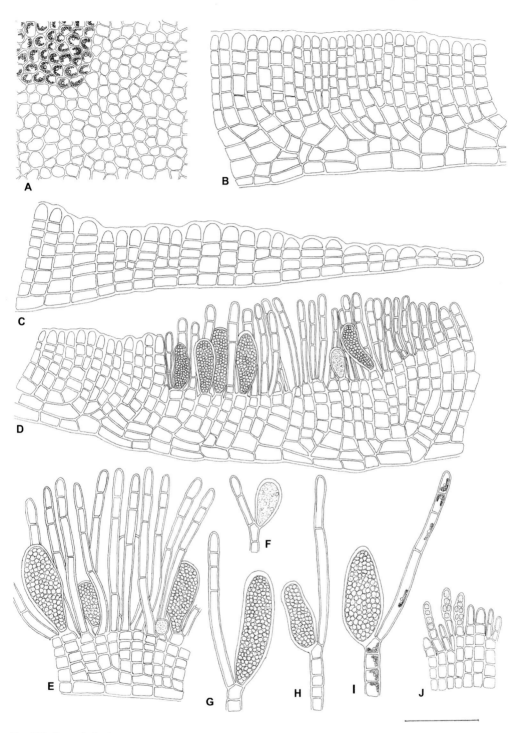

Fig. 228. *Stragularia clavata*
A. Surface view of crust cells. Note single plate-like plastid with pyrenoid drawn in some cells. B. VS of thallus showing slightly curved erect filaments. C. RVS of thallus margin showing terminal apical cell. D, E. VS of fertile thallus showing unilocular sporangial sorus. F–I. Unilocular sporangia and associated paraphyses. J. VS of fertile thallus showing terminal plurilocular sporangia. Bar = 50 μm.

S. clavata is based on *Myrionema clavatum* described by Harvey in Hooker (1833). Harvey had based his description on *Linckia clavata* originally described for Loch Linnhe on the west coast of Scotland by Captain Carmichael in his manuscripts 'Algae Appinenses' and 'Cryptogam Appinens'. Later P. Crouan & H. Crouan (1852) moved *Myrionema clavatum* to the genus *Ralfsia* quoting Harvey and thus establishing *R. clavata* (Harvey in Hooker) P. Crouan & H. Crouan accepted by various authors such as Taylor (1937a) and Parke & Dixon (1976). However, according to Reinke (1889b) and Kylin (1947) this combination is illegitimate as the Crouans' material is identical with *R. verrucosa* (see Batters 1890, p. 288 footnote). Farlow's (1881) later description of *R. clavata* collected on the coast of New England, USA appears to agree with our present-day concept of this species, and he was thus often associated with the binomial e.g. *R. clavata* Farlow in Newton (1931) and Jaasund (1965) and *R. clavata* (Carmichael) Crouan sensu Farlow in Taylor (1957) and South & Hooper (1980). A further development was initiated by Kylin (1947). He cast doubt on the conspecificity of Farlow's material and European material of *R. clavata* as Farlow did state his material was identical to Crouan brothers No. 56 of their *Algues marines du Finistère*; this as stated above, was considered by Reinke and Kylin to be *R. verrucosa*. Kylin also adopted Batters's (1890, p. 288) attitude that as the description given by Carmichael was very brief and none of the original material has been detected there must always remain some degree of doubt as to whether his specimen is identical with the presently recognized species. Because of this doubt, Kylin (1947, p. 5) proposed a new name *R. tenuis* based on *R. clavata* sensu Reinke (1889b, p. 48) as a replacement for the questionable *R. clavata*. This has been accepted by a number of authors including Lund (1959), Pankow (1971) and Rueness (1977). This binomial is not, however, adopted here as it is the author's preference to continue to accept, Harvey's *Myrionema clavatum* as the basis for our present-day concept of this taxon, pending the results of further studies. The species is also transferred here to the genus *Stragularia* following Hamel (1931–39, p. xxxi).

Material resembling *Ralfsia bornetii*, originally described for Helgoland in the North Sea by Kuckuck (1894), is herein included within the circumscription of *S. clavata*. *Ralfsia bornetii* is distinguished from *S. clavata sensu stricto* by the possession of extremely long paraphyses (approx. 140–180 μm) with elongate basal cells and associated elongate unilocular sporangia (approx. 80–125 μm). Material of *S. clavata sensu stricto* is usually described as having smaller paraphyses (approx. 70–100 μm) and smaller, more rounded unilocular sporangia (approx. 40–85 μm). Opinion is divided as to whether these represent distinct species. Newton (1931) and Hamel (1931–39) included *R. bornetii* within the synonymy of *R. clavata* while it was recognized as a separate species by Sundene (1953), Taylor (1957), Jaasund (1965), Edelstein *et al.* (1970), Rueness (1977), South & Hooper (1980) and Tanaka & Chihara (1980c). Indeed, according to Edelstein *et al.* (1970) some of the distinctive features of *R. bornetii* and *R. clavata* were retained in laboratory culture. Following examination of British and Irish material, there appears to be complete integration with respect to the lengths of paraphyses and sporangia. Therefore, pending further studies, *R. bornetii* is here included within the synonymy of *S. clavata*.

A number of laboratory culture studies have revealed Atlantic isolates of *S. clavata* (includes *R. bornetii*) to be connected in life history with *Petalonia fascia* (Edelstein *et al.* 1970; Rhodes & Connell 1973; Roeleveld *et al.* 1974; Fletcher 1974b, 1978; Yoneshigue-Valentin & Pupo 1994). The relationship between the crustose and erect bladed expressions appears to be asexual and facultative and determined by environmental conditions rather than differences in ploidy level. There is no evidence of a sexual cycle between an erect bladed (i.e. *Petalonia*-like) gametophyte and a crustose (i.e. *Stragularia*-like) sporophyte. However, this requires investigation especially given reports of sexual life histories in other members of the Scytosiphonaceae in the North Pacific. It is possible that the crustose and erect bladed expressions are linked by a sexual process in the North Atlantic and this has either passed unnoticed or requires precise environmental conditions. The sexual process might alternatively be confined to a small number of genotypes. Certainly, the indications are that different genotypes are present, which vary in their life history and developmental patterns. Most isolates investigated in culture revealed a heteromorphic life history; however, there have also been reports of direct, monophasic life histories (Kylin 1934; Loiseaux 1968). It is also likely that geographically separated ecotypes have developed varying responses to environmental conditions (as shown for *S. lomentaria* and its crustose expression by Lüning 1980a).

Ardré (1970), pp. 247, 248, pl. 36, figs 1, 2 (as *R. bornetii*); Bartsch & Kuhlenkamp (2000), p. 169; Batters (1890), p. 288, pl. 8(22); Brodie *et al.* (2016), p. 1021; Edelstein *et al.* (1970), pp. 527–31, figs 2–15; Fletcher (1974b), p. 218, (1978), pp. 371–98, figs 3, 7, 11, 12, 17, 25, 29, 30, 34, (1987), pp. 254–7, figs 59(C), 70; Fletcher in Hardy & Guiry (2003), p. 306; Gallardo García *et al.* (2016), p. 37; Guiry (2012), pp. 156, 157; Hamel (1931–39), p. 108, fig. 26(C); Jaasund (1965), pp. 61–3, fig. 19(A) (including *R. bornetii*); Keum (2010), p. 104; Kuckuck (1894), pp. 244–6, figs 14, 15 (including *R. bornetii*); Kylin (1934), pp. 17, 18, fig. 10; Loiseaux (1968), pp. 302–8, figs 3–5; Lund (1959), pp. 78, 79 (as *R. tenuis*); Morton (1994), p. 33; Newton (1931), p. 154, fig. 95(A–B, not C); Reinke (1889a), p. 9, pl. 6(10–19); Ribera *et al.* (1992), p. 113; Rueness (1977), pp. 128–30, figs 53, 54(B, not C) (including *R. bornetii* and *R. tenuis*), (2001), p. 17; Stegenga & Mol (1983), p. 82, pl. 28(1, 2); Stegenga *et al.* (1997a), p. 21; Strömfelt (1886b), pp. 49, 50, pl. II, figs 13–15; Tanaka & Chihara (1980c), pp. 341, 342, fig. I(C) (as *R. tenuis*); Taylor (1957), pp. 135, 136, p. 11(1–2) (includes *R. bornetii*); Yoneshigue-Valentin & Pupo (1994), pp. 489–96.

Stragularia spongiocarpa (Batters) Hamel (1931–39), p. 70.   Figs 2(F), 229, 230

Ralfsia spongiocarpa Batters (1888), p. 452.

Thalli encrusting, comparatively thin, discrete and orbicular to 10 mm in diameter, becoming confluent and irregular to 2–3 cm in extent, frequently with central regions eroded away, yellow to dark brown-black, firmly adherent to substratum by underside, usually without rhizoids; in surface view, central vegetative cells rounded or polygonal, thick walled, quite closely packed, 8–16 μm in diameter; thallus pseudoparenchymatous in structure, consisting of a monostromatic basal layer of outward-spreading, frequently branched, quite firmly adjoined filaments with a marginal row of discrete, synchronously dividing apical cells, giving rise immediately behind to vertical files of simple or little branched, compacted filaments; erect filaments quite firmly adjoined although separable under pressure, to 190 μm high (–20 cells) comprising cells, 1–2 times broader than long below, quadrate or only slightly broader than long above, 7–13 × 10–19 μm each containing a single, plate-like plastid and pyrenoid occupying upper cell region; hairs infrequent, approx. 6 μm in diameter, terminal on erect filaments, arising superficially or from depressions, with basal meristem but lacking obvious sheath.

Unilocular sporangia common, borne in slightly raised yellow, spongy, turf-like, gelatinous sori, extensive and occupying greater part of crust surface, obvious in surface view; unilocular sporangia ovate, obovate, pyriform or elongate-pyriform, 21–70 × 16–35 μm, terminal on erect filaments with or without associated paraphysis-like filaments, less frequently lateral at the base of the paraphysis-like filaments, sessile or on 1–2-celled pedicels, rarely reported terminal on paraphysis-like filaments; paraphysis-like filaments not markedly different from erect vegetative filaments, multicellular, gelatinous, simple, linear, slightly tapering to the apex, 105–120(–140) μm long, 8–12 cells, comprising quadrate to rectangular cells, 1–2 diameters long throughout, 8–18 × 5–9 μm; plurilocular sporangia unknown.

Epilithic, on the sloping sides of shallow pools, lower eulittoral, frequently covered by sand.

Apparently restricted to southern and eastern shores of Britain (Fife, Northumberland, Yorkshire, Sussex, Hampshire, Dorset, South Devon); probably more widely distributed.

Probably perennial, unilocular sporangia recorded December to March.

S. spongiocarpa is common locally but appears to have a limited geographical distribution. It occurs mainly in the lower eulittoral, on the sides of sand-covered shallow pools, in association with a wide range of other crustose brown and red algae. Fertile thalli are particularly common during January and February, after which the central crust areas die away leaving the outer margin portions. In a dried state the thalli are readily distinguishable from the

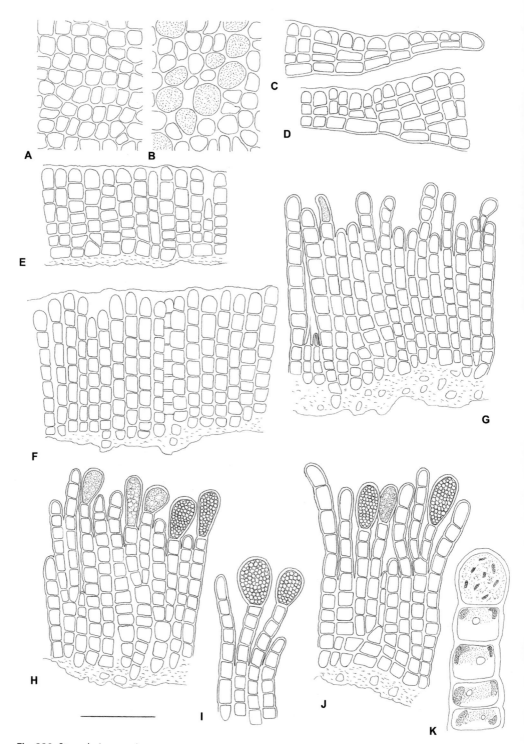

Fig. 229. *Stragularia spongiocarpa*
A. Surface view of vegetative thallus. B. Surface view of fertile thallus with unilocular sporangia. C. VS of crust margin showing single, large, apical cell. D–G. VS of vegetative thalli showing erect filaments. H–J. VS of fertile thalli showing terminal unilocular sporangia. K. Portion of erect filament showing young unilocular sporangia and cells with a single plate-like plastid and pyrenoid. Bar = 20 µm (K), = 50 µm (others).

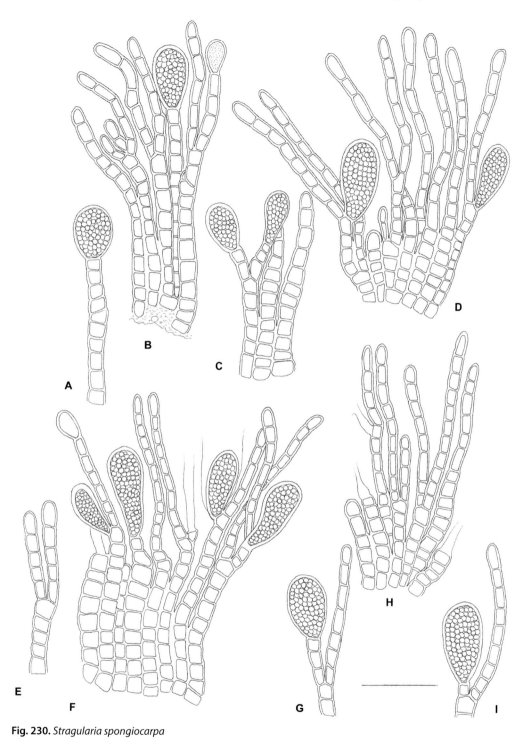

Fig. 230. *Stragularia spongiocarpa*
A–I. VS and SP of fertile thalli showing vegetative filaments giving rise to paraphyses and unilocular sporangia.
Bar = 50 µm.

morphologically similar *S. clavata* by their extensive and more prominent fructification which is yellow in colour and similar to the confluent tufts of *Compsonema saxicola*. *S. spongiocarpa* can further be distinguished from *S. clavata* by the wider vegetative cells (10–19 μm compared to 5–13 μm), the greater number of cells in the paraphyses (8–12 compared to 2–7) and the absence of elongated cells in the paraphyses (1–2 compared to 1–7 diameters long).

Initially, unilocular sporangia are formed terminally, in crowded sori on the thallus surface, unaccompanied by the paraphyses. Later, after spore release has occurred, many terminal cells either continue sporangia production within the old sporangial wall husks or continue growth to form the paraphyses. The paraphyses are not clearly distinguishable from the erect filaments of the crustose thallus.

Culture studies by Fletcher (1981b) indicate that a direct, monomorphic life history is probably operating in this species. Spores from the unilocular sporangia behaved asexually and developed directly into fertile, crustose thalli similar to the parental thallus. There was no indication that *S. spongiocarpa* was connected in life history with an erect bladed member of the Scytosiphonaceae.

Batters (1888), pp. 452, 453, pl. XVIII(17–21), (1890), p. 289, pl. VIII(17–19); Brodie *et al.* (2016), p. 1021; Dizerbo & Herpe (2007), p. 76; Fletcher (1974b), p. 218, (1975), pp. 534, 535, figs 1–3, (1978), pp. 371–98, (1981b), pp. 323–30, figs 1–19, (1987), pp. 257–61, figs 59(D), 71, 72; Fletcher in Hardy & Guiry (2003), p. 306; Hamel (1931–39), pp. 109, 110; Newton (1931), p. 153.

Excluded species

Stragularia disciformis (P. Crouan & H. Crouan) Fletcher, *comb. nov.*

Ralfsia disciformis P. Crouan & H. Crouan (1867), p. 166.

This species was described by P. Crouan & H. Crouan (1867) from material dredged at Brest. The description was inadequate and, as Batters (1896) pointed out, comes very near to that of *Ralfsia clavata* (= *Stragularia clavata* in the present text), from which it differs by the shorter, 1–2-celled, less clavate paraphyses which are hardly any longer than the unilocular sporangia. The record for Britain is based on material dredged from deep water near the mouth of the River Yealm, Plymouth Sound, south coast of England (Batters 1896). Examination of material deposited in the BM reveal 2–4-celled paraphyses and unilocular sporangia with size ranges within those given for *Stragularia clavata* in the present work. *Stragularia disciformis* is, therefore, excluded from the flora of Britain.

P. Crouan & H. Crouan (1867), p. 166; Batters (1896), p. 385; Fletcher (1987), pp. 243–4; Newton (1931), p. 153; Ribera *et al.* (1992), p. 113; Dizerbo & Herpe (2007), p. 76.

Stragularia pusilla Strömfelt (1888), p. 382.

Ralfsia pusilla (Strömfelt) Foslie (1892), p. 2.

Stragularia pusilla was originally described by Strömfelt (1888) for the coast of Norway. It was first reported for Britain in Cumbrae, Argyll by Batters (1892a) and subsequently recorded for Dorset, Northumberland and Orkney. Examination of collections of the above material, deposited in the BM confirmed the identity of *Stragularia*-like thalli (except for the Cumbrae specimen) but they appeared to show a marked similarity to *S. clavata*. The crusts were also much larger than those of *S. pusilla* described by Strömfelt and different in both the number and shape of cells in the paraphyses. For example, Strömfelt's (1888) figure 4 showed paraphyses with 9–13 cells which were approx. 1–2 diameters long, compared to only 4–7 cells, which were markedly elongate below, in material identified as this species collected in Britain. Given this discrepancy, this species is excluded from the British and Irish flora pending further studies.

Batters (1892a), p. 174; Brattegard & Holthe (2001), p. 17; Dizerbo & Herpe (2007), p. 76; Fletcher (1987), p. 244; Foslie (1892), p. 264; Jaasund (1965), p. 63, fig. 19(B); Mathieson & Dawes (2017), p. 230, pl. LX(9, 10); Newton (1931), p. 153; Norum (1913), pp. 131–60; Printz (1926), p. 139; Rueness (1997), pp. 128, 129; Strömfelt (1888), p. 382, pl. 3(4); Taylor (1957), pp. 136, 137.

Symphyocarpus Rosenvinge

SYMPHYOCARPUS Rosenvinge (1893), p. 896.

Type species: *S. strangulans* Rosenvinge (1893), p. 896.

Thalli epiphytic, epizoic, encrusting, pseudoparenchymatous consisting of a monostromatic basal layer, the cells of which give rise to loosely united and gelatinous erect filaments; cells with a single plate-like plastid and pyrenoid; ascocyst-like cells common arising from basal layer or terminating erect filaments.
   Plurilocular sporangia terminal on erect filaments, multiseriate, clustered.

One species in Britain and Ireland.

Symphyocarpus strangulans Rosenvinge (1893), p. 896, fig. 5.   Fig. 231

Thalli forming small, thin, closely adherent crusts, 1–5 mm in diameter, light brown in colour; thallus structure pseudoparenchymatous, comprising a monostromatic base of laterally adjoined, firmly attached, outward-spreading filaments of cells, with or without downward-growing rhizoids, giving rise to numerous compacted, erect filaments; in surface view, peripheral vegetative cells rectangular, 11–24 × 7–161 µm, central vegetative cells irregular or rounded 11–22 × 7–17 µm, all cells with a single, large plate-like plastid, without obvious pyrenoid; in vertical section, erect filaments straight, little branched, loosely united and easily separable under pressure, slightly gelatinous in texture, to 11 cells, 85 µm long, basal cells rectangular in section usually much wider than high 10–16 × 15–37 µm, upper cells usually subquadrate or quadrate, rarely rectangular, 4–21 × 9–20 µm, each with a large, single, plate-like plastid occupying upper cell region; hairs infrequent, usually terminal on erect filaments, approx. 8 µm in diameter, with basal meristem but lacking obvious sheath; ascocysts common, either colourless or dark yellow pigmented, borne either on the basal cells or terminating erect filaments, in surface view, rounded in shape 12–32 × 18–25 µm, discrete or in sori, in section, elongated and cylindrical or slightly clavate, 23–52 × 16–26 µm.
   Plurilocular sporangia common, usually in sori, terminal on erect filaments; in surface view, rounded or irregular in shape, slightly raised above vegetative cells, often with 4 loculi, sometimes multilocular, 14–27 × 15–24 µm; in section, multiseriate, arranged in 2 or 4 vertical columns, rarely irregularly clustered, 4–7 loculi, 14–40 × 17–24 µm, loculi 4–5 × 8–11 µm; unilocular sporangia unknown.
   Epiphytic on *Laminaria* and maerl, also epizoic on dogfish cases, lower eulittoral and sublittoral to 20 m depth.
   Only known in Britain for Berwick, Argyll and the Outer Hebrides, and in Ireland for Galway.
   Data inadequate to comment on seasonal behaviour.

Bartsch & Kuhlenkamp (2000), p. 169; Batters (1895a), p. 275; Brodie *et al.* (2016), p. 1021; Dizerbo & Herpe (2007), p. 76; Fletcher (1987), p. 92, fig. 5; Fletcher in Hardy & Guiry (2003), p. 311; Gallardo García *et al.* (2016), p. 37; Guiry (2012); Hamel (1931–39), p. XXX, fig. 62(12–15); Jaasund (1963), pp. 6, 7, fig. 4, (1965), p. 61; Lund (1959), pp. 70–3, fig. 11; Mathieson & Dawes (2017), pp. 203, 204, pls LVII(18), LVIII(1); Newton (1931), p. 126, fig. 74; Pedersen (1976), pp. 39, 40, pl. 4(d–f); Rosenvinge (1893), p. 896, figs 28, 29, (1910), p. 117; Rueness (1977), p. 130, (2001), p. 17.

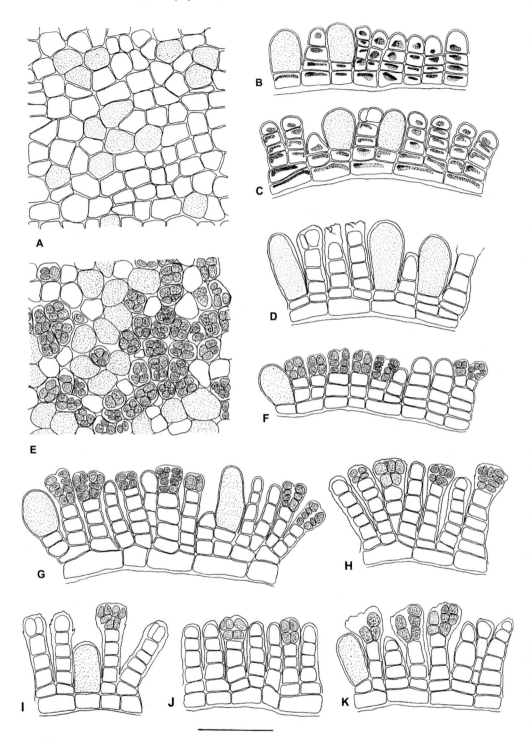

Fig. 231. *Symphyocarpus strangulans*

A. Surface view of vegetative crust showing scattered, hyaline ascocysts. B–D. VS of vegetative crusts showing erect filaments, large, hyaline ascocysts and cells with a single plate-like plastid. E. Surface view of fertile crust showing ascocysts and plurilocular sporangia with 2–4 loculi. F–K. VS of fertile crusts showing erect filaments, ascocysts and terminal plurilocular sporangia. Bar = 50 μm.

Fucales Bory

FUCALES Bory (1827), p. 62.

Thalli macroscopic, from a few cm to several metres in length, erect, yellow-brown to dark brown-black, sometimes loose-lying but mainly found attached to hard substrata by a rounded to irregularly shaped, flattened or conical holdfast, more rarely by a rhizome system; thalli largely parenchymatous (except for the filamentous medulla) and structurally complex, ranging in form from cup-shaped, cylindrical, to more commonly dorsiventrally flattened and strap-like, sometimes a mixture on the same thallus, and often morphologically differentiated into distinct holdfast, stipe and frond regions, with or without further elaboration into leaf-like, stem-like portions; thalli rarely unbranched, more commonly much branched with branching dichotomous, pseudodichotomous, radial or irregularly pinnate, margin entire or serrated, with or without air vesicles and pseudo-vesicles; structurally comprising a central, filamentous, medullary tissue of elongated cells, with or without accompanying hyphae, an outer cortical tissue of more or less isodiametric cells, and an outer, active, single-layered meristoderm; sieve plates present in medulla and cortex cells; growth monopodial, by one or several apical cells sunken in apical pit, rarely by diffuse growth, usually with supplementary growth in thickness by the peripheral meristoderm; surface cells with numerous discoid plastids without pyrenoids, and with fucosan vesicles; cryptostomata (sterile cavities) often present, with or without, emergent tufts of hairs.

Reproductive organs borne on thalli in specialized terminal or lateral, simple or branched receptacles; sexual reproduction oogamous, with oogonia and antheridia borne in sunken, flask-shaped conceptacles within the receptacles, accompanied by simple, or branched, multi-cellular paraphyses; thalli dioecious or monoecious, conceptacles of the latter with either oogonia or antheridia; the three-layer walled oogonia sessile on conceptacle wall, dividing first meiotically and then mitotically, to produce eight nuclei, all or some of which become functional to form eggs, the number varying according to genera, e.g. eight eggs formed in *Fucus*, four in *Ascophyllum nodosum*, two in *Pelvetia canaliculata* and one in *Sargassum muticum*, with the non-functional nuclei either degenerating or extruded; the two-layer walled antheridia either sessile on conceptacle wall, or lateral/terminal on the paraphyses, dividing meioti-cally and then mitotically to produce up to 64 pear-shaped, biflagellate antherozoids; prior to, or following release from the conceptacles, processes of rupturing, gelatinization and dissolution release the eggs and antherozoids into the water column; rarely in some genera (e.g. *Sargassum*) the oogonia are 'incubated' on the receptacle surface and fertilized in situ. Life history dimorphic and diplohaplontic, the sporophyte representing the large, diploid thallus and the oogonia and antheridia representing highly reduced gametophytes.

A large and structurally diverse order, globally distributed and characterized by a combi-nation of the following features: (1) a monomorphic, diplontic life history which lacks an alternation of generations or production of spores, in which the young oogonium is considered to represent a unilocular sporangium meiotically producing highly reduced, non-free-living gametophytes, (2) a wide range in morphological forms, branching patterns and differentiation of branches, (3) an often basic differentiation into a holdfast, stipe and frond, (4) a complex, differentiated, anatomical structure, with central medullary tissue, an enclosing cortical tissue and an outer meristoderm layer responsible for growth in girth, (5) growth by a single (or more) apical cell, (6) the occurrence of sterile cavities (cryptostomata) in some genera, and (7) the formation of gametes in flask-shaped cavities (conceptacles) in modified vegetative branches or specialized receptacles.

In this book, three families – Fucaceae, Himanthaliaceae and Sargassaceae – are recognized for Britain and Ireland. Support for these families was provided by molecular studies, based on sequences of partial nrSSu and nrlSu genes, by Rousseau & de Reviers (1999b), Rousseau *et al.* (2001) and Draisma *et al.* (2003), in which they emerged as distinct clades. Rousseau & de Reviers (1999b) also proposed the merger of the Cystoseiraceae within the Sargassaceae, which found support in molecular studies by Saunders & Kraft (1995) and later by Cho *et al.* (2006), Harvey & Goff (2006) and Phillips *et al.* (2008a) and which is accepted here too.

Rocky shore fucoids

Fucoid algae are a dominant component of the eulittoral on rocky shores all around Britain and Ireland with many occurring in more or less well-defined subzones under sheltered conditions. From high water downwards, these fucoid belts occur in the following order: *Pelvetia canaliculata, Fucus spiralis, F. macroguiryi, Ascophyllum nodosum/F. vesiculosus* and *F. serratus*. In some sheltered localities in the far north, *F. distichus* subsp. *evanescens* forms a belt between *F. spiralis* and *F. serratus*, replacing *F. vesiculosus* and *A. nodosum*, while *F. ceranoides* represents the dominant eulittoral fucoid in areas such as estuaries and tidal inlets which are under the influence of freshwater. With increasing wave exposure, this zonation pattern often becomes less clear, with the elimination of some fucoid species and their replacement by other algae and/or animals, or the fucoids undergoing some form of modification. For example, with increasing exposure, *Ascophyllum nodosum* is eliminated, followed by the vesiculate form of *F. vesiculosus* and then *F. serratus*. On exposed rocky shores, *F. vesiculosus* is much reduced and evesiculate (f. *linearis*) and *F. spiralis* occurs in its dwarf form (f. *nana*). On very exposed rocky shores, the fucoids are usually completely absent, or represented by dwarf *F. spiralis* (*F. spiralis* f. *nana*), sometimes either accompanied with, or replaced by *F. distichus* on extreme northern shores. All the remaining fucoid algae (*Bifurcaria bifurcata, Cystoseira* spp., *Ericaria* sp., *Gongolaria* spp., *Halidrys siliquosa, Himanthalia elongata* and *Sargassum muticum*) are primarily lower eulittoral tidal pool occupants, except for *Bifurcaria bifurcata* which occurs in mid to high eulittoral tidal pools; only *Himanthalia elongata* is mainly emergent.

Saltmarsh fucoids

Although fucoid algae are primarily epilithic, they may also occur as unattached thalli in sheltered saltmarsh habitats. They can be found embedded in the mud, loose-lying and interlaced with each other on the surface of the mud or wrapped around the stems or rhizomes of accompanying angiosperms, and thus be less vulnerable to being removed by the daily tides. Such forms are a common and conspicuous feature of the extreme shelter of saltmarshes, first recognized and reported at the beginning of the last century (Oliver in Tansley 1911; Cotton 1912; Baker 1912; Baker & Bohling 1916) and act as pioneers on the lower marshes or occur as an undergrowth to the phanerogams. All the fucoids, except *Bifurcaria bifurcata, Cystoseira* spp., *Ericaria* sp., *Gongolaria* spp., *Sargassum muticum, Himanthalia elongata, Fucus distichus* and *F. distichus* subsp. *evanescens*, have been reported with these growth forms, which are considered to be derived from populations growing attached on neighbouring rocky shores (Chapman 1960; Oliveira & Fletcher 1980; Naylor 1928), although it has been suggested that they may arise by germination of the zygotes (Baker 1912; Baker & Bohling 1916; Moss 1956). However, once established on the saltmarshes, it is likely that they can perennate by means of vegetative propagation and this is the only means of reproduction. Saltmarsh populations are characterized by:

1. a dwarf habit;
2. absence of an attaching disc;
3. a bending, spiralling, curling and increased branching of the thallus;
4. absence of sexual reproduction, with vegetative reproduction only through proliferation and fragmentation.

Saltmarsh forms are considerably smaller when compared to their rocky-shore counterparts, sometimes up to half the size (Naylor 1928), have a greater fat content (Russell-Wells 1932; Hass & Hill 1933), a much firmer thallus and a reduced mucilaginous coating (Naylor 1928), a lower water content on a dry-weight basis (Oliveira & Fletcher 1977), retain the juvenile three-sided apical cell (Baker 1950; Moss 1956a), have marginally placed cryptostomata, and the megasporangia undergo only partial division or even fail to divide. Hypotheses put forward to explain their dwarf habit and reduced width include the fact that the saltmarsh thalli grow at a higher level and have a greater degree of exposure and are also, therefore, subject to nutrient deficiency due to the shorter period of immersion in often diluted seawater (Baker 1912; Baker & Bohling 1916). Spirality and curling of the thallus have been considered to be due to the procumbent habit, and subsequent contact with the mud, one side lying on the mud growing more rapidly benefiting from increased moisture and nutrients. The lack of conceptacles has been considered to be attributable to either the high humidity, lack of attachment (Baker & Bohling 1916) or to the permanent retention of the juvenile, three-sided apical cell keeping the thallus in a juvenile state (Moss 1956a).

The dwarf habit of the populations, the similarity in their morphological expressions and the lack of reproductive organs have often led to difficulties in establishing their true affinities. However, several authors have attempted to categorized them and link their identity and derivation to local attached populations. Baker & Bohling (1916) proposed the infraspecific rank 'ecad', the latter term originally proposed by Clements (1905) indicating variability in form as a result of differences in habitat, collectively placing all the marsh ecads under the 'megecad' *limicola* to define their general characteristics due to their habitat, notably their prevalence for vegetative reproduction, absence of an attachment disc, dwarf habit and, when applicable to a saltmarsh existence, a curling and spirality of the thallus. While not accepting their taxonomic status, with the possible exception of the form now recognized as *Fucus cottonii*, and accepting the occurrence of intermediate forms, Baker & Bohling's (1916) classification system does provide a useful framework to call attention to these distinct and recognizable saltmarsh fucoids. Table 7 lists the megecads and ecads that have been described for Britain and Ireland by Baker & Bohling (1916), along with some additions, and they will be considered separately in sections following the treatment of their rocky shore counterparts.

Table 7. Checklist of fucoid saltmarsh ecads described for Britain and Ireland (as recognised by Baker & Bohling (1916) and the present author).

Fucoid species	Fucoid ecads
Ascophyllum nodosum	*Ascophyllum nodosum* megecad *limicola*
	Ascophyllum nodosum megecad *limicola* ecad *mackayi*
	Ascophyllum nodosum megecad *limicola* ecad *minor*
	Ascophyllum nodosum megecad *limicola* ecad *scorpioides*
Fucus ceranoides	*Fucus ceranoides* megecad *limicola*
	Fucus ceranoides megecad *limicola* ecad *glomerata*
	Fucus ceranoides megecad *limicola* ecad *proliferatus*
	Fucus ceranoides megecad *limicola* ecad *ramosissima*
Fucus serratus	*Fucus serratus* megecad *limicola*
	Fucus serratus megecad *limicola* ecad *limicola*
Fucus spiralis	*Fucus spiralis* megecad *limicola*
	Fucus spiralis megecad *limicola* ecad *nana*
Fucus vesiculosus	*Fucus vesiculosus* megecad *limicola*
	Fucus vesiculosus megecad *limicola* ecad *caespitosus*
	Fucus vesiculosus megecad *limicola* ecad *volubilis*
Pelvetia canaliculata	*Pelvetia canaliculata* megecad *limicola*
	Pelvetia canaliculata megecad *limicola* ecad *coralloides*
	Pelvetia canaliculata megecad *limicola* ecad *libera*
	Pelvetia canaliculata megecad *limicola* ecad *muscoides*

With respect to attempting a classification system for these saltmarsh entities, particular problems occur with some of the dwarf, embedded *Fucus* which have few characteristics to enable their true affinities to be determined. Such embedded dwarfs include *Fucus cottonii*, originally described as var. *muscoides* of *Fucus vesiculosus* (Cotton 1912), and now recognized as a distinct species, *Pelvetia canaliculata* ecad *muscoides* (Skrine 1929), a *muscoides* form of *Fucus spiralis* (Mathieson & Dawes 2001) and an alga designated as *F. cottonii* from Alaska (Ruiz *et al.* 2000). Several ecological studies (Baker & Bohling 1916; Lynn 1935a,b; Niell *et al.* 1980) have referred to a continuum of morphological expression extending between the dwarf embedded *Fucus* and the attached *F. vesiculosus* in Europe, while Mathieson *et al.* (2006) also reported a morphological continuity between a dwarf *Fucus*, *F. spiralis* ecad *lutarius* (Kützing) Sauvageau and *F. spiralis* in the north-west Atlantic Ocean. Additional transplantation experiments by Mathieson *et al.* (2006) removing the dwarf forms from the high eulittoral zone to the mid eulittoral zone also produced thalli resembling *F. spiralis* ecad *lutarius*, indicating that vertical transplantation can modify fucoid morphology and result in varying ecads. Molecular studies using microsatellite markers on dwarf *Fucus* populations in the north-west and north-east Atlantic Ocean and north-east Pacific Ocean have also thrown light on the taxonomy and systematic relationships of several of these saltmarsh *Fucus* entities (Mathieson & Dawes 2001; Wallace *et al.* 2004, 2005; Mathieson *et al.* 2006; Neiva *et al.* 2012b). Mathieson *et al.* (2006), for instance, working on populations in the north-west Atlantic similar to Irish material of *F. cottonii*, have revealed them to be transformed *F. spiralis*, as well as transformed hybrids of *F. vesiculosus* and *F. spiralis* and they thus might differ genetically from the European material which could be derived from *F. vesiculosus* and should not be designated as the same taxon, *F. cottonii*, but as a distinct taxonomic unit. Molecular studies by Neiva *et al.* (2012b) on *F. cottonii* from both the north-east Pacific and north-east Atlantic similarly concluded that this species was not a 'coherent genetic entity, but an artificial grouping of evolutionarily independent populations that converged into similar morphologies in different salt-marsh habitats'. In summary, the dwarf *muscoides*-like fucoids may well be derived from different attached parental species. Clearly, as Mathieson *et al.* (2006) note, more research – particularly molecular studies – needs to be carried out in order to determine whether all the ecads described for saltmarsh habitats truly represent phenotypic variants or whether there are underlying genetic differences.

Fucaceae Adanson

FUCACEAE Adanson (1763), pl. 12.

Thalli macroscopic, from a few cm to several metres in length, erect, branched, yellow olive to brown to dark brown-black, sometimes forming loose-lying populations, but more commonly attached to hard substrata, and often morphologically differentiated into a distinct rounded to irregularly shaped, flattened or conical holdfast, a rounded stipe and flattened frond regions; thalli largely parenchymatous (except for the filamentous medulla), strap to channel-shaped, with or without central midrib; branching dichotomous, pseudodichotomous, with or without additional short subopposite lateral branches, and in the plane of flattening, margin entire or serrated, with or without air vesicles and pseudo-vesicles; structurally complex and comprising a central filamentous medullary tissue of elongated cells, with or without accompanying hyphae, an outer cortical tissue of more isodiametric cells, and an outer, active, single-layered meristoderm contributing to the girth of the thallus; sieve plates present in medullary and cortical cells; growth monopodial by a three-sided, later four-sided apical cell sunk in an apical pit, usually with supplementary growth by the peripheral meristoderm; surface cells with numerous discoid plastids without pyrenoids, and with fucosan vesicles; cryptostomata (sterile cavities) often present, with or without, emergent tufts of hairs.

Reproductive receptacles terminal or lateral on thalli, rounded, cylindrical or dorsiventrally flattened, simple or branched; sexual reproduction oogamous, with oogonia and antheridia borne in sunken, flask-shaped conceptacles within the receptacles, accompanied by simple, or branched, multicellular paraphyses; thalli dioecious or monoecious, the latter with either unisexual or bisexual conceptacles; the three-layer walled oogonia sessile on conceptacle wall, dividing first meiotically, and then mitotically, to produce eight nuclei, all or some of which become functional to form eggs, the number of the latter varying according to genera, e.g. eight eggs formed in *Fucus* spp., four in *Ascophyllum nodosum* and two in *Pelvetia canaliculata*.

This family is represented by three genera in Britain and Ireland, *Ascophyllum*, *Fucus* and *Pelvetia* which are clearly morphologically distinguishable.

Ascophyllum Stackhouse

ASCOPHYLLUM Stackhouse (1809), pp. 54, 66.

Type species: *A. laevigatum* Stackhouse (1809), nom. illeg. (= *Ascophyllum nodosum* (Linnaeus) Le Jolis 1863, p. 96).

Thalli perennial, olive-brown, firm and cartilaginous, up to several metres in height, with usually several primary axes arising from a large discoid base; erect axes at first cylindrical, later linear and flattened, dichotomously branched, with intermittent, annual, large, ovoid to spherical, thick-walled vesicles, and often with adventitious, lateral branches.

Thalli dioecious, with oogonia and antheridia borne on separate thalli, on laterally produced oval receptacles, pinnately borne on short pedicels; oogonium with four functional eggs.

One species in Britain and Ireland.

Ascophyllum nodosum Linnaeus (Le Jolis) (1863) p. 96.   Figs 232, 233

Fucus nodosus Linnaeus (1753), p. 1159; *Fucus nodosus* var. *siliquatus* Turner (1802), p. 253; *Ascophylla laevigatum* Stackhouse (1809), p. 54; *Fistularia nodosa* (Linnaeus) Stackhouse (1816), p. xi; *Halicoccus nodosus* (Linnaeus) Lyngbye (1819), p. 37; *Chordaria scorpioides* (Hornemann) Lyngbye (1819), p. 50; *Halidrys siliquosa* var. *minor* Lyngbye (1819), p. 37; *Fucus nodosus* var. *denudatus* C. Agardh (1820), p. 86; *Fucus nodosus* var. *evesiculosus* J. Agardh (1836), p. 12; *Physocaulon nodosum* (Linnaeus) Kützing (1843), p. 352; *Ozothallia vulgaris* Decaisne & Thuret (1845), p. 13; *Ozothallia nodosa* (Linnaeus) Decaisne & Thuret (1845), p. 13; *Halicoccus nodosus* var. *furcatus* Areschoug (1847), p. 254; *Fucodium nodosum* (Linnaeus) J. Agardh (1848), p. 206; *Fucodium nodosum* var. *scorpioides* (Hornemann) J. Agardh (1848), p. 207; *Ozothallia nodosa* f. *furcata* (Areschoug) Kjellman (1880), p. 8; *Ascophyllum mackayi* f. *robertsonii* Batters (1892b), p. 14; *Ascophyllum robertsonii* (Batters) Batters (1892b), p. 14; *Ascophyllum nodosum* var. *furcatum* (Areschoug) Reinke (1892b), p. 8; *Ascophyllum nodosum* var. *typicum* Reinke (1892b), p. 7, nom. illeg.; *Ascophylla nodosa* (Linnaeus) Kuntze (1891), p. 884; *Ascophyllum nodosum* var. *minor* Batters (1902), p. 51; *Ascophyllum nodosum* var. *siliquatum* (Turner) Batters (1902), p. 51; *Ascophyllum nodosum* var. *lusitanicum* Lami (1939), p. 181; *Ascophyllum nodosum* f. *denudatum* (C. Agardh) Athanasiadis (1996), p. 218.

Thalli erect, macroscopic, usually gregarious, to 1–2(–3) m in length, light/grey-brown to olive-brown, arising from a large, flattened, discoid holdfast; thalli leathery and smooth in texture, somewhat flattened, lacking a midrib, 5–10 mm wide, irregularly dichotomously branched in one plane to several orders, with large air vesicles formed annually along the length, with a slightly serrated margin, and lacking cryptostomata on the surface; main axes often bearing adventitious branches, usually in subopposite clusters, arising from slightly raised, narrow, marginal slits; in cross section, thallus with a large central medulla of firmly aggregated filaments of thick-walled longitudinally elongated, colourless cells giving rise peripherally to a narrow cortex of 4–6, dark-pigmented and photosynthetic, radially elongated cells enclosed within an outer meristoderm of photosynthetic, palisade-like cells that in turn are enclosed by a surface cuticle which is occasionally shed.

Fig. 232. *Ascophyllum nodosum*
A. Habit of vegetative thalli showing large vesicles. B. Habit of reproductive thalli showing terminal receptacles. Darker filamentous alga is epiphytic *Vertebrata lanosa*.

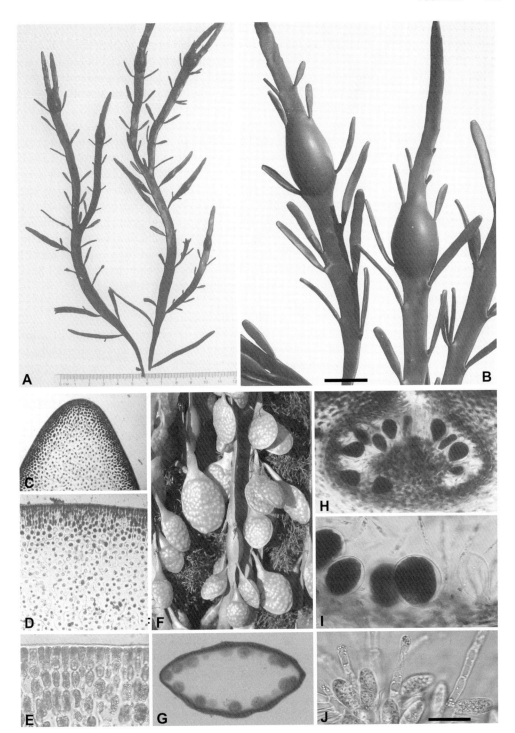

Fig. 233. *Ascophyllum nodosum*
A. Habit of young thalli. B. Terminal regions of thalli showing large vesicles and adventitious branches. C. TS of thallus margin. D. TS of thallus, mid region. E. TS of thallus edge, showing radially elongated cortical cells and outer meristoderm layer. F. Terminal branch region with receptacles. G. TS of receptacles showing immersed conceptacles. H. TS of female conceptacles, showing darkly pigmented oogonia, some of which seem to originate from a central placenta-like growth. I. Part of TS of female conceptacle showing oogonia and associated paraphyses. J. Part of TS of a male conceptacle showing granular antheridia. Bar in B – 15 mm, bar in J = 15 mm (F), = 250 μm (G), = 470 μm (D), = 240 μm (C), = 180 μm (H), = 90 μm (I), = 45 μm (E, J).

Dioecious, with receptacles arising from the tips of the lateral branches; receptacles swollen, ovate, simple, with or without a small sterile margin, 1–2 cm in length, 1–2 cm in width, highly mucilaginous in content and spongy in texture with a central tissue of interconnecting, loosely associated filaments; receptacles bearing conceptacles with either antheridia or oogonia, enclosed within paraphyses; paraphyses arising directly from conceptacle wall, extending into the central cavity, to 180(–260) μm in length, branched, comprising colourless, cylindrical to ovate cells, 2–6 diameters long, 30–72 × 8–16 μm; male receptacles yellow in colour when mature, bearing ovate, sunken conceptacles distributed around the edge, each bearing small, 27–45 × 13–20 μm, slightly pigmented, densely packed, ovate to cigar-shaped antheridia borne laterally on branched paraphyses arising from the walls of the conceptacle; female receptacles grey/green brown in colour bearing densely packed conceptacles containing large, dark-pigmented oogonia dominating conceptacle cavity arising directly from conceptacle wall; oogonia pyriform at first, later more ovate, 82–145 × 53–90 μm with four eggs when mature.

Epilithic, upper to mid eulittoral, covering the surface of substrata, forming a distinct horizontal band above or associated with *Fucus vesiculosus*, absent from exposed localities and usually forming characteristic extensive beds of thalli under sheltered conditions; tolerant of some reduced salinity and recorded in estuaries; frequently epiphytized by the red alga *Vertebrata lanosa*.

Common and widely distributed all around Britain and Ireland.

Perennial; receptacles developing in late summer, developing throughout winter and reaching maximum maturity and biomass in January and February respectively and releasing gametes throughout March and April, occasionally to May and June, depending on geographical locality (see Blackler 1956).

Åberg (1989), pp. 183–90, (1990), pp. 281–7, (1992a), pp. 1473–87, (1992b), pp. 1488–501; Ang *et al.* (1993), pp. 321–6, (1996), pp. 179–84; Araujo *et al.* (2009), pp. 81–92; Ardré (1970), pp. 311–13; Baardseth (1955), pp. 1–67, (1970), pp. 1.1–8.5; Bacon & Vadas (1991), pp. 166–73; Bartsch & Kuhlenkamp (2000), p. 165; Bertocci *et al.* (2011), pp. 73–82; Brackenbury *et al.* (2006), pp. 245–51; Brodie *et al.* (2016), p. 1022; Bunker *et al.* (2017), p. 262; Cabioc'h *et al.* (1992), pp. 72, 73, fig. 49; Cho *et al.* (2006), pp. 512–19; Chock & Mathieson (1976), pp. 171–90, (1978), pp. 21–6; Chopin *et al.* (1996), pp. 543–52; Coppejans & Kling (1995), p. 214, pl. 75(A, B); Cotton (1912), p. 97; Cousens (1981), pp. 253–8, (1982a), p. 231, (1982b), pp. 191–5, (1984), pp. 217–27, (1985), pp. 231–49; Damant (1936), pp. 198–209; David (1943), pp. 178–98; Deckert & Garbary (2005a), pp. 225–32, (2005b), pp. 363–8; Dizerbo & Herpe (2007), p. 125, pl. 41(2); Eckersley & Garbary (2007), pp. 107–16; Engel *et al.* (2003), pp. 180–2; Filion-Myklebust & Norton (1981a), p. 135, (1981b), pp. 45–51; Fletcher in Hardy & Guiry (2003), p. 225; Flothe & Molis (2013), pp. 468–74; Fries (1979), pp. 117–21, (1988), pp. 333–7, (1991), pp. 310–13, (1993), pp. 5–7; Gallardo García (2001b), pp. 29–32, fig. 5; Gallardo García *et al.* (2010), p. 37; Garbary (2017), pp. 297–301; Garbary & London (1995), pp. 529–33; Garbary *et al.* (2005), pp. 353–61, (2009), pp. 239–48, (2014), pp. 321–31; Gayral (1966), p. 317, pl. LXV; Gibb (1957), pp. 49–83; Gómez Garreta (2001), pp. 29–32, fig. 5; Guillaumont *et al.* (1993), p. 297; Guiry & Morrison (2013), pp. 1823–30; Guiry (2012), pp. 167, 168; Halat *et al.* (2015), pp. 599–608; Hamel (1931–39), pp. 375–7, fig. 59(3); Harlin & Craigie (1975), pp. 109–13, (1981), p. 193; Harvey (1846–51), pl. 158; Hession *et al.* (1998), pp. 1–74; Jaasund (1965), p. 111; John (1974), pp. 243, 244; Josselyn & Mathieson (1978), pp. 258–61; Kelly *et al.* (2001), pp. 1–51; Keser & Larson (1984), pp. 83–7; Keser *et al.* (1981), pp. 29–38; Kim *et al.* (1997), pp. 1133–6, (2006), pp. 331–6, (2009), pp. 93–104; Kingman & Moore (1982), pp. 149–53; Kohlmeyer & Kohlmeyer (1972), pp. 109–12; Kornmann & Sahling (1977), p. 168, pl. 92(A–D); Kraberg & Norton (2007), pp. 17–24; Kylin (1947), pp. 84, 85, pl. 18(56); Larsen & Haug (1958), p. 73, (1964), pp. 338–43, (1969), pp. 521–8; Lining & Garbary (1992), pp. 341–9; Lobban & Baxter (1983), pp. 533–8; Longtin & Scrosati (2009), pp. 535–9; MacFarlane (1932), pp. 27–33; Margalet & Navarro (1992), pp. 117–32; Mathieson & Dawes (2017), pp. 205–6, pl. XXXVIII(3–5); Mathieson *et al.* (1982), pp. 331–6; Miller & Vadas (1984), p. 198; Miller *et al.* (2004), pp. 1028–31; Molloy & Hills (1996), pp. 305–10; Morrison *et al.* (2008), pp. 293–303; Morton (1994), p. 44; Moss (1970), pp. 253–60, (1971), pp. 187–93, (1975), pp. 75–80; Müller *et al.* (1982), pp. 1119, 1120; Munda (1964a), pp. 123–6, (1964b), pp. 76–89, (1964c), pp. 158–88; Myklestad & Melsom (1979), pp. 143–51; Newton (1931), pp. 220–2, fig. 139(A–E); Niell & Soneira (1976), pp. 105–10; Olsen *et al.* (2010), pp. 842–56; Parys *et al.* (2009), pp. 331–8; Pearson & Evans (1990), pp. 597–603; Peckol *et al.* (1988), pp. 192–8; Pedersen (2001), pp. 110–12, figs 120–2; Penot (1974), pp. 125–31; Penot & Penot (1976), pp. 1–7, (1977), pp. 339–47, (1981), pp. 217–23; Printz (1926), p. 21, (1950), pp. 3–15, (1956a), pp. 194–7, (1956b), pp. 198–202, (1959a), pp. 1–15, (1959b), pp. 3–28; Ragan & Jensen (1978), pp. 245–58; Rawlence

(1972), pp. 279–90, (1973), pp. 17–28; Rees (1932) pp. 1063, 1064; Rheault & Ryther (1983), pp. 252–4; Rintz (1959), pp. 1–28; Rueness (1973), pp. 446–54, (1977), p. 190, pl. XXV(3), (2001), p. 14; Schneider & Searles (1991), p. 168, fig. 203; Seeley & Schlesinger (2012), pp. 84–103; Sharp (1987), pp. 3–44; Sharp & Bodiguel (2003), pp. 107–14; Sharp *et al.* (1994), pp. 1632–44; Sheader & Moss (1975), pp. 125–32; Sieburth & Tootle (1981), pp. 57–64; Soneira & Niell (1975), pp. 43–59; South & Hill (1970), pp. 1697–701; Stegenga & Mol (1983), p. 119, pl. 46(1); Stegenga *et al.* (1997a), p. 24; Stengel & Dring (1992), p. 101, (1997), pp. 193–202, (1998), pp. 259–68, (2000), pp. 145–61; Strömgren (1986), pp. 311–19; Sundene (1953), pp. 176, 177, (1973), pp. 249–55; Tammes (1954), pp. 114–23; Taylor (1957), pp. 195, 196, pl. 27(1–2); Terry & Moss (1980), pp. 291–301; Toth & Pavia (2003), pp. 411–16; Ugarte (2011a), pp. 401–7, (2011b), pp. 409–16; Ugarte *et al.* (2007), pp. 125–33, (2016), pp. 114, 115; Vadas & Wright (1986), pp. 101–13; Vadas *et al.* (1990), pp. 263–72; White *et al.* (2011), pp. 253–63.

Ascophyllum nodosum (L.) Le Jolis megecad *limicola* ecad *mackayi* Baker & Bohling (1916), p. 350. Fig. 234

Fucus mackayi Turner (1808), pl. 52; *Fistularia mackayi* (Turner) Stackhouse (1816), p. xi ('mackaei'); *Fucus nodosus* γ. *mackayi* C. Agardh (1824), p. 275; *Physocaulon mackayi* (Turner) Kützing (1843), p. 352; *Fucodium nodosum* L. var. β *mackayi* J. Agardh (1848), p. 206; *Ozothallia mackayi* ('*mackaji*') (Turner) Kützing (1849), p. 592; *Ascophyllum mackayi* (Turner) Holmes & Batters (1891a), p. 128; *Ascophyllum mackayi* Holmes & Batters f. *robertsoni* Batters (1892a), p. 175; *Ascophyllum nodosum* var. *mackayi* Cotton (1912), pp. 128, 129; *Ascophyllum nodosum* f. *mackayi* (Turner) Mathieson & Dawes (2017), p. 206, nom. inval.

Thalli macroscopic, free growing, loose-lying and unattached, forming either globular tufts, to 10–25(–50) in diameter or turf-like growths; thalli olive-brown to dark brown, sometimes yellow, slender, cylindrical or pseudo-compressed to flat, linear, margin entire, much branched, attenuate, branching irregularly or regularly dichotomous, radiating outwards from a central point when globular in habit, with no attaching holdfast; branching both apical and lateral; small vesicles usually present and often frequent, sometimes absent, their presence bearing no relationship to thallus age, considerably smaller than those on the fixed form, solitary, usually occurring below the dichotomies, sometimes above, elliptical, 0.5–1 cm long, 0.2–0.7 cm wide; occasionally with the red algal epiphyte *Vertebrata lanosa*; in cross section, thallus with a large central medulla of fairly closely aggregated filaments of thick-walled, longitudinally elongated, colourless cells, surrounded by a narrow cortical band of 4–6, more pigmented cells, enclosed within an outer meristoderm layer of photosynthetic cells.

Receptacles present or absent, if present terminal, more usually lateral near the base of branches, long and tapering, stalked, lanceolate to ovate, simple, sometimes forked, quite mucilaginous; thalli dioecious or monoecious, sometimes with male receptacles more numerous than female receptacles.

Common and locally abundant ('to several acres' – Gibb 1957) in saltmarshes or on muddy or sandy shores in land-locked bays, lying between just above high water neaps to low water springs, 90–140 cm just below the *Pelvetia* zone (Naylor 1928; Gibb 1957), usually on patches of muddy shingle between small boulders on flat parts of the beach, in extreme sheltered conditions near a freshwater source (beach form) or forming turf-like growths in very sheltered localities near a freshwater source.

Frequent at various localities on the west coast of Scotland and on islands lying off the coast (apparently absent from the east and north coasts of Scotland), and reported for Loch Duich (Ross-shire), Lochs Ainort, Beag, Eishort, na Dal, Slapin and Strollamus (Skye), Loch Seaforth (Outer Hebrides), Loch Spelve (Mull), Lochs Glencoul and Laxford (Sutherland), Loch Ranza (Arran), Lochs Eil, Arisaig and Rhue Pier (Inverness-shire), Lochs Feochan and Riddon and Atlantic Bridge (Argyllshire) and Shetland; in Ireland, at various times reported for Strangford Lough (Down), Mullroy Bay (Donegal), Birterbui Bay, Roundstone Bay, Ardbear Bay, Cashel Bay, Ballynakill bay, Costelloe and the Ardfry inlet (Galway) and Bantry Bay (Cork).

Perennial. With respect to reproductive periodicity, Gibb (1957) reported that receptacles make their first appearance in August and continue to appear in an irregular fashion until

Fig. 234. *Ascophyllum nodosum* (L.) Le Jolis megecad *limicola* ecad *mackayi*
A–D. Habit of thalli. E. Part of TS of thallus showing large inner medulla and outer, more pigmented-celled cortex. Bar in D = 2 cm, bar in E = 135 μm.

January, with maximum maturity occurring in the third week of April, with eggs being released up to the middle of June. Cotton (1912) similarly reported the 'ectomorph' (i.e. ecad) to be fertile during winter and early spring. Most the receptacles are then shed during May and June.

This free-living form was first described as *Fucus mackayi* by Turner (1808) based on material collected by Mr J.T. Mackay at Birterbui Bay, near Cashel, County Galway, Ireland. In 1824, C.A. Agardh reduced it to a variety of *Ascophyllum nodosum*, while later Holmes & Batters (1890) gave it specific rank (*Ascophyllum mackayi*), with a distinctive and less common form of it (*A. mackayi* f. *Robertson*) later additionally described by Batters (1892c). Cotton (1912) again reduced its status to an infraspecific rank (*A. nodosum* var. *mackayi*) and this seems to have received universal acceptance. It was suggested by Gibb (1957) that this variety was produced when a whole thallus or parts of thalli of *Ascophyllum nodosum* were subjected to the frequent alternation of high and low salinity. Extreme shelter from wave action is also an absolute necessity (Gibb 1957) and tidal movements should be gentle. It is considered to arise directly from broken pieces of the attached *Ascophyllum* (Gibb 1957), possibly at only certain times of the year (Cotton 1912), and Naylor (1928) described intermediate stages between the two forms, with broken pieces of *Ascophyllum* showing vigorous vegetative growth, and frequent dichotomies producing the tufted growth form. Gibb (1957) described two growth forms – a 'beach' form, as essentially described above, and a newly described 'turf' form, the latter differing from the former in being much shorter and dwarf-like, about 6 cm long, more dense, upright and with uniform growth, the lower parts of the thalli being slightly buried in mud or entangled in the roots of grass and halophytes. The frond was also described as infrequently branched, with both apical and lateral branches, without air-bladders, mainly terete, bent and twisted with the apices irregularly orientated, and usually sterile, or with occasional apical, sometimes intercalary, small, ovoid, simple or divided receptacles in which the conceptacles remained immature. In some localities, this 'turf' form formed a dense continuous growth over areas up to 2–3 m² and was considered by Gibb (1957) to have originated from fragments of the beach form of ecad *mackayi*.

Agardh (1824), p. 275, (1848), pp. 206–97; Baker & Bohling (1916), pp. 350, 351; Bárbara *et al.* (2012), pp. 5–32; Batters (1892a), pp. 13–15, pl. 183(1–5), (1892c), p. 175, (1902), p. 51; Brinkhuis (1976), pp. 325–38, figs 5(a–c), 6(a–b); Brodie *et al.* (2016), p. 1023; Bunker *et al.* (2017), p. 262; Chapman (1941), pp. 324, 325; Cotton (1912), pp. 128, 129; Fletcher in Hardy & Guiry (2003), p. 225; Gibb (1950), p. 256, (1957), pp. 54–78, pls 1(2–3), 2(4, 5), 3(6–8), 4(9, 11); Guiry (2012), p. 168; Harvey (1846–51), pl. 52; Holmes & Batters (1890), p. 85; Lynn (1935a), pp. 2–4, fig. 1; Mathieson & Dawes (2017), p. 206, pl. XXXVIII(8, 9); Mathieson & Hehre (1986), pp. 1–139; Mathieson *et al.* (2001), pp. 1–46, (2006), pp. 283–303; Morton (1994), p. 44; Moss (1971), pp. 187–93; Naylor (1928), pp. 61–8, figs 3, 4(A–C); Newton (1931), pp. 221, 222; Norton & Mathieson (1983), pp. 333–86; Parkes (1958), pp. 325–7; Schneider & Searles (1991), p. 168; South and Hill (1970), pp. 1697–701; Taylor (1957), p. 196, pl. 26(2); Turner (1808), pp. 115, 116, pl. 52.

Ascophyllum nodosum (L.) Le Jolis megecad *limicola* ecad *scorpioides* Hauck (1883), p. 289.

Fucus scorpioides Hornemann (1813), pl. 1479, nom. illeg.; *Chordaria scorpioides* (Hornemann) Lyngbye (1819), p. 50; *Fucus nodosus* β *denudatus* C.A. Agardh (1820), p. 86; *Fucodium nodosum* (L.) var. Y *scorpioides* J.A. Agardh (1848), p. 207; *Ozothallia vulgaris* Decaisne β *denudata* Kützing (1849), p. 592; *Ozothallia vulgaris* var. *scorpioides* Kützing (1860), vol. X, p. 8, pl. 20, fig. γ; *Ascophyllum nodosum* f. *scorpioides* Hauck (1883), p. 289; *Ascophyllum nodosum* var. *scorpioides* (Hauck) Reinke (1889b), p. 33.

Thalli generally similar to that of *Ascophyllum nodosum* megecad *limicola* ecad *mackayi*, but differs in some small respects, notably: thallus macroscopic, light yellowish brown in colour, dwarf-like, forming rounded, loose-lying clumps, to 6(–30) cm across, unattached by disc or with lower parts of thallus embedded in mud; thallus irregularly curled and twisted, almost cylindrical, frequently and unevenly branched, this almost entirely laterally, with branches narrow, cylindrical and elongated; air-bladders very rare or absent; commonly with the algal epiphyte *Vertebrata lanosa*.

Receptacles very rare, spherical or ovoid, 1–2 mm in diameter, on long drooping branches.

Loose-lying on turf, saltmarsh, mud or in tidal ditches above highwater neaps, in saltmarshes and estuaries, sometimes locally abundant. This ecad was considered by Gibb (1957) to have originated from fragments of typical *Ascophyllum nodosum*.

Recorded on the east coast of England, from Northumberland to Kent; in Ireland, recorded in Roundstone Bay, Co. Galway and Strangford Lough, Co. Down.

Perennial. Gibb (1957) observed that production of receptacles seemed to vary from year to year.

This free-living, loose-lying and non-floating form was first described by Hornemann (1813) as *Fucus scorpioides*, based on material collected from the north coast of the Danish island of Fyen. Later, C. Agardh reduced it to a variety of attached *Fucus nodosus* and it has subsequently been generally referred to as *Ascophyllum nodosum* (L.) Le Jolis ecad *scorpioides* (Hornemann) Hauck. It was first described for Britain from the Blackwater marshes, Essex, by Holmes (1888). It arises from detached pieces of the typical attached form, with long periods of contact with water of reduced salinity being primarily responsible for producing the change of form (Gibb 1957). Note that Gibb cast doubt on the record for Strangford Lough and considered that it probably represented a form of the ecad *mackayi*. Lynn (1935a) also referred to a marsh 'var.' of this ecad in Strangford Lough, Co. Down, which she described as 'deeply embedded in mud' with the clumps being 'much smaller, upright and closely packed together to form a close "turf"'. She likened it with a similar marsh form of var. *scorpioides* described for the Blackwater marshes in Essex by Baker (1916).

Agardh (1820), p. 86, (1824), p. 275; Baker & Bohling (1916), p. 350, pl. 29; Batters (1894c), p. 13, (1902), p. 51; Brinkhuis (1976), pp. 325–38, figs 7(a–b), 8(a–b); Brinkhuis & Jones (1976), pp. 339–48; Chapman (1941), p. 325; Chock & Mathieson (1976), pp. 171–90, figs 1–18, (1978), pp. 21–6, (1983), pp. 87–97; Cotton (1912), pp. 129, 130; Gallardo García *et al.* (2016), p. 37; Gibb (1957), pp. 50, 51, 79–81; Hauck (1883), p. 289; Holmes & Batters (1890), p. 85; Kylin (1947), pp. 84, 85, pl. 18(57); Lynn (1935a), pp. 4–6, fig. 2; Mathieson (1989), pp. 419–38; Mathieson & Dawes (2017), p. 206, pl. XXXVIII(6, 7); Mathieson *et al.* (2006), pp. 283–303; Morton (1984), p. 44; Norton & Mathieson (1983), pp. 333–86; Oltmanns (1922) p. 234, fig. 530; Polderman (1979), pp. 256, 257; Rabenhorst (1863), p. 289, fig. 120(c); Reinke (1889b), pp. 33, 34; South & Hill (1970), pp. 1697–701; Taylor (1957), p. 19, pl. 27(3); Webber (1968), pp. 1–253; Webber & Wilce (1971), p. 282.

Ascophyllum nodosum megecad *limicola* ecad *minor*

Fucus nodosus β *minor* Turner (1802), p. 253; *Ascophyllum nodosum* var. *minor* (Turner) Batters (1902), p. 51.

Unattached thalli 10–20 cm long, tufted from the base, and slightly embedded in mud, ovate in section, narrower than the attached form, often curled or bent, branching infrequent and almost entirely lateral, air-bladders scarce, small and slightly dilated; receptacles small and poorly formed, and remaining immature in reproductive development.

Found in very sheltered localities, under the influence of water of reduced salinity, just above highwater neaps. It is considered to arise from small pieces of attached *Ascophyllum nodosum* with which it is usually in association.

Reported for three localities on the west coast of Scotland; Loch Leven, Loch Melfort and Loch Ranza, and one locality on the south coast of England (Hurst Castle saltmarshes, Hampshire); in Ireland, Lough Larne, Co. Antrim.

Gibb (1957) referred this entity to a variety of *Ascophyllum nodosum* rather than an ecad because it is not always free-living. She accepted Clements (1905) terminology in regarding ecads as a form which results from adaptation or a change in morphology due to a new habitat.

Baker & Bohling (1916), p. 350; Batters (1902), p. 51; Gibb (1957), pp. 50, 51, 81; Lynn (1949), pp. 301–4, (1960), pp. 161, 162; Morris (1914); Morton (1994), p. 44; Turner (1802), p. 253.

Addendum to *Ascophyllum nodosum* ecads

Ascophyllum nodosum (L.) Le Jolis megecad *limicola* ecads 1 & 2   Figs 235, 236

Unpublished field studies by Ian Evans in a number of sheltered lochs in West Sutherland, Scotland have revealed two distinct, previously undescribed, possible ecads of *Ascophyllum nodosum*, in addition to the more widespread and common ecad *mackayi* (see Figs 235 and 236 respectively).

1. This ecad is similar to *mackayi* but forms much larger clumps. The olive-brown thalli extend up to 40 cm or more in length, 2–3 mm in width, tapering to 1 mm at the tips, are terete in section, regularly dichotomously branched and are commonly vesiculate, the small ovate to heart-shaped air vesicles measuring 7–15 × 5–9 mm, usually occurring just before a fork. No receptacles have been observed to date. Small quantities of this ecad were found in the *Fucus vesiculosus* zone in Lochs Glencoul and Nedd alongside a freshwater stream at the former (see Fig. 235).

2. This ecad is much larger than ecad 1 and more strongly resembles the morphology of attached *Ascophyllum nodosum*, forming beds of prostrate, unattached, superimposed, olive-brown to yellow thalli. The main axes are monopodial, distinctly flattened, measuring up to 40–50 cm in length, 6(–10) mm in width, dichotomously branched in one plane, with smaller subopposite adventitious branchlets. No air vesicles or receptacles have been observed to date. This ecad has only been observed at one locality, on the south side of Loch Glencoul above the *Fucus vesiculosus* zone (Ian Evans, pers. obs.) (see Fig. 236).

It is likely, therefore, that a wide range of loose-lying, unattached ecads of *Ascophyllum nodosum* can occur in extreme sheltered habitats with characteristics defined by the local environmental conditions, particularly tidal height.

Fucus Linnaeus

FUCUS Linnaeus (1753), p. 1158.

Type species: *F. vesiculosus* Linnaeus (1753), p. 1158.

Thalli perennial, olive-brown to dark brown, erect, arising from a conical disc; erect axes strap-shaped, sometimes spirally twisted, dichotomously or irregularly branched, with distinct midrib, smooth or serrated margins, with or without, innate air-filled bladders, and irregular pseudobladders; cryptostomata common; caecostomata present or absent.

Monoecious or dioecious with antheridia and oogonia borne among paraphyses in conceptacles which are immersed in terminal, variously shaped, simple or branched swollen receptacles which are entire or with wing-like lateral extensions.

Seven species of *Fucus* are currently recognized for Britain and Ireland viz. *F. ceranoides*, *F. cottonii*, *F. distichus*, *F. macroguiryi*, *F. serratus*, *F. spiralis* and *F. vesiculosus*. With the exception of the rare, embedded dwarf species, *F. cottonii*, which is restricted to the upper eulittoral zone in sandy saltmarsh environments, all the remaining *Fucus* species are locally common, macroscopic components of the eulittoral, wrack communities occurring on rocky shores, although four of them (*F. ceranoides*, *F. spiralis*, *F. serratus* and *F. vesiculosus*) do additionally occur in saltmarsh forms. While *F. distichus* is restricted to the north and north-west shores, *F. cottonii*, *F. macroguiryi*, *F. serratus*, *F. spiralis* and *F. vesiculosus* have a much wider distribution pattern.

Species of the genus *Fucus* exhibit great morphological plasticity, as exemplified by the large number of varieties and formae attributed to them, and this has resulted in considerable taxonomic confusion at both the intraspecific and interspecific levels (Baker & Bohling 1916; Burrows & Lodge 1951; Marsden *et al.* 2006; Mathieson *et al.* 2006; Norton & Mathieson 1976;

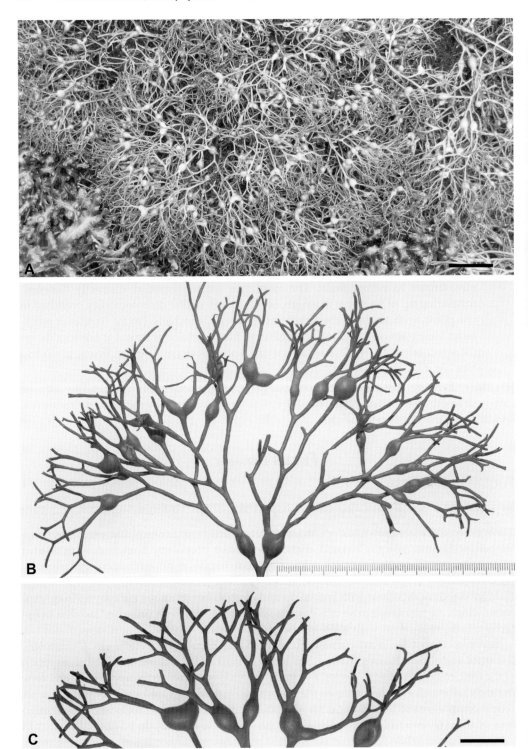

Fig. 235. *Ascophyllum nodosum* (L.) Le Jolis megecad *limicola* ecad 1
West Sutherland, Scotland, collected by Ian Evans. A–C. Habit of thalli showing prominent air vesicles. Bar in A = 50 mm, bar in C = 18 mm.

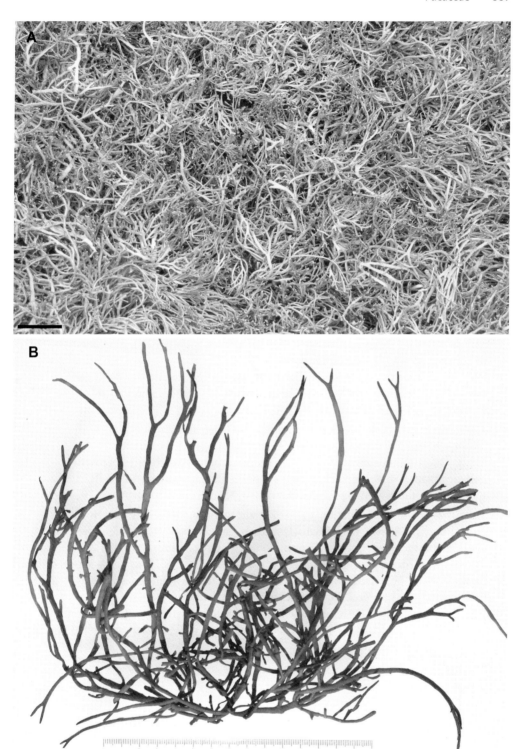

Fig. 236. *Ascophyllum nodosum* (L.) Le Jolis megecad *limicola* ecad 2
West Sutherland, Scotland, collected by Ian Evans. A–B. Habit of thalli. Bar in A = 90 mm.

Perez-Ruzafa & Garcia 2000; Powell 1963a). Such plasticity and phenotypic variability has been shown to be both environmentally and genetically based (see Mathieson *et al.* 2006 for examples and literature). At the interspecific level, hybrids between *Fucus* species, particularly the adjacent *F. spiralis* and *F. vesiculosus*, have resulted in thalli with 'intermediate' morphologies (Burrows & Lodge 1951; Evans *et al.* 1982; Hardy *et al.* 1998; Mathieson *et al.* 2006; Scott & Hardy 1994). Similar hybridization has been reported between *F. serratus* and *F. distichus* subsp. *evanescens* (Coyer *et al.* 2002a). Support for hybridization has also been provided by both pyrolysis mass spectrometry (Hardy *et al.* 1998) and the use of microsatellite markers (Coyer *et al.* 2002b). Hybrids are usually formed at the boundaries between species and are less likely to occur between more spatially distributed species. They also occur commonly in disturbed habitats (Mathieson *et al.* 2006).

Six species in Britain and Ireland, one species in Britain only.

KEY TO ROCKY SHORE SPECIES

1. Thallus edge serrated; receptacles very flat . . . . *F. serratus*
   Thallus edge smooth; receptacles slightly distended to swollen. 2

2. Thallus estuarine, membranous, thin and translucent . . . . *F. ceranoides*
   Thallus predominantly marine, leathery, non-translucent. 3

3. Receptacles globular to ovoid, swollen, entire, usually bearing sterile outer margin rim; caecostomata absent; thallus sometimes spirally twisted; widely distributed around Britain and Ireland 4
   Receptacles compressed elongate, bifid, fusiform, rarely swollen, emarginate; caecostomata present; thallus untwisted; restricted to northern and north-western shores of Britain and/or Ireland. 5

4. Thallus regularly dichotomous throughout and often spirally twisted; receptacles with or without small sterile margin rim . . . . *F. spiralis*
   Thallus dichotomous throughout, often with one branch restricted in growth and terminating in a receptacle giving a monopodial appearance to the thallus; receptacles with a prominent sterile outer margin rim. . . . . *F. macroguiryi*

5. Thallus 2–12(–20) cm in length, < 0.4 cm in width; brown to yellow-brown; stipe thin or thick, stiff; pseudovesicles absent, caecostomata rare, receptacles short, to 4 cm max; on open rock, or in pools, in upper eulittoral zone or littoral fringe on very exposed shores . . . . *F. distichus* subsp. *distichus*
   Thallus 20–30(–50) cm in length, to 0.5–2 cm in width; dark brown; stipe thick; pseudovesicles occasional, caecostomata usually common, receptacles long; on rocks, in pools throughout eulittoral zone, in relative shelter
   . . . . *F. distichus* subsp. *evanescens*

KEY TO SALTMARSH SPECIES/ECADS

1. Thallus edge serrated, receptacles absent. . . . . *F. serratus* ecad *limicola*
   Thallus edge smooth, receptacles present or absent 2

2. Thallus erect, to 6(–10) cm in height, attached at the base 3
   Thallus loose lying, commonly exceeding 10 cm in height, unattached or partially embedded in mud. 5

3. Thallus attached to the sides of steep mud and sand banks, up to 10 cm in height, > 3 mm in width, with distinct midrib, slightly spirally twisted, and usually solitary . . . . *F. spiralis* ecad *nanus*
   Thallus attached to mud flats, to 3(–6) cm in height, to 1(–3) mm in width, without distinct midrib, forming a moss-like turf. 4

4. Thallus irregularly branched, untwisted. . . . . *F. cottonii*
    Thallus dichotomously branched, sometimes spirally twisted
    . . . . *F. vesiculosus* ecad *caespitosus*
5. Thallus wings thin and papery, translucent. 6
    Thallus wings leathery, non-translucent. 8
6. Thallus forming loose lying, unattached, round, globular, tufted clumps on
    shingle beaches; receptacles abundant . . . . *F. ceranoides* ecad *glomerata*
    Thallus forming procumbent, outward-radiating, loose-lying, unattached or
    partly embedded clumps on mud; receptacles rare or infrequent 7
7. Thallus 0.5–1.5 cm in width, forming solitary radiating clumps
    . . . .*F. ceranoides* ecad *proliferatus*
    Thallus to 2.8 cm in width, forming gregarious clumps to 30 cm or more in
    diameter. . . . . *F. ceranoides* ecad *ramosissima*
8. Thallus to 80 cm in length, 1–15 mm broad, much spirally twisted, distinct
    midrib, vesicles numerous. . . . . *F. vesiculosus* ecad *volubilis*
    Thallus to 3(–6) cm long, to 1 mm broad, spirally untwisted or slightly so,
    indistinct midrib, vesicles absent . . . . *F. vesiculosus* ecad *caespitosus*

Fucus ceranoides Linnaeus (1753), p. 1158.

Figs 237, 238

Chondrus ceranoides (Linnaeus) Stackhouse (1797), p. xxiv; *Fucus harveyanus* Decaisne (1868), p. 10; *Fucus divergens* J. Agardh (1872), p. 28; *Fucus ceranoides* f. *divergens* (J. Agardh) Kjellman (1880), p. 8; *Fucus ceranoides* f. *harveyanus* (Decaisne ex J. Agardh) Kjellman (1880), p. 8; *Virsodes ceranoides* (Linnaeus) Kuntze (1891).

Thalli erect, macroscopic, usually gregarious, to 65 cm in length, 14–20(–24) mm in width, olive-brown to brown, arising from a rounded or irregularly shaped, discoid to conical-shaped holdfast; thalli flattened, sometimes terminally twisted, strap-shaped, membranous and delicate, fairly smooth-textured, thin and translucent; margin entire, regularly and repeatedly dichotomously to irregularly branched in one plane to several orders, tapering towards the apex, with a prominent, fine, raised midrib and thinner outer wings which may be worn towards the base leaving only the stalk-like midrib; thalli with or without adventitious laterals and evesiculate, although sometimes with terminal, elongate tube-shape, gas-filled, pseudovesicles each side of the midrib; scattered cryptostomata common, with or without clumps of emergent colourless hairs; caecostomata not observed; in cross section, thalli with a central medulla of densely packed filaments of irregularly shaped and contorted, longitudinally elongated cells, vertically orientated and rounded in cross section in the midrib, more loosely packed, horizontally orientated and longitudinal in cross section in the wing region, giving rise peripherally to a relatively narrow cortex of several layers of irregularly shaped, less obviously longitudinally arranged, darker cells, enclosed by an outer meristoderm layer of photosynthetic, palisade-like cells containing large numbers of discoid plastids without pyrenoids; medulla of basal holdfast, midrib and stalk regions only embedded in hyphal filaments produced from the inner cortical cells.

Thalli mostly dioecious, occasionally monoecious, with terminally positioned corymbose arranged receptacles containing sunken conceptacles bearing oogonia and/or antheridia borne among hyaline or slightly pigmented paraphyses; receptacles not much wider than the supporting branches, somewhat swollen and slightly spongy and mucilaginous in content, dorsiventrally flattened, simple, occasionally bifid, slender, fusiform, linear to lanceolate/ spindle shaped, acuminate, 8.5–18 × 4–7 mm, lacking a midrib; conceptacles prominently surrounded by a sheath of several layers of compacted tissue, hemispherical to teardrop shaped in cross section, 450–600 × 380–500 µm; oogonia large, oval to pyriform, densely pigmented, 44–70(–100) × 38–47(–58) µm, sessile, arising directly from conceptacle wall, and enclosed

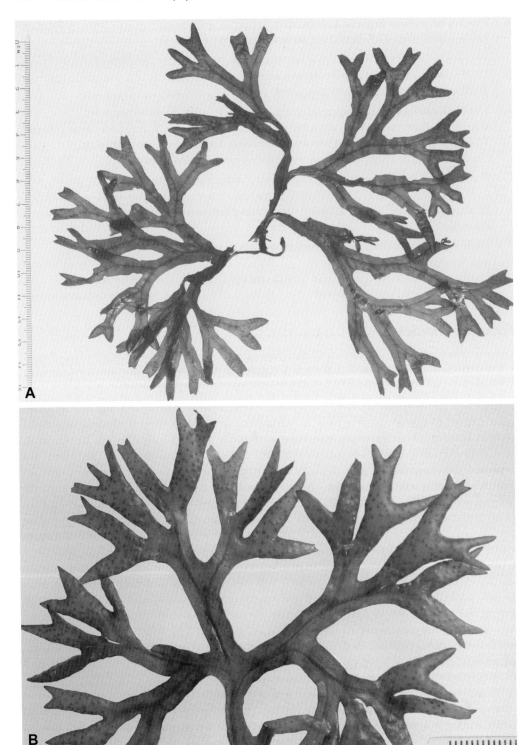

Fig. 237. *Fucus ceranoides*
A. Habit of thallus. B. Upper regions of thallus with terminal receptacles.

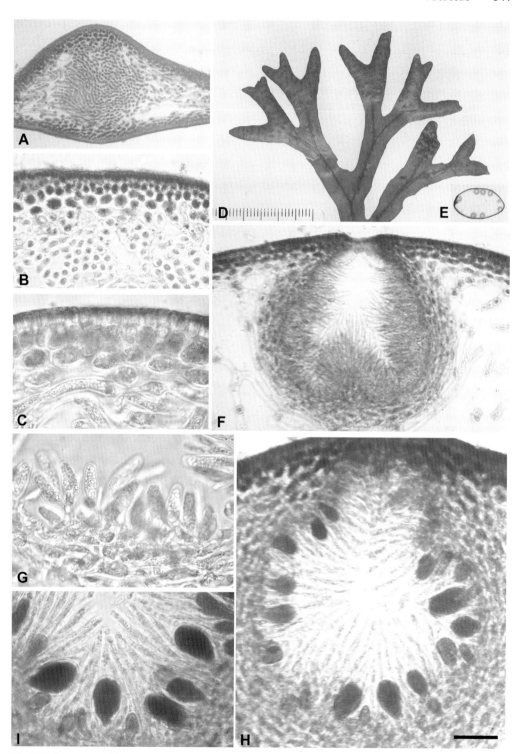

Fig. 238. *Fucus ceranoides*

A. TS of thallus, midrib region. B. TS of thallus edge, midrib region. C. TS of thallus edge, wing region. D. Upper region of branches, with terminal receptacle. E. TS of receptacle with immersed conceptacles. F. TS of male conceptacle. G. TS of male conceptacle, showing antheridia. H. TS of female conceptacle, showing darkly pigmented oogonia with associated filamentous paraphyses. I. TS of part of female conceptacle showing oogonia and paraphyses. Bar in H = 240 µm (A), = 180 µm (F), = 90 µm (B, H), = 45 µm (C, G, I).

within paraphyses; paraphyses to 170 μm, 5–8 cells in length, simple to rarely dichotomously/ pseudodichotomously branched, straight or slightly curved, somewhat acuminate, comprising cylindrical cells, 2–6 diameters long, 18–48(–62) × 6–12(–15) μm; antheridia slightly pigmented and granular, fusiform to elongate-oval in shape, 19–46 × 9–16 μm, borne in quite dense clusters, terminally and laterally, on much-branched paraphyses up to 120 μm, 3–6 cells long, comprising cells 2–6 diameters long, 10–50 × 4–9 μm.

Epilithic on stones, rocks, walls and various hard substrata, upper eulittoral, in areas of brackish water, such as estuaries and tidal inlets under the influence of fresh water; locally common.

Common and widely distributed all around Britain and Ireland.

Perennial with fertile receptacles observed all year round, although mainly April–September.

Ardré (1970), pp. 303, 304; Bäck et al. (1992a), pp. 53–9; Bartsch & Kuhlenkamp (2000), pp. 166, 178; Billard et al. (2005), pp. 900–5; Brodie et al. (2012), p. 1023; Bunker et al. (2017), pp. 21, 264; Cairrao et al. (2009), pp. 205–15; Chater (1927), pp. 362–80; Davy de Virville (1944), pp. 421–52; De Mesquita Rodrigues (1963), pp. 90–2; Dizerbo & Herpe (2007), pp. 125, 126, pl. 41(3); Fletcher in Hardy & Guiry (2003), p. 251; Gallardo García et al. (2016), p. 37; Gard (1923), pp. 294–6; Gayral (1966), p. 315, pl. LXIV; Gómez Garreta (2001), pp. 36–8, fig. 6; Guiry (2012), p. 168; Hamel (1931–39), pp. 373–5, fig. 58(IX); Harvey (1846–51), pl. 271; Khfaji & Norton (1979), pp. 433–9; Lein (1984), pp. 75–81; Margalet et al. (1993), pp. 267–90; Morton (1994), pp. 44, 45; Munda (1964a), pp. 123, 124, a (1964c), pp. 158–88; Naylor (1936), p. 437; Neiva et al. (2010), pp. 4812–22, (2012a), no. 78; Newton (1931), p. 217; Pankow (1970), p. 208, pl. 30(2); Perez-Ruzafa (2001), pp. 36–8, fig. 6; Powell (1963a), p. 74; Rueness (1977), p. 191, pl. xxvi(1), (2001), p. 14; Stegenga & Mol (1983), p. 121, pl. 47(1); Stegenga et al. (1997a), p. 24; Stomps (1911), pp. 326–77.

Fucus ceranoides Linnaeus megecad *limicola* ecad *proliferatus*

Fucus ceranoides var. *limicola* S.M. Baker & M.H. Bohling (1916), p. 354, nom. inval.

Thalli macroscopic, usually much dwarfed, procumbent and radiating over the mud, much curled and interwoven, unattached by discs, but lower parts embedded in ground, 5–10(–15) cm in length, 0.5–1(–1.5) cm in width, papery in texture, surface wrinkled and with prominent midrib; hair-pits conspicuous, marginal and superficial, with projecting tufts of hairs; proliferations common, arising from the hair-pits, produced on the wings and margins of the thallus, and also on the sides of the midrib; adventitious shoots and budding sometimes common, arising from the lower, buried parts of the thallus; pseudovesicles occasional.

Receptacles reported to be either very rare or frequent, and occurring throughout the year, arranged in lateral corymbs; conceptacles variable in their contents, sometimes hermaphrodite, sometimes either male or female and usually with the oogonia remaining undivided.

Occurring on the edges of shallow muddy creeks and in wet places amongst *Puccinellia maritima* and *Juncus maritimus* (Mochas marshes), and in the upper zones of the marsh with *Eleocharis palustris* (Keyhaven, Hampshire).

Reported for a few widely distributed localities: Mochas and Llanbedr saltmarshes on the Artro Estuary, Merioneth in Wales, Keyhaven, Hampshire, Laira Lake, Plymouth, Devon, Berwick-upon-Tweed, Northumberland in England, Loch Creran in Scotland (but see Polderman & Polderman-Hall 1980) and Strangford Lough, County Down in Ireland.

Baker & Bohling (1916), pp. 340–5, 354, figs 9(A, B), 10, 11, 12; Bárbara et al. (2012), pp. 5–32; Chapman (1960); Gallardo García et al. (2016), pp. 7–52; Lynn (1935b), pp. 275–83, (1937), pp. 193, 194; Naylor (1936), pp. 436–8; Polderman & Polderman-Hall (1980), p. 65; Skrine et al. (1932), pp. 769–79, text figs 1–5, pl. XXX(1–8).

Fucus ceranoides Linnaeus megecad *limicola* ecad *glomerata*

Fucus ceranoides (Linnaeus) var. *glomerata* (Batters) Lynn (1935b), pp. 278, 279.

Thalli forming round compact, globular, tufted clumps 15–20 cm in diameter, often with vegetative budding occurring; cryptostomata fairly numerous and superficial.

Receptacles abundant, terminal, spindle-shaped, simple or bifurcating; conceptacles monoecious, either with oogonia or antheridia.

In original report, forming a dense carpet on a shingle beach, over which a small stream flowed, a few thalli attached to stones but the majority loose-lying.

Reported for Berwick-upon-Tweed in Britain and for Strangford Lough, Co. Down in Ireland.

First reported for Britain at Berwick-upon-Tweed, Northumberland, based on a specimen deposited in the Natural History Museum, dated 1895, but not mentioned subsequently in any publication by Batters. It is not clear whether the Berwick specimen was attached or unattached.

Lynn (1935b), pp. 278, 279, pl. 19, fig. 3a, (1937), pp. 193, 194; Morton (1994), p. 45.

Fucus ceranoides Linnaeus megecad *limicola* ecad *ramosissima*

Fucus ceranoides Linnaeus f. *ramosissima* Lynn (1935b), p. 278.

Thalli procumbent, forming loose, unattached masses, to 30 cm or more in diameter, lying on the surface and slightly embedded in mud, between hummocks of *Puccinellia maritima* in brackish saltmarshes; thallus profusely branched, mainly dichotomous, with vegetative budding occurring from broken branches; wings thin and papery, to 28 mm wide, missing from lower regions; cryptostomata common, superficial, sometimes marginal; receptacles rare, and only antheridial conceptacles reported.

Reported for Strangford Lough, County Down, Ireland.

Lynn (1935b), pp. 277, 278, text fig. 2, pl. 19(2a); Morton (1994), p. 45.

Fucus cottonii M.J. Wynne & Magne (1991), p. 59.   Fig. 239

Fucus vesiculosus var. *muscoides* Cotton (1912), pp. 96, 127; *Fucus muscoides* (Cotton) Feldmann & Magne (1964), p. 16, nom. illeg.; *Fucus vesiculosus* megecad *limicola* ecad *muscoides* Baker & Bohling (1916), p. 353.

Thalli macroscopic, very short and dwarf-like, upright, to 5–6 cm in height, 1–3 mm wide, often densely crowded and forming a continuous, thick, moss-like turf, the lower parts embedded in mud; fronds cylindrical or slightly compressed, not spirally twisted, branched, branching fastigiate and irregular, occasionally with tufts of lateral proliferations; cryptostomata marginal.

Dioecious, receptacles on larger thalli, rare, minute, terminal, 2–4 mm in diameter.

Forming a dense, mossy turf on firm, usually well-drained, sandy saltmarshes, often mixed with saltmarsh flowering plants such as *Puccinellia maritima*, *Salicornia europaea*, *Armeria maritima*, *Limonium humile* and *Plantago maritima*.

Reported in Britain for Norfolk, Cardiganshire, Merioneth, Ross and Cromarty, Argyllshire, Orkney and Shetland; in Ireland, Counties Clare, Cork, Limerick, Mayo, Kerry, Donegal, Galway and Down.

This species was originally described by Cotton (1912), as *Fucus vesiculosus* var. *muscoides* for Clare Island, Ireland in view of its turf-forming habit, forming extensive patches on peaty saltmarshes. It was later given specific rank as *Fucus muscoides* (Cotton) Feldmann & Magne but as the name was illegitimate, it was renamed *F. cottoni* by Wynne & Magne. Similar material has subsequently been described throughout the world in saltmarshes. However, a number of transplantation and molecular studies have cast doubt on its taxonomic status as a single genetic

Fig. 239. *Fucus cottonii*

A. Surface view of a population of thalli embedded in mud. B. VS of a population of thalli with lower regions buried in mud. C. Habit of thalli. Bar in A = 3.8 mm, bar in B = 2.7 mm, bar in C = 3 mm.

entity and it might well represent juvenile, dwarf, phenotypic variants of other attached species or their hybrids (Coyer *et al.* 2006b; Mathieson & Clinton 2001; Mathieson *et al.* 2006; Neiva *et al.* 2012a; Sjøtun *et al.* 2017). For example, the study by Coyer *et al.* (2006b) revealed material from Co. Galway, Ireland represented polyploid *Fucus vesiculosus*, while material from Iceland represented *Fucus vesiculosus/F. spiralis* hybrids; the occurrence of dwarf *Fucus cottonii*-like hybrids in saltmarshes was also revealed in studies by Neiva *et al.* (2012b) and Sjøtun *et al.* (2017).

Baker & Bohling (1916), p. 353; Brodie *et al.* (2016), p. 1023; Bunker *et al.* (2017), p. 267; Cabioc'h *et al.* (1992), p. 70, fig. 44; Carter (1933b), p. 397; Chapman (1937), pp. 244, 260, (1938), p. 151; Cotton (1912), pp. 80, 81, 96, 127, 128, pl. 6(1, 2); Coyer *et al.* (2006b), pp. 405–8; De Valéra *et al.* (1979), pp. 259–69; Den Hartog (1959), pp. 1–241; Dizerbo & Herpe (2007), p. 126; Feldmann & Magne (1964), p. 16; Fletcher in Hardy & Guiry (2003), p. 251; Fritsch (1945), p. 385, fig. 137; Fuentes & Niell (1985), pp. 435–8; Guiry (2012), pp. 168, 169; Loisaux-de Goër & Noailles (2008), pp. 1–215; Lynn (1935a), pp. 205, 206, fig. 4; Mathieson & Dawes (2001), pp. 172–201, (2017), p. 210, pl. XXXIX(2); Mathieson *et al.* (2006), pp. 283–303; Morton (1994), p. 45, (2003), p. 131; Neiva *et al.* (2012b), pp. 461–8; Newton (1931), p. 219; Niell *et al.* (1980), pp. 303–7; Polderman & Polderman-Hall (1980), pp. 59–71; Rees (1935), p. 117; Rueness *et al.* (2001), p. 15; Serrão *et al.* (2006), p. 75; Sheehy Skeffington & Curtis (2000), pp. 179–96; Sjøtun *et al.* (2017), pp. 360–70; Skrine (1928), pp. 152, 153; Skrine *et al.* (1932), p. 772; Wallace *et al.* (2004), pp. 1013–27, (2005), pp. 15–17; Wynne & Magne (1991), pp. 55–65.

Fucus distichus Linnaeus (1767), p. 716. Figs 240, 241

Fucus linearis Oeder (1767), pl. 351, nom. illeg.; *Fucus furcatus* C. Agardh (1820), p. 97, nom. illeg.; *Fucus microphyllis* Bachelot de la Pylaie (1829), p. 95; *Fucus miclonensis* Bachelot de la Pylaie (1830), p. 90; *Fucus fuecii* Bachelot de la Pylaie (1830), p. 87; *Fucus anceps* Harvey & Ward ex Carruthers (1864), p. 54; *Fucus distichus* var. *robustior* J. Agardh (1868c), p. 37; *Fucus distichus* var. *tenuior* J. Agardh (1868c), p. 37; *Fucus distichus* var. *miclonensis* (Bachelot de la Pylaie) Kleen (1874), p. 30; *Fucus distichus* f. *robustior* (J. Agardh) Kjellman (1883b), p. 262; *Fucus distichus* f. *tenuior* (J. Agardh) Kjellman (1883b), p. 262; *Virsodes distichum* (Linnaeus) Kuntze (1891); *Virsodes furcatum* Kuntze (1891); *Fucus inflatus* f. *linearis* (Oeder) Rosenvinge (1893), p. 834; *Fucus inflatus* f. *distichus* (Linnaeus) Børgesen (1902), p. 465; *Fucus inflatus* f. *linearis* (Oeder) Børgesen (1902), p. 834; *Fucus fuecii* f. *typicus* E.S. Zinova (1922), p. 132, nom. inval.; *Fucus fuecii* f. *elenkini* E.S. Zinova (1922), p. 133; *Fucus fuecii* f. *ruprechti* E.S. Zinova (1922), p. 133; *Fucus nitens* N.L. Gardner (1922), p. 26; *Fucus evanescens* var. *cuneatus* Setchell in Gardner (1922), p. 39; *Fucus evanescens* var. *ecostatus* Setchell in Gardner (1922), p. 39; *Fucus evanescens* var. *oregonensis* Setchell in Gardner (1922), p. 40; *Fucus furcatus* f. *angustus* Setchell in Gardner (1922), p. 40; *Fucus furcatus* f. *elongatus* Setchell in Gardner (1922), p. 21; *Fucus furcatus* f. *luxurians* Setchell in Gardner (1922), p. 22; *Fucus furcatus* f. *typicus* Setchell in Gardner (1922), p. 16; *Fucus gardneri* P.C. Silva (1953), p. 227; *Fucus distichus* subsp. *anceps* (Harvey & Ward ex Carruthers) H.T. Powell (1957a), p. 421; *Fucus distichus* subsp. *distichus* (Linnaeus) Powell (1957a), p. 420.

Thalli erect, macroscopic, usually gregarious, 2–12(–20) cm in length and 2–3(–6) mm in width, yellow to dark brown, much branched, with the fronds arising from a short, thick, to 2 mm in width, erect-standing stipe attached at the base by a discoid holdfast up to 2 cm in diameter; thalli quite sturdy, coriaceous, flattened, strap-shaped with an entire margin, linear, evesiculate, with a slightly raised midrib becoming indistinct above, and with very narrow lateral wings which are absent towards the base; branching of fronds dichotomous, in one plane, to eight orders, with axils acute and branches decreasing in width above, with the terminal branches characteristically arching over; cryptostomata and caecostomata rare; in cross section, thalli comprising an extensive inner medullary region of anastomosing filaments of longitudinally elongated cells, quite densely packed, vertically orientated and rounded in cross section in the midrib region, more loosely packed, horizontally orientated and longitudinal in cross section in the wing region, giving rise peripherally to a relatively narrow band of 3–6, slightly larger, pigmented cortical cells enclosed within an outer meristoderm layer of photosynthetic palisade-like cells.

Monoecious, with terminally positioned receptacles containing conceptacle bearing both oogonia and antheridia borne among hyaline, or slightly pigmented paraphyses; receptacles only slightly wider than the supporting branches, brown to slightly yellow in colour, dorsiventrally flattened to slightly distended and quite firm throughout, simple, linear to lanceolate,

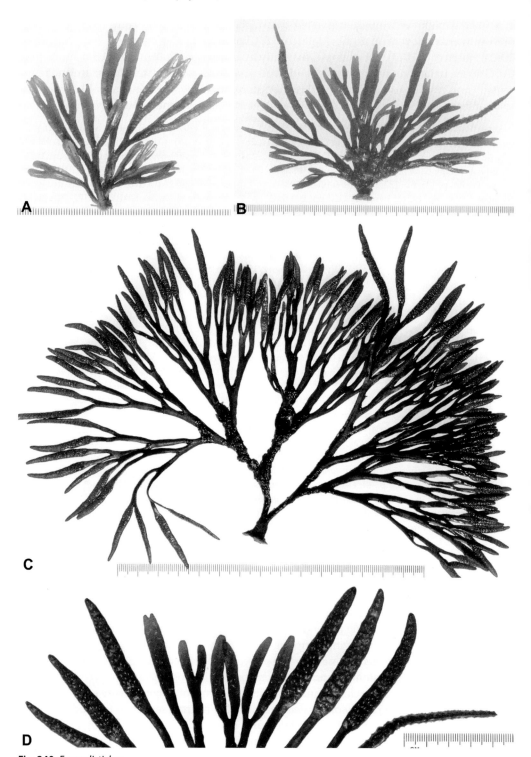

Fig. 240. *Fucus distichus*

A. Habit of small vegetative thallus. B. Habit of small fertile thallus. C. Habit of mature thallus with terminal receptacles. D. Upper regions of mature thallus with terminal receptacles.

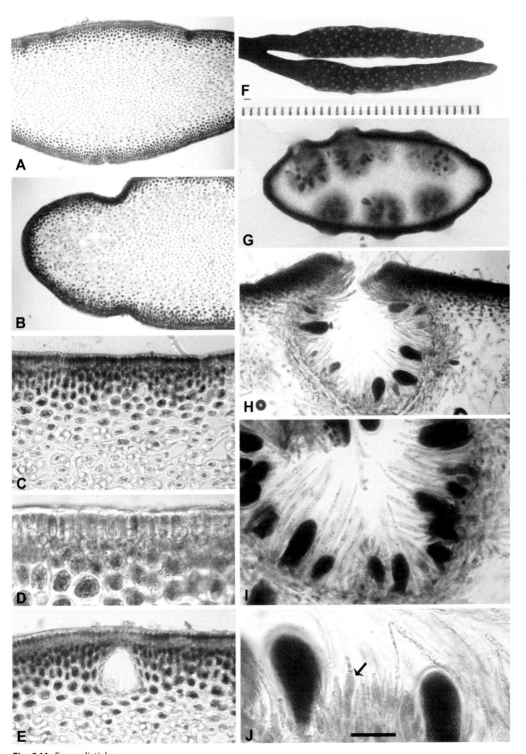

Fig. 241. *Fucus distichus*
A. TS of thallus, midrib region. B. TS of thallus margin. C, D. TS of thallus edge, showing central medulla, outer darker celled cortex with peripheral meristoderm. E. TS of thallus edge, with caecostomata. F. Terminal receptacles. G. TS of receptacle showing the sunken conceptacles. H–J. TS of conceptacles showing the large oogonia, filamentous paraphyses and antheridia (arrow in J). Bar in J = 213 μm (H), = 190 μm (Λ, B), = 120 μm (I), = 63 μm (C, E, J), = 38 μm (D).

acuminate, 14–30(–40) × 3–4.8 mm, lacking a midrib, smooth at first, becoming somewhat warty with slightly protruding conceptacles, each with emergent tuft of hyaline hairs.

Conceptacles hemispherical in cross section, 400–700 × 560–670 µm, comprising a central cavity, densely filled with oogonia, antheridia and paraphyses arising from all over the conceptacle wall; oogonia large, oval to pyriform, densely pigmented, 60–120(–160) × 32–50(–82) µm, sessile, arising directly from conceptacle wall, and enclosed within longer hyaline to slightly pigmented paraphyses; paraphyses to 500 µm, to 5–8 cells, in length, simple to rarely dichotomously branched, straight or slightly curved, acuminate with acute apices, comprising cylindrical to somewhat swollen/misshapen cells, 2.5–10 diameters long, 35–65(–85) × 8–18(–25) µm; antheridia slightly pigmented and granular, elongate-oval in shape, 28–39 × 10–13 µm, borne in dense clusters, terminally and laterally, on short, much-branched, colourless paraphyses.

Epilithic, usually emersed, more rarely in pools, forming a distinct narrow-banded community coinciding with the upper limit of barnacles, on very exposed shores, often on steep ledges/reefs facing north-west or north, and often being either the only fucoid present or mixed with and/or forming narrow patches below *F. spiralis* f. *nana*. In Britain and Ireland, it is the *anceps* morph that is emergent and the *distichus* morph that is submerged in rock pools.

Fairly widespread on exposed northern and north-western shores of Britain and Ireland; Shetland, Fair Isle, Orkney, Caithness, Lewis and Harris, North Rona and Sula Sgeir (Inner Hebrides), St Kilda; in Ireland, Counties Clare, Donegal, Kerry.

Thalli perennial, with fertile receptacles occurring from April to August.

Originally recorded for Ireland from Kilkee, Co. Clare, by Harvey (1864), it was later reported (Powell & Lewis 1952) along 'a limited and very exposed stretch of the north coast of Caithness', and subsequently reported more widely on northern and north-western shores.

Ang (1991), pp. 71–85; Ang & De Wreede (1993), pp. 253–65; Batters (1902), p. 49; Becker *et al.* (2009), pp. 609–16; Bird & McLachlan (1976), pp. 79–84; Bisgrove *et al.* (1997), pp. 823–9; Brawley & Quatrano (1979), pp. 266–72; Brodie *et al.* (2016), p. 1023; Bunker *et al.* (2017), p. 268; Burrows *et al.* (1954), pp. 283–8; Carruthers (1863), pp. 353–5, (1864), p. 54; Coleman & Brawley (2005), pp. 1110–19; Edelstein & McLachlan (1975), pp. 305–24; Fletcher in Hardy & Guiry (2003), p. 252; Gardner (1922), pp. 2–59; Guiry (2012), p. 169; Harvey (1864), p. 389, (1866), pp. 53, 54; Jónsson (1903), pp. 184–91; Kim & Garbary (2009), pp. 93–104; Kloareg & Quatrano (1987), pp. 123–9; Lewis (1954), p. 700; Mathieson & Dawes (2017), pp. 207, 208, pl. XXXVIII(10, 11, 13); McLachlan (1974), pp. 943–51; Morton (1994), p. 8; Motomura (1995), pp. 108–13; Motomura & Nagasato (2009), pp. 140–9; Nagasata *et al.* (2015), pp. 229–38, (2000), pp. 163–6; Newton (1931), p. 217; Pearson & Davison (1994), pp. 257–67; Pedersen (1976), p. 52, (2011), p. 113; Petrov (1965), pp. 64–70; Pollock (1969), pp. 1073–5; Powell (1957a), pp. 407–32, (1957b), pp. 666–84, pls I(1, 2), II(1, 2), III(1, 2), (1958), p. 1246, (1963a), pp. 63–77, (1963b), pp. 247–51, (1981), p. 139; Powell & Lewis (1952), pp. 508, 509; Quatrano *et al.* (1979), pp. 113–23; Rice & Chapman (1985), pp. 433–59; Rice *et al.* (1985), pp. 207–15; Rueness (1977), pp. 191, 192, pl. XXVII(1), (2001), p. 15; Rueter & Robinson (1986), pp. 243–6; Russell (1974), pp. 679–83; Saunders & McDevit (2013), pp. 1–23; Schonbeck & Norton (1981), pp. 475–83; Sideman & Mathieson (1983), pp. 111–27, (1985), pp. 250–7; Spencer *et al.* (2009), pp. 237–60; Strömgren (1985), pp. 263–9; Thom (1983), pp. 471–86; Thomas & Turpin (1980), pp. 479–81; Thomas *et al.* (1985), pp. 267–74; Van Alstyne (1990), pp. 412–16; Vreeland (1972), pp. 258–367.

Fucus distichus subsp. *evanescens* (C. Agardh) H.T. Powell (1957), p. 426.   Figs 242, 243

Fucus vesiculosus var. *inflatus* (Linnaeus) C. Agardh (1810), p. 9; *Fucus evanescens* C. Agardh (1820), p. 92; *Fucus edentatus* Bachelot de la Pylaie (1829), p. 84; *Fucus edentatus* f. *angustior* Bachelot de la Pylaie (1829), p. 84; *Fucus microphyllus* Bachelot de la Pylaie (1829), p. 95; *Fucus bursigerus* J. Agardh (1868c), pp. 35, 41, pl. III; *Fucus evanescens* f. *bersigerus* (J. Agardh) Kjellman (1877b), p. 4; *Fucus evanescens* f. *nanus* Kjellman (1877b), p. 4; *Fucus evanescens* f. *pergrandis* Kjellman (1877b), p. 3; *Fucus evanescens* f. *angustus* Kjellman (1877a), p. 27; *Fucus evanescens* f. *typicus* Kjellman (1877b), p. 3, nom. inval.; *Fucus edentatus* f. *contractus* Kjellman (1883b), p. 256; *Fucus evanescens* f. *norvegicus* Strömfelt (1886), p. 35; *Fucus evanescens* f. *arcticus* Strömfelt (1886), p. 35; *Fucus evanescens* f. *dendroides* Strömfelt (1886), pp. 35, 36; *Fucus evanescens* f. *cornutus* Kjellman (1889), 34; *Fucus evanescens* f. *macrocephalus* Kjellman (1889), p. 34; *Fucus evanescens* f. *rudis* Kjellman (1889), p. 34; *Fucus evanescens* f. *contractrus* Kjellman (1889), p. 35; *Fucus inflatus* var. *finmarkicus* Kjellman (1890), p. 11; *Fucus inflatus* f. *principalis* Kjellman (1890), p. 11; *Fucus inflatus* f. *densus* Kjellman (1890), p. 12; *Fucus inflatus* f. *dilutus* Kjellman (1890), p. 13; *Fucus inflatus* f. *humilis* Kjellman (1890), p. 13; *Fucus inflatus* f. *latifrons* Foslie (1890), p. 67; *Fucus inflatus* f. *latifrons* Kjellman (1890), p. 12; *Fucus inflatus* f. *pygmaeus*

Foslie (1890), p. 67; *Fucus inflatus* f. *reductus* Kjellman (1890), p. 12; *Fucus inflatus* f. *nanus* Kjellman (1890), p. 12; *Fucus inflatus* f. *gracilis* Kjellman (1890), p. 13; *Fucus inflatus* f. *nordlandicus* Kjellman (1890), p. 12; *Virsodes evanescens* (C. Agardh) Kuntze (1891), p. 930; *Fucus inflatus* var. *evanescens* (C. Agardh) Rosenvinge (1893), p. 834; *Fucus inflatus* var. *edentatus* (Bachelot de la Pylaie) Rosenvinge (1893), p. 834; *Fucus inflatus* f. *membranaceus* Rosenvinge (1898), p. 45; *Fucus inflatus* f. *edentatus* (Bachelot de la Pylaie) Børgesen (1902), p. 465; *Fucus inflatus* f. *expositus* Jónsson (1903), p. 190; *Fucus inflatus* f. *murmanicus* E.S. Zinova (1914), p. 299; *Fucus evanescens* f. *magnificus* N.L. Gardner (1922), p. 48; *Fucus distichus* subsp. *evanescens* (C. Agardh) H.T. Powell (1957a), p. 426; *Fucus distichus* subsp. *edentatus* (Bachelot de la Pylaie) H.T. Powell (1957a), p. 424; *Fucus distichus* f. *latifrons* (Foslie) Petrov (1965), p. 68.

Thalli erect, macroscopic, usually gregarious, quite large, 18–30(–50) cm in length, and up to 9–17(–25) mm in width, olive to dark brown, much branched, with the fronds arising from a discoid/conical base; thalli coriaceous, flattened, strap-shaped, linear with an entire margin, evesiculate, although sometimes with simple or forked pseudovesicle-like apical regions, midrib slightly raised and prominent below, to 5(–8) mm in width, gradually flattening and disappearing above, with broad membranous margins throughout, becoming denuded below in aged thalli and stalk-like in appearance to 7 mm in width; branching of fronds regularly dichotomous, sometimes in part alternate, in one plane, to several orders, axils acute; fronds usually with cryptostomata and caecostomata, the latter seen in great abundance in the wings and midrib in the upper regions viewed against the light; in cross section, thalli comprising an extensive inner medullary region of anastomosing filaments of longitudinally elongated cells, quite densely packed, vertically orientated and rounded in cross section in the midrib region, more loosely packed, horizontally orientated and longitudinal in cross section in the wing region, giving rise peripherally to a relatively narrow band of 3–6, slightly larger, pigmented cortical cells enclosed within an outer meristoderm layer of photosynthetic palisade-like cells.

Monoecious, with terminally positioned receptacles containing conceptacles bearing both oogonia and antheridia borne among hyaline, or slightly pigmented paraphyses; receptacles brown to slightly yellow in colour, slightly distended and firm, sometimes becoming swollen and spongy in texture and even inflated and pseudovesicle-like terminally, with or without terminal, sterile region, simple or forked and pincer-like, linear to lanceolate, acuminate, 20–40(–58) × 8–13(–16) mm, lacking a midrib, smooth at first, becoming somewhat verrucose and warty with protruding conceptacles, each usually with emergent tuft of hyaline hairs.

Conceptacles hemispherical in cross section, 330–560 × 430–680 µm, comprising a central cavity, densely filled with oogonia, antheridia and paraphyses arising from all over the conceptacle wall; oogonia large, oval to pear-shaped, densely pigmented, 120–150(–172) × 70–90(–120) µm, sessile, arising directly from conceptacle wall, and enclosed within longer hyaline to slightly pigmented paraphyses; paraphyses to 350 µm, to 5–7 cells, in length, unbranched except near the base, straight or slightly curved, linear with acute apices, comprising cylindrical cells, 3–7 diameters long, 25–55(–70) × 6–9(–15) µm; antheridia slightly pigmented and granular, elongate-oval in shape, 30–46(–50) × 12–15(–17) µm, borne in dense clusters, terminally and laterally, on short, much-branched, colourless paraphyses which occasionally extend further into conceptacle lumen.

Epilithic and usually emersed, more rarely in pools, occurring either as scattered individuals or forming a distinct belt in the mid-littoral zone, above *Fucus serratus* and below *F. spiralis*, usually on north and north-east facing rocks on moderately sheltered shores. In Shetland, Powell (1957a) suggested that this alga grows best under eutrophic conditions caused by sewage or other organic pollution.

Reported for a small number of localities in the extreme north of Britain: Lerwick and Scalloway harbours and districts in Shetland, North Haven on Fair Isle, and in the north-eastern part of the peninsula of Fianuis on the island of North Rona, the latter representing the southern limit for the species. Not reported on mainland Britain and Ireland.

Thalli perennial, with fertile receptacles occurring from April to August.

Fig. 242. *Fucus distichus* subsp. *evanescens*
A. Several fertile thalli attached to rock surface. B. Terminal regions of thalli showing bifid receptacles. Bar in A = 24 mm, bar in B = 16 mm.

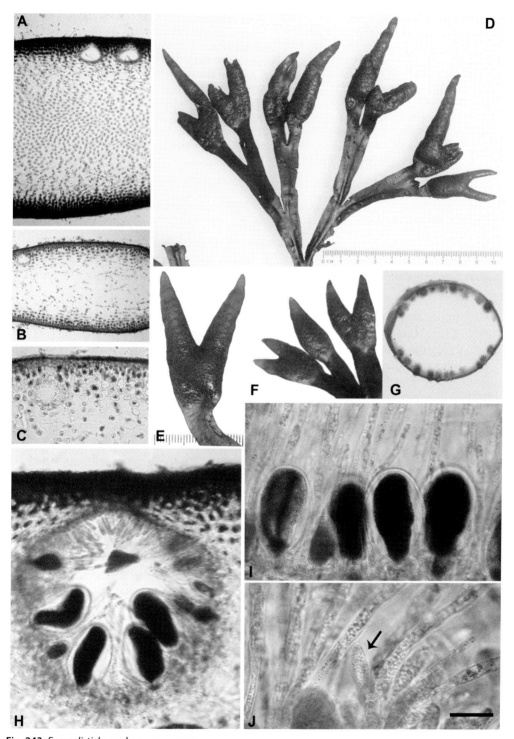

Fig. 243. *Fucus distichus* subsp. *evanescens*
A. TS of thallus, midrib region, showing two caecostomata. B. TS of thallus, wing region. C. TS of thallus, edge of wing region. D. Upper thallus region, showing terminal receptacles. E, F. Terminal receptacles. G. TS of receptacle showing peripheral conceptacles. H. TS of conceptacle. I. TS of portion of conceptacle showing oogonia among filamentous paraphyses. J. TS of portion of conceptacle, showing granular antheridia (arrow) amongst paraphyses. Bar in J = 1.5 mm (B, G), = 570 μm (A), = 90 μm (H), = 43 μm (C), = 10 mm (F), 75 μm (I), = 48 μm (J).

Børgesen (1903), p. 304; Brodie *et al.* (2016), p. 1023; Coleman & Brawley (2005), pp. 1110–19; Collén & Davidson (1997), pp. 643–8; Coyer *et al.* (2002a), pp. 35–7, (2002b), pp. 1829–34, (2006a), pp. 235–46; Fletcher in Hardy & Guiry (2003), p. 252; Gardner (1922), pp. 36, 37, pl. 1(2); Kajiwara *et al.* (1984), p. 1953; Larsen & Sand-Jensen (2005), pp. 13–23; Major & Davidson (1998), pp. 129–38; Mathieson & Dawes (2017), p. 209, pl. XXXVIII(4–16); McLachlan (1974), pp. 943–51; Pedersen (2011), p. 113, fig. 124; Petrov (1965), pp. 64–70; Powell (1957a), pp. 407–32, pl. II(2), (1958), p. 1246, (1963a), pp. 63–77, (1963b), pp. 252–4, (1981), p. 139; Rice & Chapman (1985), pp. 433–59; Rice *et al.* (1985), pp. 207–15; Rueness (1977), pp. 192, 193, fig. 113(A, B), pl. xxvii(2), (2001), p. 15; Russell (1974), pp. 679–83; Schueller & Peters (1994), pp. 471–7; Setchell & N.L. Gardner (1925); Steen & Scrosati (2004), pp. 61–70; Steen (2003a), pp. 26–30; Strömgren (1985), pp. 263–9; Taylor (1957), pp. 191, 193, 194, pls 23(3, 4), 24(2); Thom (1983), pp. 471–86; Tokida (1954), pp. 129, 130; Wikström *et al.* (2002), pp. 510–17.

Fucus macroguiryi Almeida, E.A. Serrão & G.A. Pearson 2022, p. 4802.   Figs 244, 245

Fucus platycarpus Thuret nom. illeg. 1851, p. 9 [non *Fucus platycarpus* Turner, 1809 (= *Botryoglossum platycarpum* (Turner) Kützing)]; *Fucus spiralis* var. *platycarpus* Batters 1902, p. 50.

Note: *Fucus guiryi* Zardi, Nicastro, E.S. Serrão & G.A. Pearson 2011, reported from Britain and Ireland, is a nomenclatural synonym of *Fucus limitaneus* (Montagne) Montagne [type locality: Canary Islands], but the records refer to *Fucus macroguiryi* Almeida, E.A. Serrão & G.A. Pearson.

Thalli forming erect, macroscopic, usually gregarious thalli, to 20(–30) cm in length, light brown, arising from a rounded or irregularly shaped, discoid holdfast; thalli leathery and smooth in texture, dorsiventrally flattened, strap-shaped, irregularly or regularly dichotomously branched in one plane to several orders, becoming monopodial with one branch stopping growth and developing into a receptacle; blades with a distinct prominent midrib and outer thinner wings which may be worn towards the base; branches to 15(–20) mm in width, with an entire margin, evesiculate, although sometimes with terminally positioned, elongated tube-like pseudodiscs on each side of the midrib, with scattered sterile conceptacles or cryptostomata with emergent tufts of hyaline hairs; in section, thalli comprising an extensive inner medullary region of anastomosing filaments of longitudinally elongated cells, quite densely packed, vertically orientated and rounded in cross section in the midrib region, more loosely packed, horizontally orientated and longitudinal in cross section in the wing region, giving rise peripherally to a relatively narrow band of 3–6, slightly larger, pigmented cortical cells enclosed within an outer meristoderm layer of photosynthetic palisade-like cells.

   Monoecious, with terminally positioned hermaphrodite receptacles containing conceptacles bearing both oogonia and antheridia borne among hyaline, or slightly pigmented paraphyses; receptacles firm, dorsiventrally flattened although somewhat swollen and spongy in texture, usually simple, sometimes forked, 18–28(–34) × 10–14(–23) mm, ovate, with rounded or blunt apices, lacking a midrib, smooth at first, becoming quite verrucose and warty with protruding conceptacles, each usually with emergent tuft of hyaline hairs; receptacles usually with prominent, sterile, marginal rim up to 3 mm in extent, which is usually jagged/appearing chewed in outline. Conceptacles hemispherical in cross section, 420–780 × 360–700 µm, quite closely packed, sunken at first, protruding later or when dried, comprising a central cavity, densely filled with oogonia, antheridia and paraphyses closely associated and arising from all over the conceptacle wall; oogonia large, densely pigmented, oval to pear-shaped, 90–125(–153) × 65–100 µm, sessile, arising directly from conceptacle wall, and enclosed within longer hyaline to slightly pigmented paraphyses; paraphyses to 300(–390) µm in length, unbranched except near the base, straight or slightly curved, linear with acute apices, comprising cylindrical cells, 4–6(–8) diameters long, 45–60(–82) × 9–13 µm; antheridia slightly pigmented and granular, cigar to elongate-oval in shape, 17–33 × 9–15 µm, borne in dense clusters terminally and laterally on much-branched, colourless paraphyses extending up to 180(–230) µm into conceptacle lumen.

Fig. 244. *Fucus macroguiryi*
A. Population of fertile thalli attached to rock. B. Habit of fertile thalli showing monopodial branching pattern.

Fig. 245. *Fucus macroguiryi*
A. Portion of thallus showing terminal conceptacles. B. TS of thallus, midrib region. C. TS of thallus, wing region. D. TS of thallus, edge of wing region. E. Receptacles with prominent marginal rim. F. TS of conceptacle. G. TS of portion of receptacle showing oogonia, filamentous paraphyses and clumps of antheridia (arrow). H. TS of conceptacle showing oogonia amongst paraphyses. I. Released oogonium divided into 8 eggs. J. Branched paraphyses bearing granular antheridia. Bar in A = 12 mm (A), = 15 mm (E), bar in J = 190 μm (B, C, F), = 120 μm (G), = 63 μm (D, H, I), = 38 μm (J).

Epilithic, upper culittoral, on moderately exposed shores, forming a distinct horizontal band, usually emersed, rarely in pools, occupying an intermediate position between (and overlapping with) *F. spiralis* (above) and *F. vesiculosus* (below).

Locally common, and probably widespread all around Britain and Ireland. Previously reported (as *F. spiralis* var. *platycarpus*) for the Channel Islands (Lyle 1920), and a number of counties around Britain (Batters 1902) and Ireland (Cotton 1912; Guiry 1978; Morton 1994). After 2021, reported as *F. guiryi* (see Ruiz-Medina *et al*. 2021). Atlantic coasts south to N. Portugal and Morocco (Almeida *et al*. 2022).

Perennial; receptacles developing in late winter/early spring and proceeding through spring until late summer during which gametes are shed and receptacles dehisced.

Previously separated from *Fucus spiralis* (as *F. guiryi* – Zardi *et al*. 2011), *F. macroguiryi* differs in being generally much longer, with wider blades and monopodial branching, and from *F. vesiculosus* by its monopodial branching, the absence of air vesicles, and with usually simple, ovate, rarely bifurcated, hermaphrodite receptacles possessing a very prominent sterile marginal rim. The taxonomic status of *F. macroguiryi* also found support in the molecular study of Zardi *et al*. (2011) who revealed differences with respect to the genome-wide transcriptomic sequence data and to the mtDNA haplotypes and microsatellite genotypes.

Almeida *et al*. (2022), pp. 4797–817; Bartsch & Kuhlenkamp (2000), p. 166; Batters (1902), p. 50; Børgesen (1902), pp. 472–5, fig. 95, (1909), pp. 105–19, pl. 9, (1926), pp. 96–9; Billard *et al*., 2010, pp. 163–74, fig. 6; Brodie *et al*. (2016), p. 1023; Bunker *et al*. (2017), p. 269; Burel *et al*. (2019), p. 24; Cotton (1912), p. 96; Dangeard (1949), pp. 89–189; Davy De Virville (1940), p. 230, pl. IV; De Mesquita Rodrigues (1963), pp. 97–100 pl. Xb; Dizerbo & Herpe (2007, p. 127, pl. 42(1); Fritsch (1945), pp. 324, 326, fig. 114(A); Gallardo *et al*. (2016), p. 37; Gómez Garreta (2001), pp. 46–8, fig. 10, map 7; Guiry, 1978, p. 198, (2012), p. 169; Hamel (1931–39), pp. 365–6, fig. 58(IV); Kornmann & Sahling (1977), p. 164, pl. 89(A–C); Miranda (1931), p. 35; Lyle, 1920; Morton (1994), p. 45; p. 45; Newton (1931), p. 217, fig. 138(A–D); Oltmanns (1922), fig. 438(4); Perez-Ruzafa (2001), pp. 46–8, fig. 10; Perez-Ruzafa & Gallardo (1997), map 7; Powell (1963), p. 71; Peña & Bárbara (2002), p. 46; Price *et al*. (1978), pp. 125, 126; Prinz 2020, 116 pp; Ruiz-Medina *et al*. 2021, 15–26; Sánchez de Pedro *et al*. (2014), pp. 40, 41 (2022), pp. 363–74; Sauvageau (1908b), pp. 65–224, (1909c), 291–5, (1923), p. 20; Thuret (1851), p. 9, pl. II; Zardi *et al*. (2011), pp. 1–13, fig. 2.

Fucus serratus Linnaeus (1763), p. 1626.   Figs 246, 247

Fucus serratus var. *latifolius* Turner (1809), p. 54; *Fucus serratus* var. *integer* Turner (1809), p. 54; *Halidrys serrata* (Linnaeus) Stackhouse (1809), p. 62 (as 'serratus'); *Fucus serratus* f. *typicus* Kjellman (1890), p. 6, nom. inval.; *Fucus serratus* var. *angustifrons* Stackhouse (1816), p. 1, nom. inval.; *Fucus serratus* var. *integerrimus* C. Agardh (1820), p. 95; *Fucus serratus* var. *arcticus* J. Agardh (1868b), p. 9; *Fucus serratus* f. *arcticus* (J. Agardh) Kjellman (1880), p. 8; *Fucus serratus* f. *abbreviatus* Kjellman (1883b), p. 245; *Fucus serratus* f. *angustus* Kjellman (1883b), p. 246; *Fucus serratus* f. *grandifrons* Kjellman (1883b), p. 245; *Fucus serratus* f. *laciniatus* Kjellman (1890), p. 6; *Fucus serratus* f. *elongatus* Kjellman (1890), p. 6; *Virsodes serratum* (Linnaeus) Kuntze (1891), p. 930.

Thalli forming erect, macroscopic, usually solitary, sometimes gregarious thalli, up to 1 m in length, light brown becoming dark brown, arising from a rounded or irregularly shaped, discoid or conical holdfast; thalli leathery in texture, flattened, strap-shaped, with a distinct midrib and outer thinner wings, irregularly dichotomously branched in one plane to several orders, often proliferous below from midribs; evesiculate, although sometimes with irregularly shaped, gas-filled pseudovesicles in sheltered conditions; branches to 30 mm in width, with a prominently serrated margin and scattered sterile conceptacles or cryptostomata; serrations acute, at intervals of about 5 mm; in cross section, thalli comprising an extensive inner medullary region of loosely packed anatomising filaments of longitudinally elongated cells, vertically orientated and rounded in cross section in the midrib region, more horizontally spreading and longitudinal in cross section in the wing region, enclosed by a relatively narrow band of larger, more rounded and pigmented cortical cells, and a peripheral meristoderm layer of pigmented, photosynthetic, palisade-like cells.

Fig. 246. *Fucus serratus*

A, B. Whole vegetative thalli. C. Portion of vegetative thallus. D. Terminal region of vegetative thallus showing inflated apices (pseudovesicles). E. Terminal region of vegetative thallus showing scattered cryptostomata with emergent tufts of hyaline hairs (summer material). Bar in D = 12 mm, bar in E = 12 mm.

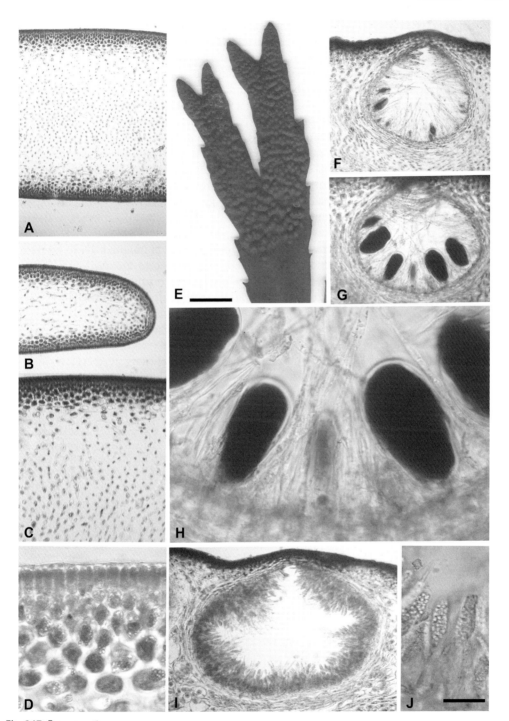

Fig. 247. *Fucus serratus*

A. TS of thallus, midrib region showing extensive inner medullary region with filaments rounded in cross section. B. TS of thallus edge, midrib region. C. TS of thallus, wing region showing filaments horizontal in cross section. D. TS of thallus, edge region, showing outer meristoderm layer. E. Branched terminal receptacles. F, G. TS of female receptacles showing darkly pigmented oogonia surrounded by paraphyses. H. TS of portion of female conceptacle, showing oogonia and associated filamentous paraphyses arising from conceptacle wall. I. TS of male conceptacle showing the clusters of antheridia arising from the paraphyses. J. Portion of male conceptacle showing the granular antheridia. Bar in E = 17 mm, bar in J = 215 μm (A, F, I), = 120 μm (B, C, G), = 63 μm (H), = 38 μm (J).

Dioecious, with reproductive receptacles developing terminally on the thallus; receptacles flattened, or very slightly swollen and usually simple, sometimes forked, with serrated margins, lanceolate to obovate, to 3 cm in length, lacking a midrib, with surface distinctly warty/verrucose in texture due to slightly raised, hemispherical pattern of the enclosed mucilaginous conceptacles; antheridia and oogonia borne in sunken, hemispherical to tear-shaped conceptacles, opening out to the receptacle surface via a small ostiole; female receptacles brown or slightly olive-green, in transverse section, the conceptacles often revealing 3–5 prominent, large, oval, dark-pigmented oogonia enclosed within extensive paraphyses development; paraphyses extending well into conceptacle cavity, to 400 μm in length, simple, or very slightly branched, comprising cylindrical, colourless cells 2.5–9 diameters long, 28–95 × 9–15 μm; oogonia 120–185 × 60–95 μm, with 8 eggs when mature; male receptacles distinctly yellow/orange in colour, in transverse section the conceptacles showing large numbers of small antheridia, arising laterally or terminally from abundantly branched paraphyses; antheridial paraphyses to 150 μm in length and comprising cylindrical colourless cells, 3–8 diameters long, 20–50 × 5–7 μm; antheridia ovate to cigar-shaped, slightly pigmented, 29–43 × 9–14 μm.

Epilithic; lower eulittoral, in pools or emersed, extending into the shallow sublittoral. In conditions of shelter from wave action blades are broad and their serratedness is very pronounced, whereas on exposed shores, serrations are small (almost absent in extreme conditions) and blades narrow.

Abundant and widely distributed all around Britain and Ireland.

Perennial; reproductive phenology has been reported to vary according to geographical locality: from July to December at Plymouth, from August to February at Aberystwyth, from October to the end of February in the Isle of Man and Portsmouth (pers. obs.), and throughout autumn, winter and spring in St Andrews, although fruiting thalli can usually be observed throughout the year in most localities.

Altamirano *et al.* (2003a), pp. 9–20, (2003b), pp. 101–6; Ardré (1970), pp. 304, 305; Armitage *et al.* (2014), pp. 85–97; Arrontes (1993), pp. 183–93, (2002), pp. 1059–67; Astudillo *et al.* (2009), p. 14; Bartsch & Kuhlenkamp (2000), p. 166; Berkaloff & Rousseau (1979), pp. 163–73; Berkaloff *et al.* (1983), pp. 96–100; Bird *et al.* (1979), pp. 521–7; Blackler (1956), pp. 158–62; Boaden *et al.* (1975), pp. 111–36; Bolwell *et al.* (1979), pp. 19–30; Brawley *et al.* (2009), pp. 8239–44; Brenchley *et al.* (1998), pp. 307–13; Brodie *et al.* (2016), p. 1023; Bunker *et al.* (2017), pp. 249, 270; Cabioc'h *et al.* (1992), p. 70, fig. 43; Callow *et al.* (1981), pp. 385–90; Chapman & Fletcher (2002), pp. 894–903; Coppejans & Kling (1995), p. 216, pl. 76; Coyer *et al.* (2002a), pp. 173–8, (2002b), pp. 35–7, (2003), pp. 1817–29, (2004), pp. 1323–8, (2006a), pp. 235–46; Creed *et al.* (1996), pp. 203–9; De Mesquita Rodrigues (1963), pp. 93, 94; Diouris (1989), pp. 504–11; Dizerbo & Herpe (2007), pp. 126, 127, pl. 41(4); Edelstein *et al.* (1971b), pp. 33–42, Engel *et al.* (2003), pp. 180–2; Fletcher in Hardy & Guiry (2003), p. 253; Fredriksen & Christie (2003), pp. 357–64; Gallardo García *et al.* (2016), p. 37; Gayral (1966), p. 311, pl. LXII; Green *et al.* (1992), p. 90; Guiry (2012), pp. 169, 170; Hamel (1931–39), pp. 372, 373, pl. 19(1); Harvey (1846–51), pl. 47; Hoarau *et al.* (2007), pp. 3606–16, (2009), pp. 621–4; Huppertz *et al.* (1990), pp. 175–82; Hurd & Dring (1988), p. 290; Isaeus *et al.* (2004), pp. 301–8; Jensen (1956), pp. 319–25, (1969), pp. 493–500; Johnston & Raven (1990), pp. 75–82; Keen *et al.* (1986), pp. 331, 332; Kim & Thomas (1997), pp. 11–16; Knight & Parke (1950), pp. 439–514; Kornmann & Sahling (1977), pp. 164, 166, pl. 91(A–D); Kremer (1975a,b), pp. 115–27; Kylin (1947), pp. 82, 83, pl. 16(51); Lemoine (1913), pp. 3–5; Liem & Laur (1976), pp. 367–76; Malm & Kautsky (2003), pp. 880–7; Malm *et al.* (2001), pp. 101–8; Margalet *et al.* (1993), pp. 267–90; Marsden *et al.* (1984), pp. 79–83; Masters *et al.* (1991), p. 92; Mathieson *et al.* (2008), pp. 730–41; McLachlan & Bidwell (1978), pp. 371–3; Morton (1994), p. 45, pl. 6; Müller & Seferiadis (1977), pp. 85–94; Newton (1931), p. 219; Pazó & Niell (1977), pp. 455–72; Pearson *et al.* (2010), pp. 195–213; Penot & Videau (1975), pp. 833, 834; Perez-Ruzafa (2001), pp. 40–2, fig. 8; Rueness (1977), p. 193, pl. XXVIII(1), (2001), p. 15; Sauvageau (1909b), pp. 833–44; Spencer *et al.* (2009), pp. 237–60; Stafford *et al.* (1991), p. 97, (1992), pp. 429–34; Steen (2003a), pp. 26–30; Steen & Scrosati (2004), pp. 61–70; Stegenga & Mol (1983), p. 121, pl. 47(2); Stegenga *et al.* (1997a), p. 24; Taylor (1957), p. 194, pl. 26(1); Taylor & Brownlee (1992), p. 101; Thélin (1981), pp. 515–19; Trembling *et al.* (1993), pp. 471–5; Wangkulangkul *et al.* (2016), pp. 63–70; Willenbrink & Kremer (1973), pp. 173–8; Williams (1996), pp. 191–7.

Fucus serratus megecad *limicola* ecad *limicola*

Fucus serratus megecad *limicola* Naylor (1928), p. 61.

Thalli small in comparison to *Fucus serratus*, average length 25 cm, with a smaller width, averaging 1.3 cm, and not so thick; bright yellow-brown when young, becoming darker in older parts, texture firm, less mucilaginous than the attached form, much branched and curled, cryptostomata fewer and smaller and arranged, as in the fixed form, in two rows on each side of the midrib; receptacles absent and thalli sterile, reproducing purely vegetatively.

Found loose-lying, or partially embedded on muddy sand, among attached thalli of *Ascophyllum nodosum, Fucus vesiculosus* and *F. serratus*, or forming extensive mats in conjunction with *Ascophyllum nodosum* var. *mackayi*.

Found in the region of Arisaig (the sheltered Arisaig Loch, and a sheltered coastal cove, Rhue Pier) on the coast of Inverness-shire; likely to be found all around the coasts of Britain and Ireland in similar sheltered regions.

This ecad was described by Naylor (1928) and she expressed little doubt that the form had arisen by vegetative growth from broken pieces of the normal form.

Naylor (1928), pp. 61–8, fig. 2.

Fucus spiralis Linnaeus (1763), p. 1159. Figs 248, 249

Fucus vesiculosus var. *spiralis* (Linnaeus) Roth (1800), p. 443; *Halidrys spiralis* (Linnaeus) Stackhouse (1809), p. 63; *Fucus vesiculosus* var. *spiralis* (Linnaeus) C. Agardh (1810), p. 9; *Fucus sherardii* f. *spiralis* (Linnaeus) Areschoug (1868), p. 106; *Fucus platycarpus* f. *spiralis* (Linnaeus) Thuret (1878), pp. 41, 42; *Fucus vesiculosus* f. *spiralis* (Linnaeus) Batters (1890), p. 304; *Fucus platycarpus* var. *spiralis* (Linnaeus) Rosenvinge in Sauvageau (1897b), p. 211; *Fucus spiralis* f. *borealis* Kjellman (1883b), p. 252; *Fucus areschougi* Kjellman (1890), p. 11; *Fucus areschougi* var. *borealis* (Kjellman) Kjellman (1890), p. 11; *Fucus spiralis* var. *typicus* Børgesen (1902), p. 472, nom. inval.; *Fucus spiralis* f. *arenicola* Hamel (1939), p. 365.

Thalli forming erect, macroscopic, usually gregarious thalli, to 20(–30) cm in length, olive-brown, arising from a rounded or irregularly shaped, discoid holdfast; thalli leathery and smooth in texture, dorsiventrally flattened, strap-shaped, irregularly or regularly dichotomously branched in one plane to several orders, often spirally twisted terminally, with a distinct midrib and thinner outer wings which may be worn towards the base; branches to 16 mm in width with an entire margin, evesiculate, although sometimes with terminal, elongated pseudovesicles, scattered cryptostomata common with emergent tufts of hyaline hairs; in cross section, thalli comprising an extensive inner medullary region of densely packed, anastomosing filaments of longitudinally elongated cells, vertically orientated and rounded in cross section in the midrib region, more horizontally orientated and longitudinal in cross section in the wing region, giving rise peripherally to a relatively narrow band of 3–6, slightly larger, pigmented cortical cells enclosed within an outer meristoderm layer of photosynthetic, palisade-like cells.

Monoecious with terminally positioned receptacles containing conceptacles bearing both oogonia and antheridia borne among hyaline paraphyses; receptacles flattened at first becoming more swollen later, mucilaginous in content and slightly spongy in texture, usually simple, sometimes forked, ovate with rounded apices, or lanceolate with pointed apices, smooth to slightly verrucose, sometimes with a small sterile margin, 11–32 mm in length × 6–13 mm in width, lacking a midrib, with scattered ostioles and emergent tufts of hyaline hairs.

Conceptacles hemispherical in cross section, quite closely packed together in parts, sunken at first, becoming more protruding later, comprising a central cavity with oogonia, antheridia and paraphyses arising from all over the conceptacle wall; oogonia large, dark-pigmented, up to 13 or more seen in cross section, spherical to oval in shape, 90–142 × 65–130 µm, sessile, arising directly from conceptacle wall, and enclosed within hyaline paraphyses; antheridia

Fig. 248. *Fucus spiralis*

A. mixture of young and mature thalli attached to a harbour wall. B. Vegetative thalli attached to wooden breakwater. C, D. Young thalli. E. Vegetative thallus showing dichotomous branching pattern. Bar in A = 26 mm, bar in B = 20 mm.

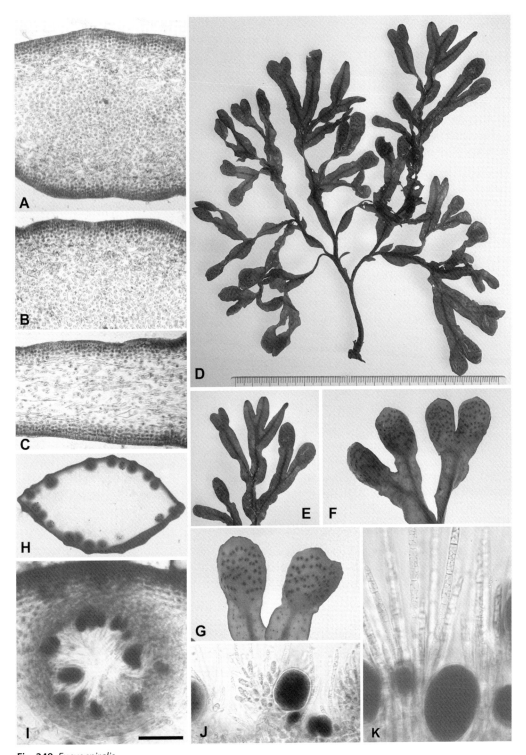

Fig. 249. *Fucus spiralis*

A, B. TS of thallus, midrib region. C. TS of thallus, wing region. D. Habit of thallus. E–G. Portions of thalli showing terminal receptacles. H. TS of receptacle showing immersed conceptacles. I. TS of conceptacle. J. TS through portion of conceptacle showing a large dark-pigmented oogonium and adjacent cluster of granular antheridia. K. TS through portion of conceptacle, showing oogonia surrounded by filamentous hyaline paraphyses. Bar = 190 μm (A, I), = 120 μm (B, C), = 26 mm (E), = 13 mm (F), = 9 mm (G), = 2.7 mm (H), = 63 μm (J, K).

sessile but more commonly borne in dense clusters on branches at the base of paraphyses, slightly pigmented, elongate-ovoid, 24–40 × 9–17 μm; paraphyses to 450 μm or more in length, hyaline multicellular, straight, sometimes slightly curved, narrow towards the apex, comprising cylindrical cells, 2–6 diameters long, 22–50(–83) × 6–12(–17) μm.

Epilithic; upper eulittoral, usually emersed, more rarely in pools, often forming a distinct horizontal band above *Fucus macroguiryi* and below *Pelvetia canaliculata*.

Abundant and widely distributed around Britain and Ireland.

Perennial; receptacles developing in late winter/early spring and proceeding through spring until late summer, during which gametes are shed and receptacles dehisced.

Afonso-Carrillo *et al.* (1988), pp. 73–6; Anderson & Scott (1998), pp. 1003–6; Ardré (1970), pp. 305–9; Bartsch & Kuhlenkamp (2000), p. 166; Billard *et al.* (2005b), pp. 397–407, (2005b), pp. 900–5, (2010), pp. 163–74; Bond *et al.* (1999), pp. 513–21; Børgesen (1909), pp. 105–19, pl. 9, (1926), pp. 96–9; Brodie *et al.* (2016), p. 1023; Bunker *et al.* (2017), p. 271; Cabioc'h *et al.* (1992) p. 68, fig. 40; Cairrar *et al.* (2009), pp. 205–15; Chapman (1989), pp. 565–72, (1990), pp. 205–9; Coleman & Brawley (2005b), pp. 753–62; Coppejans & Kling (1995), pp. 216, 218, 220, pl. 77(A–G); Cormaci *et al.* (2012), pp. 292, 294, pl. 84(1–3); Costas *et al.* (1994), pp. 11–15; Coyer *et al.* (2006b), pp. 405–8; Dangeard (1949); De Mesquita Rodrigues (1963), pp. 95–7; Dizerbo & Herpe (2007), p. 127, pl. 42(1); Engel (2005), pp. 2033–46; Engel *et al.* (2005), pp. 2033–46; Fletcher in Hardy & Guiry (2003), p. 253; Fries (1977), pp. 451–6, (1984), pp. 1616–20, (1993), pp. 5–7; Gallardo García *et al.* (2016), p. 37; Gayral (1958), p. 248, fig. 39(A), pl. XL, (1966), pp. 305, 306, fig. 39(A–C), pl. LX; Gómez Garreta (2001), pp. 42–4, 46–8, figs 9(a), 10; Guiry (2012), p. 170; Hallet *et al.* (1983), pp. 325–36; Hamel (1931–39), pp. 363–6, fig. 58(IV, V); Hazlett & Seed (1976), pp. 607–18; Hurd & Dring (1988), p. 290; Kim *et al.* (2011), pp. 193–200; Kornmann & Sahling (1977), p. 164, pl. 89(A–C); Kylin (1947), p. 83, pl. 17(55); Lopez Rodriguez & Perez-Cirera (1995a), pp. 21–9, (1995b), pp. 25–9; Madsen & Maberly (1990), pp. 24–30; Margalet *et al.* (1993), pp. 267–90; Mathieson & Dawes (2017), pp. 210, 211, pl. LXII(3); Mathieson *et al.* (2006), pp. 283–303; Morton (1994), p. 45; Munda (1986), pp. 341–9; Neiva *et al.* (2010), pp. 4812–22; Newton (1931), p. 217, fig. 138(A–D); Niell *et al.* (1987), pp. 27–32; Niemeck & Mathieson (1976), pp. 33–48; Norris & Conway (1974), pp. 79–81; Perez-Ruzafa (2001), pp. 42–4, fig. 9(a); Perez-Ruzafa & Gallardo García (1997), pp. 121–36; Perrin *et al.* (2007), pp. 219–30; Powell (1960), p. 17; Rees (1932), pp. 1063, 1064; Reyes & Sansón (1999), pp. 53–65; Ribera *et al.* (1992), p. 123; Robertson (1987), pp. 475–82; Rueness (1977), p. 193, pl. XXVIII(2), (2001), p. 15; Sancholle (1988), pp. 157–61; Schonbeck & Norton (1979a), pp. 687–96; Scott *et al.* (2001), pp. 43–50; Serrão *et al.* (1999), pp. 382–94; Spencer *et al.* (2009), pp. 237–60; Stegenga & Mol (1983), pp. 121, 124, pl. 48(1–2); Subrahmanyan (1961), pp. 335–54; Taylor (1957), pp. 191, 192, pl. 24(1); Topinka & Robbins (1976), pp. 659–64; Topinka (1978), pp. 241–7; Wallace *et al.* (2004), pp. 1013–27; Zardi *et al.* (2011), pp. 1–13.

Fucus spiralis Linnaeus megecad *limicola* ecad *nanus*

Fucus areschougii f. *nanus* Kjellman (1890), p. 11; *Fucus spiralis* f. *nanus* (Kjellman) Batters (1902), p. 50; *Fucus spiralis* (L.) var. *nanus* (Kjellman) Batters (1902), p. 50; *Fucus spiralis* (L.) var. *nanus* (Kjellman) Baker & Bohling (1916), p. 351, nom. illeg.

Thallus macroscopic, attached, dwarf, 1–10 cm long, tufted from the base; cryptostomata prominent.

Receptacles present or absent.

This marsh ecad is very similar to the normal attached form but differs in being dwarf. It has also been described as epilithic on rocky shores and phenotypically different (Scott *et al.* 2001).

On steep mud and sand banks in saltmarshes.

Reported in Britain for the Blackwater marshes, Essex and saltmarshes in the Plymouth and Devonport district, south Devon (Hamoaze Estuary), Ross and Cromarty, Argyllshire and Kirkcudbrightshire; reported in Ireland for Roundstone Bay, Clew Bay and Clare Island, County Mayo, Lough Ine, County Cork, Down and Derry.

Anderson & Scott (1998), pp. 1003–6; Baker & Bohling (1916), p. 351; Børgesen (1902), pp. 472, 475–7, figs 96, 97, (1909), p. 109, fig. 3; Chapman (1960); Cotton (1912), p. 124; De Toni (1895), pp. 206, 207; Kjellman (1890), p. 11; Lynn (1935b), pp. 279, 280, pl. 19(4); Montagne (1856), p. 139; Morton (1994), p. 45; Naylor (1936), pp. 425–39; Polderman (1979), pp. 225–66, (1980), pp. 85–95; Polderman & Polderman-Hall (1980), pp. 59–71; Rees (1935), p. 117.

Fucus vesiculosus Linnaeus (1753), p. 1158.   Figs 250, 251

Fucus inflatus Linnaeus (1753), p. 1159; *Fucus divaricatus* Linnaeus (1753), p. 1159; *Fucus excisus* Forsskål (1775), p. 193, nom. illeg.; *Fucus vesiculosus* var. *divaricatus* (Linnaeus) Goodenough & Woodward (1797), p. 144; *Halidrys vesiculosus* (Linnaeus) Stackhouse (1809), p. 62; *Fucus vesiculosus* var. *rigidus* Wahlenberg (1812), p. 490; *Fucus balticus* C. Agardh (1814), p. 29; *Fucus vesiculosus* var. *angustifolius* C. Agardh (1817), p. 5; *Fucus vesiculosus* var. *nanus* C. Agardh (1817), p. 5; *Fucus vesiculosus* var. *subecostatus* C. Agardh (1817), p. 5; *Fucus vesiculosus* f. *acutus* Lyngbye (1819), p. 3; *Fucus vesiculosus* var. *filiformis* C. Agardh (1819), pl. 516; *Fucus vesiculosus* var. *alternans* C. Agardh (1820), p. 90; *Fucus vesiculosus* var. *grandifrons* C. Agardh (1820), p. 88; *Fucus vesiculosus* var. *ceratiformis* Wahlenberg (1826), p. 887; *Fucus vesiculosus* var. *laterifructus* Greville (1828), pl. 319; *Fucus vesiculosus* var. *chondriformis* J. Agardh (1836), p. 9; *Fucus vesiculosus* var. *balticus* (C. Agardh) Areschoug (1847), p. 255; *Fucus axillaris* var. *subecostatus* J. Agardh (1868c), p. 43; *Fucus vesiculosus* subsp. *pseudoceranoides* Areschoug (1868), p. 103; *Fucus vesiculosus* f. *vadorum* Areschoug (1868), p. 102; *Fucus vesiculosus* var. *sphaerocarpus* J. Agardh (1872), p. 29; *Fucus vesiculosus* f. *vadorum* (Areschoug) Kleen (1874), p. 26; *Fucus vesiculosus* f. *pseudoceranoides* (Areschoug) Kleen (1874), p. 27; *Fucus vesiculosus* f. *sphaerocarpus* (J. Agardh) Kleen (1874), p. 27; *Fucus vesiculosus* f. *angustifrons* Gobi (1878), p. 53; *Fucus axillaris* f. *balticus* (C. Agardh) Kjellman (1880), p. 8; *Fucus vesiculosus* var. *spiralis* Farlow (1881), p. 101, nom. illeg.; *Fucus vesiculosus* f. *angustifolius* (C. Agardh) Kjellman (1890), p. 8; *Fucus vesiculosus* f. *filiformis* (C. Agardh) Kjellman (1890), p. 8; *Fucus vesiculosus* f. *nanus* (C. Agardh) Kjellman (1890), p. 8; *Fucus vesiculosus* f. *plicatus* Kjellman (1890), p. 8; *Fucus vesiculosus* f. *subecostatus* (C. Agardh) Kjellman (1890), p. 8; *Fucus vesiculosus* f. *racemosus* Kjellman (1890), p. 9; *Fucus vesiculosus* f. *tenuis* Kjellman (1890), p. 9; *Fucus vesiculosus* var. *rotundatus* Kjellman (1890), p. 6; *Fucus vesiculosus* f. *crispus* Kjellman (1890), p. 7; *Fucus vesiculosus* f. *flabellatus* Kjellman (1890), p. 7; *Fucus vesiculosus* f. *robustus* Kjellman (1890), p. 7; *Fucus vesiculosus* f. *subglobosus* Kjellman (1890), p. 7; *Fucus vesiculosus* f. *terminalis* Kjellman (1890), p. 7; *Fucus vesiculosus* f. *turgidus* (Kjellman) Kjellman (1890), p. 7; *Fucus vesiculosus* f. *vadorum* (Areschoug) Kjellman (1890), p. 7; *Fucus vesiculosus* f. *lanceolatus* Kjellman (1890), p. 9; *Fucus vesiculosus* f. *typicus* Kjellman (1890), p. 7, nom. inval.; *Fucus vesiculosus* f. *fluviatilis* Kjellman (1890), p. 7; *Fucus vesiculosus* f. *abbreviatus* Kjellman (1890), p. 9; *Fucus vesiculosus* f. *latus* Kjellman (1890), p. 9; *Fucus vesiculosus* f. *subfusiformis* Kjellman (1890), p. 8; *Fucus vesiculosus* f. *elongatus* Kjellman (1890), p. 9; *Virsodes vesiculosum* (Linnaeus) Kuntze (1891), p. 929; *Fucus vesiculosus* f. *limicola* F.S. Collins (1906), p. 109; *Fucus vesiculosus* f. *balticus* (C. Agardh) Dannenberg (1927), p. 142; *Fucus vesiculosus* var. *aestuarii* Lami (1938), p. 182; *Fucus vesiculosus* f. *balticus* Levring (1940), p. 60.

Thalli forming erect, macroscopic, usually gregarious thalli, to 50(–80) cm in length, light brown, arising from a rounded or irregularly shaped or discoid holdfast; thalli leathery and smooth in texture, dorsiventrally flattened, strap-shaped, irregularly dichotomously branched in one plane to several orders, with a distinct midrib and outer thinner wings, sometimes proliferous below from worn-down midribs; branches to 20 mm in width, with an entire margin, and with prominent, characteristic air vesicles in wings, ovate to elliptical in shape, singly or more often in opposite pairs either side on the midrib; sometimes also bearing pseudovesicles on either side of midrib, usually terminally positioned, tube-shaped, to 60 mm or more in length; blades also commonly bearing scattered sterile conceptacles or cryptostomata; in section, thallus comprising an extensive inner medullary region of quite densely packed, anastomosing filaments of longitudinally elongated cells, vertically orientated and rounded in cross section in the midrib region, more horizontally orientated and longitudinally spreading in the wing region, giving rise peripherally to a relatively narrow band of 3–6 larger, rounded in cross section, pigmented, cortical cells enclosed within an outer meristoderm of photosynthetic palisade-like cells.

Dioecious, receptacles terminal on branches, containing conceptacles bearing either oogonia or antheridia borne among hyaline, simple or branched paraphyses; receptacles swollen, highly mucilaginous in content and spongy in texture, usually simple sometimes forked, lanceolate to obovate, with acuminate apices, 1–3(–6) cm in length, 7–10(–14) mm in width, ovate in section, lacking a midrib.

Conceptacles hemispherical to tear-shaped in cross section, 360–560 μm high × 320–520 μm wide with a central cavity, closely and evenly distributed over the entire receptacle surface, sunken at first, becoming more protruding and prominent later, with tufts of hyaline hairs emerging through ostioles onto receptacle surface; female receptacles brown or slightly

Fig. 250. *Fucus vesiculosus*
A, B. Terminal regions of thalli showing dichotomous branching pattern and typical vesicles. C. Terminal regions of thalli showing tube-like pseudovesicles. Bar in B = 12 mm (A), = 10 mm (B), = 15 mm (C).

Fig. 251. *Fucus vesiculosus*

A. Terminal regions of mature thalli showing bifurcate receptacles. B. TS of thallus showing extensive inner medulla and outer dark-celled cortex. C. TS of thallus edge, midrib region, showing part of inner medulla, outer dark-celled cortex and outer meristoderm layer. D. TS of thallus edge, wing area, showing inner horizontally spreading medullary filaments, 2–3-celled cortex and outer meristoderm layer. E, F. Terminal swollen receptacles. G. TS of male conceptacle. H. Granular antheridia produced on paraphyses. I. TS. of 2 female conceptacles. J. Oogonia and filamentous paraphyses emerging from conceptacle wall. Bar in A = 23 µm (A), bar in J = 160 µm (B), = 63 µm (C), = 105 µm (D), = 29 mm (E), = 14 mm (F), = 20 µm (G), = 38 µm (H), = 225 µm (I), = 63 µm (J).

olive-green, conceptacles with up to 9 or more, prominent, darkly pigmented, oogonia arising from conceptacle wall borne among the hyaline paraphyses; oogonia club-shaped to elliptical, 80–125 × 42–70 μm; paraphyses unbranched, multicellular, to 330 μm in length, projecting into conceptacle cavity, comprising rectangular cells, 2–6 diameters long, 22–75 × 10–13 μm; male receptacles slightly yellow in colour, conceptacles with large numbers of small, 24–37 × 9–15 μm, ovate, slightly pigmented antheridia borne on much branched, hyaline paraphyses; paraphyses to 130 μm in length, comprising rectangular cells, 2–6 diameters long, 12–50 × 4–9 μm.

Epilithic; mid-eulittoral, usually emersed, more rarely in pools, often forming a distinct horizontal band below *Fucus spiralis*/*F. macroguiryi* and above *Fucus serratus*, often in conjunction/competition with *Ascophyllum nodosum*; tolerant of reduced salinity and found in estuaries, with reduced vesiculation; a non-vesicled form of this species (*F. vesiculosus* var. *evesiculosus*), often described with narrower fronds and with a more stunted growth form, is characteristically found on exposed shores.

Abundant and widely distributed around Britain and Ireland.

Perennial; mainly spring and summer fertile with ripe receptacles releasing gametes from February to July depending on geographical locality (see Blackler 1956b), although occasionally fertile thalli can be observed throughout the year.

Ardré (1970), pp. 292–9, pl. 39(1–3); Bäck *et al.* (1990), p. 84, (1992b), pp. 71–82, (1992c), pp. 39–47; Bartsch & Kuhlenkamp (2000), p. 166; Berger *et al.* (2001), pp. 265–74; Bergström *et al.* (2003), pp. 41–6; Billard *et al.* (2005b), pp. 397–407, (2005a), pp. 900–5, (2010), pp. 163–74; Blackler (1956b), pp. 158–62; Bonsdorff & Nelson (1996), pp. 129–32; Brodie *et al.* (2016), p. 1023; Bunker *et al.* (2017), p. 272; Cabioc'h *et al.* (1992), p. 68, figs 41, 42; Cairrao *et al.* (2009), pp. 205–15; Canovas (2011), pp. 342–51; Carlson (1991), pp. 447–53; Chapman (1990), pp. 205–9; Cho *et al.* (2006), pp. 512–19; Coppejans & Kling (1995), pp. 220, 221, pl. 78(A, B); Cormaci *et al.* (2012), pp. 294, pl. 85(1–6); Coudret *et al.* (1985), pp. 155–61; Coyer *et al.* (2006b), pp. 405–8; Creed *et al.* (1992), p. 87; Derenbach & Gereck (1980), pp. 61–5; Dizerbo & Herpe (2007), pp. 127, 129, pl. 42(2); Engel (2003), pp. 180–2; Engel *et al.* (2005), pp. 2033–46; Eriksson & Johansson (2003), pp. 217–22; Fischer-Piette (1961), pp. 302–16; Fletcher in Hardy & Guiry (2003), p. 254; Flothe *et al.* (2014a), pp. 564–76, (2014b), pp. 356–69; Fulcher & McCully (1969), pp. 219–22; Gallardo García *et al.* (2016), p. 37; Garbary *et al.* (2006), pp. 557–66; Gayral (1958), p. 250, pl. XLI, (1966), p. 309, pl. LXI; Gledhill *et al.* (1999), pp. 501–9; Gómez Garreta (2001), pp. 48–56, figs 11–13; Gorostiaga (1994), pp. 43, 44; Graiff *et al.* (2017), pp. 239–55; Guiry (2012), p. 171; Gylle *et al.* (2009), pp. 156–64; Hamel (1931–39), pp. 367–71, fig. 58(vii, viii); Jaasund (1965), pp. 109, 110; Johannesson *et al.* (2011), pp. 990–8; Jordan & Vadas (1972), pp. 248–52; Kalvas & Kautsky (1993), pp. 85–91, (1998), pp. 985–1001; Kangas *et al.* (1982), pp. 1–27; Kautsky *et al.* (1986), pp. 1–8, (1992), pp. 33–48; Keser *et al.* (1981), pp. 29–38; Khailov *et al.* (1978), pp. 289–311; Kiirikki (1996), pp. 61–6; Kiirikki & Ruuskanen (1996), pp. 133–9; Kim & Garbary (2009), pp. 93–104; Kim *et al.* (1997), pp. 1133–6; King & Schramm (1976), pp. 209–13; Knight & Parke (1950), pp. 439–514; Kornmann & Sahling (1977), pp. 164–6, pl. 90(A–F); Kylin (1947), pp. 83, 84, pl. 17(53–4); Lehvo *et al.* (2001), pp. 345–50; Lemoine (1913), pp. 3–5; López-Rodríguez & Pérez-Cirera (1998), pp. 581–92; López-Rodríguez *et al.* (1999), pp. 129–41; Lund (1959), pp. 168, 169; Luther (1981), pp. 187–200; Malm & Kautsky (2003), pp. 880–7; Malm *et al.* (2003), pp. 256–62; Margalet *et al.* (1993), pp. 267–90; Mathieson & Dawes (2017), pp. 211, 212, pl. XXXIX(7, 8); Mathieson *et al.* (2006), pp. 283–303; Morton (1994), pp. 45, 46; Moss (1948), pp. 267–79, (1950b), pp. 411–19, (1956a), p. 371, (1956b), pp. 387–92, (1958a), p. 21, (1964a), pp. 377–80, (1964b), pp. 117–22, (1965), pp. 387–92, (1966a), p. 154, (1966b), p. 179, (1966c), pp. 31–5, (1967a), pp. 209–12, (1967b), pp. 67–74, (1968), pp. 567–73, (1950a), pp. 395–410; Muhlin & Brawley (2009), pp. 828–37; Müller & Gassmann (1978), p. 389; Müller & Seferiadis (1977), pp. 85–94; Munda & Hudnik (1986), pp. 401–12, (1988), pp. 213–25; Neiva *et al.* (2010), pp. 4812–22; Newton (1931), pp. 217–19; Niell *et al.* (1980), pp. 303–7; Nygard & Dring (2008), pp. 253–62; Nygard & Ekelund (2007), pp. 235–41; Parusel (1991), p. 93; Pearson & Evans (1990), pp. 597–603; Pearson (2010), pp. 195–213; Pedersen (2011), p. 112, fig. 125; Perez-Ruzafa (2001), pp. 48–61, figs 11–15; Perez-Ruzafa & Gallardo García (1997), pp. 121–36; Perrin *et al.* (2007), pp. 219–30; Printz (1959b), pp. 3–28; Raven & Samuelsson (1988), p. 295; Rees (1932), pp. 1063, 1064; Ribera *et al.* (1992), p. 123; Rindi & Guiry (2004), pp. 233–52; Rintz (1959), pp. 1–128; Rohde *et al.* (2008), pp. 143–50; Rönnberg & Ruokolahti (1986), pp. 317–23; Rönnberg *et al.* (1985), pp. 231–44, (1990), pp. 388–92, (1992a), pp. 109–20, (1992b), pp. 95–9; Rothäusler *et al.* (2016), pp. 877–87; Rueness (1977), pp. 193, 194, pl. xxix, fig. 1(a–c), (2001), p. 15; Russell (1979), pp. 659–66, (1985a), pp. 87–104, (1987b), pp. 371 8; Sauvageau (1909b), pp. 833, 834; Schneider & Searles (1991), p. 169, fig. 204; Seoane-Camba (1966), pp. 561–76; Serrão *et al.* (1999), pp. 254–69; Sieburth & Tootle (1981), pp. 57–64; Spencer *et al.* (2009);

Speransky *et al.* (1999), pp. 1264–75; Spilkling *et al.* (2010), pp. 1350–5; Stegenga & Mol (1983), p. 124, pls 48(3), 49; Stegenga *et al.* (1997a), p. 24; Tatarenkov *et al.* (2007), pp. 675–85; Taylor (1957), pp. 192, 193, pl. 25(1–3); Viana *et al.* (2015), pp. 247–59; Wallace *et al.* (2004), pp. 1013–27; Zardi *et al.* (2011), pp. 1–13.

Fucus vesiculosus Linnaeus megecad *limicola* ecad *volubilis*   Fig. 252

Fucus vesiculosus var. *volubilis* Goodenough & Woodward (1797), p. 144; *Fucus vesiculosus* var. *longissimus* Clemente (1807), p. 310; *Halidrys volubilis* (Goodenough & Woodward) Stackhouse (1809), p. 63; *Fucus vesiculosus* var. *monocystus* C. Agardh (1820), p. 91; *Fucus vesiculosus* var. *tricystus* C. Agardh (1820), p. 9; *Fucus vesiculosus* var. *lutarius* Chauvin ex J. Kickx (1856), p. 23; *Fucus lutarius* (Chauvin ex J. Kickx f.) Kützing (1860), p. 7, pl. 17; *Fucus axillaris* var. *spiralis* J. Agardh (1868c), p. 43, nom. illeg.; *Fucus axillaris* var. *divaricatus* J. Agardh (1868c), p. 43, nom. illeg.; *Fucus vesiculosus* var. *axillarius* Sauvageau (1897b), p. 286; *Fucus vesiculosus* var. *axilaris* (J. Agardh) Miranda (1931), p. 35; *Fucus vesiculosus* f. *volubilis* (Goodenough & Woodward) H.T. Powell (1963a), p. 144, nom. inval.; *Fucus vesiculosus* f. *volubilis* (Goodenough & Woodward) A.C. Mathieson & Dawes (2017), p. 212.

Thalli macroscopic, loose-lying or partially embedded in mud, highly variable in size, 4–80 cm in length, 1–8(–16) mm broad and much spirally twisted; frond margin smooth or waved, vesiculate on larger fronds with numerous vesicles, few vesicles or evesiculate on smaller fronds; numerous adventitious 'buds' produced from base of thallus; cryptostomata marginal, numerous and prominent.

Receptacles rare, terminal on branches, hardly broader than the frond, 1–3 cm in length, elongated, oblong or almost acuminate, simple or rarely bifid, turgid, with or without mucilage; thalli dioecious, oogonia remaining immature and undivided.

Occurring in the lower levels of muddy saltmarshes, locally common.

Reported in Britain for Fambridge Ferry, Blackwater Estuary (Essex), Wells-next-the-Sea, Blakeney, Scolt Head, Stiffkey, Thornham (Norfolk), Isle of Wight, Hurst Castle, Keyhaven, Lymington (Hampshire), Poole Harbour (Dorset); reported in Ireland for Lough Ine, County Cork, and County Down.

Baker (1912), pp. 289, 290, figs 5(A, B), 6–8, (1916), pp. 371, 372; Baker & Bohling (1916), p. 352, figs 2–7; Batters (1902), p. 50; Lynn (1935a), pp. 201–8, 1–8 (reprint), fig. 6; Chapman (1937), p. 244; Gayral (1966), p. 313, pl. LXII; Mathieson & Dawes (2017), p. 212, pl. XXXIX(11); Morton (1994), p. 46; Niell *et al.* (1980), pp. 303–7; Polderman (1978), pp. 235–40; Rees (1935), p. 117; Sauvageau (1908b), pp. 106–60, figs 16–19.

Fucus vesiculosus Linnaeus megecad *limicola* ecad *caespitosus*

Fucus vesiculosus var. *subecostatus* C. Agardh (1817), p. 5; *Fucus vesiculosus* var. *balticus* Hooker (1833), no. 267; *Fucus balticus* Hooker (1833), no. 267; *Fucus lutarius* var. *arcassonensis* Sauvageau (1908), p. 131.

Thalli macroscopic, but short, 3–6 cm long, 1 mm broad, lower parts embedded in mud, membranaceous and flattened, dichotomously branched with proliferating branches, slightly or not at all spirally twisted, with an indistinct midrib; air vesicles absent; prominent marginal cryptostomata.

Receptacles rare, terminal, spherical or ovoid, simple or bifid, to 5–8 mm in length, dioecious with oogonia reported to divide into eggs.

Forming a saltmarsh turf mixed with *Puccinellia maritima* and *Salicornia*, very variable in length and width, the smallest thalli resembling *Fucus cottonii*, although less cylindrical, and the largest thalli resembling *F. vesiculosus* ecad *volubilis* but without the midrib and the branches not spiral.

Reported in Britain for Berwick-upon-Tweed (Northumberland), Thornham, Blakeney (Norfolk). Blackwater Estuary (Essex), Hurst Castle (Hampshire), Argyllshire, Renfrewshire, Ayrshire, Carnarvon; reported in Ireland for Clew Bay, Clare Island (Mayo), Lough Ine (Cork), Down, Galway, Leitrim, Kerry and Donegal.

Fig. 252. *Fucus vesiculosus* megecad *limicola* ecad *volubilis*
A. Extensive loose-lying population of thalli on mud surface between erect *Spartina* growths. B. Close-up of loose-lying population of thalli on mud surface. C. Vertical section of thalli emerging from mud. D. Terminal portion of thallus showing spirally twisted blades. Bar in D − 21 mm (B), = 23 mm (C), − 13 mm (D).

Baker & Bohling (1916), pp. 351, 253, fig. 2; Batters (1889), p. 84, (1890b), p. 304, (1902), p. 50; Chapman (1937), p. 244; Greville (1830), p. 12; Harvey (1846–51), pl. 24; Hooker (1833), no. 267; Lynn (1935a), p. 206, fig. 5; Niell *et al.* (1980), pp. 303–7; Rees (1935), p. 117; Sauvageau (1908), p. 131, fig. 20; Skrine *et al.* (1932), pp. 769–79.

Pelvetia Decaisne & Thuret

PELVETIA Decaisne & Thuret (1845), p. 12.

Type species: *P. canaliculata* (Linnaeus) Decaisne & Thuret (1845), p. 13.

Thalli perennial, macroscopic, erect, tufted and gregarious, subcylindrical below, flattened and channelled above, without midrib, air vesicles or cryptostomata, regularly dichotomously branched to several orders, firm and leathery, with several thalli arising from a discoid/conical holdfast.

    Thalli monoecious, with antheridia and oogonia produced in conceptacles immersed in terminal, swollen receptacles; conceptacles containing both oogonia and antheridia; oogonia with two functional eggs.

One species in Britain and Ireland.

Pelvetia canaliculata (Linnaeus) Decaisne & Thuret (1845), p. 13.   Figs 253, 254

Fucus excisus Linnaeus (1753), p. 1159; *Fucus canaliculatus* Linnaeus (1767), p. 716; *Halidrys canaliculata* (Linnaeus) Stackhouse (1809), p. 64; *Fucodium canaliculatum* (Linnaeus) J. Agardh (1848), p. 204; *Ascophyllum canaliculatum* (Linnaeus) Kuntze (1891), p. 884; *Ascophylla canaliculata* (Linnaeus) Kuntze (1891), p. 884; *Pelvetia canaliculata* f. *radicans* Foslie (1893), p. 1; *Pelvetia canaliculata* f. *minima* Simmons (1897), p. 269; *Pelvetia canaliculata* var. *libera* S.M. Baker (1912), p. 289; *Pelvetia canaliculata* var. *coralloides* S.M. Baker (1912), p. 289; *Pelvetia canaliculata* var. *acutilobata* Lami (1939), p. 179; *Pelvetia canaliculata* f. *interposita* Lami (1939), p. 181; *Pelvetia canaliculata* var. *typica* Lami (1939), p. 180, nom. illeg.; *Pelvetia canaliculata* f. *acutilobata* (Lami) Ardré (1961), p. 450, nom. inval.

Thalli forming erect, macroscopic, usually gregarious thalli, to 5–10(–20) cm in length, light/grey brown, greeny brown in colour, with several thalli arising from a rounded or irregularly shaped, discoid/conical holdfast; thalli leathery and smooth in texture, 5–10 mm wide, lacking a midrib but distinctly channelled and appearing U-shaped in cross section, regularly and dichotomously branched in one plane to several orders, evesiculate, with a smooth margin and lacking cryptostomata on the surface; in section, thalli solid or with some tissue breakdown, with an inner medullary region of branched, anastomosing, cross-connected filaments of colourless cells, enclosed by a cortex of several layers (4–6) of darker, thick-walled, longitudinally elongated cells enclosed peripherally by a cuticle-covered, single layered meristoderm of radially elongated, palisade-like cells containing large numbers of densely packed plastids; hypha formation confined to the central medulla-like region at the base of the thalli with the hyphae downward-extending 'en masse' to form the compact tissue of the attachment disc.

    Monoecious, with receptacles terminal on the erect axes; receptacles 11–21 × 3–5.5 mm, somewhat swollen, irregularly nodose, spongy in texture, highly mucilaginous in content, with large central cavity; conceptacles sunken, distributed around the edge of the receptacle but penetrating quite deeply, flask-shaped, hermaphrodite, each bearing both oogonia and antheridia variously distributed around the conceptacle wall and associated with paraphyses; paraphyses to 350 µm in length, projecting deep into central cavity of the conceptacle, trichotomously-dichotomously to pseudodichotomously branched, colourless, comprising rectangular cells, 3–10 diameters long, measuring 20–85 × 8–12 µm; oogonia comparatively large and dominating conceptacle cavity, borne directly on the conceptacle wall, at first elongate-clavate with one ovum, later more ovate and divided internally into two ova, darkly pigmented, 135–200 × 70–115 µm, surrounded by a thick gelatinous wall; antheridia

Fig. 253. *Pelvetia canaliculata*
A. Dense clump of thalli on rock surface. B. Clump of thalli. Bar = 7 cm (A), = 4 cm (B).

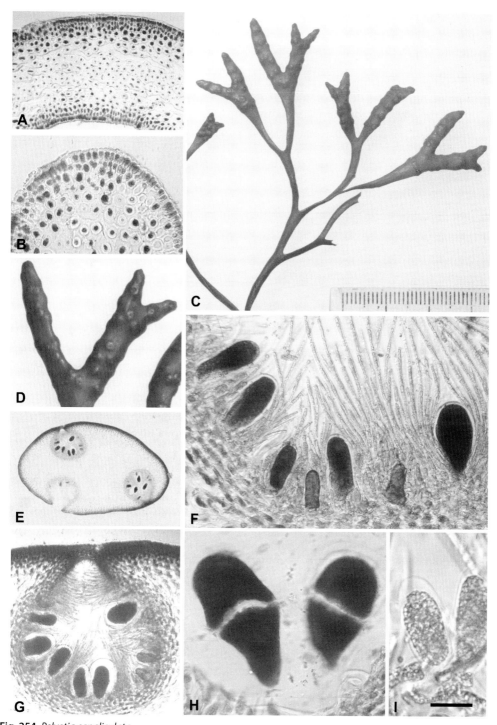

Fig. 254. *Pelvetia canaliculata*
A. TS of thallus, midrib region, showing central medulla and outer darker-celled cortex. B. TS of thallus edge, showing central medullary filaments in cross-section, and outer darker cortical cells. C. Portion of fertile branch showing terminal receptacles. D. Branched verrucose receptacles. E. TS of receptacles showing flasked-shaped conceptacles. F. TS of conceptacles showing large oogonia amongst filamentous paraphyses. G. TS of conceptacle showing oogonia and paraphyses arising from conceptacle wall. H. Two oogonia divided into two eggs. I. TS of part of conceptacle showing granular antheridia. Bar in I = 6 mm (D), = 1 mm (E), = 190 μm (G), = 120 μm (A, F), = 65 μm (B, H), = 40 μm (I).

more abundant, comparatively small, borne terminally/laterally on relatively short-branched paraphyses, ovate, lightly pigmented, 30–60 × 13–24 μm.

Epilithic, upper eulittoral, usually emersed, very rarely in pools, often forming a distinct, sometimes patchy horizontal band above *Fucus spiralis*.

Common and widely distributed around Britain and Ireland.

Perennial, receptacles developing in February, ripening in June, fertile over the summer period and then deteriorating in the autumn.

Ardré (1961), p. 450, (1969), pp. 31–40, (1970), pp. 309–11; Brodie *et al.* (2016), p. 1023; Bunker *et al.* (2017), p. 173; Cabioc'h *et al.* (1992), p. 72, fig. 465; Cho *et al.* (2006), pp. 512–19; Coppejans & Kling (1995), p. 221, pl. 79(B); De Mesquita Rodrigues (1963), pp. 104–7, fig. 9(B–D); Dizerbo & Herpe (2007), p. 129; Elliot & Moss (1953), p. 357; Fischer-Piette (1957), pp. 65–73; Fletcher in Hardy & Guiry (2003), p. 282; Gayral (1966), p. 319, pl. LXVI; Gómez Garreta (2001), pp. 63–8, fig. 16; Goodband (1973), pp. 175–9; Guiry (2012), p. 171; Hamel (1931–39), pp. 377, 378, fig. 59(4); Hardy & Moss (1979b), pp. 203–12; Harvey (1846–51), pl. 229; Isaac (1933), pp. 343–8; Jaasund (1965), p. 111; Kim *et al.* (2011), pp. 193–200; Margalet & Navarro (1992), pp. 117–32; Morton (1994), p. 46; Moss (1974), pp. 317–22; Newton (1931), p. 223, fig. 140(A–E); Oliveira Filho & Fletcher (1977), pp. 1–12, (1980), pp. 409–17; Pfetzing *et al.* (2000), pp. 399–407; Rueness (1977), p. 194, pl. xxix(2), (2001), p. 16; Rugg & Norton (1987), pp. 347–58; Schonbeck & Norton (1979b), pp. 687–96; Stegenga & Mol (1983), p. 124, pl. 46(2); Stegenga *et al.* (1997a), p. 24; Subrahmanyan (1957), pp. 373–95, (1960), pp. 614–30.

Pelvetia canaliculata (Linnaeus) Decaisne & Thuret megecad *limicola* ecad *libera*

Fig. 255

Pelvetia canaliculata var. *libera* Baker (1912), p. 289.

Thalli macroscopic, about the same size as rocky-shore *Pelvetia canaliculata*, loose-lying and unattached, or part embedded in mud, olive-green to dark-brown, to 10(–20) cm in length, forming a continuous matting of bushy clumps on mud surface; frond dorsiventrally flattened, and characteristically channelled, profusely and dichotomously branched, with terminal branches curling away from the ground, usually with numerous adventitious branches arising throughout.

Receptacles rare or absent, normal in appearance.

Locally abundant and interlaced with adjacent thalli to form extensive populations covering mud surface, or in shallow pools, among higher saltmarsh plants, such as *Salicornia annua*, *Juncus maritimus*, *Tripolium pannonicum* and *Armeria maritima*, in sheltered saltmarshes.

Reported in Britain for the sheltered saltmarshes of Blakeney Point and Burnham Overy (Norfolk), and Anglesey; reported in Ireland for Strangford Lough (Down) and Lough Ine (Cork).

Reproduction by budding from various parts of the thallus, rarely by receptacle formation.

Baker (1912), pp. 277, 289, figs 2, 3, pl. 8(1); Baker & Bohling (1916), p. 349; Carter (1932), p. 352, (1933a), p. 397; Chapman (1937), p. 245, (1939), p. 192, (1941), p. 324; Dizerbo (1960), pp. 1–3; Fritsch (1945), pp. 384, 385, fig. 138(E); Lynn (1935b), pp. 280, 281, (1937), pp. 192–5; Naylor (1928), p. 65; Oliver (1913), pp. 12, 13; Oliveira & Fletcher (1977), pp. 1–12, (1980), pp. 409–17, figs 4(A–C), 5, 6; Oliveira & Morton (1994), p. 46; Rees (1935), p. 117; Skrine (1929), pp. 241–3.

Fig. 255. *Pelvetia canaliculata* megecad *limicola* ecad *libera*
A. Large clump of thalli loose-lying on mud surface. B. Outer portion of thallus clump showing dichotomous branching.

Pelvetia canaliculata megecad *limicola* ecad *muscoides*

Thalli similar in appearance to the ecad *libera* but differing in a small number of respects. These include: positioned erect on the surface of mud but not usually entangled amongst the angiosperm vegetation, being more dwarfed in habit, with parts of the thallus embedded in the substrate, and being less proliferous at the margins.

Receptacles not observed.

Reported in Britain for Ynyslas marsh and Mochras saltmarsh at Llanbedr, Merioneth.

Carter (1932), p. 352, (1933a), pp. 204, 206, fig. 1, (1933b), pp. 397, 398; Chapman (1941), p. 324; Oliveira & Fletcher (1980), p. 416, fig. 6; Skrine (1928), p. 152, (1929), pp. 241–3, figs A–C; Skrine *et al.* (1932), p. 772.

Pelvetia canaliculata megecad *limicola* ecad *coralloides*

Pelvetia canaliculata var. *coralloides* Baker (1912), p. 289.

Thalli channelled, somewhat curled and similar in appearance to the ecad *libera* but differing in a small number of respects. These include: being more dwarfed in habit, from 1–4 cm in height, with the basal regions embedded in the mud, branching sparse and being less proliferous at the margins with branches almost cylindrical.

Receptacles unknown.

Rare, reported in Britain from the saltmarsh at Blakeney Point, Norfolk, growing amongst *Salicornia europaea*.

Baker (1912) noted the similarity in habit between this ecad collected at Blakeney and *Pelvetia canaliculata* ecad *radicans* Foslie (1894, p. 6, pl. 1, fig. 2) although the latter was larger and about the same size as normal attached *Pelvetia canaliculata*. Chapman (1960) was doubtful of the existence of the ecad *coralloides* as it could not be found in the type locality and no specimens were available for examination. Two visits by the present author to Blakeney Point also failed to find any material. However, pending further studies the ecad is retained for now.

Baker (1912), p. 289, fig. 4; Baker & Bohling (1916), p. 349; Chapman (1937), p. 245, (1941), p. 324, (1960), pp. 1–392; Lynn (1935b), p. 280; Oliveira & Fletcher (1980), p. 416, fig. 6.

Himanthaliaceae (Kjellman) De Toni

HIMANTHALIACEAE (Kjellman) De Toni (1891), p. 173.

Vegetative thalli macroscopic, small, at first pear-shaped, becoming cone-shaped and then button-shaped, to 5 cm in diameter, stalked and narrowing below to a small discoid holdfast; thalli largely parenchymatous (except for the filamentous medulla), simple, upper surface smooth, coriaceous and concave, lower surface rough and convex; in cross section, comprising an inner medulla, outer cortex and peripheral meristoderm; growth by a three-sided apical cell sunken in an apical pit; surface cells with fucosan vesicles and numerous discoid plastids without pyrenoids.

Reproductive organs borne in 1–2 m long, yellow to medium brown, dichotomously branched, linear, strap-shaped receptacles emerging from central region of vegetative thallus; receptacles bearing sunken, flask-shaped conceptacles with oogonia and antheridia; thalli dioecious and receptacles differing slightly in size; oogonia bearing a single egg.

Life history essentially diplontic and monomorphic, but with the vegetative diploid thallus acting like a sporophyte, and the young oogonium considered to represent a unilocular sporangium meiotically producing highly reduced gametophytes.

This family is represented by a single genus *Himanthalia* and a single species *H. elongata*.

Himanthalia Lyngbye

HIMANTHALIA Lyngbye (1819), p. xxix.

Type species: *H. lorea* (Linnaeus) Lyngbye (1819), p. xxix (= *Himanthalia elongata* (Linnaeus) S.F. Gray (1821), p. 389).

Thalli biennial/triennial, and clearly differentiated into a small vegetative button-shaped thallus and a large, fertile, strap-like thallus; vegetative thallus at first distinguished as pear-shaped vesicles, later becoming mushroom- and then button-shaped and up to 5 cm in diameter with a smooth, concave, upper surface and a rough convex lower surface, the centre of which extends down into a discoid holdfast. From apical pit depressions in the centre of the concave upper surface, 1–4 erect strap-shaped, dichotomously branched, quite mucilaginous receptacles emerge, to several metres in length, 1–2 cm broad, bearing, except in the proximal regions, the male and female conceptacles containing the antheridia and single-egged oogonia respectively; thalli dioecious and differing slightly in size.

One species in Britain and Ireland.

Himanthalia elongata (Linnaeus) S.F. Gray (1821), p. 389.   Figs 256, 257

Fucus elongatus Linnaeus (1753), p. 1159; *Fucus loreus* Linnaeus (1767), p. 716; *Fucus pruniformis* Gunnerus (1772), p. 89; *Fucus tomentosus* Hudson (1778), p. 584, nom. illeg.; *Ulva tomentosa* (Hudson) de Candolle in Lamarck & de Candolle, (1805), p. 6; *Funicularius tuberculatus* Roussel (1806), p. 91, nom. illeg.; *Himanthalia lorea* (Linnaeus) Lyngbye (1819), pp. xxix, 36; *Himanthalia elongata* var. *β inequalis* S.F. Gray (1821), p. 389.

Thalli erect, macroscopic, usually gregarious, at first small, pear- or trumpet-shaped, soft and filled with a watery fluid, later becoming button-like, extending to 5 cm in diameter, solid, olive-brown becoming dark shiny brown in colour, arising from a short (0.5–2 cm) cone-shaped, at first hollow, later solid, stalk, 0.5–2 cm in length attached at the base by a small, spreading disc; buttons smooth and leathery in texture, flexible to the touch, slightly inflated and usually wavy/undulating around edge, the upper surface concave, with one to several light coloured apical pits at first, later giving rise to one or more erect, strap-shaped receptacles (see below), and usually free of epiphytes, the lower surface slightly convex and frequently overgrown, along with the stalk, with epiphytes on older specimens; in cross section, thalli with an inner medullary region of branched, anastomosing, cross-connected filaments of longitudinally elongated, colourless cells enclosed by an outer cortex of several, progressively thinner walled, longitudinally elongated cells, enclosed peripherally by a meristem of longer, narrow, radially elongated, palisade-like cells; meristoderm and outer cortex cells with numerous densely packed plastids; branched, septate, narrow hyphae commonly produced from both the medullary and inner cortical cells, which grow downwards en masse to form a solid central tissue in the button and stipe, finally emerging at the base to form the attachment disc; meristoderm of the upper and lower surface of the button differing in development, that of the lower surface remaining a single layer of short, broad cells, while that of the upper surface (and button margin) becomes multicellular and tissue-like, by dividing repeatedly to give rise to vertical rows of branched, tightly packed filaments of approximately five narrow, elongated cells.

Thalli dioecious, with 1–4 receptacles arising from the upper surface of the vegetative buttons, usually in the second or third year; receptacles yellow to olive-brown at first, later dark brown, dorsiventrally flattened, strap-shaped, linear, and smooth edged in young specimens, more undulating in older specimens, attenuate, very flexible, to 2 m in length, 18 mm in width, regularly and equally dichotomously branched to several orders, smooth at first, later becoming rough and tuberculate, bearing immersed flask-shaped conceptacles over

Fig. 256. *Himanthalia elongata*
A. Dense belt of receptacles lying on rock. B. Button-like vegetative thalli. C. Vegetative button giving rise centrally to several, young, dichotomous receptacles. D. Single thallus with two erect branched receptacles arising from basal vegetative button. E. Upper portion of receptacle showing dichotomous branching pattern. Bar in E = 7 mm (B), 35 mm (E).

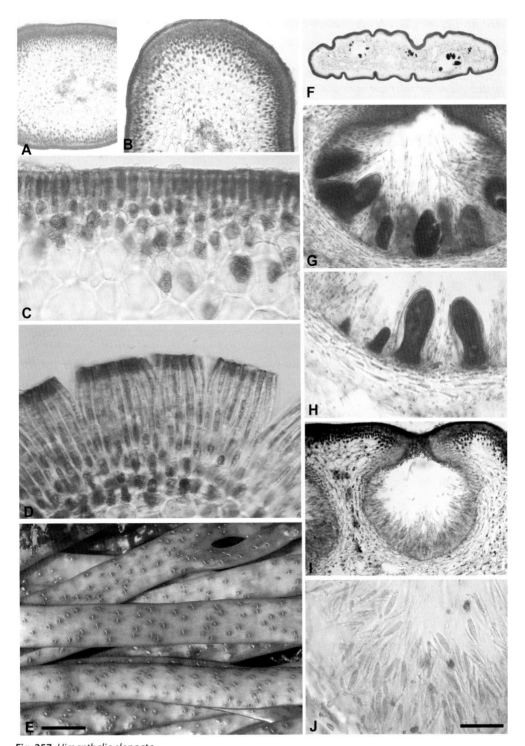

Fig. 257. *Himanthalia elongata*

A. TS through margin of vegetative button. B. VS of vegetative button, mid region. C. VS of vegetative button, lower edge. D. VS of vegetative button, upper edge, showing packed filamentous arrangement of cells. E. Mid-region portions of receptacles showing scattered conceptacle openings. F. TS of mature receptacle showing immersed conceptacles. G. TS of female conceptacle. H. Part of TS of female conceptacle showing oogonia arising from conceptacle wall. I. TS of male conceptacle. J. Part of TS of male conceptacle showing granular antheridia produced on branched paraphyses. Bar in J = 4 mm (F), = 15 mm (E), = 215 μm (A, B, G, I), = 120 μm (H), = 65 μm (C, D, J).

nearly the entire length except for a small, sterile, basal region; female conceptacles containing several large oogonia, each comprising a single egg, borne directly on the conceptacle wall, throughout the conceptacle; oogonia at first clavate, later more ovate, darkly pigmented, 70–210 × 32–125 μm, surrounded by a thick, gelatinous wall; male conceptacles containing large numbers of antheridia borne laterally and terminally on branched, multicellular paraphyses distributed throughout the conceptacle cavity; antheridia cylindrical, lightly pigmented, 30–50 × 8–13 μm; both oogonia and antheridia accompanied by densely packed multicellular paraphyses arising from conceptacle wall; paraphyses up to 500 μm or more in length, seldom branched, comprising colourless, cylindrical to irregularly swollen cells, 1.5–6(–13) diameters long, 20–93 × 5–20 μm.

In cross section, mature receptacles anatomically similar to the vegetative buttons with a central medulla of longitudinally elongated cells, a well-developed system of longitudinal and horizontal hyphae which gradually form a solid central tissue towards the base and into the button, an outer cortex, and a peripheral meristoderm, the latter well developed and multi-layered with rows of small cells elongated radially as formed on the upper surface of the buttons.

Epilithic, rarely epiphytic/epizoic; commonly in dense stands, sometimes isolated, lower eulittoral, on open rock, or in pools, in moderately exposed localities.

Common, and widely distributed around Britain and Ireland.

Perennial with receptacle developing on 1–2(–3) year old buttons; receptacle initiated in mid-summer, growing slowly over winter, more rapidly in spring/early summer, reaching maximum length/development and maturity of gametes by mid/late summer/early autumn; gamete release observed from mid/late summer (July/August) to mid-winter (Dec/January), after which the receptacles usually degenerate.

Ardré (1970), pp. 313–15; Bartsch & Kuhlenkamp (2000), p. 166; Blackler (1956b), pp. 158–62; Brodie *et al.* (2016), p. 1023; Bunker *et al.* (2017), p. 274; Cabioc'h *et al.* (1992), pp. 55–67, figs 38, 39; Cho *et al.* (2006), pp. 512–19; Coppejans & Kling (1995), pp. 222, pl. 79(a); Cox *et al.* (2011), pp. 471–80; Creed (1995), pp. 851–9, figs 1–6; De Mesquita Rodrigues (1963), pp. 107, 108, pl. 11(A); Dizerbo & Herpe (2007), pp. 130, 132, pl. 42(4); Farmer & Williams (1898), pp. 623–45; Fletcher in Hardy & Guiry (2003), p. 259; Gallardo García & Perez-Ruzafa (2001), pp. 69–71; Gallardo García *et al.* (2016), p. 37; Gayral (1966), p. 321, pl. 67; Gibb (1937), pp. 11–21; Gómez Garreta (2001), pp. 69–71, fig. 17; Guiry (2012), p. 172; Hamel (1931–39), pp. 379, 380; Harvey (1846–51), pl. 78; Jaasund (1965), p. 111; Kitching (1987), pp. 663–76; Kornmann & Sahling (1977), p. 170, pl. 94(A–C); Lagos & Cremades (2004), pp. 105–16; Margalet & Navarro (1992), pp. 117–32; Marginez *et al.* (2009), p. 11; Morton (1994), p. 46; Moss (1952), pp. 29–34, (1958b), pp. 31, 32, (1969), pp. 387–97; Moss *et al.* (1973), pp. 233–43; Naylor (1951), pp. 501–33; Newton (1931), pp. 223, 224, fig. 142; Niell (1984), pp. 71–102; Oltmanns (1889), pp. 1–100; Quéguineur *et al.* (2013), pp. 1–11; Ramon (1973), pp. 445–9; Rueness (1977), pp. 194, 195, fig. 114, (2001), p. 15; Russell & Wareing (1979), p. 128; Salgado & Neill (1974), pp. 305–24; Stegenga & Mol (1983), p. 124, pl. 50(1, 2); Stegenga *et al.* (1997a), p. 24; Stengel *et al.* (1999), pp. 213–21; Thuret & Bornet (1878); Wille (1910), pp. 495–539.

Sargassaceae Kützing

SARGASSACEAE Kützing (1843), pp. 349, 359 [including Cystoseiraceae De Toni 1891].

Thalli macroscopic, large, perennial, to 1(–5) m in length, erect, much branched, usually bushy in habit, yellow to dark brown in colour, sometimes blue/green iridescent, sometimes free-floating (in the Sargasso Sea), more commonly attached to hard substrata, arising singly or grouped from either a rounded to irregularly shaped discoid or conical holdfast, or from an expansive rhizome system; thalli largely parenchymatous (except for the filamentous medulla) and morphologically complex, representing a range of form, from cup-shaped, cylindrical, to slightly flattened; primary axis perennial and either remaining relatively short, or of indefinite length; primary axes sometimes giving rise annually to much branched, primary (secondary and tertiary) laterals of unlimited growth; branching alternate, radial, distichous and pinnate, dichotomous or pseudodichotomous; branches/laterals with or without midrib, stalked or innate air vesicles, leaf-like branchlets and spine-like appendages; structurally complex and comprising a central medulla, a surrounding cortex and an outer, single-layered meristoderm; growth monopodial by a three-sided apical cell sunken in apical pit, usually with supplementary growth by the peripheral meristem; surface cells with fucosan vesicles and numerous discoid plastids without pyrenoids; cryptostomata often present.

Reproductive organs borne in specialized receptacles which occur terminally or laterally on ultimate, specialized branchlets; receptacles of various shapes, sometimes pedicellate, simple or branched, with or without air vesicles or spines, cylindrical or dorsiventrally flattened, smooth, verrucose or spiny; sexual reproduction oogamous, with oogonia and antheridia borne in sunken, flask-shaped conceptacles within the receptacles, accompanied by simple or branched multicellular paraphyses; thalli dioecious or monoecious, conceptacles of the latter with either or both oogonia and antheridia; oogonia with a single egg; oogonia sometimes retained and fertilized on receptacle surface following release, incubated for 1–2 days and then disseminated as multicellular germlings with basally emergent rhizoids.

Six genera within this family are currently recognised for Britain and Ireland: *Bifurcaria*, *Cystoseira*, *Ericaria*, *Gongolaria*, *Halidrys* and *Sargassum* – all of which are very different in morphological and other features.

Within this family, particularly noteworthy is Orellaba *et al.*'s (2019) molecular study, using the four genetic markers *psb*A, mt23S, *cox*1 and *nad*1, of species in the genus *Cystoseira sensu lato* present in the eastern Atlantic and Mediterranean. The authors found that most of the species investigated were resolved in three clades. One clade corresponded to *Cystoseira sensu stricto*, and included the species *C. foeniculacea* and *C. humilis*, a second clade corresponded to *Treptacantha* gen. emend and included the species *T. baccata* and *T. nodicaulis*, while a third clade corresponded to *Carpodesmia* gen. emend and included the species *C. tamariscifolia*. However, the combinations for *Carpodesmia* and *Treptacantha* were relatively short-lived, as Molinari & Guiry (2020) reported that these two genera were pre-dated by *Ericaria* and *Gongolaria* respectively. This has been accepted in the present treatise.

Bifurcaria Stackhouse

BIFURCARIA Stackhouse (1809), pp. 59, 90.

Type species: *B. tuberculata* Stackhouse (1809), p. 90 (= *B. bifurcata* Ross 1958, p. 754)

Thalli macroscopic, perennial, to 20–30 (exceptionally 50) cm in height, to 4 mm in width, with several erect, branched, yellow-brown, cylindrical axes arising annually from an irregularly spreading, branched, cylindrical and somewhat gnarled/warty rhizome system attached underneath by small, adhesive disc-shaped holdfasts; erect axes irregularly dichotomously branched above to 8–9 orders in larger specimens, constricted at base, obtuse and rounded at the apex, with or without innate air vesicles; cryptostomata absent; in structure, comprising an inner anastomosing, hyphal-like, colourless medulla surrounded by layers of pigmented cortical cells with enclosing cuticle layer.

Monoecious, with slightly swollen, elongate receptacles borne terminally on erect shoots, containing sunken hermaphrodite conceptacles bearing oogonia and antheridia.

One species in Britain and Ireland.

Bifurcaria bifurcata Ross (1958), p. 754.   Figs 258, 259

Fucus rotundus Hudson (1762), p. 471; *Fucus tuberculatus* Hudson (1778), p. 588, nom. illeg.; *Fucus bifurcatus* Velley in Withering (1792), p. 257, nom. illeg.; *Bifurcaria tuberculata* Stackhouse (1809), p. 90; *Pycnophycus tuberculatus* (Stackhouse) Kützing (1843), p. 359; *Ascophylla tuberculatum* Kuntze (1891), p. 884.

Thalli forming erect, macroscopic thalli, 10–50 cm in length, olive-brown/yellow-brown to dark brown in colour, becoming black when dried, arising gregariously from a horizontally spreading, gnarled, much branched, perennating, rhizome-like base attached by adhesive discs formed on the undersides; thalli rigid and smooth in texture, cylindrical, 2–4 mm in width, sometimes commonly inflated with air vesicles, lacking in cryptostomata, unbranched below and branched towards the apices, to 5–9 orders in older specimens; branching irregularly dichotomous with branches spirally twisted, constricted at base, obtuse and rounded terminally, with newly branched tips incurled (forcipate apices); in cross section, thalli with a central medulla-like core of small, narrow, longitudinally elongated, thick-walled cells enclosed by a cortex-like region of several (8–10) layers of larger, colourless, thinner-walled, longitudinally elongated cells, enclosed peripherally by a cuticle-covered, single layered meristoderm of radially elongated, palisade-like cells containing large numbers of densely packed plastids; hyphal formation confined to protuberances on the undersurfaces of the rhizome, arising from the central cells and growing downwards to form the compact tissue of the attachment discs.

Monoecious, with receptacles terminal on the erect shoots; receptacles to 4(–6.5) cm in length, to 4(–5) mm in width, slightly wider than the branches giving rise to them, terete, simple, smooth in texture at first, later slightly tuberculate due to the protruding bumps of the enclosed reproductive conceptacles; sunken conceptacles distributed around the edge of the receptacles, flask-shaped, hermaphrodite, each bearing both oogonia and antheridia associated with paraphyses; antheridia developing before the oogonia, occurring as densely packed, greyish, ovate to elongate-cylindrical packets, 25–42 × 8–12 μm, borne laterally on richly branched paraphyses arising from the walls in the upper regions of the conceptacle; oogonia occupying the base of the conceptacle, associated with unbranched paraphyses, comparatively large and prominent, few per conceptacle, dark-coloured, sessile, ovoid, 120–190 × 50–115 μm with a single ovum.

Epilithic; mid to lower eulittoral, in pools, more rarely on open rock, in fairly exposed localities, sometimes associated with *Himanthalia elongata*, forming quite dense stands around the edge of tidal pools, frequently locally dominant, and often characteristically with blackened tips exposed above the water surface.

Restricted to the Channel Islands, Isles of Scilly, south-west coast of England (Cornwall, Devon, extending to Portland Bill, Dorset and Wales) in Britain, and to the west coast of Ireland, extending north from Co. Clare to Co. Donegal.

Perennial, receptacles observed throughout the year although the peak of fertility appears to be in winter and spring (November to April).

Ardré, F. (1970), pp. 315–17; Brodie *et al.* (2016), p. 1022; Bunker *et al.* (2017), p. 162; Cabioc'h *et al.* (1992), p. 76, fig. 56; Cho *et al.* (2006), pp. 512–19; Culioli *et al.* (2001), pp. 529–35; Daoudi *et al.* (2001), pp. 973–8; De Mesquita Rodrigues (1963), pp. 108–10; De Valéra (1962), pp. 77–101; Delf (1935), pp. 245–59, figs 1–8; Dizerbo & Herpe (2007), pp. 132, 133, pl. 43(3); Fernandez *et al.* (1983), pp. 435–55; Fletcher in Hardy & Guiry (2003), p. 228; Gallardo García (2001a), pp. 91–3; Gallardo García *et al.* (2016), p. 37; Gayral (1958), pp. 274–6, fig. 43, (1966), p. 333, pl. LXXIII(41); Gómez Garreta (2001), pp. 91–3, fig. 23; Gruet (1976), pp. 11–15; Guiry (2012), p. 165; Hamel (1931–39), pp. 381, 382, fig. 59(2); Hardy & Moss (1979a), pp. 164, 165; Harvey (1846–51), pl. 89; Le Lann *et al.* (2008), pp. 238–45; Margalet & Navarro (1990), pp. 99–107; Méndez-Sandín & Fernández (2016), pp. 46–56; Morton (1994), p. 46; Newton (1931), p. 223, fig. 141(A–D); Rees (1933), pp. 101–15, figs 1–7; Ross (1958), pp. 753, 754, (1961), p. 512; Stegenga & Mol (1983), p. 127, pl. 50(3); Stegenga *et al.* (1997a), p. 23; Valls *et al.* (1995), pp. 145–9.

Fig. 258. *Bifurcaria bifurcata*
A. Habit of thalli in eulittoral pool. B. Habit of thalli. Bar in B = 18 mm.

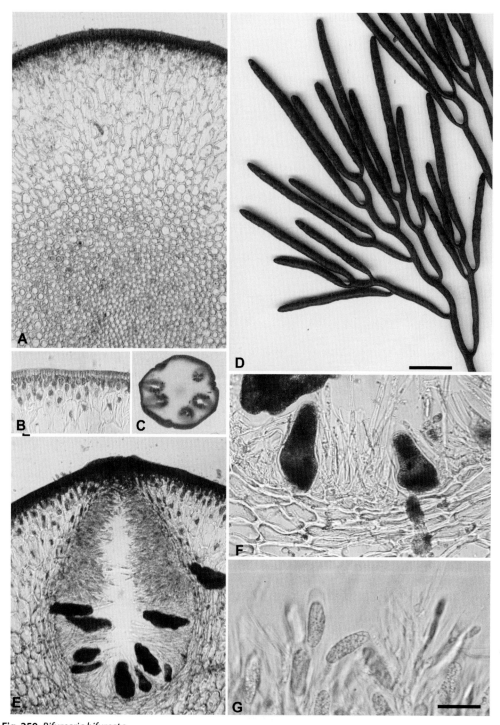

Fig. 259. *Bifurcaria bifurcata*

A. TS of thallus showing central medulla of small, tightly packed, thick-walled cells, surrounded by a cortex of larger, colourless cells. B. TS of thallus, edge region, showing outer meristoderm layer. C. TS of receptacle, showing immersed conceptacles. D. upper region of mature thallus showing the long terminal receptacles. E. TS of whole conceptacle, showing antheridial clusters distributed in the upper region, oogonia distributed in the basal region. F. TS of basal region of conceptacle showing oogonia arising from conceptacle wall. G. TS of upper region of conceptacle, showing granular antheridia borne on branched paraphyses. Bar in D = 16 mm (D), = 2.5 mm (C), bar in G = 215 μm (A, E), = 65 μm (B, F), = 38 μm (G).

Cystoseira C. Agardh

CYSTOSEIRA C. Agardh (1820), p. 50.

Type species: *C. concatenata* (Linnaeus) C. Agardh (1820), p. 57 (= *Cystoseira foeniculacea* (Linnaeus) Greville 1830, pp. xxxii, 6, 7).

Thalli large, perennial, erect, to 1(–2) m in length, caespitose, usually much branched, non-iridescent, with several axes arising from a discoid attachment disc; comprising a terete, more rarely flattened, relatively short, main perennial axis, either smooth, thorny, or covered with tubercles and sometimes basally produced spinose branches, from which arise, radially or bilaterally, usually several, deciduous, primary laterals of more unlimited growth which themselves are similarly structured and branched to several orders; tophules and spine-like appendages often inflated in parts to form internal, ovoid aerocysts in the terminal branchlets, and usually with cryptostomata scattered over the entire thallus; shedding of laterals occurs during dormant period, often leaving scars; in section, thallus parenchymatous, comprising a central medulla of small polygonal cells, surrounded by a cortex of irregularly globose, thick-walled cells progressively decreasing in size both inwards and outwards, within an enclosing, single layered, palisade-like meristoderm of photosynthetic, squared to rectangular cells.

Monoecious; receptacles borne seasonally in terminal branchlets, to several cm in length, usually clustered, cylindrical, fusiform or siliquose, verrucose, warty, often with proximal aerocysts, simple or branched, without spine-like appendages; conceptacles protruding, grouped, usually hermaphrodite, oogonia borne around base, antheridia borne above; oogonia large, sessile, containing a single egg, which upon release may be incubated on receptacle surface; antheridia small, sessile or borne on branched, multicellular paraphyses.

Two species in Britain and Ireland.

KEY TO SPECIES

Primary and secondary laterals cylindrical or slightly flattened; branching either distichous and divaricate, or alternate/zigzag and in various planes; occasional flattened, leafy laterals in adult thalli . . . . *C. foeniculacea*
Primary and secondary laterals always cylindrical, never flattened, and either pyramidal or cylindrical in appearance; branching never distichous, only sometimes alternate; no flattened, leafy laterals in adult thalli . . . *C. humilis* var. *myriophylloides*

Cystoseira foeniculacea (Linnaeus) Greville (1830), pp. xxxii, 6, 7.   Figs 260, 261, 262

Fucus foeniculaceus Linnaeus (1753), p. 1161; *Fucus abrotanifolius* Linnaeus (1753), p. 1161; *Fucus concatenatus* Linnaeus (1753), p. 1160; *Fucus barbatus* Linnaeus (1753), p. 1161; *Fucus discors* Linnaeus (1767), p. 717; *Cystoseira abrotanifolia* (Linnaeus) C. Agardh (1820), p. 63; *Cystoseira concatenata* (Linnaeus) C. Agardh (1820), p. 57; *Cystoseira discors* (Linnaeus) C. Agardh (1828), p. 62; *Phyllacantha concatenata* (Linnaeus) Kützing (1843), p. 355; *Cystoseira ercegovicii* Giaccone in Giaccone & Bruni (1973), p. 72, nom. inval.

Thalli caespitose, erect, macroscopic to 0.7 m in length, one to several fronds arising from a single, often large, irregularly shaped, discoid holdfast; primary axis of each frond usually unbranched (although sometimes producing basally positioned, divaricate, lateral branches, which can be spinose, apparently regenerating from basal stubs of dehisced primary laterals), to 15 cm in length, 6 mm in width, cylindrical and terete in section, thorny, with either a smooth apex, or especially in winter, an apex covered with small wart-like protuberances (tubercles); growing apex of primary axis dome-shaped and usually hidden beneath surrounding young primary laterals; young primary axis often densely enclosed, and considerably exceeded in length, by large numbers (up to 14 counted on one thallus) of erect primary laterals up to 60 cm in length, the longest below, produced acropetally in a radial or bilateral manner, which, upon

Fig. 260. *Cystoseira foeniculacea*
Habit of thallus showing the terminal divaricate branching of the erect primary laterals.

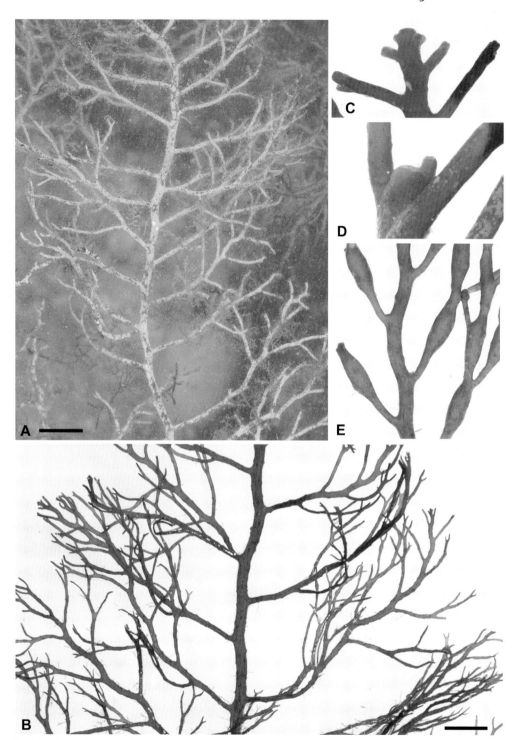

Fig. 261. *Cystoseira foeniculacea*
A. Habit of thallus in a tidal pool showing the divaricate branching of a primary lateral. Note the pale colouration of the thallus due to the extensive growth of epiphytic coralline red algae. B. Close-up view of part of primary lateral showing the alternate, divaricate branching pattern. C. Terminal region of primary lateral showing the flattened trumpet-like growing apex surrounded by young secondary branch initials. D. Terminal region of primary axis showing the dome-shaped apex and surrounding primary laterals. E. Secondary lateral with aerocysts. Bar in A = 3 cm, bar in B = 9 mm (B), = 2.6 mm (C), = 3.7 mm (D), = 3 mm (E).

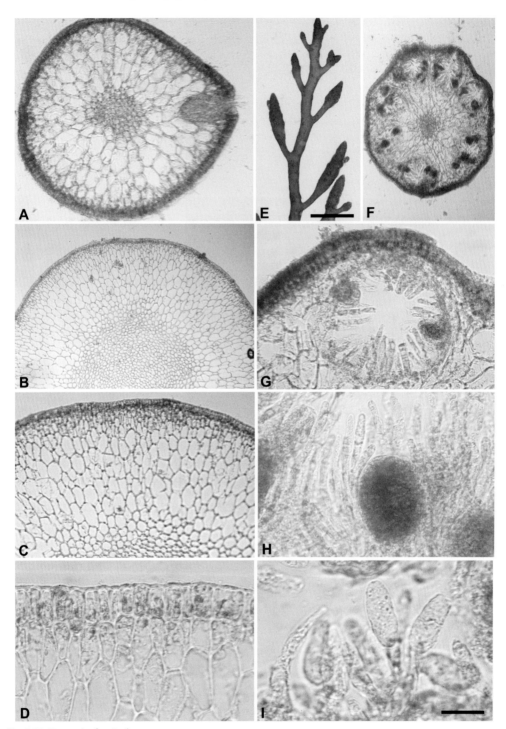

Fig. 262. *Cystoseira foeniculacea*

A. TS of young branch showing immersed cryptostomata. B, C. TS of primary lateral showing inner medulla of closely packed cells, the surrounding cortex of larger, radially elongated cells and an enclosing meristoderm. D. TS of primary lateral showing outer photosynthetic meristoderm layer. E. Receptacles on terminal branchlets. F. TS of receptacle showing immersed conceptacles. G. TS of conceptacle showing two dark-pigmented oogonia and multicellular paraphyses. H. Part of TS of conceptacle showing a central oogonium surrounded by paraphyses. I. Granular antheridia borne on branched paraphyses. Bar in E = 3.5 mm, bar in I = 215 μm (B, F), = 120 μm (A, C), = 65 μm (G), = 35 μm (D, H, I).

shedding after approximately one year, leave the basal region of older thalli quite denuded and covered with obvious circular scars; tophules absent; primary laterals usually circular in cross section throughout their length, although sometimes becoming somewhat flattened distally; young primary laterals regularly branched along their length, older primary laterals often denuded below except for short remaining stumps of dehisced secondary laterals; foliose leafy branches with toothed margin and central midrib sometimes reported at the base of the primary axes and primary laterals; secondary laterals to 10(–15) cm in length, their form varying with season, usually being spirally arranged, shorter, filiform and less branched in the winter and spring, more characteristically longer, somewhat flattened with regularly alternate and distichous branching in the late summer and autumn, giving a somewhat zigzag appearance to the parental primary lateral; secondary laterals similarly branched, usually circular in cross section, but also sometimes somewhat flattened, giving rising to tertiary branches ending in fine filiform, acuminate branches; growing apex of secondary laterals somewhat pear/trumpet shaped with a flat top and surrounding young branch initials; no leaf-like or spine like appendages observed; cryptostomata usually abundant, immersed in the terminal branchlets, somewhat nodose/papillate on the secondary and primary laterals, with or without obvious tuft of hairs emerging from the ostiole, occurring on vegetative branches, aerocysts and on receptacles interspersed with the conceptacles; aerocysts mainly associated with the fertile branches during the summer, absent in winter, elongate-ovoid, although somewhat triangular when commonly beneath a dichotomy, singly or in chains, 3–4 mm in length, 1.2–1.4 mm in width; in section, thallus parenchymatous, comprising a central medulla of small cells, surrounded by a cortex of irregularly globose cells within an enclosing, single-layered, palisade-like meristoderm of photosynthetic, squared to rectangular cells.

Monoecious, receptacles developing on the terminal branchlets of the secondary laterals, usually in an alternate/zigzag manner, with or without a subtending aerocyst, without spiny appendages, sometimes reported to be in dense yellowish clusters; receptacles simple, sometimes forked, lanceolate, often paired, 4–6.5 mm in length, 1–1.3 mm in width, with either tapered or blunt ends, sometimes pedicellate, although commonly merging basally with the aerocysts, the latter also bearing conceptacles; sunken conceptacles grouped, distributed around the edge of the receptacle, with up to 8(–10) seen in cross section, small, spherical to oval, 170–220(–270) × 130–170(–220) μm, hermaphrodite, each bearing both oogonia and antheridia associated with paraphyses, although sometimes appearing unisexual, possibly representing different timings of sexual gamete development; antheridia in large numbers, occurring throughout conceptacle cavity, borne singly or in small clusters on short, branched, paraphyses, cigar-shaped, granular, 25–34 × 10–16 μm; oogonia darkly pigmented, spherical to oval, 64–80 × 48–67 μm, borne directly on conceptacle wall; paraphyses to 100 μm, 1–5 cells in length, simple or branched towards the base, comprising colourless cylindrical cells, 1–6 diameters long, 12–24(–36) × 6–9(–11) μm.

Epilithic; lower eulittoral pools.

Infrequent, restricted to the south coast of England (Cornwall, Devon, Dorset, Hampshire, Sussex) and west coast of Ireland (Kerry, Galway, Donegal); also found in the Channel Islands.

Perennial, summer fertile with receptacles reported for June, July, August, September and October, more rarely at other times.

Brodie *et al.* (2016), p. 1022; Bunker *et al.* (2017), p. 265; Cabioc'h *et al.* (1992), p. 34, fig. 52; Cormaci *et al.* (2012), pp. 331, 339, pl. 104(1–3); Dawson (1941), pp. 316–25; De Mesquita Rodrigues (1963), pp. 114, 115; Dizerbo & Herpe (2007), p. 134, pl. 44(1); Ercegovic (1952), pp. 1–172; Feldmann (1937), p. 325, pl. 10; Fletcher in Hardy & Guiry (2003), p. 238; Gallardo García *et al.* (2016), p. 37; Gómez Garreta (2001), pp. 131–5, fig. 37(a–c), map 34; Guiry (2012), p. 166; Hamel (1931–39), pp. 408, 409, 414–16, pl. IX; Morton (1994), p. 8; Newton (1931), p. 228; Orellana *et al.* (2019), pp. 447–65; Ribera *et al.* (1992), p. 122; Roberts (1967a), pp. 345–66, figs 1, 2, 3, 13, 19, 20, (1968a), pp. 251–64, pl. 2(b), (1968b), pp. 547–64, figs 1–14, (1968c), pp. 565, 566; Sauvageau (1912), pp. 437–55, 523, 524, 533, 534, (1920b), pp. 37–42; Stegenga *et al.* (1997a), p. 24; Stegenga & Mol (1983), p. 129; Taskin *et al.* (2012), pp. 1–75.

Cystoseira humilis var. *myriophylloides* Schousboe ex Kützing (Sauvageau) Price & John in Price *et al.* (1978), p. 104.   Figs 263, 264, 265

Cystoseira myriophylloides Sauvageau (1912), pp. 455, 531.

Forming erect, macroscopic thalli, to 20(–35) cm in length – elsewhere reported to be up to 1.5 m in length – olive-brown to dark brown in colour, non-iridescent, caespitose, usually several thalli arising from a single, often large, irregularly shaped discoid holdfast; primary axis simple or branched to produce lateral axes, up to 5(–7) cm in length, 2–5 mm in width, cylindrical and terete in section, denuded of branches below and only bearing the raised, circular scars of shed primary laterals, the terminal region acropetally producing several, ephemeral, primary laterals; apex of primary axis fairly indistinct and enclosed by the enveloping, primary lateral initiatives, sometimes giving rise below to numerous, small, bump-like protuberances/tubercles; primary laterals without swollen, basal regions (tophules), up to 20–30 cm in length, greatly exceeding length of parental primary axis, often denuded below over the greater part of their length except for the short stub-like remains of shed laterals, much branched and either cylindrical or pyramidal in shape above depending on the season, giving rise to branched secondary laterals, which in turn give rise to short, terete, somewhat curved, tertiary branches up to 10(–20) mm in length; branching of laterals irregular, spiral, alternate, sometimes pseudodichotomous, with apices slightly acute; in spring, tertiary branches with oblong aerocysts, 1.5–3(–4) mm in length, to 0.8–1 mm in width, usually subtending the receptacles; no flattened, leaf-like appendages/branches seen; cryptostomata usually abundant on all laterals, sessile or sunken on the secondary and tertiary laterals, more prominent and characteristically nodose/papillate in appearance on the primary laterals; in cross section, thalli comprising an inner medulla of relatively narrow, longitudinally elongated cells, quite densely packed and rounded to angular in cross section, surrounded by a cortex of larger cells, radially elongated in cross section, enclosed within an outer meristoderm layer of photosynthetic, palisade-like cells.

Monoecious, reproductive receptacles developing terminally on the tertiary and secondary laterals, usually in dense yellowish clusters; receptacles simple, sometimes branched, fusiform, acuminate, somewhat nodose, 2–3.5(–6) mm in length, 0.5–1 mm in width, without spiny appendages, slightly wider than the subtending lateral and, therefore, somewhat pedicellate in appearance; conceptacles 270–510 × 260–410 µm, spherical to flask shaped, with 2–4(–6) seen in cross section, and closely grouped together; separate male and female conceptacles observed on each receptacle, with hermaphrodite conceptacles, sometimes reported in the literature, not observed; both antheridia and oogonia produced in association with paraphyses; antheridia produced in large numbers, borne laterally and terminally in clusters on the paraphyses, cigar-shaped, granular, 22–35 × 10–22 µm; oogonia larger, spherical to oval in shape, darkly pigmented, borne directly on conceptacle wall, 100–180 × 85–135 µm, usually up to 4–5 seen in cross section; newly released oogonia often observed adhering to receptacle wall; paraphyses, to 2–5 cells, 135 µm in length, multicellular, sometimes branched at the base, comprising mostly cylindrical cells along their length, more enlarged and barrel-shaped at their base, 1–4 diameters long, 17–50 × 8–38 µm.

Epilithic; mid to lower eulittoral pools.

Infrequent and restricted to the south-west coast of England (Channel Islands, Cornwall, South Devon); first reported for Britain by Blackler (1961).

Perennial, spring/summer fertile although usually found with receptacles throughout the year.

Ardré (1970), pp. 321, 322; Barceló *et al.* (1994), pp. 126–9; Brodie *et al.* (2016), p. 1022; Bunker *et al.* (2017), p. 266; Cabioc'h *et al.* (1992), p. 76, fig. 54; Cormaci *et al.* (2012), p. 347, pl. 110(1–3), map 38; Dizerbo & Herpe (2007), pp. 134, 135, pl. 44(2); Fletcher in Hardy & Guiry (2003), p. 238; Gallardo García *et al.* (2016), p. 38;

Fig. 263. *Cystoseira humilis* var. *myriophylloides*
Habit of thallus showing erect terminally branched primary laterals.

Fig. 264. *Cystoseira humilis* var. *myriophylloides*
A. Habit of thallus. B, C. Upper thallus region showing abundant immersed cryptostomata on the filiform branchlets (B), which become more nodular below (C). D. Terminal region of primary laterals showing the filiform aspect of the secondary and tertiary branches. E. Terminal region of primary axis showing the apex surrounded by young primary lateral initials and subterminal wart-like tubercles. Bar in D = 4 mm (D), = 2 mm (E).

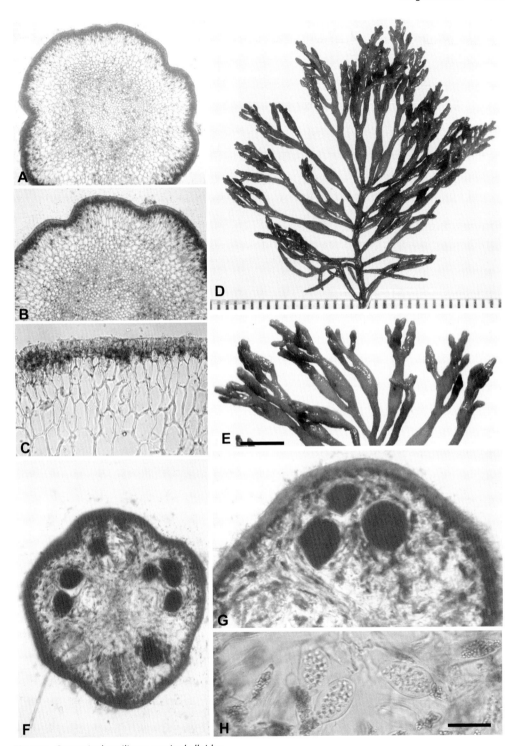

Fig. 265. *Cystoseira humilis* var. *myriophylloides*

A–C. TS of primary laterals showing inner medulla of closely packed cells, the surrounding cortex of larger, radially elongated cells and (C) an enclosing photosynthetic meristoderm. D. Fertile branch region showing aerocysts. E. Terminal fertile branch region showing young receptacles. F. TS of receptacle showing immersed conceptacles. G. TS of part of receptacle showing a single conceptacle with three darkly pigmented oogonia. H. Granular antheridia borne on branched paraphyses. Bar in H = 375 μm (A), = 215 μm (B, F), = 120 μm (G), = 65 μm (C), = 35 μm (H), bar in E = 4.5 mm.

Garcia-Fernandez & Bárbara (2016), eo35; Gayral (1966), p. 331, fig. 40(A), pl. LXXII; Gómez Garreta (2001), pp. 142–4, fig. 41(A–C), map 38; Hamel (1931–39), pp. 413, 414; Orellana *et al.* (2019), pp. 447–65; Ribera *et al.* (1992), p. 122; Roberts (1967a), pp. 345–66, figs 5, 7, 12, 17, 18, (1968a), p. 254, (1970), pp. 201–10, (1978), pp. 399–422; Sauvageau (1912), pp. 455–66, 531, 532; Taskin *et al.* (2012), pp. 1–175.

Ericaria Stackhouse

ERICARIA Stackhouse (1809), p. 56.

Type species: *Ericaria selaginoides* (Linnaeus) Molinari & Guiry (2020), p. 6.

Thalli large, perennial, erect, epilithic, caespitose or non-caespitose, arising from a discoid/conical holdfast or from prostrate axes with haptera; primary axes cylindrical, simple or branched; primary laterals annual, cylindrical and profusely radially branched; all orders of laterals decreasing in length towards the apex, the majority of species with numerous spine-like appendages; laterals sometimes iridescent, with scattered cryptostomata and inconspicuous aerocysts; tophules present or absent; in section, thallus parenchymatous, comprising a central, well-developed medulla of small polygonal cells, surrounded by a cortex of globose, thin-walled, sometimes deformed, cells within an enclosing, single layered, palisade-like meristoderm comprising cells ellipsoidal to rounded in surface view, pronouncedly rectangular in cross section.

Monoecious, receptacles formed at the end of terminal branchlets, with grouped conceptacles and spiny or filiform appendages; conceptacles hermaphrodite, with dark brown oogonia borne around base, large, with a single egg, which upon release may be incubated on receptacle surface; antheridia yellowish, small, sessile on branched multicellular paraphyses.

One species in Britain and Ireland.

Ericaria selaginoides (Linnaeus) Molinari & Guiry (2020), p. 6.   Figs 266, 267

Fucus selaginoides Linnaeus (1759), p. 1345; *Fucus tamariscifolius* Hudson (1762), p. 469; *Fucus ericoides* Linnaeus (1763), p. 1631; *Cystoseira ericoides* (Linnaeus) C. Agardh (1820), p. 52; *Cystoseira selaginoides* (Linnaeus) Bory (1832), p. 319, nom. illeg.; *Cystoseira ericoides* var. *laevis* P.L.J. Dangeard (1949), p. 130; *Cystoseira ericoides* var. *divaricata* P.L.J. Dangeard (1949), p. 130; *Cystoseira tamariscifolia* (Hudson) Papenfuss (1950), p. 185; *Carpodesmia tamariscifolia* (Hudson) Orellana & Sansón (2019), p. 461.

Thalli erect, 0.5–1 m in length, dark brown to black, often quite bushy in appearance, arising from a conical holdfast, and characteristically with a greenish or bluish iridescence under water; primary axis up to 0.6 m in length, 1 cm in width, cylindrical and copiously branched, often bearing the scars of dehisced primary laterals towards the base, tophules absent, apex inconspicuous and surrounded by developing primary lateral branch systems; primary laterals terete in section, sometimes spirally twisted around main axis, at first appearing like cylindrical mounds, densely covered with closely set whorls of bifid outgrowths (early formed appendages) giving the apex a heathery, whorled appearance, later, following rapid elongation, and the transformation of one or both of the bifid appendages into secondary laterals, comprising a basal region of bifid appendages and few or no laterals and a distal region of a mixture of closely set, spirally arranged laterals interspersed with a few appendages; secondary laterals similar in morphology and structure to the primary laterals, to 15 cm in length, but bearing simple, subulate as well as bifid appendages and prominent cryptostomata, giving rise similarly to tertiary laterals as outgrowths of one of the bifid appendages; tertiary laterals may be similarly structured and branched, forming quaternaries on large thalli but usually the tertiary laterals may only give rise to simple or bifid appendages and terminal receptacles; bifid appendages characteristically spine-like in appearance, and up to 3–4 mm in length; cryptostomata common and prominent throughout terminal branch regions; terminal branchlets with ovoid aerocysts,

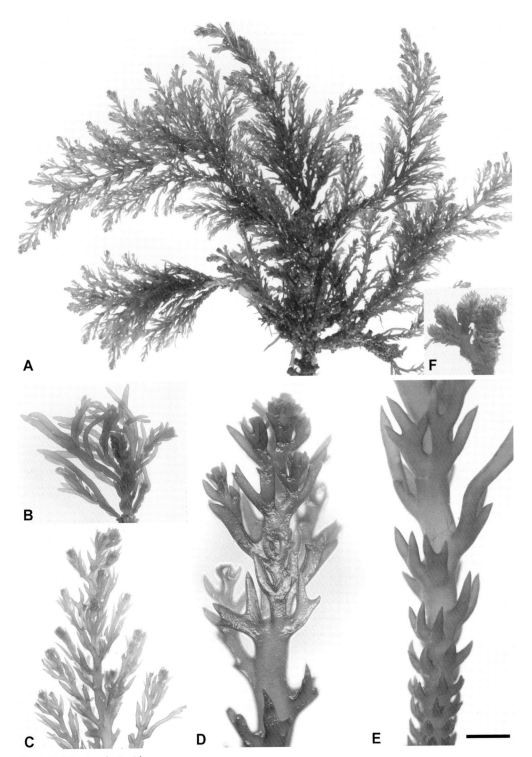

Fig. 266. *Ericaria selaginoides*

A. Habit of thallus. B. Terminal region of young primary axis showing filiform branches. C–E. Terminal region of primary laterals showing spine-like bifid appendages. F. Terminal region of a primary lateral. Bar in E = 18 mm (A), = 6 mm (B), = 7.5 mm (C), = 2.7 mm (D), = 4.5 mm (E), = 3 mm (F).

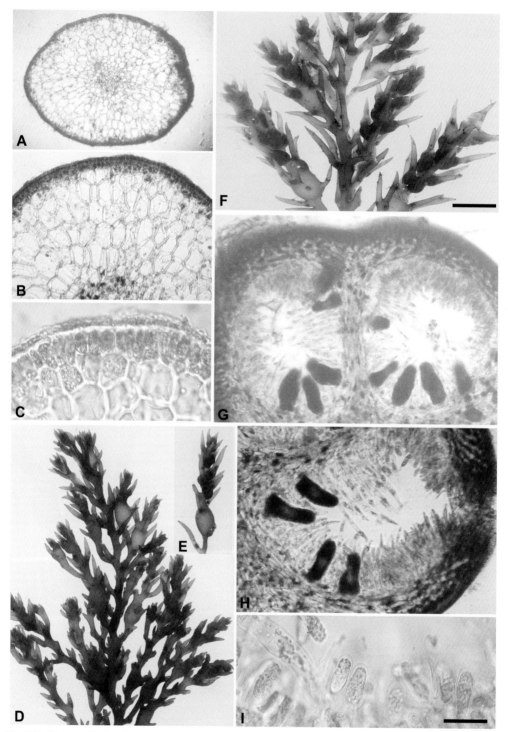

Fig. 267. *Ericaria selaginoides*

A–C. TS of primary laterals showing inner medulla of closely packed cells, the surrounding cortex of larger, radially elongated cells and (C) an enclosing photosynthetic meristoderm. D. Terminal region of primary lateral showing aerocysts. E. Terminal region of fertile primary lateral showing nodular receptacles. G. TS of receptacle showing two conceptacles. H. TS of conceptacle showing darkly pigmented oogonia surrounded by filamentous paraphyses. I. Granular antheridia borne on branched paraphyses. Bar in F = 3 mm (F), = 4.4 mm (E), = 5 mm (D), bar in I = 215 µm (G), = 120 µm (A, B, H), = 35 µm (C, I).

up to 5–6 mm in length, single or in chains, with spirally arranged, spine-like outgrowths and prominent cryptostomata; basal regions often with long cylindrical or dorsiventrally flattened leaf-like branchlets, up to 2–3 cm long; in cross section, thalli comprising an inner medulla of relatively narrow, longitudinally elongated cells, quite densely packed and rounded to angular in cross section, surrounded by a relatively larger cortex of larger, thin-walled cells, radially elongated in cross section in young axes, more rounded in older axes, enclosed within an outer meristoderm layer of photosynthetic, palisade-like, ellipsoidal to rounded cells.

Monoecious, receptacles infrequent, developing terminally on the ultimate branches of the quaternaries, tertiaries and secondaries, somewhat indistinct, 2–10(–40) mm in length, 1–2 mm wide, simple or sometimes branched, often with a subtending aerocyst, bearing short, spirally arranged, spine-like or filiform appendages, irregularly nodose in outline as the conceptacles develop within the axes and in the swollen bases of the sometimes decurrent appendages; receptacles bearing immersed flask to oval-shaped conceptacles, usually 3–5 in cross section filling most of the lumen, 500–600 × 400–560 µm, sometimes clearly adpressed and semi-circular in shape, each typically containing both oogonia and antheridia, sometimes with antheridia only, interspersed with paraphyses; if both are present, oogonia borne at the base of the conceptacle, antheridia occurring more towards the ostiole; antheridia fusiform, granular, 20–33 × 8–12 µm borne laterally in clusters on relatively short, multicellular, much-branched paraphyses, usually developing prior to the oogonia; oogonia much larger and darkly pigmented, 85–140(–160) × 40–60(–80) µm, club to oval-shaped, borne directly on conceptacle wall and enclosed by paraphyses; paraphyses multicellular, to 280 µm, 3–5 cells in length, extending well into conceptacle lumen, acuminate, simple or once branched towards the base, colourless, comprising colourless, cylindrical cells, 2–9 diameters long, 20–95(–135) × 6–12(–18) µm; released oogonia sometimes seen to be incubated on receptacle surface.

Epilithic; lower eulittoral pools or emergent, extending into the sublittoral to 5 m.

Locally common, restricted in distribution mainly to southern and western shores of Britain, reported for the Isles of Scilly, the Channel Islands, Cornwall, Devon, Dorset, Inverness and Argyll; in Ireland, Galway, Donegal and Antrim.

Perennial, with receptacles developing in spring, maturing by June/July and remaining fertile until October/November; empty receptacles can be found throughout the winter period.

Aldanondo-Aristizabal *et al.* (2004), pp. 115–28; Ardré (1970), pp. 322–5; Brodie *et al.* (2016), p. 1022; Bunker *et al.* (2017), p. 266; Cabioc'h *et al.* (1992), p. 74, fig. 51; Celis-Pla *et al.* (2014), pp. 377–88, (2016), pp. 89–97, (2017a), pp. 67–81, (2017b), pp. 157–65; Cormaci *et al.* (2012), pp. 366, pl. 123, figs 1–4; De Mesquita Rodrigues (1963), pp. 115–17, pl. 12(E); Dizerbo & Herpe (2007), pp. 136, 137, pl. 44(4); Fletcher in Hardy & Guiry (2003), p. 239; Gallardo García *et al.* (2016), p. 38; Garcia-Fernandez & Bárbara (2016); Gayral (1958), p. 256, pl. XLIV, (1966), p. 323, pl. LXVIII, fig. 40(c); Gómez Garreta (2001a), pp. 159–62, fig. 49(a–c), map 46; Gómez Garreta *et al.* (1994), pp. 110–13, 116, 117; Guiry (2012), p. 166; Hamel (1931–39), pp. 395–7; Harvey (1846–51), pl. 264; Molinari-Novoa & Guiry (2020), pp. 1–10; Morales Ayala & Viera Rodríguez (1989), pp. 107–13; Morton (1994), p. 47; Newton (1931), p. 226, fig. 144(A–E); Orellana *et al.* (2019), pp. 447–65; Ribera *et al.* (1992), p. 122; Roberts (1967a), pp. 345–66, figs 9, 10, 14, 23, (1968a), pp. 256, 257, (1970), pp. 201–10, figs 1–6, (1978), pp. 399–422; Sauvageau (1912), pp. 316–37, 513–15, (1920b), pp. 22–7; Stegenga & Mol (1983), p. 129, pl. 51(2); Stegenga *et al.* (1997a), p. 24; Taskin *et al.* (2008), p. 66, (2012), pp. 1–75.

Gongolaria Boehmer

GONGOLARIA Boehmer (1760) p. 503.

Type species: *Gongolaria abies-marina* (S.G. Gmelin) Kuntze (1891), p. 895.

Thalli large, perennial, erect, epilithic, arborescent, non-caespitose, usually arising from a discoid/conical holdfast, sometimes from prostrate axes with haptera; primary axes dorsiventrally flattened or cylindrical; primary laterals cylindrical, annual, usually exceeding length of primary axis, radially branched, mainly smooth throughout length, at least at the base, occasionally with small, widely spaced, spiny appendages upwards; higher order branches with or without spines, sometimes slightly iridescent, with scattered, inner aerocysts and cryptostomata; tophules present or absent; in section, thallus parenchymatous, comprising a central medulla of small polygonal cells, surrounded by a cortex of larger, globose, thick-walled cells within an enclosing unicellular, palisade-like meristoderm of squared cells.

Monoecious, receptacles on terminal branchlets, with spiny appendages, bearing few or clusters of conceptacles at the base of each fertile spine; conceptacles usually hermaphrodite, oogonia borne around base, large, with a single egg, which upon release may be incubated on receptacle surface; antheridia small, sessile on branched multicellular paraphyses.

Two species in Britain and Ireland.

KEY TO SPECIES

Primary axis flattened, apex surrounded by incurved young laterals; primary
laterals lacking tophules . . . . G. baccata
Primary axis cylindrical, apex smooth, rounded dome-shaped apex; primary
laterals with ovoid tophules. . . . . .G. nodicaulis

Gongolaria baccata (S.G. Gmelin) Molinari & Guiry (2020), p. 2.   Figs 268, 269, 270

Fucus abrotanoides S.G. Gmelin (1768), p. 89; *Fucus baccatus* S.G. Gmelin (1768), p. 90; *Fucus fibrosus* Hudson (1778), p. 575; *Cystoseira fibrosus* (Hudson) C. Agardh (1820), p. 65; *Mackaia fibrosus* (Hudson) S.F. Gray (1821), p. 392; *Cystoseira thesiophylla* Duby (1830), p. 937; *Phyllacantha fibrosus* (S.G. Gmelin) Kützing (1843), p. 356; *Phyllacantha thesiophylla* (Duby) Kützing (1860), p. 13; *Gongolaria fibrosa* (Hudson) Kuntze (1891), p. 895; *Cystoseira baccata* (Gmelin) Silva (1952), p. 280; *Treptacantha baccata* (S.G. Gmelin) Orellana & Sansón in Orellana *et al.* (2019), p. 456.

Thalli forming erect, usually solitary thalli, 0.5–1(–1.5) m in height, often quite bushy, arising from a discoid/conical holdfast; primary axis up to 0.5 m in height, 1 cm in width, dorsiventrally flattened and distinctly zigzagged in appearance as a result of the persistent decurrent bases of the deciduous primary laterals; apical region giving rise acropetally to the primary laterals, the youngest characteristically incurled, which are later shed gradually denuding the primary axis; tophules absent; primary laterals usually exceeding length of the primary axis, up to 1 m in length, terete in section and slightly zigzagged in appearance, much branched, with radially produced secondary laterals which are incurved, sometimes producing at their base, larger branches resembling the parental primaries which are deciduous, leaving conspicuous stumps; first formed secondary laterals flattened and leaf-like, simple or bifid and with, or without, a pronounced midrib, later shed and only seen in juvenile thalli and in the apical regions of the primary laterals; later formed secondaries are cylindrical, up to approximately 8 cm in length, with a spirally twisted axis, much spirally branched and conical in outline, giving rise spirally, or in a zigzag fashion, to tertiary laterals of both limited and unlimited growth characteristically clustered around the apex in a corymbose manner; winter developing secondaries are shorter and more compact, with tertiary laterals borne close together; spring and summer borne secondaries are more attenuate with more widely spaced tertiaries; basal tertiaries, and the apical region of secondaries, often develop into appendages, particularly in spring giving the thalli a shaggy appearance; appendages slender, filiform structures, simple, bifurcate or sometimes branched, recurved towards parental axis, tapering towards the tip, sometimes spike-like, up to 2.5 cm in length; a mixture of appendages and tertiary laterals developing over the greater length of the secondary lateral; tertiaries up to 3 cm in length, are often further branched, the basal region often inflating to form one or more ovoid aerocysts, 4–7 mm in length, and the more distal regions developing into receptacles; in cross section, thalli comprising an inner medulla of closely packed, relatively narrow, longitudinally elongated cells, surrounded by a cortex of larger thick-walled cells, somewhat radially elongated in cross section, enclosed within an outer meristoderm layer of photosynthetic, palisade-like square cells; cryptostomata scattered over entire thallus.

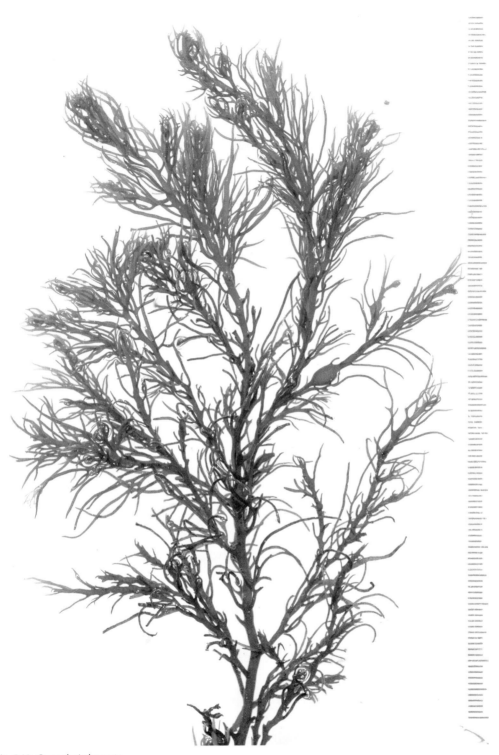

Fig. 268. *Gongolaria baccata*
Habit of upper thallus region.

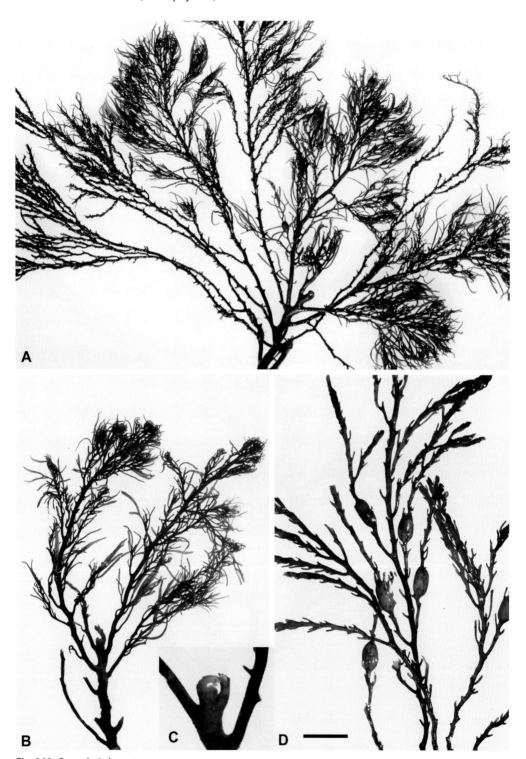

Fig. 269. *Gongolaria baccata*

A, B. Habit of thallus showing flattened primary axis giving rise, in a zigzag manner, to primary laterals. Note the denuded lower regions of the primary laterals. C. Apex of primary axis showing incurved arrangement of the young primary lateral initials. D. Terminal region of thallus showing aerocysts and spine-like appendages. Bar in D = 18 mm (A), = 17 mm (B), = 6 mm (C), = 11 mm (D).

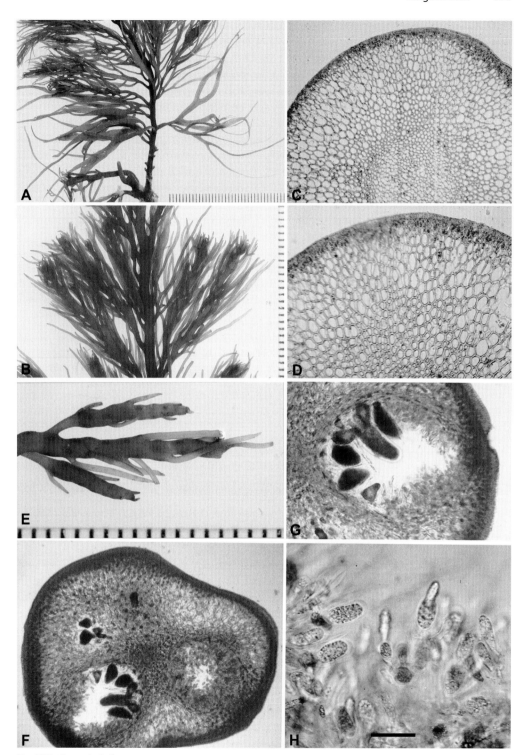

Fig. 270. *Gongolaria baccata*

A, B. Terminal region of young primary laterals. C, D. TS of primary laterals showing the central medulla of closely packed cells, surrounded by a cortex of larger, radially elongated cells and a surrounding outer meristoderm layer. E. Terminal receptacle with spine-like appendages. F. TS of receptacle showing sunken conceptacles. G. Conceptacle showing darkly pigmented oogonia arising from conceptacle wall. H. Densely crowded antheridia borne on paraphyses. Bar in H = 375 μm (F), = 215 μm (C), = 120 μm (G), = 65 μm (D), =38 μm (H).

Monoecious, receptacles developing terminally on the tertiaries, somewhat indistinct, 1–4(–5) cm in length, 1–1.5 mm wide, simple or sometimes branched, irregularly nodose in outline and sometimes appearing spirally constructed and clothed with spiny appendages; receptacles bearing a few immersed, flask to oval-shaped conceptacles at the base of each fertile spine, usually 3–6 seen in cross section, somewhat squashed together, 400–600 × 300–520 µm, each containing both oogonia and antheridia interspersed with paraphyses; oogonia borne at the base of the conceptacle, antheridia occurring more towards the ostiole, sometimes throughout conceptacle cavity; antheridia cylindrical, granular, borne laterally on relatively short, multicellular, much-branched paraphyses, measuring 18–30 × 8–12 µm, usually developing prior to the oogonia; oogonia much larger and darkly pigmented, 115–170(–220) × 50–68(–73) µm, club to oval shaped, borne directly on conceptacle wall and enclosed by multi-cellular paraphyses; paraphyses to 205 µm, 3–5 cells in length, somewhat attenuate, extending well into conceptacle lumen, simple or once branched towards the base, comprising colourless, cylindrical cells, 2.5–10 diameters long, 22–58(–72) × 8–12(–16) µm.

Epilithic, lower eulittoral pools, extending into the sublittoral to 10 m.

Restricted mainly to the Channel Islands, Isle of Man and south-west coast of England (Cornwall, Devon, Dorset, Sussex); in Ireland, Clare, Kerry, Galway, Donegal.

Perennial, receptacles reported all the year round but more actively fertile in the spring/summer period.

Ardré (1970), pp. 317, 318; Brodie *et al.* (2016), p. 1022; Bunker *et al.* (2017), p. 265; Cabioc'h *et al.* (1992), pp. 74, 76, fig. 53; Coppejans & Kling (1995), p. 212, pl. 74(A); De Mesquita Rodrigues (1963), pp. 110–12, pl. XII(d); Dizerbo & Herpe (2007), pp. 133, 134, pl. 43(4); Fletcher in Hardy & Guiry (2003), p. 237; Gallardo García *et al.* (2016), p. 37; Gayral (1958), p. 266, pls XLIX, L, fig. 42(A), (1966), p. 325, pls LXIX, LXX; Gómez Garreta (2001), pp. 110–13, fig. 28(a–c), map 25; Guiry (2012), p. 165; Hamel (1931–39), pp. 391, 392; Molinari-Novoa & Guiry, M.D. (2020), pp. 1–10; Morton (1994), p. 46; Newton (1931), p. 228; Orellana *et al.* (2019), pp. 447–65; Ribera *et al.* (1996), pp. 89–103; Roberts (1967a), pp. 345–66, figs 8, 15, 21, (1967b), pp. 367–78, figs 1–13; Rull Lluch *et al.* (1994), pp. 131–8; Sánchez *et al.* (2005), pp. 942–9; Sauvageau (1912), pp. 495–7, 510–11, (1920b), pp. 19–21; Silva (1952), p. 280; Stegenga & Mol (1983), p. 127, pl. 50(4); Stegenga *et al.* (1997a), p. 24; Taskin *et al.* (2012), pp. 1–75.

Gongolaria nodicaulis (Withering) Molinari & Guiry (2020), p. 3.   Figs 271, 272

Fucus nodicaulis Withering (1796), p. 111; *Fucus mucronatus* Turner (1802a), pp. xxxiii, 73, nom. illeg.; *Mackaia mucronata* (Hudson) S.F. Gray, (1821), p. 392; *Cystoseira nodicaulis* (Withering) M. Roberts (1967a), pp. xxxiii, 349; *Treptacantha nodicaulis* (Withering) Orellana & Sansón, in Orellana *et al.* (2019), p. 459.

Thalli erect, to 50 cm in length, dark brown to black, usually solitary in appearance, leathery, and arising from a conical, discoid holdfast; primary axis up to 24 cm in length, 1 cm in width, cylindrical, usually much branched, with a fairly conspicuous, smooth, rounded dome-shaped apex surrounded by intermittent zones of tophules (formed early on in the growing season) and the bases of developing primary lateral branches (formed later on in the growing season); tophules usually develop in spring and early summer, are ovoid, approxi-mately 1.5 cm in length, and either smooth, or covered on the surface with small tubercular outgrowths, and persistent after the dehiscence of the distal lateral system; primary laterals annual, formed as slender cylindrical outgrowths, arising from either the tips of the tophules or from the lateral branch systems in autumn and are usually shed the following summer; primary laterals terete in cross section, borne from the main axis in a radial or distichous fashion, in acropetal succession, and with a slight blue iridescence when young, to 30 cm in length, repeatedly branched in a pinnate fashion, either regularly or irregularly, with infrequent cryptostomata and bearing spine-like appendages, shed in the summer, with the secondary primaries repeating the structure; secondary and tertiary branch systems with or without aerocysts, which are often restricted to the fertile branches, isolated or in series, and sometimes with conspicuous cryptostomata; in cross section, thalli comprising an

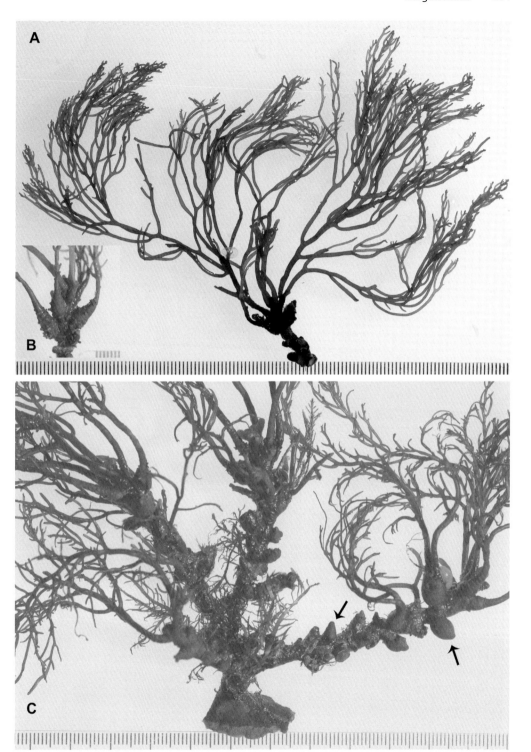

Fig. 271. *Gongolaria nodicaulis*
A–C. Habit of thalli showing characteristic tophules (arrows) on primary axes.

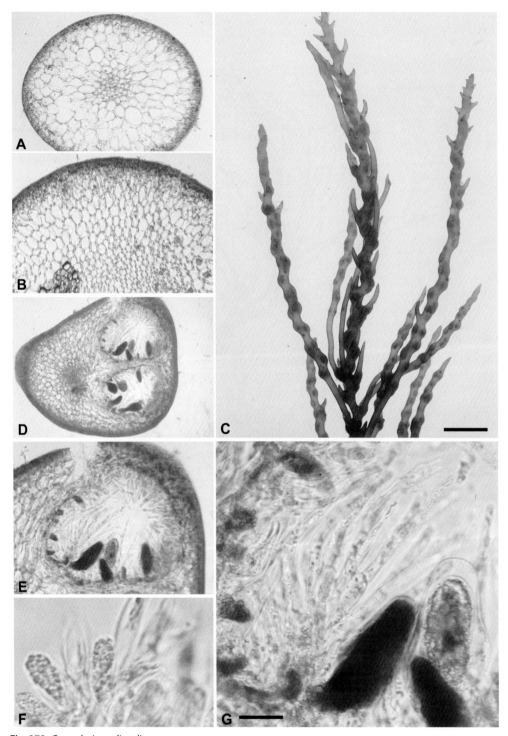

Fig. 272. *Gongolaria nodicaulis*

A, B. TS of lateral showing inner medulla of closely packed cells, the surrounding cortex of larger, radially elongated cells and an enclosing meristoderm. C. Terminal fertile branch region showing long nodular receptacles and spine-like appendages. D. TS of receptacle showing two immersed conceptacles. E. Close-up of receptacle showing darkly pigmented oogonia and surrounding filamentous paraphyses. F. Granular antheridia borne on branched paraphyses. G. Darkly pigmented oogonia surrounded by branched, multicellular, colourless paraphyses. Bar in C = 2.2 mm, bar in G = 190 μm (D), = 120 μm (A, B, E), = 70 μm (G), = 35 μm (F).

inner medulla of relatively narrow, longitudinally elongated cells, quite densely packed and rounded to angular, surrounded by a cortex of larger thick-walled cells, radially elongated in cross section in young axes, more rounded in older axes, enclosed within an outer meristoderm layer of photosynthetic, palisade-like square cells.

Monoecious, receptacles developing terminally on the ultimate branches of the tertiaries and secondaries, indistinct, to 5–6 cm in length, 1–2 mm wide, simple or sometimes branched, irregularly nodose and verrucose in outline as the conceptacles develop within the axes, with or without a proximal, ovoid aerocyst and usually with spirally arranged, stiff, spine-like appendages; receptacles bearing immersed, flask to oval shaped conceptacles, usually 3–4 seen in cross section filling most of the lumen, 380–500 × 230–430 μm, sometimes clearly adpressed and semi-circular in shape, each containing both oogonia and antheridia, sometimes with antheridia only, interspersed with paraphyses; oogonia borne at the base of the conceptacle, antheridia occurring more towards the ostiole; antheridia fusiform, granular, 22–32 × 8–12 μm, borne laterally in clusters on relatively short, multicellular, much branched paraphyses, usually developing prior to the oogonia; oogonia much larger and darkly pigmented, 63–110(–132) × 28–50(–70) μm, club to oval shaped, borne directly on conceptacle wall and enclosed by paraphyses; paraphyses multicellular, to 170 μm, 3–5 cells in length, extending well into conceptacle lumen, acuminate, simple or once branched towards the base, colourless, comprising colourless, cylindrical cells, 2–9 diameters long, 28–58(–72) × 8–12(–16) μm.

Epilithic; lower eulittoral pools, or emergent, commonly extending into the sublittoral, to 5 m.

Locally common, restricted in distribution mainly to southern and western shores, reported for the Isles of Scilly, the Channel Islands, Cornwall, Devon, Dorset, Hampshire, Sussex, Anglesey, Isle of Man, Argyll; in Ireland, Cork, Kerry, Clare, Mayo, Donegal and Antrim.

Perennial, receptacles reported all the year round but more actively fertile in the spring/summer period.

Ardré (1970), p. 321; Arenas *et al.* (1995), pp. 1–8; Brodie *et al.* (2016), p. 1022; Bunker *et al.* (2017), p. 266; Cabioc'h *et al.* (1992), p. 74, fig. 50; Cormaci *et al.* (2012), pp. 351, 352, pl. 114(1–3); Dizerbo & Herpe (2007), p. 136, pl. 44(3); Fletcher in Hardy & Guiry (2003), p. 239; Gallardo García *et al.* (2016), p. 38; Garcia-Fernandez & Bárbara (2016); Gayral (1966), p. 329, pl. LXXI; Gómez Garreta (2001), pp. 151, 152, fig. 45(a–c), map 42; Guiry (2012), p. 166; Hamel (1931–39), pp. 407, 408; Harvey (1846–51), pl. 60; Jégou *et al.* (2012), pp. 203, 204; Molinari-Novoa & Guiry (2020), pp. 1–10; Morton (1994), p. 47; Newton (1931), pp. 226, 227; Orellana *et al.* (2019), pp. 447–65; Penot *et al.* (1985), pp. 93–102; Ribera *et al.* (1992), p. 122, (1996), pp. 89–103; Rico & Fernández (1997), pp. 405–10; Roberts (1968a), pp. 254–6, (1967a), pp. 345–66, (1977), pp. 175–99, figs 1–23, (1978), pp. 399–422; Sauvageau (1912), pp. 260–74, 522–3; Stegenga & Mol (1983), p. 129, pl. 51(1); Stegenga *et al.* (1997a), p. 24; Taskin *et al.* (2012), pp. 1–75.

Halidrys Lyngbye

HALIDRYS Lyngbye (1819), p. 37.

Type species: *H. siliquosa* (Linnaeus) Lyngbye (1819), p. 37.

Thalli macroscopic, perennial, bushy, dark brown to black, leathery and smooth, much branched and feather-like in appearance, to 1(–2) m in length, and arising from a large, discoid/conical holdfast; main axes slightly flattened, profusely pinnately and distichously branched in one plane to several orders, with branching alternate and zigzag-like; terminal laterals giving rise to characteristic, siliquose, serially arranged, chambered, pod-like air-bladders, or stalked receptacles sometimes grouped in a racemose manner.

Thalli monoecious, with antheridia and oogonia borne in conceptacles within the receptacle; oogonia producing a single functional egg.

One species in Britain and Ireland.

604 Brown Seaweeds (Phaeophyceae) of Britain and Ireland

Halidrys siliquosa (Linnaeus) Lyngbye (1819), p. 37.   Figs 273, 274

Fucus siliquosus Linnaeus (1753), p. 1160; *Fucus siliculosus* Stackhouse in Withering (1796), p. 88; *Halidrys siliquosa* var. *denudata* Lyngbye (1819), p. 37; *Cystoseira siliquosa* (Linnaeus) C. Agardh (1820), p. 71; *Halidrys siliquosa* var. *evesiculosa* J. Agardh (1836), p. 12; *Halidrys siliquosa* var. *siliculosus* (Stackhouse) Batters (1902), p. 52.

Thalli erect, macroscopic, monopodially branched and bushy in habit, to 120 cm in length, seldom exceeding 10 mm in width, light olive-brown when young becoming dark brown to black later, arising from a large prominent, discoid/conical holdfast; thalli tough, leathery and smooth in texture and much branched throughout; axes and branches cylindrical below, becoming clearly dorsiventrally compressed above and alternately branched from the margins in one plane to several orders, branches variable in length, basal branches attaining length of parental axis, terminal branches more restricted and variable in length, sometimes remaining short and teeth-like, 2–3 mm in length; terminal branchlets developing into long narrow, linear, lanceolate, leaf-like branches with central vein-like midrib, air vesicles or receptacles; air vesicles characteristically siliquose, regularly septate, chambered, pod-like, to 4(–7) cm (to 12 chambers) in length, 3–5(–8) mm in width, borne on a pedicel, and usually mucronate or with terminal undifferentiated acuminate portion; in cross section, thalli with a central medullary region of small, tightly packed, polygonal, dark-pigmented cells surrounded by a cortical region of 10–15 larger, irregularly shaped, thick-walled, colourless, radially elongated cells, which become progressively smaller and pigmented towards the outside, and are enclosed within an outer meristoderm layer of palisade-like cells.

Monoecious, bearing receptacles terminally on the branches, sometimes above an air vesicle, stalked, siliquose, acuminate, linear, simple, rarely branched, solitary or in clusters around central axis, to 4 cm in length, 3–6 mm in width; in section, showing sunken, spherical to flask-shaped hermaphrodite conceptacles bearing both oogonia and antheridia distributed throughout conceptacle cavity, associated with paraphyses; paraphyses multicellular, branched, colourless or sometimes terminally pigmented, comprising cylindrical, sometimes irregularly swollen cells, 20–80(–140) × 6–11(–18) μm; oogonia large, few in number, arising directly from conceptacle wall, darkly pigmented, broadly based, club to elongate-cylindrical, 135–185(–220) × 63–80(–110) μm, with a single egg; antheridia small, abundant and densely packed, 20–33 × 11–18 μm, borne terminally and laterally on branched, multicellular paraphyses, often clustered.

Epilithic, upper to lower eulittoral, in deep, often sandy pools and channels, extending into the shallow sublittoral, rarely to 30 m.

Common and widely distributed around Britain and Ireland.

Perennial, maximum growth occurring in spring and summer, with receptacle formation and differentiation occurring from August to December, followed by gamete differentiation and discharge from approximately December to March.

Ardré (1970), p. 324; Bartsch & Kuhlenkamp (2000), p. 166; Brodie *et al.* (2016), p. 1022; Bunker *et al.* (2017), p. 171; Cabioc'h *et al.* (1992), p. 72, figs 47, 48; Coppejans & Kling (1995), p. 212, pl. 74(B); De Mesquita Rodrigues (1963), pp. 117, 118, pl. 12(B); Dizerbo & Herpe (2007), p. 138, pl. 45; Elliot & Moss (1953), p. 357; Fletcher in Hardy & Guiry (2003), p. 255; Gayral (1966), p. 335, pl. LXXIV; Gómez Garreta (2001), pp. 95–7, fig. 24, map 21; Goodband (1973), pp. 175–9; Guiry (2012), p. 166; Hamel (1931–39), pp. 382–4, fig. 59(1); Harding (1978), p. 201; Hardy & Moss (1978), pp. 69–78; Harvey (1846–51), pl. 66; Jaasund (1965), p. 112; Kornmann & Sahling (1977), p. 170, pl. 93(A–C); Kylin (1947), p. 85, pl. 18(58); Lacey & Moss (1956), p. 14; Lima *et al.* (2009), pp. 1–4; Morton (1994), p. 47; Moss (1982), pp. 185–91; Moss & Elliot (1957), pp. 143–51; Moss & Lacy (1963), pp. 67–74; Moss & Sheader (1973), pp. 63–8; Naylor (1958c), pp. 205–17; Newton (1931), pp. 225, 226, fig. 143(A–C); Pankow (1971), p. 210, fig. 271; Rueness (1977), p. 195, pl. XXIV(4), (2001), p. 15; Stegenga & Mol (1983), p. 129, pl. 52(1, 2); Stegenga *et al.* (1997a), p. 24; Wernberg *et al.* (2001), pp. 31–9.

Fig. 273. *Halidrys siliquosa*
Habit of thallus. (Herbarium specimen, NHM.)

Fig. 274. *Halidrys siliquosa*

A. Apex region of vegetative thallus. B. Upper region of thallus with air vesicles. C–E. TS of different sized axes, showing central medulla surrounded by larger celled cortex. F. TS of thallus, edge region. G. Terminal branch region with air vesicles and receptacles (top centre). H. VS of air vesicle, showing internal septate arrangement. I. Two receptacles. J. TS of receptacle showing immersed conceptacles. K. TS of conceptacle, showing three dark-pigmented oogonia. L. TS of conceptacle, showing a single oogonium surrounded by clusters of granular antheridia borne on paraphyses. Bar in B = 12 mm, bar in G = 12 mm (G), = 1.5 mm (J), bar in H/I = 5.5 mm (H), = 6.6 mm (I), bar in L = 375 μm (D), = 215 μm (C), = 140 μm (E, K), = 70 μm (L), = 35 μm (F).

Sargassum C. Agardh

SARGASSUM C. Agardh (1820), p. 1.

Type species: *S. bacciferum* (Turner) C. Agardh (1820), p. 1 (= *Sargassum natans* (Linnaeus) Gaillon (1828), p. 355).

Thalli perennial, bushy to 5 m in length, yellow to dark brown in colour, with one short perennial main axis giving rise to one to several long primary laterals, and attached at the base by a conical holdfast; sometimes free floating (Sargasso Sea); primary laterals produced seasonally, deciduous, indeterminate, branched to several orders, giving rise to secondary/ tertiary laterals of determinate growth; axis and laterals terete, with branching usually radial; laterals giving rise to flattened leaves with distinct midrib, toothed margins and cryptostomata, and to rounded, pedicellate air-bladders (pneumatocysts).

Antheridia and oogonia borne inside conceptacles within specialized, terete, swollen, linear receptacles produced in the axils of the ultimate branches; monoecious or dioecious, receptacles and conceptacles bisexual or unisexual.

One species in Britain and Ireland.

Sargassum muticum (Yendo) Fensholt (1955), p. 306.   Figs 275, 276

Sargassum kjellmanianum f. *muticum* Yendo (1907), p. 104.

Thalli erect, 0.5–1(–5) m in length, light to dark brown, arising from a discoid holdfast; primary axis perennial, formed initially as the remaining basal region of a one year old thallus, up to 0.2 m in length, 1 cm in width, cylindrical, giving rise to one to several primary laterals of varying length; apex inconspicuous and surrounded by developing primary lateral branch systems; primary laterals of indeterminate growth, terete in section, relatively short during the winter, with 2–3, sometimes more in older thalli, rapidly elongating during the spring and summer to 3(–5) m in length, highly branched, giving rise in a spiral, alternate fashion, at regular intervals, to secondary laterals of determinate growth, to 10–30 cm in length; secondary laterals similarly giving rise to tertiary laterals of determinate length, to 5–10 cm in length; primary axis and all laterals characteristically bearing leaves and air-bladders; leaves linear-lanceolate in shape in basal regions, to 10 cm in length with toothed margins, leaves in upper thallus regions to 4 cm in length with toothed or entire margins; air-bladders single or in clusters, borne in leaf axils, usually pedicellate, sometimes mucronate, to 2–3 mm in diameter; in cross section, thallus axes comprising a relatively small, central medulla of densely aggregated filaments of narrow, thick-walled, longitudinally elongated cells, which appear angular in transverse section, giving rise peripherally to a more extensive cortex of larger, thick-walled, colourless, longitudinally elongated cells which become more pigmented, less longitudinally elongated and more radially elongated towards the outside, and are enclosed within an outer meristoderm of photosynthetic, palisade-like cells.

Monoecious, with reproductive receptacles developing in the axils of the ultimate branches of the tertiaries, secondaries and primaries, to 3 cm in length, 2–3 mm in diameter, cylindrical, simple, cigar-like, irregularly nodose in outline; receptacles bisexual, bearing sunken, peripherally arranged, densely packed, globular to pyriform male and female conceptacles, which extend deeply into central receptacle region; conceptacles unisexual with oogonia and antheridia borne in separate conceptacles which are often adjoined; oogonial conceptacles more common, containing 12–16 large, densely pigmented oogonia, borne around the sides of the conceptacle and arising directly from the conceptacle wall cells; oogonia oval, 90–162 μm × 65–108 μm, extending deeply into the conceptacle and occupying most of its volume; accompanied by branched, multicellular, pigmented paraphyses, extending to 100 μm (3–6 cells) into conceptacle lumen, comprising cells

Fig. 275. *Sargassum muticum*
A. Bushy clumps of thalli dominating lower eulittoral rocky pool. B. Two young thalli with leaves. C. Young thallus with erect primary laterals. D. Young vegetative thallus with several erect primary laterals. E. Portion of primary lateral with secondary laterals. F, G. Close-ups of secondary laterals with tertiary laterals bearing vesicles and leaves. Bar in B = 16.4 mm, bar in E = 42 mm.

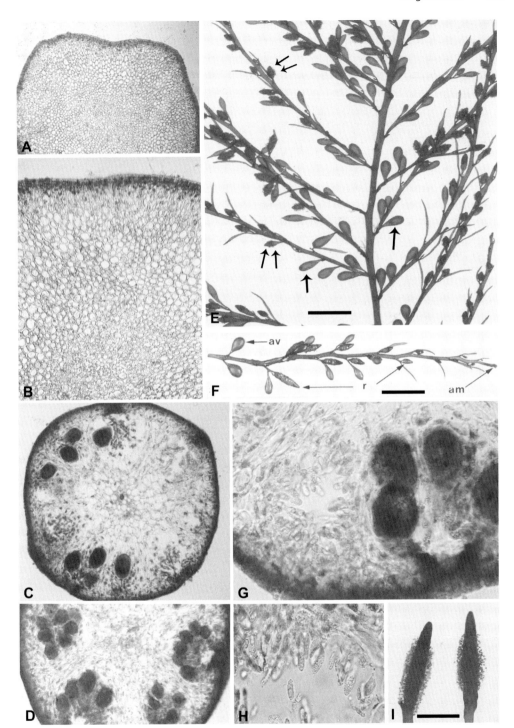

Fig. 276. *Sargassum muticum*

A, B. TS through part of primary laterals showing small central medulla and extensive outer cortex. C, D. TS of receptacles showing conceptacles. E. Portion of a mature primary lateral producing secondary laterals bearing vesicles (single arrows) and receptacles with attached zygotes (double arrow). F. Tertiary lateral showing air vesicles (av), receptacles (r) and apical meristem (am). G. TS of part of a receptacle showing adjacent female and male conceptacles. H. Male conceptacle with granular antheridia. I. Two terminal receptacles with attached 'incubating' zygotes. Bar in E = 7 mm, bar in F = 6 mm, bar in I = 3.5 mm (I), = 220 μm (A, C), = 140 μm (B, D), = 70 μm (G), = 35 μm (H).

15–30(–45) × 7–15 μm, 1.5–3(–5) diameters long, often with a bulbous terminal cell; antheridial conceptacles less common with antheridia in large numbers borne on stalk cells arising from conceptacle wall, successive divisions of the stalk cells giving rise to many antheridia borne on sympodially branched paraphyses extending up to 100 μm into conceptacle lumen; antheridia lightly pigmented, oval, 17–22 × 9–13 μm; paraphyses multicellular, branched, usually pigmented, to 100 μm (3–6 cells) in length, comprising cells 15–30(–45) × 7–15 μm, 1.5–3(–5) diameters long, sometimes with a swollen terminal cell; oogonia are released from each conceptacle all at the same time and are retained on the receptacle surface by a flexible mucilaginous stalk to form clusters around each conceptacle ostiole just visible to the unaided eye; during this approximately 48-hour tethered process, fertilization takes place and the zygotes enlarge into multicellular, pear-shaped propagules, reaching 220 × 180 μm before being released.

Thalli with a perennial primary axis, which gives rise annually to a number of long primary laterals of unlimited growth. Studies on the south coast of England have revealed that these primary laterals (and juvenile thalli) usually arise in October and slowly increase in length over winter and early spring after which growth is rapid, with maximum growth occurring in May and June, with mature thalli reaching several metres in length and giving rise to secondary and tertiary laterals of limited growth. The reproductive receptacles are first seen usually towards the end of May, reaching their maximum length and maturity by early August, but often releasing gametes into September. From late July/early August onwards, the laterals begin to senesce with decay most rapid in September and October, considerably reducing the size of mature populations with the ground covered by beds of juvenile thalli, although thalli at various stages of development can be observed throughout the year.

Epilithic, less commonly epiphytic and epizoic; lower eulittoral, emergent on rocks or more commonly in pools, lagoons, runnels etc., extending to the mid-eulittoral and higher in pools, extending into the sublittoral to 5–10 m. Thalli showing a preference for sheltered waters and little tolerance of wave action or prolonged emersion. Locally common and can form a dominant canopy in tidal pools and lagoons, the size and development of the thalli usually depending on pool/lagoon depth. Also found as a fouling alga on floating structures such as pontoons of marinas, and boat travel is likely to be a major vector contributing towards its geographical spread.

Locally common and predominantly restricted in distribution to southern and western shores, with various surveys reporting its occurrence in Britain for the Isles of Scilly, the Channel Islands, Cornwall, North and South Devon, Lundy, Dorset, Hampshire, Sussex, Kent, Essex, Suffolk, Norfolk, Lincolnshire, Pembrokeshire, Cardigan Bay, Anglesey, Lancashire, Outer Hebrides and West Lothian; in Ireland, reported for Cork, Down, Donegal, Sligo, Mayo, Galway, Clare, Kerry, Cork and Wexford.

First reported for Britain, and the North Atlantic, in the mid-south coast region of England in 1973 (Farnham et al. 1973), this alien alga quickly spread along the south coast, westwards to Cornwall and the Isles of Scilly, eastwards to Kent and south to the Channel Islands. It then spread up the west coast of Britain from North Devon to the Outer Hebrides. As a contaminant of oysters, it was introduced into Lough Stranack in Ireland (Boaden 1995) and has subsequently spread to several counties, especially on the west coast (Guiry 2012; Loughnane & Stengel 2002; Simkanin 2004; Stokes et al. 2004; Strong & Dring 2011; Winder 2002). Its spread north-eastwards has been relatively slower with reports of its discovery restricted to Norfolk and Lincolnshire. Its geographical spread on the European mainland Atlantic coast has been more rapid, with reports of attached populations reaching Norway (Rueness 1985, 1989; Rueness et al. 2001) and as far south as the Iberian Peninsula (Engelen et al. 2008, Gallardo García et al. 2016) and Morocco (Sabour et al. 2013). Following its initial introduction into the l'Etang de Thau lagoon (Gerbal et al. 1985; Gerbal 1986), probably as a contaminant on imported oysters, it is now widely distributed across the Mediterranean, with reports of its discovery as far east as the Venice Lagoon (Curiel *et al.* 1998) and Taranto, Italy (Cecere *et al.* 2000). More recently (Benali *et al.* 2019) it has been reported as drift on the Algerian coast.

It is likely that the original population discovered at Bembridge on the Isle of Wight, southern England, originated from drift material derived from across the Channel in Normandy where it had first become established as a contaminant on imported oysters (see Farnham 1980 for discussion). Despite an attempted eradication programme (Farnham & Jones 1974; Jones & Farnham 1973), this alga became firmly established on the south coast of England. Its success as an introduced species both in the North Atlantic and Mediterranean, and indeed at other localities around the world, can be attributed to its possession of a range of useful biological characteristics including: fragmentation and long distance dissemination of floating fertile branches, strong regenerative abilities, persistent apical growth of the receptacles, monoecious reproduction with hermaphrodite receptacles, incubation of zygotes on the receptacle surface and their subsequent release as young, more hardy, multicellular germlings, a rapid growth rate, high fecundity and a physiological hardiness to a wide range of environmental conditions (Engelen *et al.* 2008; Fletcher 1980b; Fletcher & Fletcher 1975 a,b; Hales & Fletcher 1989a,b, 1990, 1992; Jephson & Farnham 1974; Jephson & Gray 1977; Kjeldsen & Phinney 1972; Norton 1976; Steen 2004).

Abbott & Hollenberg (1976), pp. 275, 276, fig. 220; Aguilar-Rosas & Galindo (1990), pp. 185–90; Andrade-Sorcia *et al.* (2014), pp. 201–23, figs 1–9; Andrew & Viejo (1998), pp. 251–8; Arenas & Fernández (1998), pp. 209–16, (2000), pp. 1012–20; Arenas *et al.* (1995), pp. 1–8; Baer & Stengel (2010), pp. 185–94; Balboa *et al.* (2017), pp. 244–51; Bartsch & Kuhlenkamp (2000), p. 168; Belsher (1983–89), pp. 1–99; Belsher & Pommellec (1988), pp. 221–31; Belsher *et al.* (1985), pp. 33–6; Benali *et al.* (2019), pp. 575–81; Bjærke & Fredriksen (2003), pp. 353–64; Boaden (1995), pp. 111–13; Boalch & Potts (1977), pp. 29–31; Britton-Simmons (2004), pp. 61–78; Britton-Simmons *et al.* (2011), pp. 187–96; Brodie *et al.* (2016), p. 1022; Bunker *et al.* (2017), p. 276; Buschbaum *et al.* (2006), pp. 743–54; Cabioc'h (1981), pp. 1, 2; Cabioc'h *et al.* (1992), p. 76, fig. 55; Cacabelos *et al.* (2010), pp. 182–7; Casares Pascual *et al.* (1987), p. 151; Cecere *et al.* (2000), pp. 305–9; Chamberlain (1978), p. 198; Cheang *et al.* (2010), pp. 1063–74; Cho *et al.* (2006), pp. 512–19, (2012), pp. 473–84; Coppejans & Kling (1995), p. 224, pl. 80(A–C); Coppejans *et al.* (1980), pp. 7–13; Cormaci *et al.* (2012), pp. 385, 386, pl. 131, figs 1–3; Cosson (1999), pp. 35–42; Cosson *et al.* (1977), pp. 109–16; Critchley (1980), p. 194, (1981), p. 134, (1983a), pp. 825–31, (1983b), pp. 813–24, (1983c), pp. 617–25, (1983d), pp. 539–45, (1983e), pp. 547–52; Critchley & Dijkema (1984), pp. 211–16; Critchley *et al.* (1983), pp. 799–811, (1986), pp. 313–32, (1987), pp. 245–55, (1990a), pp. 211–17, (1990b), pp. 551–62; Curiel *et al.* (1998), pp. 17–22, (1999), pp. 17–22; De Paula & Eston (1987), pp. 405–10; De Wreede (1978), pp. 1–9, (1983), pp. 153–60; De Wreede & Vandermeulen (1988), pp. 469–76; Deysher (1984), pp. 403–7; Deysher & Norton (1982), pp. 179–95; Dizerbo & Herpe (2007), p. 132, pl. 43(2); Druehl (1973), p. 12; Engelen & Santos (2009), pp. 675–84; Engelen *et al.* (2008), pp. 275–82, (2011), pp. 157–85, (2015), pp. 81–126; Espinoza (1990), pp. 193–6; Farnham (1980), pp. 898–903; Farnham & Jones (1974), pp. 57, 58; Farnham *et al.* (1973), pp. 231, 232, (1981), pp. 277–82; Fensholt (1955), pp. 305–22; Fernández (1999), pp. 553–62; Fernández *et al.* (1990), pp. 423–8; Fletcher (1980b), pp. 425–32; Fletcher & Fletcher (1975a), pp. 157–62, (1975b), pp. 149–56; Gallardo García *et al.* (2016), p. 38; Gerbal (1986), pp. 5, 6; Gerbal *et al.* (1985), pp. 241–54; Givernaud (1984), pp. 129–32; Givernaud *et al.* (1990), pp. 293–304; Glombitza *et al.* (1982), pp. 449–54; Gómez-Garreta (2001), p. 192; Gómez Garreta *et al.* (2001b), pp. 81–3, fig. 20, map 17; Gorostiaga *et al.* (1988), pp. 437–43; Gray & Gareth Jones (1977), pp. 303–8; Guiry (2012), p. 167; Hales & Fletcher (1989a), pp. 115–25, (1989b), pp. 167–76, (1990), pp. 241–9, (1992), pp. 591–601; Haroun & Izquierdo (1991), pp. 27–48; Huang *et al.* (2017), pp. 71, 83; Hwang & Dring (2002), pp. 471–5; IARE (1992), p. 106; Incera *et al.* (2011), pp. 488–95; IUCN (2009), p. 30; Jephson & Gray (1977), pp. 367–75; Jones & Farnham (1973), pp. 394, 395; Josefsson & Jansson (2011), pp. 1–10; Kane (1978), pp. 202, 203; Karlsson (1988), pp. 199–205; Karlsson & Loo (1999), pp. 285–95; Kerckhof *et al.* (2007), pp. 243–57; Kerrison & Le (2016), pp. 481–9; Knoepffler-Péguy *et al.* (1985), pp. 291–5; Kraan (2008), pp. 825–32; Kraberg & Norton (2007), pp. 17–24; Le *et al.* (2018), pp. 2475–83; Le Lann *et al.* (2012), pp. 1–11; Lewey & Farnham (1981), pp. 388–94; Liu & Pang (2014), pp. 1129, 1130; Loughnane & Stengel (2002), pp. 70–2; Loraine (1989), p. 83; Monteiro *et al.* (2009b), pp. 1–7; Morrell & Farnham (1981), p. 138, (1982), pp. 236, 237; Nicholson *et al.* (1981), pp. 416–24; Nienhuis (1982), pp. 189–95; Norton (1976), pp. 197, 198, (1977b), pp. 41–53, (1980), pp. 197, 198, (1981a), pp. 465–70, (1981b), pp. 449–56, (1992), pp. 293–301; Norton & Deysher (1989), pp. 147–52; Oak *et al.* (2010), pp. 125–7, figs 9, 10; Ogawa (1990), pp. 357–60; Olabarría *et al.* (2009a), pp. 153–8, (2009b), pp. 194–7; Paula & Eston (1987), pp. 405–10; Peaucelle & Couder (2016), pp. 1–15; Pedroche *et al.* (2008), pp. 1–139; Pérez-Larrán *et al.* (2019), pp. 2481–95; Plouguerné *et al.* (2006), pp. 337–44, (2008), pp. 202–8; Polte & Buschbaum (2008), pp. 11–18; Prud'homme van Reine &

Nienhuis (1982), pp. 37–9; Prud'homme van Reine *et al.* (1982), p. 238; Puspita *et al.* (2017), pp. 2521–37; Ribera *et al.* (1992), p. 123; Rico & Fernández (1997), pp. 405–10; Rueness (1985), pp. 71–4, (1989), pp. 173–6, (2001), p. 16; Rull Lluch *et al.* (1994), pp. 131–8; Sabour *et al.* (2013), pp. 97–102; Sánchez & Fernández (2005), pp. 923–30; Sánchez *et al.* (2005), pp. 942–9; Scagel (1956), pp. 1–10; Schwartz *et al.* (2017), pp. 116–31; Setzer & Link (1971), pp. 5, 6; Silva *et al.* (2019), pp. 918–27; Simkanin (2004), pp. 281, 282; Sjøtun *et al.* (2007), pp. 127–38; Staehr *et al.* (2000), pp. 79–88; Steen (1992), (2003b), pp. 36–43, (2004), pp. 293–300; Steen & Rueness (2004), pp. 175–83; Stegenga & Mol (1983), pp. 129, 130, pl. 51(3, 4); Stiger *et al.* (2003), pp. 1–10; Stiger-Pouvreau & Thouzeau (2015), pp. 227–57; Stokes *et al.* (2004); Strong & Dring (2011), pp. 223–9; Strong *et al.* (2009), pp. 303–14; Tanaka & Hosoi (1967), pp. 1–11; Tanniou *et al.* (2015), pp. 451–6; Thomsen *et al.* (2006), pp. 50–8; Trowbridge (2013), pp. 323–38; Tsukidate (1984), p. 393; Ushida *et al.* (1991), pp. 2249–53; Vaz-Pinto *et al.* (2014), pp. 108–16; Viejo (1997), pp. 325–40, (1999), pp. 131–49; Viejo *et al.* (1995), pp. 437–41; Wernberg *et al.* (1998), pp. 128–32, (2001), pp. 31–9, (2004), pp. 154–61; White & Shurin (2007), pp. 1193–203, (2011), pp. 111–19; Winder (2002), p. 85; Withers *et al.* (1975), pp. 79–86; Yamauchi (1984), pp. 1115–23; Yoshida (1983), pp. 99–246, (1998), p. 394, figs 2, 31c; Zhang *et al.* (1991), pp. 87–91.

Laminariales Migula

LAMINARIALES Migula (1909), p. 243.

Thalli annual or perennial, macroscopic, 2–4(–6) m in length, typically divided into a distinct holdfast, a cylindrical stipe and one to many terminal laminae; holdfast sometimes discoid and conical, more usually a system of branched, root-like haptera which can, in some species, produce stoloniferous outgrowths; stipe varying in length, terete or flattened, smooth or roughened, giving rise to terminal laminae, secondary lateral laminae, pneumatocysts and/ or reproductive sporophylls; laminae simple or branched/segmented, usually leathery in texture, sometimes paper-like, variously shaped and patterned, with or without central midrib and buoyancy bladders; structure parenchymatous, although partly filamentous in structure internally, solid or hollow, and highly differentiated with the following integration of tissue types; an inner central medulla (usually absent in haptera), an inner and outer cortex, and a surrounding epidermis or meristoderm; medulla a relatively narrow zone, derived from transformed cortical cells, and comprising a loose network of anastomosing filaments made up of longitudinally elongated cells, characteristically thin in their middle region, expanding abruptly at the cross walls to form trumpet cells which frequently form horizontally, or obliquely spreading, cross-connections, as well as mainly radially or irregularly spreading, free, branched filaments (often termed hyphae) which penetrate the inner region of the cortex; cortical cells colourless, longitudinally elongate, with straight or oblique/pointed end walls, of variable wall thickness; medullary, cortical and hyphal septa with pits, resembling phloem sieve tubes of higher plants; meristoderm one-cell thick, pigmented, containing numerous discoid plastids without pyrenoids, meristematic and tangentially dividing to contribute cells to the cortex; some genera developing a secondary meristem from a layer of cortical cells, contributing additional layers of cortical cells, the annual production of which in perennial species results in concentric zones; some species with mucilage ducts present in the outer cortex, sometimes arranged in one or more concentric rings; growth by a narrow intercalary meristem usually positioned at junction of lamina and stipe, more rarely subterminal; hairs, if present, tufted, scattered on lamina, with basal meristem.

Life history heteromorphic with a large macroscopic sporophyte alternating with a microscopic gametophyte. Unilocular sporangia borne on the sporophyte, superficial in origin and position, usually densely packed and accompanied by unicellular colourless paraphyses, either in localized or extensive, slightly raised sori covering both sides of the lamina, or in specialized variously positioned sporophylls. Gametophytes filamentous, dioecious, oogamous; antheridia single or clustered, borne at branch tips or as lateral outgrowths of intercalary cells, each antheridium bearing a single antherozoid which is motile, pear-shaped and biflagellate;

oogonia formed from terminal or intercalary cells, bearing a single large oogonium which is usually retained with fertilization and initial sporophyte development occurring in situ.

Members of this order are mainly restricted to cool temperate and cold waters or, if in tropical/subtropical or warm temperate regions, are cooled by cold-water currents. They usually form extensive, economically important beds in the sublittoral and are harvested/maricultured throughout the world as a source of food and alginates.

Molecular evidence over recent years has revealed the need for a new classification system for the Laminariales as traditional schemes based on morphology, while useful, do not reflect their phylogeny and evolutionary relationships. Evidence suggests that the Laminariales is polyphyletic and can be split into at least two clades (Draisma *et al.* 2003). One clade contains the families Akkesiphycaceae (see Kawai & Sasaki 2001), Chordaceae, Pseudochordaceae, Alariaceae, Lessoniaceae and Laminariaceae, while the other clade comprises the families Phyllariaceae and Halosiphonaceae. Molecular studies by Kawai & Sasaki (2001) resulted in the Phyllariaceae being moved out of the Laminariales based on its close molecular affinity to the Tilopteridales, although placed temporarily in *incertae sedis* at the ordinal level. Later, based on both nuclear and chloroplast DNA sequence data (Sasaki *et al.* 2001), both the Phyllariaceae and Halosiphonaceae were placed in the Tilopteridales resulting in the latter containing the three families Tilopteridaceae, Halosiphonaceae and Phyllariaceae (Draisma *et al.* 2003). Draisma *et al.* (2003) further suggested that the Tilopteridales might become an emended order, Cutleriales, if more molecular evidence reveals *Cutleria multifida* to be closely related to *Halosiphon*, *Haplospora* and *Tilopteris*. This found support from Silberfeld *et al.* (2014) who proposed, based on molecular arguments by Phillips *et al.* (2008a) and Silberfeld *et al.* (2010), that the Cutleriaceae should also be placed within an amended Tilopteridales and that the Cutleriales should be placed as a synonym.

Molecular studies have also radically restructured the content of the so-called ALL complex of families (Alariaceae, Lessoniaceae, Laminariaceae), characterized by the presence of mucilaginous organs in the sporophyte, lack of an eyespot in the meiospores, unique flagellation of the sperm and the production of the sexual pheromone lamoxirene (Müller *et al.* 1985; Kawai & Sasaki 2001; Lane *et al.* 2006). Earlier studies by Fain *et al.* (1988) (using the chloroplast genome), by Saunders and Druehl (1992), Boo *et al.* (1999) (using the small subunit rDNA (SSU) and Saunders and Druehl (1993), Druehl *et al.* (1997) and Yoon *et al.* (2001) (using the ITS units of the nuclear ribosomal cistron) provided evidence that both the Lessoniaceae and Alariaceae were polyphyletic. Druehl *et al.* (1997) also showed that the Laminariaceae was polyphyletic, a conclusion supported by Yoon *et al.* (2001). Later molecular work by Lane *et al.* (2006) on the ALL complex, using a multi-gene approach recognized the three families Alariaceae, Laminariaceae and Lessoniaceae but with 'vastly different compositions'. Of relevance here, their data strongly supported a split of the genus *Laminaria*, and the resurrection of the genus *Saccharina* Stackhouse for the *Laminaria* clade that did not contain *L. digitata* (Hudson) J.V.V. Lamouroux, the type of the genus. Within this clade, they included *L. saccharina*, which will be removed from *Laminaria* and repositioned in *Saccharina* in the present treatise.

Five families are currently recognized for Britain and Ireland: Alariaceae, Chordaceae, Halosiphonaceae, Laminariaceae and Phyllariaceae.

Abbott & Hollenberg (1976), pp. 228–57; Boo & Yoon (2000), pp. 13–16; Boo *et al.* (1999), pp. 109–14; Christensen (1980–94), pp. 156–65; Draisma *et al.* (2003), pp. 87–102; Druehl *et al.* (1997), pp. 221–35; Fain *et al.* (1988), pp. 292–302; Fritsch (1945), pp. 192–260; Hamel (1931–39), pp. 287–314; Kawai & Sasaki (2001), pp. 416–28; Kornmann & Sahling (1977), pp. 144–52; Kraan *et al.* (2001), pp. 35–43; Lane *et al.* (2006); pp. 493–512; Maier (1995), pp. 51–102; Müller *et al.* (1985), pp. 475–7; Papenfuss (1951), pp. 145–50; Pohnert & Boland (2002), pp. 108–22; Rousseau *et al.* (2001), pp. 305–19; Sasaki *et al.* (2001), pp. 123–34; Saunders & Druehl (1993), pp. 689–97; Saunders & McDevit (2013), pp. 1–23; Setchell & N.L. Gardner (1925), pp. 590–649; Silberfeld *et al.* (2010), pp. 659–74, (2014), pp. 117–56; Phillips *et al.* (2008a), pp. 394–404; Yoon & Boo (1999), pp. 47–55; Yoon *et al.* (2001), pp. 231–43; Yotsukura *et al.* (1999), pp. 71–90.

Alariaceae Setchell and Gardner

ALARIACEAE Setchell and Gardner (1925), p. 633.

Sporophytes annual or perennial, erect, macroscopic, to several metres in length in many genera, usually clearly divided up into a distinct holdfast, stipe, and a single, terminal lamina; holdfast a flattened or conical shaped system of branched haptera; stipe simple, more rarely branched/forked, terete or compressed, solid or hollow, smooth or bullate and roughened, giving rise directly to a typically large, terminal lamina, and lateral reproductive sporophylls; terminal lamina persistent or lost, simple or branched, margin entire or undulate/split/toothed, smooth or bullate/corrugate/ridged/grooved, with or without pneumatocysts, variously shaped, although more usually lanceolate, ligulate, less often broad, thick and coriaceous or delicate, thin and paper-like, with or without a thickened central region/midrib formed as a direct extension of the stipe.

Structure typical of members of the Laminariales, parenchymatous, usually solid, and highly differentiated with an inner central medulla, an inner and outer cortex, and a surrounding pigmented meristoderm of cells containing numerous discoid plastids without pyrenoids; some species with mucilage ducts present in the outer cortex; hairs, if present, tufted, scattered on lamina, with basal meristem. Growth by an intercalary meristem situated in region between the stipe and lamina (transition zone).

Life history heteromorphic – a large macroscopic sporophyte alternating with a filamentous microscopic gametophyte. Unilocular sporangia usually borne on specialized sporophylls; sporophylls arising at the transition zone, usually borne laterally and acropetally in two rows along margins of stipe, often at the base intermingled with sterile lamina, or directly as marginal outgrowths of the terminal and/or lateral blade surface; sporophylls differing or similar to terminal blade in shape and size, lamina-like, variously shaped, smooth, without midrib, sometimes stipitate; unilocular sori borne on both sides of the sporophylls, limited or more usually extensive in development, covering the entire surface, and accompanied by unicellular paraphyses. Gametophytes branched, filamentous, microscopic, dioecious and oogamous, with the smaller-celled males producing antheridia which release a single motile antherozoid, and the larger-celled females producing oogonia-bearing a single large oogonium. Fertilization and initial sporophyte development occurring in situ.

The main determining characteristic of the Alariaceae is the production of sporophylls which arise from the transition zone as secondary lateral outgrowths of either the stipe or the lamina.

Two of the nine genera currently included in this family are present in Britain and Ireland, *Alaria* and the introduced *Undaria*, each containing a single species, *Alaria esculenta* and *Undaria pinnatifida*, respectively. Molecular evidence places both species as sibling taxa (Draisma *et al.* 2003). *Alaria* and *Undaria* are usually readily distinguishable and represent the only kelps in the flora of Britain and Ireland with a distinct midrib.

Alaria Greville

ALARIA Greville (1830), pp. xxxix, 25.

Type species: *A. esculenta* (Linnaeus) Greville (1830), pp. xxxix, 25.

Sporophytes perennial, to several metres in length and clearly differentiated into a long, dorsiventrally flattened lamina, a short, terete, flexible, unbranched stipe and an attaching root-like holdfast (haptera); lamina simple, relatively thin, membranous, smooth, with a pronounced flattened midrib, which is an upward continuation of the stalk, and usually with a wavy, torn margin; cryptostomata common on blade surface; anatomical structure typical of members of the Laminariales, with thalli comprising a central medulla, a parenchymatous cortex and an outer meristoderm; mucilage cells commonly present; annual growth by meristematic region at base of lamina.

Unilocular sporangia, accompanied by unicellular paraphyses, in extensive and continuous, dark-coloured sori borne on the surface of several, tongue-like, at first linear, later obovate, flattened, lateral, leaf-like extensions (sporophylls) borne acropetally in two opposite rows each year in the upper stipe region.

A North Atlantic and North Pacific genus, containing up to approximately 15 species (Widdowson 1971a,b; Guiry & Guiry 2023).

One species in Britain and Ireland.

Alaria esculenta (Linnaeus) Greville (1830), pp. xxxix, 25.   Figs 277, 278, 279

Fucus pinnatus Gunnerus (1766), p. 96, nom. illeg.; *Fucus esculentus* Linnaeus (1767), p. 718; *Fucus teres* Goodenough & Woodward (1797), p. 140; *Ceramium esculentum* (Linnaeus) Stackhouse (1801), p. xxiv; *Fucus esculentus* var. *β minor* Turner (1802), p. 104; *Musaefolia esculenta* (Linnaeus) Stackhouse (1809), p. 53; *Laminaria esculenta* (Linnaeus) J.V. Lamouroux (1813), p. xxx; *Orgyia esculenta* (Linnaeus) Stackhouse (1816), p. 24; *Laminaria esculenta* (Linnaeus) C. Agardh (1817), p. 16; *Phasgonon esculentum* (Linnaeus) S.F. Gray (1821), p. 385; *Phasgonon esculentum* var. *minus* S.F. Gray (1821), p. 385; *Phasgonon esculentum* var. *minus* S.F. Gray (1821), p. 385; *Laminaria polyphylla* (S.G. Gmelin) Steudel (1824), p. 230; *Agarum esculentum* (Linnaeus) Bory de Saint-Vincent (1826), p. 194; *Agarum delisei* Bory de Saint-Vincent (1826), p. 194; *Laminaria musaefolia* Bachelot de la Pylaie (1830), p. 31; *Alaria delisei* (Bory) Greville (1830), p. xxxix; *Podopteris esculentum* (Linnaeus) Bachelot de la Pylaie (1830), p. 23; *Laminaria musaefolia* var. *β remotifolia* Bachelot de la Pylaie (1830), p. 35; *Alaria esculenta* f. *latifolia* Postels & Ruprecht (1840), p. 11; *Alaria esculenta* f. *polyphylla* (S.G. Gmelin) Postels & Ruprecht (1840), p. 5; *Orgyia delisei* (Bory de Saint-Vincent) Trevisan (1845), p. 28; *Phasganon alatum* Ruprecht (1850), p. 161, nom. illeg.; *Phasganon macropterum* Ruprecht (1850), p. 161; *Alaria grandifolia* J. Agardh (1872), p. 26; *Alaria musaefolia* (Bachelot de la Pylaie) J. Agardh (1872), p. 26; *Orgyia pinnata* Gobi (1878), p. 77; *Alaria esculenta* f. *australis* Kjellman (1883b), p. 265; *Alaria esculenta* f. *musaefolia* (Pylaie) Kjellman (1883b), p. 265; *Alaria dolichorhachis* Kjellman (1883b), p. 271; *Alaria esculenta* f. *fasciculata* Strömfelt (1886), p. 38; *Alaria esculenta* f. *pinnata* (Gobi) Foslie (1886), p. 114; *Alaria linearis* Strömfelt (1886), p. 38; *Alaria esculenta* var. *β pinnata* (Gobi) Kjellman (1890), p. 20; *Alaria platyrhiza* Kjellman (1906), p. 11; *Alaria macroptera* (Ruprecht) Yendo (1919b), p. 79; *Alaria esculenta* var. *noltei* Rabenhorst (as '*noltii*'); *Alaria dolichorhachis* f. *typica* Miyabe & Nagai in Nagai (1940), p. 110, nom. inval.

Sporophytes perennial, to 1(–3) m in length, forming erect, solitary, olive to yellow-brown thalli, clearly differentiated into a basal attachment holdfast, supporting stipe and terminal lamina; holdfast prostrate, disc-shaped, not conical, comparatively small, comprising 1–3 ascending layers of laterally growing, compacted, intertwined, slender, dichotomously branched, terete to oval, finger-like haptera to 2(–3) mm in diameter, gradually narrowing towards tips and forming strongly adherent attachment pads upon surface contact; stipe comparatively small and short, dark brown, solid, simple, linear, cylindrical below, gradually becoming somewhat flattened towards base of blade, to 10(–30) cm in length, to 5(–9) mm in width, on older thalli bearing the reproductive sporophylls; lamina to 20(–25) cm in width, narrowly linear or linear-lanceolate, solid, tapering gradually above, more acutely below, simple, dorsiventrally flattened and blade-like, relatively thin, to 150(–170) μm in thickness, and translucent, smooth, flaccid, papery, almost softly coriaceous and delicate, with a wavy, often split/torn margin, with a characteristic, prominent, raised, central, somewhat flattened, solid midrib throughout, extending from the stipe, to 5(–8) mm in diameter and almost rectangular in cross-section; cryptostomata common all over thallus surface; in cross section, both lamina and stipe similar and solid, with a central medulla, a surrounding cortex and a peripheral meristoderm; in transverse section, stipe rounded to oval, the medulla forming a relatively thin, central band of fairly loosely bound tissue extending to almost the whole internal diameter of the stipe and comprising vertically elongated filaments of fairly thick-walled, transversely rounded/oval cells, traversed and entwined by a network of branched, hyphal-like, darkly pigmented

Fig. 277. *Alaria esculenta*
Habit of sporophytes in an intertidal pool, showing terminal eroded lamina with midrib and basal clump of sporophylls.

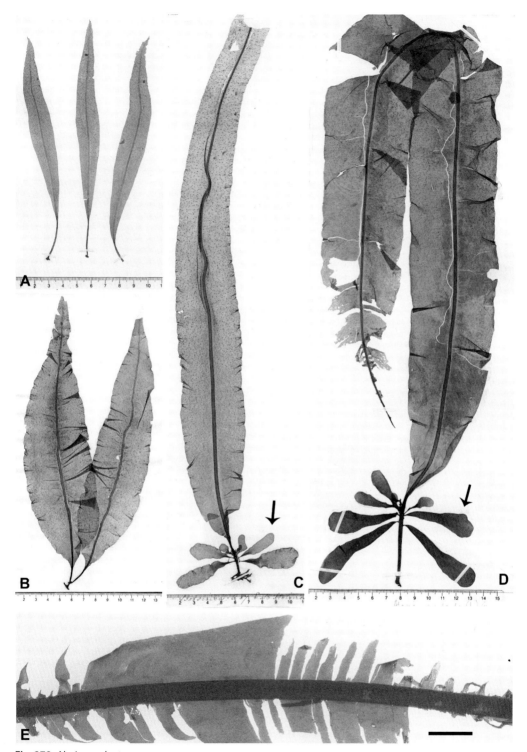

Fig. 278. *Alaria esculenta*

A, B. Habit of young sporophytes. C, D. Habit of more mature sporophytes showing sporophylls arising from stipe (arrows). E. Portion of split lamina showing central midrib. Bar in E = 6 cm. (Herbarium specimens, NHM.)

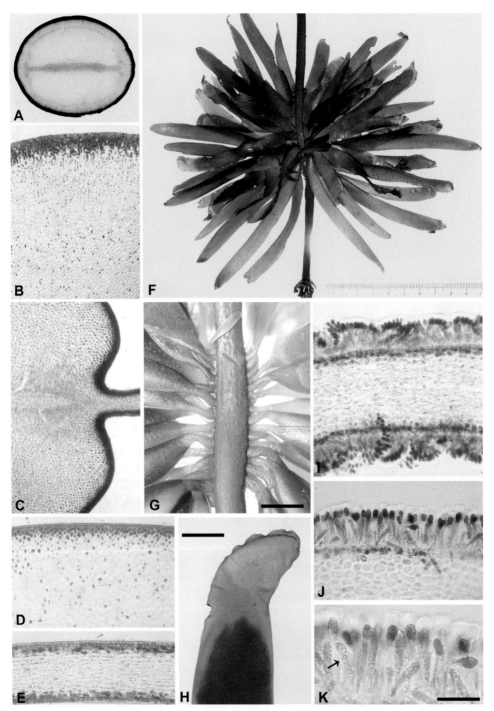

Fig. 279. *Alaria esculenta*

A. TS of stipe showing central thin band of medullary tissue. B. TS of stipe showing extensive cortex with outer cell layers noticeably more darkly pigmented. C. TS of midrib showing marginal extension into lamina. D. TS of edge of midrib in the outer cortex region. E. TS of lamina showing relatively large central medulla. F. Habit of sporophylls arising from stipe. G. Close-up of stipe showing lateral formation of sporophylls. H. Terminal region of sporophyll showing darker sorus area. I–K. TS of sporophylls showing sori with unilocular sporangia (arrow) enclosed within the paraphyses. Note the darker coloured, enlarged apex of the paraphyses and the terminal hyaline 'hat'. Bar in G = 13 mm, bar in H = 8.5 mm, bar in K = 220 μm (B–D), = 140 μm (I), = 70 μm (E, J), = 35 μm (K).

filaments derived from the inner cortical cells; inner cortex extensive and comprising closely packed, irregularly arranged, thick-walled colourless cells which become more radially aligned and elongated towards the periphery, the outer cell layers becoming noticeably more darkly pigmented; meristoderm a single layer of palisade-like photosynthetic cells enclosing large numbers of discoid plastids without pyrenoids; in transverse section, lamina midrib with a central, relatively thin, transverse band of medullary tissue similar to that of the stipe, enclosed by a wide cortex of irregularly arranged, rounded thick-walled cells, the outer ones more darkly pigmented and terminating in a single celled meristoderm layer of palisade-like photosynthetic cells containing numerous discoid plastids without pyrenoids; in transverse section, lamina relatively thin, produced marginally from the midrib with a relatively large central medulla originating from the midrib medulla, comprising a closely packed network of horizontally spreading, branched, septate filaments of cells, enclosed within a 1–2-celled cortex of rounded, dark-pigmented cortical cells situated beneath the outer palisade-like layer of photosynthetic meristoderm cells; stipe, midrib and lamina without mucilage ducts.

Reproductive sporophylls numerous, 30–60 in number on mature thalli, annually deciduous, borne in acropetal succession marginally in opposite rows in the upper regions of the stipe, light to dark brown, 55–120(–165) × 5–17(–35) mm, at first thin, membranous, oblanceolate to broadly elliptical/spatulate with rounded apices, later becoming more coriaceous and linear-lanceolate/elongate cuneate and obtuse, gradually tapering below, shortly pedicellate, solid, sometimes with central, pseudovesicle-like, air cavity; unilocular sporangia borne in slightly raised, mucilaginous, orbicular to irregularly shaped sori, to 100(–130) μm in height, covering the middle regions of both sides of the younger, thinner, more membranous sporophylls; sporangia closely packed in vertical columns, arising as extensions of the outer meristoderm layer, 45–60(–72) × 8–10(–12) μm, elongate-cylindrical, accompanied by, and enclosed under, an umbrella of unicellular, markedly elongate-clavate, terminally darkly pigmented paraphyses, 78–90(–100) × 8–12 μm, each paraphysis with a prominent, terminal, hyaline, oval to wedge-shaped 'hat', appearing densely packed and hexagonal in surface view.

Epilithic; rarely epiphytic as small specimens, lower eulittoral, emersed, occasionally in pools and shallow sublittoral.

Locally common, widely distributed although largely restricted to north and west coasts of the Britain and Ireland and on semi to fully exposed coasts where there is good wave and surf action.

Sporophyte perennial, with the lamina renewed annually in spring/early summer from a basal meristoderm, gradually replacing the old lamina. Reproductive sporophylls found throughout the year, produced in summer, in acropetal succession near the upper end of the stipe, later shed during autumn and winter, leaving scars.

Bringloe *et al.* (2018), pp. 1–8; Brodie *et al.* (2016), p. 1022; Buggeln (1974a), pp. 80–2; (1974b), pp. 283–8, (1976), pp. 439–42, (1977), pp. 212–18, (1978a), pp. 156–60, (1978b), pp. 54–6, (1981), pp. 102–4; Buggeln & Varangu (1983), pp. 205–9; Bunker *et al.* (2017), p. 253; Cabioc'h *et al.* (1992), p. 78, fig. 58; Dizerbo & Herpe (2007), p. 123, pl. 41(1); Fletcher in Hardy & Guiry (2003), p. 224; Garbary & Clarke (2002), pp. 211–16; Gayral (1966), pp. 302, 303, pl. LIX; Greville (1830), p. 25, pl. 4; Guiry (2012), p. 162; Hamel (1931–39), pp. 310, 311, fig. 51(F); Han (1994), pp. 67–75; Han & Kain (Jones) (1993), pp. 79–81; Harvey (1846–51), pl. 79; Kraan & Guiry (2000a), pp. 554–9, (2000b), pp. 190–8, (2001), pp. 1–33; Kraan *et al.* (2000), pp. 577–83, (2001), pp. 35–42; Lane & Saunders (2005), pp. 426–36; Laycock (1975), pp. 271–9; Mathieson & Dawes (2017), p. 216, pl. XL(7, 8); Minchin (1992), p. 171; Morton (1994), p. 44; Munda & Lüning (1977), pp. 311–14; Newton (1931), p. 206, fig. 130(A–E); Pedersen (2011), pp. 106, 107, fig. 114; Printz (1926), pp. 192–8, figs 16, 17; Rueness (1977), p. 177, pl. xxiv(3), (2001), p. 14; Saunders & McDevit (2013), pp. 1–23; Sauvageau (1918), p. 219, figs 82–5; Smale *et al.* (2013), pp. 4016–38; Stegenga & Mol (1983), p. 111, pl. 41(3); Stegenga *et al.* (1997a), p. 23; Sundene (1962a), pp. 155–74; Taylor (1957), pp. 186, 187, pl. 21(2); Walton (1986), p. 338; Widdowson (1971a), pp. 11–49, fig. 12(A–C), (1971b), pp. 125–43; Yendo (1919), pp. 1–145, pl. XI.

Undaria Suringar

UNDARIA Suringar (1873), p. 77.

Type species: *U. pinnatifida* (Harvey) Suringar (1873), p. 77.

Sporophyte annual, macroscopic, erect and clearly differentiated into a root-like holdfast of downward-growing, branched, finger-like haptera, a short dorsiventrally flattened stipe, and a large, dorsiventrally flattened, fairly delicate, membranous lamina, with a prominent flattened midrib, and pinnately lobed margins; in cross section, sporophytes reveal a central medulla, a parenchymatous cortex and an outer meristoderm; meristematic region at base of lamina.

Unilocular sporangia with unicellular paraphyses, borne on both sides of specialized, undulating sporophylls produced as lateral, wing-like, extensions of the stipe.

A genus of the colder waters of Japan in the North Pacific, although one adventive species (*U. pinnatifida*) has spread by human activities throughout the world in cool and warm temperate regions including Britain and Ireland.

One species in Britain and Ireland.

Undaria pinnatifida (Harvey) Suringar (1873), p. 77.   Figs 280, 281, 282

Alaria pinnatifida Harvey (1860), p. 329; *Alaria amplexicaulis* Martens (1866), p. 114; *Undaria pinnatifida* var. *vulgaris* Suringar (1872), p. 77; *Undaria pinnatifida* var. *elongata* Suringar (1872), p. 77; *Ulopteryx pinnatifida* (Harvey) Kjellman (in Kjellman & Petersen, 1885, p. 275); *Undaria pinnatifida* var. *distans* Miyabe & Okamura (in Okamura, 1902, p. 128); *Undaria pinnatifida* f. *distans* (Miyabe & Okamura) Yendo (1911), p. 708; *Undaria pinnatifida* f. *narutensis* Yendo; *Undaria pinnatifida* f. *typica* Yendo (1911), p. 708, nom. inval.; *Undaria pinnatifida* f. *subflabellata* Suringar.

Sporophytes macroscopic, to 1(–3) m in length, 65 cm in width, forming erect, solitary, light to medium golden-brown thalli, differentiated into a terminal lamina, supporting stipe and fibrous holdfast comprising several ascending layers of downward-extending, inter-twined, dichotomously branched, terete, light brown/yellow, finger-like haptera, to 5 mm in diameter, gradually narrowing towards tips and forming strongly adherent pads upon surface contact; stipe relatively short, to 14(–20) cm in length, to 20(–30) mm in diameter, dark brown, solid, firm, coriaceous, distinctly dorsiventrally flattened, almost rectangular in cross section, with projecting marginal rims, further growth of the latter giving rise to the terminal lamina and, on mature thalli, to the basal reproductive sporophylls; lamina light yellow-brown at first, becoming medium to dark brown later, solid, dorsiventrally flattened and blade-like, ranging from 220–560 μm in thickness, linear-lanceolate in shape, tapering above and below, with a prominent, dark brown, central midrib, extending from, and similar to, the stipe below, becoming more faint and less prominent towards the thallus apex; young thalli entire, simple, smooth, softly coriaceous and relatively thin, later becoming thicker, somewhat bullate/undulating, membranous, translucent, papery, delicate, easily torn, and frequently indented/ruffled to pinnately lobed marginally, with pinnae extending to 20(–30) cm in length, 25(–35) mm in width; central midrib with primary veins extending out into secondary veins over lamina surface; colourless cryptostomata with emergent tuft of hairs and smaller, light to dark coloured caecostomata common all over lamina surface; in cross section, stipe, midrib and lamina solid, with a central medulla, an enclosing cortex and a peripheral meris-toderm; in transverse section, stipe rectangular, with the medulla forming a central, thin, wide band of fairly loosely bound, septate, vertically elongated filaments of cells with some traversing hyphae-like filaments, surrounded by an inner cortex of larger, colourless, mostly rounded, irregularly arranged cells, giving rise to a more extensive outer cortex of radial rows of smaller, more radially elongated, rectangular cells, the outer cell layers enclosing a few plastids, enclosed by a peripheral meristoderm layer of small palisade-like cells with a thick outer wall and enclosing large numbers of discoid plastids without pyrenoids; in section,

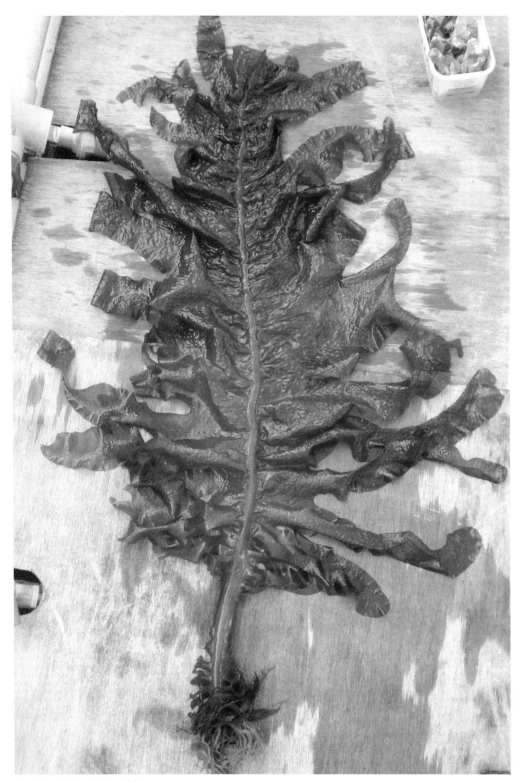

Fig. 280. *Undaria pinnatifida*
Habit of large sporophyte thallus. Specimen approximately 2 metres in length.

Fig. 281. *Undaria pinnatifida*
A–G. Habit of sporophyte thalli at different stages of development. Bar in C = 3 cm.

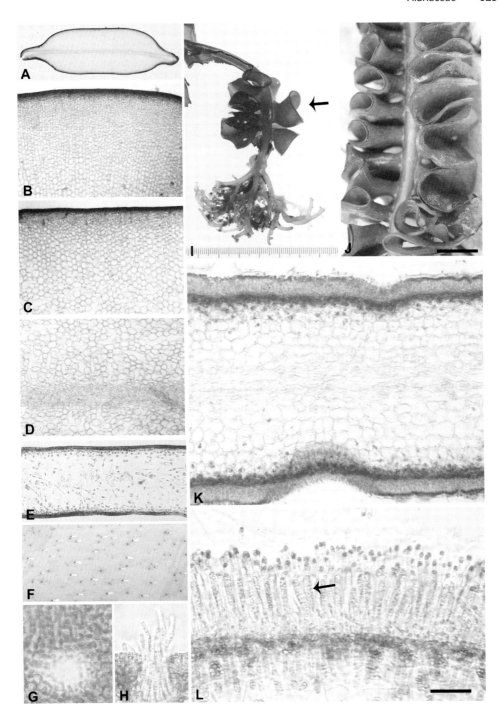

Fig. 282. *Undaria pinnatifida*
A. TS of stipe showing projecting marginal wings. B, C. TS of stipe showing outer cortex region. D. TS of stipe showing central medullary region. E. TS of lamina showing a well-developed central medulla of loosely associated filaments enclosed within a few-celled cortex and outer meristoderm layer. F. Surface view of lamina portion showing scattered cryptostomata with emergent tuft of hairs. G. Surface view of lamina portion showing a caecostomata. H. TS of lamina showing a cryptostomata with emergent tuft of hairs. I. Basal region of thallus showing the holdfast of root-like haptera and lateral sporophylls (arrow). J. Portion of stipe showing extensive sporophyll production. K. TS of sporophyll showing sorus formation on both sides. L. TS of sorus showing unilocular sporangia (arrow) and associated paraphyses. Bar in J = 13 mm (J), bar in L = 4 mm (A), = 0.5 cm (F), = 220 µm (B–D, K), = 140 µm (E), = 70 µm (L), = 35 µm (G, H).

lamina with a large central medulla of loosely associated filaments of cells within a gelatinous matrix enclosed within a few celled cortex and outer meristoderm.

Reproductive sporophylls bearing unilocular sporangia, borne as growth extensions of the marginal rim on both sides of the basal stipe region, extending to 10(–15) cm or more along the stipe, 600–1000 µm in thickness, dark brown, and fluted/deeply ruffled and slimy when mature; unilocular sporangia closely packed, in vertical columns, arising as extensions of the outer meristoderm layer, 50–60(–70) × 9–12 µm, elongate cylindrical and enclosed within unicellular, markedly elongate-clavate, terminally darkly pigmented paraphyses, 85–112 × 7–11 µm, each paraphysis with a thick, terminal hyaline layer appearing conjoined and hexagonal in surface view.

Epilithic, epizoic, less frequently epiphytic, with most populations largely restricted to the sides of floating pontoons of marinas with records on natural substrata in the littoral fringe and shallow sublittoral regions confined to a small number of localities on the south coast of England (Torquay, Plymouth (South Devon), Poole Harbour (Dorset) and the Hamble Estuary (Hampshire)). Reported from 15 m depth on the Brittany coast.

Locally abundant; since its initial arrival at a marina in Southampton Water (Hampshire) it has been recorded from several counties on the south coast (Devon, Dorset, Isle of Wight, Sussex, Kent), with reports from Suffolk and Lincolnshire on the east coast, Lancashire on the west coast and the Isle of Man; more recently it has also been reported for Belfast Lough in Northern Ireland. It is likely to continue to disperse all around Britain and Ireland, assisted by small boat travel; local spread from marina sites into the sublittoral appears to be much slower.

Sporophytes annual, appearing in March/April and reaching maximum reproductive size in June/July. Sporophytes senescing during the summer leaving only the basal regions with sporophylls, and usually becoming heavily infested with epiphytes. Fertile thalli observed all year round on the south coast of England, with a second population arising in the autumn and developing slowly overwinter.

A well-known, adventive species, introduced to several countries worldwide from its native Japan (see bibliographic section below). It was first reported for the Solent region of Britain in 1994 (Fletcher & Manfredi 1995) colonizing the sides of floating pontoons in the Hamble Estuary, Hampshire and was most probably introduced by small boats coming from Brittany and Normandy where it was already widespread. Small boats were probably also responsible for its well-documented secondary spread along the coasts of England, as it is now a common colonizer of a large number of marinas along the south coast from Ramsgate (Kent) in the east to Plymouth (South Devon) in the west, and has been found by the present author in marinas at both Lowestoft (Suffolk) and Grimsby (Lincolnshire) on the east coast of England, and at South Queensferry in the Firth of Forth on the east coast of Scotland (July 2023), with reports of its spread up the west coast of Britain (Pembrokeshire, Lancashire, Isle of Man), and in Ireland (Antrim, Louth, Wexford). However, all these records might possibly be due to secondary introductions from France. The first find of this species in the Channel Islands was also at a yachting marina (La Collette Marina, St Helier, Jersey (pers. obs., P. Farrell and R. Fletcher, April 1996)). On pontoon sides, it can be locally abundant, especially in the more wave-washed/high-current regions where it appears to replace native kelp species such as *Laminaria digitata* and, if originally present, *Saccharina latissima* and *Saccorhiza polyschides*, with quite high biomass levels. It occurs mainly just below the water line on the pontoons, but can occur at greater depths, and is often found attached to the ascidian *Styela clava*. It is also euryhaline and extends, for example, much further up the Hamble Estuary, Hampshire than *Laminaria digitata*, *Saccharina latissima* and the introduced *Sargassum muticum*. Colonization of natural shores has, however, been comparatively slow. Following its initial colonization of the pontoons at Torquay Marina, Devon, it apparently took several years to spread the comparatively short distance beyond the outer harbour walls and establish itself on the adjacent benthic rocky shore.

Aguilar-Rosas *et al.* (2004), pp. 255–8; Akiyama (1965), pp. 143–70; Akiyama & Kurogi (1982), pp. 91–100; Alexandrova & Reunov (2008), pp. 712–15; Apoya *et al.* (2002), pp. 445–52; Arnold *et al.* (2016), pp. 661–76; Baba (2008), pp. 7–15; Báez *et al.* (2010), pp. 21–31; Bollen *et al.* (2016), no. 194, (2017), pp. 1–8; Boudouresque *et al.* (1985), pp. 364–6; Brodie *et al.* (2016), p. 1022; Brown & Lamare (1994), pp. 63–70; Bunker *et al.* (2017), p. 261; Burridge & Gorski (1997); Cabioc'h *et al.* (1992), p. 180, fig. 199; Campbell & Burridge (1998), pp. 379–81; Campbell *et al.* (1999), pp. 231–42; Carvalho *et al.* (2009), pp. 245–53, (2010), pp. 1180–6; Casas & Piriz (1996), pp. 213–15; Casas *et al.* (2004), pp. 411–16, (2008), pp. 21–8; Castric-Fey *et al.* (1993), pp. 351–8, (1999a), pp. 83–96, (1999b), pp. 71–82; Cazon *et al.* (2014), pp. 100–8; Cecere *et al.* (2000), pp. 305–9, (2016), pp. 451–62; Cho (2010), pp. 181, 182; Cho *et al.* (2005), pp. 423–30; Cobacho *et al.* (2018), pp. 453–7; Cremades *et al.* (1997), pp. 29–40, (2006), pp. 169–87; Curiel *et al.* (1994), pp. 121–6, (1998), pp. 17–22, (2002), pp. 209–19; Daguin *et al.* (2005), pp. 647–50; Dan & Kato (2008), pp. 79–83; De Leij *et al.* (2017), pp. 156–71; Dean & Hurd (2007), pp. 1138–48; Dellatorre *et al.* (2014), pp. 467–78; Desmond *et al.* (2019), no. 139; Ding Min *et al.* (1984), pp. 263–5; Dizerbo & Herpe (2007), p. 123; Endo *et al.* (2009), pp. 393–400; Epstein & Smale (2017), pp. 8624–42, (2018), pp. 1049–72; Epstein *et al.* (2019), pp. 1–15; FAO (2013); Farrell & Fletcher (2006), pp. 236–43; Fletcher & Farrell (1999), pp. 259–75; Fletcher & Manfred (1995), pp. 355–8; Fleurence & LeCoeur (1994), pp. 555–9; Floc'h *et al.* (1991), pp. 379–90, (1995), pp. 157, 158, (1996a), pp. 217–22, (1996b), pp. 157, 158; Floer *et al.* (2005), pp. 765–78; Forrest & Blakemore (2006), pp. 333–45; Forrest & Hopkins (2013), pp. 317–26; Forrest & Taylor (2002), pp. 375–86; Forrest *et al.* (2000), pp. 547–52; Freire-Gago *et al.* (2006), pp. 25–32; Fukuzumi *et al.* (1999), pp. 11–17; Gao *et al.* (2013a), pp. 53–64, (2013b), pp. 269–75, (2013c), pp. 1171–8, (2013d), pp. 1331–40; Geange *et al.* (2014), pp. 1583–92; Grulois *et al.* (2011), pp. 485–98; Guiry (2012), p. 182; Hara & Akiyama (1985), pp. 47–50; Hara & Ishikawa (1988), pp. 27–35; Hay (1990), pp. 301–13; Hay & Luckens (1987), pp. 329–32; Hay & Villouta (1993), pp. 461–76; Heiser *et al.* (2014), p. e93; Henkel & Hofmann (2008), pp. 164–73; Henkel *et al.* (2009), pp. 1–13; Hewitt *et al.* (2005), pp. 251–63; Huh *et al.* (2001), pp. 429–35; Hwang *et al.* (2012), pp. 401–8, (2014), pp. 747–52; Irigoyen *et al.* (2011a), pp. 17–24, (2011b), pp. 1521–32; Ishikawa (1992), pp. 19–24, (1993), pp. 1331–6; James & Shears (2012), pp. 1–44, (2016a), no. 34, (2016b), no. 225; James *et al.* (2014), pp. 21–4, (2015), pp. 3393–408; Guzinski *et al.* (2018), pp. 1582–97; Jimenez *et al.* (2015), pp. 2521–6; Kang & Yoo (1993), pp. 77–82; Kaplanis *et al.* (2016), pp. 111–24; Kasahara (1967), pp. 279–87, figs 1–5; Kato & Dan (2010), pp. 279–87; Kato & Nakahisa (1962), pp. 998–1004; Katsuoka *et al.* (1990), pp. 3043, 3044; Kawai *et al.* (2016a), pp. 227–49; Kawashima & Tokuda (1993), pp. 385–9; Kim & Lee (1998), pp. 151–5; Kim & Nam (1997), pp. 505–10; Kim *et al.* (2007), pp. 247–52; Kimura *et al.* (1976), pp. 57–65, (2010), pp. 353–61; Kinashita & Shibuya (1944), pp. 369–73; Kito *et al.* (1976), pp. 67–73, (1981), pp. 11–18; Kraan (2017), pp. 1107–14; Kurogi & Akiyama (1957), pp. 95–117; Kusaka *et al.* (2007), pp. 17–28; Lane *et al.* (2006), pp. 493–512; Leal *et al.* (2015), pp. 165, 166, (2016), pp. 12–20, (2017a), pp. 557–66, (2017b), no. 7, (2018), no. 14763; Lee & Sohn (1993), pp. 71–87; Lee *et al.* (1995), pp. 51–7, (2006), pp. 157–60; Leliaert *et al.* (2000a), pp. 5–10; Li *et al.* (2013), pp. 774–81, (2014), pp. 365–71, (2015), pp. 953, 954, (2017), pp. 993–9; Liu *et al.* (2004), pp. 519–21; Martin & Bastida (2008), pp. 335–44; Martin & Cuevas (2006), pp. 1399–402; Matsumura *et al.* (2001), pp. 10–20; Meretta *et al.* (2012), pp. 59–63; Minchin (2007), pp. 302–13; Minchin & Nunn (2014), pp. 57–63; Minchin *et al.* (2017), pp. 53–60; Morelissen *et al.* (2013), pp. 197–206, (2016), no. 241; Morita *et al.* (2003a), pp. 154–60, (2003b), pp. 266–79; Muraoka & Saitoh (2005), pp. 1365–9; Murphy *et al.* (2016), pp. 540–5, (2017), pp. 691–702; Na *et al.* (2016), pp. 190–7; Nanba *et al.* (2011), pp. 1023–30, (2013), pp. 37–41; Nima & Harada (2016), pp. 10–18; Niwa (2015), pp. 90–7, (2016), pp. 173–82; Oh & Koh (1996), pp. 389–93; Ohno & Matsuoka (1993), pp. 41–9; Ohno *et al.* (1999), pp. 61–4; Okamura (1915), pp. 266–78; Pang & Lüning (2004), pp. 83–92; Pang & Shan (2008), pp. 1–8; Pang & Wu (1996), pp. 205–10; Pang *et al.* (2008), pp. 280–7; Park *et al.* (2008), pp. 485–90, (2009), pp. 35–40; Pereyra *et al.* (2014), pp. 65–70; Pérez *et al.* (1981), pp. 1–12, (1984), pp. 3–15, (1988), pp. 315–28; Pérez-Cirera *et al.* (1997), pp. 3–28; Pérez-Ruzafa *et al.* (2002), pp. 147–51; Peteiro (2002), pp. 8–10, (2008), pp. 413–15; Peteiro & Freire (2011a), pp. 269–76, (2012a), pp. 189–96, (2012b), pp. 1361–72, (2013a), pp. 205–13, (2013b), pp. 706–13, (2014), pp. 469–74; Peteiro & Sanchez (2012), pp. 197–200; Peteiro *et al.* (2001), pp. 23–4, (2014), pp. 518–28, (2016), pp. 9–23; Primo *et al.* (2010), pp. 3081–92; Qing *et al.* (1984), pp. 263–5; Raffo *et al.* (2009), pp. 1571–80; Reeves *et al.* (2018), pp. 1239–51; Renard *et al.* (1992), pp. 445–51; Ribera *et al.* (1992), p. 120; Rismondo *et al.* (1993), pp. 329, 330; Russell *et al.* (2008), pp. 103–15; Saito (1956a), pp. 229–34, (1956b), pp. 235–9, (1962), pp. 1–102, (1975), pp. 304–20; Saito *et al.* (2016), pp. 3447–58; Salinas *et al.* (1996), pp. 77–9, (2006), pp. 65–72; Sanderson (1990), pp. 153–7, (1997); Santiago Caamaño *et al.* (1990), pp. 1–45; Schaffelke *et al.* (2005), pp. 84–94; Schiel & Thompson (2012), pp. 25–33; Schiller *et al.* (2018), pp. 365–71; Sfriso & Facca (2013), pp. 162–72; Shan & Pang (2009), pp. 36–44, (2010), pp. 171–6; Shan *et al.* (2013), pp. 154–61, (2015), p. 902, (2019), pp. 154–61; Shibneva *et al.* (2013), pp. 1909–16; Shim *et al.* (2017), pp. 349–57; Silva *et al.* (2002), pp. 333–8; Skriptsova *et al.* (2004), pp. 17–21; Sliwa *et al.* (2006), pp. 396–405; Smale *et al.* (2013), pp. 4016–38; Sohn (1993), pp. 207–16; South & Thomsen (2016), no. 175; South *et al.* (2016), pp. 103–12, (2017), pp. 243–57; Stapleton (1988), pp. 178, 179; Stegenga (1999), pp. 71–3; Stegenga *et al.* (2007), pp. 125–43; Stiger-Pouvreau & Thouzeau (2015), pp. 227–57; Stuart & Brown (1996); Stuart *et al.* (1999), pp. 191–9; Suárez-Jiménez *et al.* (2015), pp. 2521–6, (2017a), no. 113, (2017b), pp. 45–55; Suringar (1873); Susuki *et al.* (1995), pp. 201–5; Taboada *et al.* (2013), pp. 1271–6; Tanaka &

Hosoi (1967), pp. 1–11; Taniguchi *et al.* (1981), pp. 1–9; Thomsen *et al.* (2018), no. 173; Thompson & Schiel (2012), pp. 95–105; Thornber *et al.* (2004), pp. 69–80; Tom Dieck (1993), pp. 253–64; Uwai *et al.* (2006a), pp. 687–95, (2006b), pp. 345–56, (2007), pp. 263–71; Valentine & Johnson (2003), pp. 63–90, (2004), pp. 223–30, (2005a), pp. 43–55, (2005b), pp. 106–15; Veiga *et al.* (2014), pp. 363–5; Verlaque (1994), pp. 1–23, (2001), p. 32; Verlaque *et al.* (2015), pp. 1–362; Viera-Rodriguez *et al.* (2007), p. 95; Wallentinus (1999), pp. 13–19, (2007), pp. 1–36; Wang *et al.* (2019), pp. 784–8; Watanabe & Nisizawa (1984), pp. 106–11; Watanabe *et al.* (2014), pp. 2405–15; Wotton *et al.* (2004), pp. 844–9; Wu *et al.* (2004), pp. 153–6; Yamada *et al.* (2007), pp. 222–30; Yamanaka & Akiyama (1993), pp. 249–53; Yendo (1911), pp. 691–715; Yoo *et al.* (2007), pp. 333–8; Zhang & Pang (2007), pp. 104–11; Zhang *et al.* (1984), pp. 263–5, (2005), pp. 1467–75, (2006), pp. 642–7, (2015), (2016), pp. 25, 26; Zuo-mei (1984), pp. 314–16.

Chordaceae Dumortier

CHORDACEAE Dumortier (1822), pp. 72, 102.

Sporophytes annual, erect and macroscopic, to several metres in length, unbranched, cylindrical, terete and chord/whip-like, firm to slightly spongy in texture, smooth and slimy, initially solid, becoming hollow when mature, and arising from a short, minute stipe and small attachment disc, sometimes with several thalli reported to originate from the same base; structure parenchymatous, with a central medulla of colourless, narrow, elongate, thread-like hyphae with characteristic 'trumpet'-shaped endings, an outer, largely colourless cortex of longitudinally elongated cells, and a peripheral, pigmented, single-layered meristoderm/epidermis of small isodiametric/palisade cells; inner cortical cells lining the hollow cavity producing hyphae-like lateral outgrowths which often extend inwardly and unite to form supporting cross-connections; meristoderm cells with large numbers of discoid plastids without pyrenoids and giving rise in the summer to colourless hairs which clothe the thallus of young individuals; growth by an intercalary meristem situated just above the holdfast, assisted by a secondary meristematic zone situated a short distance below the apex.

Life history heteromorphic; macroscopic sporophyte alternating with a filamentous, microscopic gametophyte. Unilocular sporangia borne in extensive, continuous sori which cover almost the entire length of the sporophyte, intermingled with unicellular paraphyses, both originating as outgrowths of the epidermal cells.

A monotypic family in the Laminariales, represented by the genus *Chorda* and a single species *C. filum* which is widely distributed in the northern hemisphere. Molecular evidence indicates that *Chorda filum* is paraphyletic (see Kawai *et al.* 2001) and may require subdivision into several species (Draisma *et al.* 2003).

Chorda Stackhouse

CHORDA Stackhouse (1797), p. xvi.

Type species: *C. filum* (L.) Stackhouse (1797), pp. xxiv, 40

Sporophytes annual, erect and macroscopic, cylindrical, unbranched, initially solid, later hollow and septate, arising from a short, minute stipe and small attachment disc; in section, thalli comprise a central, narrow, filamentous medulla, an outer large. parenchymatous cortex of longitudinally elongated cells and a peripheral meristoderm/epidermis of small, photosynthetic, palisade cells; plastids discoid, without pyrenoids; growth by a basal, intercalary meristem and a subapical secondary meristem. Life history heteromorphic, gametophytes filamentous, microscopic bearing antheridia and oogonia; unilocular sporangia accompanied by unicellular paraphyses in extensive continuous sori on sporophyte.

One species in Britain and Ireland.

Chorda filum (L.) Stackhouse (1797), pp. xxiv, 40.

Fig. 283

Fucus filum Linnaeus (1753), p. 1162; *Fucus filiformis* Strøm (1762), p. 94; *Ceramium filum* (Linnaeus) R.H. Wiggers (1780), p. 91; *Fucus thrix* Stackhouse in Withering (1796), p. 116; *Chordaria filum* (Linnaeus) C. Agardh (1817), p. 13; *Scytosiphon filum* (Linnaeus) C. Agardh (1820), p. 161; *Chondrus filum* (Linnaeus) Lamouroux (1824), p. 503; *Chorda filum* var. *thrix* (Stackhouse) Greville (1830), p. 47; *Chorda filum* var. *subtomentosa* Areschoug (1875), p. 13; *Chorda filum* f. *subtomentosa* (Areschoug) Kjellman (1883b), p. 305; *Chorda filum* f. *crassipes* Kjellman (1883b), p. 305; *Chorda filum* f. *typica* Kjellman (1883b), p. 245.

Sporophytes macroscopic, forming erect, olive to dark brown, single, or more commonly, gregarious, chord-like thalli, to 6 m in length and 5 mm wide, with one to several individuals arising from a small discoid holdfast of densely packed, branched, colourless, multicellular, irregularly shaped, rhizoidal filaments; thalli simple, terete, cylindrical, linear, sometimes spirally twisted, cartilaginous, spongy, slippery to the touch, bearing a dense growth of hyaline hairs, especially when young and during the summer, giving immersed individuals a fine white sheen over the thallus surface; thalli solid when young, and at the extreme base, becoming hollow throughout and tapering gradually to the apex and base without any clearly defined stipe region; distal regions of older thalli typically have signs of senescence; in surface view, basal cells and cells of very young thalli rectangular to quadrate in shape, 13–19 × 10–16 µm, regularly arranged in longitudinal rows, and enclosing several discoid plastids without pyrenoids; surface cells of more mature thalli replaced by paraphyses of the reproductive sorus which appear, in surface view, rounded, 14–20 µm in diameter, somewhat pigmented and irregularly arranged; in cross section, parenchymatous, young thalli and basal regions appearing solid, mature regions hollow, with a central medulla of loosely associated, anastomosing filaments, and an outer, cortex of several layers of longitudinally elongated, somewhat radially elongated, colourless cells enclosed within an outer meristoderm layer of photosynthetic isodiametric to palisade-like cells; inner cortical cells of mature, hollow thalli long and narrow with dilated ends (termed 'trumpet' cells) which regularly project into the lumen of the cavity and combine, at fairly even intervals, to form bridging diaphragms; peripheral meristoderm of mature thalli giving rise to the reproductive sori; surface hairs multicellular, with basal growth zone, unbranched and colourless.

Unilocular sporangia borne in dark-coloured, extensive, reproductive sori covering most of the surface of mature thalli except for the very basal regions; sporangia closely packed in vertical columns arising as extensions of the outer meristoderm layer, 18–25(–30) × 7–10 µm, elongate-cylindrical, enclosed under an umbrella of unicellular, elongate-clavate, rounded to flat-topped, terminally pigmented paraphyses, 40–46(–55) × 10–14(–20) µm, each containing both plastids and fucosan vesicles.

Epilithic, sometimes epiphytic; lower eulittoral, in pools, extending into the sublittoral to 15–20 m, on small stones, shells and gravel, particularly in silty, sandy areas, with holdfasts commonly buried in sand and mud, and in sheltered areas, including bays, lagoons, harbours, estuaries where it can form extensive meadows; rarely found on open and exposed coasts.

Common and widely distributed around Britain and Ireland.

Annual, with developing sporophytes observed in April/May, becoming fertile in June/July, dying off in October/November.

André (1970), p. 291; Bartsch & Kuhlenkamp (2000), p. 165; Brodie *et al.* (2016), p. 1022; Bunker *et al.* (2017), p. 254; Cabioc'h *et al.* (1992), p. 78, fig 57; Cho (2010), p. 183; Coppejans & Kling (1995), pp. 224–6, pl. 81(A–C); Cormaci *et al.* (2004), p. 177, (2012), pp. 393, 394, pl. 136(1–3); De Mesquita Rodrigues (1963), pp. 75, 76; Dizerbo & Herpe (2007), pp. 118, 119, pl. 39(3); Fletcher in Hardy & Guiry (2003), p. 231; Gallardo García *et al.* (2016), p. 38; Gayral (1966), p. 304; Guiry (2012), pp. 162, 163; Hamel (1931–39), pp. 312, 313; Harvey (1846–51), pl. 107; Izquierdo *et al.* (1996), pp. 105–15; Jaasund (1965), p. 104; Kanda (1938), pp. 87–111; Kawa *et al.* (2001), pp. 130–42, (2016), pp. 227–49; Kogame & Kawai (1996), pp. 247–60; Kornmann & Sahling (1977), p. 150, pl. 80(A–C); Kylin (1918), pp. 1–64, (1933), pp. 69–73, figs 33–5, (1947), pp. 79, 80, pl. 14(46); Lane *et al.* (2006), pp. 493–512; Lee & Oh (1998), pp. 69–77; Levring (1940), pp. 58, 59; Lund (1959), pp. 155, 156; Mathieson & Dawes (2017), p. 217, pls

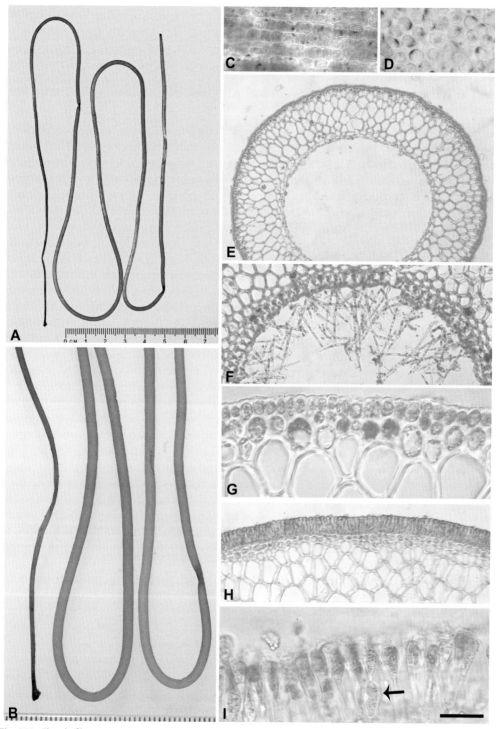

Fig. 283. *Chorda filum*

A, B. Habit of sporophyte thallus. C. Surface view of basal region showing cells arranged in longitudinal rows. D. Surface view of thallus showing closely packed paraphyses. E. TS of thallus showing central cavity. F. TS of portion of thallus showing 'diaphragm' formation. G. TS of basal thallus region edge showing outer meristoderm layer and inner large, colourless, cortical cells. H. TS of thallus showing outer reproductive sorus. I. Close-up of sorus showing unilocular sporangia (arrow) and associated taller paraphyses. Bar in I = 220 μm (E, F), = 190 μm (H), = 35 μm (C, D, G, I).

XL(9, 10), CVII(3); Morton (1994), p. 43; Newton (1931), p. 200, fig 127(A–D); Novaczek *et al.* (1986a), pp. 2414–20; Oltmanns (1922), fig. 398; Pankow (1971), pp. 206, 207, fig. 270, pl. 28; Pedersen (2011), p. 109, fig. 119; Reinke (1892a), pp. 35–41, pls 26–8; Ribera *et al.* (1992), p. 120; Riouall (1985), pp. 83–6; Rosenvinge & Lund (1947), p. 70, fig. 23; Rueness (1977), p. 173, (2001), p. 14; Russell (1985a), pp. 87–104, (1985b), pp. 343–9, (1986), pp. 309–77, (1987c), p. 311; Sasaki & Kawai (2007), pp. 10–21; Saunders & McDevit (2013), pp. 1–23; Setchell & N.L. Gardner (1903), p. 254, (1925), pp. 592, 593; South & Burrows (1967), pp. 379–402, figs 1–14; Stegenga & Mol (1983), p. 106, pl. 39(1, 2); Stegenga *et al.* (1997a), p. 23; Taskin *et al.* (2008), p. 65; Taylor (1957), p. 176, pls 14(3), 15(1); Tilden (1937), pp. 258–60, fig. 123; Tokida (1954), pp. 112, 113; Verlaque (2001), p. 32; Verlaque *et al.* (2015), pp. 1–362; Williams (1921), pp. 603–7.

Laminariaceae Bory

LAMINARIACEAE Bory de Saint-Vincent (1827), p. 63.

Sporophyte erect, macroscopic, to several metres in length, perennial, rarely annual, usually distinctly divided into a holdfast, stipe and terminal lamina; holdfast initially a rhizoidal disc which in some species remains disc-like and the only means of attachment, in others giving rise to whorled, terete, branched haptera produced annually and acropetally up the stipe; stipe, if present, simple, hollow or solid, terete to slightly flattened above, lamina flat, blade-like, simple, with or without midrib, folded or ridged, smooth or bullate, with or without perforations, entire or ruffled, usually coriaceous and leathery, slimy, entire or becoming longitudinally split and multistrap-like with increasing exposure; structure parenchymatous with central medulla, enclosing cortex and peripheral meristoderm; growth by a non-dividing, primary meristematic zone situated at the junction of the stipe and blade, with no secondary lateral outgrowths occurring; in perennial species, annual activity of the meristem produces a new blade which pushes the old blade forward which is eventually shed; hairs absent.

Life history heteromorphic with a large, macroscopic sporophyte alternating with a filamentous, microscopic gametophyte. Unilocular sporangia accompanied by unicellular paraphyses produced in extensive raised sori formed on the lamina surface of the sporophyte.

Representatives of this family are mainly distributed in the colder parts of the northern and southern hemispheres, or in areas of cold water up-welling, often dominating the sublittoral. They are also economically important, being used for food and as a source of alginic acid which has widespread industrial applications.

There are two genera, *Laminaria* and *Saccharina*, in Britain and Ireland.

Laminaria J.V. Lamouroux

LAMINARIA J.V. Lamouroux (1813), p. 40.

Type species: *L. digitata* (Hudson) J.V. Lamouroux (1813), p. 42.

Sporophytes perennial and differentiated into a root-like holdfast of downward-growing branched haptera, a terete or slightly compressed flexible or stiff stipe, solid or hollow, of variable length, which gives rise to a single broad, large, dorsiventrally flattened, coriaceous, entire or divided, smooth or ruffled, lamina; in section, thalli comprise a central medulla, a parenchymatous cortex and an outer meristoderm; growth annual, by a meristematic region situated at the base of the blade.

Unilocular sporangia, accompanied by unicellular paraphyses, borne in extensive, irregular, raised, dark brown sori on both sides of lamina surface.

Three species in Britain and Ireland.

KEY TO SPECIES

1. Mucilage ducts present in stipe and lamina; stipe narrowing/attenuate above 2
   Mucilage ducts absent in stipes; stipe linear throughout . . . . *L. digitata*

2. Stipe flexible, smooth, usually without epiphytes; lamina pale brown and
   bearing a translucent yellow patch at base; sublittoral fringe . . . . *L. ochroleuca*
   Stipe stiff, rough, usually with epiphytes; lamina deep brown and lacking a
   basal yellow patch; sublittoral zone . . . .*L. hyperborea*

Laminaria digitata (Hudson) Lamouroux (1813), p. 42   Figs 284, 285

Fucus digitatus Hudson (1762), p. 474; *Fucus bifurcatus* Gunnerus (1766), p. 96; *Ceramium digitatum* (Hudson) Stackhouse (1797), pp. xxiv, 390, 405; *Gigantea digitata* (Hudson) Stackhouse (1816), p. xi; *Laminaria latifolia* C. Agardh (1820), p. 119, nom. illeg.; *Laminaria phycodendron* Bachelot de la Pylaie (1824), p. 181; *Laminaria conica* Bory de Saint-Vincent (1826), p. 190; *Laminaria bongardiana* var. *bifurcata* (Gunnerus) Postels & Ruprecht (1840), p. 10 (as *Laminaria bongardiana β bifurcata*); *Laminaria digitata* var. *latifolia* (C. Agardh) Areschoug (1842), p. 225; *Laminaria ensifolia* Kützing (1843), p. 345; *Hafgygia digitata* (Hudson) Kützing (1843), p. 343; *Laminaria digitata* f. *stenophylla* Harvey (1851), p. 338; *Laminaria flexicaulis* Le Jolis (1855), p. 472; *Laminaria flexicaulis* var. *cucullata* Le Jolis (1856), p. 59; *Laminaria flexicaulis* f. *ovata* Le Jolis (1856), p. 59; *Laminaria flexicaulis* f. *ensifolia* Le Jolis (1856), p. 57; *Laminaria stenophylla* (Harvey) J. Agardh (1868a), p. 18; *Laminaria digitata* var. *ensifolia* (Kützing) J. Agardh (1868a), p. 24; *Laminaria digitata* var. *integrifolia* J. Agardh (1868a), p. 24; *Laminaria digitata* var. *stenophylla* (Harvey) Kleen (1874), p. 33; *Laminaria digitata* f. *latifolia* (C. Agardh) Kjellman (1877b), p. 26; *Laminaria digitata* f. *complanata* Kjellman (1877c), p. 26; *Laminaria digitata* f. *cucullata* (Le Jolis) Kjellman (1883b), p. 299; *Laminaria cucullata* f. *ovata* (Le Jolis) Foslie (1883), p. 27; *Laminaria cucullata* (Le Jolis) Foslie (1883), p. 24; *Laminaria digitata* f. *cucullata* (Foslie) Kjellman (1883b), p. 299; *Laminaria digitata* f. *ovata* (Le Jolis) Kjellman (1883b), p. 299; *Laminaria flexicaulis* f. *valida* Foslie (1883), p. 21; *Laminaria digitata* f. *valida* (Foslie) Kjellman (1883b), p. 299; *Laminaria flexicaulis* f. *latilaciniata* Foslie (1883), p. 21; *Laminaria digitata* f. *ensifolia* (Le Jolis) Kjellman (1883b), p. 299; *Laminaria digitata* f. *latilaciniata* (Foslie) Kjellman (1883b), p. 29; *Laminaria cucullata* f. *typica* Kjellman (1883b), p. 25, nom. inval.; *Laminaria intermedia* Foslie (1884), p. 81; *Laminaria intermedia* f. *cucullata* (Le Jolis) Foslie (1884), p. 82; *Laminaria digitata* f. *debilipes* Foslie (1884), p. 61; *Laminaria intermedia* f. *longipes* Foslie (1884), p. 82; *Laminaria intermedia* f. *ovata* (Le Jolis) Foslie (1884), p. 82; *Laminaria digitata* f. *grandifolia* Foslie (1884), p. 60; *Laminaria digitata* var. *valida* (Foslie) Kjellman (1890), p. 23; *Laminaria digitata* f. *cuneata* Kjellman (1890), p. 23; *Laminaria digitata* var. *debilipes* (Foslie) Kjellman (1890), p. 23; *Laminaria digitata* f. *longipes* (Foslie) Kjellman (1890), p. 24; *Laminaria digitata* var. *intermedia* (Foslie) Kjellman (1890), p. 23; *Saccharina digitata* (Hudson) Kuntze (1891), p. 915; *Laminaria cucullata* f. *longipes* (Foslie) A.D. Zinova (1950), p. 73; *Saccharina bongardiana* f. *bifurcata* Selivanova, Zhigadlova & G.I. Hansen (2007), p. 283.

Sporophytes macroscopic, to 2(–3) m in length, 60–100 cm in width, forming erect, medium to dark brown thalli, differentiated into a terminal lamina, supporting stipe and basal holdfast; lamina to 2 m in length, medium to dark brown, dorsiventrally flattened, smooth and glossy, solid, to 1.2(–1.5) mm thick, coriaceous, slimy, lacking a midrib, entire and ovate when young or in extreme sheltered conditions, commonly becoming deeply split terminally into large numbers of strap-like digits; stipe dark brown, very variable in length, to 1 m or more, to 11 mm in diameter, simple, straight or sometimes twisted, linear, very smooth, lacking epiphytes unless damaged, cylindrical or sometimes slightly flattened, gradually becoming more oval above before flattening into the lamina; stipe rigid at first and arising erect from the holdfast, very quickly becoming flaccid and flexuose and lying prostrate on the substrata; holdfast arising at stipe base, and comprising one to several superimposed layers of outward-spreading, pseudodichotomously branched, rounded, smooth, finger-like haptera, to 7 mm in diameter, becoming narrower terminally and attached on their undersides by spreading adhesive discs; in cross section, stipe and lamina solid with a central medulla, outer cortex and peripheral meristoderm; stipes with a relatively large, central medulla of vertical filaments of thick-walled, transversely rounded, longitudinally elongated, septate cells, loosely packed within a matrix centrally, more closely packed externally, which are traversed by a horizontal network of branched, hyphae-like filaments; medulla grading into an inner and then outer cortex of thin-walled, rectangular cells usually arranged in radial rows, which become progressively

Fig. 284. *Laminaria digitata*

A. Habit of sporophyte thalli in the lower eulittoral. B. Close-up of holdfast showing the root-like, dichotomously branched haptera.

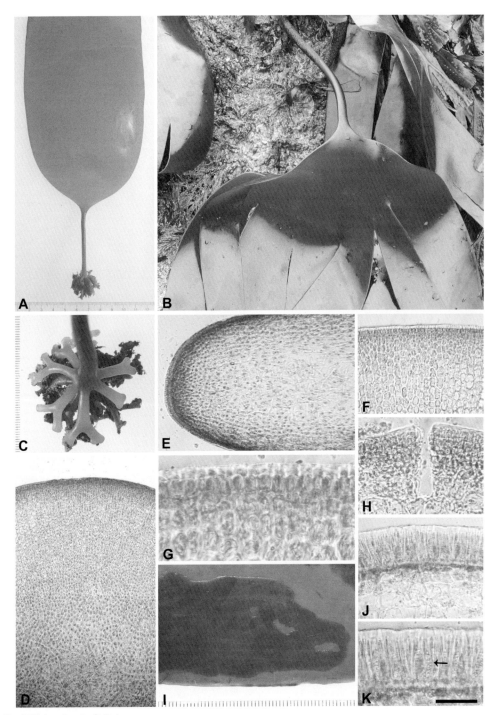

Fig. 285. *Laminaria digitata*

A. Habit of young sporophyte showing entire lamina. B. Habit of mature sporophyte showing segmented lamina. C. Close-up of young holdfast. D. TS of stipe showing the central medulla enclosed within an extensive cortex, the outer cell layers of which are arranged in radial rows. E. TS of lamina edge showing central medulla, outer cortex and peripheral meristoderm. F. TS of outer stipe region, showing the radial rows of the outer cortex and the peripheral meristoderm. G. TS close-up of F showing the peripheral meristoderm. H. TS of lamina showing mucilage canal opening out onto surface. I. Surface view of lamina showing a well-developed sorus. J, K. TS of sori showing unilocular sporangia (arrow) with associated taller paraphyses. Bar in K = 190 μm (D), = 140 μm (E), = 70 μm (F, J), = 35 μm (G, H, K).

smaller, and more photosynthetic, with discoid plastids lacking pyrenoids, up to the mucilage-covered, outer meristoderm layer of palisade-like cells; stipes lacking mucilage ducts; laminar structure similar to that of the stipe, although the cortex is less developed and comprises fewer layers of more rounded, photosynthetic cells with mucilage ducts often present, especially in young blade regions, as elongated tear-shaped cavities opening narrowly to the exterior and penetrating several cells down into the outer cortex.

Unilocular sporangia borne in extensive, irregularly shaped, mucilaginous, dark brown, slightly raised, confluent sori borne on both sides of the lamina, often in a mirror image, spreading over several cm on the laminar surface; sori produced as vertical extensions of the surface cells, to 70(–80) µm in height and comprising closely packed unilocular sporangia accompanied by paraphyses; unilocular sporangia elongate-cylindrical, 34–50 × 11–13 µm; paraphyses unicellular, elongate-clavate with a truncated, flat-topped apex and extending above and exclosing the sporangia, to 60(–70) × 13–16 µm, usually with a thin central column of cytoplasm enclosed within a thick lateral and terminal hyaline/gelatinous matrix.

Epilithic on rock, small stones and shells, occasionally epiphytic on larger, tougher algae, on a wide range of shores, mid to lower eulittoral, in pools, rarely emergent except at low water spring tides, spreading into the shallow sublittoral as a narrow, gregarious, dense or scattered band above the *Laminaria hyperborea* forest, to 6–9 m.

Locally abundant and widely distributed on rocky shores all around Britain and Ireland.

Sporophytes perennial, to 4(–6) years, with maximum growth from winter to early summer (February to July) and maximum fertile period July–August and November–December.

Ar Gall *et al.* (2004), pp. 30–7; Arseni *et al.* (2001), pp. 411–17; Bartsch & Kuhlenkamp (2000), p. 167; Bartsch *et al.* (2013), pp. 1061–73; Billot *et al.* (1999), pp. 307–14, (2003), pp. 111–21; Bringloe *et al.* (2018), pp. 1–8; Brodie *et al.* (2016), p. 1022; Bunker *et al.* (2017), p. 256; Cabioc'h *et al.* (1992), p. 79, fig. 60; Chapman (1981), pp. 307–11, (1993), pp. 263–7, (2008), pp. 307–11; Conolly & Drew (1985), pp. 181–95; Coppejans & Kling (1995), pp. 226–8, pl. 82(A, B); Corre & Prieur (1990), pp. 515–23; Cosson (1972), pp. 2501–4, (1973a), pp. 973–6, (1973b), pp. 104–12, (1975), pp. 50–4, (1976), pp. 28–34, (1977), pp. 19–26, (1999), pp. 35–42; Cosson & Gayral (1979), pp. 59–65; Cosson *et al.* (1976), pp. 1293–6; Davidson & Reed (1983), pp. 201–3, (1984), p. 192, (1985), pp. 41–50; Delebecq *et al.* (2012), pp. 503–17, (2016), pp. 71–82; Dizerbo & Herpe (2007), pp. 119, 120, pl. 39(4); Fletcher (1980), pp. 44–5, pl. 17(1); Floc'h & Penot (1976), pp. 989–92, (1981), pp. 176–87; Forster & Dring (1991), p. 85; Fries (1980), pp. 475–7; Gayral (1966), pp. 293–6, pl. LV, fig. 37; Gayral & Cosson (1973), pp. 1.1–9.11; Gómez & Lüning (2001), pp. 391–6; Gruber *et al.* (2011), pp. 603–14; Guiry (2012), p. 163; Gunnarsson (1991), p. 148; Hagen Rødde & Larsen (1997), pp. 391–5; Hagen Rødde *et al.* (1993), pp. 577–81, (1997), pp. 385–90; Hamel (1931–39), pp. 297–300, fig. 51(A, B); Haug & Jensen (1956), pp. 10–15; Hellebust & Haug (1972a), pp. 177–84, (1972b), pp. 169–76; Indergaard & Jensen (1981), pp. 411–17; Indergaard & Skjåk-Braek (1987), pp. 541–9; Jaasund (1965), pp. 105, 106; Jensen (1956); Jensen & Haug (1952), pp. 138, 139; Jordan *et al.* (1991), pp. 520–4; King & Schramm (1976), pp. 209–13; Kregting *et al.* (2015), pp. 1116–28; Klenell *et al.* (2002), pp. 1143–9; Kornmann & Sahling (1977), pp. 144–9, pl. 77(D); Kylin (1916), pp. 551–61, (1947), pp. 80, 81, pl. 15(48); Lane *et al.* (2006), pp. 493–512; Le Jolis (1855), pp. 241–312, (1856), pp. 531–91; Lubsch & Timmermans (2017), pp. 229–37; Lund (1959), pp. 164–6; Lüning *et al.* (2000), pp. 1129–34; Maier & Müller (1982), pp. 1–8, (1990), pp. 869–76; Maier *et al.* (1988), pp. 260–3; Martins *et al.* (2017), pp. 109–21; Mathieson & Dawes (2017), pp. 219, 220, pl. XLI(1–3); Mathieson *et al.* (1972), pp. 127–30; Mikelstad (1968), pp. 30–6; Molis *et al.* (2010), pp. 76–84; Morton (1994), p. 43; Moss (1977), p. 121; Moulin *et al.* (1999), pp. 1237–45; Müller *et al.* (1979), pp. 430, 431; Newton (1931), p. 204; Nitschke *et al.* (2018), pp. 114–25; P. Pedersen (1983), pp. 113–18; Penot & Videau (1975), pp. 285–93; Pérez (1969a), pp. 329–44, (1969b), pp. 117–35; Printz (1926), pp. 183–8, fig. 12, (1962), pp. 64–73; Ratcliff *et al.* (2017), pp. 1957–66; Rodrigues *et al.* (2000), pp. 97–106, (2002), pp. 939–47; Roleda *et al.* (2010), pp. 577–88; Roseder *et al.* (2005), pp. 1227–35; Rueness (1977), pp. 174, 175, (2001), p. 15; Russell (1983a), pp. 181–7, (1983b), pp. 303–4; Saunders & McDevit (2013), pp. 1–23; Schaffelke & Lüning (1994), pp. 49–56; Seoane Camba (1965b), pp. 47–55; Shaw (1957), p. 18; Sjøtun & Schoschina (2002), pp. 147–52; Smale *et al.* (2013), pp. 4016–38; Smith (1985), pp. 83–101; Stegenga & Mol (1983), p. 108, pl. 39(3); Stegenga *et al.* (1997a), p. 23; Stockton & Evans (1979), p. 128; Sundene (1958), pp. 121–8, (1962b), pp. 5–24, (1964), pp. 83–107, (1971), pp. 277–9; Taylor (1957), p. 184, pl. 18(1); Tom Dieck & Oliveira Filho (1993), pp. 151–60; Tonon *et al.* (2008), pp. 1250–6; Tovey & Moss (1978), pp. 17–22; Van Went *et al.* (1973), pp. 77, 78; Vauchel *et al.* (2008), pp. 515–17; Young *et al.* (2007), pp. 1200–8.

Laminaria hyperborea (Gunnerus) Foslie (1884), p. 42.   Figs 286, 287

Fucus scoparius Strøm (1762), p. 93; *Fucus hyperboreus* Gunnerus (1766), p. 74; *Laminaria cloustonii* Edmonston (1845), p. 54; *Hafgygia cloustoni* (Edmonston) Areschoug (1883), p. 1; *Laminaria hyperborea* f. *compressa* Foslie (1884), p. 42.

Sporophytes macroscopic, to 2(–4) m in length, 30–60 cm in width, forming erect thalli, differentiated into a terminal lamina, supporting stipe, and basal attachment holdfast; lamina light to medium brown, 30–70(–150) cm in length, more or less fan-shaped, dorsiventrally flattened, smooth, 900–1250 μm in thickness, solid, coriaceous, slightly slimy, lacking a midrib, entire and ovate at first, becoming deeply split terminally into large numbers (to 20–30) of strap-like digits; stipe light to dark brown, to 1–1.5(–3) m in length, usually greatly exceeding that of the terminal lamina, simple, solid, cylindrical throughout, erect, stiff and rigid, elongate-conical in shape, thickest at the base, to 30(–46) mm, narrowest at the apex, to 12(–15) mm, at first smooth for a short distance below the lamina, then becoming characteristically very rugose throughout due to the surface development of small, closely packed, raised, flat-topped, rounded bumps, initially laid down above in vertical rows, becoming more irregularly patterned below; stipe often heavily epiphytized, the cavities between the bumps providing habitats for various algae and invertebrates; holdfast conical, extending to 10 cm or more in height up the stipe, comprising rings of up to 12(–15) successive layers of downward-spreading, pseudodichotomously branched, sometimes with additional branches below, sturdy, thick, to 12 mm in diameter, terete, finger-like haptera attached terminally on their undersides by adhesive discs; blades and stipe solid, differentiated into a central medulla, outer cortex and peripheral meristoderm; in section, stipe with a large central medulla of vertical filaments of thick-walled, transversely rounded/oval, longitudinally elongated, septate cells, loosely distributed within a matrix which is traversed by a horizontal network of branched hyphae-like filaments; cortex comprising closely packed, irregularly arranged, vertical files of transversely rounded to angular, longitudinally elongated, thin-walled cells, becoming radially arranged towards the periphery, and enclosed by a region of several layers of small, cuboidal, darkly coloured cells, produced by the activity of the surface layer of meristematic cells, localized growth variations of which form the pattern of raised bumps seen in surface view on the stipe; transverse sections of older stipes reveal faint, concentric growth rings, produced by the meristematic activity of the outer cortex; outer cortex with a single concentric ring of mucilage canals, at first radially elongated and slit-like, later more rounded and tear-shaped extending narrowly through the outer cortex to the meristem layer and often opening out onto the surface, in older stipes, more deeply situated in the outer cortex and laterally expanded, and often fused with adjacent canals; lamina with a central medulla of closely packed filaments, a surrounding cortex of larger, thin-walled cells and an outer meristoderm layer; mucilage canals commonly occurring a few cells deep within the cortex, rounded in thinner terminal blade regions, more tear-shaped and opening out onto the surface in thicker lamina.

Unilocular sporangia borne in extensive, irregularly shaped, dark brown, slightly raised, to 80(–100) μm, mucilaginous, confluent sori spread over both sides of the lamina to several cm; sporangia closely packed in vertical columns accompanied by paraphyses, arising as extensions of the outer meristoderm layer; sporangia elongate-clavate, 38–50(–66) × 7–10(–13) μm, each enclosing relatively large zoospores; paraphyses unicellular, elongate-clavate, terminally truncated, 72–82 in length, 13–16 μm wide at the apex, extending above and enclosing the sporangia, usually with a central column of cytoplasm enclosed within a thick lateral and terminal, hyaline, gelatinous matrix.

Epilithic; on a wide range of shores, occasional in deep pools and channels in the lower, shaded eulittoral, more commonly sublittoral, forming dense, gregarious forests over bedrock below the *Laminaria digitata* belt, to 17 m and deeper in clear waters; a preference for solid, stable rock and turbulent water movement and not reported in areas of soft substrata, such as chalk, and areas influenced by sediment.

Fig. 286. *Laminaria hyperborea*

A. Small forest of sporophytes, shallow sublittoral, Berwick-upon Tweed, showing erect nature of stipe. B. Habit of sporophyte showing epiphytes on the stipe. C. Habit of two mature thalli showing the long, stout, conical stipe, basal conical holdfast and relatively shorter lamina. Bar = 81 mm (B), = 92 mm (C).

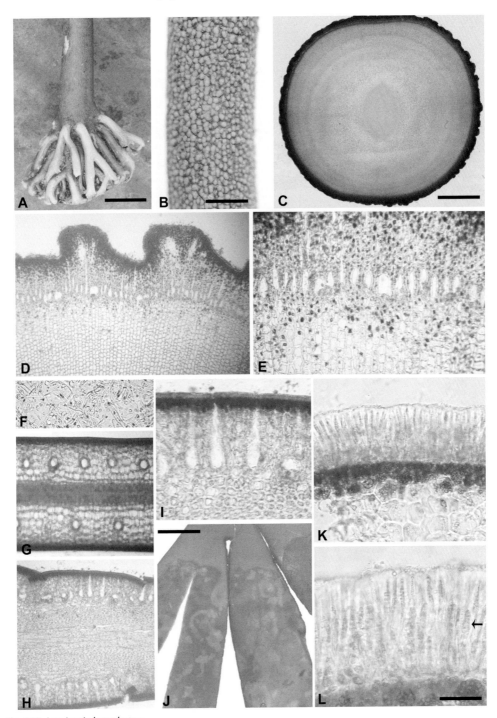

Fig. 287. *Laminaria hyperborea*
A. Conical holdfast. B. Surface view of stipe showing the characteristic raised bumps. C. TS of stipe showing the outer raised bumps, faint concentric growth rings in the cortex, and the inner medulla. D. TS through stipe edge showing the raised bumps, the single concentric ring of mucilage canals and the radial rows of outer cortical cells. E. Close-up of mucilage canals. F. TS of stipe showing central medullary tissue. G. TS of young lamina showing mucilage canals. H, I. TS of mature lamina showing mucilage canals opening out onto the surface. J. Portion of lamina segments showing darker, raised sorus areas. K, L. TS of sorus showing unilocular sporangia (arrow) with taller associated paraphyses. Bar in A = 35 mm, bar in B = 15 mm, bar in C = 6 mm, bar in J = 18 mm, bar in L = 375 μm (D), = 220 μm (H), = 190 μm (G), = 140 μm (E, I), = 70 μm (F, K), = 35 μm (L).

Abundant and widely distributed on rocky substrata around Britain and Ireland.

Sporophytes perennial, to 6–10(–20) years, with growth period from winter to early summer, November to June, and fertile during the winter, September to April.

Ardré (1970), pp. 285, 286; Assis *et al.* (2016), pp. 174–82; Bartsch & Kuhlenkamp (2016), p. 167; Black *et al.* (1959), pp. 137–49; Brodie *et al.* (2016), p. 1022; Bunker *et al.* (2017), p. 257; Cabioc'h *et al.* (1992), p. 80, fig. 61; De Mesquita Rodrigues (1963), pp. 76–8, pl. 8(D–E); Dizerbo & Herpe (2007), p. 120, pl. 40(1); Fletcher in Hardy & Guiry (2003), p. 265; Foslie (1884), pp. 1–112; Fredriksen *et al.* (1995a), pp. 47–54, (1995b), pp. 41–6; Fries (1980), pp. 475–7; Gallardo García *et al.* (2016), p. 38; Gayral (1966), pp. 296, 297, pl. LVI; Guiry (2012), pp. 163, 164; Gunnarsson (1991), pp. 1–148; Halm *et al.* (2011), pp. 16–26; Hamel (1931–39), pp. 300, 301, figs 50(VII–IX), 51(C); Han (1993), pp. 199–205, (1994a), pp. 67–75, (1994b), pp. 107–10; Han & Kain (1991), p. 87, (1992a), p. 91, (1992b), pp. 219–30, (1996), pp. 233–40; Harkin (1981), pp. 303–8; Harvey (1846–51), pl. 223; Hellebust & Haug (1972), pp. 169–76; Hopkin & Kain (1971), pp. 75–7, (1978), pp. 531–53; Izquierdo *et al.* (1993), pp. 291–304; Jaasund (1965), p. 105; Jupp & Drew (1974), pp. 185–96; Kain (1963), pp. 129–51, (1964b), pp. 207–14, (1964c), p. 390, (1969), pp. 455–73, (1971a), pp. 1–74, (1971b), pp. 387–408, (1972), p. 400, (1973), p. 213, (1974b), pp. 387–408, (1976a), pp. 603–28, (1976b), pp. 267–90, (1977), pp. 587–607, (1979), pp. 101–61; Kornmann & Sahling (1977), pp. 144–9, pl. 78(A–I); Kylin (1947), p. 81, pl. 16(50); Larkum (1972), pp. 405–18; Larsen & Haug (1960), pp. 250–4; Lobban *et al.* (1981), pp. 81–3; Longtin & Saunders (2015), pp. 440–50; Lüning (1986), pp. 269–73; Marshall (1960), pp. 18, 19; Miller *et al.* (2009), pp. 571–84; Morton (1994), p. 43; Newton (1931), p. 204, fig. 128(A–D); Oligschlager *et al.* (2012), pp. 511–25; P. Pedersen (1983), pp. 113–18; Pérez-Ruzafa *et al.* (2003), pp. 155–64; Richardson & Walker (1956), pp. 203–9; Rosenvinge & Lund (1947), p. 92, figs 31–3; Rueness (1977), p. 175, (2001), p. 15; Sauvageau (1918), pp. 183–99, figs 67–72; Schaffelke (1995), pp. 313–17; Schaffelke & Lüning (1994), pp. 49–56; Sjøtun & Fredriksen (1995), pp. 213–22; Sjøtun & Schoschina (2002), pp. 147–52; Sjøtun *et al.* (1993), pp. 215–21, (1995), pp. 525–36, (1996), pp. 1–8, (1998), pp. 337–43; Smale *et al.* (2013), pp. 4016–38, (2015), pp. 1033–44; Steinbiss & Schmitz (1974), pp. 134–52; Svendsen (1972a), pp. 448–60, (1972b), pp. 33–45; Whittick (1983), pp. 1–10; Willenbrink *et al.* (1975), pp. 161–70.

Laminaria ochroleuca Bachelot de la Pylaie (1824), p. 182.   Figs 288, 289

Laminaria lejolisii Sauvageau (1916b), p. 714; *Laminaria pallida* (Greville) J. Agardh (1848), p. 134; *Laminaria pallida* var. *iberica* G. Hamel (1928), p. 81; *Laminaria iberica* (Hamel) Lami (1934), p. 113.

Sporophytes perennial, to 2(–3) m in length, to 35 cm or more in width, forming erect, usually gregarious thalli, clearly differentiated into a terminal lamina, supporting stipe and basal attachment holdfast which is very light brown/yellow to medium brown in colour and comprises one to several, ascending layers of laterally outgrowing, branched, terete, finger-like haptera; young holdfasts with 1–2 layers of outward-spreading haptera and almost disc-like in appearance, older holdfasts giving rise to several layers of more downward-spreading, densely aggregated haptera, up to 8 cm or more up the stipe and almost conical in appearance; haptera irregularly, sometimes unilaterally and/or dichotomously branched, narrowing towards tips, forming small, strongly adherent attachment pads upon surface contact; stipe light brown, to 120 cm in length, to 27 mm in width at the base depending on thallus age, solid, terete to oval throughout, elongate-conical in shape, with a distinctly wider base which gradually narrows to 0.5–1 cm in diameter at the apex, erect and stiff over much of the length, surface very smooth and without nodular growths and epiphytes; lamina to 2(–3) m in length, 700–1100 μm in thickness, more or less fan-shaped, dorsiventrally flattened, solid, thick, coriaceous, mucilaginous, lacking a midrib, distinctly light brown/yellow throughout during spring/early summer growth season, medium brown throughout rest of year, although often remaining yellow-brown at the cordate base, entire and ovate at first, becoming deeply cleft terminally into several to large numbers (to 20–30) of strap-like digits; in cross section, both stipe and blade solid, and differentiated into a central medulla, an enclosing cortex and a peripheral meristoderm; medulla of stipe comprising vertical filaments of thick-walled, transversely rounded/oval, longitudinally elongated, septate cells loosely distributed within a matrix, which is traversed by a horizontal network of branched hyphae-like filaments derived from the inner cortex; older stipes with an extensive cortex, comprising closely packed, thin-walled, colourless

Fig. 288. *Laminaria ochroleuca*
A, D. Habit of sporophyte in the lower eulittoral, showing erect, conical, smooth stipe. B. Basal region of sporophyte showing conical stipe and basal holdfast. C. Holdfast of sporophyte.

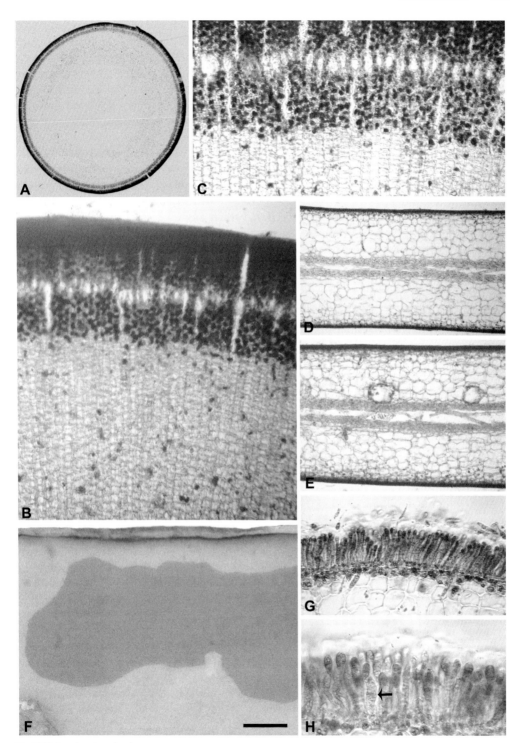

Fig. 289. *Laminaria ochroleuca*
A. TS of stipe showing smooth surface B. TS of outer stipe region showing the cortical cells in radial rows, the darkly pigmented appearance of the outer cortical cell layers, and the concentric ring of mucilage canals. C. Close-up of B. D, E. TS of lamina showing the narrow central medulla, enclosing colourless, cortical cells, outer meristoderm layer and occasional mucilage canals. F. Portion of lamina showing darker, raised sorus area. G, H. TS of sorus showing unilocular sporangia (arrow) with associated paraphyses. Bar in F = 6 mm (A), = 4.2 mm (F), = 375 μm (B), = 220 μm (C), = 190 μm (D, E), = 70 μm (G), = 35 μm (H).

cells, the inner cells appearing rounded/angular in cross section and irregularly arranged, the outer cells appearing more cuboidal/rectangular/lenticular in cross section, radially elongate and laid down in extensive radial rows; cortex surrounded by an outer, peripheral meristoderm layer of small, palisade-like, photosynthetic cells containing large numbers of enclosed discoid plastids without pyrenoids, with or without a subtending small-celled tissue several cells deep merging with the outer cortex; in transverse section, outer cortex with a single (2 in older thalli) concentric ring of radially elongated/slit-like to rounded, interconnecting mucilage ducts, with associated secretory cells, with both inwardly and outward-extending protuberances; outer cortex region, meristoderm layer and subtending small tissue usually deeply pigmented and bark-like; cross sections of the lamina reveal similar medullary, cortical and meristoderm tissues, and occasional, similar, large mucilage canals at the boundary of the medulla and inner cortex.

Unilocular sporangia borne in extensive, dark brown, slightly raised, confluent sori on the surface of both sides of the lamina, usually spread over several cm; sporangia closely packed, in vertical columns, arising as extensions of the outer meristoderm layer, 25–58 × 7–11 μm, elongate-cylindrical, accompanied by paraphyses which are unicellular, elongate-clavate, 28–65 × 7–11 μm with a thick lateral, and notably terminal, gelatinous matrix.

Epilithic, on rock; on sheltered to very exposed shores, occasional in lower eulittoral, deep shaded pools, more commonly sublittoral to 6(–8) m, reported to be much deeper, to 15(–30) m on more exposed shores, either occasional and mixed with *Saccharina latissima*, *Laminaria hyperborea* and *Saccorhiza polyschides*, or forming extensive stands.

Locally common, confined to the south-west coasts of Britain (Isles of Scilly, South Cornwall, South Devon, Lundy) and west coast of Ireland (Mayo).

Sporophytes perennial, lifespan unknown, fertile from April to November/December.

First reported for Britain in the more sheltered parts of Plymouth Sound, South Devon in 1946 (Parke 1948), occurring down to 8 m below low water springs, mixed with *Laminaria digitata* and *L. hyperborea*. Further records were then reported for South Devon (Salcombe) and West Cornwall (Helford River) by Spooner (1950). Underwater observations in the Isles of Scilly (Norton 1968) revealed it to be the dominant kelp species from 0–6 m in sheltered sites, and confined to deeper water, replacing *L. hyperborea*, in more exposed sites, disappearing at 27–30 m depth. On Lundy Island (North Devon), Irvine *et al.* (1972) reported it occurring only well below the sublittoral fringe in small but distinct stands, but in 2008 it was found at the bottom of the shore (Hiscock & Brodie 2016). It can be distinguished from *L. digitata* and *L. hyperborea* by a combination of the following features: whitish-yellow colour of the base of the frond often resembling a half-moon, smooth, conical stipe, the occurrence of an outer ring of mucilage canals seen just below the surface in transverse section of the stipe, and the pale yellow-brown colour of the whole frond during the spring and early summer growth period.

Ardré (1970), pp. 287–9, pl. XXI; Assis *et al.* (2016), pp. 174–82; Bachelot de la Pylaie (1824), pp. 174–84; Bertocci *et al.* (2014), pp. 117–22; Brodie *et al.* (2016), p. 1022; Bunker *et al.* (2017), p. 258; Cabioc'h *et al.* (1992), p. 80, fig. 62; Chemin (1934), pp. 11–13; Cormaci *et al.* (2012), p. 398, pl. 137(1–4); De Mesquita Rodrigues (1963), pp. 78–81, pl. 8; Dizerbo & Herpe (2007), pp. 120, 122, pl. 40(2); Feldmann (1934), p. 13, figs 3, 4; Fletcher in Hardy & Guiry (2003), p. 266; Gallardo García *et al.* (2016), p. 38; Gayral (1958), pp. 240, 241, figs 36, 37, (1966), p. 296, fig. 38; Gorostiaga *et al.* (1981), pp. 265–70; Guiry (2012), p. 183; Hamel (1928), p. 81, (1931–39), pp. 302–4, fig. 51(D); Hiscock & Brodie (2016), p. 41; Irvine *et al.* (1972), p. 126; Izquierdo *et al.* (1993), pp. 291–304, (1997), pp. 51–66, (2002), pp. 285–92; John (1969), pp. 175–87; Lami (1934), p. 113, (1943), pp. 75–90; Niell (1984), pp. 71–102; Parke (1948), p. 295, (1950), p. 259; Pereira *et al.* (2011), pp. 395–403; Pérez-Ruzafa *et al.* (2003), pp. 155–64; Ribera *et al.* (1992), p. 120; Roleda *et al.* (2004), pp. 603–13; Sauvageau (1916b), p. 74, (1918a), pp. 164, figs 56–66, 67(b); Schoenrock *et al.* (2019), pp. 1–8; Seoane Camba (1966), p. 432, pl. 4, fig. 1; Smale *et al.* (2013), pp. 4016–38, (2015), pp. 1033–44; Spooner (1950), pp. 261, 262, pl. I.

Saccharina Stackhouse

SACCHARINA Stackhouse (1809), nom. rejic.

Type species: *S. latissima* (Linnaeus) C.E. Lane, C. Mayes, Druehl & G.W. Saunders (2006), p. 509.

Sporophytes perennial, and differentiated into an attaching root-like holdfast of downward-growing branched haptera, a terete or slightly compressed flexible or stiff stipe, solid or hollow, of variable length, which gives rise to a single, large, usually entire, rarely split, dorsiventrally flattened, coriaceous, smooth or more commonly ruffled lamina; in cross section, lamina comprise a central medulla, a parenchymatous cortex and an outer meristoderm; growth annual, by a meristematic region situated at the base of the lamina.

Unilocular sporangia, accompanied by unicellular paraphyses, produced in raised, dark brown sori borne on both sides of lamina surface.

One species in Britain and Ireland.

Saccharina latissima (Linnaeus) C.E. Lane, C. Mayes, Druehl & G.W. Saunders (2006), p. 509.   Figs 290, 291

Ulva latissima Linnaeus (1753), p. 1163; *Fucus saccharinus* Linnaeus (1753), p. 1161; *Ulva latissima* Gunnerus (1766), p. 52, nom. illeg.; *Ulva maxima* Gunnerus (1772), p. 127; *Ulva longissima* Gunnerus (1772), p. 128; *Ulva fusca* Hudson (1778), p. 567, nom. illeg.; *Fucus phyllitis* Stackhouse in Withering (1796), p. 100; *Ceramium phyllitis* (Stackhouse) Stackhouse (1797); *Fucus saccharinus* var. *bullatus* Goodenough & Woodward (1797), p. 151; *Ceramium saccharinum* (Linnaeus) Stackhouse (1797), pp. xxiv, 390, 405; *Ulva lactuca* var. *latissima* (Linnaeus) A.P. de Candolle in Lamarck & de Candolle (1805), p. 9; *Ulva phyllitis* (Stackhouse) A.P. de Candolle in Lamarck & de Candolle (1805), p. 15; *Saccharina plana* Stackhouse (1809), p. 65, nom. illeg.; *Saccharina bullata* (Goodenough & Woodward) Stackhouse (1809), p. 65; *Fucus saccharinus* var. *latissimus* Turner (1811), pp. 69, 71 (as [*delta*] *latissimus*); *Laminaria saccharina* (Linnaeus) J.V. Lamouroux (1813), p. 42; *Laminaria phyllitis* (Stackhouse) C. Agardh (1817), p. 9; *Laminaria saccharina* var. *latissima* C. Agardh (1817), p. 18; *Laminaria saccharina* var. *bullata* C. Agardh (1817), p. 18; *Palmaria phyllitis* (Stackhouse) Nees (1820), p. 7; *Laminaria membranacea* S.F. Gray (1821), p. 384; *Laminaria caperata* Bachelot de la Pylaie (1824), p. 180; *Fucus saccharinus* var. *angusto-bullatus* Wahlenberg (1826), p. 891; *Laminaria stackhousei* Bory de Saint-Vincent (1826), p. 189 (*stackhousii*), nom. illeg.; *Laminaria saccharina* f. *linearis* J. Agardh (1848), p. 132; *Laminaria saccharina* f. *membranacea* J. Agardh (1848), p. 13; *Laminaria saccharina* f. *prima* J. Agardh (1851), p. 132; *Phycoseris latissima* (Linnaeus) Frauenfeld (1854); *Laminaria saccharina* var. *phyllitis* (Stackhouse) Le Jolis (1863), p. 91; *Laminaria hieroglyphica* J. Agardh (1868a), p. 11; *Laminaria caperata* var. *lanceolata* J. Agardh (1868a), p. 14; *Laminaria caperata* var. *obovata* J. Agardh (1868a), p. 14; *Laminaria caperata* var. *oblonga* J. Agardh (1868a), p. 14; *Laminaria caperata* var. *eliptica* J. Agardh (1868a), p. 14; *Laminaria saccharina* var. *oblonga* J. Agardh (1868a), p. 12; *Laminaria maxima* (Gunnerus) Kjellman (1877b), p. 25; *Laminaria agardhii* Kjellman (1877b), p. 18 (as *Agardhii*); *Laminaria saccharina* var. *caperata* (Bachelot de la Pylaie) Farlow (1881), p. 94; *Laminaria saccharina* f. *latissima* Kjellman (1883b), p. 287; *Laminaria saccharina* f. *oblonga* (J. Agardh) Kjellman (1883b), p. 286; *Laminaria saccharina* f. *linearis* (J. Agardh) Foslie (1884), p. 90; *Laminaria saccharina* f. *longissima* (Gunnerus) Foslie (1884), p. 90; *Laminaria saccharina* f. *borealis* Foslie (1884), p. 91; *Laminaria saccharina* var. *membranacea* (J. Agardh) Kjellman (1890), p. 25; *Laminaria saccharina* var. *sublaevis* Kjellman (1890), p. 25; *Laminaria saccharina* var. *grandis* Kjellman (1890), p. 25; *Laminaria saccharina* f. *latifolia* (Kjellman) Kjellman (1890), p. 25; *Laminaria saccharina* f. *typica* Kjellman (1890), p. 24, nom. inval.; *Saccharina phyllitis* (Stackhouse) Kuntze (1891), p. 915; *Laminaria groenlandica* Rosenvinge (1893), p. 847; *Laminaria longicruris* var. *faeroensis* Børgesen (1896), p. 403; *Laminaria saccharina* f. *caperata* (Bachelot de la Pylaie) Setchell (1900), p. 146; *Laminaria faeroensis* (Børgesen) Børgesen (1902), p. 454; *Laminaria faeroensis* f. *sacchariniformis* Børgesen (1902), p. 454; *Laminaria phyllopus* Kjellman (1906), p. 9; *Laminaria saccharina* f. *bullata* (C. Agardh) Kylin (1907), p. 98; *Laminaria saccharina* f. *sublaevis* (Kjellman) Kylin (1907), p. 98; *Saccharina groenlandica* (Rosenvinge) Lane, Mayes, Druehl & Saunders (2006), p. 509.

Sporophytes perennial, forming erect, olive to dark brown, leathery thalli, clearly differentiated into a basal holdfast, supporting stipe and terminal lamina; holdfast prostrate, not conical, light to medium brown in colour, comprising one to several (3–4), ascending layers of laterally outgrowing, dichotomously branched, terete to oval, flexible, non-rigid, finger-like

Fig. 290. *Saccharina latissima*

A–C Habit of sporophytes. D, E. Surface views of lamina showing the bullate appearance of older thalli. F. Holdfast of sporophyte attached to mussel showing the rings of haptera emerging. Bar in F = 38 mm (A), = 34 mm (B), = 40 mm (C), = 20 mm (D), = 9 mm (F).

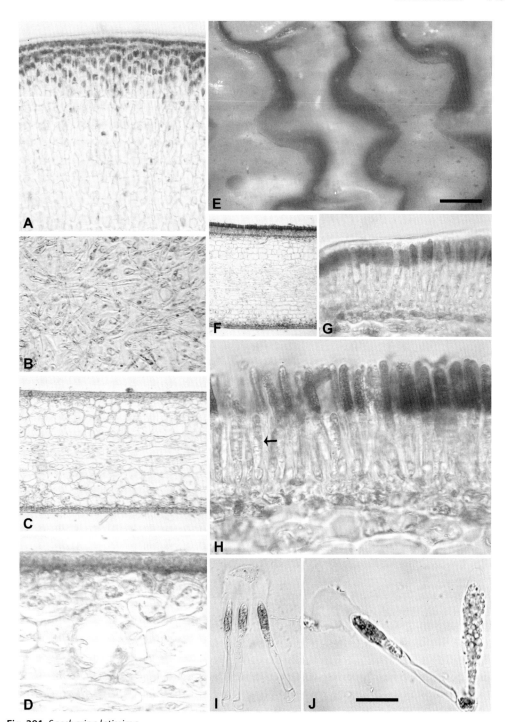

Fig. 291. *Saccharina latissima*
A. TS of stipe showing the radially elongated and arranged outer cortical cells and the surface meristoderm layer. B. TS of middle region of stipe showing the medullary tissue. C. TS of lamina showing the central medulla, outer cortex and enclosing meristoderm layer. D. TS of lamina showing a mucilage canal in the cortex. E. Surface view of fertile lamina showing the darker areas associated with sorus development. F. TS of fertile lamina showing sorus development on one side. G, H. TS of sorus showing unilocular sporangia (arrow) and taller, darkly pigmented paraphyses. I. Three paraphyses showing pronounced mucilaginous cap. J. Unilocular sporangium with associated paraphysis. Bar in E = 5 mm, bar in J = 220 μm (F), = 140 μm (B, C), = 70 μm (D), = 45 μm (A), = 35 μm (G–J).

haptera up to 5 mm in width, gradually narrowing towards the tips and forming strongly adherent, attachment pads upon surface contact; stipe solid, smooth, lacking epiphytes, cylindrical, linear, very flexible, to 20(–40) cm in length, 3(–4.5) mm in width; lamina to 2 m in length, to 30 cm in width, dorsiventrally flattened, simple, solid, lacking a midrib, at first olive-brown, cuneate, broadly rounded in shape and relatively thin, to 400 µm in thickness, later becoming medium to dark brown, linear lanceolate, strap-shaped, coriaceous, slimy, to 1 mm in thickness, narrowing abruptly at the base into the stipe; older thalli frequently bullate/verrucose, ruffled and undulating along the edge and with relatively deep depressions in rows on both sides of the lamina; young lamina commonly bearing cryptostomata on the surface, with emergent hyaline hairs; in cross section, lamina and stipe solid, with a central medulla, an enclosing cortex and a peripheral meristoderm; stipe medulla oval in shape, comprising a tangled mass of septate filaments of vertically elongated, thick-walled, transversely oval to rounded cells traversed by a network of branched hyphae-like filaments derived from the inner cortex; cortex extensive and comprising closely packed, thin-walled colourless cells, at first rounded/angular in cross section and irregularly arranged, towards the periphery becoming more radially elongated, variously shaped and laid down in radial rows, the outer few cells more darkly pigmented; meristoderm a single layer of palisade-like, photosynthetic cells with a thick cuticular covering, and enclosing large numbers of discoid plastids without pyrenoids; outer cortex without rings of mucilage canals; sections of the lamina reveal a central medulla of variously sized, fairly closely packed, predominantly longitudinally arranged, septate filaments of cells, an enclosing cortex of several, large, thin-walled, rounded to rectangular, colourless cells, occasionally with rounded mucilage-canals, and an outer photosynthetic meristoderm.

Unilocular sporangia borne in extensive, dark brown, slightly raised, confluent, mucilaginous sori which reach 100(–130) µm in height and are usually spread over several cm in the central regions of the lamina; unilocular sporangia closely packed, in vertical columns, arising as extensions of the outer meristoderm layer, a single cell dividing into a sporangial initial cell and a paraphysis initial cell; paraphyses developing before the sporangia, providing an umbrella-like covering, at first elongate-cylindrical to elongate-clavate, 62–70(–82) × 6–8(–10) µm, with a dark-pigmented, terminal region containing fucosan vesicles (and possibly plastids), later becoming truncated with a marked increase in the gelatinous thickening of the lateral walls and apex, increasing their height to 80–130 µm, and at the truncated apex, to 10–17 µm across; sporangia elongate-cylindrical to elongate-clavate, 40–60(–70) × 9–13(–15) µm.

Epilithic, solitary or aggregated, on rock, small stones, shingle and shells, often replacing *Laminaria digitata* on smaller stones, rarely epiphytic when young, on sheltered to extremely sheltered shores, lower eulittoral spreading into the shallow sublittoral, where it can be locally dominant, sometimes deeper (to 21 m); also reported to live in an unattached form in the deeper sublittoral of very sheltered sandy bays.

Common and widely distributed around Britain and Ireland.

Sporophytes biennial, lifespan rarely exceeding three years, found all year round, first appearing late winter/early spring with rapid growth January to June, slower growth from July to December, with distal lamina tissue cast continuously through the life of the sporophyte; sporophytes become fertile after approximately 8 months, and reproductive sori can be found throughout the year. According to Mathieson & Dawes (2017), the blade morphology of *Saccharina latissima* is highly variable depending on water motion. Blades with ruffled edges, which enhance nutrient uptake, occur in low energy environments, while smooth and narrow blades which help minimize drag occur in high energy environments. Blades in sheltered habitats are also reported to be more brittle and wider than those in exposed and high-salinity habitats (see original studies by Fowler-Walker *et al.* 2006 and Garbary & Tarakhovskaya 2013).

Anderson *et al.* (2013), pp. 689–700; Ardré (1970), pp. 289–91; Bartsch & Kuhlenkamp (2000), p. 167; Boden (1979), pp. 405–8; Breton *et al.* (2018), pp. 32–40; Brodie *et al.* (2016), p. 1022; Bunker *et al.* (2017), p. 259; Burrows (1961), pp. 187–9; Burrows & Pybus (1971), pp. 53–6; Cabioc'h *et al.* (1992), p. 79, fig. 59; Chapman *et al.* (1978), pp. 195–8; Conolly & Drew (1985), pp. 181–95; Coppejans & Kling (1995), p. 228, pl. 829(C); De Mesquita Rodrigues (1963), pp. 81, 82; Dizerbo & Herpe (2007), p. 122, pl. 40(3); Druehl *et al.* (2005), pp. 250–62; Dunton (1985), pp. 181–9; Fletcher (1980), p. 45, pl. 17, figs 2–6; Fletcher in Hardy & Guiry (2003), p. 266; Fowler-Walker *et al.* (2006), pp. 755–67; Freitas *et al.* (2016), pp. 377–85; Gallardo García *et al.* (2016), p. 38; Garbary & Tarakhovskaya (2013), pp. 267–80; Gayral (1966), p. 299, pl. LVII; Gerard (1987), pp. 237–44, (1988), pp. 25–6, (1990), pp. 519–28, (1997), pp. 800–10; Gerard & Du Bois (1988), pp. 575–80; Gerard *et al.* (1987), pp. 229–32; Guiry (2012), pp. 164, 165; Hamel (1931–39), pp. 295, 296, fig. 50(I); Han & Kain (Jones) (1993), pp. 79–81; Hanelt *et al.* (1997), pp. 387–95; Hagen Rødde & Larsen (1997), pp. 391–5; Harvey (1846–51), pls 192, 289; Heesch & Peters (1999), pp. 1–6; Heinrich *et al.* (2012), pp. 83–94, (2015), pp. 93–108; Hsiao & Druehl (1971), pp. 1503–8, (1973a), pp. 829–39, (1973b), pp. 989–97, (1973c), pp. 160–4; Íñiguez *et al.* (2016), pp. 248–66; Izquierdo *et al.* (1993), pp. 291–304; Kornmann & Sahling (1977), pp. 144–9, pl. 77(A–C); Kylin (1947), pp. 81, 82, pl. 15(49); Lane *et al.* (2006), pp. 493–512; Lee & Brinkhuis (1986), pp. 276–85, (1988), pp. 181–91; Longtin & Saunders (2015), pp. 440–50; Lund (1959), pp. 159–63, fig. 34; Lüning (1975), pp. 108–14, (1981a), pp. 379–93, (1988), pp. 137–44; Lüning & Dring (1972), pp. 252–6; Lüning & Markham (1979), pp. 125, 126; Lüning & Mortensen (2015), pp. 449–55; Makarov & Voskoboinikov (2001), pp. 89–94; Mathieson & Dawes (2017), pp. 221, 222, pl. XLI(11–13); McDowell *et al.* (2015), pp. 431–41; Meijer (1984), pp. 1–107; Morton (1998), p. 43; Newton (1931), pp. 202–4; Nielson *et al.* (2016), pp. 523–31; Parke (1948), pp. 651–709, pls 5–13, text figs 1–10; Paulino *et al.* (2016), pp. 3071–4; Pedersen (2017), p. 109, fig. 116; Perez Cordero *et al.* (2012), p. 19; Pérez-Ruzafa *et al.* (2003), pp. 155–64; Peteiro & Freire (2009), pp. 54–60, (2011b), p. 319, (2012a), pp. 189–96, (2012b), p. 706, (2013), pp. 205–13; Peteiro *et al.* (2006), pp. 45–52, (2014), p. 519, (2016), pp. 9–23; Peters & Schaffelke (1996), pp. 111–16; Rosenvinge & Lund (1947), p. 79, fig. 27; Rueness (1977), pp. 175, 176, (2001), p. 15; Saunders & McDevit (2013), pp. 1–23; Sauvageau (1918), p. 200, figs 73–81; Schaffelke *et al.* (1996), pp. 117–23; Setchell (1912), p. 149; Setchell & N.L. Gardner (1903), p. 261, (1925), pp. 595, 596; Sjøtun (1993), pp. 433–41; Sjøtun & Schoschina (2002), pp. 147–52; Smale *et al.* (2013), pp. 4016–38; Spurkland & Iken (2011), pp. 355–65; Stegenga & Mol (1983), p. 108, pl. 40(2); Stegenga *et al.* (1997a), p. 23; Strong & Dring (2011), pp. 223–9; Taylor (1957), p. 180; Tokida (1954), pp. 115–18, pl. X(3–7); Vilg *et al.* (2015), pp. 435–47; Wheeler & Weidner (1983), pp. 91–6.

Ralfsiales Nakamura

RALFSIALES Nakamura (1972) ex P.-E. Lim & H. Kawai in Lim *et al.* (2007).

Thallus encrusting, with or without an erect phase, of pseudoparenchymatous construction; if encrusting, circular or indefinite in outline, with or without radial and concentric ridges, with or without superimposed thalli, at first a monostromatic disc of outward-radiating, coalesced filaments, later becoming a polystromatic crust by the growth and lateral cohesion of assurgent straight, or curved cell rows of filaments arising from the basal disc cells, in some species with both assurgent and descending curved cell rows arising from a central layer; filaments loosely or tightly adjoined; disc and crust attached to the substratum by the undersurface, usually without rhizoids; surface of disc and crust with or without a pronounced cuticle; if erect, caespitose, cylindrical, multiaxial with central medullary filaments and outer cortex; plastids single, plate-like or several and discoid, without pyrenoids, often with abundant physodes; hairs usually frequent, tufted, in depressions, with basal meristem, lacking obvious sheath; growth of both basal and erect filaments by terminal apical cells or by intercalary divisions.

Reproduction by both unilocular and plurilocular sporangia, these usually occurring on separate thalli, either as small or expansive patches; unilocular sporangia borne laterally at the base of multicellular paraphyses (assimilatory filaments); plurilocular sporangia in closely packed, uniseriate or partly biseriate, vertical columns terminal on erect filaments, without paraphyses, usually with one to several terminal sterile cells; life history direct, isomorphic, or heteromorphic and apparently diplohaplontic.

Until fairly recently, the Ralfsiales in Britain and Ireland was represented by three genera, viz *Petroderma*, *Pseudolithoderma* and *Ralfsia* positioned in the Ralfsiaceae (Brodie *et al.*, 2016; Guiry, 2012), this reduced to the single genus *Ralfsia*, with general acceptance of Silberfeld *et al.*'s transfer of the former two genera to the Petrodermataceae and Sphacelariales respectively. In his treatment of *Ralfsia*, Fletcher (1989), following on from the works of Batters (1902), Newton (1931) and Parke & Dixon (1976), recognised five species: *R. clavata*, *R. disciformis*, *R. pusilla*, *R. spongiocarpa* and *R. verrucosa*. In his investigation of the marine algae of Berwick-on-Tweed, Batters (1890, pp. 286–9) identified *R. clavata*, *R. spongiocarpa* and *R. verrucosa* which he classified within two subgenera: *Euralfsia* which had delimited sori and curved erect filaments, and *Stragularia* Strömfelt (1886) which had continuous sori and vertically orientated filaments. He placed *R. verrucosa* in *Euralfsia* and *R. clavata* and *R. spongiocarpa* in *Stragularia*. This subdivision of *Ralfsia* was not widely accepted (see Abbott & Hollenberg 1976; Parke & Dixon 1976; Rueness 1977; Taylor 1957); indeed, Loiseaux (1968, p. 306) concluded on the basis of culture work that there was no justification for maintaining a distinction between *Euralfsia* and *Stragularia*.

The division of *Ralfsia* into two subgenera was, however, accepted by Newton (1931, p. 152), although unlike Batters (1902, p. 42), she included *R. clavata* and *R. spongiocarpa* with *R. verrucosa* in *Euralfsia*; the two remaining species, *R. disciformis* and *R. pusilla*, were placed in *Stragularia*.

Fletcher (1987) concluded that these five species of *Ralfsia* formed a disparate group of taxa which could be usefully separated using the characters originally proposed by Batters in his subgeneric classification. In disagreement with Newton (1931), Fletcher (1987) considered that only *R. verrucosa* possessed *Euralfsia* features while the remaining four species were better assigned to *Stragularia*. Furthermore, he proposed that these distinguishing characters were sufficient taxonomic criteria to warrant readoption of *Stragularia* as a genus, following the treatments of Hamel (1931–39, p. xxxi), Kjellman (1893) and Weber-van Bosse (1913). Additional evidence for the separation of *Stragularia* spp. from *Ralfsia sensu stricto* was provided by the results of a number of culture studies (see Fletcher 1987, p. 253 for summary) which indicated that *Stragularia*, unlike *Ralfsia*, comprised a group of crustose entities, some of which have been shown to be connected in life history with various erect thalloid members of the Scytosiphonaceae.

For this reason the genus *Stragularia* was transferred to the Scytosiphonaceae by Fletcher (1987), a position accepted by Brodie *et al.* (2016) and Guiry (2012), leaving only *Ralfsia* and the species *R. verrucosa* representing the Ralfsiaceae for Britain and Ireland.

However, the concept of this species was challenged by Parente *et al.* (2010b) working on samples collected from several localities in the north-eastern Atlantic and Mediterranean. They revealed several divergent lineages among the samples and, therefore, the presence of several cryptic species within the taxon. A follow-up to this study by Parente *et al.* (2020), using molecular analyses and morpho-anatomical comparisons to type material of North Atlantic *Ralfsia*-like species, further revealed the occurrence of a monophyletic subset of *Ralfsia*-like species containing two new genera, *Nuchella* and *Pseudoralfsia*, both of which occur in Britain. These genera did not resolve within the above-recognised families at that time, and were transferred to a new family, Pseudoralfsiaceae. Species of Pseudoralfsiaceae are morphologically similar to species of *Ralfsia sensu stricto* but are distinguished from the latter in having frequent hair pits, typically unsymmetrical thalli and in their molecular data. Pseudoralfsiaceae is, therefore, included in the present treatise.

Pseudoralfsiaceae Parente, Fletcher & G.W. Saunders

PSEUDORALFSIACEAE Parente, Fletcher & G.W. Saunders (2020), p. 16.

Thallus encrusting, light to dark brown, sometimes olive in colour, discrete and orbicular, sometimes becoming confluent and irregular in outline, firm, smooth and coriaceous sometimes becoming bullate and warty, with or without superimposed thalli, loosely or moderately attached to the substratum by the undersurface, rarely with rhizoids; undersides of crusts sometimes rust-red in colour; comprising an initial monostromatic basal disc of outward-radiating, branched filaments, with terminal, synchronously growing, apical cells, giving rise behind to assurgent, straight or slightly curved, rarely downward-curved, branched cell rows, grading into erect usually firmly adhered filaments covered by a thick surface cuticle; plastids single, plate-like lacking obvious pyrenoids; hairs frequent, tufted, arising in pits from the middle and/or lower cells of the erect filaments, with basal meristem, lacking obvious sheath; growth of both basal and erect filaments by terminal apical cells.

Reproduction by both unilocular and plurilocular sporangia, either embedded and concave or borne in slightly raised, gelatinous sori in discrete, medium or large patches; unilocular sporangia abundant, laterally positioned, sessile or on pedicels of one, rarely 2–3 cells, borne at the base of 1–2, simple, clavate, multicellular paraphyses; plurilocular sporangia in adventitious or non-adventitious sori, with synchronised or non-synchronised development, intercalary beneath one, rarely two terminal sterile cells, in closely packed, uniseriate, or at times biseriate, vertical columns terminal on the erect filaments, without paraphyses, and enclosed within a surface cuticle; life history apparently isomorphic and diplohaplontic.

A family of two genera, *Pseudoralfsia* Parente, Fletcher & G.W. Saunders and (provisionally) *Nuchella* Parente, Fletcher & G.W. Saunders, both identified for Britain. *Nuchella* is differentiated from *Pseudoralfsia* in having numerous vesicles in the erect cells, non-synchronous development of the plurilocular sporangia and hair pits arising from both middle and lower cells of the erect filaments. In the absence of sufficient field collections of *Nuchella* in Britain, only *Pseudoralfsia* is considered in the present treatise.

Pseudoralfsia Parente, Fletcher & G.W. Saunders

PSEUDORALFSIA Parente, Fletcher & G.W. Saunders (2020), pp. 15, 16.

Type species: *Pseudoralfsia verrucosa* (Areschoug) Parente, Fletcher & G.W. Saunders (2020), p. 16, figs 2–15

Thallus encrusting, discrete and orbicular, becoming confluent and irregular, without concentric zones and radial lines, sometimes superimposed in vertical section, firm, glossy, coriaceous, smooth, becoming bullate and verrucose in older thalli, loosely or moderately attached to the substratum by undersurface, with or without rhizoids; comprising a monostromatic basal layer of outward-spreading, frequently branched, firmly united filaments, with terminal, synchronously growing, apical cells, giving rise immediately behind to firmly united, branched filaments, at first almost prostrate, giving the appearance of a polystromatic base, later becoming erect, remaining fairly short and covered by a thick surface cuticle; plastids single, plate-like without obvious pyrenoids; hairs infrequent or abundant, tufted, in depressions, with basal meristem, lacking obvious sheath.

Unilocular and plurilocular sporangia borne on the same or different thalli, in raised, discrete or spreading, gelatinous sori; unilocular sporangial sori common, discrete, prominent and wart-like, very gelatinous, with sporangia pyriform, sessile or pedicellate at the base of clavate, multicellular, simple paraphyses; plurilocular sporangial sori infrequent, spreading, slightly raised and gelatinous, with sporangia in closely packed uniseriate, vertical columns terminal on erect filaments, without paraphyses, with terminal sterile cell and enclosed within a surface cuticle.

In their molecular study, Parente *et al.* (2020) further revealed a genetic group within *Pseudoralfsia* which differed from all the other groups in its vegetative construction but was common to that of the lectotype of *Ralfsia verrucosa*. This group was consequently assigned the name *Pseudoralfsia verrucosa* and material previously identified as *Ralfsia verrucosa* was transferred from the Ralfsiaceae to the new family Pseudoralfsiaceae and will be considered here. Three other genetic groups, also assigned to *Pseudoralfsia* by Parente *et al.* (2020), were reported for Britain, although they await formal description.

One species in Britain and Ireland.

Pseudoralfsia verrucosa (Areschoug) Parente, R.L. Fletcher & G.W. Saunders (2020), pp. 15, 16.
Figs 2(C,E), 292

Cruoria verrucosa Areschoug (1843), p. 264; *Ralfsia verrucosa* (Areschoug) Areschoug in Fries (1845), p. 124.

Thalli epilithic, encrusting, comparatively thick, olive to dark brown, individual and orbicular or more frequently confluent and indefinite in outline to 5(–10) cm in extent, loose to moderately well attached to substrata by underside, usually without rhizoids; peripheral region olive brown, smooth, coriaceous, quite closely adherent, but with margins usually quite easily raised intact, central regions dark brown, verrucose, bullate, brittle, loosely attached with rust-red coloured underside, frequently with young superimposed thalli; in surface view, cells rounded or polygonal, 4–9 μm in diameter, each with a single plate-like plastid without obvious pyrenoid; thalli pseudoparenchymatous in structure, comprising a monostromatic basal layer of outward-spreading, frequently branched, laterally united filaments, with terminal synchronously growing apical cells, giving rise immediately behind to strongly united, branched filaments, at first almost prostrate, giving the appearance of a polystromatic base, later becoming erect, assurgent and markedly curved, tapering slightly, covered terminally by a thick cuticle; thalli not usually exceeding 14(–20) cells, 120(–210) μm thick, superimposed thalli up to 1 mm thick; erect filament cells quadrate to rectangular, 4–11 × 5–8 μm, each with a single, terminal, plate-like plastid without obvious pyrenoid; hairs infrequent, grouped, emerging from depressions, with basal meristem, lacking obvious sheath.

Unilocular and plurilocular sporangia usually borne on separate thalli, rarely on the same thalli, terminal on erect filaments in raised, gelatinous sori; unilocular sporangial sori yellow brown, common, discrete, prominent and wart-like, spongy, very gelatinous, often concentrically zoned just behind growing margin, comprising obovate to pyriform unilocular sporangia, 60–105 × 15–37 μm, sessile or on 1–2-celled pedicels, lateral at the base of multicellular paraphyses; paraphyses simple, clavate, 7–11(–14) cells, to 115(–150) μm long, comprising rectangular cells 1–2 diameters long 7–12 × 6–7 μm in mid/upper regions usually with an enlarged subcylindrical to subpyriform apical cell, 10–20 × 6–8 μm, much narrower, 1–4(–5) diameters long, 12–18 × 3–5 μm below; plurilocular sporangial sori infrequent, slightly raised and gelatinous, spreading, with plurilocular sporangia in closely packed vertical columns terminal on erect filaments, without paraphyses, to 135 μm (–27 loculi) long, uniseriate, rarely biseriate, 6–8 μm in diameter, with subquadrate loculi, and terminal, dark-coloured sterile cell embedded in a thick cuticular layer.

Epilithic, epizoic; from the littoral fringe to the lower eulittoral, in pools and emergent, especially in damp places.

Abundant and widely distributed all around Britain and Ireland.

Thalli perennial; unilocular sporangia widely recorded throughout the year, although more common in the winter, November to March; plurilocular sporangia only recorded from scattered localities (Berwick, Northumberland, Hampshire, Dorset, Devon, Cornwall, Isle of Man), and without any obvious seasonal distribution.

Pseudoralfsia verrucosa is the most common crustose brown alga recorded in the eulittoral region around Britain and Ireland. It occurs on very exposed to very sheltered shores over the whole intertidal range, usually in pools but frequently exposed on rock, particularly in damp situations, e.g. under *Fucus* cover, in cracks and crevices, on soft, water-retaining, substrata such as chalk, etc. It occurs on a wide variety of substrata, both natural and artificial, and is commonly epizoic on barnacles, limpets, mussels and top shells; it has never, however, been observed as

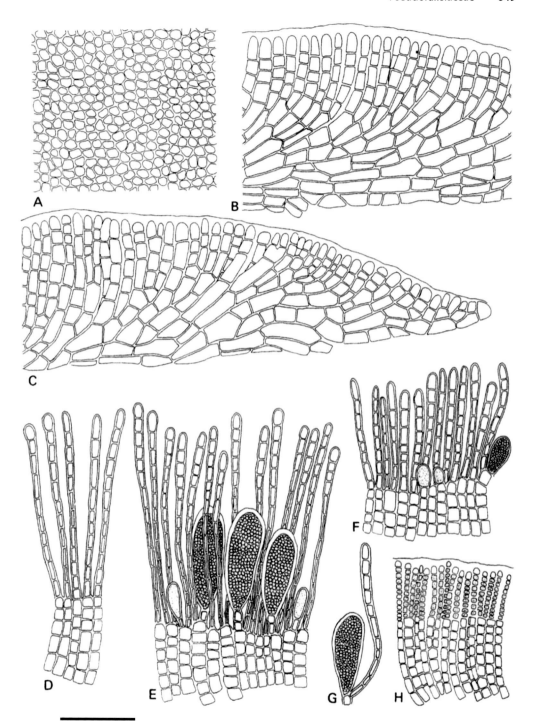

Fig. 292. *Pseudoralfsia verrucosa*
A. Surface view of crust cells. B. RVS of thallus showing upward-curving laterally adjoined filaments. C. RVS of thallus margin. D–G. VS of fertile thalli showing erect filaments giving rise terminally to paraphyses and unilocular sporangia. H. VS of fertile thallus showing terminal rows of plurilocular sporangia. Note sterile cap cell. Bar = 50 μm.

an epiphyte. It can be distinguished from the other crustose brown algae recorded for Britain and Ireland by the following combination of characteristics: it forms moderately well-attached, leathery, verrucose crusts common in the eulittoral region only, abundantly distributed all around Britain and Ireland; in radial tangential section, the erect filaments are markedly curved as they arise from the basal layer, almost prostrate at first, giving the appearance of a polystromatic base; the erect filaments are tightly bound together and can only be squashed apart with difficulty; cells contain a single plate-like plastid without obvious pyrenoid; the unilocular sporangia occur in discrete, raised sori on crust surface, and are lateral at the base of 7–14-celled, clavate paraphyses; the plurilocular sporangia occur in slightly raised, spreading sori on crust surface, and form closely packed, uniseriate, vertical columns with a sterile terminal cell.

Culture studies have only been carried out on North Atlantic isolates of *Pseudoralfsia verrucosa* (as *Ralfsia verrucosa*) by Loiseaux (1968), Edelstein *et al.* (1971) and Fletcher (1974b). The life history appears to be of the 'direct' type, with spores derived from both unilocular and plurilocular sporangia usually directly producing crusts similar to those of the parent. The cycle appears to be apomeiotic and without any indication of a sexual process. There is no evidence that *P. verrucosa* has a heteromorphic life history similar to that shown by some members of the crustose genus *Stragularia*.

Agardh (1848–76), p. 62, fig. 212, pl. 40; Areschoug (1843), p. 264, pl. 9(5–6); Bartsch & Kuhlenkamp (2000), p. 168; Branch (1971), pp. 1–38; Brodie *et al.* (2016), p. 1022; Bunker *et al.* (2017), p. 251; Coppejans (1983), pl. 66; Coppejans & Kling (1995), p. 209; Cormaci *et al.* (2012), p. 417, pl. 143(7–10); Dizerbo & Herpe (2007), p. 76, pl. 24(1); Edelstein *et al.* (1971), pp. 1249, 1250; Fletcher (1974a), pp. 67–77, figs 11, 12, pls 17–22, (1978), pp. 371–3, figs 1, 2, 5, 6, 13, 14, 24, 26, 27, (1987), pp. 241–3, figs 59(B), 66; Fletcher in Hardy & Guiry (2006), p. 291; Funk (1955), p. 35; Gallardo *et al.* (2016), p. 39; Guiry (2012), p. 158; Hamel (1931–39), pp. 106–8, fig. 26(A–B); Hauck (1884), p. 401, fig. 176; Jaasund (1964), pp. 131–3, figs 2, 3; Kain (2008), pp. 498–509; Kain *et al.* (2010), pp. 617–27; Keum (2010), pp. 102–4, figs 19, 20; Kornmann & Sahling (1977), pp. 118–20, fig. 62; León-Alvarez (2017), pp. 567–81, figs 1–6; Lindauer *et al.* (1961), pp. 211, 212, fig. 38; Loiseaux (1968), pp. 297–302, figs 1, 2; Mathieson & Dawes (2017), p. 231, pl. LX(11–14); Morton (1994), p. 33; Nakamura (1972), pp. 152, 153, fig. 3; Newton (1931), p. 153; Parente (2007), p. 137, Parente *et al.* (2010b), p. 80, (2020), pp. 12–23, figs 2–15; Pedersen (2011), p. 99; Reinke (1889a), pp. 9, 10, pl. 5(1–13); Ribera *et al.* (1992), p. 113; Rueness (1977), p. 130, fig. 55; Setchell &Gardner (1925), pp. 497, 498; Skjenja *et al.* (1997b), pp. 147–9, pl. 37(1–4); Stegenga & Mol (1983) p. 82, pl. 28(3–6); Tanaka & Chihara (1980b), pp. 227–31, figs 1(A), 2(A–B); Taylor (1957), p. 123, pl. 11(1, 2); Tokida (1954), p. 82; Womersley (1987), pp. 70–2, fig. 19(A–C).

Sporochnales Sauvageau

SPOROCHNALES Sauvageau (1926), p. 364.

Sporophyte thalli erect, terete or compressed, with or without midrib, solid, cartilaginous, much branched, arising from a fibrous holdfast; branching pseudodichotomous, radial or triradiate, each branch with a terminal tuft of pigmented, assimilatory, hair-like, unbranched filaments; surface smooth or with short, papillate branches with a similar terminal tuft of assimilatory filaments; growth trichothallic; structure pseudoparenchymatous, with a broad central region of large, longitudinally elongated, transversely rounded or irregular, colourless cells enclosed by 1–2 layers of small, pigmented cells; plastids discoid, several in each cell, without pyrenoids.

Unilocular sporangia borne laterally on branched paraphyses in slightly raised, densely packed, mucilaginous sori borne on the conical apex (*Carpomitra*) or around the apex of the short, determinate branchlets (*Sporochnus*) on the sporophyte; microscopic gametophytes filamentous, monoecious or dioecious, oogamous.

Life history diplohaplontic and heteromorphic with a macroscopic sporophyte alternating with a microscopic, filamentous gametophyte.

The Sporochnales contains a single family, Sporochnaceae.

Sporochnaceae Greville

SPOROCHNACEAE Greville (1830), p. 36 [as Order IV. Sporochnoideae]

Sporophyte thalli erect, to 20(–50) cm in length, either terete or slightly elliptical/dorsiventrally flattened with midrib, much branched, solid, arising from a fibrous holdfast; branching pseudodichotomous, alternate, or triradiate, with or without numerous short lateral branches of limited growth, all branches terminated by a tuft of pigmented, hair-like, unbranched filaments; growth trichothallic, structure pseudoparenchymatous, with a broad central region of large, longitudinally elongated, transversely rounded or irregular, colourless cells enclosed by 1–2 layers of small, pigmented cells; plastids discoid, several in each cell, without pyrenoids.

Unilocular sporangia sessile and clustered, lateral on closely packed, short, multicellular, branched paraphyses in slightly raised mucilaginous sori on terminally situated receptacles.

Life history diplohaplontic and heteromorphic with macroscopic, diploid, sporophytes alternating with microscopic, monoecious or dioecious, haploid filamentous gametophytes.

This family is represented in Britain and Ireland by two genera: *Carpomitra* and *Sporochnus*, both easily distinguished on morphological features. Members of this family have a similar life history to those of the Arthrocladiaceae and Desmarestiaceae in the Desmarestiales and for this reason many authors such as Parke & Dixon (1976) and Clayton (1981) placed the Sporochnaceae within the latter order. However, members of the Sporochnaceae differ from members of the Arthrocladiaceae and Desmarestiaceae in having a terminal tuft of unbranched hair-like filaments on all branch-apices and in possessing a multiaxial rather than a uniaxial pseudoparenchymatous mode of construction which is initiated at an early stage by parenchymatous divisions. On the basis of this unique growth mechanism a number of authors, including Abbott & Hollenberg (1976), Bold & Wynne (1978), Christensen (1980–84), Fritsch (1945), Lindauer *et al.* (1961), Rueness (1977) Schneider & Searles (1991) and Womersley (1987) adopted Sauvageau's (1926a) proposal for ordinal status of the Sporochnaceae. This was later supported by molecular studies (Draisma 2002; Draisma *et al.* 2003; de Reviers & Rousseau 1999; Silberfeld *et al.* 2014) and accepted by authors such as Brodie *et al.* (2016), Cormaci *et al.* (2014), Guiry (2012), Kim (2010) and de Reviers *et al.* (2015).

Carpomitra Kützing

CARPOMITRA Kützing (1843), p. 343.

Type species: *C. cabrerae* (Clemente) Kützing (1843) (= *Carpomitra costata* (Stackhouse) Batters 1902, p. 46).

Thalli consisting of erect, branched, dorsiventrally flattened to slightly elliptical, solid thalli arising from a fibrous holdfast; thalli with pseudodichotomous, sometimes triradiate or alternate branches, with a distinct midrib and terminal tuft of pigmented hair-like filaments; young unbranched thalli and terminal young branches often terete; growth trichothallic, structure pseudoparenchymatous, comprising a central region of axial cells surrounded by a cortex of longitudinally elongated, transversely rounded or irregular, colourless cells, enclosed by 1–2 layers of small, pigmented, peripheral cells; in surface view, cells of thalli pigmented, in longitudinal rows in midrib region, mainly irregularly arranged elsewhere; cells with numerous discoid plastids without pyrenoids.

Unilocular sporangia borne in continuous sori covering the surface of inverted conical or cylindrical receptacles which project from branch apices; receptacles usually with basal, surrounding collar and terminal annular swelling from which tufts of pigmented filaments emerge; unilocular sporangia lateral, in clusters, from branched, multicellular, slightly mucilaginous paraphyses; germination of unispores gives rise to filamentous, microscopic, monoecious gametophytes bearing oogonia and antheridia.

One species in Britain and Ireland.

Carpomitra costata (Stackhouse) Batters (1902), p. 46.   Figs 293, 294

Fucus costatus Stackhouse (1802), 109, pl. 17 (pro parte); *Fucus cabrerae* Clemente (1807), p. 313; *Carpomitra cabrerae* (Clemente) Kützing (1843), p. 343; *Carpomitra luxurians* W.R. Taylor (1945), p. 104.

Thalli forming erect, solitary, much-branched, dorsiventrally flattened, slightly elliptical thalli, 0.1–0.2 m long, 1–3 mm wide, light to olive-brown, arising from a small, terete stipe and fibrous holdfast; erect blades quite firm and cartilaginous, solid, with distinct midrib, branching pseudodichotomous, occasionally triradiate or alternate, apex obtuse or truncated, often with young, projecting, cylindrical and terete terminal branches, with or without a basal collar and/or terminal tuft of hair-like filaments; surface cells in longitudinal rows in midrib regions, irregularly arranged elsewhere, 8–21 × 6–10 μm, each containing numerous discoid plastids without pyrenoids; terminal filaments 4–6 mm long, simple, attenuate at apex, with short basal meristem, comprising quadrate to rectangular cells 30–160 × 26–34 μm, usually 1–3 diameters long below, 3–6 diameters long above, each densely packed with discoid plastids without pyrenoids; in section, thalli with thick-walled, longitudinally elongated, transversely rounded or irregularly shaped cells, distinctly smaller and more thick-walled in the centre, enclosed by a peripheral layer of 1–2, small, pigmented cells.

Unilocular sporangia borne in continuous gelatinous sori covering and encircling the surface of specialized receptacles; receptacles formed from extending, midrib tissue, conical or cylindrical, colourless, to 2 mm long, terminal on branched apices, less frequently in axes of dichotomously to trichotomously branching thalli, usually with basal, surrounding collar and with terminal annular swelling from which tufted filaments extend; unilocular sporangia elongate-clavate or elongate-pyriform, 33–48 × approx. 10–13 μm, in large numbers lateral on erect, mucilaginous, densely packed, projecting paraphyses; paraphyses dichotomously to trichotomously branched, to 140 μm, 5(–7) cells long, comprising cells 1–4 diameters long, rectangular or slightly swollen, 10–26 × 5–16 μm with enlarged, bulbous apical cell, 10–23 μm in diameter.

Epilithic on bedrock and boulders, sublittoral to 37 m; tolerant of sand cover.

South and south-western shores of Britain (Channel Islands, Hampshire, north and south Devon, Cornwall, Pembroke, Isle of Man) and Ireland (Cork, Donegal).

Probably a summer annual; recorded June to September.

C. costata was investigated in laboratory culture by Sauvageau (1926a & b) who revealed a heteromorphic life history with the erect, macroscopic sporophyte alternating with a microscopic, filamentous, monoecious gametophyte bearing oogonia and antheridia. A similar life history was later described by Motomura *et al.* (1985).

Adams (1994), p. 94, pl. 30; Brodie *et al.* (2016), p. 1022; Cormaci *et al.* (2012), pp. 438–40, pl. 150(1–5); Dizerbo & Herpe (2007), pp. 103, 104, pl. 34(2); Fletcher (1987), pp. 289–91, fig. 88; Fletcher in Hardy & Guiry (2003), p. 229; Gallardo García *et al.* (2016), p. 39; Guiry (2012), p. 159; Hainsworth (1976), p. 61; Hamel (1931–39), pp. 274–6, fig. 49(d–i); Harvey (1846–51), pl. 14; Hiscock & Maggs (1984), p. 85; Lindauer *et al.* (1961), pp. 245, 246, fig. 61; Morton (1994), p. 38; Motomura *et al.* (1985), pp. 21–31; Newton (1931), pp. 137, 138, fig. 94; Oak (2010), pp. 165, 166, fig. 47; Ribera *et al.* (1992), p. 116; Sauvageau (1926a), pp. 141–92, figs 1–17, (1926b), pp. 351–64; Stegenga *et al.* (1997b), pp. 177, 181, pl. 53(3); Taylor (1945), p. 104, pls 3(9–16), 16, 17(2); Womersley (1987), p. 270, figs 97(A), 98(A–C).

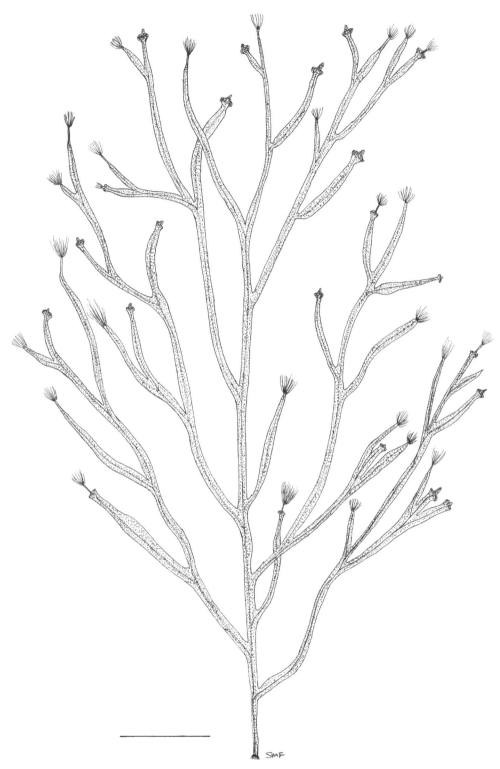

Fig. 293. *Carpomitra costata*
Habit of thallus. Bar = 20 mm.

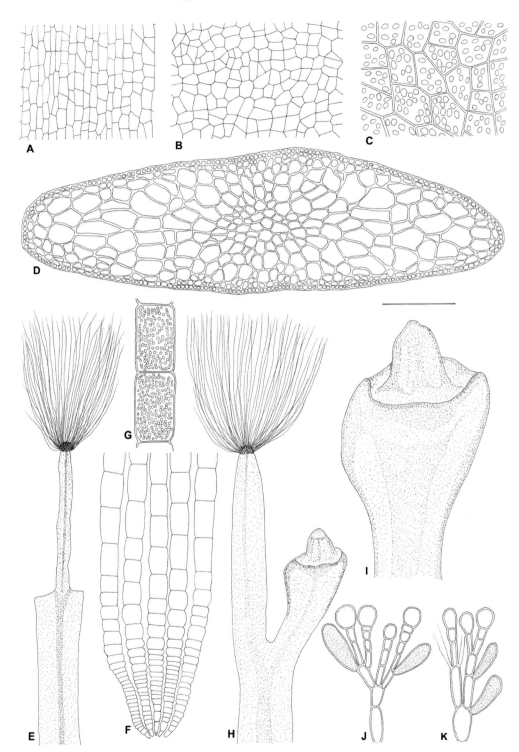

Fig. 294. *Carpomitra costata*

A. Surface view of thallus, midrib region. B, C. Surface view of thallus, blade region. Note discoid plastids in cells. D. TS of thallus. E. Branch apex showing new growth and terminal tuft of filaments. F, G. Portions of filaments. H, I. Branch apices showing receptacles. J, K. SP of unilocular sporangial sori showing unilocular sporangia associated with short paraphyses. Bar = 50 μm (A, B, G, H, J, K), = 20 μm (C), = 100 μm (D, F), = 2.5 mm (E, H), = 5 mm (I).

Sporochnus C. Agardh

SPOROCHNUS C. Agardh (1817), p. 12.

Type species: *S. pedunculatus* (Hudson) C. Agardh (1817), p. 149.

Thalli forming erect, much-branched, terete, solid thalli arising from a fibrous holdfast; main axis distinct, branching alternate or irregularly spiralled, to one order only, both primary and secondary axes beset with numerous, short, divaricate branches of limited growth, all branches terminated by a tuft of pigmented, hair-like filaments; growth trichothallic, structure pseudoparenchymatous, with a broad central region of large, fairly thick-walled longitudinally elongated, transversely rounded or irregularly shaped colourless cells, enclosed by 1–2 layers of small, pigmented, peripheral cells; surface cells in longitudinal rows, each containing a few discoid plastids without pyrenoids.

   Unilocular sporangia lateral, usually in large numbers, on short, multicellular, branched paraphyses, in slightly raised, mucilaginous sori, developing on the short lateral branches of limited growth; life history heteromorphic with unispores germinating into microscopic, either monoecious or dioecious, haploid gametophytes bearing oogonia and antheridia.

One species in Britain and Ireland.

Sporochnus pedunculatus (Hudson) C. Agardh (1817), p. xii.   Figs 295, 296

Fucus pedunculatus Hudson (1778), p. 587; *Capillaria pedunculata* (Hudson) Stackhouse (1809), p. 87.

Thalli forming erect, much-branched, terete thalli, usually solitary, occasionally gregarious, light to olive-brown, arising from a small, fibrous holdfast; erect thalli with a distinct main axis, quite firm and cartilaginous becoming soft terminally, solid, to 30(–50) cm long, 0.5(–0.65) mm wide, branching alternate, or irregularly spiralled, to one order only, with the branches at well-spaced intervals, divaricate, gradually narrowing terminally to a tuft of fine, hair-like filaments; all axes beset with numerous short, linear, clavate or pyriform divaricate branches of limited growth, to 3(–5) mm long, 0.1–0.2(–0.3) mm wide, with similar terminal tuft of filaments; filaments to 4.5 mm long, uniseriate, simple, with basal meristem, comprising cells 1–3 diameters long below, 28–75 × 16–29 µm, to 10 diameters long above, 40–70 × 5–8 µm, each containing numerous discoid plastids without pyrenoids; surface cells of axes in longitudinal rows, usually longer than wide, 1–6 diameters long, commonly rectangular, occasionally with oblique/pointed end walls, 16–52 × 5–17 µm, each with a few discoid plastids without pyrenoids; in section, thalli with a broad central region of large, fairly thick-walled, longitudinally elongate, transversely rounded or irregular, colourless cells, enclosed by 1–2 layers of small, slightly pigmented cells.

   Unilocular sporangia borne in sori on the short lateral branches of limited growth; sori slightly raised, continuous, gelatinous, completely encircling and swelling the terminal regions of the longer, older branches, appearing pedicellate, to 0.7 mm long × 0.4 mm wide, entirely covering the shorter, younger branches and shortly spreading onto the parental axes, appearing sessile; unilocular sporangia clavate, pyriform, elongate-pyriform or elongate-cylindrical, 26–43 × 8–12 µm, single or grouped, lateral and sessile on short paraphyses; paraphyses produced as lateral extensions of the outer cortical cells, simple, or more commonly much branched, clavate, multicellular, to 3(–5) cells long, 65(–90) µm, comprising mainly rectangular, colourless cells, with enlarged bulbous apical cell, 13–16 µm in diameter.

   Epilithic, on small stones, gravel, shells, etc., particularly in silty areas and often associated with *Arthrocladia villosa*, in the sublittoral to 18 m.

   Rare but generally distributed around Britain and Ireland; more records for southern and western coasts.

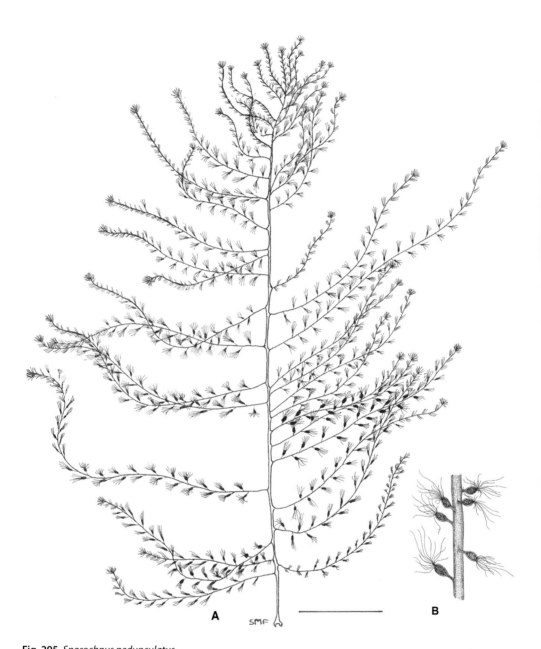

Fig. 295. *Sporochnus pedunculatus*
A. Habit of thallus. B. Portion of thallus, showing lateral branches of limited growth with terminal tuft of filaments. Bar = 2 cm (A), = 2.4 mm (B).

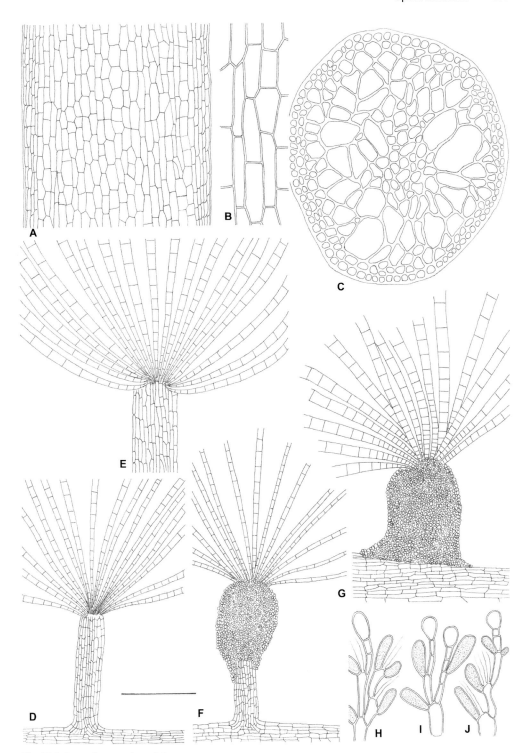

Fig. 296. *Sporochnus pedunculatus*
A. Portion of thallus. B. Surface view of thallus. C. TS of thallus. D–E. Short lateral branches of limited growth, with terminal tuft of filaments. F–G. Lateral branches clothed in unilocular sporangial sori. H–J. SP of sori showing unilocular sporangia associated with short paraphyses. Bar = 100 µm (A, C), = 50 µm (B, H–J), = 200 µm (D–G).

Annual, spring and summer, April to September.

Culture studies by Sauvageau (1931) and Caram (1965) revealed a heteromorphic life history with the macroscopic, diploid sporophyte alternating with a microscopic, either monoecious or dioecious, haploid gametophyte. Caram (1965) also noted parthenogenetic development of the antherozoids to form microthalli with unilocular sporangia; spores released from the latter recycled the male gametophyte.

Abbott & Hollenberg (1976), pp. 184, 185, fig. 149; Bartsch & Kuhlenkamp (2000), p. 168; Brodie *et al.* (2016), p. 1022; Bunker *et al.* (2017), p. 305; Caram (1965), pp. 146–65, figs 1–11, pl. VI; Coppejans (1983), pl. 91; Cormaci *et al.* (2012), pp. 443, 444, pl. 153(1–4); Dizerbo & Herpe (2007), p. 105, pl. 34(3); Fletcher (1987), pp. 292–5, figs 89, 90; Fletcher in Hardy & Guiry (2003), p. 303; Gallardo García *et al.* (2016), p. 39; Gayral (1966), p. 285, pl. 50; Guiry (2012), p. 159; Hamel (1931–39), pp. 276, 277, fig. 49(j); Hauck (1884), p. 383, fig. 165; Kylin (1947), p. 66, fig. 29; Lindauer *et al.* (1961), pp. 242, 243, fig. 59; Magne (1953), pp. 1596–8; Morton (1994), pp. 38, 39; Newton (1931), p. 137, fig. 82; Ribera *et al.* (1992), p. 116; Rosenvinge & Lund (1947), pp. 49, 50, fig. 18; Rueness (1977), p. 171, fig. 102, (2001), p. 17; Sauvageau (1931), pp. 122–6, fig. 23; Stegenga *et al.* (1997b), pp. 181, 182.

Stschapoviales H. Kawai

STSCHAPOVIALES H. Kawai (2017), p. 923.

Thalli erect, simple, terete, sometimes becoming flattened, linear throughout, or becoming thickened in the proximal regions, parenchymatous, with hair-like, multicellular, assimilatory filaments emerging in whorls from the surface cells; assimilatory filaments abundant on juvenile thalli, either remaining throughout the entire length or confined to the distal regions and absent on the proximal thickened portions; no obvious growth zones or apical cells; life history heteromorphic, with a macroscopic sporophyte with unilocular sporangia alternating with a microscopic, oogamous, monoecious gametophyte bearing oogonia and antheridia, or without an alternation of generations; reproduction also by monospore-like cells and sperm-like flagellated cells.

A recently proposed order (Kawai 2017) to accommodate three families; Halosiphonaceae, Platysiphonaceae and Stschapoviaceae, only one of which (Halosiphonaceae) is represented in the flora of Britain and Ireland.

Halosiphonaceae Christensen

HALOSIPHONACEAE Christensen (1962), p. 115.

Thalli annual, macroscopic, solitary or gregarious, medium to dark brown, to 1 metre in length, simple, cylindrical, terete and chord-like, firm but slippery, narrowing below to a small discoid holdfast, and characteristically clothed throughout by a fringe of pigmented, branched, assimilating filaments; thalli parenchymatous, at first solid, later becoming hollow, without supporting cross-connections, showing little differentiation of internal tissues with several layers of colourless, longitudinally elongated cortical cells, with oblique cross walls surrounding the central lacunae, which become progressively smaller towards the perimeter; outer perimeter cells more pigmented, containing discoid plastids without pyrenoids, and giving rise to stalked assimilating filaments, unilocular sporangia and paraphyses; growth diffuse; hairs absent.

Life history heteromorphic, with the macroscopic sporophyte alternating with a microscopic, filamentous, branched, gametophyte; unilocular sporangia borne in extensive sori covering the thallus surface, accompanied by unicellular paraphyses, releasing zoospores lacking a stigma; reproduction monoecious and oogamous.

Only one genus of the Halosiphonaceae, *Halosiphon*, occurs in Britain and Ireland.

Halosiphon Jaasund

HALOSIPHON Jaasund (1957), p. 211.

Type species: *H. tomentosa* (Lyngbye) Jaasund (1957), p. 212.

Thalli annual, solitary or gregarious, to 1 metre in length, simple, cylindrical and chord-like, and tapering below to a small discoid holdfast; thallus spongy and slippery in texture, dark brown to black, at first solid, later becoming hollow, buoyant, but not obviously gas-filled and lying flat on the substratum, and clothed, except near the base, by a dense fringe of dark brown to black, pigmented assimilating filaments; in cross section, thalli with no obvious differentiation of tissues, with a central lacuna which lacks supporting cross-connections, surrounded by 3–8 layers of large, colourless cells which are rounded or irregular in cross section, elongated in longitudinal section and often with oblique cell walls, which become progressively smaller and more radially elongated towards the periphery; peripheral cells photosynthetic, containing numerous discoid plastids without pyrenoids; assimilatory filaments branched, comprising barrel-shaped, persistent, stalk-like cells below, giving rise to a meristematic growth zone of short compacted cells from which the terminal more elongate, cylindrical, ephemeral cells arise, all cells containing numerous discoid plastids which lack pyrenoids; growth by a basally situated meristematic region and/or by an apical meristem.

Unilocular sporangia in extensive sori covering almost the entire surface of the thallus, accompanied and overgrown by cylindrical to clavate unicellular paraphyses; unilocular sporangia elliptical to subcylindrical; gametophytes microscopic, filamentous, possibly isomorphic, monoecious and oogamous; antheridia clustered, on branch tips, oogonia solitary, borne laterally on branches.

Distributed in cold temperate to polar regions in the North Atlantic.

The genus *Halosiphon* was originally described by Jaasund (1957) based on the species *Chorda tomentosa*. Up until that time, two species of *Chorda* were recognized for the North Atlantic, *C. filum* described by Stackhouse (1797), and *C. tomentosa* described by Lyngbye (1819). Although some authors (Areschoug 1847; Harvey 1846–51; Hauck 1885) later considered the latter species to be a spring form of the former, further work by Areschoug (1875) and Reinke (1891a) showed them to be separate species. The placement of both species in the Laminariales, usually by most authors within the family Chordaceae, rarely (Levring 1940) within the Laminariaceae, was based on culture studies by Kylin (1918, 1933) who revealed *C. filum* to have oogamous reproduction with microscopic, filamentous, dioecious gametophytes. The subsequent removal of *C. tomentosa* into the new genus *Halosiphon* by Jaasund (1957) was based on his erroneous interpretation of a streblonematoid epiphyte as the gametophyte. His new genus and species were consequently placed into the Dictyosiphonales. However, Sundene (1963) and Maier (1984a,b) confirmed that a *Laminaria*-type of life history was operating, although the gametophytes were monoecious which is rare in the Laminariales. Subsequently the genus *Halosiphon* was largely rejected (Kogame & Kawai 1996; Rueness 1977; South & Tittley 1986). Support for the re-recognition of Jaasund's genus *Halosiphon* was provided from molecular work by Peters (1998), with additional support from morphological and biochemical studies (see Peters 1998). The latter included differences in 1) growth mechanisms, 2) meiospore structure, 3) sexual pheromones, 4) types of reproduction (dioecious in *C. filum*, monoecious in *C. tomentosa*), and 5) the occurrence of assimilatory filaments in *C. tomentosa* rather than phaeophycean hairs as found in *C. filum*. The molecular work of Peters (1998) and Kawai & Sasaki (2000) showed that *H. tomentosum* is unlikely to be a member of the Chordaceae, so the Halosiphonaceae is retained. Peters (1998) noted that its true familial position will only be resolved by additional molecular work on other taxa, particularly those belonging to the Laminariales and the Sporochnales. A later study by Sasaki *et al.* (2001), based on both nuclear and chloroplast DNA sequence data, further placed the Halosiphonaceae within the

Tilopteridales where it was positioned until the recent work of Kawai *et al.* (2017) suggested its more appropriate placement in the Stschapoviales.

One species in Britain and Ireland.

Halosiphon tomentosus (Lyngbye) Jaasund (1957), p. 212. Figs 297, 298

Chorda tomentosa Lyngbye (1819), p. 74; *Scytosiphon filum* var. *tomentosum* (Lyngbye) C. Agardh (1820), p. 162; *Chorda filum* var. *tomentosa* (Lyngbye) Areschoug (1847), p. 365; *Chorda abbreviata* Areschoug (1875), p. 15; *Chorda tomentosa* f. *subfulva* Foslie (1890), p. 87; *Chorda tomentosa* var. *subfulva* (Foslie) Lily Newton (1931), p. 200; *Chorda filum* f. *abbreviata* (Areschoug) A.D. Zinova (1953), p. 144; *Halosiphon altae* Jaasund (1957), p. 213.

Thalli macroscopic, erect, medium to dark brown, single or more commonly, gregarious, to 1 m in length, to 1(–3) mm wide, with one to several thalli arising from a small attachment disc of densely packed and coherent rhizoidal filaments; thalli simple, terete, cylindrical, linear, firm, solid when young becoming hollow throughout later, slippery to the touch and, except in young thalli, and on the 2–3 cm long, stipe-like region at the base of older thalli, characteristically densely and prominently clothed in pigmented, brown, tapering, unbranched assimilatory filaments up to 5(–7) mm in length with a distinct basal meristem of closely packed subquadrate/lenticular cells; assimilatory filament cells quadrate to mainly rectangular in shape throughout, narrowing considerably towards the apex, 1–9 diameters long, 32–95 × 8–30 μm, each with numerous discoid plastids without pyrenoids; in surface view, thallus cells regularly arranged in longitudinal rows, less obviously in transverse rows, roughly quadrate in shape, or somewhat angular, 14–40 × 15–30 μm, each with several small discoid plastids without pyrenoids, and giving rise directly and quite densely, to the erect assimilatory filaments on more mature thalli; in cross section, parenchymatous and hollow throughout except for the extreme base of young thalli, with the central cavity lined by a 1–2-celled layer of small, thick-walled cells (medullary cells) surrounded by a well-developed cortex of longitudinally, and somewhat radially elongated, large, colourless, thick-walled cells; in young thalli, the cortical cells enclosed by an outer single layer of meristoderm-like photosynthetic cells while on mature thalli the cortical cells are enclosed by an outer 3–6-layered tissue of somewhat radially arranged, smaller, pigmented, photosynthetic, thick-walled cells, the outer cell layer giving rise to the reproductive sorus.

Unilocular sporangia borne in extensive sori, to 120 μm in height, covering the surface of mature thalli, except in the basal stipe region; sporangia closely packed in vertical columns arising as extensions of the surface cells, 80–100(–115) × 10–15(–22) μm, elongate cylindrical, accompanied by paraphyses of similar height; paraphyses unicellular, 90–118 × 9–12 μm, elongate cylindrical to elongate clavate with a somewhat bulbous, pigmented apex containing a few plastids.

Epilithic, shallow sublittoral, to 6 m, on sand covered stones and shell fragments.

Rare and mainly confined to northern shores of Britain (Yorkshire, Berwick, Fife, Caithness, Shetlands, Orkney, Hebrides, Argyll, Cumbrae, Isle of Man), rarer elsewhere (Jersey, South Cornwall, South Devon); in Ireland, reported for Down, Antrim, Derry and Clare.

Annual, March to September.

Halosiphon tomentosum has been investigated in culture by Rosenvinge & Lund (1947), Sundene (1963) and Maier (1984b), the two latter studies revealing the occurrence of a heteromorphic, *Laminaria*-type life history with monoecious gametophytes.

Bartsch & Kuhlenkamp (2000), p. 166; Batters (1892b), p. 21; Boo *et al.* (1999), pp. 109–14; Brodie *et al.* (2016), p. 1022; Bunker (2017), p. 255; Dizerbo & Herpe (2007), p. 119; Fletcher in Hardy & Guiry (2003), p. 256; Guiry (2012), p. 159; Hamel (1931–39), pp. 313, 314; Harvey (1846–51), pl. 107; Jaasund (1957), pp. 212–16, figs 1–5; (1965), pp. 100–4, figs 33, 34; Kawai & Sasaki (2000), pp. 416–28, (2004), pp. 1156–69; Kornmann & Sahling (1977), p. 150, pl. 8(A–E); Kylin (1947), p. 80, fig. 47; Levring (1940), p. 59; Lund (1947), p. 80, (1959), pp. 156, 157; Maier

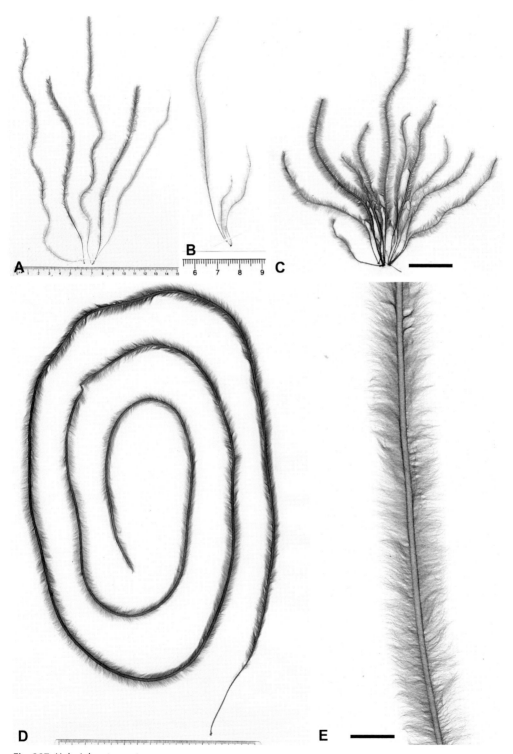

Fig. 297. *Halosiphon tomentosus*

A–D. Habit of variously sized thalli. E. Close-up of mature thallus showing fringe of assimilatory filaments. Bar in C = 16 mm, bar in E = 6 mm. (Herbarium specimens, NHM.)

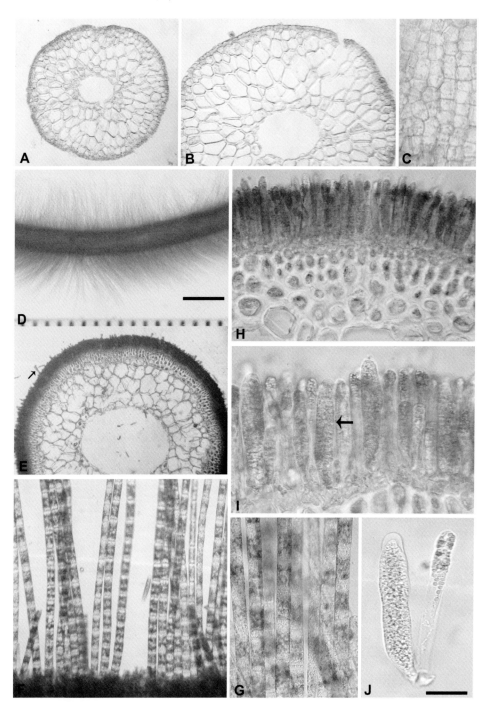

Fig. 298. *Halosiphon tomentosus*

A, B. TS of basal (stipe) region of young thallus showing central cavity, surrounding cortex of large, colourless cells and peripheral photosynthetic meristoderm. C. Surface view of basal (stipe) region showing cells in longitudinal rows, each with numerous discoid plastids. D. Close-up of mature thallus showing fringe of assimilatory filaments. E. TS of mature thallus showing central cavity, surrounding cortex of large, thin-walled, colourless cells, an enclosing 3–6-celled, darkly pigmented, radially arranged tissue, and a peripheral darkly coloured, reproductive sorus. Note remains of assimilatory filaments (arrow) lost during the sectioning process. F, G. Assimilatory filaments. H, I. TS of reproductive sorus showing unilocular sporangia (arrow) accompanied by paraphyses. J. Unilocular sporangium with accompanying paraphysis. Bar in J =440 µm (E), = 220 µm (A, F), = 140 µm (B), = 70 µm (C, H, G), = 35 µm (I, J).

(1984), pp. 95–106; Maier & Müller (1984), p. 71; Mathieson & Dawes (2017), p. 241, pl. LX(15, 17); Morton (1994), p. 43; Newton (1931), p. 200; Novaczek *et al.* (1986a), pp. 2414–20; Pankow (1971), p. 207, pl. 29; Pedersen (1976), p. 50, (2011) p. 109, fig. 118; Peters (1998), pp. 65–71; Reinke (1891a), pp. 35–7, 41–3, pl. 29(1–6); Rosenvinge & Lund (1947), pp. 73–7, figs 24, 25; Rueness (1977), pp. 173, 174, pl. XXIV(1), (2001), p. 14; Sasaki & Kawai (2007), pp. 10–21; Saunders & McDevit (2013), pp. 1–23; Sundene (1953), pp. 171, 172, (1963), pp. 159–67, text figs 1–3, pls I–III; Taylor (1957), pp. 175, 176.

Tilopteridales Bessey

TILOPTERIDALES Bessey (1907), p. 290 – see Silva & de Reviers (2000), p. 53.

Cutleriales Bessey (1907), p. 15(289).

Thalli erect, tufted or blade-like, branched, monosiphonous at first, becoming partly or fully polysiphonous and parenchymatous; growth trichothallic; cells with numerous discoid plastids without obvious pyrenoids; life history isomorphic or heteromorphic, sometimes reduced or incomplete; heteromorphic life history oogamous or anisogamous, involving a macroscopic, diploid sporophyte, alternating with a microscopic, filamentous, dioecious or monoecious gametophyte; sporophytic thalli with either large, globose, quadrinucleate, single or paired monosporangia (meiosporangia) producing monospores, or with densely packed unilocular sporangia containing large numbers of motile zoospores.

Originally created by Thuret in 1855, and based on *Tilopteris mertensii* and its unique formation of large monospores singly in globose monosporangia, the order Tilopteridales, which later included *Haplospora globosa* (Kjellman 1872), has been generally accepted (Brodie *et al.* 2016; Fritsch 1945; Guiry 2012; Hardy & Guiry 2003; Newton 1931; Papenfuss 1951; Rueness 1977; Taylor 1957), although some authors placed the two taxa within families in a broader-based Ectocarpales – for example in the Tilopteridaceae (Kornmann & Sahling 1977; Parke & Dixon 1976; Russell & Fletcher 1975) or in the Acinetosporaceae (Hamel 1931–39; Rosenvinge & Lund 1941), the latter placement probably based on Sauvageau's (1899) report of monospores in *Acinetospora pusilla*.

Molecular studies have since confirmed that the Tilopteridales *sensu stricto* form a distinct monophyletic group and are clearly separated from the larger Ectocarpales (Draisma 2002; Siemer *et al.* 1998). Later, molecular evidence, based on both nuclear and chloroplast DNA sequence data (Draisma *et al.* 2003; Kawai & Sasaki 2000; Sasaki *et al.* 2001) grouped the families Phyllariaceae and Halosiphonaceae with the Tilopteridaceae and Sasaki *et al.* (2001) proposed that all three families should be placed together in an emended Tilopteridales. Based on conventional taxonomic criteria, such as life histories and morphological/structural features, this did seem a surprising union as members of the Tilopteridales *sensu stricto* had not been considered to have a close phylogenetic relationship to these kelps. However, as Sasaki *et al.* (2001) pointed out, there is evidence that considerable and rapid changes in life history patterns and sexual reproductive structures had already been reported in some brown algal taxa. They also suggested, based on geographical distribution patterns, that it is likely that the Tilopteridales and Phyllariaceae share a common ancestor with the Halosiphonaceae and that they evolved in the Atlantic. In addition, Silberfeld *et al.* (2014) based on molecular arguments by Phillips *et al.* (2008a) and Silberfeld *et al.* (2010) proposed that the Cutleriaceae should also be placed within an emended Tilopteridales and that the Cutleriales should be placed as a synonym.

A more recent molecular study by Kawai *et al.* (2017), while supporting a Tilopteridales comprising the Cutleriaceae, Phyllariaceae and Tilopteridaceae, found that *Halosiphon* clustered with *Platysiphon* and *Stschapovia* in a clade that was paraphyletic with the Tilopteridales and that the three genera shared a number of morphological and structural features. Based on these results, they proposed a new order Stschapoviales and a new family Platysiphonaceae to accompany the Halosiphonaceae and the Stschapoviaceae. This is accepted in the present treatise.

In view of the above discussion, the Halosiphonaceae is removed from the Tilopteridales and placed into the new order Stschapoviales, leaving the following three families present in the Tilopteridales in Britain and Ireland: Cutleriaceae, Phyllariaceae and the Tilopteridaceae.

Cutleriaceae J.W. Griffith & Henfrey

CUTLERIACEAE J.W. Griffith & Henfrey (1856), p. 179.

Thalli erect or procumbent; erect fronds flattened, fan-shaped, deeply divided or dichotomously branched and tapering, all branches terminated by a fringe of hair-like, monosiphonous, unbranched, photosynthetic filaments; growth trichothallic, structure parenchymatous, medulla of large colourless thick-walled cells, cortex of small, photosynthetic thin-walled cells; plastids several, discoid without obvious pyrenoids; hairs single or grouped, arising from surface cells, with basal meristem; anisogametes on separate thalli, grouped in small, scattered, punctate sori on blade surfaces, terminal or lateral on short, branched or unbranched multicellular filaments; unilocular sporangia unknown on blade surface.

Procumbent thalli macroscopic, orbicular, irregular or fan-shaped, usually deeply divided, lobate, with or without overlapping lobes, loosely attached underneath by rhizoids, with an outer margin of large, apical cells (apical growth) or a fringe of two superimposed hair-like, monosiphonous unbranched, pigmented filaments (trichothallic growth); structure parenchymatous with inner medulla of large, colourless cells, lower cortex of small, thick-walled cells, with or without rhizoidal extensions, upper cortex of 1–2 thin-walled, photosynthetic cells, or 5–6 cells arranged in vertical, branched, files; upper cortex cells with several discoid plastids without pyrenoids; gametangia (oogonia, antheridia) similar in structure to those on erect thalli, forming mixed sori on thallus surface; unilocular sporangia densely crowded in raised, spreading, mucilaginous sori on separate thalli.

Two genera, *Cutleria* and *Zanardinia* occur in Britain and Ireland. They differ in their life histories; *Cutleria* has a heteromorphic alternation of phases involving a '*Cutleria*' gametophyte and an '*Aglaozonia*' sporophyte while *Zanardinia* has an isomorphic alternation of phases. The *Aglaozonia* phases in the life histories of species of *Cutleria* were previously described as separate entities. For example, the '*Aglaozonia*' phase of one representative of the genus *Cutleria*, *C. multifida* (Smith) Greville, was described as *Aglaozonia parvula* (Greville) Zanardini. Other reported life history associations include *Cutleria monoica* Ollivier and *Aglaozonia chilosa* Falkenberg in the Mediterranean and *Cutleria adspersa* (Mertens ex Roth) De Notaris and *Aglaozonia melanoidea* (Schousboe) Sauvageau in both the Mediterranean and North Atlantic.

Cutleria and *Zanardinia* have several features in common in their structure and reproduction. These include trichothallic growth, a parenchymatous mode of construction with an inner medulla and outer cortex, cells with several discoid plastids, an anisogamous mode of reproduction, and an 'alternation of generations' type of life history. The British and Irish representatives of these two genera, *Cutleria multifida* (Smith) Greville and *Zanardinia typus* (Nardo) Nardo are easily distinguishable on morphological features, the former producing fairly large, erect thalli, the latter producing fairly small, procumbent thalli.

Cutleria Greville

CUTLERIA Greville (1830), p. 59.

Type species: *C. multifida* (Turner) Greville (1830), p. 60.

Sexual phase (= *Cutleria*) forming erect blades, usually single, rarely gregarious, arising from a discoid holdfast, bilaterally flattened, flaccid, fan-shaped and deeply divided, becoming dichotomously branched, linear, narrowly tapering, with blunt, dissected apices fringed by a tuft of hair-like, monosiphonous, unbranched, photosynthetic filaments; surface cells irregularly placed, less obviously in longitudinal rows except near apex, with several discoid plastids without pyrenoids; in section, thallus parenchymatous with an inner medulla of 4–6, large, colourless cells and an outer cortex of 1–2 smaller, photosynthetic cells; hairs single or grouped, arising from surface cells, with basal meristem; anisogametes on separate thalli, grouped in punctate sori scattered on both sides of blade surface, terminal on short, branched, multicellular filaments, uniseriate to triseriate.

Asexual phase (= *Aglaozonia*) forming procumbent thalli, orbicular or irregularly shaped with overlapping lobes, and entire, rounded margins, smooth and membranous becoming subcoriaceous, loosely attached underneath by rhizoids; surface cells in longitudinal rows with several discoid plastids without pyrenoids; in cross section, thallus parenchymatous, arising from an outer marginal row of large apical cells, comprising an inner medulla of 2–6 large colourless cells and an upper and lower cortex of 1–3 smaller, pigmented cells; hairs single or tufted, arising superficially or from depressions, in upper cortex; unilocular sporangia in extensive, densely crowded, mucilaginous sori on thallus surface, without paraphyses.

One species in Britain and Ireland.

Cutleria multifida (Turner) Greville (1830), p. 60.   Figs 299, 300, 301

Ulva multifida Turner (1801), p. 310; *Dictyota penicillata* Lamouroux (1809), p. 13; *Zonaria multifida* C. Agardh (1820–28), p. 135; *Zonaria parvula* Greville (1828), pl. 360; *Padina reptans* P. Crouan & H. Crouan (1833), p. 398; *Cutleria laciniata* (J.V. Lamouroux) Meneghini (1842), p. 200; *Cutleria dichotoma* Kützing (1843), p. 338, pl. 25 II: fig. 2; *Padina parvula* (Greville) Zanardini (1843), p. 38; *Padinella parvula* (Greville) Areschoug (1843), p. 259; *Aglaozonia parvula* (Greville) Zanardini (1843), p. 38; *Aglaozonia reptans* (P. Crouan & H. Crouan) Kützing (1849), p. 566; *Zonaria reptans* (P. Crouan & H. Crouan) P. Crouan & H. Crouan (1852), No. 74 (exsiccatum); *Cutleria multifida* var. *dichotoma* (Kützing) Frauenfeld (1854), p. 330; *Cutleria multifida* f. *angustifrons* Holmes & Batters (1890), p. 84, nom. inval.; *Cutleria multifida* var. *confervoides* Kuckuck (1894), p. 252, fig. 20.

Sexual (*Cutleria*) phase

Thalli consisting of erect, flattened, solitary blades, light to yellow-brown, to 40 cm long, arising from a conical holdfast of fibrous, compacted, colourless, rhizoidal filaments; blades initially fan-shaped, deeply dissected, astipitate, membranous and somewhat striated at base, later extending into dichotomously branched, narrowly tapering, flaccid, linear blades, 1–7 mm (–12) mm wide; apices blunt, wedge-shaped, finely dissected and fringed terminally by a tuft of hair-like filaments; filaments simple, monosiphonous, 1–2.5 mm long, comprising cells predominantly shorter than wide 0.6–1 diameters long, 8–26 × 13–22 µm, each with numerous discoid plastids; surface cells of thallus irregularly placed, less obviously in longitudinal rows, 10–20 × 7–13 µm, each with numerous disc-shaped plastids without pyrenoids; in section, blades parenchymatous comprising an inner medulla of 4–6 large, colourless, thin-walled cells enclosed by an outer cortex of 1–2 smaller pigmented cells; hairs usually grouped, arising from surface cells, with basal meristem.

Anisogametes occurring on separate thalli, forming small, punctate sori sometimes with accompanying hairs, scattered over both sides of the blades; oogonia in small clumps, terminal

Fig. 299. *Cutleria multifida*
A. Habit of thallus (sexual *Cutleria* phase). B. Habit of thallus (asexual *Aglaozonia* phase). Bar = 15 mm.

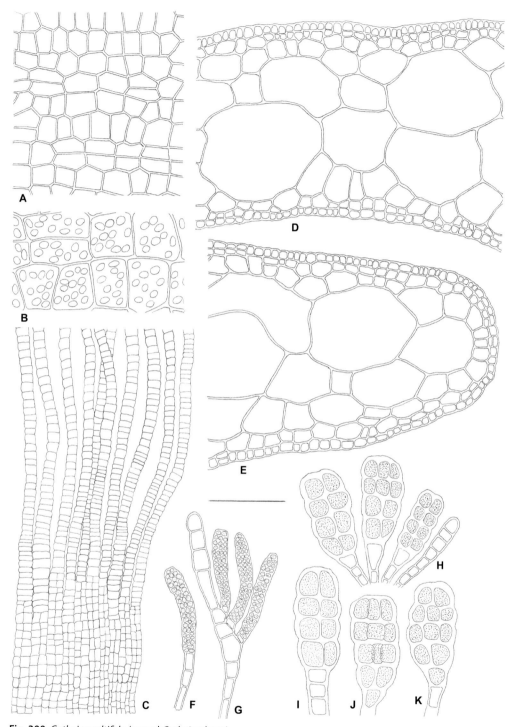

Fig. 300. *Cutleria multifida* (sexual *Cutleria* phase)

A, B. Surface view of thallus showing cells with numerous discoid plastids. C. Branch apex showing terminal fringe of hair-like filaments. D. TS of thallus. E. TS of thallus margin. F, G. Antheridia. H–K. Oogonia. Bar = 50 μm (A, F–K), = 20 μm (B), = 100 μm (C–E).

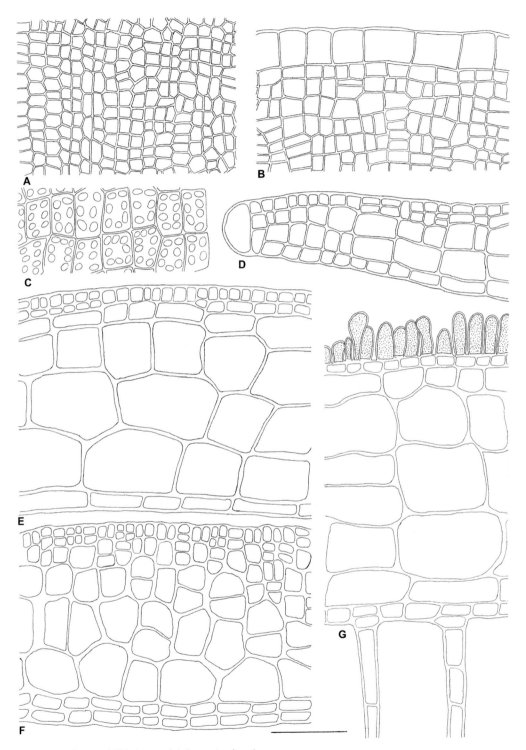

Fig. 301. *Cutleria multifida* (asexual *Aglaozonia* phase)
A. Surface view of thallus cells. B. Surface view of thallus margin showing large apical cells. C. Surface view of thallus showing cells with numerous discoid plastids. D. VS of thallus margin showing large apical cell. E, F. VS of vegetative thalli. G. VS of fertile thallus showing unilocular sporangia arising from upper surface and rhizoids produced from lower surface. Bar = 20 μm (C), = 50 μm (others).

on short, multicellular (2–4 cells) stalks, 4–5 loculi long, ovoid, usually biseriate, occasionally triseriate, 45–65 × 22–33 µm, comprising loculi 10–16 × 10–13 µm; antheridia more densely clumped, sessile, cylindrical, slightly curved and secundly arranged on short, branched, multicellular filaments, biseriate, more commonly multiseriate, 45–65 × 8–13 µm, to 16(–19) loculi long, comprising loculi 2–3 × 2–3 µm; unilocular sporangia borne on separate asexual 'Aglaozonia' thalli.

Epilithic, rarely epiphytic or epizoic; sublittoral to 10 m.

Widely distributed around Britain and Ireland, but more common on the south-west coasts and becoming rare further north and on the south-east and east coasts.

Probably annual, young thalli appearing in winter and spring and reaching maximum development in the summer; gametangia recorded June–October with oogonial thalli more frequently recorded than antheridial thalli.

Asexual (*Aglaozonia*) phase

Thalli horizontally expanded and flattened, light olive-brown to black, orbicular and discrete becoming irregular and spreading to several centimetres, with overlapping lobes, loosely attached underneath by rhizoids; surface smooth and membranous, becoming subcoriaceous, margin entire and rounded; surface cells in longitudinal rows, quadrate or rectangular, 8–16 × 5–12 µm containing densely packed discoid plastids without pyrenoids; in section, thallus parenchymatous, to 195 µm thick arising from a marginal row of large meristematic apical cells, comprising an inner medulla of 2–6 large, thick-walled, elongated, colourless cells with few plastids enclosed by an upper and lower cortex of smaller, pigmented cells; upper cortex of 2–3 cells enclosed within surface cuticle, cells subquadrate or quadrate, 8–11 high × 7–13 µm wide, lower cortex of 1–2 cells with thick, darkly pigmented walls, cells quadrate, subquadrate or usually much wider than high 14–60 µm wide × 10–16 µm high, commonly producing thick-walled, multicellular branched rhizoids, 9–23 µm wide; hairs infrequent, usually in tufts, superficial or in slight depressions in the upper cortex, with basal meristem.

Unilocular sporangia infrequent, confluent, densely packed, in rounded or irregularly shaped, spreading, slightly raised, mucilaginous sori on thallus surface, 25–37 × 11–15 µm, elongate-pyriform or cylindrical, usually enclosing 8 zoospores; gametangia borne on separate sexual 'Cutleria' thalli.

Epilithic, epiphytic (especially on haptera/stipes of *Laminaria*), less commonly epizoic, lower eulittoral pools, or more commonly sublittoral to 12 m and probably deeper.

Common and widely distributed all round Britain and Ireland.

Perennial; unilocular sporangia mainly recorded in winter and spring.

Culture studies on European isolates of *C. multifida* by various authors such as Church (1898), P. Crouan & H. Crouan (1855), Derbès & Solier (1856), Falkenberg (1879), Hartman (1950) Kuckuck (1899b, 1929), Reinke (1878b), Schlosser (1935) and Yamanouchi (1909, 1912) have all revealed the occurrence of a heteromorphic life history, with an erect bladed gametophyte (*Cutleria* phase) alternating with a prostrate sporophyte (*Aglaozonia* phase). The sexual thalli are dioecious and release anisogametes, which fuse with the aid of a sex attractant (Müller 1974; Jaenicke 1977). The resultant zygote develops into the prostrate *Aglaozonia* thallus bearing unilocular sporangia in which meiosis occurs. The released unispores then germinate directly to reform the erect bladed *Cutleria* thalli. Around Britain and Ireland the gametophyte is mainly a fairly short lived, summer annual while the sporophyte is perennial and reproduces during the winter and spring months. However, evidence from both field and laboratory studies indicates that the life history can abbreviate from such a rigid alternation of sexual phases. For instance, field studies around Britain and Ireland revealed a disproportionate number of female, compared to male, gametophytes and considerable differences in the relative abundance and geographical distribution of the *Cutleria* and *Aglaozonia* phases. The culture studies also showed that parthenogenetic development of the female gametes can occur which results in the formation of *Cutleria* and/or (haploid) *Aglaozonia* thalli. A similar diversity of development has also been reported for unispores released from the *Aglaozonia* thalli. The morphological expression appears, therefore, to be independent of ploidy level. Further, the diversity in the products of germination (e.g. filamentous forms, intermediate growth forms) described by the authors might reflect the variable culture conditions employed. Such a flexible, heteromorphic life history, markedly influenced by environmental conditions, would be analogous to that revealed for members of the Scytosiphonaceae by Nakamura & Tatewaki (1975).

Ardré (1970), p. 264; Bartsch & Kuhlenkamp (2000), p. 165; Brodie *et al.* (2016), p. 1022; Bunker *et al.* (2017), pp. 251, 282; Cabioc'h *et al.* (1992), p. 164, fig. 175; Church (1898), pp. 75–109; Cormaci *et al.* (2012), pp. 450, 452–4, pl. 155(4–10); P. Crouan & H. Crouan (1855), pp. pp. 648–50; Derbès & Solier (1856), pp. 60, 61, Dizerbo & Herpe (2007), pp. 91, 94, pl. 30(3); Falkenberg (1879), pp. 420–47; Feldmann (1937), pp. 305–8; Fletcher (1987), pp. 262–6, figs 73(A, B), 75; Fletcher in Hardy & Guiry (2003), p. 236; Fritsch (1945), pp. 157–71, figs 51–5; Funk (1955), pp. 46, 47; Gallardo García *et al.* (2016), p. 39; Gayral (1966), p. 251, pl. XXXVI; Guiry (2012), p. 158; Hartman (1950), pp. 120–8; Hamel (1931–39), pp. 321–3, figs 54(A–E, G), 55(F); Jaenicke (1977), pp. 69–75; Kuckuck (1899b), pp. 95–117, text figs 1–12, pl. 8(12–19), (1929), pp. 15–18, figs 5–7; Kylin (1947), p. 33, pl. 3, fig. 8; Lindauer *et al.* (1961), pp. 179–82, fig. 26, pl. 3; Lund (1950), pp. 65–74, figs 13, 14; Morton (1994), p. 36; Müller (1974), pp. 212–15; Newton (1931), pp. 197–9, fig. 25; Peña & Bárbara (2010), p. 54, fig. 9; Reinke (1878b), pp. 59–96; Ribera *et al.* (1992), p. 118; Rueness (1977), p. 170, pl. XXIII(la, b), (2001), p. 13; Sauvageau (1899), pp. 265–362, figs 9, 25, 26; Schlosser (1935), pp. 198–208; Stegenga *et al.* (1997a), p. 19; Turner (1801), p. 310, pl. 1(B); Womersley (1987), pp. 260–3, figs 94(A), 95; Wynne (2003), pp. 453, 454; Yamanouchi (1909), pp. 380–6, (1912), pp. 441–502.

Zanardinia Nardo ex P. Crouan & H. Crouan

ZANARDINIA Nardo ex P. Crouan & H. Crouan (1857), p. 24.

Type species: *Z. collaris* (C.A. Agardh) P. Crouan & H. Crouan (1857), p. 24 (= *Z. prototypus* Zanardini (1841), p. 236, nom. illeg.).

Thalli procumbent, smooth and membranous, becoming coriaceous, orbicular, fan-shaped or sometimes irregularly shaped, lobate, radially ridged, with entire or ruffled margins, with or without a fringe of hair-like filaments, loosely attached underneath by a felt of fibrous branched rhizoids; surface cells longitudinally arranged peripherally, more irregularly arranged centrally, quadrate or rectangular, with several discoid plastids without pyrenoids; in structure, thallus parenchymatous, arising from the trichothallic meristematic activity of a marginal row of two superimposed, monosiphonous, simple, freely extending filaments, giving rise to cells both forward and inward, the latter undergoing further divisions to produce an inner medulla of large, thick-walled, colourless, elongated cells and an upper and lower cortex of smaller, pigmented cells; upper cortex photosynthetic, at first of 1–2 cells, later 5–6 cells, in vertical, branched files, lower cortex of 1–2 cells, later becoming thick walled and dark coloured, giving rise to multicellular rhizoidal filaments; hairs single or grouped, arising superficially or from depressions in cortex.

Monoecious, life history isomorphic; gametangia (oogonia, antheridia) and unilocular sporangia in sori, arising directly from upper cortical cells; gametangia mixed in sori, elongate-cylindrical, terminal, on simple, or little branched, erect filaments; unilocular sporangia on separate thalli, crowded, elongate-pyriform enclosing 4–6 zoospores.

One species in Britain and Ireland.

Zanardinia typus (Nardo) P.C. Silva in Greuter (2000), p. 2.   Figs 302, 303

Stifftia typus Nardo (1835), p. 13; *Stifftia nardi* Zanardini (1840), p. 226; *Zanardini prototypus* Zanardini (1841), p. 236, nom. illeg.; *Stifftia prototypus* Nardo (1841), p. 184, nom. illeg.

Thalli consisting of horizontally expanded, flattened blades, spherical or fan-like, sometimes irregular in shape, to 20 cm across, olive to dark brown-black, loosely attached on the under-surface by rhizoids; surface smooth, membranous or coriaceous in texture, radially ridged comprising cells at first in longitudinal rows, later becoming irregularly arranged, quadrate or rectangular, 8–17 × 5–11 μm, each containing numerous discoid plastids without pyrenoids; margin entire or more usually ruffled with a distinct outer hair-like fringe comprising two superimposed, monosiphonous, multicellular, simple filaments, freely extending 3–4 mm, comprising cells 1–2(–2.5) diameters long, 13–36 × 14–21 μm, each with numerous discoid plastids without pyrenoids; in structure, thallus parenchymatous, 130–400 μm thick, comprising an inner medulla of 5–7(–10) large, colourless, thick-walled cells enclosed within an upper and lower cortex of smaller, pigmented, dark-coloured cells; upper cortex enclosed within a thick cuticle, at first comprising 1–2 pigmented cells, later of 5–6 cells arranged in vertical files of dichotomously branched, attenuating filaments, lower cortex of 1–2 cells, with thick, darkly

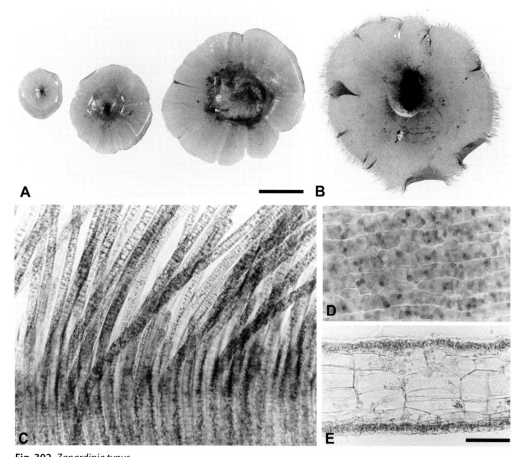

Fig. 302. *Zanardinia typus*
A. Habit of young thalli. B. Habit of mature thalli. Note outer hair-like fringe. C. Surface view of thallus margin showing hair-like fringe. D. Surface view of thallus showing cells with numerous discoid plastids. E. VS of vegetative thalli showing inner medulla of large, colourless cells enclosed within an upper and lower cortex of smaller, pigmented cells. Bar in A = 8 mm (A), = 9 mm (B), bar in E = 110 μm (C), = 43 μm (D, E).

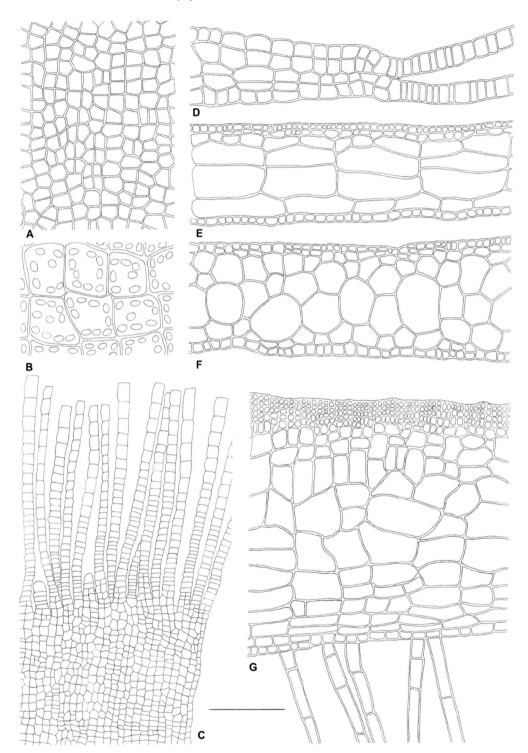

Fig. 303. *Zanardinia typus*
A, B Surface view of thallus showing cells with numerous plastids. C. Surface view of thallus margin showing outer fringe of hair-like filaments. D. VS of thallus margin showing two projecting filaments. E–G. VS of vegetative thalli, showing increased anatomical development. Bar = 50 μm (A, D), = 20 μm (B), = 100 μm (C, E–G).

staining walls, often producing free, multicellular, branched, downward-extending, thick-walled rhizoids; hairs not observed in Britain and Ireland, reported elsewhere to be single or grouped, arising superficially or from slight depressions in the upper cortex; growth trichothallic by the meristematic activity of cells at the base of the fringing filaments, giving rise to inner cells which undergo further division to produce the differentiated thallus.

Reproductive organs not observed in Britain and Ireland, reported elsewhere (see Yamanouchi 1909, 1913) to be developed in discrete, ill-defined sori on upper surface of thallus; gametangia occurring in the same sori, multiseriate, elongate-cylindrical, terminal on short, simple or little branched, multicellular filaments; oogonia reported to have 3–9 tiers of 2–4 loculi, antheridia more than 30 tiers high, each of 8 loculi; unilocular sporangia occurring on separate thalli, crowded, without accompanying paraphyses, elongate-pyriform, enclosing 4–6 zoospores.

Epilithic, rarely epiphytic, in the sublittoral to 20 m, especially on silty boulders and bedrock at the lower limit of *Laminaria hyperborea*.

Reported from southern and western shores of Britain and Ireland; Channel Islands, eastwards to Hampshire, northwards to Pembroke; recorded in Ireland from Cork.

Probably perennial, although data on seasonal behaviour too inadequate for comment.

Prior to the first report of its occurrence on the south coast of England (Jephson *et al.* 1975) the previous northern limit of *Z. prototypus* was the Channel Islands. It now appears to be widespread in south-west Britain, and in some localities quite abundant (Hiscock & Maggs 1982, 1984).

The only reported culture studies on *Z. prototypus* have been by Yamanouchi (1913) who showed the occurrence of a sexual, isomorphic lifecycle.

Ardré (1970), pp. 263, 264, pl. 37(2–30); Brodie *et al.* (2016), p. 1022; Bunker *et al.* (2007), p. 291; Cabioc'h *et al.* (1992), p. 165, fig. 176; Cormaci *et al.* (2012), pp. 454–6, pl. 156(1–7); P. Crouan & H. Crouan (1857), pp. 24–7; Dizerbo & Herpe (2007), p. 94, pl. 30(4); Feldmann (1937), p. 305; Fletcher (1987), pp. 267, 269–71, figs 76, 77; Fletcher in Hardy & Guiry (2017), p. 314; Gallardo García *et al.* (2016), p. 39; Greuter (2000), p. 2; Guiry (2012), p. 158; Hamel (1931–39), pp. 319–21, fig. 53; Hiscock & Maggs (1982), pp. 414–16, figs 1–2, (1984), pp. 84, 85; Jephson *et al.* (1975), pp. 253–5, figs 1–5; Kuckuck (1929), pp. 18, 19; Newton (1931), p. 199, fig. 128; Ribera *et al.* (1992), p. 118; Sauvageau (1899), pp. 286–96, figs 1–4; Yamanouchi (1909), pp. 280–386, (1913), pp. 1–35.

Phyllariaceae Hamel ex Petrov

PHYLLARIACEAE Hamel ex Petrov (1974), p. 153.

Thalli macroscopic, annual and clearly differentiated into a lamina, stipe and holdfast; lamina coriaceous, dorsiventrally flattened, lacking a midrib, entire or terminally split into several digits, with or without cuneate base; stipe long or short, terete or flattened, with or without a basal twist and lateral undulations; initial attachment by a small, primary, discoid holdfast or small finger-like haptera, which are later overgrown by a secondary, hollow, bulbous (*Saccorhiza polyschides*) or umbrella-like expansion (*S. dermatodea*), 5–30 cm in diameter, bearing whorls of branched or unbranched haptera; lamina with cryptostomata; in structure, comprising a central medulla, a surrounding cortex and a peripheral meristoderm layer; medullary elements distinctive and comprising thick-walled, longitudinally arranged, conducting cells (solenocysts), with lateral connecting cells (allelocysts), rather than sieve tubes as found in the kelps; cortex lacking mucilage ducts; growth by an intercalary meristem situated at junction of stipe and lamina.

Unilocular sporangia borne in continuous to discontinuous, densely packed, mucilaginous sori at the base of the lamina, additionally on the stipe frills and bulbous base in *S. polyschides*; sporangia interspersed with unicellular paraphyses which lack a hyaline cap; zoospores released from the unilocular sporangia with a stigma; life history heteromorphic with a diploid, macroscopic sporophyte alternating with a microscopic, branched, filamentous, dioecious and dimorphic (*S. polyschides*) or monecious (*S. dermatodea*) gametophyte; sexual pheromone unknown.

A family of one genus, *Saccorhiza*, containing two species, *S. dermatodea* and S. *polyschides*, the latter extending from Norway to the coasts of North Africa and the Mediterranean, the former largely confined to the Arctic Ocean. The type species of the genus is *S. polyschides* (as *S. bullosa* (Hudson) de la Pylaie), and in view of the genetic, morphological and reproductive differences between these two species, a generic assignment of *S. dermatodea* is required, as suggested by Sasaki *et al.* (2001). In view of the close molecular affinity of the Phyllariaceae to the Tilopteridales (Kawai & Sasaki 2001), *S. polyschides* is herewith removed from the Laminariales to the Tilopteridales.

Saccorhiza Bachelot de la Pylaie

SACCORHIZA Bachelot de la Pylaie (1830), p. 23.

Type species: *S. polyschides* (Lightfoot) Batters (1902), p. 48.

Sporophytic thalli macroscopic, annual, and clearly differentiated into a basal, contorted, bulbous, wart-like structure (rhizogen) attached by numerous downward-projecting finger-like haptera, a centrally emergent, at first spirally twisted, flattened, strap-like, leathery stipe with a ruffled and wavy margin, and a single, large, dorsiventrally flattened, at first entire, later digitate, coriaceous and smooth blade with a cuneate base. The rhizogen originates as a small swelling in the transition zone of young blades and then enlarges to cover the original disc/haptera; lamina with abundant cryptostomata; in section, lamina comprises a central, filamentous medulla of specialized, longitudinally arranged, conducting cells (soleno-cysts) and laterally connecting cells (allelocysts), a parenchymatous cortex lacking mucilage ducts, and an outer meristoderm; growth by an intercalary meristem situated at junction of stipe and lamina.

Unilocular sporangia, accompanied by unicellular paraphyses, formed in irregular, slightly raised, dark brown patches on laminar surface, more rarely on the rhizogen and upper stipe region; sporangia interspersed with unicellular paraphyses lacking a hyaline cap; life history heteromorphic with a diploid, macroscopic sporophyte alternating with a microscopic, branched, filamentous, dioecious and dimorphic, or monoecious gametophyte.

One species in Britain and Ireland.

Saccorhiza polyschides (Lightfoot) Batters (1902), p. 48.   Figs 304, 305

Fucus polyschides Lightfoot (1777), pp. 936–8; *Fucus bulbosus* Hudson (1778), p. 579, nom. illeg.; *Ulva bulbosa* (Hudson) de Candolle in Lamarck & de Candolle (1805), p. 16, nom. illeg.; *Laminaria bulbosa* Lamouroux (1813), p. 42, nom. illeg.; *Gigantea bulbosa* Stackhouse (1816), p. xi, nom. illeg.; *Phasgonon bulbosum* (Hudson) S.F. Gray (1821), p. 385; *Laminaria blossevillei* Bory De Saint Vincent in Chaubard & Bory de Saint-Vincent (1838), p. 74; *Haligenia bulbosa* (Hudson) Decaisne (1842), p. 346, nom. illeg.; *Phycocastanum bulbosum* (Hudson) Kützing (1843), p. 346; *Saccorhiza bulbosa* J. Agardh (1848), p. 138, nom. illeg.; *Alaria pylaiei* var. *grandifolia* (J. Agardh) Jónsson (1904), p. 20.

Sporophytic thalli macroscopic, to 3(–4) m in length, forming erect, solitary, pale olive to dark brown, leathery thalli, clearly differentiated into a terminal lamina, supporting stipe and large bulbous base attached underneath by unbranched haptera; lamina dorsiventrally flattened, solid, at first papery, later leathery in texture, smooth and often slimy to the touch, up to 1.5(–2) m in length, cuneate-cordate at base, at first entire and lanceolate, later vertically splitting into large numbers (to 30) of terminal strap-like digits, commonly bearing

Fig. 304. *Saccorhiza polyschides*
A. Habit of thallus showing bulbous base, emerging flattened stipe with basal frills and terminal segmented lamina. B–D. Stages in the development of the bulbous base. Note flattened, twisted stipe and segmented terminal lamina. Bar in D = 1.8 cm (B), = 58 mm (D).

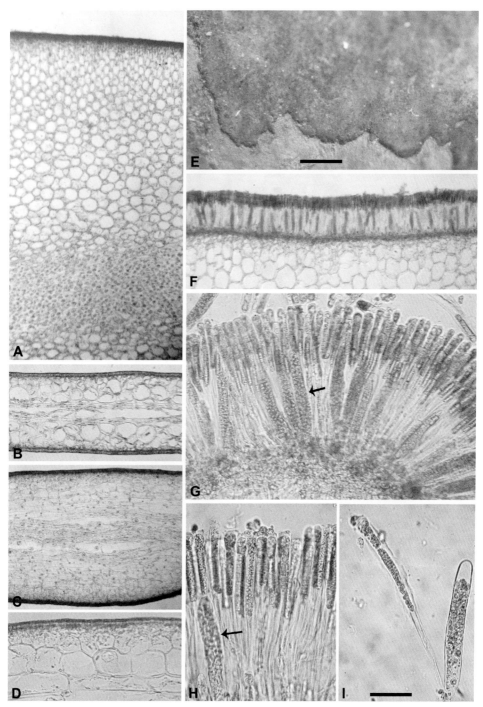

Fig. 305. *Saccorhiza polyschides*

A. TS of stipe showing central medulla of closely packed filaments, surrounding cortex of large, colourless, thin-walled cells and peripheral, photosynthetic meristoderm layer. B, C. TS of lamina of different thicknesses showing central medulla, outer cortex and peripheral meristoderm layer. D. TS of lamina, edge region, showing peripheral meristoderm enclosing a large-celled cortex. E. Surface view of lamina showing slightly raised mucilaginous reproductive sorus. F, G. TS of sorus showing unilocular sporangia (arrow) with taller associated paraphyses. H. Close-up of sorus showing unilocular sporangium (arrow) and taller, terminally pigmented paraphyses. I. Unilocular sporangia with associated paraphysis. Bar in E = 7 mm, bar in I = 220 μm (A, C), = 140 μm (B, F), = 70 μm (D, G), = 35 μm (H, I).

cryptostomata on the surface, particularly on young blades; stipe flattened and strap-like, leathery, solid, quite flexible at first, becoming stiff and inflexible on older thalli, to 210 cm in length, 8 cm in width, 2–3 mm in thickness, usually twisted towards the base and when fertile characteristically frilled with an outer fringe of undulating lateral extensions; both lamina and stipe lacking a midrib; basal, attachment region initially a small discoid swelling which gives rise to short, unbranched, finger-like protuberances (haptera) as in *Laminaria*, these later being overgrown by an enlarging discoid, circum-peripheral swelling formed higher up the stipe, which gives rise to a characteristic large, spreading, bulbous, warty, rounded to oblong, somewhat misshapen hollow structure (rhizogen), to 30 cm in diameter, covered with successive whorls of short, stubby, stiff, unbranched haptera, the first-formed basal ones extending downwards to attach to the substratum; in cross section, both stipe and blade similar and solid, with a central medulla, an enclosing cortex and a peripheral meristoderm; stipe elongate-oval in cross section, with 1) a central medullary band of fairly closely packed, vertically elongated, septate, branched filaments of small, thick-walled cells, rounded to oval in cross section, with some traversing, branched, hyphae-like filaments, all bound together in a gelatinous matrix 2) a quite extensive, surrounding cortex of closely packed, irregularly arranged, thin-walled, colourless cells, rounded to angular in cross section becoming smaller towards the periphery and 3) an outer meristoderm layer of cuboidal/ palisade-like, photosynthetic cells each enclosing large numbers of discoid plastids without pyrenoids; outer cortex without rings of mucilage canals; cross sections of the lamina edge reveal a central medulla of predominantly horizontally orientated, septate, filaments of cells, an enclosing cortex of several, large, thin-walled, colourless cells, rounded to rectangular in young lamina, more horizontally elongated and rectangular in older lamina, and an enclosing meristoderm layer of small, cuboidal cells with large numbers of discoid plastids without pyrenoids.

Unilocular sporangia borne in extensive, dark brown, slightly raised, mucilaginous, confluent sori at the base of the lamina, to a height of 130(–160) μm, less frequently on the stipe frills and bulbous base; sporangia closely packed, in vertical columns, arising as extensions of the outer meristoderm layer, 100–115(–125) × 16–18(–20) μm, elongate-oval to elongate-cylindrical in shape and enclosed under an umbrella of paraphyses; paraphyses simple, unicellular, markedly elongate-clavate with a long narrow base, 90–150(–165) long × 8–11 μm wide at the apex, darkly pigmented with plastids in the upper regions and with a marked gelatinous thickening of the lateral walls, and smaller terminal hyaline thickening of the apical cell.

Life history heteromorphic with a diploid, macroscopic sporophyte alternating with a microscopic, branched, filamentous (occasionally unicellular on female), dioecious and dimorphic gametophyte.

Epilithic, on rock, more rarely attached to small stones and shells in moderately to extremely sheltered conditions, usually associated with turbulent water habitats or areas of strong current action, sublittoral, rarely exposed during extreme low water of spring tides, to 10–15(–35) m exceptionally depending on the clarity of the water, extending from just above to below the *Laminaria hyperborea* belt; intolerant of reduced salinity.

Common and generally widely distributed around Britain and Ireland, particularly on the west coasts, more rarely distributed on the east coasts.

Sporophytic thalli annual, appearing in spring/early summer, with maximum blade growth occurring in June/July, becoming fertile by late summer/early autumn, after which decay of the blade and stipe occurs, leaving the still growing, more persistent bulbous base which usually overwinters, becoming detached in March/April; juvenile thalli sometimes observed all year round. A small proportion of late developers overwintering without decay.

Ardré (1970), pp. 282–5; Bachelot de la Pylaie (1830); Barber (1889), pp. 41–64, pls V–VI; Batters (1902), p. 48; Black (1948), pp. 165–8; Borja & Gorostiaga (1990), pp. 1–8; Brodie *et al.* (2016), p. 1022; Bunker *et al.* (2017),

p. 260; Cabioc'h *et al.* (1992), p. 80, fig. 63; Cormaci *et al.* (2012), p. 460, pl. 158(2); De Mesquita Rodrigues (1963), pp. 85, 86, pl. 8(C–F); Dizerbo & Herpe (2007), pp. 122–3, pl. 40(4); Feldmann (1934), p. 30, figs 12, 13; Fernández (1982), pp. 391–403, (2011), pp. 352–60; Fletcher in Hardy & Guiry (2003), p. 292; Gallardo García *et al.* (2016), p. 39; Gayral (1958), p. 242, pl. XXXVIII, (1966), pp. 300, 301, pl. LVIII; Guiry (2012), p. 160; Hamel (1931–39), pp. 304–7, figs 50(II–VI), 52; Harvey (1846–51), pl. 241; Henry & South (1987), pp. 9–16; Jensen *et al.* (1985), pp. 375–81; Morton (1984), p. 43; Newton (1931), p. 206, figs 128(E, F), 129(A–F); Niell (1984), pp. 71–102; Norton (1969), pp. 1025–45, figs 1–9, (1970a), pp. 1.1–9.3, figs 1–4, (1971), pp. 215–31, (1973), pp. 91–5, (1977a), pp. 625–35; (1978), pp. 527–36; Norton & Burrows (1969a), pp. 19–53, figs 1–16, (1969b), pp. 287–96, figs 1–4; Ribera *et al.* (1992), p. 120; Rueness (1977), p. 177, fig. 103, (2001), p. 16; Sauvageau (1915), pp. 445–8, (1916), pp. 396–8, (1918a), pp. 21–125, figs 3–41; Stegenga & Mol (1983), pp. 108, 110, 111, pl. 41(1–2); Stegenga *et al.* (1997a), p. 23; Svendsen (1962), pp. 11–13; Tilden (1937), pp. 261, 262, fig. 125.

Tilopteridaceae Kjellman

TILOPTERIDACEAE Kjellman (1890), p. 86.

Thalli erect, filamentous, much branched, at first monosiphonous, becoming parenchymatous, particularly in basal regions of main axes; branching regularly opposite and bipinnate, irregular, or unilateral; ultimate branchlets sometimes of limited growth and terminated by a hair; growth trichothallic; plastids discoid without obvious pyrenoids; asexual reproduction by monospores produced in large monosporangia, sexual reproduction by elongated, tubular antheridia and globose oogonia.

Until recently, the Tilopteridaceae comprised three, rare, monotypic genera, *Phaeosiphoniella*, *Haplospora* and *Tilopteris*, all restricted to the colder waters of the North Atlantic. Only *Haplospora* and *Tilopteris* have been reported for Britain and Ireland, with *Phaeosiphoniella* endemic to the south coast of Newfoundland. Following molecular evidence revealing *Phaeosiphoniella cryophila* to be sister to the Laminariales (Phillips *et al.* 2008a), but given the 'highly divergent' morpho-anatomical features between *P. cryophila* and the Laminariales and the genetic distance between them, Silberfeld *et al.* (2014) removed *P. cryophila* to a new family (Phaeosiphoniellaceae) and new order (Phaeosiphoniellales).

The main distinguishing characteristics of the Tilopteridaceae include the branching filamentous thalli being distinctly parenchymatous in parts, especially in the basal regions, trichothallic growth, oogamous sexual reproduction by large monosporangial-like oogonia and elongated, plurilocular antheridia, and asexual reproduction by large, non-motile spores formed in monosporangia, which can be uninucleate or quadrinucleate. A full life history involving a sexual process with fertilization and meiosis has not been adequately demonstrated in any species and the life history is consequently considered to be reduced, with some evidence of geographically isolated life history phases occurring. For example, in *Haplospora*, life history studies of a Newfoundland population revealed an isomorphic alternation of generations with nuclear DNA levels revealed by fluometry to be haploid in the gametophytes and diploid in the sporophytes, although the processes of fertilization and meiosis were not actually observed, and both generations displayed equal chromosome numbers (Kuhlenkamp & Müller 1985). In Helgoland, in the North Sea, only the sporophyte phase has been described. In *Tilopteris*, only the gametophyte phase is recognized, bearing monosporangia, which release large monospores considered by Kuhlenkamp (1989) to represent oogonia, and plurilocular sporangia which are interpreted as antheridia. However, fertilization has not been observed, and pheromones are not secreted by the 'oogonia', unlike the situation in *Haplospora* (Kuhlenkamp & Müller 1985). The sporophyte phase seems to have been lost.

Similar globose monosporangia, which produce non-motile monospores, have also been reported for the genus *Acinetospora* Bornet and for this reason the genus has sometimes been included within the order Tilopteridales, family Tilopteridaceae (Fritsch 1945; Newton 1931), or placed in the Ectocarpales, family Acinetosporaceae, along with *Tilopteris* and *Haplospora*

(Feldmann 1937; Hamel 1931–39, pp. xix–xx). Most authors, however, placed *Acinetospora* in the Ectocarpaceae (Ardré 1970; Cardinal 1964; Kornmann 1953a; Kornmann & Sahling 1977; Parke & Dixon 1976; Schneider & Searles 1991; Stegenga & Mol 1983; Stegenga *et al.* 1997; Womersley 1987) sometimes acknowledging that its systematic position is uncertain (Parke & Dixon 1976; Pedersen & Kristiansen 2001; Rosenvinge & Lund 1941). Molecular work has since confirmed the monophyly of the Tilopteridales/Tilopteridaceae (Draisma 2002), with *Acinetospora* grouped with *Feldmannia*, *Hincksia*, *Pogotrichum* and *Pylaiella* in a larger family Acinetosporaceae, one of five families proposed for the Ectocarpales by Peters & Ramírez (2001). This has also been accepted by Silberfeld *et al.* (2014).

Haplospora Kjellman

HAPLOSPORA Kjellman (1872), p. 3.

Type species: *H. globosa* Kjellman (1872), p. 5.

Scaphospora Kjellman (1877a), p. 29; *Capsicarpella* Bory de Saint-Vincent (1823), p. 178.

Thalli erect, tufted, monosiphonous or polysiphonous in parts, arising from a fibrous rhizoidal mat; branching radial, opposite above, irregular below with trichothallic growth; cells with numerous discoid plastids without obvious pyrenoids.

Gametangia and meiosporangia (unilocular sporangia) on separate, morphologically similar individuals; oogonia intercalary, solitary and formed by direct transformation of vegetative cells or more usually paired from (one) or both daughter cells following longitudinal division, globose; antheridia as lateral outgrowths, sessile or stalked, elongate-cylindrical, and irregular in outline, tapering to a hair-like filament; meiosporangia sessile or terminal on short branchlets, single or paired, globose.

One species in Britain and Ireland.

Haplospora globosa Kjellman (1872), p. 5.   Fig. 306

Capsicarpella speciosa Kjellman (1872), p. 26; *Scaphospora arctica* Kjellman (1877a), p. 31; *Scaphospora speciosa* (Kjellman) Kjellman (1877a), p. 32.

Thalli forming erect, densely branched, slightly matted tufts, to 15(–32) cm in length, light, golden brown in colour, attached at the base by a fibrous mat of downward growing, filamentous rhizoids which form a compact discoid holdfast; main axes with widely spreading, indeterminately growing, lateral branches, mainly uniseriate above, pluriseriate in the lower regions of axes and main branches; branching to several orders, radial, opposite, less frequently irregular below, unilateral, often widely divergent and frequently recurved above; growth trichothallic with all branches and apices with a subterminal meristem and terminal, hyaline, multicellular hair; cells variable in dimensions, 0.5–4 diameters long, 31–235 μm high × 33–62 μm wide, light brown, almost colourless, in terminal region, dark brown below, each with large numbers of discoid plastids without obvious pyrenoids.

Reproductive organs borne on separate, almost identical thalli; meiosporangia on the sporophytic (*Haplospora*) phase, oogonia and plurilocular antheridia on the gametophytic (*Scaphospora*) phase (see explanation of life history below); meiosporangia common in terminal thallus regions, sessile or terminal on short, lateral, widely divergent branches, single, rarely in groups of two, globose, 75–85 × 65–75 μm, generally more densely pigmented than accompanying vegetative cells, with 1–4 densely stained regions (representing plastid accumulation around nuclei); oogonia not unlike meiosporangia in appearance, more pigmented than vegetative cells with a single, central, densely stained region (uninucleole), intercalary, formed

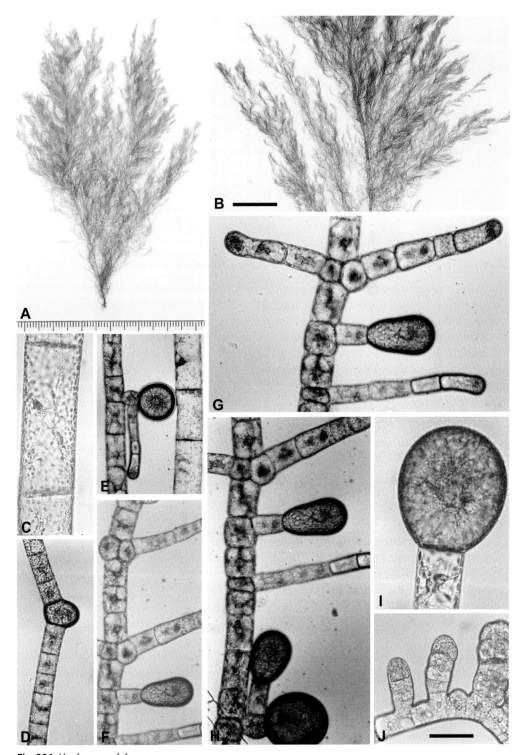

Fig. 306. *Haplospora globosa*

A, B. Habit of thalli. C. Basal filament cell showing enclosed central nucleus and numerous discoid plastids. D–H. Portions of erect thallus (sporophyte phase) showing meiosporangia. I. Terminal meiosporangum. J. Unusual branch formations. Bar in B = 12 mm, bar in J = 70 μm (D–H), = 35 μm (C, I, J). (A, B, herbarium specimens, NHM.)

by direct transformation of vegetative cells or, more usually, from one or both daughter cells following longitudinal division, globose, 50–86 × 52–70 µm; plurilocular sporangia usually closely associated with oogonia, arising as lateral, widely divergent outgrowths, sessile or shortly stalked, elongate-cylindrical, irregularly outlined, almost knot-like in appearance, hollow, colourless, tapering to a hair-like filament, 56–145 × 39–52 µm, comprising large numbers of irregularly shaped, more usually cuboidal loculi, 10–13 × 7–3 µm.

Epilithic, lower eulittoral and shallow sublittoral.

Rare and widely distributed all around Britain (Channel Islands, Cornwall, Bute, Wigtownshire, Cardiganshire) and Ireland (Clare, Galway, Mayo).

Annual, spring and early summer, with most records in March/April; erect thallus probably originating each year from prostrate microthalli (= pseudoperennial); see Kuhlenkamp & Hooper (1995, p. 237).

Life history studies undertaken by various authors such as Dammann (1930), Kornmann & Sahling (1977), Kuhlenkamp & Müller (1985), Nienburg (1923), Reinke (1889c) and Sundene (1966) revealed *Haplospora globosa* to possess an isomorphic alternation of generations. However, slight morphological differences have been noted between the two generations, with the gametophyte phase, previously recorded as a separate genus, *Scaphospora* Kjellman (Kjellman 1877a, p. 29), and species, *S. speciosa* Kjellman (1877a, p. 32), differing from the sporophyte phase in being slightly smaller, more delicate, more sparsely branched, having a greater recurvature of the upper branches and axes, and by carrying different reproductive organs – the gametophyte (*Scaphospora*-phase) bearing oogonia and antheridia, the sporophyte (*Haplospora*-phase) bearing meiosporangia. Note, however, the more detailed and complete study by Kuhlenkamp & Müller revealed the life history lacked sexuality and nuclear alteration, while Kornmann & Sahling (1977) revealed the occurrence of a reduced life history for Helgoland material with only sporophytes.

Bartsch & Kuhlenkamp (2000), p. 166; Batters (1894b), pp. 116, 117, (1902), p. 53; Brebner (1897), pp. 176–87, figs 1–8; Brodie *et al.* (2016), p. 18; Caram & Jónsson (1972); Dammann (1930), p. 19; Dizerbo & Herpe (2007), p. 94; Edelstein & McLachlan (1967), p. 207, figs 33, 36–9; Fletcher in Hardy & Guiry (2003), p. 257; Foslie (1894), pp. 139, 140; Fritsch (1945), pp. 148–57, figs 48(J), 49(E, I), 50(A, B); Guiry (2012), p. 160; Kjellman (1872), p. 5, pl. I(1), (1877a), pp. 29–35, figs 1–15; Kornmann & Sahling (1977), pp. 132, 133, fig. 70; Kuhlenkamp & Hooper (1995), pp. 229–39, figs 3, 4, 6–9, 29–31, 34; Kuhlenkamp & Müller (1985), pp. 301–12; Kuhlenkamp *et al.* (1993), pp. 377–80, figs 1–3; Kylin (1917), p. 299, (1947), pp. 33, 34, fig. 27(A–B); Lund (1959), pp. 88–94, figs 16, 17; Mathieson & Dawes (2017), p. 245, pl. LXI(1–2); Newton (1931), pp. 209, 210, fig. 131; Nienburg (1923), p. 211; Pankow (1971), pp. 165, 166, figs 192, 193; Parke & Dixon (1976), p. 561; Pedersen (2011), p. 80, fig. 86; Reinke (1889b), pp. 34, 35, (1889c), p. 108, pl. 11, figs 1–17; Rosenvinge & Lund (1941), pp. 67–72, figs 36–8; Rueness (1977), p. 178, fig. 104, (2001), p. 15; Saunders & McDevit (2013), pp. 1–23; Sauvageau (1899), pp. 107–27, (1928a), pp. 51–94; South (1975), pp. 698–700, figs 3–5; Sundene (1966), pp. 937, 938; Taylor (1957), pp. 125, 126.

Tilopteris Kützing

TILOPTERIS Kützing (1849), p. 462.

Type species: *T. mertensii* (Turner) Kützing (1849), p. 462.

Thalli erect, tufted and densely branched, uniseriate above, multiseriate below, arising from a fibrous, discoid holdfast; branching opposite, pinnate, with branches and branchlets fairly divergent, of determinate growth and often of unequal length; growth trichothalli; plastids discoid, without obvious pyrenoids; monosporangium-like oogonia and plurilocular antheridia on branchlets, terminated by a hair; oogonia globose, in series, antheridia elongated, cylindrical/elongate pyriform and hollow.

One species in Britain and Ireland.

Tilopteris mertensii (Turner) Kützing (1849), p. 462.   Figs 307, 308

Conferva mertensii Turner in Smith (1802), pl. 999; *Ectocarpus mertensii* (Turner) C. Agardh (1828), p. 47; *Trichopteris mertensii* (Turner) Kützing (1855), pl. 84, fig. 11.

Thalli forming erect, filamentous, densely branched, bushy tufts, up to 21 cm in height, light olive-brown becoming dark brown in colour, arising from a large, compact, basal, discoid holdfast; main axes dominant, of indeterminate growth, uniseriate in the upper regions, and partly/regularly multiseriate in the lower regions, clothed towards the base in downward-growing, filamentous rhizoids contributing to the fibrous, mat-like, discoid, attachment system; branches and branchlets fairly divergent, uniseriate, sometimes multiseriate in the basal regions, of determinate growth, the branchlets remaining very short; branching pinnate, somewhat flattened, distichous, opposite, to several orders, with branches often unequal in length, more rarely irregular or unilateral; growth trichothallic, all branches and apices with a distinct subterminal meristem giving rise apically to a tapered hyaline, multicellular hair; cells usually shorter than wide, 0.25–0.75 diameters long, measuring (20–)40–70(–110) × (60–)100–130(–150) μm in the main axes and branches, (8–)10–13(–18) × (30–)33–42(–50) μm in the terminal branchlets, each with numerous small discoid plastids without pyrenoids.

Monosporangium-like oogonia and plurilocular antheridia intercalary on terminal, monosiphonous branchlets, occurring separately or on same thallus; oogonia common, globose or barrel-shaped, occurring singly or in chains of 2–3(–4), (32–)52–80(–90) × (40–)45–75(–100) μm; antheridia rare, elongated, cylindrical, more rarely elongate pyriform, hollow, (80–)100–160(–210) × (38–)40–55(–66) μm.

Epilithic, lower eulittoral, or more commonly sublittoral to 21 m, particularly on mud and sand-covered rocks in sheltered to moderately exposed localities; elsewhere reported to be rarely epiphytic.

Rare and widely distributed all around Britain (Channel Islands, Jersey, Cornwall, North and South Devon, Dorset, Sussex, Norfolk, Durham, Yorkshire, Northumberland, Lothian, Fife, Aberdeenshire, Orkney, Argyllshire, Bute, Ayrshire, Isle of Man) and Ireland (Down, Antrim, Clare, Cork, Donegal, Dublin, Galway, Mayo).

Annual, spring and early summer, March to June, with most records for April and May. Erect thallus probably originating each year from prostrate microthalli (= pseudoperennial); see Parke (1932, p. 163) and Kuhlenkamp & Hooper (1995, p. 237).

Culture studies by Dammann (1930), Kuhlenkamp & Müller (1985); Moestrup *et al.* (1975) and South (1972) indicate that an asexual, 'direct' life history is operating with the 'eggs' released from the oogonia developing parthenogenetically to repeat the parental thallus. Released sperm seemed to lack any function, while an accessory direct mode of reproduction was provided by the germination of large 'spores' formed at the base of the antheridia; the reported fertilization of these spores (Dammann 1930) was not observed by Kuhlenkamp & Müller (1985).

Bartsch & Kuhlenkamp (2000), p. 169; Bornet (1891), pp. 367–71, pl. 8(6–10); Brebner (1897), pp. 176–87; Brodie *et al.* (2016), p. 18; Dammann (1930), p. 14, figs 6–9; Dizerbo & Herpe (2007), p. 94, pl. 31(1); Fletcher in Hardy & Guiry (2003), p. 312; Fritsch (1945), pp. 148–57, figs 48(A–E, K, N, O), 49(G, H, J–K), 50(C–D); Guiry (2012), p. 160; Hamel (1931–39), pp. 81–3, figs 23(5–7); Harvey (1846–51), pl. 132; Katsaros & Salla (1997), pp. 60–7, figs 1–24; Kawai *et al.* (20004), pp. 1156–69; Knight & Parke (1931), p. 75; Kornmann & Sahling (1977), pp. 130, 131, fig. 69; Kuckuck (1895), pp. 290–322; Kuhlenkamp (1989), pp. 555–62; Kuhlenkamp & Hooper (1995), pp. 229–39, figs 5, 10, 11; Kuhlenkamp & Müller (1985), pp. 301–12; Kylin (1947), p. 35, fig. 27(c); Mathieson & Dawes (2017), p. 246, pl. LXI(3–5); Morton (1984), p. 36; Newton (1931), p. 210, fig. 132; Parke (1932), p. 163; Parke & Dixon (1976), p. 561; Reinke (1889c), pp. 155–8, pl. III(21); Rosenvinge & Lund (1941), pp. 72–4; Rueness (1977), p. 179, fig. 105, (2001), p. 17; Sauvageau (1899), pp. 107–27, (1928), pp. 51–94, figs 1–4; South (1971), pp. 1027–33, (1972), pp. 83–9, figs 1–10, (1975), pp. 695–8, figs 1, 2, 5; South & Hill (1971), pp. 211–13, figs 1–10; Taylor (1957), p. 126.

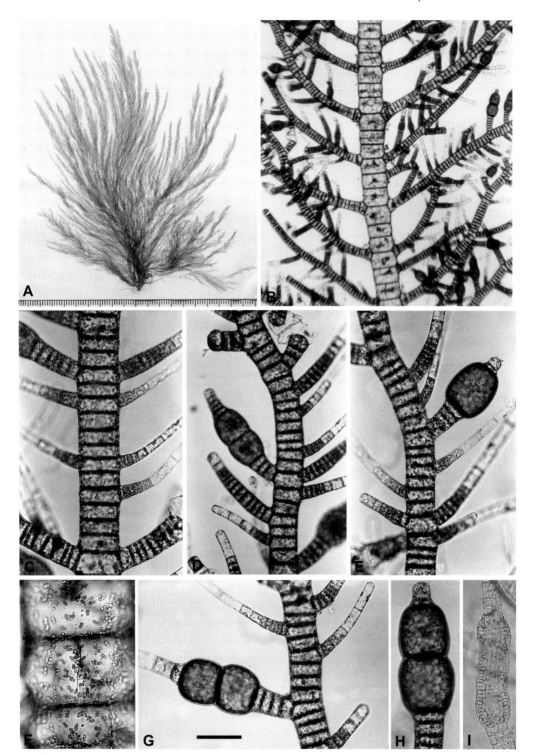

Fig. 307. *Tilopteris mertensii*
A. Habit of thallus. B–E, G, H. Portions of erect thalli showing monosporangium-like oogonia on branchlets. F. Basal filament cell showing enclosed, small, discoid plastids. I. Portion of branchlet showing intercalary antheridia. Bar in G = 140 μm (B), = 70 μm (C–E, G), = 35 μm (F, H, I). (A, herbarium specimen, NHM.)

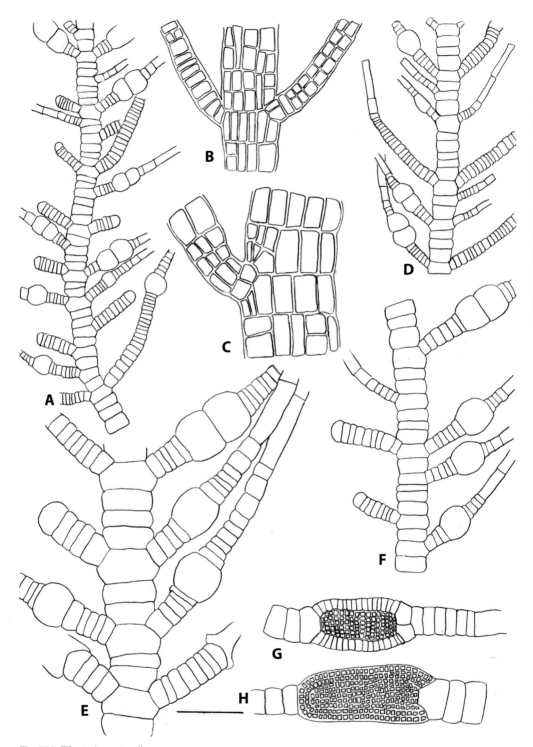

Fig. 308. *Tilopteris mertensii*
A–F. Portions of erect thallus, some showing intercalary monosporangium-like oogonia on branchlets. G, H. Portions of branchlets showing intercalary antheridia. Scale bar = 200 μm (A, E), = 125 μm (F), = 80 μm (B–D), = 50 μm (G, H).

Incertae sedis at ordinal rank

Sorapion Kuckuck

SORAPION Kuckuck (1894), p. 236.

Type species: *S. simulans* Kuckuck (1894), p. 236.

Thalli encrusting, thin, orbicular or indefinite in outline, closely adhering to substratum; thallus structure pseudoparenchymatous, consisting of a monostromatic basal layer, the cells of which give rise to simple or little-branched erect filaments, quite firmly united; vegetative cells with a single, plate-like, parietal plastid without obvious pyrenoid; hairs not observed.

Unilocular sporangia terminal on erect filaments, in sori, unaccompanied by paraphyses; plurilocular sporangia unknown.

One species in Britain.

Sorapion simulans Kuckuck (1894), p. 236.   Figs 309, 310

Lithoderma simulans (Kuckuck) Batters (1896), p. 385.

Thalli crustose, to 1–2 mm in diameter, light brown, thin, firmly attached to substratum by whole undersurface, rhizoids not observed; in central, thicker regions, surface cells polygonal, closely packed, 9–12 × 7–11 µm, each containing a single plate-like, parietal plastid, in peripheral regions more obviously one cell thick, consisting of outward-spreading, branched, firmly united filaments of cells which are mainly rectangular, 7–19 × 6–9 µm; thallus structure pseudoparenchymatous, in vertical section, the monostromatic base giving rise to straight, vertical, tightly joined filaments, to 52 µm high (6 cells), comprising cells quadrate or more often, especially at base, broader than high, 5–10 × 8–19 µm, each with a single, plate-like plastid in upper cell region, with pyrenoids uncertain; hairs not observed.

Unilocular sporangia terminal, in small groups, approximately rounded in surface view, 13–20 µm in diameter, slightly raised above vegetative cells, more distinctly pyriform in section, 20–27 × 15–18 µm; plurilocular sporangia unknown.

Epilithic on stones, in the sublittoral. Only recorded from South Devon in Britain. Insufficient data to comment on seasonal distribution.

British material of this species is represented by a single specimen collected by George Brebner in Plymouth Sound, Devon and reported by Batters (1896). Three slides are present in the Natural History Museum, London from which the above description and illustrations have been made.

The characteristic features of this species (erect filaments of cells with a single plate-like plastid and terminal, pyriform unilocular sporangia) are similar to those of an early stage in the development of another crustose alga, *Stragularia spongiocarpa* (Batters) Hamel, and it was speculated that the two might, therefore, be conspecific (Fletcher 1981b reported earlier in Parke & Dixon 1976, p. 563 note 20, as *Ralfsia spongiocarpa*). However, subsequent detailed examination of material of both species reveals this to be unlikely and they are thus described as separate entities in the present treatment.

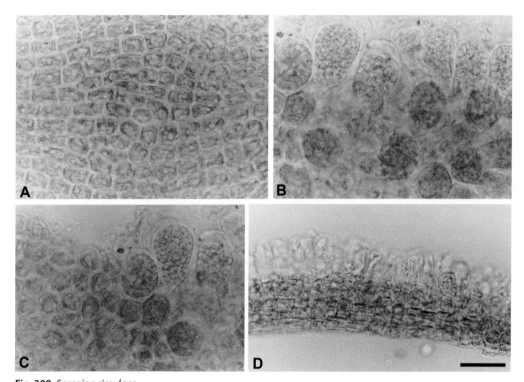

Fig. 309. *Sorapion simulans*
A. Surface view of vegetative thallus. B, C. Surface view of fertile thallus showing the unilocular sporangia. D. TS of fertile thallus showing unilocular sporangia. Bar in D = 35 μm (A–D). Photos taken from Batters's micro-slides at the NHM.

In the author's previous account of this species (Fletcher 1987), the plate-like plastids were described with a pyrenoid, which was not in agreement with Kuckuck's original description (Kuckuck 1894, p. 237; see also Mathieson & Dawes 1917). It is possible that the present author was mistaken on this point due to the poor quality of the slides, and pyrenoids may well be absent. Certainly, they are not prominent in this genus and are unlikely to be pedunculate but possibly exserted and similar to those described for *Petroderma* (Ishigeales), suggesting a possible close relationship between *Sorapion* and the latter genus. However, with its crust-like features and single plate-like plastid, *Sorapion* could also be suitably placed in the family Scytosiphonaceae, as positioned by Mathieson & Dawes (2017). This uncertainly of its position led Silberfeld *et al.* (2014) to place *Sorapion* under *Incertae sedis* – a position adopted in the present treatise.

Another species closely related to *S. simulans* that might also be present around Britain (and Ireland) is *Sorapion kjellmanii* (Wille) Rosenvinge (1898, p. 95). It has been widely reported in the North Atlantic, including from the United States (Taylor 1957; this was, however, probably *S. simulans*), Canada (Sears & Wilce 1973; South & Hooper 1980), Greenland (Lund 1959; Pedersen 1976; Rosenvinge 1898), Russia (Wille & Rosenvinge 1885), the Faeroes (Børgesen 1902), Sweden (Waern 1949) and Denmark (Kristiansen 1978). According to Rosenvinge (1898, pp. 96, 97), and as discussed by Mathieson & Dawes (2017, p. 203), *Sorapion kjellmanii* differs from *S. simulans* in having a plate-like plastid which possesses a pyrenoid, erect filaments that branch near the apices, hairs in depressions on the crust and unilocular sporangia that are poorly defined. However, agreement is expressed with Rosenvinge that these differences are not substantial and so the two species might be conspecific.

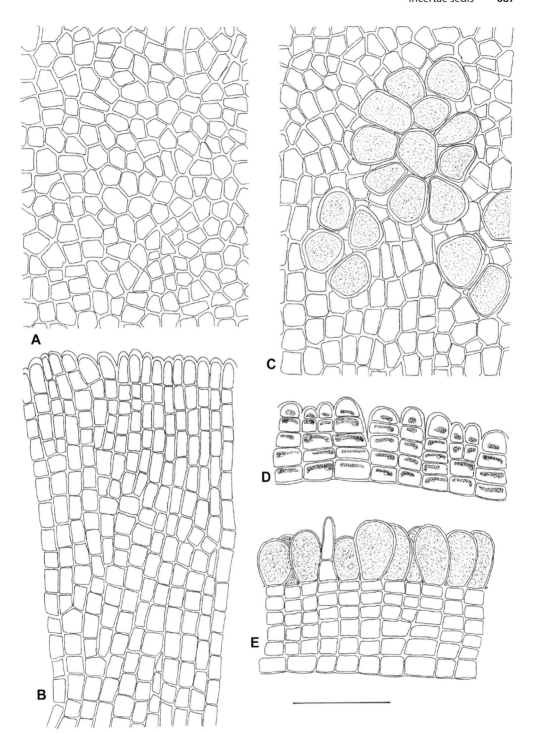

Fig. 310. *Sorapion simulans*
A. Surface view of vegetative crust. B. Surface view of crust margin showing outward-spreading, laterally united filaments with terminal apical cells. C. Surface view of fertile crust showing unilocular sporangia. D. VS of vegetative crust showing cells containing a single, large, plate-like plastid with occasional pyrenoid. E. VS of fertile crust showing large, terminal pyriform, unilocular sporangia. Note associated 2-celled filament. Bar = 50 µm. Drawings made from Batters's microslides at the NHM.

Pedersen (1981b) linked *Sorapion kjellmanii* as an alternate phase in the life history of the relatively little-known brown alga *Porterinema fluviatile* (H.C. Porter) Waern. Unispores from the unilocular sporangia in the *Sorapion* thalli germinated directly, without sexual fusion, to form new crustose *Porterinema*-like thalli with plurilocular sporangia. Plurispores from the latter also behaved asexually and germinated directly to produce thalli with plurilocular sporangia identical to the parents. The absence of cytological data, however, prevented a more precise understanding of the relationship between the two crustose phases.

Bartsch & Kuhlenkamp (2000), p. 168; Batters (1896), pp. 385, 386; Brodie *et al.* (2016), p. 1021; Dizerbo & Herpe (2007), p. 76; Fletcher (1987), pp. 88–90, fig. 4; Fletcher in Hardy & Guiry (2003), p. 295; Guiry (2012), p. 156; Hamel (1931–39), pp. 111, 112, fig. 26(F); Kuckuck (1894), pp. 236, 237, fig. 10; Lund (1959), pp. 79–81, fig. 14; Mathieson & Dawes (2017), p. 203, pl. LVII(16, 17); Newton (1931), p. 127; Pedersen (1981b), pp. 203–8, figs 1–9; Rosenvinge (1898), pp. 95–7; Waern (1949), pp. 662, 663, (1952), pp. 136–41.

Glossary

ABAXIAL: The side, usually the lowermost, of a lateral/ branch adjacent to the axis.

ACROBLASTIC: Origin of laterals/branch initials from all or part of small segments cut off from the apical cells of erect filaments (used in relation to branching).

ACROBLASTICALLY: Laterals arising in an acroblastic manner.

ACROHETEROBLASTY: Two different types of structures are formed from the small segments cut off from the apical cells of erect filaments (used in relation to branching).

ACROHOMOBLASTY: Only one type of structure is formed from the small segments cut off from the apical cells of erect filaments (used in relation to branching).

ACROPETALLY: Arising and maturing from the base to the apex, so that the youngest are near to the apex (used in relation to branching and reproductive structures).

ACUMINATE: Gradually tapering to a point.

ACUTE: Tapering sharply to a point.

ADAXIAL: The side, usually the uppermost, of a lateral/ branch adjacent to the axis.

ADHERENT: Closely attached.

ADJOINED: Joined together, weakly or tightly (used in relation to filaments of cells).

ADVENTITIOUS: Arising in an irregular manner or from an abnormal position (usually said of branching).

AEROCYSTS: A general term sometimes used for an air-bladder.

AIR-BLADDER: Air-filled vesicle serving for flotation; another commonly used term for a vesicle.

ALLELOCYSTS: Specialized cross-connecting cells associated with solenocysts.

ALTERNATE: Placed singly, with change of side.

ANASTOMOSING: A cross-connecting arrangement, usually of filaments, permitting a joining of adjacent points.

ANISOGAMETES: Similar gametes of unequal size.

ANISOGAMOUS: Characterized by anisogamy.

ANISOGAMY: The fusion of two similar motile gametes of unequal size.

ANNUAL: Living for one year only, during which the life history is completed (used in relation to the erect thallus body).

ANTHERIDIUM (pl. ANTHERIDIA): Male reproductive organ, producing male gametes.

ANTHEROZOID: A male gamete.

ANTICLINAL: At right angles to the surface.

APLANOSPORE: A non-motile spore, within a sporangium with its own wall.

APICAL CELL: Terminal initial cell of a filament or thallus.

APICAL MARGIN: Margin of thallus entirely comprising apical cells.

APOMEIOSIS: The formation of spores without meiosis.

APOMICTIC: A thallus produced by apomixis.

APOMIXIS: Asexual reproduction, without gametes or fertilization.

ARBORESCENT: Tree-like in form.

ARTICULATE: Jointed.

ASCOCYST: An abnormal enlarged, hyaline, or darkly staining cell, usually empty when old.

ASEXUAL: Not involving the process of sexual fusion.

ASSIMILATORY FILAMENTS: An erect, free, branched or unbranched, pigmented photosynthetic filament.

ASSIMILATORY HAIRS: Pigmented, photosynthetic hairs.

ASSURGENT: Curving upwards (used in relation to some horizontal filaments which turn upwards).

ASTIPITATE: Not having a stipe.

ATTENUATE: Slender, thin, tapering.

AUXOCAULIC: Growth of shoots/axes/branches derived from enlargement of secondary medullary segments and, therefore, with widths greater than the apical cells (e.g. in Sphacelariales).

AUXOCAULOUS: Characterized by auxocaulic growth.

AXIAL FILAMENT: The central filament of a thallus or its branch.

AXIS: The central portion of the thallus.

BENTHIC: Attached to or living on the sea bottom.

BIENNIAL: Living for two years only during which the life history is completed (used in relation to the erect thallus body).

BIFID: Forked, dichotomous, divided into two parts.

BIFLAGELLATE: Possessing two flagella (used in relation to gametes or spores).

BIFURCATE: Divided into approximately two equal parts.

BIPHASIC: With two ploidy levels (e.g. haploid and diploid).

BISERIATE: Consisting of two rows of cells in longitudinal series.

BLADE: The broad, flattened, leaf-like part of the thallus.

BRANCHLETS: A general term for small terminal branches.

BULBOUS: Bulb-like, swollen and rounded.

BULLATE: With a bulging or puckering of the surface.

CAECOSTOMA (pl. CAECOSTOMATA): A sterile, immersed, completely enclosed cavity lacking hairs, paraphyses and an ostiole, found in some Fucales and Laminariales.

CAESPITOSE: Growing in clusters or turfs.

CALCIFIED: Encrusted or impregnated with lime.

CARTILAGINOUS: Hard, tough and elastic.

CATENATE: In the form of a chain.

CHANNELLED: Incurved or inrolled, appearing u-shaped in section (usually said of thalli).

CIGAR-SHAPED: Elongate oval.

CLADE: A grouping of taxa which appear to have similar characteristics.

CLAVATE: Club-shaped (used in relation to a filament or sporangium, which becomes gradually wider towards the tip from a relatively narrow base).

COMPLANATE: Uniformly flattened in one plane.

COMPRESSED: Flattened from side to side.

CONCEPTACLE: A small, usually flask-shaped embedded cavity opening to the surface via an ostiole in which the reproductive organs develop.

CONFLUENT: Becoming united to form one.

CORIACEOUS: Leathery in texture.

CONICAL: With the shape of a cone.

CORDATE: Heart-shaped.

CORTEX: A term loosely used for the outer regions of a thallus, external to the medulla with or without an enclosing epidermal-like layer; usually photosynthetic.

CORTICATED/CORTICATING: Having a cortex or an enclosing layer of cells or filaments.

CORYMBOSE: Arranged in a flat-topped cluster (used in relation to branches arising at different points on an axis, but all reaching the same lengths).

CRAMPONS: Short, spine-like branches.

CRUST: A closely adherent, outwardly radiating thallus, pseudoparenchymatous in structure with laterally adjoined basal and erect filaments.

CRUSTOSE: Resembling a crust (used in relation to thalli).

CRYPTOSTOMA (pl. CRYPTOSTOMATA): A sterile, immersed conceptacle-like cavity opening to the surface and containing hairs only (common in the Fucales).

CUNEATE: Wedge-shaped.

CUTICLE: Thin sheet of extracellular material on outside of thallus.

CYTOPLASM: Living cell material.

DECIDUOUS: Not persistent, being shed/released (used in relation to blades/branches/hairs etc.).

DEHISCED: Shed/lost (used in relation to branches and hairs).

DETERMINATE: Having a limit to growth (used in relation to branching).

DICHOTOMOUS: Equally forked forming two equal dichotomies (used in relation to branching).

DICHOTOMY: Division by means of an apical cell so that forked branching occurs.

DIGITATE: Arising from a single level, like fingers on a hand, compound.

DIMORPHIC: Occurring in two morphological forms (see also = HETEROMORPHIC).

DIOECIOUS: With male and female gametangia on separate thalli.

DIPLOHAPLONTIC: A life history with separate multicellular diploid (sporophyte) and haploid (gametophyte) generations.

DIPLOID: Having double the haploid number of chromosomes.

DIPLONTIC: Thallus body consisting of a diploid phase only, with the gametophyte phase comprising only the gametes (said of life histories).

DISC: A thin, closely adherent, often circular thallus, comprising outwardly radiating, laterally adjoined, usually synchronously extending filaments, monostromatic, or at least partly distromatic in structure.

DISCOID: Like a disc, rounded and flat and closely adpressed to the surface – often with reference to a holdfast.

DISTICHOUS: Arranged in two opposite rows and lying in one plane; like a feather (usually said of branching).

DISTROMATIC: Composed of a layer two cells thick, at least in parts.

DIVARICATE: Spreading at a wide angle (used in relation to branching and sporangia).

DIVERGENT: Spreading at a moderate angle (used in relation to branching).

DORSAL: Pertaining to the upper surface of a dorsiventral thallus.

DORSIVENTRAL: Thalli with the dorsal and ventral surfaces unalike.

ECAD: An ecological variant or growth form of a species associated with a particular habitat which has no taxonomic status, e.g. salt marsh ecads of fucoids.

ECORTICATE: Without/lacking a cortex.

ECTOCARPOID: Having a structure which resembles that of *Ectocarpus* (used in relation to thalli which comprise erect, branched, uniseriate filaments, and plurilocular sporangia that are siliquose and multiseriate).

EGG: A female gamete, usually large, without flagella, non-motile and borne in an oogonium.

ELLIPSOID: Elliptical in outline or cross-section.

EMERSED: Not submerged in water (usually said of thalli on shores).

ENDOBIOTIC: Living within a plant/alga or animal.

ENDOGENOUS: Originating from below the surface of a thallus.

ENDOPHYTIC: Living within a plant/alga.

ENDOZOIC: Living within an animal.

ENTIRE: Without divisions, lobes, proliferations or indentations (used in relation to thallus margins).

EPHEMERAL: Lasting only for a short while.

EPIBIOTIC: Living on the surface of a plant/alga or animal.

EPIDERMAL: Resembling an epidermis, the outermost layer of cells of a thallus.

EPIDERMIS: General term used for the outermost layer of cells of a thallus; usage usually restricted to anatomically complex thalli (e.g. in Laminariales, Fucales).

EPILITHIC: Living on rock or stones.

EPIPHYTIC: Living on a plant/alga, but attached to the surface only.

EPI-ENDOPHYTIC: Living both within and on the surface of a plant/alga.

EPIZOIC: Living on an animal, but attached to the surface only.

EULITTORAL/LITTORAL: Applied to that portion of the shore which is alternatively exposed to the air and wetted either by the tide or by splash and spray; usually divided into upper, mid and lower regions.

EURYHALINE: Tolerant of a broad range of salinities.

EVESICULATE: Without air-bladders.

EYESPOT: A pigmented lens-like structure, functioning as a light receptor, found in some motile, reproductive cells.

EXOGENOUS: Originating from the surface of a thallus.

FALCATE: Sickle-shaped.

FASCICLE/FASCICULATE: Forming a dense cluster or bundle (used in relation to filaments, branches or sporangia).

FASTIGIATE: Erect, clustered and parallel (used in relation to branches and sporangia).

FERTILE: Bearing sporangia or gametangia.

FILAMENT: A branched or unbranched row of cells joined end to end.

FILAMENTOUS: Comprising a single row of cells.

FILIFORM: Thread-like having the form of a slender filament (used in relation to branches).

FIMBRIATE: With a fringed edge or margin.

FLABELLATE: Fan-shaped.

FLACCID: Limp.

FLAGELLUM: A thread-like outgrowth of the cytoplasm, used for locomotion.

FLEXUOSE: Full of twists and turns.

FOLIACEOUS: Leaf-like.

FORMA (pl. FORMAE): A subspecific taxonomic term to denote minor morphological differences within a species, which have no genetic basis.

FROND: That main part of the thallus, usually expanded and blade-like, other than the attachment structure and supporting stipe.

FUCOSAN: A substance that accumulates in cells, considered to be a by-product of metabolism which has tannin-like properties.

FUCOSAN VESICLES: Vesicular bodies in cells containing fucosan, a tannin-like compound.

FUCOXANTHIN: A yellow-brown polyphenolic xanthophyll pigment in the plastids of the Phaeophyceae.

FUSIFORM: Spindle-shaped, tapering at both ends.

GAMETANGIUM: A cell producing one or more gametes.

GAMETE: A sexual cell capable of uniting with another sexual cell to form a zygote.

GAMETOPHYTE: A morphological phase that bears gametangia.

GELATINOUS: Slimy and jelly-like.

GERMLING: A juvenile thallus.

GLOBOSE: Spherical or globular.

GREGARIOUS: Growing together.

HAIR: A general term for usually unbranched, hyaline deciduous filaments. Two types are distinguished in the Phaeophyceae: the common true hairs which are colourless with a distinct basal meristem and with or without a basal collar or sheath, and the less common pseudo-hairs or false hairs which possess plastids, terminate filament tips only and lack a distinct basal meristem and sheath.

HAPLOID: Having the number of chromosomes characteristic of the gamete.

HAPLONTIC: Thallus body consisting of a haploid phase only, with the diploid phase comprised only of a zygote cell (used in relation to life histories).

HAPLOSTICHOUS: Thallus constructed of free or loosely/moderately associated filaments, which lack intercalary longitudinal cell divisions.

HAPTERON (pl. HAPTERA): A specialized multicellular attachment structure, usually cylindrical and branched and present in the Laminariales.

HECATONEMOID: Said of a thallus which resembles *Hecatonema* (discoid, distromatic).

HERMAPHRODITE: Bearing both male and female sex organs on one individual; often used in relation to receptacles and conceptacles.

HETEROANTAGONISM: Antagonism between two different species.

HETEROBLASTY: The formation at two morphologically different germinating patterns from spores liberated from a single sporangium.

HETEROTHALLIC: With gametes produced from separate thalli.

HETEROMORPHIC: Having morphologically dissimilar forms (said of life histories).

HETEROTRICHOUS: With two parts to the thallus, viz. a prostrate basal portion and an erect one, both usually filamentous.

HOLDFAST: Basal attachment organ; usually multicellular, discoid or conical, comprising a mass of rhizoids.

HORNS: The short arms of propagules found in the Sphacelariales (see *Sphacelaria tribuloides*).

HYALINE: Colourless, transparent.

HYPACROBLASTIC: Origin of laterals/branch laterals from cells below the apical cell (said of branching).

HYPHA (pl. HYPHAE): Elongated, narrow filaments of cells, usually produced laterally from medullary cells in the central regions of some larger brown algae.

HYPOBLASTICALLY: Laterals arising in a hypoblastic manner.

IMMERSED: Embedded within a thallus or submerged in water.

INDETERMINATE: Having unrestricted growth (usually said of branching).

INDUSIUM: A membranous covering, usually of reproductive structures.

INFLATED: Distended with air or gases.

INITIAL: The earliest stage of a cell, tissue or structure.

INTERCALARY: Between the apex and base of a thallus (used in relation to growth zones, meristems and sporangia).

INTERCELLULAR: Between cells.

INTERTIDAL: Region of shore lying between high and low tide levels.

INTRACELLULAR: Within a cell.

INVOLUCRUM: An enclosing, arching-over structure.

IRIDESCENT: Showing rainbow-like colours.

ISODIAMETRIC: Cells of equal diameter; as broad as long.

ISOGAMETES: Similar motile gametes of equal size.

ISOGAMOUS: Characterized by isogamy.

ISOGAMY: The fusion of two similar motile gametes of equal size.

ISOMORPHIC: Having morphologically similar forms (said of life histories).

LAMINA (pl. LAMINAE): The flat, usually thin, blade-like region of a thallus.

LANCEOLATE: Narrow and tapering at both ends.

LENTICULAR: Lens-shaped.

LEPTOCAULIC: No enlargement of secondary medullary segments, with width of shoots/axes/

branches remaining equal to the apical cell (e.g. in Sphacelariales).

LEPTOCAULOUS: Characterized by leptocaulic growth.

LIGULATE: Long and narrow with parallel sides.

LINEAR: Narrow, longer than wide and parallel-sided.

LITHOPHYTE: A thallus growing on rock or stones.

LITHOPHYTIC: Living on rocks or stones or other non-biological substrata.

LITTORAL FRINGE: A term commonly applied to the upper eulittoral zone of shores.

LOCULE (pl. LOCULI): Small component cells or compartments within a structure (used in relation to plurilocular sporangia and gametangia).

LOMENTACEUS: Shaped like a pod or legume.

LOWER EULITTORAL: Applied to the lower portion of the eulittoral.

LUBRICOUS: Smooth and slippery.

LUMEN: A cavity inside a cell or thallus.

MACROSCOPIC: Visible to the unaided eye.

MACROTHALLUS (pl. MACROTHALLI): Applied to a thallus usually visible to the unaided eye.

MARGIN: The outer edge of a structure or thallus.

MASTIGONEME: Hair-like appendages on flagella.

MEDULLA: A term loosely used for the internal region of a thallus; it may be parenchymatous, pseudoparenchymatous or distinctly filamentous.

MEGA/MACROGAMETANGIUM (pl. MEGA/MACROGAMETANGIA): Plurilocular gametangium with many loculi of relatively large dimensions.

MEGAZOIDANGIA: Sporangia producing relatively large zoids.

MEGECADS: A collective term for a group of ecads.

MEIOSIS: The process of nuclear division in which the chromosome number is halved.

MEIOSPORANGIUM (pl. MEIOSPORANGIA): A sporangium in which meiosis occurs. Referred to as plurilocular sporangia with small loculi by some authors, e.g. Sauvageau.

MEIOTICALLY: By the process of meiosis.

MEISPORES: Spores produced by the process of meiosis.

MEMBRANOUS/MEMBRANACEOUS: Like a membrane, thin, skin-like and semi-transparent.

MERISTEM: A group of cells which divide and cause growth.

MERISTEMATIC: Functioning as a meristem.

MERISTODERM: An outermost layer of meristematic cells responsible for increase in girth.

MICROGAMETANGIA: Plurilocular gametangium, with many loculi of relatively small dimensions.

MICROSPORANGIUM (pl. MICROSPORANGIA): A plurilocular sporangium with many loculi of relatively small dimensions.

MICROSCOPIC: Not visible to the unaided eye.

MICROTHALLUS (pl. MICROTHALLI) Applied to a thallus not usually visible to the unaided eye.

MICROZOIDANGIA: Sporangia producing relatively small zoids.

MID-EULITTORAL: Applied to the middle portion of the eulittoral.

MIDLITTORAL: Applied to the middle portion of the littoral.

MIDRIB: The central thickened longitudinal axis of a flattened thallus.

MITOTICALLY: Produced by the process of mitosis.

MONILIFORM: Bead-like.

MONOECIOUS: With male and female gametangia on the same thallus.

MONOMORPHIC: Occurring in one morphological form (said of life histories).

MONOPHASIC: With a single ploidy level (e.g. haploid or diploid).

MONOPHYLETIC: Derived from a common ancestor.

MONOPODIAL: Growth of the main axis by an apical cell or meristem.

MONOPODIAL BRANCHING: A mode of development in which a distinct primary axis is maintained and from which secondary branches arise.

MONOSPORANGIUM: A sporangium in which a single spore (monospore) is formed.

MONOSPORE: A spore formed in a monosporangium.

MONOSTROMATIC: Composed of a single layer of cells.

MONOTYPIC: Having only one species in a genus.

MUCILAGE: A watery sticky substance.

MUCILAGE CANALS: Specialized conducting channels found in the outer cortex of some Laminariales.

MUCRONATE: Short, sharply pointed and spine-like.

MULTIAXIAL: Of a thallus containing several axial filaments.

MULTIPINNATE: Pinnate branches which are themselves pinnate.

MULTISERIATE: Consisting, wholly or in part, of several rows of cells in longitudinal series.

MYRIONEMOID: A thallus that resembles *Myrionema* (discoid, monostromatic).

NODOSE: Bearing knot-like swellings.

OBCONIC: An inverted cone, with the distal region broadest.

OBOVATE: Roughly egg-shaped, with the narrow end basal and the broader end distal.

OOGAMOUS: Characterized by oogamy.

OOGAMY: The fusion of two dissimilar gametes, usually a large non-motile egg and a small motile antherozoid.

OOGONIUM (pl. OOGONIA): Female reproductive organ in which one or more eggs are formed.

ORGANELLE: A specialized structure within a cell – e.g. plastid, nucleus.

OSTIOLE: A small pore-like opening, usually of a conceptacle.

OVAL/OVATE: Roughly egg-shaped, with a broader end at the basal region and a narrower, pointed distal region.

OVOID: Egg-shaped.

PAPILLA (pl. PAPILLAE): Small, rounded or nipple-like protuberances on the surface of a structure.

PAPILLATE/PAPILLOSE: Bearing papilla.

PARAPHYSIS (pl. PARAPHYSES): A sterile, hair-like filament or cell usually associated with reproductive organs, and occurring on the thallus surface, or within a conceptacle.

PARENCHYMA: A compact, solid, usually internal, tissue of thin-walled, isodiametric, undifferentiated cells formed by mitotic divisions in all planes and usually functioning in storage.

PARENCHYMATOUS: Composed of or containing parenchyma.

PARIETAL: Lying along the wall, peripheral to the cell.

PARTHENOGAMETES: Gametes developing without fertilisation.

PARTHENOGENETIC: Developing without fertilisation.

PEDICEL: A stalk or cell supporting a reproductive structure.

PEDICELLATE: Borne on a pedicel.

PERENNIAL/PERENNATE: Living for several years; usually said of the erect thallus body.

PERICLINAL: Parallel to the surface.

PERICYST: A distinct outermost cell which functions as a dormant initial.

PERIMEDULLARY: Situated or occurring around the medulla.

PHENOLIC: Containing phenols, usually as a chemical defence against grazing animals.

PHEROMONE: A chemical compound released from an organism that influences another or similar organisms, e.g. sexual attractant released from female eggs which induces release and/or attraction of sperm from male gametophytes; variously demonstrated in the life histories of brown algae.

PHYSODES: Vesicular bodies in cells, containing phenolic compounds.

PINNATE: Regularly arranged on each side of the axis in one plane, feather-like (used in relation to branching).

PITS: Microscopic holes in the walls of adjacent cells allowing cytoplasmic connection.

PLASMALEMMA: A very thin membrane surrounding the cytoplasm of a cell.

PLASTID: A specialized cytoplasmic organelle which may contain the photosynthetic pigments.

PLETHYSMOTHALLUS: A gametophyte or sporophyte microthallus capable of asexual production by zoospore formation.

PLURILOCULAR: Containing many loculi or cells, usually said of reproductive organs.

PLURISERIATE: More than two loculi or cells wide (also = multiseriate).

PLURISPORE: A spore produced from a plurilocular sporangium.

PLURIZOID: Zoid formed in plurilocular zoidangia.

PNEUMATOCYST: A large gas-filled floatation structure.

POLYMORPHIC: Occurring in many morphological forms.

POLYPHYLETIC: Derived from a number of ancestors.

POLYSTICHOUS: Thallus constructed of free or loosely/moderately associated filaments in which intercalary longitudinal cell divisions have occurred.

POLYSTROMATIC: Composed of more than one layer of cells.

PRIMARY ASSIMILATORS: First formed assimilatory filaments.

PRIMARY BRANCHES: The first-formed branches of the main axis.

PRIMARY RHIZOIDS: see RHIZOID.

PROCUMBENT: Lying flat on a surface.

PROLIFERATION: Development of new parts of a thallus by vegetative growth.

PROPAGULUM/PROPAGULE (pl. PROPAGULES): A modified deciduous multicellular, vegetative portion of a frond which functions as a vegetative reproductive organ (in some species of *Sphacelaria*).

PROTHALLUS (pl. PROTHALLI): A gametophyte or sporophyte microthallus capable of sexual reproduction by gamete formation.

PROTONEMATA/PROTONEMA: Early juvenile filamentous, prostrate stage in development of the thallus body.

PSEUDODICHOTOMOUS: Unequally forked, forming two unequal dichotomies (used in relation to branching).

PSEUDODISCOID: Having the appearance of a disc but comprising irregularly adjoined and associated filaments.

PSEUDOPARENCHYMA: A tissue formed by the strong or weak aggregation of branched or unbranched filaments and having the appearance of parenchyma.

PSEUDOPARENCHYMATOUS: Composed of a pseudoparenchyma (used in relation to thalli).

PSEUDOPERENNIAL: A perennial that reduces down during unfavourable periods.

PSEUDOVESICLES: Irregular swellings/dilations of a thallus, resembling vesicles; not uncommonly found in fucoids.

PYRENOID: An organelle occurring within or adjacent to a plastid, often associated with reserve food synthesis and accumulation.

PYRIFORM: Pear-shaped.

PULVINATE: Cushion-shaped.

PUNCTATE: Characterized by a dot-like appearance.

QUADRATE: Of square or cuboid shape.

QUATERNARY BRANCHES: Branches produced by the tertiary branches.

QUATERNATE: Arranged in fours, or consisting of four parts.

RACEMOSE: Raceme-like. Shaped like an elongated bunch of grapes, clustered together around an axis and borne acropetally.

RADIAL: Originating on the radii of an axis and arranged around it (usually said of branching).

RALFSIOID: Said of a thallus which resembles *Ralfsia/Pseudoralfsia* (pseudoparenchymatous and crustose).

RECEPTACLE: A specialized structure usually a modified terminal branch bearing conceptacles (Fucales) or sporangia sori.

RENIFORM: Kidney-shaped.

RETICULATE: Having the appearance of a network.

RHIZOID: A unicellular or multicellular filament formed for attachment, usually arising from the base of a thallus; primary rhizoids arise as downward-growing outgrowths, from the developing reproductive propagule, secondary rhizoids arise from the basal cells of the developing plant/alga body, and often form an enclosing sheath in some species.

RHIZOGEN: A term often used to describe the large bulbous base in *Saccorhiza polyschides*.

RHIZOME: A prostrate, horizontal axis usually associated with vegetative spread.

RHIZOMATOUS: Rhizome-like.

SACCATE: Inflated, or sac-like form.

SAXICOLOUS: Growing on rocks or stones.

SECONDARY ASSIMILATORS: Later formed assimilatory filaments.

SECONDARY BRANCHES: Branches produced by the primary branches.

SECONDARY RHIZOIDS: see RHIZOID.

SECUND: Arranged more or less in a row on one side only (usually said of branches and sporangia). See unilateral.

SERRATE: With a saw-like edge to the thallus.

SESSILE: Directly attached, lacking a stalk.

SHEATH: An enclosing enveloping structure (used in relation to true hairs).

SIEVE PLATES: Specialized end walls with pores allowing conduction between cells, found in the medulla of Laminariales and Fucales.

SIEVE TUBES: Specialized conducting filaments found in the medulla of Laminariales and Fucales.

SILICULOSE: Resembling a silicule, i.e. pod-like, short and not much longer than wide.

SILIQUOSE: Shaped like a siliqua, elongated with transverse septa.

SIMPLE: Unbranched.

SINUOUS: With a wavy outline or margin to the thallus.

SOLENOCYSTS: Specialized, elongated, medullary cells associated with translocation in some Laminariales.

SORUS (pl. SORI): An aggregation of reproductive structures, usually occurring as a raised patch.

SP: Squash preparation.

SPATULATE: Shaped like a spatula with a broad rounded tip and narrow base.

SPERM: General term used for a male gamete.

SPINOSE: Bearing spines.

SPORANGIUM (pl. SPORANGIA): A reproductive structure containing spores. Spores can be produced by mitosis (mitosporangia producing mitospores) or by meiosis (meiosporangia producing meiospores).

SPORE: An asexual reproductive body which may be motile or non-motile.

SPOROPHYLLS: Distinct, foliose, blade-like structures bearing sporangia, found in some Laminariales.

SPOROPHYTE: A morphological phase which bears sporangia. The diploid generation in a life history, which bears sporangia which produce spores usually by meiosis.

STALK CELL: A supporting cell of a structure, usually a sporangium.

STELLATE: Star-shaped.

STIGMA (pl. STIGMATA): Red eyespot found in flagellated spores or gametes associated with light perception.

STIPE: The lowermost stalk-like part of an erect frond.

STIPITATE: Marked with bands, furrows or ridges.

STOLON: A vegetative prostrate outgrowth at the base of thalli, usually multicellular, from which new shoots can arise.

STREBLONEMATOID: Said of a thallus which resembles that of *Streblonema* (prostrate, un-united branched filaments).

SUBCORIACEOUS: Approaching or roughly being coriaceous.

SUBDICHOTOMOUS: Approaching or roughly being dichotomous.

SUBFILAMENTOUS: tending towards or mostly filamentous.

SUBGLOBOSE: Nearly or approximately globose.

SUBLITTORAL: Applied to that portion of the shore below the eulittoral which is totally immersed.

SUBLITTORAL FRINGE: Applied to that intermediate portion of the shore which is between the eulittoral and sublittoral which is uncovered by the receding tide infrequently and then for very short periods.

SUBQUADRATE: Nearly or approximately square, usually applicable to cells or loculi which are slightly broader than long.

SUBSIMPLE: Partially branched.

SUBSPHERICAL: Approaching or roughly being spherical.

SUBULATE: Awl-shaped and tapering to a fine point.

SUBSTRATUM (pl. SUBSTRATA): The structure on which an alga is growing.

SWARMER: A general term used to describe a motile, reproductive spore cell.

SYMPODIA: Continuing the direction of growth of a main axis by the repeated replacement of the apical growing region by a lateral growing point from below.

SYMPODIALLY: In a sympodial manner.

SYMPODIAL BRANCHING: A mode of development in which the primary axis in continually being replaced by lateral axes which become dominant, but which are soon then replaced by their own laterals.

TERETE: Circular in transverse section.

TERTIARY BRANCHES: Branches produced by the secondary branches.

TETRASPORANGIUM (pl. TETRASPORANGIA): A sporangium in which four spores are formed.

TETRASPORE: One of the four spores formed in a tetrasporangium.

THALLUS (pl. THALLI): General term used for a body of both microscopic and macroscopic algae, to include filamentous, pseudoparenchymatous and parenchymatous forms.

THYLAKOID: Membrane-bounded flattened sacs arranged in stacks within plastids where photosynthesis takes place.

TISSUE: The structural components of the thallus body.

TOMENTOSE: Densely covered with short hairs or filaments and woolly-like in appearance.

TOPHULES: A bulbous, swollen structure situated at the base of branches in some species of *Cystoseira*, *Ericaria* and *Gongolaria*.

TORTUOUS: Twisting and bending.

TOTIPOTENT: Capable of developing into an adult thallus; said of cells.

TRICHOTHALLIC: A method of growth, with a meristematic region in an intercalary position, usually near the base of the filament or hair.

TRIRADIATE: Having three radiating branches.

TRISERIATE: Consisting, wholly or in part, of three rows of cells in longitudinal series.

TRUMPET CELLS: Specialized cells with expanded cross walls found in the medulla of Laminariales.

TRUNCATE: Ending abruptly, as if cut or broken off.

TS: Transverse section.

TUBERCLE: A small swelling or swollen axis.

TUBERCULATE: Appearing swollen like a tubercle.

TUBEROUS: Having swollen parts like a tubercle.

TURGID: Swollen, distended, inflated.

UNIAXIAL: Said of a thallus containing a single axial filament.

UNILATERAL: On one side only (see SECUND).

UNILOCULAR: Containing a single loculus or cell, said of reproductive organs.

UNISERIATE: Consisting of a single row of cells in longitudinal series (not exceeding one cell wide).

UNISPORES: A spore produced from a unilocular sporangium.

UPPER EULITTORAL: Applied to the upper portion of the eulittoral, sometimes referred to as the littoral fringe.

VENTRAL: Pertaining to the lower surface of a dorsiventral thallus.

VERRUCOSE: Warty or wart-like (usually said of a thallus surface).

VERTICILLATE: Bearing successive whorls of branches along an axis, usually at a node.

VESICLE: An air-bladder.

VESICULATE: Having air-bladders.

WHORLED: With lateral branches arising in a ring around the main axis.

ZOIDANGIUM: A structure containing zoids; unizoids are formed in unilocular zoidangia, plurizoids are formed in plurilocular zoidangia.

ZOOID: A motile reproductive cell.

ZOOSPORE: A motile spore.

ZYGOTE: Cell formed as a result of the fusion of two gametes.

References

Abe, K. 1935. Zur kenntnis der entwicklungsgeschichte von *Heterochordaria, Scytosiphon* und *Sorocarpus. Sci. Rep. Tohoku Imp. Univ. IV, biol.* 9: 329–37.

Åberg, P. 1989. Distinguishing between genetic individuals in *Ascophyllum nodosum* populations on the Swedish west coast. *British Phycological Journal* 24: 183–90.

Åberg, P. 1990. Measuring size and choosing category size for a transition matrix study of the seaweed *Ascophyllum nodosum. Marine Ecology Progress Series* 63: 281–7.

Åberg, P. 1992a. A demographic study of two populations of the seaweed *Ascophyllum nodosum. Ecology* 73: 1473–87.

Åberg, P. 1992b. Size-based demography of the seaweed *Ascophyllum nodosum* in stochastic environments. *Ecology* 73: 1488–501.

Abbott, I.A. & Hollenberg, G.J. 1976. Marine Algae of California. Stanford University Press, Stanford.

Adams, J. 1908. A synopsis of Irish algæ, freshwater and marine. *Proceedings of the Royal Irish Academy* 27B(2): 11–60.

Adanson, M. 1763. *Familles des plantes*. II. Partie [Vol. 2]. pp. [1–24], [i–iii], [1]–640. Paris: Chez Vincent, Imprimeur-Librarie de Mgr le Comte de Provence, rue S. Servin.

Afonso-Carrillo, J., Sansón, M., Gil-Rodríguez, M.C., Chacana, M. & Reyes, J. 1988. An endophytic *Streblonema* (Phaeophyta) associated with galls in *Fucus spiralis* (Phaeophyta) from the Canary Islands. *Actes del Simposi Internacional de Botanica Pius font i Quer* 1: 73–6.

Agardh, C.A. 1810. *Dispositio algarum Sueciae*, quam, publico examini subjiciunt Carl Adolf Agardh, C.A. & Gustav Sannberg Blekingus die viii decembris mdcccx. p. i. h. & l.s. pp. Pars 1: [1]–16. Lundae: Litteris Berlingianis.

Agardh, C.A. 1811. *Dispositio algarum Sueciae*, quam publico examini subjiciunt Carl Adolph Agardh, C.A. & Johannes Bruzelius, Scanus. Die xi decembris mdcccxi. p. ii. h. & l.s. pp. Pars 2: [i], 17–26. Lundae: Litteris Berlingianis.

Agardh, C.A. 1814. *Algarum decas tertia*. Londini Gothorum: Ex officina Berlingiana.

Agardh, C.A. 1817. *Synopsis algarum Scandinaviae*, adjecta dispositione universali algarum. pp. [i]–xl, [1]–135. Lundae [Lund]: Ex officina Berlingiana.

Agardh, C.A. 1819. *Fucus balticus. Svensk botanik.* 8: Table 516.

Agardh, C.A. 1820. *Species algarum rite cognitae, cum synonymis, differentiis specificis et descriptionibus succinctis*: Volumen primum. Pars prima. pp. [i–iv], [1]–168. Lundae [Lund]: ex officina Berlingiana.

Agardh, C.A. 1822. *Species algaru*: Volumen primum pars posterior. pp. [v–vi], 169–398. Lundae [Lund]: ex officina Berlingiana.

Agardh, C.A. 1824. *Systema Algarum*. pp. [i]–xxxvii, [1]–312. Lundae [Lund]: Literis Berlingianis [Berling].

Agardh, C.A. 1827. Aufzahlungen einiger in den osterreichischen Landem gefundenen neuen Gattungen und Arten von Algen. *Flora* 10: 625–56.

Agardh, C.A. 1828a. *Species algarum*: Voluminis secundi. Sectio prior. pp. [i]–lxxvi, [i]–189. Gryphiae [Greifswald]: sumptibus Ernesti Mauriti [Ernst Mauritius].

Agardh, C.A. 1828b. *Icones algarum europearum*: Fasc. 1. pp. i–xx, pls 1–20. Leipsic [Leipzig]: Leopold Voss.

Agardh, J.G. 1836. *Novitiae florae Sueciae ex Algarum familia*: pp. 1–16. Lundae [Lund]: Gleerup.

Agardh, J.G. 1841. In historiam algarum symbolae. *Linnaea* 15: 1–50, 443–57.

Agardh, J.G. 1842. *Algae maris Mediterranei et Adriatici*, observationes in diagnosin specierum et dispositionem generum. pp. [i]–x, 1–164. Parisiis [Paris]: Apud Fortin, Masson et Cie.

Agardh, J.G. 1848. *Species genera et ordines algarum*: Volumen Primum. Algas fucoideas complectens. pp. [i–vi], [i]–viii, [1]–363. Lundae [Lund]: C.W.K. Gleerup.

Agardh, J.G. 1868a. De Laminarieis symbolas offert. *Lunds Universitets Årsskrift* 4: 1–36.

Agardh, J.G. 1868b. Bidrag till kännedomen om Spetsbergens alger. *Kungliga Svenska Vetenskaps-Akademiens Handlingar, Nye Följd* 7: [1]4–12.

Agardh, J.G. 1868c. Bidrag till kännedomen om Spetsbergens alger. Tilläg till föregående afhandling. *Kungliga Svenska Vetenskaps-Akademiens Handlingar, Nye Följd* 7: [27] 28–49.

Agardh, J.G. 1872. Bidrag till Kännedomen af Grönlands Laminarieer och Fucaceer. *Kongl Svenska Vetenskaps-Akademiens Handlingar* 10: 1–31.

Agardh, J.G. 1882. Till algernes systematik. Nya bidrag. (Andra afdelningen.). *Lunds Universitets Års-Skrift, Afdelningen for Mathematik och Naturvetenskap* 17: 1–134.

Agardh, J.G. 1890. Till algernes systematik. Nya bidrag. (Sjette afdelningen.). *Lunds Universitets Års-Skrift, Andra Afdelningen, Kongl. Fysiografiska Sällskapets i Lund Handlingar* 26: 1–125.

Agardh, J.G. 1894. *Analecta algologica*, observationes de speciebus algarum minus cognitae earumque dispositione. Continuatio I. *Lunds Universitets Års-Skrift, Andra Afdelningen, Kongl. Fysiografiska Sällskapets i Lund Handlingar* 29: 1–144.

Agardh, J.G. 1841. In historiam algarum symbolae. *Linnaea* 15: 1–50, 443–57.

Agardh, J.G. 1896. *Analecta algologica*, Continuatio III. *Lunds Universitets Års-Skrift, Andra Afdelningen, Kongl. Fysiografiska Sällskapets i Lund Handlingar* 32: 1–140, 1–8 [index], 1 plate.

Aguilar-Rosas, R. & Machado-Galindo, A.M. 1990. Ecological aspects of *Sargassum muticum* (Fucales, Phaeophyta) in Baja California, Mexico: reproductive phenology and epiphytes. *Hydrobiologia* 204/5: 185–90.

Aguilar-Rosas, R., Aguilar-Rosas, L.E., Avila-Serrano, G. & Marcos-Ramírez, R. 2004. First record of *Undaria pinnatifida* (Harvey) Suringar (Laminariales, Phaeophyta) on the Pacific coast of Mexico. *Botanica Marina* 47: 255–8.

Aisha, K. & Shameel, M. 2011. Taxonomic study of the order Ectocarpales (Phaeophycota) from the coastal waters of Pakistan. *International Journal on Algae* 13: 128–48.

Ajisaka, T. 1979. The life history of *Acrothrix pacifica* Okamura et Yamada (Phaeophyta, Chordariales) in culture. *Japanese Journal of Phycology* 27: 75–81.

Ajisaka, T. & Kawai, H. 1986. The life history of *Acrothrix gracilis* Kylin (Phaeophyceae, Chondariales) in Japan. *Japanese Journal of Phycology* 34: 129–36.

Ajisaka, T. & Umezaki, I. 1978. The life history of *Sphaerotrichia divaricata* (Ag.) Kylin (Phaeophyta, Chordariales) in culture. *Japanese Journal of Phycology* 26: 53–9.

Akiyama, K. 1965. Studies of ecology and culture of *Undaria pinnatifida* (Harvey) Suringar Environmental factors affecting the growth and maturation of gametophyte. *Bulletin of the Tohoku National Fisheries Research Institute* 25: 143–70.

Akiyama, K. & Kurogi, M. 1982. Cultivation of *Undaria pinnatifida* (Harvey) Suringar, the decrease in crops from natural plants following crop increase from cultivation. *Bulletin of Tohoku Regional Fisheries Research Laboratory* 44: 91–100.

Aldanondo-Aristizabal, N., González González, R. & Gil-Rodríguez, M.C. 2004. Acerca de *Cystoseira tamariscifolia* en Tenerife y La Palma (Islas Canarias). *Revista de la Academia Canaria de Ciencias* XV: 115–28.

Alexander, W.B., Southgate, B.A. & Bassindale, B. 1935. Survey of the River Tees. Part II. The estuary – chemical and biological. D.S.I.R., London. Technical Paper Water Pollution Research No. 5.

Alexandrova, Y.N. & Reunov, A.A. 2008. The oogonia of macroalga *Undaria pinnatifida* are alkaline-phosphatase positive and contain germinal body-like structures. *Journal of Phycology* 44: 712–15.

Allender, B.M. 1980. *Dictyotopsis propagulifera* (Phaeophyta) – an algal enigma. *Phycologia* 19: 234–6.

Almeida, S.C., Neiva, J., Sousa, F., Martin, N., Cox, C.J., Melo-Ferreira, J., Guiry, M.D., Serrão, E.A. & Pearson, G.A. 2022. A low-latitude species pump: Peripheral isolation, parapatric speciation and mating-system evolution converge in a marine radiation. *Molecular Ecology* 31: 4797–817.

Altamirano, M., Bañares, E. & Zanolla, M. 2014. A new record of *Desmarestia dudresnayi* J.V. Lamouroux ex Leman (Desmarestiaceae, Heterokontophyta) represents a new southernmost limit of its distribution in the north Atlantic Ocean. *Anales de Biologia* 36: 61–3.

Altamirano, M., Flores-Moya, A. & Figueroa, F.L. 2003a. Effects of UV radiation and temperature on growth of germlings of three species of *Fucus* (Phaeophyceae). *Aquatic Botany* 75: 9–20.

Altamirano, M., Flores-Moya, A., Kuhlenkamp, R. & Figueroa, F.L. 2003b. Stage-dependent sensitivity to ultraviolet radiation on zygotes of the brown alga *Fucus serratus*. *Zygote* 11: 101–6.

Altamirano, M., Murakami, A. & Kawai, H. 2004. High light stress in the kelp *Ecklonia cava*. *Aquatic Botany* 79: 125–35.

Amsler, C.D. 1984. Culture and field studies of *Acinetospora crinita* (Carmichael) Sauvageau (Ectocarpaceae, Phaeophyceae) in North Carolina, USA. *Phycologia* 23: 377–82.

Amsler, C.D. & Neushul, M. 1989. Chemotactic effects of nutrients on spores of the kelps *Macrocystis pyrifera* and *Pterygophora californica*. *Marine Biology* 102: 557–64.

Amsler, C.D. & Neushul, M. 1990. Nutrient stimulation of spore settlement in the kelps *Pterygophora californica* and *Macrocystis pyrifera*. *Marine Biology* 107: 297–304.

Amsler, C.D., Shelton, K.L., Britton, C.J., Spencer, N.Y. & Greer, S.P. 1999. Nutrients do not influence swimming behaviour or settlement rates of *Ectocarpus siliculosus* (Phaeophyceae) spores. *Journal of Phycology* 35: 239–44.

Anand, P.L. 1937. A taxonomic study of the algae of the British chalk-cliffs. *Journal of Botany, British and Foreign* 75 (Suppl. II): 1–51.

Andersen, G.S. 2013. Growth, survival and reproduction in the kelp *Saccharina latissima*: seasonal patterns and the impact of epibionts. PhD thesis, Department of Biosciences, University of Oslo.

Andersen, G.S., Pedersen, M.F. & Nielsen, S.L. 2013. Temperature acclimation and heat tolerance of photosynthesis in Norwegian *Saccharina latissima* (Laminariales, Phaeophyceae). *Journal of Phycology* 49: 689–700.

Anderson, C.I.H. & Scott, G.W. 1998. The occurrence of distinct morphotypes within a population of *Fucus spiralis*. *Journal of the Marine Biological Association of the United Kingdom* 78: 1003–6.

Anderson, R.J. 1982. The life history of *Desmarestia firma* (C. Ag.) Skottsb. (Phaeophyceae, Desmarestiales). *Phycologia* 21: 316–22.

Anderson, R.J. & Velimirov, B. 1982. An experimental investigation of the palatability of kelp bed algae to the sea urchin *Parechinus angulosus* Leske. *Marine Ecology* 3: 357–73.

Andersson, S., Kautsky, L. & Kalvas, A. 1994. Circadian and lunar gamete release in *Fucus vesiculosus* in the tidal Baltic Sea. *Marine Ecology Progress Series* 110: 195–201.

Andrade-Sorcia, G.R., Riosmena-Rodríguez, R., Muñiz-Salazar, R., López-Vivas, J.M., Boo, G.H., Lee, K.M. & Boo, S.M. 2014. Morphological reassessment and molecular assessment of *Sargassum* (Fucales: Phaeophyceae) species from the Gulf of California, México. *Phytotaxa* 183: 201–23.

Andrew, N.L. 1989. Contrasting ecological implications of food limitation in sea urchins and herbivorous gastropods. *Marine Ecology Progress Series* 51: 189–93.

Andrew, N.L. & Viejo, R.M. 1998. Effects of wave exposure and intraspecific density on the growth and survivorship of *Sargassum muticum* (Sargassaceae: Phaeophyta). *European Journal of Phycology* 33: 251–8.

Ang, P.O. 1991. Natural dynamics of a *Fucus distichus* (*Phaeophyta, Fucales*) population: reproduction and recruitment. *Marine Ecology Progress Series* 78: 71–85.

Ang, P.O., Jr & De Wreede, R.E. 1993. Simulation and analysis of the dynamics of a *Fucus distichus* (Phaeophyceae, Fucales) population. *Marine Ecology Progress Series* 93: 253–65.

Ang, P.O., Sharp, G.J. & Semple, R.E. 1993. Changes in the population structure of *Ascophyllum nodosum* (L.) Le Jolis due to mechanical harvesting. *Proceedings of the International Seaweed Symposium* 14: 321–6.

Ang, P.O., Sharp, G.J. & Semple, R.E. 1996. Comparison of the structure of populations of *Ascophyllum nodosum* (Fucales, Phaeophyta) at sites with different harvesting histories. *Proceedings of the International Seaweed Symposium* 15: 179–84.

Anonymous. 1952. *Flora of Devon. II. 1. The Marine Algae.* Vol. 2, pp. 1–77. Torquay, Devon: The Torquay Times and Devonshire Press, Limited.

Antolic, A., Span, A., Nikolic, V., Grubelic, I., Despatalovic, M. & Cvitkovic, I. 2010. A checklist of the benthic marine macroalgae from the eastern Adriatic coast: II. Heterokontophyta: Phaeophyceae. *Acta Adriatica* 51: 9–33.

Apoya, M., Ogawa, H. & Nanba, N. 2002. Alginate content of farmed *Undaria pinnatifida* (Harvey) Suringar from the Three Bays of Iwate, Japan during harvest period. *Botanica Marina* 45: 445–52.

Ar Gall, E., Küpper, F.C. & Kloareg, B. 2004. A survey of iodine content in *Laminaria digitata*. *Botanica Marina* 47: 30–7.

Arai, A. & Miura, A. 1991. Effects of salinity and light intensity on the growth of brown alga, *Sargassum ringgoldianum*. *Suisanzoushoku* 39: 315–19.

Arakawa, H. 2005. Lethal effects caused by suspended particles and sediment load on zoospores and gametophytes of the brown alga *Eisenia bicyclis*. *Fisheries Science* 71: 133–40.

Arakawa, H. & Matsuike, K. 1990. Influence on sedimentation velocity of brown algae zoospores and their base-plate insertion exserted by suspended matters. *Nippon Suisan Gakkaishi* 56: 1741–8.

Arakawa, H. & Matsuike, K. 1992. Influence on insertion of zoospores germination, survival, and maturation of gametophytes of brown algae exserted by sediments. *Nippon Suisan Gakkaishi* 58: 619–25.

Arakawa, H. & Morinaga, T. 1994. Influence of suspended particles on dispersion of brown algal zoospores. *Nippon Suisan Gakkaishi* 60: 61–4.

Arakawa, H., Shinoda, K., Matsumoto, A., Endo, H. & Agatsuma, Y. 2014. Physical Factors Involved in the Isoyake (Seaweed Forest Depletion) at Mio, Pacific Coast of Central Japan. (Research Article). *Journal of Marine Biology and Oceanography* 3: 1–7.

Araújo, R., Bárbara, I., Tibaldo, M., Berecibar, E., Díaz-Tapia, P., Pereira, R., Santos, R. & Sousa-Pinto, I. 2009. Checklist of benthic marine algae and cyanobacteria of northern Portugal. *Botanica Marina* 52: 24–46.

Araújo, R., Vaselli, S., Almeida, M., Serrão, E. & Sousa-Pinto, I. 2009. Effects of disturbance on marginal populations: human trampling on *Ascophyllum*

nodosum assemblages at its southern distribution limit. *Marine Ecology Progress Series* 378: 81–92.

Ardissone, F. 1883. Phycologia mediterranea. Parte prima, Floridee. *Memorie della Società Crittogamologica Italiana* 1: i–x, 1–516.

Ardissone, F. 1886. Phycologia mediterranea. Parte IIa. Oosporee-Zoosporee-Schizosporee. *Memorie della Società Crittogamologica Italiana* 2: 1–128.

Ardissone, F. 1887. *Phycologia mediterranea. Parte IIa. Oosporee-Zoosporee-Schizosporee.* Vol. 2, pp. 129–320. Varese: Antica tip. ferri di Maj e Malnati.

Ardré, F. 1961. Algues Du Portugal: Liste Préliminaire. *Revue Générale de Botanique* 68: 443–56.

Ardré, F. 1969. Remarques sur le *Pelvetia canaliculata* (L.) Dec. et Thur. *Proceedings of the International Seaweed Symposium* 6: 31–40.

Ardré, F. 1970. Contribution à l'étude des algues marines du Portugal. I. La flore. *Portugaliae Acta Biologica, Série B, Sistemática, Ecologia, Biogeografia e Paleontologia* 10: 137–555.

Arenas, F. & Fernández, C. 1998. Ecology of *Sargassum muticum* (Phaeophyta) on the north coast of Spain III. Reproductive ecology. *Botanica Marina* 41: 209–16.

Arenas, F. & Fernández, C. 2000. Size structure and dynamics in a population of *Sargassum muticum* (Phaeophyceae). *Journal of Phycology* 36: 1012–20.

Arenas, F., Fernández, C., Rico, J.M., Fernández, E. & Haya, D. 1995. Growth and reproductive strategies of *Sargassum muticum* (Yendo) Fensholt and *Cystoseira nodicaulis* (Whit.) Roberts. *Scientia Marina* 59 (suppl.): 1–8.

Areschoug, J.E. 1842. Algarum minus rite cognitarum pugillus primus. *Linnaea* 16: 225–36.

Areschoug, J.E. 1843. Algarum (phycearum) minus rite cognitarum pugillus secundus. *Linnaea* 17: 257–69.

Areschoug, J.E. 1847. Enumeratio phycearum in maribus scandinaviae cresentium. Sectio prior. *Nova Acta Regiae Societatis Scientiarum Upsaliensis, Series 2*, 13: 223–382.

Areschoug, J.E. 1850. Phycearum, quae in maribus Scandinaviae crescunt, enumeratio. Sectio posterio Ulvaceas continen. *Nova Acta Regiae Societatis Scientiarum Upsaliensis* 14: 385–454.

Areschoug, J.E. 1861–72. *Algae Scandinavicae Exsiccatae (1861–1872). Serie novae. Fasc. I–IX, numbers 1–430.* Fasc. 1. 1–50. 1861. Fasc. 2–3. 51–150. 1862 Fasc. 4. 151–200. 1863. Fasc 5. 201–250. 1864. Fasc. 6. 251–300. 1866. Fasc 7–8. 301–400. 1872. Fasc. 9. 401–430. 1879.

Areschoug, J.E. 1873. Om de skandinaviska algformer, som äro närmast beslägtade med *Dictyosiphon foeniculaceus*, eller kunna med denna lättast förblandas. *Botaniska Notiser* 6: 161–71.

Areschoug, J.E. 1875. Observationes phycologicae III. *Nova Acta Regiae Societatis Scientiarum Upsaliensis, Series 3*, 10: 1–36.

Areschoug, J.E. 1876. De algis nonnullis maris Baltici et Bahusiensis. *Botaniska Notiser* 2: 33–7.

Areschoug, J.E. 1883. Observationes phycologicae. Particula quarta. De Laminariaceis nonnullis. *Nova Acta Regiae Societatis Scientiarum Upsaliensis, Series 3* 12(8): 1–23.

Armitage, C.S., Sjøtun, K. & Jensen, K.H. 2014. Correlative evidence for competition between *Fucus serratus* and the introduced chlorophyte *Codium fragile* subsp. *fragile* on the southwest coast of Norway. *Botanica Marina* 57: 85–97.

Armstrong, S.L. 1989. The behavior in flow of the morphologically variable seaweed *Hedophyllum sessile* (C. Ag.) Setchell. *Hydrobiologia* 183: 115–22.

Arnold, M., Teagle, H., Brown, M.P. & Smale, D.A. 2016. The structure and diversity of epibiotic assemblages associated with the invasive kelp *Undaria pinnatifida* in comparison to native habitat forming macroalgae on a subtidal temperate reef. *Biological Invasions* 18: 661–76.

Arrontes, J. 1993. Nature of the distributional boundary of *Fucus serratus* on the north shore of Spain. *Marine Ecology Progress Series* 93: 183–93.

Arrontes, J. 2002. Mechanisms of range expansion in the intertidal brown alga *Fucus serratus* in northern Spain. *Marine Biology* 141: 1059–67.

Asare, S.O. & Harlin, M.M. 1983. Seasonal fluctuations in tissue nitrogen for five species of perennial macroalgae in Rhode Island Sound. *Journal of Phycology* 19: 254–7.

Asensi, A. 1966. La presencia de *Pylaiella littoralis* (L.) Kjellman en la Antartida. *Contribuciones del Instituto Antártico Argentino* 101: 1–14.

Asensi, A., Ar Gall, E., Marie, D., Billot, C., Dion, P. & Kloareg, B. 2001. Clonal propagation of *Laminaria digitata* (Phaeophyceae) sporophytes through a diploid cell-filament suspension. *Journal of Phycology* 37: 411–17.

Asensi, A., Delépine, R., Rousseau, F. & de Reviers, B. 2004. Morphology and taxonomy of *Adenocystis longissima* (Skottsberg) stat. nov. (Phaeophyceae) from subantarctic South America. *Polar Biology* 28: 82–91.

Askenasy, E. 1888. Algen mit Unterstützung der Herren E. Bornet, E. Grunow, P. Hariot, M. Moebius, O. Nordstedt bearbeitet. In: *Forschungsreise S.M.S. 'Gazelle'. Theil 4: Bot.* (Engler, A., ed), pp. 1–58. Berlin.

Assali, N.-E. & Loiseaux-de Goër, S. 1992. Sequence and phylogeny of the *psaB* gene of *Pylaiella littoralis* (Phaeophyta). *Journal of Phycology* 28: 209–13.

Assali, N.-E., Mache, R. & Loiseaux-de Goër, S. 1990. Evidence for a composite phylogenetic origin of the plastid genome of the brown alga *Pylaiella littoralis* (L.) Kjellm. *Plant Molecular Biology* 15: 307–15.

Assali, N.-E., Martin, W.F., Sommerville, C.C. & Loiseaux-de Goër, S. 1991. Evolution of the Rubisco operon from prokaryotes to algae: structure and analysis of the rbcS gene of the brown alga *Pylaiella littoralis*. *Plant Molecular Biology* 17: 853–63.

Assis, J., Vaz Lucas, A., Bárbara, I. & Serrão, E.A. 2016. Future climate change is predicted to shift long-term persistence zones in the cold-temperate kelp *Laminaria hyperborea*. *Marine Environmental Research* 113: 174–82.

Astudillo, M.C. & Viejo, R.M. 2009. Recent changes in the distribution of *Fucus serratus*: effects on assemblages on rocky coast in the northern Spain. *Algas, Boletín Informativo de la Sociedad Española de Ficología* 42: 14.

Athanasiadis, A. 1985. North Aegean marine algae I. New records and observations from the Sithonia Peninsula, Greece. *Botanica Marina* 28: 453–68.

Athanasiadis, A. 1996. *Taxonomisk litteratur och biogeografi av Skandinaviska rödalger och brunalger.* Göteborg: Algologia.

Athanasiadis, A. 2021. *Phycologia Europaea Phaeophyta.* pp. [i]–xxxxvii, [1]–759. Gothenburg: Published and distributed by the author.

Aziz, K.M.S. & Humm, H.J. 1962. Additions to the algal flora of Beaufort, N.C., and vicinity. *Journal of the Elisha Mitchell Scientific Society* 78: 55–63.

Baardseth, E. 1955. *Regrowth of Ascophyllum nodosum after harvesting.* Dublin: Institute for Industrial Research and Standards.

Baardseth, E. 1970. Synopsis of biological data on knobbed wrack *Ascophyllum nodosum* (Linnaeus) Le Jolis. Rev. 1. *FAO Fisheries Synopsis* 38: 1–48.

Baba, M. 2008. Effects of temperature, irradiance and salinity on the growth of *Undaria pinnatifida* from Niigata Prefecture, Central Japan. *Report of the Marine Ecological Research Institute* 11: 7–15 (in Japanese).

Bachelot de la Pylaie, A.J.M. 1824. Quelques observations sur les productions de l'île de Terre Neuve et sur quelques algues de la côte de France, appartement au genre Laminaire. *Annales des Sciences Naturelles, Botanique* 4: 174–84.

Bachelot de la Pylaie, A.J.M. (1830 '1829'). *Flora de l'Ile Terre-Neuve et des Iles Saint Pierre et Miclon.* Livraison [Algae]. pp. 1–128. Paris: Typographie de A. Firmin Didot, rue Jacob, No. 24.

Bäck, S., Collins, J.C. & Russell, G. 1990. Reproductive biology of Baltic *Fucus vesiculosus*. *British Phycological Journal* 25: 84.

Bäck, S., Collins, J.C. & Russell, G. 1991. Effects of salinity on growth of Baltic and Atlantic *Fucus vesiculosus*. *British Phycological Journal* 26: 81–2.

Bäck, S., Collins, J.C. & Russell, G. 1992a. Recruitment of the Baltic flora: the *Fucus ceranoides* enigma. *Botanica Marina* 35: 53–9.

Bäck, S., Collins, J.C. & Russell, G. 1992b. Comparative ecophysiology of Baltic and Atlantic *Fucus vesiculosus*. *Marine Ecology Progress Series* 84: 71–82.

Bäck, S., Collins, J.C. & Russell, G. 1992c. Effects of salinity on growth of Baltic and Atlantic *Fucus vesiculosus*. *British Phycological Journal* 27: 39–47.

Bäck, S., Collins, J.C. & Russell, G. 1992b. Recruitment of the Baltic flora: the *Fucus ceranoides* Enigma. *Botanica Marina* 35: 53–9.

Bacon, L.C. & Vadas, R.L. 1991. A model for gamete release in *Ascophyllum nodosum* (Phaeophyta). *Journal of Phycology* 27: 166–73.

Baer, J. & Stengel, D.B. 2010. Variability in growth, development and reproduction of the non-native seaweed *Sargassum muticum* (Phaeophyceae) on the Irish west coast. *Estuarine, Coastal and Shelf Science* 90: 185–94.

Báez, J.C., Olivero, J., Peteiro, C., Ferri-Yáñez, F., García-Soto, C. & Real, R. 2010. Macro-environmental modelling of the current distribution of *Undaria pinnatifida* (Laminariales, Ochrophyta) in northern Iberia. *Biological Invasions* 12: 21–31.

Baker, J.R.J. & Evans, L.V. 1971. A myrionemoid variant of *Ectocarpus fasciculatus* Harv. *British Phycological Journal* 6: 73–80.

Baker, J.R.J. & Evans, L.V. 1972. Une variante myrionémoïde d'*Ectocarpus fasciculatus* Harv. *Bulletin de la Société botanique de France, Mémoires*: 99–100.

Baker, J.R.J. & Evans, L.V. 1973a. The ship-fouling alga *Ectocarpus* I. Ultrastructure and cytochemistry of plurilocular reproductive stages. *Protoplasma* 77: 1–13.

Baker, J.R.J. & Evans, L.V. 1973b. The ship-fouling alga *Ectocarpus* II. Ultrastructure of the unilocular reproductive stages. *Protoplasma* 77: 181–9.

Baker, S.M. 1912. On the brown seaweeds of the salt marsh. *Journal of the Linnean Society of London, Botany* 40: 275–91.

Baker, S.M. & Bohling, M.H. 1916. On the brown seaweeds of the salt-marsh. II. Their systematic relationships, morphology and ecology. *Journal of the Linnean Society of London, Botany* 43: 325–80.

Balboa, E.M., Gallego-Fábrega, C., Moure, A. & Domínguez, H. 2016. Study of the seasonal variation on proximate composition of oven-dried *Sargassum muticum* biomass collected in Vigo Ria, Spain. *Journal of Applied Phycology* 28: 1943–53.

Bárbara, I. & Cremades, J. 1990. *Petalonia zosterifolia* (Reinke) O. Kuntze, nuevo feófito para la Península Ibérica. *Anales del Jardín Botánico de Madrid* 47: 492–4.

Bárbara, I., Cremades, J., Calvo, S., López-Rodríguez, M.C. & Dosil, J. 2005. Checklist of the benthic marine and brackish Galician algae (NW Spain). *Anales del Jardín Botánico de Madrid* 62: 69–100.

Bárbara, I., Díaz Tapia, P., Peteiro, C., Berecibar, E., Peña, V., Sánchez, N., Tavares, A.M., Santos, R., Secilla, A., Riera Fernández, P., Bermejo, R. & García, V. 2012. Nuevas citas y aportaciones corológicas para la flora bentónica marina del Atlántico de la Península Ibérica. *Acta Botánica Malacitana* 37: 5–32.

Barceló, M.C., Gómez Garreta, A., Ribera, M.A. & Rull Lluch, J. 1998. Mapas de distribución de algas marinas de la Península Ibérica e Islas Baleares. XI. *Lobophora variegata* (Lamour.) Womersley, *Padina pavonica* (L.) Thivy y *Zonaria tournefortii* (Lamour.) Mont. (Dictyotales, Fucophyceae). *Botanica Complutensis* 22: 179–86.

Barceló, M.C., Gómez Garreta, A., Rull Lluch, J. & Ribera, M.A. 1994. Mapas de distribución de algas marinas de la Península Ibérica e Islas Baleares. VI. *Cystoseira* C. Agardh: Grupos *C. spinifero-opuntioides* y *C. discors-abrotanifolioides*. *Botanica Complutensis* 19: 119–30.

Barker, K.M. & Chapman, A.R.O. 1990. Feeding preferences of periwinkles among four species of *Fucus*. *Marine Biology* 106: 113–18.

Barnes, D.K.A. 1999. The influence of ice on polar nearshore benthos. *Journal of the Marine Biological Association of the United Kingdom* 79: 401–7.

Bartsch, I. & Kuhlenkamp, R. 2000. The marine macroalgae of Helgoland (North Sea): an annotated list of records between 1845 and 1999. *Helgoland Marine Research* 54: 160–89.

Bartsch, I., Vogt, J., Pehlke, C. & Hanelt, D. 2013. Prevailing sea surface temperatures inhibit summer reproduction of the kelp *Laminaria digitata* at Helgoland (North Sea). *Journal of Phycology* 49: 1061–73.

Bartsch, I., Wiencke, C., Bischof, K., Buchholz, C.M., Buck, B.H., Eggert, A., Feuerpfeil, P., Hanelt, D., Jacobsen, S., Karez, R., Karsten, U., Molis, M., Roleda, M.Y., Schubert, H., Schumann, R., Valentin, K., Weinberger, F. & Wiese, J. 2008. The genus *Laminaria sensu lato*: recent insights and developments. *European Journal of Phycology* 43: 1–86.

Batters, E.A.L. 1883. Notes on the marine algae of Berwick-upon-Tweed. *History of the Berwickshire Naturalists' Club* 10: 108–15.

Batters, E.A.L. 1884. Notes on the marine algae of Berwick-upon-Tweed. *History of the Berwickshire Naturalists' Club* 10: 349–55.

Batters, E.A.L. 1885. Notes on the marine algae of Berwick-on-Tweed. *History of the Berwickshire Naturalists' Club* 10: 535–8.

Batters, E.A.L. 1888. A description of three new marine algae. *Journal of the Linnean Society of London, Botany* 24: 450–3.

Batters, E.A.L. 1890a. *A list of the marine algae of Berwick-on-Tweed*. pp. 1–171, pls VII–XL. Alnwick: Henry H. Blair.

Batters, E.A.L. 1890b. A list of marine algae of Berwick-on-Tweed. *History of the Berwickshire Naturalists' Club* 12: 221–392.

Batters, E.A.L. 1892a. Additional notes on the marine algae of the Clyde sea area. *Journal of Botany, British and Foreign* 30: 170–7.

Batters, E.A.L. 1892b. New or critical British algae. *Grevillea* 21: 13–23.

Batters, E.A.L. 1892c. New or critical British algae. *Grevillea* 21: 49–53.

Batters, E.A.L. 1893a. On the necessity for removing *Ectocarpus secundus*, Kütz., to a new genus. *Grevillea* 21: 85–6.

Batters, E.A.L. 1893b. New or critical British algae. *Grevillea* 22: 20–4, 24–5.

Batters, E.A.L. 1893c. New or critical British algae. *Grevillea* 22: 50–2.

Batters, E.A.L. 1894a. New British marine algae. *Grevillea* 22: 90–2.

Batters, E.A.L. 1894b. New or critical British marine algae. *Grevillea* 22: 114–17.

Batters, E.A.L. 1894c. A provisional list of the marine algae of Essex, and the adjacent coast. *Essex Naturalist (London)* 8: 1–25.

Batters, E.A.L. 1895a. Some new British marine algae. *Journal of Botany, British and Foreign* 33: 274–6.

Batters, E.A.L. 1895b. On some new British algae. *Annals of Botany* 9: 307–21, Plate 11 (2 pages).

Batters, E.A.L. 1895c. Some new British algae. *Annals of Botany* 9: 168–9.

Batters, E.A.L. 1896. New or critical British marine algae. *Journal of Botany, British and Foreign* 34: 384–90.

Batters, E.A.L. 1897. New or critical British marine algae. *Journal of Botany, British and Foreign* 35: 433–40.

Batters, E.A.L. 1900. New or critical British marine algae. *Journal of Botany, British and Foreign* 38: 369–79, pl. 414.

Batters, E.A.L. 1902. A catalogue of the British marine algae being a list of all the species of seaweeds known to occur on the shores of the British Islands, with the localities where they are found. *Journal of Botany, British and Foreign* 40 (Supplement): 1–107.

Batters, E.A.L. 1906. New or critical British marine algae. *Journal of Botany, British and Foreign* 44: 1–3, pl. 475.

Beauchamp, K.A. & Gowing, M.M. 1982. A quantitative assessment of human trampling effects on a rocky intertidal community. *Marine Environmental Research* 7: 279–93.

Beauvois, A.M.F. & Palisot de, J. 1805. *La flore d'Oware et de Benin en Afrique*. Part 3. pp. 17–32. Paris: Fain.

Becker, S., Walter, B. & Bischof, K. 2009. Freezing tolerance and photosynthetic performance of polar seaweeds at low temperatures. *Botanica Marina* 52: 609–16.

Belsher, T. & Pommellec, S. 1988. Expansion de l'algue d'origine japonaise *Sargassum muticum* (Yendo) Fensholt, sur les côtes Français, de 1983 à 1987. *Cahiers de Biologie Marine* 29: 221–31.

Belsher, T., Boudouresque, C.F., Lauret, M. & Riouall, R. 1985. L'envahissement de l'Etang de Thau (France) par la grande Phaeophyceae *Sargassum muticum. Rapport et Procès-Verbaux des Réunions de la Commission Internationale pour l'Exploration Scientifique de la Mer Méditeranée* 29: 33–6.

Benali, M., Djebri, I., Bellouis, D., Sellam, L.N. & Rebzani-Zahaf, C. 2019. First record of drifting *Sargassum muticum* (Yendo) Fensholt thalli on the Algerian coasts of Cherchell and Sidi Fredj. *BioInvasions Records* 8: 575–81.

Benedetti-Cecchi, L. 2000. Predicting direct and indirect effects during succession in a midlittoral rocky shore assemblage. *Ecological Monographs* 70: 45–72.

Benson, E.E., Rutter, J.C. & Cobb, A.H. 1983. Seasonal variation in frond morphology and chloroplast physiology of the intertidal alga *Codium fragile* (Suringar) Hariot. *New Phytologist* 95: 569–80.

Berger, R., Malm, T. & Kautsky, L. 2001. Two reproductive strategies in Baltic *Fucus vesiculosus* (Phaeophyceae). *European Journal of Phycology* 36: 265–74.

Bergström, L., Berger, R. & Kautsky, L. 2003. Negative direct effects of nutrient enrichment on the establishment of *Fucus vesiculosus* in the Baltic Sea. *European Journal of Phycology* 38: 41–6.

Berkaloff, C. & Rousseau, B. 1979. Ultrastructure of male gametogenesis in *Fucus serratus* (Phaeophyceae). *Journal of Phycology* 15: 163–73.

Berkaloff, C., Duval, J.C., Hauswirth, N. & Rousseau, B. 1983. Freeze fracture study of thylakoids of *Fucus serratus*. *Journal of Phycology* 19: 96–100.

Berkeley, M.J. 1833. *Gleanings of British Algae; being an appendix to the supplement to English Botany.* C.E. Sowerby, London.

Berndt, M-L., Callow, J.A. & Brawley, S.H. 2002. Gamete concentrations and timing and success of fertilization in a rocky shore seaweed. *Marine Ecology Progress Series* 226: 273–85.

Berthold, G. 1881. Die geschlechtliche Fortpflanzung der eigentlichen Pheosporeen. *Mittheilungen aus der Zoologischen Station zu Neapel* 11: 401–13.

Berthold, G. 1882. Über die Verteilung der Algen im Golf von Neapel nebst einem Verzeichnis der bisher daselbst beobachteten Arten. *Mittheilungen aus der Zoologischen Station zu Neapel* 3: 393–536.

Bertness, M.D., Leonard, G.H., Levine, J.M., Schmidt, P.R. & Ingraham, A.O. 1999. Testing the relative contribution of positive and negative interactions in rocky intertidal communities. *Ecology* 80: 2711–26.

Bertocci, I., Araújo, R., Vaselli, S. & Sousa-Pinto, I. 2011. Marginal populations under pressure: spatial and temporal heterogeneity of *Ascophyllum nodosum* and associated assemblages affected by human trampling in Portugal. *Marine Ecology Progress Series* 439: 73–82.

Bessey, C.E. 1907. A synopsis of plant phyla. *University Studies of the University of Nebraska* 7: 275–373.

Bhattacharya, D. & Druehl, L.D. 1989. Morphological and DNA sequence variation in the kelp *Costaria costata* (Phaeophyta). *Marine Biology, Berlin* 102: 15–23.

Bhattacharya, D., Mayes, C. & Druehl, L.D. 1991a. Restriction endonuclease analysis of ribosomal DNA sequence variations in *Laminaria* (Phaeophyta). *Journal of Phycology* 27: 624–8.

Bhattacharya, D., Stickel, S.K. & Sogin, M.L. 1991b. Molecular phylogenetic analysis of genic regions from *Achlya bisexualis* (Oomycota) and *Costaria costata* (Chromophyta). *Journal of Molecular Evolution* 33: 525–36.

Bi, Y., Feng, M., Jiang, R., Wu, Z., Zhang, S., & Wang, W. 2016. The effects of sediment on the early settlement stages of *Sargassum horneri* on rocky subtidal reefs. *Aquatic Botany* 132: 17–23.

Biebl, R. 1937. Ökologische und zellphysiologische Untersuchungen an Rotalgen der englischen Südküste. *Beih. Bot. Z. Abt.* A. 57: 381–424.

Biebl, R. 1938. Trockenresistenz und osmotische Empfindlichkeit der Meeresalgen verschieden tiefer Standorte. *Jahrbücher für Wissenschaftliche Botanik* 86: 350–86.

Biebl, R. 1939. Uber die Temperaturresistenz von Meeresalgen verschiedener Klimazonen und verschieden tiefen Standorte. *Jahrbücher für Wissenschaftliche Botanik* 88: 389–420.

Biebl, R. 1958. Temperatur – und osmotische Resistenz von Meeresalgen der bretonischen Küste. *Protoplasma* 50: 217–42.

Billard, E., Daguin, C., Pearson, G., Serrão, E., Engel, C. & Valero, M. 2005. Genetic isolation between three closely related taxa: *Fucus vesiculosus, F. spiralis* and *F. ceranoides* (Phaeophyceae). *Journal of Phycology* 41: 900–5.

Billard, E., Serrão, E.A., Pearson, G.A., Destombe, C. & Valero, M. 2010. *Fucus vesiculosus* and *spiralis* complex: a nested model of local adaptation at the shore level. *Marine Ecology Progress Series* 405: 163–74.

Billard, E., Serrão, E.A., Pearson, G.A., Engel, C.R., Destombe, C. & Valero, M. 2006. Analysis of sexual phenotype and prezygotic fertility in natural populations of *Fucus spiralis, F. vesiculosus* (Fucaceae, Phaeophyceae) and their putative hybrids. *European Journal of Phycology* 40: 397–407.

Billot, C., Boury, S., Benet, H. & Kloareg, B. 1999. Development of RAPD markers for parentage analysis in *Laminaria digitata*. *Botanica Marina* 42: 307–14.

Billot, C., Engel, C.R., Rousvoal, S. & Kloareg, B. 2003. Current patterns, habitat discontinuities and population genetic structure: the case of the kelp *Laminaria digitata* in the English Channel. *Marine Ecology Progress Series* 253: 111–21.

Billot, C., Rousvoal, S., Estoup, A., Epplen, J.T., Saumitou-Laprade, P., Valero, M. & Kloareg, B. 1998. Isolation and characterization of microsatellite markers in the nuclear genome of the brown alga *Laminaria digitata* (Phaeophyceae). *Molecular Ecology* 7: 1778–80.

Bird, N.L. & McLachlan, J. 1976. Control of formation of receptacles in *Fucus distichus* L. subsp. *distichus* (Phaeophyceae, Fucales). *Phycologia* 15: 79–84.

Bird, N.L., Chen, L.C.M. & McLachlan, J. 1979. Effects of temperature, light and salinity on growth in culture of *Chondrus crispus*, *Furcellaria lumbricalis*, *Gracilaria tikvahiae* (Gigartinales, Rhodophyta) and *Fucus serratus* (Fucales, Phaeophyta). *Botanica Marina* 22: 521–7.

Bisalputra, T. 1974. Plastids. In: Stewart, W.D.P. (ed.), *Algal Physiology and Biochemistry*, pp. 124–60. Botanical Monographs 10. Blackwell Scientific Publications.

Bisalputra, T. & Bisalputra, A.A. 1969. The ultrastructure of chloroplast of a brown alga *Sphacelaria* sp. 1. Plastid DNA configuration – the chloroplast genophore. *Journal of Ultrastructural Research* 29: 151–70.

Bisalputra, T. & Burton, H. 1969. The ultrastructure of chloroplast of a brown alga *Sphacelaria* sp. II. Association between the chloroplast DNA and the photosynthetic lamellae. *Journal of Ultrastructural Research* 29: 224–35.

Bisgrove, S.R., Nagasato, C., Motomura, T. & Kropf, D.L. 1997. Immunolocalization of centrin during fertilization and the first cell cycle in *Fucus distichus* and *Pelvetia compressa* (Fucales, Phaeophyceae). *Journal of Phycology* 33: 823–9.

Biskupa, S., Bertocci, I., Arenas, F. & Tuya, F. 2014. Functional responses of juvenile kelps, *Laminaria ochroleuca* and *Saccorhiza polyschides*, to increasing temperatures. *Aquatic Botany* 113: 117–22.

Bitter, G. 1899. Zur Anatomieund Physiologie von *Padina pavonia*. *Berichte der Deutschen Botanischen Gesellschaft* 17: 255–74.

Bittner, L., Payri, C.E., Couloux, A., Cruaud, C., Reviers, B. de & Rousseau, F. 2008. Molecular phylogeny of the Dictyotales and their position within the Phaeophyceae, based on nuclear, plastid and mitochondrial DNA sequence data. *Molecular Phylogenetics and Evolution* 49: 211–26.

Bjærke, M.R. & Fredriksen, S. 2003. Epiphytic macroalgae on the introduced brown seaweed *Sargassum muticum* (Yendo) Fensholt (Phaeophyceae) in Norway. *Sarsia* 88: 353–64.

Black, W.A.P. 1948. The seasonal variation in chemical constitution of some of the sub-littoral seaweeds common to Scotland. Part I. *Laminaria cloustoni*. *Journal of the Society of Chemical Industry* 67: 165–8.

Black, W.A.P., Richardson, W.D. & Walker, F.T. 1959. Chemical and growth gradients of *Laminaria cloustoni* Edm. (*L. hyperborea* Fosl.). *Economic Proceedings of the Royal Dublin Society* 4 (No. 8): 137–49.

Blackler, H. 1956a. Further additions to the algal flora of St. Andrews Fife. *Transactions and Proceedings of the Botanical Society of Edinburgh* 37: 46–60.

Blackler, H. 1956b. Observations on the time of reproduction and the rate of growth of certain Fucaceae. *Proceedings of the International Seaweed Symposium* 2: 158–62.

Blackler, H. 1961. *Desmarestia dudresnayi* Lamouroux in Britain. *British Phycological Journal* 2: 87.

Blackler, H. 1964. Some observations on the genus *Colpomenia* (Endlicher) Derbès et Solier 1851. *Proceedings of the International Seaweed Symposium* 4: 50–4.

Blackler, H. 1967. The occurrence of *Colpomenia peregrina* (Sauv.) Hamel in the Mediterranean (Phaeophyta, Scytosiphonales). *Blumea* 15: 5–8.

Blackler, H. 1974. Flora. In: *Fauna and flora of St Andrews Bay*. (Laverack, M.S. & Blackler, H., Eds), pp. 167–295. Edinburgh.

Blackler, H. 1981. Some algal problems with special reference to *Colpomenia peregrina* and other members of the Scytosiphonaceae. *British Phycological Journal* 16: 133.

Blackler, H. & Jackson, D.C. 1966. The discovery of plurilocular sporangia in *Sphacelaria britannica* Sauv. *British Phycological Bulletin* 3: 85.

Blackler, H. & Katpitia, A. 1963. Observations of the life history and cytology of *Elachista fucicola*. *Transactions and Proceedings of the Botanical Society of Edinburgh* 29: 392–5.

Blanchette, C.A. 1997. Size and survival of intertidal plants in response to wave action: a case study of *Fucus gardnerii*. *Ecology* 78: 1563–78.

Blomquist, H.L. 1955. *Acinetospora* Born. new to North America. *Journal of the Elisha Mitchell Scientific Society* 71: 46–9.

Boaden, P.J.S. 1995. The adventive seaweed *Sargassum muticum* (Yendo) Fensholt in Strangford Lough, Northern Ireland. *Irish Naturalists' Journal* 25: 111–13.

Boaden, P.J.S., O'Connor, R.J. & Seed, R. 1975. The composition and zonation of a *Fucus serratus* community in Strangford Lough, Co. Down. *Journal of Experimental Marine Biology and Ecology* 7: 111–36.

Boalch, G.T. 1957. Marine algal zonation and substratum in Beer bay, south-east Devon. *Journal of the Marine Biological Association of the United Kingdom* 36: 519–28.

Boalch, G.T. 1961. Studies on *Ectocarpus* in culture. II. Growth and nutrition of a bacteria-free culture. *Journal of the Marine Biological Association of the United Kingdom* 41: 287–304.

Boalch, G.T., Holme, N.A. Jephson, N.A. & Sidwell, J.M. 1974. A resurvey of Colman's intertidal traverses at Wembury, South Devon. *Journal of the Marine Biological Association of the United Kingdom* 54: 551–3.

Boalch, G.T. & Potts, G.W. 1977. The first occurrence of *Sargassum muticum* (Yendo) Fensholt in the Plymouth area. *Journal of the Marine Biological Association of the United Kingdom* 57: 29–31.

Boden, G.T. 1979. The effect of depth on summer growth of *Laminaria saccharina* (Phaeophyta, Laminariales). *Phycologia* 18: 405–8.

Boland, W., Marner, F.-J. & Jaenicke, L. 1983. Comparative receptor study in gamete chemotaxis of the seaweeds *Ectocarpus siliculosus* and *Cutleria multifida*. *European Journal of Biochemistry* 134: 97–103.

Bold, H.C. & Wynne, M.J. 1978. *Introduction to the Algae*. Prentice-Hall, New Jersey.

Bold, H.C. & Wynne, M.J. 1985. *Introduction to the Algae: Structure and Reproduction*. 2nd edn. pp. Englewood Cliffs, NJ: Prentice Hall, Inc.

Bollen, M., Battershill, C.N., Pilditch, C.A. & Bischof, K. 2017. Desiccation tolerance of different life stages of the invasive marine kelp *Undaria pinnatifida*: potential for overland transport as invasion vector. *Journal of Experimental Marine Biology and Ecology* 496: 1–8.

Bollen, M., Pilditch, C.A., Battershill, C.N. & Bischof, K. 2016. Salinity and temperature tolerance of the invasive alga *Undaria pinnatifida* and native New

Zealand kelps: Implications for competition. *Marine Biology* 163: 194.

Bolton, J.J. 1979. Estuarine adaptations in populations of *Pilayella littoralis* (L.) Kjellm. (Phaeophyta, Ectocarpales). *Estuarine and Coastal Marine Science* 9: 273–80.

Bolton, J.J. 1983. Ecoclinal variation in *Ectocarpus siliculosus* (Phaeophyceae) with respect to temperature optima and survival limits. *Marine Biology* 73: 131–8.

Bolton, J.J. & Anderson, R.J. 1987. Temperature tolerances of two southern African *Ecklonia* species (Alariaceae: Laminariales) and of hybrids between them. *Marine Biology* 96: 293–7.

Bolton, J.J. & Levitt, G.J. 1985. Light and temperature requirements for growth and reproduction in gametophytes of *Ecklonia maxima* (Alariaceae: Laminariales). *Marine Biology* 87: 131–5.

Bolton, J.J. & Lüning, K. 1992. Optimal growth and maximal survival temperatures of Atlantic *Laminaria* species (Phaeophyta) in culture. *Marine Biology* 66: 89–94.

Bolwell, G.P., Callow, J.A., Callow, M.E. & Evans, L.V. 1979. Fertilization in brown algae. II. Evidence for lectin-sensitive complementary receptors involved in gamete recognition in *Fucus serratus*. *Journal of Cell Science* 36: 19–30.

Bond, P.R., Brown, M.T., Moate, R.M., Gledhill, M., Hill, S.J. & Nimmo, M. 1999. Arrested development in *Fucus spiralis* (Phaeophyceae) germlings exposed to copper. *European Journal of Phycology* 34: 513–21.

Bonsdorff, E. & Nelson, W.G. 1996. Apical growth-measurements of *Fucus vesiculosus* L.: Limited value in monitoring. *Botanica Marina* 39: 129–32.

Boo, S.M., Lee, K.M., Cho, G.Y. & Nelson, W. 2011a. *Colpomenia claytonii* sp. nov. (Scytosiphonaceae, Phaeophyceae) based on morphology and mitochondrial cox3 sequences. *Botanica Marina* 54: 159–67.

Boo, G.H., Lindstrom, S.C., Klochkova, N.C., Yotsukura, N., Yang, E.C., Kim, H.G., Waaland, J.R., Cho, G.Y., Miller, K.A. & Boo, S.M. 2011b. Taxonomy and biogeography of *Agarum* and *Thalassiophyllum* (Laminariales, Phaeophyceae) based on sequences of nuclear, mitochondrial, and plastid markers. *Taxon* 60: 831–40.

Boo, S.-M., Lee, W.J., Hwang, II. Ki., Keum, Y-S., Oak, J.H. & Cho, G.Y. 2010. *Algal flora of Korea. Volume 2, Number 2. Heterokontophyta: Phaeophyceae: Ishigeales, Dictyotales, Desmarestiales, Sphacelariales, Cutleriales, Ralfsiales, Fucales, Laminariales. Marine Brown Algae II*, pp. 1–203, figs 1–48, pls 1–2. Incheon: National Institute of Biological Resources.

Boo, S.M. & Yoon, H.S. 2000. Molecular relationships of giant kelp (Phaeophyceae). *Algae* 15: 13–16.

Boo, S.M., Lee, W.J., Yoon, H.S., Kato, A. & Kawai, H. 1999. Molecular phylogeny of Laminariales (Phaeophyceae) inferred from small subunit ribosomal DNA sequences. *Phycological Research* 47: 109–14.

Børgesen, F. 1896. En for Faerøerne ny *Laminaria*. *Botanisk Tidsskrift* 20: 403–5.

Børgesen, F. 1902. Marine algae. Part II. In: Warming, E. (ed.), *Botany of the Faeroes based upon Danish investigations*, pp. 339–532. Copenhagen: Carlsbergfondet.

Børgesen, F. 1909. *Fucus spiralis* Linné or *Fucus platycarpus* Thuret: a question of nomenclature. *Journal of the Linnean Society, Botany* 39: 105–19.

Børgesen, F. 1914. The marine algae of the Danish West Indies. Part 2. Phaeophyceae. *Dansk Botanisk Arkiv* 2: 1–68.

Børgesen, F. 1926. Marine algae from the Canary Islands especially from Teneriffe and Gran Canaria II. Phaeophyceae. *Det Kgl Danske Videnskabernes Selskab, Biologiske Meddelelser* 6: 1–112.

Børgesen, F. 1935. A list of marine algae from Bombay. *Kongelige Danske Videnskabernes Selskab, Biologiske Meddelelser* 12: 1–64.

Børgesen, F. 1939. Marine algae from the Iranian Gulf especially from the innermost part near Bushire and the Island Kharg. In: Jessen, K. & Spärck, R. (eds), *Danish Scientific Investigations in Iran, Part 1*, pp. 47–141. Copenhagen: Ejnar Munksgaard.

Børgesen, F. 1941. Some marine algae from Mauritius. II. Phaeophyceae. *Kongelige Danske Videnskabernes Selskab, Biologiske Meddelelser* 16: 1–81.

Borja, A. & Gorostiaga, J.M. 1990. Distribución geográfica de *Saccorhiza polychides* (Lightf.) Batt. en la costa vasca. Su posible relación con la temperatura. *Bentos* 6: 1–8.

Bornet, E. 1892a. Note sur quelques *Ectocarpus*. *Bulletin de la Société Botanique de France* 38: 353–72.

Bornet, E. 1892b. Les algues de P.-K.-A. Schousboe. *Mémoires de la Société des Sciences Naturelles et Mathématiques de Cherbourg* 28: 165–376.

Bornet, E. & Thuret, G. 1876–80. *Notes Algologiques*. Paris.

Bory de Saint-Vincent, J.B.G.M. 1822–31. *Dictionnaire classique d'histoire naturelle* (Audouin, I. *et al.* eds) 17 vols. Paris: Rey et Gravier; Baudoin frères.

———. 1822. *Botrytella* 2: 425.

———. 1823. Confervées. Vol. 4, pp. 391–3.

———. 1826. Laminaire, *Laminaria*. Vol. 9, pp. 187–94.

Bory de Saint-Vincent, J.B.G.M. 1827. Cryptogamie. In: *Voyage autour du monde, exécuté par ordre du Roi, sur la corvette du Sa Majesté, 'La Coquille', pendant les années 1822, 1823, 1824 et 1825*. (Duperrey, L.I. Eds), pp. 1–96. Paris: Bertrand.

Bory de Saint-Vincent, J.B. 1828. Botanique, Cryptogamie. In: *Voyage autour du monde, exécuté par ordre du Roi, sur la corvette de Sa Majesté, la Coquille, pendant les années 1822, 1823, 1824 et 1825*. (Duperrey, L.I. eds), pp. 97–200. Paris: Bertrand.

Bory de Saint-Vincent, J.B.G.M. 1832. Hydrophytes. In: *Expédition scientifique de Morée. Section des sciences physiques. Tome III. 2e partie. Botanique*, pp. 1–367 [368]. (Bory de Saint-Vincent, J.B.G.M. eds), pp. 316–37. Paris & Strasbourg.

Bothwell, J.H., Marie, D., Peters, A.F., Cock, J.M. & Coelho, S.M. 2010. Role of endoreduplication and apomeiosis during parthenogenetic reproduction in the model brown alga *Ectocarpus*. *New Phytologist* 188: 111–21.

Bouck, G.B. 1965. Fine structure and organelle associations in brown algae. *Journal of Cell Biology* 26: 523–37.

Bouck, G.B. 1969. Extracellular microtubules. The origin, structure and attachment of flagellar hairs in *Fucus* and *Ascophyllum* antherozoids. *Journal of Cell Biology* 4: 65–86.

Bouck, G.B. 1970. The development and postfertilization fate of the eyespot and the apparent photoreceptor in *Fucus* sperm. *Annals of the New York Academy of Sciences* 175: 673–85.

Boudouresque, C.F., Gerbal, M. & Knoepffler-Péguy, M. 1985. L'algue japonnaise *Undaria pinnatifida* (Phaeophyceae, Laminariales) en Méditerranée. *Phycologia* 24: 364–6.

Brackenbury, A.M., Kang, E.J. & Garbary, D.J. 2006. Air pressure regulation in air bladders of *Ascophyllum nodosum* (Fucales, Phaeophyceae). *Algae* 21: 245–51.

Braun, A. 1855. *Algarum unicellularium* genera nova et minus cognita praemissis observationibus de algis unicellularibus in genere. pp. [1]–114, 6 pls. Lipsiae [Leipzig]: apud W. Engelmann.

Bräutigam, M., Klein, M., Knippers, R. & Müller, D.G. 1995. Inheritance and meiotic elimination of a virus genome in the host *Ectocarpus siliculosus* (Phaeophyceae). *Journal of Phycology* 31: 823–7.

Brawley, S.H. & Quatrano, R.S. 1979. Effects of microtubule inhibitors on pronuclear migration and embryogenesis in *Fucus distichus* (Phaeophyta). *Journal of Phycology* 15: 266–72.

Brawley, S.H., Coyer, J.A., Blakeslee, A.M., Hoarau, G., Johnson, L.E., Byers, J.E., Stam, W.T., & Olsen, J.L. 2009. Historical invasions of the intertidal zone of Atlantic North America associated with distinctive patterns of trade and emigration. *Proceedings of the National Academy of Sciences of the United States of America* 106: 8239–44.

Brawley, S.H. 1992. Fertilization in natural populations of the dioecious brown alga *Fucus ceranoides* and the importance of the polyspermy block. *Marine Biology* 113: 145–57.

Brawley, S.H. & Johnson, L.E. 1991. Survival of fucoid embryos in the intertidal zone depends upon developmental stage and microhabitat. *Journal of Phycology* 27: 179–86.

Brawley, S.H. & Wetherbee, R. 1981. Cytology and ultrastructure. In: Lobban, C. & Wynne, M.J. (eds), *The Biology of Seaweeds*, pp. 248–99. Botanical Monographs 17. Blackwell Scientific Publications, Oxford.

Brawley, S.H., Johnson, L.E., Pearson, G.A., Speransky, V., Li, R. & Serrão, E. 1999. Gamete release at low tide in fucoid algae: maladaptive or advantageous? *American Zoologist* 39: 218–29.

Brebner, G. 1897. On the classification of the Tilopteridaceae. *Proceedings of the Bristol Naturalists' Society* II, 8: 176–87.

Breeman, A.M. 1988. Relative importance of temperature and other factors in determining geographic boundaries of seaweeds: Experimental and phenological evidence. *Helgoland Marine Research* 42: 199–241.

Brenchley, J.L., Raven, J.A. & Johnston, A.M. 1998. Carbon and nitrogen allocation patterns in two intertidal fucoids: *Fucus serratus* and *Himanthalia elongata* (Phaeophyta). *European Journal of Phycology* 33: 307–13.

Breton, T.S., Nettleton, J.C., O'Connell, B. & Bertocci, M. 2018. Fine-scale population genetic structure of sugar kelp, *Saccharina latissima* (Laminariales, Phaeophyceae), in eastern Maine, USA. *Phycology* 57: 32–40.

Bringloe, T.T., Bartlett, C.A.B., Bergeron, E.S., Cripps, K.S.A., Daigle, N.J., Gallagher, P.O., Gallant, A.D., Giberson, R.O.J., Greenough, S.J., Lamb, J.M., Leonard, T.W., MacKay, J.A., McKenzie, A.D.,

Persaud, S.M., Sheng, T., Mills, A.M.E.S., Moore, T.E. & Saunders, G.W. 2018. Detecting *Alaria esculenta* and *Laminaria digitata* (Laminariales, Phaeophyceae) gametophytes in red algae, with consideration of distribution patterns in the intertidal zone. *Phycologia* 57: 1–8.

Brinkhuis, B.H., Tempel, N.R. & Jones, R.F. 1976. Photosynthesis and respiration of exposed salt-marsh fucoids. *Marine Biology* 34: 349–59.

Britton-Simmons, K.H. 2004. Direct and indirect effects of the introduced alga *Sargassum muticum* on benthic, subtidal communities of Washington State, USA. *Marine Ecology Progress Series* 277: 61–78.

Britton-Simmons, K.H., Pister, B., Sánchez, I. & Okamoto, D. 2011. Response of a native, herbivorous snail to the introduced seaweed *Sargassum muticum*. *Hydrobiologia* 661: 87–196.

Brodie, J., Wilbraham, J., Pottas, J. & Guiry, M.D. 2016. A revised check-list of British seaweeds. *Journal of the Marine Biological Association of the United Kingdom* 96: 1005–29.

Brophy, T.C. & Murray, S.N. 1989. Field and culture studies of a population of *Endarachne binghamiae* (Phaeophyta) from southern California. *Journal of Phycology* 25: 6–15.

Brosnan, D.M. & Crumrine, L.L. 1994. Effects of human trampling on marine rocky shore communities. *Journal of Experimental Marine Biology and Ecology* 177: 79–97.

Brown, M.T. & Lamare, M.D. 1994. The distribution of *Undaria pinnatifida* (Harvey) Suringer within Timaru Harbour, New Zealand. *Japanese Journal of Phycology* 42: 63–70.

Buffham, T.H. 1891. The plurilocular zoosporangia of *Asperococcus bullosus* and *Myriotrichia clavaeformis*. *Journal of Botany, British and Foreign* 29: 321–3.

Buffham, T.H. 1893. Algological notes. *Grevillea* 21: 86–93.

Buggeln, R.G. 1974a. Negative phototropism of the haptera of *Alaria esculenta* (Laminariales). *Journal of Phycology* 10: 80–2.

Buggeln, R.G. 1974b. Physiological investigations on *Alaria esculenta* (L.) Grev. (Laminariales) I. Elongation of the blade. *Journal of Phycology* 10: 283–8.

Buggeln, R.G. 1976. The rate of translocation in *Alaria esculenta* (Laminariales, Phaeophyceae). *Journal of Phycology* 12: 439–42.

Buggeln, R.G. 1977. Physiological investigations on *Alaria esculenta* (Laminariales, Phaeophyceae) II. Role of translocation in blade growth. *Journal of Phycology* 13: 212–18.

Buggeln, R.G. 1978a. Physiological investigations on *Alaria esculenta* (Laminariales, Phaeophyceae) III. Exudation by the blade. *Journal of Phycology* 14: 54–6.

Buggeln, R.G. 1978b. Physiological investigations on *Alaria esculenta* (Laminariales, Phaeophyceae). IV. Inorganic and organic nitrogen in the blade. *Journal of Phycology* 14: 156–60.

Buggeln, R.G. 1981. Note: Source-sink relationship in the blade of *Alaria esculenta* (Laminariales, Phaeophyceae). *Journal of Phycology* 17: 102–4.

Buggeln, R.G. & Varangu, L.K. 1983. The cross-wing translocation pathway in the blade of *Alaria esculenta* (Laminariales, Phaeophyta). *Phycologia* 22: 205–9.

Bunker, F.StP.D., Brodie, J.A., Maggs, C.A. & Bunker, A.R. 2017. *Seaweeds of Britain and Ireland*, 2nd edn. Wild Nature Press: Plymouth.

Bunning, V.E. & Muller, D. 1961. Wie messen organismen lunare Zyklen. *Zeitschrift für Naturforschung* 16B: 391–5.

Burel, T., Le Duff, M. & Ar Gall, E. 2019. Updated checklist of the seaweeds of the French coasts, Channel and Atlantic Ocean. *An aod. Les Cahiers Naturalistes de l'Observatoire Marin Brest* 7(1): pp. 1–38.

Burkhardt, E. & Peters, A.F. 1998. Molecular evidence from nrDNA ITS sequences that *Laminariocolax* (Phaeophyceae, Ectocarpales *sensu lato*) is a worldwide clade of closely related kelp endophytes. *Journal of Phycology* 34: 682–91.

Burridge, T.R. & Gorski, J. 1997. The use of biocidal agents as potential control mechanisms for the exotic kelp *Undaria pinnatifida*. CRIMP Technical Rep. 16, CSIRO, Australia.

Burrowes, R., Rousseau, F., Muller, D.G. & de Reviers, B. 2003. Taxonomic placement of *Microzonia* (Phaeophyceae) in the Syringodermatales based on the rbcL and 28S nrDNA sequences. *Cryptogamie, Algologie* 24: 63–73.

Burrows, E.M. & Lodge, S. 1951. Autecology and the species problem in *Fucus*. *Journal of the Marine Biological Association of the UK* 30: 161–76.

Burrows, E.M. & Pybus, C. 1971. *Laminaria saccharina* and marine pollution in north-east England. *Marine Pollution Bulletin* 2: 53–6.

Burrows, E.M. 1961. Experimental ecology with particular reference to the ecology of *Laminaria saccharina* (L.) Lamour. *Recent Advances in Botany*: 187–9.

Burrows, E.M. 1991. *Seaweeds of the British Isles. Volume 2. Chlorophyta*. London: Natural History Museum Publications.

Burrows, E.M., Conway, E., Lodge, S.M. & Powell, H.T. 1954. The raising of intertidal algal zones on Fair Isle. *Journal of Ecology* 42: 283–8.

Burrows, E.M. 1958. Sublittoral algal populations in Port Erin Bay, Isle of Man. *Journal of the Marine Biological Association of the United Kingdom* 37: 687–703.

Burrows, E.M. 1964. An experimental assessment of some of the characters used for specific delimitation in the genus *Laminaria*. *Journal of the Marine Biological Association of the United Kingdom* 44: 137–43.

Burrows, E.M. & Lodge, S.M. 1950. A note on the inter-relationships of *Patella, Balanus* and *Fucus* on a semi-exposed coast. *Report of the Marine Biology Station, Port Erin* 62: 30–4.

Busch, S. & Schmid, R. 2001. Enzymes associated with beta-carboxylation in *Ectocarpus siliculosus* (Phaeophyceae): Are they involved in net carbon acquisition? *European Journal of Phycology* 36: 61–70.

Buschbaum, C., Chapman, A.S. & Saier, B. 2006. How an introduced seaweed can affect epibiota diversity in different coastal systems. *Marine Biology* 148: 743–54.

Cabioc'h, J. 1976. Sur la presence dans la region de Roscoff, de deux Pheophycees non encore mentionnees sur les cotes de France. *Travaux Station Biologique de Roscoff, Nouvelle Série* 23: 23–6.

Cabioc'h, J. 1981. Premières observations de l'algue Japonaise *Sargassum muticum* (Yendo) Fensholt dans la région de Roscoff. *Travaux Station Biologique de Roscoff, Nouvelle Série* 27: 1–2.

Cabioc'h, J., Floc'h, J.Y., Le Toquin, A., Boudouresque, C.F., Meinesz, A. & Verlaque, M. 1992. *Guide des Algues des Mers d'Europe*. pp. p. 231. Lausanne: Delachaux et Niestlé.

Cacabelos, E., Olabarria, C., Incera, M. & Troncoso, J.S. 2010. Do grazers prefer invasive seaweeds? *Journal of Experimental Marine Biology and Ecology* 393: 182–7.

Cairrao, E., Pereira, M.J., Morgado, F., Nogueira, A.J.A., Guilhermino, L. & Soares, A.M.V.M. 2009. Phenotypic variation of *Fucus ceranoides, F. spiralis* and *F. vesiculosus* in a temperate coast (NW Portugal). *Botanical Studies* 50: 205–15.

Callow, J.A., Bolwell, G.P., Evans, L.V. & Callow, M.E. 1981. Isolation and preliminary characterization of receptors involved in egg-sperm recognition in *Fucus serratus*. *Proceedings of the International Seaweed Symposium* 10: 385–90.

Callow, M.E., Coughlan, S.J. & Evans, L.V. 1978. The role of the golgi bodies in polysaccharide sulphation in *Fucus* zygotes. *Journal of Cell Science* 32: 337–56.

Calvo, S., Bárbara, I. & Cremades, J. 1999. Benthic algae of salt-marshes (Corrubedo Natural Park, NW Spain): the flora. *Botanica Marina* 42: 343–53.

Camacho, O., Mattio, L., Draisma, S., Fredericq, S. & Diaz-Pulido, G. 2014. Morphological and molecular assessment of *Sargassum* (Fucales, Phaeophyceae) from Caribbean Colombia, including the proposal of *Sargassum giganteum* sp. nov., *Sargassum schnetteri* comb. nov. and *Sargassum* section *Cladophyllum* sect. nov. *Systematics and Biodiversity* 13: 105–30.

Cambiè, G., Fernández-Marquéz, D. & Muiño, R. 2017. Modelling the distribution and density of the invasive seaweed *Sargassum muticum* (Fucales, Sargassaceae) in shallow subtidal areas. *Marine and Freshwater Research* 68: 244–51.

Campbell, S.J. & Burridge, T.R. 1998. Occurrence of *Undaria pinnatifida* (Phaeophyta: Laminariales) in Port Phillip Bay, Victoria, Australia. *Marine and Freshwater Research* 49: 379–81.

Campbell, S.J., Bité, J.S. & Burridge, T.R. 1999. Seasonal patterns in the photosynthetic capacity, tissue pigment and nutrient content of different development stages of *Undaria pinnatifida* (Phaeophyta: Laminariales) in Port Phillip Bay, South-Eastern Australia. *Botanica Marina* 42: 231–42.

Camus, C., Meynard, A.P., Faugeron, S., Kogame, K. & Correa, J.A. 2005. Differential life history phase expression in two coexisting species of *Scytosiphon* (Phaeophyceae) of northern Chile. *Journal of Phycology* 41: 931–41.

Canovas, F., Mota, C., Ferreira-Costa, J., Serrão, E., Coyer, J., Olsen, J. & Pearson, G. 2011. Development and characterization of 35 single nucleotide polymorphism markers for the brown alga *Fucus vesiculosus*. *European Journal of Phycology* 46: 342–51.

Caram, B. 1955. Sur l'alternance de générations chez *Chordaria flagelliformis*. *Botanisk Tidsskrift* 52: 18–36.

Caram, B. 1957. Sur la sexualite et le developpement d'une Pheophycee: *Cylindrocarpus berkeleyi* (Grev.) Crouan. *Compte Rendu Hebdomadaire des Séances de l'Académie des Sciences. Paris* 245: 440–3.

Caram, B. 1961. Sur l'alternance de générations et de phase cytologiques chez le *Sauvageaugloia griffithsiana*

(Greville) Hamel. *Compte Rendu Hebdomadaire des Séances de l'Académie des Sciences. Paris* 252: 594–6.

Caram, B. 1964. Sur la sexualité et l'alternance de générations d'une Phéophycée: le *Striaria attenuata*. Note. *Compte Rendu Hebdomadaire des Séances de l'Académie des Sciences. Paris* 259: 2495–7.

Caram, B. 1965. Recherches sur la reproduction et le cycle sexuel de quelques Pheophycees. *Vie et Milieu* 16: 21–221.

Caram, B. 1966. Sur la reproduction de deux Striaricées des eaux danoises. *Compte Rendu Hebdomadaire des Séances de l'Académie des Sciences. Paris. Série D* 262: 2333–5.

Caram, B. 1968. Sur la reproduction du *Spermatochnus paradoxus* en Suède. *Compte Rendu Hebdomadaire des Séances de l'Académie des Sciences. Paris. Série D* 266: 1828–30.

Caram, B. 1972. Le cycle de reproduction des Pheophycees-Pheosporees et ses modifications. *Mémoires de la Société Botanique de France* 1972: 151–60.

Caram, B. & Jónsson, S. 1972. Nouvelle inventaire des algues marines de l'Islande. *Acta Botanica Islandica* 1: 5–31.

Caram, B. & Nygren, S. 1970. A propos de la reproduction comparée en France et en Suède d'une Phéophycée-Phéosporée: le *Striaria attenuata*. *Helgoländer Wissenschaftliche Meeresuntersuchungen* 20: 130–5.

Cardinal, A. 1964. Étude sur les Éctocarpacées de la Manche. *Beihefte zur Nova Hedwigia* 15: 1–86.

Carlson, L. 1991. Seasonal variation in growth, reproduction and nitrogen content of *Fucus vesiculosus* L. in the Öresund, Southern Sweden. *Botanica Marina* 34: 447–53.

Carney, L.T. 2011. A multispecies laboratory assessment of rapid sporophyte recruitment from delayed kelp gametophytes. *Journal of Phycology* 47: 244–51.

Carney, L.T. & Edwards, M.S. 2010. Role of nutrient fluctuations and delayed development in gametophyte reproduction by *Macrocystis pyrifera* (Phaeophyceae) in southern California. *Journal of Phycology* 46: 987–96.

Carruthers, W. 1863. *Fucus distichus* Linn. as an Irish plant. 1: 353–5.

Carruthers, W. 1864. Note on the Kilkee *Fucus*. *Journal of Botany, British and Foreign* 2: 54.

Carter, N. 1932. A comparative study of the algal flora of two salt marshes Part I. *Journal of Ecology* 20: 341–70.

Carter, N. 1933a. A comparative study of the algal flora of two salt marshes. Part II. *Journal of Ecology* 21: 128–208.

Carter, N. 1933b. A comparative study of the algal flora of two salt marshes. Part III. *Journal of Ecology* 21: 385–403.

Carter, P.W. 1927. The life history of *Padina pavonia*. 1. The structure and cytology of the tetrasporangial plant. *Annals of Botany* 41: 139–59.

Carvalho, M.C., Hayashizaki, K. & Ogawa, H. 2009. Environment determines nitrogen content and stable isotope composition in the sporophyte of *Undaria pinnatifida* (Harvey) Suringar. *Proceedings of the International Seaweed Symposium* 19: 245–53.

Casares Pascual, C., Gómez Garreta, A., Ribera Siguán, M.A. & Seoane Camba, J.A. 1987. *Sargassum muticum*

(Yendo) Fensholt, nueva cita para la Península Ibérica. *Collectanea Botanica (Barcelona)* 17: 151.

Casas, G.N. & Piriz, M.L. 1996. Surveys of *Undaria pinnatifida* (Laminariales, Phaeophyta) in Golfo Nuevo, Argentina. *Proceedings of the International Seaweed Symposium* 15: 213–15.

Casas, G.N., Piriz, M.L. & Parodi, E.R. 2008. Population features of the invasive kelp *Undaria pinnatifida* (Phaeophyceae: Laminariales) in Nuevo Gulf (Patagonia, Argentina). *Journal of the Marine Biological Association of the United Kingdom* 88: 21–8.

Casas, G., Scrosati, R. & Piriz, M.L. 2004. The invasive kelp *Undaria pinnatifida* (Phaeophyceae, Laminariales) reduces native seaweed diversity in Nuevo Gulf (Patagonia, Argentina). *Biological Invasions* 6: 411–16.

Caspary, R. 1871. Die Seealgen bei Neukuhren an der samlandischen Kiiste in Preuben nach Hensche's sammlung. *Schr. kgl. physik.-ok. Ges. Konigsberg* 12: 138–46.

Castagne, L. 1851. *Supplément au catalogue des plantes qui croissent naturellement aux environs de Marseille*. pp. [1]–125, pls VIII–XI. Aix: Nicot & Pardigon.

Castric-Fey, A., Girard, A. & L'Hardy-Halos, M.T. 1993. The distribution of *Undaria pinnatifida* (Phaeophyceae, Laminariales) on the coast of St. Malo (Brittany, France). *Botanica Marina* 36: 351–8.

Castric-Fey, A., Beaupoil, C., Bouchain, J., Pradier, E. & L'Hardy-Halos, M. Th. 1999a. The introduced alga *Undaria pinnatifida* (Laminariales, Alariaceae) in the rocky shore ecosystem of the St Malo area: growth rate and longevity of the sporophyte. *Botanica Marina* 42: 83–96.

Castric-Fey, A., Beaupoil, C., Bouchain, J., Pradier, E. & L'Hardy-Halos, M. Th. 1999b. The introduced alga *Undaria pinnatifida* (Laminariales, Alariaceae) in the rocky shore ecosystem of the St. Malo area: morphology and growth of the sporophyte. *Botanica Marina* 42: 71–82.

Cavalier-Smith, T. 1986. The kingdom Chromista: origin and systematics. In: Round, F.E. & Chapman, D.J. (eds), *Progress in Phycological Research. Vol. 4*. pp. 309–47. Bristol: Biopress Ltd.

Cavalier-Smith, T. 1995a. Membrane heredity, symbiogenesis, and the multiple origins of algae. In: Arai, R., Kato, M. & Doi, Y. (eds), *Biodiversity and Evolution*. The National Science Museum Foundation, Tokyo, pp. 75–114.

Cavalier-Smith, T. 1995b. Zooflagellate phylogeny and classification. *Cytology* 37: 1010–29.

Cavalier-Smith, T. & Chao, E.E.-Y. 2006. Phylogeny and Megasystematics of Phagotrophic Heterokonts (Kingdom Chromista). *Journal of Molecular Evolution* 62: 388–420.

Cazon, J.P., Viera, M., Sala, S. & Donati, E. 2014. Biochemical characterization of *Macrocystis pyrifera* and *Undaria pinnatifida* (Phaeophyceae) in relation to their potentiality as biosorbents. *Phycologia* 53: 100–8.

Cecere, E., Petrocelli, A. & Saracino, O.D. 2000. *Undaria pinnatifida* (Fucophyceae, Laminariales) spread in the central Mediterranean: its occurrence in the Mar Piccolo of Taranto (Ionian Sea, southern Italy). *Cryptogamie, Algologie* 21: 305–9.

Cecere, E., Alabiso, G., Carlucci, R., Antonella, P. & Verlaque, 2016. Fate of two invasive or potentially invasive alien seaweeds in a central Mediterranean transitional water system: failure and success. *Botanica Marina* 59: 451–62.

Celan, M. 1964. *Ectocarpus lebelii* var. *agigensis* p. 33, pl. I [1], B.

Celia-Plá, P.S.M., Korbee, N., Gómez-Garreta, A. & Figueroa, F.L. 2014. Seasonal photoacclimation patterns in the intertidal macroalga *Cystoseira tamariscifolia* (Ochrophyta). *Scientia Marina* 78: 377–88.

Celis-Plá, P.S.M., Bouzon, Z.L., Hall-Spencer, J.M., Schmidt, E.C., Korbee, N. & Figueroa, F.L. 2016. Seasonal biochemical and photophysiological responses in the intertidal macroalga *Cystoseira tamariscifolia* (Ochrophyta). *Marine Environmental Research* 115: 89–97.

Celis-Plá, P.S.M., Martínez, B., Korbee, N., Hall-Spencer, J.M. & Figueroa, F.L. 2017a. Ecophysiological responses to elevated CO2 and temperature in *Cystoseira tamariscifolia* (Phaeophyceae). *Climatic Change* 142: 67–81.

Celis-Plá, P.S.M., Martínez, B., Korbee, N., Hall-Spencer, J.M. & Figueroa, F.L. 2017b. Photoprotective responses in a brown macroalgae *Cystoseira tamariscifolia* to increases in CO2 and temperature. *Marine Environmental Research* 130: 157–65.

Chalon, J. 1905. *Liste des algues marines observées jusqu'à ce jour entre l'Embouchure de l'Escaut et la Corogne (incl. Iles Anglo-Normandes).* pp. [1]–259. Anvers: Imprimerie J.-E. Buschmann, rempart de la Porte du Rhin.

Chamberlain, A.H.L. 1978. Preliminary observations of apical dominance effects in *Sargassum muticum* (Yendo) Fensholt. *British Phycological Journal* 13: 198.

Chapman, A.R.O. 1969. An experimental approach to the autecology of *Desmarestia aculeata* (L.) Lamour. PhD thesis, University of Liverpool.

Chapman, A.R.O. 1972a. Species delimitation in the filiform, oppositely branched members of the genus *Desmarestia* Lamour. (Phaeophyceae, Desmarestiales) in the northern Hemisphere. *Phycologia* 11: 225–31.

Chapman, A.R.O. 1972b. Morphological variation and its taxonomic implications in the ligulate members of the genus *Desmarestia* occurring on the west coast of North America. *Syesis* 5: 1–20.

Chapman, A.R.O. 1973. Phenetic variability of stipe morphology in relation to season, exposure and depth in the non-digitate complex of *Laminaria* Lamour. (Phaeophyta, Laminariales) in Nova Scotia. *Phycologia* 12: 43–57.

Chapman, A.R.O. 1974. The ecology of macroscopic marine algae. *Annual Review of Ecology, Evolution, and Systematics* 5: 65–80.

Chapman, A.R.O. 1981. Stability of sea urchin dominated barren grounds following destructive grazing of kelp in St. Margaret's Bay, Eastern Canada. *Marine Biology* 62: 307–11.

Chapman, A.R.O. 1989. Abundance of *Fucus spiralis* and ephemeral seaweeds in a high eulittoral zone: effects of grazers, canopy and substratum type. *Marine Biology, Berlin* 102: 565–72.

Chapman, A.R.O. 1990. Competitive interactions between *Fucus spiralis* L. and *Fucus vesiculosus* L. (Fucales, Phaeophyta). *Proceedings of the International Seaweed Symposium* 13: 205–9.

Chapman, A.R.O. 1993. Hard data for matrix modelling of *Laminaria digitata* (Laminariales, Phaeophyta) populations. *Proceedings of the International Seaweed Symposium* 14: 263–7.

Chapman, A.R.O. & Burrows, E.M. 1970. Experimental investigations into the controlling effects of light conditions on the development and growth of *Desmarestia aculeata* (L.) Lamour. *Phycologia* 9: 103–8.

Chapman, A.R.O. & Burrows, E.M. 1971. Field and culture studies of *Desmarestia aculeata* (L.) Lamour. *Phycologia* 10: 63–76.

Chapman, A.R.O. & Craigie, J.S. 1977. Seasonal growth in *Laminaria longicruris*: relations with dissolved inorganic nutrients and internal reserves of nitrogen. *Marine Biology* 40: 197–205.

Chapman, A.R.O. & Craigie, J.S. 1978. Seasonal growth in *Laminaria longicruris*: relations with reserve carbohydrate storage and production. *Marine Biology* 46: 209–13.

Chapman, A.R.O. & Johnson, C.R. 1990. Distribution and organization of macroalgal assemblages in the northwest Atlantic. *Hydrobiologia* 192: 77–121.

Chapman, A.R.O. & Markham, J. 1990. Size-specific concentrations of phlorotannins (anti-herbivore compounds) in three species of *Fucus*. *Marine Ecology Progress Series* 65: 103–4.

Chapman, A.R.O., Markham, J.W. & Lüning, K. 1978. Effects of nitrate concentration on the growth and physiology of *Laminaria saccharina* (Phaeophyta) in culture. *Journal of Phycology* 14: 195–8.

Chapman, A.S. & Fletcher, R.L. 2002. Differential effects of sediments on survival and growth of *Fucus serratus* embryos (Fucales, Phaeophyceae). *Journal of Phycology* 38: 894–903.

Chapman, D.J. & Chapman, V.J. 1961. Life histories in the algae. *Annals of Botany* N.S. 25: 547–61.

Chapman, V.J. 1937. A revision of the marine algae of Norfolk. *Journal of the Linnean Society of London, Botany* 51: 205–63.

Chapman, V.J. 1939. Studies in salt-marsh ecology. *Journal of Ecology* 27: 185–201.

Chapman, V.J. 1941. *An introduction to the study of algae.* pp. 1–387, Cambridge University Press.

Chapman, V.J. 1960. Saltmarshes and salt deserts of the world. In: Polunin, N. (ed.), *Plant Science Monographs*, pp. 1–392. L. Hill, London.

Chapman, V.J. 1963. The marine algae of Jamaica. Part 2. Phaeophyceae and Rhodophyceae. *Bulletin of the Institute of Jamaica: Science Series* 12: 1–201.

Charrier, B., Coelho, S.M., Le Bail, A., Tonon, T., Michel, G., Potin, P., Kloareg, B. & Boyen, C. 2008. Development and physiology of the brown alga *E. siliculosus*; two centuries of research. *New Phytologist* 178: 319–22.

Chater, E.H. 1927. On the distribution of the larger brown algae in Aberdeenshire estuaries. *Transactions and Proceedings of the Botanical Society of Edinburgh* 29: 362–80.

Chaubard, L.A. & Bory de Saint-Vincent, J.B.J.M. 1838. *Nouvelle flore du Péloponnèse et des Cyclades.* Paris.

Chauvin, F.J.L. 1842. Recherches sur l'organisation, la fructification et la classification de plusieurs genres d'Algues. Caen.

Cheang, C.C., Chu, K.H., Fujita, D., Yoshida, G., Hiraoka, M., Critchley, A., Choi, H.G., Duan, D.,

Serisawa, Y. & Ang, P.O., Jr. 2010. Low genetic variability of *Sargassum muticum* (Phaeophyceae) revealed by a global analysis of native and introduced populations. *Journal of Phycology* 46: 1063–74.

Chemin, E. 1922. Sur la parasitisme de *Sphacelaria bipinnata* Sauvageau. *Compte Rendu Hebdomadaire des Séances de l'Académie des Sciences. Paris.* 174: 244.

Chemin, E. 1934. Une excursion algologique aux îles Anglo-Normandes. *Bulletin du Laboratoire maritime de Saint-Servan* 12: 1–22.

Chi, E.Y. 1971. Brown algal pyrenoids. *Protoplasma* 72: 101–4.

Chiaje, S. delle 1829. *Hydrophytologiae Regni Neapolitani icones*. II. Vol. Fasc. 2 pp. 1–11, pls LI–C [51–100]. Neapoli [Naples]: Ex typographia Cataneo et Fernandes.

Cho, G.Y. 2010. Laminariales. In: Anon. (ed.), *Algal flora of Korea. Volume 2, Number 2. Heterokontophyta: Phaeophyceae: Ishigeales, Dictyotales, Desmarestiales, Sphacelariales, Cutleriales, Ralfsiales, Laminariales*, pp. 175–96. Incheon: National Institute of Biological Resources.

Cho, G.Y. & Boo, S.M. 2006. Phylogenetic position of *Petrospongium rugosum* (Ectocarpales, Phaeophyceae): insights from the protein-coding plastid *rbc*L and *psa*A gene sequences. *Cryptogamie, Algologie* 27: 3–15.

Cho, G.Y., Yoon, H.S., Choi, H.G., Kogame, K. & Boo, S.M. 2001. Phylogeny of the family Scytosiphonaceae (Pheophyta) from Korea based on sequences of plastid-encoded RuBisCo spacer region. *Algae* 16: 145–50.

Cho, G.Y., Yang, E.C., Lee, S.H. & Boo, S.M. 2002. First description of *Petalonia zosterifolia* and *Scytosiphon gracilis* (Scytosiphonaceae, Phaeophyceae) from Korea with special reference to nrDNA ITS sequence comparisons. *Algae* 17: 135–44.

Cho, G.Y., Lee, S.H. & Boo, S.M. 2004. A new brown algal order, Ishigeales (Phaeophyceae), established on the basis of plastid protein-coding *rbc*L, *psa*A, and *psb*A region comparisons. *Journal of Phycology* 40: 921–36.

Cho, G.Y., Boo, S.M., Nelson, W. & Clayton, M.N. 2005a. Genealogical partitioning and phylogeography of *Colpomenia peregrina* (Scytosiphonaceae), Phaeophyceae) based on plastid rbcL and nuclear ribosomal DNA internal transcribed spacer sequences. *Phycologia* 44: 101–11.

Cho, G.Y., Kim, M.S. & Boo, S.M. 2005b. Phylogenetic relationships of *Soranthera ulvoidea* (Chordariaceae, Phaeophyceae) on the basis of morphology and molecular data. *Algae* 20: 91–7.

Cho, G.Y., Klochkova, N.G., Krupnova, T.N. & Boo, S.M. 2006a. The reclassification of *Lessonia laminarioides* (Laminariales, Phaeophyceae): *Pseudolessonia* gen. nov. *Journal of Phycology* 42: 1289–99.

Cho, G.Y., Kogame, K. & Boo, Y.M. 2006b. Molecular phylogeny of the family Scytosiphonaceae (Phaeophyceae). *Algae* 21(2): 175–83.

Cho, G.Y., Rousseau, F., Reviers, B. de. & Boo, S.M. 2006c. Phylogenetic relationships within the Fucales (Phaeophyceae) assessed by the photosystem I coding *psa*A sequences. *Phycologia* 45: 512–19.

Cho, G.Y., Kogame, K., Kawai, H. & Boo, S.M. 2007. Genetic diversity of *Scytosiphon lomentaria*

(Scytosiphonaceae, Phaeophyceae) from the Pacific and Europe, based on the ITS of nrDNA, rbcL and rbc spacer regions. *Phycologia* 46: 657–65.

Cho, S.M., Lee, S.M., Ko, Y.D., Mattio, L. & Boo, S.M. 2012. Molecular systematic reassessment of *Sargassum* (Fucales, Phaeophyceae) in Korea using four gene regions. *Botanica Marina* 55: 473–84.

Cho, T.O., Cho, G.Y., Yoon, H.S., Boo, S.M. & Lee, W.J. 2003. New records of *Myelophycus cavus* (Scytosiphonaceae, Phaeophyceae) in Korea and the taxonomic position of the genus on the basis on a plastid DNA phylogeny. *Nova Hedwigia*. 76: 381–97.

Choat, J.H. & Black, R. 1979. Life histories of limpets and the limpet-laminarian relationship. *Journal of Experimental Marine Biology and Ecology* 41: 25–50.

Chock, J.S. & Mathieson, A.C. 1976. Ecological studies of the salt marsh ecad *scorpioides* (Hornemann) Hauck of *Ascophyllum nodosum* (L.) Le Jolis. *Journal of Experimental Marine Biology and Ecology* 23: 171–90.

Chock, J.S. & Mathieson, A.C. 1978. Physiological ecology of *Ascophyllum nodosum* (L.) Le Jolis and its detached ecad *scorpioides* (Hornemann) Hauck (Fucales, Phaeophyta). *Botanica Marina* 22: 21–6.

Chock, J.S. & Mathieson, A.C. 1983. Variations of New England estuarine seaweed biomass. *Botanica Marina* 26: 87–97.

Choi, H.G., Kim, Y.S., Lee, S.J., Park, E.J. & Nam, K.W. 2005. Effects of daylength, irradiance and settlement density on the growth and reproduction of *Undaria pinnatifida* gametophytes. *Journal of Applied Phycology* 17: 423–30.

Chopin, T., Marquis, P.A. & Belyea, E.P. 1996. Seasonal dynamics of phosphorus and nitrogen contents in the brown alga *Ascophyllum nodosum* (L) Le Jolis, and its associated species *Polysiphonia lanosa* (L.) Tandy and *Pilayella littoralis* (L.) Kjellman, from the Bay of Fundy, Canada. *Botanica Marina* 39: 543–52.

Christensen, T. 1958. Unilocular sporangia in *Ascocyclus orbicularis*. *Revue Algologique, Nouvelle Série* 4: 129–32.

Christensen, T. 1962. *Botanik. Bind II. Systematisk botanik. Nr. 2*. Alger. pp. [1]–178. København: I kommission hos Munksgaard.

Christensen, T. 1980–84. *Algae: A Taxonomic Survey*. AiO Tryk, Odense.

Christensen, T. & Thomsen, H.A. 1974. Algeforteg-nelse. Universitetsbogladen/Naturfagsbog-laden, Kobenhavn.

Christie, A.O. 1973. Spore settlement in relation to fouling by *Enteromorpha*. In: Acker, R.F., Floyd Brown, B., De Palma, J.R. & Iverson, W.P. (eds), *Proceedings of the 3rd International Congress on Marine Corrosion and Fouling* pp. 674–81. Northwestern University Press, Evanston.

Chuck, J.S. & Mathieson, A.C. 1976. Ecological studies of the salt marsh ecad scorpioides (Hornemann) Hauck of *Ascophyllum nodosum* (L.) Le Jolis. *Journal of Experimental Marine Biology and Ecology* 23: 171–90.

Church, A.H. 1898. The polymorphy of *Cutleria multifida* (Grev.). *Annals of Botany* 12: 75–109.

Cie, D.K. & Edwards, M.S. 2008. The effect of high irradiance on the settlement competency and viability of kelp zoospores *Journal of Phycology* 44: 495–500.

Clayton, M.N. 1972. The occurrence of variant forms in cultures of species of *Ectocarpus* and *Giffordia*. *British Phycological Journal* 7: 101–8.

Clayton, M.N. 1974. Studies on the development, life history and taxonomy of the Ectocarpales (Phaeophyta) in Southern Australia. *Australian Journal of Botany* 22: 743–813.

Clayton, M.N. 1975. A study of variation in Australian species of *Colpomenia* (Phaeophyta, Scytosiphonales). *Phycologia* 14: 187–95.

Clayton, M.N. 1976a. Complanate *Scytosiphon lomentaria* (Lyngbye) J. Agardh (Scytosiphonales: Phaeophyta) from Southern Australia: the effects of season, temperature and daylength on the life history. *Journal of Experimental Marine Biology and Ecology* 25: 187–98.

Clayton, M.N. 1976b. The morphology, anatomy and life history of a complanate form of *Scytosiphon lomentaria* (Scytosiphonales, Phaeophyta) from Southern Australia. *Marine Biology, Berlin* 28: 201–8.

Clayton, M.N. 1979. The life history and sexual reproduction of *Colpomenia peregrina* (Scytosiphonaceae, Phaeophyta) in Australia. *British Phycological Journal* 14: 1–10.

Clayton, M.N. 1980a. Sexual reproduction – a rare occurrence in the life history of the complanate form of *Scytosiphon* (Scytosiphonaceae, Phaeophyta) from Southern Australia. *British Phycological Journal* 15: 105–18.

Clayton, M.N. 1980b. Observations on the factors controlling the reproduction of two common species of brown algae, *Colpomenia peregrina* and *Scytosiphon* sp. (Scytosiphonaceae), in Victoria. *Proceedings of the Royal Society of Victoria* 92: 113–18.

Clayton, M.N. 1981. Correlated studies on seasonal changes in the sexuality, growth rate and longevity of complanate *Scytosiphon* (Scytosiphonaceae: Phaeophyta) from Southern Australia growing in situ. *Journal of Experimental Marine Biology and Ecology* 51: 87–98.

Clayton, M.N. 1984a. An electron microscope study of gamete release and settling in the complanate form of *Scytosiphon* (Scytosiphonaceae, Phaeophyta). *Journal of Phycology* 20: 276–85.

Clayton, M.N. 1984b. Evolution of the Phaeophyta with particular reference to the Fucales. In: Round, F.E. & Chapman, D.J. (eds), *Progress in Phycological Research* 3, pp. 11–46. Biopress Ltd, Bristol.

Clayton, M.N. 1985. A critical investigation of the vegetative anatomy, growth and taxonomic affinities of *Adenocystis*, *Scytothamnus*, and *Splachnidium* (Phaeophyta). *British Phycological Journal* 20: 285–96.

Clayton, M.N. 1986. Culture studies on the life history of *Scytothamnus australis* and *Scytothamnus fasciculatus* (Phaeophyta) with electron microscope observations on sporogenesis and gametogenesis. *British Phycological Journal* 21: 371–86.

Clayton, M.N. 1987. Isogamy and a fucalean type of life history in the Antarctic brown alga *Ascoseira mirabilis* (Ascoseirales, Phaeophyta). *Botanica Marina* 30: 447–54.

Clayton, M.N. 1990. 35. Phaeophyta In: Margulist, L., Corliss, J.L., Melkonian, M., Chapman, D.J. & MacKamm (editorial coordinators). *Handbook of Protocista*. Boston, Jones & Bartlett Publ., pp. 698–714.

Clayton, M.N. & Ducker, S.C. 1970. The life history of *Punctaria latifolia* Greville (Phaeophyta) in Southern Australia. *Australian Journal of Botany* 18: 293–300.

Clayton, M.N. & King, R.J. (eds). 1981. *Marine Botany: An Australasian Perspective*. Melbourne: Longman Cheshire.

Clayton, M.N. & King, R.J. (eds). 1990. *Biology of Marine Plants*. Melbourne: Longman Cheshire.

Clements, F.E. 1905. *Research Methods in Ecology*. Lincoln, Nebraska: The University Publishing Company.

Clint, H.B. 1927. The life history and cytology of *Sphacelaria bipinnata* Sauv. *Publications of the Hartley Botanical Laboratories, Univ. of Liverpool* 3: 5–23.

Clokie, J.J.P. & Boney, A.D. 1980. The assessment of changes in intertidal ecosystems following major reclamation work: framework for interpretation of algal-dominated biota and the use and misuse of data. In: Price, J.H., Irvine, D.E.G. & Farnham, W.F. (eds), *The Shore Environment 2: ecosystems*, pp. 609–75. Systematics Association Special Volume 17(b). Academic Press, London and New York.

Clokie, J.J.P. & Norton, T.A. 1974. The effects of grazing on the algal vegetation of pebbles from the Firth of Clyde. *British Phycological Journal* 9: 216.

Cobacho, S.P., Navarro, L., Pedrol, N. & Sánchez, J.M. 2018. Shading by invasive seaweeds reduces photosynthesis of maerl from the Ría de Vigo (NW Spain). *Botanica Marina* 61: 453–7.

Cock, J.M., Sterck, L., Rouze, P., Scornet, D., Allen, A.E., Amoutzias, G., Anthouard, V., Artiguenave, F., Aury, J.-M. & Badger, J.M. *et al.* 2010. The *Ectocarpus* genome and the independent evolution of multicellularity in the brown algae. *Nature* 465: 617–21.

Cole, K. 1967. The cytology of *Eudesme virescens* (Carm.) J. Ag. I. Meiosis and chromosome numbers. *Canadian Journal of Botany* 45: 665–73.

Cole, K. 1969. The cytology of *Eudesme virescens* (Carm.) J. Ag. II. Ultrastructure of the cortical cells. *Phycologia* 8: 101–8.

Coleman, M.A. & Brawley, S.H. 2005a. Variability in temperature and historical patterns in reproduction in the *Fucus distichus* complex (Heterokontophyta, Phaeophyceae): implications for speciation and the collection of herbarium specimens. *Journal of Phycology* 41: 1110–19.

Coleman, M.A. & Brawley, S.H. 2005b. Are life history characteristics good predictors of genetic diversity and structure? A case study of the intertidal alga *Fucus spiralis* (Heterokontophyta, Phaeophyceae). *Journal of Phycology* 41: 753–62.

Coleman, R.A., Ramchunder, S.J., Moody, A.J. & Foggo, A. 2007. An enzyme in snail saliva induces herbivore-resistance in a marine alga. *Functional Ecology* 21: 101–6.

Collén, J. & Davison, I.R. 1997. *In vivo* measurement of active oxygen production in the brown alga *Fucus evanescens* using 2′,7′-dichlorohydrofluorescein diacetate. *Journal of Phycology* 33: 643–8.

Collin, F. & van den Hoek, C. 1971. The life history of *Sphacelaria furcigera* Kütz. (Phaeophyceae) II. The influence of daylength and temperature on sexual and vegetative reproduction. *Nova Hedwigia* 11: 899–922.

Collins, F.S. 1896. Notes on New England marine algae-VII. *Bulletin of the Torrey Botanical Club* 23: 458–62.

Collins, F.S. 1906. New species, etc., issued in the Phycotheca Boreali-Americana. *Rhodora* 1908: 104–13.

Collins, F.S., Holden, I. & Setchell, W.A. 1899. *Phycotheca boreali-americana. A collection of dried specimens of the algae of North America.* Vol. Fasc. XII–XIII, Nos. 551–650. Malden, Massachusetts.

Colman, J. 1933. The nature of the intertidal zonation of plants and animals. *Journal of the Marine Biological Association of the United Kingdom* 18: 435–76.

Connell, J.H. & Slatyer, R.O. 1977. Mechanisms of succession in natural communities and their role in community stability and organization. *American Naturalist* 111: 1119–44.

Conolly, N.J. & Drew, E.A. 1985. Physiology of *Laminaria* III. Effect of a coastal eutrophication gradient on seasonal patterns of growth and tissue composition in *L. digitata* Lamour. and *L. saccharina* (L.) Lamour. *Marine Ecology* 6: 181–95.

Conover, J.T. 1968. The importance of natural diffusion gradients and transport of substances related to benthic marine plant metabolism. *Botanica Marina* 11: 1–9.

Contreras, L., Dennett, G., Moenne, A., Palma, R.E. & Correa, J.A. 2007. Molecular and morphologically distinct *Scytosiphon* species (Scytosiphonales, Phaeophyceae) display similar antioxidant capacities. *Journal of Phycology* 43: 1320–8.

Coppejans, E. 1983a. *Spongonema tomentosum* (Huds.) Kütz. (Phaeophyta, Ectocarpales), nouveau pour la flore du Boulonnais (Pas-de-Calais, France). *Dumortiera* 27: 1–5.

Coppejans, E. 1983b. Iconographie d'algues Méditerranées. Chlorophyta, Phaeophyta, Rhodophyta. *Bibliotheca Phycologica* 63: 1–28, 367 plates.

Coppejans, E. & Boudouresque, C.F. 1976. Présence de *Choristocarpus tenellus* (Kuetzing) Zanardini à Port-Cros. *Travaux Scientifiques du Parc national de Port Cros* 2: 195–7.

Coppejans, E. & Dhondt, F. 1976. Vegetation de l'ile de Port-Cros (Parc National) XIV. *Myrionema liechtensternii* Hauck (Phaeophyta-Chordariales), espece nouvelle pour la flora algologique de France. *Biologisch Jaarboek Dodonaea* 44: 112–17.

Coppejans, E. & Kling, R. 1995. *Flora algologique des côtes du Nord de la France et de la Belgique.* pp. [1]–454. Meise: Jardin Botanique National de la Belgique.

Coppejans, E., Rappe, G., Podoor, N. & Asperges, M. 1980. *Sargassum muticum* (Yendo) Fensholt ook langs de Belgische kust Aangespoeld. *Dumortiera* 16: 7–13.

Corguille, G. Le., Pearson, G., Valente, M., Viegas, C., Gschloessl, B., Corre, E., Bailly, X., Peters, A.F., Jubin, C., Vacherie, B., Cock, J.M. & Leblanc, C. 2009. Plastid genomes of two brown algae, *Ectocarpus siliculosus* and *Fucus vesiculosus*; further insights on the evolution of red-algal derived plastids. *BMC Evolutionary Biology,* 9: no. 253.

Cormaci, M. & Furnari, G. 1979. Flora algale marina della Sicilia orientale: 'Rhodophyceae', 'Phaeophyceae' e 'Chlorophyceae'. *Informatore Botanico Italiano* 11: 221–50.

Cormaci, M. & Furnari, G. 1988. On the occurrence in southern Italy of some benthic marine algae rare to the Mediterranean Sea. *Giornale Botanico Italiano* 122: 215–26.

Cormaci, M., Furnari, G., Giaccone, G. & Serio, D. 2004. Alien macrophytes in the Mediterranean Sea: a review. *Recent Research Developments in Environmental Biology* 1: 153–202.

Cormaci, M., Furnari, G., Catra, M., Alongi, G. & Giaccone, G. 2012. Flora marina bentonica del Mediterraneo: Phaeophyceae. *Bollettino dell'Accademia Gioenia* 45: 1–510.

Corre, S. & Prieur, D. 1990. Density and morphology of epiphytic bacteria on the kelp *Laminaria digitata.* *Botanica Marina* 33: 515–23.

Cosson, J. 1972. Action de la temperature et de la lumière sur l'émission des spores de la *Laminaria digitata* (L.) Lamouroux (Phéophycée, Laminaliela). *Compte Rendu Hebdomadaire des Séances de l'Académie des Sciences. Paris. Série D* 275: 2501–4.

Cosson, J. 1973a. Influence des conditions de culture sur le développement de *Laminaria digitata* (L.) Lam. (Phéophycées, Laminariales). *Bulletin de la Société Phycologique de France* 18: 104–12.

Cosson, J. 1973b. Action de la température et de la lumière sur le développement du gamétophyte de la *Laminaria digitata* (L.) Lam. (Phéophycée, Laminariale). *Compte Rendu Hebdomadaire des Séances de l'Académie des Sciences. Paris. Série D* 276: 973–6.

Cosson, J. 1975. Action des conditions d'éclairement sur la croissance des gamétophytes de *Laminaria digitata* (L.) Lamouroux (Phéophycée, Laminariale). *Bulletin de la Société Phycologique de France* 20: 50–4.

Cosson, J. 1976. Evolution de la fertilité des populations de *Laminaria digitata* (L.) Lamouroux (Phéophycée, Laminariale) au cours de l'année. *Bulletin de la Société Phycologique de France* 21: 28–34.

Cosson, J. 1977. Action de la durée d'éclairement sur la morphogénèse des gamétophytes de *Laminaria digitata* (L.) Lam. (Phéophycée, Laminariale). *Bulletin de la Société Phycologique de France* 22: 19–26.

Cosson, J. 1999. Sur la disparition progressive de *Laminaria digitata* sur les côtes du Calvados (France). *Cryptogamie, Algologie* 20: 35–42.

Cosson, J. & Gayral, P. 1979. Optimal conditions for growth and fertility of *Laminaria digitata* (Phaeophyceae) Gametophytes. *Proceedings of the International Seaweed Symposium* 9: 59–65.

Cosson, J., Duglet, A. & Billard, C. 1977. Sur la végétation algale de l'étage littoral dans la région de Saint-Vaast-la-Hougue et la présence d'une espèce japonaise nouvelle pour les côtes françaises: *Sargassum muticum* (Yendo) Fensholt (Pheophycée, Fucale). *Bulletin de la Société Linnéenne de Normandie* 105: 109–16.

Cosson, J., Gayral, P. & Jacques, R. 1976. Action de la composition spectrale de la lumière sur la croissance et la reproduction des gamétophytes de la *Laminaria digitata* (L.) Lam (Phéophycée, Laminariale). *Compte Rendu Hebdomadaire des Séances de l'Académie des Sciences. Paris. Série D* 283: 1293–6.

Costas, E., Aguilera, A., González-Gil, S. & López-Rodas, V. 1994. Early development in *Fucus spiralis*: analysis of surface glycan moieties and cell differentiation using lectins. *Botanica Marina* 37: 11–15.

Cotton, A.D. 1906. On some endophytic algae. *Journal of the Linnean Society of London, Botany* 37: 288–97.

Cotton, A.D. 1907. Some British species of Phaeophyceae. *Journal of Botany, London* 45: 368–73.

Cotton, A.D. 1908a. *Leathesia crispa* Harv. *Journal of Botany, London* 46: 329–31.

Cotton, A.D. 1908b. The appearance of *Colpomenia sinuosa* in Britain. *Bulletin of Miscellaneous Information (Royal Botanic Gardens, Kew)* 1908: 73–7.

Cotton, A.D. 1908c. *Colpomenia sinuosa* in Britain. *Journal of Botany, London* 46: 82–3.

Cotton, A.D. 1911. On the increase of *Colpomenia sinuosa* in Britain. *Bulletin of Miscellaneous Information (Royal Botanic Gardens, Kew)* 1911: 153–7.

Cotton, A.D. 1912. Marine Algae. In: Praeger, R.L. (ed.), *A biological survey of Clare Island in the County of Mayo, Ireland and of the adjoining district. Proceedings of the Royal Irish Academy* 31 sect. 1: 1–178.

Coudret, A., Feron, F. & Giraud, G. 1985. Effets d'une émersion prolongée sur le métabolisme photosynthétique de *Fucus vesiculosus* en culture expérimentale. *Physiologie végétale* 23: 155–61.

Cousens, R. 1981. Variation in annual production by *Ascophyllum nodosum* (L.) Le Jolis with degree of exposure to wave action. *Proceedings of the International Seaweed Symposium* 10: 253–8.

Cousens, R. 1982a. Popular misconceptions about *Ascophyllum nodosum* and their effect on the interpretation of its population dynamics from existing data. *British Phycological Journal* 17: 231.

Cousens, R. 1982b. The effect of exposure to wave action on the morphology and pigmentation of *Ascophyllum nodosum* (L.) Le Jolis in south-eastern Canada. *Botanica Marina* 25: 191–5.

Cousens, R. 1984. Estimation of annual production by the intertidal brown alga *Ascophyllum nodosum* (L.) Le Jolis. *Botanica Marina* 27: 217–27.

Cousens, R. 1985. Frond size distributions and the effects of the algal canopy on the behaviour of *Ascophyllum nodosum* (L.) Le Jolis. *Journal of Experimental Marine Biology and Ecology* 92: 231–49.

Cox, S., Gupta, S. & Abu-Ghannam, N. 2011. Application of response surface methodology to study the influence of hydrothermal processing on phytochemical constituents of the Irish edible brown seaweed *Himanthalia elongata*. *Botanica Marina* 54: 471–80.

Coyer, J.A., Olsen, J.L. & Stam, W.T. 1997. Genetic variability and spatial separation in the sea palm kelp *Postelsia palmaeformis* (Phaeophyceae) as assessed with M13 fingerprints and RAPD. *Journal of Phycology* 33: 561–8.

Coyer, J.A., Smith, G.J. & Andersen, R.A. 2001. Evolution of *Macrocystis* spp. (Phaeophyceae) as determined by ITS1 and ITS2 sequences. *Journal of Phycology* 37: 574–85.

Coyer, J.A., Hoarau, G., Stam, W.T. & Olsen, J.L. 2004. Geographically specific heteroplasmy of mitochondrial DNA in the seaweed, *Fucus serratus* (Heterokontophyta: Phaeophyceae, Fucales). *Molecular Ecology* 13: 1323–6.

Coyer, J.A., Peters, A.F., Stam, W.T. & Olsen, J.L. 2003. Post-Ice Age recolonization and differentiation of *Fucus serratus* L. (Fucaceae: Phaeophyta) populations in Northern Europe. *Molecular Ecology* 12: 1817–29.

Coyer, J.A., Peters, A.F., Hoarau, G., Stam, W.T. & Olsen, J.L. 2002a. Inheritance patterns of ITS1, chloroplasts, and mitochondria in artificial hybrids of the marine rockweeds, *Fucus serratus* and *F. evanescens* (Heterokontophyta; Fucaceae). *European Journal of Phycology* 37: 173–8.

Coyer, J.A., Peters, A.F., Hoarau, G., Stam, W.T. & Olsen, J.L. 2002b. Hybridization of the marine seaweeds, *Fucus serratus* and *Fucus evanescens* (Heterokontophyta: Phaeophyceae) in a 100-year-old zone of secondary contact. *Proceedings of the Royal Society of London, Ser. B – Biol. Sci.* 269: 1829–34.

Coyer, J.A., Veldsink, J.H., Stam, W.T. & Olsen, J.L. 2002c. Characterization of microsatellite loci in the marine rockweeds, *Fucus serratus* and *F. evanescens* (Heterokontophyta; Fucaceae). *Molecular Ecology Notes* 2: 35–7.

Coyer, J.A., Hoarau, G., Skage, M., Stam, W.T. & Olsen, J.L. 2006a. Origin of *Fucus serratus* (Heterokontophyta: Fucaceae) populations in Iceland and the Faroes: a microsatellite-based assessment. *European Journal of Phycology* 41: 235–46.

Coyer, J.A., Hoarau, G., Pearson, G.A., Serrão, E.A., Stam, W.T. & Olsen, J.L. 2006b. Convergent adaptation to a marginal habitat by homoploid hybrids and polyploid ecads in the seaweed genus *Fucus*. *Biology Letters* 2: 405–8.

Coyer, J.A., Hoarau, G., Oudot-Le Secq, M.P., Stam, W.T. & Olsen, J.L. 2006c. A mtDNA based phylogeny of the brown algal genus *Fucus* (Heterokontophyta; Phaeophyta). *Molecular Phylogenetics and Evolution* 39: 209–22.

Craigie, J.S. 1974. Storage products. In: Stewart, W.D.P. (ed.) *Algal physiology and biochemistry*, pp. 206–35. Botanical Monographs 10. Blackwell Scientific Publications, Oxford.

Creed, J.C. 1995. Spatial dynamics of a *Himanthalia elongata* (Fucales, Phaeophyta) population. *Journal of Phycology* 31: 851–9.

Creed, J.C., Kain (Jones), J.M. & Norton, T.A. 1992. The importance of density and self-thinning in a population of *Fucus vesiculosus*. *British Phycological Journal* 27: 87.

Creed, J.C., Norton, T.A. & Harding, S.P. 1996. The development of size structure in a young *Fucus serratus* population. *European Journal of Phycology* 31: 203–9.

Creese, R.G. 1988. Ecology of molluscan grazers and their interactions with marine algae in north-eastern New Zealand: a review. *New Zealand Journal of Marine and Freshwater Research* 22: 427–44.

Cremades, J. & Pérez-Cirera, J.L. 1990. Nuevas combinaciones de algas bentónicas marinas, como resultado del estudio del herbario de Simón de Rojas Clemente y Rubio (1777–1827). *Anales del Jardín Botánico de Madrid* 47: 489–92.

Cremades, J., Freire-Gago, Ó. & Peteiro, C. 2006. Biología, distribución e integración del alga alóctona *Undaria pinnatifida* (Laminariales, Phaeophyta) en las comunidades bentónicas de las costas de Galicia (NW de la Península Ibérica). *Anales del Jardín Botánico de Madrid* 63: 169–87.

Cremades, J., Salinas, J.M., Granja, A., Bárbara, I., Veiga, A.J., Pérez-Cirera, J.L. & Fuertes, C. 1997. Factores que influyen en la viabilidad y crecimiento de *Undaria pinnatifida* en cultivo: *fouling*, tamaño de plántula y periodos de aclimatación presiembra. *Nova Acta Cientifica Compostelana (Bioloxia)* 7: 29–40.

Cribb, A.B. 1965. An ecological and taxonomic account of the algae of a semi-marine cavern, Paradise Cave, Queensland. *Papers from the Department of Botany University of Queensland* 4: 259–82.

Critchley, A.T. 1980. The further spread of *Sargassum muticum*. *British Phycological Journal* 15: 194.

Critchley, A. 1981. Age determination of *Sargassum muticum* (Yendo) Fensholt. *British Phycological Journal* 16: 134.

Critchley, A.T. 1983a. Experimental observations on variability of leaf and air vesicle shape of *Sargassum muticum*. *Journal of the Marine Biological Association of the United Kingdom* 63: 825–31.

Critchley, A.T. 1983b. *Sargassum muticum*: a morphological study of European material. *Journal of the Marine Biological Association of the United Kingdom* 63: 813–24.

Critchley, A.T. 1983c. *Sargassum muticum*: a taxonomic history including world-wide and western Pacific distributions. *Journal of the Marine Biological Association of the United Kingdom* 63: 617–25.

Critchley, A.T. 1983d. The establishment and increase of *Sargassum muticum* (Yendo) Fensholt populations within the Solent area of southern Britain. I. An investigation of the increase in number of population individuals. *Botanica Marina* 26: 539–45.

Critchley, A.T. 1983e. The establishment and increase of *Sargassum muticum* (Yendo) Fensholt populations within the Solent area of southern Britain. II. An investigation of the increase in canopy cover of the alga at low water. *Botanica Marina* 26: 547–52.

Critchley, A.T. & Dijkema, R. 1984. On the presence of the introduced brown alga *Sargassum muticum*, attached to commercially imported *Ostrea edulis* in the S.W. Netherlands. *Botanica Marina* 27: 211–16.

Critchley, A.T., De Visscher, P.R.M. & Nienhuis, P.H. 1990a. Canopy characteristics of the brown alga *Sargassum muticum* (Fucales, Phaeophyta) in Lake Grevelingen, southwest Netherlands. *Proceedings of the International Seaweed Symposium* 13: 211–17.

Critchley, A.T., Farnham, W.F. & Morrell, S.L. 1983. A chronology of new European sites of attachment for the invasive brown alga, *Sargassum muticum*, 1973–1981. *Journal of the Marine Biological Association of the United Kingdom* 63: 799–811.

Critchley, A.T., Farnham, W.F. & Morrell, S.L. 1986. An account of the attempted control of an introduced marine alga, *Sargassum muticum*, in southern England. *Biological Conservation* 35: 313–32.

Critchley, A.T., Nienhuis, P.H. & Verschuure, K. 1987. Presence and development of populations of the introduced brown alga *Sargassum muticum* in the southwest Netherlands. *Proceedings of the International Seaweed Symposium* 12: 245–55.

Critchley, A.T., Farnham, W.F., Yoshida, T. & Norton, T.A. 1990b. A bibliography of the invasive alga *Sargassum muticum* (Yendo) Fensholt (Fucales; Sargassaceae). *Botanica Marina* 33: 551–62.

Crouan, P.L. & Crouan, H.M. 1833. Sur le *Padina reptans*. *Archives de Botanique, Paris* 2: 398–9.

Crouan, P.L. & Crouan, H.M. 1851. Études microscopiques sur quelques algues nouvelles ou peu connues constituant un genre nouveau. *Annales des Sciences Naturelles, Botanique, Troisième série* 15: 359–66, pls 16, 17.

Crouan, P.L. & Crouan, H.M. 1852. *Algues marines du Finistère*. Vol. 1–3 pp. Premier volume, Fucoïdees: 1 Apr 1852 (signed on p. [8]), p. [1]–[12], 1–112 specimens with detailed labels, [vii, index]. Deuxième volume, Floridées: 1852, p. [1]–[12], 113–322, id., [iv–xi, index]. Troisième volume, Zoospermées: 1852, pp. [l]–[8], 323–4. Brest: chez Crouans frères, pharmaciens.

Crouan, P.L. & Crouan, H.M. 1855. Observations microscopiques sur l'organisation, la fructification et la dissemination de plusieurs genres d'Algues appartenant a la famille des Dictyotees. *Bulletin de la Société Botanique de France* 2: 439–45, 644–52.

Crouan, P.L. & Crouan, H.M. 1857. Observations microscopiques sur l'organisation, la fructification et la dissémination de plusieurs genres d'Algues appartenant a la famille des Dictyotées. *Bulletin de la Société Botanique de France* 4: 24–9.

Crouan, P.L. & Crouan, H.M. 1867. *Florule du Finistère*. Paris and Brest.

Culioli, G., Daoudi, M., Ortalo-Magne, A., Valls, R. & Piovetti, L. 2001. (S)-12-hydroxygeranylgeraniol-derived diterpenes from the brown alga *Bifurcaria bifurcata*. *Phytochemistry* 57: 529–35.

Curiel, D., Guidetti, P., Bellemo, G. & Scattolin, M. 2002. The introduced alga *Undaria pinnatifida* (Laminariales, Alariaceae) in the lagoon of Venice. *Hydrobiologia* 477: 209–19.

Curiel, D., Bellemo, G., Marzocchi, M., Scattolin, M. & Parisi, G. 1998. Distribution of introduced Japanese macroalgae *Undaria pinnatifida, Sargassum muticum* (Phaeophyta) and *Antithamnion pectinatum* (Rhodophyta) in the Lagoon of Venice. *Hydrobiologia* 385: 17–22.

Daguin, C., Voisin, M., Engel, C. & Viard, F. 2005. Microsatellites isolation and polymorphism in introduced populations of the cultivated seaweed *Undaria pinnatifida* (Phaeophyceae, Laminariales). *Conservation Genetics* 6: 647–50.

Dalmon, J. & Loiseaux, S. 1981. The deoxyribonucleic acids of two brown algae: *Pylaiella littoralis* (L.) Kjellm. and *Sphacelaria* sp. *Plant Science Letters* 21: 241–51.

Daly, M.A. & Mathieson, A.C. 1977. The effects of sand movement on intertidal seaweeds and selected invertebrates at Bound Rock, New Hampshire, USA. *Marine Biology* 43: 45–55.

Damant, G.C.C. 1936. Storage of oxygen in the bladders of the seaweed *Ascophyllum nodosum* and their adaptation to hydrostatic pressure. *British Journal of Experimental Biology* 14: 198–209.

Dammann, H. 1930. Entwicklungsgeschichtliche und zytologische Untersuchungen an Helgolander Meeresalgen. *Helgoländer Wissenschaftliche Meeresuntersuchungen N.F.* 18: 1–36.

Dan, A. & Kato, S. 2008. Differences of morphology and growth between the two culture varieties originating from *Undaria pinnatifida* f. *distans* and *U. pinnatifida* f. *typica* in Naruto Strait. *Bulletin of the Tokushima Prefecture Fisheries Research Institute* 6: 79–83 (in Japanese with English abstract).

Dangeard, P. 1934. Un *Ectocarpus* nouveau du Croisic (*E. maculans*). *Bulletin de la Société Botanique de France* 81: 98–102.

Dangeard, P. 1949. Les algues marines de la côte occidentale du Maroc. *Le Botaniste* 34: 89–189.

Dangeard, P. 1962a. Sur la reproduction et le developpement de *Petalonia zosterifolia* (Reinke) Kuntze. *Compte Rendu Hebdomadaire des Séances de l'Académie des Sciences. Paris* 254: 1895–6.

Dangeard, P. 1962b. Sur le developpement du *Petalonia fascia* (Mueller) Kuntze et du *Scytosiphon lomentaria* (Lyngbye) Endlicher. *Compte Rendu Hebdomadaire des Séances de l'Académie des Sciences. Paris* 254: 3290–2.

Dangeard, P. 1963a. Sur le developpement du *Punctaria latifolia* Greville. *Le Botaniste* 46: 205–24.

Dangeard, P. 1963b. Recherches sur le cycle evolutif de quelques Scytosiphonacees. *Le Botaniste* 46: 5–129.

Dangeard, P. 1965a. Sur le cycle evolutif de *Litosiphon pusillus* (Carm.) Harvey. *Le Botaniste* 49: 47–62.

Dangeard, P. 1965b. Une algue nouvelle pour Roscoff: le *Myriactula clandestina* Crouan. *Mémoires de la Société Botanique de France* 11: 16–18.

Dangeard, P. 1965c. Recherches sur le cycle evolutif de '*Leathesia difformis*' (L.) Areschoug. *Le Botaniste* 48: 5–43.

Dangeard, P. 1965d. Sur un *Myriotrichia* Harvey recolte a Saint-Vaast-la-Hougue (Cotentin). *Le Botaniste* 49: 79–98.

Dangeard, P. 1965e. Sur la presence du *Stictyosiphon tortilis* (Rupr.) Reinke. *Bulletin de la Société Botanique de France* No. 11.

Dangeard, P. 1966a. Sur le '*Punctaria crouani*' (Thuret) Bornet, recolte a Soulac et sur son developpement. *Le Botaniste* 49: 157–67.

Dangeard, P. 1966b. Sur la presence d'un plethysmothalle chez *Leptonematella fasciculatum* (Reinke) Silva. *Compte Rendu Hebdomadaire des Séances de l'Académie des Sciences. Paris* 263: 1692–4.

Dangeard, P. 1966c. Sur quelques algues vertes marines nouvelles observées en culture. *Le Botaniste* 49: 5–45.

Dangeard, P. 1968a. Sur la presence d'*Elachista stellaris* Areschoug [{'*Areschougia stellaris*' (Aresch.) Menegh.] pres d'Erquy (Cotes-du-Nord). *Le Botaniste* 51: 87–94.

Dangeard, P. 1968b Sur la presence a Roscoff du '*Myriactula clandestina*' (Crouan) P. Dangeard. *Le Botaniste* 51: 81–6.

Dangeard, P. 1968c. Recherches sur le cycle evolutif de deux '*Asperococcus*'. *Le Botaniste* 51: 59–80.

Dangeard, P. 1968d. Etude du '*Leptonematella fasciculata*' (Reinke) Silva et de son developpement enculture. *Le Botaniste* 51: 117–30.

Dangeard, P. 1968e. Nouvelles Observations sur les '*Stictyosiphon*'. *Le Botaniste* 51: 131–9.

Dangeard, P. 1969. A propos des travaux recents sur le cycle evolutif de quelques Pheophycees, Pheosporees. *Le Botaniste* 52: 59–102.

Dangeard, P. 1970. Sur le *Componema minutum* (Ag.) Kuckuck et sur son developpement en culture. *Compte Rendu Hebdomadaire des Séances de l'Académie des Sciences. Paris* 280: 63–5.

Dannenberg, W. 1927. Vorarbeiten zu einer Algenflora des ostbaltischen Gebiete. *Naturforscher-Vereins zu Riga* 59: 129–44.

Daoudi, M., Bakkas, S., Culioli, G., Ortalo-Magné, A., Piovetti, L. & Guiry, M.D. 2001. Acyclic diterpenes and sterols from the genera *Bifurcaria* and *Bifurcariopsis* (Cystoseiraceae, Phaeophyceae). *Biochemical Systematics and Ecology* 29: 973–8.

Davidson, I.R. & Reed, R.H. 1984. The osmotic significance of seasonal changes in the internal solute content of *Luminaria digitata*. *British Phycological Journal* 19: 192.

Davison, I.R. & Reed, R.H. 1985. Osmotic adjustment in *Laminaria digitata* (Phaeophyta) with particular references to seasonal changes in internal solute concentrations. *Journal of Phycology* 21: 41–50.

Davison, I.R. & Stewart, W.D.P. 1983. Seasonal variation of nitrate reductase activity in *Laminaria digitata*. *British Phycological Journal* 18: 201–2.

Davison, I.R. 1987. Adaptation of photosynthesis in *Laminaria saccharina* (Phaeophyta) to changes in growth temperature. *Journal of Phycology* 23: 273–83.

Davison, I.R. 1991. Environmental effects on algal photosynthesis: temperature. *Journal of Phycology* 27: 2–8.

Davison, I.R., Greene, R.M. & Podolak, E.J. 1991. Temperature acclimation of respiration and photosynthesis in the brown alga *Laminaria saccharina*. *Marine Biology* 110: 449–54.

Davy de Virville, A. 1940. Les zones de végétation sur le littoral Atlantique. *Mémoires, Société de Biogéographie* 7: 205–51.

Davy de Virville, A. 1944. Les *Fucus* des cotes de France. *Revue Scientifique* 7: 421–52.

Dawes, C.J. & Mathieson, A.C. 2008. *The Seaweeds of Florida*. pp. [i]–viii, [1]–591, [592], pls I–LI. Gainesville, Florida: University Press of Florida.

Dawson, A.E. 1941. Some observations of *Cystoseira foeniculacea* Grev. emend Sauvageau. *New Phytologist* 40: 316–25.

Dawson, E.Y., Acleto, O.C. & Foldvik, N. 1964. The Seaweeds of Peru. *Beihefte Nova Hedwigia* 13: 1–111, 81 plates.

Dayton, P.K. 1975. Experimental evaluation of ecological dominance in a rocky intertidal algal community. *Ecological Monographs* 45: 137–59.

De Candolle, A.P. 1805. Note sur la mousse de Corse. *Bulletin des Sciences par la Société Philomathique de Paris* 3: 263–4.

de Carvalho, M.C., Hayashizaki, K. & Ogawa, H. 2010. Temperature effect on carbon isotopic discrimination by *Undaria pinnatifida* (Phaeophyta) in a closed experimental system. *Journal of Phycology* 46: 1180–6.

De Clerck, O. 2003. The genus *Dictyota* in the Indian Ocean. *Opera Botanica Belgica* 13: 1–205.

De Clerck, O., De Vos, P., Gillis, M. & Coppejans, E. 2001. Molecular systematics in the genus *Dictyota* (Dictyotales, Phaeophyta): a first attempt based on restriction patterns of the Internal Transcribed Spacer 1 of the rDNA (ARDRA-ITS1). *Systematics and Geography of Plants* 71: 25–35.

De Clerck, O. & Coppejans, E. 2003. Morphology and systematics of two aberrant species of *Dictyota* (Dictyotaceae, Phaeophyta), including a discussion on the generic boundaries in the tribe Dictyoteae. *Proceedings of the International Seaweed Symposium* 17: 275–84.

De Clerck, O., Leliaert, F., Verbruggen, H., Lane, C.E., De Paula, J.C., Payo, D.I. & Coppejans, E. 2006. A revised classification of the Dictyoteae (Dictyotales, Phaeophyceae) based on *rbc*L and 26S ribosomal DNA sequence data analyses. *Journal of Phycology* 42: 1271–88.

De Haas-Niekerk, T. 1965. The genus *Sphacelaria* Lyngbye (Phaeophyceae) in the Netherlands. *Blumea* 13: 145–61.

De Leij, R., Epstein, G., Brown, M.P. & Smale, D.A. 2017. The influence of native macroalgal canopies on the distribution and abundance of the nonnative kelp *Undaria pinnatifida* in natural reef habitats. *Marine Biology* 164: 156–71.

De Mesquita Rodrigues, J.E. 1963. Contribuição para o conhecimento das Phaeophyceae da Costa Portuguesa. *Memórias da Sociedade Broteriana* 16: 5–124.

De Notaris, G. 1846. *Prospetto della flora ligustica e dei zoofiti del Mare Ligustico.* pp. 1–80. Genova [Geneva]: Tipografia Ferrando.

De Paula, É.J. & Eston, V.R. 1987. Are there other *Sargassum* species potentially as invasive as *S. muticum*? *Botanica Marina* 30: 405–10.

De Toni, G.B. 1891. Systematische Übersicht der bisher bekannten Gattungen der echten Fucoideen. *Flora* 74: 171–82.

De Toni, G.B. 1895. *Sylloge algarum omnium hucusque cognitarum. Vol. III. Fucoideae.* Vol. 3, pp. [i]–xvi, [1]–638. Patavii [Padua]: Sumptibus auctoris.

De Valéra, M. 1962. Some aspects of the problem of the distribution of *Bifurcaria bifurcata* (Velley) Ross on the shores of Ireland, north of the Shannon Estuary. *Proceedings of the Royal Irish Academy* 62B: 77–101.

De Valéra, M., Pybus, C., Casley, B. & Webster, A. 1979. Littoral and benthic investigations on the west coast of Ireland X. Marine algae of the northern shores of the Burren, Co. Clare. *Proceedings of the Royal Irish Academy* 79B: 259–69.

De Wreede, R.E. 1983. *Sargassum muticum* (Fucales, Phaeophyta): regrowth and interaction with *Rhodomela larix* (Ceramiales, Rhodophyta). *Phycologia* 22: 153–60.

De Wreede, R.E. 1978. Phenology of *Sargassum muticum* (Phaeophyta) in the Strait of Georgia, British Columbia. *Syesis* 11: 1–9.

De Wreede, R.E. & Vandermeulen, H. 1988. *Lithothrix aspergillum* (Rhodophyta): regrowth and interaction with *Sargassum muticum* (Phaeophyta) and *Neorhodomela larix* (Rhodophyta). *Phycologia* 27: 469–76.

Deal, M.S., Hay, M.E., Wilson, D. & Fenical, W. 2003. Galactolipids rather than phlorotannins as herbivore deterrents in the brown seaweed *Fucus vesiculosus*. *Oecologia* 136: 107–14.

Dean, P.R. & Hurd, C.L. 2007. Seasonal growth, erosion rates, and nitrogen and photosynthetic ecophysiology of *Undaria pinnatifida* (Heterokontophyta) in southern New Zealand. *Journal of Phycology* 43: 1138–48.

Deboer, J.A. 1981. Nutrients. In: Lobban, C. & Wynne, M.J. (eds), *The Biology of Seaweeds* pp. 356–92. Botanical Monographs 17. Blackwell Scientific Publications, Oxford.

Decaisne, J. 1842. Essais sur une classification des algues et des polypiers calcifères de Lamouroux. *Annales des Sciences Naturelles, Botanique, Seconde série* 17: 297–380.

Decaisne, J. & Thuret, G. 1845. Recherches sur les anthéridies et les spores de quelques *Fucus*. *Annales des Sciences Naturelles, Botanique, Troisième série* 3: 5–15.

Deckert, R.J. & Garbary, D.J. 2005a. *Ascophyllum* and its symbionts. VI. Microscopic characterization of the *Ascophyllum nodosum* (Phaeophyceae), *Mycophycias ascophylli* (Ascomycetes) symbiotum. *Algae* 20: 225–32.

Deckert, R.J. & Garbary, D.J. 2005b. *Ascophyllum* and its symbionts. VIII. Interactions among *Ascophyllum nodosum* (Phaeophyceae), *Mycophycias ascophylli* (Ascopmycetes) and *Elachista fucicola* (Phaeophyceae). *Algae* 20: 363–8.

Delebecq, G., Davoult, D., Menu, D., Janquin, M., Dauvin, J. & Gevaert, F. 2012. Influence of local environmental conditions on the seasonal acclimation process and the daily integrated production rates of *Laminaria digitata* (Phaeophyta) in the English Channel. *Marine Biology* 160: 503–17.

Delebecq, G., Davoult, D., Janquin, M.-A., Oppliger, L.V., Menu, D., Dauvin, J.-C. & Gevaert, F. 2016. Photosynthetic response to light and temperature in *Laminaria digitata* gametophytes from two French populations. *European Journal of Phycology* 51: 71–82.

Delebecq, G., Davoult, D., Menu, D., Janquin, M.-A., Migné, A., Dauvin, J.-C. & Gevaert, F. 2011. In situ photosynthetic performance of *Laminaria digitata* (Phaeophyceae) during spring tides in Northern Brittany. *Cahiers de Biologie Marine* 52: 405–14.

Delf, E.M. 1935. Liberation of oogonia in *Bifurcaria* and other members of the Fucaceae. *New Phytologist* 34: 245–59.

Dellatorre, F.G., Amoroso, R., Saravia, J. & Orensanz, J.M. 2014. Rapid expansion and potential range of the invasive kelp *Undaria pinnatifida* in the Southwest Atlantic. *Aquatic Invasions* 9: 467–78.

Den Hartog, C. 1959. *The Epilithic Algal Communities Occurring Along the Coast of the Netherlands.* pp. xi + 241. Amsterdam, North-Holland: Academisch proefschrift.

Den Hartog, C. 1967. Brackish water as an environment for algae. *Blumea* 15: 31–43.

Den Hartog, C. 1972. Substratum. In: Kinne, O. (ed.) *Marine Ecology*, Vol. 1, pp. 1277–89. Wiley-Interscience, London.

Denicola, D.M. & McIntire, C.D. 1990. Effects of substrate relief on the distribution of periphyton in laboratory streams. 1. Hydrology. *Journal of Phycology* 26: 624–33.

Denny, M.W., Hunt, L.J.H., Miller, L.P. & Harley, C.D.G. 2009. On the prediction of extreme ecological events. *Ecological Monographs* 79: 397–421.

Denton, A.B. & Chapman, A.R.O. 1991. Feeding preferences of gammarid amphipods among four species of *Fucus*. *Marine Biology* 109: 503–6.

Denton, A., Chapman, A.R.O. & Markham, J. 1990. Size-specific concentrations of phlorotannins (anti-herbivore compounds) in three species of *Fucus*. *Marine Ecology Progress Series* 65: 103–4.

Derbès, F. & Solier, A.J.J. 1850. Sur les organes reproducteurs des algues. *Annales des Sciences Naturelles, Botanique, Troisième série* 14: 261–82.

Derbès, F. & Solier, A.J.J. 1851. In: J.L.M. Castagne, *Supplement au cataloge des plantes qui croissent naturellement aux environs de Marseille*, pp. 93–121. Aix.

Derbès, A. & Solier, A. 1856. Mémoire sur quelques points de la physiologie des algues. *Supplément aux Comptes Rendus Hebdomadaires des Séances de l'Académie des Sciences* 1: 1–120.

Derenbach, J.B. & Gereck, M.V. 1980. Interference of petroleum hydrocarbons with the sex pheromone reaction of *Fucus vesiculosus*. *Journal of Experimental Marine Biology and Ecology* 44: 61–5.

Desfontaines, R. 1799. *Flora atlantica, sive historia plantarum, quae in Atlante, agro Tunetano et Algeriensi crescunt*. Tomus secundus. pp. 161–458, pls 181–261. Parisiis [Paris]: L.G. Desgranges.

Desmond, M.J., Pritchard, D.W., Hurd, C.L., Richards, D.K., Schweikert, K., Wing, S. & Hepburn, C.D. 2019. Superior photosynthetic performance of the invasive kelp *Undaria pinnatifida* may contribute to continued range expansion in a wave-exposed kelp forest community. *Marine Biology* 166: 139.

Dethier, M.N. 1981. Heteromorphic algal life histories: the seasonal pattern and response to herbivory of the brown crust, *Ralfsia californica*. *Oecologia (Berl.)* 49: 333–9.

Devinny, J.S. & Volse, L.A. 1978. Effects of sediments on the development of *Macrocystis pyrifera* gametophytes. *Marine Biology* 48: 343–8.

Deysher, L.E. & Dean, T.A. 1986. Interactive effects of light and temperature on sporophyte production in the giant kelp *Macrocystis pyrifera*. *Marine Biology, Berlin* 93: 17–20.

Deysher, L.E. 1984. Reproductive phenology of newly introduced populations of the brown alga, *Sargassum muticum* (Yendo) Fensholt. *Proceedings of the International Seaweed Symposium* 11: 403–7.

Deysher, L. & Norton, T.A. 1982. Dispersal and colonization in *Sargassum muticum* (Yendo) Fensholt. *Journal of Experimental Marine Biology and Ecology* 56: 179–95.

Dillwyn, L.W. 1802–09. *British Confervae; or colored figures and descriptions of the British plants referred by botanists to the genus Conferva*. London: W. Phillips.

———. 1802, pls 1–20 (with text).

———. 1803, pls 21–38 (with text).

———. 1804, pls 39–44 (with text).

———. 1805, pls 45–56 (with text).

———. 1806, pls 57–68, 70–81 (with text).

———. 1807, pls 82–93 (with text).

———. 1808, pls 94–9 (with text).

———. 1809, pp. 1–87, 1–6 (index and errata), pls 69, 100–9, A–G.

Dimitriadis, I., Katsaros, C. & Galatis, B. 2001. The effect of taxol on centrosome function and microtubule organization in apical cells of *Sphacelaria rigidula* (Phaeophyceae). *Phycological Research* 49: 23–34.

Ding Min, Z., Guo Rong, M. & Lu Qing, P. 1984. Studies on *Undaria pinnatifida*. *Proceedings of the International Seaweed Symposium* 11: 263–5.

Diouris, M. 1989. Long-distance transport pf 14C-labelled assimilates in the Fucales: nature of translocated substances in *Fucus serratus*. *Phycologia* 28: 504–11.

Dixon, P.S. 1963. The Rhodophyta: some aspects of their biology. *Oceanography Marine Biology Annual Review* 1: 177–96.

Dixon, P.S. 1970. The Rhodophyta: some aspects of their biology. II. *Oceanography Marine Biology Annual Review* 8: 307–52.

Dixon, P.S. 1973. *Biology of the Rhodophyta*. Hafner Press, New York.

Dixon, P.S. & Irvine, L.M. 1977. *Seaweeds of the British Isles. Vol. 1. Rhodophyta. Part 1, Introduction, Nemaliales, Gigartinales*. British Museum (Natural History), London.

Dixon, R.R.M., Huisman, J.M., Buchanan, J., Gurgel, C.F.D. & Spencer, P. 2012. A morphological and molecular study of austral *Sargassum* (Phaeophyceae) supports the recognition of *Phyllotricha* at genus level, with further additions to the genus *Sargassopsis*. *Journal of Phycology* 48: 1119–29.

Dixon, N.M., Leadbeater, B.S.C. & Wood, K.R. 2000. Frequency of viral infection in a field population of *Ectocarpus fasciculatus* (Ectocarpales, Phaeophyceae). *Phycologia* 39: 258–63.

Dixon, P.S. & Russell, G. 1964. Miscellaneous notes on algal taxonomy and nomenclature. I. *Botaniska Notiser* 117: 279–84.

Dizerbo, A.H. 1960. *Pelvetia canaliculata* (L.) Dcne et Thur. var *libera* S.M. Baker (algues, Fucales), sa présence dans le Nord Finistère, sa valeur taxonomique. *Bulletin de la Société Phycologique de France* 6: 1–3.

Dizerbo, A.H. 1965. *Desmarestia dresnayi* Lamour. ex Leman in France and Spain. *British Phycological Bulletin* 2: 504.

Dizerbo, A.H. & Herpe, E. 2007. *Liste et répartition des algues marines des côtes françaises de la Manche et de l'Atlantique, Iles Normandes incluses*. Landerneau: Éditions Anaximandre.

Dodge, J.D. 1969. A review of the fine structure of algal eyespots. *British Phycological Journal* 4: 199–210.

Dodge, J.D. 1973. *The Fine Structure of Algal Cells*. Academic Press, London and New York.

Doty, M.S. 1946. Critical tide factors that are correlated with the vertical distribution of marine algae and other organisms along the Pacific coast. *Ecology* 27: 315–28.

Doty, M.S. 1947. The marine algae of Oregon. Part 1. Chlorophyta and Phaeophyta. *Farlowia* 3: 1–65.

Draisma, S.G.A. 2002. Calling a Class to Order-Phylogenetic studies in the Phaeophyceae. PhD thesis, Leiden University, Netherlands.

Draisma, S.G.A. & Prud'homme van Reine, W.F. 2001. Onslowiaceae fam. nov. (Phaeophyceae). *Journal of Phycology* 37: 647–9.

Draisma, S.G.A. & Prud'homme van Reine, W.F. 2010. Proposal to conserve the name *Sphacelaria* (Sphacelariales, Phaeophyceae) with a conserved type. *Taxon* 59: 1891–2.

Draisma, S.G.A., Peters, A. & Fletcher, R.L. 2003. Evolution and taxonomy in the Phaeophyceae; effects of the molecular age on brown algal systematic. In: Norton, T.A. (ed.), *Out of the Past: Collected Reviews to Celebrate the Jubilee of the British Phycological Society*, pp. 87–102. British Phycological Society.

Draisma, S.G.A., Ballesteros, E., Rousseau, F. & Thibaut, T. 2010a. DNA sequence data demonstrate the polyphyly of the genus *Cystoseira* and other Sargassaceae genera (Phaeophyceae). *Journal of Phycology* 46: 1329–45.

Draisma, S.G.A., Prud'homme van Reine, W.F. & Kawai, H. 2010b. A revised classification of the Sphacelariales (Phaeophyceae) inferred from a *psb*C and *rbc*L based phylogeny. *European Journal of Phycology* 45: 308–26.

Draisma, S.G.A., Prud'homme van Reine, W.F. & Kawai, H. 2014. A revised classification of the Sphacelariales (Phaeophyceae) inferred from a *psb*C and *rbc*L based phylogeny (Corrigendum). *European Journal of Phycology* 49: 576.

Draisma, S.G.A., Olsen, J.L., Stam, W.T. & Prud'homme van Reine, W.F. 2002. Phylogenetic relationships within the Sphacelariales (Phaeophyceae): *rbc*L, RUBISCO spacer and morphology. *European Journal of Phycology* 37: 385–402.

Draisma, S.G.A., Prud'homme van Reine, W.F., Stam, W.T. & Olsen, J.A. 2001. A reassessment of phylogenetic relationships within the Phaeophyceae based on RUBISCO large subunit and ribosomal DNA sequences. *Journal of Phycology* 37: 586–603.

Dreuhl, L.D. & Saunders, G.W. 1992. Molecular explorations in kelp evolution. *Phycological Research* 8: 47–83.

Drew, E.A. & Robertson, W.A.A. 1974. Direct observations of *Desmarestia dresnayi* Lamour. ex Leman in the British Isles and in the Mediterranean. *British Phycological Journal* 9: 195–200.

Drew, E.A. 1983. Light. In: Earll, R. & Erwin, D.G. (eds), *Sublittoral Ecology: The Ecology of the Shallow Sublittoral Benthos*, pp. 10–57. Clarendon Press, Oxford.

Drew, K.M. 1955. Life histories in the algae with special reference to the Chlorophyta, Phaeophyta and Rhodophyta. *Biological Reviews* 30: 343–90.

Dring, M.J. 1982. The Biology of Marine Plants. Edward Arnold, London.

Dring, M.J. 1984. Photoperiodism and phycology. In: Round, F.E. & Chapman, D.J. (eds), *Progress in Phycological Research Vol. 3*, pp. 159–92. Biopress Ltd, Bristol.

Dring, M.J. 1987. Stimulation of light-saturated photosynthesis in brown algae by blue light. *British Phycological Journal* 22: 302.

Dring, M.J. 1988. Photocontrol of development in algae. *Annual Review of Plant Physiology and Plant Molecular Biology* 39: 157–74.

Dring, M.J. & Lüning, K. 1975a. Induction of two-dimensional growth and hair formation by blue light in the brown alga *Scytosiphon lomentaria*. *Zeitschrift für Pflanzenphysiologie* 75: 107–17.

Dring, M.J. & Lüning, K. 1975b. A photoperiodic response mediated by blue light in the brown alga *Scytosiphon lomentaria*. *Planta* 125: 25–32.

Dring, M.J. & Lüning, K. 1983. Photomorphogenesis of marine macro algae. In: Shropshire, W. & Mohr, H. (eds), *Encyclopedia of plant physiology, Vol. 16B, Photomorphogenesis*, pp. 545–68. Heidelberg.

Druehl, L.D. 1967. Distribution of two species of *Laminaria* as related to some environmental factors. *Journal of Phycology* 3: 103–8.

Druehl, L.D. 1973. Marine transplantations. *Science* 179: 12.

Druehl, L.D. 1992. Molecular evolution in the Laminariales: A review. *NATO ASI ser G* 22: 205–17.

Druehl, L.D., Collins, J.D., Lane, C.E. & Saunders, G.W. 2005. An evaluation of methods used to assess intergeneric hybridization in kelp using Pacific Laminariales (Phaeophyceae). *Journal of Phycology* 41: 250–62.

Druehl, L.D., Mayes, C., Tan, I.H. & Saunders, G.W. 1997. Molecular and morphological phylogenies of kelp and associated brown algae. In: Bhattacharya, D. (ed.), *Origins of the Algae and their Plastids*, pp. 221–35. New York: Springer.

Du Rietz, G.E. 1940. On the identity of *Dictyosiphon chordaria* Areschoug and *Gobia baltica* (Gobi) Reinke. *Svensk Botanisk Tidsskrift* 34: 35–46.

Duby, J.É. 1830. *Aug. Pyrami de Candolle Botanicon gallicum sen synopsis plantarum in flora gallica descriptarum. Editio secunda. Ex herbariis et schedis Candollianis propriisque digestum a J.É. Duby V.D.M. Pars secunda plantas cellulares continens.* pp. [i–vi], [545]–1068, [i]–lviii. Paris: Ve Desray, Rue Hautefueille, No. 4.

Ducreux, G. 1977. Étude expérimentale des corrélations et des possibilités de régénération au niveau de l'apex de *Sphacelaria cirrhosa* Agardh. *Annales des Sciences Naturelles, Botanique, Paris*, 12ème serie, tome 18: 163–84.

Ducreux, G. 1983. Isolement expérimental des cellules terminales de l'apex de *Sphacelaria cirrosa* (Roth) C. Agardh (Sphacelariales, Phéophycées) et analyse comparée de leurs potentialités morphogénétiques. *Phycologia* 22: 415–29.

Ducreux, G. 1984. Experimental modifications of the morphogenetic behavior of the isolated sub-apical cell of the apex of *Sphacelaria cirrosa* (Phaeophyceae). *Journal of Phycology* 20: 447–54.

Dumortier, B.C. 1822. *Commentationes botanicae. Observations botaniques, dédiées à la Société d'Horticulture de Tournay.* pp. [i], [1]–116, [1, tabl., err.]. Tournay: Imprimerie de Ch. Casterman-Dieu, Rue de pont No. 10.

Dunton, K.H. 1985. Growth of dark-exposed *Laminaria saccharina* (L.) Lamour. and *Laminaria solidungula* J. Ag. (Laminariales: Phaeophyta) in the Alaskan Beaufort Sea. *Journal of Experimental Marine Biology and Ecology* 94: 181–9.

Eckersley, L.K. & Garbary, D.J. 2007. Developmental and environmental sources of variation on annual growth increments of *Ascophyllum nodosum* (Phaeophyceae). *Algae* 22: 107–16.

Edelstein, T. & McLachlan, J. 1967. Investigations of the marine algae of Nova Scotia, III. Species of Phaeophyceae new or rare to Nova Scotia. *Canadian Journal of Botany* 45: 203–10.

Edelstein, T. & McLachlan, J. 1969a. Investigations of the marine algae of Nova Scotia. VI. Some species new to North America. *Canadian Journal of Botany* 47: 555–60.

Edelstein, T. & McLachlan, J. 1969b. *Petroderma maculiforme* on the coast of Nova Scotia. *Canadian Journal of Botany* 47: 561–3.

Edelstein, T. & McLachlan, J. 1975. Autecology of *Fucus distichus* ssp. *distichus* (Phaeophyceae: Fucales) in Nova Scotia, Canada. *Marine Biology, Berlin* 30: 305–24.

Edelstein, T., Chen, L.C.M. & McLachlan, J. 1970. The life cycle of *Ralfsia clavata* and *R. bornetii*. *Canadian Journal of Botany* 48: 527–31.

Edelstein, T., Chen, L.C.M. & McLachlan, J. 1971a. On the life histories of some brown algae from eastern Canada. *Canadian Journal of Botany* 49: 1247–51.

Edelstein, T., Greenwell, M., Bird, C.J. & McLachlan, J. 1971b. Investigations of the marine algae of Nova Scotia. X. Distribution of *Fucus serratus* L. and some other species of *Fucus* in the Maritime Provinces. *Proceedings of the Nova Scotian Institute of Science* 27: 33–42.

Edmonston, Th. 1845. *A Flora of Shetland*. Aberdeen.

Edwards, P. 1969. Field and cultural studies on the seasonal periodicity of growth and reproduction of selected Texas benthic marine algae. *Contributions in Marine Science* 14: 59–114.

Edwards, P. 1976. *Illustrated Guide to the Seaweeds and Sea Grasses in the Vicinity of Port Aransas, Texas*. University of Texas Press, Austin and London.

Edwards, P. & Baalen, C. van 1970. An apparatus for the culture of benthic marine algae under varying regimes of temperature and light intensity. *Botanica Marina* 13: 42–3.

Egan, B. & Yarish, C. 1990. Productivity and life history of *Laminaria longicruris* at its southern limit in the Western Atlantic Ocean. *Marine Ecology Progress Series* 67: 263–73.

Ejtek, S.M., Edwards, M.S. & Kim, K.Y. 2011. Elk kelp, *Pelagophycus porra*, distribution limited due to susceptibility of microscopic stages to high light. *Journal of Experimental Marine Biology and Ecology* 396: 194–201.

Ellertsdóttir, E. & Peters, A.F. 1997. High prevalence of infection by endophytic brown algae in populations of *Laminaria* spp. (Phaeophyceae). *Marine Ecology Progress Series* 146: 135–43.

Elliot, E. & Moss, B. 1953. Incidence of meiosis in the life-cycle of *Halidrys siliquosa* Lyngb. *Nature, London* 171: 357.

Endo, H., Park, E.-J., Sato, Y. & Mizuta, H. 2009. Intraspecific diversity of *Undaria pinnatifida* (Harvey) Suringar (Laminariales, Phaeophyta) from Japan, China and Korea, based on the cox1 gene and ITS2 sequences. *Fisheries Science* 75: 393–400.

Enge, S., Sagerman, J., Wikström, S.A. & Pavia, H.A. 2017. Review of herbivore effects on seaweed invasions. In: Hawkins, S.J., Hughes, D.J., Smith, I.P., Dale, A.C., Firth, L.B. & Evans, A.J. (eds), *Oceanography and Marine Biology: An Annual Review* 55: 421–40. Taylor & Francis.

Engel, C.R., Brawley, S., Edwards, K.J. & Serrão, E. 2003. Isolation and cross-species amplification of microsatellite loci from the fucoid seaweeds *Fucus vesiculosus*, *F. serratus*, and *Ascophyllum nodosum* (Heterokontophyta, Fucaceae). *Molecular Ecology Notes* 3: 180–2.

Engel, C.R., Daguin, C. & Serrão, E. 2005. Genetic entities and mating system in hermaphroditic *Fucus spiralis* and its close dioecious relative *F. vesiculosus* (Fucaceae, Phaeophyceae). *Molecular Ecology* 14: 2033–46.

Engelen, A.H., Olsen, J.L., Breeman, A.M. & Stam, W.T. 2001. Genetic differentiation in *Sargassum polyceratium* (Fucales: Phaeophyceae) around the island of Curaçao (Netherlands Antilles). *Marine Biology* 139: 267–77.

Engelen, A.H., Åberg, P., Olsen, J.L., Stam, W.T. & Breeman, A.M. 2005. Effects of wave exposure and depth on biomass, density and fertility of the fucoid seaweed *Sargassum polyceratium* (Phaeophyta, Sargassaceae). *European Journal of Phycology* 40: 149–58.

Engelen, A. & Santos, R. 2009. Which demographic trails determine population growth in the invasive brown seaweed *Sargassum muticum*? *Journal of Ecology* 97: 675–84.

Engelen, A.H., Henriques, N., Monteiro, C. & Santos, R. 2011. Mesograzers prefer mostly native seaweeds over the invasive brown seaweed *Sargassum muticum*. *Hydrobiologia* 669: 157–65.

Engelen, A.H., Espirito-Santo, C., Simões, T., Monteiro, C., Serrão, E.A., Pearson, G.A. & Santos, R.O.P. 2008. Periodicity of propagule expulsion and settlement in the competing native and invasive brown seaweeds, *Cystoseira humilis* and *Sargassum muticum* (Phaeophyta). *European Journal of Phycology* 43: 275–82.

Engelen, A.H., Serebryakova, A., Ang, P., Britton-Simmons, K., Mineur, F., Pedersen, M.F., Arenas, F., Fernández, C., Steen, H., Svenson, R., Pavia, H., Toth, G., Viard, F. & Santos, R. 2015. Circumglobal invasion by the brown seaweed *Sargassum muticum*. *Oceanography and Marine Biology* 53: 81–126.

Engkvist, R., Malm, T. & Tobiasson, S. 2000. Density-dependent grazing effects of the isopod *Idotea baltica* Pallas on *Fucus vesiculosus* L in the Baltic Sea. *Aquatic Ecology* 34: 253–60.

Epstein, G. & Smale, D.A. 2017. *Undaria pinnatifida*: a case study to highlight challenges in marine invasion ecology and management. *Ecology and Evolution* 7: 8624–42.

Epstein, G. & Smale, D.A. 2018. Environmental and ecological factors influencing the spillover of the non-native kelp, *Undaria pinnatifida*, from marinas into natural rocky reef communities. *Biological Invasions* 20: 1049–72.

Epstein, G., Foggo, A. & Smale, D.A. 2019. Inconspicuous impacts: Widespread marine invader causes subtle but significant changes in native macroalgal assemblages. *Ecosphere* 10: 1–15. https://doi.org/10.1002/ecs2.2814.

Ercegovic, A. 1952. *Fauna i Flora Jadrana. Jadranske cistozire. Njihova morfologija, ekologija i razvitak/Fauna et Flora Adriatica. Sur les cystoseira adriatiques. Leur morphologie, écologie et évolution*. Vol. 2, pp. 1–172 (Croatian), 173–210 (French), 211–12 (references), Map. Institut za Oceanografiju i Ribarstvo Split/Institut d'Océanographie et de Pêche, Split [in Croatian, French].

Ercegovic, A. 1955a. Contribution à la connaissance des Éctocarpes (*Ectocarpus*) de l'Adriatique moyenne. *Acta Adriatica* 7: 1–74.

Ercegovic, A. 1955b. Contribution à la connaissance des Phéophycées de l'Adriatique moyenne. *Acta Adriatica* 7: 1–49.

Eriksson, B.K. & Johansson, G. 2003. Sedimentation reduces recruitment success of *Fucus vesiculosus* (Phaeophyceae) in the Baltic Sea. *European Journal of Phycology* 38: 217–22.

Esper, E.J.C. 1800. *Icones fucorum* cum characteribus systematicis, synonimis (sic) auctorum et descriptionibus novarum specierum. Abbildungen der Tange mit beygefügten systematischen Kennzeichen, Anführungen der Schriftsteller, und Beschribungen der neuen Gattungen. Vol. Erster Theil. Part 4 pp. 167–217, pls LXXXVIII–CXI. Nürnberg: Raspe.

Espinoza, J. 1990. The southern limit of *Sargassum muticum* (Yendo) Fensholt (Phaeophyta, Fucales) in the Mexican Pacific. *Botanica Marina* 33: 193–6.

Etherington, J. 1964. Rhizoid formation and fragmentation in *Feldmannia globifera* (Kiitz.) Hamel. *British Phycological Bulletin* 2: 373–5.

Etherington, J. 1965. The morphology of *Giffordia hincksiae* (Harv.) Hamel on the south coast of Britain. *British Phycological Bulletin* 2: 472–7.

Evans, L.V. 1966. Distribution of pyrenoids among some brown algae. *Journal of Cell Science* 1: 449–54.

Evans, L.V. 1968. Chloroplast morphology and fine structure in British fucoids. *New Phytologist* 67: 173–8.

Evans, L.V. 1974. Cytoplasmic organelles. In: Stewart, W.D.P. (ed.), *Algal Physiology and Biochemistry*, pp. 86–123. *Botanical Monographs* 10. Blackwell Scientific Publications, Oxford.

Evans, L.V. & Callow, M.E. 1976. Secretory processes in seaweeds. In: Sunderland, N. (ed.), *Perspectives in Experimental Biology* 2, pp. 487–99. Pergamon, Oxford and New York.

Evans, L.V. & Holligan, M.S. 1972a. Correlated light and electron microscope studies on brown algae. I. Localization of alginic acid and sulphated polysaccharides in *Dictyota*. *New Phytologist* 71: 1161–72.

Evans, L.V. & Holligan, M.S. 1972b. Correlated light and electron microscopic studies on brown algae. II. Physode production in *Dictyota*. *New Phytologist* 71: 1173–80.

Evans, L.V., Simpson, M. & Callow, M.E. 1973. Sulphated polysaccharide synthesis in brown algae. *Planta* 110: 237–52.

Fagerberg, W.R. & Dawes, C.J. 1973. An electron microscopic study of the sporophytic and gametophytic plants of *Padina vickersiae* Hoyt. *Journal of Phycology* 9: 199–204.

Fain, S.R., Druehl, L.D. & Baillie, D.L. 1988. Repeat and single copy sequences are differentially conserved in the evolution of kelp chloroplast DNA. *Journal of Phycology* 24: 292–302.

Fain, S.R. & Murray, S.N. 1982. Effects of light and temperature on net photosynthesis and dark respiration of gametophytes and embryonic sporophytes of *Macrocystis pyrifera*. *Journal of Phycology* 18: 92–8.

Falkenberg, P. 1878. Ueber *Discosporangium*, ein neues Phaeosporeen-Genus. *Mittheilungen aus der Zoologischen Station zu Neapel* 1: 54–66.

Falkenberg, P. 1879. Die Befruchtung und der Generationswechsel von *Culteria*. *Mittheilungen aus der Zoologischen Station zu Neapel* 1: 420–47.

FAO. 2013. *Undaria pinnatifida* (Harvey) Suringar, Fisheries and Aquaculture Department, http://www.fao.org/fishery/species/2777/en.

Farlow, W.G. 1881. The marine algae of New England. In: *Report of the U.S. Commissioner of Fish and Fisheries for 1879*, pp. 1–210. Washington.

Farlow, W.G. 1889. On some new or imperfectly known algae of the United States, I. *Bulletin of the Torrey Botanical Club* 16: 1–12.

Farmer, J.B. & Williams, J.Ll. 1898. Contributions to our knowledge of the Fucaceae: their life-history and cytology. *Philosophical Transactions of the Royal Society of London B Biological Sciences* 190: 623–45.

Farnham, W.F. 1980. Studies on aliens in the marine flora of southern England. In: Price, J.H., Irvine, D.E.G. & Farnham, W.F. (eds), *The Shore Environment 2: Ecosystems*, pp. 875–914. Systematics Association Special Volume 17(b). Academic Press, London and New York.

Farnham, W.F. & Jones, E.B.G. 1974. The eradication of the seaweed *Sargassum muticum* from Britain. *Biological Conservation* 6: 57–8.

Farnham, W.F., Fletcher, R.L. & Irvine, L.M. 1973. Attached *Sargassum* found in Britain. *Nature, London* 243: 231–2.

Farnham, W., Murfin, C., Critchley, A. & Morell, S. 1981. Distribution and control of the brown alga *Sargassum muticum*. *Proceedings of the International Seaweed Symposium* 10: 277–82.

Farrell, T.M. 1991. Models and mechanisms of succession: an example from a rocky intertidal community. *Ecological Monographs* 61: 95–113.

Farrell, P. & Fletcher, R.L. 2006. An investigation of dispersal of the introduced brown alga *Undaria pinnatifida* (Harvey) Suringar and its competition with some species on the man-made structures of Torquay Marina (Devon, UK). *Journal of Experimental Marine Biology and Ecology* 334: 236–43.

Fejtek, S.M., Edwards, M.S. & Kim, K.Y. 2011. Elk kelp, *Pelagophycus porra*, distribution limited due to susceptibility of microscopic stages to high light. *Journal of Experimental Marine Biology and Ecology* 396: 194–201.

Feldmann, G. & Gugliemi, G. 1972. Les physodes et les corps irisants du *Dictyota dichotoma* (Hudson) Lamouroux. *Compte Rendu Hebdomadaire des Séances de l'Académie des Sciences. Paris. Série D* 275: 751–4.

Feldmann, J. 1934. Les Laminariées de la Méditerranée et leur répartition géographique. *Bull. Trav. publiés par la Stat. d'Aquicult. et de Pêche de Castiglione 1932* 2: 141–84.

Feldmann, J. 1937. Les algues marines de la côte des Albères. I–III. Cyanophycées, Chlorophycées, Phaéophycées. *Revue Algologique* 9: 141(bis)–148(bis); 149–335.

Feldmann, J. 1945. Une nouvelle espece de *Myriactula* parasite du *Gracilaria armata* J. Ag. *Bulletin de la Société d'Histoire Naturelle de l'Afrique du Nord* 34: 222–9.

Feldmann, J. 1949. L'ordre des Scytosiphonales. *Mémoires hors-série de la Société d'Histoire Naturelle de l'Afrique du Nord* 2 (Travaux botaniques dédiés a René Maire): 103–15.

Feldmann, J. 1954. Inventaire de la flore marine de Roscoff. Algues, champignons, lichens et spermatophytes. *Travaux Station Biologique de Roscoff, Nouvelle Série* Suppl. 6: [1]–152.

Feldmann, J. & Magne, M.F. 1964. Additions a l'inventaire de la flore marine de Roscoff algues, champignons, lichens. *Travaux Station Biologique* de Roscoff, *Nouvelle Série* 15 (New supplement): 1–23 (+ 5).

Fensholt, D.E. 1955. An emendation of the genus *Cystophyllum* (Fucales). 42: 305–22.

Fernández, C. 1982. Estudio cuantitativo de la fase microscópica de *Saccorhiza polyschides* (Lightf.) Batt.: demografía y crecimiento. *Actas 1º Simp. Ibérico Est. Bentos Marino*: 391–403.

Fernández, C. 1999. Ecology of *Sargassum muticum* (Phaeophyta) on the north coast of Spain: IV. Sequence of colonization on a shore. *Botanica Marina* 42: 553–62.

Fernández, C. 2011. The retreat of large brown seaweeds on the north coast of Spain: the case of *Saccorhiza polyschides*. *European Journal of Phycology* 46: 352–60.

Fernández, C., Gutiérrez, L.M. & Rico, J.M. 1990. Ecology of *Sargassum muticum* on the north coast of Spain. Preliminary observations. *Botanica Marina* 33: 423–8.

Fernández, C., Niell, F.X. & Anadón, R. 1983. Comparación de dos comunidades de horizontes intermareales con abundancia de *Bifurcaria bifurcata* Ros. en las costas N y NO de España. *Investigación Pesquera* 47: 435–55.

Figari, A. & De Notaris, G. 1853. Nuovi materiali per l'algologia del mar Rosso. *Memorie della Reale Accademia delle Scienze di Torino* Ser. 2, 13 (Cl. Sc. Fis. e Mat.): 133–69.

Filion-Myklebust, C. & Norton, T.A. 1981a. An antifouling mechanism in *Ascophyllum nodosum*. *British Phycological Journal* 16: 135.

Filion-Myklebust, C. & Norton, T.A. 1981b. Epidermis shedding in the brown seaweed *Ascophyllum nodosum* (L.) Le Jolis, and its ecological significance. *Marine Biology Letters* 2: 45–51.

Fischer-Piette, E. 1957. *Pelvetia canaliculata* examinée de proche en proche de la manche au Portugal. *Annales du Muséum d'Histoire Naturelle, Paris*: 65–73.

Fischer-Piette, E. 1961. Sur l'ecologie de la non-vesiculisation de *Fucus vesiculosus* L. *Revue Générale de Botanique* 68: 302–16.

Fletcher, R.L. 1974a. Studies on the life history and taxonomy of some members of the Phaeophycean families Ralfsiaceae and Scytosiphonaceae. PhD thesis, University of London.

Fletcher, R.L. 1974b. Studies on the brown algal families Ralfsiaceae and Scytosiphonaceae. *British Phycological Journal* 9: 218.

Fletcher, R.L. 1975. Heteroantagonism observed in mixed algal cultures. *Nature, London* 253: 534–5.

Fletcher, R.L. 1976. Post-germination attachment mechanisms in marine fouling algae. In: Sharpley, J.M. & Kaplan, A.M. (eds), *Proceedings of the 3rd International Symposium on Biodegradation*, pp. 443–64. Applied Science Publishers, London.

Fletcher, R.L. 1977a. Observations on secondary attachment mechanisms in marine fouling algae. In: C.R.E.O. (ed.), *Proceedings of the 4th International Congress on Marine Corrosion and Fouling*, pp. 169–77. C.R.E.O., Paris.

Fletcher, R.L. 1977b. Seaweeds of Thanet. In: *Thanet Panorama: A Modern Guide to the Island* (21st Anniversary Issue) 15: 10–14.

Fletcher, R.L. 1978. Studies on the family Ralfsiaceae (Phaeophyta) around the British Isles. In: Irvine,

D.E.G. & Price, J.H. (eds), *Modern approaches to the taxonomy of red and brown algae* pp. 371–98. Systematics Association Special Volume 10. Academic Press, London and New York.

Fletcher, R.L. 1980a. *Catalogue of main marine fouling organisms. Vol. 6. Algae*. ODEMA Brussels.

Fletcher, R.L. 1980b. Studies on the recently introduced brown alga *Sargassum muticum* (Yendo) Fensholt III. Periodicity in gamete release and 'incubation' of early germling stages. *Botanica Marina* 23: 425–32.

Fletcher, R.L. 1981a. Studies on the ecology, structure and life history of the brown alga *Petalonia filiformis* (Batt.) Kuntze (Scytosiphonaceae) around the British Isles. *Phycologia* 20: 103–4.

Fletcher, R.L. 1981b. Observations on the ecology and life history of *Ralfsia spongiocarpa* Batt. *Proceedings of the International Seaweed Symposium* 8: 323–30.

Fletcher, R.L. 1981c. Studies on the marine fouling brown alga *Giffordia granulosa* (Sm.) Hamel in the Solent (South Coast of England). *Botanica Marina* 24: 211–21.

Fletcher, R.L. 1983. The occurrence of the brown alga *Streblonema oligosporum* Strömfelt in Britain. *British Phycological Journal* 18: 415–23.

Fletcher, R.L. 1984. Observations on the life history of the brown alga *Hecatonema maculans* (Coll.) Sauv. (Ectocarpales, Myrionemataceae) in laboratory culture. *British Phycological Journal* 19: 193.

Fletcher, R.L. 1987. *Seaweeds of the British Isles. Vol. 3. Fucophyceae (Phaeophyceae). Part 1*. London: British Museum (Natural History).

Fletcher, R.L. & Callow, M.E. 1992. The settlement, attachment and establishment of marine algal spores. *British Phycological Journal* 27: 303–29.

Fletcher, R.L. & Fletcher, S.M. 1975. Studies on the recently introduced brown alga *Sargassum muticum* (Yendo) Fensholt I. Ecology and reproduction. *Botanica Marina* 18: 149–56.

Fletcher, R.L., Baier, R.E. & Fornalik, M.S. 1984. The influence of surface energy on spore development in some common marine fouling algae. In: *Proceedings of the 6th International Congress on Marine Corrosion and Fouling*, pp. 129–44. Athens.

Fletcher, R.L., Baier, R.E. & Fornalik, M.S. 1985. The effects of surface energy on the development of some marine macroalgae. *British Phycological Journal* 20: 184–5.

Fletcher, R.L., Jones, A.M. & Jones, E.B.G. 1984. The attachment of fouling macroalgae. In: Costlow, J.D. & Tipper, R.C. (eds), *Marine Biodeterioration: An Interdisciplinary Study*, pp. 172–82. Naval Institute Press, Annapolis.

Fletcher, R.L. & Farrell, P. 1999. Introduced brown algae in the north east Atlantic, with particular respect to *Undaria pinnatifida* (Harvey) Suringar. *Helgoländer Wissenschaftliche Meeresuntersuchungen* 52: 259–75.

Fletcher, R.L. & Fletcher, S.M. 1975a. Studies on the recently introduced brown alga *Sargassum muticum* (Yendo) Fensholt II. Regenerative ability. *Botanica Marina* 18: 157–62.

Fletcher, R.L. & Fletcher, S.M. 1975b. Studies on the recently introduced brown alga *Sargassum muticum* (Yendo) Fensholt I. Ecology and reproduction. *Botanica Marina* 18: 149–56.

Fletcher, H. & Frid, C.L.J. 1996. Impact and management of visitor pressure on rocky intertidal algal communities. *Aquatic Conservation: Marine and Freshwater Ecosystems* 6: 287–97.

Fletcher, R.L. & Maggs, C.A. 1985. Two crustose marine brown algae new to Ireland. *Irish Naturalists' Journal* 21: 523–6.

Fletcher, R.L. & Manfredi, C. 1995. The occurrence of *Undaria pinnatifida* (Phaeophyceae, Laminarales) on the south coast of England. *Botanica Marina* 38: 355–8.

Fletcher, R.L., Munda, I.M. & Vukovic, A. 1988. *Compsonema saxicolum* (Kuckuck) Kuckuck and *Microspongium gelatinosum* Reinke (Scytosiphonaceae, Fucophyceae): Two new records from the Mediterranean. *Botanica Marina* 31: 1–8.

Fleurence, J. & Le Coeur, C. 1994. Influence of digestion procedures on the determination of lead and cadmium levels in the Laminariale *Undaria pinnatifida* (Wakame) by flame atomic absorption spectrophotometry. *Botanica Marina* 37: 555–9.

Floc'h, J.Y. & Penot, M. 1976. Étude comparative du transport à longue distance de différents radioéléments dans le thalle de *Laminaria digitata* (Linné) Lamouroux. *Compte Rendu Hebdomadaire des Séances de l'Académie des Sciences. Paris. Série D* 282: 989–92.

Floc'h, J.Y. & Penot, M. 1981. Translocation of ^{32}p and ^{86}Rb in some brown algae (*Laminaria digitata* and *Fucus vesiculosus*): influence of metabolism and sinks. *Proceedings of the International Seaweed Symposium* 8: 176–87.

Floc'h, J.Y., Pajot, R. & Mouret, V. 1995. Propagation of the Japanese brown alga *Undaria pinnatifida* (Harvey) Suringar along the French Atlantic coasts. Cryptogam *Cryptogamie, Algologie* 16: 157–8.

Floc'h, J.Y., Pajot, R. & Mouret, V. 1996a. *Undaria pinnatifida* (Laminariales, Phaeophyta) 12 years after its introduction into the Atlantic Ocean. *Proceedings of the International Seaweed Symposium* 15: 217–22.

Floc'h, J.-Y., Pajot, R. & Mouret, V. 1996b. Propagation of the Japanese brown alga *Undaria pinnatifida* (Harvey) Suringar along the French Atlantic coasts. *Cryptogamie, Algologie* 16: 157–8.

Floc'h, J.Y., Pajot, R. & Wallentinus, I. 1991. The Japanese brown alga *Undaria pinnatifida* on the coast of France and its possible establishment in European waters. *Journal du Conseil International pour l'Exploration de la Mer* 47: 379–90.

Floerl, O., Inglis, G.J. & Hayden, B.J. 2005. A risk-based predictive tool to prevent accidental introductions of non-indigenous marine species. *Environmental Management* 35: 765–78.

Flores-Moya, A., Hanelt, D., Figueroa, F.L., Altamirano, M., Viñegla, B. & Salles, S. 1999. Involvement of solar UV-B radiation in recovery of inhibited photosynthesis in the brown alga *Dictyota dichotoma* (Hudson) Lamouroux. *Journal of Photochemistry and Photobiology B: Biology* 49: 129–35.

Flores-Moya, A., Posudin, Y.I., Fernández, J.A., Figueroa, F.L. & Kawai, H. 2002. Photomovement of the swarmers of the brown algae *Scytosiphon lomentaria* and *Petalonia fascia*: effect of photon irradiance, spectral composition and UV dose. *Journal of Photochemistry and Photobiology B* 66: 134–40.

Flothe, C.R. & Molis, M. 2013. Temporal dynamics of inducible anti-herbivory defenses in the brown seaweed *Ascophyllum nodosum* (Phaeophyceae). *Journal of Phycology* 49: 468–74.

Flothe, C.R., Molis, M. & John, U. 2014a. Induced resistance to periwinkle grazing in the brown seaweed *Fucus vesiculosus* (Phaeophyceae): molecular insights and seaweed-mediated effects on herbivore interactions. *Journal of Phycology* 50: 564–76.

Flothe, C.R., Molis, M., Kruse, I., Weinberger, F. & John, U. 2014b. Herbivore-induced defence response in the brown seaweed *Fucus vesiculosus* (Phaeophyceae): temporal pattern and gene expression. *European Journal of Phycology* 49: 356–69.

Fong, P. 2008. Macroalgal-dominated ecosystems. In: Capone, D.G., Bronk, D.A., Mulholland, M.R. & Carpenter, E.J. (eds), *Nitrogen in the Marine Environment*, 2nd edn, pp. 917–47. Academic Press, San Diego, California.

Forrest, M. & Blakemore, A. 2006. Evaluation of treatments to reduce the spread of a marine plant pest with aquaculture transfers. *Aquaculture* 257: 333–45.

Forrest, B.M. & Hopkins, G.A. 2013. Population control to mitigate the spread of marine pests: insights from management of the Asian kelp *Undaria pinnatifida* and colonial ascidian *Didemnum vexillum*. *Management of Biological Invasions* 4: 317–26.

Forrest, B. & Taylor, M.D. 2002. Assessing invasion impact: Survey design considerations and implications for management of an invasive marine plant. *Biological Invasions* 4: 375–86.

Forrest, B.M., Brown, S.N., Taylor, M.D., Hurd, C.L. & Hay, C.H. 2001. The role of natural dispersal mechanisms in the spread of *Undaria pinnatifida* (Laminariales, Phaeophyceae). *Phycologia* 39: 547–53.

Forsskål, P. 1775. *Flora Aegyptiaca-Arabica* sive descriptiones plantarum, quas per Aegyptum inferiorem et Arabium delicem detexit illustravit Petrus Forskål. Prof. Haun. Post mortem auctoris edidit Carsten Niebuhr. pp. [1]–32, [i]–xxxvi, ... [1]–219, [220, err.], map. Hauniæ [Copenhagen]: ex officina Mölleri.

Forslund, H., Wikström, S. & Pavia, H. 2010. Higher resistance to herbivory in introduced compared to native populations of a seaweed. *Oecologia* 164: 833–40.

Forster, R.M. & Dring, M.J. 1991. Blue light increases carbon supply rate in *Laminaria digitata*. *British Phycological Journal* 26: 85.

Fortes, M.D. & Lüning, K. 1980. Growth rates of North Sea macroalgae in relation to temperature, irradiance and photoperiod. *Helgoländer Meeresuntersuchungen* 34: 15–29.

Foster, M.S. 1975. Regulation of algal community development in a *Macrocystis pyrifera* forest. *Marine Biology* 32: 331–42.

Forward, S.G. & South, G.R. 1985. Observations on the taxonomy and life history of North Atlantic *Acrothrix* Kylin (Phaeophyceae, Chordariales). *Phycologia* 24: 347–59.

Foslie, M. 1881. Om nogle nye arctiske havalger. *Christiania Videnskabers Selskabs Forhandlinger* 14: 1–14.

Foslie, M. 1883. Bidrag til kundskab om de til gruppen Digitatae hørende Laminarier. *Christiania Videnskabers Selskabs Forhandlinger* 2: 1–32.

Foslie, M. 1884. Ueber die Laminarien Norwegens. *Christiania Videnskabers Selskabs Forhandlinger* 14: 1–112.

Foslie, M. 1886. Kritisk Fortegnelse over Norges Havsalger efter aeldre botaniske Arbeider indtil Aar 1850. *Tromsø Museums Aarshefter* 9: 85–137.

Foslie, M. 1887. Nye hausalger. *Tromsø Museums Aarshefter* 10: 175–95.

Foslie, M. 1890. Contribution to knowledge of the marine algae of Norway. I. East-Finmarken. *Tromsø Museums Aarshefter B* 13: 1–186.

Foslie, M. 1891. Remarks on forms of *Ectocarpus* and *Pylaiella*. *Tromsø Museums Aarshefter* 14: 123–8.

Foslie, M. 1894. New or critical Norwegian algae. *Kongelige Norske Videnskabers Selskabs Skrifter* 1893: 114–44.

Fowler-Walker, M.J., Wernberg, T. & Connell, S.D. 2006. Differences in kelp morphology between wave sheltered and exposed localities: morphologically plastic or fixed traits? *Marine Biology* 148: 755–67.

Frauenfeld, G. 1854. Aufzählung der Algen der dalmatischen Küste. *Verhandlungen des Zoologisch-Botanischen Vereins in Wein* 4 (Abh): 317–50.

Fredersdorf, J., Müller, R., Becker, S., Wiencke, C. & Bischof, K. 2009. Interactive effects of radiation, temperature and salinity on different life history stages of the Arctic kelp *Alaria esculenta* (Phaeophyceae). *Oecologia* 160: 483.

Fredriksen, S. & Christie, H. 2003. *Zostera marina* (Angiospermae) and *Fucus serratus* (Phaeophyceae) as habitat for flora and fauna – seasonal and local variation. *Proceedings of the International Seaweed Symposium* 17: 357–64.

Fredriksen, S., Sjøtun, K., Lein, T.E. & Rueness, J. 1995a. Spore dispersal in *Laminaria hyperborea* (Laminariales, Phaeophyceae). *Sarsia* 80: 47–54.

Fredriksen, S., Sørlie, A.C. & Kjøsterud, A.-B. 1995b. *Titanoderma pustulatum* (Lamouroux) Nägeli and *Lithophyllum crouanii* Foslie (Corallinales, Rhodophyta): two common epiphytes on *Laminaria hyperborea* (Gunnerus) Foslie stipes in Norway. *Sarsia* 80: 41–6.

Freire-Gago, O., Peteiro, C. & Cremades, J. 2006. La integración del alga alóctona *Undaria pinnatifida* (Laminariales, Phaeophyta) en la flora de las costas atlánticas peninsulares. *Algas* 35 (Junio): 25–32.

Freitas, J.R.C., Salinas Morrondo, J.M. & Cremades Ugarte, J. 2016. *Saccharina latissima* (Laminariales, Ochrophyta) farming in an industrial IMTA system in Galicia (Spain). *Journal of Applied Phycology* 28: 377–85.

Fries, E.M. 1835. *Corpus florarum provincialium Sueciae*. I. Floram scanicam. pp. 1–192. Upsaliae [Uppsala]: Typis Palmbead, Serell & C.

Fries, E.M. 1836. *Corpus florarum provincialium Sueciae*. I. Floram scanicam. pp. i–xxiv, 193–346. Upsaliae [Uppsala]: excudebant Regiae Acad. Typographi.

Fries, E.M. 1845. *Summa vegetabilium Scandinaviae*: Sectio prior: [i], [iii], [i]–258, signatures 1–16 in fours. Holmiae & Lipsiae [Stockholm & Leipzig]: A. Bonnier.

Fries, L. 1977. Growth regulating effects of phenylacetic acid and *p*-hydroxy-phenylacetic acid on *Fucus spiralis* L. (Phaeophyceae, Fucales) in axenic culture. *Phycologia* 16: 451–6.

Fries, L. 1980. Axenic tissue cultures from the sporophytes of *Laminaria digitata* and *Laminaria hyperborea* (Phaeophyta). *Journal of Phycology* 16: 475–7.

Fries, L. 1984. Induction of plantlets in axenically cultivated rhizoids of *Fucus spiralis*. *Canadian Journal of Botany* 62: 1616–20.

Fries, L. 1988. *Ascophyllum nodosum* (Phaeophyta) in axenic culture and its response to the endophytic fungus *Mycosphaerella ascophylli* and epiphytic bacteria. *Journal of Phycology* 24: 333–7.

Fries, L. 1991. Formation of filaments and single cells by the activity of the meristoderm of axenic *Ascophyllum nodosum* (Fucaceae, Phaeophyta) (Research Note). *Phycologia* 30: 310–13.

Fries, L. 1993. Vitamin B_{12} heterotrophy in *Fucus spiralis* and *Ascophyllum nodosum* (fucales, Phaeophyta) in axenic cultures. *Botanica Marina* 36: 5–7.

Fries, N. 1979. Physiological characteristics of *Mycosphaerella ascophylli*, a fungal endophyte of the marine brown alga *Ascophyllum nodosum*. *Physiologia Plantarum* 45: 117–21.

Friess-Klebl, A.-K., Knippers, R. & Müller, D.G. 1994. Isolation and characterization of a DNA virus infecting *Feldmannia simplex* (Phaeophyceae). *Journal of Phycology* 30: 653–8.

Fritsch, F.E. 1945. *The Structure and Reproduction of the Algae. Volume II. Foreword, Phaeophyceae, Rhodophyceae, Myxophyceae*. Cambridge: Cambridge University Press.

Frye, T.C. & Phifer, M.W. 1930. Some questions in the life histories of the Phaeophyceae with particular reference to *Scytosiphon lomentarius*. In: *Contributions to Marine Biology*, pp. 234–45. Stanford University Press, Stanford.

Fu, G., Kinoshita, N., Nagasato, C. & Motomura, T. 2014. Fertilization of brown algae: flagellar function in phototaxis and chemotaxis. In: Sawada, H., Inoue, N. & Iwano, M. (eds), *Sexual Reproduction in Animals and Plants*, pp. 359–67. Springer, Tokyo.

Fuentes, J.M. & Niell, F.X. 1985. La entidad taxonómica de *Fucus muscoides* (Cotton) Feldmann: interpretación de un experimento de trasplante. *Investigación Pesquera* 49: 435–8.

Fukumoto, R., Borlongan, I.A., Nishihara, G.N., Endo, H. & Terada, R. 2018. Effect of photosynthetically active radiation and temperature on the photosynthesis of two heteromorphic life history stages of a temperate edible brown alga, *Cladosiphon umezakii* (Chordariaceae, Ectocarpales), from Japan. *Journal of Applied Phycology*. https://doi.org/10.1007/s10811-018-1655-3.

Fukuzumi, K., Tachiyama, T. & Fukagawa, A. 1999. Differences of the growth and morphological characteristics, *Undaria pinnatifida*, cultured in Fukuoka Bay. *Bulletin of Fukuoka Fisheries and Marine Technology Research Center* 9: 11–17 (in Japanese).

Fulcher, R.G. & McCully, M.E. 1969. Laboratory culture of the intertidal brown alga *Fucus vesiculosus*. *Canadian Journal of Botany* 47: 219–22.

Funk, G. 1927. Die Algenvegetation des Golfs von Neapel. *Publicazioni della Stazione Zoologica di Napoli* 7: 1–507.

Funk, G. 1955. Beiträge zur Kenntnis der Meeresalgen von Neapel: Zugleich mikrophotographischer Atlas.

Pubblicazioni della Stazione Zoologica di Napoli 25 (Suppl.): i–x, 1–178.

Furnari, G., Cormaci, M. & Serio, D. 1999. Catalogue of the benthic marine macroalgae of the Italian coast of the Adriatic Sea. *Bocconea* 12: 1–214.

Gachon, C.M.M., Küpper, H., Küpper, F.C. & Šetlík, I. 2007. Single-cell chlorophyll fluorescence kinetic microscopy of *Pylaiella littoralis* (Phaeophyceae) infected by *Chytridium polysiphoniae* (Chytridiomycota). *European Journal of Phycology* 41: 395–403.

Gagnon, P., Himmelman, J.H. & Johnson, Ladd, E. 2003. Algal colonization in urchin barrens: defense by association during recruitment of the brown alga *Agarum ribrosum*. *Journal of Experimental Marine Biology and Ecology* 290: 179–96.

Gaillard, J. 1963. Sur le mode de croissance des plantules de *Taonia atomaria* (Woodw) J. Ag. (Dictyotales). *Compte Rendu Hebdomadaire des Séances de l'Académie des Sciences, Paris* 257: 725–6.

Gaillard, J. 1968. Quelques caractères différentiels des genres *Dictyota* et *Dilophus* des Côtes de France. *Bulletin du Laboratoire Maritime De Dinard, Nouvelle série.* 1: 109–15.

Gaillard, J. 1972a. L'iridescence chez deux Dictyotales, *Dictyota dichotoma* (Huds.) Lamouroux et *Zonaria tournefortii* (Lamour.) Montagne. Cytologie des cellules iridescentes. *Botaniste* 55: 71–9.

Gaillard, J. 1972b. Quelques remarques sur le cycle reproducteur des Dictyotales et sur ses variations. *Mémoires de la Société Botanique de France* 1972: 145–50.

Gaillard, J. & L'Hardy-Halos, M.-T. 1976. Corrélations morphogènes et comportement des deux initiales filles chez le *Dictyota dichotoma* (Huds.) Lamouroux (Dictyotales-Phéophycée). *Compte Rendu Hebdomadaire des Séances de l'Académie des Sciences. Paris. Série D* 282: 2167–70.

Gaillard, J. & L'Hardy-Halos, M.T. 1977. A propos de la morphogenèse du *Dictyota dichotoma* (Huds.) Lamouroux (Phéophycée, Dictyotale; phénomènes corrélatifs mis en évidence sur les tronçons apicaux isolés expérimentalement. *Revue Algologique, Nouvelle Série* 12: 101–10.

Gaillard, J. & L'Hardy-Halos, M.T. 1979. Corrélations de croissance chez le *Dictyota dichotoma* (Huds.) Lamouroux (Phóphycée, Dictyotale); contrôles mutuels de l'apex et la base du thalle au cours du développement des gamétophytes juvéniles. *Revue Algologique, Nouvelle Série* 14: 149–62.

Gaillard, J. & L'Hardy-Halos, M.T. 1980. Croissance et ramification chez le *Dictyota dichotoma* (Hudson) Lamouroux (Phéophycée, Dictyotale). *Phycologia* 19: 159–67.

Gaillard, J. & L'Hardy-Halos, M.T. 1990 Morphogenèse du *Dictyota dichotoma* (Dictyotales, Phaeophyta). III. Ontogenèse et croissance des frondes adventives. *Phycologia* 29: 39–53.

Gaillard, J., L'Hardy-Halos, M.T. & Pellegrini, L. 1986. Morphogenèse du *Dictyota dichotoma* (Huds.) Lamouroux (Phaeophyta. II. Ontogenèse du thalle et cytologie ultrastructurale de différents types de cellules. *Phycologia* 25: 340–57.

Gaillon, B. 1828. Résumé méthodique des classifications des Thalassiophytes. *Dictionnaire des Sciences Naturelles* [*Levrault*] 53: 350–406.

Gaillon, B. 1833. *Aperçu d'histoire naturelle* et observations sur les limites qui séparent le règne végétal du règne animal [suivi des] tableaux synoptiques et méthodiques des genres des Nématozoaires (Lu à la Société d'Agriculture, du Commerce et des Arts, de Boulogne-sur-mer, dans sa séance publique du 19 Septembre 1832). pp. [1]–24, [11]. Boulogne: Imprimerie de Le Roy-Mabille, Grande Rue, no. 51.

Galatis, B., Katsaros, C. & Mitrakos, K. 1977. Fine structure of vegetative cells of *Sphacelaria tribuloides* Menegh. (Phaeophyceae, Sphacelariales) with special reference to some unusual proliferations of the plasmalemma. *Phycologia* 16: 139–51.

Gallardo Garciá, T. 1992. Nomenclatural notes on some Mediterranean algae, I: Phaeophyceae. *Taxon* 41: 324–6.

Gallardo Garciá, T. 2001a. *Bifurcaria* Stackh. In: Gómez Garreta, A. (ed.), *Flora phycologica iberica Vol. 1 Fucales*, pp. 91–3. Murcia: Universidad de Murcia.

Gallardo Garciá, T. 2001b. *Ascophyllum* Stackh. In: Gómez Garreta, A. (ed.), *Flora phycologica iberica Vol. 1 Fucales*, pp. 29–32. Murcia: Universidad de Murcia.

Gallardo Garciá, T. & Pérez-Ruzafa, I.M. 2001a. *Himanthalia* Lyngb. In: Gómez Garreta, A. (ed.), *Flora phycologica iberica Vol. 1 Fucales*, pp. 69–71. Murcia: Universidad de Murcia.

Gallardo Garciá, T. & Pérez-Ruzafa, I.M. 2001b. *Pelvetia* Decne. & Thur. In: Gómez Garreta, A. (ed.), *Flora phycologica iberica Vol. 1 Fucales*, pp. 63–6. Murcia: Universidad de Murcia.

Gallardo Garciá, T., Bárbara, I., Afonso-Carrillo, J., Bermejo, R., Altamirano, M., Gómez Garreta, A., Barceló Martí, M.C., Rull Lluch, J., Ballesteros, E. & De la Rosa, J. 2016. Nueva lista crítica de las algas bentónicas marinas de España. A new checklist of benthic marine algae of Spain. *Algas, Boletín Informativo de la Sociedad Española de Ficología* 51: 7–52.

Gao, K. & Nakahara, H. 1990. Effects of nutrients on the photosynthesis of *Sargassum thunbergii*. *Botanica Marina* 33: 375–83.

Gao, X. 2017. Interactive effects of nutrient availability and temperature on growth and survival of different size classes of *Saccharina japonica* (Laminariales, Phaeophyceae). *Phycologia* 56: 253–60.

Gao, X., Choi, H.G., Park, S.K., Lee, J.R., Kim, J.H., Hu, Z.-M. & Nam, K.W. 2017. Growth, reproduction and recruitment of *Silvetia siliquosa* (Fucales, Phaeophyceae) transplants using polyethylene rope and natural rock methods. *Algae: An International Journal of Algal Research* 324: 337–47.

Gao, X., Agatsuma, Y. & Taniguchi, K. 2013a. Effect of nitrate fertilization of gametophytes of the kelp *Undaria pinnatifida* on growth and maturation of the sporophytes cultivated in Matsushima Bay, northern Honshu, Japan. *Aquaculture International* 21: 53–64.

Gao, X., Endo, H., Taniguchi, K. & Agatsuma, Y. 2013b. Combined effects of seawater temperature and nutrient condition on growth and survival of juvenile sporophytes of the kelp *Undaria pinnatifida* (Laminariales; Phaeophyta) cultivated in northern Honshu, Japan. *Journal of Applied Phycology* 25: 269–75.

Gao, X., Endo, H., Yamana, M., Taniguchi, K. & Agatsuma, Y. 2013c. Compensation of the brown alga *Undaria pinnatifida* (Laminariales; Phaeophyta) after thallus excision under cultivation in Matsushima Bay, northern Japan. *Journal of Applied Phycology* 25: 1171–8.

Gao, X., Endo, H., Yamana, M., Taniguchi, K. & Agatsuma, Y. 2013d. Compensatory abilities depending on seasonal timing of thallus excision of the kelp *Undaria pinnatifida* cultivated in Matsushima Bay, northern Japan. *Journal of Applied Phycology* 25: 1331–40.

Garbary, D.J. 2017. Harvesting *Ascophyllum nodosum* (Phaeophyceae) reduces the abundance of its host-specific epiphyte *Vertebrata lanosa* (Rhodophyta). *Botanica Marina* 60: 297–301.

Garbary, D.J. & Clarke, B. 2002. Intraplant variation in nuclear DNA content in *Laminaria saccharina* and *Alaria esculenta* (Phaeophyceae). *Botanica Marina* 45: 211–16.

Garbary, D.J. & London, J.F. 1995. The *Ascophyllum/ Polysiphonia/Mycosphaerella* symbiosis. V. Fungal infection protect *A. nodosum* from desiccation. *Botanica Marina* 38: 529–33.

Garbary, D.J. & Tarakhovskaya, E.R. 2013. Marine macroalgae and associated flowering plants from the Keret Archipelago, White Sea, Russia. *Algae. An International Journal of Algal Research* 28: 267–80.

Garbary, D.J., Brackenbury, A., McLean, A.M. & Morrison, D. 2006. Structure and development of air bladders in *Fucus* and *Ascophyllum* (Fucales, Phaeophyceae). *Phycologia* 45: 557–66.

Garbary, D.J., Deckert, R.J. & Hubbard, C.B. 2005. *Ascophyllum* and its symbionts. VII. Three-way interactions among *Ascophyllum nodosum* (Phaeophyceae), *Mycophycias ascophylli* (Ascomycetes) and *Vertebrata lanosa* (Rhodophyta). *Algae* 20: 353–61.

Garbary, D.G., Lawson, J., Clement, K. & Galway, M.E. 2009. Cell division in the absence of mitosis: the unusual case of the fucoid *Ascophyllum nodosum* (L.) Le Jolis (Phaeophyceae). *Algae* 24: 239–48.

Garbary, D.J., Miller, A.G. & Scrosati, R.A. 2014. *Ascophyllum nodosum* and its symbionts: XI. The epiphyte *Vertebrata lanosa* performs better photosynthetically when attached to *Ascophyllum* than when alone. *Algae. An International Journal of Algal Research* 29: 321–31.

García-Fernández, A. & Bárbara, I. 2016. Studies of *Cystoseira* assemblages in Northern Atlantic Iberia. *Anales del Jardín Botánico de Madrid* 73: e035.

Gard, M. 1923. Sur l'etat sexuel du *Fucus ceranoides* L. *Bulletin de la Société Botanique de France* 70: 294–6.

Gardner, N.L. 1922. The genus *Fucus* on the Pacific coast of North America. *University of California Publications in Botany* 10: 1–180.

Gatty, M. 1863. *British Seaweeds*. London.

Gayral, P. 1958. *Algues de la Côte atlantique marocaine*. pp. 523. Rabat: Société des Sciences naturelles et physiques du Maroc.

Gayral, P. 1966. *Les algues des côtes Françaises (Manche et Atlantique). Notions fondementales sur l'ecologie, la biologie et la systématique des algues marines.* Paris: Éditions Doin.

Gayral, P. & Cosson, J. 1973. Exposé synoptique des données biologiques sur la Laminaria digitée *Laminaria digitata*. *FAO Fisheries Synopses* 89: 1.1–9.11.

Geange, S.W., Powell, A., Clemens-Seely, K. & Cárdenas, C.A. 2014. Sediment load and timing of sedimentation affect spore establishment in *Macrocystis pyrifera* and *Undaria pinnatifida*. *Marine Biology* 161: 1583–92.

Geiselman, J.A. & McConnell, O.J. 1981. Polyphenols in brown algae *Fucus vesiculosus* and *Ascophyllum nodosum*: chemical defenses against the marine herbivorous snail, *Littorina littorea*. *Journal of Chemical Ecology* 7: 1115–33.

Geller, A. & Müller, D.G. 1981. Analysis of the flagellar beat pattern of male *Ectocarpus siliculosus* gametes (Phaeophyta) in relation to chemotactic stimulation by female cells. *Journal of Experimental Biology* 92: 53–66.

Geoffroy, A., Mauger, S., De Jode, A., Le Gall, L. & Destombe, C. 2015. Molecular evidence for the coexistence of two sibling species in *Pylaiella littoralis* (Ectocarpales, Phaeophyceae) along the Brittany coast. *Journal of Phycology* 51: 480–9.

Georgevitch, P. 1918. Génération asexuée du *Padina pavonia* Lamour. *Compte Rendu Hebdomadaire des Séances de l'Académie des Sciences. Paris* 167: 536–7.

Gepp, E.S. 1904. Chinese marine algae. *Journal of Botany, London* 42: 161–6.

Gerard, V.A. 1982. Growth and utilization of internal nitrogen reserves by the giant kelp *Macrocystis pyrifera* in a low-nitrogen environment. *Marine Biology* 66: 27–35.

Gerard, V.A. 1986. Photosynthetic characteristics of giant kelp (*Macrocystis pyrifera*) determined in situ. *Marine Biology* 90: 473–82.

Gerard, V.A. 1987. Hydrodynamic streamlining of *Laminaria saccharina* Lamour. in response to mechanical stress. *Journal of Experimental Marine Biology and Ecology* 107: 237–44.

Gerard, V.A. 1988. Ecotypic differentiation in light-harvesting traits of the kelp *Laminaria saccharina*. *Marine Biology, Berlin* 97: 25–36.

Gerard, V.A. 1990. Ecotypic differentiation in the kelp *Laminaria saccharina*: phase-specific adaptation in a complex life cycle. *Marine Biology* 107: 519–28.

Gerard, V.A. 1997. The role of nitrogen nutrition in high-temperature tolerance of the kelp, *Laminaria saccharina* (Chromophyta). *Journal of Phycology* 33: 800–10.

Gerard, V.A. & Du Bois, K.R. 1988. Temperature ecotypes near the southern boundary of the kelp *Laminaria saccharina*. *Marine Biology, Berlin* 97: 575–80.

Gerard, V.A., DuBois, K. & Greene, R. 1987. Growth responses of two *Laminaria saccharina* populations to environmental variation. *Proceedings of the International Seaweed Symposium* 12: 229–32.

Gerard, V.A. & Mann, K.H. 1979. Growth and production of *Laminaria longicruris* (Phaeophyta) populations exposed to different intensities of water movement. *Journal of Phycology* 15: 33–41.

Gerbal, M. 1986. *Sargassum muticum* (Phaeophyceae, Fucales) dans l'Étang de Thau (Mediterranee, France): Presence de thalles cespiteux. *Rapport et Procès-Verbaux des Réunions de la Commission Internationale pour l'Exploration Scientifique de la Mer Méditeranée* 30: 5–6.

Gerbal, M., Maiz, N.B. & Boudouresque, C.-F. 1985. Les peuplements à *Sargassum muticum* de l'Étang de Thau: données préliminaires sur la flore algale. *Actes Congrès national des sociétés savantes* 110: 241–54.

Gessner, F. 1970. Temperature. In: Kinne, O. (ed.), *Marine Ecology*, Vol. 1, Part 1, pp. 363–406. Wiley & Sons, Chichester.

Gessner, F. & Schramm, W. 1971. Salinity: plants. In: Kinne, O. (ed.), *Marine Ecology*, Vol. 1, Part 2, pp. 705–820. Wiley & Sons, Chichester.

Gevaert, F., Créach, A., Davoult, D., Migné, A., Levavasseur, G., Arzel, P., Holl, A.-C. & Lemoine, Y. 2003. *Laminaria saccharina* photosynthesis measured *in situ*: photoinhibition and xanthophyll cycle during a tidal cycle. *Marine Ecology Progress Series* 247: 43–50.

Geyler, T. 1866. Zur Kenntniss der Sphacelarieen. *Jahrbücher für Wissenschaftliche Botanik* 4: 479–535.

Giaccone, G. & Bruni, A. 1973. Le Cistoseire e la vegetazione sommersa del Mediterraneo. *Atti del Reale Istituto Veneto di Scienze, Lettere ed Arti* 131: 59–103.

Gibb, D.C. 1937. Observations on *Himanthalia lorea* (L.) *Lyngb. Journal of the Linnean Society of London, Botany* 51: 11–21.

Gibb, D.C. 1950. A survey of some of the commoner fucoid algae on Scottish shores. *Journal of Ecology* 38: 253–69.

Gibb, D.C. 1957. The free living forms of *Ascophyllum nodosum* (L.) Le Jol. *Journal of Ecology* 45: 49–83.

Givernaud, T. 1984. Recherches sur l'algue brune *Sargassum muticum* (Yendo) Fensholt en Basse-Normandie. In: Elliott, M. & Ducrotoy, J.P. (eds), *Estuaries and coasts: Spatial and temporal inter-comparisons*, pp. 129–32. ECSA 19 Symposium. Olsen & Olsen, Fredensborg, for ECSA – Estuarine and Coastal Sciences Association, Caen, France.

Givernaud, T., Cosson, J. & Givernaudmouradi, A. 1990. Regeneration of the brown seaweed *Sargassum muticum* (Phaeophyceae, Fucale). *Cryptogamie, Algologie* 114: 293–304.

Glombitza, K.W., Forster, M. & Eckhardt, G. 1982. Polyhydroxyphenyl ethers from *Sargassum-muticum* (Phaeophyceae). *Botanica Marina* 25: 449–54.

Gledhill, M., Nimmo, M., Hill, S.J. & Brown, M.T. 1999. The release of copper-complexing ligands by the brown alga *Fucus vesiculosus* (Phaeophyceae) in response to increasing total copper levels. *Journal of Phycology* 35: 501–9.

Gmelin, S.G. 1768. *Historia fucorum*. pp. [i–xii], [i]–239, [i]–6 expl. tab., 35 plates [IA, IB, IIA, IIB, III–XXXIII]. Petropoli [St. Petersburg]: Ex typographia Academiae scientiarum.

Gobi, C. 1874. Die Brauntange (Phaeosporeae und Fucaceae) des finnischen Meerbusens. *Mémoires de l'Académie Impériale des Sciences de St. Petersburg* 21: 1–21.

Gobi, C. 1878. Die Algenflora des Weissen Meeres und der demselben zunächstliegenden Theile des nördlichen Eismeeres. *Mémoires de l'Académie Impériale des Sciences de St. Petersburg, VIIé série* 26: 1–92.

Gobi, C. 1879. Berichte über die algologische Forschungen in finnischen Meerbusen im Sommer 1877 ausgeführt. *Trudy Leningr. Obshch. Estest.* 10: 83–92.

Goecker, M. & Kåll, S.E. 2003. Grazing preferences of marine isopods and amphipods on three prominent

algal species of the Baltic Sea. *Journal of Sea Research* 50: 309–14.

Gómez Garreta, A. 2001. *Flora phycologica iberica* [Phycological flora of the Iberian Peninsula], Volume 1: *Fucales*. Universidad de Murcia, 192 pp.

Gómez Garreta, I. & Lüning, K. 2001. Constant short-day treatment of outdoor-cultivated *Laminaria digitata* prevents summer drop in growth rate. *European Journal of Phycology* 36: 391–6.

Gómez Garreta, A. & Ribera, M.A. 2002. Lectotypification of several species of *Cystoseira* (Cystoseiraceae, Fucales) described by Sauvageau. *Cryptogamie, Algologie* 23: 291–300.

Gómez Garreta, A., Barcelo i Martí, M.C., Ribera Siguán, M.A. & Rull Lluch, J. 2001a. *Cystoseira* C. Agardh. In: Gómez Garreta, A. (ed.), *Flora Phycologica Iberica Vol. 1 Fucales*. pp. 99–166. Murcia: Universidad de Murcia.

Gómez Garreta, A., Barceló i Martí, M.C., Ribera Siguán, M.A. & Rull Lluch, J. 2001b. *Sargassum* C. Agardh. In: Gómez Garreta, A. (ed.), *Flora Phycologica Iberica Vol. 1 Fucales*, pp. 75–87. Murcia: Universidad de Murcia.

Gómez Garreta, A., Barceló Martí, M.C., Gallardo García, T., Pérez-Ruzafa, I.M., Ribera Siguán, M.A. & Rull Lluch, J. 2001c. *Flora Phycologica Iberica. Vol. 1. Fucales*. pp. 192. Universidad de Murcia.

Gómez Garreta, A., Ribera, M.A., Barceló, M.C. & Rull Lluch, J. 1994. Mapas de distribución de algas marinas de la Península Ibérica e Islas Baleares. V. *Cystoseira* C. Agardh: Grupos *C. ericaefolia* y *C. crinito-selaginoides*. *Botanica Complutensis* 19: 109–18.

Gómez Garreta, A., Ribera, M.A., Barceló, M.C. & Rull Lluch, J. 2002. Mapas de distribución de algas marinas de la Península Ibérica e Islas Baleares. XVI. *Dictyopteris polypodioides* (DC.) Lamour. Y *Spatoglossum solieri* (Chauv. ex Mont.) Kütz. (Dictyotales, Fucophyceae). *Botanica Complutensis* 26: 153–60.

Gómez Garreta, A., Rull Lluch, J., Barceló Martí, M.C. & Ribera Siguán, M.A. 2007. On the presence of fertile gametophytes of *Padina pavonica* (Dictyotales, Phaeophyceae) from the Iberian coasts. *Anales del Jardín Botánico de Madrid* 64: 27–33.

Goodband, S.J. 1971. The taxonomy of *Sphacelaria cirrosa* (Roth) Ag., *Sphacelaria fusca* (Huds.) Ag., and *Sphacelaria furcigera* (Kütz.) Sauv. A comparison statistical approach. *Annals of Botany* 35: 957–80.

Goodband, S.J. 1973. Observations on the development of endophytic filaments of *Sphacelaria bipinnata* (Kütz.) Sauv. on *Halidrys siliquosa* (l.) Lyngb. *British Phycological Journal* 8: 175–9.

Goodenough, S. & Woodward, T.J. 1797. Observations on the British Fuci, with particular descriptions of each species. *Transactions of the Linnean Society of London* 3: 84–235.

Goodwin, T.W. 1974. Carotenoids and biliproteins. In: Stewart, W.D.P. (ed.), *Algal Physiology and Biochemistry*, pp. 176–205. Botanical Monographs 10. Blackwell Scientific Publications, Oxford.

Gorostiaga, J.M. 1994. *Fucus chalonii* or *Fucus vesiculosus* var. *linearis*? *The Phycologist* 37: 43–4.

Gorostiaga, J.M., Angulo, R. & Ibáñez, M. 1981. Nueva cita de *Saccorhiza polyschides* y *Laminaria ochroleuca* en la Costa Vasca. *Lurralde* 4: 265–70.

Gorostiaga, J.M., Casares, C., Fernandez, J.A., Perez, B. & Sarasua, A. 1988. Sobre la Expansion de *Sargassum muticum* (Yendo) Fensholt en la Costa Atlantica Europea. *Lurralde* 11: 437–43.

Graham, L.E. & Wilcox, L.W. 2000. *Algae*. 640 pp. Upper Saddle River, New Jersey: Prentice-Hall.

Graham, M.H. 1996. Effect of high irradiance on recruitment of giant kelp *Macrocystis* (Phaeophyta) in shallow water. *Journal of Phycology* 32: 903–6.

Graham, M.H. 1999. Identification of kelp zoospores from *in situ* plankton samples. *Marine Biology* 135: 709–20.

Graiff, A., Dankworth, M., Wahl, M., Karsten, U. & Bartsch, I. 2017. Seasonal variations of *Fucus vesiculosus* fertility under ocean acidification and warming in the western Baltic. *Botanica Marina* 60: 239–55.

Gran, H.H. 1893. Algevegetationen i Tonsbergfjorden. *Forh. VidenskSelsk. Krist.* 7: 1–38.

Gran, H.H. 1897. Kristianiafjordens algeflora, I. Rhodophyceae og Phaeophyceae. *Skrifter udgivne af Videnskabsselskabet i Christiania. I, Mathematisknaturvidenskabelig klasse*. 1896: 1–56.

Grateloup, J.P.A.S. 1806. *Descriptiones aliquorum Ceramiorum novorum, cum iconum explicationibus Observations sur la constitution de l'été de 1806.* pp. [1], [1 pl]. Montpellier.

Gray, J.E. 1864. *Handbook of British water-weeds or algae. The Diatomaceae by W. Carruthers*. pp. i–iv, 1–123. London: R. Hardwicke, Piccadilly.

Gray, P.W.G. & Gareth Jones, E.B. 1977. The attempted clearance of *Sargassum muticum* from Britain. *Environmental Conservation* 4: 303–8.

Gray, S.F. 1821. *A natural arrangement of British plants*, vol. 1: [i]–xxviii, [1]–824, pls I–XXI plates. London: Printed for Baldwin, Cradock & Joy, Paternoster-Row.

Green, J.R., Stafford, C.J., Wright, P.J. & Callow, J.A. 1992. Analysis of *Fucus serratus* gamete surfaces and vegetative tissue using monoclonal antibodies. *British Phycological Journal* 27: 90.

Green, L.A., Mathieson, A.C., Neefus, C.D., Traggis, H.M. & Dawes, C.J. 2012. Southern expansion of the brown alga *Colpomenia peregrina* Sauvageau (Scytosiphonales) in the Northwest Atlantic Ocean. *Botanica Marina* 55: 643–7.

Greuter, W. 2000. Notices of publications. Cryptogamae, 2. *Optima Newsletter* 35: 1–38.

Greville, R.K. 1824a. *Scottish Cryptogamic Flora*, or coloured figures and descriptions of cryptogamic plants, belonging chiefly to the order Fungi; and intended to serve as a continuation of English Botany. Vol. 2 (fasc. 19–24), pls 91–120 [with text]. Edinburgh & London: MacLachlan & Stewart; Baldwin, Craddock & Joy.

Greville, R.K. 1824b. *Flora edinensis*: or a description of plants growing near Edinburgh, arranged according to the Linnean system, with a concise introduction to the natural orders of the class Cryptogamia, and illustrated plates. pp. i–lxxxi, 1–478. Edinburgh: Printed for William Blackwood, Edinburgh; and T. Cadell, Strand, London.

Greville, R.K. 1827a. *Scottish Cryptogamic Flora, or coloured figures and descriptions of cryptogamic plants, belonging chiefly to the order Fungi; and intended to serve as a continuation of English Botany*. Vol. 5 (fasc. 55–60).

pls 271–300 (with text). Edinburgh & London: MacLachlan & Stewart; Baldwin, Craddock & Joy.

Greville, R.K. 1827b. *Scottish Cryptogamic Flora, or coloured figures and descriptions of cryptogamic plants, belonging chiefly to the order Fungi; and intended to serve as a continuation of English Botany*. Vol. 6 (fasc. 61–66), pls 301–330. Edinburgh & London: MacLachlan & Stewart; Baldwin, Craddock & Joy.

Greville, R.K. 1828. *Scottish Cryptogamic Flora, or coloured figures and descriptions of cryptogamic plants, belonging chiefly to the order Fungi; and intended to serve as a continuation of English Botany*. Vol. 6 (fasc. 67–72), pls 331–360. Edinburgh & London: MacLachlan & Stewart; Baldwin, Cradock & Joy.

Greville, R.K. 1830. *Algae britannicae*: pp. [i*–iii*], [i]–lxxxviii, [1]–218, pls 1–19. Edinburgh & London: McLachlan & Stewart; Baldwin & Cradock.

Griffith, J.W. & Henfrey, A. 1856. *The Micrographic Dictionary: a guide to the examination and investigation of the structure and nature of microscopic objects*; illustrated by forty-one plates and eight hundred and sixteen woodcuts. pp. [i]–v, [1]–696, 41 pls. London: John van Voorst.

Gross, V.A. & Cheney, D.P. 1993. Some conclusions on beach-fouling *Pilayella littoralis* in Nahant Bay. In: *Program and Abstracts, 33rd Northeast Algal Symposium, April 24–25, 1993*. Marine Biological Laboratory, Woods Hole, MA.

Gross, V.A. & Cheney, D.P. 1994. Biological and Oceanographic causes of a long-sanding algal bloom of *Pilayella littoralis* in Nahant Bay, Massachusetts. In: *Program and Abstracts, 33rd Northeast Algal Symposium, April 23–24, 1994*. Marine Biological Laboratory, Woods Hole, MA, p. 13.

Gruber, A., Roleda, M.Y., Bartsch, I., Hanelt, D. & Wiencke, C. 2011. Sporogenesis under ultraviolet radiation in *Laminaria digitata* (Phaeophyceae) reveals protection of photosensitive meiospores within soral tissue: physiological and anatomical evidence. *Journal of Phycology* 47: 603–14.

Grulois, D., Lévêque, L. & Viard, F. 2011. Mosaic genetic structure and sustainable establishment of the invasive kelp *Undaria pinnatifida* within a bay (Bay of St. Malo, Brittany). *Cahiers De Biologie Marine* 52: 485–98.

Grunow, A. 1874. Algen der Fidschi-, Tonga- und Samoa-Inseln, gesammelt von Dr. E. Graeffe. *Journal des Museum Godeffroy* 3: 23–50.

Guillaumont, B., Callens, L. & Dion, P. 1993. Spatial distribution and quantification of *Fucus* species and *Ascophyllum nodosum* beds in intertidal zones using spot imagery. *Proceedings of the International Seaweed Symposium* 14: 297.

Guiry, M.D. 1977. Notes on Irish marine algae 1. New records from the west Waterford coast. *Irish Naturalists' Journal* 19: 80–85.

Guiry, M.D. 1978. *A Consensus and Bibliography of Irish Seaweeds*. pp. [1]–287, 1 map. Vaduz: J. Cramer.

Guiry, M.D. 2012. *A Catalogue of Irish Seaweeds*. pp. [1]–250, 1 pl. Ruggell: A.R.G. Gantner Verlag KG.

Guiry, M.D. & Guiry, G.M. 2023 (continuously updated). AlgaeBase. World-wide electronic publication, National University of Ireland, Galway. http://www.algaebase.org.

Guiry, M.D. & Morrison, L. 2013. The sustainable harvesting of *Ascophyllum nodosum* (Fucaceae, Phaeophyceae) in Ireland, with notes on the collection and use of some other brown algae. *Journal of Applied Phycology* 25: 1823–30.

Gunnarsson, K. 1991. Populations de *Laminaria hyperborea* et *Laminaria digitata* (Phéophycées) dans la Baie de Breidifjördur, Islande. *Journal of the Marine Institue Reykjavik* 12: 148.

Gunnarsson, K. & Jónsson, S. 2002. Benthic marine algae of Iceland: revised checklist. *Cryptogamie, Algologie* 23: 131–58.

Gunnerus, J.E. 1766. *Flora norvegica*, observationibus praesertim oeconomicis panosque norvegici locupletata. Vol. 1, pp. [i]–viii, 1–96, [4], 3 pls. Nidrosia [Trondheim] & Hafnia [Copenhagen]: Vinding & F.C. Pelt.

Gunnerus, J.E. 1772. *Flora norvegica*, observationibus praesertim oeconomicis panosque norvegici locupletata. Pars posterior, cum iconibus. pp. [i]–viii, [1]–148, indices, pls I–IX. Nidrosia [Trondheim]; Hafnia [Copenhagen]: Vinding; F.C. Pelt.

Gutt, J. 2001. On the direct impact of ice on marine benthic communities, a review. *Polar Biology* 24: 553–64.

Guzinski, J., Ballenghien, M., Daguin-Thiébaut, C., Lévêque, L. & Viard, F. 2018. Population genomics of the introduced and cultivated Pacific kelp *Undaria pinnatifida*: Marinas – not farms – drive regional connectivity and establishment in natural rocky reefs *Evolutionary Applications* 11: 1582–97.

Gylle, A.M., Nygård, C.A. & Ekelund, N.G.A. 2009. Desiccation and salinity effects on marine and brackish *Fucus vesiculosus* L. (Phaeophyceae). *Phycologia* 48: 156–64.

Haas, S.P. & Hill, T.G. 1933. Observations on the metabolism of certain seaweeds. *Annals of Botany* 47: 55–67.

Haavisto, F., Koivikko, R. & Jormalainen, V. 2017. Defensive role of macroalgal phlorotannins: benefits and trade-offs under natural herbivory. *Marine Ecology Progress Series* 566: 79–90.

Haavisto, F., Välikangas, T. & Jormalainen, V. 2010. Induced resistance in a brown alga: phlorotannins, genotypic variation and fitness costs for the crustacean herbivore. *Oecologia* 162: 685–95.

Hagen Rødde, R.S. & Larsen, B. 1997. Protoplasts of *Laminaria digitata* and *Laminaria saccharina* (Phaeophyta) – Cultivation and Biosynthesis of Alginate. *Botanica Marina* 40: 391–5.

Hagen Rødde, R.S., Østgaard, K. & Larsen, B.A. 1993. Mannuronan C-5 epimerase activity in protoplasts of *Laminaria digitata*. *Proceedings of the International Seaweed Symposium* 14: 577–81.

Hagen Rødde, R.S., Østgaard, K. & Larsen, B. 1997. Chemical composition of protoplasts of *Laminaria digitata* (Phaeophyceae). *Botanica Marina* 40: 385–90.

Hainsworth, S. 1976. Some interesting additions to the marine fauna of Lundy. *Report of the Lundy Field Society* 26: 61–2.

Halat, L., Galway, M.E., Gitto, S. & Garbary, D.J. 2015. Epidermal shedding in *Ascophyllum nodosum* (Phaeophyceae): seasonality, productivity and relationship to harvesting. *Phycologia* 54: 599–608.

Hales, J.M. & Fletcher, R.L. 1989a. Aspects of the ecology of *Sargassum muticum* (Yendo) Fensholt in the Solent region of the British Isles: II. Reproductive phenology and senescence. In: *Reproduction, Genetics and Distributions of Marine Organisms*. 23rd European Marine Biology Symposium, pp. 115–25.

Hales, J.M. & Fletcher, R.L. 1989b. Studies on the recently introduced brown alga *Sargassum muticum* (Yendo) Fensholt. IV. The effect of temperature, irradiance and salinity on germling growth. *Botanica Marina* 32: 167–76.

Hales, J.M. & Fletcher, R.L. 1990. Studies on the recently introduced brown alga *Sargassum muticum* (Yendo) Fensholt. 5. Receptacle initiation and growth, and gamete release in laboratory culture. *Botanica Marina* 33: 241–9.

Hales, J.M. & Fletcher, R.L. 1992. Receptacle regeneration in *Sargassum muticum* (Phaeophyta). *Phycologia* 31: 591–601.

Hall, A. 1980. Heavy metal co-tolerance in a copper-tolerant population of the marine fouling alga, *Ectocarpus siliculosus* (Dillw.) Lyngbye. *New Phytologist* 85: 73–8.

Hall, A. 1981. Copper accumulation in copper-tolerant and non-tolerant populations of the marine fouling alga, *Ectocarpus siliculosus* (Dillw.) Lyngbye. *Botanica Marina* 24: 223–8.

Hall, A., Fielding, A.H. & Butler, M. 1979. Mechanisms of copper tolerance in the marine fouling alga *Ectocarpus siliculosus* – evidence for an exclusion mechanism. *Marine Biology, Berlin* 54: 195–9.

Hallet, J.N., Guinel, F. & Lecocq, F-M. 1983. Variations de l'activité mitotique au cours d'un cycle journalier et évolution du contenu en DNA nucléaire dans les territoires meristématiques et différenciés du *Fucus spiralis* Linné (Fucales, Pheophyceae). *Phycologia* 22: 325–36.

Halm, H., Lüder, U.H. & Wiencke, C. 2011. Induction of phlorotannins through mechanical wounding and radiation conditions in the brown macroalga *Laminaria hyperborea*. *European Journal of Phycology* 46: 16–26.

Hamel, C. 1928. Les algues de Vigo. *Revue Algologique* 4: 81–95.

Hamel, G. 1931–39. *Pheophycees de France*, Vans. Paris [1931. *Fasc. I.* Ectocarpacees pp. 1–80; 1935. *Fasc. II.* Myronematacees-Spermatochnacees pp. 81–176; 1937. *Fasc. III.* Spermatochnacees-Sphacelariacees pp. 177–240; 1938. *Fasc. IV.* Sphacelariacees-Dictyotacees pp. 241–336; 1939. *Fasc. V.* Dictyotacees-Sargassacees pp. i–xlvii + 337–432].

Hamel, G. 1939. Sur la classification des Ectocarpales. *Botaniska Notiser* 1939: 65–70.

Han, T. 1993. Wavelength dependent effect of high irradiance on early sporophytes of *Laminaria hyperborea* (Phaeophyta). *Korean Journal of Phycology* 8: 199–205.

Han, T. 1994a. Growth responses of early sporophytes of *Alaria esculenta* and *Laminaria hyperborea* to light qualities. *Korean Journal of Phycology* 9: 67–75.

Han, T. 1994b. Sample size determination for estimating growth rates of early sporophytes of *Laminaria hyperborea* (Phaeophyta). *Korean Journal of Phycology* 9: 107–10.

Han, T.J. & Kain, J.M. 1991. Blue light reactivation of UV-irradiated young sporophytes of *Laminaria hyperborea*. *British Phycological Journal* 26: 87.

Han, T.J. & Kain, J.M. 1992a. Growth responses of early sporophytes of *Laminaria hyperborea* to light. *British Phycological Journal* 27: 91.

Han, T.J. & Kain, J.M. 1992b. Blue light sensitivity of UV-irradiated young sporophytes of *Laminaria hyperborea*. *Journal of Experimental Marine Biology and Ecology* 158: 219–30.

Han, T. & Kain (Jones), J.M. 1993. Note: Blue light photoreactivation in ultraviolet-irradiated young sporophytes of *Alaria esculenta* and *Laminaria saccharina* (Phaeophyta). *Journal of Phycology* 29: 79–81.

Han, T. & Kain (Jones), J.M. 1996. Effect of photon irradiance and photoperiod on young sporophytes of four species of the Laminariales. *European Journal of Phycology* 31: 233–40.

Hanelt, D., Wiencke, C., Karsten, U. & Nultsch, W. 1997. Photoinhibition and recovery after high light stress in different developmental and life-history stages of *Laminaria saccharina* (Phaeophyceae). *Journal of Phycology* 33: 387–95.

Hanisak, M.D. 1983. The nitrogen relationships of marine macroalgae. In: Carpenter, E.J. & Capone, D.G. (eds), *Nitrogen in the Marine Environment*, pp. 699–730. Academic Press, New York.

Hanna, H. 1899. The plurilocular sporangia of *Petrospongium berkeleyi*. *Annals of Botany* 13: 461–4.

Hara, M. & Akiyama, K. 1985 Heterosis in growth of *Undaria pinnatifida* (Harvey) Suringar. *Bulletin of Tohoku Regional Fisheries Research Laboratory* 47: 47–50 (in Japanese with English abstract).

Hara, M. & Ishikawa, Y. 1988. Differences of the growth and shapes among four local forms, *Undaria pinnatifida*, cultured in same grounds. *Fish Genetics and Breeding Science* 13: 27–35 (in Japanese).

Harding, F.G. 1978. Attachment and development of *Halidrys siliquosa*. *British Phycological Journal* 13: 201.

Hardy, F.G & Guiry, M.D. 2006. *A Check-list and Atlas of the Seaweeds of Britain and Ireland*. Revised edn. London: British Phycological Society.

Hardy, F.G. & Moss, B.L. 1978. The attachment of zygotes and germlings of *Halidrys siliquosa* (L.) Lyngb. (Phaeophyceae, Fucales). *Phycologia* 17: 69–78.

Hardy, F.G. & Moss, B.L. 1979a. Notes on the attachment of zygotes and germlings of *Bifurcaria bifurcata* Ross (Phaeophyceae, Fucales). *Phycologia* 18: 164–70.

Hardy, F.G. & Moss, B.L. 1979b. The effects of the substratum on the morphology of the rhizoids of *Fucus* germlings. *Estuarine, Coastal and Marine Science* 9: 577–84.

Hardy, F.G. & Moss, B.L. 1979c. Attachment and development of the zygotes of *Pelvetia canaliculata* (L.) Dcne et Thur. (Phaeophyceae, Fucales). *Phycologia* 18: 203–12.

Hardy, F.G., Scott, G.W., Sisson, P.R., Lightfoot, N.F. & Mulyadi, 1998. Pyrolysis mass spectrometry as a technique for studying inter- and intraspecific relationships in the genus *Fucus*. *Journal of the Marine Biological Association of the United Kingdom* 78: 35–42.

Harkin, E. 1981. Fluctuations in epiphyte biomass following *Laminaria hyperborea* canopy removal. *Proceedings of the International Seaweed Symposium* 10: 303–8.

Harley, C.D.G. & Helmuth, B.S.T. 2003. Local- and regional-scale effects of wave exposure, thermal stress, and absolute versus effective shore level on patterns of intertidal zonation. *Limnology and Oceanography* 48: 1498–508.

Harlin, M.M. 1974. The surfaces seaweeds grow on may be a clue to their control. Maritimes (*Univ. R.I. Grad. Sch. Ocean*) 18: 7–8.

Harlin, M.M. & Lindbergh, J.M. 1977. Selection of substrata by seaweeds: optimal surface relief. *Marine Biology* 40: 33–40.

Harlin, M.M. & Craigie, J.S. 1975. The distribution of photosynthate in *Ascophyllum nodosum* as it relates to epiphytic *Polysiphonia lanosa*. *Journal of Phycology* 11: 109–13.

Harlin, M.M. & Craigie, J.S. 1981. The export of organic carbon from *Ascophyllum nodosum* as it relates to epiphytic *Polsiphonia lanosa*. *Proceedings of the International Seaweed Symposium* 8: 193.

Hartman, M. 1950. Beitrage zur Kenntnis der Befruchtung und Sexualitatt mariner Algen. I. Uber die Befruchtung von *Cutleria multifida*. *Pubblicazioni della Stazione Zoologica di Napoli* 22: 120–8.

Haroun, R. & Izquierdo, M.S. 1991. Distribución de *Sargassum muticum* (Yendo) Fensholt en Europa. Peligros de su penetración en la Península Ibérica. *Actas V Symposium Iberico Estudios Bentos Marino Tomo* 1: 27–48.

Harris, J.E. 1943. First report of the marine corrosion sub-committee. Section C. Antifouling investigations. *Journal of the Iron and Steel Institute* 147: 405–20.

Harris, J.E. 1946. Report on antifouling research, 1942–44. *Journal of the Iron and Steel Institute* 154: 297–333.

Harries, R. 1932. An investigation by cultural methods of some of the factors influencing the development of the gametophytes and early stages of the sporophytes of *Laminaria digitata*, *L. saccharina* and *L. cloustoni*. *Annals of Botany, Lond.* 46: 893–928.

Harvey, J.B.J. & Goff, L.J. 2006. A reassessment of species boundaries in *Cystoseira* and *Halidrys* (Phaeophyceae, Fucales) along the North American west coast. *Journal of Phycology* 42: 707–20.

Harvey, W.H. 1833. Div. II. Confervoideae; Div. III. Gloiocladeae. In: *The English Flora of Sir James Edward Smith. Class XXIV. Cryptogamia. Vol. V (or Vol. II of Dr. Hooker's British flora). Part I. Comprising the Mosses, Hepaticae, Lichens, Characeae and Algae*. (Hooker, W.J. Eds), pp. 263–5, 265–6, 326–89, 389–405. London: Longman, Brown, Green & Longmans Paternoster-Row.

Harvey, W.H. 1834. Algological illustrations. No. 1 Remarks on some British algae and descriptions of new species recently added to our flora. *Journal of Botany* [*Hooker*] 1: 296–305.

Harvey, W.H. 1838. *The genera of South African plants*, arranged according to the natural system. pp. [i]–lxvi, [i]–429. Cape Town: A.S. Robertson.

Harvey, W.H. 1841. *A manual of the British algae*: containing generic and specific descriptions of the known British species of sea-weeds and of *Confervae* both marine and fresh-water. pp. [i–v]–lvii, [1]–229. London: John Van Voorst.

Harvey, W.H. 1846–51. *Phycologia britannica*, or, a history of British sea-weeds: containing coloured figures, generic and specific characters, synonyms, and descriptions of all the species of algae inhabiting the shores of the British Islands. Text with plates,

London: Reeve & Benham. [1846, pls 1–78; 1847, pls 79–144; 1848, pls 145–216; 1849, pls 217–70; 1850, pls 271–318; 1851, pls 319–260].

Harvey, W.H. 1849. *A Manual of the British Marine Algae:* containing generic and specific descriptions of all the known British species of sea-weeds. With plates to illustrate the genera. London: John Van Voorst.

Harvey, W.H. 1852. *Nereis boreali-americana;* or, contributions towards a history of the marine algae of the Atlantic and Pacific coasts of North America. Part I. Melanospermeae. *Smithsonian Contributions to Knowledge* 3: 1–150.

Harvey, W.H. 1857. Short descriptions of some new British algae, with two plates. *Natural History Review* 4: 201–4.

Harvey, W.H. 1860. Characters of new algae, chiefly from Japan and adjacent regions, collected by Charles Wright in the North Pacific Exploring Expedition under Captain James Rodgers. *Proceedings of the American Academy of Arts and Sciences* 4: 327–35.

Harvey, W.H. 1862. Notice of a collection of algae made on the northwest coast of North America, chiefly at Vancouver Island, by David Lyall, Esq., MD, RN, in the years 1859–1861. *Journal of the Proceedings of the Linnean Society. Botany* 6: 157–77.

Harvey, W.H. 1864. The Kilkee *Fucus. Popular Science Review* 3: 389.

Harvey, W.H. 1866. Notice of the discovery of *Fucus distichus* L. at Duggerna, County Clare, Ireland. *Transactions of the Botanical Society of Edinburgh* 8: 53–4.

Hauck, F. 1877. Beitrage zur Kenntniss der adriatischen Algen. II. *Myrionema liechtensternii* n. sp. *Österreichische Botanische Zeitschrift* 6: 185–6.

Hauck, F. 1879. Beitrage zur Kenntniss der adriatischen Algen. X. *Österreichische Botanische Zeitschrift* 29: 151–4.

Hauck, F. 1883–85. Die Meeresalgen Deutschlands und Oesterreichs. In: Rabenhorst, L. *Kryptogamen.-Flora von Deutschland, Oesterreich und der Schweiz,* ed. 2, 2, Leipzig, pp. 1–320 (1883), 321–512 (1884), 513–75 (1885).

Hauck, F. 1887. *Choristocarpus tenellus. Hedwigia* 26: 122–4.

Haug, A. & Jensen, A. 1956. Seasonal variation in chemical composition of *Laminaria digitata* from different parts of the Norwegian Coast. *Proceedings of the International Seaweed Symposium* 2: 10–15.

Hawkins, S.J. 1981. The influence of season and barnacles on the algal colonization of *Patella vulgata* exclusion areas. *Journal of the Marine Biological Association of the United Kingdom* 61: 1–15.

Hawkins, S.J. & Hartnoll, R.G. 1980. Small-scale relationship between species number and area on a rocky shore. *Estuarine, Coastal and Marine Science* 10: 201–14.

Hawkins, S.J. & Hartnoll, R.G. 1983. Grazing of intertidal algae by marine-invertebrates. *Oceanography and Marine Biology: An Annual Review* 21: 195–282.

Hawkins, S.J. & Hartnoll, R.G. 1985. Factors determining the upper limits of intertidal canopy-forming algae. *Maine Ecology progress Series* 20: 265–71.

Hawkins, S.J., Hartnoll, R.G., Kain, J.M. & Norton, T.A. 1992. Plant–animal interactions on hard substrata in the north-east Atlantic. In: John, D.M., Hawkins, S.J.

& Price, J.H. (eds), *Plant–Animal Interactions in the Marine Benthos,* pp. 1–32. Clarendon Press, Oxford.

Hay, C.H. 1990. The dispersal of sporophytes of *Undaria pinnatifida* by coastal shipping in New Zealand, and implications for further dispersal of *Undaria* in France. *British Phycological Journal* 25: 301–13.

Hay, C.H. & Luckens, P.A. 1987. The Asian kelp *Undaria pinnatifida* (Phaeophyta: Laminariales) found in a New Zealand harbour. *New Zealand Journal of Botany* 25: 329–32.

Hay, C.H. & Villouta, E. 1993. Seasonality of the adventive Asian kelp *Undaria pinnatifida* in New Zealand. *Botanica Marina* 36: 461–76.

Häyrén, E. 1940. Über die Meeresalgen der Insel Hogland im Finnischen Meeresbusen. *Acta Phytogeographica Suecica* 13: 52–60.

Hazlett, A. & Seed, R. 1976. A study of *Fucus spiralis* and its associated fauna in Strangford Lough, Co. Down. *Proceedings of the Royal Irish Academy* 76B: 607–18.

Heesch, S. & Peters, A.F. 1999. Scanning electron microscopy observation of host entry by two brown algae endophytic in *Laminaria saccharina* (Laminariales, Phaeophyceae). *Phycological Research* 47: 1–6.

Heesch, S., Rindi, F., Guiry, M.D. & Nelson, W.A. 2020. Molecular phylogeny and taxonomic reassessment of the genus *Cladostephus* (Sphacelariales, Phaeophyceae). *European Journal of Phycology* 55: 426–43. https://doi.org/10.1080/09670262.2020.1740947

Heil, H. 1924. Die Basalzelle des Tetrasporangiums von *Dictyota dichotoma* und einiges uber die Zellwandstruktur. *Berichte der Deutschen Botanischen Gesellschaft* 42: 119–25.

Heinrich, S., Frickenhaus, S., Glochner, G. & Valentin, K. 2012. A comprehensive cDNA library of light- and temperature-stressed *Saccharina latissima* (Phaeophyceae). *European Journal of Phycology* 47: 83–94.

Heinrich, S., Valentin, K., Frickenhaus, S. & Wiencke, C. 2015. Temperature and light interactively modulate gene expression in *Saccharina latissima* (Phaeophyceae). *Journal of Phycology* 51: 93–108.

Heiser, S., Hall-Spencer, J.M. & Hiscock, K. 2014. Assessing the extent of establishment of *Undaria pinnatifida* in Plymouth Sound Special Area of Conservation, UK. Marine *Biodiversity Records* 7: e93.

Hellebust, J.A. 1970. Light. In: Kinne, O. (ed.), *Marine Ecology* Vol. 1, Part 1, pp. 125–58. Wiley & Sons, Chichester.

Hellebust, J.A. & Haug, A. 1972a. In situ studies on alginic acid synthesis and other aspects of the metabolism of *Laminaria digitata. Canadian Journal of Botany* 50: 177–84.

Hellebust, J.A. & Haug, A. 1972b. Photosynthesis, translocation, and alginic acid synthesis in *Laminaria digitata* and *Laminaria hyperborea. Canadian Journal of Botany* 50: 169–76.

Hellebust, J.A. 1974. Extracellular products. In: Stewart, W.D.P. (ed.), *Algal Physiology and Biochemistry,* pp. 838–63. *Botanical Monographs* 10. Blackwell Scientific Publications, Oxford.

Henkel, S.K. & Hofmann, G.E. 2008. Thermal ecophysiology of gametophytes cultured from invasive *Undaria pinnatifida* (Harvey) Suringar in coastal California harbors. *Journal of Experimental Marine Biology and Ecology* 367: 164–73.

Henkel, S.K., Kawai, H. & Hofmann, G.E. 2009. Inter-specific and interhabitat variation in hsp70 gene expression in native and invasive kelp populations. *Marine Ecology Progress Series* 386: 1–13.

Henry, E.C. 1987a. The life history of *Phyllariopsis brevipes* (*Phyllaria reniformis*) (Phyllariaceae, Laminariales, Phaeophyceae), a kelp with dioecious but sexually monomorphic gametophytes *Phycologia* 26: 17–22.

Henry, E.C. 1987b. Primitive reproductive characters and a photoperiodic response in *Saccorhiza dermatodea* (Laminariales, Phaeophyceae). *British Phycological Journal* 22: 23–31.

Henry, E.C. 1987c. Morphology and life histories of *Onslowia bahamensis* sp. nov. and *Verosphacela ebrachia* gen. et sp. nov., with a reassessment of the Choristocarpaceae (Sphacelariales, Phaeophyceae). *Phycologia* 26: 182–91.

Henry, E.C. & Cole, K.M. 1982a. Ultrastructure of swarmers in the Laminariales (Phaeophyceae). I. Zoospores. *Journal of Phycology* 18: 550–69.

Henry, E.C. & Cole, K.M. 1982b. Ultrastructure of swarmers in the Laminariales (Phaeophyceae). II. Sperm. *Journal of Phycology* 18: 570–9.

Henry, E.C. & South, G.R. 1987. *Phyllariopsis* gen. nov. and a reappraisal of the Phyllariaceae Tilden 1935 (Laminariales, Phaeophyceae). *Phycologia* 26: 9–16.

Hepburn, C.D., Holborow, J.D., Wing, S.R., Frew, R.D. & Hurd, C.L. 2007. Exposure to waves enhances the growth rate and nitrogen status of the giant kelp *Macrocystis pyrifera*. *Marine Ecology Progress Series* 339: 99–108.

Herbert, R.J.H., Ma, L., Marston, A., Farnham, W.F., Tittley, I. & Cornes, R.C. 2016. The calcareous brown alga *Padina pavonica* in southern Britain: population change and tenacity over 300 years. *Marine Biology* 163: 1–15.

Herbert, R.J.H., Marston, A., Manley, H. & Farnham, W.F. 2018. Distribution of the 'Peacocks Tail' seaweed *Padina pavonica* on the Isle of Wight. *Proceedings of the Isle of Wight Natural History and Archaeological Society* 32: 112–19.

Hession, C.C., Guiry, M.D., McGarvey, S. & Joyce, D. 1998. Mapping and assessment of the seaweed resources (*Ascophyllum nodosum*, *Laminaria* spp.) off the west coast of Ireland. *Marine Resource Series* [*Ireland*] 5: 1–74.

Hewitt, C.L., Campbell, M.L., McEnnulty, F., Moore, K.M., Murfet, N.B., Robertson, B. & Schaffelke, B. 2005. Efficacy of physical removal of a marine pest: the introduced kelp *Undaria pinnatifida* in a Tasmanian marine reserve. *Biological Invasions* 7: 251–63.

Higgins, E.M. 1931. A cytological investigation of *Stypocaulon scoparium* (L.) Kutz., with special reference to the unilocular sporangia. *Annals of Botany* 45: 345–53.

Hillrichs, S. & Schmid, R. 2001. Activation by blue light of inorganic carbon acquisition for photosynthesis in *Ectocarpus siliculosus*: organic acid pools and short-term carbon fixation. *European Journal of Phycology* 36: 71–80.

Hiscock, K. 1983. Water movement. In: Earll, R. & Erwin, D.G. (eds), *Sublittoral Ecology: The Ecology of the Shallow Sublittoral Benthos*, pp. 58–96. Clarendon Press, Oxford.

Hiscock, K. & Brodie, J. (2016). The character and status of rocky shore communities at Lundy: historic and recent surveys. *Journal of the Lundy Field Society* 5: 35–54.

Hiscock, S. & Maggs, C.A. 1982. Notes on Irish marine algae-6. *Zanardinia prototypus* (Nardo) Nardo (Phaeophyta). *Irish Naturalists' Journal* 20: 414–16.

Hiscock, S. & Maggs, C.A. 1984. Notes on the distribution and ecology of some new and interesting seaweeds from south-west Britain. *British Phycological Journal* 19: 73–87.

Hoarau, G., Coyer, J.A. & Olsen, J.L. 2009. Paternal leakage of mitochondrial DNA in a *Fucus* (Phaeophyceae) hybrid zone (Note). *Journal of Phycology* 45: 621–4.

Hoarau, G., Coyer, J.A., Veldsink, J.H., Stam, W.T. & Olsen, J.L. 2007. Glacial refugia and recolonization pathways in the brown seaweed *Fucus serratus*. *Molecular Ecology* 16: 3606–16.

Hoffmann, A.J. & Malbrán, M.E. 1989. Temperature, photoperiod and light interactions on growth and fertility of *Glossophora kunthii* (Phaeophyta, Dictyotales) from central Chile. *Journal of Phycology* 25: 129–33.

Hoffmann, A.J. & Santelices, B. 1982. Effects of light intensity and nutrients on gametophytes and gametogenesis of *Lessonia nigrescens* Bory (Phaeophyta). *Journal of Experimental Marine Biology and Ecology* 60: 77–89.

Hoffmann, A.J., Avila, M. & Santelices, B. 1984. Interactions of nitrate and phosphate on the development of microscopic stages of *Lessonia nigrescens* Bory (Phaeophyta). *Journal of Experimental Marine Biology and Ecology* 78: 177–86.

Holden, H.S. 1913. On some abnormal specimens of *Dictyota dichotoma*. *Memoirs and Proceedings of the Manchester Literary and Philosophical Society* 57 (No. 9) (II): 1–6.

Hollenberg, G.J. 1969. An account of the Ralfsiaceae (Phaeophyta) of California. *Journal of Phycology* 5: 290–301.

Hollenberg, G.J. & Abbott, I.A. 1966. *Supplement to Smith's marine algae of the Monterey Peninsula*. pp. [i]–ix [xii], [1]–130, 53 figs. Stanford: Stanford University Press.

Hollenberg, G.J. & Abbott, I.A. 1968. New species of marine algae from California. *Canadian Journal of Botany* 46: 1235–51.

Holmes, E.M. 1883. New British marine algae. *Grevillea* 11: 140–2.

Holmes, E.M. 1885. Review of G.W. Traill: A monograph of the Algae of the Firth of Forth. *Journal of Botany, British and Foreign* 23: 61–2.

Holmes, E.M. 1887a. Remarks on *Sphacelaria radicans* Harv. and *Sphacelaria olivacea* J. Ag. *Transactions and Proceedings Of The Botanical Society Of Edinburgh* 17: 79–82.

Holmes, E.M. 1887b. Two new British Ectocarpi. *Journal of Botany, British and Foreign* 25: 161–2.

Holmes, E.M. 1901. Alg. Brit. Rare. *Exc. Fasc. Xi*.

Holmes, E.M. & Batters, E.A.L. 1890. A revised list of the British marine algae. *Annals of Botany* 5: 63–107.

Holmes, E.M. & Batters, E.A.L. 1891. Appendix to the revised list of British marine algae. *Annals of Botany* 5: 518–26.

Hooker, W.J. 1833. Cryptogamia Algae [pp. 264–322] in. Hooker, W.J. *The English Flora of Sir James Edward Smith*. Class XXIV, Cryptogamia. Vol. V, Part 1.

Hooper, R.G. 1987. Benthic algal communities of an extremely isolated offshore shoal – Virgin Rocks, Grand Banks of Newfoundland. *British Phycological Journal* 22: 316.

Hooper, R.G., South, G.R. & Nielsen, R. 1987. Transfer of *Pilinia* Kuetzing from Chlorophyceae with *Waerniella* Kylin in synonymy. *Taxon* 36: 439–40.

Hooper, A., Ten Bos, S. & Bieeman, A.M. 1983. Photoperiodic response in the formation of gametangia of the long-day plant *Sphacelaria rigidula* (Phaeophyceae) *Marine Ecology Progress Series* 13: 285–9.

Hopkin, R. & Kain, J.M. 1971. The effect of marine pollutants on *Laminaria hyperborea*. *Marine Pollution Bulletin* 2: 75–7.

Hopkin, R. & Kain, J.M. 1978. The effects of some pollutants on the survival, growth and respiration of *Laminaria hyperborea*. *Estuarine, Coastal and Marine Science* 7: 531–53.

Hori, T. 1971. Survey of pyrenoid distribution in brown algae. *Botanical Magazine, Tokyo* 84: 231–42.

Hori, T. 1972. Further survey of the pyrenoid distribution in Japanese brown algae. *Botanical Magazine, Tokyo* 85: 125–34.

Horiguchi, T. & Yoshida, T. 1998. The phylogenetic affinities of *Myagropsis myagroides* (Fucales, Phaeophyceae) as determined from 18S rDNA sequences. *Phycologia* 37: 237–45.

Hornemann, J.W. 1813. *Flora danica*. Vol. 9, fasc. 25 pp. 1–8, pls MCCCCXLI–MD. Havniae [Copenhagen].

Hornemann, J.W. 1816. *Flora danica*. Vol. 9, fasc. 26 pp. 1–8, pls MDI–MDLX. Havniae [Copenhagen].

Hornemann, J.W. 1818. *Flora danica*. Vol. 9, fasc. 27. Havniae [Copenhagen].

Hornemann, J.W. 1837. *Dansk økonomisk Plantelaere*. Del. 2. Kjøbenhavn.

Hoshina, R., Hasegawa, K., Tanaka, J. & Hara, Y. 2004. Molecular phylogeny of the Dictyotaceae (Phaeophyceae) with emphasis on their morphology and its taxonomic implication. *Japanese Journal of Phycology* 52 (Suppl.): 189–94.

Howe, M.A. 1911. Phycological studies – V. Some marine algae of Lower California, Mexico. *Bulletin of the Torrey Botanical Club* 38: 489–514.

Howe, M.A. 1914. The marine algae of Peru. *Memoirs of the Torrey Botanical Club* 15: 1–185.

Hoyt, W.D. 1907. Periodicity in the production of the sexual cells of *Dictyota dichotoma*. *Botanical Gazette* 43: 383–92.

Hoyt, W.D. 1910. Alternation of generations and sexuality in *Dictyota dichotoma*. *Botanical Gazette* 49: 55–7.

Hoyt, W.D. 1927. The periodic fruiting of *Dictyota* and its relation to the environment. *American Journal of Botany* 14: 592–619.

Hsiao, S.I.C. 1969. Life history and iodine nutrition of the marine brown alga, *Petalonia fascia* (O.F. Mull.) Kuntze. *Canadian Journal of Botany* 47: 1611–16.

Hsiao, S.I.C. 1970. Light and temperature effects on the growth, morphology and reproduction of *Petalonia fascia*. *Canadian Journal of Botany* 48: 1359–61.

Hsiao, S.I.C. & Druehl, L.D. 1971. Environmental control of gametogenesis in *Laminaria saccharina*. I. The effects of light and culture media. *Canadian Journal of Botany* 49: 1503–8.

Hsiao, S.I. & Druehl, L.D. 1973a. Environmental control of gametogenesis in *Laminaria saccharina*. II. Correlation of nitrate and phosphate concentrations with gametogenesis and selected metabolites. *Canadian Journal of Botany* 51: 829–39.

Hsiao, S.I.C. & Druehl, L.D. 1973b. Environmental control of gametogenesis in *Laminaria saccharina*. III. The effects of different iodine concentrations, and chloride and iodide ratios. *Canadian Journal of Botany* 51: 989–97.

Hsiao, S.I.C. & Druehl, L.D. 1973c. Environmental control of gametogenesis in *Laminaria saccharina*. IV. *In situ* development of gametophytes and young sporophytes. *Journal of Phycology* 9: 160–4.

Huang, C.H., Sun, Z.M., Dao, D.H., Yao, J.T., Hu, Z.M., Li, Y.H., Yang, Y.Q., Xu, K.D. & Chen, W.D. 2017. Molecular analysis of *Sargassum* from the northern China seas. *Phytotaxa* 319: 71–83.

Hudson, [W.] 1762. *Flora anglica*: pp. [i]–viii, [1–8], 1–506, [1–22, ind.]. Londini [London]: Impensis auctoris: Prostant venales apud J. Nourse in the Strand, et C. Moran in Covent-Garden.

Hudson, [W.] 1778. *Flora anglica*: Tomus II. Editio altera, emendata et aucta. pp. [i], 335–690. Londini [London]: impensis auctoris: prostant venales apud J. Nourse, in the Strand.

Huh, M.K., Huh, H.W., Moon, S.G., Choi, C.M. & Lee, H.Y. 2001. Genetic diversity of cultivated *Undaria pinnatifida* populations from Korea based on allozyme variation analysis. *Algae* 16: 429–35.

Hunger, F.W.T. 1903. Ueber das Assimilationsproduct der Dictyotaceen. *Jahrbücher für wissenschaftliche Botanik* 38: 70–82.

Huppertz, K., Hanelt, D. & Nultsch, W. 1990. Photoinhibition of photosynthesis in the marine brown alga *Fucus serratus* as studied in field experiments. *Marine Ecology Progress Series* 66: 175–82.

Hurd, C.L. & Dring, M.J. 1988. Phosphate uptake by *Fucus spiralis* and *Fucus serratus*. *British Phycological Journal* 23: 290.

Hurd, C.L., Harrison, P.J. & Druehl, L.D. 1996. Effect of seawater velocity on inorganic nitrogen uptake by morphologically distinct forms of *Macrocystis integrifolia* from wave-sheltered and exposed sites. *Marine Biology* 126: 205–14.

Hurd, C.L., Galvin, R.S., Norton, T.A. & Dring, M.J. 1993. Production of hyaline hairs by intertidal species of *Fucus* (Fucales) and their role in phosphate-uptake. *Journal of Phycology* 29: 160–5.

Hutchins, L.W. 1947. The basis for temperature zonation in geographical distribution. *Ecological Monographs* 17: 325–35.

Hwang, E.K. & Dring, M.J. 2002. Quantitative photoperiodic control of erect thallus production in *Sargassum muticum*. *Botanica Marina* 45: 471–5.

Hwang, E.K., Gong, Y.G. & Park, C.S. 2012. Cultivation of a hybrid of free-living gametophytes between *Undariopsis peterseniana* and *Undaria pinnatifida*: morphological aspects and cultivation period. *Journal of Applied Phycology* 24: 401–8.

Hwang, E.K., Ha, D.S. & Park, C.S. 2017. The influences of temperature and irradiance on thallus length of *Saccharina japonica* (Phaeophyta) during the early stages of cultivation. *Proceedings of the 6th Congress of the International Society for Applied Phycology*, pp. 2875–82.

Hwang, E.K., Hwang, I.K., Park, E.J., Gong, Y.G. & Park, C.S. 2014. Development and cultivation of F2 hybrid between *Undariopsis peterseniana* and *Undaria pinnatifida* for abalone feed and commercial mariculture in Korea. *Journal of Applied Phycology* 26: 747–52.

Hwang, I.K., Kim, H.-S. & Lee, W.J. 2004a. Confirmation on taxonomic status of *Spatoglossum pacificum* Yendo (Dictyotaceae, Phaeophyceae) based on morphology and plastid protein coding rbcL, rbcS, psaA, and psbA gene sequences. *Algae* 19: 161–74.

Hwang, I.K., Kim, H.-S. & Lee, W.J. 2004b. Evidence for taxonomic status of *Pachydictyon coriaceum* (Holmes) Okamura (Dictyotales, Phaeophyceae) based on morphology and plastid protein encoding rbcL, psaA and psbA gene sequences. *Algae* 19: 175–90.

Hwang, I.K., Kim, H.-S. & Lee, W.J. 2005. Polymorphism in the brown alga *Dictyota dichotoma* (Dictyotales, Phaeophyceae) from Korea. *Marine Biology, Berlin* 147: 999–1015.

Hygen, G. 1934. Über den Lebenszyklus und die entwicklungsgeschichte der Phaeosporeen. Versuche an *Nemacystus divaricatus* (Ag.) Kuch. *Nyt Magazin for Naturvidenskaberne* 74: 187–268.

Hygen, G. & Jorde, I. 1935. Beitrag zur kenntnis der algenflora der norwegischen Westküste – Bergens Museums Årbok 1934. *Naturvidensk. Rekke* 9: 1–60.

Hörnig, I. & Schnetter, R. 1988. Notes on *Dictyota dichotoma*, *D. menstrualis*, *D. indica* and *D. pulchella* spec. nova (Phaeophyta). *Phyton (Horn)* 28: 277–91.

Hörnig, I., Schnetter, R. & Prud'homme van Reine, W.F. 1992b. The genus *Dictyota* (Phaeophyceae) in the North Atlantic. II. Key to the species. *Nova Hedwigia* 54: 397–402.

Hörnig, I., Schnetter, R. & Prud'homme van Reine, W.F. 1993. Additional note to 'The genus *Dictyota* (Phaeophyceae) in the North Atlantic. I. A new generic concept and new species.' Correction and validation of new combinations in the genus *Dictyota*. *Nova Hedwigia* 56: 169–71.

Hörnig, I., Schnetter, R., Prud'homme van Reine, W.F., Coppejans, E., Achenbach-Wege, K. & Over, J.M. 1992a. The genus *Dictyota* (Phaeophyceae) in the North Atlantic. I. A new generic concept and new species. *Nova Hedwigia* 54: 45–62.

IARE. 1992. Proposition en vue de la définition d'un schéma de développement de l'aquaculture sur les côtes de Corse. [Proposal of an aquaculture development plan in Corsica coast]. Rapport. IARE/Service Etudes & Aides Economiques, Collectivité Territoriale de Corse.

Igic, L. 1968. The fouling on ships as the consequence of their navigation in the Adriatic and other world seas. In: *Proceedings of the 2nd International Congress on Marine Corrosion and Fouling*, pp. 571–7. Athens.

Inagaki, K.I. 1954. Contributions to the knowledge of the Chordariales from Japan I. *Scientific Papers of the Institute of Algological Research, Faculty of Science, Hokkaido Imperial University* 4: 1–14.

Incera, M., Olabarria, C., Cacabelos, E., César, J. & Troncoso, J.S. 2011. Distribution of *Sargassum muticum* on the North West coast of Spain. Relationships with urbanization and community diversity. *Continental Shelf Science* 31: 488–95.

Indergaard, M. & Jensen, A. 1981. Nitrate and phosphate uptake in small populations of *Laminaria digitata* (Phaeophyceae). *Proceedings of the International Seaweed Symposium* 10: 411–17.

Indergaard, M. & Skjåk-Braek, G. 1987. Characteristics of alginate from *Laminaria digitata* cultivated in a high-phosphate environment. *Proceedings of the International Seaweed Symposium* 12: 541–9.

Íñiguez, C., Carmona, R., Lorenzo, M.R., Wiencke, C. & Gordillo, F.J.L. 2016. Increased temperature, rather than elevated CO2, modulates the carbon assimilation of the Arctic kelps *Saccharina latissima* and *Laminaria solidungula*. *Marine Biology* 163: 248–66.

Irigoyen, A.J., Eyras, C. & Parma, A.M. 2011a. Alien algae *Undaria pinnatifida* causes habitat loss for rocky reef fishes in north Patagonia. *Biological Invasions* 13: 17–24.

Irigoyen, A.J., Trobbiani, G., Sgarlatta, M.P. & Raffo, M.P. 2011b. Effects of the alien algae *Undaria pinnatifida* (Phaeophyceae, Laminariales) on the diversity and abundance of benthic macrofauna in Golfo Nuevo (Patagonia, Argentina): potential implications for local food webs. *Biological Invasions* 13: 1521–32.

Irvine, D.E.G. 1956. Notes on British species of the genus *Sphacelaria* Lyngb. *Transactions and Proceedings of the Botanical Society of Edinburgh* 37: 24–45.

Irvine, D.E.G. 1974. The marine vegetation of the Shetland Isles. In: Goodier, R. (ed.), *The Natural Environment of Shetland*, pp. 107–13. The Nature Conservancy Council, Edinburgh.

Irvine, D.E.G. 1982. Seaweeds of the Faroes. I. The flora. *Bulletin of the British Museum (Natural History) Botany* 10: 109–31.

Irvine, D.E.G., Guiry, M.D., Tittley, I. & Russell, G. 1975. New and interesting marine algae from the Shetland Islands. *British Phycological Journal* 10: 57–71.

Isaeus, M., Malm, T., Persson, S. & Svensson, A. 2004. Effects of filamentous algae and sediment on recruitment and survival of *Fucus serratus* (Phaeophyceae) juveniles in the eutrophic Baltic sea. *European Journal of Phycology* 39: 301–8.

Isaac, W.E. 1933. Some observations and experiments on the drought resistance of *Pelvetia canaliculata*. *Annals of Botany* 47: 343–8.

Ishikawa, Y. 1992. Morphological characters of the thalli obtained from a clone in *Undaria pinnatifida*. *Fish Genetics and Breeding Science* 18: 19–24 (in Japanese).

Ishikawa, Y. 1993. A simple method for growth estimation of blades in *Undaria pinnatifida*. *Nippon Suisan Gakkaishi* 59: 1331–6 (in Japanese with English abstract).

IUCN. 2009. *Marine Menace. Alien invasive species in the marine environment*. IUCN Global Marine Programme. Gland, Switzerland, pp. 30.

Izquierdo, J.L., Gallardo, T. & Pérez-Ruzafa, I. 1996. Mapas de distribución de algas marinas de la Península Ibérica e Islas Baleares. IX. *Saccorhiza polyschides* (Lightf.) Batt. y *Chorda filum* (L.) Stackh. (Laminariales, Fucophyceae). *Botanica Complutensis* 20: 105–15.

Izquierdo, J.L., Navarro, M.J. & Gallardo, T. 1993. Mapas de distribución de algas marinas de la Península Ibérica. IV. *Laminaria ochroleuca* Pylaie, *L. hyperborea*

(Gunner) Foslie y *L. saccharina* (L.) Lamour. (Laminariales, Fucophyceae). *Botanica Complutensis* 18: 291–304.

Izquierdo, J.L., Pérez-Ruzafa, I.M. & Gallardo, T 1997. An anatomical study of *Laminaria ochroleuca* Pylaie (Laminariales, Phaeophyta). *Nova Hedwigia* 66: 51–66.

Izquierdo, J.L., Pérez-Ruzafa, I. & Gallardo, T. 2002. Effect of temperature and photon fluence rate on gametophytes and young sporophytes of *Laminaria ochroleuca* Pylaie. *Helgoland Marine Research* 55: 285–92.

Jaasund, E. 1951. Marine algae from northern Norway, I. *Botaniska Notiser* 1951: 128–42.

Jaasund, E. 1957. Marine algae from northern Norway, II. *Botaniska Notiser* 110: 205–31.

Jaasund, E. 1960a. *Elachista lubrica* Ruprecht and *Elachista fucicola* (Velley) Areschoug. *Botanica Marina* 1: 101–7.

Jaasund, E. 1960b. *Fosliea curta* (Fosl.) Reinke and *Isthmoplea sphaerophora* (Carm.) Kjellman. *Botanica Marina* 2: 174–81.

Jaasund, E. 1961a. A note on *Ectocarpus fasciculatus* (Griff.) Harv. *Botaniska Notiser* 114: 239–41.

Jaasund, E. 1961b. Further studies of *Isthmoplea spaerophora* (Carm) Kjellman. *Botanica Marina* 2: 215–22.

Jaasund, E. 1963. Beitrage zur Systematik der norwegischen Braunalgen. *Botanica Marina* 5: 1–8.

Jaasund, E. 1964. Marine algae from northern Norway, III. *Botanica Marina* 6: 129–33.

Jaasund, E. 1965. Aspects of the marine algal vegetation of North Norway. *Botanica Gothoburgensia. Acta Universitatis Gothoburgensis* 4: 1–174.

Jaenicke, L. 1977. Sex hormones of brown algae. *Naturwissenschaften* 64: 69–75.

Jaenicke, L., Müller, D.G. & Moore, R.E. 1974. Multifidene and aucantene, C. hydrocarbons in the male-attracting essential oil from the gynogametes of *Cutleria multifida* (Smith) Grey. (Phaeophyta). *Journal of the American Chemical Society* 96: 3324–5.

Jaffe, L. 1954. Stimulation of the discharge of gametangia from a brown alga by a change from light to darkness. *Nature, London* 174: 743.

James, K. & Shears, N. 2012. Spatial distribution and seasonal variation in *Undaria pinnatifida* populations around the Coromandel Peninsula. *Waikato Regional Council Technical Report* 2013/15.

James, K. & Shears, N.T. 2016a. Proliferation of the invasive kelp *Undaria pinnatifida* at aquaculture sites promotes spread to coastal reefs. *Marine Biology* 163: 34.

James, K. & Shears, N.T. 2016b. Population ecology of the invasive kelp *Undaria pinnatifida* towards the upper extreme of its temperature range. *Marine Biology* 163: 225.

James, K., Kibele, J. & Shears, N.T. 2015. Using satellite-derived sea surface temperature to predict the potential global range and phenology of the invasive kelp *Undaria pinnatifida*. *Biological Invasions* 17: 3393–408.

James, K., Middleton, I., Middleton, C. & Shears, N. 2014. Discovery of *Undaria pinnatifida* (Harvey) Suringar, 1873 in northern New Zealand indicates increased invasion threat in subtropical regions. *BioInvasions Records* 3: 21–4.

Jarman, N.G. & Carter, R.A. 1981. The primary producers of the inshore regions of the Benguela. *Transactions of the Royal Society of South Africa* 44: 321–6.

Jégou, C., Culioli, G. & Stiger-Pouvreau, V. 2012. Meroditerpene from *Cystoseira nodicaulis* and its taxonomic significance. *Biochemical Systematics and Ecology* 44: 202–4.

Jenkins, S.R., Arenas, F., Arrontes, J., Bussell, J., Castro, J. & Coleman, R.A. 2001. A European scale analysis of seasonal variability in limpet grazing activity and microalgal abundance. *Marine Ecology Progress Series* 211: 193–203.

Jenks, A. & Gibbs, S.P. 2000. Immunolocalization and distribution of form II Rubisco in the pyrenoid and chloroplast stroma of *Amphidinium carterae* and form I Rubisco in the symbiont-derived plastids of *Peridinium foliaceum* (Dinophyceae). *Journal of Phycology* 36: 127–38.

Jensen, A. & Haug, A. 1952. Fargereaksjon til adskillelse av Stortare (*Laminaria cloustonii*) og Fingertare (*Laminaria digitata*). *Tidsskr. Kjemi, Bergvesen oc Metallurgi* 8: 138–9.

Jensen, A. 1956. Preliminary investigations of the carbohydrates of *Laminaria digitata* and *Fucus serratus*. *Reports of the Norwegian Institute of Seaweed Research* No. 10.

Jensen, A. 1964. Ascorbic acid in *Ascophyllum nodosum*, *Fucus serratus* and *Fucus vesiculosus*. *Proceedings of the International Seaweed Symposium* 4: 319–25.

Jensen, A. 1969. Seasonal variations in the content of individual tocopherols in *Ascophyllum nodosum*, *Pelvetia canaliculata* and *Fucus serratus* (Phaeophyceae). *Proceedings of the International Seaweed Symposium* 6: 493–500.

Jensen, A., Indergaard, M. & Holt, T.J. 1985. Seasonal variation in the chemical composition of *Saccorhiza polyschides* (Laminariales, Phaeophyceae). *Botanica Marina* 28: 375–81.

Jensen, J.B. 1974. Morphological studies in Cystoseiraceae and Sargassaceae (Phaeophyceae) with special reference to apical organisation. *University of California Publications in Botany* 68: vi + 61 pp.

Jephson, N.A. & Gray, P.W.G. 1977. Aspects of the ecology of *Sargassum muticum* (Yendo) Fensholt in the Solent region of the British Isles. I. The growth cycle and epiphytes. In: Keegan, B.F., Ceidigh, P.O. & Boaden, P.J.S. (eds), *Biology of benthic organisms. Proceedings of the XIth European Symposium in Marine Biology, Galway, 1976)*, pp. 367–75. Pergamon Press, Oxford.

Jephson, N.A., Fletcher, R.L. & Berryman, J. 1975. The occurrence of *Zanardinia prototypus* on the south coast of England. *British Phycological Journal* 10: 253–5.

Jernakoff, P. 1985. Interactions between the limpet *Patelloida latistrigulata* and algae on an intertidal rock platform. *Marine Ecology Progress Series* 23: 71–8.

Jimenez, R.S., Hepburn, C.D., Hyndes, G.A., McLeod, R.J., Taylor, R.B. & Hurd, C.L. 2015. Do native subtidal grazers eat the invasive kelp *Undaria pinnatifida*? *Marine Biology* 162: 2521–6.

Johannesson, K., Johansson, D., Larsson, K.H., Huenchunir, C.J. Perus, J., Forslund, H., Kautsky, L. & Pereyra, R.T. 2011. Frequent clonality in fucoids

(*Fucus radicans* and *Fucus vesiculosus*; Fucales, Phaeophyceae) in the Baltic Sea. *Journal of Phycology* 47: 990–8.

John, D.M. 1974. New records of *Ascophyllum nodosum* (L.) Le Jol. from the warmer parts of the Atlantic Ocean. *Journal of Phycology* 10: 243–4.

John, D.M., Prud'homme van Reine, W.F., Lawson, G.W., Kostermans, T.B. & Price, J.H. 2004. A taxonomic and geographical catalogue of the seaweeds of the western coast of Africa and adjacent islands. *Beihefte zur Nova Hedwigia* 127: 1–339.

Johnson, L.E. & Brawley, S.H. 1998. Dispersal and recruitment of a canopy-forming intertidal alga: the relative roles of propagule availability and post-settlement process. *Oecologia* 117: 517–26.

Johnson, T. 1891. On the systematic position of the Dictyotaceae, with special reference to the genus *Dictyopteris* Lamour. *Journal of the Linnean Society of London, Botany* 27: 463–70.

Johnson, T. 1892. Seaweeds from the west coast of Ireland. *Irish Naturalist* 1: 4–6.

Johnston, A.M. & Raven, J.A. 1990. Effects of culture in high CO_2 on the photosynthetic physiology of *Fucus serratus*. *British Phycological Journal* 25: 75–82.

Joly, A.B. 1965. Flora marinha do litoral norte do estado de São Paulo e regiões circunvizinhas. *Boletim da Faculdade de Filosofia, Ciências e Letras da Universidade de São Paulo, Botânica* 21: 5–393.

Jones, E. & Long, J. 2017. The relative strength of an herbivore-induced seaweed defence varies with herbivore species. *Marine Ecology Progress Series* 581: 33–44.

Jones, G. & Farnham, W. 1973. Japweed: new threat to British coasts. *New Scientist* 60: 394–5.

Jones, J.P. & Kingston, J.F. 1829. *Flora devoniensis*: pp. i–xlvii, 1–162 (part 1); i–lxvii, 1–217 [218, err] (part 2). London: Longman, Rees, Orme, Brown and Green.

Jones, W.E. 1974. Changes in the seaweed flora of the British Isles. In: Hawksworth, D.L. (ed.), *The changing flora and fauna of Britain*, pp. 97–113. Systematics Association Special Volume 6. Academic Press, London and New York.

Jones, W.E. & Demetropoulos, A. 1968. Exposure to wave action: measurements of an important ecological parameter on rocky shores of Anglesey. *Journal of Experimental Marine Biology and Ecology* 2: 46–63.

Jones, W.E., Bennell, S., Beveridge, C., McConnell, B., Mack-Smith, S. & Mitchell, J. 1980. Methods of data collection and processing in rocky intertidal monitoring. In: Price, J.H., Irvine, D.E.G. & Farnham, W.F. (eds), *The shore environment 1: methods*, pp. 137–70. Systematics Association Special Volume 17(a). Academic Press, London and New York.

Jónsson, D. 1903a. The marine algae of Iceland. II. Phaeophyceae. *Botanisk Tidsskrift* 25: 141–95.

Jónsson, H. 1904. The marine algae of East Greenland. *Meddelelser om Groenland* 30: 1–73.

Jónsson, S. 1977. Existence d'une race haploïde agame dans le complex polyploïde intraspécifique de l'*Isthmoplea sphaerophora* (Carm.) Kjellm., Phéophycée. *Compte Rendu Hebdomadaire des Séances de l'Académie des Sciences. Paris. Série D* 284: 433–5.

Jordan, A.J. & Vadas, R.L. 1972. Influence of environmental parameters on infraspecific variation in *Fucus vesiculosus*. *Marine Biology, Berlin* 14: 248–52.

Jordan, P., Kloareg, B. & Vilter, H. 1991. Detection of vanadate-dependent bromoperoxidases in protoplasts from the brown algae *Laminaria digitata* and *L. saccharina*. *Journal of Plant Physiology* 137: 520–4.

Jormalainen, V., Honkanen, T. & Heikkilä, N. 2001. Feeding preferences and performance of a marine isopod on seaweed hosts: cost of habitat specialization. *Marine Ecology Progress Series* 220: 219–30.

Josefsson, M. & Jansson, K. 2011. Invasive Alien Species Fact Sheet, *Sargassum muticum*. Online Database of the European Network on Invasive Alien Species. NOBANIS. www.nobanis.org.

Josselyn, M.N. & Mathieson, A.C. 1978. Contribution of receptacles from the fucoid *Ascophyllum nodosum* to the detrital pool of a North temperate estuary. *Estuaries* 1: 258–61.

Jupp, B.P. & Drew, E.A. 1974. Studies on the growth of *Laminaria hyperborea* (Gunn.) Fosl. I. Biomass and productivity. *Journal of Experimental Marine Biology and Ecology* 15: 185–96.

Kain (Jones), J.M. 1963. Aspects of the biology of *Laminaria hyperborea* II. Age, weight and length. *Journal of the Marine Biological Association of the United Kingdom* 43: 129–51.

Kain, J.M. 1964a. Aspects of the biology of *Laminaria hyperborea* III. Survival and Growth of Gametophytes. *Journal of the Marine Biological Association of the United Kingdom* 44: 415–33.

Kain, J.M. 1964b. A study on the ecology of *Laminaria hyperborea* (Gunn.) Fosl. *Proceedings of the International Seaweed Symposium* 4: 207–14.

Kain, J.M. 1964c. The growth of gametophytes and young sporophytes of *Laminaria hyperborea* under various artificial conditions. *British Phycological Bulletin* 2: 390.

Kain, J.M. 1965. Aspects of the biology of *Laminaria hyperborea* IV. Growth of early sporophytes. *Journal of the Marine Biological Association of the UK* 45: 129–43.

Kain, J.M. 1969. The biology of *Laminaria hyperborea*. Comparison with early stages of competitors. *Journal of the Marine Biological Association of the United Kingdom* 49: 455–73.

Kain (Jones), J.M. 1971a. Synopsis of biological data on *Laminaria hyperborea*. *FAO Fisheries Synopsis* No 87 (Rev. 1): 1–74.

Kain (Jones), J.M. 1971b. The biology of *Laminaria hyperborea* VI. Some Norwegian populations. *Journal of the Marine Biological Association of the United Kingdom* 51: 387–408.

Kain, J.M. 1972. Reproduction and colonization in *Laminaria hyperborea*. *Proceedings of the International Seaweed Symposium*: 400.

Kain, J.M. 1973. Growth rates of stipes and fronds of *Laminaria hyperborea* (Gunn.) Fosl. *British Phycological Journal* 8: 213.

Kain (Jones), J.M. 1975. Algal recolonisation of some cleared subtidal areas. *Journal of Ecology* 63: 739–65.

Kain (Jones), J.M. 1976a. The biology of *Laminaria hyperborea* IX. Growth pattern of fronds. *Journal of the Marine Biological Association of the United Kingdom* 56: 603–28.

Kain (Jones), J.M. 1976b. The biology of *Laminaria hyperborea* VIII. Growth on cleared areas. *Journal of the Marine Biological Association of the United Kingdom* 56: 267–90.

Kain, J.M. 1976c. New and interesting marine algae from the Shetland Isles II. Hollow and solid stiped *Laminaria* (Simplices). *British Phycological Journal* 11: 1–11.

Kain (Jones), J.M. 1977. The biology of *Laminaria hyperborea* X. The effect of depth on some populations. *Journal of the Marine Biological Association of the United Kingdom* 57: 587–607.

Kain (Jones), J.M. 1979. A view of the genus *Laminaria*. *Marine Biology Annual Review* 17: 101–61.

Kajiwara, T., Katayama, S., Abe, M. & Hatanaka, A. 1984. Male gamete-attracting activities of hermaphrodite brown algae *Pelvetia wrightii* and *Fucus evanescens*. *Bulletin of the Japanese Society of Scientific Fisheries* 50: 1953.

Kalvas, A. & Kautsky, L. 1993. Geographical variation in *Fucus vesiculosus* morphology in the Baltic and North Seas. *European Journal of Phycology* 28: 85–91.

Kalvas, A. & Kautsky, L. 1998. Morphological variation in *Fucus vesiculosus* populations along temperate and salinity gradients in Iceland. *Journal of the Marine Biological Association of the United Kingdom* 78: 985–1001.

Kanda, T. 1938. On the gametophytes of some Japanese species of Laminariales. II. *Scientific Papers of the Institute of Algological Research, Faculty of Science, Hokkaido Imperial University* 2: 87–111.

Kang, R.S. & Yoo, S. 1993. The acute toxicity of three oils to the early life stages of *Undaria pinnatifida* (Harvey) Suringar. *Korean Journal of Phycology* 8: 77–82.

Kangas, P., Autio, H., Hällfors, G., Luther, H., Niemi, Å. & Salemaa, H. 1982. A general model of the decline of *Fucus vesiculosus* at Tvärminne, south coast of Finland in 1977–81. *Acta Botanica Fennica* 118: 1–27.

Kaplanis, N.J., Harris, J.L. & Smith, J.E. 2016. Distribution patterns of the non-native seaweeds *Sargassum horneri* (Turner) C. Agardh and *Undaria pinnatifida* (Harvey) Suringar on the San Diego and Pacific coast of North America. *Aquatic Invasions* 11: 111–24.

Karlsson, J. 1988. *Sargassum muticum*, a new member of the algal flora of the Swedish West Coast. *Svensk Botanisk Tidskrift* 82: 199–205.

Karlsson, J. & Loo, L.O. 1999. On the distribution and continuous expansion of the Japanese seaweed – *Sargassum muticum* – in Sweden. *Botanica Marina* 42: 285–95.

Karsakoff, N. 1892. Quelques remarques sur le genre *Myriotrichia. Journale de Botany, Paris* 6: 433–44.

Karyophyllis, D., Katsaros, C. & Galatis, B. 2000b. F-actin involvement in apical cell morphogenesis of *Sphacelaria rigidula* (Phaeophyceae): mutual alignment between cortical actin filaments and cellulose microfibrils. *European Journal of Phycology* 35: 195–203.

Karyophyllis, D., Katsaros, C., Dimitriadis, I. & Galatis, B. 2000a. F-Actin organization during the cell cycle of *Sphacelaria rigidula* (Phaeophyceae). *European Journal of Phycology* 35: 25–33.

Kasahara, K. 1967. On the development of the mucilage gland of *Undaria pinnatifida. Botanical Magazine, Tokyo* 80: 279–87.

Kato, S. & Dan, A. 2010. Growth and morphological variations of *Undaria pinnatifida* cultivated in Naruto Strait using seedling derived from eight areas in *Japanese Algal Resources* 3: 19–26 (in Japanese with English abstract).

Kato, T. & Nakahisa, Y. 1962. A comparative study on two local forms of *Undaria pinnatifida* Sur., a brown alga, grown in a culture ground. *Bulletin of the Japanese Society of Scientific Fisheries* 28: 998–1004 (in Japanese with English abstract).

Katpitia, A.D. & Blackler, H. 1962. Plurilocular sporangia on *Elachista scutulata* (Sm.) Duby in Britain. *British Phycological Bulletin* 2: 173–4.

Katsaros, C. & Galatis, B. 1985. Ultrastructural studies on thallus development in *Dictyota dichotoma* (Phaeophyta, Dictyotales). *British Phycological Journal* 20: 263–76.

Katsaros, C. & Salla, C. 1997. Ultrastructural changes during development, release, and germination of monospores of *Tilopteris mertensii* (Tilopteridales, Phaeophyta). *Phycologia* 36: 60–7.

Katsaros, C., Galatis, B. & Mitrakos, K. 1983. Fine structural studies on the interphase and dividing apical cells of *Sphacelaria tribuloides* (Phaeophyta). *Journal of Phycology* 19: 16–30.

Katsaros, C., Kreimer, G. & Melkonian, M. 1991. Localization of tubulin and a centrin-homologue in vegetative cells and developing gametangia of *Ectocarpus siliculosus* (Dillw.) Lyngb. (Phaeophyceae, Ectocarpales) – a combined immunofluorescence and confocal laser scanning microscope study. *Botanica Acta* 104: 87–92.

Katsaros, C.I., Maier, I. & Melkonian, M. 1993. Immunolocalization of centrin in the flagellar apparatus of male gametes of *Ectocarpus siliculosus* (Phaeophyceae) and other brown algal motile cells. *Journal of Phycology* 29: 787–97.

Katsuoka, M., Ogura, C., Etoh, H., Sakata, K. & Ina, K. 1990. Galactosyldiacylglycerols and Sulfoquinovosyldiacyl-Glycerols isolated from the brown algae, *Undaria pinnatifida* and *Costaria costata* as repellents of the blue mussel, *Mytilus edulis. Agricultural and Biological Chemistry, Tokyo* 54: 3043–4.

Kautsky, N., Kautsky, H., Kautsky, U. & Waern, M. 1986. Decreased depth penetration of *Fucus vesiculosus* (L.) since the 1940s indicates eutrophication of the Baltic Sea. *Marine Ecology Progress Series* 28: 1–8.

Kautsky, H., Kautsky, L., Kautsky, N., Kautsky, U. & Lindblad, C. 1992. Studies on the *Fucus vesiculosus* community in the Baltic Sea. In: Sjögren, E., Wallentinus, I. & Snoeijs, P. (eds), *Phycological Studies of Nordic Coastal Waters – a festschrift dedicated to Prof. Mats Wærn on his 80th birthday*, 78, pp. 33–48. Uppsala: Opulus Press AB.

Kawai, H. 1983. Morphological observations on *Acrothrix gracilis* Kylin (Chordariales, Phaeophyta) newly found in Japan. *Japanese Journal of Phycology* 31: 167–72.

Kawai, H. 1986. The life history of Japanese *Eudesme virescens* (Carm.) J. Ag. (Phaeophyceae, Chordariales) in culture. *Japanese Journal of Phycology* 34: 203–8.

Kawai, H. 1991. Critical review of the taxonomy and life history of *Kjellmania arasakii* (Dictyosiphonales, Phaeophyceae). *Japanese Journal of Phycology* 39: 319–28.

Kawai, H. & Prud'homme van Reine, W.F. 1998. Life history of Japanese *Stypocaulon durum*

(Sphacelariales, Phaeophyceae). *Phycological Research* 46: 263–70.

Kawai, H. & Sasaki, H. 2000. Molecular phylogeny of the brown algal genera *Akkesiphycus* and *Halosiphon* (Laminariales), resulting in the circumscription of the new families Akkesiphycaceae and Halosiphonaceae. *Phycologia* 39: 416–28.

Kawai, H. & Sasaki, H. 2004. Morphology, life history and molecular phylogeny of *Stschapovia flagellaris* (Tilopteridales, Phaeophyceae), and the erection of the Stschapoviaceae fam. nov. *Journal of Phycology* 40: 1156–69.

Kawai, H., Hanyuda, T., Ahibata, K., Kamiya, M. & Peters, A.F. 2019. Proposal of a new algal species, *Mesogloia japonica* sp. nov. (Chordariaceae, Phaeophyceae), and transfer of *Sauvageaugloia ikomae* to *Mesogloia*. *Phycologia* 58(1): 63–69.

Kawai, H., Hanyuda, T. & Uwai, S. 2016a. Evolution and Biogeography of Laminarialean Kelps. In: Hu, Z.M. & Fraser, C. (eds), *Seaweed Phylogeography*, pp. 227–49. Springer, Dordrecht. https://doi.org/10.1007/978-94-017-7534-2_9

Kawai, H., Hanyuda, T., Bolton, J. & Anderson, R. 2016b. Molecular phylogeny of *Zeacarpa* (Ralfsiales, Phaeophyceae) proposing a new family Zeacarpaceae and its transfer to Nemodermatales (Note). *Journal of Phycology* 52: 682–6.

Kawai, H., Hanyuda, T., Yamagishi, T., Kai, A., Lane, C.E., McDevit, D., Küpper, F.C. & Saunders, G.W. 2015a. Reproductive morphology and DNA sequences of the brown alga *Platysiphon verticillatus* support the new combination *Platysiphon glacialis*. *Journal of Phycology* 51: 910–17.

Kawai, H., Hanyuda, T., Draisma, S.G.A., Wilce, R.T. & Andersen, R.A. 2015b. Molecular phylogeny of two unusual brown algae, *Phaeostrophion irregulare* and *Platysiphon glacialis*, proposal of the Stschapoviales ord. nov. and Platysiphonaceae fam. nov., and a re-examination of divergence times for brown algal orders. *Journal of Phycology* 51: 918–28.

Kawai, H., Hanyuda, T., Draisma, S.G.A. & Müller, D.G. 2007. Molecular phylogeny of *Discosporangium mesarthrocarpum* (Phaeophyceae) with a reinstatement of the order Discosporangiales. *Journal of Phycology* 43: 186–94.

Kawai, H., Hanyuda, T., Lindeberg, M. & Lindstrom, S.C. 2008. Morphology and molecular phylogeny of *Aureophycus aleuticus* gen. et sp. nov. (Laminariales, Phaeophyceae) from the Aleutian Islands. *Journal of Phycology* 44: 1013–21.

Kawai, H., Kubota, M., Kondo, T. & Watanabe, M. 1991. Action spectra for phototaxis in zoospores of the brown alga *Pseudochorda gracilis*. *Protoplasma* 161: 17–22.

Kawai, H., Muto, H., Fujii, T. & Kato, A. 1995. A linked 5S rRNA gene in *Scytosiphon lomentaria* (Scytosiphonales, Phaeophyceae). *Journal of Phycology* 31: 306–11.

Kawai, H., Müller, D.G., Fölster, E. & Häder, D.P. 1990. Phototactic responses in the gametes of the brown alga, *Ectocarpus siliculosus*. *Planta* 182: 292–7.

Kawai, H., Nakayama, T., Inouye, I. & Kato, A. 1997. Linkage of 5S ribosomal DNA to other rDNAs in the chromophytic algae and related taxa. *Journal of Phycology* 33: 505–11.

Kawai, H., Sasaki, H., Maeda, Y. & Arai, S. 2001. Morphology, life history, and molecular phylogeny of *Chorda rigida*, sp. nov. (Laminariales, Phaeophyceae) from the Sea of Japan and the genetic diversity of *Chorda filum*. *Journal of Phycology* 37: 130–42.

Kawai, H., Sasaki, H., Maeba, S. & Henry, E.C. 2005. Morphology and molecular phylogeny of *Phaeostrophion irregulare* (Phaeophyceae) with a proposal for Phaeostrophiaceae fam. nov., and a review of Ishigeaceae. *Phycologia* 44: 169–82.

Kawashima, Y. & Tokuda, H. 1993. Regeneration from callus of *Undaria pinnatifida* (Harvey) Suringar (Laminariales, Phaeophyta). *Hydrobiologia* 260/1: 385–9.

Keen, J.N. & Evans, L.V. 1986. Partial amino-acid sequence of the small subunit of Rubisco from the brown alga *Fucus serratus*. *British Phycological Journal* 21: 331–2.

Kelly, L., Collier, L., Costello, M.J., Diver, M., McGarvey, S., Kraan, S., Morrissey, J. & Guiry, M.D. 2001. Impact assessment of hand and mechanical harvesting of *Ascophyllum nodosum* on regeneration and biodiversity. *Marine Resource Series [Ireland]* 19: [i]–vi, 1–51.

Keough, M.J. & Quinn, G.P. 1998. Effects of periodic disturbances from trampling on rocky intertidal algal beds. *Ecological Applications* 8: 141–61.

Kerby, N.W. & Evans, L.V. 1978. Isolation and partial characterization of pyrenoids from the brown alga *Pilayella littoralis* (L.) Kjellm. *Planta* 142: 91–5.

Kerckhof, F., Haelters, J. & Gollasch, S. 2007. Alien species in the marine and brackish ecosystem: the situation in Belgian waters. *Aquatic Invasions* 3: 243–57.

Kerrison, P. & Le, H.N. 2016. Environmental factors on egg liberation and germling production of *Sargassum muticum*. *Journal of Applied Phycology* 28: 481–9.

Keser, M. & Larson, B.R. 1984. Colonization and growth of *Ascophyllum nodosum* (Phaeophyta) in Maine. *Journal of Phycology* 20: 83–7.

Keser, M., Vadas, R.L. & Larson, B.R. 1981. Regrowth of *Ascophyllum nodosum* and *Fucus vesiculosus* under various harvesting regimes in Maine, USA. *Botanica Marina* 24: 29–38.

Keum, Y.S. 2010. Sphacelariales, Cutleriales, Ralfsiales. In: Anon. (ed.), *Algal flora of Korea. Volume 2, Number 2. Heterokontophyta: Phaeophyceae: Ishigeales, Dictyotales, Desmarestiales, Sphacelariales, Cutleriales, Ralfsiales, Laminariales*, pp. 19–69. Incheon: National Institute of Biological Resources.

Keum, Y.-S., Oak, J.H., Prud'homme van Reine, W.F. & Lee, I.K. 2003. Comparative morphology and taxonomy of *Sphacelaria* species with tribuliform propagules (Sphacelariales, Phaeophyceae). *Botanica Marina* 46: 113–24.

Keum, Y.S., Oak, J.H., Draisma, S.G.A., Prud'homme van Reine, W.F. & Lee, I.K. 2005. Taxonomic reappraisal of *Sphacelaria rigidula* and *S. fusca* (Sphacelariales, Phaeophyceae) based on morphology and molecular data with special reference to *S. didichotoma*. *Algae* 20: 1–13.

Khailov, K.M., Kholodov, V.I., Firsov, Y.K. & Prazulin, A.V. 1978. Thalli of *Fucus vesiculosus* in ontogenesis: changes in morpho-physiological parameters. *Botanica Marina* 21: 289–311.

Khfaji, A.K. & Norton, T.A. 1979. The effects of salinity on the distribution of *Fucus ceranoides*. *Estuarine, Coastal and Shelf Science* 8: 433–9.

Kickx, J. 1856. Essai sur les variétés indigènes du *Fucus vesiculosus*. *Bulletins de l'Académie royale des sciences, des lettres et des beaux-arts de Belgique*. 23 (Ier Partie): 477–526.

Kiirikki, M. 1996. Experimental evidence that *Fucus vesiculosus* (Phaeophyta) controls filamentous algae by means of the whiplash effect. *European Journal of Phycology* 31: 61–6.

Kiirikki, M. & Ruuskanen, A. 1996. How does *Fucus vesiculosus* survive ice scraping? *Botanica Marina* 39: 133–9.

Kim, B.Y., Park, S.K., Norton, T.A. & Choi, H.G. 2011. Effects of temporary and periodic emersion on the growth of *Fucus spiralis* and *Pelvetia canaliculata* germlings. *Algae* 26: 193–200.

Kim, H.-S. 2010. Ectocarpaceae, Acinetosporaceae, Chordariaceae. In: Kim, H.-S. & Boo, S.-M. (eds), *Algal Flora of Korea. Volume 2, Number 1. Heterokontophyta: Phaeophyceae: Ectocarpales. Marine brown algae I*, pp. [3]–137. Incheon: National Institute of Biological Resources.

Kim, H.-S. & Boo, S.-M. 2010. *Algal flora of Korea. Volume 2, Number 1. Heterokontophyta: Phaeophyceae: Ectocarpales. Marine brown algae I*, pp. [1–6], 1–195, figs 1–19. Incheon: National Institute of Biological Resources.

Kim, H.S. & Lee, I.K. 1992a. Morphotaxonomic studies on the Korean Ectocarpaceae (Phaeophyta) I. Genus *Ectocarpus* Lyngbye. *Korean Journal of Phycology* 7: 225–42.

Kim, H.S. & Lee, I.K. 1992b. Morphotaxonomic studies on the Korean Ectocarpaceae (Phaeophyta) II. Genus *Hincksia* J.E. Gray. *Korean Journal of Phycology* 7: 243–56.

Kim, H.S. & Lee, I.K. 1994. Morphotaxonomic studies on the Korean Ectocarpaceae (Phaeophyta). III. Genus *Feldmannia* Hamel, specially referred to morphogenesis and phylogenetic relationship among related genera. *Korean Journal of Phycology* 9: 153–68.

Kim, K.Y. & Garbary, D.J. 2009. Form, function and longevity in fucoid thalli: chlorophyll a fluorescence differentiation of *Ascophyllum nodosum*, *Fucus vesiculosus* and *F. distichus* (Phaeophyceae). *Algae* 24: 93–104.

Kim, K.Y., O'Leary, S.J. & Garbary, D.J. 1997. Artificial hybridization between *Ascophyllum nodosum* and *Fucus vesiculosus* (Phaeophyceae) from Nova Scotia, Canada. *Canadian Journal of Botany* 75: 1133–6.

Kim, K.Y., Jeong, H.J., Main, H.P. & Garbary, D.J. 2006. Fluorescence and photosynthetic competency in single eggs and embryos of *Ascophyllum nodosum* (Phaeophyceae). *Phycologia* 45: 331–6.

Kim, M.-K. & Lee, H.-W. 1998. Changes in B-carotene content in fresh and dry thalli of *Undaria pinnatifida* and *Enteromorpha compressa* from Korea. *Algae* 13: 151–5.

Kim, M.-K. & Thomas, J.-C. 1997. Growth of phaeophyte (*Fucus serratus*) in culture and changes of the amounts of triacylglycerols and fatty acids. *Algae* 11: 11–16.

Kim, S.-H. & Kawai, H. 2002. Taxonomic revision of *Chordaria flagelliformis* (Chordariales, Phaeophyceae) including novel use of the intragenic spacer region of rDNA for phylogenetic analysis. *Phycologia* 41: 328–39.

Kim, S.-H., Peters, A.F. & Kawai, H. 2003. Taxonomic revision of *Sphaerotrichia divaricata* (Ectocarpales, Phaeophyceae), with a reappraisal of *S. firma* from the north-west Pacific. *Phycologia* 42: 183–92.

Kim, W.-J., Kim, S.-M., Kim, H.-G., Oh, H.-R., Lee, K.B., Lee, Y.-K. & Park, Y.-I. 2007. Purification and anticoagulant activity of a fucoidan from Korean *Undaria pinnatifida* sporophyll. *Algae* 22: 247–52.

Kim, Y.S. & Nam, K.W. 1997. Temperature and light responses on the growth and maturation of gametophytes of *Undaria pinnatifida* (Harvey) Suringar in Korea. *Journal of the Korean Fisheries Society* 30: 505–10.

Kimura, K., Nagasato, C., Uwai, S. & Motomura, T. 2010a. Sperm mitochondrial DNA elimination in the zygote of the oogamous brown alga *Undaria pinnatifida* (Laminariales, Phaeophyceae). *Cytologia* 75: 353–61.

Kimura, K., Nagasato, C., Kogame, K. & Motomura, T. 2010b. Disappearance of male mitochondrial DNA after the four-cell stage in sporophytes of the isogamous brown alga *Scytosiphon lomentaria* (Scytosiphonaceae, Phaeophyceae). *Journal of Phycology* 46: 143–52.

Kimura, T., Ezura, Y. & Tajima, K. 1976. Microbiological study of a disease of wakame (*Undaria pinnatifida*) and of the marine environments of wakame culture sites in Kesennuma Bay. *Bulletin of Tohoku Regional Fisheries Research Laboratory* 36: 57–65.

King, R.J. & Schramm, W. 1976. Determination of photosynthetic rates for the marine algae *Fucus vesiculosus* and *Laminaria digitata*. *Marine Biology, Berlin* 37: 209–13.

Kingman, A.R. & Moore, J. 1982. Isolation, purification, and quantitation of several growth-regulating substances in *Ascophyllum nodosum* (Phaeophyta). *Botanica Marina* 25: 149–53.

Kinoshita, T. & Shibuya, S. 1944. Study in optimal temperature on development of *Undaria pinnatifida*. *Bulletin of the Fisheries Experimental Station of the Hokkaido Prefecture* 1: 369–73.

Kinoshita, N., Shiba, K., Inaba, K., Fu, G., Nagasato, C. & Motomura, T. 2016. Flagellar waveforms of gametes in the brown alga *Ectocarpus siliculosus*. *European Journal of Phycology* 51: 139–48.

Kinoshita, N., Nagasato, C. & Motomura, T. 2017a. Chemotactic movement in sperm of the oogamous brown alga, *Saccharina japonica* and *Fucus distichus*. *Protoplasma* 254: 547–55.

Kinoshita, N., Nagasato, C. & Motomura, T. 2017b. Phototaxis and chemotaxis of brown algal swarmers. *Journal of Plant Research* 130: 443.

Kitayama, T. 1994. A taxonomic study of the Japanese *Sphacelaria* (Sphacelariales, Phaeophyceae). *Bulletin of the National Science Museum, Tokyo, Ser. B (Botany)* 20: 37–141.

Kitching, J.A. 1987. The flora and fauna associated with *Himanthalia elongata* (L.) S.F. Gray in relation to water current and wave action in the Lough Hyne marine nature reserve. *Estuarine, Coastal and Shelf Science* 25: 663–76.

Kito, H., Akiyama, K. & Sasaki, M. 1976. Electron microscopic observations on the diseased thalli of *Undaria pinnatifida* (Harvey) Suringar, caused by parasitic bacteria. *Bulletin of Tohoku Regional Fisheries Research Laboratory* 36: 67–73.

Kito, H., Taniguchi, K. & Akiyama, K. 1981. Morphological variation of *Undaria pinnatifida* (Harvey) Suringar – II. Comparison of the thallus morphology of cultivated F_1 plants originated from parental types of two different morphologies. *Bulletin of Tohoku Regional Fisheries Research Laboratory* 42: 11–18 (in Japanese with English abstract).

Kjeldsen, C.K. & Phinney, H.K. 1972. Effects of variations in salinity and temperatures on some estuarine macro-algae. *Proceedings of the International Seaweed Symposium* 7: 301–8.

Kjellman, F.R. 1872. *Bidrag till kännedomen om Skandinaviens Ectocarpeer och Tilopterider*. Stockholm: Akademisk afhandling Upsala.

Kjellman, F.R. 1877a. Über die Algenvegetationen des Murmanschen Meeres av der Westküste von Nowaja Semlja und Wajgatsch. *Nova Acta Regiae Societatis Scientiarum Upsaliensis, Series 3* Vol. ext. ord. (Art. XII): 1–86.

Kjellman, F.R. 1877b. Om Spetsbergens marina, klorogyllförande Thallophyter. II. *Kungliga Svenska Vetenskapsakademiens Handlingar* 4: 1–61.

Kjellman, F.R. 1879. Bidrag till Kännedomen om Islands Hafsalgflora. *Botanisk Tidsskrift* 11: 77–80.

Kjellman, F.R. 1880. *Points-förteckning öfver Skandinaviens växter. Enumerantur plantae scandinaviae* Vol. 1B: Algae. pp. 1–85. Lund: C.W.K. Gleerup.

Kjellman, F.R. 1883a. The algae of the Arctic Sea. A survey of the species, together with an exposition of the general characters and the development of the flora. *Kongliga Svenska Vetenskaps-Akademiens Handlingar* 20: 1–351.

Kjellman, F.R. 1883b. Norra Ishafvets algflora. *Vega-expeditionens Vetenskapliga Iakttagelser* 3: 1–431.

Kjellman, F.R. 1889. Om Beringhafvets algflora. *Kongl. Svenska Vetenskaps-Akademiens Handlingar* 23: 1–58.

Kjellman, F.R. 1890. *Handbok i Skandinaviens hafsalgflora. I. Fucoideae.* pp. [i]–vi, 1–103, 17 figs. Stockholm: Oscar L. Lamms Förlag.

Kjellman, F.R. 1891a. En för Skandinaviens flora ny Fucoidé, *Sorocarpus uvaeformis* Pringsheim. *Botaniska Notiser* 1891: 177–8.

Kjellman, F.R. 1891b. Choristocarpaceae. In: Engler, A. & Prantl, K. (eds), *Die natürlichen Pflanzenfamilien nebst ihren Gattungen und wichtigeren Arten insbesondere den Nutzpflanzen unter Mitwirkung zahlreicher hervorragender Fachgelehrten, Teil 1, Abteilung 2*, pp. 190–1. Leipzig: verlag von Wilhelm Engelmann.

Kjellman, F.R. 1891c. Phaeophyceae (Fucoideae). In: Engler, A. & Prantl, K. (eds), *Die natürlichen Pflanzenfamilien nebst ihren Gattungen und wichtigeren Arten insbesondere den Nutzpflanzen unter Mitwirkung zahlreicher hervorragender Fachgelehrten, Teil 1, Abteilung 2*, pp. 176–81. Leipzig: verlag von Wilhelm Engelmann.

Kjellman, F.R. 1893. Encoeliaceae. In: Engler, A. & Prantl, K., Die naturlichen Pflanzenfamilien. Nachtrage 2 Teil I, Abt. 2 Leipzig.

Kjellman, F.R. 1906. Zur Kenntnis der marinen Algenflora von Jan Mayen. *Arkiv für Botanik, Uppsala* 5: 1–30.

Kjellman, F.R. & Petersen, J.V. 1885. Om Japans Laminariaceer. *Vega-expeditionens Vetenskapliga Iakttagelser, Stokholm* 4: 255–80.

Kjellman, F.R. & Svedelius, N.L. 1910. Lithodermataceae. In: Engler, A. & Prantl, K., Die naturlichen Pflanzenfamilien. *Nachtrage 2 Teil I, Abt. 2. Leipzig.*

Kleen, E.A.G. 1874. Om Nordlandens högre hafsalger. *Öfversigt af Kongl. Vetenskaps-Akademiens Förhandlingar, Stockholm* 31: 1–46.

Klenell, M., Snoeijs, P. & Pedersén, M. 2002. The involvement of a membrane H+-ATPase in the blue-light enhancement of photosynthesis in *Laminaria digitata* (Phaeophyta). *Journal of Phycology* 38: 1143–9.

Klimova, A.V., Klochkova, N.G., Klochkova, T.A. & Kim, G.H. 2018. Morphological and molecular identification of *Alaria paradisea* (Phaeophyceae, Laminariales) from the Kurile Islands. *Algae. An International Journal of Algal Research* 33: 37–48.

Klinger, T. & DeWreede, R.E. 1988. Stipe rings, age, and size in populations of *Laminaria setchellii* Silva (Laminariales, Phaeophyta) in British Columbia, Canada. *Phycologia* 27: 234–40.

Kloareg, B. & Quatrano, R.S. 1987. Enzymatic removal of the cell walls from zygotes of *Fucus distichus* (L.) Powell (Phaeophyta). *Proceedings of the International Seaweed Symposium* 12: 123–9.

Klochkova, T.A., Klochkova, N.G., Yotsukura, N. & Kim, G.H. 2017. Morphological, molecular, and chromosomal identification of dwarf haploid parthenosporophytes of *Tauya basicrassa* (Phaeophyceae, Laminariales) from the sea of Okhotsk. *Algae* 32: 15–28.

Klochkova, T.A., Klochkova, N.G., Belij, M.N., Hyung-Seop, K. & Kim, G.W. 2012. Morphology and Molecular Phylogeny of *Chordaria okhotskensissp.* nov. (Ectocarpales, Phaeophyceae) from the Sea of Okhotsk. *Cryptogamie, Algologie* 33: 3–20.

Knight, M. 1923. Studies in the Ectocarpaceae. I. The life-history and cytology of *Pylaiella littoralis* Kjellm. *Transactions of the Royal Society of Edinburgh* 53: 343–60.

Knight, M. 1929. Studies in the Ectocarpaceae. II. The life history and cytology of *Ectocarpus siliculosus* Dillw. *Transactions of the Royal Society of Edinburgh* 56: 307–32.

Knight, M. & Parke, M.W. 1931. Manx algae. An algal survey of the south end of the Isle of Man. *Proceedings and Transactions of the Liverpool Biological Society* 45 (Appendix II): 1–155.

Knight, M. & Parke, M. 1950. A biological study of *Fucus vesiculosus* L. and *F. serratus* L. *Journal of the Marine Biological Association of the United Kingdom* 26: 439–514.

Knight, M., Blackler, M.C.H. & Parke, M.W. 1935. Notes on the life cycle of species of *Asperococcus*. *Proceedings and Transactions of the Liverpool Biological Society* 48: 79–97.

Knoepffler-Péguy, M. 1970. Quelques *Feldmannia* Hamel, 1939 (Phaeophyceae-Ectocarpales) des côtes d'Europe. *Vie et Milieu* Ser. A, 21: 137–88.

Knoepffler-Péguy, M. 1972. Comportement de deux espèces suédoises de *Feldmannia* cultivées en diverses conditions de température et de salinité. *Mémoires de la Société Botanique de France* 1972: 101–4.

Knoepffler-Péguy, M. 1973. Étude experimentale du polymorphisme chez quelques Ectocarpales (*Feldmannia* et *Acinetospora*). I. Methodes d'etude experimentale. *Vie et Milieu* Ser. A, 23: 171–89.

Knoepffler-Péguy, M. 1974. Le genre *Acinetospora* Bornet 1891 (Phaeophyceae, Ectocarpales). *Vie et Milieu* 24: 43–72.

Knoepffler-Péguy, M. 1977. Polymorphisme et environnement chez les *Feldmannia* (Ectocarpacées). *Revue Algologique, Nouvelle Serie* 12: 111–28.

Knoepffler-Péguy, M., Belsher, T., Boudouresque, C.-F. & Lauret, M. 1985. *Sargassum muticum* begins to invade the Mediterranean. *Aquatic Botany* 23: 291–5.

Koehl, M.A.R. 1986. Seaweeds in moving water: Form and mechanical function. In: Gavinish, T.J. (ed.), *On the Economy of Plant Form and Function*, pp. 603–34. Cambridge University Press, Cambridge.

Koehl, M.A.R. & Alberte, R.S. 1988. Flow, flapping, and photosynthesis of *Nereocystis luetkeana*: a functional comparison of undulate and flat blade morphologies. *Marine Biology* 99: 435–44.

Koeman, R.P.T. & Cortel-Breeman, A.M. 1973. The life-history of *Elachista fucicola* (Vell.) Aresch. *British Phycological Journal* 8: 213.

Koeman, R.P.T. & Cortel-Breeman, A.M. 1976. Observations on the life history of *Elachista fucicola* (Veil.) Aresch. (Phaeophyceae) in culture. *Phycologia* 15: 107–17.

Kogame, K. 1996. Morphology and life history of *Scytosiphon canaliculatus* comb. nov. (Scytosiphonales, Phaeophyceae) from Japan. *Phycological Research* 44: 85–94.

Kogame, K. 1997a. Life histories of *Colpomenia sinuosa* and *Hydroclathrus clathratus* (Scytosiphonaceae, Phaeophyceae in culture). *Phycological Research* 45: 227–1.

Kogame, K. 1997b. Sexual reproduction and life-history of *Petalonia fascia* (Scytosiphonales, Phaeophyceae). *Phycologia* 36: 389–94.

Kogame, K. 1998. A taxonomic study of Japanese *Scytosiphon* (Scytosiphonales, Phaeophyceae), including two new species. *Phycological Research* 46: 39–46.

Kogame, K. 2001. Life history of *Chnoospora implexa* (Chnoosporaceae, Phaeophyceae) in culture. *Phycological Research* 49: 123–8.

Kogame, K. & Boo, S.M. 2006. Molecular phylogeny of the family Scytosiphonaceae (Phaeophyceae). *Algae* 21(2): 175–83.

Kogame, K. & Kawai, H. 1993. Morphology and life history of *Petalonia zosterifolia* (Reinke) O. Kuntze (Scytosiphonales, Phaeophyceae) from Japan. *Japanese Journal of Phycology* 41: 29–37.

Kogame, Y. & Kawai, H. 1996. Development of the intercalary meristem in *Chorda filum* (Laminariales, Phaeophyceae) and other primitive Laminariales. *Phycological Research* 44: 247–60.

Kogame, K. & Masuda, M. 2001. Crustose sporophytes of *Colpomenia bullosa* (Scytosiphonaceae, Phaeophyceae) in nature. *Cryptogamie, Algologie* 22: 201–8.

Kogame, K. & Yamagishi, Y. 1997. The life history and phenology of *Colpomenia peregrina* (Scytosiphonales, Phaeophyceae) from Japan. *Phycologia* 36: 337–44.

Kogame, K., Horiguchi, T. & Masuda, M. 1999. Phylogeny of the order Scytosiphonales (Phaeophyceae) based on DNA sequences of rbcL, partial rbcS, and partial LSU nrDNA. *Phycologia* 38: 496–502.

Kogame, K., Uwai, S., Shimada, S. & Masuda, M. 2005. A study of sexual and asexual populations of *Scytosiphon lomentaria* (Scytosiphonaceae, Phaeophyceae) in Hokkaido, northern Japan, using molecular markers. *European Journal of Phycology* 40: 313–22.

Kogame, K., Nagasato, C. & Motomura, T. 2006. Inheritance of mitochondrial and chloroplast genomes in the isogamous brown alga *Scytosiphon lomentaria* (Phaeophyceae). *Phycological Research* 54: 65–71.

Kogame, K., Rindi, F., Peters, A.F. & Guiry, M.D. 2015a. Genetic diversity and mitochondrial introgression in *Scytosiphon lomentaria* (Ectocarpales, Phaeophyceae) in the north-eastern Atlantic Ocean. *Phycologia* 54: 367–74.

Kogame, K., Ishikawa, S., Yamauchi, K., Uwai, S., Kurihara, A. & Masuda, M. 2015b. Delimitation of cryptic species of the *Scytosiphon lomentaria* complex (Scytosiphonaceae, Phaeophyceae) in Japan, based on mitochondrial and nuclear molecular markers. *Phycological Research* 63: 167–77.

Kohlmeyer, J. & Kohlmeyer, E. 1972. Is *Ascophyllum nodosum* lichenized? *Botanica Marina* 15: 109–12.

Komazawa, I., Sakanishi, Y. & Tanaka, J. 2015. Temperature requirements for growth and maturation of the warm temperate kelp *Eckloniopsis radicosa* (Laminariales, Phaeophyta). *Phycological Research* 63: 64–71.

Kornmann, P. 1953. Der Formenkreis von *Acinetospora crinita* (Carm.) nov. comb. *Helgoländer Wissenschaftliche Meeresuntersuchungen* 4: 205–24.

Kornmann, P. 1954. *Giffordia fuscata* (Zan.) Kuck. nov. comb. eine Ectocarpaceae mit heteromorphen, homophasischen Generationen. *Helgoländer Wissenschaftliche Meeresuntersuchungen* 5: 41–52.

Kornmann, P. 1956. Uber die entwicklung einer *Ectocarpus confervoides*-form. *Pubblicazioni della Stazione Zoologica di Napoli* 28: 32–43.

Kornmann, P. 1962a. Der Lebenszyklus von *Desmarestia viridis*. *Helgoländer Wissenschaftliche Meeresuntersuchungen* 8: 287–92.

Kornmann, P. 1962b. Plurilokulare sporangien bei *Elachista fucicola*. *Helgoländer Wissenschaftliche Meeresuntersuchungen* 8: 293–7.

Kornmann, P. 1962c. Die Entwicklung von *Chordaria flagelliformis*. *Helgoländer Wissenschaftliche Meeresuntersuchungen* 8: 276–9.

Kornmann, P. & Sahling, P.H. 1973. *Striaria attenuata* (Phaeophyta, Dictyosiphonales), neu bei Helgoland: Entwicklung und Aufbau. *Helgoländer Wissenschaftliche Meeresuntersuchungen* 25: 14–25.

Kornmann, P. & Sahling, P.H. 1977. Meeresalgen von Helgoland. Benthische Grün-, Braun und Rotalgen. *Helgoländer Wissenschaftliche Meeresuntersuchungen* 29: 1–289.

Kornmann, P. & Sahling, P.H. 1983. Meeresalgen von Helgoland: Erganzung. *Helgoländer Wissenschaftliche Meeresuntersuchungen* 36: 1–65.

Kornmann, P. & Sahling, P.-H. 1984. Der *Sorocarpus*-complex (Ectocarpaceae, Phaeophyta). *Helgoländer Meeresuntersuchungen* 38: 87–101.

Kornmann, P. & Sahling, P.H. 1988. The disentanglement of the *Botrytella* (*Sorocarpus*)-complex

(Ectocarpaceae, Phaeophyta). *Helgoländer Wissenschaftliche Meeresuntersuchungen* 42: 1–12.

Korpinen, S. & Jormalainen, V. 2008a. Grazing and nutrients reduce recruitment success of *Fucus vesiculosus* L. (Fucales: Phaeophyceae). *Estuarine, Coastal and Shelf Science* 78: 437–44.

Korpinen, S. & Jormalainen, V. 2008b. Grazing effects in macroalgal communities depend on timing of patch colonization, *Journal of Experimental Marine Biology and Ecology* 360: 39–46.

Korpinen, S., Jormalainen, V. & Honkanen, T. 2007. Effects of nutrients, herbivory and depth on the macroalgal community in a rocky sublittoral. *Ecology* 88: 839–52.

Korpinen, S., Jormalainen, V. & Ikonen, J. 2008. Selective consumption and facilitation by mesograzers in adult and colonizing macroalgal assemblages. *Marine Biology* 154: 787–94.

Kraan, S. 2008. *Sargassum muticum* (Yendo) Fensholt in Ireland: an invasive species on the move. *Journal of Applied Phycology* 20: 825–32.

Kraan, S. 2017. *Undaria* marching on; late arrival in the Republic of Ireland. *Journal of Applied Phycology* 29: 1107–14.

Kraan, S. & Guiry, M.D. 2000a. Molecular and morphological character inheritance in hybrids of *Alaria esculenta* and *A. praelonga* (Alariaceae, Phaeophyceae). *Phycologia* 39: 554–9.

Kraan, S. & Guiry, M.D. 2000b. Sexual hybridization experiments and phylogenetic relationships as inferred from Rubisco spacer sequences in the genus *Alaria* (Phaeophyceae). *Journal of Phycology* 36: 190–8.

Kraan, S. & Guiry, M.D. 2001. Phase II: Strain hybridisation field experiments and genetic fingerprinting of the edible brown seaweed *Alaria esculenta*. *Marine Resource Series* [*Ireland*] 18: [i]–vi, 1–33.

Kraan, S., Rueness, J. & Guiry, M.D. 2001. Are North Atlantic *Alaria esculenta* and *A. grandifolia* (Alariaceae, Phaeophyceae) conspecific? *European Journal of Phycology* 39: 35–42.

Kraan, S., Tramullas, A.V. & Guiry, M.D. 2000. The edible brown seaweed *Alaria esculenta* (Phaeophyceae, Laminariales): hybridization, growth and genetic comparisons of six Irish populations. *Journal of Applied Phycology* 12: 577–83.

Kraberg, A.C. & Norton, T.A. 2007. Effect of epiphytism on reproductive and vegetative lateral formation in the brown, intertidal seaweed *Ascophyllum nodosum* (Phaeophyceae). *Phycological Research* 55: 17–24.

Krauss, F. [Krauss, C.F.F. von] 1846. Pflanzen des Cap- und Natal-Landes, gesammelt und zusammengestellt von Dr. Ferdinand Krauss. (Schluss.). *Flora* 29: 209–15.

Kregting, L.T., Hepburn, C.D. & Savidge, G. 2015. Seasonal differences in the effects of oscillatory and uni-directional flow on the growth and nitrate-uptake rates of juvenile *Laminaria digitata* (Phaeophyceae). *Journal of Phycology* 51: 1116–26.

Kreimer, G., Kawai, H., Müller, D.G. & Melkonian, M. 1991. Reflective properties of the stigma in male gametes of *Ectocarpus siliculosus* (Phaeophyceae) studied by confocal laser scanning microscopy. *Journal of Phycology* 27: 268–76.

Kremer, B.P., 1973. Isolation of mannitol from *Desmarestia viridis*. *Phytochemistry* 12: 609–10.

Kremer, B.P. 1975a. Physiological and chemical characteristics of different thallus regions of *Fucus serratus*. *Helgoländer Wissenschaftliche Meeresuntersuchungen* 27: 115–27.

Kremer, B.P. 1975b. Mannitmetabolismus in der marinen braunalge *Fucus serratus*. *Zeitschrift für Pflanzenphysiologie* 74: 255–63.

Kristiansen, A. 1960. *Microspongium globosum*, en for Danmark ny brunalge. *Botanisk Tidsskrift* 56: 251–4.

Kristiansen, A. 1978. Marine algal vegetation in shallow water around the Danish Island of Saltholm, The Sound. *Botanisk Tidsskrift* 72: 203–26.

Kristiansen, A. 1984. Experimental field studies on the ecology of *Scytosiphon lomentaria* (Fucophyceae, Scytosiphonales) in Denmark. *Nordic Journal of Botany* 4: 719–24.

Kristiansen, A. & Pedersen, P.M. 1979. Studies on the life history and seasonal variation of *Scytosiphon lomentaria* (Fucophyceae, Scytosiphonales) in Denmark. *Botanisk Tidsskrift* 74: 31–56.

Kristiansen, A., Pedersen, P.M. & Mosholm, L. 1991. Growth and reproduction of *Scytosiphon lomentaria* (Fucophyceae) in relation to temperature in two populations from Denmark. *Nordic Journal of Botany* 11: 375–83.

Kucera, H. & Saunders, G.W. 2008. Assigning morphological variants of *Fucus* (Fucales, Phaeophyceae) in Canadian waters to recognised species using DNA barcoding. *Botany* 86: 1065–79.

Kuckuck, P. 1891. Beiträge zur Kenntniss der *Ectocarpus*-Arten der Kieler Föhrde. *Botanisches Centralblatt* 48: 1–6, 33–41, 65–71, 97–104, 129–41.

Kuckuck, P. 1892. *Ectocarpus siliculosus* Dillw. sp. forma *varians* n. f., ein Beispiel für ausserordentliche Schwankungen der pluriloculären Sporangienform. *Berichte der deutschen Botanischen Gesellschaft* 10: 256–9.

Kuckuck, P. 1894. Bemerkungen zur marinen Algenvegetation von Helgoland. *Wissenschaftliche Meeresuntersuchungen. Neue Folge* 1: 223–63.

Kuckuck, P. 1895a. Ueber. Schwarmsporenbildung bei den Tilopterideen und uber *Choristocarpus tenellus*. *Jahrbucher fur wissenschaftliche Botanik.* 28: 290–322.

Kuckuck, P. 1895b. Ueber einige neue Phaeosporeen der westlichen Ostsee. *Botanische Zeitung* 8: 175–87.

Kuckuck, P. 1897a. Bemerkungen zur marinen Algenvegetation von Helgoland. *Wissenschaftliche Meeresuntersuchungen. Neue Folge* 2: 371–400.

Kuckuck, P. 1897b. Beiträge zur Kenntnis der Meeresalgen, 3. Die Gattung *Mikrosyphar* Kuckuck. *Wissenschaftliche Meeresuntersuchungen. Neue Folge* 2: 349–58.

Kuckuck, P. 1897c. Beiträge zur Kenntnis der Meeresalgen, 4. Über zwei hohlenbewohnende Phaeosporeen. *Wissenschaftliche Meeresuntersuchungen. Neue Folge* 2: 359–69.

Kuckuck, P. 1898. Ueber die Paarung von Schwarmsporen bei *Scytosiphon*. *Berichte der Deutschen Botanischen Gesellschaft* 16: 35–7.

Kuckuck, P. 1899a. Beiträge zur Kenntnis der Meeresalgen. 5–9. *Helgoländer Wissenschaftliche Meeresuntersuchungen* 3: 11–81, 4 + 21 + 15 figs, pls II (8)–VIII (14).

Kuckuck, P. 1899b. Ueber den Generationswechsel von *Culteria multifida* (Engl. Bot.) Grev. *Helgoländer*

Wissenschaftliche Meeresuntersuchungen N.F. 3: 95–116.

Kuckuck, P. 1912. Beitrage zur Kenntnis der Meeresalgen. XI. Zur Fortpflanzung der Phaeosporeen. *Helgoländer Wissenschaftliche Meeresuntersuchungen* N.F. 5: 155–88.

Kuckuck, P. 1917. Über zwerggenerationen bei *Pogotrichum* und über die Fortpflanzung von *Laminaria*. *Berichte der Deutschen Botanischen Gesellschaft* 35: 557–78.

Kuckuck, P. 1929. Fragmente einer Monographie der Phaeosporeen. *Helgoländer Wissenschaftliche Meeresuntersuchungen* 17: 1–93.

Kuckuck, P. 1953. Ectocarpaceen-Studien I: *Hecatonema, Chilionema, Compsonema*. *Helgoländer Wissenschaftliche Meeresuntersuchungen* 4: 316–52.

Kuckuck, P. 1955. Ectocarpaceen-Studien III. *Protectocarpus* nov. gen. (Kuckuck). *Helgoländer Wissenschaftliche Meeresuntersuchungen* 5: 119–40.

Kuckuck, P. 1956. Ectocarpaceen-Studien IV. *Herponema, Kützingiella* nov. gen., *Farlowiella* nov. gen. *Helgoländer Wissenschaftliche Meeresuntersuchungen* 5: 292–325.

Kuckuck, P. 1958. Ectocarpaceen-Studien V. *Kuckuckia, Feldmannia*. *Helgoländer Wissenschaftliche Meeresuntersuchungen* 6: 171–92.

Kuckuck, P. 1960. Ectocarpaceen-Studien VI. *Spongonema*. *Helgoländer Wissenschaftliche Meeresuntersuchungen* 7: 93–113.

Kuckuck, P. 1961. Ectocarpaceen-Studien VII. *Giffordia*. *Helgoländer Wissenschaftliche Meeresuntersuchungen* 8: 119–52.

Kuckuck, P. 1963. Ectocarpaceen-Studien VIII. Einige Arten aus warmen Meeren. *Helgoländer Wissenschaftliche Meeresuntersuchungen* 9: 361–82.

Kuckuck, P. 1964. Ectocarpaceen-Studien IX. Untersuchung von Herbarmaterial. *Helgoländer Wissenschaftliche Meeresuntersuchungen*: 1–23.

Kuckuck, V.P. & Kornmann, H.V.P. 1954. Ectocarpaceen-Studien II. *Streblonema*. *Helgoländer Wissenschaftliche Meeresuntersuchungen* 5: 103–17.

Kuhlenkamp, R. 1989. Photomorphogenesis in early development of *Tilopteris mertensii* (Tilopteridales: Phaeophyceae). *Journal of the Marine Biological Association of the United Kingdom* 69: 555–62.

Kuhlenkamp, R. & Hooper, R.G. 1995. New observations on the Tilopteridaceae (Phaeophyceae). I. Field studies of *Haplospora* and *Phaeosiphoniella* with implications for survival, perennation and dispersal. *Phycologia* 34: 229–39.

Kuhlenkamp, R. & Müller, D.G. 1985. Culture studies on the life history of *Haplospora globosa* and *Tilopteris mertensii* (Tilopteridales, Phaeophyceae). *British Phycological Journal* 20: 301–12.

Kuhlenkamp, R. & Müller, D.G. 1994. Isolation and regeneration of protoplasts from healthy and virus-infected gametophytes of *Ectocarpus siliculosus* (Phaeophyceae). *Botanica Marina* 37: 525–30.

Kuhlenkamp, R., Müller, D.G. & Whittick, A. 1993. Genotypic variation and alternating DNA levels at constant chromosome numbers in the life history of the brown alga *Haplospora globosa* (Tilopteridales). *Journal of Phycology* 29: 377–80.

Kumagae, N., Inoh, S. & Nishibayashi, T. 1960. Morphogenesis in Dictyotales. II. On the meiosis of the tetraspore mother cell in *Dictyota dichotoma* (Huds.) Lamour. and *Padina japonica* Yamada. *Biological Journal of Okayama University* 6: 91–102.

Kumke, J. 1973. Beiträge zur Periodizität der Oogon-Entleerung bei *Dictyota dichotoma* (Phaeophyta). *Zeitschrift für Pflanzenphysiologie* 70: 191–210.

Kunieda, H. & Arasaki, S. 1947. On the life history of *Ilea fascia* and *Punctaria* sp. *Seibutsu* 2: 185–8.

Kunieda, H. & Suto, S. 1938. The life history of *Colpomenia sinuosa* (Scytosiphonaceae) with special reference to the conjugation of anisogametes. *Botanical Magazine, Tokyo* 52: 539–46.

Kuntze, O. 1891–98. *Revisio generum plantarum*. Leipzig, London, Milano, New York, Paris: Arthur Felix, Dulau & Co., U. Hoepli, Gust. A. Schechert, Charles Klincksierck. [Pars I. pp. i–clvi, 1–376, 1891; Pars II. pp. 377–1011, 1891; Pars III: (1): I–VI + 1–202 + 1–576. 1898].

Kurata, K., Taniguchi, K., Shiraishi, K. & Suzuki, M. 1990. Feeding deterrent diterpenes from the brown alga *Dilophus okamurai*. *Phytochemistry* 29: 3453–5.

Kurogi, M. & Akiyama, K. 1957. Studies of ecology and culture of *Undaria pinnatifida* (Sur.) Hariot. *Bulletin of Tohoku Regional Fisheries Research Laboratory* 10: 95–117.

Kusaka, K., Sasaki, R., Tsukada, T. & Oikawa, H. 2007. Growth and morphological characteristics of native wakame (*Undaria pinnatifida*) seed collected from eight locations around Japan coast. *Miyagi Prefecture Report Fisheries Science* 7: 17–28 (in Japanese).

Kuylenstierna, M. 1990. *Benthic algal vegetation in the Nordre Älve Estuary (Swedish west coast). Vol. 1. Text*: Univ. of Göteborg.

Kylin, H. 1907. *Studien über die Algenflora der schwedischen Westküste*. Akademische Abhandlung. pp. [i–iii]–iv, 1–287, 7 pls. Uppsala: K.W. Appelbergs Buchdruckerei.

Kylin, H. 1910. Zur Kenntnis der Algenflora der norwegischen Westküste. *Arkiv för Botanik* 10 No. 1.

Kylin, H. 1916. Über die Generationswechsel bei *Laminaria digitata*. *Svensk Botanisk Tidskrift* 10: 551–61.

Kylin, H. 1917. Über die Entwicklungsgeschichte und die systematische Stellung der Tilopterideen. *Berichte der deutsche botanischen Gesellschaft* 35: 298–310.

Kylin, H. 1918. Studien über die Entwicklungsgeschichte der Phaeophyceen. *Svensk Botanisk Tidskrift* 12: 1–64.

Kylin, H. 1933. Über die Entwicklungsgeschichte der Phaeophyceen. *Acta Universitatis Lundensis* 29: 1–102.

Kylin, H. 1934. Zur Kenntnis der Entwicklungsgeschichte einiger Phaeophyceen. *Acta Universitatis Lundensis* 30: 1–18.

Kylin, H. 1937. Bemerkungen über die Entwicklungsgeschichte einiger Phaeophyceen. *Acta Universitatis Lundensis* 33: 1–34.

Kylin, H. 1938. Bemerkungen über die Fucosanblasen der Phaeophyceen. *Kungliga Fysiografiska Sällskapets i Lund Förhandlingar* 8: 1–10.

Kylin, H. 1940. Die Phaeophyceenordnung Chordariales. *Acta Universitatis Lundensis* 36: 1–67.

Kylin, H. 1943. *Ectocarpus siliculosus* mit unilokulären und plurilokulären Sporangien in kultur aufgezogen. *Botaniska Notiser* 1943: 295–8.

Kylin, H. 1947. Die Phaeophyceen der schwedischen Westküste. *Acta Universitatis Lundensis* 43: 1–99.

Kützing, F.T. 1833. Algologische Mittheilungen. *Flora* 16: 513–21.

Kützing, F.T. 1843. *Phycologia generalis* oder Anatomie, Physiologie und Systemkunde der Tange. Mit 80 farbig gedruckten Tafeln, gezeichnet und gravirt vom Verfasser. pp. [part 1]: [i]–xxxii, [1]–142, [part 2:] 143–458, 1, err.], pls 1–80. Leipzig: F.A. Brockhaus.

Kützing, F.T. 1845. *Phycologia germanica*, d. l. Deutschlands Algen in bündigen Beschreibungen. Nebst einer Anleitung zum Untersuchen und Bestimmen dieser Gewächse für Anfänger. pp. i–x, 1–340. Nordhausen: W. Köhne.

Kützing, F.T. 1845–71. *Tabulae phycologicae*. I–XIX Index Nordhausen. 1845–49 (V. 1); 1850–52 (V. 2); 1853–69 (V. 3–19 respectively); 1871 (index).

Kützing, F.T. 1847. Diagnosen und Bemerkungen zu neuen oder kritischen Algen. *Botanische Zeitung* 5: 1–5, 22–5, 33–8, 52–5, 164–7, 177–80, 193–8, 219–23.

Kützing, F.T. 1849. *Species algarum*. pp. [i]–vi, [1]–922. Lipsiae [Leipzig]: F.A. Brockhaus.

Kützing, F.T. 1855. *Tabulae phycologicae*; oder, Abbildungen der Tange. Vol. V pp. i–ii, 1–30, 100 plates. Nordhausen: Gedruckt auf kosten des Verfassers (in commission bei W. Köhne).

Kützing, F.T. 1856. *Tabulae phycologicae*; oder, Abbildungen der Tange. Vol. VI pp. i–iv, 1–35, 100 pls. Nordhausen: Gedruckt auf kosten des Verfassers (in commission bei W. Köhne).

Kützing, F.T. 1859. *Tabulae phycologicae*; oder, Abbildungen der Tange. Vol. IX pp. i–vii, 1–42, 100 pls. Nordhausen: Gedruckt auf kosten des Verfassers (in commission bei W. Köhne).

Kützing, F.T. 1860. *Tabulae phycologicae*; oder, Abbildungen der Tange. Vol. X pp. i–iv, 1–39, 100 pls. Nordhausen: Gedruckt auf kosten des Verfassers.

La Claire, J.W. 1982. Light and electron microscopic studies of growth and reproduction in *Cutleria* (Phaeophyta). III. Nuclear division in the trichothallic meristem of *Cutleria cylindrica*. *Phycologia* 21: 273–87.

La Claire, J.W. II & West, J.A. 1978. Light and EM studies of growth and reproduction in *Cutleria*. I. Gametogenesis in the female plant of *C. hancockii*. *Protoplasma* 97: 93–110.

La Claire, J.W. II & West, J.A. 1979. Light and EM studies of growth and reproduction in *Cutleria*. II. Gametogenesis in the male plant of *C. hancockii*. *Protoplasma* 101: 247–67.

Lacey, A.J. & Moss, B. 1956. The autoecology of *Halidrys siliquosa* (L.) Lyngb. *British Phycological Bulletin* 1: 14.

Ladah, L.B. & Zertruche-González, J.A. 2007. Survival of microscopic stages of a perennial kelp (*Macrocystis pyrifera*) from the centre and the southern extreme of its range in the northern hemisphere after exposure to simulated El Nino stress. *Marine Biology* 152: 677–86.

Lagos, V. & Cremades, J. 2004. Contribución al conocimiento de la biología del alga parda alimentaria *Himanthalia elongata* (Fucales, Phaeophyta) en las costas de Galicia. *Anales de Biología* 26: 105–16.

Laing, R.M. 1927. A reference list of New Zealand marine algae. *Transactions and Proceedings of the New Zealand Institute* 57: 126–85.

Lakowitz, K. 1929. *Die Algenflora der gesamten Ostsee (ausschl. Diatomeen)*. pp. [i]–viii, [1]–474, illus. Danzig: Kommissions-Verlag von R. Friedländer.

Lamarck, J.B. de & De Candolle, A.P. 1805. *Flore française*: Troisième Édition. Tome second. pp. i–xii, 1–600, 1 folded map. Parisiis [Paris]: Chez H. Agasse, rue de Poitevine, No. 6 de l'Imprimerie de Stoupe, An XIII.

Lami, R. 1934. Sur une nouvelle espèce de Laminaire de la région ibérico-marocaine: *Laminaria ibérica* (Hamel) Lami. *Comptes Rendus Hebdomadaires des Seances de l'Academie des Sciences* 198: 113–14.

Lami, R. 1939. Sur quelques Fucacées de la côte du Portugal et leur répartition. *Boletim da Sociedade Broteriana, II série* 13: 177–86.

Lami, R. 1943. Notules d'Algologie marine (Suite) IX. Sur l'écologie et la répartition dans la Manche de *Laminaria ochroleuca* De La Pylaie. *Bulletin du Laboratoire Maritime De Dinard* 25: 75–90.

Lamouroux, J.V.F. 1805. *Dissertations sur plusieurs espèces de* Fucus, peu connues ou nouvelles; avec leur description en latin et en français. pp. xxiv + 85, 36 plates. Agen & Paris: de l'Imprimerie de Raymond Nouvel & Chez Treuttel et Würtz.

Lamouroux, J.V.F. 1809a. Exposition des caractères du genre *Dictyota*, et tableau des espèces qu'il renferm. *Journal de Botanique* [Desvaux] 2: 38–44.

Lamouroux, J.V.F. 1809b. Observations sur la physiologie des algues marines, et description de cinq nouveaux genres de cette famille. *Nouveau Bulletin des Sciences, par la Société Philomathique de Paris* 1: 330–3.

Lamouroux, J.V.F. 1813. Essai sur les genres de la famille des thalassiophytes non articulées. *Annales du Muséum d'Histoire Naturelle, Paris* 20: 21–47, 115–39, 267–93.

Lamouroux, J.V.F. 1816. *Histoire des polypiers coralligènes flexibles*, vulgairement nommés zoophytes. pp. [i]–lxxxiv, chart, [1]–560, [560, err], pls I–XIX, uncol. by author. Caen: De l'imprimerie de F. Poisson.

Lamouroux, J.V., Bory de Saint-Vincent, J.B.G.M. & Deslongchamps, E. 1824. *Encyclopédie méthodique ou par ordre de matières. Histoire naturelle des zoophytes, ou animaux rayonnés, faisant suite à l'histoire naturelle des vers de Bruguière*. pp. [i*–iii*], [i]–viii, [i]–869. Paris: Mme veuve Agasse.

Lane, C.E. & Saunders, G.W. 2005. Molecular investigation reveals epi/endophytic extrageneric kelp (Laminariales, Phaeophyceae) gametophytes colonizing *Lessoniopsis littoralis* thalli. *Botanica Marina* 48: 426–36.

Lane, C.E., Mayes, C., Druehl, L.D. & Saunders, G.W. 2006. A multi-gene molecular investigation of the kelp (Laminariales, Phaeophyceae) resolves competing phylogenetic hypotheses and supports substantial taxonomic re-organization. *Journal of Phycology* 42: 493–512.

Lane, C.E., Lindstrom, S.C. & Saunders, G.W. 2007. A molecular assessment of northeast Pacific *Alaria* species (Laminariales, Phaeophyceae) with reference to the utility of DNA barcoding. *Molecular Phylogenetics and Evolution* 44: 634–48.

Larkum, A.W.D. 1972. Frond structure and growth in *Laminaria hyperborea*. *Journal of the Marine Biological Association of the United Kingdom* 52: 405–18.

Larsen, A. & Sand-Jensen, K. 2005. Salt tolerance and distribution of estuarine benthic macroalgae in the Kattegat-Baltic Sea area. *Phycologia* 45: 13–23.

Larsen, B. & Haug, A. 1958. The influence of habitat on the chemical composition of *Ascophyllum nodosum*. *Proceedings of the International Seaweed Symposium* 3: 73.

Larsen, B. & Haug, A. 1960. The distribution of Iodine and other constituents in stipe of *Laminaria hyperborea* (Gunn.) Foslie. *Botanica Marina* 2: 250–4.

Larsen, B. & Haug, A. 1964. Polysaccharides occurring in the sodium carbonate extract of *Ascophyllum nodosum*. *Proceedings of the International Seaweed Symposium* 4: 338–43.

Larsen, B. & Haug, A. 1969. Fucose-containing Polysaccharides in *Ascophyllum nodosum*. *Proceedings of the International Seaweed Symposium* 6: 521–8.

Lawrence, J.M. 1975: On the relationships between marine plants and sea urchins. *Oceanography and Marine Biology Annual Review* 13: 213–86.

Lawson, G.W. & John, D.M. 1982. The marine algae and coastal environment of tropical west Africa. *Beihefte zur Nova Hedwigia* 70: 1–455.

Laycock, M.V. 1975. The amino acid sequence of cytochrome f from the brown alga *Alaria esculenta* (L.) Grev. *Biochemical Journal* 149: 271–9.

Le, H.N., Hughes, A.D. & Kerrison, P.D. 2018. Early development and substrate twine selection for the cultivation of *Sargassum muticum* (Yendo) Fensholt under laboratory conditions. *Journal of Applied Phycology* 30: 2475–83.

Le Bail, A., Billoud, B., Maisonneuve, C., Peters, A.F., Cock, J.M. & Charrier, B. 2008. Early development pattern of the brown alga *Ectocarpus siliculosus* (Ectocarpales, Phaeophyceae) sporophyte. *Journal of Phycology* 44: 1269–81.

Le Jolis, A. 1855. Examen des espèces confondues sous le nom de *Laminaria digitata* auct., suivi de quelques observations sur le genre *Laminaria*. *Mémoires de la Société Impériale des Sciences Naturelles de Cherbourg* 3: 241–312.

Le Jolis, A. 1856. Examen des espèces confondues sous le nom de *Laminaria digitata*, suivi de quelques observations sur le genre *Laminaria*. *Novorum actorum Academie Casareae Leopoldino-Carolinae naturae curiosorum* 25: 531–91.

Le Jolis, A. 1861. On the synonymy of *Ectocarpus brachiatus*. *Transactions of the Botanical Society [Edinburgh]* 7: 36–7.

Le Jolis, A. 1863. Liste des algues marines de Cherbourg. *Mémoires de la Société Impériale des Sciences Naturelles de Cherbourg* 10: 5–168.

Le Lann, K., Connan, S. & Stiger-Pouvreau, V. 2012. Phenology, TPC and size-fractioning phenolics variability in temperate Sargassaceae (Phaeophyceae, Fucales) from Western Brittany: Native versus introduced species. *Marine Environmental Research* 80: 1–11.

Le Lann, K., Jégou, C. & Stiger-Pouvreau, V. 2008. Effect of different conditioning treatments on total phenolic content and antioxidant activities in two Sargassacean species: Comparison of the frondose *Sargassum muticum* (Yendo) Fensholt and the cylindrical *Bifurcaria bifurcata* R. Ross. *Phycological Research* 56: 238–45.

Leal, P.P., Hurd, C.L., Fernández, P.A. & Roleda, M.Y. 2017a. Ocean acidification and kelp development: reduced pH has no negative effects on meiospore germination and gametophyte development of *Macrocystis pyrifera* and *Undaria pinnatifida*. *Journal of Phycology* 53: 557–66.

Leal, P.P., Hurd, C.L., Fernandez, P.A. & Roleda, M.Y. 2017b. Meiospore development of the kelps *Macrocystis pyrifera* and *Undaria pinnatifida* under ocean acidification and ocean warming: independent effects are more important than their interaction. *Marine Biology* 164: 7.

Leal, P.P., Hurd, C.L., Sander, S.S. & Armstrong, E. 2015. Effects of seawater pH temperature and copper exposure on the development of meiospores of the kelps *Macrocystis pyrifera* and *Undaria pinnatifida* (Laminariales, Ochrophyta). *Poster, 6th European Phycological Congress (EPC6)*, 50: 165–6.

Leal, P.P., Hurd, C.L., Sander, S.G., Kortner, B. & Roleda, M.Y. 2016. Exposure to chronic and high dissolved copper concentrations impedes meiospore development of the kelps *Macrocystis pyrifera* and *Undaria pinnatifida* (Ochrophyta). *Phycologia* 55: 12–20.

Leal, P.P., Hurd, C.L., Sander, S.G., Armstrong, E., Fernández, P.A., Suhrhoff, T.J. & Roleda, M.Y. 2018. Copper pollution exacerbates the effects of ocean acidification and warming on kelp microscopic early life stages. *Scientific Reports* 8: 14763.

Leclerc, M.C., Barriel, V., Lecointre, G., de Reviers, B. 1998. Low divergence in rDNA ITS sequences among five species of *Fucus* (Phaeophyceae) suggests a very recent radiation. *Journal of Molecular Evolution* 46: 115–20.

Lee, J.A. & Brinkhuis, B.H. 1986. Reproductive phenology of *Laminaria saccharina* (L.) Lamour. (Phaeophyta) at the southern limit of its distribution in the northwestern Atlantic. *Journal of Phycology* 22: 276–85.

Lee, J.A. & Brinkhuis, B.H. 1988. Seasonal light and temperature effects on development of *Laminaria saccharina* (Phaeophyta) gametophytes and juvenile sporophytes. *Journal of Phycology* 24: 181–91.

Lee, J.-A., Sunwoo, Y.-I., Lee, H.-J., Park, I.-H. & Chung, I.-K. 1990. The effects of copper on the early stages of *Undaria pinnatifida* (Harv.) Suringar (Laminariales, Phaeophyta) under temperature-irradiance gradient. *Korean Journal of Phycology* 4: 41–53.

Lee, J.C., Soh, W.Y. & Lee, J.H. 1995. Effects of plant growth regulators on growth in the various explants of *Undaria pinnatifida* (Harvey) Suringar. *Korean Journal of Phycology* 10: 51–7.

Lee, E.-Y., Lee, I.K. & Choi, H.-G. 2003. Morphology and nuclear small-subunit rDNA sequences of *Ishige* (Ishigeaceae, Phaeophyceae) and its phylogenetic relationship among selected brown algal orders. *Botanica Marina* 46: 193–201.

Lee, E.-Y., Pedersen, P.M. & Lee, I.K. 2002. *Neoleptonema yongpillii* E.-Y. Lee & I.K. Lee, gen. et sp. nov. (Phaeophyceae), based on morphological characters and RuBisCO spacer sequences. *British Phycological Journal* 37: 227–36.

Lee, K.M., Boo, G.H., Riosmena-Rodriguez, R., Shin, J.-A. & Boo, S.M. 2009. Classification of the genus *Ishige* (Ishigeales, Phaeophyceae) on the North

Pacific Ocean with recognition of *Ishige foliacea* based on plastid rbcL and mitochondrial cox3 gene sequences. *Journal of Phycology* 45: 906–13.

Lee, J.W. & Hwang, I.K. 2010. Dictyotales, Desmarestiales. In: Anon. (ed.), *Algal flora of Korea. Volume 2, Number 2. Heterokontophyta: Phaeophyceae: Ishigeales, Dictyotales, Desmarestiales, Sphacelariales, Cutleriales, Ralfsiales, Laminariales*, pp. 19–69. Incheon: National Institute of Biological Resources.

Lee, Y.K., Yoon, H.S., Motomura, T., Kim, Y.J. & Boo, S.M. 1999. Phylogenetic relationships between *Pelvetia* and *Pelvetiopsis* (Fucaceae, Phaeophyta) inferred from sequences of the RuBisCo spacer region. *European Journal of Phycology* 34: 205–11.

Lee, W.J. & Bae, K.S. 2002. Phylogenetic relationship among several genera of Dictyotaceae (Dictyotales, Phaeophyceae) based on 18S rRNA and partial rbcl gne sequences. *Marine Biology* 140: 1107–15.

Lee, W.J. & King, R.J. 1996. The molecular characteristics of five genera of Dictyotaceae (Phaeophyta) from Australia: based on DNA sequence of nuclear rDNA internal transcribed spacer (ITS) and 5.8S. *Algae* 11: 381–8.

Lee, K.Y. & Sohn, C.H. 1993. Morphological characteristics and growth of two forms of sea mustard, *Undaria pinnatifida* f. *distans* and *U. pinnatifida* f. *typica*. *Journal of Aquaculture* 6: 71–87.

Lee, Y. & Kang, S. 2001. *A catalogue of the seaweeds in Korea*. pp. [8], 1–662. Jeju: Cheju National University Press.

Lee, Y.P. & Oh, Y.S. 1998. Morphology and anatomy of *Chorda filum* (Linnaeus) Stackhouse (Chordaceae, Phaeophyta) in Korea. *Algae (The Korean Journal of Phycology)* 13: 69–77.

Lee, Y.K., Lim, D.-J., Lee, Y.-H. & Park, Y.-I. 2006. Variation in fucoidan contents and monosaccharide compositions of Korean *Undaria pinnatifida* (Harvey) Suringar (Phaeophyta. *Algae* 21: 157–60.

Lehvo, A., Bäck, S. & Kiirikki, M. 2001. Growth of *Fucus vesiculosus* L. (Phaeophyta) in the Northern Baltic Proper: Energy and Nitrogen storage in seasonal environment. *Botanica Marina* 44: 345–50.

Lein, T.E. 1984. Distribution, reproduction, and ecology of *Fucus ceranoides* L. (Phaeophyceae) in Norway. *Sarsia* 69: 75–81.

Leliaert, F., Kerckhof, F. & Coppejans, E. 2000a. Eerste waarnemingen van *Undaria pinnatifida* (Harvey) Suringar (Laminariales, Phaeophyta) en de epifyt *Pterothamnion plumula* (Ellis) Nägeli (Ceramiales, Rhodophyta) in Noord Frankrijk en België. *Dumortiera* 75: 5–10.

Leman, D.S. 1819. *Annales des Sciences Naturelles*. Tom XIII, DEA-DZW. Paris, 104–6.

León-Alvarez, D., Reyes-Gómez, V.P., Wynne, M.J., Ponce-Márquez, M.E. & Quiróz-González 2017. Morphological and molecular characterization of *Hapalospongidion gelatinosum*, Hapalospongidiaceae fam. nov. (Ralfsiales, Phaeophyceae) from Mexico. *Botanica Marina* 60: 567–81.

Levring, T. 1935a. Über einige meeresalgen bei Kristineberg an der schwedischen westküste. *Botaniska Notiser* 1935: 454–63.

Levring, T. 1935b. Undersökningar över Öresund … XIX. Zur kenntnis der algenflora von Kullen an

der Schwedischen Westküste. *Lunds Universitets Årsskrift*. 31: 1–64.

Levring, T. 1937. Zur Kenntnis der Algenflora der Norwegischen Westküste. *Lunds Universitets Årsskrift. N.F. avd. 2* 33: 1–147.

Levring, T. 1938. Verzeichnis einiger Chlorophyceen und Phaeophyceen von Südafrika. *Lunds Universitets Årsskrift, Ny Följd, Andra Afdelningen* 34: 1–25.

Levring, T. 1940. *Studien über die Algenvegetation von Blekinge, Südschweden*. pp. i–vii, 1–178, [1]. Lund: Håkan Ohlssons Buchdruckerei.

Levring, T. 1945. Zur Kenntnis zweier endophytischen Phaeophyceen. *Acta Horti Gothoburgensis. Meddelanden från Götëborgs botaniska trädgard.* 16: 184–90.

Lewey, S. & Farnham, W.F. 1981. Observations on *Sargassum muticum* in Britain. *Proceedings of the International Seaweed Symposium* 8: 388–94.

Lewis, I.F. & Taylor, W.R. 1933. Notes from the Woods Hole Laboratory. *Rhodora* 35: 147–54.

Lewis, I.F. 1910. Periodicity in *Dictyota* at Naples. *Botanical Gazette* 50: 59–64.

Lewis, J.R. 1954. The ecology of exposed rocky shores of Caithness. *Transactions of the Royal Society of Edinburgh* 62: 695–723.

Lewis, J.R. 1964. The Ecology of Rocky Shores. English Universities Press, London.

Lewis, R.J., Green, M.K. & Afzal, M.E. 2012. Effects of chelated iron on oo-genesis and vegetative growth of kelp gametophytes (Phaeophyceae). *Phycological Research* 61: 46–51.

Li, J., Pang, S. & Shan, T. 2017. Existence of an intact male life cycle offers a novel way in pure-line cross-breeding in the brown alga *Undaria pinnatifida*. *Journal of Applied Phycology* 29: 993–9.

Li, J., Pang, S., Shan, T. & Liu, F. 2014. Zoospore-derived monoecious gametophytes in *Undaria pinnatifida* (Phaeophyceae). *Chinese Journal of Oceanology and Limnology* 32: 365–71.

Li, J., Pang, S., Liu, F., Shan, T. & Gao, S. 2013. Spermatozoid life-span of two brown seaweeds, *Saccharina japonica* and *Undaria pinnatifida*, as measured by fertilization efficiency. *Chinese Journal of Oceanology and Limnology* 31: 774–81.

Li, T-Y., Qu, J.-Q., Feng, Y.-J., Cui, L., Shan, C. & Liu, T. 2015. Complete mitochondrial genome of *Undaria pinnatifida* (Alariaceae, Laminariales, Phaeophyceae). *Mitochondrial DNA* 26: 953–4.

Lichtenberg, M., Nørregaard, R.D. & Kühl, M. 2017. Diffusion or advection? Mass transfer and complex boundary layer landscapes of the brown alga *Fucus vesiculosus*. *Journal of the Royal Society Interface* 14: 1–10.

Liem, P.Q. & Laur, M.H. 1976. Teneur, composition et répartition cytologique des lipides polaires soufrés et phosphorés de *Pelvetia canaliculata* (L.), Decn. et Thur., *Fucus vesiculosis* (L.) et *Fucus serratus* (L.). *Phycologia* 15: 367–76.

Lightfoot, J. 1777. *Flora scotica*: or, a systematic arrangement, in the Linnaean method, of the native plants of Scotland and the Hebrides. Vol. II. pp. 545–1151 [1–24], pls 1–35. London: printed for B. White at Horace's Head, in Fleet-Street.

Lilley, S.A. & Schiel, D.R. 2006. Community effects following the deletion of a habitat-forming alga from rocky marine shores. *Oecologia* 148: 672–81.

Lim, B.-K., Kawai, H., Hori, H. & Osawa, S. 1986. Molecular evolution of 5S ribosomal RNA from red and brown algae. *Japanese Journal of Genetics* 61: 169–76.

Lim, P.E., Sakaguchi, M., Hanyuda, T., Kogame, K., Phang, S.M. & Kawai, H. 2007. Molecular phylogeny of crustose brown algae (Ralfsiales, Phaeophyceae) inferred from rbcL sequences resulting in the proposal for Neoralfsiaceae fam. nov. *Phycologia* 46: 456–66.

Lima, F.P., Queiroz, N., Ribeira, P., Xavier, R., Hawkins, S.J. & Santos, A.M. 2009. First record of *Halidrys siliquosa* on the Portuguese coast: counter-intuitive range expansion? *JMBA Biodivesity Records* 2: 1–4.

Lindauer, V.W., Chapman, V.J. & Aiken, M. 1961. The marine algae of New Zealand. Part II. Phaeophyceae. *Nova Hedwigia* 3: 129–350.

Lining, T. & Garbary, D.J. 1992. The *Ascophyllum/Polysiphonia/Mycosphaerella* symbiosis. III. Experimental studies on the interactions between *P. lanosa* and *A. nodosum*. *Botanica Marina* 35: 341–9.

Link, H.F. 1833. Handbuch zur Erkennung der nulzbarsten und am haufigsten vorkommenden Gewachse. Bd 3. Berlin.

Linnaeus, C. 1753. *Species plantarum*: Vol. 2, pp. [i], 561–1200, [1–30, index], [i, err.]. Holmiae [Stockholm]: Impensis Laurentii Salvii.

Linnaeus, C. 1759. *Systema naturae*: Tomus I. Editio decima, reformata. Editio decima revisa. Vol. 2, pp. 825–1384. Holmiae [Stockholm]: impensis direct. Laurentii Salvii.

Linnaeus, C. 1763. *Species plantarum, exhibentes plantas rite cognitas, ad genera relatas, cum differentiis specificis, nominibus trivialibus, synonymis selectis, locis natalibus, secundum systema sexuale digestas*. Tomus II. Editio secunda, aucta. pp. 785–1684, [1–64, Indices]. Homiae [Stockholm]: Impensis direct. Laurentii Salvii.

Linnaeus, C. 1767. *Genera Plantarum*: Editio novissima. pp. i–xx, 1–580, 1–44 (index). Viennae [Vienna]: Typis Joan. Thomae nob. de Trattnern.

Linnaeus, C. 1755. *Flora Svecica*. Editio Secunda, Stockholmiae.

Linskens, H.F. 1966. Adhasion von Fort pflanzungszellen Benthontischer. *Planta* 68: 99–110.

Littlauer, R. 2010. Mung! (or *Pilayella* and macroalgal blooms). *The Merecat Crossing*, 2 pp.

Littler, M.M. & Arnold, K.E. 1982. Primary productivity of marine macroalgal functional-form groups from southwestern North America. *Journal of Phycology* 18: 307–11.

Littler, M.M. & Kauker, B.J. 1984. Heterotrichy and survival strategies in the red alga *Corallina officinalis* L. *Botanica Marina* 27: 37–44.

Littler, M.M. & Littler, D.S. 1980. The evolution of thallus form and survival strategies in benthic marine macroalgae: field and laboratory tests of a functional form mode. *The American Naturalist* 116: 25–44.

Littler, M.M. & Littler, D.S. 1983. Heteromorphic life history strategies in the brown alga *Scytosiphon lomentaria* (Lyngb.) Link. *Journal of Phycology* 19: 425–31.

Littler, M.M., Littler, D.S. & Taylor, P.R. 1983a. Evolutionary strategies in a tropical barrier reef system: functional-form groups of marine macroalgae. *Journal of Phycology* 19: 229–37.

Littler, M.M., Martz, D.R. & Littler, D.S. 1983a. Effects of recurrent sand deposition on rocky intertidal organisms: importance of substrate heterogeneity in a fluctuating environment. *Marine Ecology Progress Series* 11: 129–39.

Littler, M.M., Taylor, P.R. & Littler, D.S. 1983b. Algal resistance to herbivory on a Caribbean barrier reef. *Coral Reefs* 2: 111–18.

Liu, J., Dong, L., Shen, Y. & Wu, C. 2004. Effect of light period on egg-discharge of gametophyte clones of *Undaria pinnatifida* (Phaeophyta). *Journal of Applied Phycology* 16: 519–21.

Liu, F. & Pang, S.J. 2009. Performances of growth, photochemical efficiency, and stress tolerance of young sporophytes from seven populations of *Saccharina japonica* (Phaeophyta) under short-term heat stress. *Journal of Applied Phycology* 22: 221–9.

Liu, S.X., Zou, D.H. & Xu, J.T. 2010. Response of the young sporophytes of *Hizikia fusiformis* to different N growth conditions and the solar radiation. *Acta Ecologica Sinica* 30: 5562–8 (in Chinese with English abstract).

Liu, F. & Pang, S. 2014. Complete mitochondrial genome of the invasive brown alga *Sargassum muticum* (Sargassaceae, Phaeophyceae). *Mitochondrial DNA*. https://doi.org/10.3109/19401736.2014.933333

Lizumi, H. & Sakanishi, Y. 1994. Temperature dependence of photosynthesis–irradiation (P–I) relationship of gametophytes of *Laminaria religiosa* Miyabe. *Bulletin of Hokkaido National Fisheries Research Institute* 58: 45–51.

Lobban, C.S. & Baxter, D.M. 1983. Distribution of the red algal epiphyte *Polysiphonia lanosa* on its brown algal host *Ascophyllum nodosum* in the Bay of Fundy, Canada. *Botanica Marina* 26: 533–8.

Lobban, C.S., Weidner, M. & Lüning, K. 1981. Photoperiod affects enzyme activities in the kelp, *Laminaria hyperborea*. *Zeitschrift für Pflanzenphysiologie* 105: 81–3.

Lobban, C.S. & Harrison, P.J. 1997. Seaweed Ecology and Physiology. Cambridge University Press, Cambridge, 366 pp.

Lobban, C.S. & Wynne, M.J. 1981. *The Biology of Seaweeds*. Vol. 17. Oxford.

Lockhart, J.C. 1979. Factors determining various forms in *Cladosiphon zosterae* (Phaeophyceae). *American Journal of Botany* 66: 836–44.

Lockhart, J.C. 1982. Influence of light, temperature and nitrogen on morphogenesis of *Desmotrichum undulatum* (J. Agardh) Reinke (Phaeophyta, Punctariaceae). *Phycologia* 21: 264–72.

Lodge, S.M. 1948a. Additions to algal records for the Manx Region. *Report of the Marine Biological Station at Port Erin* 58/60: 59–62.

Lodge, S.M. 1948b. Algal growth in the absence of *Patella* on an experimental strip of foreshore, Port St Mary, Isle of Man. *Proceedings and Transactions of the Liverpool Biological Society* 56: 78–85.

Loffler, Z., Graba-Landry, A., Kidgell, J.T., McClure, E.C., Pratchett, M.S. & Hoey, A.S. 2018. Holdfasts of *Sargassum swartzii* are resistant to herbivory and resilient to damage. *Coral Reefs* 37: 1075–84.

Lofthouse, P.F. & Capon, B. 1975. Ultrastructural changes accompanying mitosporogenesis. *Protoplasma* 84: 83–99.

Loiseaux, S. 1964. Sur l'heteroblastie et le cycle de deux *Ascocyclus* de la region Roscoff. *Compte Rendu Hebdomadaire des Séances de l'Académie des Sciences. Paris* 259: 2903–5.

Loiseaux, S. 1966. Sur le cycle de developpement de l'*Ascocyclus hispanicus* (Pheophycees, Myrionematacees) et la formation en culture de stades coccoides. *Compte Rendu Hebdomadaire des Séances de l'Académie des Sciences. Paris*, ser 3, 262: 68–71.

Loiseaux, S. 1967a. Morphologie et cytologie des Myrionemacees. Criteres taxonomiques. *Revue Générale de Botanique* 74: 329–47.

Loiseaux, S. 1967b. Recherches sur les cycles de developpement des Myrionematacees (Pheophycees) I–II Hectanonematees et Myrionematees. *Revue Générale de Botanique* 74: 529–76.

Loiseaux, S. 1968. Recherches sur les cycles de developpement des Myrionematacees (Pheophycees) III Tribu des Ralfsiees IV Conclusions generates. *Revue Générale de Botanique* 75: 295–318.

Loiseaux, S. 1969. Sur une espèce de *Myriotrichia* obtenue en culture à partir de zoïdes d'*Hecatonema maculans* Sauv. *Phycologia* 8: 11–15.

Loiseaux, S. 1970a. Notes on several Myrionemataceae from California using culture studies. *Journal of Phycology* 6: 248–60.

Loiseaux, S. 1970b. *Streblonema anomalum* S. et G. and *Compsonema sporangiiferum* S. et G., stages in the life history of a minute *Scytosiphon*. *Phycologia* 9: 185–91.

Loiseaux, S. & West, J.A. 1970. Brown algal mastigonemes: comparative ultrastructure. *Transactions of the American Microscopical Society* 89: 524–32.

Loiseaux, S. & Mache, R. 1977. Caractères particuliers de l'ARN ribosomal de *Pylaiella littoralis* (L.) Kjellm. (Ectocarpacées). *Bulletin de la Société Phycologique de France* 22: 110.

Loiseaux, S. & Mache, R. 1980. Cytoplasmic rRNA synthesis is probably controlled by plastids in the brown alga: *Pylaiella littoralis* (L.) Kjellm. *Physiologie végétale* 18: 199–206.

Loiseaux, S. & Rozier, C. 1978. Culture axénique de *Pylaiella littoralis* (L.) Kjellm. (Phéophycées). *Revue Algologique, Nouvelle Serie* 13: 333–40.

Loiseaux, S., Mache, R. & Rozier, C. 1979. Heterogeneity of the 23 S ribosomal RNA in phaeoplasts of *Pylaiella littoralis* (L.) Kjellm., Phaeophyta. *Physiologie végétale* 17: 619–29.

Loiseaux, S., Rozier, C. & Dalmon, J. 1980. Plastidial origin of a large ribosomal precursor molecule in the brown alga *Pylaiella littoralis* (L.) Kjellm. *Plant Science Letters* 18: 381–8.

Loiseaux-de Goër, S. & Noailles, M.-C. 2008. *Algues de Roscoff.* pp. [1]–215, col. figs. Roscoff: Editions de la Station Biologique de Roscoff.

Longtin, C.M. & Saunders, G.W. 2015. On the utility of mucilage ducts as a taxonomic character in *Laminaria* and *Saccharina* (Phaeophyceae) – the conundrum of *S. groenlandica. Phycologia* 54: 440–50.

Longtin, C.M. & Scrosati, R.A. 2009. Role of surface wounds and brown algal epiphytes in the colonization of *Ascophyllum nodosum* (Phaeophyceae) fronds by *Vertebrata lanosa* (Rhodophyta) (Note). *Journal of Phycology* 45: 535–9.

Lopez-Garcia, M., Masters, N., O'Brien, H.E., Lennon, J., Atkinson, G., Cryan, M.J., Oulton, R., & Whitney, H.M. 2018. Light-induced dynamic structural color by intracellular 3D photonic crystals in brown algae. *Science Advances* 4(4): eaan8917. https://doi.org/10.1126/sciadv.aan8917

López Rodríguez, M.C. & Pérez-Cirera, J.L. 1995a. Aportación al conocimiento de la influencia de la contaminación industrial en *Fucus vesiculosus* y *Fucus spiralis* en el N.O. de la Península Ibérica. *Studia Botanica* 13: 21–9.

López Rodríguez, M.C. & Pérez-Cirera, J.L. 1995b. Estudio de la colonización y sucesión de la comunidad de *Fucus spiralis* L. (Fucaceae, Phaeophyta) en las cercanías de un colector de aguas residuales de la Ría de Pontevedra (N.O. de España). *Nova Acta Científica Compostelana (Bioloxía)* 5: 25–9.

López-Rodríguez, M.C. & Pérez-Cirera, J.L. 1998. The effects of different environmental conditions on colonization and succession in the *Fucus vesiculosus* community on the Galician coasts (Northwestern Iberian Peninsula). *Botanica Marina* 41: 581–92.

López-Rodríguez, M.C., Bárbara, I. & Pérez-Cirera, J.L. 1999. Effects of pollution on *Fucus vesiculosus* communities on the northwest Iberian Atlantic coast. *Ophelia* 51: 129–41.

Loraine, I. 1989. L'algue japonaise *Sargassum muticum* (Yendo) Fensholt. Caractéristiques et répartition. *Rapport Archive Institutionnelle de l'Ifremer.*

Lotze, H.K., Schramm, W., Schories, D. & Worm, B. 1999. Control of macroalgal blooms at early developmental stages: *Pilayella littoralis* versus *Enteromorpha* spp. *Oecologia* 119: 46–54.

Loughnane, C. & Stengel, D.B. 2002. Attached *Sargassum muticum* (Yendo) Fensholt found on the west coast of Ireland. *Irish Naturalists' Journal* 27: 70–2.

Lu, T.T. & Williams, S.L. 1994. Genetic diversity and genetic structure in the brown alga *Halidrys dioica* (Fucales: Cystoseiraceae) in southern California. *Marine Biology* 121: 363–71.

Lubchenco, J. 1978. Plant species diversity in a marine intertidal community: importance of herbivore food preference and algal competitive abilities. *American Naturalist* 112: 23–39.

Lubchenco, J. 1982. Effects of grazers and algal competitors on fucoid colonisation in tide pools. *Journal of Phycology* 18: 544–50.

Lubchenco, J. 1983. *Littorina* and *Fucus*: effects of grazers, substratum heterogeneity, and plant escapes during succession. *Ecology* 64: 1116–23.

Lubchenco, J. 1986. Relative importance of competition and predation: early colonization by seaweeds in New England. In: Diamond, J. & Case, T.J. (eds), *Community Ecology*, pp. 537–55. Harper & Row, New York.

Lubchenco, J. & Cubit, J. 1980. Heteromorphic life histories of certain marine algae as adaptations to variations in herbivory. *Ecology* 61: 676–87.

Lubchenco, J. & Gaines, S.D. 1981. A unified approach to marine plant-herbivore interactions. I. Populations and communities. *Annual Review of Ecology, Evolution, and Systematics* 12: 405–37.

Lubchenco, J. & Menge, B.A. 1978. Community development and persistence in a low rocky intertidal zone. *Ecological Monographs* 59: 67–94.

Lubsch, A. & Timmermans, K. 2017. Texture analysis of *Laminaria digitata* (Phaeophyceae) thallus reveals trade-off between tissue tensile strength and toughness along lamina. *Botanica Marina* 60: 229–37.

Lüder, U.H. & Clayton, M.N. 2004. Induction of phlorotannins in the brown macroalga *Ecklonia radiata* (Laminariales Phaeophyta) in response to simulated herbivory – the first microscopic study. *Planta* 218: 928–37.

Ludwig, C.G. 1760. *Definitiones generum plantarum olim in usum auditorum collectas nunc auctas et emendates edidit* D. Georgius Rudolphus Boehmer Medic. Prof. Publ. Ord. Wittebergensis [Ed. 3]. pp. xlviii, index, 516, 1 Erratum, numerous woodcuts. Lipsiae [Leipzig]: Ex officina Joh. Frideric Gleditschii.

Lund, S. 1938. On *Lithoderma fatiscens* Areschoug and *L. fatiscens* Kuckuck. *Meddelser om Grønland* 116: 1–18.

Lund, S. 1940. Om *Dictyota dichotoma* (Huds.) Lamour. og andre nye arter for floraen i Nissum Bredning. *Botanisk Tidsskrift* 45: 180–94.

Lund, S. 1950. The marine algae of Denmark. Contributions to their natural history Vol II. Phaeophyceae IV. Sphacelariaceae, Cutleriaceae, Dictyotaceae. *Det Kongelige Danske Videnskabernes Selskabs. Biologiske Skrifter* 6: 1–80.

Lund, S. 1959. The marine algae of East Greenland. I. Taxonomical part. *Meddelser om Grønland* 156: 1–248.

Lund, S. 1966. On a sporangia-bearing microthallus of *Scytosiphon lomentaria* from nature. *Phycologia* 6: 67–78.

Lüning, K. 1971. Seasonal growth of *Laminaria hyperborea* under recorded underwater light conditions near Helgoland. In: Crisp, D.J. (ed.), *Proceedings of the 4th European Marine Biology Symposium*, pp. 347–61. Cambridge University Press, Cambridge.

Lüning, K. 1975. Crossing experiments in *Laminaria saccharina* from Helgoland and from the Isle of Man. *Helgoländer Wissenschaftliche Meeresuntersuchungen* 27: 108–14.

Lüning, K. 1980a. Control of algal life history by daylength and temperature. In: Price, J.H., Irvine, D.E.G. & Farnham, W.F. (eds), *The Shore Environment 2: Ecosystems*, pp. 915–45. Systematics Association Special Volume 17(b). Academic Press, London and New York.

Lüning, K. 1980b. Critical levels of light and temperature regulating the gametogenesis of three *Laminaria* species (Phaeophyceae). *Journal of Phycology* 16: 1–15.

Lüning, K. 1981a. Egg release in gametophytes of *Laminaria saccharina*: induction by darkness and inhibition by blue light and u.v. *British Phycological Journal* 16: 379–94.

Lüning, K. 1981b. Light. In: Lobban, C.S. & Wynne, M.K. (eds), *The Biology of Seaweeds*, pp. 326–55. Blackwell. Oxford.

Lüning, K. 1981c. Photomorphogenesis of reproduction in marine macroalgae. *Berichte der Deutschen Botanischen Gesellschaft* 94: 401–17.

Lüning, K. 1986. New frond formation in *Laminaria hyperborea* (Phaeophyta): a photoperiodic response. *British Phycological Journal* 21: 269–73.

Lüning, K. 1988. Photoperiodic control of sorus formation in the brown alga *Laminaria saccharina*. *Marine Ecology Progress Series* 45: 137–44.

Lüning, K. 1990. *Seaweeds: Their Environment, Biogeography and Rcophysiology* 2nd edn. New York: Wiley-Interscience.

Lüning, K. 1991. Circannual growth rhythm in a brown alga, *Pterygophora californica*. *Acta Botanica* 104: 157–62.

Lüning, K. & Tom Dieck, I. 1989. Environmental triggers in algal seasonality. Environmental triggers in algal seasonality. *Botanica Marina* 32: 389–97.

Lüning, K. & Dring, M.J. 1972. Reproduction induced by blue light in female gametophytes of *Laminaria saccharina*. *Planta* 104: 252–6.

Lüning, K. & Dring, M.J. 1973. The influence of light quality on the development of the brown algae *Petalonia* and *Scytosiphon*. *British Phycological Journal* 8: 333–8.

Lüning, K. & Dring, M.J. 1975. Reproduction, growth and photosynthesis of gametophytes of *Laminaria saccharina* grown in blue and red light. *Marine Biology* 29: 195–200.

Lüning, K. & Kadel, P. 1993. Daylength range for circannual rhythmicity in *Pterygophora californica* (Alariaceae, Phaeophyta) and synchronization of seasonal growth by daylength cycles in several other brown algae. *Phycologia* 32: 379–87.

Lüning, K. & Markham, J.W. 1979. Morphogenetic responses of *Laminaria saccharina* sporophytes to red and blue light. *British Phycological Journal* 14: 125–6.

Lüning, K. & Müller, D.G. 1978. Chemical interaction in sexual reproduction of several Laminariales (Phaeophyceae): release and attraction of spermatozoids. *Zeitschrift für Pflanzenphysiologie* 89: 333–41.

Lüning, K. & Mortensen, L. 2015. European aquaculture of sugar kelp (*Saccharina latissima*) for food industries: iodine content and epiphytic animals as major problems. *Botanica Marina* 58: 449–55.

Lüning, K. & Neushul, M. 1978. Light and temperature demands for growth and reproduction of Laminarian gametophytes in Southern and Central California. *Marine Biology* 45: 297–309.

Lüning, K., Wagner, A. & Buchholz, C. 2000. Evidence for inhibitors of sporangium formation in *Laminaria digitata* (Phaeophyceae) during the season of rapid growth (Note). *Journal of Phycology* 36: 1129–34.

Luther, G. 1976. Bewuchsuntersuchungen auf Natursteinsubstraten im Gezeitenbereich des Nordsylter Wattenmeeres: Algen. *Helgoländer Wissenschaftliche Meeresuntersuchungen* 28: 318–51.

Luther, H. 1981. Occurrence and ecological requirements of *Fucus vesiculosus* in semi-enclosed inlets of the Archipelago Sea, SW Finland. *Annales Botanici Fennici* 18: 187–200.

Lyle, L. 1920. The marine algae of Guernsey. *Journal of Botany, British and Foreign* 58(Supplement): 1–53.

Lyngby, J.E. & Mortense, S.M. 1996. Effects of Dredging Activities on Growth of *Laminaria saccharina*. *Marine Ecology* 17: 345–54.

Lyngbye, H.C. 1819. *Tentamen hydrophytologiae danicae*: pp. [i]–xxxii, [1]–248, 70 pls. Hafniae

[Copenhagen]: typis Schultzianis, in commissis Librariae Gyldendaliae.

Lynn, M.J. 1935a. Rare algae from Strangford Lough Part I. *Irish Naturalists' Journal* 5: 1–8.

Lynn, M.J. 1935b. Rare algae from Strangford Lough Part II. *Irish Naturalists' Journal* 5: 275–83.

Lynn, M.J. 1937. Notes on the algae of the district of Whiterock, Strangford Lough. *Irish Naturalists' Journal* 6: 192–5.

Lynn, M.J. 1949. A rare alga from Larne Lough. *Irish Naturalists' Journal* 9: 301–5.

Lynn, M.J. 1960. Coastal Survey X (new series) southern end of Larne Lough, Co. Antrim. *Irish Naturalists' Journal* 13: 159–63.

MacFarlane, C. 1932. Observations on the annual growth of *Ascophyllum nodosum*. *Proceedings of the Nova Scotian Institute of Science*: 27–33.

Machalek, K.M., Davison, I.R. & Falkowski, P.G. 1996. Thermal acclimation and photoacclimation of photosynthesis in the brown alga *Laminaria saccharina*. *Plant, Cell and Environment* 19: 1005–16.

Madsen, T.V. & Maberly, S.C. 1990. A comparison of air and water as environments for photosynthesis by the intertidal alga *Fucus spiralis* (Phaeophyta). *Journal of Phycology* 26: 24–30.

Maegawa, M., Yokohama, Y. & Aruga, Y. 1987. Critical light conditions for young *Ecklonia cava* and *Eisenia bicyclis* with reference to photosynthesis. *Hydrobiologia* 151–2: 447–55.

Maggi, E., Bertocci, I., Vaselli, S. & Benedetti-Cecchi, L. 2011. Connell and Slatyer's models of succession in the biodiversity era. *Ecology* 92: 1399–406.

Maggs, C.A., Freamhainn, M.T. & Guiry, M.D. 1983. A study of the marine algae of subtidal cliffs in Lough Hyne (Ine), Co. Cork. *Proceedings of the Royal Irish Academy* 83B: 251–66.

Magnus, P. 1873. Zur Morphologie der Sphacelarien. *Gesellschaft Naturforschender Freunde zu Berlin*. 1873: 129–56.

Magnus, P. 1875. Die botanischen Ergebnisseder Nordseefahrt 1872. *Jahresberichte der Kommission zur Wissenschaftlichen Untersuchung der Deutschen Meere in Kiel* 2: 59–79.

Maier, I. 1984. Culture studies of *Chorda tomentosa* (Phaeophyta, Laminariales). *British Phycological Journal* 19: 95–106.

Maier, I. 1995. Brown algal pheromones. *Progress in Phycological Research* 11: 51–102.

Maier, I. 1997a. The fine structure of the male gamete of *Ectocarpus siliculosus* (Ectocarpales, Phaeophyceae). I. General structure of the cell. *European Journal of Phycology* 32: 241–53.

Maier, I. 1997b. The fine structure of the male gamete of *Ectocarpus siliculosus* (Ectocarpales, Phaeophyceae). II. The flagellar apparatus. *European Journal of Phycology* 32: 255–66.

Maier, I. & Müller, D.G. 1982. Antheridium fine structure and spermatozoid release in *Laminaria digitata* (Phaeophyceae). *Phycologia* 21: 1–8.

Maier, I. & Müller, D.G. 1984. Pheromone-triggered gamete release in *Chorda tomentosa*. *Naturwissenschaften* 71: 48–9.

Maier, I. & Müller, D.G. 1990. Chemotaxis in *Laminaria digitata* (Phaeophyceae). I. Analysis of spermatozoid movement. *Journal of Experimental Botany* 41: 869–76.

Maier, I., Müller, D.G. & Katsaros, C. 2002. Entry of the DNA virus, *Ectocarpus fasciculatus* virus type 1 (Phycodnaviridae), into host cell cytosol and nucleus. *Phycological Research* 50: 227–32.

Maier, I., Müller, D.G., Schmid, C., Boland, W. & Jaenicke, L. 1988. Pheromone receptor specificity and threshold concentrations for spermatozoid release in *Laminaria digitata*. *Naturwissenschaften* 75: 260–3.

Maier, I., Rometsch, E., Wolf, S., Kapp, M. & Müller, D.G. 1997. Passage of a marine brown algal DNA virus from *Ectocarpus fasciculatus* (Ectocarpales, Phaeophyceae) to *Myriotrichia clavaeformis* (Dictyosiphonales, Phaeophyceae): infection symptoms and recovery. *Journal of Phycology* 33: 838–44.

Maier, I. & Schmid, C.E. 1995. An immunofluorescence study on lectin binding sites in gametes of *Ectocarpus siliculosus* (Ectocarpales, Phaeophyceae). *Phycological Research* 43: 33–42.

Maier, I., Wenden, A. & Clayton, M.N. 1992. The movement of *Hormosira banksii* (Fucales, Phaeophyta) spermatozoids in response to sexual pheromone. *Journal of Experimental Botany* 43: 1651–7.

Maier, I., Müller, D.G., Gassman, G., Boland, W. & Jaenicke, L. 1987. Sexual pheromones and related egg secretions in Laminariales (Phaeophyta). *Zeitschrift für Naturforschung* 42e: 948–54.

Maier, I., Wolf, S., Delaroque, N., Müller, D.G. & Kawai, H. 1998. A DNA virus infecting the marine brown alga *Pilayella littoralis* (Ectocarpales, Phaeophyceae) in culture. *European Journal of Phycology* 33: 213–20.

Major, K.M. & Davison, I.R. 1998. Influence of temperature and light on growth and photosynthetic physiology of *Fucus evanescens* (Phaeophyta) embryos. *European Journal of Phycology* 33: 129–38.

Makarov, M.V. & Voskoboinikov, G.M. 2001. The influence of Ultraviolet-B radiation on spore release and growth of the kelp *Laminaria saccharina*. *Botanica Marina* 44: 89–94.

Malm, T., Engkvist, R. & Kautsky, L. 1999. Grazing effects of two freshwater snails on juvenile *Fucus vesiculosus* in the Baltic Sea. *Marine Ecology Progress Series* 188: 63–71.

Malm, T. & Kautsky, L. 2003. Differences in life-history characteristics are consistent with the vertical distribution pattern of *Fucus serratus* and *Fucus vesiculosus* (Fucales, Phaeophyceae) in the Central Baltic Sea. *Journal of Phycology* 39: 880–7.

Malm, T., Kautsky, L. & Claesson, T. 2003. The density and survival of *Fucus vesiculosus* L. (Fucales, Phaeophyta) on different bedrock types on a Baltic Sea moraine coast. *Botanica Marina* 46: 256–62.

Malm, T., Kautsky, L. & Engkvist, R. 2001. Reproduction, recruitment and geographical distribution of *Fucus serratus* L. in the Baltic Sea. *Botanica Marina* 44: 101–8.

Manton, I. 1957. Observations with the electron microscope on the internal structure of the zoospore of a brown alga (*Scytosiphon lomentarius*). *Journal of Experimental Botany* 8: 294–303.

Manton, I. 1959. Observations on the internal structure of the spermatozoid of *Dictyota*. *Journal of Experimental Botany* 10: 448–61.

Manton, I. 1964a. A contribution towards understanding of 'The Primitive Fucoid'. *The New Phycologist* 63: 244–52.

Manton, I. 1964b. The possible significance of some details of flagellar bases in plants. *Journal of the Royal Microscopical Society* 82: 279–85.

Manton, I. & Clarke, B. 1951a. An electron microscope study of the spermatozoid of *Fucus serratus*. *Annals of Botany* N.S. 15: 461–71.

Manton, I. & Clarke, B. 1951b. Electron microscope observations on the zoospores of *Pylaiella* and *Laminaria*. *Journal of Experimental Botany* 2: 242–6.

Manton, I. & Clarke, B. 1956. Observations with the electron microscope on the internal structure of the spermatozoid of *Fucus*. *Journal of Experimental Botany* 7: 416–32.

Manton, I., Clarke, B. & Greenwood, A.D. 1953. Further observations with the electron microscope on spermatozoids in the Brown algae. *Journal of Experimental Botany* 4: 319–29.

Margalet, J.L. & Navarro, M.J. 1990. Mapas de distribución de algas marinas de la Península Ibérica. I. *Bifurcaria bifurcata* R. Ross y *Halydris siliquosa* (L.) Lyngb. *Botanica Complutensis* 16: 99–107.

Margalet, J.L. & Navarro, M.J. 1992. Mapas de distribución de algas marinas de la Península Ibérica. II. *Ascophyllum nodosum* (L.) Le Jolis, *Pelvetia canaliculata* (L.) Decne. et Thur e *Himanthalia elongata* (L.) S.F. Gray. *Botanica Complutensis* 17: 117–32.

Margalet, J.L., Almaraz, T., Navarro, M.J. & Pérez-Ruzafa, I.M. 1993. Mapas de distribución de algas marinas de la Península Ibérica. III. *Fucus ceranoides* L., *F. serratus* L., *F. spiralis* L. y *F. vesiculosus* L. (Fucales, Fucophyceae). *Botanica Complutensis* 18: 267–90.

Markey, D.R. & Wilce, R.T. 1975. The ultrastructure of reproduction in the brown alga *Pylaiella littoralis* I. Mitosis and cytokinesis in the plurilocular gametangia. *Protoplasma* 85: 219–41.

Markey, D.R. & Wilce, R.T. 1976a. The ultrastructure of reproduction in the brown alga *Pylaiella littoralis*. II. Zoosporogenesis in the unilocular sporangia. *Protoplasma* 88: 147–73.

Markey, D.R. & Wilce, R.T. 1976b. The ultrastructure of reproduction in the brown alga *Pylaiella littoralis*. III. Later stages of gametogenesis in the plurilocular gametangia. *Protoplasma* 88: 175–86.

Maron, J.L. & Vila, M. 2001. When do herbivores affect plant invasion? Evidence for the natural enemies and biotic resistance hypotheses. *Oikos* 95: 361–73.

Marsden, W.J.N., Evans, L.V., Callow, J.A. & Keen, J.N. 1984. A preliminary electrophoretic comparison of *Fucus serratus* and *Fucus vesiculosus*. *Botanica Marina* 27: 79–83.

Marshall, W. 1960. An underwater study of the epiphytes of *Laminaria hyperborea* (Gunn.) Fosl. *British Phycological Bulletin* 2: 18–19.

Martens, G. von. 1868. *Die Tange. Die Preussische Expedition nach Ost-Asien. Nach amtlichen Quellen. Botanischer Theil.* pp. 1–152, pls I–VII. Berlin: Verlag de Königlichen Geheimen Ober-Hofbuchdruckerei (R.Y. Decker).

Martin, J.P. & Bastida, R. 2008. The invasive seaweed *Undaria pinnatifida* (Harvey) Suringar in Ria Deseado (southern Patagonia, Argentina): Sporophyte cycle and environmental factors determining its distribution. *Revista De Biologia Marina Y Oceanografia* 43: 335–44.

Martin, J.P. & Cuevas, J.M. 2006. First record of *Undaria pinnatifida* (Laminariales, Phaeophyta) in Southern Patagonia, Argentina. *Biological Invasions* 8: 1399–402.

Martín, M.J., Sansón, M. & Reyes, J. 2002. Morfología, anatomía y distribución de *Liebmannia leveillei* en las islas Canarias (Phaeophyceae, Chordariales). *Vieraea* 30: 87–98.

Martin, W.F., Jouannic, S. & Loiseaux-de Goër, S. 1993. Molecular phylogeny of the *atp*B and *atp*E genes of the brown alga *Pylaiella littoralis*. *European Journal of Phycology* 28: 111–13.

Martínez, B., Viejo, R.M., Calvo, S. & Carreño, F. 2009. Habitat model of *Himanthalia elongata*: Predictions in a warming up scenario. *Algas, Boletín Informativo de la Sociedad Española de Ficología* 42: 11.

Martínez, E.A. 1996. Micropopulation differenciation in phenol content and susceptibility to herbivory in the Chilean kelp *Lessonia nigrescens* (Phaeophyta, Laminariales). *Proceedings of the International Seaweed Symposium* 15: 205–11.

Martínez, E.A. 1999. Latitudinal differences in thermal tolerance among microscopic sporophytes of the kelp *Lessonia nigrescens* (Phaeophyta: Laminariales). *Pacific Science* 53: 74–81.

Martins, N., Tanttu, H., Pearson, G.A., Serrão, E.A. & Bartsch, I. 2017. Interactions of daylength, temperature and nutrients affect thresholds for life stage transitions in the kelp *Laminaria digitata* (Phaeophyceae). *Botanica Marina* 60: 109–21.

Masters, A.K., Sheeran, E., Shirras, A.D., Brownlee, C. & Hetherinton, A.M. 1991. *In vitro* translation of mRNA from the eggs of *Fucus serratus*. *British Phycological Journal* 26: 92.

Mathias, W.T. 1935. The life history and cytology of *Phloeospora brachiata* Born. *Publications of the Hartley Botanical Laboratories, University of Liverpool* 13: 1–24.

Mathieson, A.C. 1989. Phenological patterns of Northern New England Seaweeds. *Botanica Marina* 32: 419–38.

Mathieson, A.C. & Dawes, C.J. 2001. A muscoides-like *Fucus* from a Maine salt marsh: its origin, ecology and taxonomic implications. *Rhodora* 103(914): 172–201.

Mathieson, A.C. & Dawes, C.J. 2017. *Seaweeds of the Northwest Atlantic.* pp. [i]–x, 1–798, CIX pls. Amherst & Boston: University of Massachusetts Press.

Mathieson, A.C. & Hehre, E.J. 1986. A synopsis of New Hampshire seaweeds. *Rhodora* 88: 1–139.

Mathieson, A.C., Dawes, C.J., Anderson, M.L. & Hehre, E.J. 2001. Seaweeds of the Brave Boat Harbor salt marsh and adjacent open coast of southern Maine. *Rhodora* 103: 1–46.

Mathieson, A.C., Dawes, C.J., Green, L.A. & Traggis, H. 2016. Distribution and ecology of *Colpomenia peregrina* (Phaeophyceae) within the Northwest Atlantic. *Rhodora* 118: 276–305.

Mathieson, A.C., Dawes, C.J., Wallace, A.L. & Klein, A.S. 2006. Distribution, morphology, and genetic affinities of dwarf embedded *Fucus* populations from the Northwest Atlantic Ocean. *Botanica Marina* 49: 283–303.

Mathieson, A.C., Pedersen, J.R., Neefus, C.D., Dawes, C.J. & Bray, T.L. 2008. Multiple assessments of introduced seaweeds in the Northwest Atlantic. *ICES Journal of Marine Science* 65: 730–41.

Mathieson, A.C., Penniman, C.A., Busse, P.K. & Tveter-Gallagher, E. 1982. Effects of ice on *Ascophyllum nodosum* within the Great Bay estuary system of New Hampshire-Maine. *Journal of Phycology* 18: 331–6.

Mathieson, A.C., Swan, E.F. & Fralick, R.A. 1972. An abnormal specimen of *Laminaria digitata* (L.) Lamouroux. *Rhodora* 74: 127–30.

Matson, P.G. & Edwards, M.S. 2007. Effects of ocean temperature on the southern range limits of two understory kelps, *Pterygophora californica* and *Eisenia arborea*, at multiple life-stages. *Marine Biology* 151: 1941–9.

Matsumura, W., Yasui, H. & Yamamoto, H. 2001. Successful sporophyte regeneration from protoplasts of *Undaria pinnatifida* (Laminariales, Phaeophyceae). *Phycologia* 40: 10–20.

Mattio, L. & Payri, C.E. 2009. Taxonomic revision of *Sargassum* species (Fucales, Phaeophyceae) from New Caledonia based on morphological and molecular analyses. *Journal of Phycology* 45: 1374–88.

Mattio, L., Payri, C. & Stiger-Pouvreau, V. 2008. Taxonomic revision of *Sargassum* (Fucales, Phaeophyceae) from French Polynesia based on morphological and molecular analyses. *Journal of Phycology* 44: 1541–55.

Mattio, L., Payri, C. & Verlaque, M. 2009. Taxonomic revision and geographic distribution of subgen. *Sargassum* (Fucales, Phaeophyceae) in the western and central Pacific islands based on morphological and molecular analyses. *Journal of Phycology* 45: 12131227.

Mattio, L., Payri, C.E., Verlaque, M. & de Reviers, B. 2010. Taxonomic revision of *Sargassum* sect. *Acanthocarpicae* (Fucales, Phaeophyceae). *Taxon* 59: 896–904.

May, D.I. & Clayton, M.N. 1991. Oogenesis, the formation of oogonial stalks and fertilization in *Sargassum vestitum* (Fucales, Phaeophyta) in southern Australia. *Phycologia* 30: 243–56.

Mayes, C., Saunders, G.W., Tan, I.H. & Druehl, L.D. 1992. DNA extraction methods for kelp (Laminariales) tissue. *Journal of Phycology* 28: 712–16.

McCauley, L.A.R. & Wehr, J.D. 2007. Taxonomic reappraisal of the freshwater brown algae *Bodanella*, *Ectocarpus*, *Heribaudiella*, and *Pleurocladia* (Phaeophyceae) on the basis of *rbc*L sequences and morphological characters. *Phycologia* 46: 429–39.

McDevit, D.C. & Saunders, G.W. 2009. On the utility of DNA barcoding for species differentiation among brown macroalgae (Phaeophyceae) including a novel extraction protocol. *Phycological Research* 57: 131–41.

McDevit, D.C. & Saunders, G.W. 2010. A DNA barcode examination of the Laminariaceae (Phaeophyceae) in Canada reveals novel biogeographical and evolutionary insights. *Phycologia* 49: 235–48.

McDevit, D.C. & Saunders, G.W. 2017. A molecular investigation of Canadian Scytosiphonaceae (Phaeophyceae) including descriptions of *Planosiphon* gen. nov. and *Scytosiphon promiscuus* sp. nov. *Botany* 95: 653–71.

McDowell, R.E., Amsler, M.O., Li, Q., Lancaster, J.R., Jr & Amsler, C.D. 2015. The immediate wound-induced oxidative burst of *Saccharina latissima* depends on light via photosynthetic electron transport. *Journal of Phycology* 51: 431–41.

McLachlan, J. 1974. Effects of temperature and light on growth and development of embryos of *Fucus edentatus* and *F. distichus* ssp. *distichus*. *Canadian Journal of Botany* 52: 943–51.

McLachlan, J. & Bidwell, R.G.S. 1978. Photosynthesis of eggs, sperm, zygotes, and embryos of *Fucus serratus*. *Canadian Journal of Botany* 56: 371–3.

McLachlan, J. & Bidwell, R.G.S. 1983. Effects of colored light on the growth and metabolism of *Fucus* embryos and apices in culture. *Canadian Journal of Botany* 61: 1993–2003.

McLachlan, J., Greenwell, M., Bird, C.J. & Holmsgaard, J.E. 1987. Standing stocks of seaweeds of commercial importance on the north shore of Prince Edward Island, Canada. *Botanica Marina* 30: 277–89.

McLachlan, J., Chen, L.C.M. & Edelstein, T. 1971. The life history of *Microspongium* sp. *Phycologia* 10: 83–7.

McPeak, R.H. 1981. Fruiting in several species of Laminariales from southern California. *Proceedings of the 8th International Seaweed Symposium*, pp. 404–9.

Meeks, J.C. 1974. Chlorophylls. In: Stewart, W.D.P. (ed.), *Algal Physiology and Biochemistry*, pp. 161–75. Botanical Monographs 10. Blackwell Scientific Publications, Oxford.

Meijer, A.J.M. 1984. Literatuurstudie aan *Laminaria saccharina*. *Bureau Waardenburg Bv*: 1–107.

Méndez-Sandín, M. & Fernández, C. 2016. Changes in the structure and dynamics of marine assemblages dominated by *Bifurcaria bifurcata* and *Cystoseira* species over three decades (1977–2007). *Estuarine, Coastal and Shelf Science* 175: 46–56.

Meneghini, G. 1840. *Lettera del Prof. Giuseppe Meneghini al Dott. Iacob Corinaldi a Pisa*. pp. [1–4]. Pisa: Tipografia Prosperi.

Meneghini, G. 1841. Algologia dalmatica. *Atti della terza Riunione degli scienziati italiani tenuta in Firenze* 3: 424–31.

Meneghini, G. 1842a. *Alghe Italiane e Dalmatiche*. Vol. Fascicle I pp. 1–80. Padova [Padua]: Angelo Sicca.

Meneghini, G. 1842b. *Alghe Italiane e Dalmatiche*. Vol. Fascicle II pp. 81–160. Padova [Padua]: Angelo Sicca.

Meneghini, G. 1842c. *Alghe Italiane e Dalmatiche*. Vol. Fascicle III pp. 161–255. Padova [Padua]: Angelo Sicca.

Meneghini, G. 1842d. Monographia Nostochinearum italicarum addito specimine de Rivulariis. *Memorie della Reale Accademia delle Scienze di Torino, ser. 2* 5 (Cl. Sc. Fis. e Mat): 1–143, pls I–XVII.

Meneghini, G. 1843. *Alghe Italiane e Dalmatiche*. Vol. Fascicle IV pp. 256–352. Padova [Padua]: Angelo Sicca.

Meneghini, G. 1844. Observazioni su alcuni generi della famiglia delle Cordariee. *Giornale Botanico Italiano* 1: 291–5.

Menge, B.A. 1976. Organization of the New England rocky intertidal community: role of predation, competition, and environmental heterogeneity. *Ecological Monographs* 46: 355–93.

Menge, B.A. & Sutherland, J.P. 1976. Species diversity gradients: synthesis of the roles of predation, competition and temporal heterogeneity. *American Naturalist* 110: 351–69.

Meretta, P.E., Matula, C.V. & Casas, G. 2012. Occurrence of the alien kelp *Undaria pinnatifida* (Laminariales,

Phaeophyceae) in Mar del Plata, Argentina. *BioInvasions Records* 1: 59–63.

Micheli, F., Heiman, K.W., Kappel, C.V., Martone, R.G., Sethi, S.S., Osio, G.C., Fraschetti, S., Shelton, A.O. & Tanner, J.M. 2016. Impacts of natural and human disturbances on rocky shore communities. *Ocean & Coastal Management* 126: 42–50.

Migula, W. 1908–09. *Kryptogamen-Flora von Deutschland, Deutsch-Österreich und der Schweiz im Anschluß an Thomé's Flora von Deutschland. Band II. Algen. 2. Teil. Rhodophyceae, Phaeophyceae, Characeae.* Gera, R.: Verlag Friedrich von Zezschwitz.

Mikelstad, S. 1968. Ion-exchange properties of brown algae. I. Determination of ratio mechanism for calcium-hydrogen ion-exchange for particles for *Laminaria hyperbora* and *Laminaria digitata. Journal of Applied Chemistry* 18: 30–6.

Miller, A.W., Chang, A.L., Consentino-Manning, N. & Ruiz, G.M. 2004. A new record and eradication of the Northern Atlantic alga *Ascophyllum nodosum* (Phaeophyceae) from San Francisco Bay, California, USA. *Journal of Phycology* 40: 1028–31.

Miller, D.D. & Connell, M.U. 2012. Photoperiodic mediation of differential gene expression in *Scytosiphon lomentaria* (Phaeophyceae). *Phycological Research* 60: 105–13.

Miller, H.L., III, Neale, P.J. & Dunton, K.H. 2009. Biological weighting functions for UV inhibition of photosynthesis in the kelp *Laminaria hyperborea* Phaeophyceae). *Journal of Phycology* 45: 571–84.

Miller, S.L. & Vadas, R.L. 1984. The population biology of *Ascophyllum nodosum*: biological and physical factors affecting survivorship of germlings. *British Phycological Journal* 19: 198.

Minchin, D. 1992. Extensive grazing of the prosobranch *Lacuna vincta* (Montagu) on the kelp *Alaria esculenta* (L.) Grev. *Irish Naturalists' Journal* 24: 171.

Minchin, D. 2007. Aquaculture and transport in a changing environment: overlap and links in the spread of alien biota. *Marine Pollution Bulletin* 55: 302–13.

Minchin, D. & Nunn, J. 2014. The invasive brown alga *Undaria pinnatifida* (Harvey) Suringar, 1873 (Laminariales: Alariaceae), spreads northwards in Europe. *BioInvasions Records* 3: 57–63.

Minchin, D., Nunn, J., Murphy, J., Edwards, H. & Downie, A. 2017. Monitoring temporal changes in the early phase of an invasion: *Undaria pinnatifida* (Harvey) Suringar using the abundance and distribution range method. *Management of Biological Invasions* 8: 53–60.

Minchinton, T.E., Scheibling, R.E. & Hunt, H.L. 1997. Recovery of an intertidal assemblage following a rare occurrence of scouring by sea ice in Nova Scotia, Canada. *Botanica marina* 40: 139–48.

Miranda, F. 1931. Sobre las algas y cianofíceas del Cantábrico, especialmente de Gijón. *Trabajos del Museo de Ciencias Naturales. Serie Botánica* 25: 1–106.

Miranda, F. 1936. Nuevas localidades de algas de las costas septentrionales y occidentales de España y otras contribuciones ficológicas. *Boletín de la Real Sociedad Española de Historia Natural* 36: 367–81.

Misra, J.N. 1966. *Phaeophyceae in India.* pp. (x +) 203, 100 figs, 5 tables, 6 plates. New Delhi: Indian Council of Agricultural Research.

Mizuta, H., Kai, T., Tabuchi, K. & Yasui, H. 2007. Effects of light quality on the reproduction and morphology of sporophytes of *Laminaria japonica* (Phaeophyceae). *Journal of Applied Phycology* 38: 1323–9.

Moe, R.L. & Silva, P.C. 1977. Sporangia in the brown algal genus *Desmarestia* with special reference to Antarctic *D. ligulata. Bulletin of the Japanese Society for Phycology* 25: 159–67.

Mohring, M.B., Kendrick, G.A., Wernberg, T., Rule, M.J. & Vanderklift, M.A. 2013. Environmental Influences on Kelp Performance across the Reproductive Period: An Ecological Trade-Off between Gametophyte Survival and Growth? *PLoS One* 8: e65310.

Mohring, M., Wernberg, T., Kendrick, G. & Rule, M. 2013. Reproductive synchrony in a habitat-forming kelp and its relationship with environmental conditions. *Marine Biology* 160: 119–26.

Molinari-Novoa, E.A. & Guiry, M.D. (2020). Reinstatement of the genera *Gongolaria* Boehmer and *Ericaria* Stackhouse (*Sargassaceae*, Phaeophyceae). *Notulae Algarum* 172: 1–10.

Molis, M., Enge, A. & Karsten, U. 2010. Grazing impact of, and indirect interactions between mesograzers associated with kelp (*Laminaria digitata*). *Journal of Phycology* 46: 76–84.

Molloy, R.J. & Bolton, J.J. 1996. The Effects of Wave Exposure and Depth on the Morphology of Inshore Populations of the Namibian Kelp, *Laminaria schinzii* Foslie. B *Botanica Marina* 39: 525–32.

Molloy, F.J. & Hills, J.M. 1996. Long-term changes in heavy metal loadings of *Ascophyllum nodosum* from the Firth of Clyde, UK. *Proceedings of the International Seaweed Symposium* 15: 305–10.

Montagne, C. 1840. Plantae cellulares. In: Barker-Webb, P. & Berthelot, S. (eds), *Histoire naturelle des Iles Canaries.* Vol. 3, part 2, sect. ultima [4], pp. 137–92. Paris: Mellier.

Montagne, C. 1856. *Sylloge generum specierumque cryptogamarum*: pp. [i]–xxiv, [1]–498. Parisiis [Paris] & Londini [London]: sumptibus J.-B. Baillière …; H. Baillière.

Montagne, J.F.C. 1846. Flore d'Algérie. Ordo I. Phyceae Fries. In: *Exploration scientifique de l'Algérie pendant les années 1840, 1841, 1842… Sciences physiques. Botanique. Cryptogamie.* (Durieu De Maisonneuve, M.C., eds) Vol. 1, pp. 1–197. Paris: Imprimerie Royale, publiée par ordre du Gouvernement et avec le concours d'une Commission Académique.

Montecinos, A.E., Couceiro, L., Peters, A.F., Desrut, A., Valero, M. & Guillemin, M.L. 2017. Species delimitation and phylogeographic analyses in the *Ectocarpus* subgroup *siliculosi* (Ectocarpales, Phaeophyceae). *Journal of Phycology* 53: 17–31.

Monteiro, C.A., Engelen, A.H. & Santos, R.O.P. 2009a. Macro- and mesoherbivores prefer native seaweeds over the invasive brown seaweed *Sargassum muticum*: a potential regulating role on invasions. *Marine Biology* 156: 2505–15.

Monteiro, C., Engelen, A.H., Serrão, E.A. & Santos, R. 2009b. Habitat differences in the timing of reproduction of the invasive alga *Sargassum muticum* (Phaeophyta, Sargassaceae) over tidal and lunar cycles. *Journal of Phycology* 45: 1–7.

Moore, P.G. 1983. Biological interactions. In: Earll, R. & Erwin, D.G. (eds), *Sublittoral Ecology: The Ecology of* the Shallow Sublittoral Benthos, pp. 125–43. Clarendon Press, Oxford.

Morelissen, B., Dudley, B.D. & Phillips, N.E. 2016. Recruitment of the invasive kelp *Undaria pinnatifida* does not always benefit from disturbance to native algal communities in low-intertidal habitats. *Marine Biology* 163: 241.

Morelissen, B., Dudley, B.D., Geange, S.W. & Phillips, N.E. 2013. Gametophyte reproduction and development of *Undaria pinnatifida* under varied nutrient and irradiance conditions. *Journal of Experimental Marine Biology and Ecology* 448: 197–206.

Moris, G. & De Notaris, G. 1839. Florula caprariae sive enumeratio plantarum in insula Capraria vel sponte nascentium vel ad utilitatem latius excultarum. *Memorie della Reale Accademia delle Scienze di Torino, ser. 2*: 59–300, pls I–VI.

Morita, T., Kurashima, A. & Maegawa, M. 2003a. Temperature requirements for the growth and maturation of the gametophytes of *Undaria pinnatifida* and *U. undarioides* (Laminariales, Phaeophyceae). *Phycological Research* 51: 154–60.

Morita, T., Kurashima, A. & Maegawa, M. 2003b. Temperature requirements for the growth of young sporophytes of *Undaria pinnatifida* and *Undaria undarioides* (Laminariales, Phaeophyceae). *Phycological Research* 51: 266–70.

Morrell, S. & Farnham, W.F. 1981. The effects of mechanical clearance on Solent populations of *Sargassum muticum*. *British Phycological Journal* 16: 138.

Morrell, S.L. & Farnham, W.F. 1982. Some effects of substratum on *Sargassum muticum*. *British Phycological Journal* 17: 236–7.

Morrison, L., Baumann, H.A. & Stengel, D.B. 2008. An assessment of metal contamination along the Irish coast using the seaweed *Ascophyllum nodosum* (Fucales, Phaeophyceae). *Environmental Pollution* 152: 293–303.

Morton, O. 1994. *Marine Algae of Northern Ireland*. Belfast: Ulster Museum, Botanic Gardens, Belfast.

Morton, O. 2003. The marine macroalgae of County Donegal, Ireland. *Bulletin of the Irish Biogeographical Society* 27: 3–165.

Moss, B.L. 1950a. Studies in the genus *Fucus* II. The anatomical structure and chemical composition of receptacles of *Fucus vesiculosus* from three contrasting habitats. *Annals of Botany* 14: 395–410.

Moss, B.L. 1950b. Studies in the genus *Fucus* III. Structure and development of the attaching discs of *Fucus vesiculosus*. *Annals of Botany* 14: 411–19.

Moss, B. 1952. Variations in chemical composition during the development of *Himanthalia elongata* (L.) S.F. Gray. *Journal of the Marine Biological Association of the United Kingdom* 31: 29–34.

Moss, B. 1956. Apical cell of salt marsh fucoids. *Nature, London* 171: 371.

Moss, B. 1958a. Growth substances and the regeneration of adventitious branches of *Fucus vesiculosus*. *Proceedings of the International Seaweed Symposium* 3: 21.

Moss, B. 1958b. Observations on the development and cytology of *Himanthalia elongata* (L.) S.F. Gray. *British Phycological Bulletin* 1: 31–2.

Moss, B. 1964a. Growth and regeneration of *Fucus vesiculosus* in culture. *British Phycological Bulletin* 2: 377–80.

Moss, B. 1964b. Wound healing and regeneration in *Fucus vesiculosus* L. *Proceedings of the International Seaweed Symposium* 4: 117–22.

Moss, B. 1965. Apical dominance in *Fucus vesiculosus*. *New Phytologist* 64: 387–92.

Moss, B. 1966a. Apical dominance in *Fucus vesiculosus*. *British Phycological Bulletin* 3: 154.

Moss, B. 1966b. Apical dominance in *Fucus vesiculosus*. *Proceedings of the International Seaweed Symposium* 5: 179.

Moss, B. 1966c. Polarity and apical dominance in *Fucus vesiculosus*. *British Phycological Bulletin* 3: 31–5.

Moss, B. 1967a. The culture of fertile tissue of *Fucus vesiculosus*. *British Phycological Bulletin* 3: 209–12.

Moss, B. 1967b. The apical meristem of *Fucus*. *New Phytologist* 66: 67–74.

Moss, B. 1968. The transition from vegetative to fertile tissue in *Fucus vesiculosus*. *British Phycological Bulletin* 3: 567–73.

Moss, B. 1969. Apical meristems and growth control in *Himanthalia elongata* (S.F. Gray). *New Phytologist* 68: 387–97.

Moss, B. 1970. Meristems and growth control in *Ascophyllum nodosum* (L.) Le Jol. *New Phytologist* 69: 253–60.

Moss, B. 1971. Meristems and morphogenesis in *Ascophyllum nodosum* ecad *mackaii* (Cotton). *British Phycological Journal* 6: 187–93.

Moss, B. 1974. Attachment and germination of the zygotes of *Pelvetia canaliculata* (L.) Dcne. et Thur. (Phaeophyceae, Fucales). *Phycologia* 13: 317–22.

Moss, B. 1975. Attachment of zygotes and germlings of *Ascophyllum nodosum* (L.) Le. Jol. (Phaeophyceae, Fucales). *Phycologia* 14: 75–80.

Moss, B. 1977. Attachment of the holdfasts of *Laminaria digitata* (Huds.) Lamour. *British Phycological Journal* 12: 121.

Moss, B.L. 1982. The control of epiphytes by *Halidrys siliquosa* (L.) Lyngb. (Phaeophyta, Cystoseiraceae). *Phycologia* 21: 185–91.

Moss, B.L. & Elliot, E. 1957. Observations on the cytology of *Halidrys siliquosa* (L.) Lyngb. *Annals of Botany* 21: 143–51.

Moss, B. & Lacy, A. 1963. The development of *Halidrys siliquosa* (L.) Lyngb. *New Phytologist* 62: 67–74.

Moss, B., Mercer, S. & Sheader, A. 1973. Factors affecting the distribution of *Himanthalia elongata* (L.) S.F. Gray on the north-east coast of England. *Estuarine, Coastal and Shelf Science* 1: 233–43.

Moss, B. & Sheader, A. 1973. The effect of light and temperature upon the germination and growth of *Halidrys siliquosa* (L.) Lyngb. (Phaeophyceae, Fucales). *Phycologia* 12: 63–8.

Motomura, T. 1995. Premature chromosome condensation of the karyogamy-blocked sperm pronucleus in the fertilization of *Fucus distichus* (Fucales, Phaeophyceae). *Journal of Phycology* 31: 108–13.

Motomura, T. & Nagasato, C. 2009. Functional and non-functional spindle formation affecting mitosis and cytokinesis in *Fucus distichus* zygotes: the role of the centrosome. *Botanica Marina* 52: 140–9.

Motomura, T., Nagasato, C. & Kimura, K. 2010. Cytoplasmic inheritance of organelles in brown algae. *Journal of Plant Research* 123: 185–92.

Motomura, T., Kawaguchi, S. & Sakai, Y. 1985. Life history and ultrastructure of *Carpomitra cabrerae* (Clemente) Kützing (Phaeophyta, Sporochnales). *Japanese Journal of Phycology* 33: 21–31.

Moulin, P., Crépineau, F., Kloareg, B. & Boyen, C. 1999. Isolation and characterization of six cDNAs involved in carbon metabolism in *Laminaria digitata* (Phaeophyceae). *Journal of Phycology* 35: 1237–45.

Muhlin, J.F. & Brawley, S.H. 2009. Recent versus relic: discerning the genetic signature of *Fucus vesiculosus* (Heterokontophyta; Phaeophyceae) in the northwestern Atlantic. *Journal of Phycology* 45: 828–37.

Munda, I. 1964a. The influence of salinity on the chemical composition, growth and fructification of some Fucaceae. *Proceedings of the International Seaweed Symposium* 4: 123–6.

Munda, I. 1964b. The quantity and chemical composition of *Ascophyllum nodosum* (L) Le Jol. along the coast between the Rivers Ölfusá and Thjorsá (Southern Iceland). *Botanica Marina* 7: 76–89.

Munda, I. 1964c. Water and electrolyte exchange in the brown algae *Ascophyllum nodosum* (L.) Le Jol., *Fucus ceranoides* L. *Botanica Marina* 6: 158–88.

Munda, I. 1964d. A found of *Sorocarpus uvaeformis* on the western coast of Norway. *Nova Hedwigia* 7: 535–6.

Munda, I.M. 1964e. The quantity of alginic acid in some Adriatic brown algae. *Acta Adriatica* 11: 205–13.

Munda, I.M. 1979a. Additions to the check-list of benthic marine algae from Iceland. *Botanica Marina* 22: 459–63.

Munda, I.M. 1979b. A note on the ecology and growth forms of *Chordaria flagelliformis* (O.F. Müll.) C. Ag. of Icelandic waters. *Nova Hedwegia.* 31: 567–91.

Munda, I.M. 1986. Differences in heavy metal accumulation between vegetative parts of the thalli and receptacles in *Fucus spiralis* L. *Botanica Marina* 29: 341–9.

Munda, I.M. & Hudnik, V. 1986. Growth responses of *Fucus vesiculosus* to heavy metals, singly and in dual combinations, as related to accumulation. *Botanica Marina* 29: 401–12.

Munda, I.M. & Hudnik, V. 1988. The effects of Zn, Mn, and Co accumulation on growth and chemical composition of *Fucus vesiculosus* L. under different temperature and salinity conditions. *Pubblicazioni della Stazione Zoologica di Napoli Marine Ecology* 9: 213–25.

Munda, I.M. & Lüning, K. 1977. Growth performance of *Alaria esculenta* off Helgoland. *Helgoländer Wissenschaftliche Meeresuntersuchungen* 29: 311–14.

Muraoka, D. & Saitoh, K. 2005. Identification of *Undaria pinnatifida* and *Undaria undarioides* Laminariales, Phaeophyceae using mitochondrial 23S ribosomal DNA sequences. *Fisheries Science* 71: 1365–9.

Murúa, P., Küpper, F.C., Muñoz, L.A., Bernard, M. & Peters, A.F. 2018. *Microspongium alariae* in *Alaria esculenta*: a widely-distributed non-parasitic brown algal endophyte that shows cell modifications within its host. *Botanica Marina* 61: 343–54.

Myklestad, S. & Melsom, S. 1979. Heavy metal exchange by *Ascophyllum nodosum* (Phaeophyceae) plants

in situ. *Proceedings of the International Seaweed Symposium* 9: 143–51.

Müller, D.G. 1962. Über jahres – und lunarperiodische Erscheinungen bei einigen Braunalgen. *Botanica Marina* 4: 140–55.

Müller, D.G. 1963. Die Temperaturabhängigkeit der Sporangienbildung bei *Ectocarpus siliculosus* von verschiedenen standorten. *Pubblicazioni della Stazione Zoologica di Napoli* 33: 310–14.

Müller, D.G. 1964a. Die Beteiligung eines Berührungsreizes beim Festsetzen von Algenschwärmen auf dem Substrat. *Zeitschrift für Botanik* 52: 193–8.

Müller, D.G. 1964b. Life cycle of *Ectocarpus siliculosus* from Naples, Italy. *Nature* 203: 1402.

Müller, D.G. 1967. Generationswechsel, Kernphasenwechsel und Sexualität der Braunalge *Ectocarpus siliculosus* im Kulturversuch. *Planta* 75: 39–54.

Müller, D.G. 1969. Anisogamy in *Giffordia* (Ectocarpales). *Naturwissenschaften* 56: 220.

Müller, D.G. 1972a. Life cycle of the brown alga *Ectocarpus fasciculatus* var. *refractus* (Kütz.) Ardis. (Phaeophyceae, Ectocarpales) in culture. *Phycologia* 11: 11–14.

Müller, D.G. 1972b. Studies on reproduction in *Ectocarpus siliculosus*. *Mémoires de la Société Botanique de France* 1972: 87–98.

Müller, D. 1974. Sexual reproduction and isolation of a sex attractant in *Cutleria multifida* (Smith) Grev. (Phaeophyta). *Biochemie und Physiologie der Pflanzen* 165: 212–15.

Müller, D.G. 1975. Experimental evidence against sexual fusions of spores from unilocular sporangia of *Ectocarpus siliculosus* (Phaeophyta). *British Phycological Journal* 10: 315–21.

Müller, D.G. 1976a. Relative sexuality in *Ectocarpus siliculosus*. A scientific error. *Archives for Microbiology* 109: 89–94.

Müller, D.G. 1976b. Sexual isolation between a European and an American population of *Ectocarpus siliculosus* (Phaeophyta). *Journal of Phycology* 12: 252–4.

Müller, D.G. 1977. Sexual reproduction in British *Ectocarpus siliculosus* (Phaeophyta). *British Phycological Journal* 12: 131–6.

Müller, D.G. 1978. Locomotive responses of male gametes to the species specific sex attractant in *Ectocarpus siliculosus* (Phaeophyta). *Archiv für Protistenkunde* 120: 371–7.

Müller, D.G. 1979. Genetic affinity of *Ectocarpus siliculosus* (Dillw.) Lyngbye from the Mediterranean, North Atlantic and Australia. *Phycologia* 18: 312–18.

Müller, D.G. 1980. Development of *Ectocarpus siliculosus* (Phaeophyta). *Publikationen zu Wissenschaftlichen Filmen Biologie* No 13: 3–15.

Müller, D.G. 1981a. Culture studies on reproduction of *Spermatochnus paradoxus* (Phaeophyceae, Chordariales). *Journal of Phycology* 17: 384–9.

Müller, D.G. 1981b. Sexuality and sex attraction. In: Lobban, C.S. & Wynne, M.J. (eds), *The Biology of Seaweeds*, pp. 661–74. Blackwell, Oxford.

Müller, D.G. 1986. Apomeiosis in *Acinetospora* (Phaeophyceae, Ectocarpales). *Helgoländer Meeresuntersuchungen* 40: 219–24.

Müller, D.G. 1988. Studies on sexual compatibility between *Ectocarpus siliculosus* (Phaeophyceae) from Chile and the Mediterranean Sea. *Helgoländer Meeresuntersuchungen* 42: 469–76.

Müller, D.G. 1991a. Marine virioplankton produced by infected *Ectocarpus siliculosus* (Phaeophyceae. *Marine Ecology Progress Series* 76: 101–2.

Müller, D.G. 1991b. Mendelian segregation of a virus genome during host meiosis in the marine brown alga *Ectocarpus siliculosus*. *Journal of Plant Physiology* 37: 739–43.

Müller, S. & Clauss, H. 1976. Aspects of photomorphogenesis in the brown alga *Dictyota dichotoma*. *Zeitschrift für Pflanzenphysiologie* 78: 461–5.

Müller, D.G. & Eichenberger, W. 1994. Betaine lipid content and species delimitation in *Ectocarpus, Feldmannia* and *Hincksia* (Ectocarpales, Phaeophyceae). *European Journal of Phycology* 29: 219–25.

Müller, D.G. & Eichenberger, W. 1995. Note: Crossing experiments, lipid composition, and the species concept in *Ectocarpus siliculosus* and *E. fasciculatus* (Phaeophyceae, Ectocarpales). *Journal of Phycology* 31: 173–6.

Müller, D.G. & Eichenberger, W. 1997. Mendelian genetics in brown algae: inheritance of a lipid defect mutation and sex alleles in *Ectocarpus fasciculatus* (Ectocarpales, Phaeophyceae). *Phycologia* 36: 79–81.

Müller, D.G. & Falk, H. 1973. Flagellar structure of the gametes of *Ectocarpus siliculosus* (Phaeophyta) as revealed by negative staining. *Archiv für Hydrobiologie* 91: 313–22.

Müller, D.G. & Frenzer, K. 1993. Virus infections in three marine brown algae: *Feldmannia irregularis, F. simplex,* and *Ectocarpus siliculosus. Proceedings of the International Seaweed Symposium* 14: 37–44.

Müller, D.G. & Gassmann, G. 1978. Identification of the sex attractant in the marine brown alga *Fucus vesiculosus. Naturwissenschaften* 65: 389.

Müller, D.G. & Gassmann, G. 1980. Sexual hormone specificity in *Ectocarpus* and *Laminaria* (Phaeophyceae). *Naturwissenschaften* 67: 462.

Müller, D.G. & Jaenicke, L. 1973. Fucoserraten, the female sex attractant of *Fucus serratus* L. (Phaeophyta). *F.E.B.S. Lett.* 30: 127–39.

Müller, D.G. & Kawai, H. 1991. Sexual reproduction of *Ectocarpus siliculosus* (Ectocarpales, Phaeophyceae) in Japan. *Japanese Journal of Phycology* 39: 151–5.

Müller, D.G. & Lüthe, N.M. 1981. Hormonal interaction in sexual reproduction of *Desmarestia aculeata* (Phaeophyceae). *British Phycological Journal* 16: 351–8.

Müller, D.G. & Meel, H. 1982. Culture studies on the life history of *Arthrocladia villosa* (Desmarestiales, Phaeophyceae). *British Phycological Journal* 17: 419–25.

Müller, D.G. & Seferiadis, K. 1977. Specificity of sexual chemotaxis in *Fucus serratus* and *Fucus vesiculosus* (Phaeophyceae). *Zeitschrift für Pflanzenphysiologie* 84: 85–94.

Müller, D.G. & Schmid, C.E. 1988. Qualitative and quantitative determination of pheromone secretion in female gametes of *Ectocarpus siliculosus. Biological Chemistry Hoppe-Seyler* 369: 647–53.

Müller, D.G. & Schmidt, C.U. 1988. Culture studies on the life history of *Elachista stellaris* Aresch. (Phaeophyceae, Chordariales). *British Phycological Journal* 23: 153–8.

Müller, D.G. & Seferiadis, K. 1977. Specificity of sexual chemotaxis in *Fucus serratus* and *Fucus vesiculosus*

(Phaeophyceae). *Zeitschrift für Pflanzenphysiologie* 84: 85–94.

Müller, D.G. & Stache, B. 1989. Life history studies on *Pilayella littoralis* (L.) Kjellman (Phaeophyceae, Ectocarpales) of different geographical origin. *Botanica Marina* 32: 71–8.

Müller, D.G., Bräutigam, M. & Knippers, R. 1996. Virus infection and persistence of foreign DNA in the marine brown alga *Feldmannia simplex* (Ectocarpales, Phaeophyceae). *Phycologia* 35: 61–3.

Müller, D.G., Gassmann, G. & Lüning, K. 1979. Isolation of spermatozoid-releasing and -attracting substance from female gametophytes of *Laminaria digitata. Nature, London* 279: 430–1.

Müller, D.G., Maier, I. & Gassmann, G. 1985. Survey on sexual pheromone specificity in Laminariales (Phaeophyceae). *Phycologia* 24: 475–7.

Müller, D.G., Kawai, H., Stache, B. & Lanka, S. 1990. A virus infection in the marine brown alga *Ectocarpus siliculosus* (Phaeophyceae). *Botanica Acta* 103: 72–82.

Müller, D.G., Gassmann, G., Boland, W., Marner, F. & Jaenicke, L. 1981a. *Dictyota dichotoma* (Phaeophyceae): identification of the sperm attractant. *Science* 212: 1040–1.

Müller, D.G., Gassmann, G., Marner, F.J., Boland, W. & Jaenicke, L. 1982. The sperm attractant of the marine brown alga *Ascophyllum nodosum* (Phaeophyceae). *Science* 218: 1119–20.

Müller, D.G., Marner, F.J., Boland, W., Jaenicke, L. & Gassmann, G. 1981b. Identification of a volatile gamete secretion in *Spermatochnus paradoxus. Naturwissenschaften* 68: 478–80.

Müller, D.G., Parodi, E.R. & Peters, A.F. 1998. *Asterocladon lobatum* gen. et sp. nov., a new brown alga with stellate chloroplast arrangement, and its systematic position judged from nuclear rDNA sequences. *Phycologia* 37: 425–32.

Müller, D.G., Jaenicke, L., Donike, M. & Akintobi, T. 1971. Sex attractant in a brown alga: chemical structure. *Science* 171: 815–17.

Müller, O.F. 1771–82. *Flora Danica* 4 & 5. Havniae.

Müller, R., Wiencke, C. & Bischof, K. 2008. Interactive effects of UV radiation and temperature on microstages of Laminariales (Phaeophyceae) from the Arctic and North. *Journal of Sea Climate Research* 37: 203–13.

Müller, S. & Clauss, H. 1976. Aspects of photomorphogenesis in the brown alga *Dictyota dichotoma*. *Zeitschrift für Pflanzenphysiologie* 78: 461–5.

Munda, I. 1964. The influence of salinity on the chemical composition, growth and fructification of some Fucaceae. *Proceedings of the International Seaweed Symposium* 4: 123–6.

Munda, I.M. 1967a. Observations on the benthic marine algae in a land-locked fjord (Nordàsvatnet) near Bergen, western Norway. *Nova Hedwigia* 14: 519–48.

Munda, I.M. 1967b. Changes in the algal vegetation of a part of the Deltaic area in the southern Netherlands (Veerse Meer) after its closure. *Botanica Marina* 10: 141–57.

Munda, I.M. 1977. Combined effects of temperature and salinity on growth rates of germlings of three *Fucus* species from Iceland, Helgoland and the North Adriatic Sea. *Helgoländer Wissenschaftliche Meeresuntersuchungen* 29: 302–10.

Munda, I.M. 1978. Salinity dependent distribution of benthic algae in estuarine areas of Icelandic fjords. *Botanica Marina* 21: 451–68.

Munda, I.M. 1979. A note on the ecology and growth forms of *Chordaria flagelliformis* (O.F. Muell.) C. Ag. in Icelandic waters. *Nova Hedwigia* 31: 567–91.

Murase, N., Kito, H., Mizukami, Y. & Maegawa, M. 2002. Relationships between critical photon irradiance for growth and daily compensation point of juvenile *Sargassum macrocarpum. Fisheries Science* 66: 1032–8.

Murphy, J.T., Johnson, M.P. & Viard, F. 2017. A theoretical examination of environmental effects on the life cycle schedule and range limits of the invasive seaweed *Undaria pinnatifida. Biological Invasions* 19: 691–702.

Murphy, J.T., Voisin, M., Johnson, M. & Viard, F. 2016. Abundance and recruitment data for *Undaria pinnatifida* in Brest harbour, France: Model versus field results. *Data in Brief* 7: 540–5.

Murúa, P., Küpper, F.C., Muñoz, L.A., Bernard, M. & Peters, A.F. 2018. *Microspongium alariae* in *Alaria esculenta*: a widely-distributed non-parasitic brown algal endophyte that shows cell modifications within its host. *Botanica Marina* 61: 343–54.

Na, Y.J., Jeon, D.V., Han, S.J., Maranguy, C.A.O., An, D.S., Cha, H.K., Lee, J.B., Yang, J.H., Lee, H.W.L. & Choi, H.G. 2016. Crossed effects of light and temperature on the growth and maturation of gametophytes in *Costaria costata* and *Undaria pinnatifida. Korean Journal of Fisheries and Aquatic Sciences* 49: 190–7.

Nagai, M. 1940. Marine algae of the Kurile Islands, I. *Journal of the Faculty of Agriculture, Hokkaido Imperial University* 46: 1–137.

Nagasato, C., Motomura, T. & Ichimura, T. 2000. Parthenogenesis and abnormal mitosis in unfertilized eggs of *Fucus distichus* (Fucales, Phaeophyceae) (Research Note). *Phycologia* 39: 163–6.

Nagasato, C., Terauchi, M., Tanaka, A. & Motomura, T. 2015. Development and function of plasmodesmata in zygotes of *Fucus distichus. Botanica Marina* 58: 229–38.

Nägeli, C. 1847. Die neuern Algensysteme. *Neue Denkschr. allg. schweiz. Ges. ges. Naturw. 9* (unnumbered art. no. 2): 1–275.

Nägeli, M. 1932. On a new species of *Sphacelaria. Transactions of the Sapporo Natural History Society* 12: 142–7.

Nakahara, H. 1984. Alternation of generations of some brown algae in unialgal and axenic cultures. *Scientific Papers of the Institute of Algological Research, Faculty of Science, Hokkaido Imperial University* 7: 77–194.

Nakahara, H. & Nakamura, Y. 1971. The life history of *Desmarestia tabacoides* Okamura. *Botanical Magazine, Tokyo* 84: 69–75.

Nakamura, Y. 1965. Development of zoospores in *Ralfsia*-like thallus, with special reference to the life cycle of the Scytosiphonales. *Botanical Magazine, Tokyo* 78: 109–10.

Nakamura, Y. 1972. A proposal on the classification of the Phaeophyta. In: Abbott, I.A. and Kurogi, M. (eds), *Contributions to the Systematics of Marine Algae of the North Pacific*, pp. 147–56. Japanese Society of Phycology. Kobe.

Nakamura, Y. & Tatewaki, M. 1975. The life history of some species of Scytosiphonales. *Scientific Papers of the Institute of Algological Research, Faculty of Science, Hokkaido Imperial University* 6: 57–93.

Nanba, N., Fujiwara, T., Kuwano, K., Ishikawa, Y. & Kado, R. 2013. Morphological variation of local *Undaria pinnatifida* strains (Laminariales, Phaeophyceae) from Iwate Prefecture, northeastern Japan. *Sessile Organisms* 30: 37–41 (in Japanese with English abstract).

Nanba, N., Fujiwara, T., Kuwano, K., Ishikawa, Y., Ogawa, H. & Kado, R. 2011. Effect of water flow velocity on growth and morphology of cultured *Undaria pinnatifida* sporophytes (Laminariales, Phaeophyceae) in Okirai Bay on the Sanriku coast, northeast Japan. *Journal of Applied Phycology* 23: 1023–30.

Nardo, G.D. 1834. De novo genere *Stifftia* noncupando. *Isis*, Jena. Heft VI & VII, col. 677–8. Leipzig.

Nardo, G.D. 1835. *Considerazioni generali sulle Alghe* Loro caratteri, classificazione composizione chimica e applicazioni alla medicina, all'arti, all'agricoltura ec. seguite da brevi cenni storici sui progressi dell'algologia e da due articoli riguardanti la formazione degle algaei vivi e secchi, ec. e da un'Appendice sulle Alighe. Lette al Veneto ateneo il giorno 3 agosto 1835. pp. 1–46. Venezia [Venice]: Dai tipi di Guiseppe Antonelli.

Nardo, G.D. 1841. Nuovo osservazioni sulla struttura, abituoline e valore dei generi *Stifftia, Hildenbrandtia* et *Agardhina. 2nd Ruin. Sci. Ital* Torino.

Nasr, A.H. 1941. Some new and little known algae from the Red Sea. *Revue Algologique* 12: 57–76.

Naylor, G.L. 1928. Some observations on free-growing fucoids. *New Phytologist* 27: 61–8.

Naylor, G.L. 1936. The Fucoids of St. John's Lake, Plymouth, etc. *Revue Algologique* 7: 425–39.

Naylor, M. 1951. The structure and development of *Himanthalia lorea* (L.) Lyngb. *Annals of Botany* 15: 501–33.

Naylor, M. 1958a. Observations on the taxonomy of the genus *Stictyosiphon* Kütz. *Revue Algologique, Nouvelle Serie* 4: 1–24.

Naylor, M. 1958b. Some aspects of the life history and cytology of *Stictyosiphon tortilis* (Rupr.) Reinke. *Acta Adriatica* 8: 1–22.

Naylor, M. 1958c. The cytology of *Halidrys siliquosa* (L.) Lyngb. *Annals of Botany* 22: 205–17.

Nees, C.G. 1820. *Horae physicae Berolinenses* collectae ex symbolis virorum doctorum H. Linkii …; edicuravit Christianus Godof. Nees ab Esenbeck. pp. [i–xii], 1–123, [4], 27 pls. Bonnae [Bonn]: Sumtibus Adolphi Marcus.

Neiva, J.N., Pearson, G.A., Valero, M. & Serrão, E.A. 2010. Surfing the wave on a borrowed board: range expansion and spread of introgressed organellar genomes in the seaweed *Fucus ceranoides* L. *Molecular Ecology* 19: 4812–22.

Neiva, J., Pearson, G.A., Valero, M. & Serrão, E.A. 2012a. Fine-scale genetic breaks driven by historical range dynamics and ongoing density-barrier effects in the estuarine seaweed *Fucus ceranoides* L. *BMC Evol. Biol.* 12 Article No. 78.

Neiva, J., Hansen, G.I., Pearson, G.A., Van de Vliet, S., Maggs, C. & Serrão, E.A. 2012b. *Fucus cottonii*

(Fucales, Phaeophyceae) is not a single genetic entity but a convergent salt-marsh morphotype with multiple independent origins. *European Journal of Phycology* 47: 461–8.

Nelson, W.A. 1982. A critical review of the Ralfsiales, Ralfsiaceae and the taxonomic position of *Analipus japonicus* (Harv.) Wynne (Phaeophyta). *British Phycological Journal* 17: 311–20.

Nelson, W.A. 2013. *New Zealand Seaweeds: An Illustrated Guide*. Wellington: Te Papa Press.

Nelson, W.A. 2005. Life history and growth in culture of an endemic New Zealand kelp *Lessonia variegata* J. Agardh in response to different regimes of temperature, photoperiod and light. *Journal of Applied Phycology* 17: 23–8.

Nelson, W.A. & Maggs, C.A. 1996. Records of adventive marine algae in New Zealand: *Antithamnionella ternifolia*, *Polysiphonia senticulosa* (Ceramiales, Rhodophyta) and *Striaria attenuata* (Dictyosiphonales, Phaeophyta). *New Zealand Journal of Marine and Freshwater Research* 30: 449–53.

Neushul, M. 1972. Functional interpretation of benthic marine algal morphology. In: Abbott, I.A. & Kurogi, M. (eds), *Contributions to the Systematics of Benthic Marine Algae of the North Pacific*, pp. 47–73. Japanese Society of Phycology, Kobe, Japan.

Neushul, M., Foster, M.S., Coon, D.A., Woessner, J.W. & Harger, B.W.W. 1976. An *in situ* study of recruitment, growth, and survival of subtidal marine algae: techniques and preliminary results. *Journal of Phycology* 12: 397–408.

Newton, L. 1931. *A Handbook of the British Seaweeds*. British Museum (Natural History), London.

Ni-Ni-Win, Hanyuda, T., Arai, S., Uchimura, M., Prathep, A., Draisma, S.G.A., Phang, S.M., Abbott, I.A., Millar, A.J.K. & Kawai, H. 2011. A taxonomic study of the genus *Padina* (Dictyotales, Phaeophyceae) including the description of four new species from Japan, Hawaii, and the Andaman Sea. *Journal of Phycology* 47: 1193–209.

Nicholson, N., Hosmer, H., Bird, K., Hart, L., Sandlin, W., Shoemaker, C. & Sloan, C. 1981. The biology of *Sargassum muticum* (Yendo) Fensholt at Santa Catalina Island, California. *Proceedings of the International Seaweed Symposium* 8: 416–24.

Niell, F.X. 1984. Variación estacional de la concentración de clorofila *a* en los tejidos de tres macrófitos intermareales de la ría de Vigo (NO de España): *Himanthalia elongata* (L.) S.F. Gray, *Saccorhiza polyschides* (Lightf.) Batt. y *Laminaria ochroleuca* Pyl. *Investigación Pesquera* 48: 71–102.

Niell, F.X. & Soneira, A. 1976. On the biology of *Ascophyllum nodosum* (L.) in Galicia. Standing crop in the Vigo Bay. *Investigación Pesquera* 40: 105–10.

Niell, F.X., Jiménez, C. & Fernández, J.A. 1987. The forms of *Fucus spiralis* L. in the Canary Islands: discriminant and canonical analysis applied to define a new form. *Botanica Marina* 30: 27–32.

Niell, F.X., Miranda, A. & Pazó, J.P. 1980. Studies on the morphology of the Megaecade *limicola* of *Fucus vesiculosus* L. with taxonomical comments. *Botanica Marina* 23: 303–7.

Nielsen, M.M., Kumar, J.P., Soler-Vila, A., Johnson, M.P. & Bruhn, A. 2016. Early stage growth responses of *Saccharina latissima* spores and gametophytes. Part 1: inclusion of different phosphorus regimes. *Journal of Applied Phycology* 28: 387–93.

Nielsen, M.M., Paulino, C., Neiva, J., Krause-Jensen, D., Bruhn, A. & Serrão, E.A. 2016. Genetic diversity of *Saccharina latissima* (Phaeophyceae) along a salinity gradient in the North Sea-Baltic transition zone. *Journal of Phycology* 52: 523–31.

Nielsen, R. & Gunnarsson, K. 2001. Seaweeds of the Faroe Islands: an annotated checklist. *Fródskaparrit* 49: 45–108.

Niemeck, R.A. & Mathieson, A.C. 1976. An ecological study of *Fucus spiralis* L. *Journal of Experimental Marine Biology and Ecology* 24: 33–48.

Nienburg, W. 1923. Zur Entwicklungsgeschichte der Helgolander *Haplospora*. *Berichte der Deutschen Botanischen Gesellschaft* 41: 211–17.

Nienhuis, P.H. 1969. The significance of the substratum for intertidal algal growth on the artificial rocky shore of the Netherlands. *Internationale Revue der gesamten Hydrobiologie und Hydrographie* 54: 207–15.

Nienhuis, P.H. 1982. Attached *Sargassum muticum* found in the south-west Netherlands. *Aquatic Botany* 12: 189–95.

Nieuwland, J.A. 1917. Critical notes on new and old genera of plants, X. *American Midland Naturalist* 5: 50–2.

Nigan, N., Dethier, M. & Mumford, T. 2014. Response of haptera growth to different frequencies of light in deep water *Agarum fimbriatum* and shallow water *Alaria marginata*. Student Research paper, Friday Harbor Laboratories.

Nitschke, U., Walsh, P., McDaid, J. & Stengel, D.B. 2018. Variability in iodine in temperate seaweeds and iodine accumulation kinetics of *Fucus vesiculosus* and *Laminaria digitata* (Phaeophyceae, Ochrophyta). *Journal of Phycology* 54: 114–25.

Niwa, K. 2015. Experimental cultivation of *Undaria pinnatifida* for double cropping in *Pyropia* farms around the Akashi Strait, Hyogo Prefecture. *Japanese Journal of Phycology* 63: 90–7 (in Japanese with English abstract).

Niwa, K. 2016. Seedling production of *Undaria pinnatifida* using free-living gametophytes in a large indoor tank. *Aquaculture Research* 64: 173–82 (in Japanese with English abstract).

Niwa, K. & Harada, K. 2016. Experiment on forcing cultivation of *Undaria pinnatifida* sporophytes in the Seto Inland Sea by using free-living gametophytes cultured in laboratory. *Japanese Journal of Phycology* 64: 10–18 (in Japanese with English abstract).

Nizamuddin, M. 1981. Contribution to the marine algae of Libya Dictyotales. *Bibliotheca Phycologica* 54: 1–122.

Noda, M. 1969. The species of Phaeophyta from Sado Island in the Japan Sea. *Science Reports of Niigata University Series D (Biology)* 6: 1–64.

Noda, M. 1970. Some marine algae collected on the coast of Iwagasaki, Prov. Echigo facing the Japan Sea. *Science Reports of Niigata University, Series D (Biology)* 7: 27–35.

Noda, M. 1973. Some marine algae collected on the coast of Kashiwazaki Province facing the Japan Sea (2). *Scientific Reports Niigata University, Ser. D. (Biology)* 10: 1–10.

Norris, J.N. 2010. Marine algae of the Northern Gulf of California: Chlorophyta and Phaeophyceae. *Smithsonian Contributions to Botany* 94: i–x, 1–276.

Norris, R.E. & Conway, E. 1974. *Fucus spiralis* L. in the northeast Pacific. *Syesis* 7: 79–81.

Norton, T.A. 1968. Underwater observations on the vertical distribution of algae at St Mary's, Isles of Scilly. *British Phycological Bulletin* 3: 585–8.

Norton, T.A. 1969. Growth form and environment in *Saccorhiza polyschides*. *Journal of the Marine Biological Association of the United Kingdom* 49: 1025–45.

Norton, T.A. 1970a. Synopsis of biological data on *Saccorhiza polyschides*. *FAO Fisheries Synopsis* No 83: 11–93.

Norton, T.A. 1970b. The marine algae of County Wexford, Ireland. *British Phycological Journal* 5: 257–66.

Norton, T.A. 1973. Orientated growth of *Membranipora membranacea* (L.) on the thallus of *Saccorhiza polyschides* (Lightf.) Batt. *Journal of Experimental Marine Biology and Ecology* 13: 91–5.

Norton, T.A. 1976. Why is *Sargassum muticum* so invasive? *British Phycological Journal* 11: 197–8.

Norton, T.A. 1977a. Experiments on the factors influencing the geographical distributions of *Saccorhiza polyschides* and *Saccorhiza dermatodea*. *New Phytologist* 78: 625–35.

Norton, T.A. 1977b. The growth and development of *Sargassum muticum*. *Journal of Experimental Marine Biology and Ecology* 26: 41–53.

Norton, T.A. 1978. The factors influencing the distribution of *Saccorhiza polyschides* in the region of Lough Ine. *Journal of the Marine Biological Association of the United Kingdom* 58: 527–36.

Norton, T.A. 1980. Sink, swim or stick: the fate of *Sargassum muticum* propagules. *British Phycological Journal* 15: 197–8.

Norton, T.A. 1981a. Gamete expulsion and release in *Sargassum muticum*. *Botanica Marina* 24: 465–70.

Norton, T.A. 1981b. *Sargassum muticum* on the Pacific coast of North America. *Proceedings of the International Seaweed Symposium* 8: 449–56.

Norton, T.A. 1992. Dispersal by macroalgae. *British Phycological Journal* 27: 293–301.

Norton, T.A., Ed. 2003. *Out of the Past. Collected reviews to celebrate the Jubilee of the British Phycological Society*. pp. 180. British Phycological Society.

Norton, T.A. & Burrows, E.M. 1969a. Studies on marine algae of the British Isles 7. *Saccorhiza polyschides* (Lightf.) Batt. *British Phycological Journal* 4: 19–53.

Norton, T.A. & Burrows, E.M. 1969b. The environmental control of the seasonal development of *Saccorhiza polyschides* (Lightf.) Batt. *Proceedings of the International Seaweed Symposium* 6: 287–96.

Norton, T.A. & Deysher, L.E. 1989. The reproductive ecology of *Sargassum muticum* at different latitudes. In: Ryland, J.S. & Tyler, P.A. (eds), *Reproduction, Genetics and Distributions of Marine Organisms*, pp. 147–52. 23rd European Marine Biology Symposium, School of Biological Sciences, University of Wales, Swansea, 5–9 September 1988. International Symposium Series.

Norton, T.A. & Fetter, R. 1981. The settlement of *Sargassum muticum* propagules in stationary and flowing water. *Journal of the Marine Biological Association of the United Kingdom* 61: 929–40.

Norton, T.A., Hiscock, K. & Kitching, J.A. 1977. The ecology of Lough Ine. XX. The *Laminaria* forest at Carrigathorna. *Journal of Ecology* 65: 919–41.

Norton, T.A., Mathieson, A.C. & Neushul, M. 1981. Morphology and environment. In: Lobban, C. & Wynne, M.J. (eds), *The Biology of Seaweeds*, pp. 421–51. Botanical Monographs 17. Blackwell Scientific Publications, Oxford.

Norton, T.A. & Mathieson, A.C. 1983. The biology of unattached seaweeds. In: Round, F.E. & Chapman, A.R.O. (eds), *Progress in Phycological Research Vol. 2*, pp. 333–86. Elsevier Science Publishers.

Norum, E. 1913. Brunalger fra Haugesund og omegn. *Nyt Magazin for Naturvidenskaberne* 51: 131–60.

Novaczek, I., Bird, C.J. & McLachlan, J. 1986a. The effect of temperature on development and reproduction in *Chorda filum* and *C. tomentosa* (Phaeophyta, Laminariales) from Nova Scotia. *Canadian Journal of Botany* 64: 2414–20.

Novaczek, I., Bird, C.J. & McLachlan, J.L. 1986b. Culture and field study of *Stilophora rhizodes* (Phaeophyceae, Chordariales) from Nova Scotia, Canada. *British Phycological Journal* 21: 407–16.

Novaczek, I. & McLachlan, J. 1987. Correlation of temperature and daylength response of *Sphaerotrichia divaricata* (Phaeophyta, Chordariales) with field phenology in Nova Scotia and distribution in eastern North America. *British Phycological Journal* 22: 215–19.

Novaczek, I., Breeman, A.M. & van den Hoek, C. 1989. Thermal tolerance of *Stypocaulon scoparium* (Phaeophyta, Sphacelariales) from eastern and western shores of the North Atlantic Ocean. *Helgoländer Meeresuntersuchungen* 43: 183–93.

Nultsch, W., Rüffer, U. & Pfau, J. 1984. Circadian rhythms in the chromatophore movements of *Dictyota dichotoma*. *Marine Biology, Berlin* 81: 217–22.

Nygard, C.A. & Dring, M.J. 2008. Influence of salinity, temperature, dissolved inorganic carbon and nutrient concentration on the photosynthesis and growth of *Fucus vesiculosus* from the Baltic and Irish Seas. *European Journal of Phycology* 43: 253–62.

Nygard, C.A. & Ekelund, N.G.A. 2007. Photosynthesis and UV-B tolerance of the marine alga *Fucus vesiculosus* at different sea water salinities. *Proceedings of the International Seaweed Symposium* 18: 235–41.

Nygren, S. 1975a. Life history of some Phaeophyceae from Sweden. *Botanica Marina* 18: 131–41.

Nygren, S. 1975b. Influence of salinity on the growth and distribution of some Phaeophyceae on the Swedish west coast. *Botanica Marina* 8: 143–7.

Nygren, S. 1979. Life histories and chromosome numbers in some Phaeophyceae from Sweden. *Botanica Marina* 22: 371–3.

Nylund, G.M.N., Pereyra, R.T., Wood, H.L. & Johannesson, K. 2012. Increased resistance towards generalist herbivory in the new range of a habitat-forming seaweed. *Ecosphere* 3: 1–13.

Oak, J.H. 2010. Fucales. In: Anon. (ed.), *Algal flora of Korea. Volume 2, Number 2. Heterokontophyta: Phaeophyceae: Ishigeales, Dictyotales, Desmarestiales, Sphacelariales, Cutleriales, Ralfsiales, Laminariales*, pp.

111–72. Incheon: National Institute of Biological Resources.

Oak, J.H., Suh, Y. & Lee, I.K. 2002. Phylogenetic relationships of *Sargassum* subgenus *Bactrophycus* (Sargassaceae, Phaeophyceae) inferred from rDNA ITS sequences. *Algae* 17: 235–47.

Oates, B.R. 1985. Photosynthesis and amelioration of desiccation in the intertidal saccate alga *Colpomenia peregrina*. *Marine Biology, Berlin* 89: 109–19.

Oates, B.R. 1988. Water relations of the intertidal saccate alga *Colpomenia peregrina* (Phaeophyta, Scytosiphonales). *Botanica Marina* 31: 57–63.

Oates, B.R. 1989. Articulated coralline algae as a refuge for the intertidal saccate species, *Colpomenia peregrina* and *Leathesia difformis* in Southern California. *Botanica Marina* 32: 475–78.

Ogata, H. 1990. Effects of temperature on the maturation and early development of *Sargassum muticum* (Yendo) Fensholt, Phaeophyta. In: *Proceedings of the Second Asian Fisheries Forum, Tokyo, Japan, 1989*, pp. 357–60.

Oh, S.H. & Koh, C.H. 1996. Growth and photosynthesis of *Undaria pinnatifida* (Laminariales, Phaeophyta) on a cultivation ground in Korea. *Botanica Marina* 39: 389–93.

Ohmori, T. 1973. Sporangium development in *Striaria attenuata* (Ag.) Grev., *Asperococcus bullosus* Lam. and *Punctaria* sp. *Bulletin of the Sanyo Gakuen Women's Junior College* 4: 87–95.

Ohno, M. 1969. A physiological ecology of the early stage of some marine algae. *Report of the Usa Marine Biology Station, Kochi University* 16: 1–46.

Ohno, M. & Matsuoka, M. 1993. *Undaria* cultivation 'wakame'. In: Ohno, M. & Critchley, A.T. (eds), *Seaweed cultivation and marine ranching*, pp. 41–9. Japan International Cooperation Agency (JICA), Yokosuka.

Ohno, M., Matsuoka, M., Dan, N., Pand, S. & Wu, C. 1999. Morphological characteristics of the wild *Undaria pinnatifida* strain from Qindao, China grown in Naruto, Japan, as compared with wild plants in Naruto. *Aquaculture Research* 47: 61–4 (in Japanese with English summary).

Okamura, K. 1902. *Nippon Sorui-meii [Book listing Japanese Algae]*. pp. i–vi, 1–276. Tokyo: Keigyosha. (in Japanese).

Okamura, K. 1915. *Undaria* and its species. *Botanical Magazine, Tokyo* 29: 266–78.

Okamura, K. 1916. *Icones of Japanese Algae*. Vol. IV pp. 1–40, pls CLI–CLX. Tokyo.

O'Kelly, C.J. 1989. Preservation of cytoplasmic ultrastructure in dried herbarium specimens: the lectotype of *Pilinia rimosa* (Phaeophyta, formerly Chlorophyta). *Phycologia* 28: 369–74.

O'Kelly, C.J. & Floyd, G.L. 1984. The absolute configuration of the flagellar apparatus in zoospores from two species of Laminariales (Phaeophyceae). *Protoplasma* 123: 18–25.

Okuda, T. 1981. Egg Liberation in Some Japanese Sargassaceae (Phaeophyceae). In: *Proceedings of the Xth International Seaweed Symposium*, pp. 197–202, Berlin.

Olabarría, C., Rodil, I.F., Incera, M. & Troncoso, J.S. 2009a. Limited impact of *Sargassum muticum* on native algal assemblages from rocky intertidal shores. *Marine Environmental Research* 67: 153–8.

Olabarría, C., Rossi, F., Rodil, I.F., Quintas, P. & Troncoso, J.S. 2009b. Uso de diseños jerárquicos en la detección de escalas de heterogeneidad en la especie invasora *Sargassum muticum*. *Scientia Marina* 98: 194–7.

Oligschlager, M., Bartsch, I., Gutow, L. & Wiencke, C. 2012. Effects of ocean acidification on different life-cycles stages of the kelp *Laminaria hyperborea* (Phaeophyceae). *Botanica Marina* 55: 511–25.

Oliveira Filho, E.C. de & Fletcher, A. 1977. Comparative observations on some physiological aspects of rocky-shore and salt marsh populations of *Pelvetia canaliculata* (Phaeophyta). *Boletim de Botânica da Universidade de São Paulo* 5: 1–12.

Oliveira Filho, E.C. de & Fletcher, A. 1980. Taxonomic and ecological relationships between rocky-shore and saltmarsh populations of *Pelvetia canaliculata* (Phaeophyta) at Four Mile Bridge, Anglesey, UK. *Botanica Marina* 23: 409–17.

Oliveira, L. & Bisalputra, T. 1978. A virus infection in the brown alga *Sorocarpus uvaeformis* (Lyngbye) Pringsheim (Phaeophyta, Ectocarpales). *Annals of Botany* 42: 439–45.

Olsen, J.L., Sadowski, G., Stam, W.T., Veldsink, J.H. & Jones, K. 2002. Characterization of microsatellite loci in the marine seaweed *Ascophyllum nodosum* (Phaeophyceae; Fucales). *Molecular Evolution Notes* 2: 33–4.

Olsen, J.L., Zechman, F.W., Hoarau, G., Coyer, J.A., Stam, W.T., Valero, M. & Åberg, P. 2010. The phylogeographic architecture of the fucoid seaweed *Ascophyllum nodosum*: an intertidal 'marine tree' and survivor of more than one glacial-interglacial cycle. *Journal of Biogeography* 37: 842–56.

Oltmanns, F. 1889. Beiträge zur Kenntnis der Fucaceen. *Bibliotheca Botanica* 3: 94.

Oltmanns, F. 1892. Ober die kultur – and Lebensbedingungen der Meeresalgen. *Jahrbücher für wissenschaftliche Botanik* 23: 349–440.

Oltmanns, F. 1894. Über einige parasitische Meeresalgen. *Botanische Zeitung* 52: 207–16.

Oltmanns, F. 1922. *Morphologie und biologie der algen*. Zweite, umgearbeitete Auflage. Zweiter Band. Phaeophyceae-Rhodophyceae. Ed. 2. pp. [i]–iv, [1]–439, figs 288–612. Jena: Gustav Fischer.

Oppliger, L.V., Correa, J.A., Faugeron, S., Beltrán, J., Tellier, F., Valero, M. & Destombe, C. 2011. Sex ratio variation in the *Lessonia nigrescens* complex (Laminariales, Phaeophyceae): effect of latitude, temperature, and marginality. *Journal of Phycology* 47: 5–12.

Oppliger, L.V., Correa, J.A., Engelen, A.H., Tellier, F., Vieira, V., Faugeron, S., Valero, M., Gomez, G. & Destombe, C. 2012. Temperature effects on gametophyte life-history traits and geographic distribution of two cryptic kelp species. *PLoS One* 7: e39289.

Orellana, S., Hernández, M. & Sansón, M. 2019. Diversity of *Cystoseira sensu lato* (Fucales, Phaeophyceae) in the eastern Atlantic and Mediterranean based on morphological and DNA evidence, including *Carpodesmia* gen. emend. and *Treptacantha* gen. emend. *European Journal of Phycology* 54: 447–65.

Oróstica, M.H., Aguilera, M.A., Donoso, G.A., Vásquez, J.A. & Broitman, B.R. 2014. Effect of grazing on distribution and recovery of harvested stands of *Lessonia berteroana* kelp in northern Chile. *Marine Ecology Progress Series* 511: 71–82.

Oteng'o, A.O., Won, B.Y. & Cho, T.O. 2022. Proposal for the Sungminiaceae fam. nov. (Ralfsiales, Phaeophyceae) for *Sungminia* gen. nov. with three new species based on molecular and morphological analyses. *Journal of Phycology* 58: 719–28.

Oudot-Le-Secq, M.-P., Kloareg, B. & Loiseaux-de Goër, S. 2002. The mitochondrial genome of the brown alga *Laminaria digitata*: a comparative analysis. *European Journal of Phycology* 37: 163–72.

Oudot-Le-Secq, M.-P., Loiseaux-de Goër, S., Stam, W.T. & Olsen, J.I. 2006. Complete mitochondrial genomes of the three brown algae (Heterokonta: Phaeophyceae) *Dictyota dichotoma*, *Fucus vesiculosus* and *Desmarestia viridis*. *Current Genetics* 42: 47–58.

Oudot-Le-Secq, M.-P., Fontaine, J.-M., Rousvoal, S., Kloareg, B. & Loiseaux-de Goër, S. 2001. The complete sequence of a brown algal mitochondrial genome, the ectocarpale *Pylaiella littoralis* (L.) Kjellm. *Journal of Molecular Evolution* 53: 80–8.

Pang, S. & Lüning, K. 2004. Photoperiodic long-day control of Sporophyll and hair formation in the brown alga *Undaria pinnatifida*. *Journal of Applied Phycology* 16: 83–92.

Pang, S. & Shan, T. 2008. Zoospores of *Undaria pinnatifida*: their efficiency to attach under different water velocities and conjugation behavior during attachment. *Acta Oceanologica Sinica* 27: 1–8.

Pang, S. & Wu, C.Y. 1996. Study on gametophyte vegetative growth of *Undaria pinnatifida* and its application. *Chinese Journal of Oceanology and Limnology* 14: 205–10.

Pang, S.J., Shan, T.F. & Zhang, Z.H. 2008. Responses of vegetative gametophytes of *Undaria pinnatifida* to high irradiance in the process of gametogenesis. *Phycological Research* 56: 280–7.

Pankow, H. 1971. *Algenflora der Ostsee I. Benthos*. 419 pp. Gustav Fischer Verlag, Stuttgart.

Papenfuss, G.F. 1933. Note on the life-cycle of *Ectocarpus siliculosus* Dillw. *Science* 77: 390–1.

Papenfuss, G.F. 1934. Alternation of generations in *Sphacelaria bipinnata* Sauv. *Bot. Notiser* 1934: 437–44.

Papenfuss, G.F. 1935a. Alternation of generations in *Ectocarpus siliculosus*. *Botanical Gazette* 96: 421–46.

Papenfuss, G.F. 1935b. The development of the gametophyte of *Spermatochnus paradoxus*. *Kungliga Fysiografiska Sällskapets i Lund Förhandlingar* 5: 1–4.

Papenfuss, G.F. 1950. Review of the genera of algae described by Stackhouse. *Hydrobiologia* 2: 181–208.

Papenfuss, G.F. 1951. Phaeophyta. In: Smith, G.M. (ed.), *Manual of Phycology*, pp. 119–58. Waltham, MA: Chronica Botanica.

Papenfuss, G.F. 1955. Classification of the algae. In: Kessel, E.L. (ed.), *A Century of Progress in the Natural Sciences, 1853–1953*, pp. 115–224. San Francisco: California Academy of Sciences.

Parente, M.I. 2007. *Life history studies and the systematic status of selected brown algae (Phaeophyceae) in the North Atlantic*. A thesis presented in candidature for the degree of Doctor of Philosophy. University of Portsmouth, Portsmouth.

Parente, M.I., Costa, F.O., Fletcher, R.L. & Saunders, G.W. 2011a. DNA barcode library for crustose Brown algal species. In: Abstracts, *Proceedings of the 5th European Phycological Congress, Rhodes, Greece*, 4–9 Sept., p. 66.

Parente, M.I., Costa, F.O. & Saunders, G.W. 2011b. DNA barcoding representative Scytosiphonaceae (Phaeophyceae) from the Northeast Atlantic emphasizing the Azores. In: *Abstracts of the 50th Northeast Algal Symposium*, April 15–17, 2011. Marine Biological Laboratory, Woods Hole, MA, p. 43.

Parente, M.I., Neto, A.I. & Fletcher, R.L. 2003a. Morphology and life history of *Scytosiphon lomentaria* (Scytosiphonaceae, Phaeophyceae) from the Azores. *Journal of Phycology* 39: 353–9.

Parente, M.I., Neto, A.I. & Fletcher, R.L. 2003b. Morphology and life history studies of *Endarachne binghamiae* (Scytosiphonaceae, Phaeophyta) from the Azores. *Aquatic Botany* 76: 106–9.

Parente, M.I., Rousseau, F., Fletcher, R.L., Neto, A.I. & de Reviers, B. 2005. Should we recognize an order Ralfsiales within the Phaeophyceae?. *Phycologia* 44 (Suppl.): 78.

Parente, M.I., Fletcher, R.L., Neto, A.I., Tittley, I., Sousa, A.F., Draisma, S. & Gabriel, D. 2010a. Life history and morphological studies of *Punctaria tenuissima* (Chordariaceae, Phaeophyceae), a new record for the Azores. *Botanica Marina* 53: 223–31.

Parente, M.I., Rousseau, F., de Reviers, B., Fletcher, R.L., Costa, F. & Sanders, G.W. 2010b. Molecular divergence within *Ralfsia verrucosa* (Ralfsiales, Phaeophyceae) indicates cryptic species. In: Abstracts, 49th Annual Northeast Algal Symposium, April 16–18, Roger Williams University, Bristol, RI, p. 18.

Parente, M.I., Fletcher, R.L., Costa, F.O. & Saunders, G.W. 2020 (2021). Taxonomic investigation of Ralfsia-like (Ralfsiales, Phaeophyceae) taxa in the North Atlantic Ocean based on molecular and morphological data, with descriptions of Pseudoralfsiaceae fam. nov., *Pseudoralfsia azorica* gen. et sp. nov. and *Nuchella vesicularis* gen. et sp. nov. *European Journal of Phycology* 56: 12–23.

Park, C.S., Hwang, E.K., Yi, Y.H. & Sohn, C.H. 1995. Effects of daylength on the differentiation and receptacle formation of *Hizikia fusiformis* (Harvey) Okamura. *Korean Journal of Phycology* 10: 45–9.

Park, C.S., Park, K.Y. & Baek, J.M. 2008. The occurrence of pinhole disease in relation to developmental stage in cultivated *Undaria pinnatifida* (Harvey) Suringar (Phaeophyta) in Korea. *Journal of Applied Phycology* 20: 485–90.

Park, C.S., Park, K.Y., Baek, J.M. & Hwang, E.K. 2009. The occurrence of pinhole disease in relation to developmental stage in cultivated *Undaria pinnatifida* (Harvey) Suringar (Phaeophyta) in Korea. *Proceedings of the International Seaweed Symposium* 19: 35–40.

Parke, M. 1933. A contribution to the knowledge of the Mesogloiaceae and associated families. *Publications of the Hartley Botanical Laboratories, Univ. of Liverpool* 9: 1–43.

Parke, M. 1953. A preliminary check list of British marine algae. *Journal of the Marine Biological Association of the United Kingdom* 32: 497–520.

Parke, M. & Dixon, P.S. 1964. A revised check list of British marine algae. *Journal of the Marine Biological Association of the United Kingdom* 44: 499–542.

Parke, M. & Dixon, P.S. 1968. Check list of British marine algae – second revision. *Journal of the Marine Biological Association of the United Kingdom* 48: 783–832.

Parke, M. & Dixon, P.S. 1976. Check list of British marine algae – third revision. *Journal of the Marine Biological Association of the United Kingdom* 56: 527–94.

Parke, M. 1948. *Laminaria ochroleuca* De La Pylaie growing on the coast of Britain. *Nature (London)* 162: 295–6.

Parker, J.D. & Hay, M.E. 2005. Biotic resistance to plant invasions? Native herbivores prefer non-native plants. *Ecology Letters* 8: 959–67.

Parker, J. & Philpott, D.E. 1960. EM studies of *F. vesiculosus* cytoplasm in summer and winter. *Biological Bulletin of the Marine Biology Laboratory, Woods Hole* 119: 330–1.

Parkes, H.M. 1958. A general survey of marine algae of Mulroy Bay, Co. Donegal I & II. *Irish Naturalists' Journal* 12: 277–83, 324–30.

Parodi, E.R. & Müller, D.G. 1994. Field and culture studies on virus infections in *Hincksia hincksiae* and *Ectocarpus fasciculatus* (Ectocarpales, Phaeophyceae). *European Journal of Phycology* 29: 113–17.

Parsons, M.J. 1982. *Colpomenia* (Endlicher) Derbès et Solier (Phaeophyta) in New Zealand. *New Zealand Journal of Botany* 20: 289–301.

Parusel, E.S. 1991. Brown algae (*Fucus vesiculosus* f. *mytili*) entangled by blue mussels (*Mytilus edulis*) – a beneficial status of nutrient supply? *British Phycological Journal* 26: 93.

Parys, S., Kehraus, S., Pete, R., Küpper, F.C., Glombitza, K.-W. & König, G.M. 2009. Seasonal variation of polyphenolics in *Ascophyllum nodosum* (Phaeophyceae). *European Journal of Phycology* 44: 331–8.

Paula, E.J. & Eston, V.R. 1987. Are there other *Sargassum* species potentially as invasive as *S. muticum*? *Botanica Marina* 30: 405–10.

Paula, R.J. de & Oliveira, E.C. de 1982. Wave exposure and ecotypical differentiation in *Sargassum cymosum* (Phaeophyta-Fucales). *Phycologia* 21: 145–52.

Paulino, C., Neiva, J., Coelho, N.C., Aires, T., Marbà, N., Krause-Jensen, D. & Serrão, E.A. 2016. Characterization of 12 polymorphic microsatellite markers in the sugar kelp *Saccharina latissima*. *Journal of Applied Phycology* 28: 3071–4.

Pavia, H. & Toth, G.B. 2000. Inducible chemical resistance to herbivory in the brown seaweed *Ascophyllum nodosum*. *Ecology* 81: 3212–25.

Pavia, H., Toth, G.B., Lindgren, A. & Åberg, P. 2003. Intraspecific variation in the phlorotannin content of the brown alga *Ascophyllum nodosum*. *Phycologia* 42: 378–83.

Pavia, H., Baumgartner, F., Cervin, G., Enge, S., Kubanek, J., Nylund, G.M., Selander, E., Svensson, J.R. & Toth, G.B. 2012. Chemical defences against herbivores. In: Brönmark, C. & Hansson, L.-A (eds), *Chemical Ecology in Aquatic Systems*, pp. 210–35. Oxford University Press, Oxford.

Pazó, J.P. & Niell, F.X. 1977. Distribución y características de *Fucus serratus* L. en las Rías Bajas Gallegas. *Investigación Pesquera* 41: 455–72.

Pearson, G.A. & Brawley, S.H. 1996. Reproductive ecology of *Fucus distichus* (Phaeophyceae): an intertidal alga with successful external fertilization. *Marine Ecology Progress Series* 143: 211–23.

Pearson, G.A. & Davison, I.R. 1994. Freezing stress and osmotic dehydration in *Fucus distichus* (Phaeophyta): evidence for physiological similarity. *Journal of Phycology* 30: 257–67.

Pearson, G.A. & Evans, L.V. 1990. Settlement and survival of *Polysiphonia lanosa* (Ceramiales) spores on *Ascophyllum nodosum* and *Fucus vesiculosus* (Fucales). *Journal of Phycology* 26: 597–603.

Pearson, G.A. & Serrão, E.A. 2006. Revisiting synchronous gamete release by fucoid algae in the intertidal zone: fertilization success and beyond? *Integrative and Comparative Biology* 46: 587–97.

Pearson, G.A., Serrão, E.A. & Brawley, S.H. 1998. Control of gamete release in fucoid algae: sensing hydrodynamic conditions via carbon acquisition. *Ecology* 79: 1725–39.

Peaucelle, A. & Couder, Y. 2016. Fibonacci spirals in a brown alga [*Sargassum muticum* (Yendo) Fensholt] and in a land plant [*Arabidopsis thaliana* (L.) Heynh.]: a case of morphogenetic convergence. *Acta Societatis Botanicorum Poloniae* 85: 1–15.

Peckol, P., Harlin, M.M. & Krumscheid, P. 1988. Physiological and population ecology of intertidal and subtidal *Ascophyllum nodosum* (Phaeophyta). *Journal of Phycology* 24: 192–8.

Pedersén, M. 1983. Ultrastructural localization of bromine and iodine in the stipes of *Laminaria digitata* (Huds.) Lamour., *Laminaria saccharina* (L.) Lamour. and *Laminaria hyperborea* (Gunn.) Fosl. *Botanica Marina* 26: 113–18.

Pedersen, P.M. 1974a. The life history of *Sorocarpus micromorus* (Phaeophyceae, Ectocarpaceae) in culture. *British Phycological Journal* 9: 57–61.

Pedersen, P.M. 1974b. On the systematic position of *Delamarea attenuata* (Phaeophyceae). *British Phycological Journal* 9: 313–18.

Pedersen, P.M. 1975. Culture studies on marine algae from West Greenland I. Chromosomal information relating to the life history of *Isthmoplea sphaerophora* (Phaeophyceae, Dictyosiphonales). *British Phycological Journal* 10: 165–8.

Pedersen, P.M. 1976. Marine, benthic algae from southernmost Greenland. *Meddelelser om Grønland* 199: 1–80.

Pedersen, P.M. 1978a. Culture studies on marine algae from West Greenland. III. The life histories and systematic positions of *Pogotrichum filiforme* and *Leptonematella fasciculata* (Phaeophyceae). *Phycologia* 17: 61–8.

Pedersen, P.M. 1978b. Culture studies on the pleomorphic brown alga *Myriotrichia clavaeformis* (Dictyosiphonales, Myriotrichiaceae). *Norwegian Journal of Botany* 25: 281–91.

Pedersen, P.M. 1979a. Culture studies on marine algae from West Greenland IV. *Giffordia ovata* (Fucophyceae, Ectocarpales). *Botanisk Tidsskrift* 74: 57–65.

Pedersen, P.M. 1979b. Culture studies on the brown algae *Halothrix lumbricalis* and *Elachista fucicola* (Elachistaceae). *Botaniska Notiser* 132: 151–9.

Pedersen, P.M. 1980a. Culture studies on complanate and cylindrical *Scytosiphon* (Fucophyceae,

Scytosiphonales) from Greenland. *British Phycological Journal* 15: 391–8.

Pedersen, P.M. 1980b. *Giraudyopsis stellifer* (Chrysophyceae) and *Streblonema immersum* (Phaeophyceae), additions to the British marine algae. *British Phycological Journal* 15: 247–8.

Pedersen, P.M. 1981a. The life histories in culture of the brown algae *Gononema alariae* sp. nov. and *G. aecidioides* comb. nov. from Greenland. *Nordic Journal of Botany* 1: 263–70.

Pedersen, P.M. 1981b. *Porterinema fluviatile* as a stage in the life history of *Sorapion kjellmanii* (Fucophyceae, Ralfsiaceae). *Proceedings of the International Seaweed Symposium* 10: 203–8.

Pedersen, P.M. 1981c. Life histories of brown algae. In: Lobban, C. & Wynne, M.J. (eds), *The biology of seaweeds*, pp. 194–217. Botanical Monographs 17. Blackwell Scientific Publications, Oxford.

Pedersen, P.M. 1981d. Culture studies on the rare brown alga *Phaeostroma longisetum* comb. nov. and its common relative *P. pustulosum* from Greenland. *Nordic Journal of Botany* 1: 271–6.

Pedersen, P.M. 1983. Notes on marine, benthic algae from Madeira in nature and in culture. *Bocagiana* 70: 1–8.

Pedersen, P.M. 1984. Studies on primitive brown algae (Fucophyceae). *Opera Botanica* 74: 1–76.

Pedersen, P.M. 2011. *Grønlands havalger.* pp. [1] 7–208. Copenhagen: Forlaget Epsilon. DK.

Pedersen, P.M. & Kristiansen, A. 2001. On the enigmatic brown alga *Acinetospora crinita* (Ectocarpales, Fucophyceae). *Cryptogamie, Algologie* 22: 209–18.

Pedroche, P.F., Silva, P.C., Aguilar Rosas, L.E., Dreckmann, K.M. & Aguilar Rosas, R. 2008. *Catálogo de las algas benthónicas del Pacífico de México II. Phaeophycota.* pp. [i–viii], i–vi, 15–146. Mexicali & Berkeley: Universidad Autónoma Metropolitana; Universidad Autónoma de Baja California; University of California Berkeley.

Pellegrini, L. 1974. Origine et modifications ultrastructurales du materiel osmiophile contenu dans les physodes et dans certains corps iridescents des cellules vegetatives apicales chez *Cystoseira stricta* Sauvageau (Pheophycee, Fucale). *Compte Rendu Hebdomadaire des Séances de l'Académie des Sciences, Paris ser. D* 279: 903–6.

Peña, V. & Bárbara, I. 2002. Caracterización florística y zonación de las algas bentónicas marinas del puerto de A Coruña (NO Península Ibérica). *Nova Acta Científica Compostelana (Bioloxía)* 12: 35–66 [in Spanish, English abstract].

Peña, V. & Bárbara, I. 2010. New records of crustose seaweeds associated with subtidal maërl beds and gravel bottoms in Galicia (NW Spain). *Botanica Marina* 53: 41–62.

Penot, M. 1974. Ionic exchange between tissue of *Ascophyllum nodosum* (L.) Le Jolis and *Polysiphonia lanosa* (L.) Tandy. *Zeitschrift für Pflanzenphysiologie* 73: 125–31.

Penot, M. & Penot, M. 1976. Nouvel aspect des relations physiologiques entre l'*Ascophyllum nodosum* (L.) Le Jolis. (Fucacées) et ses épiphytes. *Bulletin de la Société Phycologique de France* 21: 1–7.

Penot, M. & Penot, M. 1977. Quelques aspects originaux des transports à ongue distance dans le thalle de

Ascophyllum nodosum (L.) Le Jolis (Phaeophyceae, Fucales). *Phycologia* 16: 339–47.

Penot, M. & Penot, M. 1981. Ions transport and exchange between *Ascophyllum nodosum* (L.) Le Jolis and some epiphytes. *Proceedings of the International Seaweed Symposium* 8: 217–23.

Penot, M. & Videau, C. 1975. Adsorption du 86Rb et du 99Mo par deux algues marines: Le *Laminaria digitata* et le *Fucus serratus*. *Zeitschrift für Pflanzenphysiologie* 76: 285–93.

Penot, M., Dumay, J. & Pellegrini, M. 1985. Contribution à l'étude de la fixation et du transport du ^{14}C chez *Cystoseira nodicaulis* (Fucales, Cystoseiraceae). *Phycologia* 24: 93–102.

Pereira, T., Engelen, A., Pearson, G., Serrão, E., Destombe, C. & Valero, M. 2011. Temperature effects on the microscopic haploid stage development of *Laminaria ochroleuca* and *Saccorhiza polyschides*, kelps with contrasting life histories. *Cahiers de biologie marine* 52: 395–403.

Pereira, T.R., Engelen, A.H., Pearson, G.A., Valero, M. & Serrão, E.A. 2015a. Contrasting timing of life stages across latitudes – a case study of a marine forest-forming species. *European Journal of Phycology* 50: 361–9.

Pereira, T., Engelen, H., Pearson, G., Valero, M. & Serrão, E. 2015b. Response of kelps from different latitudes to consecutive heat shock. *Journal of Experimental Marine Biology and Ecology* 463: 57–62.

Pérez Cordero, S., Ribamar Da Cruz Freitas Junior, J. & Cremades Ugarte, J. 2012. Optimización del cultivo de *Saccharina latissima* (Laminariales, Phaeophyta) en sistemas IMTA en la Ría de Ares y Betanzos (A Coruña). *Algas, Boletín Informativo de la Sociedad Española de Ficología* 46: 19.

Pérez, R. 1969a. Le repeuplement des champs de *Laminaria digitata*. Influence comparée de la coupe et de l'arrachage. *Science et Peche*: 329–44.

Pérez, R. 1969b. Etude biometrique d'une population de *Laminaria digitata* Lamouroux de l'étage infralittoral profond. I. Résultats relatifs aux dimensions des thalles. *Revue des travaux de l'Institut des pêches maritimes* 33: 117–35.

Pérez, R. 1969c. Croissance de *Laminaria digitata* (L.) Lamouroux etudiée sur trois années consécutives. *Proceedings of the International Seaweed Symposium* 6: 329–44.

Pérez, R. 1971. Écologie, croissance et régénération, teneurs en acide alginique de *Laminaria digitata* sur les cotes de la Manche. *Revue des Travaux de l'Institut des Peches Maritimes* 35: 287–346.

Pérez, R., Kaas, R. & Barbaroux, O. 1984. Culture expérimentale de l'algue *Undaria pinnatifida* sur les côtes de France. *Science et Pêche, Bulletin Institutionnel des Pêches Marines* 343: 3–15.

Pérez, R., Lee, J.Y. & Juge, C. 1981. Observations sur la biologie de l'algue japonaise *Undaria pinnatifida* (Harvey) Suringar introduite accidentellement dans l'étang de Thau. *Science et Pêche, Bulletin Institutionnel des Pêches Marines* 315: 1–12.

Pérez, R., Durand, P., Kaas, R., Barbaroux, O., Barbier, V., Vinot, C., Bourgeau-Causse, M., Leclerq, M. & Moigne, J.Y. 1988. *Undaria pinnatifida* on the French coasts. Cultivation method, biochemical composition of the sporophyte and the gametophyte. In:

Staedler, T., Mollion, J., Verdus, M.C., Karamanos, Y., Morvan, H. & Christiaen, D. (eds), *Algal Biotechnology*, pp. 315–28. Elsevier, London.

Pérez-Cirera, J.L., Cremades, J. & Bárbara, I. 1991. Consideraciones sobre *Scytosiphon dotyi* Wynne (Scytosiphonaceae, Fucophyceae), novedad para las costas de la Península Ibérica. *Anales del Jardín Botánico de Madrid* 49: 135–8.

Pérez-Cirera, J.L., Salinas, J.M., Cremades, J., Bárbara, J., Granja, A., Veiga, A.J. & Fuertes, C. 1997. Cultivo de *Undaria pinnatifida* (Laminariales, Phaeophyta) en Galicia. *Nova Acta Científica Compostelana (Bioloxía)* 7: 3–28.

Pérez-Larrán, P., Torres, M.D., Flórez-Fernández, N., Balboa, E.M., Moure, A. & Domínguez, H. 2019. Green technologies for cascade extraction of *Sargassum muticum* bioactives. *Journal of Applied Phycology* 31: 2481–95.

Pérez-Ruzafa, I.M. 2001. *Fucus* L. In: Gómez Garreta, A. (ed.), *Flora phycologica iberica Vol. 1 Fucales*, pp. 33–61. Murcia: Universidad de Murcia.

Pérez-Ruzafa, M. & Gallardo, T. 1997. Mapas de distribución de algas marinas de la Península Ibérica e Islas Baleares. X. Variedades de *Fucus spiralis* L. y de *F. vesiculosus* L. (Fucales, Fucophyceae). *Botanica Complutensis* 21: 121–36.

Pérez-Ruzafa, I., Izquierdo, J.L., Araújo, R., Sousa-Pinto, I., Pereira, L. & Bárbara, I. 2003. Distribution maps of marine algae from the Iberian Peninsula and the Balearic Islands. XVII. *Laminaria rodriguezii* Bornet and additions to the distribution maps of *L. hyperborea* (Gunner.) Foslie, *L. ochroleuca* Bach. pyl. and *L. saccharina* (L.) Lamour. (*Laminariales, Fucophyceae*). *Botanica Complutensis* 27: 155–64.

Pereyra, P.J., Arias, M., Gonzalez, R. & Narvarte, M. 2014. Moving forward: the Japanese kelp *Undaria pinnatifida* (Harvey) Suringar, 1873 expands in northern Patagonia, Argentina. *BioInvasions Records* 3: 65–70.

Pérez-Ruzafa, I., Menéndez, J.L. & Salinas, J.M. 2002. Mapas de distribución de algas marinas de la Península Ibérica y las Islas Baleares. XV. *Undaria pinnatifida* (Harvey) Suringar (Laminariales, Fucophyceae). *Botanica Complutensis* 26: 147–51.

Perrin, C., Daguin, C., Van de Vliet, M., Engel, C.R., Pearson, G.A. & Serrão, E.A. 2007. Implications of mating system for genetic diversity of sister algal species: *Fucus spiralis* and *Fucus vesiculosus* (Heterokontophyta, Phaeophyceae). *European Journal of Phycology* 42: 219–30.

Peteiro, C. 2002. Explotación y cultivo de macroalgas en México. El instituto de investigaciones Oceanológicas de Universidad Autónoma de Baja California. *Algas* 28: 8–10 (Biog).

Peteiro, C. 2008. A new record of the introduced seaweed *Undaria pinnatifida* (Laminariales, Phaeophyceae) from the Cantabrian Sea (northern Spain) with comments on its establishment. *Aquatic Invasions* 3: 413–15.

Peteiro, C. & Freire, Ó. 2009. Effect of outplanting time on the commercial cultivation of the kelp *Laminaria saccharina* at the southern limit in the Atlantic Coast (NW Spain). *Chinese Journal of Oceanology and Limnology* 27: 54–60.

Peteiro, C. & Freire, Ó. 2011a. Effect of water motion on the cultivation of the commercial seaweed *Undaria pinnatifida* in a coastal bay of Galicia, Northwest Spain. *Aquaculture* 314: 269–76.

Peteiro, C. & Freire, Ó. 2011b. Offshore cultivation methods affects blade features of the edible seaweed *Saccharina latissima* in a bay of Galicia, Northwest Spain. *Russian Journal of Marine Biology* 37: 319.

Peteiro, C. & Freire, Ó. 2012a. Observations on fish grazing of the cultured kelps *Undaria pinnatifida* and *Saccharina latissima* (Phaeophyceae, Laminariales) in Spanish Atlantic waters. *Aquaculture, Aquarium, Conservation & Legislation, International Journal of the Bioflux Society* 5: 189–96.

Peteiro, C. & Freire, Ó. 2012b. Outplanting time and methodologies related to mariculture of the edible kelp *Undaria pinnatifida* in the Atlantic coast of Spain. *Journal of Applied Phycology* 24: 1361–72.

Peteiro, C. & Freire, Ó. 2013a. Biomass yield and morphological features of the seaweed *Saccharina latissima* cultivated at two different sites in a coastal bay in the Atlantic coast of Spain. *Journal of Applied Phycology* 25: 205–13.

Peteiro, C. & Freire, Ó. 2013b. Epiphytism on blades of the edible kelps *Undaria pinnatifida* and *Saccharina latissima* farmed under different abiotic conditions. *Journal of the World Aquaculture Society* 44: 706–13.

Peteiro, C. & Freire, Ó. 2014. Morphological traits of wild and selected strains of cultured *Undaria pinnatifida* from Galicia (NW Spain). *Journal of the World Aquaculture Society* 45: 469–74.

Peteiro, C. & Sánchez, N. 2012. Comparing salinity tolerance in early stages of the sporophytes of a non-indigenous kelp (*Undaria pinnatifida*) and a native kelp (*Saccharina latissima*). *Russian Journal of Marine Biology* 38: 197–200.

Peteiro, C., Cremades, J. & Salinas, J.M. 2001. Cultivo experimental con fines industriales de *Undaria pinnatifida* (Laminariales, Phaeophyta) en la ría de Ares y Betanzos (A Coruña) (Resumen). *Algas* 26: 23–4.

Peteiro, C., Salinas, J.M., Freire, Ó. & Fuertes, C. 2006. Cultivation of the autoctonous seaweed *Laminaria saccharina* off the Galician coast (NW): production and features of the sporophytes for an annual and biennial harvest. *Thalassas* 22: 45–52.

Peteiro, C., Sánchez, N. & Martínez, V. 2016. Mariculture of the Asian kelp *Undaria pinnatifida* and the native kelp *Saccharina latissima* along the Atlantic coast of Southern Europe: An overview. *Algal Research* 16: 9–23.

Peteiro, C., Sánchez, N., Dueñas-Liaño, C. & Martínez, B. 2014. Open-sea cultivation by transplanting young fronds of the kelp *Saccharina latissima*. *Journal of Applied Phycology* 26: 519–28.

Peters, A.F. 1987. Reproduction and sexuality in the Chordariales (Phaeophyceae). *Progress in Phycological Research* 5: 223–64.

Peters, A.F. 1989. Sexual reproduction in the crustose brown alga *Pseudolithoderma extensum*. *Helgoländer Wissenschaftliche Meeresuntersuchungen N.F.* 43: 195–205.

Peters, A.F. 1991. First record of *Striaria attenuata* (Phaeophyceae, Dictyosiphonales) in South America, and its life history in laboratory cultures. *Revista Chilena de Historia Natural* 64: 261–9.

Peters, A.F. 1998. Ribosomal DNA sequences support taxonomic separation of the two species of *Chorda*: reinstatement of *Halosiphon tomentosus* (Lyngbye) Jaasund (Phaeophyceae, Laminariales). *European Journal of Phycology* 33: 65–71.

Peters, A.F. 2003. Molecular identification, distribution and taxonomy of brown algal endophytes, with emphasis on species from Antarctica. *Proceedings of the International Seaweed Symposium* 17: 293–302.

Peters, A.F. & Burkhardt, E. 1998. Systematic position of the kelp endophyte *Laminarionema elsbetiae* (Ectocarpales *sensu lato*, Phaeophyceae) inferred from nuclear ribosomal DNA sequences. *Phycologia* 37: 114–20.

Peters, A.F. & Clayton, M.N. 1998. Molecular and morphological investigations of three brown algae genera with stellate plastids: evidence for Scytothamnales ord. nov. (Phaeophyceae). *Phycologia* 37: 106–13.

Peters, A.F. & Moe, R.L. 2001. DNA sequences confirm that *Petroderma maculiforme* (Phaeophyceae) is the brown algal phycobiont of the marine lichen *Verrucaria tavaresiae* (Verrucariales, Ascomycota) from Central California. *Bulletin of the California Lichen Society* 8: 41–3.

Peters, A.F. & Müller, D.G. 1986a. Sexual reproduction of *Stilophora rhizodes* (Phaeophyceae, Chordariales) in culture. *British Phycological Journal* 21: 417–23.

Peters, A.F. & Müller, D.G. 1986b. Critical re-examination of sexual reproduction in *Tinocladia crassa*, *Nemacystus decipiens* and *Sphaerotrichia divaricata* (Phaeophyceae, Chordariales). *Japanese Journal of Phycology* 34: 69–73.

Peters, A.F. & Ramírez, M.E. 2001. Molecular phylogeny of small brown algae, with special reference to the systematic position of *Caepidium antarcticum* (Adenocystaceae, Ectocarpales). *Cryptogamie, Algologie* 22: 187–200.

Peters, A.F., Ramírez, M.E. & Rülke, A. 2000. The phylogenetic position of the subantarctic marine macroalga *Desmarestia chordalis* (Phaeophyceae) inferred from nuclear ribosomal ITS sequences. *Polar Biology* 23: 95–9.

Peters, A.F. & Schaffelke, B. 1996. *Streblonema* (Ectocarpales, Phaeophyceae) infection in the kelp *Laminaria saccharina* (Laminariales, Phaeophyceae) in the western Baltic. *Proceedings of the International Seaweed Symposium* 15: 111–16.

Peters, A.F., Couceiro, L., Tsiamis, K., Küpper, F.C. & Valero, M. 2015. Barcoding of cryptic stages of marine brown algae isolated from incubated substratum reveals high diversity in Acinetosporaceae (Ectocarpales, Phaeophyceae). *Cryptogamie, Algologie* 36: 3–29.

Peters, A.F., Van Oppen, M.J.H., Wiencke, C., Stam, W.T. & Olsen, J.L. 1997. Phylogeny and historical ecology of the Desmarestiaceae (Phaeophyceae) support a southern hemisphere origin. *Journal of Phycology* 33: 294–309.

Peters, A.F., Kawai, H. & Novaczek, I. 1992. Intraspecific sterility barrier provides evidence that introduction of *Sphaerotrichia divaricata* (Phaeophyceae, Chordariales) into Mediterranean was from Japan. *Proceedings of the International Seaweed Symposium* 14, abstract 249, 115.

Peters, A.F., Kawai, H. & Novaczek, I. 1993. Intraspecific sterility barrier confirms that introduction of *Sphaerotrichia divaricata* (Phaeophyceae, Chordariales) into the Mediterranean was from Japan. *Proceedings of the International Seaweed Symposium* 14: 31–6.

Peters, A.F., Novaczek, I., Müller, D.G. & McLachlan, J. 1987. Culture studies on reproduction of *Sphaerotrichia divaricata* (Chordariales, Phaeophyceae). *Phycologia* 26: 457–66.

Peters, A.F., Marie, D., Scornet, D., Kloareg, B. & Cock, J.M. 2004. Proposal of *Ectocarpus siliculosus* (Ectocarpales, Phaeophyceae) as a model organism for brown algal genetics and genomics. *Journal of Phycology* 40: 1079–88.

Peters, A.F., Scornet, D., Müller, D.G., Kloareg, B. & Cock, J.M. 2004a. Inheritance of organelles in artificial hybrids of the isogamous multicellular chromist alga *Ectocarpus siliculosus* (Phaeophyceae). *European Journal of Phycology* 39: 235–42.

Peters, A.F., Van Wijk, S.J., Scornet, D., Kawai, H., Schroeder, D.S., Cock, J.M. & Boo, S.M. 2010. Reinstatement of *Ectocarpus crouaniorum* Thuret in Le Jolis as a third common species of *Ectocarpus* (Ectocarpales, Phaeophyceae) in Western Europe, and its phenology at Roscoff, Brittany. *Phycological Research* 58: 157–70.

Petersen, J.B., Caram, B. & Hansen, J.B. 1958. Observations sur les zoïdes du *Chordaria flagelliformis* au microscope électronique. *Botanisk Tidsskrift* 54: 57–60.

Petrov, J.E. 1965. *Fucus distichus* L. emend. Powell et *Fucus evanescens* C. Ag. *Novitates Systematicae Plantarum non Vascularium* 2: 64–70.

Petrov, Y.E. 1974. Clavis synoptica Laminarialium et Fucalium e maribus URSS [Synoptical Key to the Laminariales and Fucales of the Seas of the USSR]. *Novosti Sistematiki Nizshikh Rastenii (Novitates Systematicae Plantarum non Vascularium)* 11: 153–69.

Pfetzing, J., Stengel, D.B., Cuffe, M.M., Savage, A.V. & Guiry, M.D. 2000. Effects of temperature and prolonged emersion on photosynthesis, carbohydrate content and growth of the brown intertidal alga *Pelvetia canaliculata*. *Botanica Marina* 43: 399–407.

Phillips, J.A., Clayton, M.N., Maier, I., Boland, W. & Müller, D.G. 1990a. Multifidene, the spermatozoid attractant of *Zonaria angustata* (Dictyotales, Phaeophyta). *British Phycological Journal* 25: 295–8.

Phillips, J.A., Clayton, M.N., Maier, I., Boland, W. & Müller, D.G. 1990b. Sexual reproduction in *Dictyota diemensis* (Dictyotales, Phaeophyta). *Phycologia* 29: 367–79.

Phillips, N. & Fredericq, S. 2000. Biogeographic and phylogenetic investigation of the panpacific genus *Sargassum* (Fucales, Phaeophyceae) with respect to the Gulf of Mexico species. *Gulf of Mexico Science* 18: 1–11.

Phillips, N.E., Clifford, C.M. & Morden, C.W. 2005. Testing systematic concepts of *Sargassum* (Fucales, Phaeophyceae) using portions of the rbcLS operon. *Phycological Research* 53: 1–10.

Phillips, N., Smith, C.M. & Morden, C.W. 2001. An effective DNA extraction protocol for brown algae. *Phycological Research* 49: 97–102.

Phillips, N., Burrowes, R., Rousseau, F., Reviers, B. de, Saunders, G.W. 2008a. Resolving evolutionary relationships among the brown algae using chloroplast and nuclear genes. *Journal of Phycology* 44: 394–404.

Phillips, N., Calhoun, S., Moustafa, A., Bhattacharya, D. & Braun, E.L. 2008b. Genomic insights into evolutionary relationships among heterokont lineages emphasizing the Phaeophyceae. *Journal of Phycology* 44: 15–18.

Piccone, A. 1884. *Crociera del Corsaro alle Isole Madera e Canarie del Capitano Enrico d'Albertis. Alghe.* pp. [3]–60, 1 pl. Genova [Genoa]: Tipografia del r. Istituto Sordo-Muti.

Plouguerné, E., Hellio, C., Deslandes, E., Véron, B. & Stiger-Pouvreau, V. 2008. Anti-microfouling activities in extracts of two invasive algae: *Grateloupia turuturu* and *Sargassum muticum. Botanica Marina* 51: 202–8.

Plouguerné, E., Le Lann, K., Connan, S., Jechoux, G., Deslandes, E. & Stiger-Pouvreau, V. 2006. Spatial and seasonal variations in density, maturity, length and phenolic content of the invasive brown macroalga *Sargassum muticum* along the coast of Western Brittany (France). *Aquatic Botany* 85: 337–44.

Plouguerné, E., Georgantea, P., Ioannou, E., Vagias, C., Roussis, V., Hellio, C., Kraffe, E. & Stiger-Pohnert, G. & Boland, W. 2002. The oxylipin chemistry of attraction and defense in brown algae and diatoms. *Natural Product Reports* 19: 108–22.

Polderman, P.J.G. 1978. Algae of the saltmarshes on the south and southwest coasts of England. *British Phycological Journal* 13: 235–40.

Polderman, P.J.G. 1979. The saltmarsh algal communities in the Wadden area, with reference to their distribution and ecology in NW Europe. I. The distribution and ecology of the algal communities. *Journal of Biogeography* 6: 225–66.

Polderman, P.J.G. & Polderman-Hall, R.A. 1980. Algal communities in Scottish saltmarshes. *British Phycological Journal* 15: 59–71.

Pollock, E.G. 1969. Effect of 17 B-estradiol on early cleavage patterns in the embryo of *Fucus distichus. Experientia* 25: 1073–5.

Polte, P. & Buschbaum, C. 2008. Native pipefish *Entelurus aequoreus* promoted by the introduced seaweed *Sargassum muticum* in the northern Wadden Sea, North Sea. *Aquatic Biology* 33: 11–18.

Poore, A.G., Gutow, L., Pantoja, J.F., Tala, F., Jofré Madariaga, D. & Thiel, M. 2014. Major consequences of minor damage: impacts of small grazers on fast-growing kelps. *Oecologia* 174: 789–801.

Poore, A.G.B., Campbell, A.H., Coleman, R.A., Edgar, G.J., Jormalainen, V., Reynolds, P.L., Sotka, E.E., Stachowicz, J.J., Taylor, R.B., Vanderklift, M.A. & Duffy, J.E. 2012. Global patterns in the impact of marine herbivores on benthic primary producers. *Ecology Letters* 15: 912–22.

Postels, A. & Ruprecht, F. 1840. *Illustrationes algarum:* pp. [i–vi], [i]–iv, 1–28 [1–2, index], [Latin:] [–iv], [1]–22, [1–2, index], 40 pls. Petropoli [St. Petersburg]: Typis Eduardi Pratz.

Pouvreau, V. 2010. Anti-microfouling activity of lipidic metabolites from the invasive brown alga *Sargassum muticum* (Yendo) Fensholt. *Marine Biotechnology* 12: 52–61.

Povey, A. & Keough, M.J. 1991. Effects of trampling on plant and animal populations on rocky shores. *Oikos* 61: 355–68.

Powell, H.T. 1957a. Studies in the genus *Fucus* L.I. *Fucus distichus* L. emend. Powell. *Journal of the Marine Biological Association of the United Kingdom* 36: 407–32.

Powell, H.T. 1957b. Studies in the genus *Fucus* L. II. Distribution and ecology of forms of *Fucus distichus* L. emend. Powell in Britain and Ireland. *Journal of the Marine Biological Association of the United Kingdom* 36: 663–93.

Powell, H.T. 1958. Occurrence of forms of *Fucus distichus* L. *emend.* Powell on North Rona and Sula Sgeir. *Nature, London* 182: 1246.

Powell, H.T. 1960. The typification of *Fucus spiralis* L. *British Phycological Bulletin* 2: 17.

Powell, H.T. 1963a. Speciation in the genus *Fucus* L., and related genera. *Systematics Association Publication* 5: 63–77.

Powell, H.T. 1963b. New records of *Fucus distichus* subspecies for the Shetland and Orkney Islands. *British Phycological Bulletin* 2: 247–54.

Powell, H.T. 1981. The occurrence of *Fucus distichus* subsp. *edentatus* in Macduff harbour, Scotland – the first record for mainland Britain. *British Phycological Journal* 16: 139.

Powell, H.T. & Lewis, J.R. 1952. Occurrence of *Fucus inflatus* L. forma *distichus* Börgesen on the North Coast of Scotland. *Nature, London* 169: 508, 509.

Poza, A.M., Gauna, M.C., Escobar, J.F. & Parodi, E.R. 2017. Heteromorphic phases of *Leathesia marina* (Ectocarpales, Ochrophyta) over time from northern Patagonia, Argentina. *Phycologia* 56: 579–89.

Prescott, G.W. 1969. *The Algae: A Review.* Thomas Nelson and Sons Ltd, London.

Price, J.H., John, D.M. & Lawson, G.W. 1978. Seaweeds of the Western coast of tropical Africa and adjacent islands: a critical assessment. II. Phaeophyta. *Bulletin of the British Museum (Natural History) Botany* 6: 87–182.

Price, J.H., Tittley, I. & Richardson, W.D. 1979. The distribution of *Padina pavonica* (L.) Lamour. (Phaeophyta: Dictyotales) on British and adjacent European shores. *Bulletin of the British Museum (Natural History) Botany* 7: 1–67.

Primo, C., Hewitt, C.L. & Campbell, M.L. 2010. Reproductive phenology of the introduced kelp *Undaria pinnatifida* (Phaeophyceae, Laminariales) in Port Phillip Bay (Victoria, Australia). *Biological Invasions* 12: 3081–92.

Prince, J.S. & LeBlanc, W.G. 1992. Comparative feeding preference of *Strongylocentrotus droebachiensis* (Echinoidea) for the invasive seaweed *Codium fragile* ssp. *tomentosoides* (Chlorophyceae) and four other seaweeds. *Marine Biology* 113: 159–63.

Pringsheim, N. 1862. Beiträge zur Morphologie der Meeres-Algen. *Physikalische Abhandlungen der Königlichen Akademie der Wissenschaften zu Berlin* 1862: 1–37, pls I–VIII.

Pringsheim, N. 1873. Ueber den Gang der morphologischen Differenzierung in der Sphacelarien – Reihe. *Physikalische Abhandlungen der Königlichen Akademie der Wissenschaften zu Berlin* 137–91, 1873 (Gesammelte Abhandl., 1, 359–414, 1895).

Pringsheim, N. 1874. Ueber den Gang der morphologischen Differenzierung in de Sphacelarien-Reihe. *Physikalische Abhandlungen der Königlichen Akademie der Wissenschaften in Berlin, Phys. Kl.* 1873: 137–91.

Printz, H. 1926. *Die Algenvegetation des Trondhjemsfjordes.* pp. [1]–273, [274], pls 1–10, map. Oslo: i kommision hos Jacob Dybwad.

Printz, H. 1950. Seasonal growth and production of dry matter in *Ascophyllum nodosum* (L.) Le Jol. *Det Norske Videnskaps Akademi i Oslo, I. Mat. Naturv. Klasse* 4: 3–15.

Printz, H. 1956a. Recuperation and recolonisation in *Ascophyllum*. *Proceedings of the International Seaweed Symposium* 2: 194–7.

Printz, H. 1956b. Phenology of *Ascophyllum nodosum*. *Proceedings of the International Seaweed Symposium* 2: 198–202.

Printz, H. 1959a. Investigations of the failure of recuperation and repopulation in cropped *Ascophyllum* areas. *Det Norske Videnskaps-Akademi i Oslo I. Mat.-Naturv. Klasse* 3: 1–15.

Printz, H. 1959b. Phenological studies of marine algae along the Norwegian Coast. I. *Ascophyllum nodosum* (L.) Le Jol. II. *Fucus vesiculosus* L. *Det Norske Videnskaps-Akademi i Oslo I. Mat.-Naturv. Klasse* 4: 3–28.

Printz, H. 1962. Phykochronologische Untersuchungen in Assoziationen von *Laminaria digitata* (L.) Lamour. f. *stenophylla* (Kütz.) Harv. an der norwegischen Westküste. *Archiv für Mikrobiologie* 42: 64–73.

Prinz, A. 2020. Evolutionary History of the *Fucus spiralis/ Fucus guiryi* Complex Large-scale Analysis of the Distribution Range and Genetic Structure across the Northern Hemisphere. Algarve: Universidade do Algarve.

Probyn, T.A. 1981. Aspects of the light and nitrogenous nutrient requirement for growth of *Chordaria flagelliformis* (O.F. Mull) C. Agardh. *Proceedings of the International Seaweed Symposium* 10: 339–44.

Probyn, T.A. 1984. Nitrate uptake by *Chordaria flagelliformis* (Phaeophyta). *Botanica Marina* 27: 271–5.

Probyn, T.A. & Chapman, A.R.O. 1982. Nitrogen uptake characteristics of *Chordaria flagelliformis* (Phaeophyta) in batch mode and continuous mode experiments. *Marine Biology, Berlin* 71: 129–33.

Probyn, T.A. & Chapman, A.R.O. 1983. Summer growth of *Chordaria flagelliformis* (O.F. Muell.) C. Ag.: physiological strategies in a nutrient stressed environment. *Journal of Experimental Marine Biology and Ecology* 73: 243–71.

Prud'homme van Reine, W.F. 1968. Einge gegevens over de veervormige *Sphacelaria* soorten in Nederland (Bruinwieren). *Gorteria* 4: 114–21.

Prud'homme van Reine, W.F. 1972. Notes on Sphacelariales (Phaeophyceae) II. On the identity of *Cladostephus setaceus* Suhr and remarks on European *Cladostephus*. *Blumea* 20: 138–44.

Prud'homme van Reine, W.F. 1974. Geographic distribution of European *Sphacelaria* species in the world (Phaeophyceae, Sphacelariales). *Bulletin de la Société Phycologique de France* 19: 171–7.

Prud'homme van Reine, W.F. 1978. Criteria used in systematic studies in the Sphacelariales. In: Irvine, D.E.G. & Price, J.H. (eds), *Modern approaches to the taxonomy of red and brown algae*, pp. 301–23.

Systematics Association Special Vol. 10. London: Academic Press.

Prud'homme van Reine, W.F. 1982a. A taxonomic revision of the European Sphacelariaceae (Sphacelariales, Phaeophyceae). *Leiden Botanical Series* 6: [i–x], 1–293, 660.

Prud'homme van Reine, W.F. 1982b. List of Localities. In: *Supplement to: A taxonomic revision of European Sphacelariaceae (Sphacelariales, Phaeophyta).* (Eds), pp. 1–86. Leiden: Rijksherbarium.

Prud'homme van Reine, W.F. 1991. *Stypocaulon* and *Halopteris* (Stypocaulaceae, Sphacelariales, Phaeophycophyta), reasons to separate these genera again. *Journal of Phycology* 27 (suppl.): 60.

Prud'homme van Reine, W.F. 1993. Sphacelariales (Phaeophyceae) of the world, a new synthesis. *Korean Journal of Phycology* 8: 145–60.

Prud'homme van Reine, W.F. & Nienhuis, P.H. 1982. Occurrence of the brown alga *Sargassum muticum* (Yendo) Fensholt in The Netherlands. *Botanica Marina* 25: 37–9.

Prud'homme van Reine, W.F. & Star, W. 1981. Transmission electron microscopy of apical cells of *Sphacelaria* spp. (Sphacelariales, Phaeophyceae). *Blumea* 27: 523–46.

Prud'homme van Reine, W.F., van der Wiele, P. & Bom, H. 1982. Studies on *Sargassum muticum* in the Netherlands. *British Phycological Journal* 17: 238.

Puspita, M., Déniel, M., Widowati, I., Radjasa, O.K., Douzenel, P., Marty, C., Vandanjon, L., Bedoux, G. & Bourgougnon, N. 2017. Total phenolic content and biological activities of enzymatic extracts from *Sargassum muticum* (Yendo) Fensholt. Proceedings of the 22nd International Seaweed Symposium (ISS) Location: Copenhagen, Denmark, June 19–24, 2016 Sponsor(s): International Seaweed Association Vol. 29: 2521–37.

Pybus, C. 1974. Some observations on the behaviour of excised segments from *Dictyota dichotoma* (Huds.) Lamour. and *Dilophus spiralis* (Mont.) Hamel in culture. *Irish Naturalists' Journal* 18: 25–7.

Pybus, C. 1975. A new record of *Corynophlaea crispa* (Harv.) Kuck. *Irish Naturalists' Journal* 18: 153–5.

Qing, P.L., Rong, M.G. & Min, Z.D. 1984. Studies on *Undaria pinnatifida*. *Proceedings of the International Seaweed Symposium* 11: 263–5.

Quatrano, R.S., Hogsett, W.S. & Roberts, M. 1979. Localization of a sulfated polysaccharide in the rhizoid wall of *Fucus distichus* (Phaeophyceae) zygotes. *Proceedings of the International Seaweed Symposium* 9: 113–23.

Quéguineur, B., Goya, L., Ramos, S., Angeles Martín, M., Mateos, R., Guiry, M.D. & Bravo, L. 2013. Effect of phlorotannin-rich extracts of *Ascophyllum nodosum* and *Himanthalia elongata* (Phaeophyceae) on cellular oxidative markers in human HepG2 cells. *Journal of Applied Phycology* 25: 1–11.

Rabenhorst, L. 1863. *Kryptogamen-Flora von Sachsen*: pp. i–xx, 1–653, Un-numbered figs. Leipzig: Verlag von Eduard Kummer.

Racault, M.-F.L.P., Fletcher, R.L., de Reviers, B., Cho, G.Y., Boo, S.M., Parente, M.I. & Rousseau, F. 2009. Molecular phylogeny of the brown algal genus *Petrospongium* Nägeli ex Kütz. (Phaeophyceae) with evidence for Petrospongiaceae fam. nov. *Cryptogamie, Algologie* 30: 111–23.

Raffo, M.P, Eyras, M.C. & Iribarne, O.O. 2009. The invasion of *Undaria pinnatifida* to a *Macrocystis pyrifera* kelp in Patagonia (Argentina, south-west Atlantic). *Journal of the Marine Biological Association of the United Kingdom* 89: 1571–80.

Ragan, M.A. 1976. Physodes and the phenolic compounds of brown algae. Composition and significance of physodes in vivo. *Botanica Marina* 19: 145–54.

Ragan, M.A. & Jensen, A. 1978. Quantitative studies in brown algal phenols. II. Seasonal variation in polyphenol content of *Ascophyllum nodosum* (L.) Le Jol. and *Fucus vesiculosus* (L.). *Journal of Experimental Marine Biology and Ecology* 34: 245–58.

Ramon, E. 1973. Germination and attachment of zygotes of *Himanthalia elongata* (L.) S.F. Gray. *Journal of Phycology* 9: 445–9.

Ramon, E. & Friedmann, I. 1966. The gametophyte of *Padina* in the Mediterranean. *Proceedings of the International Seaweed Symposium* 5: 183–96.

Ratcliff, J.J., Soler-Vila, A., Hanniffy, D., Johnson, M.P. & Edwards, M.D. 2017. Optimisation of kelp (*Laminaria digitata*) gametophyte growth and gametogenesis: effects of photoperiod and culture media. *Journal of Applied Phycology* 29: 1957–66.

Rautenberg, E. 1960. Zur Morphologie und Ökologie einiger epiphytischer und epi-endophytischer Algen. *Botanica Marina* 2: 133–45.

Ravanko, O. 1970. Morphological, developmental and taxonomic studies in the *Ectocarpus* complex (Phaeophyceae). *Nova Hedwigia* 20: 179–252.

Raven, J.A. & Samuelsson, G. 1988. Ecophysiology of *Fucus vesiculosus* L. at its northern limit in the Gulf of Bothnia. *British Phycological Journal* 23: 295.

Rawlence, D.J. 1972. An ultrastructural study of the relationship between rhizoids of *Polysiphonia lanosa* (L.) Tandy (Rhodophyceae) and tissue of *Ascophyllum nodosum* (L.) Le Jolis (Phaeophyceae). *Phycologia* 11: 279–90.

Rawlence, D.L. 1973. Some aspects of the ultrastructure of *Ascophyllum nodosum* (L.) Le Jolis (Phaeophyceae, Fucales), including observations on cell plate formation. *Phycologia* 12: 17–28.

Reed, D.C. 1987. Factors affecting the production of sporophylls in the giant kelp *Macrocystis pyrifera* (L.) C. Ag. *Journal of Experimental Marine Biology and Ecology* 113: 61–9.

Reed, D.C. & Foster, M.S. 1984. The effects of canopy shading on algal recruitment and growth in a giant kelp forest. *Ecology* 65: 937–48.

Reed, D.C., Ebeling, A.W., Anderson, T.W. & Anghera, M. 1996. Differential reproductive responses to fluctuating resources in two seaweeds with different reproductive strategies. *Ecology* 77: 300–16.

Reed, R.H. & Barron, J.A. 1983. Physiological adaptation to salinity change in *Pilayella littoralis* from marine and estuarine sites. *Botanica Marina* 26: 409–16.

Reed, R.H., Davison, I.R., Chudek, J.A. & Foster, R. 1985. The osmotic role of mannitol in the Phaeophyta: an appraisal. *Phycologia* 24: 35–47.

Reed, R.H., Wright, P.J., Chudek, J.A. & Hunter, G. 1995. Turnover of hexitols in the marine macroalga

Himanthalia elongata (Phaeophyta, Fucales). *European Journal Phycology* 30: 169–77.

Rees, E.M. 1933. Some observations on *Bifurcaria bifurcata* Stackh. *Annals of Botany* 47: 101–15.

Rees, T.K. 1932. A note on the longevity of certain species of the Fucaceae. *Annals of Botany* 46: 1062–4.

Rees, T.K. 1935. The marine Algae of Lough Ine. *Journal of Ecology* 23: 69–133.

Reinbold, T. 1893. Die Phaeophyceen (Brauntange) der Kieler Föhrde. *Schriften des Naturwissenschaftlichen Vereins für Schleswig-Holstein* 10: 21–59.

Reeves, S.E., Kriegisch, N., Johnson, C.R. & Ling, S.D. 2018. Reduced resistance to sediment-trapping turfs with decline of native kelp and establishment of an exotic kelp *Oecologia* 188: 1239–51.

Reinke, J. 1878a. Ueber die Entwicklung von *Phyllitis, Scytosiphon* und *Asperococcus*. *Jahrbücher für wissenschaftliche Botanik* 11: 262–73.

Reinke, J. 1878b. Entwicklungsgeschichtliche Untersuchungen über die Culteriaceen des Golfs von Neapel. *Nova Acta Academiae Caesareae Leopoldino-Carolinae Germanicae Naturae Curiosorum* 40: 59–96.

Reinke, J. 1878c. Entwicklungsgeschichtliche Untersuchungen über die Dictyotaceen des Golfs von Neapel. *Nova Acta Academiae Caesareae Leopoldino-Carolinae Germanicae Naturae Curiosorum* 40: 1–56.

Reinke, J. 1888a. Die braunen Algen (Fucaceen und Phaeosporeen) der Kieler Bucht. *Berichte der deutsche botanischen Gesellschaft* 6: 14–20.

Reinke, J. 1888b. Einige neue braune und grüne algen der Kieler Bucht. *Berichte der deutsche botanischen Gesellschaft* 6: 240–1.

Reinke, J. 1889a. *Atlas deutscher Meeresalgen*: Vol. 1, pp. [i–iv], 1–34, pls 1–4, 5/6, 7–11, 12/13, 14–25. Berlin: Paul Parey.

Reinke, J. 1889b. Algenflora der westlichen Ostsee deutschen Antheils. Eine systematisch-pflanzengeographische Studie. *Bericht der Kommission zur Wissenschaftlichen Untersuchung der deutschen Meere in Kiel* 6: [i–] iii–xi, 1–101.

Reinke, J. 1889c. Ein Fragment aus der Naturgeschichte der Tilopterideen. *Botanische Zeitung* 47: 101 and sequence.

Reinke, J. 1890. Übersicht der bisher bekannten Sphacelariaceen. *Berichte der Deutsche Botanischen Gesellschaft* 8: 201–15.

Reinke, J. 1891a. Beiträge zur vergleichenden anatomie und morphologie der Sphacelariaceen. *Bibliotheca Botanica* 5: 1–40, XIII plates.

Reinke, J. 1891b. *Atlas deutscher Meeresalgen*: Vol. 2 (1, 2), pp. 35–54, pls 26–35. Berlin: Paul Parey.

Reinke, J. 1892a. *Atlas deutscher Meeresalgen*: Vol. 2, pp. [i–iv], 55–70, pls 36–50. Berlin: Paul Parey.

Reinke, J. 1892b. Über Gaste der Ostserflors. *Berichte der deutsche botanischen Gesellschaft* 10: 4–12.

Reinsch, P.F. 1875. *Contributiones ad algologiam et fungologiam*. Vol. 1, pp. [i]–xii, [1]–103, [104, err.], 131 plates [I–III, IIIa, IV–VI, VIa, VII–XII, XIIa, XIII–XX, XXa, XXI–XXXV, XXXVa, XXXVI (Melanophyceae); I–XLII, XLIIa, XLIII–XLVII, XLVIIa, XLVIII–LXI (Rhodophyceae); I–XVIII (Chlorophyllophyceae); I–IX (Fungi)]. Norimbergae [Nürnberg]: Typis Theodor Haesslein.

Renard, P., Arbault, S., Kaas, R. & Perez, R. 1992. A method for the cryopreservation of the gametophytes of the food alga *Undaria pinnatifida* (Laminariales). *Comptes rendus de l'Académie des sciences. Série III*, 315: 445–51.

Reviers, B. de 2003. *Biologie et phylogénie des algues*. 2. Berlin, Paris.

Reviers, B. de & Rousseau, F. 1999. Towards a new classification of the brown algae. *Progress in Phycological Research* 13: 107–201.

Reviers, B. de, Rousseau, F. & Draisma, S.G.A. 2007. Classification of brown algae from past to present and current challenges. In: Brodie, J. & Lewis, J. (eds), *Unravelling the Algae: The Past, Present and Future of Algal Molecular Systematics*, pp. 267–84. The Systematics Association, London.

Reviers, B. de, Rousseau, F. & Silberfeld, T. 2015. 19. Class Phaeophyceae Kjellman. In: Frey, W. (ed.), *Syllabus of Plant Families, Adolf Engler's syllabus de Pflanzenfamilien, 13th edition, Part 2/1 Phototrophic eukaryotic Algae*, pp. 139–76.

Reyes, J. & Sansón, M. 1999. Estudio fenológico de dos poblaciones de *Fucus spiralis* en Tenerife, Islas Canarias (Fucales, Phaeophyta). *Vieraea* 27: 53–65.

Rheault, R.B. & Ryther, J.H. 1983. Note: Growth, yield and morphology of *Ascophyllum nodosum* (Phaeophyta) under continuous and intermittent seawater spray culture regimens. *Journal of Phycology* 19: 252–4.

Rhodes, R.G. 1970. Relation of temperature to development of the macrothallus of *Desmotrichum undulatum*. *Journal of Phycology* 6: 312–14.

Rhodes, R.G. & Connell, M.V. 1973. The biology of brown algae on the Atlantic coast of Virgina. II. *Petalonia fascia* and *Scytosiphon lomentaria*. *Chesapeake Science* 14: 211–15.

Ribamar da Cruz Freitas Junior, J. 2015. Desarrollo y aplicaciones del cultivo de Saccharina latissima (Laminariales, Ochrophyta) en sistemas de acuicultura multitrófica integrada (AMTI). Doctoral thesis.

Ribera, M.A., Gómez Garreta, A., Barceló, M.A. & Rull Lluch, J. 1996. Mapas de distribución de algas marinas de la Península Ibérica e Islas Baleares. VIII. *Cystoseira* C. Agardh y *Sargassum* C. Agardh. *Botanica Complutensis* 20: 89–103.

Ribera, M.A., Gómez Garreta, A., Gallardo, T., Cormaci, M., Furnari, G. & Giaccone, G. 1992. Check-list of Mediterranean seaweeds. I. Fucophyceae (Warming 1884). *Botanica Marina* 35: 109–30.

Rice, E.L. & Chapman, A.R.O. 1982. Net productivity of two cohorts of *Chordaria flagelliformis* (Phaeophyta) in Nova Scotia, Canada. *Marine Biology, Berlin* 71: 107–11.

Rice, E.L. & Chapman, A.R.O. 1985. A numerical taxonomic study of *Fucus distichus* (Phaeophyta). *Journal of the Marine Biological Association of the United Kingdom* 65: 433–59.

Richardson, J.P. 1979. Overwintering of *Dictyota dichotoma* (Phaeophyceae) near its northern distribution limit on the east coast of North America. *Journal of Phycology* 15: 22–6.

Richardson, W.D. & Walker, F.T. 1956. Perennial changes of *Laminaria cloustoni* Edm. (*L. hyperborea* (Gunn.) Fosl.) around Scotland. *Proceedings of the International Seaweed Symposium* 2: 203–9.

Rico, J.M. & Fernández, C. 1997. Ecology of *Sargassum muticum* on the North Coast of Spain II. Physiological differences between *Sargassum muticum* and *Cystoseira nodicaulis*. *Botanica Marina* 40: 405–10.

Rietema, H. & Hoek, C. van den. 1981. The life history of *Desmotrichum undulatum* (Phaeophyceae) and its regulation by temperature and light conditions. *Marine Ecology Progress Series* 4: 321–35.

Rindi, F. & Guiry, M.D. 2004. Composition and spatio temporal variability of the epiphytic macroalgal assemblage of *Fucus vesiculosus* Linnaeus at Clare Island, Mayo, western Ireland. *Journal of Experimental Marine Biology and Ecology* 311: 233–52.

Rintz, H. 1959. Phenological studies of marine algae along the Norwegian coast. I. *Ascophyllum nodosum* (L.) Le Jol. II. *Fucus vesiculosus* L. *Det Norske Videnskaps-Akademi i Oslo I. Mat.-Naturv. Klasse* 1959: 1–28.

Riouall, R. 1985. Sur la présence dans l'Etange de Thau (Hérault-France) de *Sphaerotrichia divaricata* (C. Ag.) Kylin et *Chorda filum* (L.) Stackhouse. *Botanica Marina* 28: 83–6.

Rismondo, A., Volpe, S., Curiel, D. & Solazzi, A. 1993. Segnalazione di *Undaria pinnatifida* (Harvey) Suringar a Chioggia (Laguna Veneta). *Lavori della Società veneziana di scienze naturali* 18: 329–30.

Ritter, A., Cabioch, L., Brillet-Guéguen, L., Corre, E., Cosse, A., Dartevelle, L., Duruflé, H., Fasshauer, C., Goulitquer, S., Thomas, F., Correa, J.A., Potin, P., Faugeron, S., Leblanc, C. & Gaquerel, E. 2017. Herbivore-induced chemical and molecular responses of the kelps *Laminaria digitata* and *Lessonia spicata*. *PLoS ONE* 12. https://doi.org/10.1371/journal.pone.0173315

Roberts, M. 1967a. Studies on marine algae of the British Isles 3. The genus *Cystoseira*. *British Phycological Bulletin* 3: 345–66.

Roberts, M. 1967b. Studies on marine algae of the British Isles 4. *Cystoseira baccata* (Gmelin) Silva. *British Phycological Bulletin* 3: 367–78.

Roberts, M. 1968a. Taxonomic and nomenclatural notes on the genus *Cystoseira* C. Ag. *Journal of the Linnean Society of London, Botany* 60: 251–64.

Roberts, M. 1968b. Studies on marine algae of the British Isles. 6. *Cystoseira foeniculacea* (Linnaeus) Greville. *British Phycological Bulletin* 3: 547–64.

Roberts, M. 1968c. Studies on marine algae of the British Isles: an amendment to 'The Genus *Cystoseira*'. *British Phycological Bulletin* 3: 565–6.

Roberts, M. 1970. Studies of the marine algae of the British Isles 8. *Cystoseira tamariscifolia* (Hudson) Papenfuss. *British Phycological Journal* 5: 201–10.

Roberts, M. 1977. Studies on marine algae of the British Isles. 9. *Cystoseira nodicaulis* (Withering) M. Roberts. *British Phycological Journal* 12: 175–99.

Roberts, M. 1978. Active speciation in the taxonomy of the genus *Cystoseira* C. Ag. In: Irvine, D.E.G. & Price, J.H. (eds) *Modern approaches to the taxonomy of Red and Brown Algae*, pp. 399–422. Systematics Association Special Vol. 10. London, New York. Academic Press.

Roberts, M. & Ring, F.M. 1972. Preliminary investigations into conditions affecting the growth of the microscopic phase of *Scytosiphon lomentaria* (Lyngb.) Link. *Mémoires de la Société Botanique de France* 1972: 117–28.

Robertson, B.L. 1987. Reproductive ecology and canopy structure of *Fucus spiralis* L. *Botanica Marina* 30: 475–82.

Robinson, W. 1932. Observations on the development of *Taonia atimaria* Ag. *Annals of Botany* 46: 113–20.

Robledo, D.R., Sosa, P.A., Garcia-Reina, G. & Müller, D.G. 1994. Photosynthetic performance of healthy and virus-infected *Feldmannia irregularis* and *F. simplex* (Phaeophyceae). *European Journal of Phycology* 29: 247–51.

Robuchon, M., Couceiro, L., Peters, A.K., Destombe, C. & Valero, M. 2014. Examining the bank of microscopic stages in kelps using culturing and barcoding. *European Journal of Phycology* 49: 128–33.

Rodrigues, M.A., Pereira dos Santos, C., Yoneshigue Valentin, Y., Strbac, D. & Hall, D.O. 2000. Photosynthetic light-response curves and photoinhibition of the deep-water *Laminaria abyssalis* and the intertidal *Laminaria digitata* (Phaeophyceae). *Journal of Phycology* 36: 97–106.

Rodrigues, M.A., Pereira dos Santos, C., Young, A.J., Strbac, D. & Hall, D.O. 2002. A smaller and impaired xanthophyll cycle makes the deep sea macroalgae *Laminaria abyssalis* (Phaeophyceae) highly sensitive to daylight when compared with shallow water *Laminaria digitata*. *Journal of Phycology* 38: 939–47.

Roeder, V., Collén, J., Rousvoal, S., Corre, E., Leblanc, C. & Boyen, C. 2005. Identification of stress gene transcripts in *Laminaria digitata* (Phaeophyceae) protoplast cultures by expressed sequence tag analysis. *Journal of Phycology* 41: 1227–35.

Roeleveld, J.G., Duisterhof, M. & Vroman, M. 1974. On the year cycle of *Petalonia fascia* in the Netherlands. *Netherlands Journal of Sea Research* 8: 410–26.

Rohde, S., Hiebenthal, C., Wahl, M., Karez, R. & Bischof, K. 2008. Decreased depth distribution of *Fucus vesiculosus* (Phaeophyceae) in Western Baltic: effects of light deficiency and epibionts on growth and photosynthesis. *European Journal of Phycology* 43: 143–50.

Rohde, S., Molis, M. & Wahl, M. 2004. Regulation of anti-herbivore defence by *Fucus vesiculosus* in response to various cues. *Journal of Ecology* 92: 1011–18.

Roleda, M.Y. 2009. Photosynthetic response of Arctic kelp zoospores exposed to radiation and thermal stress. *Photochemical and Photobiological Sciences* 8: 1302–12.

Roleda, M. & Dethleff, D. 2011. Storm-generated sediment deposition on rocky shores: simulating burial effects on the physiology and morphology of *Saccharina latissima* sporophytes. *Marine Biology Research* 7: 213–23.

Roleda, M., Hanelt, D., Kräbs, G. & Wiencke, C. 2004. Morphology, growth, photosynthesis and pigments in *Laminaria ochroleuca* (Laminariales, Phaeophyta) under ultraviolet radiation. *Phycologia* 43: 603–13.

Roleda, M.Y., Hanelt, D. & Wiencke, C. 2006. Exposure to ultraviolet radiation delays photosynthetic recovery in Arctic kelp zoospores. *Photosynthesis Research* 88: 311–22.

Roleda, M., Lüder, U. & Wiencke, C. 2010. UV-susceptibility of zoospores of the brown macroalga *Laminaria digitata* from Spitsbergen. *Polar Biology* 33: 577–88.

Rönnberg, O., Ådjers, K., Ruokolahti, C. & Bondestam, M. 1990. *Fucus vesiculosus* as an indicator of heavy metal availability in a fish farm recipient in the Northern Baltic Sea. *Marine Pollution Bulletin* 21: 388–92.

Rosenvinge, L.K. 1893. Grønlands Havalger. *Meddelelser om Grønland* 3: 763–981.

Rosenvinge, L.K. 1894. Les algues marines du Groenland. *Annales des Sciences Naturelles, Botanique Ser. 7*, 19: 53–164.

Rosenvinge, L.K. 1899. Deuxième mémoire sur les algues marines du Groenland. *Meddelelser om Grønland* 20: 1–128.

Rosenvinge, L.K. & Lund, S. 1935. On some Danish Phaeophyceae. *Det Kongelige Danske Videnskabernes Selskabs. Biologiske Skrifter, Naturv. og Mathem. Afd.* 6: 1–40.

Rosenvinge, L.K. & Lund, S. 1941. The marine algae of Denmark. Vol. II. Phaeophyceae. I. Ectocarpaceae and Acinetosporaceae. *Det Kongelige Danske Videnskabernes Selskabs. Biologiske Skrifter 7 Raekke, Afd.* 1: 1–79.

Rosenvinge, L.K. & Lund, S. 1943. The marine algae of Denmark. Vol. II. Phaeophyceae, II. Corynophlaeaceae, Chordariaceae, Acrothricaceae, Spermatochnaceae, Sporochnaceae, Desmarestiaceae, Arthrocladiaceae with supplementary comments on Elachistaceae. *Det Kongelige Danske Videnskabernes Selskabs. Biologiske Skrifter* 2: 1–59.

Rosenvinge, L.K. & Lund, S. 1947. The marine algae of Denmark. Vol. II. Phaeophyceae. III. Encoeliaceae, Myriotrichiaceae, Giraudiaceae, Striariaceae, Dictyosiphonaceae, Chordaceae and Laminariaceae. *Det Kongelige Danske Videnskabernes Selskabs. Biologiske Skrifter* 4: 1–99.

Ross, R. 1958. The type species of *Bifurcaria* Stackhouse. *Journal of the Linnean Society of London, Botany* 55: 753–4.

Ross, R. 1961. The type species of *Bifurcaria* Stackhouse – a correction. *Journal of the Linnean Society of London, Botany* 56: 512.

Rostafinski, J. 1877. *Ueber das Spitzenwachstum von Fucus vesiculosus und Himanthalia lorea*. Leipzig, 1876 (see also *Bot. Zeit.* 35, 613–16, 1877).

Roth, A.W. 1797. *Catalecta botanica*: Fasc. 1. pp. [i]–viii, [1]–244 [1–2, index, pls], pls I–VIII. Lipsiae [Leipzig]: in Bibliopolo I.G. Mülleriano.

Roth, A.W. 1800a. *Catalecta botanica*: Fasciculus secundus cum tabulis aeneis. IX. pp. [i–x], [i]–258, [1–2, add.], [1–2 index icon.], [1–5, index, 6–7 err., 8 note], IX pls. Lipsiae [Leipzig]: in Bibliopolio Io. Fr. Gleditschiano.

Roth, A.W. 1800b. *Tentamen florae Germanicae*: Tomus III. Pars prior. pp. [i]–vii, 1–578. Lipsiae [Leipzig]: In Bibliopolio Gleditschiano.

Roth, A.G. 1806. *Catalecta botanica*: Fasciculus tertius cum tabulis aenaeis. XII. pp. [i–viii], [1]–350, [1–2, index pi.], [1–6, index] [1, err.], pls I–XII. Lipsiae [Leipzig]: in Bibliopolio Io. Fr. Gleditschiano.

Rothäusler, E., Haavisto, F. & Jormalainen, V. 2017. Is the future as tasty as the present? Elevated temperature and hyposalinity affect the quality of *Fucus* (Phaeophyceae, Fucales) as food for the isopod *Idotea balthica*. *Marine Biology*, 164: 11. https://doi.org/10.1007/s00227-017-3237-3

Rothäusler, E., Sjöroos, J., Heye, K. & Jormalainen, V. 2016. Genetic variation in photosynthetic performance and tolerance to osmotic stress (desiccation, freezing, hyposalinity) in the rocky littoral foundation species *Fucus vesiculosus* (Fucales, Phaeophyceae). *Journal of Phycology* 52: 877–87.

Round, F.E. 1973. *The Biology of the Algae*. Edward Arnold, New York.

Rousseau, F. & de Reviers, B. 1999a. Circumscription of the order Ectocarpales (Phaeophyceae): bibliographical synthesis and molecular evidence. *Cryptogamie, Algologie* 20: 5–18.

Rousseau, F. & de Reviers, B. 1999b. Phylogenetic relationships within the Fucales (Phaeophyceae) based on combined partial SSU þ LSU rDNA sequence data. *European Journal of Phycology* 34: 53–64.

Rousseau, F., de Reviers, B., Leclerc, M.-C., Asensi, A. & Delépine, R. 2000. Adenocystaceae fam. nov. (Phaeophyceae) based on morphological and molecular evidence. *European Journal of Phycology* 35: 35–43.

Rousseau, F., Burrowes, R., Peters, A.F., Kuhlenkamp, R. & de Reviers, B. 2001. A comprehensive phylogeny of the Phaeophyceae based on nrDNA sequences resolves the earliest divergences. *Compte Rendu Hebdomadaire des Séances de l'Académie des Sciences, Paris, Sciences de la Vie* 324: 305–19.

Rousseau, F., Leclerc, M.C. & de Reviers, B. 1997. Molecular phylogeny of European Fucales (Phaeophyceae) based on partial large-subunit rDNA sequence comparisons. *Phycologia* 36: 438–46.

Roussel, H.F.A. 1806. *Flore du Calvados*: pp. [1]6–340, [2]. A Caen: De l'imprimerie de F. Poisson, rue Froide-Rue.

Roxas Clemente y Rubio, S. de 1807. *Ensayo sobre las variedades de la vid común que vegetan en Andalucía*: pp. [i]–xviii + [1]–324, 1 pl. Madrid: En la imprenta de Villalpando.

Rubin, E., Rodriguez, P., Herrero, R., Cremades, J., Barbara, I. & de Vicente, M.E.S. 2005. Removal of Methylene Blue from aqueous solutions using as biosorbent *Sargassum muticum*: an invasive macroalga in Europe. *Journal of Chemical Technology and Biotechnology* 80: 291–8.

Rueness, J. 1973. Pollution effects on littoral algal communities in the inner Oslofjord, with special reference to *Ascophyllum nodosum*. *Helgoländer Wissenschaftliche Meeresuntersuchungen* 24: 446–54.

Rueness, J. 1974. Life history in culture and chromosome number in *Isthmoplea sphaerophora* (Phaeophyceae) from southern Scandinavia. *Phycologia* 13: 323–8.

Rueness, J. 1977. *Norsk Algeflora*. pp. 1–266. Oslo: Universitetsforlaget.

Rueness, J. 1985. Japweed – *Sargassum muticum* – biological pollution of European waters. *Blyttia* 43: 71–4.

Rueness, J. 1989. *Sargassum muticum* and other introduced Japanese macroalgae: biological pollution of European coasts. *Marine Pollution Bulletin* 20: 173–6.

Rueness, J. 1997. Algae. In: Brattegard, T. & Holthe, T. (eds) *Distribution of marine, benthic macro-organisms in Norway. A tabulated catalogue. preliminary Edition*. Research Report No. 1997–1. Trondheim: Direktoratet for Naturforvaltning.

Rueness, J., Brattegard, T., Lein, J.E., Küfner, R., Pedersen, A. & Sørlie, A.C. 1997. Class Phaeophyceae (division Chromophyta). In: Brattegard, T. & Holthe, T. (eds), *Distribution of marine, benthic macro-organisms in Norway*, pp. 46–52. Research Report for DN 1997–1. Directorate for Nature Management.

Rueness, J., Brattegard, T., Lein, J.E., Küfner, R., Pedersen, A. & Sørlie, A.C. 2001. Distribution of marine, benthic macroorganisms in Norway. A tabulated catalogue. Class Phaeophyceae (division Chromophyta). In: Brattegard, T. & Holthe, T. (eds) *Distribution of Marine, Benthic Macro-organisms in Norway: A Tabulated Catalogue*. Preliminary Edition. Research Report No. 1997–1: 1–394.

Rueness, J., Mathisen, H.A. & Tananger, T. 1987. Culture and field observations on *Gracilaria verrucosa* (Huds.) Papenf. (Rhodophyta) from Norway. *Botanica Marina* 30: 267–76.

Rueter, J.G. & Robinson, D.H. 1986. Inhibition of carbon uptake and stimulation of nitrate uptake at low salinities in *Fucus distichus* (Phaeophyceae). *Journal of Phycology* 22: 243–6.

Rugg, D.A. & Norton, T.A. 1987. *Pelvetia canaliculata*, a high-shore seaweed that shuns the sea. In: Crawford, R.M.M. (ed.), *Plant Life in Aquatic and Amphibious Habitats*, pp. 347–58. Oxford: Blackwell Scientific Publications.

Ruggiero, M.A., Gordon, D.P., Orrell, T.M., Bailly, N., Bourgoin, T., Brusca, R.C., Cavalier-Smith, T.C., Guiry, M.D. & Kirk, P.M. 2015. A higher level classification of all living organisms. *PloS One* 10(4): e0119248.

Ruiz-Medina, M.A., Gonzales-Rodríguez, Á.M. & Sansón, M. 2021. Distinctive morphological characteristics of *Fucus guiryi* (Fucales, Phaeophyceae) from Canary Islands, subtropical eastern Atlantic Ocean. *Phytotaxa* 52: 15–26.

Rull Lluch, J., Gómez Garreta, A., Barceló, M.C. & Ribera, M.A. 1994. Mapas de distribución de algas marinas de la Península Ibérica e Islas Baleares. VII. *Cystoseira* C. Agardh (Grupo *C. baccata*) y *Sargassum* C. Agardh (*S. muticum* y *S. vulgare*). *Botanica Complutensis* 19: 131–8.

Rull Lluch, J., Ribera, M.A., Barceló, M.C. & Gómez Garreta, A. 2005. Mapas de distribución de algas marinas de la Península Ibérica e Islas Baleares. XVIII. *Dictyota dichotoma*, *D. linearis* y *D. mediterranea* (Dictyotales, Fucophyceae). *Botanica Complutensis* 29: 61–70.

Ruprecht, F.J. 1850. *Algae ochotenses*. Die ersten sicheren Nachrichten über die Tange des Ochotskischen Meeres. pp. 1–243, 10 pls. St. Petersburg: Buchdruckerei der Kaiserlichen Akademie der Wissenschaften.

Russell-Wells, B. 1932. Fats of brown seaweeds. *Nature, London* 14: 33–42.

Russell, G. 1958. The autecology of *Pylaiella littoralis* (L.) Kjellm. *Proceedings of the International Seaweed Symposium* 3: 34–5.

Russell, G. 1961a. The autecology and life-history of *Pylaiella littoralis* (L.) Kjellman. *Reports of the Challenger Society* 3: 30–1.

Russell, G. 1961b. The taxonomic status of *Pylaiella rupincola*. *British Phycological Bulletin* 2: 101.

Russell, G. 1963. A study in populations of *Pylaiella littoralis*. *Journal of the Marine Biological Association of the United Kingdom* 43: 469–83.

Russell, G. 1964a. Systematic position of *Pilayella littoralis* and status of the order Dictyosiphonales. *British Phycological Bulletin* 2: 322–6.

Russell, G. 1964b. *Laminariocolax tomentosoides* on the Isle of Man. *Journal of the Marine Biological Association of the United Kingdom* 44: 601–12.

Russell, G. 1966. The genus *Ectocarpus* in Britain I. The attached forms. *Journal of the Marine Biological Association of the United Kingdom* 46: 267–94.

Russell, G. 1967a. The genus *Ectocarpus* in Britain II. The free-living forms. *Journal of the Marine Biological Association of the United Kingdom* 47: 233–50.

Russell, G. 1967b. The ecology of some free-living Ectocarpaceae. *Helgoländer Wissenschaftliche Meeresuntersuchungen* 15: 155–62.

Russell, G. 1970. Rhizoid production in excised *Dictyota dichotoma*. *British Phycological Journal* 5: 243–5.

Russell, G. 1971. Marine algal reproduction in two British estuaries. *Vie et milieu. Suppl.* 22: 219–30.

Russell, G. 1972. Phytosociological studies on a two-zone shore I. Basic pattern. *Journal of Ecology* 60: 539–45.

Russell, G. 1973a. The Phaeophyta: a synopsis of some recent developments. *Oceanography Marine Biology Annual Review* 11: 45–88.

Russell, G. 1973b. The litmus line; a reassessment. *Oikos* 24: 158–61.

Russell, G. 1973c. Phytosociological studies on a two-zone shore II. Community structure *Journal of Ecology* 61: 525–53.

Russell, G. 1974. *Fucus distichus* communities in Shetland. *Journal of Applied Ecology* 11: 679–84.

Russell, G. 1977. Vegetation on rocky shores at some North Irish sea sites. *Journal of Ecology* 65: 485–95.

Russell, G. 1979. Heavy receptacles in estuarine *Fucus vesiculosus* L. *Estuarine, Coastal and Marine Science* 9: 659–61.

Russell, G. 1980. Applications of simple numerical methods to the analysis of intertidal vegetation. In: Price, J.H., Irvine, D.E.G. & Farnham, W.F. (eds), *The shore environment 1: methods*, pp. 171–92. Systematics Association Special Volume 17(a). Academic Press, London and New York.

Russell, G. 1983a. Formation of an ectocaroid epiflora on blades of *Laminaria digitata*. *Marine Ecology Progress Series* 11: 181–7.

Russell, G. 1983b. Parallel growth patterns in algal epiphytes and *Laminaria* blades. *Marine Ecology Progress Series* 13: 303–4.

Russell, G. 1985a. Recent evolutionary changes in the algae of the Baltic Sea. *British Phycological Journal* 20: 87–104.

Russell, G. 1985b. Some anatomical and physiological differences in *Chorda filum* from coastal waters of Finland and Great Britain. *Journal of the Marine Biological Association of the United Kingdom* 65: 343–9.

Russell, G. 1986. Variation and natural selection in marine macroalgae. *Oceanography Marine Biology Annual Review* 24: 309–77.

Russell, G. 1987a. Salinity and seaweed vegetation. In: Crawford, R.M.M. (ed.), *Plant Life in Aquatic and Amphibious Habitats*, pp. 35–52. Blackwell, Oxford.

Russell, G. 1987b. Spatial and environmental components of evolutionary change: interactive effects of salinity and temperature on *Fucus vesiculosus* as an example. *Helgoländer Meeresuntersuchungen* 41: 371–6.

Russell, G. 1987c. Note on Baltic *Chorda filum*. *British Phycological Journal* 22: 311.

Russell, G. 1988. The seaweed flora of a young semi-enclosed sea: The Baltic. Salinity as a possible agent of flora divergence. *Helgoländer Meeresuntersuchungen* 42: 243–50.

Russell, G. & Bolton, J.J. 1975. Euryhaline ecotypes of *Ectocarpus siliculosus* (Dillw.) Lyngb. *Estuarine, Coastal and Marine Science* 3: 91–4.

Russell, G. & Fielding, A.H. 1974. The competitive properties of marine algae in culture. *Journal of Ecology* 62: 689–98.

Russell, G. & Fielding, A.H. 1981. Individuals, populations and communities. In: Lobban, C.S. & Wynne, M.J. (eds), *The Biology of Seaweeds*, pp. 393–420. Botanical Monographs 17. Blackwell Scientific Publications, Oxford.

Russell, G. & Fletcher, R.L. 1975. A numerical taxonomic study of the British Phaeophyta. *Journal of the Marine Biological Association of the United Kingdom* 55: 763–83.

Russell, G. & Morris, O.P. 1970. Copper tolerance in the marine fouling alga *Ectocarpus siliculosus*. *Nature, London* 228: 288–9.

Russell, G. & Morris, O.P. 1971. A ship model in anti-fouling research. *Sea Breezes* July: 512–13.

Russell, G. & Pedersen, P.M. 1994. *Microspongium globosum* (Algae: Fucophyceae). An addition to the marine flora of Finland. *Annals of Botany, Fennici* 31: 143–6.

Russell, G. & Veltkamp, C.J. 1984. Epiphyte survival on skin-shedding macrophytes. *Marine Ecology progress Series* 18: 149–53.

Russell, G. & Wareing, A.-M. 1979. The *Herponema velutinum–Himanthalia elongata* coincidence. *British Phycological Journal* 14: 128.

Russell, L.K., Hepburn, C.D., Hurd, C.L. & Stuart, M.D. 2008. The expanding range of *Undaria pinnatifida* in southern New Zealand: distribution, dispersal mechanisms and the invasion of wave exposed environments. *Biological Invasions* 10: 103–15.

Rönnberg, O. & Ruokolahti, C. 1986. Seasonal variation of algal epiphytes and phenolic content of *Fucus vesiculosus* in a northern Baltic archipelago. *Annales Botanici Fennici* 23: 317–23.

Rönnberg, O., Lehto, J. & Haahtela, I. 1985. Recent changes in the occurrence of *Fucus vesiculosus* in the Archipelago Sea, SW Finland. *Annales Botanici Fennici* 22: 231–44.

Rönnberg, O., Östman, T. & Ådjers, K. 1992b. Effects of ferry traffic on the metal content of *Fucus vesiculosus* in the Åland archipelago, northern Baltic Sea. In: Sjögren, E., Wallentinus, I. & Snoeijs, P. (eds), *Phycological studies of Nordic coastal waters – a festschrift dedicated to Prof. Mats Wærn on his 80th birthday*, Vol. 78, pp. 95–9. Uppsala: Opulus Press AB.

Rönnberg, O., Ådjers, K., Ruokolahti, C. & Bondestam, M. 1990. *Fucus vesiculosus* as an indicator of heavy metal availability in a fish farm recipient in the northern Baltic Sea. *Marine Pollution Bulletin* 21: 388–92.

Rönnberg, O., Ådjers, K., Ruokolahti, C. & Bondestam, M. 1992a. Effects of fish farming on growth, epiphytes and nutrient content of *Fucus vesiculosus*

L. in the Åland Archipelago, northern Baltic Sea. *Aquatic Botany* 42: 109–20.

Rothäusler, E., Sjöroos, J., Heye, K. & Jormalainen, V. 2016. Genetic variation in photosynthetic performance and tolerance to osmotic stress (desiccation, freezing, hyposalinity) in the rocky littoral foundation species *Fucus vesiculosus* (Fucales, Phaeophyceae). *Journal of Phycology* 52: 877–87.

Sabour, B., Reani, A., El-Magouri, H. & Haroun, R. 2013. *Sargassum muticum* (Yendo) Fensholt (Fucales, Phaeophyta) in Morocco, an invasive marine species new to the Atlantic coast of Africa. *Aquatic Invasions* 8: 97–102.

Sáez, C.A., Ramesh, K., Greco, M., Bitonti, M.B. & Brown, M.T. 2015. Enzymatic antioxidant defences are transcriptionally regulated in Es524, a copper-tolerant strain of *Ectocarpus siliculosus* (Ectocarpales, Phaeophyceae). *Phycologia* 54: 425–9.

Saito, Y. 1956a. An ecological study of *Undaria pinnatifida* Sur. I: on the influence of environmental factors upon the development of gametophytes. *Bulletin of the Japanese Society of Scientific Fisheries* 22: 229–34 (in Japanese).

Saito, Y. 1956b. An ecological study of *Undaria pinnatifida* Sur. II: on the influence of environmental factors upon the maturity of gametophytes and early development of sporophytes. *Bulletin of the Japanese Society of Scientific Fisheries* 22: 235–9 (in Japanese).

Saito, Y. 1962. Fundamental studies on the propagation of *Undaria pinnatifida* (Harv.) *Contributions of the Fisheries Laboratory of the Faculty of Agriculture of the University of Tokyo* 3: 1–102 (in Japanese with English summary).

Saito, Y. 1975. *Undaria*. In: Tokida, J. & Hirose, H. (eds), *Advance of Phycology in Japan*, pp. 304–20.

Saito, Y., Hirano, T., Niwa, K., Suzuki, T., Fukunishi, N., Abe, T. & Kawano, S. 2016. Phenotypic differentiation in the morphology and nutrient uptake kinetics among *Undaria pinnatifida* cultivated at six sites in Japan. *Journal of Applied Phycology* 28: 3447–58.

Salgado, J.M. & Niell, F.X. 1974. The structure of a population of epiphytes on *Himanthalia elongata*. *Investigación Pesquera* 38: 305–24.

Salinas, J.M., Llera, E.M. & Fuertes, C. 1996. Sobre la presencia de *Undaria pinnatifida* (Harvey) Suringar (Laminariales, Phaeophyta) en Asturias (mar Cantábrico). *Boletín del Instituto Español de Oceanografía* 12: 77–9.

Salinas, J.M., Cremades, J., Peteiro, C. & Fuertes, C. 2006. Influencia de las características del hilo de semilla en el cultivo industrial de *Undaria pinnatifida* y *Laminaria saccharina* (Laminariales, Phaeophyta). *Boletín del Instituto Español de Oceanografía* 22: 65–72.

Sánchez, I. & Fernández, C. 2005. Impact of the invasive seaweed *Sargassum muticum* (Phaeophyta) on an intertidal macroalgal assemblage. *Journal of Phycology* 41: 923–30.

Sánchez, I., Fernández, C. & Arrontes, J. 2005. Long-term changes in the structure of intertidal assemblages following invasion by *Sargassum muticum* (Phaeophyta). *Journal of Phycology* 41: 942–9.

Sánchez de Pedro, R., Fernández, A., García-Sánchez, M.J., Flores-Moya, A. & Bañares-España, E. 2019. When the microclimate does matter: Differences in the demographic, morphometric and reproductive variables of *Fucus guiryi* from two nearby populations. *Algas* 55: 40–41.

Sánchez de Pedro, R., Fernández, A.N., García-Sanchez, M.J., Flores-Moya, A. & Bañares-España, E. 2022. Seasonal and ontogenetic variability in the photosynthetic thermal tolerance of early-life stages of *Fucus guiryi* (Phaeophyceae, Fucales). *Phycologia* 61: 363–74.

Sancholle, M. 1988. Présence de *Fucus spiralis* (Phaeophyceae) en Méditerranee occidentale. *Cryptogamie, Algologie* 9: 157–61.

Sanderson, J.C. 1990. A preliminary survey of the distribution of the introduced macroalga, *Undaria pinnatifida* (Harvey) Suringar on the East Coast of Tasmania, Australia. *Botanica Marina* 33: 153–7.

Sanderson, J.C. 1997. Survey of *Undaria pinnatifida* in Tasmanian coastal waters, January–February 1997. Report to the Tasmanian Department of Marine Resources, Hobart, Australia.

Sansón, M., Martín, M.J. & Reyes, J. 2006. Vegetative and reproductive morphology of *Cladosiphon contortus*, *C. occidentalis* and *C. cymodoceae* sp. nov. (Ectocarpales, Phaeophyceae) from the Canary Islands. *Phycologia* 45: 529–45.

Santelices, B. 1990. Patterns of reproduction, dispersal and recruitment in seaweeds. *Oceanography Marine Biology Annual Review* 28: 177–276.

Santelices, B., Ramírez, M.E. & Abbott, I.A. 1989. A new species and new records of marine algae from Chile. *British Phycological Journal* 24: 73–82.

Santiago Caamaño, J., Durán Neira, C. & Acuña Castro-viejo, R. 1990. Aparición de *Undaria pinnatifida* en las costas de Galicia (España). Un nuevo caso en la problemática de introducción de especies foráneas. *Informes técnicos del Centro de Investigaciones Submarinas* 3: 1–45.

Santiañez, W.J.E. & Kogame, K. 2017. Transfer of *Petalonia filiformis* (Batters) Kuntze to the genus *Planosiphon* McDevit & G.W. Saunders (Scytosiphonaceae, Phaeophyceae). *Notulae Algarum* 40: 1–3.

Santiañez, W.J.E., Macaya, E.C., Lee, K.M., Cho, G.Y., Boo, S.M. & Kogame, K. 2018a. Taxonomic reassessment of the Indo-Pacific Scytosiphonaceae (Phaeophyceae): *Hydroclathrus rapanuii* sp. nov. and *Chnoospora minima* from Easter Island, with proposal of *Dactylosiphon* gen. nov. and *Pseudochnoospora* gen. nov. *Botanica Marina* 61: 47–64.

Santiañez, W.J.E., Lee, K.M., Uwai, S., Kurihara, A., Geraldino, P.J.L., Ganzon-Fortes, E.T., Boo, S.M. & Kogame, K. 2018b. Untangling nets: elucidating the diversity and phylogeny of the clathrate brown algal genus *Hydroclathrus*, with the description of a new genus *Tronoella* (Scytosiphonaceae, Phaeophyceae). *Phycologia* 57: 61–78.

Sasaki, H. & Kawai, H. 2007. Taxonomic revision of the genus *Chorda* (Chordaceae, Laminariales) on the basis of sporophyte anatomy and molecular phylogeny. *Phycologia* 46: 10–21.

Sasaki, H., Flores-Moya, A., Henry, E.C., Müller, D.G. & Kawai, H. 2001. Molecular phylogeny of Phyllariaceae, Halosiphonaceae and Tilopteridales (Phaeophyceae). *Phycologia* 40: 123–34.

Saunders, De A. 1898. Phycological memoirs. *Proceedings of the California Academy of Sciences. Series 3, Botany* 1: 147–68, pls XII–XXXII.

Saunders, G.W. & Druehl, L.D. 1992. Nucleotide sequences of the small-subunit ribosomal RNA genes from selected Laminariales (Phaeophyta): Implications for kelp evolution. *Journal of Phycology* 28: 544–9.

Saunders, G.W. & Druehl, L.D. 1993. Revision of the kelp family Alariaceae and the taxonomic affinities of *Lessoniopsis* Reinke (Laminariales, Phaeophyta). *Proceedings of the International Seaweed Symposium* 14: 689–97.

Saunders, G.W. & Kraft, G.T. 1995. The phylogenetic affinities of *Notheia anomala* (Fucales, Phaeophyceae) as determined from partial small-subunit rRNA gene sequences. *Phycologia* 34: 382–9.

Saunders, G.W. & McDevit, D.C. 2013. DNA barcoding unmasks overlooked diversity improving knowledge on the composition and origins of the Churchill algal flora. *BMC Ecology* 13: 1–23.

Saunders, G.W. & McDevit, D.C. 2014. A DNA barcode survey of Haida Gwaii kelp (Laminariales, Phaeophyceae) reveals novel ecological and distributional observations and *Saccharina druehlii* sp. nov. *Botany*. https://doi.org/10.1139/cjb-2014-0119.

Sauvageau, C. 1892. Sur quelques algues phéosporées parasites. *Journal de Botanique* [Morot] 6: 36–43, 55–9, 76–80, 90–6, 97–106, [1–48 (reprint nos)], 4 plates.

Sauvageau, C. 1895a. Note sur l'*Ectocarpus battersii* Bornet. *Journal de Botanique* [Morot] 9: 351–64.

Sauvageau, C. 1895b. Note sur l'*Ectocarpus pusillus* Griffiths. *Journal de Botanique* [Morot] 9: 274–80, 281–91, 307–18.

Sauvageau, C. 1895c. Note sur l'*Ectocarpus tomentosus* Lyngbye. *Journal de Botanique* [Morot] 9: 153–6, 157–66.

Sauvageau, C. 1895d. Sur les sporanges pluriloculaires de l'*Asperococcus compressus* Griff. *Journal de Botanique* [Morot] 9: 336–8.

Sauvageau, C. 1896a. Remarques sur la reproduction des Phéosporées et en particulier des *Ectocarpus*. *Annales des Sciences Naturelles, Botanique, série 8*, 2: 223–74.

Sauvageau, C. 1896b. Observations relatives à la sexualité des Phéosporées. *Journal de Botanique* [Morot] 10: 357–67, 388–98.

Sauvageau, C. 1896c. Sur l'*Ectocarpus virescens* Thuret et ses deux sortes des sporanges pleuriloculaires. *Journal de Botanique* [Morot] 10: 98–107, 113–26.

Sauvageau, C. 1896d. Note sur le *Strepsithalia*. *Journal de Botanique* [Morot] 10: 53–65.

Sauvageau, C. 1896e. Sur la conjugaison des zoospores de l'*Ectocarpus siliculosus*. *Compte Rendu Hebdomadaire des Séances de l'Académie des Sciences, Paris* 123: 431–3 (see also *Mémoires de la Société Impériale des Sciences Naturelles de Cherbourg?* 30, 294–304, 1897).

Sauvageau, C. 1897a. Sur quelques Myrionémacées. *Annales des Sciences Naturelles, Botanique, série 8*, 5: 161–288.

Sauvageau, C. 1897b. Note préliminaire sur les algues marines du golfe de Gascogne. *Journal de Botanique* [Morot] 11: 166–79, 202–14, 252–7, 263–88, 301–11.

Sauvageau, C. 1897c. Observations relative a la sexualite des Pheosporees. III. *Ectocarpus lebelii*. *Journal de Botanique* [Morot] 11: 5–14.

Sauvageau, C. 1897d. Sur les antheridies du *Taonia atomaria*. *Journal de Botanique* [Morot] 11: 86–90.

Sauvageau, C. 1897e. Observations relative a la sexualite des Pheosporees. II. *Ectocarpus padinae* (*Giffordia* Buffh.). *Journal de Botanique* [Morot] 11: 24–33.

Sauvageau, C. 1898a. Sur l'*Acinetospora pusilla* et la sexualité des Tilopteridées. *Compte Rendu Hebdomadaire des Séances de l'Académie des Sciences, Paris* 126: 1581–3.

Sauvageau, C. 1898b. Sur la sexualite et les affinities des Sphacelariacees. *Compte Rendu Hebdomadaire des Séances de l'Académie des Sciences, Paris* 126: 1672–5.

Sauvageau, C. 1899a. Les *Acinetospora* et la sexualité des Tiloptéridacées. *Journal de Botanique* [Morot] 13: 107–27.

Sauvageau, C. 1899b. Les Cutleriacees et leur alternance der generations. *Annales des Sciences Naturelles, Botanique, série 8 Bot.* 10: 265–362.

Sauvageau, C. 1900. Remarques sur les Sphacélariacées. *Journal de Botanique* [Morot] 14: 213–34, 247–59, 304–12, 313–22.

Sauvageau, C. 1901. Remarques sur les Sphacélariacées. *Journal de Botanique* [Morot] 15: 22–36, 50–62, 105–16, 137–49, 222–36, 237–55, 368–80, 408–18.

Sauvageau, C. 1902. Remarques sur les Sphacélariacées (suite). *Journal de Botanique* [Morot] 16: 325–49, 379–92, 393–416.

Sauvageau, C. 1903. Remarques sur les Sphacélariacées. *Journal de Botanique* [Morot] 17: 45–56, 69–95, 332–53, 378–422.

Sauvageau, C. 1904. Remarques sur les Sphacélariacées. *Journal de Botanique* [Morot] 18: 321–480.

Sauvageau, C. 1906. Sur les pousses indefinites dresses du *Cladostephus verticillatus*. *Actes de la Société linnéenne de Bordeaux* 61: 69–94.

Sauvageau, C. 1907a. Sur la germination et les affinities des *Cladostephus*. *Compte Rendu Hebdomadaire des Séances de l'Académie des Sciences, Paris* 62: 921–2.

Sauvageau, C. 1907b. 'Sur La sexualite de l'*Halopteris* (*Stypocaulon*) *scoparia*. *Compte Rendu Hebdomadaire des Séances de l'Académie des Sciences, Paris* 62: 506–7.

Sauvageau, C. 1908a. Nouvelles observations sur la germination du *Cladostephus verticillatus*. *Comptes Rendus Hebdomadaire des Séances de l'Académie des Sciences, Paris* 64: 695–7.

Sauvageau, C. 1908b. Sur deux *Fucus* récoltés à Arcachon (*Fucus platycarpus* et *Fucus lutarius*). *Bulletin de la Station Biologique d'Arcachon* 11: 65–224.

Sauvageau, C. 1909a. Sur le developpement echelonne de l'*Halopteris* (*Stypocaulon* Kutz.) *scoparia* Sauv. et remarques sur le *Sphacelaria radicans* Harv. *Journal de Botanique* [Morot] II, 2, 44–71.

Sauvageau, C. 1909b. Sur l'hybride des *Fucus vesiculosus* et *F. serratus*. *Compte Rendu Hebdomadaire des Séances de l'Académie des Sciences, Paris* 62: 833–4.

Sauvageau, C. 1909c. Une question de nomenclature botanique, Fucus platycarpus ou *Fucus spiralis*. *Bulletin de la Station Biologique d'Arcachon* 12: 291–95.

Sauvageau, C. 1912. A propos des *Cystoseira* de Banyuls et Guéthary. *Bulletin de la Station Biologique d'Arcachon* 14: 133–556.

Sauvageau, C. 1914. *Remarques sur les Sphacélariacées*. Vol. 3, pp. iii–xii, 481–634, 128 figs. Bordeaux.

Sauvageau, C. 1915. Sur le developpement et la biologie d'une Laminaire (*Saccorhiza bulbosa*). *Comptes Rendus Hebdomadaire des Séances de l'Académie des Sciences, Paris* 160: 445–8.

Sauvageau, C. 1916a. Sur les plantules de quelques Laminaires. *Compte Rendu Hebdomadaire des Séances de l'Académie des Sciences, Paris* 163: 522–4.

Sauvageau, C. 1916b. Sur les variations biologiques d'une Laminaire (*Saccorhiza bulbosa*). *Compte Rendu Hebdomadaire des Séances de l'Académie des Sciences, Paris* 163: 396–8.

Sauvageau, C. 1916c. Sur les gametophytes de deux Laminaires (*L. flexicaulis* et *L. saccharina*). *Compte Rendu Hebdomadaire des Séances de l'Académie des Sciences, Paris* 162: 601–4.

Sauvageau, C. 1916d. Sur la sexualite heterogamique d'une Laminaire (*Alaria esculenta*). *Compte Rendu Hebdomadaire des Séances de l'Académie des Sciences, Paris* 162: 840–2.

Sauvageau, C. 1918a. Recherches sur les Laminaires des côtes de France. *Mémoires de l'académie royale des sciences de Paris* 56.

Sauvageau, C. 1918b. Sur le dissemination et la naturalisation de quelques Algues marines. *Bulletin de l'Institut Océanographique Monaco* 342: 1–28.

Sauvageau, C. 1920a. Nouvelles observations sur l'*E. Padinae. Comptes Rendus Hebdomadaire des Séances de l'Académie des Sciences, Paris* 171: 1041–4.

Sauvageau, C. 1920b. A propos des *Cystoseira. Bulletin de la Station Biologique d'Arcachon* 19–51.

Sauvageau, C. 1923. A propos de quelques *Fucus* du bassin d'Arcachon. *Bulletin de la Station Biologique d'Arcachon* 20: 21–137.

Sauvageau, C. 1924. Sur le curieux développement d'une Algue phéosporée, *Castagnea Zosterae* Thuret. *Comptes Rendus Hebdomadaire des Séances de l'Académie des Sciences, Paris* 179: 1381–4.

Sauvageau, C. 1924. Sur quelques exemples d'heteroblastie dans le development des algues pheosporees. *Compte Rendu Hebdomadaire des Séances de l'Académie des Sciences. Paris ser. D.* 179: 1576–9.

Sauvageau, C. 1925. Sur le developpement d'une Algue Pheosporee, *Leathesia difformis* Aresch. *Compte Rendu Hebdomadaire des Séances de l'Académie des Sciences, Paris* 180: 1632–5.

Sauvageau, C. 1926a. Sur un nouveau type d'alterance des generations chez les Algues brunes; les Sporochnales. *Compte Rendu Hebdomadaire des Séances de l'Académie des Sciences, Paris* 182: 361–4.

Sauvageau, C. 1926b. Sur l'alternance des generations chez le *Carpomitra cabrerae* Kütz. *Bulletin de la Station Biologique d'Arcachon* 23: 141–92.

Sauvageau, C. 1927a. Sur le *Colpomenia sinuosa* Derb. et Sol. *Bulletin de la Station Biologique d'Arcachon* 24: 309–53.

Sauvageau, C. 1927b. Sur les problèmes du *Giraudia. Bulletin de la Station Biologique d'Arcachon* 24: 1–74.

Sauvageau, C. 1927c. Sur le *Castagnea zosterae. Bulletin de la Station Biologique d'Arcachon* 24: 1–74.

Sauvageau, C. 1928a. Sur la vegetation et la sexualite des Tilopteridales. *Bulletin de la Station Biologique d'Arcachon* 25: 51–94.

Sauvageau, C. 1928b. Sur les Algues Pheosporees a eclipse ou Eclipsiophycees. *Recueil des travaux botaniques néerlandais A*, 25: 260–70.

Sauvageau, C. 1928c. Seconde note sur l'*Ectocarpus tomentosus* Lyngbye. *Bulletin de la Station Biologique d'Arcachon* 25: 121–35.

Sauvageau, C. 1929. Sur le developpement de quelques Pheosporees. *Bulletin de la Station Biologique d'Arcachon* 26: 253–420.

Sauvageau, C. 1931a. Sur la troisième sorte d'organes pluriloculaires de l'*Ectocarpus secundus* Kutz. *Compte Rendu Hebdomadaire des Séances de l'Académie des Sciences, Paris* 193: 971–2.

Sauvageau, C. 1931b. Sur quelques algues Pheosporees de Guethary (Basses-Pyrenees). *Bulletin de la Station Biologique d'Arcachon* 30: 1–128.

Sauvageau, C. 1931c. Sur quelques Algues phéosporées de la rade de Villefranche (Alpes-Maritimes). *Bulletin de la Station Biologique d'Arcachon* 28: 7–168.

Sauvageau, C. 1932. Le plethysmothalle. *Bulletin de la Station Biologique d'Arcachon* 29: 1–16.

Sauvageau, C. 1933b. Un genre *Symphoriococcus* Reinke est-il justifie? *Bulletin de la Station Biologique d'Arcachon* 30: 179–88.

Sauvageau, C. 1936. Second mémoire sur les algues Phéosporées de Villefranche-sur-Mèr. *Bulletin de la Station Biologique d'Arcachon* 33: 107–214.

Sauvageau, C. 1971. *Remarques sur les Sphacelariacees*. Wheldon & Wesley: J. Cramer (this publication represents a reprint of all Sauvageau's 'Remarques' papers on the Sphacelariaceae).

Sawai, Y., Fujita, Y., Sakata, K. & Tamashiro, E. 1994. 20-Hydroxy-4,8,13,17-tetramethyl-4,8,12,16-eicosatetraenoic acid, a new feeding deterrent against herbivorous gastropods, from the subtropical brown alga *Turbinaria ornata. Fish Science* 60: 199–201.

Scagel, R.F. 1966. The Phaeophyceae in perspective. *Oceanography Marine Biology Annual Review* 4: 123–94.

Schaffelke, B. 1995. Storage carbohydrates and abscisic acid contents in *Laminaria hyperborea* are entrained by experimental daylengths. *European Journal of Phycology* 30: 313–17.

Schaffelke, B. & Lüning, K. 1994. A circannual rhythm controls seasonal growth in the kelps *Laminaria hyperborea* and *L. digitata* from Helgoland (North Sea). *European Journal of Phycology* 29: 49–56.

Schaffelke, B., Campbell, M.L. & Hewitt, C.L. 2005. Reproductive phenology of the introduced kelp *Undaria pinnatifida* (Phaeophyceae, Laminariales) in Tasmania, Australia. *Phycologia* 44: 84–94.

Schaffelke, B., Evers, D. & Walhorn, A. 1995. Selective grazing of the isopod *Idotea baltica* between *Fucus evanescens* and *F. vesiculosus* from Kiel Fjord (western Baltic). *Marine Biology* 124: 215–18.

Schaffelke, B., Peters, A.F. & Reusch, T.B.H. 1996. Factors influencing depth distribution of soft bottom inhabiting *Laminaria saccharina* (L.) Lamour. in Kiel Bay, Western Baltic. *Proceedings of the International Seaweed Symposium* 15: 117–23.

Scheibling, R. & Anthony, S. 2001. Feeding, growth and reproduction of sea urchins (*Strongylocentrotus droebachiensis*) on single and mixed diets of kelp (*Laminaria* spp.) and the invasive alga *Codium fragile* ssp. *tomentosoides. Marine Biology* 139: 139–46.

Schiel, D.R. & Lilley, S.A. 2011. Impacts and negative feedbacks in community recovery over eight years following removal of habitat-forming macroalgae. *Journal of Experimental Marine Biology and Ecology* 407: 108–15.

Schiel, D.R., Taylor, D.I. 1999. Effects of trampling on a rocky intertidal algal assemblage in southern New Zealand. *Journal of Experimental Marine Biology and Ecology* 235: 213–35.

Schiel, D.R. & Thompson, G.A. 2012. Demography and population biology of the invasive kelp *Undaria pinnatifida* on shallow reefs in southern New Zealand. *Journal of Experimental Marine Biology and Ecology* 434–5: 25–33.

Schiffner, V. 1916. Studien über Algen des Adriatischen Meeres. *Helgoländer Wissenschaftliche Meeresuntersuchungen* (Helgol.), N.F. 11: 129–98.

Schiffner, V. 1933. Meeresalgen aus Sud-Dalmatien, gesammelt von Franz Berger. *Österreichische Botanische Zeitschrift* 82: 283–304.

Schiffner, V. & Vatova, A. 1938. Le alghe della Laguna di Venezia. Sezione I. Chlorophyceae – Phaeophyceae – Rhodophyceae. (Appendix: Myxophyceae). In: Brunelli, G., Magrini, G., Milani, L. & Orsi, P. (eds), *La Laguna di Venezia*. Vol. III, Parte V, Tomo IX, Fascicolo i, pp. [83]–250. Venezia: C. Ferrari.

Schiller, J., Lackschewitz, D., Buschbaum, C., Reise, K., Pang, S. & Bischof, K. 2018. Heading northward to Scandinavia: *Undaria pinnatifida* in the northern Wadden Sea. *Botanica Marina* 61: 365–71.

Schlosser, L.A. 1935. Zur Entwicklungsphysiologie des Generationswechsels von *Cutleria*. *Biol. Centralbl.* 55: 198–208.

Schmid, C.E. 1993. Cell–cell-recognition during fertilisation in *Ectocarpus siliculosus* (Phaeophyceae). *Proceedings of the International Seaweed Symposium* 14: 437–43.

Schmid, R. & Dring, M.J. 1992. Stimulation of red light saturated photosynthesis and shifting of the circadian photosynthetic rhythm by blue light in *Ectocarpus siliculosus*. *British Phycological Journal* 27: 100.

Schmid, R. & Hillrichs, S. 2001. Uptake and accumulation of inorganic carbon in *Ectocarpus siliculosus* and its relation to blue light stimulation of photosynthesis. *European Journal of Phycology* 36: 257–64.

Schmid, C.E., Schroer, N. & Müller, D.G. 1994. Female gamete membrane glycoproteins potentially involved in gamete recognition in *Ectocarpus siliculosus* (Phaeophyceae). *Plant Science* 102: 61–7.

Schmidt, O.C. 1937. Choristocarpaceen und Discosporangiaceen. *Hedwigia* 77: 1–4.

Schmidt, O.C. 1938. Beiträge zur Systematik der Phaeophyten I. *Hedwigia* 77: 213–30.

Schmidt, P. 1940. Ueber *Acinetospora pusilla* (Bornet) Sauvageau, etc. *Berichte der Deutsche Botanischen Gesellschaft* 58: 23–8.

Schmidt, P. 1942. *Krobylopteris oltmansii* n.g., n. sp., die neue Tilopteridee der Helgoländer algenflora. *Botanische Zeitung* 37: 321–4.

Schneider, C.W. & Searles, R.B. 1991. *Seaweeds of the Southeastern United States: Cape Hatteras to Cape Canaveral*. Duke University Press, Durham, NC, & London.

Schnetter, R., Hörnig, I. & Weber-Peukert, G. 1987. Taxonomy of some North Atlantic *Dictyota* species (Phaeophyta). *Proceedings of the International Seaweed Symposium* 12: 193–7.

Schoenrock, K., O'Callaghan, T., O'Callaghan, R. & Krueger-Hadfield, S.A. 2019. First record of *Laminaria ochroleuca* Bachelot de la Pylaie in Ireland in Béal an Mhuirthead, county Mayo. *Marine Biodiversity Records* 12: 1–8.

Schoenwaelder, M.E.A. 2002. The occurrence and cellular significance of physodes in brown algae (Phycological Reviews 21). *Phycologia* 41: 125–39.

Schonbeck, M. & Norton, T.A. 1978. Factors controlling the upper limits of fucoid algae on the shore. *Journal of Experimental Marine Biology and Ecology* 31: 303–13.

Schonbeck, M. & Norton, T.A. 1979a. The effects of diatoms on the growth of *Fucus spiralis* germlings in culture. *Botanica Marina* 22: 233–6.

Schonbeck, M. & Norton, T.A. 1979b. Drought-hardening in the upper-shore seaweeds *Fucus spiralis* and *Pelvetia canaliculata*. *Journal of Ecology* 67: 687–96.

Schonbeck, M. & Norton, T.A. 1981. Growth forms of *Fucus distichus* in the San Juan Islands of Washington State. *Proceedings of the International Seaweed Symposium* 8: 475–83.

Schramm, A. & Mazé, H. 1865. *Essai de classification des algues de la Guadeloupe*. pp. 1–52. Basse Terre.

Schreiber, E. 1931. Ueber die geschlechtliche Fortpflanzung der *Sphacelariales*. *Berichte der Deutsche Botanischen Gesellschaft* 49: 235–40.

Schreiber, E. 1932. Ober die Entwicklungsgeschichte und die systematische stellung der Desmarestiaceen. *Zeitschrift für Botanik* 25: 561–82.

Schreiber, E. 1935. Ueber Kultur und Geschlechtsbestimmung von *Dictyota dichotoma*. *Planta* 24: 266–75.

Schueller, G.H. & Peters, A.F. 1994. Arrival of *Fucus evanescens* (Phaeophyceae) in Kiel Bight (western Baltic). *Botanica Marina* 37: 471–7.

Schuh, R.E. 1900. *Rhadinocladia*, a new genus of brown algae. *Rhodora* 2: 111–12.

Schultz, N.E., Lane, C.E., Le Gall, L., Gey, D., Bigney, A.R., de Reviers, B., Rousseau, F. & Schneider, C.W. 2015. A barcode analysis of the genus *Lobophora* (Dictyotales, Phaeophyceae) in the western Atlantic Ocean with four novel species and the epitypification of *L. variegata* (J.V. Lamouroux) E.C. Oliveira. *European Journal of Phycology* 50: 481–500.

Schussnig, B. 1930. Phykologische Beiträge II. *Österreichische Botanische Zeitschrift* 79: 171–8.

Schwartz, N., Dobretsov, S., Rohde, S. & Schupp, P.J. 2017. Comparison of antifouling properties of native and invasive *Sargassum* (Fucales, Phaeophyceae) species. *European Journal of Phycology* 52: 116–31.

Scott, G.W., Hull, S.L., Hornby, S.E., Hardy, F.G. & Owens, N.J.P. 2001. Phenotypic variation in *Fucus spiralis* (Phaeophyceae): morphology, chemical phenotype and their relationship to the environment. *European Journal of Phycology* 36: 43–50.

Scrosati, R. & Heaven, C. 2006. Field technique to quantify intensity of scouring by sea ice in rocky intertidal habitats. *Marine Ecology Progress Series* 320: 293–5.

Scrosati, R. & Heaven, C. 2007. Spatial trends in community richness, diversity, and evenness across rocky intertidal environmental stress gradients in eastern Canada. *Marine Ecology Progress Series* 342: 1–14.

Scrosati, R. & Heaven, C. 2008. Trends in abundance of rocky intertidal seaweeds and filter feeders across gradients of elevation, wave exposure, and ice scour in eastern Canada. *Hydrobiologia* 603: 1–14.

Sears, J.R. 1998. List of taxa. In: Sears, J.R. (ed.), *NEAS Keys to benthic marine algae of the northeastern coast of North America from Long Island Sound to the start of Belle Isle.*

Sears, J.R. & Wilce, R.T. 1973. Sublittoral benthic marine algae of southern Cape Cod and adjacent islands: *Pseudolithoderma paradoxum* sp. nov. (Ralfsiaceae, Ectocarpales). *Phycologia* 12: 75–82.

Seeley, R.H. & Schlesinger, W.H. 2012. Sustainable seaweed cutting? The rockweed (*Ascophyllum nodosum*) industry of Maine and the Maritime Provinces. *Annals of the New York Academy of Sciences* 1249: 84–103.

Segawa, S. 1941. New or noteworthy algae from Izu. *Scientific Papers of the Institute of Algological Research, Faculty of Science, Hokkaido Imperial University* 2: 251–71.

Segi, N. & Kita, W. 1957. Studies on the development of *Undaria undarioides* (Yendo) Okamura. On the development of gametophytes and influence of light intensity on it. *Report of the Faculty of Fisheries, Prefectural University of Mie* 2: 517–26 (in Japanese).

Selivanova, O.N., Zhigadlova, G.G. & Hansen, G.I. 2007. Revision of the systematics of algae in the order Laminariales (Phaeophyta) from the Far-Eastern Seas of Russia on the basis of molecular-phylogenetic data. *Russian Journal of Marine Biology* 33: 278–89.

Sengco, M.R., Bräutigam, M., Kapp, M. & Müller, D.G. 1996. Detection of virus DNA in *Ectocarpus siliculosus* and *E. fasciculatus* (Phaeophyceae) from various geographic areas. *European Journal of Phycology* 31: 73–8.

Seoane-Camba, J. 1965a. Estudios sobre las algas bentónicas en la costa sur de la Península Ibérica (litoral de Cádiz). *Investigación Pesquera* 29: 3–216.

Seoane-Camba, J. 1965b. Las Laminarias de España, su distribución y el problema de *Laminaria digitata* Lamour. V Reunión sobre Productividad y Pesquerías, Instituto de Investigaciones Pesqueras de Barcelona. 47–55.

Seoane-Camba, J. 1966. Sobre la variabiliad morfológica de *Fucus vesiculosus* en las rías gallegas. *Investigación Pesquera* 30: 561–76.

Seoane-Camba, J.A. 1989. On the possibility of culturing *Gelidium sesquipedale* by vegetative propagation. In: Kain, J.M., Andrews, J.W. & McGregor, B.J. (eds), *COST 48, Aquatic Primary Biomass, Marine Macroalgae Outdoor Seaweed Cultivation*, pp. 59–68. EEC, Brussels.

Serisawa, Y., Yokohama, Y., Aruga, Y. & Tanaka, J. 2001. Photosynthesis and respiration in bladelet of *Ecklonia cava* Kjellman (Laminariales, Phaeophyta) in two localities with different temperature conditions. *Phycological Research* 49: 1–11.

Serrão, E.A., Alice, L.A. & Brawley, S.H. 1999. Evolution of the Fucaceae (Phaeophyceae) inferred from nrDNA-ITS. *Journal of Phycology* 35: 382–94.

Serrão, E.A., Brawley, S.H., Hedman, J., Kautsky, L. & Samuelsson, G. 1999. Reproductive success of *Fucus vesiculosus* (Phaeophyceae) in the Baltic Sea. *Journal of Phycology* 35: 254–69.

Serrão, E.A., Pearson, G.A., Kautsky, L. & Brawley, S.H. 1996. Successful external fertilization in turbulent environments. *Proceedings of the National Academy of Sciences of the United States of America* A 93: 5286–90.

Serrão, E., Vliet, M., Hansen, G.I., Maggs, C. & Pearson, G. 2006. Molecular characterization of the '*cottonii*' form of Fucus in the northeastern Pacific versu the Atlantic. In: Meeting Program, 60th Annual meeting. Phycological Society of America, July 6–12, 2006, University of Alaska Southeast, Juneau, AK, p. 75.

Setchell, W.A. 1912. The kelps of the United States and Alaska. In: Cameron, F.K.A. (ed.), *Report on the Fertilizer Resources of the United States, Appendix K*, pp. 130–78. Washington: United States Senate Document No. 190.

Setchell, W.A. & Gardner, N.L. 1903. Algae of northwestern America. *University of California Publications in Botany* 1: 165–418.

Setchell, W.A. & Gardner, N.L. 1922. Phycological contributions. VI. New species of *Ectocarpus*. *University of California Publications in Botany* 7: 403–26.

Setchell, W.A. & Gardner, N.L. 1924a. Phycological contributions. VII. *University of California Publications in Botany* 13: 1–13.

Setchell, W.A. & Gardner, N.L. 1924b. New marine algae from the Gulf of California. *Proceeding of the California Academy of Science, Series* 4 12: 695–949.

Setchell, W.A. & Gardner, N.L. 1925. The marine algae of the Pacific coast of North America. Ill Melanophyceae. *University of California Publications in Botany* 8: 383–898.

Setchell, W.A. & Gardner, N.L. 1930. Marine algae of the Revillagigedo Islands Expedition in 1925. *Proceedings of the California Academy of Science, Series* 4 19: 109–215.

Sfriso, A. 2011. *Ochrophyta (Phaeophyceae e Xanthophyceae). Ambiente di transizione italiani e litorali adiacenti.* pp. [1]–234, pls 1–51. Bologna: Arpa Emilia-Romagna.

Sfriso, A. & Facca, C. 2013. Annual growth and environmental relationships of the invasive species *Sargassum muticum* and *Undaria pinnatifida* in the lagoon of Venice. *Estuarine, Coastal and Shelf Science* 129: 162–72.

Shan, T.F. & Pang, S.J. 2009. Assessing genetic identity of sporophytic offspring of the brown alga *Undaria pinnatifida* derived from mono-crossing of gametophyte clones by use of amplified fragment length polymorphism and microsatellite markers. *Phycological Research* 57: 36–44.

Shan, T.F. & Pang, S.J. 2010. Sex-linked microsatellite marker detected in the female gametophytes of *Undaria pinnatifida* (Phaeophyta). *Phycological Research* 58: 171–6.

Shan, T.F., Pang, S.J. & Gao, S.Q. 2013. Novel means for variety breeding and sporeling production in the brown seaweed *Undaria pinnatifida* (Phaeophyceae): crossing female gametophytes from parthenosporophytes with male gametophyte clones. *Phycological Research* 61: 154–61.

Shan, T., Pang, S., Li, J., Li, X. & Su, L. 2015. Construction of a high-density genetic map and mapping of a sex-linked locus for the brown alga *Undaria pinnatifida* (Phaeophyceae) based on large scale marker development by specific length amplified fragment (SLAF) sequencing. *BMC Genomics* 16: 902.

Shan, T., Pang, S., Wang, X., Li, J., Su, L., Schiller, J., Lackschewitz, D., Hall-Spencer, J.M. & Bischof, K. 2019. Genetic analysis of a recently established

Undaria pinnatifida (Laminariales: Alariaceae) population in the northern Wadden Sea reveals close proximity between drifting thalli and the attached population. *European Journal of Phycology* 54: 154–61.

Shanab, S., Jacques, R. & Magne, F. 1988. Croissance et ramification du thalle de *Bachelotia antillarum* cultivé en éclairement monochromatiques. *Plant Physiology and Biochemistry* 26: 303–11.

Shannon, R.K., Crow, G.E. & Mathieson, A.C. 1988. Evidence for sexual reproduction in *Petalonia fascia* (O.F. Mueller) Kuntze. *Botanica Marina* 6: 511–13.

Sharp, G. 1987. *Ascophyllum nodosum* and its harvesting in Eastern Canada. *FAO Fisheries Technical Paper* 281: 3–44.

Sharp, G. & Bodiguel, C. 2003. Introducing integrated management, ecosystem and precautionary approaches in seaweed management: the *Ascophyllum nodosum* (rockweed) harvest in New Bruswick, Canada and implications for industry. *Proceedings of the International Seaweed Symposium* 17: 107–14.

Sharp, G.J., Ang, P.O., Jr & MacKinnon, D. 1994. Rockweed (*Ascophyllum nodosum* (L.) Le Jolis) harvesting in Nova Scotia: its socioeconomic and biological implications for coastal zone management. *Proceedings Coastal Zone Canada Association* 92: 1632–44.

Sharp, G.J. & Carter, J.A. 1986. Biomass and population structure of kelp (*Laminaria* spp.) in southwestern Nova Scotia. *Canadian Manuscript Report of Fisheries and Aquatic Sciences no. 1907:* 42 pp.

Shaw, T.I. 1957. Iodide and the respiration of *Laminaria digitata*. *British Phycological Bulletin* 1: 18.

Sheader, A. & Moss, B.L. 1975. Effects of light and temperature on germination and growth of *Ascophyllum nodosum* (L.) Le Jol. *Estuarine, Coastal and Marine Science* 3: 125–32.

Sheehy Skeffington, M. & Curtis, T.G.F. 2000. The Atlantic element in Irish salt marshes. In: Rushton, B. (ed.), *Biodiversity: The Irish Dimension*, pp. 179–96. Dublin: Royal Irish Academy.

Shibneva, S., Skriptsova, A., Shan, T.F. & Pang, S.J. 2013. The different morphs of *Undaria pinnatifida* (Phaeophyceae, Laminariales) in Peter the Great Bay (Sea of Japan) are phenotypic variants: direct evidence. *Journal of Applied Phycology* 25: 1909–16.

Shim, J.H., Kim, J.B., Hwang, D.-W., Choi, H.-G. & Lee, Y. 2017. Variations in carbon and nitrogen stable isotopes and in heavy metal contents of mariculture kelp *Undaria pinnatifida* in Gijang, southeastern Korea. *Algae. An International Journal of Algal Research* 32: 349–57.

Shinmura, I. 1974. Studies on the cultivation of an edible brown *alga Cladosiphon okamuranus* III Development of zoospores from plurilocular sporangium. *Bulletin of the Japanese Society of Scientific Fisheries* 40: 1213–22.

Shinmura, I. 1977. Life history of *Cladosiphon okamuranus* Tokida from southern Japan. *Bulletin of the Japanese Society for Phycology* 25: 333–40.

Sideman, E.J. & Mathieson, A.C. 1983. The growth, reproductive phenology, and longevity of non-tide-pool *Fucus distichus* (L.) Powell in New England. *Journal of Experimental Marine Biology and Ecology* 68: 111–27.

Sideman, E.J. & Mathieson, A.C. 1985. Morphological variation within and between natural populations of non-tide pool *Fucus distichus* (Phaeophyta) in New England. *Journal of Phycology* 21: 250–7.

Sieburth, J.M. & Tootle, J.L. 1981. Seasonality of microbial fouling on *Ascophyllum nodosum* (L.) Lejol., *Fucus vesiculosus* L., *Polysiphonia lanosa* (L.) Tandy and *Chondrus crispus* Stackh. *Journal of Phycology* 17: 57–64.

Siemer, B.L. & Pedersen, P.-M. 1995. The taxonomic status of *Pilayella littoralis*, *P. varia* and *P. macrocarpa* (Pilayellaceae, Fucophyceae). *Phycologia* 34: 257–66.

Siemer, B.L., Stam, W.T., Olsen, J.L. & Pedersen, P.M. 1998. Phylogenetic relationships of the brown algal orders Ectocarpales, Chordariales, Dictyosiphonales, and Tilopteridales (Phaeophyceae) based on RuBisCo large subunit and spacer sequences. *Journal of Phycology* 34: 1038–48.

Silar, P. 2016, Protistes Eucaryotes: Origine, Evolution et Biologie des Microbes Eucaryotes, *HAL Archives-ouvertes*: 1–462.

Silberfeld, T. 2010. Contributions des phylogenies moleculaires a la systematique et a la comprehension de l'evolution des algues brunes (Ochrophyta, Phaeophyceae). Thèse doctorale, Muséum National d'histoire Naturelle, Paris, Oct. 2010.

Silberfeld, T., Leigh, J.W., Verbruggen, H., Cruaud, C., Reviers, B. de & Rousseau, F. 2010. A multilocus time-calibrated phylogeny of the brown algae (Heterokonta, Ochrophyta, Phaeophyceae): Investigating the evolutionary nature of the 'brown algal crown radiation'. *Molecular Phylogenetics and Evolution* 56: 659–74.

Silberfeld, T., Bittner, L., Fernández-Garcia, C., Cruaud, C., Rousseau, F., de Reviers, B., Leliaert, F., Payri, C.E. & De Clerck, O. 2013. Species diversity, phylogeny and largescale biogeographic patterns of the genus *Padina* (Phaeophyceae, Dictyotales). *Journal of Phycology* 49: 130–42.

Silberfeld, T., Rousseau, F. & de Reviers, B. 2014. An updated classification of brown algae (Ochrophyta, Phaeophyceae). *Cryptogamie, Algologie* 35: 117–56.

Silberfeld, T., Racault, M.F.L.P., Fletcher, R.L., Couloux, A., Rousseau, F. & de Reviers, B. 2011. Systematics and evolutionary history of pyrenoid-bearing taxa in brown algae (Phaeophyceae). *European Journal of Phycology* 46: 361–77.

Silva, L.D., Bahcevandziev, K. & Pereira, L. 2019. Production of bio-fertilizer from *Ascophyllum nodosum* and *Sargassum muticum* (Phaeophyceae). *Journal of Oceanology and Limnology* 37: 918–27.

Silva, P.C. 1952. A review of nomenclatural conservation in the algae from the point of view of the type method. *University of California Publications in Botany* 25: 241–323.

Silva, P.C. 1953. The identity of certain *Fuci* of Esper. *Wasmann Journal of Biology* 11: 221–32.

Silva, P.C. 1959. Remarks on algal nomenclature II. *Taxon* 8: 60–4.

Silva, P.C. 1980. *Names of classes and families of living algae Regnum Vegetabile vol. 103.* Utrecht & The Hague.

Silva, P.C. & de Reviers, B. 2000. Ordinal names in the phaeophyceae. *Cryptogamie, Algologie* 21: 49–58.

Silva, P.C., Basson, P.W. & Moe, R.L. 1996. Catalogue of the benthic marine algae of the Indian Ocean. *University of California Publications in Botany* 79: 1–1259.

Silva, P.C., Meñez, E.G. & Moe, R.L. 1987. Catalog of the benthic marine algae of the Philippines. *Smithsonian Contributions to Marine Sciences* 27: [i–ii] iii–iv, 1–179.

Silva, P.C., Woodfield, R.A., Cohen, A.N., Harris, L.H. & Goddard, J.H.R. 2002. First report of the Asian kelp *Undaria pinnatifida* in the Northeastern Pacific Ocean. *Biological Invasions* 4: 333–8.

Simberloff, D., Abele, L.G. & Thistle, A.B. (eds), *Ecological Communities: Conceptual Issues and the Evidence.* University Press, Princeton, pp. 151–80.

Simkanin, C.M. 2004. The invasive seaweed *Sargassum muticum* (Yendo) Fensholt in Lough Hyne Marine Nature Reserve, Co. Cork. *Irish Naturalists' Journal* 27: 481–2.

Simmons, H. 1897. Zur Kenntniss der Meeresalgen der Färöer. *Hedwigia* 36: 247–76.

Simon, M.F. 1954. Recherches sur les pyrenoides des Phaeophycees. *Revue de Cytolologie et de Biologie vegetales* 15: 74–106.

Sinclair, C. & Whitton, B.A. 1977. Influence of nutrient deficiency on hair formation in the Rivulariaceae. *British Phycological Journal* 12: 297–313.

Sjøtun, K. 1993. Seasonal lamina growth in two age groups of *Laminaria saccharina* (L.) Lamour. in Western Norway. *Botanica Marina* 36: 433–41.

Sjøtun, K. & Fredriksen, S. 1995. Growth allocation in *Laminaria hyperborea* (Laminariales, Phaeophyceae) in relation to age and wave exposure. *Marine Ecology Progress Series* 126: 213–22.

Sjøtun, K. & Schoschina, E.V. 2002. Gametophytic development of *Laminaria* spp. (Laminariales, Phaeophyta) at low temperature. *Phycologia* 41: 147–52.

Sjøtun, K., Eggereide, S.F. & Høisaeter, T. 2007. Grazer-controlled recruitment of the introduced *Sargassum muticum* (Phaeophyceae, Fucales) in northern Europe. *Marine Ecology Progress Series* 342: 127–38.

Sjøtun, K., Fredriksen, S. & Rueness, J. 1996. Seasonal growth and carbon and nitrogen content in canopy and first-year plants of *Laminaria hyperborea* (Laminariales, Phaeophyceae). *Phycologia* 35: 1–8.

Sjøtun, K., Fredriksen, S. & Rueness, J. 1998. Effect of canopy biomass and wave exposure on growth in *Laminaria hyperborea* (Laminariaceae: Phaeophyta). *European Journal of Phycology* 33: 337–43.

Sjøtun, K., Fredriksen, S., Rueness, J. & Lein, T.E. 1995. Ecological studies of the kelp *Laminaria hyperborea* (Gunnerus) Foslie in Norway. In: Skjoldal, H.R., Hopkins, C., Erikstad, K.E. & Leinaas, H.P. (eds), *Ecology of fjords and coastal waters*, pp. 525–36. Elsevier Science BV.

Sjøtun, K., Fredriksen, S., Lein, T.E., Rueness, J. & Sivertsen, K. 1993. Population studies of *Laminaria hyperborea* from its northern range of distribution in Norway. *Proceedings of the International Seaweed Symposium* 14: 215–21.

Sjøtun, K., Heesch, S., Rull Lluch, J., Martín, R.M., Gómez Garreta, A., Brysting, A.K. & Coyer, J.A. 2017. Unravelling the complexity of salt marsh 'Fucus cottonii' forms (Phaeophyceae, Fucales). *European Journal of Phycology* 52: 360–70.

Skinner, S. & Womersley, H.B.S. 1984. Southern Australian taxa of Giraudiaceae (Dictyosiphonales, Phaeophyta). *Phycologia* 23: 161–81.

Skottsberg, C. 1911. Beobachtungen über einige Meeresalgen aus der Gegend von Tvärminne im südwestlichen Finnland. *Acta Societatis pro Fauna et Flora Fennica* 34: 3–18.

Skrine, P.M. 1929. A member of the Fucaceae from the Dovey salt-marshes. *Journal of Botany* 67: 241–3.

Skrine, P.M., Newton, L. & Chater, E.H. 1932. A salt-marsh form of *Fucus ceranoides* L. From Llanbedr, Merioneth. *Annals of Botany* 46: 769–79.

Skriptsova, A., Khomenko, V. & Isakov, I.V. 2004. Seasonal changes in growth rate, morphology and alginate content in *Undaria pinnatifida* at the northern limit in the Sea of Japan (Russia). *Journal of Applied Phycology* 16: 17–21.

Skuja, H. 1928. Vorarbeiten zur einer Algenflora von Lettland. IV. *Acta Horti Botanici Universitatis Latviensis* 3: 103–218.

Sliwa, C., Johnson, C.R. & Hewitt, C.L. 2006. Mesoscale dispersal of the introduced kelp *Undaria pinnatifida* attached to unstable substrata. *Botanica Marina* 49: 396–405.

Slocum, C.J. 1980. Differential susceptibility to grazers in two phases of an intertidal alga: advantages of heteromorphic generations. *Journal of Experimental Marine Biology and Ecology* 46: 99–110.

Smale, D.A., Burrows, M.T., Moore, P., O'Connor, N. & Hawkins, S.J. 2013. Threats and knowledge gaps for ecosystem services provided by kelp forests: a northeast Atlantic perspective. *Ecology and Evolution* 3: 4016–38.

Smale, D.A., Wernberg, T., Yunnie, A.L.E. & Vance, T. 2015, The rise of *Laminaria ochroleuca* in the Western English Channel (UK) and comparisons with its competitor and assemblage dominant *Laminaria hyperborea*. *Marine Ecology* 36: 1033–44.

Smith, B.D. 1985. Recovery following experimental harvesting of *Laminaria longicruris* and *L. digitata* in southwestern Nova Scotia. *Helgöländer Meeresuntersuchungen* 39: 83–101.

Smith, G.M. 1944. *Marine Algae of the Monterey Peninsula.* Stanford, CA: Stanford University Press.

Smith, G.M. 1955. *Cryptogamic Botany.* Vol. 1. McGraw-Hill, New York.

Smith, J.E. 1790–1814. *English Botany:* Vols 1–36, 2,592 plates. London: printed for the author by J. Davis.

Smith, J.E. & Sowerby, J. 1843. Supplement to the English Botany of the late Sir J.E. Smith and Mr Sowerby. Vol. Ill [Tabs 2797–867]. J. Sowerby; Longman & Co. Sherwood & Co. London.

Snirc, A., Silberfeld, T., Bonnet, J., Tillier, A., Tuffet, S. & Sun, J.-S. 2010. Optimization of DNA extraction from brown algae (Phaeophyceae) based on a commercial kit (Note). *Journal of Phycology* 46: 616–21.

Sohn, C., H. 1993. *Porphyra, Undaria* and *Hizikia* cultivation in Korea. *Korean Journal of Phycology* 8: 207–16.

Sommerfelt, C. 1826. *Supplementum Florae lapponicae quam edidit Dr. Georgius Wahlenberg auctore.* pp. [i*–iii*], [i]–xii, [1]–331, [332, err.], 3 pls. Christianiae [Oslo]: typis Borgianis et Gröndahlianis.

Soneira, A. & Niell, F.X. 1975. On the biology of *Ascophyllum nodosum* in the coasts of Galicia. *Investigación Pesquera* 39: 43–59.

Sousa, W.P. 1979. Experimental investigations of disturbance and ecological succession in a rocky intertidal algal community. *Ecological Monographs* 49: 227–54.

South, G.R. 1972. On the life history of *Tilopteris mertensii* (Turn. in Sm.) Kütz. *Proceedings of the International Seaweed Symposium* 7: 83–9.

South, G.R. 1975. Contributions to the flora of marine algae of eastern Canada III. Order Tilopteridales. *Naturaliste Canadien* 102: 693–702.

South, G.R. 1976. A check list of marine algae of Eastern Canada – first revision. *Journal of the Marine Biological Association of the United Kingdom* 56: 871–43.

South, G.R. 1980. Observations on the life histories of *Punctaria plantaginea* (Roth) Greville and *Punctaria orbiculata* Jao (Punctariaceae, Phaeophyta). *Phycologia* 19: 266–72.

South, G.R. & Burrows, E.M. 1967. Studies on marine algae of the British Isles. 5. *Chorda filum* (L.) Stackh. *British Phycological Bulletin* 3: 379–402.

South, G.R. & Hill, R.D. 1970. Studies on marine algae of Newfoundland. I. Occurrence and distribution of free-living *Ascophyllum nodosum* in Newfoundland. *Canadian Journal of Botany* 48: 1697–701.

South, G.R. & Hill, R.D. 1971. Studies on marine algae of Newfoundland. II. On the occurrence of *Tilopteris mertensii*. *Canadian Journal of Botany* 49: 211–13.

South, G.R. & Hooper, R. 1976. *Stictyosiphon soriferus* (Phaeophyta, Dictyosiphonales) from eastern North America. *Journal of Phycology* 12: 24–9.

South, G.R. & Hooper, R.G. 1980. A catalogue and atlas of the benthic marine algae of the Island of Newfoundland. *Memorial University of Newfoundland Occasional Papers in Biology* 3: 1–136.

South, G.R. & Tittley, I. 1986. *A Checklist and Distributional Index of the Benthic Marine algae of the North Atlantic Ocean.* London and St. Andrews, New Brunswick, Canada: British Museum (Natural History) and Huntsman Marine Laboratory.

South, G.R. & Whittick, A. 1987. *Introduction to Phycology.* pp. vii +341. Oxford: Blackwell Scientific Publications.

South, P.M. & Thomsen, M.S. 2016. The ecological role of invading *Undaria pinnatifida*: an experimental test of the driver-passenger models. *Marine Biology* 163: 175.

South, P.M., Floerl, O., Forrest, B.M. & Thomsen, M.S. 2017. A review of three decades of research on the invasive kelp *Undaria pinnatifida* in Australasia: an assessment of its success, impacts and status as one of the world's worst invaders. *Marine Environmental Research* 131: 243–57.

South, P.M., Lilley, S.A., Tait, L.W., Alestra, T., Hickford, M.J.H., Thomsen, M.S. & Schiel, D.R. 2016. Transient effects of an invasive kelp on the community structure and primary productivity of an intertidal assemblage. *Marine and Freshwater Research* 67: 103–12.

Southward, A.J. 1956. The population balance between limpets and seaweeds on wave-beaten rocky shores. *Report of the Marine Biological Station at Port Erin* 68: 20–9.

Spencer, M.A., Irvine, L.M. & Jarvis, C.E. 2009. Typification of Linnaean names relevant to algal nomenclature. *Taxon* 58: 237–60.

Sprengel, K. [P.J.] 1827. *Systema vegetabilium* Editio decima sexta. Voluminis IV. Pars I. Classis 24. Vol. 4, pp. [i]–iv, [1]–592. Gottingae [Göttingen]: sumtibus Librariae Dieterichianae.

Spurkland, T. & Iken, K. 2011. Salinity and irradiance effects on growth and maximum photosynthetic quantum yield in subarctic *Saccharina latissima* (Laminariales, Laminaraceae). *Botanica Marina* 54: 355–65.

Stache, B. 1989. Sexual compatibility and species concept in *Ectocarpus siliculosus* (Ectocarpales, Phaeophyceae) from Italy, North Carolina, Chile, and New Zealand. In: Garbary, D.J. & South, R.G. (eds). Evolutionary biogeography of the marine algae of the North Atlantic. Berlin, Germany: Springer Verlag, pp. 173–86.

Stache-Crain, B., Müller, D.G. & Goff, L.J. 1997. Molecular systematics of *Ectocarpus* and *Kuckuckia* (Ectocarpales, Phaeophyceae) inferred from phylogenetic analysis of nuclear- and plastid-encoded DNA sequences. *Journal of Phycology* 33: 152–68.

Stackhouse, J. 1795. *Nereis Britannica*: Fasc. 1. pp. i–viii, 1–30, pls I–VIII. Bathoniae [Bath] & Londini [London]: S. Hazard; J. White.

Stackhouse, J. 1797. *Nereis Britannica*: Fasc. 2. pp. ix–xxiv, 31–70, pls IX–XIII. Bathoniae [Bath] & Londini [London]: S. Hazard; J. White.

Stackhouse, J. 1801. *Nereis Britannica.* ed. 1.3. Bathoniae and Londini.

Stackhouse, J. 1809. Tentamen marino-cryptogamicum, ordinem novum; in genera et species distributum, in Classe XXIVta Linnaei sistens. *Mémoires de la Société Imperiale des Naturalistes de Moscou* 2: [50]–97.

Stackhouse, J. 1816. *Nereis britannica* Editio altera. Nova addita classificatione cryptogamiarum [sic] respectu generis Fuci. pp. [i]–xii, [i]–68, 20 pls. Oxonii [Oxford]: excudebat S. Collingwood.

Staehr, P.A. & Wernberg, T. 2009. Physiological responses of *Ecklonia radiata* (Laminariales) to a latitudinal gradient in ocean temperature. *Journal of Phycology* 45: 91–9.

Stafford, C.J., Callow, J.A. & Green, J.R. 1992. Isolation and characterization of plasma membranes from *Fucus serratus* eggs (Short Note). *British Phycological Journal* 27: 429–34.

Stafford, C.J., Green, J.R. & Callow, J.A. 1991. Characterization of plasma membranes from *Fucus serratus* eggs. *British Phycological Journal* 26: 97.

Stam, W.T., Bot, P.V.M., Boele-Bos, S.A., van Rooij, J.M. & van den Hoek, C. 1988. Single copy DNA-DNA hybridization among five species of *Laminaria* (Phaeophyceae): phylogenetic and biogeographic implications. *Helgoländer Wissenschaftliche Meeresuntersuchungen* 42: 251–67.

Stapleton, J.C. 1988. Occurrence of *Undaria pinnatifida* (Harvey) Suringar in New Zealand. *Japanese Journal of Phycology* 36: 178–9.

Steen, H. 1992. Sargassum muticum i Norge: Årssyklus og Utbrebelse i Relasjon til Toleranse Overfor Regulerende Miljøfaktorer. MSc. thesis, University of Oslo, Oslo (in Norwegian).

Steen, H. 2003a. Apical hair formation and growth of *Fucus evanescens* and *F. serratus* (Phaeophyceae) germlings under various nutrient and temperature regimes. *Phycologia* 42: 26–30.

Steen, H. 2003b. Intraspecific competition in *Sargassum muticum* (Phaeophyceae) germlings under various density, nutrient and temperature regimes. *Botanica Marina* 46: 36–43.

Steen, H. 2004. Effects of reduced salinity on reproduction and germling development in *Sargassum muticum* (Phaeophyceae, Fucales). *European Journal of Phycology* 39: 293–300.

Steen, H. & Rueness, J. 2004. Comparison of survival and growth in germlings of six fucoid species (Fucales, Phaeophyceae) at two different temperature and nutrient levels. *Sarsia* 89: 175–83.

Steen, H. & Scrosati, R. 2004. Intraspecific competition in *Fucus serratus* and *F. evanescens* (Phaeophyceae, Fucales) germlings: effects of settlement density, nutrient concentration, and temperature. *Marine Biology, Berlin* 144: 61–70.

Stefanov, K., Dimitrova-Konaklieva, St., Frette, X., Christova, D., Nikolova, Ch. & Popov, S. 2000. Sterols and acylglycerols in the brown algae *Zanardinia prototypus* Nardo and *Striaria attenuata* (Grev.) Grev. from the Black Sea. *Botanica Marina* 43: 141–5.

Stegenga, H. 1996. Recente veranderingen in de Nederlanss zeewierflora. I Additionele soorten van *Hincksia, Herponema* en *Kuetzingiella*. (Ectocarpaceae, Phaeophyta). *Gorteria* 21: 198–204.

Stegenga, H. 1999. *Undaria pinnatifida* in Nederland gearriveerd. *Het Zeepaard* 59: 71–3.

Stegenga, H. & Mol, I. 1983. *Flora van de'Nederlandsezeewieren*. nr 33. Koninklijke Nederlandse Natuurhistorische Vereniging. Bibliotheek. Uitgave.

Stegenga, H. & Mol, I. 1996. Recente veranderingen in de Nederlandse Zeewierflora II. Additionele soorten bruinwieren (Phaeophyta) in de genera *Botrytella* en *Feldmannia* (Ectocarpaceae), *Leptonematella* (Elachistaceae) en *Stictyosiphon* (Striariaceae). *Gorteria* 22: 103–10.

Stegenga, H., Bolton, J.J. & Anderson, R.J. 1997b. *Seaweeds of the South African West Coast*. Cape Town: Bolus Herbarium, University of Cape Town.

Stegenga, H., Karremans, M. & Simons, J. 2007. Zeewieren van de voormalige oesterputten bij Yerseke. *Gorteria* 32: 125–43.

Stegenga, H., Mol, I., Prud'homme van Reine, W.F. & Lokhorst, G.M. 1997a. Checklist of the marine algae o the Netherlands. *Gorteria* Supplement 4: 3–57.

Steinberg, P.D. 1985. Feeding preferences of *Tegula funebralis* and chemical defences of marine brown algae. *Ecological Monographs* 55: 333–49.

Steinbiss, H.-H. & Schmitz, K. 1974. Zur Entwicklung und funktionellen anatomie des phylloids von *Laminaria hyperborea. Helgoländer Wissenschaftliche Meeresuntersuchungen* 26: 134–52.

Steneck, R.S. & Dethier, M.N. 1994. A functional group approach to the structure of algal-dominated communities. *Oikos* 69: 476–98.

Stengel, D.B. & Dring, M.J. 1992. *Ascophyllum nodosum* as an indicator of copper and iron levels in Strangford Lough and Co. Sligo. *British Phycological Journal* 27: 101.

Stengel, D.B. & Dring, M.J. 1997. Morphology and *in situ* growth rates of plants of *Ascophyllum nodosum* (Phaeophyta) from different shore levels and responses of plants to vertical transplantation. *European Journal of Phycology* 32: 193–202.

Stengel, D.B. & Dring, M.J. 1998. Seasonal variation in the pigment content and photosynthesis of different thallus regions of *Ascophyllum nodosum* (Fucales, Phaeophyta) in relation to position in the canopy. *Phycologia* 37: 259–68.

Stengel, D.B. & Dring, M.J. 2000. Copper and iron concentrations in *Ascophyllum nodosum* (Fucales, Phaeophyta) from different sites in Ireland and after culture experiments in relation to thallus age and epiphytism. *Journal of Experimental Marine Biology and Ecology* 246: 145–61.

Stengel, D.B., Wilkes, R.J. & Guiry, M.D. 1999. Seasonal growth and recruitment of *Himanthalia elongata* (Fucales, Phaeophycota) in different habitats on the Irish west coast. *European Journal of Phycology* 34: 213–21.

Stephens, T.A. & Hepburn, C.D. 2014. Mass-transfer gradients across kelp beds influence *Macrocystis pyrifera* growth over small spatial scales. *Marine Ecology Progress Series* 515: 97–109.

Stephenson, T.A. & Stephenson, A. 1949. The universal features of zonation between tide-marks on rocky coasts. *Journal of Ecology* 37: 289–305.

Stephenson, T.A. & Stephenson, A. 1972. *Life between Tidemarks on Rocky Shores*. Freeman & Company, San Francisco.

Stephenson, W. 1961. Experimental studies on the ecology of intertidal environments at Heron Island. II. The effect of substratum. *Australian Journal of Marine and Freshwater Research* 12: 164–76.

Stephenson, W. & Searles, R.B. 1960. Experimental studies on the ecology on intertidal environments at Heron Island I. Exclusion of fish from beach rock. *Australian Journal of Marine and Freshwater Research* 11: 241–67.

Steudel, E. 1824. *Nomenclator botanicus* enumerans ordine alphabetico nomina atque synonyma tum generica tum specifica et a Linnaeo et recentioribus de re botanica scriptoribus plantis phanerogamis imposita. Vol. 2, pp. [i]–xviii, [i]–450. Stuttgardtiae et Tubingae [Stuttgart & Tübingen]: sumtibus I.G. Cottae.

Stiger, V., Horiguchi, T., Yoshida, T., Coleman, A.W. & Masuda, M. 2000. Phylogenetic relationships of *Sargassum* (Sargassaceae, Phaeophyceae) with reference to a taxonomic revision of the section Phyllocystae based on ITS-2 nrDNA sequences. *Phycological Research* 48: 251–60.

Stiger, V., Horiguchi, T., Yoshida, T., Coleman, A.W. & Masuda, M. 2003. Phylogenetic relationships within the genus *Sargassum* (Fucales, Phaeophyceae), inferred from it ITS nrDNA, with an emphasis on the taxonomic revision of the genus. *Phycological Research* 51: 1–10.

Stiger-Pouvreau, V. & Thouzeau, G. 2015. Marine species introduced on the French Channel–Atlantic coasts: A review of main biological invasions and impacts. *Open Journal of Ecology* 5: 227–57.

Stimson, J., Cunha, T. & Philippoff, J. 2007. Food preferences and related behavior of the browsing sea urchin *Tripneustes gratilla* (Linnaeus) and its potential for use as a biological control agent. *Marine Biology* 151: 1761–72.

Stockton, B. & Evans, L.V. 1979. Characterization of alginates from *Laminaria digitata*. *British Phycological Journal* 14: 128.

Stokes, K., O'Neill, K. & McDonald, R.A. 2004. *Invasive species in Ireland*. Unpublished report to Environment & Heritage Service and National Parks & Wildlife Service. pp. [i], 1–152. Belfast: Quercus.

Stomps, T.J. 1911. Etudes topographyques sur la variabilite des *Fucus vesiculosus* L., *platycarpus* Thur., et *ceranoides* L. *Recueil de l'Institut Botanique [Universite Libre de Bruxelles]* 8: 326–77.

Strong, J.A. & Dring, M.J. 2011. Macroalgal competition and invasive success: testing competition in mixed canopies of *Sargassum muticum* and *Saccharina latissima*. *Botanica Marina* 54: 223–9.

Strong, J.A., Maggs, C.A. & Johnson, M.P. 2009. The extent of grazing release from epiphytism for *Sargassum muticum* (Phaeophyceae) within the invaded range. *Journal of the Marine Biological Association of the United Kingdom* 89: 303–14.

Strøm, H. 1762. *Physisk og oeconomisk beskrivelse over fogderiet Søndmør, beliggende i Bergens Stift i Norge. Oplyst med landkort og kobberstykker. Første part*. pp. [1–18], 1–572, pls I–IV, Kort over Söndmør [= 1 map]. Sorøe, Kjøbenhavn: Rothe.

Strøm, H. 1788. Fortegnelse over Norske Søevexter. *Nye Samling af det Kongelige Norske Videnskabers Selskabs Skrifter* 2: 345–56.

Strömfelt, H.F.G. 1884. Om algvegetationen i Finlands sydvestra skärgård. *Fin. Vet.-Soc.* 39: 1–22.

Strömfelt, H.F.G. 1886a. Einige fur die Wissenschaft neue Meeresalgen aus Island. *Bot. Zbl.* 26: 172–3.

Strömfelt, H.F.G. 1886b. *Om algvegetationen vid Islands Kuster*. pp. 1–89. Göteborg: Akademisk Afhandling.

Strömfelt, H.F.G. 1888. Algae novae quas ad litora Scandinaviae indagavit. *Notarisia* 3: 381–4.

Strömfelt, T. 1978. The effect of photoperiod on the length growth of five species of intertidal Fucales. *Sarsia* 63: 155–8.

Strömgren, T. 1985. Effect of light on apical length growth of *Fucus distichus* ssp. *edentatus* (Phaeophyta). *Marine Biology, Berlin* 86: 263–9.

Strömgren, T. 1986. Comparative growth ecology of *Ascophyllum nodosum* from Norway (lat. 63°N) and Wales (lat. 53°N). *Aquatic Botany* 24: 311–19.

Stuart, M.D. & Brown, M.T. 1996. Phenology of *Undaria pinnatifida* (Harvey) Suringar in Otago Harbour, New Zealand. http://nzmss.rsnz.govt.nz/conf96/abstrs2s.html.

Stuart, M.D., Hurd, C.L. & Brown, M.T. 1999. Effects of seasonal growth rate on morphological variation of *Undaria pinnatifida* (Alariaceae, Phaeophyceae). *Proceedings of the International Seaweed Symposium* 16: 191–9.

Stæhr, P.A., Pedersen, M.F., Thomsen, M.S., Wernberg, T. & Krause-Jensen, D. 2000. Invasion of *Sargassum muticum* in Limfjorden (Denmark) and its possible impact on the indigenous macroalgal community. *Marine Ecology Progress Series* 207: 79–88.

Suárez-Jiménez, R., Hepburn, C.D., Hyndes, G.A., McLeod, R.J., Taylor, R.B. & Hurd, C.L. 2015. Do native subtidal grazers eat the invasive kelp *Undaria pinnatifida*? *Marine Biology* 162: 2521–6. https://doi.org/10.1007/s00227-015-2757-y

Suárez-Jiménez, R., Hepburn, C.D., Hyndes, G.A., McLeod, R.J., Taylor, R.B. & Hurd, C.L. 2017a. Importance of the invasive macroalga *Undaria pinnatifida* as trophic subsidy for a beach consumer. *Marine Biology* 164: 113. https://doi.org/10.1007/s00227-017-3140-y

Suárez-Jiménez, R., Hepburn, C.D., Hyndes, G.A., McLeod, R.J., Taylor, R.B. & Hurd, C. 2017b. The invasive kelp *Undaria pinnatifida* hosts an epifaunal assemblage similar to native seaweeds with comparable morphologies. *Marine Ecology Progress Series* 582: 45–55.

Subrahmanyan, R. 1957. Observations on the anatomy, cytology, development of the reproductive structures of *Pelvetia canaliculata* Dcne. *et* Thur. Pert III. The liberation of reproductive bodies, fertilization and embryology. *Journal of the Indian Botanical Society* 36: 373–95.

Subrahmanyan, R. 1960. Ecological studies on the Fucales I. *Pelvetia canaliculata* Dcne. *et* Thur. *Journal of the Indian Botanical Society* 39: 614–30.

Subrahmanyan, R. 1961. Ecological studies on the Fucales II. *Fucus spiralis* L. *Journal of the Indian Botanical Society* 40: 335–54.

Sundene, O. 1953. The algal vegetation of Oslofjord. *Det Norske Videnskaps Akademi i Oslo, I. Mat. Naturv. Klasse* 2: 1–244.

Sundene, O. 1958. Interfertility between forms of *Laminaria digitata*. *Nytt Magasin for Botanikk* 6: 121–18.

Sundene, O. 1962a. The implications of transplant and culture experiments on the growth and distribution of *Alaria esculenta*. *Nytt Magasin for Botanikk* 9: 155–74.

Sundene, O. 1962b. Growth in the sea of *Laminaria digitata* sporophytes from culture. *Nytt Magasin for Botanikk* 9: 5–24.

Sundene, O. 1963. Reproduction and Ecology of *Chorda tomentosa*. *Nytt Magasin for Botanikk* 10: 159–67.

Sundene, O. 1964. The ecology of *Laminaria digitata* in Norway in view of transplant experiments. *Nytt Magasin for Botanikk* 11: 83–107.

Sundene, O. 1966. *Haplospora* Kjellm. and *Scaphospora speciosa* in culture. *Nature, London* 209: 937–8.

Sundene, O. 1973. Growth and reproduction in *Ascophyllum nodosum* (Phaeophyceae). *Norwegian Journal of Botany* 20: 249–55.

Suneson, S. 1939. On *Ectocarpus fasciculatus* growing on the finspines of fishes. *Botaniska Notiser* 1939: 53–6.

Suneson, S. 1971. A bifurcate frond of *Laminaria digitata* from Sweden. *Botaniska Notiser* 124: 277–9.

Suringar, W.F.R. 1873. Illustrationes des algues du Japon. *Musée Botanique de Leide* 1: 77–90, pls 26–33.

Susuki, Y., Kuma, K., Kudo, I. & Matsunaga, K. 1995. Iron requirement of the brown macroalgae *Laminaria japonica*, *Undaria pinnatifida* (Phaeophyta) and the crustose coralline alga *Lithophyllum yessoense* (Rhodophyta), and their competition in the northern Japan sea. *Phycologia* 34: 201–5.

Suzuki, Y., Maruyama, T., Takami, T. & Miura, A. 1998. Inhibition effects of suspended and accumulated particles on adhesion and development of *Undaria pinnatifida* zoospores. *Journal of Japan Society on Water Environment* 21: 670–5.

Svedelius, N. 1901. *Studier ofver Ostersjons Hafsalgflora*. Akad. Afh. Uppsala.

Svedelius, N. 1915. Zytologisch-entwicklungsgeschichtliche Studien über *Scinaia furcellata*. *Nova Acta Regiae Societatis Scientiarum Upsaliensis ser. 4*, 4: 1–55.

Svedelius, N. 1928. On the number of chromosomes in the two different kinds of plurilocular sporangia of *Ectocarpus virescens* Thur. *Svensk Botanisk Tidskrift Bd.* 22, Stockholm.

Svedelius, N. 1931. Nuclear phases and alternation in the Rhodophyceae. *Beihefte zum Botanischen Centralblatt* XLVIII: 38–59.

Svendsen, P. 1962. Some observations on *Saccorhiza polyschides* (Lightf.) Batt. (Phaeophyceae). *Sarsia* 7: 11–13.

Svendsen, P. 1972a. Noen observasjoner over taretrålingog gjenvekst av stortare, *Laminaria hyperborea*. *Fiskets Gang* 22: 448–60.

Svendsen, P. 1972b. Some observations on commercial harvesting and regrowth of *Laminaria hyperborea*. In: *Fisken og Havet. Fiskeridirektoratets Havforskningsinstitutt* 2: 33–45.

Svendsen, P. & Kain, J.M. 1971. The taxonomic status, distribution and morphology of *Laminaria cucullata* sensu Jorde and Klavestad. *Sarsia* 46: 1–22.

Taboada, M.C., Millán, R. & Miguez, M.I. 2013. Nutritional value of the marine algae wakame (*Undaria pinnatifida*) and nori (*Porphyra purpurea*) as food supplements. *Journal of Applied Phycology* 25: 1271–6.

Tahara, M. 1909. On the periodical liberation of oospheres in *Sargassum* (prelim). *Botanical Magazine, Tokyo* 23: 151–3.

Takamatsu, M. 1936. *Sorocarpus* aus der Matsushima-Bucht (Prov. Miyagi, Japan), auch morphologisch untersucht. *Research Bulletin Saito Ho-on Kai Museum of Natural History* 8: 71–99.

Takamatsu, M. 1943. The species of *Sphacelaria* from Japan I. *Journal. Sigenkagaku Kenkyusyo* 1: 153–87.

Tala, F., Véliz, K., Gomez, I. & Edding, M. 2007. Early life stages of the South Pacific kelps *Lessonia nigrescens* and *Lessonia trabeculata* (Laminariales, Phaeophyceae) show recovery capacity following exposure to UV radiation (Research Note). *Phycologia* 46: 467–70.

Tammes, P.M.L. 1954. Gas-exchange in the vesicles (air bladders) of *Ascophyllum nodosum*. *Acta Botanica Neerlandica* 3: 114–23.

Tamura, H., Mine, I. & Okuda, K. 1996. Cellulose-synthesizing terminal complexes and microfibril structure in the brown alga *Sphacelaria rigidula* (Sphacelariales, Phaeophyceae). *Phycological Research* 44: 63–8.

Tan, I.H.A. 1995. Ribosomal DNA phylogeny of the plant division *Phaeophyta*. Simon Fraser University, British Columbia, Canada, Ph.D. thesis.

Tan, I.H. & Druehl, L.D. 1993. Phylogeny of the Northeast Pacific brown algal (Phaeophycean) orders as inferred from 18S rRNA gene sequences. *Hydrobiologia* 260/1: 699–704.

Tan, I.H. & Druehl, L.D. 1994. A molecular analysis of *Analipus* and *Ralfsia* (Phaeophyceae) suggests the order Ectocarpales is polyphyletic. *Journal of Phycology* 30: 721–9.

Tan, I.H. & Druehl, L.D. 1996. A ribosomal DNA phylogeny supports the close evolutionary relationships among the Sporochnales, Desmarestiales, and Laminariales (Phaeophyceae). *Journal of Phycology* 32: 112–18.

Tanaka, J. 1986. The taxonomy of *Protectocarpus speciosus* (Børgesen) Kornmann (Myrionemataceae, Phaeophyceae). *Japanese Journal of Phycology* 34: 287–92.

Tanaka, J. & Chihara, M. 1980a. Taxonomic study of the Japanese crustose brown algae (1) General account and the order Ralfsiales. *Japanese Journal of Botany* 56: 193–202.

Tanaka, J. & Chihara, M. 1980b. Taxonomic study of the Japanese crustose brown algae (2) *Ralfsia* (Ralfsiaceae, Ralfsiales) (Part 1). *Japanese Journal of Botany* 55: 225–36.

Tanaka, J. & Chihara, M. 1980c. Taxonomic study of the Japanese crustose brown algae (3) *Ralfsia* (Ralfsiaceae, Ralfsiales) (Part 2). *Japanese Journal of Botany* 55: 337–42.

Tanaka, J. & Chihara, M. 1982. Morphology and taxonomy of *Mesospora schmidtii* Weber van Bosse, Mesosporaceae fam. nov. (Ralfsiales, Phaeophyceae). *Phycologia* 21: 382–9.

Tanaka, T. & Hosoi, T. 1967. The electron microscopic observation on the zoospore of *Undaria pinnatifida* Sur. *Memoirs of the Faculty of Fisheries, Kagoshima University. Kagoshima* 16: 1–11.

Taniguchi, K., Kito, H. & Akiyama, K. 1981. Morphological variation of *Undaria pinnatifida* (Harvey) Suringer – I. On the difference of growth and morphological characteristics of two types at Matsushima Bay, Japan. *Bulletin of the Tohoku National Fisheries Research Institute* 42: 1–9 (in Japanese with English abstract).

Tanniou, A., Vandanjon, L., Gonçalves, O., Kervarec, N. & Stiger-Pouvreau, V. 2015. Rapid geographical differentiation of the European spread brown algae *Sargassum muticum* using HRMAS NMR and Fourier-Transform Infrared spectroscopy. *Talanta* 132: 451–6.

Targett, N.M. & Arnold, T.M. 1998. Predicting the effects of brown algal phlorotannins on marine herbivores in tropical and temperate oceans. *Journal of Phycology* 34: 195–205.

Taskin, E. 2006. First report of *Corynophlaea crispa* (Harvey) Kuckuck (Phaeophyceae, Corynophlaeaceae) in the Mediterranean Sea. *Nova Hedwigia* 82: 217–25.

Taskin, E. 2012. First report of the alien brown alga *Scytosiphon dotyi* M.J. Wynne (Phaeophyceae, Scytosiphonaceae) in Turkey. *Mediterranean Marine Science* 13: 33–5.

Taskin, E., Ozturk, M. & Wynne, M.J. 2006. First report of *Microspongium globosum* Reinke (Phaeophyceae, Myrionemataceae) in the Mediterranean Sea. *Nova Hedwigia* 82: 135–42.

Taskin, E., Öztürk, M., Kurt, O. & Öztürk, M. 2008. *The check-list of the marine algae of Turkey*. pp. [i–ii]–[1]–87. Manisa, Turkey: Ecem Kirtasiye.

Taskin, E., Öztürk, M., Kurt, O. & Uclay, S. 2013. Benthic marine algae in Northern Cyprus (Eastern Mediterranean Sea). *Journal of Black Sea/Mediterranean Environment* 19: 143–61.

Taskin, E., Jahn, R., Öztürk, M., Furnari, G. & Cormaci, M. 2012. *The Mediterranean Cystoseira (with photographs)*. pp. [2] 1–75, photographs. Manisa, Turkey: Celar Bayar University.

Taskin, E., Kurt, O., Cormaci, M., Furnari, G. & Öztürk, M. 2010. Two brown algae from the eastern

Mediterranean Sea: *Microcoryne ocellata* Strömfelt and *Corynophlaea flaccida* (C. Agardh) Kützing. *Fresenius Environmental Bulletin* 19: 892–6.

Tatarenkov, A., Jönsson, R.B., Kautsky, L. & Johannesson, K. 2007. Genetic structure in populations of *Fucus vesiculosus* (Phaeophyceae) over spatial scales from 10 m to 800 km. *Journal of Phycology* 43: 675–85.

Tatewaki, M. 1966. Formation of a crustaceous sporophyte with unilocular sporangia in *Scytosiphon lomentaria*. *Phycologia* 6: 62–6.

Taylor, A.R. & Brownlee, C. 1992. Plasmalemma currents in unfertilized and fertilized eggs of *Fucus serratus*. *British Phycological Journal* 27: 101.

Taylor, P.R. & Littler, M.M. 1982. The roles of compensatory mortality, physical disturbance, and retention substrate in the development and organization of a sand-influenced, rocky-intertidal community. *Ecology* 63: 135–46.

Taylor, W.R. 1929. A species of *Acrothrix* on the Massachusetts coast. *American Journal of Botany* 15: 577–83.

Taylor, W.R. 1937a. Marine algae of the northeastern coast of North America. *University of Michigan Studies, Scientific Series* 13: 1–427.

Taylor, W.R. 1937b. Notes on North Atlantic marine algae. I. *Papers of the Michigan Academy of Science* 22: 225–33.

Taylor, W.R. 1945. Pacific marine algae of the Allan Hancock Expeditions to the Galapagos Islands. *Allan Hancock Pacific Expeditions* 12: i–iv, 1–528, 3 figs, 100 pls.

Taylor, W.R. 1957. *Marine Algae of the Northeastern Coast of North America*. Ann Arbor: The University of Michigan Press.

Taylor, W.R. 1960. *Marine Algae of the Eastern Tropical and Subtropical Coasts of the Americas*. Ann Arbor: The University of Michigan Press.

Teixeira, V.L., Da Silva Almeida, S.A. & Kelecom, A. 1990. Chemosystematic and biogeographic studies of the diterpenes from the marine brown alga *Dictyota dichotoma*. *Biochemical Systematics and Ecology* 18: 87–92.

Ten Hoopen, A. 1983. Effects of daylength and irradiance on the formation of reproductive organs in two algae: *Acrosymphyton purpuriferum* (J. Ag) Sjöst. (Rhodophyceae) and *Sphacelaria rigidula* Kütz. (Phaeophyceae). Thesis, University of Groningen: 1–87.

Ten Hoopen, A., Bos, S. & Breeman, A.M. 1983. Photoperiodic response in the formation of gametangia of the long-day plant *Sphacelaria rigidula* (Phaeophyceae). *Marine Ecology Progress Series* 13: 285–9.

Terry, L.A. & Moss, B.L. 1980. The effect of photoperiod on receptacle initiation in *Ascophyllum nodosum* (L.) Le Jol. *British Phycological Journal* 15: 291–301.

Thélin, I. 1981. Effets, en culture, de deux pétroles bruts et d'un dispersant pétrolier sur les zygotes et les plantules de *Fucus serratus* Linnaeus (Fucales, Phaeophyceae). *Botanica Marina* 24: 515–19.

Thom, R.M. 1983. Spatial and temporal patterns of *Fucus distichus* ssp. *edentatus* (de la Pyl.) Pow. (Phaeophyceae: Fucales) in central Puget Sound. *Botanica Marina* 26: 471–86.

Thomas, D.N. & Kirst, G.O. 1990a. Salinity tolerance of *Ectocarpus siliculosus* – influence of life-cycle. *British Phycological Journal* 25: 97.

Thomas, D.N. & Kirst, G.O. 1990b. Salt tolerance of *Ectocarpus siliculosus* (Dillw.) Lyngb.: Comparison of gametophytes, sporophytes and isolates of different geographic origin. *Botanica Acta* 104: 26–36.

Thomas, D.N. & Kirst, G.O. 1991. Differences in osmoacclimation between sporophytes and gametophytes of the brown alga *Ectocarpus siliculosus*. *Physiologia Plantarum* 82: 1–9.

Thomas, T.E. & Turpin, D.H. 1980. Desiccation enhanced nutrient uptake rates in the intertidal alga *Fucus distichus*. *Botanica Marina* 23: 479–81.

Thomas, T.E., Harrison, P.J. & Taylor, E.B. 1985. Nitrogen uptake and growth of the germlings and mature thalli of *Fucus distichus*. *Marine Biology, Berlin* 84: 267–74.

Thomsen, M.S., Wernberg, T., Staehr, P.A. & Pedersen, M.F. 2006. Spatio-temporal distribution patterns of the invasive macroalga *Sargassum muticum* within a Danish *Sargassum*-bed. *Helgoland Marine Research* 60: 50–8.

Thomsen, M.S., Alestra, T., Brockerhoff, D., Lilley, S.A., South, P.M. & Schiel, D.R. 2018. Modified kelp seasonality and invertebrate diversity where an invasive kelp co-occurs with native mussels. *Marine Biology* 165: 173.

Thompson, G. & Schiel, D. 2012. Resistance and facilitation by native algal communities in the invasion success of *Undaria pinnatifida*. *Marine Ecology Progress Series* 468: 95–105.

Thornber, C.S., Kinlan, B.P., Graham, M.H. & Stachowicz, J.J. 2004. Population ecology of the invasive kelp *Undaria pinnatifida* in California: environmental and biological controls on demography. *Marine Ecology Progress Series* 268: 69–80.

Thuret, G. 1850. Recherches sur les zoospores des Algues et les anthéridies des Cryptogames. *Annales des Sciences Naturelles, Troisième série, Botanique* 14: 214–60.

Thuret, G. 1851. Recherches sur les zoospores des algues et les anthéridies des cryptogames. Suite. *Annales des Sciences Naturelles, Troisième série, Botanique* 16: 5–39.

Thuret, G. & Bornet, É. 1878. *Études phycologiques*. Analyses d'algues marines, pp. [i–v], i–iii, 1–105, pls I–LI. Paris: G. Masson.

Tilden, J.E. 1937. *The Algae and their Life Relations*. Minnesota: The University of Minnesota Press Minneapolis.

Tittley, I. & Price, J.H. 1977a. The marine algae of the tidal Thames. *London Naturalist* 56: 10–17.

Tittley, I. & Price, J.H. 1977b. An atlas of the seaweeds of Kent. *Transactions of the Kent Field Club* 7: 1–80.

Tittley, I. & Price, J.H. 1978. The benthic marine algae of the eastern English Channel: a preliminary floristic and ecological account. *Botanica Marina* 21: 499–512.

Tittley, I. 1985a. Zonation and seasonality of estuarine benthic algae: artificial embankments in the River Thames. *Botanica Marina* 28: 1–8.

Tittley, I. 1985b. Seaweed communities on the artificial coastline of south eastern England. I Reclaimed saline wetland and estuaries. *Transactions of the Suffolk Naturalists Society* 21: 54–64.

Tittley, I. 1986. Seaweed communities on the artificial coastline of south eastern England. 2. Open-Sea Shores. *Transactions of the Kent Field Club* 10: 55–67.

Tittley, I., Irvine, D.E.G. & Jephson, N.A. 1977. The infralittoral marine algae of Sullom Voe, Shetland. *Transactions and Proceedings of the Botanical Society of Edinburgh* 42: 397–419.

Tittley, I. & Shaw, K.M. 1980. Numerical and field methods in the study of the marine flora of chalk cliffs. In: Price, J.H., Irvine, D.E.G. & Farnham, W.F. (eds), *The shore environment 1: methods*, pp. 213–40. Systematics Association Special Volume 17(a). Academic Press, London and New York.

Tokida, J. 1931. On two species of Sphacelariales new to Japan. *Transactions of the Sapporo Natural History Society* 11: 215–20.

Tokida, J. 1954. The marine algae of Southern Saghalien. *Memoirs Faculty Fisheries Hokkaido University* 2: 1–264.

Tom Dieck, I. 1987. Temperature tolerance and daylength effects in isolates of *Scytosiphon lomentaria* (Phaeophyceae) of the North Atlantic and Pacific Ocean. *Helgoland Marine Research* 41: 307–21.

Tom Dieck, I. 1991. Circannual growth rhythm and photoperiodic sorus induction in the kelp *Laminaria setchellii* (Phaeophyta). *Journal of Phycology* 27: 341–50.

Tom Dieck, I. 1992. North Pacific and North Atlantic digitate *Laminaria* species (Phaeophyta): hybridization experiments and temperature responses. *Phycologia* 31: 147–63.

Tom Dieck, I. 1993. Temperature tolerance and survival in darkness of kelp gametophytes (Laminariales, Phaeophyta): ecological and biogeographical implications. *Marine Ecology Progress Series* 100: 253–64.

Tom Dieck, I. & Oliveira Filho, E.C. de. 1993. The section Digitatae of the genus *Laminaria* (Phaeophyta) in the northern and southern Atlantic: crossing experiments and temperature response. *Marine Biology, Berlin* 115: 151–60.

Tonon, T., Rousvoal, S., Roeder, V. & Boyen, C. 2008. Expression profiling of the mannuronan C5-epimerase multigenic family in the brown alga *Laminaria digitata* (Phaeophyceae) under biotic stress conditions. *Journal of Phycology* 44: 1250–6.

Topinka, J.A. 1978. Nitrogen uptake by *Fucus spiralis* (Phaeophyceae). *Journal of Phycology* 14: 241–7.

Topinka, J.A. & Robbins, J.V. 1976. Effects of nitrate and ammonium enrichment on growth and nitrogen physiology in *Fucus spiralis*. *Limnology and Oceanography* 21: 659–64.

Toste, M.F., Parente, M.I., Neto, A.I. & Fletcher, R.L. 2003a. Life history of *Colpomenia sinuosa* (Scytosiphonaceae, Phaeophyceae) in the Azores. *Journal of Phycology* 39: 1268–74.

Toste, M.F., Parente, M.I., Neto, A.I. & Fletcher, R.L. 2003b. Life history of *Hydroclathrus clathratus* (Boryex C. Agardh) M. Howe (Scytosiphonaceae, Phaeophyta) in the Azores. *Cryptogamie, Algologie* 24: 209–18.

Toth, G.B. & Pavia, H. 2002. Lack of phlorotannin induction in the kelp *Laminaria hyperborea* in response to grazing by two gastropod herbivores. *Marine Biology* 140: 403–9.

Toth, G.B. & Pavia, H. 2003. Fertilization and germination tolerance to copper in *Ascophyllum nodosum* (Phaeophyceae). *Proceedings of the International Seaweed Symposium* 17: 411–16.

Toth, R. 1976. The release, settlement and germination of zoospores in *Chorda tomentosa* Phaeophyceae, Laminariales. *Journal of Phycology* 12: 222–33.

Tovey, D.J. & Moss, B.L. 1978. Attachment of the haptera of *Laminaria digitata* (Huds.) Lamour. *Phycologia* 17: 17–22.

Traill, G.W. 1885. *A Monograph of the Algae of the Firth of Forth*. pp. 1–16, Appendix, 8 pressed herbarium specimens. Edinburgh: Edinburgh Co-operative Printing Company, Limited, Bristo Place.

Tremblin, G., Jolivet, P. & Coudret, A. 1993. Light quality effects on subsequent dark $^{14}CO_2$-fixation in *Fucus serratus*. *Proceedings of the International Seaweed Symposium* 14: 471–5.

Trevisan [de Saint-Léon], V.B.A. (1849). De *Dictyoteis adumbratio*. *Linnaea* 22: 421–64.

Trevisan, V.B.A. 1845. *Nomenclator algarum*, ou collection des noms imposées aux plantes de la famille des algues. pp. 1–80. Padoue [Padua]: Imprimerie du Seminaire.

Tronholm, A., Sansón, M., Afonso-Carrillo, J. & De Clerck, O. 2008. Distinctive morphological features, file-cycle phases and seasonal variations in subtropical populations of *Dictyota dichotoma* (Dictyotales, Phaeophyceae). *Botanica Marina* 51: 132–44.

Tronholm, A., Steen, F., Tyberghein, L., Leliaert, F., Verbruggen, H., Siguan, M.A.R. & De Clerck, O. 2010. Species delimitation, taxonomy, and biogeography of *Dictyota* in Europe (Dictyotales, Phaeophyceae). *Journal of Phycology* 46: 1301–21.

Trowbridge, C.D. 2013. Changes in brown seaweed distributions in Lough Hyne, SW Ireland: a long-term perspective. *Botanica Marina* 56: 323–38.

Tsiamis, K., Panayotidis, P., Economou-Amilli, A. & Katsaros, C. 2013. Seaweeds of the Greek coasts. I. Phaeophyceae. *Mediterranean Marine Science* 14: 141–57.

Tsukidate, J. 1984. Studies on the regenerative ability of the brown algae, *Sargassum muticum* (Yendo) Fensholt and *Sargassum tortile* C. Agardh. *Proceedings of the International Seaweed Symposium* 11: 393.

Tugwell, S. & Branch, G. 1989. Differential polyphenolic distribution among tissues in the kelps *Ecklonia maxima*, *Laminaria pallida* and *Macrocystis angustifolia* in relation to plant-defence theory. *Journal of Experimental Marine Biology and Ecology* 129: 219–30.

Turner, D. 1801. *Ulva furcellata* et *multifida*, descriptae. *Journal für die Botanik, Schrader* 1: 300–2.

Turner, D. 1802a. *A Synopsis of the British Fuci*. 2 vols. London: Sold by J. White, Fleet-Street; and T. Longman and O. Rees, Paternoster-Row (Printed by F. Bush, Yarmouth).

Turner, D. 1802b. Descriptions of four new species of *Fucus*. *Transactions of the Linnean Society of London* 6: 125–36, pls VIII–X.

Turner, D. 1807–08. *Fuci* sive plantarum fucorum generi a botanicis ascriptarum icones descriptiones et historia. Fuci, or coloured figures and descriptions of the plants referred by botanists to the genus *Fucus*. Vol. I pp. [i, iii], [1]–164, [1]–2, pls 1–71 (col. copp. W.J. Hooker). Londini [London]: typis J. M'Creery, impensis J. et A. Arch.

Turner, D. 1809–11. *Fuci* sive plantarum fucorum generi a botanicis ascriptarum icones descriptiones et historia. Fuci, or colored figures and descriptions of the plants referred by botanists to the genus *Fucus*. Vol. III pp. [i], [1]–148, [1–2], pls 135–96 (col. copp. by W.J. Hooker and others). Londini [London]: typis J. M'Creery, impensis J. et A. Arch.

Turner, D. 1811–19. *Fuci* sive plantarum fucorum generi a botanicis ascriptarum icones descriptiones et historia. Fuci, or coloured figures and descriptions of the plants referred by botanists to the genus *Fucus*. Vol. IV pp. [i, iii], [1]–153, [1–2], [1–7], pls 197–258 (col. copp. W.J. Hooker). Londini [London]: typis J. M'Creery, impensis J. et A. Arch.

Turner, T. 1983. Facilitation as a successional mechanism in a rocky intertidal community. *American Naturalist* 121: 729–38.

Ubisch, G. 1928. Zur Entwicklungsgeschichte von *Taonia atomaria* Ag. *Berichte der Deutschen Botanischen Gesellschaft* 46: 457–63.

Ubisch, G. 1931. Zur Entwicklungsgeschichte von *Taonia atomaria* Ag. II. Weibliche Geschlechts – und Tetrasporenpflanzen. *Pubblicazioni della Stazione Zoologica di Napoli* 11: 361–6.

Uchida, T. 1993. The life cycle of *Sargassum horneri* (Phaeophyta) in laboratory culture. *Journal of Phycology* 29: 231–5.

Uchida, T., Yoshikawa, K., Arai, A. & Arai, S. 1991. Life cycle and its control of *Sargassum muticum* (Phaeophyta) in batch cultures. *Nippon Suisan Gakkaishi* 57: 2249–53.

Ueda, S. 1929. On the temperature in relation to the development of the gametophytes of *Laminaria religiosa* Miyabe. *Journal of the Fisheries Institute, Tokyo* 24: 138–9.

Ugarte, R.A. 2011a. An evaluation of the mortality of the brown seaweed *Ascophyllum nodosum* (L.) Le Jol. produced by cutter rake harvests in southern New Brunswick, Canada. *Journal of Applied Phycology* 23: 401–7.

Ugarte, R.A. 2011b. Management and production of the brown algae [sic] *Ascophyllum nodosum* in the Canadian maritimes. *Journal of Applied Phycology* 24: 409–16.

Ugarte, R., Lauzon-Guay, J.-S. & Critchley, A.T. 2016. Comments on Halat, L., Galway, M.E., Gitto, S. & Garbary, D.J. 2015. Epidermal shedding in *Ascophyllum nodosum* (Phaeophyceae): seasonality, productivity and relationship to harvesting. *Phycologia* 54: 599–608. *Phycologia* 56: 114–15.

Ugarte, R.A., Sharp, G. & Moore, B. 2007. Changes in the brown seaweed *Ascophyllum nodosum* (L.) Le Jol. plant morphology and biomass produced by cutter rake harvests in southern New Brunswick, Canada. *Proceedings of the International Seaweed Symposium* 18: 125–33.

Underwood, A.J. 1979. The ecology of intertidal gastropods. *Advances in Marine Biology* 16: 111–210.

Underwood, A.J. & Denley, E.J. 1984. Paradigms, explanations and generalizations in models for the structure of intertidal communities on rocky shores. In: Strong, D.R., Simberloff, D., Abele, L.G. & Thistle, A.B. (eds), *Ecological Communities: Conceptual Issues and the Evidence*, pp. 151–80. University Press, Princeton.

Underwood, A.J., Denley, E.J. & Moran, H.J. 1983. Experimental analyses of the structure and dynamics

of mid-shore rocky intertidal communities in New South Wales. *Oecologia* 56: 202–19.

Uwai, S., Kogame, K. & Masuda, M. 2000. Morphology and life history of four Japanese species of *Elachista* (Elachistaceae, Phaeophyceae), including a new record of *E. fucicola*. *Phycological Research* 48: 267–80.

Uwai, S., Kogame, K. & Masuda, M. 2001a. Reassessment of the taxonomic status of *Elachista tenuis* and related species (Elachistaceae, Phaeophyceae), based on culture studies and molecular phylogenetic analyses. *European Journal of Phycology* 36: 103–11.

Uwai, S., Kogame, K. & Masuda, M. 2001b. Conspecificity of *Elachista nigra* and *Elachista orbicularis* (Elachistaceae, Phaeophyceae). *Phycological Research* 50: 217–25.

Uwai, S., Arai, S., Morita, T. & Kawai, H. 2007. Genetic distinctness and phylogenetic relationships among *Undaria* species (Laminariales, Phaeophyceae) based on mitochondrial cox3 gene sequences. *Phycological Research* 55: 263–71.

Uwai, S., Nagasato, C., Motomura, T. & Kogame, K. 2005. Life history and molecular phylogenetic relationships of *Asterocladon interjectum* sp. nov. (Phaeophyceae). *European Journal of Phycology* 40: 179–94.

Uwai, S., Nelson, W., Neill, K., Wang, W.D., Aguilar-Rosas, L.E., Boo, S.M., Kitayama, T. & Kawai, H. 2006a. Genetic diversity in *Undaria pinnatifida* (Laminariales, Phaeophyceae) deduced from mitochondria genes – origins and succession of introduced populations. *Phycologia* 45: 687–95.

Uwai, S., Yotsukura, N., Serisawa, Y., Muraoka, D., Hiraoka, M. & Kogame, K. 2006b. Intraspecific genetic diversity of *Undaria pinnatifida* in Japan, based on the mitochondrial *cox3* gene and the ITS1 of nrDNA. *Hydrobiologia* 553: 345–56.

Vadas, R.L. 1972. Ecological implications of culture studies on *Nereocystis luetkeana*. *Journal of Phycology* 8: 196–203.

Vadas, R.L. 1977. Preferential feeding: an optimization strategy in sea urchins. *Ecological Monographs* 47: 337–71.

Vadas, R.L. & Wright, W.A. 1986. Recruitment, growth and management of *Ascophyllum nodosum*. *Actas Congr. Algas Marinas Chilenas* 2: 101–13.

Vadas, R.L., Wright, W.A. & Miller, S.L. 1990. Recruitment of *Ascophyllum nodosum*: wave action as a source of mortality. *Marine Ecology Progress Series* 61: 263–72.

Valentine, J.P. 2003. Establishment and persistence of dense stands of the introduced kelp *Undaria pinnatifida*. PhD thesis, University of Tasmania.

Valentine, J.P. & Johnson, C.R. 2003. Establishment of the introduced kelp *Undaria pinnatifida* in Tasmania depends on disturbance to native algal assemblages. *Journal of Experimental Marine Biology and Ecology* 295: 63–90.

Valentine, J.P. & Johnson, C.R. 2004. Establishment of the introduced kelp *Undaria pinnatifida* following dieback of the native macroalga *Phyllospora comosa* in Tasmania, Australia. *Marine and Freshwater Research* 55: 223–30.

Valentine, J.P. & Johnson, C.R. 2005a. Persistence of the exotic kelp *Undaria pinnatifida* does not depend on sea urchin grazing. *Marine Ecology Progress Series* 285: 43–55.

Valentine, J.P. & Johnson, C.R. 2005b. Persistence of sea urchin (*Heliocidaris erythrogramma*) barrens on the east coast of Tasmania: Inhibition of macroalgal recovery in the absence of high densities of sea urchins. *Botanica Marina* 48: 106–15.

Valls, R., Piovetti, L., Banaigs, B., Archavlis, A. & Pellegrini, M. 1995. (S)-13-hydroxygeranylgeraniol-derived furanoditerpenes from *Bifurcaria bifurcata*. *Phytochemistry* 39: 145–9.

van Alstyne, K.L. 1988. Herbivore grazing increases polyphenolic defenses in the intertidal brown alga *Fucus distichus*. *Ecology* 69: 655–63.

van Alstyne, K.L. 1990. Effects of wounding by the herbivorous snails *Littorina sitkana* and *L. scutulata* (Mollusca) on growth and reproduction of the intertidal alga *Fucus distichus* (Phaeophyta). *Journal of Phycology* 26: 412–16.

van den Hoek, C. 1958a. *Sphacelaria britannica* Sauv. nouveau pour la côte Française et quelques algues marines nouvelles ou rares pour la region de Roscoff. *Blumea, Supplement* 4: 188–95.

van den Hoek, C. 1958b. Observations on the algal vegetation of the northern pier at Hoek van Holland, made from October 1953 till August 1954. *Blumea* 9: 187–205.

van den Hoek, C. 1975. Phycological Reviews 3. Phytogeographic provinces along the coasts of the Northern Atlantic Ocean. *Phycologia* 14: 317–30.

van den Hoek, C. & Flinterman, A. 1968. The life-history of *Sphacelaria furcigera* Kütz. (Phaeophyceae). *Blumea* 16: 193–242.

van den Hoek, C., Cortel-Breeman, A.M., Rietema, H. & Wanders, J.B.W. 1972. L'interpretation des donnees obtenues par des cultures unialgales, sur les cycles evolutifs des algues. Quelques examples tires des recherches conduites au laboratoire de Groningue. *Mémoires de la Société Botanique de France* 1972: 45–66.

van den Hoek, C., Breeman, A.M. & Stam, W.T. 1990. The geographic distribution of seaweed species in relation to temperature – present and past. In: Beukema, J.J., Wolff, W.J. & Brouns, J.J.W. (eds), *Expected Effects of Climatic Change on Marine Coastal Ecosystems*, pp. 55–67. Kluwer Academic, Dordrecht, The Netherlands.

van den Hoek, C., Mann, D.G. & Jahns, H.M. 1995. *Algae. An introduction to phycology.* pp. i–xiv, 1–623. Cambridge: Cambridge University Press.

van den Hoek, C. & Jahns, H.M. 1978. *Algen*. Einführung in die Phykologie. Unter Mirwirkung von Hand Martin Jahns. Ed. 1. Stuttgart: Thieme.

van Heurck, H. 1908. *Prodrome de la flore des algues marines des Iles Anglo-Normandes et des cotes nord-ouest de la France.* pp. iii–xii, 1–120. Jersey: Labey et Blampied, Beresford Library, St.-Hélier.

Van Tamelen, P.G. 1987. Early successional mechanisms in the rocky intertidal: the rôle of direct and indirect interactions. *Journal of Experimental Marine Biology and Ecology* 112: 39–48.

van Went, J.L., van Aelst, A.C. & Tammes, P.M.L. 1973. Transverse connections between cortex and trans-locating medulla in *Laminaria digitata*. *Acta Botanica Neerlandica* 22: 77–8.

Vandermeulen, H., DeWreede, R.E. & Cole, K.M. 1984. Nomenclatural recommendations for three species of *Colpomenia* (Scytosiphonales, Phaeophyta). *Taxon* 33: 324–9.

Varvarigos, V., Galatis, B. & Katsaros, C. 2007. Radial endoplasmic reticulum arrays co-localize with radial F-actin in polarizing cells of brown algae. *European Journal of Phycology* 42: 253–62.

Vauchel, P., Arhaliass, A., Legrand, J., Kaas, R. & Baron, R. 2008. Decrease in dynamic viscosity and average molecular weight of alginate from *Laminaria Digitata* during alkaline extraction. *Journal of Phycology* 44: 515–17.

Vaz-Pinto, F., Olabarria, C. & Arenas, F. 2014. Ecosystem functioning impacts of the invasive seaweed *Sargassum muticum* (Fucales, Phaeophyceae). *Journal of Phycology* 50: 108–16.

Veiga, A.J., Cremades, J. & Bárbara, I. 1997. *Gononema aecidioides* (Ectocarpaceae) un nuevo feófito para la Península Ibérica. *Anales del Jardín Botánico de Madrid* 55: 155–6.

Veiga, P., Torres, A.C., Rubal, M., Troncoso, J. & Sousa-Pinto, I. 2014. The invasive kelp *Undaria pinnatifida* (Laminariales, Ochrophyta) along the north coast of Portugal: Distribution model versus field observations. *Marine Pollution Bulletin* 84: 363–5.

Velley, T. 1795. *Coloured figures of marine plants found on the southern coast of England*. Bath.

Verlaque, M. 1988. Végétation marine de la Corse (Méditerranée) VII. Documents pour la flore des algues. *Botanica Marina* 31: 187–94.

Verlaque, M. 1994. Inventaire des plantes introduites en Méditerranée origines et répercussions sur l'environnement et les activités humaines. *Oceanologica Acta* 17: 1–23.

Verlaque, M. 2001. Checklist of the macroalgae of Thau Lagoon (Hérault, France), a hot spot of marine species introduction in Europe. *Oceanologica Acta* 24: 29–49.

Verlaque, M. & Boudouresque, C.F. 1981. Vegetation marine de la Corse (Mediterranee) V. Documents pour la flore des algues. *Biol. Ecol. Medit.* 8: 139–56.

Verlaque, M., Ruitton, S., Mineur, F. & Boudouresque, C.-F. 2015. *CIESM Atlas of Exotic Species of the Mediterranean. Macrophytes.* Monaco: CIESM Publishers.

Viana, I.G., Bode, A. & Fernández, C. 2015. Ecology of *Fucus vesiculosus* (Phaeophyceae) at its southern distributional limit: growth and production of early developmental stages. *European Journal of Phycology* 50: 247–59.

Viejo, R.M. 1997. The effects of colonization by *Sargassum muticum* on tidepool macroalgal assemblages. *Journal of the Marine Biological Association of the United Kingdom* 77: 325–40.

Viejo, R.M. 1999. Mobile epifauna inhabiting the invasive *Sargassum muticum* and two local seaweeds in northern Spain. *Aquatic Botany* 64: 131–49.

Viejo, R.M., Arrontes, J. & Andrew, N.L. 1995. An experimental evaluation of the effect of wave action on the distribution of *Sargassum muticum* in Northern Spain. *Botanica Marina* 38: 437–41.

Viera-Rodríguez, M.A., Polifrone, M. & Haroun, R. 2007. *Undaria pinnatifida* (Harvey) Suringar a new introduced species in warm-temperate waters of the

Canary Islands: IV European Phycological Congress, Oviedo, p. 95.

Vilg, J.V., Nylund, G.M., Werner, T., Qvirist, L., Mayers, J.A., Pavia, H., Undeland, H. & Albers, E. 2015. Seasonal and spatial variation in biochemical composition of *Saccharina latissima* during a potential harvesting season for Western Sweden. *Botanica Marina* 58: 435–47.

Villalard-Bohnsack, M. 2002. Non-indigenou benthic algal specis introduced into the Northeastern coast of America. In: Sears, J. (ed.), *NEAS keys to benthic marine algae of the Northeastern coast of North America from Long island Sound to the Strait of Belle Isle*. 2nd edn. NEAS contribution No. 2. Northeast Algal Society, Dartmouth, MA, pp. 130–2.

Villalard-Bohnsack, M. & Harlin, M.M. 2001. *Grateloupia doryphora* (Halymeniaceae, Rhodophyta) in Rhode Island waters (USA): geographical expansion, morphological variations and associated algae. *Phycologia* 40: 372–80.

Vreeland, V. 1972. Immunocytochemical localization of the extracellular polysaccharide alginic acid in the brown seaweed, *Fucus distichus*. *Journal of Histochemistry & Cytochemistry* 20: 358–67.

Välikangas, L. 1928. Neuer fundort für *Eudesme virescens* J.G. Ag. im Finnischen Meerbusen. *Memoranda Societatis pro Fauna et Flora Fennica*. 4: 59–60.

Waaland, J.R., Cho, G.Y., Miller, K.A. & Boo, S.M. 2011. Taxonomy and biogeography of *Agarum* and *Thalassiophyllum* (Laminariales, Phaeophyceae) based on sequences of nuclear, mitochondrial, and plastid markers. *Taxon* 60: 831–40.

Waern, M. 1936. *Leptonema lucifugum*, en för Sverige ny brunalg i Hygrohalina grottor. *Svensk Botanisk Tidskrift* 30: 329–42.

Waern, M. 1940. *Cladophora pygmaea* und *Leptonema lucifugum* an der schwedischen Westküste. *Acta Phytogeographica Suecica* 13: 1–6.

Waern, M. 1945. Remarks on some Swedish Sphacelariaceae. *Svensk Botanisk Tidskrift* 39: 396–418.

Waern, M. 1949. Remarks on Swedish *Lithoderma*. *Svensk Botanisk Tidskrift* 43: 633–70.

Waern, M. 1952. Rocky shore algae in the Oregrund Archipelago. *Acta Phytogeographica Suecica* 30: 1–298.

Wahlenberg, G. 1812. *Flora lapponica*: XXX. pp. i–lxvi, 1–550, 30 pls. Berolini: in taberna libraria scholae realis.

Wahlenberg, G. 1826. *Flora suecica*: pp. 429–1117. Upsaliae [Uppsala]: suis impensis excudebant Palmblad & C.

Wallace, A.L., Klein, A.S. & Mathieson, A.C. 2004. Determining the affinities of salt marsh fucoids using microsatellite markers: evidence of hybridization and introgression between two species of *Fucus* (Phaeophyta) in a Maine estuary. *Journal of Phycology* 40: 1013–27.

Wallentinus, I. 1999. *Undaria pinnatifida* (Harvey) Suringar. In: Gollasch, S., Minchin, D., Rosenthal, H. & Voight, M. (eds), *Case histories on introduced species: their general biology, distribution, range expansion and general impact*, pp. 13–19. Department of Fishery Biology, Institute for Marine Science, University of Kiel, Germany.

Wallentinus, I. 2007. Alien species alert: *Undaria pinnatifida* (wakame or Japanese kelp). ICES Co-operative Research Report No 283, 36 pp.

Walton, A.J. 1986. Maturation of gametophytes of *Alaria esculenta*. *British Phycological Journal* 21: 338.

Wanders, J.B.W., van den Hoek, C. & Schillern-van Nes, E.N. 1972. Observations on the life history of *Elachista stellaris* (Phaeophyceae) in culture. *Netherlands Journal of Sea Research* 5: 458–91.

Wang, L., Zhang, X., Zou, D., Chen, W. & Jiang, H. 2018. Growth and F v/F m in embryos of *Hizikia fusiformis* (Harvey) Okamura (Sargassaceae, Phaeophyta) cultured under different temperature and irradiance conditions. *Journal of Oceanology and Limnology* 36: 1798–805.

Wang, W., Sun, X., Wang, G., Xu, P., Wang, X., Lin, Z. & Wang, F. 2010. Effect of blue light on indoor seedling culture of *Saccharina japonica* (Phaeophyta). *Journal of Applied Phycology* 22: 737–44.

Wang, X., Shan, T. & Pang, S. 2019. Effects of Cobalt on Spore Germination, Gametophyte Growth and Gametogenesis of *Undaria pinnatifida* (Phaeophyceae). *Bulletin of Environmental Contamination and Toxicology* 102: 784–8.

Wangkulangkul, K., Hawkins, S.J. & Jenkins, S.R. 2016. The influence of mussel-modified habitat on *Fucus serratus* L. a rocky intertidal canopy-forming macroalga. *Journal of Experimental Marine Biology and Ecology* 481: 63–70.

Warming, E. 1884. *Haandbog i den systematiske botanik: naermest til brug for laerere og universitets-studerende*. pp. iv, 1–434. Kjøbenhavn: P.G. Philipsens.

Watanabe, T. & Nisizawa, K. 1984. The utilization of wakame (*Undaria pinnatifida*) in Japan and manufacture of 'haiboshi wakame' and some of its biochemical and physical properties. *Proceedings of the International Seaweed Symposium* 11: 106–11.

Watanabe, H., Ito, M., Matsumoto, A. & Arakawa, H. 2016. Effects of sediment influx on the settlement and survival of canopy-forming macrophytes. *Scientific Reports* 6: no. 18677.

Watanabe, Y., Nishihara, G.N., Tokunaga, S. & Terada, R. 2014. The effect of irradiance and temperature responses and the phenology of a native alga, *Undaria pinnatifida* (Laminariales), at the southern limit of its natural distribution in Japan. *Journal of Applied Phycology* 26: 2405–15.

Webber, E.E., 1968. Systematics and ecology of benthic salt marsh algae at Ipswich, Massachusetts. PhD thesis, University of Massachusetts.

Webber, E.E. & Wilce, R.T. 1971. Benthic salt marsh algae at Ipswich, Massachusetts. *Rhodora* 73: 262–91.

Weber, F. & Mohr, D.M.H. 1805. Einige Worte über unsre bisherigen, hauptsächlich carpologischen Zergliederungen von kryptogamischen Seegewächsen. *Beiträge zur Naturkunde Niedersachsens* 1: 204–329.

Weber, F. & Mohr, D.M.H. 1810. *Beiträge zur Naturkunde*. In Verbindung mit meinen Freunden verfasst und herausgegeben. Zweiter Band. Mit vier Kupfertafeln. pp. [i]–vi, [1]–400, 5 pls. Kiel: bei August Schmidt.

Weber-Van Bosse, A. 1913. Liste des algues du Siboga. I. Myxophyceae, Chlorophyceae, Phaeophyceae. *Siboga Expeditie Monographic* 59a: 1–186. Leiden.

Wenderoth, H. 1933. Einige Ergänzungen zur Kenntnis des Aufbaues von *Dictyota dichotoma* Lamour. und *Padina pavonia* Lamour. *Flora* 127: 185–9.

Wernberg, T., Thomsen, M.S., Staehr, P.A. & Pedersen, M.F. 2001. Comparative phenology of *Sargassum muticum* and *Halidrys siliquosa* (Phaeophyceae: Fucales) in Limfjorden, Denmark. *Botanica Marina* 44: 31–9.

Wernberg, T., Thomsen, M.S., Staehr, P.A. & Pedersen, M.F. 2004. Epibiota communities of the introduced and indigenous macroalgal relatives *Sargassum muticum* and *Halidrys siliquosa* in Limfjorden (Denmark). *Helgoland Marine Research* 58: 154–61.

Wernberg, T. & Thomsen, M.S. 2005. The effect of wave exposure on the morphology of *Ecklonia radiata*. *Aquatic Botany* 83: 61–70.

West, J.A., Zuccarello, G.C., Pedroche, F.F. & Loiseaux-de Goër, S. 2010. *Rosenvingea orientalis* (Scytosiphonaceae, Phaeophyceae) from Chiapas, Mexico: life history in culture and molecular phylogeny. *Algae* 25: 187–95.

Wheeler, A. 1980. Fish–algal relations in temperate waters. In: Price, H., Irvine, D.E.G. & Farnham, W.F. (eds), *The Shore Environment, Vol. 2: Ecosystems*, pp. 677–98. Systematics Association Special Volume 17(b). Academic Press, London and New York.

Wheeler, W.N. & Weidner, M. 1983. Effects of external inorganic nitrogen concentration on metabolism, growth and activities of key carbon and nitrogen assimilatory enzymes of *Laminaria saccharina* (Phaeophyceae) in culture. *Journal of Phycology* 19: 91–6.

White, K.L., Kim, J.K. & Garbary, D.J. 2011. Effects of land-based fish farm effluent on the morphology and growth of *Ascophyllum nodosum* (Fucales, Phaeophyceae) in southwestern Nova Scotia. *Algae* 26: 253–63.

White, L. & Shurin, J.B. 2007. Diversity effects on invasion vary with life history stage in marine macroalgae. *Oikos* 116: 1193–203.

White, L. & Shurin, J.B. 2011. Density dependent effects of an exotic marine macroalga on native community diversity. *Journal of Experimental Marine Biology and Ecology* 405: 111–19.

Whittick, A. 1983. Spatial and temporal distributions of some dominant epiphytes on the stipes of *Laminaria hyperborea* (Gunn.) Fosl. (Phaeophyta, Laminariales) in SE Scotland. *Journal of Experimental Marine Biology and Ecology* 73: 1–10.

Widdowson, T.B. 1971a. A taxonomic revision of the genus *Alaria* Greville. *Syesis* 4: 11–49.

Widdowson, T.B. 1971b. A statistical analysis of variation in the brown alga *Alaria*. *Syesis* 4: 125–43.

Widdowson, T.B. 1971c. Changes in the intertidal algal flora of the Los Angeles area since the survey by E.Y. Dawson in 1956–1959. *Bulletin of the Southern California Academy of Sciences* 70: 2–16.

Wiggers, F.H. 1780. *Primitiae florae holsaticae*; quas praeside D. Joh. Christiano Kerstens. pp. [i–viii], 1–112. Kiliae [Kiel]: litteris Mich. Frider. Bartschii Acad. Typogr.

Wikström, S.A., von Wachenfeldt, T. & Kautsky, L. 2002. Establishment of the exotic species *Fucus evanescens* C. Ag. (Phaeophyceae) in Öresund, Southern Sweden. *Botanica Marina* 45: 510–17.

Wikström, S.A., Steinarsdóttir, M.B., Kautsky, L. & Pavia, H. 2006. Increased chemical resistance explains low herbivore colonization of introduced seaweed. *Oecologia* 148: 593–601.

Wilce, R.T. 1959. The marine algae of the Labrador Peninsula and northwest Newfoundland (ecology and distribution). *Bulletin National Museum of Canada* 158: 1–103.

Wilce, R.T. 1966. *Pleurocladia lacustris* in arctic America. *Journal of Phycology* 2: 57–66.

Wilce, R.T., Webber, E.E. & Sears, J.R. 1970. *Petroderma* and *Porterinema* in the New World. *Marine Biology, Berlin* 5: 119–35.

Wilce, R.T., Schneider, C.W., Quinlan, A.V. & van den Bosch, K. 1982. The life history and morphology of free-living *Pilayella littoralis* (L.) Kjellm. (Ectocarpaceae, Ectocarpales) in Nahant Bay, Massachusetts. *Phycologia* 21: 336–54.

Wilkinson, M. 1980. Estuarine benthic algae and their environment: a review. In: Price, J.H., Irvine, D.E.G. & Farnham, W.F. (eds), *The Shore Environment, Vol. 2: Ecosystems*, pp. 425–86. Systematics Association Special Volume 17(b). Academic Press, London and New York.

Wille, N. 1910. Der anatomische Bau bei *Himanthalia lorea* (.) Lyngbe. *Jahrbücher für wissenschaftliche Botanik* 47: 495–538.

Wille, N. & Rosenvinge, L.K. 1885. Alger fra Novaia-Zemlia og Kara-Havet, samlede paa Dijmphna – Expeditionen 1882–83 af Th. Holm. Dijmphna-Togtets zoologisck-botaniske Udbytte. Kjobenhavn.

Willenbrink, J. & Kremer, B.P. 1973. Localization of mannitol biosynthesis in the marine brown alga *Fucus serratus*. *Planta (Berl)* 113: 173–8.

Willenbrink, J., Rangoni-Kübbeler, M. & Tersky, B. 1975. Frond development and CO_2-fixation in *Laminaria hyperborea*. *Planta* 125: 161–70.

Williams, G.A. 1996. Seasonal variation in a low shore *Fucus serratus* (Fucales, Phaeophyta) population and its epiphytic fauna. *Proceedings of the International Seaweed Symposium* 15: 191–7.

Williams, J.L. 1897a. Mobility of antherozoids of *Dictyota* and *Taonia*. *Journal of Botany, British and Foreign* 35: 361–2.

Williams, J.L. 1897b. The antherozoids of *Dictyota* and *Taonia*. *Annals of Botany* 11: 545–53.

Williams, J.L. 1898. Reproduction in *Dictyota dichotoma*. *Annals of Botany* 12: 559–60.

Williams, J.L. 1903. Alternation of generations in the Dictyotaceae. *New Phytologist* 2: 184–6.

Williams, J.L. 1904a. Studies in the Dictyotaceae. I. The cytology of the tetrasporangium and the germinating tetraspore. *Annals of Botany* 18: 141–60.

Williams, J.L. 1904b. Studies in the Dictyotaceae. II. The cytology of the gametophyte generation. *Annals of Botany* 18: 183–204.

Williams, J.L. 1905. Studies in the Dictyotaceae. III. The periodicity of the sexual cells in *Dictyota dichotoma*. *New Phytologist* 19: 531–60.

Williams, J.L. 1921. The gametophytes and fertilisation in *Laminaria* and *Chorda*. *Annals of Botany* 35: 603–7.

Winder, F. 2002. *Sargassum muticum* in Kenmare River, Co. Kerry. *Irish Naturalists' Journal* 27: 85.

Withering, W. 1792. *A botanical arrangement of British plants*; Vol. 3, part 2 pp. [i–xxiv] 1–503, pls XIII–XIX.

Birmingham: Printed by Swinney & Walker; for G.G.J. & J. Robinson, Paternoster-Row, and J. Robson, New Bond-Street, London; J. Balfour, and G. Elliot, Edinburgh.

Withering, W. 1796. *An Arrangement of British Plants*; Vol. IV pp. [i–iii], [i]–418 [p. 418 in some copies err. numbered 420], pls XVII, XVIII, XXXI [sic]. Birmingham & London: Printed for the Author, by M. Swinney; Sold by C.G. & J. Robinson [etc.].

Withers, R.G., Farnham, W.F., Lewey, S., Jephson, N.A., Haythorn, J.M. & Gray, P.W.G. 1975. The epibionts of *Sargassum muticum* in British waters. *Marine Biology, Berlin* 31: 79–86.

Wolfe, J.J. 1918. Alternation and parthenogenesis in *Padina*. *Journal of the Elisha Mitchell Scientific Society* 34: 78–109.

Wollny, R. 1881. Die Meeresalgen von Helgoland. *Hedwigia* 20: 1–32.

Wollny, R. 1886. Algologische Mittheilungen. *Hedwigia* 25: 125–33.

Womersley, H.B.S. 1987. *The Marine Benthic Flora of Southern Australia. Part II.* Adelaide: South Australian Government Printing Division.

Wood, W.F. 1987. Effect of solar ultra-violet radiation on the kelp *Ecklonia radiata*. *Marine Biology* 96: 143–50.

Woodward, T.J. 1797. Observations upon the generic character of *Ulva*, with descriptions of some new species. *Transactions of the Linnean Society of London* 3: 46–58.

Woolery, M.L. & Lewin, R.A. 1973. Influence of iodine on growth and development of the brown alga *Ectocarpus siliculosus* in axenic cultures. *Phycologia* 12: 131–8.

Worm, B. & Chapman, A.R.O. 1998. Relative effects of elevated grazing pressure and competition from a red algal turf on two post-settlement stages of *Fucus evanescens* C. Ag. *Journal of Experimental Marine Biology and Ecology* 220: 247–68.

Wotton, D.M., O'Brien, C., Stuart, M.D. & Fergus, D.J. 2004. Eradication success down under: heat treatment of a sunken trawler to kill the invasive seaweed *Undaria pinnatifida*. *Marine Pollution Bulletin* 49: 844–9.

Wright, P.J., Chudek, J.A., Foster, R., Davison, I.R. & Reed, R.H. 1985. The occurrence of altritol in the brown alga *Himanthalia elongata*. *British Phycological Journal* 20: 191–2.

Wu, C.Y., Li, D.L., Liu, H.H., Peng, G.P. & Liu, J.X. 2004. Mass culture of *Undaria* gametophyte clones and their use in sporeling culture. *Hydrobiologia* 512: 153–6.

Wulfen, F.X. 1803. Cryptogama aquatica. *Archives de Botanique* 3: 1–64.

Wynne, M.J. 1969. Life history and systematic studies of some Pacific North American Phaeophyceae (brown algae). *University of California Publications in Botany* 50: 1–88.

Wynne, M.J. 1972a. Studies on the life forms in nature and in culture of selected brown algae. In: Abbott, I.A. & Kurogi, M. (eds), *Contributions to the Systematics of Benthic Marine Algae of the North Pacific*, pp. 133–45. Japanese Society of Phycology, Japan.

Wynne, M.J. 1972b. Culture studies of Pacific coast Phaeophyceae. *Mémoires de la Société Botanique de France* 1972: 129–44.

Wynne, M.J. 1981. Phaeophyta: morphology and classification. In: Lobban, C. & Wynne, M.J. (eds), *The Biology of Seaweeds*, pp. 52–85. Botanical Monographs 17. Blackwell Scientific Publications, Oxford.

Wynne, M.J. 2003a. The identity of *Laminaria ensiformis* Delle Chiaje (Phaeophyceae). *Webbia* 58: 471–6.

Wynne, M.J. 2003b. The original author of *Ulva multifida* is Dawson Turner (1801) not J.E. Smith (1808) (Notas Breves). *Anales del Jardín Botánico de Madrid* 60: 453–4.

Wynne, M.J. 2011. A checklist of benthic marine algae of the tropical and subtropical western Atlantic: third revision. *Nova Hedwigia Beihefte* 140: 7–166.

Wynne, M.J. & Loiseaux, S. 1976. Recent advances in the life history studies of the Phaeophyta. *Phycologia* 15: 435–52.

Wynne, M.J. & Magne, F. 1991. Concerning the name *Fucus muscoides* (Cotton) J. Feldmann et Magne. *Cryptogamie, Algologie* 12: 55–65.

Xu Gao, X., Lee, J.R., Park, S.K., Kim, N.G. & Choi, H.G. 2019. Detrimental effects of sediment on attachment, survival and growth of the brown alga *Sargassum thunbergii* in early life stages. *Phycological Research* 67: 77–81.

Yamada, M., Yamamoto, K., Ushihara, Y. & Kawai, H. 2007. Variation in metal concentrations in the brown alga *Undaria pinnatifida* in Osaka Bay, Japan. *Phycological Research* 55: 222–30.

Yamada, Y. 1935. The marine algae of Urup, Middle Kuriles, especially from the vicinity of Ioma Bay. *Scientific Papers of the Institute of Algological Research, Faculty of Science, Hokkaido Imperial University* 1: 1–26.

Yamada, Y. 1936. Notes on some Japanese algae VII. *Scientific Papers of the Institute of Algological Research, Faculty of Science, Hokkaido Imperial University* 1: 135–40.

Yamagishi, Y. & Kogame, K. 1998. Female dominant population of *Colpomenia peregrina* (Scytosiphonales, Phaeophyceae). *Botanica Marina* 41: 217–22.

Yamanaka, R. & Akiyama, K. 1993. Cultivation and utilization of *Undaria pinnatifida* (wakame) as food. *Journal of Applied Phycology* 5: 249–53.

Yamanouchi, S. 1909. Cytology of *Cutleria* and *Aglaozonia*. A preliminary paper. *Botanical Gazette* 48: 380–6.

Yamanouchi, S. 1912. The life history of *Cutleria*. *Botanical Gazette* 54: 441–502.

Yamanouchi, S. 1913. The life history of *Zanardinia*. *Botanical Gazette* 56: 1–35.

Yamauchi, K. 1984. The formation of *Sargassum* beds on artificial substrata by transplanting seedlings of *S. horneri* (Turner) C. Agardh and *S. muticum* (Yendo) Fensholt. *Bulletin of the Japanese Society of Scientific Fisheries* 50: 1115–23.

Yang, E.C., Peters, A.F., Kawai, H., Stern, R., Hanyuda, T., Bárbara, I., Müller, D.G., Strittmatter, M., Prud'homme van Reine, W.F. & Küpper, F.C. 2014. Ligulate *Desmarestia* (Desmarestiales, Phaeophyceae) revisited: *D. japonica* sp. nov. and *D. dudresnayi* differ from *D. ligulata*. *Journal of Phycology* 50: 149–66.

Yarish, C., Penniman, C.A. & Egan, B. 1990. Growth and reproductive responses of *Laminaria longicruris* (Laminariales, Phaeophyta) to nutrient enrichment. *Hydrobiologia* 204/5: 505–11.

Yee, N.R. 2007: Phylogenetic studies of the marine brown algal order Sporochnales (Phaeophyceae). PhD dissertation, University of Melbourne.

Yendo, K. 1907. The Fucaceae of Japan. *Journal of the College of Science, Tokyo Imperial University* 21 (Article 12): 1–174, folded table, pls I–XVIII.

Yendo, K. 1911. The development of *Costaria, Undaria,* and *Laminaria. Annals of Botany* 25: 691–715.

Yendo, K. 1919a. The germination and development of some marine algae. *Botanical Magazine, Tokyo* 33: 171–84.

Yendo, K. 1919b. A monograph of the genus *Alaria. Journal of the college of Science, Tokyo Imperial University* 43: 1–145.

Yoneshigue-Valentin, Y. & Pupo, D. 1994. Estudos *in situ* e *in vitro* de fase Ralfsióide *Stragularia clavata* (Harv. in Hook.) Hamel de *Petalonia fascia* (O.F. Müller) Kuntze (Scytosiphonales – Phaeophyta). *Revista Brasileira de Biologia* 54: 489–96.

Yoo, Y.-C., Kim, W.-J., Kim, S.-Y., Kim, S.-M., Chung, M.-K., Park, J.-W., Suh, H.-H., Lee, K.-B. & Park, Y.-I. 2007. Immunomodulating activity of a fucoidan isolated from Korean *Undaria pinnatifida* sporophyll. *Algae* 22: 333–8.

Yoon, H.S. & Boo, S.M. 1999. Phylogeny of Alariaceae (Phaeophyta) with special reference to *Undaria* based on sequences of the RuBisCo spacer region. *Proceedings of the International Seaweed Symposium* 16: 47–55.

Yoon, H.S., Lee, J.Y., Boo, S.M. & Bhattacharya, D. 2001. Phylogeny of Alariaceae, Laminariaceae, and Lessoniaceae (Phaeophyceae) Based on plastid-encoded RuBisCo Spacer and nuclear-encoded ITS sequence comparisons. *Molecular Phylogenetics and Evolution* 21: 231–43.

Yoshida, T. & Akiyama, K. 1979. *Streblonema* (Phaeophyceae) infection in the frond of cultivated *Undaria. Proceedings of the International Seaweed Symposium* 9: 219–23.

Yoshida, T. 1983. Japanese species of *Sargassum* subgenus *Bactrophycus* (Phaeophyta, Fucales). *Journal of the Faculty of Science, Hokkaido University, Series V (Botany)* 13: 99–246.

Yoshida, T., Nakajima, Y. & Nakata, Y. 1990. Check-list of marine algae of Japan (revised in 1990). *Japanese Journal of Phycology* 38: 269–320.

Yoshida, T., Stiger, V. & Horiguchi, T. 2000. *Sargassum boreale* sp. nov. (Fucales, Phaeophyceae) from Hokkaido, Japan. *Phycological Research* 48: 125–31.

Yoshida, T., Stiger, V., Ajisaka, T. & Noro, T. 2002. A molecular study of section-level classification of *Sargassum* subgenus Bactrophycus (Sargassaceae, Phaeophyta). In: Abbott, I.A. & McDermid, K.J. (eds), *Taxonomy of Economic Seaweed with Reference to some Pacific Species,* vol. 8, pp. 89–94. La Jolla: California Sea Grant College Program.

Yotsukura, N., Denboh, T., Motomura, T., Horiguchi, T., Coleman, A.W. & Ichimura, T. 1999. Little divergence in ribosomal DNA internal transcribed spacer-1 and -2 sequences among non-digitate species of *Laminaria* (Phaeophyceae) from Hokkaido, Japan. *Phycological Research* 47: 71–80.

Yoshida, T., Yoshinaga, K. & Nakajima, Y. 1995. Check-list of marine algae of Japan *Japanese Journal of Phycology* 43: 115–71 (update of a 1990 publication).

Young, E.B., Dring, M.J. & Berges, J.A. 2007. Distinct patterns of nitrate reductase activity in brown algae: light and ammonium sensitivity in *Laminaria digitata* is absent in *Fucus* species. *Journal of Phycology* 43: 1200–8.

Younger, J. 2011. What is controlling the distribution of *Undaria pinnatifida* (Phaeophyceae, Laminariales) in Victoria, Australia. Honours thesis, Central Queensland University, Gladstone, Australia.

Zanardini, G. 1840. Sopra le alghe del mare Adriatico. Lettera seconda di Giovanni Zanardini, medico fisico in Venezia, alla Direzione della Biblioteca Italiana. *Biblioteca Italiana Ossia Giornale di Letteratura Scienze ed Arti* 99: 195–229.

Zanardini, G. 1841. Synopsis algarum in mari Adriatico hucusque collectarum, cui accedunt monographia siphonearum nec non generales de algarum vita et structura disquisitiones cum tabulis auctoris manu ad vivum depictis. *Memorie della Reale Accademia delle Scienze di Torino, ser. 2* 4: 105–255.

Zanardini, G. 1843. *Saggio di classificazione naturale delle Ficee*: pp. 1–64, 2 folded tables, 1 plate. Venezia [Venice]: Dallo Stabilimento tipographico enciclopedico di Girolamo Tasso.

Zanardini, G. 1860. *Iconographia phycologica adriatica*: Vol. 1, pp. [i*–ii*], [i]–viii, [1]–175, [176], pls I–XL (col. liths., auct.). Venezia: nel priv. stabil. di G. Antonelli.

Zanardini, G. 1862. Scelta di Ficee nuove o più rare del mare Adriatico. Decade terza. *Memorie del Reale Istituto Veneto di Scienze, Lettere ed Arti* 10: [447]–484.

Zanardini, G. 1864. Scelta di Ficee nuove o più rare del mare Adriatico. *Memorie del Reale Istituto Veneto di Scienze, Lettere ed Arti* 12: [7]–43.

Zanardini, G. 1865. *Iconographia phycologica adriatica*: Vol. II pp. [i]–viii, [ix, err.], [1]–168, pls XLI–LXXX. Venezia: nel priv. stabil. di G. Antonelli edit.

Zardi, G.I., Nicastro, K.R., Canovas, F., Costa, J.F., Serrão, E.A. & Pearson, G.A. 2011. Adaptive traits are maintained on steep selective gradients despite gene flow and hybridization in the intertidal zone. *PLoS ONE* 6: 1–13.

Zhang, D.M., Miao, G.R. & Pei, L.Q. 1984. Studies on *Undaria pinnatifida. Proceedings of the Eleventh International Seaweed Symposium,* Qingdao, People's Republic of China, June 19–25, 1983, pp. 263–5.

Zhang, L., Wang, X., Liu, T., Wang, G., Chi, S., Liu, C. & Wang, H. 2015. Complete Plastid Genome Sequence of the Brown Alga *Undaria pinnatifida. PLoS ONE* 10: e0139366.

Zhang, W., Chapman, D.J., Phinney, B.O., Spray, C.R., Yamane, H. & Takahashi, N. 1991. Identification of cytokinins in *Sargassum muticum* (Phaeophyta) and *Porphyra perforata* (Rhodophyta). *Journal of Phycology* 27: 87–91.

Zhang, L., Cui, C., Li, X., Zhang, Z., Luo, S., Liang, G., Liu, Y. & Yang, G. 2013. Effect of temperature on the development of *Saccharina japonica* gametophytes. *Journal of Applied Phycology* 25: 261–7.

Zhang, X., Hu, H. & Tan, T. 2006. Photosynthetic inorganic carbon utilization of gametophytes and sporophytes of *Undaria pinnatifida* (Phaeophyceae). *Phycologia* 45: 642–7.

Zhang, X., Li, D., Hu, H. & Tan, T. 2005. Growth promotion of vegetative gametophytes of *Undaria pinnatifida* by blue light. *Biotechnology Letters* 27: 1467–75.

Zhang, Y., Guo, Y.-M., Li, T.J., Ching-Hung, C., Kang-Ning, S. & Cung-der, H. 2016. The complete chloroplast genome of Wakame (*Undaria pinnatifida*), an important economic macroalga of the family Alariaceae. *Mitochondrial DNA Part B* 1: 25–6.

Zhang, Z. & Pang, S. 2007. Circadian rhythms in the growth and reproduction of the brown alga *Undaria pinnatifida* and gametogenesis under different photoperiods. *Acta Oceanologica Sinica* 26: 104–11.

Zhao, S.F., Li, H.J., Sun, H.Q., Li, J.P. & Li, G.R. 2013. Effects of light intensity on sexual reproduction and early development of *Sargassum cinereum* (Fucales, Phaeophyta) germlings. *Journal of Shanghai Ocean University* 22: 563–70 (in Chinese with English abstract).

Zhao, S.F., Yao, W.L., Guo, X.Z., He, T.T., Sun, H.Q., Guo, S.D. & Huang, G.H. 2015. Combined effects of temperature and light on the growth rate of *Hizikia fusiformis* young sporophyte. *Journal of Aquaculture* 36: 42–7 (in Chinese with English abstract).

Zhu, Z.J. & Chen, P.M. 1997. The relationship between water temperature, light intensity and the photosynthetic rates of *Sargassum fusiforme*. *Journal of Fisheries of China*, 21: 165–70 (in Chinese with English abstract).

Zinova, A.D. 1950. *Laminaria apoda* Post. et Rupr. *Botanical Magazine* 31(1).

Zinova, A.D. 1953. *Opredelitel' burykh vodoroslej severnykh morej SSSR* [*Manual for identification of the brown algae of the northern seas of the USSR*]. pp. 1–224. Moscow & Leningrad: Akad. Nauk SSSR.

Zinova, E.S. 1914. Algae of murman. Brown algae. Trudysankt-peterburgskogo obschestva estest-voispytatelei. *Proceedings of the St. Petersburg Society of Naturalists* 44–45: 212–326.

Zinova, E.C. 1922. *Fucus fueci* De la Pyl. *Bot. materials, section cryptogamic plants* 1.

Zoega, J. 1772. *Flora islandica*. Copenhagen & Leipzig.

Zou, D.H., Liu, S.X., Du, H. & Xu, J.T. 2012. Growth and photosynthesis in seedlings of *Hizikia fusiformis* (Harvey) Okamura (Sargassaceae, Phaeophyta) cultured at two different temperatures. *Journal of Applied Phycology* 24: 1321–7.

Zou, D., Ji, Z., Chen, W. & Li, G. 2018. High temperature stress might hamper the success of sexual reproduction in *Hizikia fusiformis* from Shantou, China: a photosynthetic perspective. *Phycologia* 57: 394–400.

Zuo-mei, Y. 1984. Studies on tissue culture of *Laminaria japonica* and *Undaria pinnatifida*. *Proceedings of the International Seaweed Symposium* 11: 314–16.

Taxonomic index

Subclasses, orders and families are shown in capitals. All genera, species and infraspecific taxa are given in roman type. Taxa present in Britain and Ireland, and their main page numbers, are in **bold**. All other entries, including all synonyms, taxa not known to occur within the geographical area and those other than brown algae are shown in *italics*. Page numbers in italics denote an illustration.